STATISTICS IN ACTION

Case: Managing Financial Risk

INVESTMENT RISK
THE DICE GAME
UNDERSTANDING WHAT HAPPENS
CASE SUMMARY

Financial markets can feel overwhelming. Markets offer so many choices with different promises. How can you make sense of it all and come away feeling that you've made good choices? The right place to start is by understanding the role of random variation. Luck and chance play a large role in deciding which investment wins and which investment loses. This case uses a simulation with dice to follow the performance of several investments. Even though the roll of a die offers only six outcomes, dice provide enough randomness to simulate important features of real investments.

INVESTMENT RISK

When it comes to investments, **risk** is the variance of the percentage changes in value (returns) over time. The two histograms in Figure 1 summarize monthly percentage changes of two investments.

The histogram on the top shows monthly percentage changes on the stock market from 1926 through 2005; the histogram on the bottom shows US Treasury Bills (basically, a short-term savings bond). There's almost no month-to-month variation (on this scale) in Treasuries. The stock market is much riskier. Some times stocks they go up a lot (the largest monthly increase is 38%), and other times they go down. This variation in what happens from month to month is risk.

If stocks are risky, why does anyone buy them? The answer is simple: stocks offer the potential for higher gains as compensation for taking risks. Stocks over these years grew on average 12% annually, compared to 4% for Treasuries. Is this enough reward for the risk? If those were your retirement dollars, would you put the money into stocks?

It's not an easy question to answer. The following simulation shows you how to understand the relationship between risk and reward. Several questions

risk   The variance of the returns on an investment.

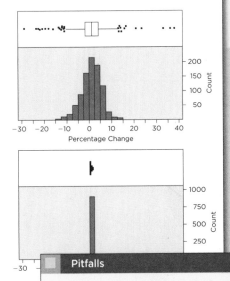

FIGURE
Bills (bo

along t
and the
the obs

THE

To learn
financia
can do
interest
dice tha
Three o
all three
market
One per
dice an
the outc

STATISTICS IN ACTION

Case studies follow each Part, extending the statistical methods and delving into substantive aspects of real-world scenarios.

Pitfalls

- *Avoid elaborate plots that may be deceptive.* Modern software makes it easy to add an extra dimension to punch up the impact of a plot. For instance, this 3-D pie chart shows the top six hosts from the Amazon example.

Referring Sites

Showing the pie on a slant violates the area principle and makes it harder to compare the shares, the principal feature of data that a pie chart ought to show.

- *Do not show too many categories.* A bar chart or pie chart with too many categories might conceal the more important categories. In these cases, group other categories together and be sure to let your audience know how you have done this.
- *Do not put ordinal data in a pie chart.* Because of the circular arrangement, pie charts are not well suited to ordinal data because the order gets lost. The pie chart shown at the top of the next column conceals the ordinal nature of the data. The slice indicating companies who expect revenue to increase more than 50% is next to the slice for those who expect revenue to fall the most.[9]

Upbeat

U.S. companies' responses when asked to forecast changes in their 2006 China revenue from 2005

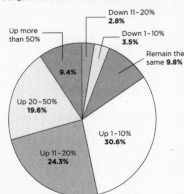

Source: AmCham Shanghai 2006 China Business Report

- *Do not carelessly round data.* Do you see the problem with the following pie chart? Evidently, the author rounded the shares to integers at some point, but then forgot to check that the rounded values add up to 100%. Few things hurt your credibility more than these little blunders.

U.S. Wireless Phone Market

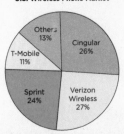

BEST PRACTICES and PITFALLS

Best Practices demonstrate how to successfully and ethically apply concepts, and **Pitfalls** help you avoid making common mistakes.

Statistics for Business

DECISION MAKING AND ANALYSIS

ROBERT STINE
Wharton School of the University of Pennsylvania

DEAN FOSTER
Wharton School of the University of Pennsylvania

Addison-Wesley

Boston Columbus Indianapolis
New York San Francisco Upper Saddle River
Amsterdam Cape Town Dubai London
Madrid Milan Munich Paris Montréal Toronto
Delhi Mexico City São Paulo Sydney
Hong Kong Seoul Singapore Taipei Tokyo

Editor in Chief: *Deirdre Lynch*
Project Editor: *Elizabeth Bernardi*
Associate Editor: *Christina Lepre*
Editorial Assistant: *Dana Jones*
Executive Director of Development: *Carol Trueheart*
Senior Managing Editor: *Karen Wernholm*
Senior Production Project Manager: *Kathleen A. Manley*
Digital Assets Manager: *Marianne Groth*
Media Producer: *Lin Mahoney*
Software Development: *Edward Chappell, MyStatLab and Marty Wright, TestGen*
Marketing Manager: *Alex Gay*
Marketing Assistant: *Kathleen DeChavez*
Senior Author Support/Technology Specialist: *Joe Vetere*
Senior Prepress Supervisor: *Caroline Fell*
Senior Manufacturing Manager: *Evelyn Beaton*
Senior Media Buyer: *Ginny Michaud*
Text Design: *Ellen Pettengell Design*
Production Coordination, Composition, Illustrations: *Pre-PressPMG*
Art Director: *Heather Scott*
Cover Design: *Tamara Newnam*
Cover Photo: *Shutterstock*

For permission to use copyrighted material, grateful acknowledgment has been made to the copyright holders listed on C-1, which is hereby made part of the copyright page.

Many of the designations used by manufacturers and sellers to distinguish their products are claimed as trademarks. Where those designations appear in this book, and Addison-Wesley was aware of a trademark claim, the designations have been printed in initial caps or all caps.

Library of Congress Cataloging-in-Publication Data

Stine, Robert A.
    Statistics for business: decision making and analysis/Robert Stine,
Dean Foster.—1st ed.
        p. cm.
    ISBN 0-321-12391-3
    1. Commercial statistics. 2. Statistics. I. Foster, Dean P. II.
Title.
    HF1017.S74 2011
    519.5024′65—dc22

                                                      2009018725

1 2 3 4 5 6 7 8 9 10—WC—13 12 11 10 09

Addison-Wesley
is an imprint of

ISBN-13: 978-0-321-12391-6
ISBN-10: 0-321-12391-3

ABOUT THE AUTHORS

Robert Stine holds a Ph.D. from Princeton University. He has taught at the Wharton School since 1983, during which time he has regularly taught business statistics. During his tenure, Bob has received a variety of teaching awards, including regularly winning the MBA Core Teaching Award, which is presented to faculty for outstanding teaching of the required curriculum at Wharton. He also received the David W. Hauck Award for Outstanding Teaching, awarded to the most highly rated faculty member teaching in the Wharton undergraduate program. Bob actively consults for industry. His clients include the pharmaceutical firms Merck and Pfizer, and he regularly works with the Federal Reserve Bank of Philadelphia on models for retail credit risk. This collaboration has produced three well-received conferences held at Wharton. His areas of research include computer software, time series analysis and forecasting, and general problems related to model identification and selection. Bob has published numerous articles in research journals, including the *Journal of the American Statistical Association*, *Journal of the Royal Statistical Society*, *Biometrika*, and *The Annals of Statistics*.

Dean Foster holds a Ph.D. from the University of Maryland. He has taught at the Wharton School since 1992 and previously taught at the University of Chicago. Dean teaches courses in introductory business statistics, probability and Markov chains, statistical computing and advanced statistics for managers. Dean's research areas are statistical inference for stochastic processes, game theory, machine learning, and variable selection. He is published in a wide variety of journals, including *The Annals of Statistics*, *Operations Research*, *Games and Economic Behaviour*, *Journal of Theoretical Population Biology*, and *Econometrica*.

Bob Stine and Dean Foster have co-authored two casebooks: *Basic Business Statistics* (Springer-Verlag) and *Business Analysis Using Regression* (Springer-Verlag). These casebooks offer a collection of data analysis examples that motivate and illustrate key ideas of statistics, ranging from standard error to regression diagnostics and time series analysis. They also have collaborated on a number of research articles.

CONTENTS

PART TWO   Probability

PART FOUR   Regression Models

PREFACE

Statistics are a great asset to the business world, but only when they are used to answer the right questions. Choosing the right question and solving the right problem have more to do with understanding the business in question than the details of how to perform the statistical analysis. First, understand the business issue from a finance, marketing, management, or accounting perspective, and then figure out how statistics can help solve the problem. Performing the statistical analysis must wait until you have grasped the issue facing the business.

Solving Business Problems. This approach shapes our examples. We open each chapter by framing a business question that motivates the contents of the chapter. For extra practice, worked-out examples within each chapter follow our 4M (Motivation, Method, Mechanics, Message) problem-solving strategy. The *motivation* sets up the problem and explains the relevance of the question at hand. We then identify the appropriate statistical *method* and work through the *mechanics* of its calculation. Finally, the *message* answers the question in language suitable for a business presentation or report. Through the 4Ms, we'll show you how a business context guides the statistical procedure and how the results determine a course of action. Motivation and Message are critical. If you do not convey the relevance of the problem at the start of the problem in the Motivation and express the Message in suitable language at the end of the problem, it won't matter how well you do the statistical analysis. Understand the business first, then use statistics to help formulate your conclusion. Notice that we said "help." A statistical analysis by itself is not the final answer. You must frame that analysis in terms that others in the business will understand and find persuasive.

Our emphasis on the substantive use of statistics in business shapes our view that the ideal reader for this text is someone with an interest in learning how statistical thinking improves the ability of a manager to run or contribute to a business. Whether you're an undergraduate with an interest in business, an MBA looking to improve your skills, or a business owner looking for another way to get ahead of the competition, the key is a desire to learn how statistics can produce better decisions and insights from the growing amount of data generated in modern businesses.

We don't assume that readers have mastered the domains of a business education, such as economics, finance, marketing, or accounting. We do assume, though, that you care about how ideas from these areas can improve a business. If you're interested in these applications—and we think you will be—then our examples provide the background you will need in order to appreciate why we want to solve the challenges that we present in each chapter. Readers with more experience will discover that we've simplified the technical details of some applications, such as those in finance or marketing. Even so, we think that the examples offer those with substantive experience a new perspective on problems that may already be familiar. We hope that you will agree that the examples are realistic and get to the heart of quantitative applications of statistics in business.

Technology. The growing power of computer software has had a dramatic impact on the field of statistics, and it's our intention to take advantage of this progress in our textbook. A casual glance at the table of contents of a recent research journal in statistics shows that most of this research relies on computers. You simply cannot do research in modern applied statistics without computing. Data sets have grown in size and complexity, making it impossible to work out the calculations by hand. Rather than dwell on routine calculations, we rely on software (often referred to as a statistics package) to compute the results. That said, we do not treat this software as a black box; we give the formulas and illustrate the calculations introduced in each chapter so you will always know what is being done by software. It is essential to appreciate what happens in the calculations, and it is also a crucial part of decision-making: You need to understand how the calculations are done in order to recognize when they are appropriate and when they fail. That does not mean, however, that you need to spend hours doing routine calculations. Your time is precious and there's only so much of it to go around. We think it makes good economic sense to take advantage of the relatively inexpensive cost of computation in order to give us more time to think harder and more thoroughly about the motivating context for an application and successfully present the business message. To help you learn how to use software, each chapter includes hints on using Excel®, Minitab®, and JMP®, for calculations. These hints won't replace the help provided by your software, but they will point you in the right direction so that you don't spin your wheels figuring out how to get started with an analysis.

Data.   Statistical analysis uses data, and we've provided lots of that to give you the opportunity to have some real hands-on experience. As you read through the chapters, you'll discover a variety of data sets that include real estate markets, stocks and bonds, technology, retail sales, human resource management, and fundamental economics. These data come from a range of sources, and each chapter includes a discussion about where we collected the data used in examples. We hope you'll use our suggestions and find more.

Prerequisite Knowledge.   To appreciate the illustrative calculations and formulas, readers will need to be familiar with basic algebra. Portions of chapters that introduce a statistical method often include some algebra to show where a formula comes from. Usually, we only use basic algebra (up through topics such as exponents and square roots). Several chapters make extensive use of the logarithm function. If you're interested in business and economics, this is a function worth getting to know a lot better. The applications we've provided, such as modeling sales or finding the best price, show why the logarithm is so important. Occasionally, we give credit to calculus for solving a problem, but we don't present derivations using calculus. You'll do fine if you are willing to accept that calculus is a branch of more advanced mathematics that provides, among other things, the ability to derive formulas that have special properties. If you do know calculus, you'll be able to see where these expressions come from.

Coverage & Organization

We have organized the chapters of this book into four parts:

1. Variation
2. Probability
3. Inference
4. Regression Models

Part 1.   Readers who have worked with data will be able to skip portions of Part 1 or move through it fairly quickly. These chapters introduce summary statistics such as the mean and important graphical summaries such as bar charts, histograms, and scatterplots. Even if you are familiar with these methods, we encourage you to skim the examples in these chapters. These examples introduce important terminology that appears in subsequent chapters. A quick review will introduce the notation that we use (which is rather standard) as well as give you a chance to look at some interesting data. If you do skip past these, take advantage of the index of Key Terms in each chapter to find definitions and examples.

Part 2.   The contents of Part 2 also will be familiar to some readers; many courses in mathematics now include topics from probability. Even if you have seen basic probability, you might benefit from reviewing how methods, such as Bayes rule, can be used to improve business processes (Chapter 8). If you plan to skip or move briskly through the rest of the chapters in Part 2, be sure that you're familiar with the concept of a random variable (Chapter 9). Statistical models use random variables to present an idealized description of the data in applications. Unless you're familiar with random variables, you won't appreciate the important assumptions that come with their use in practice. Chapter 11 describes special random variables used to model counts, and Chapter 12 defines normal random variables that appear so often in statistical models.

Part 3.   This part presents the foundations for statistical inference, the process of inferring properties of an entire population from those of a subset known as a sample. Even if you are not interested in quality control, we encourage you to read Chapter 14. Chapter 14 uses quality control to introduce a fundamental concept of inferential statistics, the sampling distribution and standard error. You can get by in statistics with a basic understanding of the concept of a sampling distribution, but the more you know about sampling distributions, the better. If you are in a hurry, you can also skip Chapter 17; it offers methods for situations in which the standard procedures don't work. How will you know that you need these? Each inferential procedure comes with a checklist of conditions that tell you whether your data and situation match up to the various inferential techniques in these chapters.

Part 4.   The chapters in Part 4 describe regression modeling. Regression modeling allows us to associate how differences in data that describe one phenomenon are related to differences in others. Regression models are among the most powerful ways to use statistics in business, providing methods for assessing profitability, setting prices, identifying anomalies, and generating forecasts. We encourage you to slow down and take your time studying these chapters. Even if you don't see yourself doing statistics in your career in business, you can be sure that you will be presented with the results of regression models. Because the examples in these chapters allow us to describe the interconnectedness of several business processes at once, they become even more interesting than those in prior chapters. Be careful if you plan to skip Chapter 20. The material in this chapter shows how to model a richer set of patterns and is less common in business textbooks, but we think these ideas are an essential component of every manager's tool set.

Case Studies.   Each of these parts of the book comes with two supplemental case studies that are called Statistics in Action. These provide additional examples of the methods introduced in several chapters and so review many of the ideas. We've found that it is easy to have a "chapter-centric" view of any subject; you know how to approach a problem if the question identifies a chapter. Executing the right approach is more difficult without that sort of clue. Statistics in Action cases take on a business problem using a variety of methods presented in several chapters, occasionally in a novel way that is not described within the text itself. We selected the data in these case studies to be interesting. Data sets include prices of stock, executive compensation, defaults on corporate bonds, retail sales management, and process control. The case studies in Part 2 (Probability) do not include data and instead provide experiments that illustrate random processes. Rolling dice simulate fundamental characteristics of risky investments, and opening packages of M&M candies reveals both sampling distributions and normality.

Features

Motivating Examples.   Each chapter opens with a business example that frames a question and motivates the contents of the chapter. We return to the example throughout the chapter, as we present the statistical methods that provide answers to the question posed in the opening example.

4M Worked Examples.   The 4M (Motivation, Method, Mechanics, Message) problem-solving strategy gives students a clear outline for solving any business problem. Each 4M example first expresses a business question in context, then guides students along the path to determine the best statistical method for working the problem using statistical software, framing the analysis in terms that others in the business world will understand.

What Do You Think?   We've included short question sets throughout each chapter to give students the opportunity to check their understanding of what they've just read. These questions are intended to be a quick check of key concepts and ideas presented in the chapter; most questions involve very little calculation. Answers are located in a footnote so that students can easily check their answers before moving on in the text.

Tips.   We highlight useful hints for applying statistical methods within the exposition so that students don't miss them.

Caution.   You'll see the caution icon in the margin next to material that might be confusing.

Best Practices.   At the end of each chapter, we include a collection of tips for applying the chapter's concepts successfully and ethically.

Pitfalls.   Most of the unintentional mistakes people make when learning statistics are avoidable and usually come from using the wrong method for the situation or misinterpreting the results. This feature at the end of each chapter provides useful tips for avoiding common mistakes.

Software Hints.   Each chapter includes hints on using Excel (2003 and 2007), Minitab, and JMP for calculations. These hints give students a jumping off point for getting started doing statistical analysis with software.

Behind the Math.   At the end of most chapters, we've included a Behind the Math section that provides interesting technical details that explain important results, such as the justification or interpretation for an underlying formula. If you are so inclined, they will help you appreciate the subtleties and logic behind the mechanics, but they are not necessary for using statistics.

Chapter Summary.   These chapter-ending summaries provide a complete review of the content.

- **Key Terms** We provide an index of the chapter's key terms at the end of each chapter to give students a quick and easy way to return to important definitions in the text.
- **Formulas** Important formulas introduced within the chapter are restated. Many include a descriptions.
- **About the Data** This feature describes and provides sources for the data used throughout the chapter.

Exercises. Each chapter contains a variety of exercises at escalating levels of difficulty in order to give students a full complement of practice in problem solving using the skills they've learned in the chapter. Types of exercises include Matching, True/False, Think About It, You Do It, and 4M Exercises.

- **Matching and True/False** exercises test students' ability to recognize the basic mathematical symbols and terminology they have learned in the chapter.
- **Think About It** exercises ask students to pull together the chapter's concepts in order to solve basic, conceptual problems.
- **You Do It** exercises give students practice solving problems that reinforce the mechanics they've learned in the chapter.

- **4M** exercises are rich, challenging applications rooted in real business situations that ask students to apply the statistical knowledge they've developed in the chapter to a set of questions about a particular business problem.

Statistics in Action Case Studies. After each of the book's four parts, we've included two in-depth case studies that use real data and ask students to use and extend the statistical concepts introduced in the preceding chapters to investigate a longer, more complex business case. Data sets include prices of stock, executive compensation, defaults on corporate bonds, retail sales management, and process control.

Supplements

Student Supplements

Student's Solutions Manual, by Sarah Streett

This manual provides detailed, worked-out solutions to all odd-numbered text exercises. (ISBN-10: 0-321-28614-6; ISBN-13: 978-0-321-28614-7)

Excel Technology Manual, by Zhiwei Zhu, University of Louisiana, Lafayette

This manual provides tutorial instruction and worked-out text examples for Excel. (ISBN-10: 0-321-64534-0; ISBN-13: 978-0-321-64534-0)

Minitab Manual, by Dorothy Wakefield, University of Connecticut Health Center, and Kathleen McLaughlin, University of Connecticut

This manual provides tutorial instruction and worked-out text examples for Minitab. (ISBN-10: 0-321-64565-0; ISBN-13: 978-0-321-64565-4)

Business Insight Videos This series of ten 5–7 minute videos, each about a well-known business and the challenges it faces, focuses on statistical concepts as they pertain to the real world. The videos can be downloaded to Video iPods® from within MyStatLab. Contact your Pearson representative for details.

Study Cards for Business Statistics Software This series of study cards provide students with easy step-by-step instructions for using statistics software. Available for Excel® (0-321-64191-4), Minitab (0-321-64421-2), JMP (0-321-64423-9), SPSS®/PASW® (0-321-64422-0), and R (0-321-64469-7).

Instructor Supplements

Instructor's Edition This version of the text contains short answers to all of the exercises within the exercise sets. (ISBN-10: 0-321-28616-2; ISBN-13: 978-0-321-28616-1)

Instructor's Solutions Manual, by Sarah Streett

This manual provides detailed, worked-out solutions to all of the book's exercises. Careful attention has been paid to ensure that all methods of solution and notation are consistent with those used in the core text. (ISBN-10: 0-321-28612-X; ISBN-13: 978-0-321-28612-3)

Online Test Bank, by Dave Bregenzer, Utah State University

This test bank contains several ready-to-use tests for each chapter in the text. The testbank is available for download from Pearson Education's online catalog (http://www.pearsonhighered.com/irc).

TestGen® (www.pearsonhighered.com/testgen) enables instructors to build, edit, print, and administer tests using a computerized bank of questions developed to cover all the objectives of the text. TestGen is algorithmically based, allowing instructors to create multiple but equivalent versions of the same question or test with the click of a button. Instructors can also modify test bank questions or add new questions. The software is available for download from Pearson Education's online catalog (http://www.pearsonhighered.com/irc).

Technology Resources

- A companion CD is bound in new copies of this text. The CD holds a number of supporting materials, including:
  - **Data sets** formatted for Minitab 14 and 15, SPSS/PASW, Excel, JMP, and text files.
  - **DDXL™**, an Excel add-in, adds sound statistics and statistical graphics capabilities to Excel. Among other capabilities, DDXL adds boxplots, histograms, statistical scatterplots, normal probability plots, and statistical inference procedures not available in Excel's Data Analysis pack.

- **ActivStats® for Business Statistics (Mac and PC)** by Data Description and Paul Velleman, is an award-winning multimedia program that supports learning chapter-by-chapter with the book. It complements the book with videos of real-world stories, worked examples, animated expositions of all major statistics topics, and tools for performing simulations, visualizing inference, and learning to use statistics software. ActivStats includes 15 short video clips; 183 animated activities and teaching applets; 260 data sets; interactive graphs, simulations, visualization tools, and much more. ActivStats (Mac and PC) is available in an all-in-one version for Data Desk, Excel, JMP, Minitab, and SPSS. (ISBN-10: 0-321-57719-1; ISBN-13: 978-0-321-57719-1)

○ **MathXL® for Statistics Online Course (access code required)** MathXL for Statistics is a powerful online homework, tutorial, and assessment system that accompanies Pearson textbooks in statistics. With MathXL for Statistics, instructors can:

- Create, edit, and assign online homework and tests using algorithmically generated exercises correlated at the objective level to the textbook.

- Create and assign their own online exercises and import TestGen tests for added flexibility.

- Maintain records of all student work, tracked in MathXL's online gradebook.

With MathXL for Statistics, students can:

- Take chapter tests in MathXL and receive personalized study plans based on their test results.

- Use the student plan to link directly to tutorial exercises for the objectives they need to study and retest.

- Students can also access supplemental animations and video clips directly from selected exercises.

MathXL for Statistics is available to qualified adopters. For more information, visit our Web site at www.mathxl.com, or contact your Pearson representative.

○ **MyStatLab™ Online Course (access code required)** MyStatLab—part of the MyMathLab® product family—is a text-specific, easily customizable online course that integrates interactive multimedia instruction with textbook content. MyStatLab gives you the tools you need to deliver all or a portion of your course online, whether your students are in a lab setting or working from home.

- **Interactive homework exercises**, correlated to your textbook at the objective level, are algorithmically generated for unlimited practice and mastery. Most exercises are free-response and provide guided solutions, sample problems, and learning aids for extra help. StatCrunch, an online data analysis tool, is available with online homework and practice exercises.

- **Personalized Study Plan,** generated when students complete a test or quiz, indicates which topics have been mastered and links to tutorial exercises for topics students have

not mastered. You can customize the Study Plan so that the topics available match your course contents or so that students' homework results also determine mastery.

- **Multimedia learning aids**, such as video lectures and podcasts, animations, and a complete multimedia textbook, help students independently improve their understanding and performance. You can assign these multimedia learning aids as homework to help your students grasp the concepts. In addition, applets are also available to display statistical concepts in a graphical manner for classroom demonstration or independent use.

- **Statistics Tools:** MyStatLab includes built-in tools for Statistics, including StatCrunch, a full online data analysis engine. For those who use technology in their course, technology manual PDFs are included.

- **Homework and Test Manager** lets you assign homework, quizzes, and tests that are automatically graded. Select just the right mix of questions from the MyStatLab exercise bank, instructor-created custom exercises, and/or TestGen test items.

- **Gradebook,** designed specifically for mathematics and statistics, automatically tracks students' results, lets you stay on top of student performance, and gives you control over how to calculate final grades. You can also add offline (paper-and-pencil) grades to the gradebook.

- **MathXL Exercise Builder** allows you to create static and algorithmic exercises for your online assignments. You can use the library of sample exercises as an easy starting point or use the Exercise Builder to edit any of the course-related exercises.

- **Pearson Tutor Center** (www.pearsontutor services.com) access is automatically included with MyStatLab. The Tutor Center is staffed by qualified math instructors who provide textbook-specific tutoring for students via toll-free phone, fax, email, and interactive Web sessions.

Students do their assignments in the Flash®-based MathXL Player which is compatible with almost any browser (Firefox®, Safari™, or Internet Explorer®) on almost any platform (Macintosh® or Windows®). MyStatLab is powered by CourseCompass™, Pearson Education's online teaching and learning environment, and by MathXL, our online homework, tutorial, and assessment system. MyStatLab is available to qualified adopters. For more information, visit www.mystatlab.com or contact your Pearson representative.

○ **PowerPoint Lecture Slides** provide an outline to use in a lecture setting, presenting definitions, key concepts, and figures from the text. These slides are available within MyStatLab or at www.pearsonhighered.com/irc.

○ **StatCrunch** is an online statistical software Web site that allows users to perform complex analyses, share data sets, and generate compelling reports of their data. Developed by programmers and statisticians, StatCrunch already has more than ten thousand data sets available for students to analyze, covering almost any topic of interest. Interactive graphics are embedded to help users understand statistical concepts and are available for export to enrich reports with visual representations of data. Additional features include:

- A full range of numerical and graphical methods that allow users to analyze and gain insights from any data set.

- Flexible upload options that allow users to work with their .txt or Excel files, both online and offline.

- Reporting options that help users create a wide variety of visually-appealing representations of their data.

StatCrunch is available to qualified adopters. For more information, visit our Web site at www.statcrunch.com, or contact your Pearson representative.

○ **The Student Edition of Minitab** is a condensed edition of the Professional release of Minitab statistical software. It offers the full range of statistical methods and graphical capabilities, along with worksheets that can include up to 10,000 data points. Individual copies of the software can be bundled with the text. (ISBN-10: 0-321-11313-6; ISBN-13: 978-0-321-11313-9)

○ **JMP Student Edition** is an easy-to-use, streamlined version of JMP desktop statistical discovery software from SAS Institute, Inc., and is available for bundling with the text. (ISBN-10: 0-321-67212-7; ISBN-13: 978-0-321-67212-4)

○ **SPSS (PASW),** a statistical and data management software package, is also available for bundling with the text. (ISBN-10: 0-321-67537-1; 978-0-321-67537-8)

Acknowledgments

We didn't develop our approach to business statistics in isolation. Our colleagues at Wharton have helped shape our approach to teaching statistics in business. Many of the ideas and examples that you'll find here arose from suggestions made by colleagues, including Andreas Buja, Sasha Rakhlin, Paul Shaman, and Adi Wyner. Over the years, members of our department have come to share a common attitude toward the use of statistics in business, and this text reflects that shared perspective. Most of the examples and many exercises from the text have been tried in other classes and improved using that feedback. We owe these friends a debt of gratitude for their willingness to talk about the fundamental use of statistics in business and to explore alternative explanations and examples.

Many thanks to the following reviewers for their comments and suggestions during the development of this text.

Kunle Adamson, *DeVry University*
Elaine Allen, *Babson College*
Randy Anderson, *California State University—Fresno*
Djeto Assane, *University of Nevada, Las Vegas*
Dipankar Basu, *Miami University*
David Booth, *Kent State University, Main Campus*
John E. Boyer, Jr., *Kansas State University*
Daniel G. Brick, *University of St. Thomas*
Nancy Burnett, *University of Wisconsin - Oshkosh*
Richard Cleary, *Bentley College*
Ismael Dambolena, *Babson College*
Anne Davey, *Northeastern State University*
Dr. Michael Deis, *Clayton University*
Frederick W. Derrick, *Loyola University Maryland*
Neil Desnoyers, *Drexel University*
Joan Donohue, *University of South Carolina*
Steve Erikson, *Babson College*
Nancy Freeman, *Shelton State Community College*
Daniel Friesen, *Midwestern State University*
Deborah J. Gougeon, *University of Scranton*
Christian Grandzol, *Bloomsburg University*
Betsy Greenberg, *University of Texas—Austin*
John Grout, *Berry College*
Warren Gulko, *University Of North Carolina, Wilmington*
Clifford B. Hawley, *West Virginia University*
Bob Hopfe, *California State University—Sacramento*
Max Houck, *West Virginia University*
David Hudgins, *University of Oklahoma*
Jeffrey Jarrett, *University of Rhode Island*
Chun Jin, *Central Connecticut State University*
Christopher K. Johnson, Ph.D. *University of North Florida*
Morgan Jones, *University of North Carolina*
Ronald K. Klimberg, *Saint Joseph's University*

David Kopcso, *Babson College*
Supriya Lahiri, *University of Massachusetts, Lowell*
Mark T. Leung, *University of Texas—San Antonio*
John McKenzie, *Babson College*
Mark R. Marino, *Niagara University*
Dennis Mathaisel, *Babson College*
Sherryl May, *University of Pittsburgh*
Bruce McCullough, *Drexel University*
Richard McGowan, *Boston College*
Constance McLaren, *Indiana State University*
Robert Meeks, *Pima Community College*
Jeffrey Michael, *Towson University*
Prakash Mirchandani, *University of Pittsburgh*
Jason Molitierno, *Sacred Heart University*
Carolyn H. Monroe, *Baylor University*
Patricia Ann Mullins, *University of Wisconsin—Madison*
Quinton J. Nottingham, *Virginia Polytechnic & State University*
Keith Ord, *Georgetown University*
Michael Parzen, *Emory University*
M. Patterson, *Midwestern State University*
Leonard Presby, *William Paterson University*
Darrell Radson, *Drexel University*
Farhad Raiszadeh, *University of Tennessee—Chattanooga*
Ranga Ramasesh, *Texas Christian University*
Deborah Rumsey, *The Ohio State University*
John Saber, *Babson College*
Dr. Subarna Samanta, *The College of New Jersey*
Hedayeh Samavati, *Indiana University—Purdue Fort Wayne*
Gary Smith, *Florida State University*
Erl Sorensen, *Bentley College*
J. H. Sullivan, *Mississippi State University*
Dr. Kathryn A. Szabat, *LaSalle University*
Rajesh Tahiliani, *University of Texas—El Paso*
Patrick A. Thompson, *University of Florida*
Denise Sakai Troxell, *Babson College*
Bulent Uyar, *University of Northern Iowa*
John Wang, *Montclair State University*
Dr. William D. Warde, *Oklahoma State University, Main Campus*
Elizabeth Wark, *Worcester State University*
James Weber, *University of Illinois—Chicago*
Fred Wiseman, *Northeastern University*
Roman Wong, *Barry University*
Zhiwei Zhu, *University of Louisiana at Lafayette*

We also thank our developmental editor, Anne Scanlan-Rohrer and accuracy checkers Ann Cannon and Frederick Derrick. Thanks also to our Pearson Education team for their help and support, especially: Greg Tobin, Deirdre Lynch, Alex Gay, Kathy Manley, Elizabeth Bernardi, Christina Lepre, Dana Jones, Kathleen DeChavez, Carol Trueheart, and Lin Mahoney.

INDEX OF APPLICATIONS

CO = Chapter Opener; **IE** = In-Text Example; **WT** = What Do You Think? **4M** = Motivation, Method, Mechanics, Message; **P** = Pitfalls; **BP** = Best Practices; **AD** = About the Data; **BTM** = Behind the Math; **TAI** = Think About It; **YDI** = You Do It; **SA** = Statistics in Action; **QT** = Questions for Thought

Science

Service Industries

Sports

Surveys and Opinion Polls

Technology

Transportation

PART I

Variation

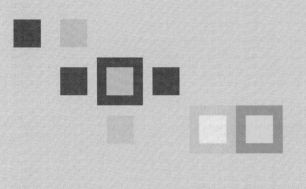

1.1 | WHAT IS STATISTICS?

What do you think statistics is all about?

Many people believe that statistics combines lots of numbers with formulas and tedious calculations. You've probably heard the famous quote, "There are three kinds of lies: lies, damn lies, and statistics." This statement suggests that statisticians are dishonest and use statistics to perpetuate dishonesty. That's not true, and it's not the way you should think about statistics.

Statistics answers questions using data, or information about the situation. A **statistic** is a property of data, be it a number such as an average or a graph that displays information. Statistics—the discipline—is the science and art of extracting answers from data. Some of these answers do require numbers and formulas, but you can also do every statistical analysis with pictures—graphs and tables. It's hard to lie or be fooled when the answer stares you in the face.

statistic A property of data.

Think of statistics as art. An artist must choose the right subject, and a good statistician finds the right picture. Rather than learning to paint, in this course you'll learn how to use statistics to interpret data and answer interesting questions.

Which questions are interesting? The answer is simple: those you care about. However, what interests one person may be of no interest to someone else. In this text we apply statistics to a mix of topics, ranging from finance and marketing to personal choices. Most of the questions in this book concern business, but statistics applies more generally. We'll help you appreciate the generality of statistics by solving problems from health and science, too.

Variation

What kinds of questions can be answered with statistics? Let's start with an example. In November 2006, Microsoft launched its MP3 player called the Zune to challenge the market-leading Apple iPod. The question facing Microsoft was, What's the right price for the Zune?

That's a hard question. To find an answer, you need to know basic economics, particularly the relationship among price, supply, and demand. A little finance and accounting help in determining the cost of development and production. Then come questions about the customers. Which customers are interested in an MP3 player? How much are they willing to pay? Does the Zune need to cost less than the iPod? How much less?

Suddenly, the initial pricing question branches into several questions, and the answers depend on whom you ask. There's **variation** among customers; customers react differently. One customer might be willing to pay $300 whereas another would pay only $200. Once you recognize these differences among customers, how are you going to set *one* price? Statistics shows how to use your data—what you know about your product and your customers—to set a price that will attract business and earn a profit.

Here's another interesting question: Why does a shopper choose a particular box of cereal? Modern grocers have become information-rich retailers, tracking every item purchased by each patron. That's why they give out personalized shopping cards; they're paying customers with discounts in return for tracking purchases. Customers keep retailers off balance because they don't buy the same things every time they shop. Did the customer buy that box of cereal because it was conveniently positioned at the end of an aisle, because he or she had a discount coupon, or simply because a six-year-old just saw a commercial while watching *Sponge Bob*? Again, variation makes the question hard to answer. If they find that coupons improve sales, store managers might decide to place more advertising in the local newspaper.

Patterns and Models

Statistics helps you answer questions by providing methods designed to handle variation. These methods filter out the clutter by revealing patterns. A **pattern** in data is a systematic, predictable feature. If customers who receive coupons buy more cereal than customers without coupons, there's a pattern.

Patterns form one part of a **statistical model**. A statistical model describes the variation in data as the combination of a pattern plus a background of remaining, unexplained variation. The pattern in a statistical model describes the variation that we claim to understand. The pattern tells us what we can anticipate in new data and thus goes beyond describing the data we observe. Often, an equation can summarize the pattern in a precise mathematical form. Background variation represents variation due to factors we cannot explain because we lack enough information to do so. For instance, retail sales increase during holiday seasons. Retailers recognize this pattern and prepare by increasing inventories and hiring extra employees. It's impossible, though, for retailers to know exactly which items customers will want and how much they will spend. The pattern does not explain everything.

Good statistical models simplify reality to help us answer questions. Indeed, the word *model* once meant the blueprints, the plans, for a building. Plans answer some questions about the building. How large is the building? Where are the bathrooms? The model isn't the building, but we can learn a lot from the model. A model of an airplane in a wind tunnel provides insights about flight even though it doesn't mimic every detail of flight. Models of data provide answers to questions even though those answers may not be entirely right. A famous statistician, George Box, once said, "All models are wrong, but some are useful."

A simple model that we understand may be better than a sophisticated model that we do not understand or have the means to test. A challenge in learning statistics is to recognize when a model can be trusted. Models based on physics and engineering often look impressively complex, but don't confuse being complex with being correct. Complex models fail when the science does not mimic reality. For example, NASA used the following elaborate equation to estimate the chance of foam insulation breaking off during take-off and damaging the space shuttle:

$$p = \frac{0.0195(L/d)^{0.45}(d)\rho_F^{0.27}(V - V^*)^{.67}}{S_T^{.25}\rho_T^{.16}}$$

This equation that represents an underlying model failed to anticipate the risk of damage from faulty insulation. Damage from insulation caused the space shuttle *Columbia* to break apart on reentry in 2003.

Models often fail because we've imagined them. People are great at finding patterns. Ancient people looked into the sky and found patterns among the stars. Psychiatrists use the Rorschach ink blot test to probe deep feelings. People even find patterns in clouds, imagining shapes or faces floating in the sky. This tendency to find patterns when nothing is there has a technical name: *pareidolia* or *patternicity* (from the book *Why People Believe Weird Things* by Michael Shermer).

A key task in statistics is deciding whether the pattern we think we see is real or something that we've imagined. Finding a pattern allows us to anticipate what is most likely to happen next, to understand the data in order to plan for the future and make better decisions. But if we imagine a pattern when there is none, we become overconfident and react inappropriately to the data.

1.2 | PREVIEWS

The following two examples preview statistics as theaters preview movies; they show the highlights, with lots of action and explosions, and save the character development for later. These examples introduce a couple of recurring themes and showcase several methods that are fully developed in later chapters. The point is to illustrate the types of analyses you will be able to do after you finish this book.

Each example begins with a question motivated by a story in the news, and then uses a statistical model to formulate an answer to the question. The first example uses a model to predict the future, and the second uses a model to fill in for an absence of data. These are previews, so we emphasize the results and skip the details.

Predicting Employment

> ## Job growth weaker than expected in October Reuters, November 4, 2005 _____
>
> WASHINGTON — Payrolls grew a smaller-than-expected 56,000 in October despite the fading impact of Hurricane Katrina, a Labor Department report Friday showed. Wall Street economists had forecast that 100,000 jobs would be created last month.

Financial markets don't react well to surprises. If everyone on Wall Street expects the Labor Department to report large numbers of new jobs, the stock market can tumble if the report announces only modest growth. What should we have expected? What made Wall Street economists expect 100,000 jobs to be created in October? Surely they didn't expect *exactly* 100,000 jobs to be created. Was the modest growth a fluke? These are serious questions. If the shortfall in jobs is the start of a downward trend, it could indicate the start of an economic recession. Businesses need to anticipate where the economy is headed in order to schedule production and supplies.

Was the weather responsible for the modest growth? On August 29, 2005, Hurricane Katrina slammed into Louisiana, devastating the Gulf Coast (see Figure 1.1). Packing sustained winds of 175 miles per hour, Katrina overwhelmed levees in New Orleans, flooded the city, and wrecked the local economy. Estimates of damages reached $130 billion, the highest ever attributed

FIGURE 1.1 Hurricane Katrina on August 29, 2005.

to a hurricane in the United States, with more than 1,000 deaths. Katrina and the hurricanes that followed during the 2005 season devastated the oil industry concentrated along the Gulf of Mexico and disrupted energy supplies around the country. Did Katrina wipe out the missing jobs?

Let's see if we can build our own forecast. Back in September 2005, how could you forecast employment in October?

We need two things to get started: relevant data and a method for using these data to address the question at hand. Virtually every statistical analysis proceeds in this way. Let's start with data. At a minimum, we need the number employed before October. For example, if the number of jobs had been steady from January through the summer of 2005, our task would be easy; it's easy to forecast something that doesn't change.

The problem is that employment does change. Table 1.1 shows the number of thousands employed each month since 2003. These are the data behind the story.

TABLE 1.1 Nonfarm employment in the United States, in thousands on payrolls.

	2003	2004	2005
Jan	130,247	130,372	132,573
Feb	130,125	130,466	132,873
Mar	129,907	130,786	132,995
Apr	129,853	131,123	133,287
May	129,827	131,373	133,413
Jun	129,854	131,479	133,588
Jul	129,857	131,562	133,865
Aug	129,859	131,750	134,013
Sep	129,953	131,880	134,005
Oct	130,076	132,162	134,061
Nov	130,172	132,294	
Dec	130,255	132,449	

Each column gives the monthly counts for a year. The first number in the table represents 130,247,000 jobs on payrolls in January 2003. The following number shows that payrolls in February 2003 fell by 122,000. At the bottom of the third column, we can see that employment increased by 56,000 from September to October 2005, as reported by Reuters. This variation complicates the task of forecasting. We've got to figure out how we expect employment to change next month.

We won't replicate the elaborate models used by Wall Street economists, but we can go a long way toward understanding their models by plotting the data. Plots are among the most important tools of statistics. Once we see the plot, we can decide what we need to do to make a forecast.

timeplot A chart of values ordered in time, usually with the values along the *y*-axis and time along the *x*-axis.

The graph in Figure 1.2 charts employment over time, a common type of display that we'll call a **timeplot**. To keep the vertical axis nicely scaled and avoid showing extraneous digits, we labeled the employment counts in millions rather than thousands. A good sense of rounding produces a better presentation.

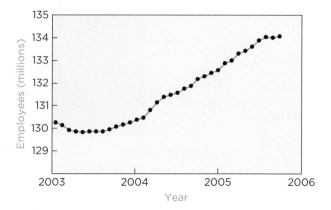

FIGURE 1.2 Timeplot of employment.

Once you've seen this timeplot, you can probably come up with your own forecast. Employment grew steadily during 2004 into 2005. This steady growth is a pattern. Employment varies from month to month, but the steady growth suggests a way to extrapolate the pattern into the future. The line drawn in the next figure summarizes the pattern we see in the data and suggests a forecast for October.

Models help us simplify complicated phenomena, but that does not make them correct. The past data on employment follow along a line after 2004, but that does not mean that this pattern will continue. It's not as though the economy knows about the line in Figure 1.3 and must follow this trend. This line allows us to anticipate where employment is headed, but reality may not match our forecast. Consider what would happen if we were to extend this pattern farther back in time. The line would underpredict employment in 2003 by a huge margin.

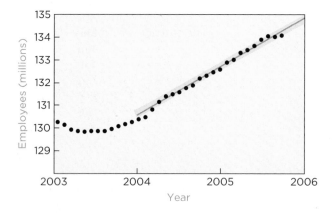

FIGURE 1.3 Linear pattern and region of uncertainty.

We use ranges to convey the uncertainty associated with using a statistical model. Ranges are used in many fields to indicate uncertainty. For example, Figure 1.4 shows the uncertainty around the projected path for Katrina before it made landfall. The wider the cone, the more doubt about the path of the storm.

Similarly, the shaded region around the line in Figure 1.3 indicates how closely the historic data stick to this pattern. In addition, this region suggests the amount of unexplained variation around the pattern. The more closely the

FIGURE 1.4 Projected path for Katrina.

data follow the pattern, the narrower this region becomes. The employment data track the line very closely during the period summarized by the line.

To get a forecast, we extend the pattern. It is easy to extend the line beyond the data, as shown in Figure 1.3. The line passes above the count for October. This model predicts employment to be near 134,150,000 in October 2005, about 90,000 more than the reported count. That's close to the value claimed in the Reuters article at the beginning of this subsection. We can also extend the region of uncertainty around the line. We should not expect counts of employment in the future to be closer to the line than those in the past. The line—our pattern—forecasts employment to be near that predicted by Wall Street economists. On the basis of this pattern, we would forecast employment in October to lie between 134,130,000 and 134,533,000 jobs.

Our simple model confirms that the level of employment in October 2005 *is* surprising. Not only is the reported count of 134,061,000 employed for October less than expected, but it's below the anticipated range. Even allowing for variation around the pattern, employment is smaller than expected. Something happened in October that reduced the number of jobs. This is a large break from the pattern and demands our attention. Do we know *why* the employment was less than expected?

Anticipating the impact of weather on employment is an ongoing concern. Again, in the fall of 2008, Hurricanes Ike and Gustav were blamed for putting 50,000 Americans out of work. The only way to arrive at these estimates is to predict what would have happened had these hurricanes not struck.

Pricing a Car

For our second preview, let's take a look at this story from the Web site of *Car and Driver* magazine.

10Best Cars: Best Luxury Sports Sedan
BMW 3-Series Car and Driver, January 2009

What are the key elements of automotive perfection? From our perspective, the list of qualifications includes eager responses, supple ride quality, smooth power, supportive seats, athletic proportions with limited front overhang, attractive styling with familial features that endure through the generations, a car that is always entertaining to drive.

For us, the sum of those attributes is epitomized by the *BMW 3-series*. Not only is this true for 2009, the addition has been coming out the same way now for 18 years: a string of consecutive *10Best Cars* appearances that's unique in the 27-year history of these awards.

You may not believe the advertising slogan that touts a BMW as the "ultimate driving experience," but you've probably heard of BMWs. The design has been so successful that *Car and Driver* chose this model as one of its 10 Best for 18 consecutive years.

What's it going to cost to get behind the wheel of a BMW 3-series? The manufacturer's suggested retail price for a basic 2009 BMW 328i is $33,400, excluding options like leather seats that add $1,500 to the bottom line. That's a lot to spend, so let's see what a used BMW costs. For example, a classified ad in the *Philadelphia Inquirer* offered to sell a 2002 BMW 325i for $22,000. The car has 53,000 miles, an automatic transmission, and a variety of performance options.

It sounds like a lot to spend $22,000 on a 2002 BMW 325i with 53,000 miles. On the other hand, this price is more than $10,000 less than the cost of a stripped-down new car that lacks options found on this car.

Companies face similar decisions: Should a company spend more to buy new equipment, or should it look for a deal on used substitutes? As for cars, auto dealerships face these questions every day. What should the dealership charge per year on a three-year lease? The ultimate resale value has a big effect on the annual cost. If, when the car comes back, the dealership can sell it for a good price, then it can offer a better deal on a lease.

Once again, we need to identify data and decide how to use them. To get data on the resale value of used BMWs, we downloaded prices for certified used cars from BMW dealers within 50 miles of Philadelphia in November 2005. We found 218 cars in the 3-series, ranging from the 2001 through the 2006 model years and categorized by style in Table 1.2. Of these, 29 are 325i models from 2002, just like the advertised car. On average these were priced at $26,323; other styles have higher average prices.

TABLE 1.2 Average prices vary by style.

Style	Number	Average Price
325ci	14	$28,466
325i	**29**	**26,323**
325xi	21	28,291
330ci	8	32,294
330i	14	30,729
330xi	12	31,296

Are the 29 cars in the 325i style similar enough to the car in the classified ad? You can probably name other things that affect the price. We've taken into account the age (2002 model year) and style (325i), but not the mileage. We'd guess you'd be willing to pay more for a car with 10,000 miles than an otherwise similar car with 100,000 miles.

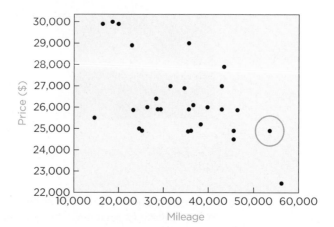

FIGURE 1.5 Scatterplot of price versus mileage.

scatterplot A graph that shows pairs of values (*x*, *y*) laid out on a two-dimensional grid.

The **scatterplot** in Figure l.5 shows that mileage is related to price. Each point shows the mileage and price of one of these 29 cars. None of these cars has exactly the same mileage as the advertised car. Only one, shown circled in red, has mileage close to that of the advertised car. It costs $25,000. By the time we find cars that match the age, style, and mileage of the car in the classified ad, we're down to one.

We can use a statistical model to compensate for our lack of matching cars. The plot in Figure 1.6 shows a line that relates mileage to price. The plot includes the region of uncertainty around the line. The region of uncertainty is so wide because other factors, such as the condition of the car and its options, affect the price. There's much more unexplained variation around this pattern than around the line shown in Figure 1.3.

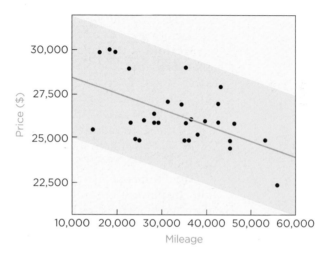

FIGURE 1.6 A line relates mileage to price for these cars.

borrowing strength The use of a model to allow all of the data to answer a specific question.

In our first example, we extrapolated a pattern in historic data. In this example, the pattern serves a different purpose. This line allows us to borrow strength. **Borrowing strength** refers to using a model to glean more data for answering a question. Without a model, only one car seems relevant to this pricing question. By using a line rather than matching, we can use data for 29 cars. Even though the mileage varies among the 29 cars, the pattern allows us to "borrow" some of the information about each car to estimate the price of a car with specific mileage. The estimated value, reading off the height of the line, is $24,600. That's $400 less than what we found by matching. As an added bonus, the negative slope of the line in this graph shows how mileage reduces the value of a car. On average, cars of this type lose about 10 cents of value per mile.

Having seen this analysis, do you think the car in the classified ad is a bargain? Its $22,000 price is less than the predicted price from our model, but well within the range for cars of this age and mileage. It might be worth a further look, but don't forget that cars from dealers generally come with warranties that might reduce the variation in your future expenses!

1.3 | HOW TO USE THIS BOOK

Each chapter in this book starts with a question. This initial question illustrates an issue that arises in business situations, such as, Who are our best customers? What is the quality of our manufacturing process? What are the chances that we will not have enough inventory on hand to meet demand? Following this motivating question, an introductory paragraph summarizes how the methods presented in the chapter will help us answer this question and others like it. The rest of the chapter develops these statistical methods and presents an answer to the initial question.

The Four Ms

The only way to learn statistics is to do it. To help you learn how to do statistics, every chapter includes a variety of examples. One or two examples run throughout the chapter, introducing a new method or illustrating a concept. Other, shorter examples allow us to review the use of the new method in a different context. To convey the structure of an analysis using the methods of a chapter, we present these examples as a four-step process: *motivation, method, mechanics,* and *message*.

Motivation is the reason for solving a problem or answering a question. In business, understanding the problem usually involves money and the underlying economics. Most companies are in business to make a profit, and decisions that affect profits require choices. How will your choice affect sales, profits, or costs? You must appreciate the implications of your decision. Motivation often reduces to estimating how much money you might make or lose.

Method is your plan for addressing the problem. Think of this as a map that describes how you are going to get from the question stage to the answer stage. Your method has two key steps: identifying the relevant data and picking the best statistical technique, or tool, to analyze the data. Often, the *method* reduces to choosing the right picture, or graph. If you show the right picture of the data to someone familiar with the context, she or he will be able to make a decision. Often the method requires identifying a model.

Mechanics means performing the calculations and generating the graphs and displays. Computers now perform most of the calculations that once dominated a statistics course. We'll do some calculations by hand so you can appreciate how the concepts work, but let software handle the details for more extensive examples.

Message is the explanation of your conclusions. Until you've explained your results in a way that others can appreciate, your job is not done. You need to explain your conclusions to coworkers so that they appreciate the implications. Displays often provide the essence of a compelling story.

Getting your message across requires awareness of what your audience knows and expects. When your results fly in the face of conventional wisdom, you should also be prepared to defend your analysis and conclusions. If you can't get your message across, your colleagues won't recognize your contribution.

Other Features

You'll find sections called **What Do You Think?** in each chapter. These sections help you check that you understand the important ideas; these sections pose questions that you should be able to answer before you continue in the chapter. The answers are provided in a footnote so you can see how you are doing. If you have trouble with these questions, you may have missed a key idea and should review the material before going on.

Some statistics presume that the information presented satisfies several conditions or assumptions. For example, certain statistics only detect patterns that resemble lines. You would not want to use these if you were looking for a curve. To help you keep track of the assumptions, the conditions are collected in a **Checklist**.

Each chapter concludes with a list of **Best Practices** that you ought to be doing and a list of **Pitfalls** that you should avoid. Most of the unintentional mistakes people make are avoidable and come from using the wrong method for the situation or misinterpreting the results.

Although we show you all the formulas required for the calculations in this book, you'll most often use software to perform the mechanics of an analysis. The easiest way to calculate statistics with a computer is with a specialized program called a statistics package. There are a number of packages available, and they differ widely in how they are used and in how they present their results. But they all work from the same basic information and find the same results. Rather than adopt one package for this book, we'll present generic output and point out features that you should look for in programs.

After the lists of Best Practices and Pitfalls, each chapter has a section called **Software Hints**. This section introduces relevant commands used by three popular tools for doing statistics on the computer (Excel, Minitab, and JMP). These hints might be all you need, but they are just hints that point in the right direction. If you are unfamiliar with the software, chances are that you will want to read more of its documentation to get the most out of it. To navigate the documentation with greater ease, search for the names of key commands that are mentioned in the Software Hints. Because Excel, unlike Minitab and JMP, is a general-purpose spreadsheet program rather than a statistics package, the hints for Excel include directions for using DDXL. DDXL supplements Excel by adding specialized commands for statistics.

A **Chapter Summary** follows the Software Hints. This summary includes a list of **Key Terms** from the chapter. Each of these key terms is shown in **boldface** when it is introduced in the chapter and accompanied by a definition in the margin. You've seen a few of these already in this chapter. The concluding Chapter Summary also includes formulas that appear in the chapter.

Along with the formulas in each chapter, we've included a section called **Behind the Math**. The text refers you to these sections with a remark and a marginal icon like the one shown here. These sections collect technical details that explain important results, such as the mathematical justification for an underlying formula or comment. These sections will help you appreciate the mathematics behind the methods, but they are not necessary for using statistics.

caution

tip

Two other icons appear in the margins. The caution sign indicates that you should be extra careful to make sure you understand the material being discussed. The tip icon **highlights a subtle idea or hint that you might otherwise overlook.**

A final section in each chapter called **About the Data** describes the data used in the chapter and provides source information (e.g., Web sites) so you can learn more about the data.

Finally, each chapter (except this one) concludes with **Exercises**. You cannot learn statistics without doing statistics. We've grouped the exercises into five stages. The first stage consists of *Mix and Match* exercises, and the second stage consists of *True/False* questions. Both groups of exercises test whether you recognize symbols and important steps of calculations. We avoid unnecessary formulas, but certain symbols and terminology show up so often that you'll be well served to recognize them. These details are important, but not so important that you need to memorize long lists. It's more important that you are able to recognize symbols and match concepts to problems. Questions in the *Think About It* section encourage you to pull together concepts and perhaps extend the ideas of the chapter. You don't need a computer or calculator for most of these questions.

You Do It exercises do require a computer unless you're really good with pencil and paper. These exercises apply the methods of the chapter, often to data related to a business application. Working through the steps of these questions helps you practice the mechanics. We expect you to use a statistics software package for many of these exercises.

4M questions have a richer, more substantive orientation. These exercises come closer to real applications of statistics in business than do the other exercises. You'll find the data for the 4M exercises and the You Do It exercises on the CD in the back of the book.

Questions in life do not come with a chapter label. You have to figure out for yourself which ideas are relevant to a particular statistical problem. That can be the hardest part of solving a problem: Which method should I use here? A **Review Test** at the end of each part of the book poses questions similar to those you have to deal with in the real world. These questions will help you determine when to use which method and to identify weak spots that you may have missed on your first reading.

Statistics in Action cases follow each of the four main parts of this book. These applications combine and sometimes expand on the methods you've learned and use them to examine an important topic. Statistics in Action cases delve into substantive aspects of the problem. Rather than emphasize methods, each case study focuses on the problem itself. These cases show other ways to use the methods that we've introduced in the chapters, but the plots and summaries are the same as those presented in the chapters.

Calculations

This book teaches you how to use data to answer questions that arise in business. The key to success involves choosing and using the correct method for each problem and then explaining your results clearly. You've got to find the right data, pick the appropriate method, and then communicate the results.

A part of learning these skills requires calculation. Often the calculation confirms that you understand what you're doing. Other times, it's an essential part of the answer. When we present results obtained with a calculator or computer, we typically round them. You don't need to know that the profits from a projected sale are $123,234.32529. It's usually better to round such a number to $123 thousand. To let you know when we've rounded a calculation, we say *about* or *approximately*. In expressions, we denote rounding with the symbol $\approx$, as in $1/6 \approx 0.167$.

Data

RUNNING A BUSINESS IS HARD WORK. One way to succeed in retail is to offer something missing from the big-box stores like Wal-Mart or Target, such as a personal touch. Store clerks once knew each customer, but few such stores survive because most of us prefer the lower prices and wider selections that we find at discount centers or on the Internet.

Paul Molino, owner of Bike Addicts in Philadelphia, loves to ride bicycles and hang out with riders who share his passion for cycling. What could be better than running his own bike shop? At first, it was easy to keep up with his customers; he talked with them every weekend when the friends met to ride.

That got harder as the business grew. Without that personal touch, why would a customer with plans to spend $6,000 for an Italian bike come to him instead of shopping online? Perhaps because Paul knows what each customer wants—favorite brands, sizes, and styles—just like those old-fashioned clerks.

Databases allow modern businesses to keep track of vast amounts of information about customers. With a database, Paul can keep track of preferences for brands, sizes, and even colors. A database of purchases also comes in handy at the end of the year when it's time for inventory. After all, Paul would rather be riding than counting parts.

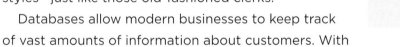

Statistics relies on data, information that describes the world around us. If these data are to be useful, however, we need to organize them. That's what this chapter is about: organizing data in a way that allows us to perform a statistical analysis. Most statistical data are organized into tables in which the rows and columns have specific meaning. Part of making sense of data involves recording what those values mean.

2.1 | DATA TABLES

Data are a collection of numbers, labels, or symbols and the context of those values. Often these values are a subset of a much larger group. The data might consist of the purchases of a few customers, the amounts of several invoices, or the performance ratings of several employees. Data can also arrive sequentially, such as the closing price of a stock each day or the monthly sales at a Web site.

Whatever the source, data are more than numbers or names. For example, one value in Paul's records is 61505, but without a frame of reference, a context for this value, this number is useless: 61505 might stand for a purchase of $615.05 or perhaps $6,150.50; some bicycles are that expensive. Figure 2.1 shows a few more values from Paul's shop.

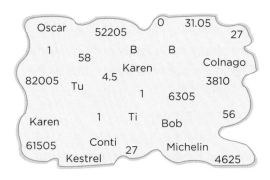

FIGURE 2.1 Disorganized data hide their meaning.

When they are scattered about with no structure, we cannot tell what these numbers and names mean. They aren't going to be useful for helping Paul to run his business. He won't be able to figure out which sizes are most popular until the data are organized.

When organized into a table with helpful column titles, as in Table 2.1, the values begin to take on meaning. This simple arrangement makes the data interpretable. The values in Table 2.1 describe four items bought at Paul's shop. Each row describes a purchased item, and each column collects a property of the items. The columns identify *who* made the purchase, *what* was bought, and *when* the purchase occurred. The first row, for example, tells us that Oscar bought a size 58 Colnago bike on May 22, 2005. Tables such as this are often called **data tables**, data sets, or data frames. Most statistical software organizes data into a table like this one. Spreadsheet programs like Excel do the same.

data table A rectangular arrangement of data, as in a spreadsheet. Rows and columns carry specific meaning.

TABLE 2.1 Same data as in Figure 2.1, but in a table.

Customer	Club Member	Date	Type	Brand	Size	Amount
Oscar	0	52205	B	Colnago	58	4625
Karen	1	6305	Tu	Conti	27	4.50
Karen	1	61505	Ti	Michelin	27	31.05
Bob	1	82005	B	Kestrel	56	3810

This table is a huge improvement over the jumble in Figure 2.1, but it omits important properties of the data. What does it mean that Oscar has a 0 in the second column? We can guess, but only Paul would know for sure. To remove the ambiguity, we should use more meaningful names, improve the formatting, and add units, as in Table 2.2.

TABLE 2.2 Units and formatting make data self-explanatory.

Customer	Club Member	Date	Type	Brand	Size	Amount
Oscar	No	5-22-05	Bike	Colnago	58 cm	$4,625.00
Karen	Yes	6-3-05	Tube	Conti	27 in	$4.50
Karen	Yes	6-15-05	Tire	Michelin	27 in	$31.05
Bob	Yes	8-20-05	Bike	Kestrel	56 cm	$3,810.00

The meaning of the second column is clear. A 0 in the second column of Table 2.1 means that the customer who made the purchase is not a club member; a 1 means that the customer is a club member. Similarly, a B in the fourth column of Table 2.1 stands for Bike. Table 2.2 also formats dates and dollar amounts. Presenting dates and dollars in a familiar format makes the values self-explanatory. Now that we're looking at data, we can begin to learn about these customers. For example, it seems that Karen is having problems with flat tires.

Rows and Columns

The rows of a data table describe a collection of things, with one row for each thing. Each row in the data for Bike Addicts denotes a purchased item. The rows are often called **observations** (as in "*observed* items purchased at Bike Addicts") or **cases**. The cases in Paul's data are the four items bought by Bob, Oscar, and Karen. Other names for the rows signify special situations: People who participate in a survey are respondents; people in a medical study are subjects or participants. In a database, rows are called records—in this example, purchase records or transaction records.

observation, case Names given to the rows in a data table.

Each column of a data table describes a *common attribute* shared by the items. Columns in a data table are usually called **variables** since we expect to find different values within a column. Consider the amount column in Paul's data. Each row describes a different item, and the amount paid varies from item to item.

variable A column in a data table that holds a common attribute of the cases.

The name of a variable should describe the attribute. Avoid abbreviated names like V1, V2, V3, and so forth. An acronym like DOLPTIS (for Dollars Per Transaction In Store) might make sense today but be puzzling in a month.

The number of rows in a data table is almost universally denoted by the symbol n. For the small data table in Table 2.1, $n = 4$. There's no consensus on a symbol for the number of columns, but you will see the symbol n for the number of rows in a data table over and over.

tip A universal convention among statistics packages that handle computing is that **the observations form the rows and the variables form the columns.** If your data table is arranged the reverse way, most software includes a function called transpose that turns rows into columns and columns into rows.

2.2 | CATEGORICAL AND NUMERICAL DATA

Variables come in several flavors, depending on the data in the column. Most variables consist of either **categorical** or **numerical** data. Categorical variables are sometimes called qualitative or nominal variables, and numerical variables are sometimes called quantitative or continuous variables. You need

categorical variable Column of values in a data table that identifies cases with a common attribute (also called nominal variable).

numerical variable Column of values in a data table that records numerical properties of cases (also called continuous variable).

measurement unit Scale that defines the meaning of numerical data, such as weights measured in kilograms or purchases measured in dollars.

to be able to recognize the measurement scale of a variable in order to choose the right method of analysis.

Categorical variables identify group membership. The labels within the column name the categories. Most of the columns in Table 2.2 are categorical variables. The first column names the customer making the purchase. The fourth column names the item.

Numerical variables describe quantitative attributes of the rows. Familiar examples include incomes, heights, weights, ages, and counts. Numerical variables allow us to perform calculations that don't make sense with categorical variables. Paul certainly wants to know the total amount of purchases made each day at his shop, but not the "average name" of his customers.

Numerical variables have **measurement units** that tell how the variable has been measured. The most important units in business denote currency, such as dollars $, euros €, or yen ¥. Lacking units, the values of a numerical variable have no meaning. The data that make up a numerical variable in a data table must share a common unit. While this convention seems routine, mixing numbers with mismatched units can have disastrous consequences. The $200 million *Mars Climate Orbiter* crashed on its way to Mars because some of the numbers it processed were in a metric scale but others were in English units.

Measurement Scales

The classical system for labeling the data held by a variable refines categorical and numerical data into four categories: *nominal, ordinal, interval,* and *ratio*. Nominal and ordinal variables are both categorical. Nominal variables name categories without implying an ordering, whereas the categories of an ordinal variable can be ordered. Interval and ratio variables are both numerical. It makes sense to add or subtract interval variables; ratio variables allow multiplication and division as well. For instance, temperature in degrees Fahrenheit is an interval variable. We can add and subtract temperatures, but ratios don't make sense. Consider two days, one at 40° and the other at 80°. The difference in temperature is 40°, but it would not make sense to say that the warmer day is twice as hot as the cooler day. The point 0° does not mean the total absence of heat. In this text, we avoid the distinction between interval and ratio scales, and we call both numerical.

Some variables seem to mix both numerical and categorical data. The data define categories, but the categories have a natural order. For example, a customer survey might ask, How would you rate our service? 5 = Great, 4 = Very good, 3 = Typical, 2 = Fair, and 1 = Poor. Higher numbers indicate greater satisfaction. You would probably prefer doing business with a firm that earns an average rating of 3.5 over another firm with an average rating of 2.5.

ordinal variable A categorical variable whose labels have a natural order.

Because the labels that define this rating can be ordered, such variables are called **ordinal variables**. Even though it is common to treat ordinal variables as numerical and do things like find the average rating, be careful about how you interpret the results. Does the difference between ratings of 4 and 5 mean the same thing as the difference between ratings of 2 and 3? Is a hotel with average rating 4 twice as good as a hotel rated 2?

Many of the questions in surveys produce ordinal variables. Surveys frequently ask respondents to rate how much they agree with a statement. The result is an ordinal variable. For example, a question might look like this.

This MP3 player has all of the features that I want.

Strongly disagree		Undecided				Strongly agree
1	2	3	4	5	6	7

Likert scale A measurement scale that produces ordinal data, typically with 5 to 7 categories.

The question usually offers 5 to 7 choices. This approach to gauging attitudes was developed by Rensis Likert in 1932.[1] A **Likert scale** produces an ordinal variable even though numbers label the choices.

What Do You Think?

The second preview example in Chapter 1 considered several variables that describe a used BMW. These factors included the price, mileage, model year, and model style (denoted by the letters i, ci, and xi) of 218 used BMWs.

a. If we organize these four characteristics into a data table, what are the rows and what are the columns?[2]
b. What's n?[3]
c. Which variables are numerical and which are categorical? Do any seem to be both numerical and categorical?[4]
d. What are the units of the numerical variables?[5]

2.3 | RECODING AND AGGREGATION

Sometimes we have to rearrange data into a more convenient table. For instance, consider a data table that records purchases of clothing at a retail store. The variable column that describes each item might include detailed labels such as Polo shirt, white, M; Polo shirt, red, L; or Polo shirt, pink, S; and so on. Such a column is a categorical variable, but one with hundreds of distinct categories, one for each combination of color and size.

These distinctions are essential for managing inventory, but managers are usually more interested in a column with fewer categories. Rather than distinguish Polo shirts by size and color, for example, we could **recode** this column by using it to define another column that only distinguishes sizes. For a categorical variable, recoding produces a new column that consolidates some of the labels to reduce the number of categories. A manager could then use this new column to see which sizes are most popular overall.

recode To build a new variable from another, such as by combining several distinct values into one.

Recoding can also be applied to numerical variables. Recoded numerical variables are typically ordinal and associate a range of values with a single label. Rather than look at the prices of individual items, for instance, Paul at Bike Addicts might want to distinguish expensive items from others. Starting from the purchase prices, he could define an ordinal variable that indicates if the item costs more than $1,000.

Paul at Bike Addicts tracks *every* purchased item in order to maintain his inventory. At times, however, he'd rather see summaries, such as how many tires were sold, rather than a list of each transaction. To answer such

---

[1] "A Technique for the Measurement of Attitudes," *Archives of Psychology*, 1932, Volume 140, 1–55.
[2] The rows identify cars, and the columns identify the four variables: price, mileage, age, and style.
[3] $n = 218$, the number of cars.
[4] The price, mileage, and age are numerical. The style is categorical. We might think of the age as categorical if we use it as in Chapter 1 to group the cars. BMW might treat the style as ordinal if some styles generate higher profits!
[5] Price is measured in dollars, mileage is measured in miles, and age is measured in years.

aggregate To reduce the number of rows in a data table by counting or summing values within categories.

questions, he can **aggregate** his data into a smaller table by summing values or counting cases within categories. Aggregation generates a new data table with *fewer* rows. A row in an aggregated table, for instance, could count the number of purchases and sum the total value of the purchases made day by day. Table 2.3 shows a portion of the original transaction data table and the aggregated data built from this portion (the actual data table of transactions has many more rows).

TABLE 2.3 Aggregating the data table of transactions produces a new table with fewer rows that summarizes the transactions for each day.

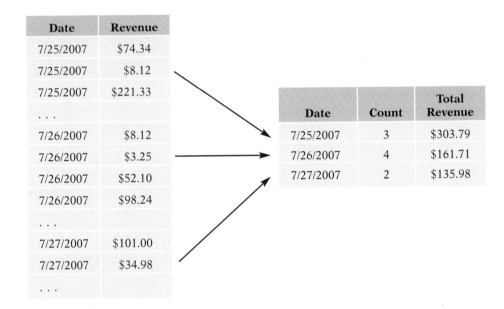

Date	Revenue
7/25/2007	$74.34
7/25/2007	$8.12
7/25/2007	$221.33
. . .	
7/26/2007	$8.12
7/26/2007	$3.25
7/26/2007	$52.10
7/26/2007	$98.24
. . .	
7/27/2007	$101.00
7/27/2007	$34.98
. . .	

Date	Count	Total Revenue
7/25/2007	3	$303.79
7/26/2007	4	$161.71
7/27/2007	2	$135.98

MEDICAL ADVICE

MOTIVATION

STATE THE QUESTION

Private doctors often contract with health maintenance organizations (HMOs) that provide health care benefits to employees of large firms. Typically, HMO plans require a copayment from the patient at the doctor's office. This copayment, along with fees paid by the insurer, keeps the doctor in business. To reduce paperwork, the insurer pays the doctor a fixed amount based on the number of patients in the care of that doctor, regardless of whether they visit the office.

This business model presents problems. To be profitable, the doctor (or office manager) needs to answer a few simple questions such as: Are patients from one HMO more likely to visit the doctor than those from another HMO? If so, the doctor may negotiate higher fees for covering patients from the plan with patients who are more likely to visit.

Many physicians still track their patients' records on paper. We can advise a doctor's office about setting up the data to compare the patients from different HMOs. There's not much that we can do with the data yet, but we can think about the big picture so that we have the data ready.

METHOD | **DESCRIBE THE DATA AND SELECT AN APPROACH**

Let's describe the data table. We start by conceiving a plan to collect and organize the data for these comparisons. We may eventually need help from computer programmers, but we first have to choose which data to gather. To organize the data, we'll put them into a table. The cases that make up the rows of the table will be office visits. The variables, or properties of these cases, will define the columns.

MECHANICS | **DO THE ANALYSIS**

We cannot show all of the data, but we can figure out how to organize portions of them. Regardless of what else gets added, the data table for this example has at least three columns: a record of a patient identifier, the name of the HMO plan, and the length of a patient's visit in minutes. The length of the visit is a column of numerical data, and the other two columns are categorical.

The following table lists 8 cases obtained from the data collection:

Patient ID	HMO Plan	Duration
287648723	Aetna	15
174635497	Blue Cross	20
173463524	Blue Cross	15
758547648	Aetna	25
837561423	Aetna	20
474739201	Blue Cross	20
463638200	AARP	30
444231452	AARP	25

Although the patient ID looks numerical, these numbers label the cases. This variable is categorical (a category identifies each patient). The HMO plan is also categorical. The durations in the third column are numerical but appear to have been rounded to multiples of five minutes. We might need to check that the staff actually recorded the duration rather than taking it from a preliminary schedule.

MESSAGE | **SUMMARIZE THE RESULTS**

The data table for answering questions about the length of time spent with patients should have at least three columns: one identifying the patient, one identifying the HMO, and one recording the length of the visit. Once these data become available, we can aggregate the counts to learn whether patients from one plan are consuming most of the doctor's office time. We will see several ways to compare these numbers in the next few chapters.

It's good to include concerns about your data in your summary message so that others are aware of possible limitations. We should contact the administrator to see how the durations of visits were measured. Our data seem to have been rounded. We should also find out how these records were chosen. Is the office going to supply us with data for every patient visit or just some? If they're only going to give us data that describe a portion of the visits (called a sample), then we need to find out how they picked this portion. Otherwise, our results might turn out to be an artifact of how the subset of records was selected. The right way to identify a subset requires some ideas that we'll come to in Chapter 13.

Gathering and organizing data is an important step of any statistical analysis. Don't be too hasty with this task. A statistical analysis is no more reliable than the data it is based on. You don't want to spend hours analyzing data and writing up your report only to discover that your data do not measure what you thought they measured.

2.4 | TIME SERIES

time series A sequence of data recorded over time.

A **time series** is a sequence of data that records an attribute at different times. Examples of time series are the daily prices of Microsoft stock during the last year, the dollar value of purchases of a regular customer at Wal-Mart, and the monthly employment counts reported by the Bureau of Labor Statistics. A **timeplot** graphs a time series by showing the values arranged in chronological order. This timeplot in Figure 2.2 from the Federal Reserve Bank of St. Louis shows the percentage unemployed monthly in the United States.

timeplot A graph of a time series showing the values in chronological order.

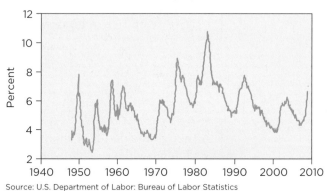

Civilian Unemployment Rate

Source: U.S. Department of Labor: Bureau of Labor Statistics
2008 Federal Reserve Bank of St. Louis: research.stlouisfed.org

FIGURE 2.2 Timeplot of the monthly unemployment rate in the United States.

Time series data are arranged in a table just like other types of data. Time series track stock prices, dollars in sales, employment, and other economic and social variables. The variables can be categorical or numerical.

The rows of time series data, however, carry a different meaning from those in previous examples. In place of different transactions, people, or countries, a time series records the changing values of a variable over time. The rows measure the variable over and over at different points in time. Each day, for instance, we can record the price of Microsoft's stock. Every row describes this same stock but at different points in time.

Some time series come from aggregating. For example, the purchases at Bike Addicts in Table 2.2 include the date. If we aggregate the data table by summing the purchased amounts for each day, we can form a new data table that has a time series of the daily total purchases. When produced by aggregation, a time series has a regular spacing known as its **frequency**. Each

frequency The time spacing of data recorded in a time series.

purchase at Bike Addicts happens at a unique time, and there's no reason for purchases to happen regularly. Once aggregated to a daily total, however, the rows progress regularly, one day at a time. The time that separates the rows tells us the frequency of measurement.

cross-sectional data Data that measure attributes of different objects observed at the same time.

If data measured over time are time series, do we have a special name for data *not* recorded over time? Not usually. If we want to be explicit, however, we can describe data like those stored for Paul's items as **cross sectional data**. His data record properties of a cross section of purchased items.

2.5 | FURTHER ATTRIBUTES OF DATA

We must understand the rows and columns of the tables to analyze data. Other characteristics put this information into context. For example, it can be useful to know *when* and *where* the data were collected. A time or date can be relevant even if data are cross sectional. The prices of the used BMWs in the second preview example of Chapter 1 are not a time series, but it's important to know when these data were collected. The price of a 2004 BMW offered for sale in 2006 differs from the asking price of the same car four years later. Location also matters; prices in the Northeast differ from those in the South.

The *source* of data may be useful to know as well, particularly if the data are available online. Unless you gathered the measurements yourself, it's good to give credit (or assign blame) to those who did.

It is also useful to know *why* data are collected. The reason behind a survey might lead us, for example, to question the validity of the information. If we know that Levi Strauss conducted a survey, we may be skeptical when we hear that 95% of students expect Levi's 501 jeans to be the most popular clothing item on campus.

It might even be important to know *how* data were collected. For example, after a steep rise in oil prices, Congress created the U.S. Department of Energy (DOE) in 1977 and tasked this agency with keeping an eye on the supply and price of gasoline. At one point, DOE collected three time series, all claiming to measure the amount of gasoline used each month. When an analyst looked at the data, he found that the series occasionally differed by 600,000 gallons of gasoline a day! The explanation was simple: Each series defined gasoline differently. One originated in a survey of gas stations, another came from reports from refiners, and the third was derived from state taxes on gasoline.

Sources of Data

Whether you're searching for a new suit or the latest numbers on retail sales at Wal-Mart, it's hard to beat Google. Common Web sites that we used in gathering the data that appear in this book include a data repository run by the Federal Reserve Bank in St. Louis (FRED2) and the sources at the Bureau of Labor Statistics. Look for files labeled "ascii" or "csv" for data that you can import into Excel or a statistics package. There's also a project known as Data Ferret to simplify access to a variety of federal data sources. (You can download this software from the Census Web site.) You won't even need to download the data; this application can build tables and compute various summaries.

Once you exhaust sites run by the government, the going gets harder, particularly if you want to find data related to a specific company. Data used for strategic decision making are closely held corporate secrets. To assemble data about a company, look in the business portions of Yahoo or Hoover's online. Unless you are prepared to pay for data, commercial sites might make you work to get all the data you want. After all, data are valuable!

4M EXAMPLE 2.2 CUSTOMER FOCUS

MOTIVATION ▸ STATE THE QUESTION

Marketing studies use a focus group to test the reaction of consumers to a new product. A company shows its product to a few selected customers—the focus group—and measures their reactions. For example, an electronics company could ask a focus group to rate features of a prototype for a cellular telephone and say how much they would pay for such a phone. How would you organize these data? Take a moment and think about the rows, columns, and anything else that is relevant.[6]

Our data record how customers in a focus group react to a new design. We plan to use this information to decide which type of advertising is most likely to reach interested customers, such as whether younger customers give it a higher rating.

METHOD ▸ DESCRIBE THE DATA AND SELECT AN APPROACH

Organize the data from the focus group as a table. Each row of this table will represent a participant of the focus group, and the columns will record the variables associated with a participant, particularly the rating given to the product.

MECHANICS ▸ DO THE ANALYSIS

Keep track of the details, particularly the units, of the variables in the data table. In addition to recording the rating assigned by each participant to the product, the columns in this data table should include characteristics of the participants, such as the name, age (in years), sex, and income (in dollars). The name and sex are categorical, and the age and income are numerical. Participants' ratings of the product are likely going to be ordinal data, such as values from a Likert scale.

MESSAGE ▸ SUMMARIZE THE RESULTS

Once we have the data, we will be able to find out if, for example, younger members of the focus group like the design or whether more affluent customers like it more than those with less to spend. Once we know who likes the design, we will be able to pick the form of advertising that is most likely to appeal to that audience.

---

[6] Not every company is enamored with focus groups. Have a look at "Shoot the focus group" from the November 14, 2005 issue of *Business Week*. A large group may not mean much if one or two personalities dominate the group.

Best Practices

- *Provide a context for your data.* At a minimum identify the rows and columns and include units for numerical variables. Verify that all of the cases (or rows) share the same units. Ideally, embellish the data with other information such as the name of the person who collected the data or the source of the data, when and where the data were gathered, and why the data were collected.
- *Use clear names for your variables.* Use names that you will understand later and that will make sense to your audience. If you include units for numerical variables as part of the name, then the data become easier to use later on.
- *Distinguish numerical data from categorical data.* Data do not have to be numerical to be useful. Information like names and addresses

can help us learn a lot about customers. Label categories with names rather than numbers.
- *Track down the details when you get the data.* Even if you are told a context for the data, it may turn out that the true situation is a bit (or even a lot) different. The context colors our interpretation of the data, so those who want to influence what you think may slant the story. If you don't learn these details as you gather the data, you probably won't be able to fill them in later.
- *Keep track of the source of data.* If you later discover a strange value or question the meaning of a variable, it will be helpful to have a record of where the data came from. Also, some data change over time or become more extensive as time passes. It will be easier to update your data if you know where they came from.

Pitfalls

- *Do not assume that a list of numbers provides numerical data.* Categories often masquerade as numbers. Don't let that fool you into thinking they have quantitative meaning. Pay attention to the context. We could add phone numbers to Paul's database, but it wouldn't be useful to average these numbers.
- *Don't trust all of the data that you get from the Internet.* The Internet is the place to get data, but that does not mean that every piece of information that you find there is accurate. If you can, find out how the data were collected and dig into the pedigree of that information before you put too much trust in the numbers.
- *Don't believe every claim based on survey data.* A survey that claims to represent consumers

may in fact report just the opinions of those who visit a Web site. The question that respondents answered may be slanted to favor a particular answer. Think about how people would respond to the question, Should the government raise taxes so that it can waste money? Not too many would favor raising taxes, would they? Now suppose the question were asked as, Should the government raise taxes to protect our citizens and improve health care? We might expect this question to find that more are in favor of higher taxes than we thought. If you don't know how the question was asked, try to get a copy of the questionnaire. We return to these issues in Chapter 13.

Software Hints

Most often we use computer software to handle the mechanics of calculating statistics from data. There are many different statistics packages that all do essentially the same things, but each speaks a different language. You have to figure out how to tell your package what you want it to do. That's easy once you become familiar with your software.

The first step gets your data ready for analysis. No matter which software program you use, you need to get your data into the program. This typically re-

quires that you type the data into the program (as in a spreadsheet) or tell the software (in a dialogue of some sort):
- *Where to find the data.* This usually means identifying the name of a file on your computer. Most software expects the data to already be laid out as a data table with one row per case. The values for the variables in each row need to be separated, or delimited, by some common character known as the delimiter. Typical choices for the delimiter are spaces, com-

mas (as in common csv format), or tabs. The return character marks the end of a row (or case).

- *Where to put the data.* After the package reads the data, it converts the data into an internal format that you can save as another file (usually identified in the name by a suffix such as .xls for Excel files). You'll want to save this file so you don't have to read the original data each time you look at the data. Plus, this internal file will hold changes you make to the data, such as recoding or aggregating cases.
- *Names for the variables.* Most packages name variables automatically, but names like Column 1 aren't very helpful. Most packages allow you to include names for the variables in the first line of your external data file. If not, you'll have to enter these manually within your package.

The following sections describe commands for specific packages: Excel, Minitab, and JMP. The symbol > denotes following a menu tree down to the next level. The software hints given here will get you started, but you'll certainly want to read the documentation for your package in order to exploit its full range of capabilities.

EXCEL

To read data into Excel from a text file laid out like a data table, follow the sequence of menu commands

Data > Get External Data > Import text file

Then fill in the dialog boxes. Excel will show you how it interprets the items along the way so you can interactively change the way that it distinguishes columns, for example. As a convention, put the variable names in the first row of the worksheet. (You will need to do that with DDXL as well.) Then freeze these cells so that they remain visible if you scroll down the rows. (Use the Window > Freeze panes command after selecting the row below the labels. This option is only available in the normal view of the worksheet.)

Excel is particularly flexible for recoding and aggregation. For example, in the retail example, you want to recode a column of numerous labels such as Polo shirt, red, M. For this task, follow the menu items

Data > Text to columns

and fill in the resulting dialog box. For aggregating tables, use pivot tables. You can find details about the use of pivot tables in Excel's online help.

MINITAB

Minitab can read data from text files or Excel files. In either case, it's best to arrange the data as a table to begin with. Variables define the columns and cases define each row. Identify the variables with names in the top row as described previously for Excel, and then add information on each observation in the remaining rows. Be sure that each row describes the same variables.

If you have your data in a text file, follow the menu commands

File > Open worksheet. . .

and then identify your file from the menu. Be sure to click the Options button in this dialog to make sure that Minitab understands how to read your data (such as whether values are separated by spaces or commas).

Recoding operations, such as changing numerical data into categories, are handled by the Calc > Calculator or Data > Code commands.

JMP

JMP will also read data directly from Excel if the data are arranged in the natural way with names of the variables in the first row and data for these variables in following rows. To import data from a text file, use the

File > Open

menu command to get to a file dialog box. In this dialog box, enable the feature "All text documents," pick the file, and open the file with the option Data (Using Preview). JMP also reads data from text files laid out in this format. It is usually able to recognize the layout of the file.

For recoding in JMP, you'll need to master its formula calculator (for example, the Match command is very useful for grouping data). For aggregating a table (combining rows into a smaller table), use the Tables > Summary command and fill in the resulting dialog box.

CHAPTER SUMMARY

Data are most often organized, or framed, as a **table**. The rows of the table denote **cases** or **observations**. These are the things that we have observed and measured. The columns of the data denote **variables**, with each column containing the measured values of some attribute of the cases. Variables may be numerical or categorical. **Numerical variables** have common **units**, and it makes sense to average these values.

Categorical variables identify cases with a shared label. **Ordinal variables**, such as those from **Likert scales**, order the cases and may assign numerical labels; these are often treated as numerical data. Variables measured over time are **time series**. The rows of a time series identify the times of the measurements, and the columns denote what was measured. Data that are not a time series are often called **cross sectional**.

◼ Key Terms

aggregate, 18	Likert scale, 17	ordinal variable, 16
case, 15	measurement unit, 16	recode, 17
categorical variable, 16	n (number of rows in data	timeplot, 20
cross-sectional data, 21	table), 15	time series, 20
data table, 14	numerical variable, 16	variable, 15
frequency, 20	observation, 15	

◼ About the Data

Bike Addicts is a privately held bike shop in Philadelphia owned by Paul Molino. The customer data are real, but we altered the names. The increasing use of databases that track customers has made privacy and data security important to both customers and businesses.

EXERCISES

Mix and Match

The following items describe a column in a data table. On the basis of the description, assign a one- or two-word name to the variable, indicate its type (categorical, ordinal, or numerical), and describe the cases. If the data are a time series, identify the frequency of the data. The first item is done as an example.

	Variable Name	Type			Cases
		Cat	Ord	Num	
Weights of people, measured in pounds	Weight			✓	People
1. Brand of car owned by drivers					
2. Income of households, measured in dollars					
3. Color preference of consumers in a focus group					
4. Counts of customers who shop at outlets of a chain of retail stores					
5. Size of clothing items given as Small, Medium, Large, and Extra large					
6. Shipping costs in dollars for deliveries from a catalog warehouse					
7. Prices of various stocks, in dollars per share					
8. Number of employees absent from work each day					
9. Sex (male or female) of respondents in a survey					
10. Education of customer, recorded as High school, Some college, or College Graduate					

True/False

Mark each statement as True or False. If you believe that a statement is false, briefly say why you think it is false.

11. Zip codes are an example of numerical data.

12. Quantitative data consist of numbers along with their units.

13. *Cases* is another name for the columns in a data table.

14. The number of rows in a data table is indicated by the symbol n.

15. The frequency of a time series refers to the time spacing between rows of data.

16. A column in a data table holds the values associated with an observation.

17. A Likert scale represents numerical data.

18. Recoding adds further columns to a data table.

19. Aggregation adds further rows to a data table.

20. Aggregation can convert cross-sectional data into a time series.

Think About It

For each situation described in Exercises 21–30,
   (a) Identify whether the data are cross sectional or a time series.
   (b) Give a name to each variable and indicate if the variable is categorical, ordinal, or numerical (if a variable is numerical, include its units if possible).
   (c) List any concerns that you might have for the accuracy of the data.

21. The debate about the future of Social Security has renewed interest in the amount saved for retirement by employees at a company. While it is possible to open an Individual Retirement Account (IRA), many employees may not have done so. To learn about its staff, a company asked its 65 employees whether they have an IRA and how much they have contributed.

22. A bank is planning to raise the fees that it charges customers for writing checks. It is concerned that the increase will cause some customers to move to other banks. It surveyed 300 customers to get their reaction to the possible increase.

23. A hotel chain sent 2,000 past guests an email asking them to rate the service in the hotel during their most recent visit. Of the 500 who replied, 450 rated the service as excellent.

24. A supermarket mailed 4,000 uniquely identifiable coupons to homes in local residential communities. During the next week, it counted the number of coupons that were redeemed and the size of the purchase.

25. A Canadian paper manufacturer sells much of its paper in the United States. The manufacturer is paid in U.S. dollars but pays its employees in Canadian dollars. The manufacturer is interested in the fluctuating exchange rate between these two currencies. Each trading day in 2005 the exchange rate fluctuated by several basis points. (A basis point is 1/100 of a percent.) A year of data is collected.

26. Car manufacturers measure the length of time a car sits on the dealer's lot prior to sale. Popular cars might only sit on average for 10 days, whereas overstocked models sit for months. This company collected data each month for the length of time cars sat on dealer lots for 10 models from 2002 to 2005.

27. A video game was previewed to a group of 30 teenagers. The teens were asked to rate the quality of the graphics and indicate if the game was too violent.

28. A major bank collected data on 100,000 of its customers (income, sex, location, number of cards, etc.) and then computed how much money it made from the account of each customer during 2005.

29. A start-up company built a database of customers and sales information. For each customer, it recorded the customer's name, zip code, region of the country (East, South, Midwest, West), date of last purchase, amount of purchase, and item purchased.

30. A retail chain lines its parking lots depending on the types of vehicles driven by local shoppers. Large trucks and SUVs require bigger spots than compact cars. Each day, while collecting shopping carts, an employee records a list of the types of vehicles in the lot.

4M Economic Time Series

Data don't always arrive in the most convenient form. You may have the data, but they're not in the same place or the same format. You'll have to combine the separate data tables from various sources into one. Unless you have data organized in a common data table, you'll find it hard or impossible to get the answers you need.[7]

For this exercise, you must prepare the data that are needed to explore the relationship between the sales of a company, retailer Best Buy, and the health of the economy. You'd like to understand if there's a relationship between the amount of money available to be spent, called the disposable income, and the net sales of Best Buy. Best Buy has done well, but is it just riding a wave of spending or has it been more successful?

The economic data come from the online repository hosted by the Federal Reserve Bank of St. Louis. The data are collected monthly, are reported in the number of billions of dollars (at an annual rate), and date back to 1959. The data for 2004 are as shown in the following table:

Month	Disp Inc
2004-01-01	8432.6
2004-02-01	8478.3
2004-03-01	8515.0
2004-04-01	8545.4
2004-05-01	8592.0
2004-06-01	8603.5
2004-07-01	8640.1
2004-08-01	8682.2
2004-09-01	8690.3
2004-10-01	8778.4
2004-11-01	8830.9
2004-12-01	9182.1

The data for Best Buy come from another source. *CompuStat* maintains a database of company information gleaned from reports that are required of all publicly traded firms. For a company to have its stock bought and sold, it must report data such as these quarterly

---

[7] We'll see these data again in Part 4 when we look at methods for forecasting.

net sales figures, given in millions of dollars. The sales data extend back to 1998 (only one year is shown).

Year	Quarter	Net Sales
2004	1	$5,475.00
2004	2	$6,085.00
2004	3	$6,646.00
2004	4	$9,227.00

Motivation

a) Explain why it would it be useful to merge these two data tables. What questions do you think would be interesting to answer based on the merged information?

Method

b) Describe the difference in interpretation of a row in the two tables. Do the tables have a common frequency?

c) The separate data tables each have a numerical column of sales or income. Are the units of these comparable, or should they be expressed with common scales?

Mechanics

d) What should you do if you want to arrange the data in a table that has a quarterly time frequency? Can you copy the data columns directly, or do you have to perform some aggregation or recoding first?

e) Suggest improved names for the columns in the merged data table. How do you want to represent the information about the date?

f) Show the merged data table for 2004.

Message

g) With the data merged, do you see anything interesting happening in 2004 that was not so obvious before?

4M Textbooks

This exercise requires research on how books are priced on the Internet. We will divide books into three broad categories: popular books, school books or textbooks, and recreational books. Some books may fall in two or even three of these categories.

Motivation

a) If you were able to reduce the cost of the books that you buy each year by 5 or 10%, how much do you think you might be able to save in a year?

Method

b) Create a data table of prices for the three types of books. You will compare prices of books at two Internet stores. Your data table will have four columns: one for the name of the book, one for the type of the book, and two more columns for the prices. (If you don't have two Internet stores at which you usually shop, use amazon.com and barnesandnoble.com.) What types of data are these?

c) Identify five books of each type. These form the rows of the data table.

Mechanics

d) Fill in your table from prices on the Web. Did you find the prices of the books that you started with, or did you have to change the choice of books to find the books at both stores?

e) Do your prices include costs for shipping and taxes, if any? Should they?

Message

f) Summarize the price difference that you found for each of the three types of books. Describe any clear patterns you found. For example, if one store is *always* cheaper, then you have an easy recommendation. Alternatively, if one store is cheaper for textbooks but more expensive for popular books, you have a slightly more complex story to tell.

3 Describing Categorical Data

CONSUMERS SPEND BILLIONS OF DOLLARS ONLINE, and the market is growing. Amazon, one of the first e-tailers, remains a leading shopping destination on the Web. Every day, more than 100 million shoppers visit its Web site, either by typing amazon.com into their browser or by clicking through from another Web site, a host with an ad that directs shoppers to Amazon.

Table 3.1 describes six visits to Amazon in the fall of 2002. Rows describe sessions in which a person visited Amazon. Columns indicate the date, whether the customer made a purchase, the host, and demographics of the shopper (number of people in the household, geographic location, and income). If the customer typed amazon.com into a browser, as in the second row, then the host field is blank. The data are a mix of categorical and numerical variables. For instance, the name of the host is categorical, whereas the household size is numerical.

Which hosts generate the most business? Amazon might be willing to pay more to keep links on busy sites. Ad space on busy Web sites has gotten more expensive. Spending for Internet ads doubled from $10 billion in 2004 to $20 billion in 2007. Space on an Internet portal that went for $100,000 in 2002 cost more than $300,000 in 2006, if space was even available on the site. A banner ad on Yahoo or MSN goes for

TABLE 3.1 Six visits to amazon.com.

Date	Purchase	Host	Household Size	Region	Income Range
24Oct2002	No	msn.com	5	North Central	35–50k
05Dec2002	No		5	North Central	35–50k
05Dec2002	No	google.com	4	West	50–75k
05Oct2002	Yes	dealtime.com	3	South	<15k
19Nov2002	No	mycoupons.com	2	West	50–75k
18Oct2002	No	yahoo.com	5	North Central	100k+

about $500,000 per day, about the cost of a 30-second ad on a popular TV show. Revenue from online advertising is growing so fast that AOL dropped its $26 monthly member fee in 2006 to attract visitors to its Web portal. More visitors mean more eyes see its ads, and those eyes generate more income than membership fees.[1]

This chapter shows how to describe a categorical variable using two types of charts: bar charts and pie charts. You've probably seen these before, so this chapter concentrates on the important choices that are made when building these graphs. The chapter also introduces principles that guide the construction of any graph that shows data.

3.1 | LOOKING AT DATA

Let's get specific about our *motivation*. You need a clear goal when you look at data, particularly a large data table. Our task in this chapter is to answer the following question:

> Which hosts send the most visitors to Amazon's Web site?

The data we have track a small percentage of activity at Amazon, but that still amounts to 188,996 visits. Table 3.1 lists six visits that we chose at random to illustrate the contents of each column. You would not want to read through the rest of these data, much less tabulate by hand the number of visits from various hosts.

That's the problem with huge data tables: You cannot *see* what's going on, and seeing is what we want to do. How else can we find patterns and answers to questions in data unless we can see the data?

Variation

variation in data Differences among the values within a column of a data table.

The difficulty in answering questions from data arises from the case-to-case variation. **Variation in data** appears in the form of differences among the values in the columns of a data table. For a categorical variable, variation measures the number of different values and how often each occurs. If every visitor came from Google, then the host column in every row would identify Google and there would be no variation. That's not the case with these data. The column that identifies the host in this data table has 11,142 different values.

Variation is also related to predictability. No variation would mean every one of these visitors came from the same host. In that case, where would you guess the next customer came from? If every prior visitor came from Google, we would expect the next to come from Google as well. Variation reduces our ability to anticipate what will happen next. With 11,142 different hosts, we cannot be sure of the next host or the most common host.

Frequency Table

distribution The collection of values of a variable and how often each occurs.

frequency table A tabular summary that shows the distribution of a variable.

To describe the variation in categorical data, statistics reduces to counting. Once we have counts, we'll graph them. The trick is to decide what to count and how to graph the counts.

That's the point of a clear motivation. Because we want to identify the hosts that send the most visitors, we know to count the number from each host. **The distribution** of a categorical variable is a list of the values of the variable along with the count or frequency of each. A **frequency table** summarizes

[1] "Wiser about the Web," *Business Week*, March 27, 2006.

the distribution of a categorical variable as a table. Each row of a frequency table lists a category along with the number of cases in this category.

A statistics package can list the visits from every host, but we need to be choosy. These data include 11,142 different hosts, and we only want to identify the hosts that send the most visitors. In fact, among all of these hosts, only 20 delivered 500 or more visits. Table 3.2 lists these individually and combines hosts with fewer visits into a large category labeled Other. This frequency table summarizes the recoded categorical variable.

TABLE 3.2 Frequency table of hosts.

Host	Frequency	Proportion
Typed "amazon.com"	89,919	0.47577
msn.com	7,258	0.03840
yahoo.com	6,078	0.03216
google.com	4,381	0.02318
recipesource.com	4,283	0.02266
aol.com	1,639	0.00867
iwon.com	1,573	0.00832
atwola.com	1,289	0.00682
bmezine.com	1,285	0.00680
daily-blessings.com	1,166	0.00617
imdb.com	886	0.00469
couponmountain.com	813	0.00430
earthlink.net	790	0.00418
popupad.net	589	0.00312
overture.com	586	0.00310
dotcomscoop.com	577	0.00305
netscape.com	544	0.00288
dealtime.com	543	0.00287
att.net	533	0.00282
postcards.org	532	0.00281
24hour-mall.com	503	0.00266
Other	63,229	0.33455
Total	**188,996**	**1.00**

This frequency table reveals several interesting characteristics of the hosts. First, the most popular way to get to Amazon in 2002 was to type amazon.com into a browser. You probably recognize the next three hosts: msn.com, yahoo.com, and google.com. More than 70% of Americans online visit these portals. The surprise is how many visits originate from less familiar hosts like recipesource.com. The last column of Table 3.2 shows that other hosts that provide fewer than 500 visits supply about a third of all visitors to Amazon.

The last column in Table 3.2 shows the proportion of the visits from each host. A proportion is also known as a **relative frequency**, and a table that shows these is known as a relative frequency table. Proportions are sometimes multiplied by 100 and given as percentages. We prefer tables that show both counts and proportions side by side as in Table 3.2. With the column of proportions added, it is easy to see that nearly half of these visitors typed amazon.com and that a third came from small hosts. The choice between proportions and percentages is a matter of style. Just be clear when you label the table.

relative frequency The frequency of a category divided by the number of cases; a proportion or percentage.

3.2 | CHARTS OF CATEGORICAL DATA

Let's concentrate on the top 10 hosts. A frequency table is hard to beat as a summary of several counts, but it becomes hard to compare the counts as the table grows. Unless you need to know the exact counts, a chart or plot beats a table as a summary of the frequencies of more than five categories. The two most

tip

common displays of a categorical variable are a bar chart and a pie chart. Both describe a categorical variable by displaying its frequency table.

Bar Chart

bar chart A display using horizontal or vertical bars to show the distribution of a categorical variable.

A **bar chart** displays the distribution of a categorical variable using a sequence of bars. The bars are positioned along a common baseline, and the length of each bar is proportional to the count of the category. Bar charts have small spaces between the bars to indicate that the bars could be rearranged into any order. The bar chart in Figure 3.1 uses horizontal bars to show the counts for each host, ordered from largest to smallest.

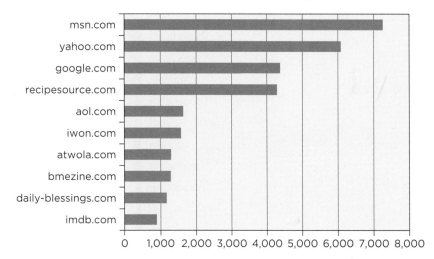

FIGURE 3.1 Bar chart of top 10 hosts, with the bars drawn horizontally.

By showing the hosts in order of frequency, this bar chart emphasizes the bigger hosts. You can also orient the bars vertically, as in Figure 3.2.

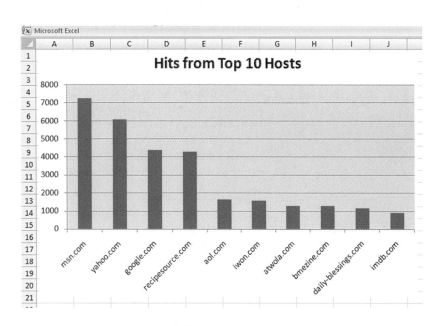

FIGURE 3.2 Bar chart of top 10 hosts, with the bars drawn vertically.

We find it easier to compare the lengths of horizontal bars than vertical bars. Vertical bars also limit the number of categories in order to show the labels. That said, orientation is a matter of taste. Bar charts may be drawn either way.

Pareto chart A bar chart with categories sorted by frequency.

tip

When the categories in a bar chart are sorted by frequency, as in Figures 3.1 and 3.2, the bar chart is sometimes called a **Pareto chart**. Pareto charts are popular in quality control to identify problems in a business process.

We could have shown the hosts in alphabetical order to help readers find a particular host, but with 10 labels, that's not necessary. **If the categorical variable is ordinal, however, you must preserve the ordering.** To show a variable that measures customer satisfaction, for instance, it would not make sense to arrange the bars alphabetically as Fair, Great, Poor, Typical, and Very good.

Bar charts become cluttered when drawn with too many categories. A bar chart that showed all 20 hosts in Table 3.2 would squish the labels, and the smallest bars would become invisible. There's also a problem if the frequency of one category is much larger than those of the rest. The bar chart in Figure 3.3 adds a bar that counts the visits from every other host.

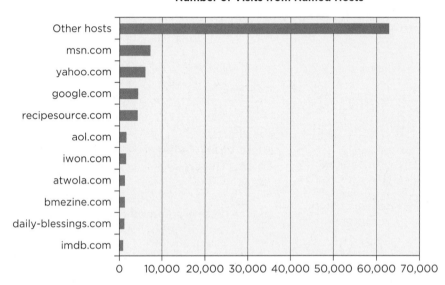

FIGURE 3.3 One category dominates this bar chart.

The long bar that accumulates small hosts dwarfs those representing the most popular hosts. That might be a point worth making; most visits don't come from the top 10 hosts. The prevalence of small hosts is an example of what's called the long tail in online retailing; see About the Data for another example.

Which of these bar charts provides the best answer to the motivating question? We prefer the chart in Figure 3.1, or perhaps Figure 3.2. The accumulated number of hits from the other hosts in Figure 3.3 obscures the differences among the top 10. It's impossible to see that AOL generates almost twice as many visitors as imdb.com. The *message* from Figure 3.1 is clear: Among hosts, MSN sends the most visitors to Amazon, more than 7,000 during this period, followed by Yahoo, Google, and RecipeSource. A variety of other hosts make up the balance of the top 10.[2]

caution We use bar charts only to show frequencies of a categorical variable. Recognize that you will sometimes see bar charts used to graph other types of data, such as a time series of counts. For example, the charts from the front page of the *Wall Street Journal* in Figure 3.4 look like bar charts, but neither shows the distribution of a categorical variable. Both charts use bars to show a sequence of counts over time. It is fine to use bars in this way to emphasize the changing level of a time series, but these are not displays of a frequency table.[3]

[2] Edward Tufte offers several elegant improvements to the bar chart, but most of these have been ignored by software packages. Look at his suggestions in his book *The Visual Display of Quantitative Information* (Cheshire, CT: Graphics Press, 1983), pp. 126–128.
[3] *Wall Street Journal*, August 21, 2006, and September 11, 2006.

Scaling Back

As Lucent Technologies has restructured, Bell Labs' work force has shrunk:

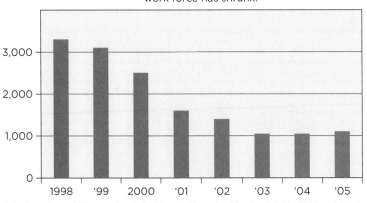

Note: Lucent says 500 researchers left with spinoff or sale of 3 units: Avaya in 2000 and Agere and Optical Fiber Solutions in 2001.

Source: The company

Big Footprint

Wal-Mart's total U.S. retail space in fiscal 2006 was 498.5 million square feet. Net square footage added through Wal-Mart's new and expanded U.S. stores:

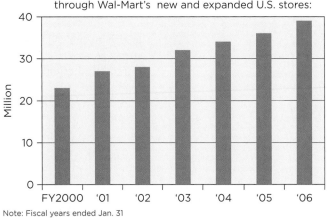

Note: Fiscal years ended Jan. 31

Source: The company

FIGURE 3.4 Not every chart with bars shows the distribution of a categorical variable.

Pie Chart

pie chart A display that uses wedges of a circle to show the distribution of a categorical variable.

Pie charts are another way to draw the distribution of a categorical variable. A **pie chart** shows the distribution of a categorical variable as wedges of a circle. The area of each wedge is proportional to the count in a category. Large wedges indicate categories with large relative frequencies.

Pie charts convey immediately how the whole divides into shares, which makes these a common choice when illustrating, for example, market shares or sources of revenue within a company, as in Figure 3.5.

Tribune's Content

2005 top holdings and operating revenue:

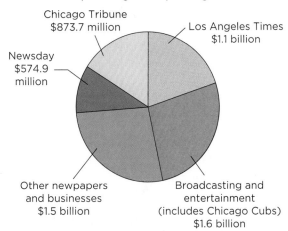

Source: The company

Big Business

Dell revenue by product category in the fiscal first quarter, in billions

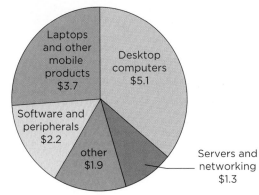

Note: The first quarter of fiscal 2007 ended May 5
Source: The company

FIGURE 3.5 Pie charts of the composition of two companies.

Pie charts are also good for showing whether the relative frequency of a category is near $1/2$, $1/4$, or $1/8$ because we're used to cutting pies into 2, 4, or 8 slices. In the pie chart in Figure 3.6, you can tell right away that msn.com generates about one-quarter of the visitors among the top 10 hosts.

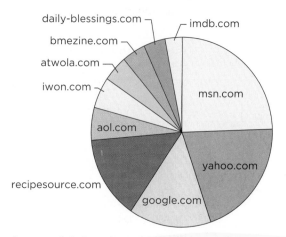

FIGURE 3.6 Pie chart of the top 10 hosts.

Pie charts are less useful than bar charts if we want to compare the actual counts. People are better at comparing lengths of bars than comparing the angles of wedges.[4] Unless the slices align conveniently, it's hard to compare the sizes of the categories. Can you tell from Figure 3.6 whether RecipeSource or Google generates more visitors? Comparisons such as this are easier in a bar chart.

Because they slice the whole thing into portions, pie charts make it easy to overlook what is being divided. It is easy to look at the pie chart in Figure 3.6 and come away thinking that 25% of visitors to Amazon come from MSN. Sure, MSN makes up one-quarter of the pie. It's just that this pie summarizes fewer than 16% of the visits. The pie chart looks very different if we add back the Other category or the typed-in hits.

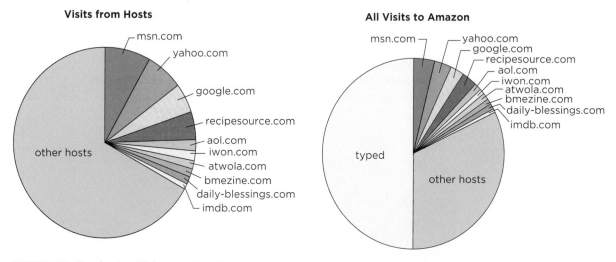

FIGURE 3.7 Pie charts with large categories.

In Figure 3.7 the chart on the left shows that the top 10 hosts generate slightly more than 25% of the host-generated visitors to Amazon. More come from small sites. Any marketing plan author who forgets this fact might be in for a big surprise. The pie chart on the right makes you realize that any one of the large hosts generates only a small share of the traffic at Amazon.

Which pie chart is best? As with choosing a bar chart, the motivation determines the best choice. Figure 3.6 emphasizes the relative sizes of the major hosts; it's best suited to our motivating question. What about the choice between a bar chart and a pie chart? For this analysis, we could use either the bar chart in Figure 3.1 or the pie chart in Figure 3.6. Because it's easier to

[4] See W. S. Cleveland and R. McGill, "Graphical Perception: Theory, Experimentation and Application to the Development of Graphical Methods," *Journal of the American Statistical Association* (1984), 531–554.

compare the sizes of categories using a bar chart, we prefer bar charts. That said, pie charts are popular, so make sure you can interpret these figures.

| **What Do You Think?** | Consider this situation: A manager has partitioned the company's sales into six districts: North, East, South, Midwest, West, and international. What graph or table would you use to make these points in a presentation for management? |

a. A figure that shows that slightly more than half of all sales are made in the West district

b. A figure that shows that sales topped $10 million in every district[5]

3.3 | THE AREA PRINCIPLE

area principle The area of a plot that shows data should be proportional to the amount of data.

There's flexibility in the mechanics of a bar chart, from picking the categories to laying out the bars. One aspect of a bar chart, however, is not open to choice. Displays of data must obey a fundamental rule called the **area principle**. The area principle says that the area occupied by a part of the graph should correspond to the amount of data it represents. Violations of the area principle are a common way to mislead with statistics.

The bar charts we've shown obey the area principle. The bars have the same width, so their areas are proportional to the counts from each host. By presenting the relative sizes accurately, bar charts make comparisons easy and natural. Figure 3.1 shows that MSN generates almost twice as many visits as RecipeSource and reminds you that RecipeSource generated almost as many hits as Google.

How do you violate the area principle? It is more common than you might think. News articles often decorate charts to attract attention. Often, the decoration sacrifices accuracy because the plot violates the area principle. For instance, the chart in Figure 3.8 shows the value of wine exports from the United States.[6]

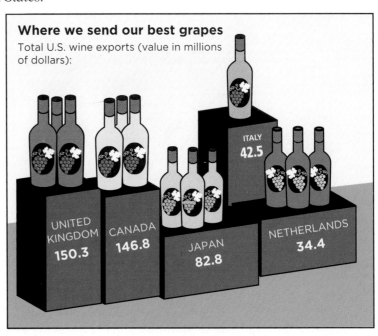

FIGURE 3.8 Decorated graphics may ignore the area principle.

[5] Use a table (there are only two numbers) or a pie chart for part a. Pie charts emphasize the breakdown of the total into pieces and make it easy to see that more than half of the total sales are in the western region. For part b, use a bar chart because bar charts show the values rather than the relative sizes. Every bar would be long enough to reach past a grid line at $10 million.

[6] Data from *USA Today*, November 1, 2006. With only five numbers to show, it might make more sense to list the values, but readers tend to skip over articles that are dense with numbers.

Where's the baseline for this chart? Is the amount proportional to the number of wine bottles? The chart shows bottles on top of labeled boxes of various sizes and shapes. The bar chart in Figure 3.9 shows the same data but obeys the area principle.

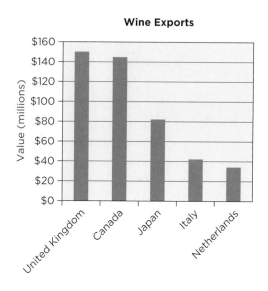

Wine Exports

FIGURE 3.9 Bar charts must respect the area principle.

The bar chart in Figure 3.9 is less attractive than its counterpart in Figure 3.8, but it is accurate. For example, you can tell that exports to Italy are slightly larger than exports to the Netherlands.

You can respect the area principle and still be artistic. Just because a chart is not a bar chart or a pie chart doesn't mean that it's dishonest. For example, the plots in Figure 3.10 divide net sales at Lockheed Martin into its main lines of business.

 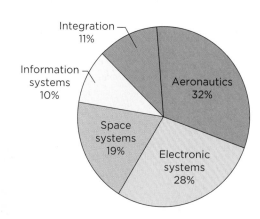

FIGURE 3.10 Alterative graphs of Lockheed earnings.

The artist who prepared the chart on the left divided a grid with photos associated with the five divisions of Lockheed.[7] This chart obeys the area principle. Each square in the grid represents $100 million in sales during 2003. The pie chart shows the same data. Which grabs your attention?

---

[7] This chart appeared in the business section of the November 28, 2004 issue of the *New York Times*.

4M EXAMPLE 3.1 ROLLING OVER

MOTIVATION ▸ STATE THE QUESTION

Authorities enter a description of every fatal automobile accident into the Fatality Analysis Reporting System (FARS). Once obscure, FARS became well known when reporters from the *New York Times* discovered something unusual. Their tools: a question, some data, and a bar chart.

The question arose anecdotally. Numerous accidents seemed associated with vehicles rolling over. Were these reports coincidental, or was something systematic and dangerous happening on the highways? For this example, we'll put you in the place of a curious reporter following up on a lead that has important implications for the businesses involved. We'll use data for the year 2000 because that's when this issue came to the public's attention.

News reports suggested some types of cars were more prone to roll-over accidents than others. Most of the reported incidents involved SUVs, but are all SUVs dangerous? If some types of cars are more prone to these dangerous accidents, the manufacturer is going to have a real problem.

METHOD ▸ DESCRIBE THE DATA AND SELECT AN APPROACH

Identify the data and come up with a plan for how you will use the data. For this example, we will extract from FARS those accidents for which the primary cause was a roll-over. We'll also stick to accidents on interstate highways.

The *rows* of the data table are accidents resulting in roll-overs in 2000. The one column of interest is the model of the car involved in the accident. Once we have the data, we plan to use a frequency table that includes percentages and a bar chart to show the results.

MECHANICS ▸ DO THE ANALYSIS

FARS reports 1,024 fatal accidents on interstates in which the primary event was a roll-over. The accidents include 189 different types of cars. Of these, 180 models were involved in fewer than 20 accidents each. We will combine these into a category called Other. Here's the frequency table, with the names sorted alphabetically. (As in the example of Web visits, most of the rows land in the Other category and we have not shown it in this bar chart.)

Model	Count	Percentage
4-Runner	34	3.3
Bronco/Bronco II/Explorer	122	11.9
Cherokee	25	2.4
Chevrolet C, K, R, V	26	2.5
E-Series Van/Econoline	22	2.1
F-Series Pickup	47	4.6
Large Truck	36	3.5
Ranger	32	3.1
S-10 Blazer	40	3.9
Other	640	62.5
Total	1,024	

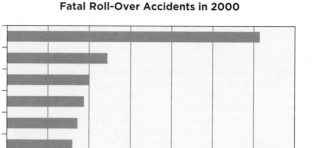

Fatal Roll-Over Accidents in 2000

The bar chart sorted by count from largest to smallest shows that the Ford Bronco (or Explorer) had the most fatal roll-overs on interstates in 2000.

MESSAGE | SUMMARIZE THE RESULTS

Data from FARS in 2000 show that Ford Broncos (and the comparable Explorer) were involved in more fatal roll-over accidents on interstate highways than any other model. Ford Broncos were involved in more than twice as many roll-overs as the next-closest model (the Ford F-series truck) and more than three times as many as the similar Chevy Blazer.

Be honest about any caveats or assumptions you have made. These results could also be explained by popularity. Perhaps more Broncos show up in roll-over accidents simply because there are more Broncos being driven on interstate highways. Perhaps Broncos are driven more aggressively. We cannot resolve these questions from the data in FARS alone, but this bar chart begs for an explanation.

The explanation of the bar chart in Example 3.1 came several years later and brought with it a bitter dispute between Ford and Bridgestone, maker of the Firestone tires that were standard equipment on Ford Broncos and Explorers. Ford claimed that the Firestone tires caused the problem, not the vehicle. The dispute led to a massive recall of more than 13 million Firestone tires.

4M EXAMPLE 3.2 | **COMPARING CHIP SALES**

MOTIVATION | STATE THE QUESTION

The U.S. Department of Justice investigated several manufacturers of computer memory for price fixing. The government alleged that the companies agreed to set price targets to avoid competing on price. The alleged price fixing occurred from 1999 through 2002. In September 2004, Infineon pled guilty to participating in a conspiracy to fix prices for dynamic random access memory (DRAM) and paid a $160 million fine.[8] The market shares also changed, as shown in this table.

Company	1999 Share (%)	2002 Share (%)
Hynix	19	13
Infineon	6	13
Micron	16	19
Samsung	23	32
Others	36	23

Did Infineon actually gain a larger share of the market for chips during this period?

METHOD

DESCRIBE THE DATA AND SELECT AN APPROACH

This task requires us to compare categorical variables. Often, the categorical variables measure the same thing at different points in time.

Our data show shares of the market for DRAM chips produced by major vendors in 1999 and 2002. The columns are the numbers of chips sold in 1999 and 2002. The row labels identify the manufacturers. We plan to use both pie charts and bar charts to contrast the shares, and pick the plot that most clearly shows the differences in the two years.

MECHANICS

DO THE ANALYSIS

You often need to consider several plots in order to decide which style presents the data in a way that best addresses the problem at hand.

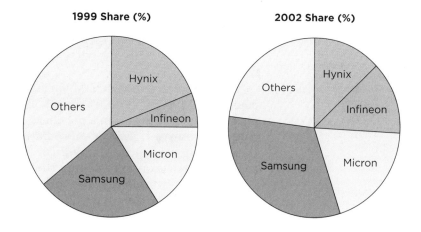

For example, these two pie charts emphasize market shares within each year. We can see that Samsung has a larger share in 2002, but the amounts are hard to judge.

[8] "Price Fixing in the Memory Market," *IEEE Spectrum*, December 2004, 18–20.

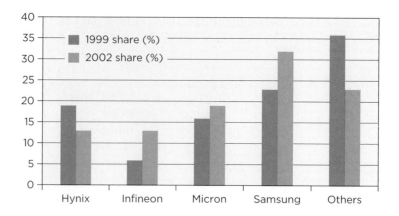

In contrast, this paired bar chart combines two bar charts of the shares together and simplifies the comparison. The bars representing market share are next to each other, making it easier to draw comparisons.

Now it's easy to see that Infineon's share more than doubled.

We prefer this graph because it compares the shares for each company in the two years rather than comparing the proportions within each year.

SUMMARIZE THE RESULTS

Comparison of the market share for chips shows that Infineon and Samsung increased their shares from 1999 to 2002. Most of the gain in share appears to have been at the expense of smaller companies (the Others).

3.4 | MODE AND MEDIAN

Plots are the best summaries of categorical data, but you may need a more compact summary. Compact summaries are handy when you need a quick way to compare categorical variables, such as variables measured at two points in time. The mode tells you which category is most common, and the median tells you what is the middle category, assuming you can order the values.

The **mode** of a categorical variable is the most common category, the category with the highest frequency. The mode labels the longest bar in a bar chart or the widest slice in a pie chart. In a Pareto chart, the mode is the first category shown. If two or more categories tie for the highest frequency, the data are said to be *bimodal* (in the case of two) or *multimodal* (more than two). For example, among the visitors to Amazon, the modal behavior (the most common) is to type amazon.com into the browser. Changes in the mode over time tell you about a shift in the data. In Example 3.2, for example, the mode is the category representing chips made by smaller manufacturers in 1999, but by 2002 the mode is the category representing chips made by Samsung.

Ordinal data offer another summary, the **median**, that is not available unless the data can be put into order. The median of an ordinal variable is the label of the category of the middle observation when you sort the values. If you have an even number of items, choose the category on either side of the middle of the sorted list as the median. For instance, letter grades in courses

mode The mode of a categorical variable is the most common category.

median The median of an ordinal variable is the category of the middle observation of the sorted values.

are ordinal; someone who gets an A did better than someone who got a B. In a class with 13 students, the sorted grades might look like

A A A A A A **B** B B B C C C

The most common grade, the mode, is an A but the median grade is a B. The median does not match the mode in this example. Most students got an A, with fewer getting a B and fewer still getting a C. The median lies in the middle of the list of grades, but that does not make it the most common.

The bar chart in Figure 3.11 shows another use of letter grades. These grades were assigned to summarize and contrast the performance of 80 real estate investment trusts (REITs) that own office properties.

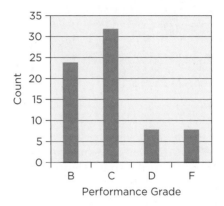

FIGURE 3.11 Ratings of the performance of 80 real estate investment trusts.

As published by Forbes, none of these REITs rated an A grade. The most common grade, the mode, is a C. The median grade, in the middle of the ordered grades, is also a C. Investors in these trusts would certainly like to learn that both the median and the mode of the performance ratings have increased.

Best Practices

- *Use a bar chart to show the frequencies of a categorical variable.* Order the categories either alphabetically or by size. The bars can be oriented either horizontally or vertically.
- *Use a pie chart to show the proportions of a categorical variable.* Arrange the slices (if you can) to make differences in the sizes more recognizable. A pie chart is a good way to show that one category makes up more than half of the total.
- *Preserve the ordering of an ordinal variable.* Arrange the bars on the basis of the order implied by the categories, not the frequencies.
- *Respect the area principle.* The relative size of a bar or slice should match the count of the associated category in the data relative to the total number of cases.
- *Show the best plots to answer the motivating question.* You may have looked at several plots when you analyzed your data, but that does not mean you have to show them all to someone else. Choose the plot that makes your point.
- *Label your chart to show the categories and indicate whether some have been combined or omitted.* Name the bars in a bar chart and slices in a pie chart. If you have omitted some of the cases, make sure the label of the plot defines the collection that is summarized.

Pitfalls

- *Avoid elaborate plots that may be deceptive.* Modern software makes it easy to add an extra dimension to punch up the impact of a plot. For instance, this 3-D pie chart shows the top six hosts from the Amazon example.

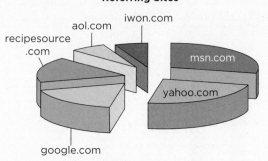

Referring Sites

Showing the pie on a slant violates the area principle and makes it harder to compare the shares, the principal feature of data that a pie chart ought to show.

- *Do not show too many categories.* A bar chart or pie chart with too many categories might conceal the more important categories. In these cases, group other categories together and be sure to let your audience know how you have done this.

- *Do not put ordinal data in a pie chart.* Because of the circular arrangement, pie charts are not well suited to ordinal data because the order gets lost. The pie chart shown at the top of the next column conceals the ordinal nature of the data. The slice indicating companies who expect revenue to increase more than 50% is next to the slice for those who expect revenue to fall the most.[9]

Upbeat

U.S. companies' responses when asked to forecast changes in their 2006 China revenue from 2005

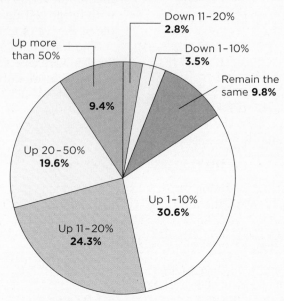

Source: AmCham Shanghai 2006 China Business Report

- *Do not carelessly round data.* Do you see the problem with the following pie chart? Evidently, the author rounded the shares to integers at some point, but then forgot to check that the rounded values add up to 100%. Few things hurt your credibility more than these little blunders.

U.S. Wireless Phone Market

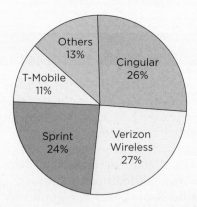

Software Hints

EXCEL

To build your own bar charts starting from the raw data, use pivot tables to create a frequency table, following the menu commands (see the online help for assistance with pivot tables)

Insert > Pivot Table (report)

[9] Data from *Wall Street Journal*, November 2, 2006.

Once you have the frequency distribution, use the commands

Insert > Chart

to build a bar chart or a pie chart. The categories shown in the bar chart appear in the order listed in the spreadsheet. If you'd like some other arrangement of the categories in the chart, just move the rows around in the spreadsheet and the chart will be redrawn. DDXL makes this process much easier because it avoids the need to build a pivot table. Start by selecting the column that contains the categorical variable. (Make sure that the name of the variable appears in the first row of the spreadsheet.) Then use the command

DDXL > Charts and Plots

to obtain a dialog for building the bar chart (or pie chart). Pick the function from the pop-up menu (labeled "Click Me") and then select the name of the variable from the list of variable names.

MINITAB

If your data use numerical codes for categories, you might first want to convert the numbers into text. That will keep you from averaging zip codes! Use the Data > Change data type... commands to convert the data in a column. Columns of text data have a T in the column headers.

The commands for bar charts and pie charts are next to each other on the Graph menu. To get the frequency table of a categorical variable, follow the sequence of commands

Stat > Tables > Tally individual variables...

For a bar chart, look in the graph menu and follow the commands

Graph > Bar chart

indicate the type of chart, and then fill the dialog with the name of the variable. (Notice that you can produce stacked and clustered versions of bar charts using this dialog as well.) Options allow the plot to show percentages in place of counts. To make a pie chart, follow the commands

Graph > Pie chart

JMP

If you have used numbers to represent categorical data, you can tell JMP that these numbers are categories, not counts. A panel at the left of JMP's spreadsheet lists the columns. If you click on the symbol next to the column name, a pop-up dialog allows you to change numerical data (called continuous by JMP) to categorical data (either ordinal or nominal).

To get the frequency table of a categorical variable, use the

Analyze > Distribution

command and fill the dialog with the categorical variable of interest. You also will see a bar chart, but the bars are drawn next to each other without the sort of spacing that's more natural.

To obtain a better bar chart and a pie chart, follow the menu items

Graph > Chart

place the name of the variable in the Categories, X, Levels field, and click the OK button. That gets you a bar chart. To get a pie chart, use the option pop-up menu (click on the red triangle above the bar chart) to change the type of plot. You can rearrange the chart by using the pop-up menu under the red triangle.

Version 6 (and later) of JMP includes an interactive table-building function. Access this command by following the menu items

Tables > Tabulate

This procedure becomes more useful when summarizing two categorical variables at once.

CHAPTER SUMMARY

Frequency tables display the **distribution** of a categorical variable by showing the counts associated with each category. **Relative frequencies** show the proportions or percentages. **Bar charts** summarize graphically the counts of the categories, and **pie charts** summarize the proportions of data in the categories. Both charts obey the **area principle**. The area principle requires that the share of the plot region devoted to a category is proportional to its frequency in the data. A bar chart arranged with the categories ordered by size is sometimes called a **Pareto chart**. When showing the bar chart for an ordinal variable, keep the labels sorted in their natural order rather than by frequency. The **mode** of a categorical variable is the most frequently occurring category. The **median** of an ordinal variable is the value in the middle of the sorted list of all the groups.

▇ Key Terms

▇ About the Data

The data on Internet shopping in this chapter come from ComScore, one of many companies that gather data on Web surfing. They get the data from computers that people receive for free in return for letting ComScore monitor how they surf the Internet. These visits cover the time period from September through December 2002. An interesting research project would be to see whether the distribution has changed since then.

An important aspect of these data is the prevalence of visits from small, specialized Web sites. Imagine how the bar charts of hosts shown in this chapter would look if we tried to show all 11,142 hosts! This long tail (a term coined by Chris Anderson in 2004) is prominent in online sales of books, music, and video. For example, a typical Barnes & Noble carries 130,000 titles. That's a lot, but more than half of Amazon's book sales come from books that are not among its top 130,000 titles. A physical bookstore lacks the space to carry so many titles, but an online, virtual bookstore has plenty of room to stock them.

The data on real estate investment trusts (REITs) comprise a portion of a data table posted on forbes.com in 2008. Sources of other examples are noted in the text.

EXERCISES

Mix and Match

Match these descriptions of variables to a bar chart, Pareto chart, pie chart, or frequency table. Some are counts and others are shares. If not given, indicate whether you plan to show the frequencies or relative frequencies.

1. Proportion of automobiles sold during the last quarter by eight major manufacturers

2. Number of different types of defects found in computer equipment

3. Number of cash-back coupons returned by purchasers of cameras, computers, and stereo equipment

4. Counts of the type of automobile involved in police traffic stops

5. Destinations for travelers leaving the United States and heading abroad

6. Reason for customers hanging up when calling a computer help desk

7. Excuses given for not finishing homework assignments on time

8. Brand of computer chosen by new students of a large university

9. Share of software purchases devoted to games, office work, and design

10. Customer choices of compact, zoom, and high-end digital cameras

11. Number of cellular telephone customers rating their service as Poor, OK, Good, and Great

12. Share of bank loans that are in good standing, 30 days past due, and in default

True/False

Mark True or False. If you believe that a statement is false, briefly say why you think it is false.

13. Charts are better than frequency tables as summaries of the distribution of a categorical variable.

14. If all of the bars in a bar chart have the same length, then the categorical variable shown in the bar chart has no variation.

15. The frequency of a category is the dollar value of the observations in that group.

16. A relative frequency is the number of observations belonging to a category.

17. Use a bar chart to show frequencies and a pie chart to show shares of a categorical variable that is not ordinal.

18. The area principle says that the proportion of the area of a graph devoted to a category must match the number of cases found in the category.

19. Put the labels of the categories in order in a bar chart when showing the frequencies of an ordinal variable.

20. The bar chart of a recoded categorical variable has the same number of bars as the bar chart of the original categorical variable.

21. A Pareto chart puts the modal category first.

22. The median of a categorical variable is the category with the most items.

Think About It

23. These two pie charts show the origin of cars bought by customers who traded in domestic and Asian models. For example, among those trading in a car produced by a domestic manufacturer, 29% purchased a car made by an Asian manufacturer. What's the message of these two pie charts?

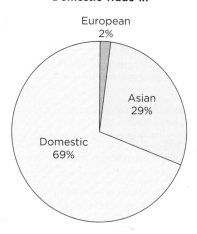

Domestic Trade-in

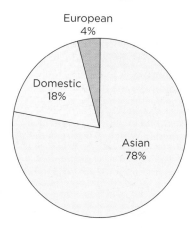

Asian Trade-in

24. To reduce energy consumption, many states and utilities have encouraged customers to replace incandescent light bulbs by more efficient compact fluorescent lamps. Can this really make much difference? The California Energy Commission reported the following summary of household electricity use. Based on this chart, can more efficient lighting have much impact on household energy use?

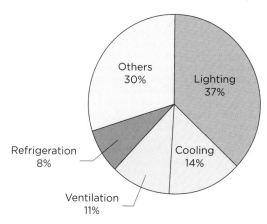

Household Uses of Electricity

25. This plot shows the holdings, in billions of dollars, of U.S. Treasury bonds in five Asian countries. Is this a bar chart in the sense of this chapter or a chart of a table of five numbers that uses bars?

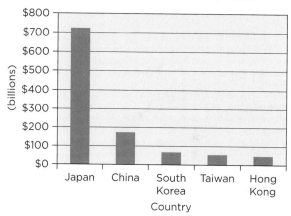

Holdings of U.S. Treasury Bonds

26. Concern over an aging product line and withdrawal of a cholesterol drug led the *Wall Street Journal* to publish this graphic with a cover-page article on future profits of drug-maker Pfizer.[10] Is it a bar chart in the sense of this chapter or a timeplot that uses bars?

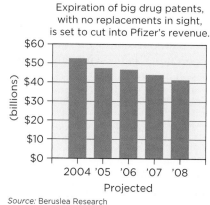

Cold Turkey

Expiration of big drug patents, with no replacements in sight, is set to cut into Pfizer's revenue.

Source: Beruslea Research

---

[10] "Demise of a Blockbuster Drug Complicates Pfizer's Revamp," *Wall Street Journal*, December 4, 2006.

27. This table summarizes results of a survey of the purchasing habits of 1,800 13- to 25-year-olds in the United States in 2006, a much sought-after group of consumers.[11] Each row of the table indicates the percentage of participants in the survey who say the following:

Participants Say That . . .	Percentage
They consider a company's social commitment when deciding where to shop.	69
Given equal price and quality, they are likely to switch brands to support a cause.	89
They are more likely to pay attention to messages from companies deeply committed to a cause.	74
They consider a company's social commitment when recommending products to friends.	66

(a) Would it be appropriate to summarize these percentages in a single pie chart?

(b) What type of chart would you recommend for these four percentages?

28. The following table summarizes a survey done in December 2005.[12] Each of the participating 1,011 adults was asked whether he or she was likely or somewhat likely to perform the activity listed in the Item column of the table.

Item	Percentage
Reduce his or her credit card debt.	84
Check his or her credit score.	48
Cut back on spending for entertainment.	45
Cut down on the number of credit cards he or she has.	35

(a) Would it be appropriate to summarize these responses together in a single bar chart?

(b) What sort of chart would you recommend to summarize these results?

29. These data summarize how 360 executives responded to a question that asked them to list the factors that impede the flow of knowledge within their company. (From the Data Warehousing Institute; *Business Week*, Decmber 6, 2004.) Do these percentages belong together in a pie chart? State why or why not.

Problem	Percentage Reporting
Delivering trustworthy data	53
Selecting key measures of performance	39
Coordinating departments	35
Resistance to sharing data	33
Getting users to use data	29

30. This table summarizes shares of the U.S. wireless telephone market in 2005. To summarize these percentages together in a pie chart, what needs to be added?

Company	Share (%)
Cingular	27
Verizon	26
Sprint	24
T-Mobile	11

31. What would happen if we were to generate a bar chart for a column that measures the amount spent by the last 200 customers at a convenience store?

32. Suppose that the purchase amounts in the previous question were coded as Small if less than $5, Typical if between $5 and $20, and Large if more than $20. Would a bar chart be useful? How should the categories be displayed?

33. Describe the bar chart of a categorical variable for which 900 of the 1,000 rows of data are in one category and the remaining 100 are distributed evenly among five categories.

34. Sketch by hand the pie chart of the data in the previous question.

35. A data table with 100 rows includes a categorical variable that has five levels. Each level has 20 rows. Describe the bar chart.

36. Compare the bar chart of the variable described in the previous question with the pie chart of the same data. Which makes it easier to see that each category has the same number of cases?

37. A categorical variable has two values, male and female. Would you prefer to see a bar chart, a pie chart, or a frequency table? Why?

38. A categorical variable denotes the state of residence for customers to a Web site. Would you prefer to see a bar chart, a pie chart, or a frequency table? Would it be useful to combine some of the categories? Why?

39. This pie chart shows the type of college attended by the CEOs of America's 50 largest companies.[13] Which category is the mode? Which category is the median?

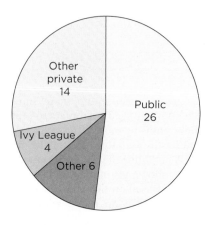

[11] "Up Front: The Big Picture," *Business Week*, November 6, 2006.
[12] "Sure You Will," *New York Times*, January 22, 2006.
[13] "Any College Will Do," *Wall Street Journal*, September 18, 2006.

40. The next two charts show other features of the colleges attended by CEOs of the 50 largest companies in the United States. Should the enrollment data be shown in a pie chart? What is the modal location? What is the median size?

Academic Diversity

Alma maters of CEOs of the 50 biggest U.S. companies by revenue:

By current enrollment size

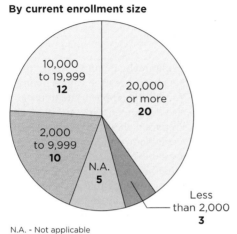

N.A. - Not applicable

By geography

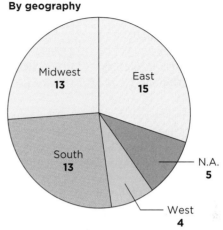

41. Auto manufacturers are sensitive to the color preferences of consumers. Would they like to know the modal preference or the median preference?

42. A consumer group surveyed customers and asked them to rate the quality of service delivered by their wireless phone company. A publication of this group summarizes the rating provided by each customer as Excellent, Good, Fair, or Poor. What does it mean for a provider to get a modal rating of Excellent or a median rating of Excellent?

43. The *New York Times Magazine* (November 27, 2005) reported the following answers to the question, Do you favor or oppose allowing cloning or genetically altering humans?

Response	Percentage
Strongly favor	3
Somewhat favor	12
Somewhat oppose	18
Strongly oppose	63
No answer	4

Is this variable ordinal or categorical? Should this distribution be shown in a pie chart?

44. Suggest a plot of your own design to summarize the data shown in Exercise 43.

You Do It

Bold names shown with a question identify the data table for the problem.

45. **Soft drinks** This table summarizes the number of cases (a case has 24 bottles or cans) of different types of soft drinks sold in the United States in 2004. (From *Beverage Digest*, March 4, 2005.)

Brand	Company	Cases (millions)
Coke Classic	Coke	1833
Pepsi-Cola	Pepsi	1180
Diet Coke	Coke	998
Mt Dew	Pepsi	648
Diet Pepsi	Pepsi	625
Sprite	Coke	580
Dr Pepper	Cadbury	574
Caffeine-free Diet Coke	Coke	170
Diet Dr Pepper	Cadbury	140
Sierra Mist	Pepsi	139

(a) Describe your impression of the underlying data table. Do you think that it has a row for every case of soft drinks that was sold?

(b) Prepare a chart of these data to be used in a presentation. The plot should emphasize the *share* of the market held by the different brands.

(c) Prepare a chart of these data that compares the *share* of the market held by each of the three companies. Is this chart different from the chart used in part (b)? If not, explain how you could use the chart in part (a).

(d) Prepare a chart that contrasts the *amounts* of diet and regular soft drinks sold.

46. **Breakfast bars** This table summarizes retail sales of breakfast bars for 2004. The data record sales from most retail outlets, except for Wal-Mart, which keeps this sort of information confidential. Total sales in this product category was $513,209,024. (*Manufacturing Confectioner*, April 2005.)

Manufacturer	Brand	Sales
Kelloggs	Nutri Grain Bars	$102,551,232
Kelloggs	Special K Bar	$50,757,024
General Mills	Milk N Cereal Bars	$49,445,124
Quaker	Fruit & Oatmeal	$32,836,244
Quaker	Oatmeal Breakfast Sq.	$29,525,036
Atkins	Morning Start	$27,970,354
Kelloggs	Cereal & Milk Bars	$25,565,480
Retailer	Private Label	$23,880,972
Quaker	Fruit & Oatmeal Bites	$19,356,590
Slim Fast	Snack Options	$18,917,658

(a) Do you believe that the underlying data table has a row for every bar or package of breakfast bars that was sold in 2004? How would you expect the underlying data to be organized?

(b) Show a chart of the shares of the total market by brand. Be sure to account for other brands that did not make the top 10.

(c) Prepare a chart that shows the amounts sold by the six named manufacturers. The audience for this chart expects to see the dollar amounts and can figure out the shares for themselves.

47. **Teen research** These data summarize results of a survey that asked teens and adults to identify the single top government research priority. (*Business Week*, December 6, 2004.)

Research Area	Adults (%)	Teens (%)
Alternative energy	33	23
Stem-cell research	12	4
Water purification	11	20
Space exploration	3	15

(a) Compare the responses of adults to those of teens using two pie charts. Be sure to add a category for the Other responses. How does the presence of a large group of Other responses affect the use of these charts?

(b) Compare the responses using a bar chart. Do you have to show the other category when using the bar chart?

(c) If each respondent in the survey listed several areas for research, would it then be appropriate to form a pie chart of the values in a column of the table?

48. **Media** A report of the Kaiser Foundation in the *New York Times* shows the presence of different types of media in the bedrooms of children living at home.

Medium	Age (years) 2–4 (%)	5–7 (%)	8–13 (%)	14–18 (%)
TV	26	39	65	65
Radio	14	18	34	38
Video game	7	18	47	42
Cable TV	10	18	28	32
Computer	4	9	23	19

(a) Explain why it would *not* be appropriate to form a pie chart using the values in any row of this table. Would it be appropriate to use a pie chart to summarize the percentages within a column?

(b) Prepare a chart that compares the types of media available to children from 5 to 7 years old to those available to adolescents from 14 to 18 years old. What seems to be a key point made by the chart?

(c) For a company that makes video game consoles, what age range do these data suggest marketing should target for new sales?

49. **Sugar Share** These data summarize the market share for artificial sweeteners held by the major brands in 2002 and 2005. Prepare a chart that shows the shares each year and the brands that gained the most share from 2002 to 2005.

Brand	2002 (%)	2005 (%)
Splenda	22	52
Sweet N Low	21	15
Equal	30	18
Others	27	15

50. **Internet** This table summarizes results from a survey of more than 30,000 North American households that asks, "What kind of Internet connection do you use at home?" (State of Consumers and Technology: Benchmark 2007 from Forrester.)

Connection Type	2003 (%)	2007 (%)
Cable modem	16	33
DSL	8	33
Satellite	0	1
Fixed wireless	0	2
Dial-up	68	26

(a) Prepare a chart that shows the shares in both years as well as identifies the technologies that are growing or fading.

(b) Does this table prove that fewer households were using dial-up in 2007 than in 2003?

51. **Women** The US Census tracks the number of women-owned businesses that have a payroll. In 1992, women owned more than 800,000 such firms in the

United States. The data identify the industry for all but 9,000 of these. This table also shows the number of businesses in the different industries, from a study in 1995. (These data are taken from a census study in 1992 and reported in the 2000 *Statistical Abstract of the United States*, Table 873.)

Industry	Women Owned	All
Agriculture	15	108
Mining	3	27
Construction	65	634
Manufacturing	41	390
Transportation	31	285
Wholesale trade	46	518
Retail trade	244	1,568
Finance and insurance	56	628
Services	307	2,386

(a) Make a bar chart to display the counts of businesses owned by women and a second that shows the counts for all businesses, and label them correctly. Do these plots suggest that women-owned business concentrate in some industries?

(b) Construct a single plot that directly shows which industries have the highest concentration of women-owned businesses. Would a pie chart of the second column in the table be appropriate?

(c) One data table tracks firms owned by women in 1992. The other tracks all firms and was collected three years later in 1995. Is this time gap a problem?

52. **Gift Card** A survey in the fall of 2004 by Deloitte & Touche found that 64% of responding shoppers planned to buy gift cards as part of their holiday giving. Here's a breakdown of the reported types of gift cards.

Type	Percentage
Specific store or product	50
Restaurant	34
Service	27

(a) Would it be appropriate to show these percentages in a pie chart?

(b) If you receive a gift card, according to these data, is it more likely to be for a product or for a meal?

(c) Is it necessary to summarize these data with a chart, or do you suggest leaving them in a table?

53. **Absences** A company reported the following summary of the days during the last year that were missed by employees who work on its assembly line.

Reason	Days
Illness	4,463
Planned medical leave	2,571
Vacation	15,632
Training	5,532

(a) The only absences that cannot be anticipated are those due to illness; the others are scheduled absences. Provide a figure, table, or chart that shows that most absences are anticipated.

(b) Show a Pareto chart of these data.

54. **Browser** The first popular Internet browser was Netscape. In 1995, Microsoft added Web browsing to Windows by licensing features from Netscape. A year later, the two companies competed head-on for domination of the browser market. As late as 1997, Netscape held 72% of the browser market. On the basis of Web traffic at popular servers, the safalra.com Web site reported these shares for September 2004.

Browser	Share (%)
Internet Explorer	73
Mozilla (Netscape, Firefox)	22
Konquerer (Safari)	2.6
Opera	2.1
Others	0.3

(a) List two questions that you need to resolve in order to use these data properly.

(b) Present a pie chart of these data. What does the chart have to say about the current status of competition among browsers?

(c) Identify the modal choice of Web surfers.

55. **Outsourcing** A survey of 1,000 large U.S. companies conducted by Forrester Research asked their plans for outsourcing high-level white-collar jobs to countries outside the United States. This table shows the percentage that indicated a certain action. The data for 2003 are real; the data for 2008 are speculative. (*Business Week*, December 12, 2004.)

Plan to	2003 (%)	2008 (%)
Do virtually no white-collar offshoring	63	46
Offshore white-collar work to some extent	33	44
Offshore any white-collar work possible	4	10

(a) Summarize the responses to these questions using two pie charts. Are pie charts appropriate for summarizing these percentages?

(b) Summarize the data using a bar chart with side-by-side columns.

(c) Compare the impression conveyed by these plots of any anticipated trend in outsourcing of white-collar jobs.

(d) Does the mode differ from the median for either distribution?

56. The charts in the second 4M example of this chapter compare the shares in the market held by the four major DRAM manufacturers. These changed from 1999 to 2002, as did the size of the market. Total revenues in this industry were $20.8 billion in 1999 and $15.25 billion in 2002.

(a) If the pie charts in that example are to reflect dollars as well as shares, how would the diameters of the pies have to be chosen?

(b) Micron's share grew from 1999 to 2002. Use a paired bar chart for the dollars sold to see whether the DRAM sales of Micron and the other companies increased.

57. **Mathsci** The Bureau of Labor Statistics reported the following employment counts for occupations in the mathematical sciences in 2006.

Occupation	Employment
Computer and information scientists, research	27,650
Computer programmers	396,020
Computer software engineers, applications	472,520
Computer software engineers, systems software	329,060
Computer support specialists	514,460
Computer systems analysts	446,460
Database administrators	109,840
Network and computer systems administrators	289,520
Network systems and data communications analysts	203,710
Computer specialists (other)	180,270
Actuaries	16,620
Mathematicians	2,840
Operations research analysts	56,170
Statisticians	19,660
Mathematical technicians	1,210
Mathematical science occupations (other)	10,190
Total	3,076,200

(a) Create a pie chart of these data showing the share of people that are employed by job classification. What question does this chart answer? What does it conceal?

(b) Which is the modal category?

(c) Improve the pie chart by altering the data to make the figure more interpretable.

58. **Handheld** A survey of the competition between two new handheld games produced the following data.

Otonafami magazine in Japan surveyed 1,000 of its readers; 140 reported owning a Sony PSP and 250 reported that they owned a Nintendo DS. In addition, the two groups reported owning the following number of games. (Because systems like the PSP can be used to play a certain type of DVD, it is possible that someone might have one and not own any games!)

Number of Games	Sony PSP (%)	Nintendo DS (%)
None	2	2
1	27	16
2	25	20
3	14	19
4	8	12
5	10	8
6 or more	14	23

(a) Is a pie chart of either column appropriate?

(b) In order to show a bar chart of the number of owners rather than percentages, what must be done to the table first?

(c) Use a single chart to compare the number of games owned by these respondents. Are percentages or counts more natural for this comparison?

(d) What is the median number of games bought by an owner of the Sony? Of the Nintendo?

(e) How do you interpret the differences in game ownership between those who own a PSP and those who own a DS?

4M Growth Industries

The U.S. Department of Commerce tracks the number of workers employed in various industries in the United States. It summarizes these data in the *Statistical Abstract of the United States*. This regularly appearing volume contains a rich collection of tables and charts that describe the country.

The data shown in the following table appear in Table 886 in the 2000 edition, available on line at www.census.gov. This table summarizes a categorical variable from each of two data tables. These categorical variables identify the type of industry that employs the worker. Each row in the underlying data tables represents an employee, either in 1980 or 1997.

Industry	1980	1997
Agriculture	290	727
Mining	994	586
Construction	4,473	5,513
Manufacturing	21,165	18,633
Transportation	4,623	6,247
Wholesale trade	5,211	6,810
Retail trade	15,047	22,003
Finance and insurance	5,295	7,367
Services	17,186	37,380

Each value shown in this frequency table is a rounded count given in thousands. For example, about 290,000 were employed in agriculture in 1980.

Motivation

a) A firm that sells business insurance that covers the health of employees is interested in how it should allocate its sales force over the different industries. Why would this firm be interested in a table such as this?

Method

b) What type of plot would you recommend to show the distribution of the workforce over industries in each year?

c) Management is particularly interested in knowing which industries have experienced the most growth in the number of employees. What single graph do you recommend to show changes in the size of the workforce in the different industries?

Mechanics

d) Prepare the chart that you chose in part (c). Explain the order you chose to show the categories.

e) How does the chart highlight industries that have smaller employment in 1997 than in 1980?

f) Do you think you should show all nine categories, or should some rows be combined or set aside?

Message

g) What message does your chart convey about the changing nature of employment?

h) By focusing on the growth and using a single chart, what's hidden?

CHAPTER 4

Describing Numerical Data

THE APPLE IPOD™ HAS BEEN A HUGE SUCCESS, carrying Apple through a period of slow computer sales and remaking the music industry. Even the tiniest model in the iPod lineup, the Shuffle™, claims to let you bring 500 songs with you, wherever you go. To get that capacity, you have to get a Shuffle that has 2 gigabytes (GB) of storage capacity.

Let's think about that claim. Two gigabytes translates into 2,048 megabytes (MB) of space for your songs. Do you believe that, no matter what your taste in music is, 500 songs you pick out will fit on a Shuffle?

Questions like this are easy to answer when there's no variation. The answer is a simple yes or no. In fact, the fine print at the Apple Web site answers the question by assuming every song is the same size. If every song that is stored on an iPod uses the same amount of space, say 4 MB, then 500 will fit, with a little room left to spare. If each song takes up 5 MB, however, then you're not getting 500 of them on a Shuffle.

It is a lot harder to answer this question with real songs. Digital recordings of real songs don't take up the same amount of space. The longer the song, the more space it requires. That means there's variation in the amount of space needed to store the songs. Now, do you think 500 different real songs will fit on a Shuffle?

It is fun to see how statistics provides answers to questions such as this, but beyond the fun, the presence of variation affects most business decisions as well. Customers with the same type of loan pay back different amounts each month. Packages labeled to weigh a certain amount in fact weigh different amounts. Sales at the same store vary from day to day. Managing loans, manufacturing, sales, and just about every other business activity requires that we understand variation.

This chapter surveys methods used to display and summarize numerical data. These methods characterize what is typical about data as well as how the data spread out around this typical value. The description of a categorical variable in Chapter 3 essentially amounts to a bar chart that shows the categories and their frequencies. The description of a numerical variable offers many more possibilities. This chapter considers graphical approaches and the use of numerical summaries. One type of numerical summary is based on putting the observed data in order; another type relies on averaging.

4.1 | SUMMARIES OF NUMERICAL VARIABLES

Variation complicates answering questions about the capacity of an iPod, and there's a lot of variation in the size of files that store digital music. Table 4.1 describes five digital recordings of songs randomly picked from a friend's computer. These songs vary in length, with longer songs taking up more room than shorter songs.

TABLE 4.1 Description of several digital recordings of songs.

Song	Artist	Genre	Size (MB)	Length (sec)
My Friends	D. Williams	Alternative	3.83	247
Up the Road	E. Clapton	Rock	5.62	378
Jericho	k.d. lang	Folk	3.48	225
Dirty Blvd.	L. Reed	Rock	3.22	209
Nothingman	Pearl Jam	Rock	4.25	275

To answer our question about fitting 500 songs on a Shuffle, we need to understand the typical length of a song as well as the variation in the sizes around this typical value. You can see that there is variation from reading the few rows of data in Table 4.1, but to understand this variation, we need to consider all of the data.

We'll first collapse all of the data on sizes into a few summary numbers. These summary numbers that characterize a variable are known as **summary statistics**. We emphasize two types of summary statistics that describe the typical value and variation of numerical data. One type of summary statistic comes from putting the data in order, and the second type comes from averaging. We'll then relate these summary statistics to graphs that show a more complete view of a numerical variable.

summary statistic Numerical summary of a variable, such as its median or average.

Percentiles

Several percentiles are commonly used to summarize a numerical variable. Of these, the median is the best-known percentile. We met the median in Chapter 3 as a summary of an ordinal variable. It's also useful as a summary of a numerical variable. The **median** is the 50th percentile, the value that falls in the middle when we sort the observations of a numerical variable. Half of the observed values are smaller than the median, and half are larger. If we have an even number of cases (n is even), the median is the average of the two values in the middle.

median Value in the middle of a sorted list of numbers; the 50th percentile.

Two other percentiles are also commonly used. The 25th percentile and 75th percentile are known as the lower and upper **quartiles**. You can think of the lower and upper quartiles as medians of the lower half and the upper half of the data. One-quarter of the data are smaller than the lower quartile, and one-quarter of the data are larger than the upper quartile. Hence, half of the data fall between the quartiles. The distance from the lower quartile to the upper quartile is known as the **interquartile range** (IQR). Because it's based on

quartile The 25th or 75th percentile.

interquartile range (IQR) Distance between the 25th and 75th percentiles.

percentiles, the IQR is a natural summary of the amount of variation to accompany the median.

The minimum and the maximum of the data are also percentiles. The maximum is the 100th percentile, and we'll call the minimum the 0th percentile. The **range** of a variable is the difference between its minimum and maximum values.

range Distance between the smallest and largest values.

We can illustrate these percentiles using the lengths of the 3,984 songs on our friend's computer. The sorted sizes of all of these songs are:

0.148, 0.246, 0.396, . . . , 3.501, **3.501**, **3.502**, 3.502, . . . 17.878, 21.383, 21.622

1st   2d   3d   1991st 1992d 1993d 1994th   3,982d 3,983d 3,984th

Because $n = 3,984$ is even, the median is the average of the middle two sizes, 3.5015 MB. Half of these songs require less than 3.5015 MB and half need more. The lower quartile of the songs is 2.85 MB and the upper quartile is 4.32 MB: 25% of the songs require less than 2.85 MB, and 25% of the songs require more than 4.32 MB. The middle half of the songs between the quartiles use between 2.85 and 4.32 MB. The difference between these quartiles is the interquartile range.

$$\text{IQR} = 75\text{th percentile} - 25\text{th percentile}$$
$$= 4.32 \text{ MB} - 2.85 \text{ MB} = 1.47 \text{ MB}$$

The range is considerably larger than the interquartile range. These songs range in size from 0.148 MB all the way up to 21.622 MB. The largest song is almost 150 times bigger than the smallest. Because the range depends on only the most extreme values of a numerical variable, it is less often used as a measure of variation. The collection of five summary percentiles,

(minimum, lower quartile, median, upper quartile, and maximum)

five-number summary The minimum, lower quartile, median, upper quartile, and maximum.

is known as the **five-number summary** of a numerical variable.

Averages

mean The average, found by dividing the sum of the values by the number of values. Shown as a symbol with a line over it, as in $\bar{y}$.

The most familiar summary statistic is the **mean**, or average, calculated by adding up the data and dividing by n, the number of values. The mean, like the median, has the same units as the data. Be sure to include the units whenever you discuss the mean.

The mean gives us an opportunity to introduce notation that you will see on other occasions. Throughout this book, the symbol y stands for the variable that we are summarizing, the variable of interest. In this example, our interest lies in the sizes of songs, so the symbol y stands for the size of a song in megabytes. To identify the sizes of specific songs, we add subscripts. For instance, y_1 denotes the size of the song in the first row of the data table, and y_2 denotes the size of the song in the second row. Because n stands for the number of rows in the data table, the size of the song in the last row is denoted by y_n. With this convention, the formula for the mean is

$$\bar{y} = \frac{y_1 + y_2 + \cdots + y_n}{n}$$

Those three dots stand for the sizes of songs between the second and the last. The y with a line over it is pronounced "y bar." In general, a bar over a symbol or name denotes the mean of a variable. The average size of the 3,984 songs works out to be

$$\bar{y} = \frac{3.0649 + 3.9017 + \cdots + 3.5347}{3,984} = 3.7794 \text{ MB}$$

On average, each song needs about 3.8 MB, which is larger than the median size 3.5 MB.

Averaging also produces a summary statistic that measures the amount of variation in a numerical variable. Consider how far each observation is from $\bar{y}$. These differences are called *deviations*. The average squared deviation from the mean is called the **variance**. We use the symbol s^2 for the variance, with the exponent 2 to remind you that we square the deviations *before* averaging them.

variance The average of the squared deviations from the mean, denoted by the symbol s^2.

$$s^2 = \frac{(y_1 - \bar{y})^2 + (y_2 - \bar{y})^2 + \cdots + (y_n - \bar{y})^2}{n - 1}$$

If the data spread out far from $\bar{y}$, then s^2 is large. If the data pack tightly around $\bar{y}$, then s^2 is small. It's natural to wonder why the divisor is $n - 1$ rather than n. The complete explanation appears in Chapter 15, but notice that dividing by $n - 1$ makes s^2 a little bit bigger than dividing by n. That turns out to be just the right adjustment. For now, even though the divisor is not n, think of s^2 as the average squared deviation from $\bar{y}$. (With $n = 3{,}984$, the difference between dividing by n or $n - 1$ is small.)

(p. 69)

It is important that we square the deviations before adding them. Regardless of our data, the sum of the deviations is exactly zero. Positive and negative deviations cancel. (In this chapter's Behind the Math section, a unique property of the mean explains why.) To avoid the cancellation, we square each deviation before adding them. Squared deviations cannot be negative, so their sum won't be zero unless all of the data are the same. Squaring also emphasizes larger deviations, so we'll have to watch for the effects of large deviations. (Absolute values would also avoid the cancellation, but they produce a less common measure of variation.)

Whatever the units of the original data, the variance has *squared* units: s^2 is the average *squared* deviation from the mean. The variance of the size of the songs is 2.584 MB^2, about $2\frac{1}{2}$ "squared megabytes." How should we interpret that?

To obtain a summary we can interpret, we'll convert s^2 into a statistic that does have the same units as the data. To measure variation in the original units, take the square root of the variance. The result is called the **standard deviation**. In symbols, the formula for the standard deviation is

standard deviation A measure of variability found by taking the square root of the variance. It is abbreviated as SD in text and identified by the symbol s in formulas.

$$s = \sqrt{s^2}$$

The square root puts s on the scale of the data. The standard deviation of the songs is $s = \sqrt{2.584\ \text{MB}^2} \approx 1.607$ MB. We use the standard deviation so often that we'll frequently abbreviate it SD. The SD has the right units, but we need to introduce an important concept to appreciate its value in summarizing variation.

4M EXAMPLE 4.1 | **MAKING M&Ms**

MOTIVATION | STATE THE QUESTION

Highly automated manufacturing processes often convey the impression that every item is the same. Think about the last time that you ate a handful of M&M chocolate candies. They all seem to be the same size, but are they?

Mars, the manufacturer, would like to know how many pieces are needed to fill a bag that is labeled to weigh 1.6 ounces. M&Ms are packaged by a highly automated process. If every piece weighs the same amount, the system can count them out. If there's variation, a different type of packaging is necessary. (See Statistics in Action: Modeling Sampling Variation.)

METHOD

DESCRIBE THE DATA AND SELECT AN APPROACH

For this question, we need to know the amount of variation relative to the typical size. If the SD is tiny relative to the mean, then the candies will appear identical even though there's actually some variation present. The ratio of the SD to the mean is known as the **coefficient of variation**.

coefficient of variation The ratio of the SD to the mean, $s/\bar{y}$.

$$c_v = \frac{s}{\bar{y}}$$

The coefficient of variation c_v has no units. The coefficient of variation is typically more useful for data with a positive mean that's not too close to 0. Values of c_v larger than 1 suggest a distribution with considerable variation relative to the typical size. Values of c_v near 0 indicate relatively small variation. In these cases, even though the SD may be large in absolute terms, it is small relative to the mean.

The data for this example are weights of 72 plain chocolate M&Ms taken from several packages. The data table has just one column. Each observation is one piece of candy, and the single data column is the weight (in grams). We are interested in the coefficient of variation c_v of the weights relative to their size.

MECHANICS

DO THE ANALYSIS

The mean weight is $\bar{y}$ = 0.86 gram, and the median nearly matches the mean. The SD is s = 0.04 gram, so the coefficient of variation is small.

$$c_v = 0.04/0.86 \approx 0.0465$$

To determine how many M&Ms are needed to fill the package, we need to convert the package weight into grams. One ounce is equivalent to 28.35 grams, so a package that weighs 1.6 ounces weighs 1.6 × 28.35 = 45.36 grams. If every M&M were to weigh exactly 0.86 gram, it would take 45.36/0.86 ≈ 52.74, about 53, pieces to obtain the sought package weight. The small size of c_v suggests this rule ought to work pretty well.

MESSAGE

SUMMARIZE THE RESULTS

The average weight of chocolate M&Ms in a sample of 72 is 0.86 gram, with SD 0.04 gram. The weights of M&Ms are not identical, but the SD is quite small compared to the mean, producing a coefficient of variation of about 5% (0.04/0.86 ≈ 5%). (No wonder they look identical when I'm eating them.)

Be sure to answer the question that motivated the analysis. These data suggest that 53 pieces are usually enough to fill the bag labeled to contain 1.6 ounces. Because of the variation, however, the packaging system may need to add a few more to allow for pieces that weigh less. As a result, the presence of variation adds to the cost of packaging. That's typical: Variation generally adds to the cost of manufacturing and explains why so many manufacturers work hard to remove variation from their processes. (That's the whole objective of Six Sigma manufacturing developed at Motorola and widely implemented at GE and other companies.)

What Do You Think?

A manager recorded the number of days employees were out sick from the office during the last year. For $n = 7$ employees, the counts are 0, 0, 4, 7, 11, 13, and 35 days.

a. What are the units of $\bar{y}$ and s? Are these different from the units of the median and interquartile range?[1]

b. Find $\bar{y}$ and the median. Which is larger? Why?[2]

c. Find s and the IQR. [*Hint:* The five-number summary is (0, 2, 7, 12, 35) and $s^2 = 146.67$.][3]

4.2 | HISTOGRAMS AND THE DISTRIBUTION OF NUMERICAL DATA

Summary statistics are concise but omit other important properties of numerical variables. A plot can show more about the variable and also help us appreciate what summary statistics like the mean and standard deviation tell us about data.

As a first step to building such a plot, we begin by grouping songs of similar length. We do this by partitioning the sizes into adjacent, equal-width intervals. Once the songs have been grouped, we count the number in each interval. These intervals and counts summarize the **distribution** of the numerical variable. We could show the distribution in a table as we did for categorical data, but it's much more useful to see a plot.

distribution The distribution of a numerical variable is the collection of possible values and their frequencies (just as for categorical variables).

histogram A plot of the distribution of a numerical variable as counts of occurrences within adjacent intervals.

The most common plot of the distribution of a numerical variable is a **histogram**. A histogram shows the counts as the heights of bars that satisfy the area principle. The area of each bar is proportional to the fraction of the total number of cases that fall within its interval. The intervals slice up *all possible values* of a numerical variable. Unlike a bar chart, a histogram positions the bars next to each other, with no gaps. A *relative frequency histogram* displays the *proportion* or *percentage* of cases in each interval instead of the count.

The histogram in Figure 4.1 shows the distribution of the sizes of all 3,984 songs on our friend's computer. Tall bars identify sizes that occur frequently; shorter bars identify sizes that seldom occur.

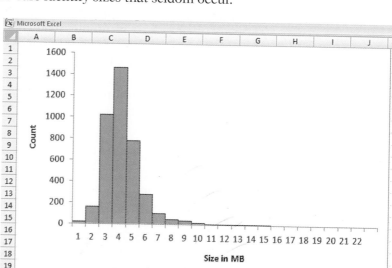

FIGURE 4.1 Histogram of song sizes.

---

[1] The unit of all of these summary statistics is that of the data: days.
[2] The mean is 10 days and the median is 7. The large number of days missed by one employee makes the mean larger.
[3] $s = \sqrt{146.67} \approx 12.1$ days and IQR $= 12 - 2 = 10$ days.

The histogram also suggests both the mean and the SD. The clump of tall bars tells us that most songs require 2 to 6 MB. The tallest bar indicates that nearly 1,500 of the 3,984 songs require between 3 and 4 MB. That's about where we may have expected the songs to cluster because the mean $\bar{y} \approx 3.8$ MB and the median is 3.5 MB.

Histograms versus Bar Charts

It is worthwhile to compare a histogram to a bar chart. The two are similar because they both display the distribution of a variable. Histograms are designed for numerical data, such as weights, that can take on any value. Bar charts show counts of discrete categories and are poorly suited for numerical data.

An explicit example makes the differences more apparent. The histogram on the left of Figure 4.2 gives the weights, in grams, of 48 individual packages of M&M candies. Five of the packages, for example, weigh more than 48 grams and are less than or equal to 49 grams. Although the packages are all labeled to contain 1.69 ounces, about 48 grams, you can see that the weights vary. The bars of the histogram are adjacent to suggest that any weight within the limits of the intervals is possible. (The bar for the highest category is separated from the others since none of the packages weighs between 52 and 53 grams.)

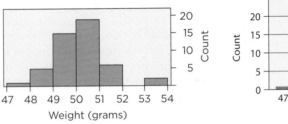

FIGURE 4.2 Histogram and bar chart of M&M package weights.

The right side of Figure 4.2 shows a bar chart derived from the same data. To make this bar chart, we truncated the package weights to two digits, keeping only the integer weights and discarding any fraction of a gram. The bar chart describes the resulting ordinal variable.

The two descriptions of these data are clearly similar. Both show a peak (mode) around 50 grams, for instance. The differences, however, are important. The bars in the bar chart are not placed next to each other as in the histogram; because of the way we constructed the ordinal variable, you cannot have a category between 49 and 50. Also notice that there's no extra space to remind you that the last two categories (labeled 51 and 53) are not so close as the others. For an ordinal variable, we only use the order of the categories, not the size of any differences.

The White Space Rule

Recall an important attribute of the recorded songs: A few songs are evidently much larger than the others. The bars toward the far right of the histogram in Figure 4.1 become so short that you cannot see them. Two songs require between 21 and 22 MB, but you need to trust the *x*-axis to know they are here. (Length aside, these two songs are *very* different: a selection from Gershwin's *Porgy and Bess* and "Whipping Post" by the Allman Brothers.)

The very long songs are outlying values, or **outliers**. Outliers affect almost every statistic. An outlier can be the most informative part of your data, or it can be an error. To hide an outlier conceals an important part of the data, but to let one dominate a plot is not the best choice either. The graph in Figure 4.1 devotes more than half of its area to show the 36 songs that are larger than 10 MB. More than half of the plot shows less than 1% of the songs.

outlier A data value that is isolated far away from the majority of cases.

The **white space rule** says that if your plot is mostly white space—like the histogram in Figure 4.1—then you can improve it by refocusing to emphasize data, not white space. When most of the data occupy a small part of a plot, there's more going on than meets the eye. In this example, we get a better look at the distribution by focusing the plot on the songs below 10 MB. We cannot see what's happening among the data and see the outliers at the same time.

Width of Histogram Intervals

What might be happening that we cannot see in Figure 4.1? Figure 4.3 shows three histograms of songs smaller than 10 MB. The histograms use intervals of different lengths to group the songs.

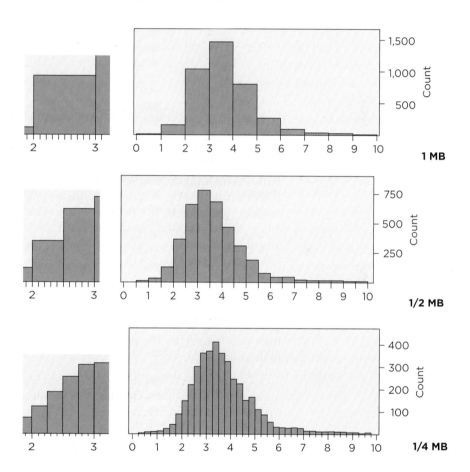

FIGURE 4.3 Histograms on finer scales show more detail.

The intervals that group songs are 1 MB long in the histogram at the top of Figure 4.3. In the second histogram, the intervals are half as long, and in the third histogram, one-quarter as long.

Narrow intervals expose details that are smoothed over by wider intervals. For example, consider songs between 2 and 3 MB. A histogram with 1-MB intervals piles these together and shows a flat rectangle. Narrower intervals reveal that these songs are not evenly spread from 2 to 3 MB; instead, they concentrate toward 3 MB. You can, however, take this process too far. Very narrow intervals that hold just a few songs produce a histogram that looks like a wild forest. At that point, we're back to a bar chart with too many categories.

What's the right length for the intervals in a histogram? Most software packages have a built-in rule that determines the length. The rule takes into account the range of the data and the number of observations. You should take such rules as suggestions and try several as in Figure 4.3.

4.3 | BOXPLOT

boxplot A graphic consisting of a box, whiskers, and points that summarize the distribution of a numerical variable using the median and quartiles.

Sometimes it's handy to have a more compact summary of the distribution of a numerical variable, particularly when we have several variables to see at once. A **boxplot** is a graphical summary of a numerical variable that shows the five-number summary of a variable in a graph. This schematic of a boxplot identifies the key features.

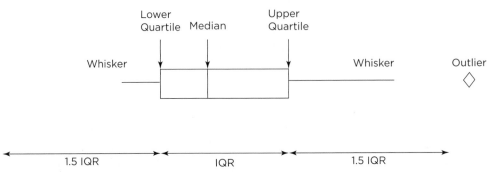

Vertical lines locate the median and quartiles. Joining these with horizontal lines forms a box. A vertical line within the box locates the median, and the left and right sides of the box mark the quartiles. Hence, the span of the box locates the middle half of the data, and the length of the box is equal to the IQR. Horizontal lines known as *whiskers* locate observations that are outside the box, but nearby. A whisker joins any value within $1.5 \times$ IQR of the upper or lower quartile to the box. Distinct symbols identify cases that are farther away. Most boxplots show roughly 1 to 5% of the observations outside the whiskers. Because we have almost 4,000 songs, we can expect to find quite a few outliers in a boxplot of the sizes of the songs.

Combining Boxplots with Histograms

A boxplot is a helpful addition to a histogram. It locates the quartiles and median and highlights the presence of outliers. Figure 4.4 adds a boxplot to the histogram of the sizes.

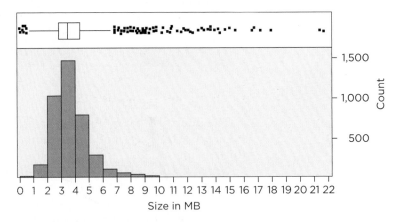

FIGURE 4.4 Histogram with boxplot of the sizes of songs.

The addition of the boxplot makes it easy to locate the median; the vertical line inside the box indicates that the median lies between 3 and 4 MB. (It is exactly 3.5015 MB.) The relatively short length of the box shows that the data concentrate near the median. The majority of the outliers (songs more than $1.5 \times$ IQR away from the box and shown as individual points in the boxplot) lie to the right of the box. Whereas the short bars of the histogram make it easy to miss outliers, the boxplot draws our attention to them.

The combination of boxplot and histogram reinforces the connection between the median and the distribution of the data. The median is the 50th percentile, in the middle of the sorted values. That means that one-half of the area of the histogram is to the left of the median and one-half of the area is to the right, as shown in Figure 4.5. The shaded half of the histogram to the left of the median holds the smaller songs.

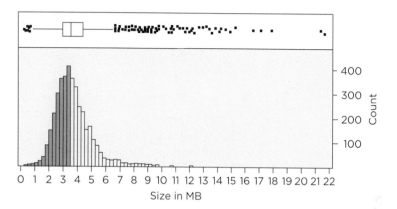

FIGURE 4.5 The median splits the area of the histogram in half.

The mean has a different connection to the histogram. Imagine putting the histogram in Figure 4.5 on a balance beam. Where should you put the fulcrum to make the histogram balance? The answer is at the mean. Think of two children on a seesaw. If one child is a heavier than the other, then the heavier child has to sit near the fulcrum unless he wants to sit there with his friend lifted in the air. Values in the histogram that lie far from the center are like children sitting far from the fulcrum; they exert more leverage. The data cannot move, so the fulcrum—the mean—slides over to make the histogram balance.

The big outliers in this distribution are long songs. To accommodate these outliers, the fulcrum shifts to the right. Consequently, the mean length 3.8 MB is larger than the median 3.5 MB. Outliers have less effect on the median. Had the Beatles let "Hey Jude" run on for several hours, imagine what such an outlier would do to the mean. The mean would chase after this one song, growing larger and larger. The median would remain where it was, unaffected by the ever-longer song. This *robustness* of the median, or resistance to the effect of an outlier, becomes a liability when it comes to dollars. **For variables measured in dollars, we will frequently want to know the mean regardless of outliers.** Would you really want a summary statistic that ignores your most profitable customers or your highest costs?

tip

We can also visually connect the interquartile range to the histogram. The middle half of the data lie within the box of the boxplot, as shown in Figure 4.6.

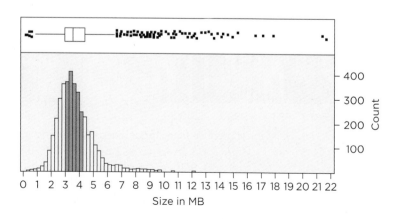

FIGURE 4.6 The inner 50% of the values lie in the box.

Half of the area of the histogram lies within the span of the box.

There is also a connection among means, standard deviations, and histograms, but that connection requires that we first understand the shape of the distribution of a numerical variable.

4.4 | SHAPE OF A DISTRIBUTION

Summary statistics describe the center of the distribution of a numerical variable (mean, median) and its variation (standard deviation, IQR). Additional statistics describe other features of the distribution, collectively known as the shape of a distribution.

Modes

mode Position of an isolated peak in the histogram. A histogram with one mode is unimodal, two is bimodal, and three or more is multimodal.

The distribution of the sizes of the songs shown in these histograms has a common shape. Most data produce a histogram with a single, rounded peak. This peak identifies where the data cluster. A **mode** of a numerical variable identifies the position of an isolated cluster of values in the distribution. The data become sparser as you move away from a mode, and so the heights of the bars in the histogram steadily decline. The tallest bar in Figure 4.4 shows that the mode lies between 3 and 4 MB.

For a categorical variable, the mode is the most common category. A numerical variable can have several modes because the data may concentrate in several locations. The histograms of such variables have several distinct peaks. A histogram such as shown in Figure 4.4 with one main peak is said to be **unimodal**; histograms with two distinct peaks are **bimodal**. Those with three or more are called **multimodal**. A histogram in which all the bars are approximately the same height is called **uniform**.

uniform A uniform histogram is a flat histogram with bars of roughly equal height.

Modes are tricky to identify because the number of modes depends on the width of the bars in the histogram. For example, look back at the histograms in Figure 4.3. With intervals that are 1 MB long, the distribution is unimodal. With one-quarter MB intervals, another mode emerges near 5 MB. Narrower intervals show even more. If the bins in a histogram are too narrow, you will find modes all over the place. As an illustration, let's consider the year in which rock and roll songs in this collection were released. The histogram on the left in Figure 4.7 has narrow intervals and many modes; the histogram on the right uses longer intervals and shows two well-defined modes (bimodal).

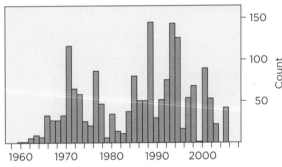
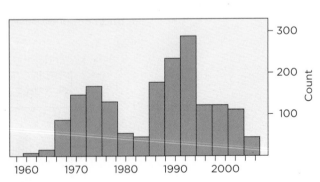

FIGURE 4.7 Narrow intervals produce many modes, and wide intervals produce fewer modes.

The histogram on the right is smoother because it uses wider intervals. Its bimodal shape begs for an explanation: It seems this listener isn't fond of music produced near 1980.

Symmetry and Skewness

symmetric The histogram of the distribution matches its mirror image.

Another aspect of the shape of a distribution of a numerical variable is symmetry. A distribution is **symmetric** if you can fold the histogram along a vertical line through the median and have the edges match up, as if the two sides are mirror images as in Figure 4.8. It won't be a perfect match, but to be symmetric, the two sides should appear similar.

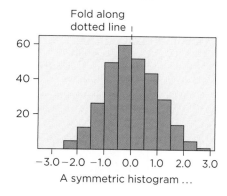
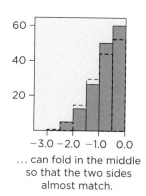

Fold along dotted line

A symmetric histogram …

… can fold in the middle so that the two sides almost match.

FIGURE 4.8 Checking for symmetry.

tails The left and right sides of a histogram where the bars become short.

The extremes at the left and right of a histogram where the bars become short locate the **tails** of the distribution. If one tail of the distribution stretches out farther than the other, the distribution is **skewed** rather than symmetric.

skewed An asymmetric histogram is skewed if the heights of the bins gradually decrease toward one side.

The distribution of the sizes of songs in Figure 4.4 is right skewed. The right tail of the distribution reaches out further from the single mode than the left tail. The boxplot confirms this skewness. The whisker and outliers extend more to the right than to the left of the box. The left side of the distribution is compact with just a few outliers below the left whisker. The half of the songs from the minimum size to the median ranges from near 0 to 3.5 MB, but the right half extends all the way out to 22 MB.

Data encountered in business are often skewed, usually to the right. Variables measured in dollars, such as payrolls, income, costs, and sales, are frequently right skewed because the values cannot be negative and because a few are occasionally much larger than the others. Skewness can be more extreme than in the example of the sizes of songs, as in the example of executive compensation below.

4M EXAMPLE 4.2

EXECUTIVE COMPENSATION

MOTIVATION ▶ STATE THE QUESTION

The press devotes a lot of attention to the compensation packages of chief executive officers (CEOs) of major companies. Total compensation includes much more than salary, such as bonuses tied to company performance. Let's exclude those bonuses and consider *salaries* of CEOs in 2003.

METHOD ▶ DESCRIBE THE DATA AND SELECT AN APPROACH

For data, we will examine the salaries of 1,501 CEOs reported (in thousands of dollars) by Compustat. (Compustat obtains these data from official financial reports that include salary information.) Each row gives the salary of a CEO. We only need one column, the column of salaries.

The data are numerical and we're interested in how they are distributed. A graph showing a histogram along with a boxplot gives a complete summary.

MECHANICS | DO THE ANALYSIS

The middle of the distribution, covered by the box shown in the following histogram, is unimodal and nearly symmetric around the median salary near $650,000. The average salary is a bit larger at $697,000. The middle half of salaries (those inside the box of the boxplot) ranges from $450,000 to $900,000. The data are right skewed. A few CEOs (shown as individual points above $1.5 million) earn much higher salaries. The largest salary, $4,000,000 (4,000 × $1,000), is that of Sumner Redstone, CEO of Viacom.

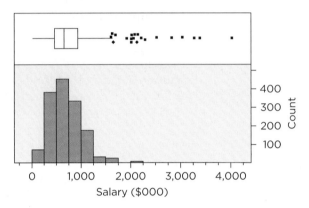

MESSAGE | SUMMARIZE THE RESULTS

When summarizing a distribution, describe its shape, center, and spread without using many technical terms. Also report the symmetry, number of modes, and any gaps or outliers. The annual base salary of CEOs ranges from less than $100,000 into the millions. The median salary is $650,000, with half of the salaries in the range of $450,000 to $900,000. The distribution of salaries is right skewed: Several CEOs make considerably more than most of the others. Even though the majority of CEOs have salaries below $1,000,000, a few salaries exceed $3,000,000.

What Do You Think?

The histogram and boxplot below summarize tuition (in dollars) charged by 100 public colleges around the United States.

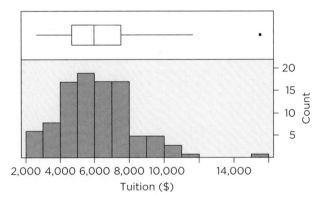

a. Estimate the median and the interquartile range.[4]
b. Describe the shape of the distribution of tuition. Is it symmetric? Unimodal? What about any outliers?[5]
c. If someone claims that 25% of these schools charge less than $5,000 for tuition, can that be right?[6]

---

[4] The median is about $6,000 (centerline of the boxplot) and the IQR (the length of the box) is near 7,200 − 4,700 = $2,500.
[5] The histogram is unimodal and a bit skewed to the right. There's a large outlier at the right.
[6] The claim is reasonable, because the lower edge of the box, the lower quartile, is a bit less than $5,000.

Bell-Shaped Distributions and the Empirical Rule

bell-shaped A bell-shaped distribution represents data that are symmetric and unimodal.

> **caution** The interquartile range is the length of the box that covers the middle half of the data. If the IQR of daily sales is $10,000, then we should not be surprised to learn that sales today were $2,000 more than sales yesterday. The standard deviation *s* has a similar property, but only in some cases. If data are symmetric and unimodal with a histogram that slowly tails off from the mode, we say that the distribution is **bell shaped**. For these, a special relationship known as the empirical rule connects the mean, standard deviation, and histogram.

empirical rule 68% of data within 1 SD of mean, 95% within 2 SDs, and almost all within 3 SDs.

The **empirical rule**, shown in Figure 4.9, uses the standard deviation *s* to describe how data that have a bell-shaped distribution cluster around the mean. According to the empirical rule, 68% (about 2/3) of the data lie within one standard deviation of the mean, and 95% of the data lie within 2 standard deviations. Reaching out farther, the empirical rule says that almost all of the data fall within three standard deviations of the mean.

Interval	Percentage of Data
$\bar{y} - s$ to $\bar{y} + s$	68
$\bar{y} - 2s$ to $\bar{y} + 2s$	95
$\bar{y} - 3s$ to $\bar{y} + 3s$	99.7

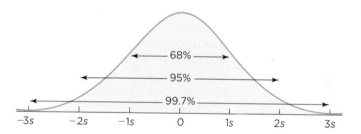

FIGURE 4.9 The empirical rule.

The catch to using the empirical rule is that this relationship among $\bar{y}, s$, and a histogram works well *only* when the distribution is unimodal and symmetric. Does the empirical rule describe the variation in the sizes of songs? We can anticipate that the match will be loose at best because the histogram is not exactly bell shaped; it's skewed to the right. Even so, the range

$$\begin{array}{ccc} \bar{y} - 2s & \text{to} & \bar{y} + 2s \\ 3.8 - 2 \times 1.6 = 0.6 \, \text{MB} & & 3.8 + 2 \times 1.6 = 7 \, \text{MB} \end{array}$$

holds 96% of the data. That's close to what the empirical rule suggests. The empirical rule is less accurate for the shorter interval. According to the empirical rule, the sizes of 68% of the songs should fall between

$$\begin{array}{ccc} \bar{y} - s & \text{and} & \bar{y} + s \\ 3.8 - 1.6 = 2.2 \, \text{MB} & & 3.8 + 1.6 = 5.4 \, \text{MB} \end{array}$$

In the data, however, 84% land in this interval instead.

Although it's not a perfect description of the distribution of the sizes of these songs, consider what the empirical rule allows us to do with just a mean and standard deviation. We can describe the distribution of the sizes of songs like this: The distribution of the sizes of songs is right skewed with a bell-shaped distribution near the mean of the data. The mean size is 3.8 MB with standard deviation 1.6 MB. If someone knows the empirical rule, she or he can imagine what the histogram looks like from two summary statistics

($\bar{y}$ and s) and our hint about skewness. That's a lot to get from two summary statistics.

The empirical rule also explains why the IQR is usually bigger than the SD. The IQR holds half of the data. If the data have a bell-shaped distribution, then ± 1 SD holds about 68% of the data. One interquartile range holds 50% of the data, but it takes $2s$ to hold 68% of the data. Consequently, the IQR is about 35% larger than the SD for bell-shaped distributions. The relationship between the IQR and SD changes for other shapes. If the distribution is skewed, for example, the SD may be larger than the IQR.

Standardizing

The empirical rule lets you judge the relative position of an observation in the distribution of a variable with a bell-shaped distribution *without* needing to see the histogram. For instance, what does it mean when the news reporter announces, "The Dow Jones index of the stock market is down by 200 points today, and the S&P index is down 20 points"? Why should one measure of the stock market change so much more than the other? It helps to know that the SD of day-to-day changes in the Dow Jones index is about 10 times larger than the SD of changes in the S&P index. Relative to their variation, both changed by similar amounts.

A related use of the SD happens in some courses. For example, if you score 91 on a test, did you do well? If the scores on the test are bell shaped with mean $\bar{y} = 95$ and $s = 2$ points, don't expect an A if your professor grades on the curve. A score of 91 on that test is two standard deviations *below* the mean. In other words, only about 2.5% (1/2 of 5%) of the grades lie below.

This notion of measuring size relative to variation is so common that there's a symbol that frequently appears in this context. A **z-score** is the ratio of a deviation from the mean divided by the standard deviation.

z-score The distance from the mean, counted as a number of standard deviations.

$$z = \frac{y - \bar{y}}{s}$$

In place of real-world units such as scores on a test or dollars of income, z-scores measure the distance from the mean in standard deviations; z-scores have no measurement units. The numerator has the same measurement units as the data, but so does the denominator. Dividing one by the other cancels the units. Converting data to z-scores is known as **standardizing** the data. If $z = 2$, then y is two standard deviations above the mean. Data below the mean have negative z-scores; $z = -1.6$ means that y lies 1.6 standard deviations below the mean. Investors who own stocks or students in a class like positive z-scores, but golfers prefer negative z-scores.

standardizing Converting deviations into counts of standard deviations from the mean.

4.5 | EPILOG

What about the motivating question? Did our friend fit 500 of her songs onto her brand-new iPod Shuffle?

Because of variation, not every collection of 500 songs will fit. The longest 500 songs on her computer don't make it, but those were not the songs she chose to download onto her Shuffle. In fact, she squeezed 549 songs onto her Shuffle, including that long song by the Allman Brothers! It looks like she won't be angry about the iPod claim. Steve Jobs, CEO of Apple, can relax!

Best Practices

- *Be sure that data are numerical when using histograms and summaries such as the mean and standard deviation.* Your telephone number and Social Security number are both numbers, but how would you interpret the bars if you put values such as these into a histogram? How would you interpret the average Social Security number?

- *Summarize the distribution of a numerical variable with a graph.* A histogram is most common, and a boxplot shows a more concise summary. If you have to use numerical summaries, be sure to note the shape of the histogram along with measures of its center and spread of the distribution. Also report on the symmetry, number of modes, and any gaps or outliers.

- *Choose interval widths appropriate to the data when preparing a histogram.* Computer programs do a pretty good job of automatically choosing histogram widths. Often, there's an easy way to adjust the width, sometimes interactively. Since it's easy to do, explore how the width of the bars affects the appearance of the shape of the histogram. If the shape seems consistent over several widths, choose the one you like best. If modes appear, you need to figure out why the data cluster in two or more groups.

- *Scale your plots to show the data, not empty space.* In other words, don't forget the white space rule. You should not hide outliers, but it's also not a good idea to show the extremes and ignore the details that might be hidden in the majority of your data.

- *Anticipate what you will see in a histogram.* Think about what the histogram should look like before you build the plot. If you see something that you did not expect, then confirm whether your data measure what you think they do.

- *Label clearly.* Variables should be identified clearly and axes labeled so a reader knows what the plot displays. Show labels with summary statistics like the mean, median, and standard deviation.

- *Check for gaps.* Large gaps often mean that the data are not simple enough to be put into a histogram together. The histogram in Figure 4.10 shows the number of calls handled daily at a call center operated by a bank. This histogram has two modes with a large gap between them.

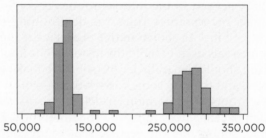

FIGURE 4.10 Number of calls handled daily at a corporate call center.

You might be able to guess why there's a gap: The histogram mixes data for weekdays with data for weekends. It's not surprising to find two modes; fewer calls arrive over the weekend. You ought to separate the cases and show two histograms.

Pitfalls

- *Do not use the methods of this chapter for categorical variables.* This one is worth repeating several times. Some believe that it's okay to use these methods with ordinal data (such as those gained from Likert scales), but be careful if you do this.

- *Do not assume that all numerical data have a bell-shaped distribution.* If data are summarized by only a mean and standard deviation, you may not be seeing the whole picture. Don't assume that the data have a bell-shaped distribution.

- *Do not ignore the presence of outliers.* If you do not look at your data and instead rely on summary statistics, particularly measures like the mean and standard deviation, you can be fooled. These statistics are very sensitive to the presence of extremely unusual observations.

■ *Do not remove outliers unless you have a good reason.* It is easy to get the impression that outliers are bad because of the way they pull the mean, inflate the SD, and make it hard to see the rest of the data. An outlier might be a coding error, but it might also represent your most valuable customer. Think hard before you exclude them.

■ *Do not forget to take the square root of a variance.* The units of the variance, or average squared deviation around the mean, are the squares of the units of the data. These squared units make the variance hard to interpret. By taking the square root, we get the standard deviation, which is a more useful summary.

Software Hints

Software handles most of the details of building a histogram. For example, what should be done for cases that fall on the boundary between two bins? (For the examples in this chapter, those on the boundary get put in the interval to the right.) Software also incorporates rules for determining the number of bins. As shown in this chapter, the number of intervals used to build the histogram is important. You should generally let the software make the first plot, and then try some variations.

Histograms are common in statistics software, but each package does things a little differently. For example, some label the vertical scale with counts whereas others use relative frequencies. Some don't label the axes at all! Software generally chooses the number and placement of the bins by default. You should explore several widths for the bins to see how this choice affects the appearance of the histogram and any conclusions that you might draw.

EXCEL

Excel prefers to make bar charts. You have to work a bit harder to get a proper histogram. Once you have counts for the intervals, you can put these into a bar chart. To remove the gaps between the bars, some versions of Excel allow you to right-click on the plot and change the formatting to set the width between the bars to zero. DDXL makes this process simpler. After selecting the columns with your data (variable names in the first row, please), follow the menu commands

DDXL > Charts and Plots

Then pick the Histogram function from the pop-up Click-Me list in the DDXL dialog box and choose the numerical variable from the list of variable names. You can get a boxplot in the same way.

To obtain summary statistics, you can easily use the formulas that are built into Excel. For example,

the formula average (*range*) produces the mean value of the data in the indicated cell range. To find functions that are related to statistics, follow the menu commands

Formulas > Function > Statistical

to get a list of the available relevant functions. In addition to the function average, Excel includes

median, mode, max, min, percentile, quartile, and stdev

The last finds s, the standard deviation. DDXL shows these summary statistics (and a few others) along with the histogram, or you can use the Summaries command to get them separately.

MINITAB

Follow the menu commands

Graph > Histogram

and fill in the dialog box with the name of a numerical variable. Once the graph has been produced, you can change the width of the intervals by right clicking on any bin and selecting "edit bins," for example. To change these characteristics, you'll need to create another histogram and modify the default options.

To obtain a boxplot, follow the menu sequence

Graph > Boxplot

and enter the names of the variables from the list shown in the dialog. Each boxplot appears vertically oriented in a separate window.

To obtain summary statistics, follow the menu items

Stat > Basic statistics > Display descriptive statistics

Enter the name of the variable and click the **OK** button.

JMP

Follow the menu commands

Analyze > Distribution

and fill in the name of one or more numerical variables; then click OK. (This command produces a bar chart without spaces between the bars if you supply a categorical variable.) The Hand tool allows you to interactively vary the width and position of the bins within the plot itself. Click the red triangle beside the name of the variable in the histogram and choose the item Histogram options to change other aspects of the display.

By default, the command that produces a histogram also shows a boxplot of the same data, aligned on the same scale as the histogram for easy comparison. (You can easily find the median, for instance, in the histogram from the boxplot.) Your boxplot might contain a few extra bells and whistles, such as a diamond and red bar, like this one:

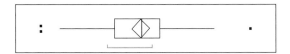

If you right-click on the boxplot and uncheck the options Mean Confid Diamond and Shortest Half Bracket, then the boxplot will look more like you expect.

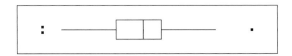

A table that shows the basic summary statistics, including the median, quartiles, mean, and standard deviation, appear below the histogram.

 BEHIND the Math

A Unique Property of the Mean

The mean $\bar{y}$ has a special property that explains why it is the balance point of the histogram. The deviations show how far each value lies from the mean.

$$y_1 - \bar{y}, y_2 - \bar{y}, \ldots, y_n - \bar{y}$$

If you add these *deviations* from the mean, you get zero. To see that, rearrange the sum of the deviations like this.

$$(y_1 - \bar{y}) + (y_2 - \bar{y}) + \cdots + (y_n - \bar{y}) = \sum_{i=1}^{n} y_i - n\bar{y}$$

$$= n\bar{y} - n\bar{y} = 0$$

Deviations to the right of the mean (the y's that are bigger than $\bar{y}$) cancel those to the left (the y's that are less than $\bar{y}$). Because the sum is zero, the average deviation from $\bar{y}$ is also zero. The mean is the *unique* value with this property. By subtracting $\bar{y}$ from each value, we've converted our original data into a new set of values with mean zero.

CHAPTER SUMMARY

Histograms and **boxplots** display the distributions of numerical variables. The bars of the histogram partition the range of the variable into disjoint intervals. The distribution can be either **symmetric** or **skewed** with a long **tail** reaching out to one side or the other. The histogram can be flat, called **uniform**; **bell shaped** with one **mode**; or split into two (*bimodal*) or more (*multimodal*) peaks. Stragglers on the edges of the distribution are known as **outliers**. The **white space rule** reminds you to think about focusing your plot on the data rather than empty space.

Summary statistics quantify various attributes of the distribution. The **median** is the middle value of the distribution, and the **mean** is the average. Both identify the center of a histogram. The **interquartile range** and **standard deviation** summarize the spread of the distribution. The **coefficient of variation** is the ratio of the standard deviation to the mean. The boxplot shows the median and **quartiles** of the data.

The **empirical rule** connects the mean and standard deviation to the bell-shaped distributions. When the histogram is bell shaped, about two-thirds of the data fall in the range $\bar{y} - s$ to $\bar{y} + s$, and 19/20 in the range $\bar{y} - 2s$ up to $\bar{y} + 2s$. If you know the mean and SD, you can find the relative location of a value by computing its **z-score**, the number of SDs above or below the mean.

■ Key Terms

■ Formulas

Many formulas in statistics involve averaging, as in the calculation of the mean and variance. To make these formulas more compact, it is common to use summation notation. **Summation notation** replaces a sum of similar terms by an expression starting with Σ followed by a generic summand. Subscripts and superscripts on Σ define a range that customizes the generic summands. For example, the sum that defines the mean can be abbreviated as

$$y_1 + y_2 + \cdots + y_n = \sum_{i=1}^{n} y_i$$

Subscripts denote a row in the data table. The generic summand is y_i. The subscript i stands for the cases to include in the sum, and the subscript and superscript on Σ indicate that the cases run from the first row ($i = 1$) to the last row (n).

MEAN (AVERAGE)

$$\bar{y} = \frac{y_1 + y_2 + \cdots + y_n}{n} = \frac{\sum_{i=1}^{n} y_i}{n}$$

VARIANCE

$$s^2 = \frac{(y_1 - \bar{y})^2 + (y_2 - \bar{y})^2 + \cdots + (y_n - \bar{y})^2}{n - 1}$$

$$= \frac{\sum_{i=1}^{n} (y_i - \bar{y})^2}{n - 1}$$

We generally use software to compute s^2, but a small example reviews the calculations and reinforces the

point that you have to square the values before adding them. Table 4.2 shows the calculations needed to find s^2 for a data table with $n = 5$ values.

TABLE 4.2 The calculation of the variance s^2 from five values.

	Value	Deviation	Squared
y_1	4	$4 - 7 = -3$	9
y_2	0	$0 - 7 = -7$	49
y_3	8	$8 - 7 = 1$	1
y_4	11	$11 - 7 = 4$	16
y_5	12	$12 - 7 = 5$	25
Sum	35	0	100

The sum of the values is 35, so $\bar{y} = 35/5 = 7$. Deviations from the mean fill the third column. The sum of the deviations equals 0, which is a handy way to check the calculations. The last column holds the squared deviations, which sum to 100. So, $s^2 = 100/(5 - 1) = 25$.

STANDARD DEVIATION (SD)

$$s = \sqrt{s^2}$$

Z-SCORE

$$z = \frac{y - \bar{y}}{s}$$

COEFFICIENT OF VARIATION

$$c_v = \frac{s}{\bar{y}}$$

summation notation Compact notation for a sum of simular terms.

About the Data

The data on digital recordings used in this chapter does indeed come from a friend's computer. The application iTunes™ offers an option to export a summary of the stored music as a text file. We used that option to obtain these data. We should also note that all of these songs are recorded using a standard format (identified as AAC in iTunes). The weights of M&M packages are from an experiment done by children of the authors for showing bar charts in school; their project was to identify the most common color.

EXERCISES

Mix and Match

Match the brief description to the item or value.

1. Position of a peak in the histogram	(a) median
2. Half of the cases are smaller than this value	(b) mean
3. Length of the box in the boxplot	(c) standard deviation
4. Proportion of cases lying within the box of the boxplot	(d) variance
5. A histogram with a long right tail	(e) z-score
6. The average of the values of a numerical variable	(f) two-thirds
7. The average squared deviation from the average	(g) mode
8. The square root of the variance	(h) interquartile range
9. The number of standard deviations from the mean	(i) one half
10. Proportion of a bell-shaped distribution within 1 SD of $\bar{y}$	(j) skewed

True/False

Mark each statement True or False. If you believe that a statement is false, briefly say why you think it is false.

11. The boxplot shows the mean plus or minus one standard deviation of the data.

12. Any data outside the box of the boxplot are outliers and should be removed from the data.

13. Wider bins produce a histogram with fewer modes than would be found in a histogram with very narrow bins.

14. If a histogram is multimodal, the bin widths should be increased until only one bin remains.

15. If data are right skewed, the mean is larger than the median.

16. The histogram balances on the mean.

17. The empirical rule indicates that the range from $\bar{y} - s$ up to $\bar{y} + s$ holds two-thirds of the distribution of any numerical variable.

18. If a distribution is bell shaped, then about 5% of the z-scores are larger than 1 or less than -1.

19. The removal of an outlier with $z = 3.3$ causes both the mean and the SD of the data to decrease.

20. The interquartile range of a distribution is half the range from the smallest to largest value.

21. If the standard deviation of a variable is 0, then the mean is equal to the median.

22. The variance of a variable increases as the number of observations of the variable increases.

Think About It

23. If the median size used by 500 songs is 3.5 MB, will these all fit on a Shuffle that has 2 GB of storage? Can you tell?

24. (a) If the average size of 500 songs is 5.1 MB with standard deviation 0.5 MB, will these fit on a Shuffle that has 2 GB of storage?
    (b) Does the size of the standard deviation matter in answering part (a)?

25. Suppose you were looking at the histogram of the incomes of all of the households in the United States. Do you think that the histogram would be bell shaped? Skewed to the left or right?

26. For the incomes of all of the households in the United States, which do you expect to be larger, the mean or the median?

27. An adjustable rate mortgage allows the rate of interest to fluctuate over the term of the loan, depending on economic conditions. A fixed rate mortgage holds the rate of interest constant. Which sequence of monthly payments has smaller variance, those on an adjustable rate mortgage or those on a fixed rate mortgage?

28. Many atheletes use interval training to improve their strength and conditioning. Interval training requires

working hard for short periods, say a minute, followed by longer intervals, say five minutes, that allow the body to recover. Does interval training increase or decrease the variation in effort?

29. The range measures the spread of a distribution. It is the distance from the smallest value to the largest value. Is the range sensitive to the presence of outliers?

30. What happens to the range as the number of observations increases? If you have a data table that shows the balance of 100 loans, does the range get smaller, get larger, or stay the same if the number of loans is increased to 250?

31. Which has larger standard deviation, the distribution of weekly allowances to 12-year-olds or the distribution of monthly household mortgage payments? Would the same distribution also have the larger coefficient of variation?

32. Which has larger variance, the prices of items in a grocery or the prices of cars at a new car dealer? Which has the larger coefficient of variation?

33. In a study of how the environment of shopping malls affects spending, researchers varied the presence of music and scents in a shopping mall. Interviews with shoppers who made unplanned purchases revealed the following average purchase amounts. Each group has about 100 shoppers.[7]

Condition	Mean Spending ($)
None	64.00
Scent	55.34
Music	96.89
Scent and music	36.70

(a) If each group includes 100 shoppers, which group spent the most money in toto?
(b) If medians summarized each group's spending rather than means, would you be able to identify the group with the largest spending?
(c) Did every shopper who was exposed to scent and music spend less than every shopper who was only exposed to music?

34. The following table summarizes sales by day of the week at a convenience store operated by a major domestic gasoline retailer. The data cover about one year, with 52 values for each day of the week.

	Mean ($)	Std.Dev. ($)
Mon	2,824.00	575.587
Tue	2,768.85	314.924
Wed	3,181.32	494.785
Thur	3,086.00	712.135
Fri	3,100.82	415.291
Sat	4,199.33	632.983
Sun	3,807.27	865.858

(a) Which consecutive two-day period produces the highest total level of sales during the year?
(b) Do the distributions of the sales data grouped by day (as summarized here) overlap, or are the seven groups relatively distinct?
(c) These data summarize sales over $52 \times 7 = 364$ consecutive days. With that in mind, what important aspect of these data is hidden by this table?

35. Suppose the mean time sales representatives spend on the road averages 45 hours per week with standard deviation 5. Would it be surprising to meet a representative who spent 51 hours on the road last week? Do you need to make any assumptions?

36. If a variable, such as income, is very right skewed, would you expect one-sixth of the values to be less than one SD below the mean, or would you expect more or less than one-sixth?

37. Would you expect distributions of these variables to be uniform, unimodal, or bimodal? Symmetric or skewed? Explain why.
(a) The number of songs that each student in your class has downloaded online
(b) Cost of each order from a mail-order catalog for clothing
(c) Weights of bags of M&Ms that are labeled to contain 6 ounces
(d) Heights of students in your class

38. Would you expect distributions of these variables to be uniform, unimodal, or bimodal? Symmetric or skewed? Explain why.
(a) Ages of shoppers in a convenience store near a university late Saturday night
(b) Number of children of shoppers in a toy store
(c) Amount of cash taken in by retail cashiers during a two-hour shift
(d) Number of packages processed each day by Federal Express in their hub location in Memphis, Tennessee, during August and the four weeks before Christmas

39. This histogram shows the distribution of the amount of sales tax (as percentages) in the 50 states and District of Columbia.

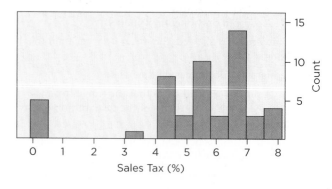

(a) About how many states charge between 4% and 5% sales tax?
(b) Estimate the mean from the histogram. Attach the appropriate unit to your estimate.

[7] Adapted from M. Morrin and J.-C. Chebat (2005), "Person-Place Congruency: The Interactive Effects of Shopper Style and Atmospherics on Consumer Expenditures," *Journal of Service Research*, 8, 181–191.

(c) Which do you think is larger, the mean or median? Why?

(d) Suggest an explanation for the multiple modes that are evident in the histogram.

(e) Is the SD of these data closer to 2, 5, or 10? (You don't need to calculate the SD exactly to answer this question. Think about what the SD tells you about how data cluster around the mean.)

40. This histogram shows the average annual premium (in dollars) for homeowner's insurance paid in the 50 states and District of Columbia.[8]

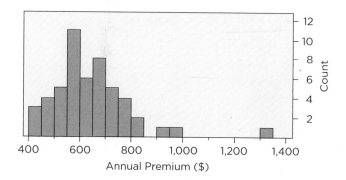

Annual Premium ($)

(a) Which holds more states, the interval from $600 to $700 or the interval from $700 to $800?

(b) Estimate the mean and standard deviation of the rates from the histogram. Be sure to attach the appropriate units to your estimates.

(c) Which do you think is larger, the mean or median? Why?

(d) Suggest an explanation for the multiple modes that are evident in the histogram. If the intervals in the histogram were wider, do you think there would still be two modes?

(e) Is the state with the highest rates slightly larger than the others or an outlier? Which state do you think earns this honor?

(f) Is the SD of these data closer to $100 or $500? (Use what you know about the relationship between the SD and the concentration of the data around the mean rather than trying to calculate the SD from the counts.)

41. The following histogram and boxplot summarize the distribution of income for 1,000 households in Colorado.

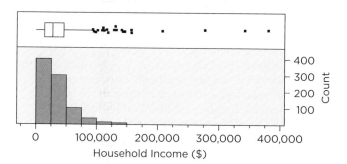

Household Income ($)

(a) Estimate the mean and median. How do you know that the mean is larger than the median for this distribution?

(b) Estimate the interquartile range from the figure.

(c) Which is larger, the IQR or the SD? (*Hint:* These data are *very* skewed.)

(d) How would the figure change if income were expressed in thousands of dollars rather than in dollars?

42. The following display summarizes the percentage of household income that goes to rent for 1,000 families that live in Denver, Colorado.

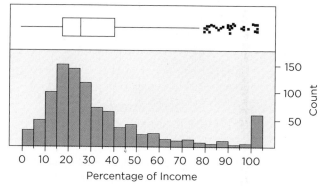

Percentage of Income

(a) Estimate the mean and median from the figure.

(b) Estimate the IQR from the figure.

(c) What's the effect of the cluster of outliers on the right of the SD?

(d) What's your explanation for the tall bin at the right side of the histogram?

43. Price scanners at the checkout register sometimes make mistakes, charging either more or less than the labeled price.[9] Suppose that a scanner is wrong 4% of the time, with half of the errors in favor of the store and half in favor of the customer. Assume that an error in favor of the store means that the scanned price is 5% above the listed price and an error in favor of the customer means that the scanned price is 5% below the listed price. How would these errors affect the distribution of the prices of sold items?

44. The Energy Policy Act of 2005 mandates that gasoline sold in the United States must average at least 2.78% ethanol in 2006. Does this mean that every gallon of gas sold has to include ethanol?

45. The Federal Reserve Survey of Consumer Finances in 2006 reported that the median household net worth in the United States was $93,100 in 2004. In contrast, the mean household net worth was $448,200. How is it possible for the mean to be so much larger than the median?

46. The Federal Reserve Survey of Consumer Finances in 2006 also reported that median household net worth was $91,700 and mean household net worth was $421,500 in 2001. Explain how it is possible for the

---

[8] From the Insurance Information Institute for 2003 (Web at www.iii.org.).

[9] "What, You Got a Problem Paying $102.13 for Tomatoes?" *New York Times*, January 28, 2006.

median net worth to rise only 1.5% but mean net worth to rise by 6.3%.

47. The data in this chapter concern the amount of space needed to hold various songs. What about the length, in minutes? Suppose you get 60 seconds of music from every megabyte of recorded songs. Describe the histogram that shows the number of seconds of each of the 3,984 songs in Figure 4.1.

48. This histogram shows the price in dollars of 218 used BMWs.

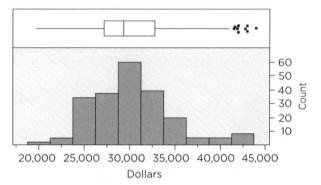

(a) Suppose we converted the prices of these cars from dollars to euros, with an exchange rate of 1.2 dollars per euro. Describe the histogram of the prices if denominated in euros.

(b) Cars are often sold with an additional dealer preparation fee that gets added on to the listed price. If the actual cost of each of the cars shown in this histogram goes up by $500, how does the histogram change?

49. (a) Convert the following summary values given in megabytes (MB) to seconds. Fill in the column for the times in seconds assuming that 1 MB of space provides 60 seconds of music.

Summary	File Size (MB)	Song Length (sec)
Mean	3.8	
Median	3.5	
IQR	1.5	
Standard deviation	1.6	

(b) If the size of each song were increased by 2 MB, how would the summary statistics given in MB change?

(c) If we were to add a very long song to this collection, one that ran on for 45 minutes, how would the median and interquartile range change?

50. (a) The following table summarizes the prices of a collection of used BMWs (introduced in Chapter 1). What are the comparable summary statistics when

Summary	Price ($)	Price (€)
Mean	30,300	
Median	29,200	
IQR	5,700	
Standard deviation	4,500	

measured in euros? Fill in the column for the prices in euros if the exchange rate is $1.20 per euro.

(b) If the price of each car were increased by $500, how would the summary statistics given in dollars change?

(c) If we were to include a very expensive BMW that sells for $125,000 to the collection considered in this exercise, how would the median and interquartile range change?

51. This figure shows the histogram of the annual tuition at 61 top undergraduate business schools, as rated by *Business Week* (May 8, 2006).

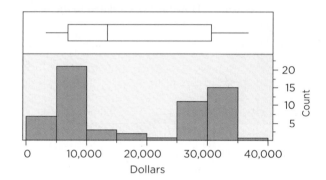

(a) Estimate from the figure the center and spread of the data. Are the usual notions of center and spread useful for these data?

(b) Describe the shape of the histogram.

(c) If you were only shown the boxplot, would you be able to identify the shape of the distribution of these data?

(d) Can you think of an explanation for the shape of the histogram?

52. This histogram shows the winning times in seconds for the Kentucky Derby since 1875.

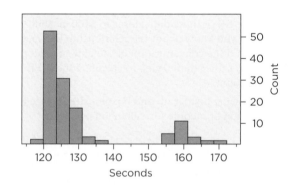

(a) Estimate the center and scale of these data from the histogram. Are the usual notions of center and spread useful summaries of these data?

(b) Describe the shape of the distribution of these data.

(c) What do you think the boxplot of these data would look like?

(d) Can you think of an explanation for the shape of the histogram? What plot of these data would help you resolve whether you have the right explanation?

You Do It

The title of each of the following exercises identifies the data table to be used for the problem.

53. **Cars** A column in this data file gives the rated horsepower for 233 cars sold in the United States during 2003 and 2004.
    (a) Produce a histogram of these data. Describe and interpret the histogram.
    (b) Compare the histogram to the boxplot. What does the histogram tell you that the boxplot does not, and vice versa?
    (c) Find the mean and standard deviation of the horsepower values. How are these related to the histogram, if at all?
    (d) Find the coefficient of variation and briefly interpret its value.
    (e) Identify any unusual values (outliers). Do you think that these are coding errors?
    (f) How does a car with 200 horsepower stack up against these models? Is it typical or one of the more under- or overpowered models?

54. **Cars** Another column in this data file cars gives the rated highway gasoline mileage (in miles per gallon) for 233 cars sold in the United States during 2003 and 2004.
    (a) Produce a histogram of these data. Describe and interpret the histogram.
    (b) Compare the histogram to the boxplot. What does the histogram tell you that the boxplot does not, and vice versa?
    (c) Find the mean and standard deviation of the rated mileages. How are these related to the histogram, if at all?
    (d) Find the coefficient of variation and briefly interpret its value.
    (e) Identify any unusual values (outliers). Do you think that these are coding errors?
    (f) How does a car that gets 20 miles per gallon stack up against these models? Is it typical or does it get relatively low or high mileage?

55. **Beatles** Outliers have a more dramatic effect on smaller data sets. For this example, the data consist of the sizes (in seconds and MB) of the 27 #1 hits on the Beatles' album 1.
    (a) Generate the boxplot and histogram of the sizes of these songs.
    (b) Identify any outliers. What is the size of this song, in minutes and megabytes?
    (c) What is the effect of excluding this song on the mean and median of the sizes of the songs?
    (d) Which summary, the mean or median, is the better summary of the center of the distribution of sizes?
    (e) Which summary, the mean or median, is the more useful summary if you want to know if you can fit this album on your iPod?

56. **Beatles** Outliers affect more than the statistics that measure the center of a distribution. Outliers also affect statistics that measure the spread of the distribution. Use the sizes of the songs in Exercise 55 for this exercise.
    (a) Find the standard deviation and interquartile range of the sizes of the songs (in megabytes).

(b) Which of these summary statistics is most affected by the presence of the outlier? How do you know?
(c) Exclude the most extreme outlier from the data and find the mean and SD without this song. Which summary changes more when the outlier is excluded: the mean or SD?

57. **DP Industry** This data table includes various financial measurements of 196 companies in the data-processing (DP) industry in 2003. One column gives the net sales of the company, in millions of dollars.
    (a) Find the median, mean, and standard deviation of net sales of these companies. What units do these summary statistics share?
    (b) Describe the shape of the histogram and boxplot. What does the white space rule have to say about the histogram?
    (c) Do the data have any extreme outliers? Identify the company if there's an extreme outlier.
    (d) What do these graphs of the distribution of net sales tell you about the DP industry? Is the industry dominated by a few companies, or is there a level playing field with many comparable rivals?

58. **DP Industry** Using the data from Exercise 57, consider the level of sales *per employee*. Sales are in millions of dollars, and the count of employees is given in the data table in thousands.
    (a) Why are there fewer than 196 observations available for making a histogram?
    (b) Find the median, mean, and standard deviation of the ratio of sales per employee. Be sure to include units with these.
    (c) Describe the shape of the histogram. Would the empirical rule be useful for describing how observations cluster near the mean of this ratio?
    (d) Which company has the largest ratio? Is it the sales or the number of employees that makes the ratio so large for this company?
    (e) A start-up DP firm has 15 employees. What net sales does the company need to achieve in order to be comparable to other firms in this industry?

59. **Tech Stocks** These data give the monthly returns on stocks in three technology companies: Dell, IBM, and Microsoft. For each month from January 1990 through the end of 2005 (192 months), the data give the return earned by owning a share of stock in each company. The return is the percentage change in the price, divided by 100.
    (a) Describe and contrast histograms of the three companies. Be sure to use a common scale for the data axes of the histograms to make the comparison easier and more reliable.
    (b) Find the mean, SD, and coefficient of variation for each set of returns. Are means and SDs useful summaries of variables such as these?
    (c) What does comparison of the coefficients of variation tell you about these three stocks?
    (d) Investors prefer stocks that grow steadily over time. In that case, what values are ideal for the mean and SD of the returns? For the coefficient of variation?
    (e) It is common to find that stocks that have a high average return also tend to be more volatile, with larger swings in price. To earn the higher average

rate of return, the investor has to tolerate more volatility. Is that true for these three stocks?

60. **Familiar Stocks** These data give monthly returns for stocks of three familiar companies: Disney, Exxon, and McDonald's from January 1990 through December 2005.

(a) Describe and contrast histograms of the three companies. Be sure to use common axes to scale the histograms to make the comparison easier and more reliable.

(b) Find the mean, SD, and c_v for each set of returns. Can these means and SDs be combined with the empirical rule to summarize the distributions of these returns?

(c) What does comparison of the coefficients of variation tell you about these three stocks?

(d) Typically, stocks that generate high average returns also tend to be volatile, with larger upward and downward swings in price. To get a higher average rate of return, an investor has to tolerate more volatility. Is that true for these three stocks?

61. **Tech Stocks** (See Exercise 59.) Some investors use the Sharpe ratio as a way of comparing the benefits of owning shares of stock in a company to the risks. The Sharpe ratio of a stock is defined as the ratio of the difference between the mean return on the stock and the mean return on government bonds (called the risk-free rate r_f) to the SD of the returns on the stock.

$$\text{Sharpe ratio} = \frac{\bar{y} - r_f}{s}$$

The mean return on government bonds is $r_f = 0.0033$ per month (that is, about 1/3 of 1% per month, or 4% annually).

(a) Find the Sharpe ratio of stock in these three companies. Which looks best from this investment point of view?

(b) Form a new column by subtracting r_f from the return each month on Dell. Then divide this column of differences by the SD for Dell. What's the mean value for this column?

(c) How does the Sharpe ratio differ from the type of standardizing used to form z-scores?

62. **Familiar Stocks** (See Exercise 60 regarding the data, and Exercise 61 for the Sharpe ratio.)

(a) Find the Sharpe ratio of stock in these three companies. Which looks best from this investment point of view?

(b) Form a new column by subtracting r_f from the return each month on Exxon. Next divide this column of differences by the SD for Exxon. What is the mean value for this column?

(c) Look at the returns for December 2005. Do the returns in this month match up to the performance suggested by the Sharpe ratio? Explain briefly what happens.

4M Financial Ratios

Financial ratios like the book-to-market ratio are common in comparisons of the performance of companies. The vast differences in the size of companies make it hard to compare the performance of managment. Which is better: $10 million net income from $100 million in sales or $20 million on $1 billion in sales? To compensate for the differences in magnitude, we can look at the net income divided by the level of sales. How much of each dollar in sales is kept as net income?

Financial ratios also have advantages for data analysis. The distributions of net income and sales are highly skewed. Some companies are a lot larger than others. The ratio of the income to sales, however, is more likely to have a distribution that is closer to bell shaped and more revealing of the variation in performance.

For this exercise, consider the net income and sales (both in billions of dollars) for the 78 energy companies in the Compustat data for 2003. Energy companies include big oil and gas producers like Exxon-Mobil as well as smaller refiners.

Motivation

a) Why would it be helpful to describe management's performance using a single ratio of *Net Income* to *Sales* rather than use the two separate variables?

b) What is the interpretation of the ratio of *Net Income* to *Sales*? What are its units?

c) Describe how a firm might use the ratio of *Net Income* to *Sales* as a yardstick for the performance of its management.

d) Are there other ratios that might serve a similar purpose, or is this the only ratio that will be useful?

Method

e) What plots do you plan to use to see whether the distribution of the ratio of *Net Income* to *Sales* is more bell shaped than either the distribution of *Net Income* or the distribution of *Sales*?

f) Why is it useful to have a bell-shaped distribution for the variation rather than one that is very skewed?

Mechanics

g) Produce histograms of the two variables *Net Income* and *Sales*. Are these skewed or bell shaped?

h) Describe the distributions.

i) Identify any exceptionally large or small outliers in these two distributions.

j) Produce a histogram with a boxplot for the ratio of *Net Income* to *Sales*. Describe the shape of this distribution. Are there features reminiscent of the distribution of songs in this chapter?

k) Is it necessarily true that the ratio of two skewed variables will be less skewed than either of the originals? Under what conditions is the ratio less skewed?

Message

l) Summarize the distribution of the ratio of *Net Income* to *Sales*.

m) On the basis of the distribution of *Net Income* to *Sales*, can you suggest guidelines for performance?

Association between Categorical Variables

BUSY PORTALS LIKE GOOGLE AND YAHOO CHARGE PLENTY FOR THE PRIVILEGE OF ADVERTISING ON THEIR PAGES. If you are a retailer, how are you to decide which locations deliver buyers?

The answer comes from understanding the variation in the variable that indicates whether a visitor makes a purchase. This variation arises because some visitors buy, but others only browse. Suppose that it were the case that every visitor from Yahoo was a buyer, but none of the visitors from Google were. Knowing the link that attracted the shopper explains variation in behavior and reveals the better location for ads.

Real data are more complicated, with less consistent, more variable behavior. Let's focus on the choices faced by an advertising manager at Amazon. She has a budget for advertising to allocate among three busy hosts: MSN, RecipeSource, and Yahoo. Together, these three sites delivered 17,619 visits to Amazon during the fall of 2002, summarized in Table 5.1.

5.1 **CONTINGENCY TABLES**

5.2 **LURKING VARIABLES AND SIMPSON'S PARADOX**

5.3 **STRENGTH OF ASSOCIATION**

CHAPTER SUMMARY

TABLE 5.1 Frequency table of the categorical variable that identifies shoppers from three hosts.

Host	Visits
MSN	7,258
RecipeSource	4,283
Yahoo	6,078
Total shoppers	17,619

MSN generates the most visits, but more visits do not automatically mean more sales. More visits do translate into higher costs, however. Amazon pays a fee to the originating Web site for every visit, whether the shopper buys anything or not. Hosts that generate many visits but few sales are costly to Amazon. Should Amazon pay some hosts more than others for each shopper sent to Amazon?

Understanding which host delivers the best shoppers amounts to understanding the association, or relationship, between two categorical variables. In this example, the variables identify the host and whether a purchase was made. The presence of association means that we can anticipate purchasing behavior once we know the host. This chapter shows how to measure the amount of association by arranging data from these variables into tables. This chapter also presents methods for graphing the information in the table to display the association visually.

5.1 | CONTINGENCY TABLES

To discover whether some hosts yield more purchasers than others, we have to consider the categorical variable that identifies those visits that result in a sale. Figure 5.1 shows the bar charts.

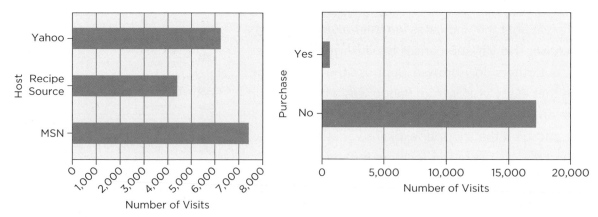

FIGURE 5.1 Bar charts of hosts and purchase actions.

The categorical variable *Host* identifies the originating site summarized in Table 5.1. The categorical variable *Purchase* indicates whether the session produced a sale. There's precious little variation in *Purchase*. Only 516 visits, less than 3%, result in a purchase. If every one of these sales came from one host, Amazon would know where to place its ads.

The bar charts in Figure 5.1 summarize each categorical variable separately, but we need to consider them simultaneously to discover the origin of the sales. For instance, we need to separate visitors from MSN into those who made a purchase and those who did not.

The most common arrangement of such counts organizes them in a table. The rows of the table identify the categories of the one variable, and the columns of the table identify the categories of the other. Such a table is called a **contingency table**. Table 5.2 shows the variable *Purchase* (along the rows) and the variable *Host* (columns).

contingency table A table that shows counts of the cases of one categorical variable contingent on the value of another.

TABLE 5.2 Contingency table of Web shopping.

		Host			
		MSN	**RecipeSource**	**Yahoo**	**Total**
Purchase	**No**	6,973	4,282	5,848	17,103
	Yes	285	1	230	516
	Total	7,258	4,283	6,078	17,619

cell Intersection of a row and a column of a table.

mutually exclusive The cells of a contingency table are mutually exclusive if each case appears in only one cell of the table.

The **cells** of this contingency table count the visits for every combination of *Host* and *Purchase*. The cells of the contingency table are **mutually exclusive**; each case appears in exactly one cell. For example, the column labeled MSN shows that 285 of the 7,258 visits from MSN generated a purchase. By comparison, only 1 of the 4,283 visits from Recipe Source led to a purchase.

Marginal and Conditional Distributions

The margins of Table 5.2 (shaded gray) give the total counts in each row and column. Because the cells of the table are mutually exclusive, the sum of the counts in the cells of the first column equals the total number of visits from MSN. The sum for each column appears in the bottom margin of the contingency table; these total counts match the frequency distribution of *Host* shown in Table 5.1. The right margin shows the frequency distribution of *Purchase*. Because these counts of the totals in each row and column are typically placed along the margins of a contingency table, the frequency distributions of the variables in the table are also called **marginal distributions**. The bar charts in Figure 5.1 show the marginal distributions.

marginal distribution The frequency distribution of a variable in a contingency table given by counts of the total number of cases in rows (or columns).

Percentages help us interpret a contingency table, but we've got to choose which percentage to show. For example, 285 shoppers from MSN made a purchase. To show this count as a percentage, we have three choices.

$$285 \text{ is } \begin{cases} 1.62\% \text{ of all } 17{,}619 \text{ visits.} \\ 3.93\% \text{ of the } 7{,}258 \text{ visits from MSN.} \\ 55.23\% \text{ of the } 516 \text{ visits that made a purchase.} \end{cases}$$

All are potentially interesting. Some statistics packages embellish a contingency table with every percentage, as in Table 5.3.

TABLE 5.3 Too many percentages clutter this contingency table.

Count Total % Col % Row %		Host			
		MSN	**Recipe Source**	**Yahoo**	**Total**
Purchase	**No**	6,973 39.58 96.07 40.77	4,282 24.30 99.98 25.04	5,848 33.19 96.22 34.19	17,103 97.07
	Yes	285 1.62 3.93 55.23	1 0.01 0.02 0.19	230 1.31 3.78 44.57	516 2.93
	Total	7,258 41.19	4,283 24.31	6,078 34.50	17,619

Each cell lists the count along with its percentage of the total, the column, and the row. Tables like this one are enough to give percentages a bad reputation. The table shows too many percentages. To decide which percentage to show, it's best to choose the percentage that answers the relevant question.

conditional distribution The distribution of a variable restricted to cases that satisfy a condition, such as those in a row or column of a contingency table.

Because the account manager at Amazon is interested in which host produces the largest proportion of purchasers, a better table shows only the counts and column percentages. Let's start with MSN. For the moment, we're interested in *only* the 7,258 visits from MSN in the first column of Table 5.2. The distribution of a variable that is restricted to cases satisfying a condition is called a **conditional distribution**. In a table, a conditional distribution refers to counts within a row or column. By limiting our attention to visits from MSN, we see the distribution of *Purchase* conditional on the host being MSN.

Table 5.4 shows the counts and the column percentages. The percentages within each column show the conditional distribution of *Purchase* for each host.

TABLE 5.4 Contingency table with relevant percentages.

Count Col %		Host			
		MSN	**Recipe Source**	**Yahoo**	**Total**
Purchase	**No**	6,973 96.07%	4,282 99.98%	5,848 96.22%	17,103 97.07%
	Yes	285 3.93%	1 0.02%	230 3.78%	516 2.93%
	Total	7,258	4,283	6,078	17,619

Compare this table with Table 5.3. Without the distraction of extraneous percentages, we can quickly see that visitors from MSN and Yahoo yield similar percentages of purchases (3.93% and 3.78%, respectively). In comparison, the percentage of purchases among visitors from Recipe Source is much smaller (0.02%).

associated Two categorical variables are associated if the conditional distribution of one variable depends on the value of the other.

We've just discovered that *Host* and *Purchase* are **associated**. Categorical variables are associated if the column percentages vary from column to column (or if the row percentages vary from row to row). In this case, the proportion of visits that produce a purchase differs among hosts. The association between *Host* and *Purchase* means that knowing the host changes your impression of the chance of a purchase.

Variables can be associated to different degrees. The least association occurs when the column percentages are identical. Overall, 516/17,619 = 2.93% of the visits made a purchase. If 2.93% of visits from *every* host made a purchase, then the rate of purchases would not depend on the host. Each conditional distribution of *Purchase* given *Host* would match the marginal distribution of *Purchase,* and it would not matter which host was chosen.

That's not what happens here: *Host* and *Purchase* are associated. Visitors from some hosts are more likely to make a purchase. Because *Host* and *Purchase* are associated, the account manager at Amazon might be willing to pay more for visits from MSN or Yahoo and less for those from Recipe Source. The value of the visit *depends* on the host. (The account manager would also be interested in the dollar value of these purchases. It turns out that the dollar value of these purchases is comparable for the three hosts, averaging about $50. We'll revisit that aspect of these data later in Chapter 18.)

Segmented Bar Charts

Bar charts that show the marginal distributions (Figure 5.1) don't reveal association, but charts that reach inside the table do. For example, managers who control the distribution channels for Amazon locate warehouses near large concentrations of shoppers to reduce shipping costs. Being close makes it cheaper to offer free shipping as well as rapid delivery. Table 5.5 shows the counts of *Purchase* by *Location* over a wider range of hosts (23,709 visits compared to 17,619).

segmented bar chart A bar chart that divides the bars into shares based on a second categorical variable.

Because we're interested in discovering where those who make a purchase live, this table shows row percentages. With four percentages in each conditional distribution, it becomes helpful to have a plot. A **segmented bar chart** divides the bars in a bar chart proportionally into segments corresponding to

TABLE 5.5 Contingency table of purchases organized by region.

		Location				
		North Central	**North East**	**South**	**West**	**Total**
Purchase	**No**	5,640 24.46	4,450 19.30	8,321 36.09	4,645 20.15	23,056
	Yes	161 24.66	146 22.36	177 27.11	169 25.88	653
	Total	5,801	4,596	8,498	4,814	23,709

the percentage in each group. If the bars look identical, then the variables are not associated.

Although the South sends the largest number of visitors, visitors from the South are more likely to browse than buy. About 36% of the browsers come from the South, but only 27% of the buyers. If *Purchase* and *Region* were not associated, then these percentages should be about the same. Because they differ, *Region* is associated with making a purchase. The segmented bar chart in Figure 5.2 shows these differences more clearly than Table 5.5. The yellow segment that identifies visits from the South makes up a larger share among those who don't purchase (the top bar) than among those who do make a purchase.

Conditional Distribution of Region

FIGURE 5.2 A segmented bar chart shows the presence of association.

caution	Be careful interpreting a segmented bar chart. This chart compares relative frequencies of two conditional distributions. Because these are relative frequencies rather than counts, the bars do not represent the same number of cases. The bar on the top in Figure 5.2 summarizes 23,056 cases whereas the bar on the bottom summarizes 653 purchases. The chart obeys the area principle, but the area is proportional to the percentages within each row of the table.

Segmented bar charts frequently appear in news items. For example, the chart and text below appeared in the *New York Times* at a time when the

But Will They Spend it?

In 2001, many households received tax rebates of up to $600 as part of President Bush's $1.35 trillion tax cut package. At the time, the University of Michigan conducted a survey asking people what they intended to do with the payments and found only about 22% planned to spend them.

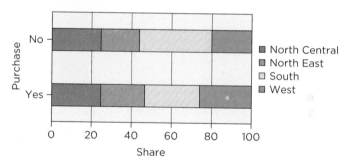

People expecting to receive the 2001 rebate said they would

HOUSEHOLD INCOME	Spend It	Save It	Pay Off Debt
$20,000 and less	18%	21	62
20,001 to 35,000	19	33	48
35,001 to 50,000	19	34	47
50,001 to 75,000	27	28	45
More than $75,000	24	33	43

New York Times

government was debating the use of tax cuts to stimulate consumer spending and avoid a recession.[1] The conditional distribution of behavior changes with income, so there's association. In this example, households with smaller incomes are more likely to use tax rebates to pay down debt than households with higher incomes.

Mosaic Plots

mosaic plot A tiled plot in which the size of each tile is proportional to the count in a cell of a contingency table.

A **mosaic plot** is an alternative to the segmented bar chart. A mosaic plot displays colored tiles, rectangular regions that represent the counts in each cell of a contingency table. The size of a tile is proportional to the count rather than the percentage within a row or column (as is done with segmented bar charts). The layout of the tiles matches the layout of the cells in a contingency table. The tiles within a column have the same width, but possibly different heights. The widths of the columns are proportional to the marginal distribution of the variable that defines the columns. For example, Figure 5.3 shows the mosaic plot of Table 5.5. The tiny height of the red tiles shows the counts of purchases; the tiny sizes remind you how rare it is to find a purchase among the visits.

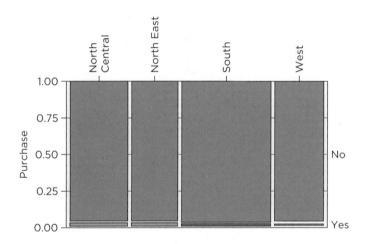

FIGURE 5.3 Mosaic plot of the purchases by region.

The widest tiles are in the column for the South because the South contributes the most visits. Because purchases are so rare, however, it's hard to see in the mosaic plot that the share of purchases is smaller for visitors from the South.

Mosaic plots are more useful for seeing association in data for which the relative frequencies do not get so small. As an example, Table 5.6 shows counts of sales of shirts at a men's clothing retailer. Do *Size* and *Style* appear associated? If the two are not associated, managers should order the same proportion of sizes in every style. If the two are associated, the distribution of sizes varies from style to style and managers should order in different proportions.

TABLE 5.6 Sales of shirts at a men's clothing retailer.

		Style			
		Button Down	**Polo**	**Small Print**	**Total**
Size	**Small**	18	27	36	81
	Medium	65	82	28	175
	Large	103	65	22	190
	Total	186	174	86	446

[1] "Economists Debate the Quickest Cure," *New York Times*, January 19, 2008.

It's hard to see the association quickly in this table of counts, but a mosaic plot as shown in Figure 5.4 makes the association very clear.

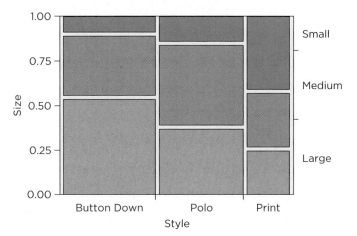

FIGURE 5.4 Mosaic plot of the shirt sales shows association between Size and Style.

The tiles would line up in the absence of association. In this example, the proportions of sizes vary across the styles, causing the tiles to vary in height. The irregular heights indicate that these variables are associated. Small sizes are much more prevalent among beach prints than button-down shirts. Because the mosaic plot respects the area principle, we can also see that the button-down style is the biggest seller overall (these tiles are wider than the others) and the beach-print style is the smallest seller. A segmented bar chart loses track of the marginal distribution, contrasting only the conditional distributions within rows or columns.

4M EXAMPLE 5.1 | CAR THEFT

MOTIVATION ▸ STATE THE QUESTION

Auto theft costs owners and insurance companies billions of dollars. The FBI estimates that 1.2 million cars worth $8.4 billion were stolen in 2002.

Should insurance companies charge the same premium for every car model for theft insurance or should they vary the premium? Obviously, a policy that insures a $90,000 Porsche should cost more than one for a $15,000 Hyundai. But should the premium be a fixed percentage of the car's value, or should the percentage vary from model to model? To answer this question, we need to know whether some cars are more likely to be stolen than others. It comes down to whether there is an association between car theft and car model.

For this example, it's up to you to decide whether or not an insurance company should charge a fixed percentage of the price to insure against theft. The company can either charge a fixed percentage of the replacement cost or charge a variable percentage for models of cars that are stolen more often. Are there large differences in the rates of theft?

METHOD ▸ DESCRIBE THE DATA AND SELECT AN APPROACH

Identify your data and verify that the categories are mutually exclusive. Then identify how you plan to use these data. The two categorical variables needed in this analysis identify the type of car and whether the car was stolen. The following data come from the National Highway Traffic Safety Administration (NHTSA). We picked seven popular models. We will organize these data into a table that shows the rates of theft for the different models.

MECHANICS | **DO THE ANALYSIS**

Make an appropriate display or table to see whether there is a difference in the relative proportions. For example, this table shows the counts along with the percentage of each model that is stolen. Among these models, the Dodge Intrepid has the highest percentage stolen (1.486%), followed by the Dodge Neon (0.804%). The Honda Accord has the lowest percentage stolen (0.167%).

Model	Stolen	Made	% Stolen
Chevrolet Cavalier	1,017	259,230	0.392
Dodge Intrepid	1,657	111,491	1.486
Dodge Neon	959	119,253	0.804
Ford Explorer	1,419	610,268	0.233
Ford Taurus	842	321,556	0.262
Honda Accord	702	419,398	0.167
Toyota Camry	1,027	472,030	0.218

Notice that we did not add the extra column that shows the number that were not stolen. If you do that, you'll see that the number made is the marginal (row) total. For example, for the Dodge Intrepid we have

Model	Stolen	Not Stolen	Total
Intrepid	1,657	109,834	111,491

We think that the table looks better, with more emphasis on the number stolen, without adding the additional column.

MESSAGE | **SUMMARIZE THE RESULTS**

Discuss the patterns in the table and displays. The following chart shows the percentage of cars within each category that were stolen. Some models (e.g., Dodge Intrepid) are more likely to be stolen than others (e.g., Honda Accord). About 1.5% of 2002 Intrepids were stolen, compared to less than 0.17% of 2002 Accords.

A lot of Accords get stolen, but that count can be explained by the sheer number of Accords sold each year. As in the example of Web shopping, a segmented bar chart or mosaic chart is less useful because the percentages stolen are so small.

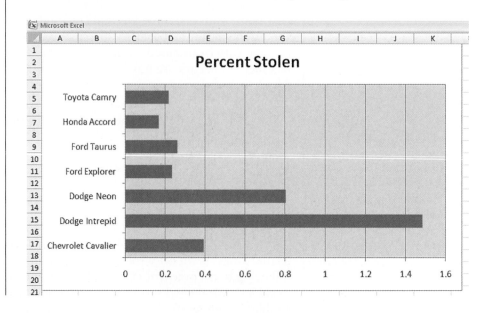

Discuss the real-world consequences and address the motivating question. These data suggest that the insurer should charge higher premiums for theft insurance for models that are most likely to be stolen. A Dodge Intrepid is seven times as likely to be stolen as a Toyota Camry. Customers who buy an Intrepid (which costs about the same as a Camry) should pay a higher premium for theft insurance.

What Do You Think?

An online questionnaire asked visitors to a retail Web site if they would like to join a mailing list. This contingency table summarizes the counts of those who joined as well as those who made a purchase.

		Mailing List	
		Join	**Decline**
Purchase	**Yes**	52	12
	No	343	3,720

The columns indicate whether the visitor signed up (*Mailing List* = Join or Decline), and the rows indicate whether the visitor made a purchase (*Purchase* = Yes or No). For example, 52 visitors joined the mailing list and made a purchase.

a. Find the marginal distribution of *Purchase*.[2]
b. Find the conditional distributions of *Purchase* given whether the customer signed up or not. Do the conditional distributions differ? How can the retailer exploit this property of the data?[3]
c. Does a segmented bar chart provide a helpful plot for these data?[4]
d. Is the variable *Purchase* associated with the variable *Mailing List*?[5]

5.2 | LURKING VARIABLES AND SIMPSON'S PARADOX

Association often gets confused with causation. This mistake can lead to serious errors in judgment. Consider Table 5.7, which compares the performance of two shipping services.

TABLE 5.7 Counts of damaged packages.

Count Column %		Service		
		Orange Arrow	**Brown Box**	**Total**
Status	**Damaged**	45 15%	66 33%	111 22.2%
	OK	255 85%	134 67%	389 77.8%
	Total	300	200	500

[2] The row totals determine the marginal distribution, 64 Yes who made a purchase and 4,063 No.
[3] Among the 395 who joined the list, 52 made a purchase (13%). Among those who declined, 12 out of 3,732 made a purchase (0.32%). Customers who joined were more likely to make a purchase. The retailer could use this distinction by encouraging customers who haven't made a purchase to join the mailing list and thus be converted.
[4] Not really, because one percentage is so small. You could show a figure like that in the prior 4M.
[5] The two are associated because the percentages within the rows (or within the columns) differ. The fraction of customers who make a purchase depends on whether they sign up.

This contingency table shows the number of cartons that were damaged when shipped by two delivery services. The percentages in each cell are column percentages. Overall, 22.2% of the 500 cartons that were shipped arrived with visible damage. Among these, 15% of cartons shipped via Orange Arrow arrived damaged compared to 33% for Brown Box. There's definitely association; neither conditional distribution matches the marginal distribution of *Status*.

Table 5.7 suggests that Orange Arrow is the better shipper, and we might be tempted to believe that cartons are more likely to arrive okay *because* they are shipped on Orange Arrow. If we believe that, we might decide to ship everything on Orange Arrow. Before we do that, however, we'd better make sure that this table offers a fair comparison. Maybe there's another explanation for why packages shipped on Brown Box arrive damaged more often.

To think of an alternative explanation, we have to know more about these packages. In this instance, the cartons hold car parts. Some cartons hold engine parts whereas others hold plastic molding. Guess which cartons are heavier and more prone to damage. The two tables in Table 5.8 separate the counts in Table 5.7 into those for heavy cartons (left table) and those for light cartons (right table).

TABLE 5.8 Separate tables for heavy and light packages.

Count Column %		Heavy			Light		
		Service			Service		
		Orange Arrow	Brown Box	Total	Orange Arrow	Brown Box	Total
Status	Damaged	20 67%	60 40%	80 44.4%	25 9%	6 12%	31 9.7%
	OK	10 33%	90 60%	100 55.6%	245 91%	44 88%	289 90.3%
	Total	30	150	180	270	50	320

Orange Arrow is no longer clearly the better shipper. Among heavy packages, 67% of those shipped on Orange Arrow arrived damaged compared to 40% for Brown Box. For light cartons, 9% of those shipped by Orange Arrow arrived damaged compared to 12% for Brown Box.

The initial comparison favors Orange Arrow because it handles a higher share of light packages. Brown Box seems more likely to damage packages because it handles a greater proportion of heavy cartons. Heavy cartons more often arrive with some damage (44.4% versus 9.7%). Table 5.7 presents a misleading comparison; it compares how well Orange Arrow ships light cartons to how well Brown Box ships heavy cartons. The weight of the cartons is a hidden, **lurking variable**. Table 5.8 adjusts for the lurking variable by separating the data into heavy cartons and light cartons.

Such reversals often go by the name **Simpson's paradox**. It can seem surprising—downright paradoxical—that one service looks better overall, but the other looks better when we restrict the comparison. The explanation lies in recognizing the presence of a lurking variable. Before you act on association (like sending all the business to Orange Arrow), be sure to identify the effects of lurking variables.

One of the best-known examples of Simpson's paradox occurred when the University of California at Berkeley was sued for bias against women applying to graduate school. When data from all of the graduate programs at Berkeley were pooled, the admission rate for men was much higher than that for women. However, it was discovered that the rate was very similar in every department. In fact, most departments had a small bias *in favor of women*.

lurking variable A concealed variable that affects the apparent relationship between two other variables.

Simpson's paradox An abrupt change in the association between two variables that occurs when data are separated into groups defined by a third variable.

The explanation for the apparent overall bias was that women tended to apply to departments that had many applicants and therefore low rates of admission. Men tended to apply to departments such as mathematics that had fewer applicants and higher admission rates.[6]

4M EXAMPLE 5.2	AIRLINE ARRIVALS

MOTIVATION ▸ STATE THE QUESTION

A corporate CEO regularly flies to meetings with branch managers in Boston, Orlando, Philadelphia, and San Diego. The CEO is very pressed for time and must schedule flights that arrive just in time for meetings. A delay in the arrival might cause the CEO to miss the meeting and have to reschedule future meetings—a major headache.

The corporation has arranged deals with two airlines, Delta and US Airways, to provide travel services. Does it matter which of these two airlines the CEO chooses?

METHOD ▸ DESCRIBE THE DATA AND SELECT AN APPROACH

This analysis requires at least two categorical variables that describe flights into these four destinations. One variable identifies the airline (US Airways or Delta) and the second variable identifies whether the flight arrived on time. We will use data from the U.S. Bureau of Transportation Statistics, the agency that monitors the status of all domestic flights within the United States.

MECHANICS ▸ DO THE ANALYSIS

The following table summarizes 10,906 arrivals on Delta and US Airways at these four destinations during a recent month.

Arrivals to Four Destinations				
Count Column %		**Airline**		
		Delta	**US Airways**	**Total**
Arrival	**Delayed**	659 20%	1,685 22%	2,344
	On Time	2,596 80%	5,966 78%	8,562
	Total	3,255	7,651	10,906

This table suggests that the two airlines perform comparably, with a slight edge in favor of Delta. Eighty percent of flights on Delta arrived on time compared to 78% for US Airways.

Before we tell the CEO to book a flight on Delta, however, we should think about whether there's a lurking variable. That requires knowing a bit more about these flights.

[6] P. J. Bickel, E. A. Hammel, and J. W. O'Connell (1975), "Sex Bias in Graduate Admissions: Data from Berkeley," *Science* 187:4175, pp. 398–404.

Arrivals to Orlando			
Count Col %	**Delta**	**US Airways**	**Total**
Delayed	228 19.5%	150 15.5%	378
On Time	940 80.5%	820 84.5%	1,760
Total	1,168	970	2,138

For example, consider the preceding contingency table, which isolates flights into Orlando. For flights to Orlando, US Airways is the better choice. In fact, no matter which destination, US Airways has a higher percentage of on-time arrivals.

On Time %	Delta	US Airways
Boston	80.1%	81.7%
Orlando	80.5%	84.5%
Philadelphia	70.5%	74.3%
San Diego	84.2%	85.4%

That's Simpson's paradox.

MESSAGE | SUMMARIZE THE RESULTS

We recommend booking the flight for the CEO on US Airways. No matter which destination, US Airways is more likely to arrive on time.

It is worthwhile to review why Delta appears better overall, even though US Airways arrives on time more often for each destination. The initial 2-by-2 table in Example 5.2 masks a lurking variable: destination. The destination matters: Delays are more common at Philadelphia, as shown in Table 5.9.

TABLE 5.9 Delayed arrivals by destinations.

Count Col %		Destination				
		Boston	**Orlando**	**Philadelphia**	**San Diego**	**Total**
Arrival	**Delayed**	615 19%	378 18%	1,230 26%	121 15%	2,344
	On Time	2,620 81%	1,760 82%	3,505 74%	677 85%	8,562
	Total	3,235	2,138	4,735	798	10,906

In addition, most flights on US Airways go to Philadelphia, whereas most on Delta go to Boston, as shown in Table 5.10.

TABLE 5.10 Airlines by destinations.

Count Row %		Destination				
		Boston	**Orlando**	**Philadelphia**	**San Diego**	**Total**
Airline	**Delta**	1,409 43%	1,168 36%	312 10%	366 11%	3,255
	US Airways	1,826 24%	970 13%	4,423 58%	432 6%	7,651
	Total	3,235	2,138	4,735	798	10,906

Table 5.9 answers the question, "Am I more likely to arrive on time flying to Boston on Delta or arrive on time flying to Philadelphia on US Airways?" The answer: Take Delta to Boston. There's nothing wrong with that answer; it's just an odd question. By focusing the analysis on flights into a specific destination, we control for this lurking variable and answer the right question.

Once you identify a lurking variable, you can remove its effects as we did in this example. But here's the hard part. How do you know whether there is a lurking variable in this example? It's easy to imagine other possible lurking variables, too. Maybe it's the type of airplane, the day of the week, or the time of day. Make no mistake about it. You need to understand the context of your data to find a lurking variable.

5.3 | STRENGTH OF ASSOCIATION

In the first example in this chapter, we concluded that *Purchase* and *Host* are associated because the proportion of visitors who make purchases differs from host to host. How different are they? Would you describe the association as weak or strong?

Chi-Squared

To answer this question, it is useful to have a statistic that quantifies the amount of association. Instead of saying "There's some association" or "There's a lot of association," we can quantify the degree of association with the statistic called **chi-squared** (pronounced "kī squared"). (Some authors and software call the statistic chi-square, without the final "d.") The larger chi-squared becomes, the greater the amount of association. This statistic also offers a preview of an approach frequently taken in statistics. To quantify the degree of association between variables, we compare the data we observe to artificial data that have no association. Chi-squared does this by comparing the observed contingency table to an artificial table with the same marginal totals but no association. If the tables are similar, then there's not much association. The greater the difference between the tables becomes, the greater the association.

We'll illustrate the use and calculation of chi-squared with an example. A recent poll asked 200 people at a university about their attitudes toward sharing copyrighted music. Half of the respondents were students and the others were staff (administrators or faculty). Table 5.11 summarizes the counts.

chi-squared A statistic that measures association in a contingency table; larger values of chi-squared indicate more association.

TABLE 5.11 Attitudes toward sharing copyright materials.

		Attitude toward Sharing		
		OK	**Not OK**	**Totals**
Group	**Staff**	30	70	100
	Student	50	50	100
	Total	80	120	200

Overall, 40% (80 of 200) of those questioned thought it was OK to share copyrighted music. That's the marginal percentage. Each row determines a conditional distribution of the attitude, one for staff and one for students. Only 30% of the staff thought it was OK to share, compared to 50% of students. Because the row percentages differ, *Group* and *Attitude* are associated.

To quantify the strength of association, we need a benchmark for comparison, a point of reference. For that, consider what Table 5.11 would look like if there were *no* association. To figure this out, pretend that we know the marginal totals, but not the counts within the table, as shown in Table 5.12.

		Attitude toward Sharing		
		OK	**Not OK**	**Total**
Group	**Staff**	?	?	100
	Student	?	?	100
	Total	80	120	200

Overall, one-half of the respondents are staff and one-half are students. Were *Group* and *Attitude* not associated, then one-half of the cases in each column would be staff and one-half would be students. The table would look like Table 5.13.

TABLE 5.13 Artificial table with *Group* and *Attitude* not associated.

		Attitude toward Sharing		
		OK	**Not OK**	**Total**
Group	**Staff**	40	60	100
	Students	40	60	100
	Total	80	120	200

Chi-squared measures the difference between the cells in the real table and those in the artificial table. We first subtract the values in the cells (the margins must match). The differences in the counts are as shown in Table 5.14.

TABLE 5.14 Deviations from the original counts.

Real Data			Artificial			Difference	
30	70	−	40	60	=	−10	10
50	50		40	60		10	−10

Next, we combine the differences. If we add them, we get zero because the negative and positive values cancel. We had this problem with cancellation when we defined the variance s^2 in Chapter 4. We'll solve the problem as we did then: square the differences *before* we add them. Chi-squared also requires another step before we add them.

Chi-squared assigns some of these differences a larger contribution to the total. Look at the differences in the first row. Both are 10, but the difference in the first column is larger relative to what we expected than the difference in the second column (10 out of 40 compared to 10 out of 60). Rather than treat these the same, chi-squared assigns more weight to the first. After all, saying 40 and finding 30 is a larger proportional error than saying 60 and finding 70. To give more weight to larger proportional deviations, chi-squared divides the squared deviations by the expected values in the artificial table.

The **chi-squared** statistic is the sum of these weighted, squared differences. For this table, chi-squared, denoted in formulas as χ^2, is

$$\chi^2 = \frac{(30-40)^2}{40} + \frac{(70-60)^2}{60} + \frac{(50-40)^2}{40} + \frac{(50-60)^2}{60}$$
$$= \frac{(-10)^2}{40} + \frac{(10)^2}{60} + \frac{(10)^2}{40} + \frac{(-10)^2}{60}$$
$$= 2.5 + 1.67 + 2.5 + 1.67$$
$$= 8.33$$

We generally use software to compute chi-squared, but it is useful to know what the program is doing in order to understand how it measures association. To summarize, here are the steps to follow in calculating chi-squared.

1. Make an artificial table with the same margins as the original table, but without any association.
2. Subtract the cells of the artificial table from those of the original table, and then square the differences.
3. Divide the squared differences in each cell by the count in the artificial table.
4. Sum the normalized, squared deviations over all of the cells.

The formula used to compute chi-squared shows that we get the same value if we exchange the rows and columns. It does not matter which variable defines the rows and which defines the columns because the artificial table only uses the product of the marginal frequencies.

What Do You Think?

Here's the contingency table from the prior "what do you think?" section, including the marginal totals.

TABLE 5.15 Contingency table of *Purchase* by *Mailing List*

		Mailing List		
		Join	**Decline**	**Total**
Purchase	**Yes**	52	12	64
	No	343	3,720	4,063
	Total	395	3,732	4,127

a. Chi-squared requires an artificial table of counts. What count would be expected in the highlighted cell for those who join the mailing list and make a purchase if *Purchase* and *Mailing List* are not associated?[7]
b. What is the contribution to chi-squared from the cell for those who join the mailing list and make a purchase?[8]
c. The value of chi-squared for this table is $\chi^2 \approx 385.9$. Does your answer in part b reveal which cell produces the largest contribution to chi-squared?[9]

Cramer's *V*

caution   Chi-squared has another similarity to s^2: It's hard to interpret. The value of chi-squared depends on the total number of cases and the size of the table. The larger the table and the larger the number of observations, the larger chi-squared becomes. A more interpretable measure of association allows comparison of the association across tables. Chi-squared for the example of music sharing is 8.33 whereas chi-squared for Table 5.15 is 385.9. Is there that much more association in the exercise, or is chi-squared larger because $n = 4{,}127$ in Table 5.15 compared to $n = 200$ in Table 5.11?

---

[7] If the two variables are not associated, then the percentage who make a purchase among those who join ought to be the same as the percentage in the margin of the table, which is 64/4,127, or about 1.55%. The expected count in the first cell is then $395 \times 64/4{,}127 \approx 6.126$.

[8] Subtract the expected count from the observed count in part a to get the deviation. Then square the deviation and divide by the expected count. The contribution is $(52 - 6.126)^2/6.126 \approx 343.5$.

[9] Each summand that goes into χ^2 is positive, so the largest contribution comes from the first cell. The big deviation from the artificial table is the large count in the first cell.

Cramer's V A statistic derived from chi-squared that measures the association in a contingency table on a scale from 0 to 1.

Cramer's V handles this comparison by adjusting chi-squared so that the resulting measure of association lies between 0 and 1. **If $V = 0$, the variables are not associated. If $V = 1$, they are perfectly associated.** If $V < 0.25$, we will say that the association is weak. If $V > 0.75$, we will say that the association is strong. In between, we will say there is moderate association.

To find Cramer's V, divide χ^2 by the number of cases times the smaller of the number of rows minus 1 or the number of columns minus 1, and take the square root. The formula for Cramer's V is simpler than writing out the definition. As usual, n stands for the total number of cases, and let r be the number of rows and c the number of columns. The formula for Cramer's V is

$$V = \sqrt{\frac{\chi^2}{n \min(r - 1, c - 1)}}$$

If $V = 0$, the two categorical variables are not associated. If $V = 1$, the two variables are perfectly associated. If variables are perfectly associated, you know the value of one once you know the value of the other. For the survey of file sharing, $\chi^2 = 8.33$ and both r and c are 2 with $n = 200$. Hence,

$$V = \sqrt{\frac{\chi^2}{200 \min(2 - 1, 2 - 1)}} = \sqrt{\frac{8.33}{200}} \approx 0.20$$

There's association, but it's weak. Staff and students have different attitudes toward file sharing, but the differences are not very large. For Table 5.15, $\chi^2 = 385.9, n = 4{,}127$, and $r = c = 2$. In this case,

$$V = \sqrt{\frac{\chi^2}{n \min(r - 1, c - 1)}} = \sqrt{\frac{385.9}{4{,}127}} \approx 0.31$$

There is indeed more association in this case than in the example of file sharing, but not that much more. The huge difference between the values of chi-squared is a consequence of the difference in sample sizes, not the strength of the association.

What does a table look like when there is strong association? Strong association implies very large differences among row or column percentages of a table. Suppose the survey results had turned out as shown in Table 5.16.

TABLE 5.16 A table with strong association.

	OK to Share	Not OK	Total
Staff	0	100	100
Students	80	20	100
Total	80	120	200

No staff thought it was okay to share files, compared to 80% of the students. You'd expect arguments between staff and students about sharing materials on this campus. Let's find χ^2 and Cramer's V for this table. The margins of Table 5.16 are the same as those in Table 5.11, so the calculation of χ^2 is similar. We just need to replace the original counts by those in Table 5.16.

$$\chi^2 = \frac{(0 - 40)^2}{40} + \frac{(100 - 60)^2}{60} + \frac{(80 - 40)^2}{40} + \frac{(20 - 60)^2}{60}$$

$$= 40 + 26.67 + 40 + 26.67$$

$$= 133.33$$

Cramer's V indicates strong association between the variables.

$$V = \sqrt{\frac{133.33}{200}} = 0.816$$

The size of Cramer's V indicates that you can almost predict what a respondent will say if you know whether the respondent is on the staff or is a student. If you know a person is a staff member, then you know her or his attitude toward sharing files. Every member of the staff says that file sharing is not okay. Among students, 80% say that it's OK to share.

> **Checklist: Chi-squared and Cramer's V.** Chi-squared and Cramer's V measure association between two *categorical* variables that define a contingency table. Before you use these, verify that your data meet these prerequisites.
>
> ✓ **Categorical variables.** If a variable is numerical, there are better ways to measure association.
> ✓ **No obvious lurking variables.** A lurking variable means that the association you've found is the result of some other variable that's not shown.

4M EXAMPLE 5.3   REAL ESTATE

MOTIVATION ▸ STATE THE QUESTION

A developer needs to pick heating systems and appliances for newly built single-family homes. If the home has electric heat, it's cheaper to install electric appliances in the kitchen. If the home has gas heat, gas appliances make more sense in the kitchen. If the developer is limited to gas or electric heating, how many of each should he offer? Does everyone who heats with gas prefer to cook with gas as well?

METHOD ▸ DESCRIBE THE DATA AND SELECT AN APPROACH

The relevant categorical variables are the type of fuel used for heating (gas or electric) and the type of fuel used for cooking (gas or electric). Now that we have χ^2 and Cramer's V, we can quantify the association as well. Strong association indicates that buyers have strong preferences, and weak association indicates vague preferences.

The developer wants to configure homes that match the demand for gas or electric heat. The developer also has to decide the types of appliances customers want in kitchens. If there's little association, then the developer needs a mix of configurations.

MECHANICS ▸ DO THE ANALYSIS

The developer obtained the preferences of residents in 447 homes in the area. For each, his data give the type of fuel used for cooking and the type used for heating. The contingency table at the top of the next page shows column percentages, which give the conditional distributions of cooking fuel given the type of fuel used for heating. About two-thirds heat with natural gas (298/447) and one-third with electricity.

The contingency table shows association between these variables. Among homes with electric heat, 91% cook with electricity. Among homes with gas heat, 46% cook with electricity. To quantify the strength of the association, we calculated that $\chi^2 = 98.62$ and $V = \sqrt{(\chi^2/n \min(r-1, c-1)} = \sqrt{(98.62/447 \times 1)} \approx 0.47$. That's moderate association.

Count Column %	Home Heat Fuel		
	Electricity	Gas	Total
Cooking Fuel			
Electricity	136 91.28%	136 45.64%	272
Gas	10 6.71%	162 54.36%	172
Other	3 0.20%	0 0.00%	3
Total	149	298	447

MESSAGE | SUMMARIZE THE RESULTS

Homeowners contacted by the developer prefer natural gas to electric heat by about 2 to 1. These findings suggest building about two-thirds of the homes with gas heat and the rest with electric heat. Of those with electric heat, keep it simple and install an electric kitchen. For those with gas heat, put an electric kitchen in one half and gas in the other half.

Be up front about your concerns. If you have some reservations, mention them. There's a big caveat, however. This analysis assumes that buyers who are looking for new homes have the same preferences that these residents have—a big if.

Best Practices

- *Use contingency tables to find and summarize association between categorical varibles.* You cannot see the association in the separate bar charts. It only becomes evident when you look at the table and compare the conditional distributions with the marginal distributions.
- *Be on the lookout for lurking variables.* Before you interpret the association you find between two variables, think about whether there is some other variable that offers a different explanation for your table. Are the data in the columns or rows of your table really comparable, or might some other factor that's not evident explain the association that you see?

- *Use plots to show association.* Segmented bar charts and mosaic plots are useful for comparing relative frequencies in larger tables. Adjacent pie charts are another choice, but these can make it hard to compare percentages unless the differences are large.
- *Exploit the absence of association.* If the two categorical variables are not associated, the variation of each is self-contained in the variables, and you do not need the complexity of a table to summarize the variables. You can study each, one at a time.

Pitfalls

- *Don't interpret association as causation.* You might have found association, but that hardly means that you know why values fall in one category rather than another. Think about the possibility of lurking variables.
- *Don't display too many numbers in a table.* Computers make it easy to decorate a table

with too many percentages. Choose just the ones that you need, those that help you answer the question at hand. Numerical tables with lots of rows and columns are also overwhelming; summarize these with a chart.

Software Hints

EXCEL

Excel has a powerful feature for producing contingency tables, but you need to master the concept of pivot tables. If you want to stay with Excel for all of your computing, then it's probably worth the effort. Start by reading the Help files that come with Excel by searching for "pivot tables" from the Help menu. DDXL allows you to get right to the table; it will do the counting for you. Select the region of your spreadsheet that has the variables (with names in the first row), and then follow the commands

DDXL > Tables

to get the tables dialog box. Choose the Contingency Table option from the Click-Me menu and then choose the categorical variables from the list shown at the right of the dialog box. DDXL does the rest; to see chi-squared, click the button in the window that shows the table.

If you are working without DDXL, it is not too hard to compute the value of chi-squared once you have the contingency table. Cramer's V is easy to compute once you have chi-squared. We find it easiest to build a table of expected counts (the reference table in which there is no association), then subtract this table from the observed table and square each cell. Adding up the squared deviations divided by the expected counts gives chi-squared.

MINITAB

To obtain the contingency table, follow the menu items

Stat > Tables > Cross-Tabulation and chi-squared

and fill in the dialog box with the names of two categorical variables. Pick one variable to identify the rows of the table and the other for the columns. (Layers allow you to produce tables such as Table 5.8 that show a separate table for each value of a third variable.) Options also produce intermediate steps in the calculation of chi-squared, such as the contribution from each cell to the total. It's an easy calculation to convert chi-squared to Cramer's V.

JMP

Follow the menu commands

Analyze > Fit Y by X

and pick one categorical variable for the Y variable and one for X. The variable chosen for Y identifies the columns of the contingency table and the variable chosen for X identifies the rows of the contingency table. By default, the output from JMP shows the mosaic plot. The pop-up menu produced by clicking on the red triangle beside the header Contingency Table in the output window allows you to modify the table by removing, for instance, some of the shown percentages.

The value of chi-squared appears below the table in the section of the output labeled Tests. The value of chi-squared is in the output in the row labeled Pearson. (There are variations on how to compute the chi-squared statistic.)

Test	Chi-Squared	Prob > Chi-Sq
Likelihood Ratio	16.570	0.0054
Pearson	16.056	0.0067

Once you have chi-squared, use the formula given in the text to obtain Cramer's V.

CHAPTER SUMMARY

A **contingency table** displays counts and may include selected percentages. The totals for rows and columns of the table give the **marginal distributions** of the two variables. Individual rows and columns of the table show the **conditional distribution** of one variable given a label of the other. If the conditional distribution of a variable differs from its marginal distribution, the two variables are **associated**. **Segmented** bar charts and **mosaic plots** are useful for seeing association in a contingency table. A **lurking variable** offers another explanation for the association found in a table. A lurking variable can produce **Simpson's paradox**; the association in the table might be the result of a lurking variable rather than the two that define the rows and columns. **Chi-squared** and **Cramer's V** are statistics that quantify the degree of association.

Key Terms

associated, 80
chi-squared, 89
contingency table, 78
cell, 79
Cramer's V, 92
distribution, conditional, 79
marginal, 79
lurking variable, 86

■ Formulas

CHI-SQUARED

Begin by forming a table with the same marginal counts as in the data, but no association. A formula shows how to compute the cells of the artificial table from the marginal counts. Let row_i denote the marginal frequency of the ith row (the total number of observations in this row), and let col_j denote the marginal frequency of the jth column (the number in this column). The count for the cell in row i and column j if there is no association is

$$\text{expected}_{i,j} = \frac{\text{row}_i \times \text{col}_j}{n}$$

A spreadsheet is helpful to organize the calculations. To find χ^2, sum the weighted, squared deviations

between $\text{expected}_{i,j}$ and $\text{observed}_{i,j}$. Using the summation notation introduced in Chapter 4, the formula for chi-squared is

$$\chi^2 = \sum_{i,j} \frac{(\text{observed}_{i,j} - \text{expected}_{i,j})^2}{\text{expected}_{i,j}}$$

The sum extends over all of the cells of the table.

CRAMER'S V

$$V = \sqrt{\frac{\chi^2}{n \min(r - 1, c - 1)}}$$

for a table with r rows and c columns that summarizes n cases.

■ About the Data

The Amazon data in this chapter (and Chapter 3) come from ComScore, a firm that monitors the Web-browsing habits of a sample of consumers around the country. The data on airline arrivals in the 4M example of Simpson's paradox is from the Web site of the Bureau of Transportation Statistics. (From the main page, follow the links to data that summarize information about various types of travel

in the United States.) We used arrival data for January 2006. The data for kitchen preferences is a subset of the Residential Energy Consumption Survey (RECS), performed by the Department of Energy. The example of attitudes toward file sharing is from a story in the *Daily Pennsylvanian*, the student newspaper at the University of Pennsylvania.

EXERCISES

Mix and Match

Match the description of the item in the first column to the term in the second column.

1. Table of cross-classified counts	(a) chi-squared
2. Counts cases that match values of two categorical variables	(b) expected
3. Shown in the bar chart of a categorical variable	(c) associated
4. Shown by a segmented bar chart	(d) Simpson's paradox
5. Measure of association between two categorical variables that grows with increased sample size	(e) marginal distribution
6. Measure of association between two categorical variables that lies between 0 and 1	(f) cell
7. Conditional distribution matches marginal distribution	(g) conditional distribution
8. Percentages within a row differ from marginal percentages	(h) Cramer's V
9. Produced by a variable lurking behind a table	(i) not associated
10. Cell counts produced by assuming no association	(j) contingency table

True/False

Mark each statement True or False. If you believe that a statement is false, briefly say why you think it is false.

11. We can fill in the cells of the contingency table from the marginal counts alone if the two categorical variables are not associated.

12. We can see association between two categorical variables by comparing their bar charts.

13. The percentages of cases in the first column within each row of a contingency table are the same if the variables are not associated.

14. A large chi-squared tells us that there is strong association between two categorical variables.

15. The value of chi-squared depends on the number of observations in a contingency table.

16. Cramer's V is 0 if the categorical variables are not associated.

17. The value of chi-squared depends on which of two categorical variables defines the rows and which of the two variables defines the columns of the contingency table.

18. If variable X is associated with variable Y, then Y is caused by X.

19. If the categorical variable that identifies the supervising manager is associated with the categorical variable that indicates a problem with processing orders, then the manager is causing the problems.

20. A small chi-squared statistic suggests that a lurking variable conceals the association between two categorical variables.

21. If the percentage of female job candidates who are hired is larger than the percentage of male candidates who are hired, then there is association between the categorical variables *Sex* (male, female) and *Hire* (yes, no).

22. If the percentage of defective items produced by a manufacturing process is about the same on Monday, Tuesday, Wednesday, Thursday, and Friday, then the day of the week is associated with defective items.

Think About It

23. This table shows counts from a consumer satisfaction survey of 2,000 customers who called a credit card company to dispute a charge. One thousand customers were retired and the remaining were employed.
    (a) What would it mean to find association between these variables?
    (b) Does the table show association? (You shouldn't need to do any calculation.)

	Employed	Retired
Satisfied	697	715
Unsatisfied	303	285

24. This table summarizes the status of 1,000 loans made by a bank. Each loan either ended in default or was repaid. Loans were divided into large (more than $50,000) or small size.
    (a) What would it mean to find association between these variables?
    (b) Does the table show association? (You shouldn't need to do much calculation.)

	Repaid	Default
Large	40	10
Small	930	20

25. A marketing study asked potential customers for their preferences for two attributes of a new product, the color and the packaging. The color of the final product is to be chosen by one manager and the packaging by another. Is this division of labor simpler if the two variables are associated or if they are not associated?

26. Executives at a retailer are comparing items chosen by shoppers who either saw or did not see its most recent television advertising. Do the managers in charge of advertising hope to find association between viewing ads and the choice of items?

27. This chart summarizes explanations given for missing work. The data are the explanations given for 100 absences by employees on the assembly line, administration, and supervising managers. The explanations are classified as medical, family emergency, or other.

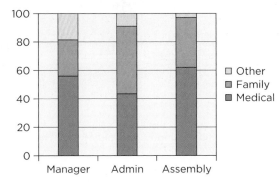

Reason for Absence

(a) For which group are absences due to family emergencies most common?
(b) Are the two variables associated? How can you tell?

28. The following chart summarizes the complaints received at a catalog sales call center by day of the week. The chart shows percentages for each day. The complaints are classified as related to delivery (the order has not arrived), fulfillment (an item was missing from an order or not what was ordered), size (an item did not fit), and damage.

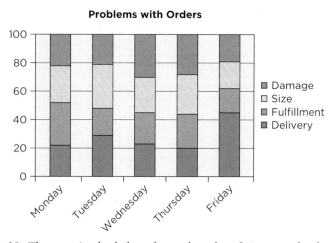

Problems with Orders

Legend: Damage, Size, Fulfillment, Delivery

(a) Is the day of the week associated with the nature of the complaint? Explain why you think that these variables are or are not associated.

(b) What supplemental information that is not shown by the chart is important to managers in charge of dealing with complaints?

29. The mosaic plot below shows the sales of cigarettes by the three leading manufacturers, divided into six international regions. (Total sales in 2006 were approximately 2 trillion.)
    (a) In which region are total cigarette sales the largest?
    (b) In which two regions does British-American Tobacco command more than half of the market?
    (c) Which share leads to higher sales: 80% of the market in the U.S. and Canada (like that of Philip Morris) or 80% of the market in Asia and the Pacific?
    (d) Can you easily tell what percentage of Philip Morris's sales is in western Europe?
    (e) Are brand and region associated?

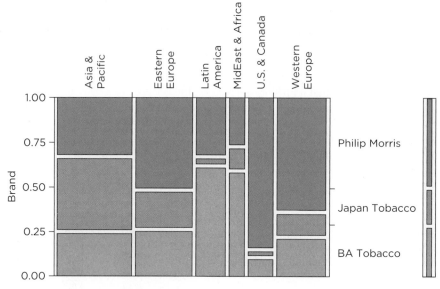

30. The following table compares market shares of several companies in the computer industry in fourth quarters of 2007 and 2008 as reported by Gartner, an industry monitor. The table gives the number of computer systems shipped during each quarter, in thousands.

Company	4th quarter, 2008	4th quarter, 2007
Dell	4,465.8	5,344.6
HP	4,288.3	4,439.5
Acer	2,373.9	1,527.3
Apple	1,255.0	1,159.3
Toshiba	1,007.7	900.0
Others	2,219.2	3,992.6

(a) Summarize these data using both a stacked bar chart (of market share) and a mosaic plot. What message do these plots convey about the changes in market share from 2007 to 2008?

(b) Explain the differences between the two charts. One plot shows a feature of the data that the other does not show. What is it, and why is the difference subtle in this example?

31. A survey of car buyers found that the favorite color of all buyers overall is silver. In this case, should a new car dealer order most pick-up trucks to stock on its lot in silver?

32. A chain of stores suspects that coupons will increase sales to customers in its frequent shopper program. To test this theory, the chain mails some customers

coupons and does not send these coupons to others. A manager builds a contingency table of two variables: whether the customer was sent a coupon and whether the customer used the coupon in the store. Both variables have Yes/No values. Must the manager find association?

33. A study of purchases at a 24-hour supermarket recorded two categorical variables: the time of the purchase (8 A.M to 8 P.M vs. late night) and whether the purchase was made by someone with children present. Would you expect these variables to be associated?

34. Among other services, Standard and Poor's (S&P) rates the risk of default for corporate bonds. AAA bonds are top rated (smallest chance of default), followed by AA, A, BBB, BB, B, CCC, CC, and last D. In a table of bond ratings crossed with the presence or absence of default, would you expect to see association?

35. Numerous epidemiological studies associate a history of smoking with the presence of lung cancer. If a study finds association (cancer rates are higher among smokers), does this mean that smoking *causes* cancer?

36. Incidences of flu are much more prevalent in the winter months than in the summer. Does this association between respiratory illness and outside temperature imply that cold weather causes the flu?

37. (a) What is Cramer's *V* for this contingency table, which shows choices for paint colors and finishes at a hardware store? (You should not need to do any calculations.)

		Color		
		Red	**Green**	**Blue**
	Low	20	30	40
Gloss	**Medium**	10	15	20
	High	40	60	80

(b) What is the value of Cramer's *V* for this contingency table, which reverses the order of the counts in the rows?

		Color		
		Red	**Green**	**Blue**
	Low	40	60	80
Gloss	**Medium**	10	15	20
	High	20	30	40

(c) What is the implication of the association between these variables for stocking paint at this hardware store?

38. (a) What is the value of Cramer's *V* for this contingency table? (You should not need to do any calculations.)

		Color		
		Red	**Green**	**Blue**
	Low	20	0	0
Gloss	**Medium**	0	45	0
	High	0	0	10

(b) What is the value of Cramer's *V* for this contingency table, which reverses the position of the nonzero values?

		Color		
		Red	**Green**	**Blue**
	Low	0	0	20
Gloss	**Medium**	0	45	0
	High	10	0	0

(c) What is the implication of the association between these variables for stocking paint at this hardware store?

You Do It

Bold names shown with a question identify the data table for the problem.

39. **Gasoline sales** A service station near an interstate highway sells three types of gasoline: regular, plus, and premium. During the last week, the manager counted the number of cars that purchased these types of gasoline. He kept the counts separate for weekdays and the weekend. Here are the counts.

	Weekday	**Weekend**
Premium	126	62
Plus	103	28
Regular	448	115

(a) Complete the table by finding the marginal totals for the rows and columns.
(b) Find the conditional distribution of purchase type for weekday purchases.
(c) Find the conditional distribution of premium purchases during the week and weekend.
(d) Does the fact that your answers to parts (b) and (c) are different indicate association?
(e) The owner of the station would like to develop better ties to customers who buy premium gas. (The owner expects these customers to be more affluent and likely to purchase other services from the station.) If the owner wants to meet more of these customers, when should the owner be around the station: on weekdays or weekends?

40. **Mens Shirts** A men's clothing retailer stocks a variety of men's shirts. On the basis of sales during the last three months, a manager prepared the following table of the sizes and styles. (You should recognize these data!)

	Button Down	Polo	Beach Print
Small	18	27	36
Medium	65	82	28
Large	103	65	22

(a) Complete the table by finding the marginal distributions for the two variables.
(b) Find the conditional distribution of sizes of polo shirts.
(c) Find the conditional distribution of the styles of large shirts.
(d) How should the manager in charge of stocking the store use this table?

41. The contingency table in Exercise 39 shows counts of the types of gasoline bought during the week and during weekends.
(a) Find the value of chi-squared and Cramer's V for this table.
(b) Interpret these values. What do these tell you about the association in the table?

42. The contingency table in Exercise 40 shows counts of the sizes of shirts bought in three styles in a retail store.
(a) Find the value of chi-squared and Cramer's V for this table.
(b) Interpret these values. What do these tell you about the association in the table?

43. **Owner Satisfaction** A marketing survey of car owners was presented in two ways.[10] One way asked the customer whether he was satisfied with his car. The other way asked the customer whether he was dissatisfied. This table summarizes the results.

	Question Uses *Satisfied*	Question Uses *Dissatisfied*
Very satisfied	139	128
Somewhat satisfied	82	69
Somewhat dissatisfied	12	20
Very dissatisfied	10	23

(a) Complete the table by adding the marginal row and column that give the totals.
(b) Which group reported being more satisfied (either very satisfied or somewhat satisfied) with their cars?
(c) If a company wants to produce an ad citing this type of survey of its customers, how should it word the question?

44. **Poll** After the collapse of the stock market in October 1987, *Business Week* polled its readers and asked whether they expected another big drop in the market during the next 12 months.[11] This table summarizes what *Business Week* found.

	Stockholders	Nonstockholders
Very likely	18	26
Somewhat likely	41	65
Not very likely	52	68
Not likely at all	19	31
Unsure	8	13

(a) Expand the table by adding row and column marginal totals.
(b) Did stockholders think that a drop was either somewhat likely or very likely?
(c) Did nonstockholders think that a drop was either somewhat likely or very likely?
(d) Do the differences between parts (b) and (c) make sense? Do these differences imply that the two variables summarized in the table are associated?

45. Exercise 43 compares the responses of customers in a survey depending on how a question about their satisfaction with their car was phrased.
(a) Quantify the amount of association between the type of question and the degree of satisfaction.
(b) Combine the counts of very satisfied with somewhat satisfied, and very dissatisfied with somewhat dissatisfied, so that the table has only two rows rather than four. What happens to the degree of association if the table merges the satisfied categories into only two rows rather than four?

46. **Employment** Exercise 44 summarizes results from a survey of stockholders following the October 1987 stock market crash.
(a) Quantify the amount of association between the respondents' stock ownership and expectation about the chance for another big drop in stock prices.
(b) Reduce the table by combining the counts of very likely and somewhat likely and the counts of not very likely and not likely at all, so that the table has three rows: likely, not likely, and unsure. Compare the amount of association in this table to that in the original table.

	Men	Women
Advertising	34%	66%
Book publishing	40%	60%
Law firms	38%	62%
Investment banking	60%	40%

[10] R. A. Peterson and W. R. Wilson (1992), "Measuring Customer Satisfaction: Fact and Artifact," *Journal of the Academy of Marketing Science, 20,* 61–71.

[11] November 9, 1987, p. 36.

47. This table shows percentages of men and women employed in four industries.[12]
    (a) Is there association between the gender of the employee and the industry? How can you tell?
    (b) Interpret the association (or lack thereof).
    (c) Find the chi-squared statistic and Cramer's V if these data were derived from $n = 400$ employees, with 100 in each industry. Repeat the process for data derived from $n = 1,600$ employees, with 400 in each industry. Which statistic changes? Which remains the same?

48. **Credit Rating** The best-known credit rating for consumers in the United States is the FICO score (named for the source, a company called Fair Isaac). The score ranges from 300 to 850, with higher scores indicating that the consumer is a better credit risk. For the following table, consumers with scores below 620 are labeled Risky. Those having scores between 620 and 660 are labeled Uncertain. Those between 660 and 720 have an Acceptable credit rating, and consumers with scores over 720 have Perfect credit. A department store kept track of loans (in the form of a store credit card) given to customers with various ratings. This table shows the proportion of customers within each risk category that did not repay the loan (defaulted).

	Defaulted	**Repaid**
Risky	30%	70%
Uncertain	22%	78%
Acceptable	2%	98%
Perfect	2%	98%

   (a) Is the credit score (as defined by these four categories) associated with default? How can you tell?
   (b) What would it mean for the use of the FICO score as a tool for stores to rate consumers if there were no association between the category and default?
   (c) If virtually all customers at this store have acceptable or perfect credit scores, will the association be strong or weak? Suppose that the store has 10,000 loans, with 50 made to risky customers, 100 to uncertain customers, 9,000 to acceptable customers, and the rest to perfect customers.
   (d) If a higher proportion of the loans had been made to customers who were risky or uncertain, would the association have remained the same, increased, or decreased?

49. **Trust** The National Opinion Research Center at the University of Chicago reported the following results of a survey in fall of 2004. The survey asked 2,812 adults to state how much they trust different figures in the news. This table summarizes the amount of confidence the public puts in each group.

Group	Great Deal of Trust	Some Trust	Hardly Any Trust	No Answer
Scientists	42%	49%	6%	3%
Banks	29%	57%	13%	1%
Organized religion	24%	52%	22%	2%
Executive branch of government	22%	47%	31%	0%

   (a) Does this table show any association between the group and the amount of trust placed in the group? Interpret the category No Answer as simply another category. (*Hint:* You might want to think about your answer to part (b) first.)
   (b) Should these data be interpreted as a contingency table?

50. **Student Loan** This table summarizes the effects of student loan debt found in two surveys of college graduates, one in 1987 and the second in 2002.[13] Respondents were asked if they had encountered each of the following problems:

	1987	**2002**
Changed career plans	11%	17%
Delayed buying a home	23%	38%
Delayed getting married	9%	14%
Delayed having children	12%	21%

   (a) What does this table suggest is happening to the effects of student loans on college graduates?
   (b) Does this table show association between the nature of the problem and the year of the survey? Is this a contingency table?

51. **Article tone** Does the fact that medical researchers get money from drug manufacturers affect their results? The *New England Journal of Medicine* published a paper that included the following results.[14] The table at the top of the next page summarizes the tone of articles published by 69 authors during 1996–1997. The publication was labeled either supportive, neutral, or critical. For each author, the data show whether the researchers received support from a pharmaceutical company.
   (a) Do the data support the claim that by supporting research, drug companies influence the results of scientists?
   (b) Does the table show a plain relationship, or might another variable lurk in the background? What might be its impact? Explain.

[12] *New York Times*, November 21, 2005.

[13] *Business Week*, November 14, 2005.
[14] H. T. Stelfox, G. Chua, K. O'Rourke, and A. S. Detsky (1998), "Conflict of Interest in the Debate Over Calcium Channel Antagonists," *New England Journal of Medicine*, January 8, 1998, pp. 101–106.

	Tone of Article		
	Supportive	**Neutral**	**Critical**
Supported by pharmaceutical	24	10	13
Not supported by pharmaceutical	0	5	17

52. **Reaction** To gauge the reactions of possible customers, the manufacturer of a new type of cellular telephone displayed the product at a kiosk in a busy shopping mall. The following table summarizes the results for the customers who stopped to look at the phone:

	Male	**Female**
Favorable	36	18
Ambivalent	42	7
Unfavorable	29	9

(a) Is the reaction to the new phone associated with the sex of the customer? How strong is the association?
(b) How should the company use the information from this study when marketing its new product?
(c) Can you think of an underlying lurking variable that might complicate the relationship shown here? Justify your answer.

53. These data compare the on-time arrival performance of United and US Airways. The table shows the status of 13,511 arrivals during January 2006.

	United	**US Airways**
On time	6,291	4,366
Delayed	1,466	1,388

(a) On the basis of this initial summary, find the percentages (row or column) that are appropriate for comparing the on-time arrival rates of the two airlines. Which arrives on time more often?
(b) The next two tables organize these same flights by destination. The first also shows arrival time and the second shows airline. Does it appear that a lurking variable might be at work here? How can you tell?

	Dallas	**Denver**	**Minneapolis**	**Philadelphia**
On time	829	5,661	518	3,649
Delayed	205	1,205	160	1,284

	Dallas	**Denver**	**Minneapolis**	**Philadelphia**
United	450	6,323	474	510
US Airways	584	543	204	4,423

(c) Each cell of the following table shows the number of on-time arrivals for each airline at each destination. Is *Destination* a lurking factor behind the original 2 × 2 table?

	Dallas	**Denver**	**Minneapolis**	**Philadelphia**
United	359	5,208	360	364
US Airways	470	453	158	3,285

54. These data compare the on-time arrival performance of flights on American and Delta. The table shows the status of 17,064 arrivals during January 2006.

	American	**Delta**
On time	1,536	1,1769
Delayed	416	3,343

(a) On the basis of this initial summary, find the percentages (row or column) that are appropriate for comparing the on-time arrival rates of the two airlines. Which arrives on time more often?
(b) The next two tables organize these same flights by destination. The first shows arrival time and the second shows airline. Does it appear that a lurking variable might be at work here? How can you tell?

	Atlanta	**Las Vegas**	**San Diego**
On time	11,512	1,007	786
Delayed	3,334	244	181

	Atlanta	**Las Vegas**	**San Diego**
American	653	698	601
Delta	14,193	553	366

(c) Each cell of the following table shows the number of on-time arrivals for each airline at each destination. Is *Destination* a lurking factor behind the original 2 × 2 table?

	Atlanta	**Las Vegas**	**San Diego**
American	497	561	478
Delta	11,015	446	308

4M Discrimination in Hiring

Contingency tables appear frequently in legal cases, such as those that allege that a company has discriminated against a protected class. The following table gives the number of employees of different ages who were laid off when a company anticipated a decline in business:[15]

		Laid Off	Retained	Total
Age	**< 40**	18	787	805
	40–49	14	632	646
	50–59	18	374	392
	60 or more	18	107	125

Several long-time employees who were laid off filed a suit against the company for wrongful termination. All were over 50, and they claimed the company discriminated on the basis of age.

To use data like these in a trial requires a more complete analysis than we are ready to do. For now, we'll settle for a descriptive analysis that lays the foundation.

Motivation

a) *From the employees' point of view.* On the basis of the claim of their lawsuit, would the laid-off employees expect to find association in this table?

b) *From the company's point of view.* If the company does not discriminate on the basis of age, would you expect to find association in this table?

Method

c) *From either point of view.* Would it be more useful to find the percentages within the columns or the percentages within the rows?

d) What statistic would you choose to represent the presence or absence of evidence of discrimination?

Mechanics

e) Are age and employment status associated? Compute the appropriate summary statistic and interpret its value.

Message

f) Summarize what you find in your analysis of the table.

g) How would the presence of a lurking factor compromise the use of data such as these in a legal case?

4M Picking a Hospital

One effort to improve the efficiency of the health care system provides patients and doctors with more information about the quality of care delivered by hospitals.

The idea is to help consumers find the best hospital to treat their illness.

These data describe patients with breast cancer. Breast cancer is the most frequently diagnosed cancer among women in the United States, with 211,300 invasive and 55,700 in situ cases in 2003, accounting for nearly one in every three cancers diagnosed. The data indicate outcomes at two hospitals, Community Hospital (CH) and University Hospital (UH), and the type of breast cancer. Early-stage cancers are more easily treated than later-stage cancers. Each cell of the table gives the number of deaths due to breast cancer and the number of surgeries. For example, at the Community Hospital (first row), there were 3 deaths among the 30 surgeries to treat early-stage cancer.

		Stage of Cancer		
		Early	**Late**	**Both**
Hospital	**CH**	3/30	12/21	15/51
	UH	5/98	180/405	185/503

Motivation

a) In general, should patients be allowed to see only the marginal totals for each hospital rather than all of the information in this table? Give the advantages of releasing just the marginal information (the summary of both types of cancers) versus releasing the full table.

Method

b) If a woman is diagnosed with breast cancer at an early stage, what probabilities should she focus on when comparing these two hospitals: marginal or conditional?

Mechanics

c) Create a table that shows the percentages of patients who die from breast cancer in each hospital at each stage.

d) What proportion of cases at the community hospital are late stage? At the university hospital?

e) Which hospital has the lower death rate for early-stage cancers? Late-stage cancers?

f) Which hospital has the lower overall death rate from breast cancer?

Message

g) How should a woman diagnosed with breast cancer interpret the results of these data? Write a few sentences giving guidance in how to interpret these results.

h) In order to help patients choose a good hospital, is it sufficient to release marginal statistics like the overall death rate from breast cancer? Or is more information needed for patients to interpret health outcomes?

6 Association between Quantitative Variables

CHAPTER

WHETHER MOTIVATED BY RISING PRICES OR GLOBAL WARMING, many homeowners who heat with natural gas would like to use less. Businesses have noticed the interest in conservation and found ways to use conservation as a marketing tool. For example, a recent government study found that homeowners could reduce their energy use by up to 30% by upgrading the insulation of their homes (plus a few other improvements). Manufacturers of insulation such as Owens-Corning quoted these potential savings in their advertisements.

Saving 30% sounds great, but most consumers would rather hear how much money they'll save. An ad that promises savings in dollars might be more attractive. A homeowner is more willing to spend $400 for insulation if it will save that much in a year. Let's design an ad to motivate homeowners to insulate.

The potential savings from insulating a home vary widely because homes differ in their use of natural gas. Figure 6.1 summarizes the annual consumption of 1,658

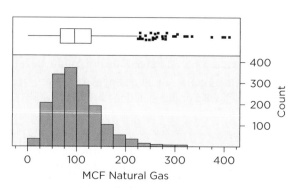

FIGURE 6.1 Annual use of natural gas per household, in thousands of cubic feet (MCF).

homes that heat with natural gas (in thousands of cubic feet per year, abbreviated MCF).

The distribution in Figure 6.1 is right skewed. Values accumulate near the median at 90 MCF and tail off to the right. The considerable variation translates into a wide range of costs. To convert these amounts to dollars, add a zero to the scale: Natural gas sells for about $10 per MCF. For instance, the home at the far right used 409 MCF. At $10 per MCF, its annual heating bill is $4,090. By comparison, homeowners at the far left spent less than $100. In order to anticipate how much a homeowner might save, we need to understand why there's so much variation in consumption.

This chapter shows that we can understand variation by measuring the association between numerical variables. For example, an obvious explanation for the variation in gas usage is that these homes are located in different climates. Winters are a lot colder in Minnesota than in Florida, so we can expect homes in Minnesota to need more fuel for heating. This association also allows us to predict gas usage on the basis of the climate. The methods in this chapter resemble those for categorical variables discussed in Chapter 5 but take advantage of the properties of numerical data.

January Average Temperatures		
	Duluth, MN	**Miami, FL**
High	16.3	75.2
Low	−2.1	59.2
Daily	7.2	67.1

6.1 | SCATTERPLOTS

The National Weather Service and Department of Energy use heating degree-days (HDD) to measure the severity of winter weather. The larger the number of heating degree-days becomes, the colder the weather. For example, a typical winter in Florida produces 700 HDD compared to 9,000 in Minnesota. Figure 6.2 shows the distribution of the number of HDD experienced by the homes in Figure 6.1.[1]

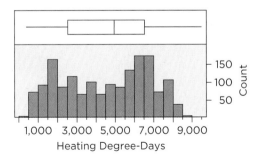

FIGURE 6.2 Heating degree-days around the United States.

The homes on the left are evidently in Florida whereas those at the right appear to be in Minnesota (or other locations as warm or cold, respectively).

---

[1] To calculate HDD for a day, average the high and low temperatures. If the average is less than 65, then the number of heating degree-days for that day is 65 minus the average. For example, if the high temperature is 60 and the low is 40, the average is 50. This day has 65 − 50 = 15 HDD. If the average is above 65, HDD = 0. According to HDD, you don't need heat on such days!

scatterplot A graph that displays pairs of values as points on a two-dimensional grid.

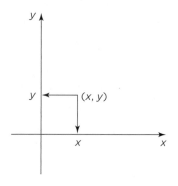

response Name for the variable that has the variation we'd like to understand, placed on the y-axis in scatterplots.

explanatory variable Name for a variable that is to be used to explain variation in the response; placed on the x-axis in scatterplots.

For heating degree-days to explain variation in the use of natural gas, these variables must be associated. Association between numerical variables resembles association between categorical variables (Chapter 5). Two categorical variables are associated if the distribution of one depends on the value of the other. That's also true for numerical variables, but the methods for seeing and quantifying association differ. In place of a contingency table, we use a **scatterplot** to look for association between numerical variables.

A scatterplot displays *pairs* of numbers. In this example, we have 1,658 pairs, one for each home. One element of each pair is the number of heating degree-days and the other is the amount of natural gas used. The scales of these variables determine the axes of the scatterplot. The vertical axis of the scatterplot is called the *y*-axis, and the horizontal axis is the *x*-axis. The variable that defines the *x*-axis specifies the horizontal location of a specific observation, and the variable that defines the *y*-axis specifies the vertical location of the same observation. Together, a pair of values defines the coordinates of a point, which is usually written (x, y).

tip

To decide which variable to put on the *x*-axis and which to put on the *y*-axis, display the variable you would like to explain or predict along the *y*-axis. We will call the variable on the *y*-axis the **response**. The variable that we use to explain variation in the response is the **explanatory variable**; it goes on the *x*-axis. In this example, the amount of natural gas used is the response, and the number of heating degree-days is the explanatory variable.

Figure 6.3 shows the completed scatterplot. Each point in the plot represents one of the 1,658 households that make up these data. For example, the point marked with an *x* represents a household for which $(x, y) = (2776 \text{ HDD}, 229 \text{ MCF})$.

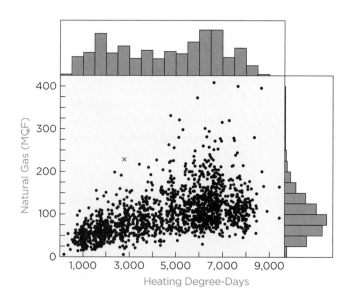

FIGURE 6.3 Scatterplot with histograms along the axes.

Figure 6.3 adds the histogram of the response on the right and the histogram of the explanatory variable at the top. These histograms show the marginal distributions of the variables in a scatterplot because they summarize the variation in one variable without taking account of the other. These histograms are analogous to the marginal counts of the categorical variables in a contingency table. As with contingency tables, we cannot judge the presence of association from marginal distributions. For a contingency table, we considered patterns in the cells; for a scatterplot, we look for patterns in the display of points in the scatterplot.

6.2 | ASSOCIATION IN SCATTERPLOTS

Is the amount of natural gas used associated with heating degree-days? Can we explain some of the differences in consumption by taking account of differences in climate? It appears that we can: The points in Figure 6.3 generally shift upward as we move from left to right, from warmer to colder climates. This upward drift means that homes in warm climates typically use less natural gas than homes in cold climates.

Visual Test for Association

Before we conclude that there's association between degree-days and gas use, we need to decide whether this pattern is real. Is there a pattern that relates degree-days to gas use, or are we imagining it?

Here's a simple way to decide. Recall how we measured association in a table (Chapter 5). The chi-squared statistic compares the counts in a contingency table to those in an artificial table that forces the variables to be unrelated. The cells of the artificial table remove any association between the variables. We can do the same with scatterplots: Compare the scatterplot that we see to a scatterplot that has the same marginal distributions but no association. To construct numerical variables that are not associated, randomly pair x and y values. Pair each value of y with a randomly chosen value of x. If the scatterplot of the randomly paired data looks like the scatterplot of the original data, then there's little or no association.

As an example, one of the scatterplots in Figure 6.4 is the scatterplot of gas use versus heating degree-days (Figure 6.3). The other three scatterplots

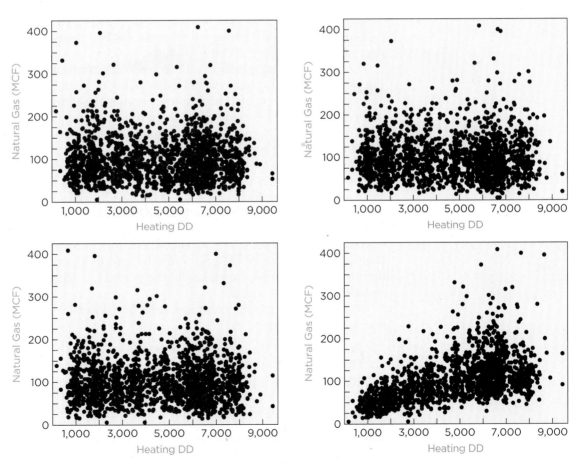

FIGURE 6.4 Do you recognize the original scatterplot?

randomly pair the number of heating degree-days with the amount of gas used. Do you recognize the original?

The original scatterplot is at the lower right. Unlike in the others, in this frame the amount of gas used appears lower at the left (less in warm climates) and higher at the right (more in cold climates). In the other plots, the variation in gas use is the same regardless of the climate. This comparison of the original plot to several artificial plots in which the variables are unrelated is the **visual test for association**. If you can pick out the original, there's association. Otherwise, the variability in the response looks the same everywhere. Because the original plot stands out from the artificial plots in Figure 6.4, there's a pattern that indicates association between HDD and the amount of gas used.

visual test for association A method for identifying a pattern in a plot of numerical variables. Compare the original scatterplot to others that randomly match the coordinates.

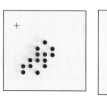

Describing Association in a Scatterplot

Once you decide that a scatterplot shows association, then you need to describe the association. To describe the association, start with its *direction*. In this example, the colder the winter, the larger the gas consumption tends to be. This pattern has a *positive direction* because the points in the scatterplot tend to concentrate in the lower left and upper right corners. As the explanatory variable increases, so does the response. A pattern running the other way has a *negative direction*. As *x* increases, *y* tends to decrease.

Another property of the association is its *curvature*. Does the pattern resemble a line or does it bend? The scatterplot of the natural gas use versus HDD appears *linear*. The points roughly concentrate along an upward-sloping line that runs from the lower left corner to the upper right. Linear patterns have a consistent direction. Linear patterns with positive direction follow a line with positive slope; linear patterns with negative direction follow a line with negative slope. Curved relationships are harder to describe because the direction changes.

The third property of the association is the amount of *variation* around the pattern. In this case, there's quite a bit of variation among homes in similar climates. For instance, among homes in a climate with 6,000 HDD, some use 40 cubic feet of gas compared to others that use six times as much. Plus, some homes in cold climates use less gas than others in warm climates. The variation around the linear pattern in gas use also appears to increase with the amount used. The points stick closer to the linear pattern in the warm climates at the left of Figure 6.3 than in the colder climates on the right.

Finally, look for *outliers* and other surprises. Often the most interesting aspect of a scatterplot is something unexpected. **An outlying point is almost always interesting and deserves special attention.** Clusters of several outliers raise questions about what makes the group so different. In Figure 6.3, we don't see outliers so much as an increase in the variation that comes with higher consumption of natural gas.

Let's review the questions to answer when describing the association that you see in a scatterplot.

1. *Direction.* Does the pattern trend up, down, or both?
2. *Curvature.* Does the pattern appear to be linear or does it curve?
3. *Variation.* Are the points tightly clustered along the pattern? Strong association means little variation around the trend.
4. *Outliers and surprises.* Did you find something unexpected?

Don't worry about memorizing this list. The key is to look at the scatterplot and think. We'll repeat these steps over and over again, and soon the sequence will become automatic.

What Do You Think?   One of the following scatterplots shows the price of diamonds versus their weights. The other shows the box-office gross in dollars of movies versus a critic's rating on a 5-point Likert scale. Identify which is which and describe any association.[2]

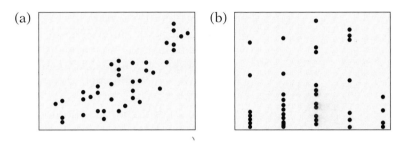

(a)   (b)

6.3 | MEASURING ASSOCIATION

Rather than rely on a visual impression from a scatterplot, we need a way to measure the strength of the association. We will use two statistics that allow us to convey precisely the degree of association, as we did with categorical variables. The second statistic converts the first statistic onto a scale that is more easily interpreted, analogous to how Cramer's V converts the chi-squared statistics for the association between categorical variables in Chapter 5.

Covariance

covariance A statistic that measures the amount of linear association between two numerical variables.

Covariance quantifies the strength of the linear association between two numerical variables. It measures the degree to which data concentrate along a diagonal line in a scatterplot. To see how covariance works, let's continue with the energy data. In the scatter plot in Figure 6.5, we've colored the points that show the amount of natural gas versus the number of heating degree-days (Figure 6.3).

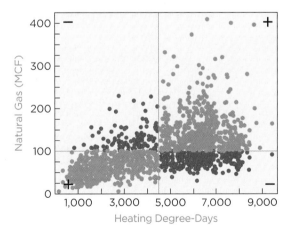

FIGURE 6.5 Scatterplot of the amount of natural gas versus HDD with reference lines at the means.

[2] The diamonds are on the left, with positive association. The pattern is vaguely linear, with quite a bit of variation around the linear trend. The movies are on the right with little or no association. The rating assigned by the critic on the *x*-axis has little relationship with the popularity at the box office.

The horizontal and vertical gray lines locate the means of the two variables and divide the plot into four quadrants. Green points in the upper right and lower left quadrants identify households for which both energy use and heating degree-days are either larger or smaller than average. For instance, a household in the upper right quadrant is in a colder than average climate and uses more than the average amount of gas. Red points in the other two quadrants identify households for which one variable is larger than its mean and the other is smaller. These include, for instance, homes in relatively warm locations that use more gas than average.

Points in the green quadrants indicate positive association. Positive association implies that the variables vary together. For example, cases that are relatively large on one axis are also large on the other. If there is positive association, most of the points should be in the upper right and lower left quadrants. That's the case in this example. Points in the other two quadrants (red) suggest negative association: One variable is relatively large whereas the other is relatively small. Negative association implies that most of the data should be in the upper left and lower right quadrants.

Rather than count the points in the quadrants, covariance takes into account the distances from the means. Cases that lie farther from the central point $(\bar{x}, \bar{y})$ contribute more to the covariance than cases that are close to $(\bar{x}, \bar{y})$. To see how we determine the contribution of each point, look at Figure 6.6, which zooms in on one point in Figure 6.5. The rectangle in this figure has one corner at the point $(\bar{x}, \bar{y})$, and the opposite corner lies at the point (x, y) associated with a specific home. The area of this rectangle is $(x - \bar{x})(y - \bar{y})$.

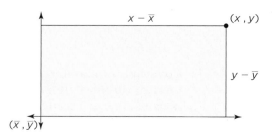

FIGURE 6.6 Deviations from the means in the scatterplot.

Since both deviations $x - \bar{x}$ and $y - \bar{y}$ in Figure 6.6 are positive, their product $(x - \bar{x})(y - \bar{y})$ is positive. That's true for all of the points in the two green quadrants because both deviations have the same sign.

$$(x - \bar{x})(y - \bar{y}) > 0$$

For points in the two red quadrants, the product is negative. One variable is above its mean (positive deviation) and the other is below its mean (negative deviation). Hence, the area $(x - \bar{x})(y - \bar{y})$ is positive for points indicating positive association (green) and negative for points indicating negative association (red). Points with either deviation equal to zero don't contribute to the covariance.

The covariance is (almost) the average of the signed areas of the rectangles defined by the data. The formula for the covariance is

$$\text{cov}(x, y) = \frac{(x_1 - \bar{x})(y_1 - \bar{y}) + (x_2 - \bar{x})(y_2 - \bar{y}) + \cdots + (x_n - \bar{x})(y_n - \bar{y})}{n - 1}$$

As in Chapter 4, subscripts identify rows of the data table; $x_1 - \bar{x}$ denotes the difference between the heating degree-days of the home in the first row of the data table (x_1) and the mean HDD $(\bar{x})$. Similarly, $y_1 - \bar{y}$ denotes the difference between the gas used by the first home (y_1) and the mean gas use $(\bar{y})$. The divisor $n - 1$ matches the divisor in s^2.

Because the energy example has a large number of cases $(n = 1,658)$, we used a computer to compute the covariance between heating degree-days and gas use. To illustrate how the data enter into the formula, the next expression plugs several values from the data table (a portion of the data table and the two averages are shown in the left margin) into the formula for the covariance.

$$
\text{cov}(HDD, Gas)
$$

$$
= \frac{(HDD_1 - \overline{HDD})(Gas_1 - \overline{Gas}) + (HDD_2 - \overline{HDD})(Gas_2 - \overline{Gas}) + \cdots + (HDD_{1658} - \overline{HDD})(Gas_{1658} - \overline{Gas})}{1658 - 1}
$$

$$
= \frac{(4{,}080 - 4{,}547.5)(72.5 - 99.9) + (2{,}068 - 4{,}547.5)(35.6 - 99.9) + \cdots + (8{,}896 - 4{,}547.5)(107.5 - 99.9)}{1657}
$$

$$
= 63{,}357 \; HDD \times MCP
$$

Case	HDD	Gas
1	4,080	72.5
2	2,068	35.6
...	...	...
1,658	8,896	107.5
Average	4,547.5	99.9

The covariance confirms the presence of positive linear association, but it is difficult to interpret the specific value. Positive covariance is consistent with our intuition and the scatterplot of gas consumption versus heating degree-days. On average, homes in colder climates use more than the average amount of natural gas. The size of the covariance, however, is difficult to interpret. This difficulty occurs because the covariance has units: Those of the x-variable times those of the y-variable. In this case, the covariance is 63,357 HDD × MCF. To make sense of the covariance, we convert it into a statistic that more easily conveys the strength of the association.

Correlation

correlation (r) A standardized measure of linear association between two numerical variables; the correlation is always between −1 and +1.

Correlation is a more easily interpreted measure of linear association derived from the covariance. The correlation between two numerical variables is easy to determine from the covariance. Just divide the covariance by the product of the standard deviations.

$$
\text{corr}(x, y) = \frac{\text{cov}(x, y)}{s_x s_y}
$$

Because the standard deviations of two variables appear in this calculation, we use subscripts that identify the variables to distinguish them. For example, the correlation between heating DD and gas usage is

$$
\text{corr}(HDD, Gas) = \frac{\text{cov}(HDD, Gas)}{s_{HDD}\, s_{Gas}}
$$

$$
= \frac{63{,}357 \; HDD \times MCF}{2{,}235.4 \; HDD \times 51.26 \; MCF}
$$

$$
\approx 0.55
$$

The units of the two standard deviations cancel the units of the covariance. The resulting statistic, the correlation, does not have units and can be shown *always to lie between −1 and +1*,

$$
-1 \leq \text{corr}(x, y) \leq +1
$$

The correlation is usually denoted by the letter r.

The correlation in this example is well below its upper limit at 1. Because $r = 0.55$ is much less than 1, the data show considerable variation around the linear pattern. Homes in the same climate vary considerably in how much gas they use. Other factors beyond climate, such as thermostat settings, affect consumption.

Because the correlation has no units, it is unaffected by the scale of measurement. For instance, had we measured gas use in cubic meters and climate in degrees Celsius, the correlation would be $r = 0.55$ in both cases. Figure 6.7 shows two scatterplots, one with the original units and one with metric units.

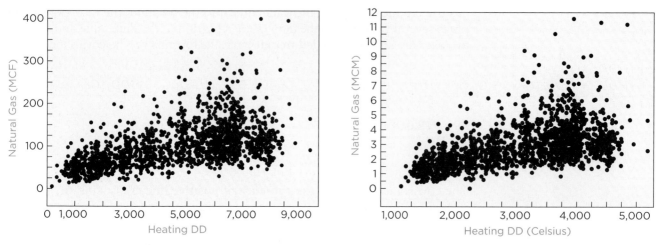

FIGURE 6.7 Scales affect the axes but not the content of a scatterplot.

If you cover the axes, these scatterplots are the same. Only the labels on the axes differ. These differences don't affect the direction, curvature, or variation in the scatterplot, so they don't change the strength of the association either. The correlation is the same in both cases.

The correlation *can* reach −1.0 or +1.0, but these extremes are unusual. They happen only if *all* the data fall exactly on a diagonal line. For example, Figure 6.8 shows the temperature at 50 locations on both Fahrenheit and Celsius scales.

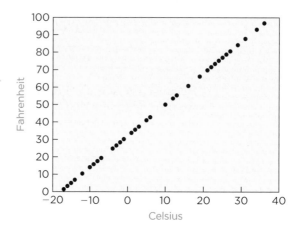

FIGURE 6.8 Celsius temperatures are perfectly associated with Fahrenheit temperatures.

The correlation $r = 1$ in this case because there is a line that gives one temperature *exactly* in terms of the other (Fahrenheit = 32 + 1.8 × Celsius). Data on a line with negative slope implies $r = -1$. Keep these properties of the correlation in mind:

1. r measures the strength of linear association.
2. r is always between −1 and +1, $-1 \leq r \leq 1$.
3. r does not have units.

To help you appreciate the relationship between r and the strength of linear association, Figure 6.9 shows nine scatterplots and their correlations. The values of r run from $r = \pm 0.95$ in the first row down to $r = 0$ at the bottom. The pattern in a scatterplot tilts up when the correlation is positive (plots on the left) and tilts down when the correlation is negative (right). The larger the magnitude of r, the tighter the points cluster along the diagonal line. Each plot shows 200 points.

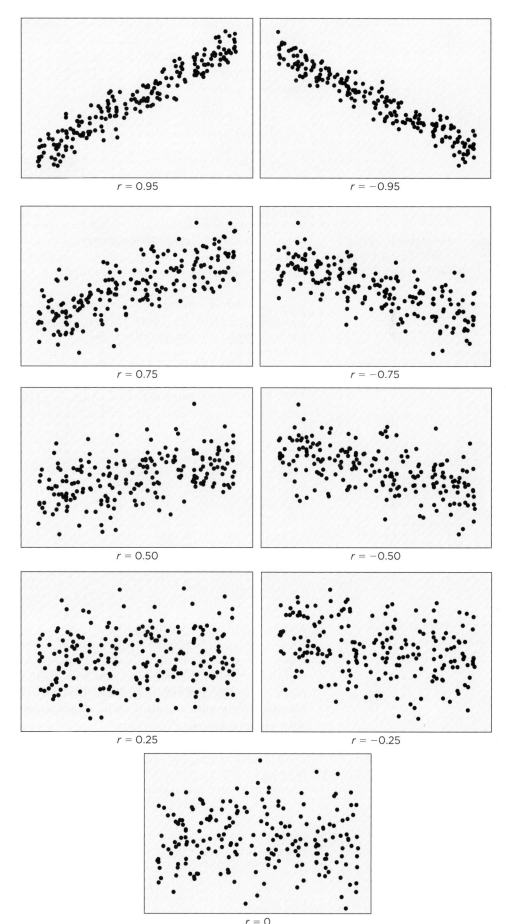

FIGURE 6.9 Correlation measures the tendency of data to concentrate along a diagonal of the scatterplot.

The pattern associated with the correlation is subtle when r gets close to 0. Indeed, you may not recognize any pattern in the two scatterplots with $r = \pm 0.25$. That's because when r gets close to zero, the variation around the pattern obscures the pattern itself. In these cases, most of the variation in y is unrelated to x.

Correlation Matrix

It is common in some fields to report the correlation between every available pair of numerical variables and arrange these in a table. The rows and columns of the table name the variables, and the cells of the table hold the correlations.

A table that shows all of the correlations among a collection of numerical variables is known as a **correlation matrix.** Correlation matrices are compact and convey a lot of information at a glance. They can be an efficient way to explore a large set of variables because they summarize the pairwise relationships.

correlation matrix A table showing all of the correlations among a set of numerical variables.

caution Before relying on a large table of correlations, be sure to review the checklist for these correlations. Because a correlation matrix does not show plots, you cannot tell whether the results conceal outliers or miss bending patterns.

As an example, Table 6.1 shows the correlation matrix for several characteristics reported in *Forbes* magazine for large companies.

TABLE 6.1 Correlation matrix of the characteristics of large companies.

	Assets	Sales	Market Value	Profits	Cash Flow	Employees
Assets	1.000	0.746	0.682	0.602	0.641	0.594
Sales	0.746	1.000	0.879	0.814	0.855	0.924
Market value	0.682	0.879	1.000	0.968	0.970	0.818
Profits	0.602	0.814	0.968	1.000	**0.989**	0.762
Cash flow	0.641	0.855	0.970	**0.989**	1.000	0.787
Employees	0.594	0.924	0.818	0.762	0.787	1.000

For example, the correlation between profits and cash flow is quite high, 0.989 (shown in boldface). Notice that this correlation appears twice in the table, once above the diagonal and once below the diagonal. The table is symmetric about the diagonal that runs from the top left cell to the bottom right cell. The correlation between profits and cash flow is the same as the correlation between cash flow and profits, so the values match on either side of the diagonal. The correlation (and covariance) is the same regardless of which variable you call x and which you call y. The diagonal cells of a correlation table are exactly 1. (Can you see why? Think about where the points lie in a scatterplot of assets on assets, for instance. The plot would resemble Figure 6.8 and look like a single line.)

What Do You Think?

The scatterplot in Figure 6.10 shows the amount of natural gas used versus another variable, the size of the home measured in the number of square feet:

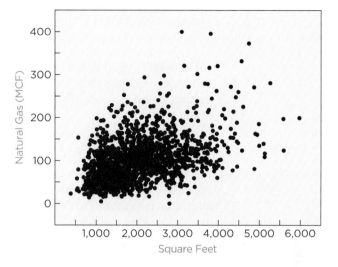

FIGURE 6.10 Scatterplot of natural gas used versus size of home in square feet.

a. Describe the association between the two variables.[3]
b. The covariance between these variables is 19,096. What are the units of the covariance?[4]
c. The SD of natural gas is $s_y = 51.26$ MCF, and the SD of the size is $s_x = 876.5$ sq ft. What is the correlation between the use of natural gas and the size of the home?[5]
d. Explain the size of the correlation. Is there a reason that the correlation is not larger?[6]

6.4 | SUMMARIZING ASSOCIATION WITH A LINE

Correlation measures the strength of linear association between variables. The larger $|r|$ becomes, the more closely the data cluster along a line. We can use r to find the equation of this line. Once we have this equation, it is easy to predict one variable from the other.

The simplest expression for this equation uses z-scores of the two variables. A z-score (Chapter 4) is a deviation from the mean divided by the standard deviation. The correlation converts a z-score of one variable (say, heating degree-days) into a z-score of the other (gas usage). If we know that a home is, for instance, located in a climate that is 1 SD above the mean, then we expect to find its use of natural gas r SDs above the mean. For homes in a climate that is 1 SD below the mean, we expect their gas use to average r SDs below the mean. Since $|r| \leq 1$, the line associated with the correlation pulls values toward the mean.

We can express this relationship as a simple equation. Consider a home in a climate with x degree-days that uses y MCF of natural gas. Let z_x

---

[3] There's weak positive association, vaguely linear. It looks like the variation around the pattern increases with the size of the home.
[4] The units of the covariance are those of x times those of y, so that's 19,096 (square feet $\times$ MCF).
[5] The correlation is $19,096/(876.5 \times 51.26) \approx 0.425$.
[6] The correlation confirms that there is some linear association (it's about halfway between 0 and its largest positive value $+1$), but it's not as strong as the correlation in the text example. These homes are in very different climates; a large house does not require much heating if it's in Florida.

denote the z-score for its climate, and let z_y denote the z-score for its use of natural gas:

$$z_x = (x - \bar{x})/s_x \qquad z_y = (y - \bar{y})/s_y$$

The line determined by the correlation is

$$\hat{z}_y = r\,z_x$$

The ˆ over z_y distinguishes what this formula predicts from the real thing. Unless $r = \pm 1$, the data are not exactly on a single line. If $r = 1$, then we could predict gas usage exactly using this formula. If $r = 0$, this line is flat; we'd predict $\hat{z}_y = 0$ for *any* value of x.

Slope-Intercept Form

The equation $\hat{z}_y = rz_x$ is compact, but it is also common to see this equation expressed directly in terms of x and y. We will explore this approach to association much more in Part 4 of this book, but it may help to see a formula for the line defined by the correlation so that you can add this line to a plot. A bit of algebra shows that we can express the line associated with the correlation as

$$\hat{y} = a + bx$$

with $\qquad a = \bar{y} - b\bar{x} \qquad$ and $\qquad b = rs_y/s_x$

y-intercept The location at which a line intersects the *y*-axis in a graph.

slope Rate of change of a line; steepness of the line.

where a is known as the **y-intercept** and b is the **slope**. This way of writing the line is hence called slope-intercept form. Once we have a and b, we simply plug in a value for x and the equation predicts y. It is also easy to add a line expressed in slope-intercept form to the scatterplot of the associated data. The intercept a shows where the line intersects the y-axis, and the slope b determines how rapidly the line rises or falls.

As an example, let's find the line that describes the association between gas use (y) and climate (x). The underlying statistics for gas usage are $\bar{y} = 99.9$ MCF and $s_y = 51.26$ MCF; those for climate are $\bar{x} = 4{,}547.5$ HDD and $s_x = 2{,}235.4$ HDD. Hence the slope of the line is

$$b = rs_y/s_x = 0.55(51.26/2{,}235.4) \approx 0.0126 \text{ MCF/HDD}$$

The slope has units: those of y (from s_y) divided by those of x (from s_x). The intercept is

$$a = \bar{y} - b\bar{x} = 99.9 - 0.0126 \times 4{,}547.5 \approx 42.6 \text{ MCF}$$

The intercept takes its units from $\bar{y}$ (b converts the units of $\bar{x}$ into the units of y). Figure 6.11 adds the correlation line to the scatterplot of natural gas usage on heating degree-days.

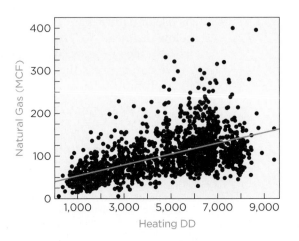

FIGURE 6.11 The correlation defines a line that summarizes the association between HDD and usage.

The intercept tells us that on average homes with no need for heat (zero heating degree-days) evidently use some gas ($a = 42.6\,\text{MCF}$) for other purposes, such as heating water or cooking. The slope says that the amount of gas used goes up by about 12.6 MCF (1,000 times $b = 0.0126$ MCF/HDD) when we compare homes in a climate with, say, 2,000 HDD to those in a colder climate with 3,000 HDD.

Lines and Prediction

Let's return to the problem suggested at the beginning of the chapter and use the correlation line to build a customized ad. We set out to design an ad that would convince homeowners to insulate by telling them their potential savings. Once we know where someone lives, we can use the equation of this line to guess his or her gas usage and the cost of that gas. (Few homeowners know offhand how much they spend annually for heating.) For example, we could replace the ad shown in the introduction with an interactive Web ad: "Click to find typical savings from insulating!"

As an example, suppose that a homeowner who lives in a cold climate with 8,800 HDD clicks on the ad. Let's estimate the possible savings from insulating. This climate is about 1.9 SDs above the mean number of heating degree-days:

$$z_x = \frac{x - \bar{x}}{s_x} = \frac{8,800 - 4,547.5}{2,235.4} \approx 1.9$$

The positive association between heating degree-days and gas usage suggests that gas use at this home is about $r \times 1.9 \approx 1.05$ SDs above the mean of natural gas used, at

$$\bar{y} + 1.9s_y = 99.9 + 1.05(51.26) \approx 154\,\text{MCF}$$

[Using the line in slope-intercept form gets the same answer; gas use in this climate averages $\hat{y} = a + bx = 42.6 + 0.0126(8,800) \approx 154$ MCF.] At \$10 per MCF, this gas would cost \$1,540. If the savings from insulating were 30% of the predicted costs, then the homeowner would save $0.3 \times \$1,540 = \$462$.

Let's do another example, this time for a warmer climate. For a home in a climate 1 SD *below* the mean, we expect its gas use to be about 0.55 SD *less* than the average gas used, or $\bar{y} - \hat{r}s_y = 99.9 - 0.55 \times 51.26 = 71.707$ MCF. Because we expect less to be used, the predicted savings are smaller, only $0.3 \times \$717.07 \approx \$215$.

We've now got the ingredients for an effective interactive ad. Customers at our Web site could click on their location on a map like this one or enter their zip code. Our software would look up the climate in that area and use the correlation line to estimate the possible dollar savings.

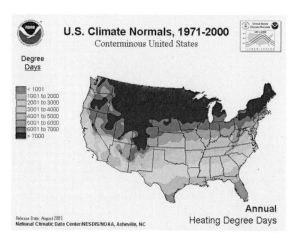

Nonlinear Patterns

caution If the association between numerical variables is not linear, a line may be a poor summary of the pattern. Linear patterns are common but do not apply in every situation. Chi-squared (Chapter 5) measures any type of association between categorical variables. Covariance and correlation measure only *linear* association. If the pattern in a scatterplot bends, then covariance and correlation miss some of the association. Be sure that you inspect the scatterplot before relying on these statistics to measure association.

For example, the scatterplot in Figure 6.12 shows data on employees of a small firm. The *x*-axis gives the age of each employee and the *y*-axis shows the cost in dollars to the employer to provide life insurance benefits to each employee.

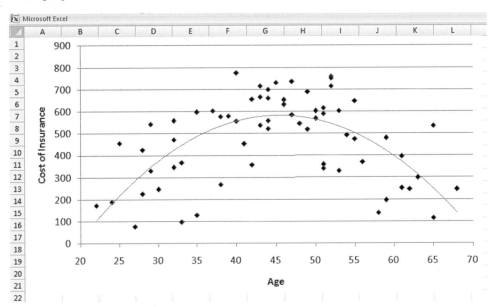

FIGURE 6.12 The cost of insurance does not continue to increase.

You can see a strong pattern, but it's not linear. The direction of the pattern changes as suggested by the curve in the figure. The cost of providing life insurance initially grows as young employees pass through their family years and opt for more insurance. As employees age, fewer want life insurance, and the cost to the firm to provide this benefit shrinks.

Correlation does not measure this association because the pattern is not linear. In fact, the correlation between age and cost in Figure 6.12 is near zero ($r = 0.09$). When you think about the correlation and its associated line, this value for r makes sense. In a way, the overall slope *is* zero; the positive slope on the left cancels the negative slope on the right.

A problem occurs if benefit managers only compute a correlation and do not look at the scatterplot. They might conclude from the small value of r that these two variables are unrelated. That mistake could produce poor decisions when it comes time to anticipate the cost to the firm to provide this benefit.

6.5 | SPURIOUS CORRELATION

A scatterplot of the damage (in dollars) caused to homes by fire would show a strong correlation with the number of firefighters who tried to put out the blaze. Does that mean that firefighters cause damage? No, and a lurking variable, the size of the blaze, explains the superficial association. A correlation that results from an underlying, lurking variable rather than the shown *x*- and *y*-variables is often called a **spurious correlation**.

spurious correlation Correlation between variables due to the effects of a lurking variable.

Scatterplots and correlation reveal association, not causation. You have to be cautious interpreting association in a contingency table (recall Simpson's paradox), and the same warning applies to scatterplots. Lurking variables can affect the relationship in a scatterplot in the same way that they affect the relationship between categorical variables in a contingency table.

Interpreting the association in a scatterplot requires deeper knowledge about the variables. For the example concerning energy use in this chapter, it seems sensible to interpret the scatterplot of gas usage and heating degree-days as meaning that "colder weather causes households to use more natural gas." We can feel pretty good about this interpretation because we've all lived in homes and had to turn up the heat when it gets cold outside. Though this interpretation makes sense, it's not proof.

What other factors might affect the relationship between heating degree-days and the amount of gas that gets used? We already mentioned two: thermostat settings and the size of the home. How might these factors affect the relationship between heating degree-days and gas use? Maybe homes in colder climates are bigger than those in warmer climates. That might explain the larger amount of gas used for heating. It's not the climate; it's the size of the home. These alternatives may not seem plausible, but we need to rule out other explanations before we make decisions on the basis of association.

Checklist: Covariance and Correlation. Covariance and correlation measure the strength of *linear* association between *numerical* variables. Before you use correlation (and covariance), verify that your data meet the prerequisites in this checklist.

✓ **Numerical variables.** Correlation applies only to quantitative variables. Don't apply correlation to categorical data.
✓ **No obvious lurking variables.** There's a spurious correlation between math skills and shoe size for kids in an elementary school, but there are better ways to convey the effect of age.
✓ **Linear.** Correlation measures the strength of *linear* association. A correlation near zero implies no linear association, but there could be a nonlinear pattern.
✓ **Outliers.** Outliers can distort the correlation dramatically. An outlier can enlarge an otherwise small correlation or conceal a large correlation. If you see an outlier, it's often helpful to report the correlation with and without the point.

caution   These conditions are easy to check in a scatterplot. Be cautious if a report gives a correlation without the associated scatterplot. The concerns in this checklist are important even if you don't see the plot.

4M EXAMPLE 6.1 | **LOCATING A NEW STORE**

MOTIVATION | STATE THE QUESTION

When retailers like Target look for locations for opening a new store, they take into account how far the location is from the nearest competition. It seems foolish to open a new store right next to a Wal-Mart, or does it? Is it really the case that it's better to locate farther from the competition? We'll consider data from a regional chain of retail stores.

METHOD

DESCRIBE THE DATA AND SELECT AN APPROACH

Plan your approach before you get wrapped up in the details. Our interest lies in the association between sales at the retail outlets and the distance to the nearest competitor. Do stores located far from the competition have higher sales? If there's positive association between distance and sales, then we have support for maintaining some separation from rivals.

Summarize how you'll approach the analysis. This chain operates 55 stores. For each store, our data include the total retail sales over the prior year (in dollars). The data also include a numerical variable that gives the distance in miles from the nearest competitor. The first step of the analysis is to inspect the scatterplot of sales versus distance. Once we see the scatterplot, we will be able to check the conditions for using a correlation to quantify the strength of the association. If the association is linear, we can use the correlation line to summarize the relationship.

MECHANICS

DO THE ANALYSIS

The scatterplot of sales versus distance for these stores shows positive association. Visually, we'd say this was linear and moderately strong. The scatterplot includes the line determined by the correlation.

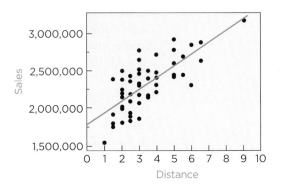

Let's run through the checklist for the correlation.

✓ **Numerical variables:** Both variables are numerical.

✓ **No obvious lurking variable:** Without more background, we cannot be sure. Perhaps stores that are farther from competitors are also larger. This chain operates some superstores that combine a traditional department store with a grocery market. If large stores tend to be farther from the competition, the association may be due to size, not distance.

✓ **Linear:** The pattern of the data in the scatterplot seems linear, with substantial variation around the pattern.

✓ **Outliers:** Some stores sell a lot and others a little, but that's to be expected. One store seems a bit of an outlier; this store is about 9 miles from the nearest competition and has the largest annual sales. Though separated from the others, this store falls in line with the others.

Using software, we computed the correlation between sales and distance to be $r = 0.741$. To draw the line as in the scatterplot, express the equation in slope-intercept form. The relevant summary statistics are $\bar{y} = \$2.308$ million and $s_y = \$340,000$ for sales and $\bar{x} = 3.427$ miles and $s_x = 1.582$ miles for distance. Hence the slope and intercept of the line are

$$b = rs_y/s_x = 0.741(340,000/1.582) = \$159,254.1/\text{mile}$$
$$a = \bar{y} - b\bar{x} = 2,308,000 - 159,254.1 \times 3.427 = \$1,762,237 \approx \$1.76\,\text{million}$$

Though not part of the motivating task, this line anticipates sales levels at new locations. For example, sales at a location adjacent to a competitor ($Distance = 0$) are estimated to be $\hat{y} = a + b(0) = a \approx \$1.76$ million. If the distance were, say, 5 miles, then the expected sales grow by five times the slope, $\hat{y} = a + b(5) = 1,762,237 + 159,254.1 \times 5 = 2,558,508 \approx \$2.56$ million. **It's a good idea when doing these calculations to look back at the plot to make sure that you have not made a mistake.**

tip

MESSAGE | SUMMARIZE THE RESULTS

Sales and distance are positively related. The data show a strong, positive linear association between distance to the nearest competitor and annual sales. The correlation between sales and distance is 0.74. As distance grows, the data suggest sales grow on average as well, rising by about $160,000 per mile of added separation. There is moderate variation around the linear pattern.

If you are unsure or suspect a possible lurking variable, mention your concerns in the summary message. In this example, we have lingering concerns that there may be other lurking variables (such as the size of the store) behind this relationship. If stores that are farther from competitors are larger, we may be mistaking distance from the competition for size.

Best Practices

- *To understand the relationship between two numerical variables, start with a scatterplot.* If you see the plot, you won't be fooled by outliers, bending patterns, or other features that can mislead the correlation.
- *Look at the plot, look at the plot, look at the plot.* It's the most important thing to do. Be wary of a correlation if you are not familiar with the underlying data.
- *Use clear labels for the scatterplot.* Many people don't like to read; they only look at the pictures. Use labels for the axes in your scatterplots that are self-explanatory.
- *Describe a relationship completely.* If you do not show a scatterplot, make sure to convey the direction, curvature, variation, and any unusual features.
- *Consider the possibility of lurking variables.* Correlation shows association, not causation. There might be another factor lurking in the background that's a better explanation for the pattern that you have found.
- *Use a correlation to quantify the association between two numerical variables that are linearly related.* Don't use correlation to summarize the association unless you are sure that the pattern is linear. Verify the correlation checklist before you report the correlations.

Pitfalls

- *Don't use the correlation if the data are categorical.* Keep categorical data in tables where they belong.
- *Don't treat association and correlation as causation.* The presence of association does not mean that changing one variable *causes* a change in the other. The apparent relationship may in fact be due to some other, lurking variable.
- *Don't assume that a correlation of zero means that the variables are not associated.* If the rela-

tionship bends, the correlation can miss the pattern. A single outlier can also hide an otherwise strong linear pattern or give the impression of a pattern in data that have little linear association.
- *Don't assume that a correlation near −1 or +1 means near perfect association.* Unless you see the plot, you'll never know whether all that you have done is find an outlier.

Software Hints

Statistics packages generally make it easy to look at a scatterplot to check whether the correlation is appropriate. Some packages make this easier than others.

Many packages allow you to modify or enhance a scatterplot, altering the axis labels, the axis numbering, the plot symbols, or the colors used. Some options, such as color and symbol choice, can be used to display additional information on the scatterplot.

EXCEL

Use the Insert > Chart... command to produce a scatterplot of two columns. The process is simplest if you select the two columns prior to starting this command, with the *x* column adjacent and to the left of the *y* column. (If the columns are not adjacent or in this order, you'll need to fill in a data dialog later.) Select the icon that looks like a scatterplot without connected dots. Click the button labeled Next and you're on your way. Subsequent dialogs allow you to modify the scales and labels of the axes. The chart is interactive. If you change the data in the spreadsheet, the change is transferred to the chart.

To compute a correlation, select the function CORREL from the menu of statistical functions. Enter ranges for two numerical variables in the dialog. (You can type the function directly into a cell as well.)

DDXL automates this process. Again, start by selecting the portion of the spreadsheet with your data (names in the first row, please). Then follow the DDXL > Charts and Plots menu command. In the resulting dialog, choose the Scatterplot item in the Click-Me menu. Then identify the variables in the

resulting dialog. To get a scatterplot *and* the correlation, follow the DDXL > Regression menu commands and pick the Correlation item in the Click-Me menu. You can color points and change the symbols using the floating palettes in the DDXL window.

MINITAB

To make a scatterplot, follow the menu commands

Graph > Scatterplot

Select the default plot (others, for example, let you include the correlation line), then identify the numerical variable for the *y*-axis and the numerical variable for the *x*-axis.

To compute a correlation, follow the menu items

Stat > Basic Statistics > Correlation

Fill the dialog with at least two numerical variables. If you choose more than two, you get a table of correlations. The sequence Stat > Basic Statistics > Covariance produces the covariance of the variables.

JMP

To make a scatterplot, follow the menu commands

Analyze > Fit Y by X

In the dialog, assign numerical variables from the available list for the *y*-axis and the *x*-axis.

Once you have formed the scatterplot, to obtain a correlation use the pop-up menu identified by the red triangle in the upper left corner of the output window holding the scatterplot. Select the item labeled Density ellipse and choose a level of coverage (such as 0.95). The ellipse shown by JMP is thinner as the correlation becomes larger. A table below the scatterplot shows the correlation.

To obtain a correlation table, follow the menu commands

Analyze > Multivariate Methods > Multivariate

and select variables from the list shown in the dialog. By default, JMP produces the correlation table and the scatterplot matrix (a table of scatterplots, one for each correlation). Clicking on the red triangle button in the output window produces other options, including the table of covariances.

CHAPTER SUMMARY

Scatterplots show association between two numerical variables. The **response** goes on the y-axis and the **explanatory variable** goes on the x-axis. The **visual test for association** compares the observed scatterplot to artificial plots that remove any pattern. To describe the pattern in a scatterplot, indicate its *direction, curvature, variation,* and other surprising features such as *outliers*. **Covariance** measures the amount of linear association between two numerical variables. The **correlation r** scales the covariance so that the resulting measure of association is always between -1 and $+1$. The correlation is also the slope of a line that relates the standardized deviations from the mean on the x-axis to the standardized deviations from the mean on the y-axis. **Spurious correlation** results from the presence of a lurking variable, and curvature or outliers can cause correlation to miss the pattern.

■ Key Terms

correlation, 111
  matrix, 114
covariance, 109
explanatory variable (x), 106

r (correlation), 111
response (y), 106
scatterplot, 106
slope, 116

spurious correlation, 118
visual test for association, 108
y-intercept, 116

■ Formulas

COVARIANCE

$$\text{cov}(x, y)$$
$$= \frac{(x_1 - \bar{x})(y_1 - \bar{y}) + (x_2 - \bar{x})(y_2 - \bar{y}) + \cdots + (x_n - \bar{x})(y_n - \bar{y})}{n - 1}$$
$$= \frac{\sum_{i=1}^{n}(x_i - \bar{x})(y_i - \bar{y})}{n - 1}$$

CORRELATION

$$r = \text{corr}(x, y) = \frac{\text{cov}(x, y)}{s_x s_y}$$

When calculations are done by hand, the following formula is also used.

$$r = \frac{\sum_{i=1}^{n}(x_i - \bar{x})(y_i - \bar{y})}{\sqrt{\sum_{i=1}^{n}(x_i - \bar{x})^2 \sum_{i=1}^{n}(y_i - \bar{y})^2}}$$

Review this checklist when using a correlation.

✓ Numerical variables
✓ No obvious lurking variable
✓ Linear pattern
✓ No extreme outliers

EXPRESSIONS RELATING COVARIANCE AND CORRELATION

$$\text{cov}(x, y) = s_x s_y \,\text{corr}(x, y)$$
$$\text{corr}(x, y) = \text{cov}(x, y)/(s_x s_y)$$

CORRELATION LINE

Using z-scores $\hat{z}_y = r z_x$
Using x and y
$\hat{y} = a + bx, \quad a = \bar{y} - b\bar{x}, \quad b = r s_y/s_x$

▨ About the Data

The data on household energy consumption was collected by the Energy Information Agency of the U.S. Department of Energy (DoE) as part of its survey of residential energy consumption around the United States. How does the government use these data? For one, it uses the data to provide guidance to homeowners who are thinking about making their homes more energy efficient. The Home Energy Saver at DoE's Web site, hes.lbl.gov, provides a calculator that compares the energy costs for homes in your area. The calculator uses data like those in this chapter to estimate the value of modernizing appliances and adding insulation.

EXERCISES ▪ ▪

Mix and Match

1. Match the description to the scatterplot.
   (a) Negative direction, linear, moderate variation around line
   (b) Positive direction, linear, small variation around line
   (c) No pattern. A plot of scrambled pairs would look the same
   (d) Negative then positive direction, valley-shaped curve, moderate variation around curve

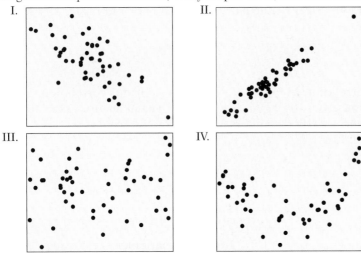

2. Match the description to the scatterplot.
   (a) No pattern. A plot of scrambled pairs would look the same
   (b) Negative direction, linear, small variation around line
   (c) Positive direction, but bending with small variation around curving pattern
   (d) Positive direction, linear, large variation around line

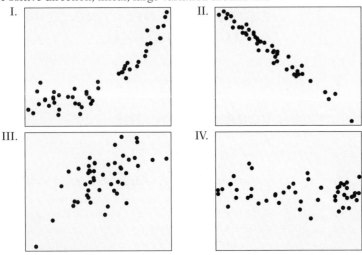

3. Match the value of the correlation to the data in the scatterplot.
   (a) $r = 0$ (b) $r = 0.5$ (c) $r = 0.8$ (d) $r = -0.6$

I.
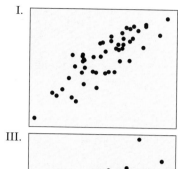

II.
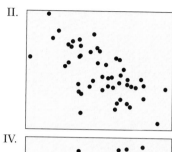

III.
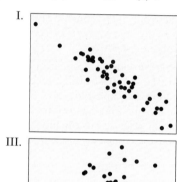

IV.
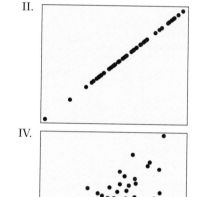

4. Match the value of the correlation to the data in the scatterplot.
   (a) $r = 0$ (b) $r = -0.9$ (c) $r = 1$ (d) $r = 0.4$

I.

II.

III.

IV.

True/False

Mark each statement True or False. If you believe that a statement is false, briefly say why you think it is false.

5. The explanatory variable defines the x-axis in a scatterplot.

6. In a plot of income (the response) versus education (the explanatory variable) for managers, managers with the lowest levels of education are at the right-hand side of the figure.

7. The presence of a pattern in a scatterplot means that values of the response increase with the values of the explanatory variable.

8. The visual test for association is used to decide whether a plot has a real or imagined pattern.

9. A line with positive slope describes a linear pattern with a positive direction.

10. A company has learned that the relationship between its advertising and sales shows diminishing marginal returns. That is, as it saturates consumers with ads, the benefits of increased advertising diminish. The company should expect to find linear association between its advertising and sales.

11. If net revenue at a firm is about 10% of gross sales, then a scatterplot of net revenue versus gross sales would show a nonlinear pattern rather than a line.

12. If two variables are linearly associated, the peak in the histogram of the explanatory variable will line up with the peak in the histogram of the other variable.

13. If the correlation between the growth of a stock and the growth of the economy as a whole is close to 1, then this would be a good stock to hold during a recession when the economy shrinks.

14. If the covariance between x and y is 0, then the correlation between x and y is 0 as well.

15. If the correlation between x and y is 1, then the points in the scatterplot of y on x all lie on a single line with slope s_y/s_x.

16. An accountant at a retail shopping chain accidentally calculated the correlation between the phone number of customers and their outstanding debt. He should expect to find a substantial positive correlation.

17. A report gives the covariance between the number of employees at an assembly factory and the total number of items produced daily. The covariance would be larger if these data were aggregated to monthly totals.

18. The correlation between sales and advertising when both are measured in millions of dollars is 0.65. The correlation remains the same if we convert these variables into thousands of dollars.

19. A gasoline station collected data on its monthly sales over the past three years along with the average selling price of gasoline. A scatterplot of sales on price of gasoline showed positive association. This association means that higher prices lure more customers.

20. Half of the numbers in a correlation matrix are redundant, ignoring the diagonal.

Think About It

21. Suppose you collect data for each of the following pairs of variables. You want to make a scatterplot. Identify the response and explanatory variables. What would you expect to see in the scatterplot? Discuss the direction, curvature, and variation.
    (a) Sales receipts: number of items, total cost
    (b) Productivity: hours worked, items produced
    (c) Football players: weight, time to run 40 yards
    (d) Fuel: number of miles since filling up, gallons left in your tank
    (e) Investing: number of analysts recommending a stock, its subsequent price change

22. Suppose you collect data for each of the following pairs of variables. You want to make a scatterplot. Identify the response and explanatory variables. What would you expect to see in the scatterplot? Discuss the likely direction, curvature, and variation.
    (a) Winter electricity: kilowatts, temperature
    (b) Long-distance calls: time (minutes), cost
    (c) Air freight: weight of load, fuel consumption
    (d) Advertising: length of commercial, number who remember ad
    (e) Health: amount of exercise per week (hours), percentage body fat

23. States in the United States are allowed to set their own rates for sales taxes as well as taxes on services, such as telephone calls. The scatterplot at the top of the next column graphs the state and local taxes charged for wireless phone calls (as a percentage) versus the state sales tax (also as a percentage).

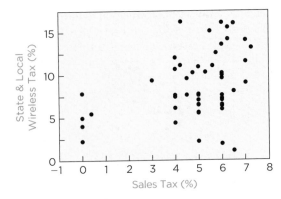

(a) Describe the association, if any, that you find in the scatterplot.
(b) Estimate the correlation between the two variables from the plot. Is it positive, negative, or zero? Is it closer to zero or to ± 0.5?
(c) The cluster of states with no sales tax in the lower left corner of the plot includes Alaska, Delaware, Montana, New Hampshire, and Oregon. What is the effect of this cluster on the association? If these were excluded from the analysis, would the correlation change?
(d) Would it be appropriate to conclude that states that have high sales tax charge more for services like wireless telephone use?

24. The Organization for Economic Cooperation and Development (OECD) is a loose affiliation of 30 nations. The data in this scatterplot show the percentage of disposable income for a single citizen that remains after paying taxes versus the top personal income tax rate (also a percentage).

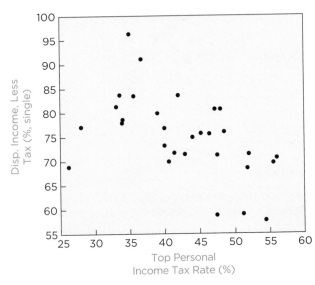

(a) Describe the association in the scatterplot.
(b) Does the association have the direction that you'd expect?
(c) Estimate the correlation between the two variables.
(d) The data for which country appears most unusual relative to the others? How does the correlation change if this country is set aside and excluded from the calculation? (It may help to think about the line associated with the correlation.)

25. If the correlation between number of customers and sales in dollars in retail stores is $r = 0.6$, then what would be the correlation if the sales were measured in thousands of dollars? In euros? (1 euro is worth about 1.2 to 1.5 dollars.)

26. A fitness center weighed 53 male athletes and then measured the amount that they were able to lift. The correlation was found to be $r = 0.75$. Can you interpret this correlation without knowing whether the weights were in pounds or kilograms or a mixture of the two?

27. To inflate the correlation, a conniving manager added $200,000 to the actual sales amount for each store in Exercise 25. Would this trickery make the correlation larger?

28. After calculating the correlation, the analyst at the fitness center in Exercise 26 discovered that the scale used to weigh the athletes was off by 5 pounds; each athlete's weight was measured as 5 pounds more than it actually was. How does this error affect the reported correlation?

29. Management of a sales force has noticed that there is a positive linear association ($r = 0.6$) between sales produced by company representatives in adjacent months. If a sales representative does really well this month (say, her z-score is $+2$), why should we expect her z-score to be lower next month?

30. The human resources department at a company noticed that there is a linear association between the number of days that an employee is absent from year to year, with correlation 0.4. The correlation of absences this year with the number of absences last year is 0.4. Managers claim that an award given to employees who are seldom absent has inspired those who are frequently absent to be present more often in the following year. What do you think?

31. The timeplot at the bottom of the page shows the values of two indices of the economy in the United States: Inflation (left axis, in red, measured as the year-over-year change in the Consumer Price Index) and the Survey of Consumer Sentiment (right axis, in blue, from the University of Michigan). Both series are monthly and cover the time period January 2004 through September 2006.

(a) From the chart, do you think that the two sequences are associated?

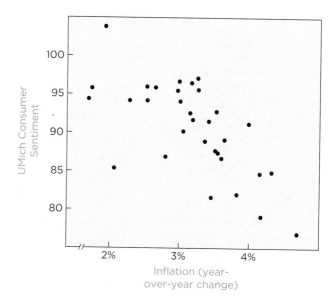

(b) The scatterplot shown above displays the same time series, with Consumer Sentiment plotted versus Inflation. Does this scatterplot change your impression of the association between the two?

(c) Estimate the correlation between these two series.

(d) For looking at the relationship between two time series, what are the advantages of these two plots? Each shows some things, but hides others. Which helps you visually estimate the correlation? Which tells you the timing of the extreme lows and highs of each series?

(e) Does either plot prove that inflation causes changes in consumer sentiment?

32. The scatterplot at the top of the next page shows monthly values of Consumer Sentiment plotted versus Inflation (see Exercise 31 for the definition of these variables). For this plot, we've shown the full history of

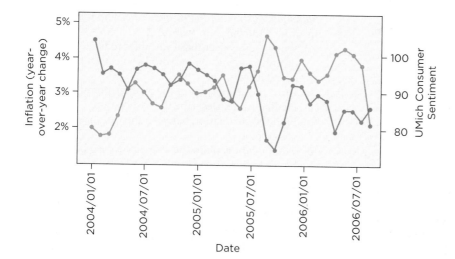

the consumer survey that goes back to January 1978. The subset of the points in black shows data since 2004.

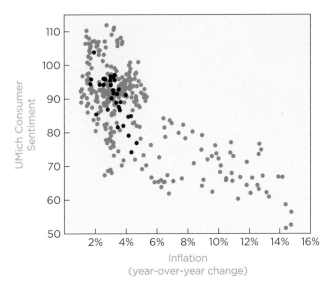

(a) Describe the association between these two variables over the full time period. Is the association strong? Is the relationship linear?

(b) Does the addition of the older data change your impression of the association between these two variables? To see a scatterplot of only the data since 2004, see Exercise 31.

(c) Do you think that the addition of the data prior to 2004 increases or decreases the size of the correlation?

33. Which do you think produces a larger correlation between the weight and the price of diamonds: using a collection of gems of various cuts, colors, and clarities or a collection of stones that have the same cut, color, and clarity?

34. To measure the association between the number of employees and the number of items produced on a small assembly line, a company recorded the number of employees each day (from 5 to 10) and the number of items produced. To make a nicer plot, a manager found the average number of items produced when 5 employees worked, the average number when 6 worked, and so forth. He then plotted these average levels of production versus the number of employees. How does this averaging affect the correlation?

35. Cramer's V (Chapter 5) measures dependence and varies from 0 to 1. The correlation also measures dependence, but varies from −1 to +1. Why doesn't it make sense for V also to take on negative values?

36. The correlation measures the strength of linear dependence. Does Cramer's V (Chapter 5) only measure linear dependence?

37. The visual test for association can be used to help distinguish whether the pattern in a plot is real or imagined. The original ought to stand out clearly. If you look at four plots, one the original and three with scrambled coordinates, and guess, what is the chance that you'll guess the original even if there is no pattern?

38. The visual test for association requires you to compare several plots. One has the original data, and the others show the data with scrambled coordinates. Would it help in finding the original if these scatterplots include the marginal histograms along the edges of the plot?

You Do It

Bold names shown with a question identify the data table for the problem. The title of each of the following exercises identifies the data table to be used for the problem.

39. **Housing and Stocks** Each month, the Department of Commerce releases its latest estimate of construction activity in the housing industry. A key measure is the percentage change in the number of new homes under construction. Does the release of this number come with a change in the stock market? These data show the percentage change in the number of new, privately owned housing units started each month, as reported by the Department of Commerce. The data also include the percentage change in the S&P 500 index on the same day the Department of Commerce releases the housing results.

(a) Before looking at these data, do you expect these variables to be associated? If so, what do you anticipate the correlation to be?

(b) Draw the scatterplot of the percentage change in the S&P 500 index on the percentage change in the number of new housing units started. Describe the association.

(c) Find the correlation between the two variables in the scatterplot. What does the size of the correlation suggest about the strength of the association between these variables?

(d) Suppose you know that there was a 5% increase in the number of new homes. From what you've seen, can you anticipate movements in the stock market?

40. **Homeowners' Insurance** These data give the median dollar value in 2003 of homes in the 50 states and District of Columbia along with the average annual premium for homeowners' insurance (from the *Statistical Abstract of the United States* and the Insurance Information Institute).

(a) Would you expect these variables to be dependent or independent? What would you expect the correlation to be?

(b) Draw the scatterplot of the average premium on the median value. Describe the association.

(c) Find the correlation between the variables in the scatterplot.

(d) What might explain the size of the correlation between these two variables?

41. **Employee Testing** The slope for the correlation line when expressed in z-scores is r. Hence when formulated in z-scores, the absolute value of the slope is less than 1 unless the data lie along a single line. The data in this example are productivity measures used to grade 15 employees who work on an assembly

line. Each employee was tested once, then again a month later.

First Test	Second Test
50	58
35	46
15	40
64	76
53	62
18	39
40	57
24	41
16	31
67	75
46	62
64	64
32	54
71	65
16	51

(a) Would you expect the scores to be associated?
(b) Make a scatterplot of these data, with the first test along the *x*-axis. Describe the relationship, if any.
(c) On the basis of the correlation, if an employee scores two SDs above the mean on the first test, do you expect him or her to be so far above the mean on the second test?
(d) Find the employee with the highest score on the first test. The score for this employee is not the best on the second test. Does this relative decline mean the employee has become less productive, or can you offer an alternative reason for the relatively lower score?

42. **Tour de France** These data give the times in minutes for 140 cyclists who competed in the 2006 Tour de France and raced in both time trial events. In a time trial, each cyclist individually rides a set course. The rider with the shortest time wins the event. The courses in the 2006 tour were each about 35 miles long. (Stage 7 was 52 kilometers long and Stage 19 was 57 kilometers.)
(a) Would you expect the times to be associated?
(b) Make a scatterplot of these data, with the first set of times (from Stage 7) along the *x*-axis. Do the variables appear to be associated? Describe the relationship, if any.
(c) Find the correlation between these variables, if appropriate.
(d) Identify the times for Floyd Landis in the scatterplot. Did he perform as well in the second race, with the championship on the line, as in the first race?

43. **Data Entry** How many mistakes get made during data entry? The following table gives the number of mistakes made by 15 data entry clerks who enter medical data from case report forms. These forms are submitted by doctors who participate in studies of the performance of drugs for treating various illnesses. The column Entered indicates the number of values entered, and the column Errors gives the number of coding errors that were detected among these.

Entered	Errors
4,434	35
4,841	42
6,280	15
1,958	28
7,749	36
2,829	42
4,239	18
3,303	54
5,706	34
3,770	40
3,363	36
1,740	23
3,404	27
1,640	26
3,803	56
1,529	20

(a) Make a scatterplot of these data. Which did you choose for the response and which for the explanatory variable? Describe any patterns.
(b) Find the correlation for these data.
(c) Suppose we were to record the counts in the table in hundreds, so 4,434 became 44.34. How would the correlation change? Why?
(d) Write a sentence or two that interprets the value of this correlation. Use language that would be understood by someone familiar with data entry rather than correlations.
(e) One analyst concluded, "It is clear from this correlation that clerks who enter more values make more mistakes. Evidently they become tired as they enter more values." Explain why this explanation is not an appropriate conclusion.

44. **Truck Weight** The Minnesota Department of Transportation would like to measure the weights of commercial trucks on its highways without the expense of maintaining weigh stations along the highway. Weigh stations are expensive to operate, delay shipments, and aggravate truckers. To see if a proposed remote weight system was accurate, transportation officials conducted a

test. They weighed trucks when stopped on a regular scale. They then used the remote system to estimate the weights. Here are the data.

Weight of a Truck (thousands of pounds)	
New System	**Scale Weight**
26	27.9
29.9	29.1
39.5	38
25.1	27
31.6	30.3
36.2	34.5
25.1	27.8
31	29.6
35.6	33.1
40.2	35.5

(a) Make a scatterplot of these data and describe the direction, form, and scatter of the plot.
(b) Find the correlation.
(c) If the trucks were weighed in kilograms, how would this change the correlation? (1 kilogram = 2.2 pounds)
(d) Interpret the correlation in the context of trucks and weights. What does the correlation tell the Minnesota Department of Transportation?

45. **Cars** These data report characteristics of 233 types of vehicles sold in the United States in 2003–2004. One column gives the official mileage (MPG City, estimated miles per gallon for city driving), and another gives the rated horsepower.
(a) Make a scatterplot of these two columns. Which variable makes the most sense to put on the x-axis, and which belongs on the y-axis?
(b) Describe any pattern that you see in the plot. Be sure to identify any outliers.
(c) Find the correlation between these two variables.
(d) Interpret the correlation in the context of these data. Does the correlation provide a good summary of the strength of the relationship?
(e) Describe the marginal distributions of MPG City and Horsepower. Include the mean and SD of both variables.
(f) Use the correlation line to estimate the mileage of a car with 125 horsepower. Does this seem like a sensible procedure?

46. **Cars** These data are described in Exercise 45.
(a) Make a scatterplot of weight and highway mileage. Which variable do you think makes the most sense to put on the x-axis, and which belongs on the y-axis?
(b) Describe any pattern that you see in the plot. Be sure to identify any outliers.
(c) Find the correlation between these two variables.

(d) Interpret the correlation in the context of these data. Does the correlation provide a good summary of the strength of the relationship?
(e) Describe the marginal distributions of MPG Highway and Weight. Include the mean and SD of both variables.
(f) Use the correlation line to estimate the mileage of a car that weighs 4,000 pounds. Does this seem like a sensible procedure?

47. **Philadelphia Housing** These data describe housing prices in the Philadelphia area. Each of the 110 rows of this data table describes a region of the metropolitan area. (Several make up the city of Philadelphia.) One column, labeled Selling Price, gives the median price for homes sold in that area during 1999 in thousands of dollars. Another, labeled Crime Rate, gives the number of crimes committed in that area, per 100,000 residents.
(a) Make a scatterplot of the selling price on the crime rate. Which observation stands out from the others? Is this outlier unusual in terms of either marginal distribution?
(b) Find the correlation using all of the data as shown in the prior scatterplot.
(c) Exclude the distinct outlier and redraw the scatterplot focused on the rest of the data. Does your impression of the relationship between the crime rate and selling price change?
(d) Compute the correlation without the outlier. Does it change much?
(e) Can we conclude from the correlation that crimes in the Philadelphia area cause a rise or fall in the value of real estate?

48. **Boston Housing** These data describe 506 census tracts in the Boston area. These data were assembled in the 1970s and used to build a famous model in economics, known as a hedonic pricing model.[7] The column Median Value is the median value of owner-occupied homes in the tract (in thousands of dollars back in the 1970s, so the amounts will seem quite small), and the column Crime Rate gives the number of crimes per 100,000 population in that tract.
(a) Make a scatterplot of the median value on the crime rate. Do any features of this plot strike you as peculiar?
(b) Find the correlation using all of the data as shown in the prior scatterplot.
(c) Tracts in 1970 with median home values larger than $50,000 were truncated so that the largest value reported for this column was 50. The column Larger Value replaces the value 50 for these tracts with a randomly chosen value above 50. Use this variable for the response, make the scatterplot, and find the correlation with Crime Rate. Does the correlation change?
(d) What can we conclude about the effects of crime on housing values in Boston in the 1970s?

[7] D. Harrison and D. L. Rubinfeld (1978), "Hedonic Housing Prices and the Demand for Clean Air," *Journal of Environmental Economics and Management,* 5, 81–102. The data are rather infamous. See the discussion in O. W. Gilley and R. Kelley Pace (1996), "On the Harrison and Rubinfeld Data," *Journal of Environmental Economics and Management, 31,* 403–405.

49. **Cash Counts** There's a special situation in which you can measure dependence using either the correlation or Cramer's V (Chapter 5). Suppose both variables indicate group membership. One variable might distinguish male from female customers, and the other whether the customer uses a credit card, as in this example. These variables define a contingency table for a sample of 150 customers at a department store.

	Sex		
	Male	**Female**	**Total**
Pay with cash	50	10	60
Use a credit card	55	35	90
Total	105	45	150

   (a) Compute the value of Cramer's V for this table. You may need to look back at Chapter 5.
   (b) Find the correlation between the numerical variables *Sex* (coded as 1 for male, 0 for female) and *Cash* (coded 1 for those paying with cash, 0 otherwise). Use the counts in the table to manufacture these two columns.
   (c) What's the relationship between Cramer's V and the correlation?

50. A scatterplot is not so useful for numerical data when the data include many copies. In Exercise 49, there are 50 cases of male customers who paid with cash. That's 50 pairs of values with *Sex* = 1 and *Cash* = 1.
   (a) What does the scatterplot of *Cash* versus *Sex* look like?
   (b) Suggest a method for improving the scatterplot that helps you see the association.
   (c) Cramer's V measures *any* type of association, but correlation only measures linear association. When you consider the association between the two 0/1 variables in this example, can there be anything other than linear association?

51. **Macro** The Bureau of Labor Statistics monitors many characteristics of the United States economy, and some of these time series are in the data table for this question. These time series are quarterly and span the time period from 1970 through the third quarter of 2004. For example, the bureau monitors hourly compensation and hourly output in the business sector. The variable *Compensation* is an index of the output per hour of all persons in the business sector. The variable *Output* is an index of the hourly output.
   (a) A timeplot is another name for a scatterplot of a sequence of values versus a variable that measures the passage of time. Generate timeplots of *Compensation* and *Output*. Do these variables seem related?
   (b) Make a scatterplot of *Output* versus *Compensation*. Does this plot suggest that the two series are related? What does the white space rule suggest about this plot?
   (c) Find the correlation between the two series as shown in the scatterplot in part (b).

   (d) Does the correlation between *Output* and *Compensation* imply that managers can get more output by increasing the pay to their employees?

52. **Macro** The government wants consumers in the United States to have money to spend to keep the economy moving, so it watches the national level of personal disposable income. The Federal Reserve wants to make sure that these same consumers do not get into too much debt. The variable *Disposable Income* tracks the money available to consumers to spend (in billions of dollars), and the variable *Debt* tracks the money owed by households in the retail credit market (not including mortgages, also in billions of dollars). Both series are quarterly and span the time period from 1970 through the third quarter of 2004.
   (a) Do you expect the two series to be positively or negatively related?
   (b) A timeplot is a scatterplot of the values in a sequence in chronological order. Generate timeplots of these two series. Do the series appear to be related?
   (c) Make a scatterplot of *Disposable Income* on *Debt*. Does this plot reveal things that you could not tell from the timeplots? Does the white space rule suggest a need for care when looking at this relationship?
   (d) Find the correlation between *Disposable Income* and *Debt*.
   (e) Is the correlation a good summary of the relationship between these two series? Does the correlation imply that the government should try to lower the level of disposable income to get consumers to save more?

53. **Flight Delays** If the airline flight that you are on is 20 minutes late departing, can you expect the pilot to make these minutes up by, say, flying faster than usual? These data summarize the status of a sample of 984 flights during August 2006. (Like the data used to illustrate Simpson's paradox in Chapter 5, these data come from the Bureau of Transportation Statistics.)
   (a) Do you expect the number of minutes that the flight is delayed departing to be associated with the arrival delay?
   (b) Make a scatterplot of the arrival delay (in minutes) on the departure delay (also in minutes). Summarize the association present in the scatterplot, if any.
   (c) Find the correlation between arrival delay and departure delay.
   (d) How is the correlation affected by the evident outlier, a flight with very long delays?
   (e) How would the correlation change if delays were measured in hours rather than minutes?

54. **CO$_2$** These data give the amount of CO$_2$ produced in 15 nations along with the level of economic activity, both in 2004. CO$_2$ emissions are given in thousands of metric tons, extracted from a study by the United Nations. Economic activity is given by the gross domestic product (GDP), a summary of overall economic output measured in billions of dollars by the International Monetary Fund.

(a) Make a scatterplot of CO_2 emissions and GDP. Which variable have you used as the response and which as the explanatory variable?

(b) Describe any association between CO_2 emissions and GDP.

(c) Find the correlation between CO_2 emissions and GDP.

(d) Find the correlation between CO_2 emissions and GDP, excluding Canada. How does the correlation change when this observation is removed?

4M Correlation in the Stock Market

Dell, IBM, and Microsoft are well known in the computer industry. If the computer industry is doing well, then we might expect the stocks of all three to increase in value. If the industry goes down, we'd expect all three to go down as well. How strong is the association among these three companies? After all, they compete in some areas. For example, Dell and IBM both sell powerful computer systems designed to be used as Web servers by Web sites like Amazon or weather.com.

This data set has monthly returns for Dell, IBM, and Microsoft for January 1990 through December 2006. The returns are calculated as

$$R_t = \frac{P_t - P_{t-1}}{P_{t-1}}$$

In this fraction, P_t is the price of the stock at the end of a month and P_{t-1} denotes the price at the end of the prior month. If multiplied by 100, the return is the percentage change in the value of the stock during the month. The returns given here have been adjusted for accounting activities (such as dividend payments and splits) that would otherwise produce misleading results. The data are from the Center for Research in Security Prices (CRSP), as are other stock returns used in this book.

Motivation

a) Investors who buy stocks often buy several to avoid putting all their eggs into one basket. Why would someone who buys stocks care whether the returns for these stocks were related to each other?

b) Would investors who are concerned about putting all their eggs into one basket prefer to buy stocks that are positively related, unrelated, or negatively related?

Method

c) How can an investor use correlations to determine whether these three stocks are related to each other? How many correlations are needed?

d) Correlations can be fooled by patterns that are not linear or distorted by outliers that do not conform to the usual pattern. Before using correlations, how can the investor check the conditions needed for using a correlation?

e) A key lurking variable anytime we look at a scatterplot of two time series is time itself. How can an investor check to see if time is a lurking factor when looking at stock returns?

Mechanics

f) Obtain all of the scatterplots needed to see whether there are patterns that relate the returns of these stocks. (A scatterplot matrix works nicely for this if your software provides one.) Does it matter which stock return goes on the x-axis and which goes on the y-axis? Do you find that the returns are associated? Is any association linear?

g) Obtain all of the correlations among these three stocks.

h) Look at timeplots of the returns on each stock. Why are these important when looking at time series?

Message

i) Summarize your analysis of the relationships among these returns for an investor who is thinking of buying these stocks. Be sure to talk about stocks, not correlations.

4M Cost Accounting

Companies that manufacture products that conform to a unique specification for each order need help in estimating their own costs. Because each order is unique, it can be hard to know for sure just how much it will cost to manufacture. Plus, the company often needs to quote a price when the customer calls rather than after the job is done. The company in this exercise manufactures customized metal blanks that are used for computer-aided machining. The customer sends a design via computer (a 3-D blueprint), and the manufacturer replies with a price per unit. Currently, managers quote prices on the basis of experience.

What factors are related to the order's cost? It's easy to think of a few. The process starts with raw metal blocks. Each block is then cut to size using a computer-guided tool. All of this work requires a person to keep an eye on the process.

The data for the analysis were taken from the accounting records of 200 orders that were filled during the last three months. The data have four variables of interest. These are the final cost (in dollars per unit manufactured), the number of milling operations required to make each block, the cost of raw materials (in dollars per unit), and the total amount of labor (hours per unit).

Motivation

a) If the manufacturer can find a variable that is highly correlated with the final cost, how can it use this knowledge to improve the process that is used to quote a price?

Method

b) What variable is the response? Which are the possible explanatory variables?

c) How can the manufacturer use the correlation to identify variables that are related to this response?

d) Explain why it is important to consider scatterplots as well.

Mechanics

e) Obtain all of the scatterplots needed to understand the relationship between the response and the three explanatory variables. Briefly describe the association in each.

f) Obtain all of the correlations among these four variables. (These are most often displayed in a correlation matrix.) Which explanatory variable is most highly correlated with the response?

g) Check the conditions for the correlation of the response with the three predictors. Do the scatterplots suggest any problems that make these three correlations appear unreliable?

h) Adjust for any problems noted in part (g) and recompute the correlations.

Message

i) Which variable is most related to the response?

j) Explain in the context of this manufacturing situation why the correlation is not perfect and what this lack of perfection means for predicting the cost of an order.

Case: Financial Time Series

STOCK PRICES

STOCK RETURNS

VALUE AT RISK

CASE SUMMARY

Time series, data that flow over time, are common in business. Among these, time series that track the flow of dollars are the most studied. The financial time series considered in this case tracks the value of stock in a company, Enron.

This case reviews many of the essential concepts covered in the chapters of this section: the importance of understanding the data, numerical summaries and distributions, plots to find patterns, and the use of models to describe those patterns. It is important to appreciate several quirks of stock prices in order to make sense of these data. The emphasis is upon data that track a stock over time, but the concerns and methods are familiar.

STOCK PRICES

At the end of the 20th century, Enron was one of the world's largest corporations. It began in 1986 as a regional supplier of natural gas and grew into a dominant player in the energy trading business. All of that came to an end when it abruptly collapsed in 2001.

The price of a share of stock in Enron mirrors the spectacular rise and fall of the company. In April 1986, each of the 45 million shares of stock in the newly formed Enron Corporation sold for $37.50. By the end of 2000, there were 740 million shares valued at $80 apiece. At that point, Enron's market value was $60 billion. Had you bought and held one of those first shares that cost $37.50, it would have grown into eight shares worth $650 by the end of 2000. Not bad. To match that, you'd need a savings account that paid 21% interest each year!

Timing in the stock market, however, is important. Had you held onto that stock in Enron a little longer, it would have been worth less than 650 cents! The price collapsed to $0.26 a share by the end of 2001.

Were there hints of trouble before the bottom dropped out? Books that describe Enron's wheeling and dealing, such as *The Smartest Guys in the Room*

(written by two reporters from *Fortune* magazine), convey the impression that a lot of people should have known something was going to happen. But what about regular investors? Could they have recognized the risks of investing in Enron? Let's see how investors could have used statistics to appreciate the risks in owning stock in Enron.

Stock Prices

To measure the risk of owning a stock, we start with data that record the price of one share. The data in this analysis record the price of Enron stock at the end of each month from April 1986 through December 2001. These prices form a time series. Each row in the data table refers to a specific month, and the variable *Price* records the price of one share of Enron stock at the end of the month. The rows are in order, so that running down the column of prices tracks Enron's history. Table 1 lists first 4 months of data.

TABLE 1 Prices of Enron stock in 1986.

Date	Price
April 1986	$37.50
May 1986	$41.38
June 1986	$44.00
July 1986	$40.00
⋮	⋮

We begin our analysis as usual: with a plot. We've only got one time series and the dates, but we'll begin with a scatterplot. How are we going to build a scatterplot from one column of numbers? Easy. The time order defines the x-axis for the scatterplot, and the prices define the y-axis forming the timeplot (Chapter 1) shown in Figure 1.

The timeplot begins in April 1986 when Enron sold for $37.50 per share. The price bounces around after that. Several times it runs up only to fall suddenly. Up until the end, the price recovered after each fall.

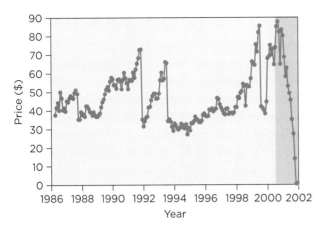

FIGURE 1 Timeplot of the price of Enron stock.

Once we've seen Figure 1, it's possible to wonder why anyone would be surprised by the collapse of Enron. The price of a share fell 50% three times in the 1990s, and it steadily recovered after the first two. Perhaps the drop at the end of 1999 was the time to buy shares while the price was low, hoping for another steady recovery. That might seem like a risky bet. Wasn't it inevitable that one of those drops would be deadly?

Some Details: Stock Splits

We have to understand the data in order to use statistics. In this case, a simple accounting practice explains the sudden collapses in the price of Enron stock. To understand these data, we have to learn more about how stocks are priced.

Figure 1 doesn't present the history of Enron in a way that's helpful to an investor. It displays the prices of a share of stock in Enron at the end of each month. Had we gone to the New York Stock Exchange back then, this plot shows what we would have paid for a share of Enron. To appreciate that Figure 1 hides something, take a look at Figure 2. This timeplot tracks

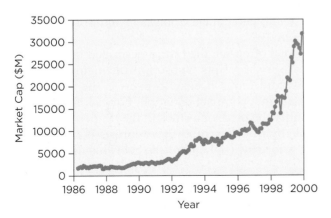

FIGURE 2 Market capitalization of Enron, in millions of dollars.

the capitalized value of Enron, in millions of dollars, through the end of 1999 before the collapse. (Capitalized value, or market cap, is the price of a share times the number of outstanding shares. It's the total value of all of the stock in a company.)

The capitalized value rises steadily, without the sudden falls seen in the value of individual shares. The explanation is that the price of stock tracks only one component of the value of Enron. When the price of a stock increases, it is common for companies to "split" the stock, converting each outstanding share into two or three. So, if we held one share of XYZ Corporation that was valued at $100 before a split, then we'd have two shares after the split, each worth $50. Our investment is still worth $100, only split between two shares. We'll leave it for a finance class to explain the logic for splitting stocks, but the common explanation is that lower prices attract more buyers.

Stock splits explain the rapid drops in the value of Enron stock seen in Figure 1. The market did not abruptly devalue Enron. Rather, these rapid drops were stock splits. For example, at the end of November 1991, Enron sold for about $70 per share. The stock split "2 for 1" in December, and not surprisingly, the price at the end of December was about $35 per share.

Dividend payments also add variation to stock prices. Let's go back to our $100 investment in one share of stock. If that's the price today and the company pays a dividend of $5 per share tomorrow, then the value of each share drops to $95. We were paid the other $5 in cash. Like splits, dividends cause prices to change but don't change the value of our investment.

We have to account for splits and dividends in order to build a useful plot. Splits and dividend payments add variation to the prices, but these do not materially affect the value of the investment. We would not have lost money in 1991 when Enron's stock split; we'd simply have had twice as many shares, each worth half of the original price. A more informative plot removes these accounting artifacts. We have to adjust for the splits and track the wealth of an investor who started with one share. Finance pros would recognize the splits right away.

Figure 3 on the next page shows the timeplot of the wealth of an investor who bought one share of Enron stock in 1986 and held onto it and all of its "children" (the stocks obtained when that original share split). These data are corrected for dividend payments.

This timeplot looks very different from the timeplot of prices in Figure 1. The stock soars. Without the jumps caused by splits, we can appreciate the appeal of Enron. The soaring growth pulled in investors.

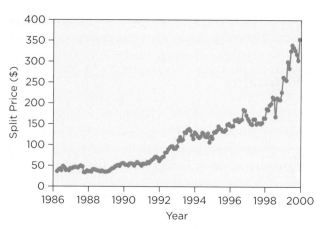

FIGURE 3 Value of investment in Enron.

STOCK RETURNS

We ended the timeplots in Figure 2 and Figure 3 at the end of 1999. We want to pretend that we're investors at the end of 1999 who are considering stock in Enron. We don't know what's coming, just what's happened so far. Figure 3 certainly makes the stock look attractive, but do other views of these data expose a different, less inviting perspective? What could we have seen in the history up to then that might have warned us to stay away?

> **caution** There's a problem with plots that track the cumulative value of an investment. Plots like Figure 3 appear throughout the financial section of the newspaper—such as in ads for mutual funds—but these plots conceal certain risks. We get a hint of the problem from the white space rule. Most of the area of these timeplots does not show data. We cannot see the variation in the first half of the data that are squeezed down near the x-axis.

A substantive insight also suggests a problem with Figure 3. The stock market fell more than 20% in October 1987. Now look at the history of Enron in Figure 3. There's hardly a blip for October 1987. It could be that the drop in markets in 1987 didn't affect Enron, but that's wishful thinking. The problem lies with this view of the data. A plot that can hide October 1987 is capable of hiding other things as well.

Let's think about what happened in October 1987. To construct a more revealing plot, we need to think more like gamblers. If we had put $100 into almost any stock on October 1, 1987, we would have had less than $100 by the end of the month. All the various "bets" on the market lost money. Bets on IBM lost money. Bets on Disney lost money. So did GM, Enron, and most others. If we were to bet on a card game, we'd probably look at the game one play at a time. It's too hard to think about how one round fits into the whole game, so most gamblers think about games of chance one play at a time.

We can think of the prices of Enron stock as telling us what happens in a sequence of bets, one wager at a time. Imagine that we start each month with $100 invested in Enron, regardless of what happened last month. Each month is a fresh bet.

We want a plot that tracks our winnings each month by converting these prices into percentage changes. Let's do one calculation by hand. On January 1, 1997, Enron's stock sold for $43.125. (Weird prices like this one come from the old convention of pricing stocks in eighths of a dollar; nowadays, stocks are priced to the nearest penny.) Our $100 would buy 100/43.125 = 2.32 shares. At the end of January, the price per share had dropped to $41.25. So, each of our shares lost 43.125 − 41.25 = $1.875. Taking into account the fraction of a share that we owned, we lost $4.35. Here's the calculation:

$$\left(\frac{\$100}{\$43.125/\text{share}} \right) \times (\$41.25 - \$43.125)/\text{share}$$
$$= -\$4.35$$

The first expression inside the parentheses tells us how many shares our $100 buys. The second term tells us what happened to the price of each. The end result tells how much we win or lose by investing $100 in Enron for one month. The amount won or lost is equivalent to the percentage change:

$$\text{Percentage change} = 100 \times \left(\frac{\text{Change in Price}}{\text{Starting Price}} \right)$$
$$= 100 \times \left(\frac{\$41.25 - \$43.125}{\$43.125} \right)$$
$$= -4.35\%$$

The ratio of the change in price to the starting price is known as the **return** on the stock. We convert these into percentage changes by multiplying the return by 100.

You'll often see the return described in a formula that uses subscripts. Subscripts distinguish time periods and make it easy to express the percentage change in a compact equation. For example, we can identify the price of a stock in the tth row of the data table by P_t. The symbol P_{t-1} represents the price in the prior row (the end of the previous month). The

---

return Ratio of the change in the value of an investment during a time period to the initial value; if multiplied by 100, the percentage change.

change in price during this month is $P_t - P_{t-1}$. The return on the stock is then

$$R_t = \frac{P_t - P_{t-1}}{P_{t-1}}$$
$$= \frac{\$41.25 - \$43.125}{\$43.125}$$
$$= -0.0435$$

The percentage change multiplies the return by 100, a loss of 4.35%.

Simple Time Series

The timeplot of the monthly percentage changes in Figure 4 conveys a different impression of the risks associated with investing in Enron.

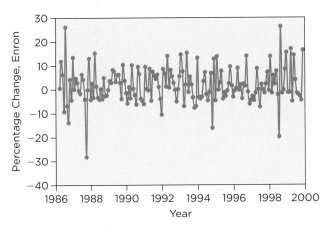

FIGURE 4 Monthly percentage changes in price of Enron stock.

These hardly look like the same data. What happened to the smooth upward growth evident in Figure 2 and Figure 3? Where did all the rough, irregular ups and downs come from?

By thinking of investing as a series of bets, we remove the compounding that happens to investments and concentrate on what happens each month anew. Most of these monthly bets made money. After all, the price went up on average. But we can also see quite a few losing bets. Bets on Enron were hardly a lock to make money. We can also appreciate the magnitude of what happened in October 1987: Stock in Enron fell nearly 30% that month. That wasn't the only big drop either. During August 1998, stock in Enron fell 20%.

There's another key difference between the timeplot of the prices (Figure 3) and the timeplot of percentage changes (Figure 4). Which plot shows a pattern? We would say that the timeplot of percentage changes in Enron stock is *simpler* than the timeplot of the prices because there's no trend in the returns. *Simple* in this sense does not mean "constant." All of that irregular, zigzagging variation in Figure 4 looks complicated,

but that's common in simple data. We say that these data are **simple** because it is easy to summarize this sort of variation. There's variation, but the appearance of the variation is consistent. Some would call this variation "random," but we will see in later chapters that random variation often has patterns. Because there's no pattern that would be hidden, we can summarize simple variation with a histogram. Simple variation is "histogrammable."

It can be hard to decide if a timeplot shows simple variation because people find it easy to imagine patterns. The visual test for association works nicely in this context. In Chapter 6, we showed several scrambled scatterplots of energy prices. If the original can easily be identified, there's a pattern. If not, the variation is simple and can be summarized with a histogram. To use the visual test for association with time series, mix the order of the data as though shuffling a deck of cards. If the timeplot of shuffled data looks different from the original timeplot, then there's a pattern and the variation is not simple.

Figure 5 compares the timeplot of the price of Enron stock to the plot of the data after shuffling.

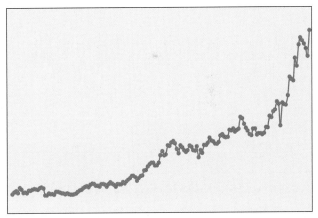

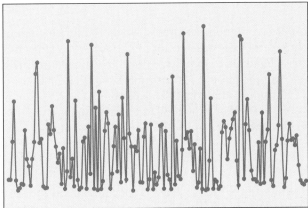

FIGURE 5 It's easy to recognize the timeplot of prices.

simple Variation in simple data lacks a pattern, making it easily summarized in a histogram.

It's easy to distinguish the actual prices from the scrambled prices. The prices of Enron stock are *not* simple and definitely show a pattern.

Now, let's do the same thing with the percentage changes. Which of the timeplots in Figure 6 is the timeplot of the percentage changes?

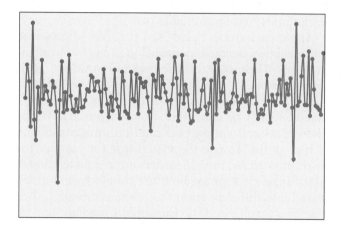

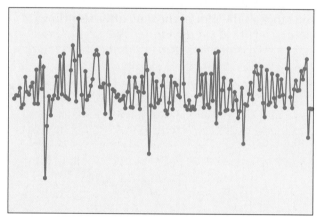

FIGURE 6 It's not so easy to recognize the timeplot of returns.

Unless we remember precisely where a big drop happens, either of these could be the original. This similarity leads us to conclude that the returns on Enron are "simple enough" to summarize with a histogram.

tip Simple data that lack a pattern allow for a simple summary, namely a histogram. Time series that show a pattern require a description of the pattern.

Histograms

Simple time series free us to concentrate on the variation and ignore the timing. To focus on the variation, we collapse the plot into a histogram that ignores the time order. For an investor, it's crucial to know that the price fell during some months in the past. The fact that the price of Enron stock bounced around a lot over its history is crucial. The precise

timing of the ups and downs isn't crucial unless you think you can predict the timing of these events.

What do we expect the distribution of the percentage changes to look like? **tip It's often a good idea to imagine the distribution before asking software to make the picture for us. That way we'll be ready if the plot shows something unexpected.** Figure 7 shows what we are doing when we make a histogram of a time series.

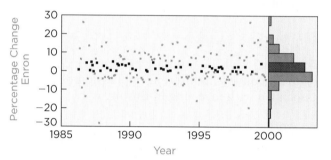

FIGURE 7 Histogram and time plot of the percentage changes.

We did not connect the dots in the time plot so that the separate points can be seen more easily. Since time order is by and large irrelevant with a simple time series, there's no point connecting the dots.

Instead of plotting the returns as a long sequence of dots, the histogram shows how many land in each interval of the *y*-axis. The histogram ignores the order of the data and counts the number of months in which the return falls in each interval. For instance, the highlighted bar of the histogram in Figure 7 counts the number of months in which the percentage change in Enron stock was between 0% and 5%. If there were a pattern in the time series, the histogram would hide it.

Figure 8 rotates the histogram to the more familiar horizontal layout and adds a boxplot to locate the center of the data.

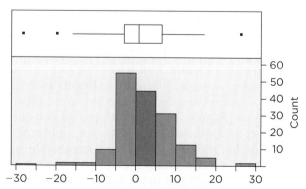

FIGURE 8 Histogram of percentage changes in the price of Enron stock.

Once we have the histogram (and boxplot), it's easy to see that the percentage changes for about one-half of the months lie between −5% and 5%. We can also see that more months show an increase of 10% or higher than show a decrease of 10% or lower. The first bin on the left, which counts months in which the stock lost between 25% and 30%, has only one month, October 1987.

VALUE AT RISK

The histogram of stock returns reveals a great deal of risk. The average return during these 165 months is 1.9%. On average, an investor made almost 2% *per month* by owning stock in Enron. That sounds great compared to a savings account. Owning a stock, however, isn't quite the same because the stock can go down in value. We need to think about the standard deviation of the percentage changes as well. For returns on Enron stock, $s = 7.4\%$.

The bell-shaped distribution of percentage changes in Figure 8 suggests that we can use the Empirical Rule. Recall from Chapter 4 that the Empirical Rule connects the mean and SD to the concentration of data around the mean. Of all the various applications of the Empirical Rule in business, one of the most common is in the analysis of stock returns. The Empirical Rule predicts that in about two-thirds of the months, Enron stock changes by $\bar{x} \pm s = 1.9 \pm 7.4\%$ from −5.5% to 9.3%. That's a wide interval. Reaching out further, the Empirical Rule suggests that returns in 19/20 months lie in the range $\bar{x} \pm 2s = 1.9 \pm 2 \times 7.4\%$, or −12.9% to 16.7%. With so much variation, a 2% average return per month seems less comforting.

In finance, **risk** is variation in the percentage change of the value of an investment. Investors considering Enron in December 1999 should have been prepared for the possibility of losing 5% or 10% of their investment in a single month. Of course, the histogram also shows that the investment is more likely to increase by 10%. This possibility was evidently enough to lure investors.

Finance professionals often quantify the risk of an investment using a concept known as the **value at risk (VaR)**. Rather than simply quote the mean and SD of returns, VaR converts these into a dollar amount that conveys possible consequences of variation. The idea is to use the Empirical Rule (or other more elaborate rules) to make a statement about how much money might be lost.

For instance, suppose we invest $1,000 in Enron at the start of a month. If we take the Empirical Rule seriously (maybe too seriously), then there's a 95% chance that the return during the month lies between −12.9% and 16.7%. Because the bell-shaped curve is symmetrical, that means there's a 2.5% chance that the percentage change could be less than −12.9%. If that were to happen, the stock would drop by more than 12.9% by the end of the month. For VaR, this investment puts $1,000 \times 0.129 = $129 "at risk." Notice that saying that the VaR is $129 doesn't mean that we cannot lose more. Instead, it means that this is the most that we would lose if we exclude the worst 2.5% of events. (VaR calculations vary in the percentage excluded. We're using 2.5% since its easy to find with the Empirical Rule.)

This calculation of VaR takes the Empirical Rule at face value. The Empirical Rule is a powerful technique that allows us to describe a histogram with the mean and standard deviation. Two summary statistics and the Empirical Rule describe the chances for all sorts of events—as long as the data really are bell shaped. (The match seems pretty good. Stock in Enron fell by 12.9% or more during 4 of the 165 months prior to 2000. That works out to about 2.4%, close to the 2.5% anticipated by the Empirical Rule.)

For comparison, let's figure out the VaR for a $1,000 investment spread over the whole stock market. The stock market as a whole grew on average 1.3% per month over this period with standard deviation 4.4%. The market as a whole has less growth and less variation. The Empirical Rule says there's a 95% chance that the return on the market lies in the range 1.3% $\pm$ 2 $\times$ 4.4% = −7.5% to 10.1% The VaR for this investment is $1,000 \times 0.075 = $75 compared to the VaR of $129 for Enron.

What's an investor to do? An investment in Enron puts more value at risk, but it offers a higher average return than investing in the market as a whole. To choose between these requires trading risk for return. Part 2 of this book offers several suggestions based on models for the data.

---

risk Variation in the returns (or percentage changes) on an investment.

value at risk (VaR) An estimate of the maximum potential losses of an investment at a chosen level of chance.

CASE SUMMARY

Returns summarize the performance of an asset such as stock in a firm. The calculation for returns on stock requires careful accounting for stock splits and dividend payments. Variation in returns defines the **risk** of the asset and allows one to anticipate the chance for the asset to rise or fall in value. **Value at risk** (VaR) measures the amount that might be lost in an investment over a chosen time horizon if we rule out a fraction of the possibilities. When viewed as a time series, a sequence of returns typically has simpler structure than the underlying prices. Returns on most stocks are simple time series that can be usefully summarized by a histogram.

■ Key Terms

return, 136
risk, 139

simple, 137
value at risk (VaR), 139

■ Formula

RETURN ON AN INVESTMENT

If the price of the investment at the end of the current period is P_t and the price at the end of the prior period is P_{t-1}, then the return is

$$R_t = \frac{P_t - P_{t-1}}{P_t}$$

100 times the return is the percentage change in the price of the asset. Returns are computed for various time periods, such as daily, monthly, or annually.

■ About the Data

The stock data used in this chapter come from the Center for Research in Security Prices (CRSP).

Analysts at CRSP carefully adjust their data for accounting adjustments and stock splits.

■ Questions for Thought

These questions consider stock in Pfizer, the pharmaceutical manufacturer. The data are monthly and span 1944 through the end of 2005.

1. The timeplot in Figure 9 shows the price at the end of the month of one share of stock in Pfizer.

What seems to be happening? How could you verify your answer?

2. Figure 10 tracks the value of an initial $100 investment made in Pfizer in 1944. For each month, we multiplied the amount invested in Pfizer at the end of the previous month by the

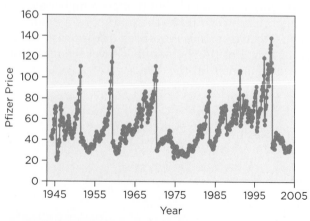

FIGURE 9 Timeplot of price of stock in Pfizer.

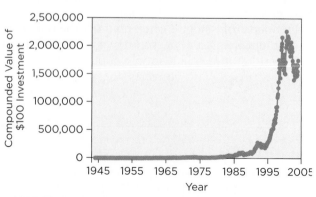

FIGURE 10 Compounded value of $100 investment made in Pfizer in 1944.

percentage change during the month. The initial $100 grows to more than $2 million by the year 2000.

   a. What does the white space rule say about this figure?

   b. Would it be appropriate to summarize these data using a histogram?

3. Figure 11 shows the timeplot of month-to-month percentage changes in the value of Pfizer, adjusting for dividend payments and stock splits. Would it be appropriate to summarize these data using a histogram?

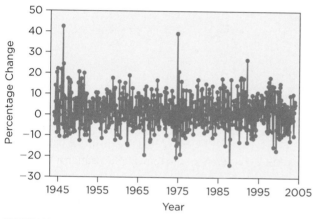

FIGURE 11 Monthly returns on Pfizer stock.

4. Explain in your own words how it is that the percentage changes (Figure 11) look so random, whereas the cumulative value of investing in Pfizer (Figure 10) grows so smoothly until the last few years.

5. The average percentage change in Pfizer stock is 1.55% with $s = 7.55$.

   a. What is the standard deviation of the monthly *returns* on Pfizer?

   b. From the shown summary statistics, how might you estimate the chance that stock in Pfizer would fall 2.5% or more next month? What would you like to check before accepting this estimate?

   c. If we invest $1,000 in Pfizer for one month, what is the Value at Risk based on the Empirical Rule? Interpret the VaR in your own words.

Case: Executive Compensation

This case study explores variation among the salaries of 1,500 business leaders. Though we hear a lot about the stratospheric salaries of a few, most executives earn quite a bit less than the superstars. To understand this variation requires that we transform the data to a new scale. Otherwise, histograms and summary statistics show us only the most extreme observations, ignoring the majority of the data. The transformation used in this example is the most common and important in business statistics, the logarithm.

INCOME AND SKEWNESS

The salaries of the Chief Executive Officer, or CEO, of large corporations have become regular news items. Most of these stories describe the salaries of top executives and ignore everyone else. It turns out that there's a surprising amount of variation in salaries across different firms, if you know where and how to look.

The data table for this case study has 1,495 rows, one for each chief executive officer at 1,495 large companies in the United States. The data table has several columns, but we begin with the most interesting: total compensation. Total compensation combines salary with other bonuses paid to reward the CEO for achievements during the year. Some bonuses are paid as cash, and others come as stock in the company. The histogram in Figure 1 shows the distribution of total compensation, in millions of dollars, in 2003. Does it look like what you expect for data that have a median of $2.5 million with interquartile range of $4.2 million?

This distribution is very skewed. A few outlying CEOs earned extraordinarily high compensation: 14 earned more than $30 million during this year. These 14 are evident in the boxplot, though virtually invisible in the histogram. The histogram leaves so much of the area blank because these 14 cases reach from $30 million to more than $70 million. It's hard

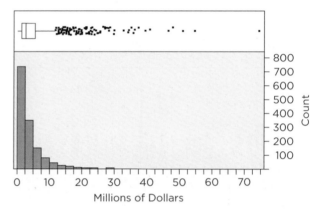

FIGURE 1 Histogram of the total compensation of CEOs.

to see the tiny bars that count these. You probably recognize the names of several of these high-paid executives. The highest is Steve Jobs of Apple Computer. He earned close to $75 million in 2003. Table 1 lists the top 5.

TABLE 1 Top-paid CEOs in 2003.

CEO	Company	Industry	Compensation
Steven Jobs	Apple	IT	$74.8 million
Sanford Weill	Citigroup	Financial	$54.8 million
Larry Glasscock	Anthem Inc	Health	$51.2 million
H. Culp, Jr.	Danaher Corp	Industrial	$48.1 million
Charles Cawley	MBNA	Financial	$46.7 million

When compared to the earnings of this elite group, the earnings of the remaining CEOs appear downright reasonable. The tallest bar at the far left of the histogram represents CEOs who made less than $2.5 million. That group accounts for almost half of the cases shown in Figure 1, but we can barely see them in the boxplot because they are squeezed together at the left margin.

The skewness evident in the histogram affects numerical summaries as well. **When data are highly skewed, we have to be very careful interpreting summary statistics such as the mean and standard deviation.** Skewed distributions are hard to summarize with one or two statistics. Neither the mean nor the median appears to be a good summary. The boxplot gives the median, but it seems so far to the left. The mean is larger, but still to the left of Figure 1 because so many cases accumulate relatively near zero. The median compensation is $2.53 million. The mean is almost twice as large, $4.63 million. The skewness pulls the mean to the right in order to balance the histogram (see Chapter 4).

For some purposes, the mean remains useful. Suppose you had to pay all of these salaries! Can you figure out the total amount of money that you would need from the median or from the mean? You need the mean. Because the mean is the sum of all of the compensations divided by the count, $4.63 \times 1495 =$ $6,922 million, about $6.9 *billion*, would be needed to pay them all.

Skewness also affects measures of variation. The interquartile range, the length of the box in the boxplot, is $4.22 million. This length is quite a bit smaller than the wide range in salaries. The standard deviation of these data is larger at $6.23 million, but still small compared to the range of the histogram. (For data with a bell-shaped distribution, the IQR is usually larger than s.)

We should not even think of using the Empirical Rule (Chapter 4). The Empirical Rule describes bell-shaped distributions and gives silly answers for data so skewed as these. When the distribution is bell shaped, we can sketch the histogram if we know the mean and standard deviation. For instance, the interval formed as the mean plus or minus one standard deviation holds about two-thirds of the data when the distribution is bell shaped. Imagine using that approach here. For the total compensation, the interval from $\bar{x} - s$ to $\bar{x} + s$ reaches from $4.63 - 6.23 = -$1.60$ million to $4.63 + 6.23 = $10.86 million. The lower endpoint of this interval is negative, but negative compensations don't happen! Skewness ruins any attempt to use the Empirical Rule to connect $\bar{x}$ and s to the histogram.

LOG TRANSFORMATION

Visually, the histogram itself seems a poor summary. While complete (it does obey the area principle, Chapter 3), the histogram confines most of the data to a small part of the plot. The span of the histogram in Figure 1 ranges from zero to more than $70 million. As a result, it crams about half of the cases into one bin that is squeezed into a corner of the histogram. The large expanse of white space on the right sends a warning that the data use only a small part of the figure. The white space rule (Chapter 4) hints that we should consider other displays that show more data and less empty space.

To see the variation among the less well-paid executives, we need to magnify the left side of the histogram and condense the right. If we could stretch the x-axis on the left and squeeze it on the right, we might be able to see the variation among all of the amounts rather than emphasize the outliers. As an added benefit, the resulting histogram would be more bell shaped, making it sensible to summarize using its mean and standard deviation.

There's a simple way to spread out the lower compensations and pull together the larger compensations. We will do this with a **transformation**. The most useful transformation of business data is a log transformation. Logs allow us to re-express highly skewed data in a manner that can be summarized with a mean and a standard deviation. We will use base 10, or common, logs.[1] The base-10 log of a number, abbreviated $\log_{10}$, is the power to which you must raise 10 in order to find the number. This relationship is easier to express with symbols. The $\log_{10}$ of 100 is 2 because $10^2 = 100$, and the $\log_{10}$ of 1,000 is 3 because $10^3 = 1,000$. Logs do not have to be integers. For instance, $\log_{10} 250 = 2.4$ because $10^{2.4} = 250$.

Logs change how we look at compensation. Table 2 shows several salaries with the base-10 logs. The two columns labeled as "Gap to Next Salary" list differences between adjacent rows.

TABLE 2 Illustration of salaries in dollars and on a log scale.

Salary	Gap to Next Salary	Log$_{10}$ Salary	Gap to Next Log Salary
$100	$900	2	1
$1,000	$9,000	3	1
$10,000	$90,000	4	1
$100,000	$900,000	5	1
$1,000,000	$9,000,000	6	1
$10,000,000		7	

---

transformation In data analysis, the use of a function such as the log to re-express the values of a numerical variable.

---

[1] There's another common choice for logs known as *natural logs*, often abbreviated as ln rather than log. The base for a natural log is a special number $e = 2.71828\ldots$. There's a reason for the popularity of natural logs, but that comes later. For dollar amounts, base-10 logs are easier to think about.

Notice that we can interpret base-10 logs as telling us the number of digits in someone's paycheck (if we add 1). So, Steve Jobs is an 8-figure CEO and Carly Fiorina, then CEO of Hewlett-Packard who made $8.3 million, is a 7-digit CEO.

As we read down the rows of Table 2, the salaries increase by a factor of 10. The gaps between the salaries grow wider and wider. Compare this growth to the fixed gaps between the logs of the salaries. In terms of dollars, there's a much bigger gap between $10,000 and $100,000 than between $100 and $1,000. As far as logs are concerned, however, these gaps are the same because each represents a 10-fold increase.

Utility for Wealth

Relative change is natural when describing differences among salaries. Think back to the last time that you heard of a salary increase, either your own or in the news. Chances are that the salary increase was described in percentage terms, not dollars. When you read about unions negotiating with companies, you hear reports of, say, a 5% increase, not a **tip** $500 increase. **Putting data on a log scale emphasizes relative differences (or percentage change) rather than absolute differences.**

When describing the value of money, economists often use an important concept known as **utility.** Is every dollar that you earn as important as every other dollar? To most of us, earning another $1,000 a year has a lot of "utility." The extra money means that we can afford to do and buy more than we would otherwise. But does the extra money affect us all in the same way? Would another $1,000 change the sort of things Steve Jobs could afford to do? Probably not. The formal way to convey the impact of the added $1,000 is through utilities.

Consider the impact of a $1,000 bonus. For Steve Jobs, going from $75,000,000 to $75,001,000 amounts to a 0.00133% increase—hardly noticeable. This bonus is a tiny fraction of a huge salary. For an engineer making $100,000, the bonus adds 1% more. In the language of utility, we'd say that the utility of the added $1,000 to the engineer is $1/0.00133 = 752$ times larger than the utility to Steve Jobs.

Logs capture this notion. Differences between logs measure percentage differences rather than actual differences. The difference between $\log_{10} 101,000$ and $\log_{10} 100,000$ is (recall that $\log a - \log b = \log a/b$)

$$\log_{10} 101,000 - \log_{10} 100,000$$
$$= \log_{10} \frac{101,000}{100,000} \approx 0.0043$$

That's small and suggests that if we make $100,000, a 1% bonus is not going to put us in a new house. For Steve Jobs, the difference on a log scale is nearly zero:

$$\log_{10} 75,001,000 - \log_{10} 75,000,000$$
$$= \log_{10} \frac{75,001,000}{75,000,000}$$
$$\approx 0.00000581$$

Logs change the way we think about differences. Differences between logs are relative differences. So, when we look at variation on a log scale, we're thinking in terms of relative comparisons.

Logs and Histograms

Let's get back to executive compensation and transform the total compensations to a log scale. Transformation using logs does more than change the labels on the histogram; it also changes the shape of the histogram. On a log scale, the distribution of total compensation is bell shaped instead of right skewed.[2]

Figure 2 labels the x-axis in both $\log_{10}$ units and **tip** dollar units. **When using an unusual scale, help the reader by showing familiar units.** For instance, rather than only show 6, the plot shows $1,000,000. When labeled with the corresponding amounts, readers can figure out right away what the histogram shows. For example, we can see from Figure 2 that a typical compensation lies between $1 million and $10 million (6 to 7 on the $\log_{10}$

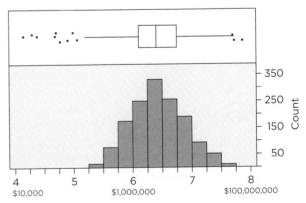

FIGURE 2. Histogram of total compensation on a log scale.

---

utility A measure of the desire for or satisfaction gained from acquiring a quantity.

---

[2] Indeed, the histogram is slightly left skewed! Richard Kinder of Kinder-Morgan (a supplier of natural gas) earned $1 in total compensation in 2003, making him a big outlier on the left (log 1 = 0). We excluded him for the sake of these plots.

scale). We can also see outliers, but this time on *both* sides of the boxplot. One of those on the left with a relatively low compensation is Jeff Bezos, CEO of Amazon. Data are full of surprises if we know where to look.

Because the histogram of the logs of the total CEO compensations is bell shaped, it is easy to summarize. The mean and median are almost identical and in the middle of the histogram. We can guess them from the plot. The average $\log_{10}$ compensation is 6.41 and the median is 6.40. The standard deviation on the log scale is 0.47, and the range $\bar{x} \pm 2s = 6.41 \pm 2 \times 0.47$ holds about 95% of the data.

caution The use of a logarithm transformation, however, does lose some information. For instance, when we say that the mean of log compensation is 6.41, what happens to the units? Should we say 6.41 log-dollars? Units are hard to define when data are transformed by logs, and we'll avoid doing so for the time being. In general, avoid converting summary statistics on the log scale back to the original units. It might be tempting, for example, to convert 6.41 log-dollars back to dollars by "unlogging" the mean of the logs, getting $10^{6.41} = \$2.57$ million. But this is not the mean of the data and instead is close to the median compensation. The mean of the logs is not equal to the log of the mean; it's much smaller and closer to the median.

Association and Transformations

If summaries like the mean are hard to interpret when the data are transformed, then why transform? The key reason to transform a skewed variable is not only to see details in its distribution that are concealed by skewness, but also to see how this variation is related to *other* variables. With the variation exposed, it becomes easier to see how relative differences in compensation are related to other variables.

Let's look for association in a scatterplot using the original units. It seems reasonable that CEOs of companies with large sales should make more than CEOs of companies with small sales. To measure sales, our data include net sales of each company. Net sales measures the total amount of goods that it sold, minus the cost of the materials used to produce those goods. Figure 3 shows a scatterplot of total compensation versus net sales.

What do we learn from this plot? Certainly, we see the outliers. Those at the right are companies with large sales. Wal-Mart, Exxon-Mobil, GM, Ford, and GE had the largest net sales in 2003. (Car manufacturers have since fallen from this list.) We can also see Steve Jobs at the top left.

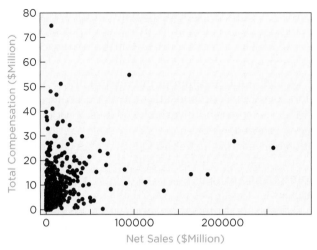

FIGURE 3 Scatterplot of total compensation and net sales, both in millions of dollars.

We'd be hard pressed, however, to guess the strength of the association in Figure 3. Intuitively, there ought to be some association between the compensation of the CEO and net sales of the company. Most of these data, however, hide in the lower left corner of the plot because *both* variables are right skewed. To see the skewness, we can add histograms to the axes.

The histogram of net sales along the *x*-axis at the top of Figure 4 is even more skewed than the histogram of executive compensation! The net sales of about two-thirds of the companies fall in the interval at the far left of the histogram. The skewness makes it impossible to see how variation in one variable is associated with variation in the other.

If all we want to know is how compensation and size are related for the superstars, then we might want to use the correlation between these variables. Even so, it is hard to guess the correlation between net sales and compensation; we might not even be

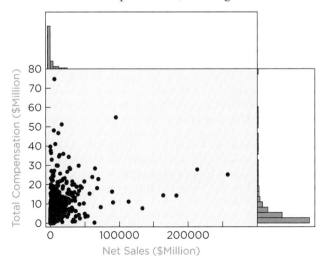

FIGURE 4 Histograms along the margins show the skewness of both variables.

sure of the direction. It turns out that the correlation between net sales and total compensation is $r = 0.39$. The direction is positive, but the size might be surprising.

We get a much different impression of the association between compensation and net sales if we **tip** re-express both variables as logs. **If variables are skewed, we often get a clearer plot of the association between them after using logs to reduce skewness.** Figure 5 shows the scatterplot with both variables transformed using $\log_{10}$. Notice the log scale on both axes.

The association is more apparent: This plot of the logs shows that percentage differences in net sales are associated with percentage differences in total compensation. We can imagine a line running along the diagonal of this plot. The direction of the association is clearly positive, and the pattern is linear. (On the left of the plot, we find a few other interesting companies. Several companies have particularly small net sales but moderate CEO salaries. These are mostly biotech or technology firms.)

The correlation between the $\log_{10}$ of total compensation and the $\log_{10}$ of net sales is $r = 0.60$, quite

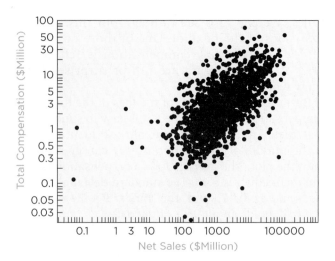

FIGURE 5 Transformation to a log scale shows evident association.

a bit larger than correlation between compensation and net sales. More importantly, the correlation is a useful summary of the association between the logs. If we told you the correlation between two variables is 0.6, you ought to be thinking of a plot that looks like the data in Figure 5.

CASE SUMMARY

Variables that measure income are typically right skewed. This skewness makes the mean and SD less useful as summaries. A **transformation** changes the scale of measurement in a fundamental way that alters the distribution of data. The log transformation is the most useful transformation for business data because it shows variation on a relative scale. Economics sometimes use logs to measure **utility**. Variation on a log scale is equivalent to variation among the percentage differences. The log transformation also allows us to measure association between income and other factors using the correlation.

■ Key Terms

transformation, 143
utility, 144

■ About the Data

These data come from Standard & Poor's Execu-Comp database. This database tracks the salaries of top executives at large firms whose stock is publicly traded. Publicly traded firms are required by law to report executive salaries. Privately owned companies are not.

Questions for Thought

1. If someone tells you the mean and SD of a variable, what had you better find out before trying to use the Empirical Rule?

2. Another approach to skewness is to remove the extreme values and work with the remaining, more typical, values. Does this approach work? Remove the top 5 or 10 compensations from these salaries. Does the skewness persist? What if you remove more?

3. You will often run into natural logs (logs to base e, sometimes written ln) rather than base-10 logs. These are more similar than you might expect. In particular, what is the association between $\log_{10}$ compensation and $\log_e$ compensation? What does the strength of the association tell you?

4. Calculate the $\log_{10}$ of net sales for these companies and find the mean of these values. Then convert the mean back to the scale of dollars by raising 10 to the power of the mean of the logs (take the antilog of the mean of the logs). Is the value you obtain closer to the mean or the median of net sales?

PART II

Probability

CHAPTER 7

Probability

EVERY DAY THOUSANDS OF CUSTOMERS DIAL TOLL-FREE TELEPHONE NUMBERS TO GET HELP SETTING UP A NEW COMPUTER. The customer service agent who answers the phone in Bangalore, India, can solve most problems, but not every one. The agent forwards hard calls to a second tier of more experienced agents back in the United States. Some days every caller seems to need more help; on other days, every call seems to get an instant reply. On busy days, callers get put on hold, and you know how that feels. Surveys report that two-thirds of customers who don't get help before the end of their call are at risk of defecting to a rival. No one likes to wait when having a problem.

What's going on? The number of calls varies from day to day, and the questions change, too. Plus, calls don't arrive at the same times every day. The timing and complexity of calls are both random. This randomness

7.1 FROM DATA TO PROBABILITY
7.2 RULES FOR PROBABILITY
7.3 INDEPENDENT EVENTS
    CHAPTER SUMMARY

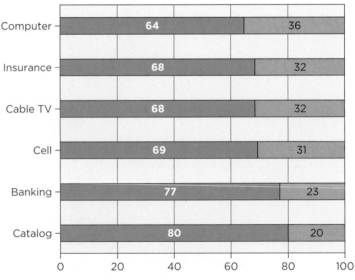

Customer satisfaction with calls to customer service varies by industry[1]

[1] From the report "Issue Resolution and Offshoring Have Major Impact on Customer Satisfaction with Call Centers," June 12, 2007, CFI Group.

complicates management's decisions. The managers of the call center must decide how many agents to have available to respond to calls, particularly to help callers who reach the second tier. Agents in the second tier are expensive. The business can't afford to pay them if they are not actively helping customers. Nor do managers want to deal with customers who are put on hold when the second tier backs up.

The accounting department knows the hourly costs for both types of agents, but these costs are not enough. Management also needs two *probabilities*: the *probability* of a call arriving in the next few minutes and the *probability* that the agent who answers the call resolves the problem. What are probabilities? How are they useful?

This chapter builds on our intuitive sense of probability, providing a framework that helps us use probability in making decisions. Part of this framework distinguishes probabilities from percentages like these that summarize customer satisfaction. Probability and data are related, but the relationship requires important assumptions that run throughout statistics. Probabilities allow us to speculate about what might happen in the future rather than just describe what happened in the past. To anticipate what might happen, we still work with counts, but counts of *possibilities* rather than data.

7.1 | FROM DATA TO PROBABILITY

Most of us have an intuitive sense for what it means to say "The probability of a tossed coin landing heads is 1/2." A coin has two sides, and either side is equally likely to be on top when it lands. With two equally likely possibilities, the probability of one or the other is one-half. The same logic applies to tossing a six-sided die. The six outcomes are equally likely, so we'd say that the probability for each is 1/6.

Away from simple games of chance, however, the notion of probability becomes subtle. Suppose we're at a call center in India and a phone starts ringing. If the first-tier agent handles the call, it's an easy call. Calls that need further help are hard calls. What is the probability that it's an easy call when the agent answers the ringing phone?

Answering the phone has one characteristic in common with tossing a coin: There are only two possible outcomes, easy calls and hard calls. There's also a big difference: Easy calls may not arrive at the same rate as hard calls. It's not as though the caller tosses a coin to decide whether to pose an easy question or a hard question. If the probability of an easy call is not one-half, then what is it? We will look to data for an answer.

Let's track a sequence of calls. As each call arrives, record a 1 for an easy call and record a 0 for a hard call. Figure 7.1 shown at the top of the following page shows the timeplot and histogram of the results for 100 consecutive calls handled by this agent.

There does not appear to be any association between the order of arrival and the outcomes. Scramble the time order, and the plot appears the same. The visual test for association (Chapter 6) finds no pattern in this sequence of calls. **The absence of a pattern in the timeplot implies that the histogram is a useful summary of these data.**

The mean of these data is 0.64; that is, the agent handled 64% of the incoming calls. Is this the probability that he handles the next call? Maybe not. It's a busy call center so we can get more data. Figure 7.2 shows the timeplot of the

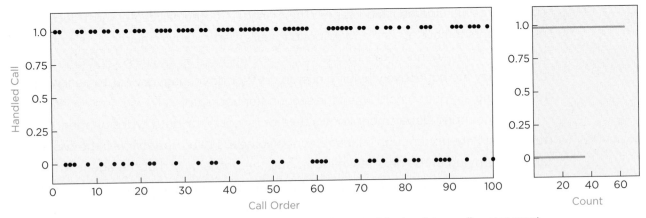

FIGURE 7.1 Timeplot with histogram of the type of 100 calls (1 for easy, 0 for hard) to a call-center agent.

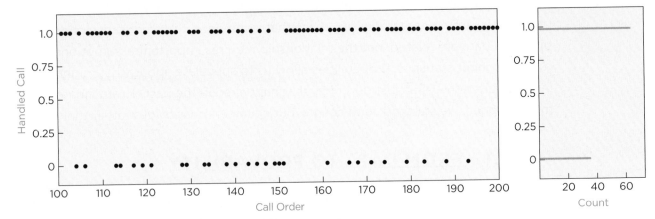

FIGURE 7.2 Timeplot of the next 100 calls to a call-center agent.

outcomes of the *next* 100 calls. Of these 100 calls, the agent handled 72, not 64. So, what's the probability that he handles the next call: 0.64, 0.72, or the average of these?

None of these is quite right. Relative frequencies describe *past* data rather than what will happen next, and the relative frequency depends on which part of the past we observe. A different view of the data, however, reveals a consistency that lurks behind the randomness. Instead of looking at each call one at a time, we can graph the accumulated relative frequency of easy calls. The accumulated relative frequency converges to a constant as we consider more calls, as shown in Figure 7.3.

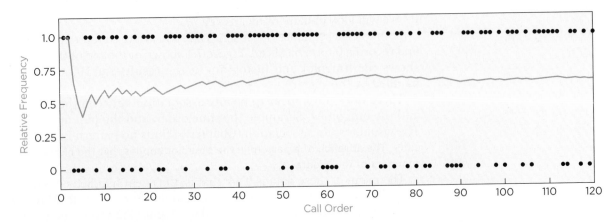

FIGURE 7.3 Accumulated relative frequency of easy calls.

Initially, the relative frequencies bounce around. The graph starts at 1 (for 100%) because the agent handled the first two calls. He forwarded the third call, so the accumulated relative frequency of easy calls drops to 2/3. He forwarded the fourth and fifth calls as well, so the accumulated relative frequency drops to 2/4 and then to 2/5. Because each new call makes up a smaller and smaller portion of the accumulated percentage, the relative frequency changes less and less as each call arrives. By the 1,000th call—more than shown here—the relative frequency settles down to 70%.

probability The long-run relative frequency of an event.

We define the **probability** of an event to be its long-run relative frequency. The probability of an easy call is the relative frequency *in the long run* as the agent handles more and more calls. If we were asked for the *probability* that the *next* call is easy, we would say 0.7 or 70%. That's based on knowing that over a long run of calls in the past, 70% of the incoming calls were easy.

The Law of Large Numbers

Knowing the probability is 0.7 does not mean that we know what will happen for any *particular* call. It only means that 70% of the calls are easy *in the long run*. The same thinking applies to tossing coins. We do not know whether the next toss of a coin will land on heads or tails. Our experience tells us that if we keep tossing the coin, the proportion of tosses that are heads will eventually be close to one-half.

The Law of Large Numbers (LLN) guarantees that this intuition is correct in these ideal examples.

Law of Large Numbers (LLN)   The relative frequency of an outcome converges to a number, the probability of the outcome, as the number of observed outcomes increases.

For calls to the agent, the relative frequency of calls that are easily resolved settles down to 70%. We say that the *probability* of an easy call is 0.7 and write $P(\text{easy}) = 0.7$.

The LLN does not apply to every situation. The LLN applies to data like those for the call center in Figures 7.1 and 7.2 because these data don't have a pattern. Weird things *can* happen for data with patterns. In such cases, the accumulated relative frequency does not settle down. For instance, suppose an agent handles the first 2 calls, forwards the next 4, handles the next 8, forwards the next 16, and so forth. The timeplot of this sequence has a very strong pattern, as shown in Figure 7.4.

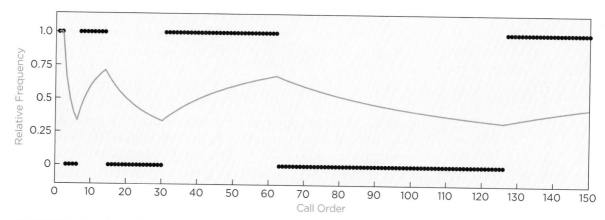

FIGURE 7.4 The relative frequency does not settle down if the data have a strong pattern.

This pattern causes the accumulated relative frequency to wander endlessly between 1/3 and 2/3, no matter how long we accumulate the results. We cannot learn a probability from such data.

The Law of Large Numbers is often misunderstood because we forget that it only applies in the long run. Consider a contractor who bids on construction projects. Over the past few years, the contractor has won about one-third of the projects that he bid on. Recently, however, the contractor has lost four bids in a row. It would be a mistake for him to think that these losses mean that he's due to win the next contract. The LLN doesn't describe what happens next. The connection to probabilities emerges *only in the long run.* A random process does not compensate for what has recently happened.

If you toss a fair coin and get five heads in a row, is the next toss more likely to be tails because the coin *owes* you a tail? No. The coin doesn't remember what's happened and attempt to balance the outcomes. In fact, if you flipped a fair coin several thousand times, you would find long streaks of heads and tails. Even so, the next toss is just as likely to be heads or tails. The LLN promises that the results *ultimately* overwhelm any drift away from what is expected, just not necessarily on the next toss.

The Law of Large Numbers establishes a strong connection between data and probabilities. For our purposes, probabilities are linked to data. You might *believe* that there's a 50% chance for rain tomorrow, but a business such as AccuWeather that sells weather forecasts to broadcasters builds its probabilities from vast amounts of meteorological data.

Beliefs often masquerade as probabilities. For example, what is the probability that the economy expands by more than 3% next year? Figure 7.5 summarizes the amount of growth anticipated by 60 economists at the start of 2007. The news article displayed these opinions as a histogram. (The footnote reveals what happened.)

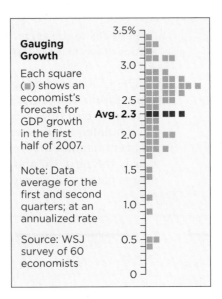

FIGURE 7.5 Expectations for the economy.[2]

It would not be appropriate to conclude from this figure that the probability that GDP will grow by more than 3% is 8/60 ≈ 0.13. This histogram summarizes opinions. Opinions such as these are often called subjective probabilities.

[2] "Economy Poised for '07 Rebound, Forecasters Say," *Wall Street Journal,* January 2, 2007. GDP stands for gross domestic product, a measure of the total output of goods and services in the U.S. economy. The Federal Reserve reports that GDP in fact grew from $13.51 trillion to $13.95 trillion in the first half of 2007—an increase of 3.1%.

In this book we do not use probabilities in this sense. A probability must be connected to data as described by the Law of Large Numbers.

This link to data is essential because chance does not always produce the results that we believe it should. Managers of a hospital, for instance, might be inclined to assign nurses in the maternity ward evenly over the week. It's "obvious" that babies are equally likely to be born on any day. Data, however, tell a different story in Figure 7.6. Babies may not care whether they are born on a weekend, but it appears that pregnant mothers and doctors who deliver the babies do!

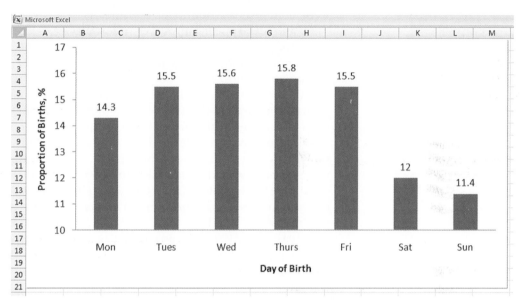

FIGURE 7.6 Births by day of week.[3]

What Do You Think?

a. Supervisors at a construction site post a sign that shows the number of days since the last serious accident. Does the chance for an accident tomorrow increase as the number of days since the last accident increases? Explain why or why not.[4]

b. Over its 40-year history, a bank has made thousands of mortgage loans. Of these, 1% resulted in default. Does the LLN mean that 1 out of the next 100 mortgages made by the bank will result in default?[5]

c. On the basis of Figure 7.6, how would you recommend staffing the maternity ward?[6]

d. Pick a number at random from 1, 2, 3, or 4. Which did you pick? How many others do you think pick the same number as you?[7]

---

[3] From "Risks of Stillbirth and Early Neonatal Death by Day of Week," by Zhong-Cheng Luo, Shiliang Liu, Russell Wilkins, and Michael S. Kramer, for the Fetal and Infant Health Study Group of the Canadian Perinatal Surveillance System. They collected data on 3,239,972 births in Canada between 1985 and 1998.

[4] If the workers continue their duties in the same manner regardless of the sign, then the chances remain the same regardless of the sign. If the sign changes attitudes toward safety, then the chances might change.

[5] The answer is no, for two reasons. First, the LLN only applies in the long run. Second, the local economy of this bank may have changed: Current borrowers are more or less likely to default than those in the past.

[6] The histogram shows births are about equally likely during the week, but about 1/4 less likely on the weekend. So we recommend equal staffing Monday to Friday, but with 1/4 fewer on Saturday and Sunday.

[7] Almost 75% of people pick 3. About 20% pick 2 or 4. If the choice were at random, you'd expect 25% to pick each number. That's not what people seem to do.

7.2 | RULES FOR PROBABILITY

In many situations, we know some probabilities, but our needs require others. Consider risk managers, who decide which loans to approve at a bank. Their experience shows that for a particular type of loan, 2% of borrowers default, 15% of borrowers fall behind, and the rest repay the debt on schedule. It is essential for the managers to know these probabilities, but they also need other probabilities that are determined by these. For example, the bank may exhaust its cash reserves if too many borrowers default or pay late. What are the chances that more than 30 of the 100 borrowers who were approved this week do not repay their loans on time?

Probability appears intuitive—it's a relative frequency—but the manipulations that convert one probability into another become confusing without rules. Fortunately, we only need a few. In fact, three intuitive rules (formally known as Kolmogorov's axioms) are sufficient. We will start with these rules, and then add two more that simplify common calculations.

Probability rules use sets to describe the outcomes of interest. Consider a single borrower at the bank. The loan made to this borrower results in three possible outcomes: default, late, or on-time. These are the only possible outcomes and define a sample space. The **sample space** is the set of *all* possible outcomes that can happen in a chance situation. The boldface letter **S** denotes the sample space (some books use the Greek letter Ω). The sample space depends on the problem. The status of the next call to the agent at the help line is either easy or hard, so the sample space for a call is **S** = {easy, hard}. For a single loan, the sample space has 3 possible outcomes, **S** = {default, late, on-time}. We use curly braces { } to enclose the list of the items in the sample space. Also notice that the events that make up a sample space don't have to be equally likely. Outcomes from tossing a fair coin or rolling a die are equally likely. Those in most business applications are not.

sample space S The sample space, denoted **S**, is the set of all possible outcomes.

The sample space can grow very large. The sample space for the status of a single loan at the bank has 3 elements; the sample space for 2 loans has $3^2 = 9$ outcomes (not 2×3). Each element of this sample space is a pair. The first item of each pair shows the outcome of the first loan, and the second item shows the outcome of the second loan, like this:

(default, default)   (default, late)   (default, on time)

(late, default)   (late, late)   (late, on time)

(on time, default)   (on time, late)   (on time, on time)

The sample space for 10 calls to the help line has $2^{10} = 1,024$ elements, and the sample space for 100 loans has $3 \times 3 \times \cdots \times 3 = 3^{100} \approx 5 \times 10^{47}$ elements.

event A portion (subset) of the sample space.

Rather than list every element of the sample space, it is generally more useful to identify subsets. Subsets of a sample space are known as **events**. Following convention, we denote events by single capital letters, such as **A**, **B**, or **C**. For example, when considering two loans, the event

A = {(default, default), (late, late), (on time, on time)}

includes the outcomes in which both loans have the same status. We can list the three outcomes in this event, but that will not be feasible when the event includes hundreds of outcomes.

Three Essential Rules

Every event **A** has a probability, denoted $P(\mathbf{A})$. To find the probability, we use three rules along with others built from these. To illustrate each rule, we use a **Venn diagram**. A Venn diagram is a graphical method for depicting the relationship among sets. Since events are sets, Venn diagrams are helpful for seeing the rules of probability. In a Venn diagram, the area of a set represents its probability. Events with larger probabilities have larger area.

The first rule of probability says that *something must happen*. Because the sample space is the collection of all possible outcomes, it is assigned probability 1. Our Venn diagrams display the sample space as a shaded rectangle such as the one shown here.

Venn diagram An abstract graphic that shows one or more sets.

Rule 1 The probability of an outcome in the sample space is 1, $P(\mathbf{S}) = 1$.

In the Venn diagram, the area of the rectangle that denotes the sample space—the probability of the sample space—is 1. The something-must-happen rule provides a useful check on a collection of probabilities. When we assign probabilities to outcomes, we must distribute *all* of the probability. If the probabilities do not add up to 1, we have forgotten something, double counted, or made an error elsewhere in our calculations.

The second rule of probability comes from the definition of a probability as a relative frequency: Every probability lies between 0 and 1.

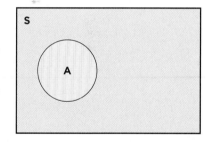

Rule 2 For any event **A**, the probability of **A** is between 0 and 1, $0 \le P(\mathbf{A}) \le 1$.

The Venn diagram makes the second rule appear reasonable. This Venn diagram shows an event **A**. Since events are subsets contained within the sample space, $P(\mathbf{A})$ cannot be larger than 1. Since we associate area with probability, then it's also clear that the probability of **A** cannot be negative. Even if you think an event is inconceivable, its probability cannot be less than zero. You might say casually, "The probability that I will loan you \$1,000 is −1," but zero is as low as a probability can go. You can imagine an event with probability zero, but you won't observe it. Events with probability zero *never* occur.

disjoint events Events that have no outcomes in common; mutually exclusive events.

The third rule of probability has wider consequences because it shows how to combine probabilities. The events described by this rule must be disjoint. **Disjoint events** have no outcomes in common; disjoint events are also said to be **mutually exclusive**. An outcome in one cannot be in the other. In a Venn diagram, disjoint events **A** and **B** have nothing in common and hence do not overlap.

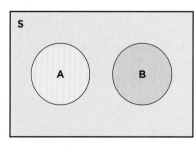

The third rule gives the probability of a union of disjoint events. The **union** of two events **A** and **B** is the collection of outcomes in **A**, in **B**, or in both. We will write unions using words as

mutually exclusive Events are mutually exclusive if no element of one is in the other.

union Collection of outcomes in **A** or **B** or both.

A *or* **B**

(Math texts typically use the symbol $\cup$ in place of *or*.) Because disjoint events have no outcomes in common, the probability of their union is the sum of the probabilities. This rule is known as the **Addition Rule for Disjoint Events**.

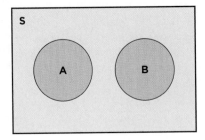

Rule 3. Addition Rule for Disjoint Events
The probability of a union of disjoint events is the sum of the probabilities. If **A** and **B** are disjoint events, then

$$P(\mathbf{A}\,or\,\mathbf{B}) = P(\mathbf{A}) + P(\mathbf{B})$$

The union (shown in green in the Venn diagram) combines **A** (which was yellow) with **B** (which was cyan). The Addition Rule for Disjoint Events states the obvious: The area (probability) of the union of **A** with **B** is the sum of the two areas.

caution Sometimes when speaking, we say the word *or* to mean one or the other, but not both. "I'm going to finish this paper or go to the gym." That usage is not correct when describing unions of events. The union (**A** *or* **B**) is the collection of outcomes in **A**, in **B**, or in both of them.

The Addition Rule for Disjoint Events extends to more than two events. So long as the events $\mathbf{E}_1, \mathbf{E}_2, \ldots, \mathbf{E}_k$ are disjoint, the probability of their union is the sum of the probabilities

$$P(\mathbf{E}_1\,or\,\mathbf{E}_2\,or \ldots or\,\mathbf{E}_k) = P(\mathbf{E}_1) + P(\mathbf{E}_2) + \cdots + P(\mathbf{E}_k)$$

This extension is intuitively reasonable. Since the events have nothing in common, the probabilities accumulate without any double counting.

That's it. Every other rule comes from these three.

Essential Rules of Probability

1. $P(\mathbf{S}) = 1$.
2. For any event **A**, $0 \leq P(\mathbf{A}) \leq 1$.
3. For disjoint events **A** and **B**, $P(\mathbf{A}\,or\,\mathbf{B}) = P(\mathbf{A}) + P(\mathbf{B})$.

The Complement and Addition Rules

Two additional rules make solving many problems easier and faster. The first gives you a way to figure out the probability of an event using what does *not* happen. Suppose that the probability of an investment increasing in value next year is 0.6. What's the probability that the investment does *not* increase in value? It's $0.4 = 1 - 0.6$. To express this intuitive notion as a rule, we need a way to denote an event that does not happen.

The outcomes that are not in the event **A** form another event called the **complement** of **A**, denoted $\mathbf{A}^c$. The Complement Rule says that the probability of an event is 1 minus the probability that it does not happen.

complement Elements not in a set.

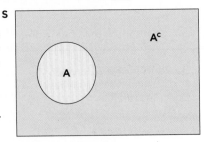

Rule 4. Complement Rule   The probability of an event is one minus the probability of its complement.

$$P(\mathbf{A}) = 1 - P(\mathbf{A}^c)$$

The complement of **A** in the Venn diagram is everything in the sample space **S** that is not in **A** (colored cyan in the Venn diagram).

The second extra rule generalizes the Addition Rule for Disjoint Events (Rule 3). It shows how to find the probability of unions of events that are *not* disjoint. In general, events share outcomes. Suppose the sample space consists of customers who regularly shop in the local mall, and define the events

A = {customers who use credit cards}

and

B = {customers who are women}

Female customers who use credit cards are in both **A** and **B**. Because they share outcomes, these events are not disjoint. Because the events are not disjoint, the probability of the union **A** *or* **B** is not equal to the sum of the probabilities of the two events.

To find the probability of the union of any events, we first identify the overlap between events. The **intersection** of two events **A** and **B** is the event consisting of the outcomes in both **A** and **B**. The intersection occurs if both **A** and **B** occur. We write the intersection of two events as **A** *and* **B**. (Math texts generally use the symbol ∩ for "intersection.")

In general, the probability of the union **A** *or* **B** is less than $P(\mathbf{A}) + P(\mathbf{B})$. You can see why by looking at the Venn diagram. Events that are not disjoint overlap; this overlap is the intersection. Adding the probabilities double counts the intersection. The Venn diagram also suggests an easy fix. Add the probabilities of the two events and then subtract the probability of the intersection to compensate for double counting. This logic leads to the Addition Rule.

Rule 5. Addition Rule For two events **A** and **B**, the probability that one or the other occurs is the sum of the probabilities minus the probability of their intersection.

$$P(\mathbf{A} \text{ } or \text{ } \mathbf{B}) = P(\mathbf{A}) + P(\mathbf{B}) - P(\mathbf{A} \text{ } and \text{ } \mathbf{B})$$

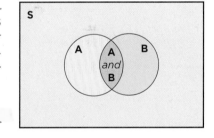

The Addition Rule for Disjoint Sets is a special case. If **A** and **B** are disjoint, then their intersection is empty and $P(\mathbf{A} \text{ } and \text{ } \mathbf{B}) = 0$. To see how these rules are derived from the three essential rules, read the section Behind the Math: Making More Rules at the end of the chapter.

(p. 164)

An Example

Let's try these rules with an example. Movies are commonly shown in cinemas that combine several theatres. That way, one cashier and snack bar can serve customers for several theatres. Here is the schedule of the next six movies that are about to start.

Movie	Time	Genre
M_1	9:00 P.M.	Action
M_2	9:00 P.M.	Drama
M_3	9:00 P.M.	Drama
M_4	9:05 P.M.	Horror
M_5	9:05 P.M.	Comedy
M_6	9:10 P.M.	Drama

---

set operations

union: A *or* **B** Outcomes in **A**, in **B**, or in both.

intersection: A *and* **B** Outcomes in both **A** and **B**.

complement Outcomes not in **A**; written $\mathbf{A}^c$.

Intersection Outcomes in both **A** and **B**.

Because all six movies start in the next few minutes, let's assume that the customers waiting to buy tickets are equally likely to buy a ticket for each of these movies. What's the probability that the next customer buys a ticket for a movie that starts at 9 o'clock *or* is a drama?

Let's practice using the notation for events. Define the events $\mathbf{A}$ = {movie starts at 9 P.M.} and $\mathbf{B}$ = {movie is a drama}. The probability sought by the question is $P(\mathbf{A}\,or\,\mathbf{B})$. We can draw the Venn diagram like this, with M's representing the movies.

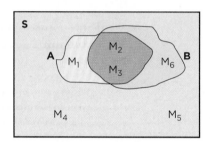

The event $\mathbf{A}$ = {M_1, M_2, M_3} and the event $\mathbf{B}$ = {M_2, M_3, M_6}. Because tickets for all movies are equally likely to be purchased, $P(\mathbf{A})$ = $P(\mathbf{B})$ = 3/6. However, $P(\mathbf{A}\,or\,\mathbf{B})$ ≠ $P(\mathbf{A})$ + $P(\mathbf{B})$. The events $\mathbf{A}$ and $\mathbf{B}$ are not disjoint; movies M_2 and M_3 belong to both.

We have two ways to find $P(\mathbf{A}\,or\,\mathbf{B})$. The first is obvious when the sample space is small and the outcomes are equally likely: Count the outcomes in the union.

$$\mathbf{A}\,or\,\mathbf{B} = \{M_1, M_2, M_3\}\,or\,\{M_2, M_3, M_6\} = \{M_1, M_2, M_3, M_6\}$$

Because the outcomes are equally likely, $P(\mathbf{A}\,or\,\mathbf{B})$ = 4/6 = 2/3.

The second way to find $P(\mathbf{A}\,or\,\mathbf{B})$ uses the Addition Rule. M_2 and M_3 form the intersection, and hence $P(\mathbf{A}\,and\,\mathbf{B})$ = 2/6. From the Addition Rule,

$$P(\mathbf{A}\,or\,\mathbf{B}) = P(9\text{ P.M. }or\text{ Drama})$$
$$= P(\mathbf{A}) + P(\mathbf{B}) - P(\mathbf{A}\,and\,\mathbf{B})$$
$$= 3/6 + 3/6 - 2/6 = 2/3$$

Both ways give the same answer. If they don't, then we've made a mistake.

What Do You Think?

Let's revisit the example of the risk manager at a bank who has just approved a loan. The loan can end in one of three states: default, late, or repaid on time. We're given that $P(\text{default})$ = 0.02 and $P(\text{late})$ = 0.15.

a. Which rules tell you how to find $P(\text{repaid on time})$?[8]
b. Is $P(\text{default }or\text{ late})$ = 0.17, or is it smaller?[9]
c. What rule tells you how to determine the probability a loan does not default?[10]

---

[8] This is a consequence of Rules 1, 3, and 4. The probabilities of the possibilities have to sum to 1, so the probability of repaid on time is 1 − (0.02 + 0.15) = 0.83.
[9] These are disjoint events, so the probabilities add. The probability of default or late is 0.02 + 0.15 = 0.17.
[10] Use the Complement Rule. The probability of not defaulting is 1 − $P(\text{default})$ = 1 − 0.02 = 0.98.

7.3 | INDEPENDENT EVENTS

Many applications of probability in business require us to estimate chances of a large number of similar events. Consider bank loans. Risk managers are concerned about the chances for default among a large number of customers, not just one or two. Similarly, insurance companies sell thousands of policies. The chances associated with each loan or each policy may be quite similar. For instance, the bank believes that $P(\text{default}) = 0.02$ for any one loan of this type. What can be said, though, about the chances for default among a large collection of loans?

independent events Events that do not influence each other; the probabilities of independent events multiply.

$$P(\mathbf{A} \, and \, \mathbf{B}) = P(\mathbf{A}) \times P(\mathbf{B}).$$

A common approach that greatly simplifies the calculation of probabilities is to treat the individual loans or policies as independent events. Two events are said to be **independent events** if the occurrence of one has no effect on the chances for the occurrence of the other. The simplest familiar example occurs when you roll two dice when playing a board game like Monopoly™. The value shown on one die doesn't influence the other. The decisions of two managers are independent if the choices made by one don't influence the choices made by the other. The precise definition of independence says that probabilities multiply.

Multiplication Rule for Independent Events   Two events **A** and **B** are independent if the probability that both **A** and **B** occur is the product of the probabilities of the two events.

$$P(\mathbf{A} \, and \, \mathbf{B}) = P(\mathbf{A}) \times P(\mathbf{B})$$

The Addition Rule applies to unions; the Multiplication Rule for Independent Events applies to intersections.

In the context of the call center, the probability that an arriving call is easy is 0.7. What's the probability that the next *two* calls are easy? If the calls are independent, then the probability is $0.7 \times 0.7 = 0.49$. Independence extends to more than two events. What is the probability that the next five calls are easy? If they are independent, the probabilities again multiply.

$$P(\text{easy}_1 \, and \, \text{easy}_2 \, and \, \text{easy}_3 \, and \, \text{easy}_4 \, and \, \text{easy}_5) = 0.7^5 \approx 0.17$$

about 1 chance in 6. For loans at the bank, what is the probability that all of 100 loans made to customers are repaid on time? The probability that any one loan repays on time is 0.83. If the events are independent, then the probability that all repay on time is $0.83^{100} \approx 8 \times 10^{-9}$. Even though each loan is likely to repay on time, the chance that *all* 100 repay on time is very small.

dependent events Events that are not independent, for which $P(\mathbf{A} \, and \, \mathbf{B}) \neq P(\mathbf{A}) \times P(\mathbf{B})$.

The assumption of independence may not match reality. At the call center, it could be that calls for assistance that arrive around the same time of day are related to each other. That could mean that an easy call is more likely to be followed by another easy call. For bank loans, the closure of a large factory could signal problems that would affect several borrowers. A default by one may signal problems with others as well. Events that are not independent are said to be **dependent events**. You cannot simply multiply the probabilities if the events are dependent. For a specific illustration, let's use the previous example of selling tickets at a movie theatre. Are the events $\mathbf{A} = \{9 \text{ P.M.}\}$ and $\mathbf{B} = \{\text{Drama}\}$ independent? We know that $P(\mathbf{A}) = 1/2$ and $P(\mathbf{B}) = 1/2$. If these events are independent, then $P(\mathbf{A} \, and \, \mathbf{B}) = 1/4$. Is it? No. The intersection contains two movies, so $P(\mathbf{A} \, and \, \mathbf{B}) = 1/3$, not 1/4. These events are dependent. Drama movies make up two-thirds of the films offered at 9 P.M., but only one-third of those offered at other times.

Much of statistics concerns deciding when it is reasonable to treat events as independent. The presence of association in a contingency table (Chapter 5) or a scatterplot (Chapter 6) indicates dependence that should be incorporated into the underlying probabilities. Data in the example of Web shopping in Chapter 5, for example, revealed that the relative frequency of making a purchase depends on the host site that pulls the shopper to Amazon. We will return to the topic of dependence in the next chapter.

4M EXAMPLE 7.1   MANAGING A PROCESS

MOTIVATION ▸ STATE THE QUESTION

A manufacturer has a large order to fulfill. To complete the order on schedule, the assembly line needs to run without a problem—a breakdown—for the next five days. *What is the probability that a breakdown on the assembly line will interfere with completing the order?*

METHOD ▸ DESCRIBE THE DATA AND SELECT AN APPROACH

Past data indicate that the assembly line runs well most of the time. On 95% of days, everything runs as designed without a hitch. That is, the probability is 0.95 that the assembly line runs the full day without a delay. Barring information to the contrary, assume that the performance is independent from one day to the next.

MECHANICS ▸ DO THE ANALYSIS

tip **All but the simplest problems require you to combine several rules to find the needed probability. State a rule as you use it so that when you look back at what you've done, you'll be able to check your thinking.**

The Multiplication Rule for Independent Events implies that the probability that the line runs well for five consecutive days is the product

$$P(\text{OK for 5 days}) = 0.95^5 \approx 0.774$$

The Complement Rule shows that the probability that the line has a breakdown is

$$P(\text{breakdown during 5 days}) = 1 - P(\text{OK for 5 days})$$
$$= 1 - 0.774 = 0.226$$

MESSAGE ▸ SUMMARIZE THE RESULTS

The probability that a breakdown disrupts production during the next five days is 0.226. There is about 1 chance in 4 that a breakdown will interfere with the schedule. It would be wise to warn the customer that the delivery may be late.

Clearly state any assumptions that you have made in nontechnical language. In this example, the calculation is based on past performance of the assembly line; if things have changed since then, the probability might be too high or too low. The calculation also relies on independence from day to day.

Boole's Inequality

Independence allows us to find the exact probability of intersections of many events, but these events must be independent. Without independence, it is often useful to be able to put an upper bound on the probability of the intersection. The following approach does not give an exact answer, but the bound that it provides is often accurate enough when the events in question have small probability. It's also very easy to use.

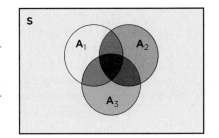

Boole's Inequality

$$P(\mathbf{A}_1 \, or \, \mathbf{A}_2 \, or \ldots or \, \mathbf{A}_k) \leq p_1 + p_2 + \cdots + p_k$$

Boole's inequality Probability of a union ≤ sum of the probabilities of the events.

Boole's inequality (sometimes called Bonferroni's inequality) is named for the English mathematician George Boole, known for creating Boolean logic. If the events $\mathbf{A}_1, \mathbf{A}_2, \ldots, \mathbf{A}_k$ have probabilities p_i, then Boole's inequality says that the probability of a union is less than or equal to the sum of the probabilities.

Look at the Venn diagram with $k = 3$ to convince yourself; the area of the union of these three events is less than the sum of the areas because of the overlap. Boole's inequality estimates the probability of the union as if the events were disjoint and did not overlap. By ignoring the double counting, it gives a probability that is too large unless the events are disjoint. If the events have the same probability $P(\mathbf{A}_i) = p$, then Boole's inequality simplifies to

$$P(\mathbf{A}_1 \, or \, \mathbf{A}_2 \, or \, \ldots \, or \, \mathbf{A}_k) \leq p + p + \ldots + p = k\,p$$

Boole's inequality is most useful when the events have small probabilities. Let's apply Boole's inequality to the previous example of breakdowns on the assembly line. The event $\mathbf{A}_1$ means that at least one breakdown occurs on Monday, $\mathbf{A}_2$ means that at least one breakdown occurs on Tuesday, and so forth up to $\mathbf{A}_5$ for Friday. Each of these events has probability 0.05. In Example 7.1, we assumed that these events were independent, as if breakdowns one day did not lead to more the next day. With Boole's inequality, we don't need that assumption.

$$P(\text{breakdown during 5 days}) = P(\mathbf{A}_1 \, or \, \mathbf{A}_2 \, or \, \mathbf{A}_3 \, or \, \mathbf{A}_4 \, or \, \mathbf{A}_5)$$
$$\leq 0.05 + 0.05 + 0.05 + 0.05 + 0.05 = 0.25$$

The exact answer if the events are independent is 0.226. Boole's inequality comes close, without requiring that we assume that the events are independent.

Best Practices

- *Make sure that your sample space includes all of the possibilities.* It may be helpful to list a few members of the sample space, as we did early in the chapter.
- *Include all of the pieces when describing an event.* As events become more complex, it becomes easy to omit an outcome and get the wrong count. What is the probability that exactly two out of three calls made to the help line in the text example are easy? There are three outcomes: EEH, EHE, or HEE. Because of independence, each of these outcomes has probability $0.7^2 \times 0.3 = 0.147$, so the probability is $3 \times 0.147 = 0.441$.
- *Check that the probabilities assigned to all of the possible outcomes add up to 1.* Once you are sure you have identified the entire sample space, check that the total probability for the possible outcomes in the sample space is 1. If the sum is less than 1, you may need another category (Other). If the sum is more than 1, check that the outcomes are disjoint. You may have double counted.
- *Only add probabilities of disjoint events.* As the name says, events must be disjoint to use the Addition Rule for Disjoint Events. If the events

are not disjoint, you'll double count the intersection. The probability that a randomly chosen customer is under 80 years old *or* female is not the probability of being under 80 *plus* the probability of being female. That sum is too large.
- *Be clear about independence* when you use probabilities. Independence means that the occurrence of one event does not affect others. That's natural for tossing coins but may not be reasonable for events that describe the economy, for instance. Unemployment, inflation, and the stock market are all linked. What happens to one almost always ripples into the others.
- *Only multiply probabilities of independent events.* Many auto insurance companies, for example, believe that the probability that you are late paying your credit card *and* you have an accident is *not* the product of the probability of paying late *times* the probability of being in an accident. Their data indicate that these events are *not* independent, and so they do not multiply the probabilities. Rather, they adjust the probabilities (and insurance rates) to reflect credit records (as well as numerous other factors, such as age and sex).

Pitfalls

- *Do not multiply probabilities of dependent events.* It's tempting and makes it easy to find the probability that several events happen together, but that does not make it right. If you are unsure about whether the events are independent, you may want to use Boole's inequality to approximate the probability rather than assume independence.
- *Avoid assigning the same probability to every outcome.* The faces of a coin or die may be

equally likely, but these are special cases. Most outcomes in business, even when you can count the possibilities, are not equally likely.
- *Do not confuse independent events with disjoint events.* Independence means that the occurrence of one event does not affect the chances for the other. That cannot happen for disjoint events. If one of two disjoint events happens, the other cannot. Disjoint events are dependent.

 BEHIND the Math

Making More Rules

The three essential rules are important because they bring mathematical precision to probability. The only way to add a new rule is to derive it from these

three. No matter how intuitive or obvious, you have to be able to derive a rule from these principles.

For example, it seems obvious that the probability of an event **A** cannot be larger than 1. We've even listed this property as part of Rule 2. Though intuitively obvious, the fact that $P(\mathbf{A}) \le 1$ is a consequence of the other rules.

To see that the remaining rules imply that $P(\mathbf{A}) \le 1$, write the sample space as the union of two disjoint events, $\mathbf{A}$ and its complement $\mathbf{A}^c$, $\mathbf{S} = \mathbf{A}\,or\,\mathbf{A}^c$. Because $\mathbf{A}$ and $\mathbf{A}^c$ are disjoint, the probability of their union is the sum of their probabilities by the Addition Rule for Disjoint Events. Hence,

$$P(\mathbf{S}) = P(\mathbf{A}\,or\,\mathbf{A}^c) = P(\mathbf{A}) + P(\mathbf{A}^c)$$

Now use the other two rules. The first rule tells us that $1 = P(\mathbf{S}) = P(\mathbf{A}) + P(\mathbf{A}^c)$. Because the second rule tells us that $P(\mathbf{A}^c) \ge 0$, it follows that

$$P(\mathbf{A}) = 1 - P(\mathbf{A}^c) \le 1$$

For a more challenging exercise, consider the general addition rule.

$$P(\mathbf{A}\,or\,\mathbf{B}) = P(\mathbf{A}) + P(\mathbf{B}) - P(\mathbf{A}\,and\,\mathbf{B})$$

To derive this rule from the three essential rules, build a little machinery first. Define the difference between two sets as

$$\mathbf{A} - \mathbf{B} = \{\text{those elements of } \mathbf{A} \text{ that are not in } \mathbf{B}\}$$

Assume $\mathbf{B}$ is a subset of $\mathbf{A}$. In this case, we can write $\mathbf{A}$ as a union of two disjoint sets, $\mathbf{A} = (\mathbf{A} - \mathbf{B})\,or\,\mathbf{B}$. Hence, the addition rule for disjoint sets (Rule 3) shows that $P(\mathbf{A}) = P(\mathbf{A} - \mathbf{B}) + P(\mathbf{B})$, from which we get $P(\mathbf{A} - \mathbf{B}) = P(\mathbf{A}) - P(\mathbf{B})$.

Now for the Addition Rule itself. Write the union of any two sets $\mathbf{A}$ and $\mathbf{B}$ as

$$\mathbf{A}\,or\,\mathbf{B} = \mathbf{A}\,or\,(\mathbf{B} - (\mathbf{A}\,and\,\mathbf{B}))$$

The sets on the right-hand side are disjoint, so

$$\begin{aligned} P(\mathbf{A}\,or\,\mathbf{B}) &= P(\mathbf{A}) + P(\mathbf{B} - (\mathbf{A}\,and\,\mathbf{B})) \\ &= P(\mathbf{A}) + P(\mathbf{B}) - P(\mathbf{A}\,and\,\mathbf{B}) \end{aligned}$$

The last step works because the intersection is a subset of $\mathbf{B}$.

CHAPTER SUMMARY

The collection of all possible outcomes forms the **sample space S**. An **event** is a subset of the sample space. The **probability** of an **event** is the long-run relative frequency of the event. The **Law of Large Numbers** guarantees that the relative frequency of an event in data that lack patterns converges to the probability of the event in the long run. Several rules allow us to manipulate probabilities. If $\mathbf{A}$ and $\mathbf{B}$ are events, then

1. Something must happen $P(\mathbf{S}) = 1$

2. Probabilities lie between 0 and 1 $0 \le P(\mathbf{A}) \le 1$

3. Addition Rule
$$P(\mathbf{A}\,or\,\mathbf{B})$$
$$= P(\mathbf{A}) + P(\mathbf{B}) - \mathbf{P}(\mathbf{A}\,and\,\mathbf{B})$$
$$= P(\mathbf{A}) + P(\mathbf{B})$$
if $\mathbf{A}$ and $\mathbf{B}$ are disjoint

4. Complement Rule $P(\mathbf{A}) = 1 - P(\mathbf{A}^c)$

5. Multiplication Rule for *Independent* Events $P(\mathbf{A}\,and\,\mathbf{B}) = P(\mathbf{A}) \times P(\mathbf{B})$

Independence is often assumed in order to find the probability of a large collection of similar events. **Boole's inequality** allows us to estimate the probability of a large collection of events without requiring the assumption of independence.

▇ Key Terms

Addition Rule for Disjoint Events, 158
Boole's inequality 163
complement, 158, 159
dependent events, 161
disjoint events, 157

event, 156
independent events, 161
intersection, 159
Law of Large Numbers, 153
Multiplication Rule for Independent Events, 161

mutually exclusive, 157
probability, 153
sample space **S**, 156
union, 157, 159
Venn diagram, 157

▨ Formulas and Notation

Complement of an event

$$\mathbf{A}^c = \{\text{outcomes not in } \mathbf{A}\}$$

Union of events

$$\mathbf{A} \text{ or } \mathbf{B} = \{\text{outcomes in } \mathbf{A}, \text{ in } \mathbf{B}, \text{ or in both}\} \quad (= \mathbf{A} \cup \mathbf{B})$$

Intersection of events

$$\mathbf{A} \text{ and } \mathbf{B} = \{\text{outcomes in both } \mathbf{A} \text{ and } \mathbf{B}\} \quad (= \mathbf{A} \cap \mathbf{B})$$

Boole's inequality

$$P(\mathbf{A}_1 \text{ or } \mathbf{A}_2 \text{ or } \ldots \text{ or } \mathbf{A}_k) \leq p_1 + p_2 + \cdots + p_k$$

EXERCISES

Mix and Match

Find the matching item from the second column.

1. Independent events	(a) $P(\mathbf{A} \text{ and } \mathbf{B}) = 2P(\mathbf{A}) \times P(\mathbf{B})$
2. Disjoint events	(b) $\mathbf{A}^c$
3. Union	(c) $\mathbf{S}$
4. Intersection	(d) $P(\mathbf{A} \text{ or } \mathbf{B}) + P(\mathbf{A} \text{ and } \mathbf{B}) = P(\mathbf{A}) + P(\mathbf{B})$
5. Complement of $\mathbf{A}$	(e) $P(\mathbf{A} \text{ or } \mathbf{B}) \leq P(\mathbf{A}) + P(\mathbf{B})$
6. Sample space	(f) $\mathbf{A} \text{ and } \mathbf{B}$
7. Addition Rule	(g) $P(\mathbf{A} \text{ and } \mathbf{B}) = 0$
8. Complement Rule	(h) $\mathbf{A} \text{ or } \mathbf{B}$
9. Boole's inequality	(i) $P(\mathbf{A}^c) = 1 - P(\mathbf{A})$
10. Dependent events	(j) $P(\mathbf{A} \text{ and } \mathbf{B}) = P(\mathbf{A}) \times P(\mathbf{B})$

True/False

Mark each statement True or False. If you believe that a statement is false, briefly say why you think it is false.

Exercises 11–16. A market research assistant watches the next five customers as they leave the store. He records whether the customer is carrying a store bag that indicates the customer made a purchase. He writes down a yes or a no for each. Define the events.

$\mathbf{A} = \{\text{first two shoppers have a bag}\}$

$\mathbf{B} = \{\text{last two shoppers have a bag}\}$

$\mathbf{C} = \{\text{last three shoppers have a bag}\}$

11. The sample space $\mathbf{S}$ for this experiment has 10 elements.

12. The assumption of independence means that each shopper has the same probability for carrying a bag.

13. $P(\mathbf{A}) + P(\mathbf{B}) = P(\mathbf{A} \text{ or } \mathbf{B})$.

14. The probability that both events $\mathbf{B}$ and $\mathbf{C}$ occur is equal to $P(\mathbf{B})$.

15. The probability that a randomly chosen customer purchases with a credit card **or** spends more than $50 is the same as or larger than the probability that the customer purchases with a credit card **and** spends more than $50.

16. If each shopper has the same chance of making a purchase and shoppers behave independently of one another, then $P(\mathbf{A} \text{ and } \mathbf{C}) = P(\mathbf{A}) \times P(\mathbf{C})$.

Exercises 17–22. The Human Resources (HR) group at a large accounting firm interviews prospective candidates for new hires. After each interview, the firm rates the candidate on a 10-point scale, with the rating 10 denoting exceptionally good candidates and 1 denoting those that the firm rates poor. The HR group rated 6 candidates on Monday and 6 candidates on Tuesday. The outcomes of these 12 ratings form the sample space.

17. The events $\mathbf{A} = \{3 \text{ candidates on Monday rate above } 7\}$ and $\mathbf{B} = \{\text{two candidates on Tuesday rate above } 7\}$ are disjoint events.

18. If 8 of the 12 candidates on Monday and Tuesday rate above 6, then the probability of a candidate rating above 6 on Wednesday is 8/12.

19. The HR group has monitored the outcome of these interviews for several years. The Law of Large Numbers assures us that HR personnel can use these data to learn the probability of a candidate scoring above 8 during an interview.

20. The probability that the ratings of the six candidates on Monday are {6, 4, 3, 8, 11, 8} is zero.

21. Define the events **A** = {6 out of the 12 candidates rate 8 or better}, **B** = {3 out of the 6 candidates on Monday rate 8 or better}, and **C** = {3 out of the 6 candidates on Tuesday rate 8 or better}. Independence implies that $P(\mathbf{A}) = P(\mathbf{B}) \times P(\mathbf{C})$.

22. Define the events **A** = {first candidate is rated 8, 9, or 10} and **B** = {first candidate is rated 5, 6, or 7}. Then $P(\mathbf{A}\ or\ \mathbf{B}) = P(\mathbf{A}) + P(\mathbf{B})$.

Think About It

23. Each of the following scatterplots shows a sequence of observations; the *x*-axis enumerates the sequence as in the example of the arrival of calls to the agent at the help desk in this chapter. For which cases does the Law of Large Numbers apply to probabilities based on these observations?

(a)
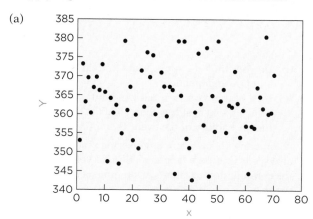

(b)
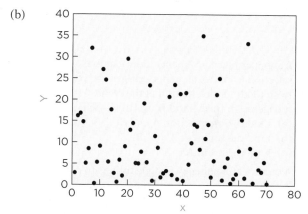

(c)
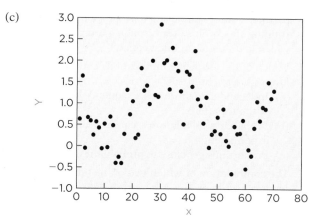

(d)
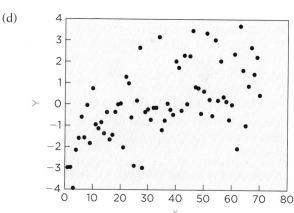

24. As in Exercise 23, these scatterplots graph a sequence of observations taken under similar conditions. If we could watch these processes longer and longer and accumulate more data, in which cases would the Law of Large Numbers apply?

(a)
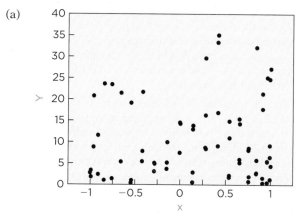

(b)
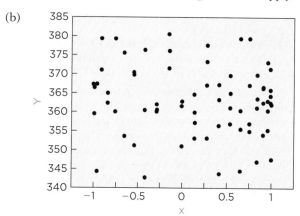

(c)

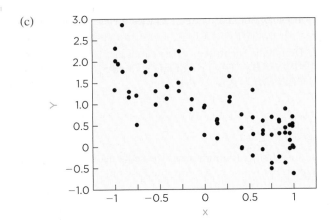

(d)

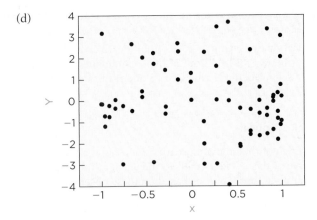

25. A shopper in a convenience store can make a food selection from frozen items, refrigerated packages, fresh foods, or deli items. Let the event **A** = {frozen, refrigerated, fresh} and **B** = {fresh, deli}.
    (a) Find the intersection **A** *and* **B**.
    (b) Find the union of **A** and **B**.
    (c) Find the event **A**ᶜ.

26. A credit-rating agency assigns ratings to corporate bonds. The agency rates bonds offered to companies that are most likely to honor their liabilities AAA. The ratings fall as the company becomes more likely to default, dropping from AAA to AA, A, down to BBB, BB, B, CCC, CC, R, and then D (for in default). Let the event **W** = {AAA, AA, A, BBB, BB, B} and **V** = {BBB, BB, B, CCC, CC}.
    (a) Find the intersection **W** *and* **V**.
    (b) Describe the union **W** *or* **V**.
    (c) Find the complement (**W** *or* **V**)ᶜ.

27. A brand of men's pants offered for sale at a clothing store comes in various sizes. The possible waist sizes are

    Waist: {24 inches, 26 inches, . . . , 46 inches}

    with inseams (length of the pant leg)

    Inseam: {28 inches, 29 inches, . . . , 40 inches}

    Define the event **B** = {waist 40 inches or larger} and **T** = {inseam 36 inches or larger}.
    (a) Describe the choice of a customer that is in the event (**B** *and* **T**).
    (b) What would it mean if $P(\mathbf{B}\ and\ \mathbf{T}) = P(\mathbf{B}) \times P(\mathbf{T})$?
    (c) Does the choice of a tall, thin customer lie in the event (**B** *and* **T**) or the event (**B** *or* **T**)?

28. An auto dealer sells several brands of domestic and foreign cars in several price ranges. The brands sold by the dealer are Saturn, Pontiac, Buick, Acura, and Ferrari. The prices of cars range from under $20,000 to well over $100,000 for a luxury car. Define the event **D** = {Saturn, Pontiac, Buick} and **C** = {price above $60,000}.
    (a) Does the sale of a Ferrari lie in the union of **D** with **C**?

    (b) Do you think it's appropriate to treat the events **D** and **C** as independent?
    (c) If all of the models of Saturn, Pontiac, and Buick vehicles sold cost less than $55,000, then what is the probability of the event **D** *and* **C**?

29. A company seeks to hire engineering graduates who also speak a foreign language. Should you describe the combination of talents as an intersection or a union?

30. Customers visit a store with several departments. A randomly chosen customer goes to the sporting goods department. After wandering for a bit, she buys a pair of running shoes. This event is most naturally viewed as which, a union or an intersection?

31. Which of the following implications comes from the Law of Large Numbers?
    (a) Independence produces a sequence of observations without patterns.
    (b) Probability calculations require a large number of experiments.
    (c) Proportions get close to probabilities in the long run.

32. The combination of which two of the following characteristics would produce the sort of data needed in order for the Law of Large Numbers to guarantee that a proportion converges to a probability?
    (a) Independent trials.
    (b) Constant chance for the event to occur on each trial.
    (c) The probability of the event must be 1/2.

33. Count the number of cars that pass by an intersection during consecutive five-minute periods on a highway leading into a city. Would these data allow us to use the Law of Large Numbers eventually to learn P(more than 50 cars in five minutes)?

34. An airline would like to know the probability of a piece of passenger's luggage weighing more than 40 pounds. To learn this probability, a baggage handler picks off every 20th bag from the conveyor and weighs it. Do you think these data allow the airline to use the Law of Large Numbers eventually to learn P(luggage weighs more than 40 pounds)?

35. A Web site recorded whether visitors click on a shown ad. The following plot shows the outcomes for a sequence of 100 visitors, with a 1 shown if the visitor clicked on the ad and a 0 otherwise.
    (a) Does it appear that the Law of Large Numbers is applicable if this sequence continues indefinitely?
    (b) Can we find the probability of clicking on the shown ad from looking at these 100 observations?

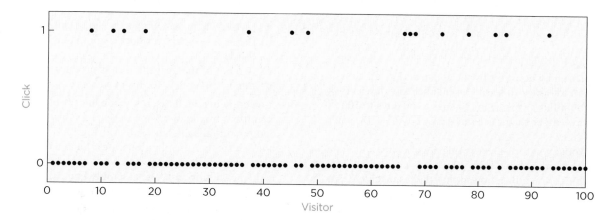

36. The manager of a beachside vacation resort tracks the local weather. Each day, she records a 1 if the day is cloudy. She records a 0 if the day is sunny. The following plot shows a sequence from her data. Would these data be well suited for estimating the probability of sunny weather?

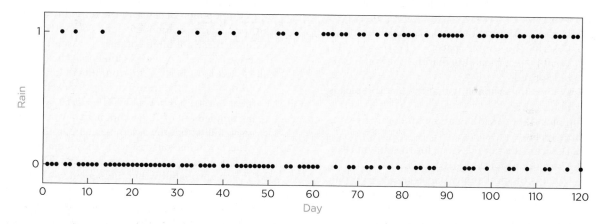

37. A basketball team is down by 2 points with only a few seconds remaining in the game. There's a 50% chance that the team will be able to make a 2-point shot and tie the game, compared to a 30% chance that it will make a 3-point shot and win. If the game ends in a tie, the game continues to overtime. In overtime, the team has a 50% chance of winning. What should the coach do, go for the 2-point shot or the 3-point shot? Be sure to identify any assumptions you make.

38. It's time for an advertising firm to renew its contract with a client. The advertising firm wants the client to increase the amount it spends to advertise. If the firm proposes to continue the current contract, there's a 75% chance that the client will accept the proposal. To increase the business will require a second proposal that has a 50% chance of approval. Alternatively, the advertising firm can begin the negotiations by proposing an elaborate advertising campaign. If it takes this approach, there's a 40% chance that the client will approve the expanded proposal without needing a second proposal. Which approach should the advertising firm take if it wants to grow the business? Identify any assumptions you've made.

39. A market analyst on TV makes predictions such as saying that there is a 25% chance that the market will move up the next week. What do you think is the meaning of such a phrase?

40. A basketball player who has missed his last seven consecutive shots then makes the game-winning shot. When speaking to reporters afterward, the player says he was very confident that last time because he knew he was "due to make a basket." Comment on his statement.

41. In the weeks following a crash, airlines often report a drop in the number of passengers, probably because people are wary of flying because they just learned of an accident.
    (a) A travel agent suggests that, since the Law of Large Numbers makes it highly unlikely to have two plane

crashes within a few weeks of each other, flying soon after a crash is the safest time. What do you think?

(b) If the airline industry proudly announces that it has set a new record for the longest period of safe flights, would you be reluctant to fly? Are the airlines due to have a crash?

42. Many construction sites post a sign that proclaims the number of days since the last accident that injured an employee.

(a) A friend who walks by the construction site tells you to expect to see an ambulance any day. The count has grown from 20 days to 60 days to 100 days. An accident is coming any day now. Do you agree?

(b) Would you feel safer visiting a work site that proclaimed it had been 100 days since the last accident, or one that showed that the last accident was 14 days ago?

You Do It

43. In January 2006, the Web site for M&M™ candies claimed that 24% of plain M&M candies are blue, 20% are orange, 16% green, 14% yellow, and 13% each red and brown.

(a) Pick one M&M at random from a package.
   1. Describe the sample space.
   2. What is the probability that the one you pick is blue or red?
   3. What is the probability that the one you pick is not green?

(b) You pick three M&M's in a row randomly from three separate packages.
   1. Describe the sample space for the outcomes of your three choices.
   2. What is the probability that every M&M is blue?
   3. What is the probability that the third M&M is red?
   4. What is the probability that at least one is blue?

44. At a large assembly line, the manufacturer reassigns employees to different tasks each month. This keeps the workers from getting bored by endlessly repeating the same task. It also lets them see how the work done in different stages must fit together. It is known that 55% of the employees have less than or equal to two years of experience, 32% have between three and five years of experience, and the rest have more than five years of experience. Assume that teams of three employees are formed randomly. Dave works at this assembly line.

(a) Consider one of Dave's teammates. What is the probability that this teammate has
   1. two or fewer years of experience?
   2. more than five years of experience?

(b) What is the probability that, considering both of Dave's two teammates,
   1. both have more than two years of experience?
   2. exactly one of them has more than five years of experience?
   3. at least one has more than five years of experience?

45. A survey found that 62% of callers in the United States complain about the service they receive from a call center if they suspect that the agent who handled the call is foreign.[11] Given this context, what is the probability that (state your assumptions)

(a) the next three consecutive callers complain about the service provided by a foreign agent?

(b) the next two calls produce a complaint, but not the third?

(c) two out of the next three calls produce a complaint?

(d) none of the next 10 calls produces a complaint?

46. A recently installed assembly line has problems with intermittent breakdowns. Recently, it seems that the equipment fails at some point during 15% of the eight-hour shifts that the plant operates. Each day contains three consecutive shifts. State your assumptions. What is the probability that the assembly line

(a) works fine throughout the three shifts on Monday?

(b) works fine on Monday but breaks down during the first shift on Tuesday?

(c) breaks down on a given day?

(d) breaks down during only one shift during a chosen day?

47. Modern manufacturing techniques rely on highly reliable components. Because of the difficulty and time to test the separate components of a computer system, for example, these are often completely assembled before the power is ever applied. Assume that the computer fails to work properly if any component fails.

(a) Suppose that the rate of defects for each component of the system is only 0.1%. If the system is assembled out of 100 components, what is the probability that the assembled computer works when switched on, assuming the components are independent?

(b) For the probability that the system works to be 99%, what is the largest allowable defect rate for the components? Assume independence.

(c) Use Boole's inequality to estimate the probability that the system described in part (a) works. Why does Boole's inequality work so well in this example?

48. A chain of fast-food restaurants offers an instant-winner promotion. Drink cups and packages of French fries come with peel-off coupons. According to the promotion, 1 in 20 packages is a winner.

(a) A group of friends buys four drinks and three orders of fries. What's the probability that they win, assuming the prizes are attached to the cups and fry containers independently?

(b) Assuming independence, how many drinks and fries would the group have to order to have a 50% chance of finding a winner?

(c) Use Boole's inequality to estimate the probability in part (a). Do you expect this approach to work well in this application?

---

[11] "Making Bangalore Sound Like Boston," *Business Week*, April 10, 2006.

49. You take a quiz with six multiple-choice questions. After you studied, you estimated that you would have an 80% chance of getting any individual question right. What are your chances of getting them all right, assuming that one hard question does not confuse you and cause you to miss others?

50. A friend of yours took a multiple-choice quiz and got all six questions right, but now claims to have guessed on every question. If each question offers four possible answers, do you believe her? Explain why you either believe her or not.

51. A down side to running a call center is the high turnover in the operators who answer calls. As an alternative, some companies have turned to "homeshoring." Rather than have operators come to a call center, the company uses technology to route calls to the operators at home. Recent analysis shows that, compared to the usual staff, home-based operators are older (presumably more mature), have 5 to 10 more years of work experience, and stay on the job longer.[12]

	Call Center	Home Based
College education	20%	75%
Annual attrition	100%	25%

(a) What is the probability that a home-based operator stays for more than a year and has a college education? Identify any assumptions that you make and offer some justification.

(b) What is the probability that a college-educated operator at a call center stays for two years? Again, note your assumptions.

52. Intel has long dominated the market for CPU chips, which are the brains in modern desktop and laptop computers. That domination was shaken in 2005 when rival AMD made inroads with a new line of processors. Intel captured 70% of the market for laptop computers in the fourth quarter of 2004, but only 42% in the same period of 2005. In some brands, the use of Intel chips fell more. For example, 70% of HP laptops in 2004 used Intel chips compared to 31% in 2005.

(a) What is the probability that someone who bought a laptop in 2004 and again in 2005 purchased both systems with Intel chips? Identify any assumptions that you make and offer some justification.

(b) Four customers in 2005 all buy HP laptops. If the customers decide independently, what is the probability that at least three of the laptops have Intel chips?

(c) Compare these changes from 2004 to 2005: the change in the probability that at least one of four customers picks an Intel-based HP laptop and the change in the probability of one customer picking an Intel-based HP laptop.

53. Text messaging is more common in Asia than in the United States. An article in the news reports the following counts of text messages that were sent in four countries in the third quarter of 2005:[13]

Country	Text Messages	Population
China	76 billion	1306 million
Philippines	21 billion	88 million
United States	19 billion	296 million
United Kingdom	8 billion	60 million

(a) If we select a message from these at random, what is the probability of getting a message sent in Asia?

(b) If we pick three messages from these at random, what is the probability that none of them were sent in the United States or the United Kingdom?

(c) Suppose messages are sent randomly to the people in each country. That is, suppose that the 76 billion messages sent in China had been randomly delivered to the 1.3 billion people in China. What's the probability that a person does not get a message?

54. A study reported in The *New England Journal of Medicine* revealed surprisingly large differences in rates of lung cancer among smokers of different races.[14] For each group, the study reported a rate among smokers per 100,000.

Race	Male	Female
Black	264	161
White	158	134
Japanese-American	121	50
Latino	79	47

(a) What is the probability that a black male smoker develops lung cancer?

(b) What is the probability that at least one of four Japanese-American women who smoke develops cancer? Do you need any assumptions for this calculation?

(c) If the four women were from the same family, would you question any assumptions used in answering the previous item?

55. To boost interest in its big sale, a retailer offers special scratch-off coupons. When the shopper makes a purchase, the clerk scratches off the covering to reveal the amount of the discount. The discount is 10%, 20%, 30%, 40%, or 50% of the initial amount. Most of the

---

[12] "Call Centers in the Rec Room," *Business Week Online*, January 23, 2006.

[13] *New York Times Magazine*, January 21, 2006. The population estimates are from the *CIA World Factbook* for July 2005.

[14] C. A. Haiman, D. O. Stram, L. R. Wilkens, M. C. Pike, L. N. Kolonel, B. E. Henderson, and L. Le Marchand (2006), "Ethnic and Racial Differences in the Smoking-Related Risk of Lung Cancer," *New England Journal of Medicine*, 354, 333–342.

coupons are for 10% off, with fewer as the amount of the discount increases. Half of the coupons give the shopper 10% off, 1/8 give 20% off, 1/8 give 30% off, 3/32 give 40% off, and 1/32 give 50% off.

(a) What are your chances of getting *more* than 30% off of your purchase?

(b) A clerk was surprised—and suspicious—when three shoppers in a row appeared with coupons that gave them half off. Should he have been suspicious?

(c) Half of the customers at a register purchase a sweater that retails for $50 and another half purchase a suit that retails for $200. What is the probability that a customer saves *more than* $20 by using one of these coupons? Be clear about any assumptions you need to make.

56. A fast-food chain randomly attaches coupons for prizes to the packages used to serve French fries. Most of the coupons say "Play again," but a few are winners. Seventy-five percent of the coupons pay nothing, with the rest evenly divided between "Win a free order of fries" and "Win a free sundae."

(a) If each member of a family of three orders fries with her or his meal, what is the probability that someone in the family is a winner?

(b) What is the probability that one member of the family gets a free order of fries and another gets the sundae? The third wins nothing.

(c) The fries normally cost $1 and the sundae $2. What are the chances of the family winning $5 or more in prizes?

57. In an NBA basketball game on January 22, 2006, Kobe Bryant of the Los Angeles Lakers scored 81 points, a total score second only to a 100-point performance by Wilt Chamberlain in 1962. The data file [kobe] contains the sequence of his attempts, with 0 indicating a miss, 2 for a made regular basket, and 3 for a 3-point basket.

(a) Does it seem to you that Kobe had a hot hand in this game, or do these data resemble a sequence of independent events?

(b) Kobe missed five out of his last six shots. Earlier in the third quarter, he made seven in a row. Explain why this difference does not necessarily mean that he cooled off at the end of the game.

4M Odds: Where's the rest of the probability?

1	2	3	4	5	6	7	9	8
MASTER DAVID 20-1	PUROE 5-1	CAIMAN 50-1	BIRDSTONE 15-1	ROCK HARD TEN 8-1	ROYAL ASSAULT 20-1	TAP DANCER 50-1	EDDINGTON 10-1	SMARTY JONES 5-2

Race tracks tell the odds for each horse winning a race rather than the probability. Bettors at the track like to see the odds to know what sort of payoff they'll get if a horse wins. Here are the odds for the 2004 Belmont Stakes.

Odds at a racetrack tell you right away the payoff that you earn if you pick the winning horse. For example, the odds for the Belmont Stakes mean that a $1 bet on Master David, horse number 1, wins you $20 if Master David wins the race.

Odds are equivalent to probabilities. If you know the probabilities, you can get the odds and vice versa. The odds for an event **E** are the ratio of the probability that it happens to the probability that it does not happen. We can use the Rule for Complements to get the probability of the event not happening.

$$Odds(\mathbf{E}) = \frac{P(\mathbf{E})}{1 - P(\mathbf{E})}$$

If you know the odds, you can get the probability from this formula (which comes from solving the previous formula for the probability).

$$P(\mathbf{E}) = \frac{Odds(\mathbf{E})}{1 + Odds(\mathbf{E})}$$

The odds shown at the track are the odds against a horse winning. For example, the odds shown for Master David are defined as

$$Odds(\text{Master David}) = \frac{P(\text{Master David loses})}{1 - P(\text{Master David loses})}$$

Motivation

a) Odds might be useful for wagering, but probabilities are handy as well. Would you rather know the probability or the odds?

b) At a popular race like the Belmont Stakes, it is likely that wagers are placed on every horse in the race. If we solve for the probabilities of every horse, what should be the sum of these probabilities, according to the rules of probability?

Method

c) If the formula gives the probability of a horse losing, then what rules can you use to find the probability of a horse winning the race?

Mechanics

d) How large can the odds against a horse be? How small?

e) Find the probabilities for each horse winning the race. Do these probabilities provide a proper description of all the possibilities?

Message

f) Interpret the results of your analysis of the probabilities. Does the mismatch between theory and practice make sense in this context?

4M Auditing a Business

Many businesses use auditors to check over their operations. With thousands of transactions, it can be easy for a crook to slip in a few bogus checks. Sales fraud occurs when managers whose pay is based on performance inflate reported sales figures in order to get a bonus or higher salary. One way to do this is to shift the dates of sales. For example, to reach a sales target for last year, a manager could change a date to make it appear as though something that was just sold yesterday got sold last year. A second type of fraud is inventory fraud. This happens when an employee overstates the level of inventory to conceal off-book transactions, sales that were made but not reported.

An auditor can spot sales fraud by contacting clients and confirming dates of transactions and their account balances. The auditor can inspect the inventory to see what is actually on hand and compare this count to the claimed levels. This is not possible, however, for a large firm. For a corporation that conducts thousands of transactions, an auditor cannot check every item. At best the auditor can check a sample of sales receipts and inventoried items.

For this exercise, you're the auditor. The directors of a large corporation suspect that some sales receipts have been incorrectly dated and some inventory levels have been overstated.

Motivation

a) Explain why an auditor need only inspect a subset of transactions rather than census every transaction.

Method

b) In order for the probability calculations to be simplified, how should the inventory items be chosen?

c) Suppose the corporation is divided into several sales divisions. The directors would like to simplify the gathering of sales records by picking one division at random. Would this be a good idea?

Mechanics

For these calculations, assume that 2% of sales receipts have been date shifted and 3% of inventory levels have been overstated.

d) What is the probability that a random sample of 25 sales receipts will find none that are fraudulent?

e) What is the probability that a random sample of 25 sales receipts and 25 inventory counts will find at least one incident of fraud?

f) You can afford to audit a total of at most 100 sales receipts and inventory items. Under these conditions, how many of each should you sample in order to maximize the chances of finding at least one fraudulent record? Are you satisfied with this result?

Message

g) Suppose that checking a random sample of 25 sales receipts and 25 inventory counts finds no evidence of fraud. Does this mean that the probability of fraud at this company is zero?

CHAPTER 8

Conditional Probability

HOW DOES EDUCATION AFFECT THE INCOME that you can expect to earn? Table 8.1 summarizes two aspects of the 98 million households surveyed in the 2000 census: the income bracket of the household and the education of the head of the household.

Each cell of Table 8.1 gives a percentage of the entire population, divided into four income brackets and seven levels of education. For instance, the leading cell indicates that 10.51% of all households in the United States earn less than $25,000 and are headed by someone who lacks a high school diploma. We can treat these percentages as probabilities of intersections. If you were to pick a household at random, there's a 0.1051 chance that the household was headed by someone who did not have a high school diploma *and* earned less than $25,000. The shaded margins give the totals for the rows and columns, defining marginal probabilities. Ignoring education, the probability that a randomly chosen household earns less than $25,000 is 0.309.

TABLE 8.1 Household income and education in the United States. (Cells show percentages.)

	A	B	C	D	E	F
1		Household Income Bracket				
2		Poor	Lower	Middle	Upper	Row Total
3	Education	Less than $25,000	$25,000 to $50,000	$50,000 to $75,000	$75,000 and up	
4	No H.S. diploma	10.51	4.26	1.32	0.68	16.77
5	High School Grad	10.93	10.57	5.84	3.87	31.21
6	Some College	6.36	8.36	5.84	5.23	25.79
7	Bachelor's Degree	2.23	4.16	4.15	6.57	17.11
8	Master's Degree	0.62	1.26	1.36	2.84	6.08
9	Professional Degree	0.16	0.22	0.29	0.99	1.66
10	Doctorate	0.09	0.21	0.29	0.8	1.39
11	Column Total	30.9	29.04	19.09	20.98	100

The percentages in Table 8.1 reflect the United States as a whole, but other percentages offer better views of the relationship between education and income. For example, about 21% of all households were in the upper income bracket in 2000. Among households with a professional degree, however, the percentage in the upper bracket is much higher, $0.99/1.66 \approx 60\%$. Similarly, households with a professional degree are rare overall but more common within the upper bracket. These households represent 1.66% of all households but comprise $0.99/20.98 \approx 5\%$ of the upper bracket. These differences imply dependent events.

Percentages computed within rows or columns of the table correspond to conditional probabilities, the topic of this chapter. Conditional probabilities allow us to find probabilities under various conditions, showing how these conditions affect the chances. Conditional probabilities are often quite different from the related marginal probabilities. If the probability depends on the situation, we have to take these conditions into account.

8.1 | FROM TABLES TO PROBABILITIES

Table 8.2 summarizes two categorical variables, *Host* and *Purchase*, that were considered in Chapter 5.

TABLE 8.2 Amazon purchase counts.

		Host			
		MSN	Recipe Source	Yahoo	Total
Purchase	No	6,973	4,282	5,848	**17,103**
	Yes	285	1	230	**516**
	Total	**7,258**	**4,283**	**6,078**	**17,619**

Rather than think of these counts as summaries of past events, we can use them to define probabilities. Think about the *next* visitor to Amazon from one of these hosts. If the next visitor behaves like a random choice from the 17,619 cases in Table 8.2, then we can use these counts to define probabilities. It's a big assumption to pretend that the next person behaves just like those who came before, but a common way to associate counts with probabilities. Table 8.3 divides each count in Table 8.2 by 17,619 and rounds the result to three decimals.

TABLE 8.3 This table converts the counts into fractions that we interpret as probabilities.

		Host			
		MSN	Recipe Source	Yahoo	Total
Purchase	No	0.396	0.243	0.332	**0.971**
	Yes	0.016	0.000	0.013	**0.029**
	Total	**0.412**	**0.243**	**0.345**	**1**

The resulting sample space has six outcomes, one for each of the six cells in Table 8.3.

{No *and* MSN}   {No *and* RecipeSource}   {No *and* Yahoo}
{Yes *and* MSN}   {Yes *and* RecipeSource}   {Yes *and* Yahoo}

These outcomes are not equally likely. The most common outcome is {No *and* MSN}, which occurs with probability 0.396. The least common outcome, {Yes *and* RecipeSource}, has virtually *zero* probability (1/17,619).

Joint Probability

joint probability Probability of an outcome with two or more attributes, as found in the cells of a table; probability of an intersection.

Each of the outcomes defined by the cells in Table 8.3 describes two attributes of a visit, the host and whether a purchase will be made. Because the probabilities in the cells of the table describe a combination of attributes, these probabilities are called **joint probabilities**. A joint probability gives the chance for an outcome having two or more attributes. In the language of Chapter 7, a joint probability is the probability of an intersection of two or more events. For example, define the event **Y** as visits that will result in a purchase, and define the event **M** as visits from MSN. The first cell in the second row of Table 8.3 tells us that the joint probability is $P(\mathbf{Y}\ and\ \mathbf{M}) = 0.016$.

Marginal Probability

marginal probability Probability that takes account of one attribute of the event; found in margins of a table.

The probabilities in the six cells of Table 8.3 give the chances for a visit with two attributes, the host and whether a purchase was made. If we care only about one attribute, we need marginal probabilities. A **marginal probability** is the probability of observing an outcome with a single attribute, regardless of its other attributes. Typically, marginal probabilities are linked to rows and columns of a table and displayed in the margins, as in Table 8.3. This common positioning is the origin of the name *marginal probability*.

As an example, what is the probability that the next visitor will make a purchase? This is the probability of the event **Y** defined in the prior section.

$$\mathbf{Y} = \{\{\text{Yes}\ and\ \text{MSN}\}\ or\ \{\text{Yes}\ and\ \text{RecipeSource}\}\ or\ \{\text{Yes}\ and\ \text{Yahoo}\}\}$$

These outcomes have the form {Yes *and* *}. We write * because we don't care which host sends the visitor; any of them will do. We care about the first attribute of the outcome, whether a purchase will be made. These outcomes form the second row of Table 8.4 (yellow).

TABLE 8.4 The second row identifies visits that result in a purchase, the marginal event **Y**.

		Host		
		MSN	**Recipe Source**	**Yahoo**
Purchase	**No**	0.396	0.243	0.332
	Yes	**0.016**	**0.000**	**0.013**

The probability of a purchase is the sum

$$P(\mathbf{Y}) = P(\{\text{Yes}\ and\ \text{MSN}\}\ or\ \{\text{Yes}\ and\ \text{RecipeSource}\}\ or\ \{\text{Yes}\ and\ \text{Yahoo}\})$$
$$= 0.016 + 0.000 + 0.013 = 0.029$$

The probabilities add because the cells define disjoint events. Though it may seem like common sense, we are using the Addition Rule for Disjoint Events defined in Chapter 7.

Analogously, the probability that the next visit comes from MSN is another marginal probability. The event $\mathbf{M} = \{*, \text{MSN}\}$ is represented by the first column of Table 8.5 (blue).

TABLE 8.5 The MSN column defines the marginal event **M**.

		Host		
		MSN	**Recipe Source**	**Yahoo**
Purchase	**No**	**0.396**	0.243	0.332
	Yes	**0.016**	0.000	0.013

The marginal probability of a visit from MSN is the sum of the probabilities in this column.

$$P(\mathbf{M}) = P(\{\text{No and MSN}\} \text{ or } \{\text{Yes and MSN}\})$$
$$= 0.396 + 0.016 = 0.412$$

These probabilities also add because the cells define disjoint events.

Conditional Probability

Table 8.3 shows the joint and marginal probabilities, but these do not directly answer an important question: Which host will deliver the best visitors, those who are more likely to make purchases? The answer to this question requires conditional probabilities.

When working with a table of probabilities, we obtain a conditional probability when we restrict the sample space to a particular row or column. To find the proportion of visitors from MSN who make purchases, we condition on the first column of the table. That is, we restrict the sample space to this one column, as in Table 8.6.

TABLE 8.6 Conditioning on the event that the visit comes from MSN limits the sample space to the first column of the table.

		Host		
		MSN	**Recipe Source**	**Yahoo**
Purchase	**No**	0.396	0.243	0.332
	Yes	0.016	0.000	0.013

This restriction makes the rest of the sample space irrelevant. It's as though we know that $\mathbf{M} = \{*, \text{MSN}\}$ must occur. Conditioning on $\mathbf{M}$ means that we confine our attention to outcomes in $\mathbf{M}$, {No and MSN} and {Yes and MSN}. Within this new sample space, what's the probability of {Yes and MSN}? If we use the joint probability 0.016, we violate Rule 1 of probability: The probabilities of the outcomes in the new sample space sum to 0.412, not 1. The solution is to scale up the probabilities so that they do sum to 1, as in Table 8.7.

TABLE 8.7 Conditional probabilities must add up to 1.

		Host		
		MSN	**Recipe Source**	**Yahoo**
Purchase	**No**	$0.396/0.412 \approx 0.961$	0.243	0.332
	Yes	$0.016/0.412 \approx 0.039$	0.000	0.013

Among visitors from MSN, the chance of a purchase is $0.016/0.412 \approx 0.039$. This ratio is the **conditional probability** of a purchase (the event $\mathbf{Y}$) given that the visitor comes from MSN (the event $\mathbf{M}$). Symbolically, the conditional probability of $\mathbf{Y}$ given $\mathbf{M}$ is

$$P(\mathbf{Y}|\mathbf{M}) = \frac{P(\mathbf{Y} \text{ and } \mathbf{M})}{P(\mathbf{M})} = \frac{0.016}{0.412} \approx 0.039$$

conditional probability The conditional probability of $\mathbf{A}$ given $\mathbf{B}$ is

$$P(\mathbf{A}|\mathbf{B}) = P(\mathbf{A} \text{ and } \mathbf{B})/P(\mathbf{B})$$

Conditional probabilities in a table refer to proportions within a row or column.

tip

The symbol | in $P(\mathbf{Y}|\mathbf{M})$ means "given." **The phrases "given that," "conditional on," or "if it is known that" signal conditional probabilities.** In general, the conditional probability of the event $\mathbf{A}$ given that the event $\mathbf{B}$ occurs is

$$P(\mathbf{A}|\mathbf{B}) = \frac{P(\mathbf{A} \text{ and } \mathbf{B})}{P(\mathbf{B})}$$

Suppose that the event $\mathbf{R} = \{*, \text{RecipeSource}\}$ is known to occur (a visit from Recipe Source, the second column of the table). What is $P(\mathbf{Y}|\mathbf{R})$, the probability of a purchase given that the visit comes from RecipeSource? By conditioning on $\mathbf{R}$, we move our focus from the first column of the table to the second column, as shown in Table 8.8.

TABLE 8.8 Conditioning on the event that the visit comes from RecipeSource (to four decimals).

		Host		
		MSN	**Recipe Source**	**Yahoo**
Purchase	**No**	0.3958	0.2430	0.3317
	Yes	0.0162	0.0001	0.0131

The conditional probability of a purchase within this column of the table is zero unless we carry out the calculations to the fourth decimal place.

$$P(\mathbf{Y}|\mathbf{R}) = \frac{P(\mathbf{Y}\,and\,\mathbf{R})}{P(\mathbf{R})} = \frac{0.0001}{0.2430} \approx 0.0004$$

In Chapter 5, we learned that *Host* and *Purchase* are associated. That's important for advertising because it means that some sites deliver a higher percentage of purchasing customers. The association between *Host* and *Purchase* is evident in the conditional probabilities of a purchase. To three decimals, these are (with $\mathbf{H}$ denoting the visits from Yahoo)

$$P(\mathbf{Y}|\mathbf{M}) = 0.039, \quad P(\mathbf{Y}|\mathbf{R}) = 0.000, \quad \text{and} \quad P(\mathbf{Y}|\mathbf{H}) = 0.038$$

Purchases occur at a much higher rate among visits from MSN or Yahoo than from RecipeSource. Rather than having a one-size-fits-all probability that applies to visits from all hosts, the probability of a purchase depends on the host.

What Do You Think?   Conditional probabilities apply to rows as well as columns.

a. In words, describe the probability $P(\mathbf{M}|\mathbf{Y})$.[1]
b. Which row or column defines the sample space for $P(\mathbf{M}|\mathbf{Y})$?[2]
c. What is $P(\mathbf{M}|\mathbf{Y})$?[3]

8.2 | DEPENDENT EVENTS

Chapter 7 defines two events $\mathbf{A}$ and $\mathbf{B}$ to be *independent* if the probability that both occur (their intersection) is the product of the probabilities of each taken separately.

$$P(\mathbf{A}\,and\,\mathbf{B}) = P(\mathbf{A}) \times P(\mathbf{B})$$

Independence makes it easy to find the probability of a combination of events because we can treat the events one at a time, multiplying their probabilities. Many events in business, however, are intentionally *dependent*. For instance, suppose event $\mathbf{A}$ identifies customers who see an advertisement for a

[1] Among purchases (or given a purchase occurs), $P(\mathbf{M}|\mathbf{Y})$ is the probability that the visit comes from MSN.
[2] By conditioning on $\mathbf{Y}$, the second row of the table (the Yes row) becomes the new sample space.
[3] $P(\mathbf{M}|\mathbf{Y}) = P(\mathbf{M}\,and\,\mathbf{Y})/P(\mathbf{Y}) = 0.016/(0.016 + 0.000 + 0.013) \approx 0.55$. More than half of the purchases from these three hosts come from MSN.

personal service and event **B** identifies customers who purchase the service. If these events are independent, then the conditional probability of a purchase given that the customer has seen the advertisement is

$$P(\mathbf{B}|\mathbf{A}) = \frac{P(\mathbf{A}\,and\,\mathbf{B})}{P(\mathbf{A})} = \frac{P(\mathbf{A})P(\mathbf{B})}{P(\mathbf{A})} = P(\mathbf{B})$$

In words, independence implies that seeing the ad has no effect on the chance that a customer will purchase the service. The managers who decided to show this ad expected dependence: They expected the chance for a purchase to *depend* on whether the customer saw the ad.

The events **Y** and **M** in the Web-hosting example are dependent. The product of the marginal probabilities does not match the probability of the intersection.

$$P(\mathbf{Y}) \times P(\mathbf{M}) = 0.029 \times 0.412 \approx 0.012 \quad \neq \quad P(\mathbf{Y}\,and\,\mathbf{M}) = 0.016$$

The probability of the intersection is 33% larger than implied by independence. Similarly, other events in Table 8.3 are also dependent. For instance, $P(\mathbf{Y}) \times P(\mathbf{R}) \neq P(\mathbf{Y}\,and\,\mathbf{R})$ and $P(\mathbf{Y}) \times P(\mathbf{H}) \neq P(\mathbf{Y}\,and\,\mathbf{H})$.

dependent events Events that are not independent, indicated by $P(\mathbf{A}\,and\,\mathbf{B}) \neq P(\mathbf{A}) \times P(\mathbf{B})$ or $P(\mathbf{A}) \neq P(\mathbf{A}|\mathbf{B})$.

Another way to recognize **dependent events** is to compare the conditional probability $P(\mathbf{A}|\mathbf{B})$ to the marginal probability $P(\mathbf{A})$. If **A** and **B** are independent, then $P(\mathbf{A}) = P(\mathbf{A}|\mathbf{B})$ and $P(\mathbf{B}) = P(\mathbf{B}|\mathbf{A})$. Independence means that the chance for **A** is the same whether or not **B** occurs, and the chance for **B** is the same whether or not **A** occurs. For the events **Y** and **M**, the marginal probability of a purchase is $P(\mathbf{Y}) = 0.029$, whereas the conditional probability of a purchase by a visitor from MSN is $P(\mathbf{Y}|\mathbf{M}) = 0.039$. Purchases are more likely among visits from MSN than in general. Hence, **Y** and **M** are dependent.

The Multiplication Rule

We multiply marginal probabilities of independent events to find the probability of intersecting events. If the events are dependent, we have to use conditional probabilities. If we rearrange the definition of the conditional probability $P(\mathbf{A}|\mathbf{B})$, we obtain a multiplication rule for intersections that applies in general.

$$P(\mathbf{A}|\mathbf{B}) = \frac{P(\mathbf{A}\,and\,\mathbf{B})}{P(\mathbf{B})} \Rightarrow P(\mathbf{A}\,and\,\mathbf{B}) = P(\mathbf{B}) \times P(\mathbf{A}|\mathbf{B})$$

We can also write this with the events **A** and **B** in the other order.

$$P(\mathbf{B}|\mathbf{A}) = \frac{P(\mathbf{A}\,and\,\mathbf{B})}{P(\mathbf{A})} \Rightarrow P(\mathbf{A}\,and\,\mathbf{B}) = P(\mathbf{A}) \times P(\mathbf{B}|\mathbf{A})$$

These expressions form the **Multiplication Rule**.

Multiplication Rule   The joint probability of two events **A** and **B** is the product of the marginal probability of one times the conditional probability of the other.

$$P(\mathbf{A}\text{ and }\mathbf{B}) = P(\mathbf{A}) \times P(\mathbf{B}|\mathbf{A})$$
$$= P(\mathbf{B}) \times P(\mathbf{A}|\mathbf{B})$$

The probability that events **A** and **B** *both* occur is the probability of **A** times the probability of **B** *given* that **A** occurs (or the probability of **B** times the probability of **A** given that **B** occurs).

Because it is so easy, it is common to see a joint probability calculated as the product of marginal probabilities, as if the events were independent.

For example, a key component of credit risk is the probability that the borrower defaults. When a bank loans money, it needs to know the chance that the borrower will not repay the loan. The higher the chance of default, the higher the risk and the higher the rate of interest charged by the bank. Consider three loans, with probabilities of default given by

$$P(\text{Loan 1 defaults}) = p_1, P(\text{Loan 2 defaults}) = p_2, P(\text{Loan 3 defaults}) = p_3$$

Assigning a probability of default to each loan is only the beginning. What is the probability that all three default? If the outcomes are independent, the joint probability is the product

$$P(\text{Loan 1 defaults } and \text{ Loan 2 defaults } and \text{ Loan 3 defaults}) = p_1 p_2 p_3$$

Is this correct, or might the failure of one borrower signal problems for others as well? Suppose, for instance, that these borrowers are companies that operate similar lines of business, such as three companies that supply auto parts to Ford. In that case, problems at Ford affect all three simultaneously, making the outcomes dependent. The probability that all three default might be much larger than the product $p_1 p_2 p_3$. **Anytime you see unconditional probabilities multiplied together, stop and ask whether the events are independent.**

Order Matters

The order in which events are listed does not matter in unions or intersections. The union **A** *or* **B** is the same as the union **B** *or* **A**, and the intersection **A** *and* **B** is the same as **B** *and* **A**. For conditional probabilities, however, order matters. The conditional probability $P(\mathbf{A}|\mathbf{B})$ usually differs from $P(\mathbf{B}|\mathbf{A})$.

For example, compare the conditional probabilities $P(\mathbf{Y}|\mathbf{M})$ and $P(\mathbf{M}|\mathbf{Y})$ in the Web-hosting example. $P(\mathbf{Y}|\mathbf{M})$ is the probability of a purchase given that the visit comes from MSN. In contrast, $P(\mathbf{M}|\mathbf{Y})$ is the probability of selecting a visit from MSN given that a purchase occurs. Both conditional probabilities involve purchases and MSN, but that's where the similarity ends. $P(\mathbf{Y}|\mathbf{M})$ restricts the outcomes to the first column of Table 8.3, whereas $P(\mathbf{M}|\mathbf{Y})$ restricts the outcomes to the second row. There's no reason for the chance of a purchase within the first column to be the same as the chance for a visit from MSN within the second row. They are indeed very different. Visits from MSN are common among purchases,

$$P(\mathbf{M}|\mathbf{Y}) = \frac{P(\mathbf{Y} \, and \, \mathbf{M})}{P(\mathbf{Y})} = \frac{P(\{\text{Yes } and \text{ MSN}\})}{P(\{\text{Yes } and \, *\})} = \frac{0.016}{0.029} \approx 0.552$$

whereas purchases are rare among visits from MSN, $P(\mathbf{Y}|\mathbf{M}) = 0.039$.

Independence in Venn Diagrams

What does independence look like in a Venn diagram? Think about it, and then draw a Venn diagram with two independent events.

caution A common, but incorrect, choice is to draw a Venn diagram with disjoint events. Disjoint (mutually exclusive) events are *never* independent. The conditional probability of one disjoint event given the other occurs is zero. If **A** and **B** are disjoint events, then

$$P(\mathbf{A}|\mathbf{B}) = \frac{P(\mathbf{A} \, and \, \mathbf{B})}{P(\mathbf{B})} = \frac{0}{P(\mathbf{B})} = 0 \neq P(\mathbf{A})$$

Since $P(\mathbf{A}) \neq P(\mathbf{A}|\mathbf{B})$, disjoint events are dependent.

How *does* the Venn diagram of independent events look? We'll use a mosaic plot (Chapter 5) to illustrate, shown in Figure 8.1.

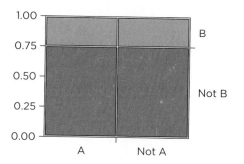

FIGURE 8.1. Mosaic plot of independent events.

Assume the total colored area of the mosaic plot is 1, serving as the sample space **S**. The entire left column (both red and blue portions) is **A**. The area in the left column is $P(\mathbf{A}) = 1/2$. The blue slice across the top of the figure identifies the event **B**, and $P(\mathbf{B}) = 1/4$. The blue slice comprises 25% of each column, meaning that $P(\mathbf{B}|\mathbf{A}) = P(\mathbf{B}|\mathbf{A}^c) = 1/4$. (The scale on the left of the plot shows the conditional probability of $\mathbf{B}^c$ within the columns.) Since these conditional probabilities match the marginal probability, **A** and **B** are independent events. Aligned, horizontal rows of tiles in a mosaic plot indicate independence.

It is easy to confuse independent events with disjoint events because

If **A** and **B** are disjoint, $P(\mathbf{A} \ or \ \mathbf{B}) = P(\mathbf{A}) + P(\mathbf{B})$.

If **A** and **B** are independent, $P(\mathbf{A} \ and \ \mathbf{B}) = P(\mathbf{A}) \times P(\mathbf{B})$.

tip **For disjoint events, probabilities add. For independent events, marginal probabilities multiply.** Try not to confuse when to add and when to multiply. Sums occur when you are looking for the probability of one event or the other. Products occur when you are looking for the probability that *both* events occur.

What Do You Think?

a. Which is larger, the probability that a customer at a department store makes a purchase or the probability that a customer who is known to have a store coupon makes a purchase? Which is a marginal probability and which is a conditional probability?[4]

b. Which is larger, the probability of a fall in the stock market or the probability of a fall in the stock market and a recession?[5]

c. Officials believe that the probability of a major failure at a nuclear plant is so small that we should not expect a failure for hundreds of years. However, two such failures occurred within about a decade of each other—Chernobyl and Three Mile Island. How could the estimates be wrong?[6]

---

[4] The probability of a customer making a purchase is probably less than the *conditional* probability of a customer with a coupon making a purchase. Given that the customer has gone to the trouble to bring a coupon, chances are that the customer has a purchase in mind. $P(\text{purchase})$ is marginal, $P(\text{purchase} \mid \text{coupon})$ is the conditional probability.

[5] Let's use set notation. Let **D** denote a drop in stocks and **R** denote a recession. Then it must be the case that $P(\mathbf{D}) > P(\mathbf{D} \ and \ \mathbf{R})$. The joint probability must be less than either marginal probability.

[6] A simple explanation lies in assuming independence of redundant safety components. Each component may be reliable, with little chance of failing. If we calculate $P(C_1 \text{ fails } and \ C_2 \text{ fails } and \dots)$ by multiplying small probabilities, the product gets very small. If failures are dependent, however, the probability may be much larger. For example, the components may share a common defect. Or, perhaps, the failure of one may lead to a cascade of failures.

8.3 | ORGANIZING PROBABILITIES

In order to find a probability that involves several events, it is useful to organize the information into either a tree or a table. These schemes help you see the information and keep track of the various pieces.

Probability Trees

Pictures help us understand data. They're also an important way to understand probabilities. Venn diagrams help visualize basic manipulations, such as unions or complements. Probability trees organize the conditional probabilities that describe a sequence of events.

Consider the success of advertisements on TV. Advertisers spend billions on television advertising. This money gets the ads broadcast, but does anyone watch? An episode of a popular television program, such as *Desperate Housewives*, might be seen in 25 million households, but do viewers watch the ads? Because a 30-second spot for a top-rated show goes for as much as $600,000, sponsors care.

If we randomly pick a viewer on a Sunday evening in the fall, let's suppose that there's a 15% chance that we catch her watching *60 Minutes*. To limit the possibilities, let's assume that there are only two other possible programs at the same time. There's a 50% chance that we catch her watching a football game and a 35% chance that she is watching *Desperate Housewives*. Of those watching *60 Minutes*, assume that 10% of the viewers use a digital video recorder to skip the ads. Half of those watching the game skip the ads, and 80% of those watching *Desperate Housewives* skip ads. What's the probability, based on these percentages, that a randomly chosen viewer watched the game *and* saw the ads?

To start, the probability of selecting someone watching the game is one-half. Now we can use the Multiplication Rule,

$$P(\text{Watch game } and \text{ See ads}) = P(\text{Watch game}) \times P(\text{See ads} \mid \text{Watch game})$$

$$= \frac{1}{2} \times \frac{1}{2} = \frac{1}{4}$$

probability tree (tree diagram) A graphical depiction of conditional probabilities that shows the probabilities as paths that join a sequence of labeled events.

That's quick and easy. For bigger problems, a probability tree organizes the information in a way that avoids formulas. A **probability tree** (or **tree diagram**) shows sequences of events as paths that suggest branches of a tree. The number of paths can grow large, so we usually draw the tree sideways, starting from the left and growing vinelike across the page. You might also see trees drawn from the bottom up or top down. Similar decision trees are used to organize a sequence of management choices.

The first branch of the probability tree for this example separates viewers of each program. That's in keeping with the information stated in the problem. We are given the percentage watching each program. Each of the initial branches takes us to a possible program. The labels on the branches show the probabilities. The branches must include *all* possible outcomes, so the probabilities that label the initial branches add to 1.

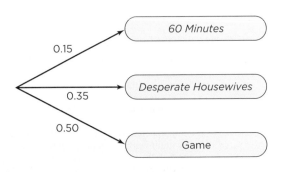

The tree grows when we show whether a viewer watches the advertisements. For each program, we add branches that identify the alternatives; *conditional* probabilities label the paths. As with the initial branches, the conditional probabilities leaving each node add to 1.

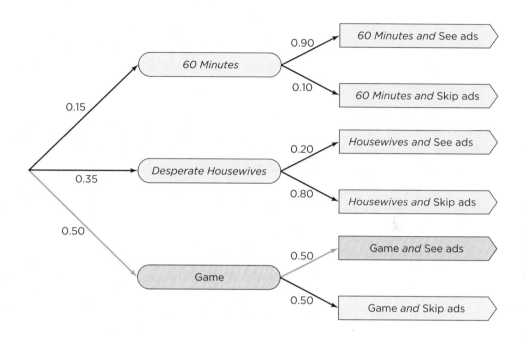

Let's use this tree to find the probability of picking a viewer that watches the game *and* sees the ads. The golden node identifies these viewers. To find the probability of this combination, follow the branches and multiply the probabilities along the way.

$$P(\text{Watch game } and \text{ See ads}) = 0.50 \times 0.50 = 0.25$$

This is exactly the calculation of this probability we got from the Multiplication Rule, only now it is laid out in a tree.

The events at the far right of the preceding probability tree are disjoint. These final nodes, or *leaves* as they are sometimes called, identify *all* the possibilities. Hence, the probabilities of the leaves of the tree sum to 1.

$$0.15 \times 0.90 + 0.15 \times 0.10 + 0.35 \times 0.20 + 0.35 \times 0.80$$
$$+ 0.50 \times 0.50 + 0.50 \times 0.50 = 1$$

We can use these final outcomes to find other probabilities. For example, what's the probability that a randomly chosen viewer sees an ad? For this, add the probabilities for each disjoint event on the far right that identifies viewers who see ads. Because the final outcomes are disjoint, we can add their probabilities. There are three of these, and the sum of the probabilities is

$$P(\text{See ads}) = 0.15 \times 0.90 + 0.35 \times 0.20 + 0.50 \times 0.50 = 0.455$$

The probability that a randomly chosen viewer sees the ads is a bit less than one-half.

Probability Tables

A probability tree works well when, as in the prior example, the context puts the information in order. We're given the program that the viewer is watching, and then the chances that a viewer of each program watches ads. So long as your questions align with this order, trees are effective. Otherwise, you have to add the probabilities for many leaves. Tables, on the other hand, are insensitive to the sequencing, plus they imbed the probability rules. Rules like $P(\mathbf{A}) = 1 - P(\mathbf{A}^c)$ become automatic in a table, making it easy to check your calculations.

For comparison, let's fill in a probability table with the information we just presented as a tree. We have three programs: *60 Minutes, Desperate Housewives,* and the game. These choices label the columns of Table 8.9; the rows identify whether the viewer sees the ads. (You could arrange it the other way; this way fits better on our page.)

TABLE 8.9 Initial layout for probability table.

	60 Minutes	*Desperate Housewives*	**Game**	
See ads				
Skip ads				

Now let's fill in the table. Cells inside Table 8.9 are meant to show joint probabilities of row and column events. The margins shaded at the right and bottom will show the marginal probabilities. We will start with the given information. The initial description gives the probability that a viewer watches each program. These are marginal probabilities and go in the bottom margin of Table 8.10.

TABLE 8.10 Probability table with marginal probabilities for the columns.

	60 Minutes	*Desperate Housewives*	**Game**	
See ads				
Skip ads				
	0.15	**0.35**	**0.50**	**1**

The 1 in the lower right corner is the sum of the three marginal probabilities. Tables are self-diagnosing: If we leave something out, the probabilities in the table do not sum to 1 and we are warned that we've made a mistake.

Now let's add the information about seeing ads. The description gives conditional probabilities of seeing ads among those watching each program. Of those watching *60 Minutes,* the conditional probability of seeing ads is 0.90. That's 90% of the probability in the first column, or $0.15 \times 0.90 = 0.135$. Of those watching *Desperate Housewives,* 20% of the probability in the second column goes to those who watch ads ($0.35 \times 0.20 = 0.07$). For the third column, half see the ads. With these cells filled, we get Table 8.11.

TABLE 8.11 The conditional probabilities determine the first row of joint probabilities.

	60 Minutes	*Desperate Housewives*	**Game**	
See ads	0.135	0.07	0.25	
Skip ads				
	0.15	**0.35**	**0.50**	**1**

With the first row done, it is straightforward to fill in the rest. The joint probabilities in each column and row must sum to the associated marginal total. For example, P(*Desperate Housewives and* Skip ads) $= 0.35 - 0.07 = 0.28$. Table 8.12 is the finished table, after we have filled in all of the cells and margins.

TABLE 8.12 Completed probability table.

	60 Minutes	*Desperate Housewives*	Game	Total
See ads	0.135	0.07	0.25	**0.455**
Skip ads	0.015	0.28	0.25	**0.545**
Total	**0.15**	**0.35**	**0.50**	**1**

Once we have filled in the table, we can answer any question about the marginal, conditional, and joint probabilities. For example, the marginal probability P(See ads) $= 0.455$. Plus, if we cannot completely fill in the table, then we are missing some information.

8.4 | ORDER IN CONDITIONAL PROBABILITIES

If we know a viewer sees the ads, then what is the probability that the viewer is watching *Desperate Housewives*? That's an interesting question because the answer describes those who see ads. Among those who watch ads, which program is most popular? Marginally, the game is the largest draw; it pulls in one-half of the audience. Does it also provide the largest audience for ads?

The tree we built is not organized to answer this question easily. Trees work well if the question we need to answer matches the direction of the tree; this one does not. The tree shows us conditional probabilities such as P(See ads | *Desperate Housewives*). It's not so easy to find P(*Desperate Housewives* | See ads). Those who watch ads are scattered among the branches.

This task is easier with the completed probability table. To find any conditional probability, divide the probability of the intersection of the events (the joint probability of the cell) by the marginal probability of the conditioning event. Visually, to find P(*Desperate Housewives* | See ads), we need to focus on the first row of the table, as in Table 8.13.

TABLE 8.13 Only the first row is relevant for the conditional probability P(Program | See ads).

	60 Minutes	*Desperate Housewives*	Game	Total
See ads	0.135	0.07	0.25	0.455
Skip ads	**0.015**	**0.28**	**0.25**	**0.545**
Total	**0.15**	**0.35**	**0.50**	**1**

The conditional probability rescales the joint probabilities in the first row so that the sum of the conditional probabilities is 1. For example,

$$P(\textit{Desperate Housewives} \mid \text{See ads}) = \frac{P(\textit{Desperate Housewives and } \text{See ads})}{P(\text{See ads})}$$

$$= \frac{0.07}{0.455} \approx 0.154$$

The conditional probability is the proportion of the probability in the first row (those who see ads) that is in the second column. Similarly,

$$P(60\ Minutes \mid \text{See ads}) = \frac{P(60\ Minutes\ and\ \text{See ads})}{P(\text{See ads})}$$

$$= \frac{0.135}{0.455} \approx 0.297$$

Even though *Desperate Housewives* draws a larger audience overall, *60 Minutes* generates a larger share of those who watch ads. The game draws over half of those who watch ads.

$$P(\text{Game} \mid \text{See ads}) = \frac{P(\text{Game}\ and\ \text{See ads})}{P(\text{See ads})}$$

$$= \frac{0.25}{0.455} \approx 0.549$$

4M EXAMPLE 8.1 | **DIAGNOSTIC TESTING**

MOTIVATION | STATE THE QUESTION

Breast cancer is the most common nonskin malignancy among women in the United States, and it is the second leading cause of death from cancer among women (lung cancer ranks first). A mammogram is a diagnostic procedure designed to quickly detect breast cancer. Such a test can exhibit two kinds of errors. One error is known as a *false positive:* The test incorrectly indicates cancer in a healthy woman. The other error (which is more serious error in this situation) is a *false negative:* The test fails to detect cancer.

The probabilities of these errors are established in clinical trials. In a clinical trial, the test is applied to two groups of people, some known to be healthy and others known to have cancer (through an established test that is more complicated, time consuming, and expensive than the proposed new method). Among women with breast cancer, the probability that mammography detects the cancer is 0.85. Among women without breast cancer, the chance for a negative result is 0.925.

If a mammogram indicates that a 55-year-old Caucasian woman tests positive for breast cancer, then what is the probability that she in fact has breast cancer?

METHOD | DESCRIBE THE DATA AND SELECT AN APPROACH

This problem requires conditional probabilities. *In such problems, be precise in stating the order of conditioning. You do not have to use formal set notation, but you must express the conditioning.* The clinical trials of this mammography exam indicate that

$$P(\text{Test negative} \mid \text{No cancer}) = 0.925$$
$$P(\text{Test positive} \mid \text{Cancer}) = 0.85$$

The motivating question asks us to find $P(\text{Cancer} \mid \text{Test positive})$. A small probability table with two rows and two columns is useful to organize the calculations. The test outcomes define the columns, and the presence or absence of cancer defines the rows.

It may not be obvious yet, but we need one more probability: the overall presence of this cancer in the population. In this case, epidemiologists claim that $P(\text{Cancer}) = 0.003$ in Caucasian women of this age.

DO THE ANALYSIS

The challenge is to fill in the probability table. Then it will be easy to find the needed conditional probability. The first step is to use the probability from the epidemiologists to fill in the marginal probabilities of the rows.

	Positive	Negative	Total
Cancer			**0.003**
No cancer			**0.997**
Total			**1**

The results from the clinical trials allocate these marginal probabilities across the columns. The rule $P(\mathbf{A}\,and\,\mathbf{B}) = P(\mathbf{A}) \times P(\mathbf{B}|\mathbf{A})$ implies that

$$P(\text{Cancer}\,and\,\text{Test positive}) = P(\text{Cancer}) \times P(\text{Test positive}|\text{Cancer})$$
$$= 0.0030 \times 0.85 = 0.00255$$

This goes in the first cell of the table. Since the probability in the first row sums to 0.00300, that leaves 0.00045 for the other cell of the first row. For the second row, the probability of a healthy person taking the test and getting a negative result is

$$P(\text{No cancer}\,and\,\text{Test negative}) = P(\text{No cancer}) \times P(\text{Test negative}|\text{No cancer})$$
$$= 0.997 \times 0.925 = 0.922225$$

That leaves $0.997 - 0.922225 = 0.074775$ in the final cell. Here's the finished table.

	Positive	Negative	Total
Cancer	0.00255	0.00045	**0.003**
Healthy	0.074775	0.922225	**0.997**
Total	**0.077325**	**0.922675**	**1**

As a check, the total probabilities for the cells inside the table sum to 1. With the complete table, it's easy to use the definition $P(\mathbf{A}|\mathbf{B}) = P(\mathbf{A}\,and\,\mathbf{B})/P(\mathbf{B})$ to find the needed conditional probability.

$$P(\text{Cancer}|\text{Test positive}) = P(\text{Cancer}\,and\,\text{Test positive})/P(\text{Test positive})$$
$$= 0.00255/0.077325 = 0.033$$

SUMMARIZE THE RESULTS

The test is hardly definitive. The chance that a woman who tests positive actually has cancer is small, a bit more than 3%.

The outcome is less certain than it might seem. With events of low probability, the result of reversing the conditioning can be surprising. That's the reason people who test positive are sent for follow-up testing. They may have only a small chance of actually having cancer. Disease testing can have unexpected consequences if people interpret *testing* positive as *being* positive.

Bayes' Rule

Instead of using a table, we can express the calculation that reverses a conditional probability algebraically. The formula for doing this is known as **Bayes' Rule**. The inputs to Bayes' Rule are the conditional probabilities in the direction *opposite* to the one that you want.

Bayes' Rule   The conditional probability of **A** given **B** can be found from the conditional probability of **B** given **A** by using the formula

$$P(\mathbf{A}\mid\mathbf{B}) = \frac{P(\mathbf{A} \, and \, \mathbf{B})}{P(\mathbf{B})} = \frac{P(\mathbf{B}\mid\mathbf{A}) \times P(\mathbf{A})}{P(\mathbf{B}\mid\mathbf{A}) \times P(\mathbf{A}) + P(\mathbf{B}\mid\mathbf{A}^c) \times P(\mathbf{A}^c)}$$

Although the rule shown here has only two events, it extends to more. The principle remains the same, but the formula gets more complicated.

The formula used in the bottom of the fraction is important as well (though we won't need it). This expression equates the marginal probability $P(\mathbf{B})$ to a weighted average of conditional probabilities: $P(\mathbf{B}) = P(\mathbf{B}\mid\mathbf{A}) \times P(\mathbf{A}) + P(\mathbf{B}\mid\mathbf{A}^c) \times P(\mathbf{A}^c)$.

4M EXAMPLE 8.2 | FILTERING JUNK MAIL

MOTIVATION | STATE THE QUESTION

Junk email consumes system resources and distracts everyone. Is there a way to help workers filter out junk mail from important messages?

A rigid set of fixed rules won't work at most businesses. The type of messages that each individual at a business receives depends upon his or her work at the company. A good filter has to adapt to the type of messages an individual sees during her or his daily routine.

METHOD | DESCRIBE THE DATA AND SELECT AN APPROACH

Bayes' Rule (really, conditional probabilities) makes it possible to build an adaptive system that can filter most junk mail. Spam filters such as Spam Assassin use Bayes' Rule to learn your preferences rather than require you to define a rigid set of rules.

A spam filter based on Bayes' Rule works by having each user train the system. As an employee works through his or her email, he or she identifies messages that are junk. The spam filter then looks for patterns in the words that appear in junk mail. For example, the system might learn conditional probabilities such as

$$P(Nigerian \, general \mid \text{Junk mail}) = 0.20$$

That is, the term *Nigerian general* appears in 20% of the junk mail of an employee. Evidently, this employee is the target of a scam that offers money for helping a "millionaire" in Nigeria move wealth to the United States. Variations on this scam have been going around since email was invented.

For a spam filter to reverse this conditional probability and find $P(\text{Junk mail} \mid Nigerian \, general)$, it tracks how often the term *Nigerian general* shows up in mail that the employee identifies as *not* junk. Let's assume this term is unusual in the normal work of this employee, say $P(\textit{Nigerian general} \mid \text{Not junk mail}) = 0.001$. As in the prior example, we need a marginal probability. Let's assume that half of the employee's email is junk. We'll again use a table to organize the information and reverse the conditional probability rather than use the algebraic formula for Bayes' Rule.

MECHANICS | DO THE ANALYSIS

Here's the completed table. The marginal information tells us that the probability in each column is one-half. The given conditional probabilities allocate the probability within the columns. For example,

$$P(\textit{Nigerian general} \text{ and Junk}) = P(\text{Junk}) \times P(\textit{Nigerian general} \mid \text{Junk})$$
$$= 0.5 \times 0.2 = 0.1$$

	Junk mail	Not junk mail	Total
Nigerian general **appears**	0.1	0.0005	**0.1005**
Nigerian general **does not appear**	0.4	0.4995	**0.8995**
Total	**0.5**	**0.5**	**1**

From the table, we find that

$$P(\text{Junk mail} \mid \textit{Nigerian general}) = 0.1/0.1005 = 0.995$$

The spam filter can then classify messages as junk if it finds this phrase, saving this employee the hassle.

MESSAGE | SUMMARIZE THE RESULTS

Email messages to this employee with the phrase *Nigerian general* have a high probability (more than 99%) of being spam. The system can classify these as junk so that the employee can keep up with the real messages.

An added benefit of a Bayes' filter for spam is the way that it adapts to the habits of each user. The rule we just found for removing spam would not be good for everyone. If an employee managed trade with a Nigerian subsidiary, then this spam filter would be a poor setup. Let's suppose that half of this employee's mail is junk, too. The difference is that the term *Nigerian general* shows up in a lot of regular messages. If the phrase appears in half of the regular messages, then his spam filter builds the probability table shown in Table 8.14.

TABLE 8.14 Alternative probability table for different user.

	Junk mail	Not junk mail	Total
Nigerian general **appears**	0.1	0.25	**0.35**
Nigerian general **does not appear**	0.4	0.25	**0.65**
Total	**0.5**	**0.5**	**1**

Now the system learns

$$P(\text{Junk mail} \mid \textit{Nigerian general}) = 0.1/0.35 = 0.286$$

It would not likely classify messages with this term as junk and would instead leave them in the inbox to be read.

Best Practices

- *Think conditionally.* Many problems that involve probability are more easily understood and solved by using conditional probability. Rather than count in haphazard ways, conditional probability provides a framework for solving complex problems.
- *Presume events are dependent and use the Multiplication Rule.* When finding the probability of two events simultaneously occurring, use the multiplication rule with a conditional probability

$$P(\mathbf{A} \ and \ \mathbf{B}) = P(\mathbf{A}) \times P(\mathbf{B} \mid \mathbf{A})$$

rather than routinely multiplying the marginal probabilities, unless it is known that the events are independent.
- *Use tables to organize probabilities.* There's no need to memorize Bayes' Rule if you organize the information that you are given in a table. Tables provide built-in checks that the probabilities that you have found make sense.
- *Use probability trees for sequences of conditional probabilities.* If the problem has a long sequence of events, particularly three or more, tables get cumbersome.
- *Check that you have included all of the events.* It is easy to lose track of some of the probabilities when several things happen at once. In a probability tree, make sure that the probabilities assigned to the leaves add to 1. In a probability table, check that the probabilities assigned to cells add up to the marginal probabilities and that the total probability assigned to the cells sums to 1.
- *Use Bayes' Rule to reverse the order of conditioning.* In general, the probability of **A** given **B** does not resemble the probability of **B** given **A** or the probability of **A**. Tables and trees can help you see what's happening.

Pitfalls

- *Do not confuse $P(\mathbf{A}|\mathbf{B})$ for $P(\mathbf{B}|\mathbf{A})$.* Be sure to condition on the event that you know has occurred, not the other way around. The order of the events for a conditional probability matters.
- *Don't think that "mutually exclusive" means the same thing as "independent."* It's just the opposite. Mutually exclusive events are disjoint; they have nothing in common. Once you know one has occurred, the other cannot.
- *Do not confuse counts with probabilities.* The difference lies in the *interpretation.* A contingency table describes two columns of observed data. Probabilities describe the chances for events that have yet to occur. You don't need data to come up with a collection of joint probabilities. You can propose any probabilities you like, so long as they satisfy the rules introduced in Chapter 7 (no negative probabilities, probabilities add to 1, and so forth). Unless you connect probabilities to data, however, there's no reason that the probabilities will be useful. You can make up a joint distribution that shows advertising generates large sales, but that's not going to boost sales in real stores.

CHAPTER SUMMARY

A **joint probability** gives the chance for an intersection of events, such as those associated with cells in a table. A **marginal probability** gives the chance of a single event. A **conditional probability** is the chance for an event to occur once it is known that another event has occurred. The **Multiplication Rule** for the probability of two events occurring simultaneously is $P(\mathbf{A} \ and \ \mathbf{B}) = P(\mathbf{A}) \times P(\mathbf{B}|\mathbf{A})$. This rule does not presume independence. For independent events, $P(\mathbf{B}|\mathbf{A}) = P(\mathbf{B})$ and $P(\mathbf{A}|\mathbf{B}) = P(\mathbf{A})$. **Probability trees** and **probability tables** organize calculations involving conditional probabilities. **Bayes' Rule** shows how to get $P(\mathbf{A}|\mathbf{B})$ from $P(\mathbf{B}|\mathbf{A})$ and $P(\mathbf{B}|\mathbf{A}^c)$ without organizing probabilities into a table or tree.

Key Terms

Bayes' Rule, 188
conditional probability, 177
dependent events, 179

joint probability, 176
marginal probability, 176
Multiplication Rule, 179

probability tree (tree diagram), 182

Formulas

Conditional Probability

$$P(\mathbf{A}|\mathbf{B}) = \frac{P(\mathbf{A}\,and\,\mathbf{B})}{P(\mathbf{B})}$$

If $P(\mathbf{B}) = 0$, the definition of $P(\mathbf{A}|\mathbf{B})$ doesn't make sense. That's not a problem in practice since events with probability 0 never occur.

Bayes' Rule

$$P(\mathbf{A}|\mathbf{B}) = \frac{P(\mathbf{B}|\mathbf{A}) \times P(\mathbf{A})}{P(\mathbf{B}|\mathbf{A}) \times P(\mathbf{A}) + P(\mathbf{B}|\mathbf{A}^c) \times P(\mathbf{A}^c)}$$

Multiplication Rule

$$P(\mathbf{A}\,and\,\mathbf{B}) = P(\mathbf{A}) \times P(\mathbf{B}|\mathbf{A}) = P(\mathbf{B}) \times P(\mathbf{A}|\mathbf{B})$$

About the Data

The initial table at the beginning of this chapter comes from the *Statistical Abstract of the United States,* available online from the Bureau of the Census. The *Statistical Abstract* summarizes hundreds of views of the United States, from tables that report economic activity to law enforcement to recreation.

We used Table 738, which gives the money income of households. We converted the counts to percentages (and made sure that the rounded percentages in Table 8.1 add to 100%). That makes the counts look more like probabilities.

EXERCISES

Mix and Match

Match the item from the first column to the best item from the second.

1. Probability of **B** given **A**	(a) $P(\mathbf{A}\,and\,\mathbf{B}) = P(\mathbf{A}) \times P(\mathbf{B}|\mathbf{A})$
2. Probability of **B**^{c} given **A**	(b) $P(\mathbf{A}|\mathbf{B}) = P(\mathbf{B}|\mathbf{A}) \times P(\mathbf{A})/P(\mathbf{B})$
3. Identifies independent events	(c) $P(\mathbf{A}) = P(\mathbf{A}|\mathbf{B})$
4. Identifies dependent events	(d) $1 - P(\mathbf{B}|\mathbf{A})$
5. Bayes' Rule	(e) $P(\mathbf{A}) \neq P(\mathbf{A}|\mathbf{B})$
6. Multiplication Rule	(f) $P(\mathbf{A}\,and\,\mathbf{B})/P(\mathbf{A})$

True/False

Mark each statement True or False. If you believe that a statement is false, briefly say why you think it is false.

Exercises 7–14. An administrator tracks absences among the staff working in an office. For each employee, define the events

$$\mathbf{A} = \{\text{employee is absent}\}$$
$$\mathbf{S} = \{\text{employee is sick}\}$$

Let $\mathbf{A}_1$ and $\mathbf{S}_1$ refer to one employee and let $\mathbf{A}_2$ and $\mathbf{S}_2$ refer to another.

7. The probability of an employee being absent is greater than the probability that the employee is absent given that the employee is sick.

8. The probability that an employee is sick when it is known that the employee is absent is equal to the probability that an employee is absent when it is known that the employee is sick.

9. If the chance for an employee to be absent is greater than the chance for an employee to be sick, then **A** and **S** are dependent events.

10. If the chance for an employee to be absent is greater than the chance for an employee to be sick, then $P(\mathbf{A}|\mathbf{S}) > P(\mathbf{S}|\mathbf{A})$.

11. If she knows that $P(\mathbf{A}) = 0.20$ and $P(\mathbf{S}) = 0.15$, the administrator can find $P(\mathbf{S}|\mathbf{A})$.

12. If the event **A** is independent of the event **S**, then **S** is independent of **A**.

13. If $\mathbf{A}_1$ is independent of $\mathbf{A}_2$, then finding out that Employee 1 is absent increases the chance that Employee 2 is absent.

14. If $\mathbf{A}_1$ is independent of $\mathbf{A}_2$, $P(\mathbf{A}_1|\mathbf{A}_2) = P(\mathbf{A}_2)$.

Exercises 15–20. The Human Resources division classifies employees of the firm into one of three categories: administrative, clerical, or management. Suppose we choose an employee at random. Define the events

$$\mathbf{A} = \{\text{administrative}\},$$

$$\mathbf{C} = \{\text{clerical}\},$$

$$\mathbf{M} = \{\text{management}\}$$

Event **A** occurs, for example, if the randomly chosen employee is an administrator. Event **S** occurs if the randomly chosen employee (from any category) makes more than $120,000 annually.

15. $P(\mathbf{A} \, and \, \mathbf{C}) = P(\mathbf{A}) \times P(\mathbf{C})$.

16. If event **A** is independent of event **S**, then $P(\mathbf{A}|\mathbf{S}) = P(\mathbf{S}|\mathbf{A})$.

17. Independence of **S** with each of **A**, **C**, and **M** implies that an equal proportion of employees within these categories makes above $120,000 annually.

18. If 20% of the employees who make more than $120,000 annually are in management and 40% of the employees who make more than $120,000 annually are administrative, then $P(\mathbf{M}) < P(\mathbf{A})$.

19. If we pick an employee at random from the clerical employees, then $P(\mathbf{S}|\mathbf{C})$ is the probability that this employee makes more than $120,000 annually.

20. If 75% of employees who work in management make more than $120,000 annually but only 30% of employees in the firm as a whole make more than $120,000 annually, then the events **S** and **M** are dependent.

Think About It

Exercises 21–26. Describe the outcomes of these random processes as either independent or dependent. Briefly explain your choice.

21. Recording whether the manufacturer is located in Europe, Asia, or North America for a sequence of cars observed on an interstate highway

22. Recording the type of accident reported to an auto insurance firm in the weeks following a major ice storm in a region where the firm sells insurance

23. Tracking the number of visits to a major Web site day by day

24. Tracking the daily price of a share of stock in Microsoft

25. Recording the amount purchased by individual customers entering a large retail store (with a zero for those who do not make a purchase)

26. Counting the number of gallons of gasoline put into cars arriving at a service station on the turnpike

In the following Venn diagrams, we have drawn the sample space **S** as a square, and you may assume that the area of the regions is proportional to the probability. Describe the events **A** and **B** as either independent or dependent, with a short explanation of your reasoning. The areas of **A** and **B** are both one-quarter of the area of **S**.

27.

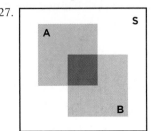

28.

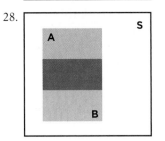

29. (a) Many companies test employees for recreational drug use. What simple diagnostic test procedure for whether the employee has used drugs has sensitivity equal to 1? Recall that the sensitivity of a test procedure is the probability that the test detects the indicated problem given that the problem is present. (*Hint:* To obtain this high sensitivity, this simple procedure requires neither messy measurements nor data.)

   (b) What is the problem with such a test procedure?

30. A pharmaceutical company has developed a diagnostic test for a rare disease. The test has *sensitivity* 0.99 (the probability of testing positive among people with the disease) and *specificity* 0.995 (the probability of testing negative among people who do not have the disease). What other probability must the company determine in order to find the probability that a person who tests positive is in fact healthy?

31. An insurer is studying the characteristics of those who buy its policies. It discovered that, among young drivers, 45% insure a foreign-made car. Among those who drive foreign-made cars, the insurer also discovered that 45% are young. Consider the events

$$\mathbf{Y} = \{\text{randomly chosen driver is young}\}$$

$$\mathbf{F} = \{\text{randomly chosen driver insures foreign-made car}\}$$

Does the equality of these percentages mean that **Y** and **F** are independent events?

32. Does the insurer described in Exercise 31 cover more drivers who are young or more drivers who insure foreign-made cars?

33. A recent study looked into the amount of debt accumulated by recent college graduates. The study found that, among those with student loans, 42% said working during college affected their grades?[7]
    (a) Convert this statement into a conditional probability, including a short description of the associated sample space.
    (b) What percentage of students working during college have student loans? Can you tell?

34. The study of recent college graduates described in Exercise 33 found that among those who had graduated with debts from student loans, 33% had sold possessions since graduating. Among those who had graduated free of debt, only 17% had sold possessions since graduating.
    (a) Express these two percentages as conditional probabilities, including a description of the relevant sample space.
    (b) What do you need to know in order to determine the proportion of recent college graduates who sold possessions after college?

You Do It

Throughout the following exercises, you can form the probability table with whichever variables you want along the rows and columns. The solutions show the tables drawn one way, but you can swap the rows and columns and still get the right answers.

35. An auto manufacturer has studied the choice of options available to customers when ordering a new vehicle. Of the customers who order a car, 25% choose a sunroof, 35% choose a leather interior, and 10% choose both. (The rest opt for neither.) Of the customers who order a truck, 20% choose an extended cab, 40% choose all-wheel drive, and 15% choose both. You cannot get a sunroof on a truck or an extended cab on a car. For this brand, half of the customers who order a vehicle choose a truck.
    (a) Organize these probabilities as a probability tree or probability table. Explain why you picked a tree or a table.
    (b) What's the probability that the customer orders a vehicle with a sunroof?

36. A vendor specializing in outdoor gear offers customers three ways to place orders for home delivery. Customers can place orders either in the store, online, or via telephone using a catalog. Of the orders, one-half are made online with one-quarter each made in stores or via telephone. Of the online orders, 40% are only for clothing, 30% are only for camping supplies, and 30% are for both. For orders placed in the store, all are for clothing (to order a

size that is not in stock). Among orders placed via telephone, 60% are for clothing alone, 20% are for camping supplies alone, and 20% are for both clothing and camping supplies.
    (a) Organize these probabilities as a probability tree or probability table. Explain why you picked a tree or a table.
    (b) What's the probability that the customer orders clothing, either by itself or in combination with camping supplies?

37. Seventy percent of customers at the snack counter of a movie theater buy drinks. Among those who buy drinks, 30% also purchase popcorn. What's the probability that a customer at the counter buys a drink and popcorn? Theaters use this type of calculation to decide which products should be bundled to appeal to customers.

38. Seventy percent of service calls regarding kitchen appliances involve a minor repair. For example, the customer did not read the instructions correctly or perhaps failed to connect the appliance correctly. These service calls cost $75. Service calls for major repairs are more expensive. More than one-half of service calls for major repairs cost the customer $200 or more. What's the probability that a randomly selected service call costs $200 or more?

39. Some electronic devices are better used than new: The failure rate is higher when they are new than when they are six months old. For example, half of the personal music players of a particular brand have a flaw. If the player has the flaw, it dies in the first six months. If it does not have this flaw, then only 10% fail in the first six months. Yours died after you had it for three months. What are the chances that it has this flaw?

40. Some of the managers at a company have an MBA degree. Of managers at the level of director or higher, 60% have an MBA. Among other managers of lower rank, 35% have an MBA. For this company, 15% of managers have a position at the level of director or higher. If you meet an MBA from this firm, what are the chances that this person is a director (or higher)?

41. A shipment of assembly parts from a vendor offering inexpensive parts is used in a manufacturing plant. The box of 12 parts contains 5 that are defective and will not fit during assembly. A worker picks parts one at a time and attempts to install them. Find the probability of each outcome.
    (a) The first two chosen are both good.
    (b) At least one of the first three is good.
    (c) The first four picked are all good.
    (d) The worker has to pick five parts to find one that is good.

42. Many retail stores open their doors early the day after Thanksgiving to attract shoppers to special bargains. A shopper wanting two medium blouses heads for the sale rack, which is a mess, with sizes jumbled together. Hanging on the rack are 4 medium, 10 large, and

6 extra-large blouses. Find the probability of each event described.

(a) The first two blouses she grabs are the wrong size.

(b) The first medium blouse she finds is the third one she checks.

(c) The first four blouses she picks are all extra-large.

(d) At least one of the first four blouses she checks is a medium.

43. After assembling an order for 12 computer systems, the assembler noticed that an electronic component that was to have been installed was left over. The assembler then checked the 12 systems in order to find the system missing the component. Assume that he checks them in a random order.

(a) What is the probability that the first system the assembler checks is the system that is missing the component?

(b) What is the probability that the second system the assembler checks is missing the component, assuming that the first system he checked was OK?

(c) Explain why the answers to parts (a) and (b) are different.

44. Among employees at a retailer, there is a 15% chance that an employee will be absent from work for some reason. The floor manager has been talking with two randomly selected employees each day to get suggestions to improve sales.

(a) What is the probability that the first employee he chooses to speak with is absent?

(b) Describe the probability that the second employee he chooses to speak with is absent, assuming that the first employee is present. Is it higher, lower, or the same as the probability in part (a)?

(c) Should your answers to parts (a) and (b) match? Explain why or why not.

45. Choice leads for developing new business are randomly assigned to 50 employees who make up the direct sales team. Half of the sales team is male, and half is female. An employee can receive at most one choice lead per day. On a particular day, five choice leads are assigned.

(a) Are the events {first lead is to a male} and {second lead is to a male} dependent or independent?

(b) If the first four leads all go to men, what is the probability that the fifth lead also goes to a man?

(c) What is the probability that all five leads go to men *if you know that at least* four of the leads go to men? (This item is harder than most. A procedure to handle these sorts of problems is given in Chapter 11, but you *can* do this exercise from principles in this chapter.)

46. Modern assembly relies on parts built to specifications (spec) so that the components fit together to make the whole. An appliance requires two components, Type A and Type B, supplied by different manufacturers. A robot randomly selects a component from a batch of Type A components and a second component from a batch of Type B components. The appliance works only if both components are within specifications. It is known that 5% of Type A components and 10% of Type B components are out of specifications.

(a) What is the probability that the appliance produced by the robot selecting one component from each batch at random works? State any assumptions clearly.

(b) The robot tests the appliance as it is assembled. It discovers that the current appliance does not work because the parts are not compatible. What is the probability that component that is out of spec is the Type A component?

47. The following table summarizes the share of the workforce in the United States in several activities in 1965 and 40 years later (according to the Bureau of Labor Statistics). Use these percentages to define a joint probability distribution of *Year* and *Occupation*. Assume that the size of the workforce is 70 million in 1965 and is 140 million in 2005, with no overlap.

	1965	2005
Durable goods manufacturing	19%	8%
Professional services	8	15
Education and health care	7	16
Other occupations	66	61

(a) Given that a randomly chosen person from the workforce is in the durable goods industry, what is the probability that the year is 1965?

(b) Given that a randomly chosen worker is in professional services, education, or health care, what is the probability that the year is 1965?

(c) You've been trapped in a time warp and suddenly reappear at a party. It's either 1965 or 2005, so you ask someone at the party what he or she does. A response of which of the four listed occupations tells you the most about the year of your reappearance?

48. A company buys components from two suppliers. One produces components that are of higher quality than the other. The high-quality supplier, call it Supplier A, has a defect rate of 2%. The low-quality supplier, Supplier B, has a defect rate of 10% but offers lower prices. This company buys in equal volume from both suppliers, with half of the orders going to each supplier.

(a) What is the probability that a component to be installed is defective?

(b) If a defective component is found, what is the probability it came from Supplier A?

49. You fly from Philadelphia to San Francisco with a connection in Dallas. The probability that your flight from Philadelphia to Dallas arrives on time is 0.8. If you arrive on time, then the probability that your luggage makes the connection to San Francisco is 0.9. If you are delayed, then the chance of your luggage making the connection with you is 0.5. In either case, you make the flight.

(a) What is the probability that your luggage is there to meet you in San Francisco?

(b) If your luggage is not there to meet you, what is the probability that you were late in arriving in Dallas?

50. A survey reports that 62% of callers to a help desk complain about the service if they think they spoke to a foreign agent, compared to 31% who complain if they think they spoke to a native agent. Suppose that 40% of all calls to service centers are handled by agents in the United States. If a caller complains, what is the probability she was dealing with a foreign agent?

51. Recent surveys report that although Internet access has grown rapidly, it's not universal: Only 75% of U.S. households have a computer with Internet access. Internet access isn't universal for households that have computers; 18% of households with a computer are not on the Internet.[8] Treat these percentages as probabilities. What is the probability that, among households not connected to the Internet, the household does not have a computer?

52. Suppose that 20% of the clerical staff in an office smoke cigarettes. Research shows that 60% of smokers and 15% of nonsmokers suffer a breathing illness by age 65.
    (a) Do these percentages indicate that smoking and this breathing illness are independent?
    (b) What's the probability that a randomly selected 65-year-old employee who has this breathing illness smokes?

4M Scanner Data

Many supermarkets offer customers discounts in return for using a shopper's card. These stores also have scanner equipment that records the items purchased by the customer. The combination of a shopping card and scanner software allow the supermarket to track which items customers regularly purchase.

The data in this table are based on items purchased in 54,055 shopping trips made by families participating in a test of scanners. As part of the study, the families also reported a few things about themselves, such as the number of children and pets. These data are the basis of the following probabilities.

This table shows the number of dogs owned by the person who applied for the shopping card. The table also shows in rows the number of dog food items purchased at the supermarket during each shopping trip in the time period of the study. The table gives the probability for each combination.

Motivation

a) Should markets anticipate the row and column events described in this table to be independent?

b) What could it mean that the probability of customers with no dogs buying dog food is larger than zero?

Method

c) Which conditional probability will be most useful here? Will it be more useful to markets to calculate row probabilities or column probabilities?

d) The smallest probabilities in the table are in the last column. Does this mean that owners of more than three dogs buy relatively little dog food?

e) These probabilities come from counts of actual data. What would it mean if the probabilities added up to a value slightly different from 1?

Mechanics

f) Expand the table by adding row and column marginal probabilities to the table.

g) Find the conditional probability of buying more than three items among customers reported to own no dogs. Compare this to the conditional probability of buying more than three dog food items among customers reported to own more than three dogs.

h) If a customer just bought eight cans of dog food, do you think that she or he owns dogs? Give a probability.

Message

i) What do you conclude about the use of this version of the scanner data to identify customers likely to purchase dog food?

	No Dogs	1 Dog	2 Dogs	3 Dogs	More than 3 Dogs
No dog items	0.0487	0.0217	0.0025	0.0002	0
1 to 3	0.1698	0.0734	0.0104	0.0004	0.0002
4 to 6	0.1182	0.0516	0.0093	0.0006	0.0002
7 to 12	0.1160	0.0469	0.0113	0.0012	0.0005
More than 12	0.2103	0.0818	0.0216	0.0021	0.0011

---

[8] "Why the Web Is Hitting a Wall," *Business Week,* March 20, 2006.

Random Variables

CHAPTER 9

DAY TRADING HAS BECOME POPULAR WITH YOUNG PEOPLE AROUND THE WORLD. The Japanese homemaker Yuka Yamamoto got bored watching television. After seeing an ad, she took $2,000 in savings and started trading stocks from home. Within a year, she turned the initial $2,000 into $1 million and became a celebrity in Japan. Her success as a day trader made her into a popular speaker and author. The 20-year-old student Yuta Mimura found similar success as a day trader on the Tokyo Stock Exchange. His first investment turned out to be a great choice. He happily watched shares that he bought for $.25 jump to $.45 in two days. His parents were not happy with his new hobby, at least not until he fixed up their home with $100,000 earned trading![1]

Day traders guess when the price of a stock is headed up or down. To do this, they've invented rules for deciding when to buy and sell stocks, such as the rule suggested by the chart on the next page. A big valley in the sequence of prices followed by a small valley, the "cup and handle" in the figure, signals a change in the trend of these prices of stock in Walt Disney, if you believe the rule. The underlying objective of these trading rules is simple: Enable the trader to consistently buy at a low price, then sell at a higher price. If you know a stock is going up, buy the shares now. Sell them after the price has risen and make a profit.

Day traders need to be right more often than wrong. They'd like it if these trading rules always worked, but that's asking for too much. Sometimes the price of stock that they bought stays the same or goes down.

This chapter defines a concise language and notation for describing processes that show random behavior, such as stock returns, weather patterns, or manufacturing. This language is essential for making decisions in the face of uncertainty because the

[1] "In Japan, Day-Trading Like It's 1999," *New York Times*, February 19, 2006.

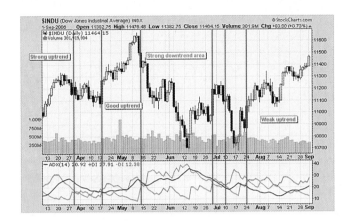

language separates what is known from what is unknown. The main component of this language is an object called a random variable. Whether we're anticipating the price of a stock, deciding among strategic plans, scheduling shipments, or forecasting economic trends, random variables allow us to concisely describe the possibilities and evaluate the alternatives.

9.1 | RANDOM VARIABLES

Our first random variable expresses the expectations of a day trader in terms of events and probabilities. This day trader is interested in stock in IBM. She can buy one share of stock in IBM for $100 at the close of the stock market today. The stock market *tomorrow* determines whether she makes money. To simplify the arithmetic, we'll restrict what can happen to three possible outcomes. The price of the stock at the close of the market tomorrow will either go up to $105, stay at $100, or fall to $95. If she buys one share, she might make $5, break even, or lose $5. To decide if IBM is a good investment, she needs to assign probabilities to these outcomes. Together, these possible values and the corresponding probabilities define a random variable.

A **random variable** describes the probabilities for an uncertain *future* numerical outcome of a random process. For a day trader considering stock in IBM, the random variable is the change in the price of IBM stock tomorrow. We can't be sure *today* how activity in the market *tomorrow* will impact the price of IBM stock, but we can assign probabilities to the possible outcomes.

Following convention, we use a capital letter X to denote the random variable that represents the change in the price of IBM stock. It takes some getting used to, but you have to remember that X does *not* stand for a number, as in algebra. A random variable represents the correspondence between the collection of all of the possibilities and their probabilities. To define a random variable, you have to list every possible outcome along with its probability.

Let's express the opinions of the day trader as a random variable. Most of the time, the day trader expects the price to stay the same; she assigns this event probability 0.80. She divides the rest of the probability almost equally between going up (with probability 0.11) and going down (with probability 0.09). There's a slight edge in favor of an increase. Because one of these outcomes has to happen, the probabilities add to 1. Table 9.1 summarizes the outcomes and probabilities.

random variable The uncertain outcome of a random process, often denoted by a capital letter.

TABLE 9.1 Probabilities that
define the random variable *X*.

Stock Price	Change x	Probability $P(X = x)$
Increases	$5	0.11
Stays same	0	0.80
Decreases	−$5	0.09

This table defines the random variable *X*. It lists the possible outcomes ($5, 0, and −$5) and assigns a probability to each. Because we can list all of the possible outcomes, *X* is called a **discrete random variable**. A **continuous random variable** can take on any value within an interval; a continuous random variable shows how probability is spread over an interval rather than assigned to specific values. We will examine continuous random variables in Chapter 12.

discrete random variable A random variable that takes on one of a list of possible values; typically counts.

continuous random variable A random variable that takes on any value in an interval.

> **caution**   The notation in Table 9.1 illustrates that we denote possible outcomes, or realizations, of a random variable by the corresponding lowercase letter. Since we're talking about the random variable *X*, we use *x* to denote a possible outcome. This convention produces possibly confusing statements, such as $P(X = x)$ in the heading of the third column. The first time you see this, it might seem as though we're saying something as useless as $P(1 = 1)$. That is not what is meant, and you must *pay attention to the capitalization*. The statement $P(X = x)$ is shorthand for the probability of any one of the possible outcomes. The possible values of *X* are $x = 5, 0,$ and $−5$, so $P(X = x)$ is a generic expression for any of the three probabilities: $P(X = 5)$, $P(X = 0)$, or $P(X = −5)$.

Graphs of Random Variables

Tables work well for random variables that have few outcomes, but plots are better when there are more outcomes. Table 9.1 would be huge if we allowed the random variable to take on all possible changes in the price of real shares of IBM stock. We'll have to handle those situations eventually, so to prepare we need to get comfortable with plots.

How do we plot a random variable? Think about the plot that we would choose if we had data that recorded the changes in price of IBM. Imagine that we had 100 observed changes in the price. What plot would you use to describe these daily changes? If the timeplot of the changes lacks a pattern, we would create a histogram. We need a comparable display for a random variable. There's a big difference, however, between the histogram of data and the display of a random variable. A histogram shows relative frequencies of what happened in the *past*. The equivalent display for a random variable shows probabilities assigned to events that have yet to happen.

probability distribution A function that assigns probabilities to each possible value of the random variable.

This graph of a random variable shows its **probability distribution** (also called its probability density function, probability distribution function, or probability mass function). The probability distribution of a random variable is often abbreviated as

$$p(x) = P(X = x)$$

The name reminds you that the graph shows probabilities, not relative frequencies in data. Whereas a histogram sets the heights of the bars by counting data, the heights of the probability distribution $p(x)$ are probabilities. Figure 9.1 graphs $p(x)$.

This graph also defines the random variable *X* because it shows all of the relevant information. It shows every possible outcome and its probability. [The gray vertical lines that connect $p(x)$ to the *x*-axis are not part of the probability distribution. We include them for discrete random variables to make it easier to identify positions on the *x*-axis.]

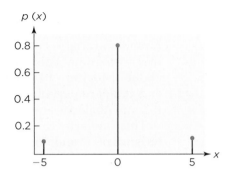

FIGURE 9.1 Plot of the probability distribution of X.

Random Variables as Models

Statistical model A breakdown of variation into a predictable pattern and the remaining variation.

A random variable is a type of **statistical model**. As defined in Chapter 1, a statistical model presents a simplified or idealized view of reality. When the day trader uses the random variable X to describe what she thinks will happen to IBM tomorrow, she hopes that her probabilities match reality. Later in this chapter we'll see that X mimics several properties of the real stock, even though it is limited to three possible outcomes.

Data affect the choice of a probability distribution for a random variable. Experiences in the past affect what we expect in the future. If a stock has always gone up or down by $5 and never stayed the same, this random variable wouldn't be a good description of what is likely to happen. This model suggests that it is common to see no change in the value of the stock. If the day trader's model for the probabilities is right, you'd expect to find 80% of the *future* changes at zero, 11% at $5, and 9% at −$5. The histogram that accumulates the outcomes *in the future* should eventually match her probability distribution in the long run, if she's right.

That is how we defined probability in Chapter 7: A probability is a long-run relative frequency. When we say that X is a random variable with these probabilities, we claim to *know* the chances for what will happen tomorrow and the days that follow. We cannot say which outcome will happen tomorrow any more than we can say whether a coin will land on heads or tails, but we can describe what will happen for lots of tomorrows. In this sense, the probability distribution of a random variable describes a histogram of possibilities rather than a histogram of data.

What Do You Think?

Customers who buy tires at an auto service center purchase one tire, two tires, three tires, or a full set of four tires. The probability of buying one tire is 1/2, the probability of buying two is 1/4, and the probability of buying three is 1/16. The random variable Y denotes the number of tires purchased by a customer who has just entered the service center.

a. What is $P(Y = 4)$?[2]
b. What is $P(Y > 1)$?[3]
c. Graph the probability distribution of Y.[4]

---

[2] Since the probability distribution must sum to 1, $P(Y = 4) = 1 − (0.5 + 0.25 + 0.0625) = 0.1875$ = 3/16.
[3] $P(Y > 1) = 1 − P(Y = 1) = 1/2$.
[4] The graph should show the points (1, 1/2), (2, 1/4), (3, 1/16), and (4, 3/16), possibly connected by vertical lines down to the x-axis.

9.2 | PROPERTIES OF RANDOM VARIABLES

A random variable conveys information that resembles what we might obtain in a histogram. This similarity allows us to exploit what we have learned about histograms to summarize random variables. For example, the mean $\bar{x}$ and standard deviation s are two important summaries of the data in a histogram. Analogous measures summarize a random variable.

This similarity can, however, cause confusion. Do not confuse the characteristics of a random variable, called **parameters**, with statistics computed from data. To distinguish parameters from statistics, we'll use Greek symbols to denote the parameters of a random variable.

parameter A characteristic of a random variable, such as its mean or standard deviation. Parameters are typically denoted by Greek letters.

Mean of a Random Variable

The probability distribution of a random variable determines its mean. Even though the mean of the random variable is not an average of data, it has a similar definition. To see the connection, imagine that we have data whose relative frequencies happen to match the probabilities for X. We have $n = 100$ observed changes in the price of IBM stock. Of these, the price fell by $5 on 9 days, stayed the same 80 times, and increased by $5 on 11 days. The histogram of these 100 changes matches the probability distribution of X in Figure 9.1.

What is the mean of these imagined data? Because of the repetition, this is an easy calculation.

$$
\begin{aligned}
\bar{x} &= \frac{\overbrace{(-5) + \cdots + (-5)}^{9} + \overbrace{0 + \cdots + 0}^{80} + \overbrace{5 + \cdots + 5}^{11}}{100} \\
&= \frac{-5(9) + 0(80) + 5(11)}{100} \\
&= -5(0.09) + 0(0.80) + 5(0.11) \\
&= \$.10
\end{aligned}
$$

The average of these 100 changes is $.10.

We define the **mean μ of a random variable** in just this way, but replace relative frequencies in imaginary data with probabilities. The mean of the random variable X is a weighted average of the possible outcomes, with probabilities $p(x)$ for the weights. The most common symbol for the mean of a random variable is the Greek letter μ (mu, pronounced "mew"). The mean of the random variable X is

mean μ of a random variable The weighted sum of possible values, with the probabilities as weights.

$$
\begin{aligned}
\mu &= -5p(-5) + 0p(0) + 5p(5) \\
&= -5(0.09) + 0(0.80) + 5(0.11) \\
&= \$.10
\end{aligned}
$$

Notice that the mean μ has the same units as the possible values of the random variable. In this example, the mean of X matches the average of the imagined data because we matched the relative frequencies to the probabilities. In general, the mean of a discrete random variable X that has k possible values $x_1, x_2, \ldots, x_k$ is

$$
\mu = x_1 p(x_1) + x_2 p(x_2) + \cdots + x_k p(x_k)
$$

The sum includes *all* of the possible outcomes of the random variable. (When dealing with several random variables, we add a subscript to μ to distinguish each mean, as in $\mu_X = \$.10$.)

The mean μ reveals the same things about the random variable X that $\bar{x}$ tells us about data in a histogram. In both cases, the mean is the balance point. To balance a histogram on a seesaw, we locate the fulcrum at $\bar{x}$. To balance the probability distribution of a random variable, we locate the fulcrum at μ. Figure 9.2 shows the probability distribution of the random variable X with a fulcrum added at μ.

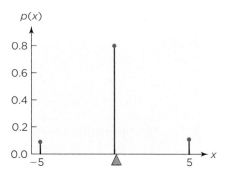

FIGURE 9.2 The mean balances the probability distribution.

The mean μ tells us that the day trader expects *on average* to make 10 cents on every share of IBM she buys. The mean is positive because there is more probability to the right of 0 than to the left.

The average gain seems small until you think about what happens over time. Each share costs $100, so a gain of $.10 amounts to an increase of $0.10/100 = 0.1\%$ *daily*. If you had a savings account that paid 0.1% daily, that would pile up more than 44% interest annually!

Expected Value

The mean of a random variable is a special case of a more general concept. A weighted average of outcomes that uses probabilities as weights is known as an **expected value**. Because the mean of a random variable is a weighted average of the possible values of the random variable, the mean of X is also known as the expected value of X. It is written in this notation as

$$E(X) = \mu = x_1 p(x_1) + x_2 p(x_2) + \cdots + x_k p(x_k)$$

expected value A weighted average that uses probabilities to weight the possible outcomes.

caution The name *expected value* can be confusing. For the stock, the expected value of *X* is not one of the possible outcomes. The price of the stock either stays the same or changes by five dollars. It never changes by 10 cents. *The expected value of a random variable may not match one of the possible outcomes.*

Only when we see the results of *many* future outcomes does the expected value emerge: The expected value is the average of these future outcomes. The expected value is long-run averages of *anything*. Usually *anything* involves a random variable, as in the following illustration.

A company uses free giveaways to increase the number of shoppers who regularly visit its retail stores. Four prize-winning tickets are hidden in randomly chosen items. The total prize of $30,000 is split among the winners. Because some tickets may not be found by the end of the contest, the amount won depends on the number of winning tickets that are claimed. Table 9.2 defines the random variable W that denotes the number of winners in a future contest. The mean, or expected value, of W is slightly less than 3.

$$\mu = E(W)$$
$$= 1(0.10) + 2(0.25) + 3(0.50) + 4(0.15)$$
$$= 2.7$$

TABLE 9.2 Probabilities of the number of winners in the future contest.

Number of Winners	$P(W = w)$
1	0.10
2	0.25
3	0.50
4	0.15

Figure 9.3 shows the probability distribution of W with a triangle locating μ.

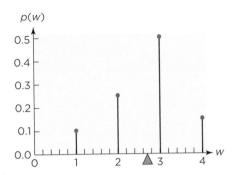

FIGURE 9.3 Plot of the probability distribution of W.

The amount won by those who find a ticket depends on the number of customers who claim winning prizes. Let's find the expected value of the amount won, $E(30,000/W)$. This expected value is a weighted average of the possible outcomes, only now the outcomes are possible values of the ratio of 30,000 to the number of winners.

We start by organizing the calculations. The first column in Table 9.3 lists the possible values $w = 1, 2, 3,$ or 4 of the random variable W. The second column shows the amount won if w customers claim prizes. The third column lists the probabilities of these outcomes, and the last column uses these probabilities to weight the outcomes. The sum of the values in the fourth column is the expected value of the amount won.

TABLE 9.3 Calculations for the expected amount won.

Number of Winners w	Amount Won 30,000/w	Probability $P(W = w)$	Amount × $P(W = w)$
1	30,000	0.10	3,000
2	15,000	0.25	3,750
3	10,000	0.50	5,000
4	7,500	0.15	1,125
		Sum	**12,875**

If you find a winning ticket in your package, you win—on average—$E(\$30,000/W) = \$12,875$. The expected value of the amount won may be larger than you anticipated. We expect $E(W) = 2.7$ winners to share the prize. The expected value $E(30,000/W)$ is *larger* than $30,000/E(W) = 30,000/2.7 \approx \$11,111$.

What Do You Think?

People who play the carnival game Lucky Ducks pick a plastic duck from among those floating in a small pool. Each duck has a value written on the bottom that determines the prize. Most say "Sorry, try again!" Otherwise, 25% pay $1, 15% pay $5, and 5% pay $20. Let the random variable Y denote the winnings from picking a duck at random.

a. Graph the probability distribution of Y.[5]
b. Estimate the mean of Y from the graph of its probability distribution.[6]
c. Find the expected value of Y. What's your interpretation of the mean if the carnival charges $2 to play this game?[7]

[5] Your graph should have four points: $p(0) = 0.55, p(1) = 0.25, p(5) = 0.15,$ and $p(20) = 0.05$.
[6] Skewness of the probabilities suggests that the mean is larger than $1 or $2, but probably not larger than $5.
[7] The mean is $0 \times 0.55 + 1 \times 0.25 + 5 \times 0.15 + 20 \times 0.05 = \$2$. This is a so-called fair game. On average customers win as much as they spend.

Variance and Standard Deviation

Day traders have to understand more than the expected value. Day traders, and anyone else who has to make decisions, also need to appreciate the variation in what might happen.

The mean change for IBM stock is positive, $E(X) = \$.10$. The fact that the mean is positive does not imply stock in IBM goes up in value every day. On average, the value increases, but an investor could lose money on any given day if the price goes down. The same goes for players of the Lucky Ducks carnival game. The expected payoff ($2) matches the cost of the game, but few break even.

The variance and standard deviation of a random variable summarize the uncertainty among the outcomes. The **variance** of a random variable is the expected value of the squared deviation from μ. We will explain this step by step; the calculations will seem familiar if you recall the calculation of the variance s^2 from data. Because it is an expected value, the variance of a random variable is a weighted sum of squared deviations from the mean, $(x - \mu)^2$. The variance of a random variable is another parameter, so we denote it by a Greek symbol σ^2 (pronounced "sigma-squared"). To avoid using Greek, we often write $\text{Var}(X)$ for the variance of the random variable X. Here's the formula.

$$
\begin{aligned}
\sigma^2 &= \text{Var}(X) \\
&= E(X - \mu)^2 \\
&= (x_1 - \mu)^2 p(x_1) + (x_2 - \mu)^2 p(x_2) + \cdots + (x_k - \mu)^2 p(x_k)
\end{aligned}
$$

As an example, we'll use the day trader's random variable. The calculations resemble those used to find s^2 from data. With data, we calculate s^2 by subtracting $\bar{x}$ from each value, squaring these deviations, adding them up, and dividing by $n - 1$. Similar steps produce the variance. First, find the deviation of each outcome from μ, then square the deviations, as in Table 9.4.

variance The variance of a random variable X is the expected value of the squared deviation from its mean μ:

$$\sigma^2 = E(X - \mu)^2$$

TABLE 9.4 Calculating the variance.

Change in Price x	Deviation $x - \mu$	Squared Deviation $(x - \mu)^2$	Probability $p(x)$
$-\$5$	$-5 - 0.10 = -5.1$	$(-5.1)^2$	0.09
0	$0 - 0.10 = -0.1$	$(-0.1)^2$	0.80
$\$5$	$5 - 0.10 = 4.9$	$(4.9)^2$	0.11

Next, multiply the squared deviations by the corresponding probabilities. Finally, add the weighted, squared deviations. The variance of the day trader's random variable is

$$
\begin{aligned}
\sigma^2 &= \text{Var}(X) \\
&= (-5 - 0.10)^2(0.09) + (0 - 0.10)^2(0.80) + (5 - 0.10)^2(0.11) \\
&= 4.99
\end{aligned}
$$

Variances have units. The variance is the expected value of the *squared* deviation from μ. Hence, the measurement units of σ^2 are the square of the units of the random variable. Squaring makes the variance hard to interpret.

We remedy this problem for random variables as we did for data, by taking the square root of the variance. The **standard deviation** of a random variable is the square root of its variance.

standard deviation The square root of the variance, whether from data or probabilities.

$$
\begin{aligned}
\sigma = \text{SD}(X) &= \sqrt{\text{Var}(X)} \\
&= \sqrt{4.99} \approx \$2.23
\end{aligned}
$$

The expected daily change in the random variable for the price of IBM stock is $\mu = \$.10$ with standard deviation $\sigma = \$2.23$. Both μ and σ have the same units as the random variable X. If we want to distinguish these parameters from those of a different random variable, we use subscripts and write them as μ_X and σ_X.

4M EXAMPLE 9.1   COMPUTER SHIPMENTS AND QUALITY

MOTIVATION   STATE THE QUESTION

CheapO Computers promptly shipped two servers to its biggest client. The company profits $5,000 on each one of these big systems. Executives are horrified, though, to learn that someone restocked 4 refurbished computers among the 11 new systems in the warehouse. The guys in shipping randomly selected the systems that were delivered from the 15 computers in stock.

If the client gets two new servers, CheapO earns $10,000 profit. If the client gets a refurbished computer, it's coming back for replacement and CheapO must pay the $400 shipping fee. That still leaves $9,600 profit. If both servers that were shipped are refurbished, however, the client will return both *and* cancel the order. CheapO will be out any profit and left with $800 in shipping costs. What are the expected value and the standard deviation of CheapO's profits?

METHOD   DESCRIBE THE DATA AND SELECT AN APPROACH

Begin by identifying the relevant random variable. In this example, the random variable is the amount of profit earned on this order. The probabilities for the alternatives come directly from the statement of the problem. We'll arrange the information as a tree. (You could use a table as well.)

MECHANICS   DO THE ANALYSIS

This tree identifies the possible outcomes and the probability for each. Notice that two paths lead to one refurbished computer and one new computer. The probability that the second system is refurbished depends on whether the first is refurbished. If X denotes the profit earned on the shipment, then this tree lists its values and probabilities. In the table on the following page, the event **R** denotes that a computer is refurbished, and the event **N** denotes a computer that is new.

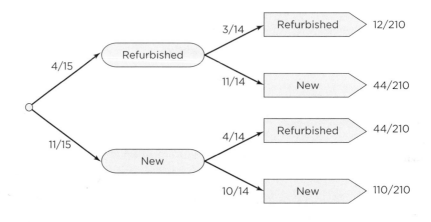

Outcome	x	p(X)
Both refurbished	−$800	$P(\mathbf{R} \text{ and } \mathbf{R}) = 12/210$
One refurbished	$9,600	$P((\mathbf{N} \text{ and } \mathbf{R}) \text{ or } (\mathbf{R} \text{ and } \mathbf{N})) = 88/210$
New/New	$10,000	$P(\mathbf{N} \text{ and } \mathbf{N}) = 110/210$

The expected value of X is a weighted sum of the possible profits, with the probabilities defining the weights.

$$\text{E}(X) = 10{,}000(110/210) + 9{,}600(88/210) - 800(12/210) = \$9{,}215$$

The variance is expected *squared* deviation of the profit from the mean.

$$\text{Var}(X) = (-800 - 9{,}215)^2(12/210) + (9{,}600 - 9{,}215)^2(88/210)$$
$$+ (10{,}000 - 9{,}215)^2(110/210) = 6{,}116{,}340 \ \$^2$$

The measurement units of the variance are unusual, so we wrote them in this odd style as a reminder. Taking the square root gives the standard deviation, which is in units of dollars and more easy to interpret.

$$\text{SD}(X) = \sqrt{\text{Var}(X)} = \sqrt{6{,}116{,}340} \approx \$2{,}473$$

MESSAGE   SUMMARIZE THE RESULTS

This is a very profitable deal on average, unless both systems that were shipped are refurbished units. The expected value of the sale is $9,215. The large SD (almost $2,500) is a reminder that the profits are wiped out if both systems are refurbished.

9.3 | PROPERTIES OF EXPECTED VALUES

We often have to add a constant to a random variable or multiply a random variable by a constant. For example, consider the computer shipper in the previous example. Suppose the company in Example 9.1 has to pay out of its profits a fixed delivery fee of, say, $400 for shipping both computers. In this case the net profits are not X dollars, but $X - \$400$. Similarly, we often multiply random variables by constants. The car shop that sells tires makes a sale of Y tires to a customer. If the shop profits $25 on each tire, then the profit from a sale of Y tires is $\$25 \times Y$.

What are the mean and standard deviation of these quantities? We could start over: Define new random variables in each case and find their expected values. There's an easier approach, however. Let's start with the effect of adding (or subtracting) a constant.

Adding or Subtracting a Constant

Adding or subtracting a constant from a random variable shifts every possible value of the random variable, changing the expected value by a fixed amount. If c is any constant, then

$$\text{E}(X \pm c) = \text{E}(X) \pm \text{E}(c) = \text{E}(X) \pm c$$

The expected value of a constant is the constant. Hence, the expected net profit to the computer shipper after paying the initial shipping fee of $400 is

$$\text{E}(X - 400) = \text{E}(X) - \$400 = \$9{,}215 - \$400 = \$8{,}815$$

| tip | Capital letters denote random variables whereas lowercase letters like *a*, *b*, or *c* identify constants. When you see a capital letter, think random. When you see a small letter, think 2 or 3.14. |

What about the standard deviation? How does shifting the outcomes affect the standard deviation? With data, adding or subtracting a constant doesn't change the variance or standard deviation. The same holds for random variables. A shift has no effect on the variance or standard deviation of a random variable.

$$\text{Var}(X \pm c) = \text{Var}(X)$$
$$\text{SD}(X \pm c) = \text{SD}(X)$$

Multiplying by a Constant

Multiplying the value of a random variable by a constant changes both the expected value and the variance. Recall from Chapter 4 that multiplying or dividing every data value by a constant changes the mean and the standard deviation by that factor. Variance, being the square of standard deviation, changes by the square of the constant. The same is true for random variables.

Multiplying a random variable by a constant *c* multiplies the mean and standard deviation by *c*.

$$\text{E}(cX) = c\text{E}(X) \qquad \text{SD}(cX) = |c|\, \text{SD}(X)$$

Because it has squared units, multiplying by a constant changes the variance by the *square* of *c*.

$$\text{Var}(cX) = c^2\text{Var}(X)$$

For example, the car shop in a previous example sells on average

$$\text{E}(Y) = 1(0.5) + 2(0.25) + 3(0.0625) + 4(0.1875) = 1.9375 \text{ tires}$$

per sale. A similar calculation shows that $\text{SD}(Y) \approx 1.144$ tires. Since it profits $25 per tire, the expected profit is $\text{E}(25Y) = 25\text{E}(Y) \approx \$48.4$ with standard deviation $\text{SD}(25Y) = 25\text{SD}(Y) \approx \$28.6$.

Let's do another example, returning to the day trader. Suppose rather than buying one share of stock, she buys two. By investing $200 and buying two shares of the same stock, the day trader doubles μ and σ. With $c = 2$, the expected change in the value of her investment is twice what we found previously:

$$\text{E}(2X) = 2\text{E}(X) = 2 \times 0.10 = \$.20$$

with standard deviation

$$\text{SD}(2X) = 2\,\text{SD}(X) = \$4.46$$

She expects to earn twice as much on average, but she also doubles the standard deviation. The variance grows by a factor of 4.

$$\text{Var}(2X) = 2^2\text{Var}(X) = 4 \times 4.99 = 19.96 \ \$^2$$

Let's do another example. Let the random variable S denote the number of windows shipped daily by a manufacturer of building materials. Some days the plant produces more and on other days less. The mean of S is 2,500 windows with standard deviation 500 windows. Each window has 8 panes of glass. What is the expected number of panes of glass shipped each day and its standard deviation?

We must use the rules for expected values to find the answer since we don't know the underlying probabilities in this example. The expected number of panes is $E(8S) = 8E(S) = 8(2,500) = 20,000$ panes with standard deviation $SD(8S) = 8SD(S) = 8(500) = 4,000$ panes.

The following equations summarize the rules for finding expected values and variances when constants are mixed with random variables. Lowercase letters denote constants.

Rules for Expected Values If a and b are constants and X is a random variable, then

$$E(a + bX) = a + bE(X)$$
$$SD(a + bX) = |b|SD(X)$$
$$Var(a + bX) = b^2Var(X)$$

Expected values of sums are sums of expected values, the expected value of a constant is that constant, and constants factor out (with squares for the variance).

What Do You Think?

a. If the day trader has to pay \$.10 for each trade of a share of stock, what is the expected change in the value of her investment? In general, what is $E(X - \mu)$? (*Hint:* The mean μ is a constant.)[8]

b. Suppose the day trader prefers to track the value of her investments in cents rather than dollars. What are the mean and variance of $100X$?[9]

9.4 | COMPARING RANDOM VARIABLES

Expressions that combine random variables with constants often arise when comparing random variables. Comparison requires transforming the initial random variables into new random variables that have a common scale or a more useful scale. Consider the following example:

An international business based in the United States is considering launching a product in two locations outside the United States. One location is in Europe, and the other is in India. Regional management in each location offers a random variable to express what it believes will happen if the product is launched. These random variables are expressed in the local currency. The division in Europe believes that the launch will produce net profits of 2.5 million euros in the first year, with standard deviation 0.75 million euros. The division in India expects that the launch will generate net profits of 125 million rupees, with standard deviation 45 million rupees.

We cannot directly compare these random variables because they are measured on different scales, one in euros and the other in rupees. To compare them, as well as to make them more familiar to managers in the United States, let's convert both into dollars using the exchange rates of 1 dollar = 0.75 euro and 1 dollar = 40 rupees. Let X denote the random variable that models the launch in Europe, and let Y denote the random variable for India. To convert X into dollars, we have to divide by 0.75. Hence, the mean and variance of a launch in Europe are $E(X/0.75) = 2.5/0.75 = \$3.33$ million and $SD(X/0.75) = 0.75/0.75 = \$1$ million. By comparison, the mean and SD of the proposed launch in India are $E(Y/40) = 125/40 = \$3.125$ million and

---

[8] Zero: $E(x - \mu) = E(X) - \mu = \mu - \mu = 0$. Trading costs are spoilers.
[9] $E(V) = E(100X) = 100E(X) = 10\text{¢}$; $Var(V) = Var(100X) = 100^2Var(X) = 100^2 \times 4.99 = 49,900\text{¢}^2$.

SD($Y/40$) = 45/40 = \$1.125 million. If both random variables accurately reflect the potential gains, then the launch in Europe seems the better choice. This option offers both a larger mean and a smaller SD.

Sharpe Ratio

Comparisons of some random variables require adjustments even if they do have the same units. In the prior example, the launch in Europe wins on two counts: It offers larger expected sales along with smaller uncertainty. Often, however, the comparison is more mixed. When the performance of two stocks is compared, for example, the stock with higher average return usually has a higher standard deviation. For instance, during the seven years 2000–2006, stock in Disney on average grew 0.61% monthly with standard deviation 8.3%. During this same period, stock in McDonald's grew 0.53% with SD = 7.6%. Stock in McDonald's grew at a slower average rate, but the growth was steadier.

We can use these characteristics to define two random variables as models for what we expect to happen in future months, as shown in Table 9.5.

TABLE 9.5 Random variables based on monthly returns on Disney and McDonald's.

Company	Random Variable	Mean	SD
Disney	D	0.61%	8.3%
McDonald's	M	0.53%	7.6%

Which random variable represents the better investment going forward? The random variables have the same units (monthly percentage changes), but the comparison is difficult because the one with the higher average also has the higher SD.

Finance offers many ways to judge the relative size of rewards in comparison to risks. One of the most popular is the **Sharpe ratio**. The higher the Sharpe ratio, the better the investment. The Sharpe ratio is a fraction. The numerator of the Sharpe ratio is the mean return *minus* what you would earn on an investment that presents no risk (such as a savings account). The denominator of the Sharpe ratio is the standard deviation of the return on the investment. Suppose X is a random variable that measures the performance of an investment, and assume that the mean of X is μ and its SD is σ. The Sharpe ratio of X is

Sharpe ratio Ratio of an investment's net expected gain to its standard deviation.

$$S(X) = \frac{\mu - r_f}{\sigma}$$

The symbol r_f stands for the return on a risk-free investment (resembling the rate of interest on a savings account). Notice that $S(X) \neq$ SD(X).

Let's use the Sharpe ratio to compare investing in Disney and McDonald's. We'll set the risk-free rate of interest to 0.4% per month (about 5% annual interest). Subscripts on μ and σ identify the random variables. The Sharpe ratio for Disney is

$$S(D) = \frac{\mu_D - r_f}{\sigma_D} = \frac{0.61 - 0.40}{8.3} \approx 0.0253$$

The Sharpe ratio for McDonald's is

$$S(M) = \frac{\mu_M - r_f}{\sigma_M} = \frac{0.53 - 0.40}{7.6} \approx 0.0171$$

An investor who judges investments with the Sharpe ratio prefers investing in Disney to investing in McDonald's because $S(D) > S(M)$. Disney offers a higher average rate of return relative to its standard deviation.

Best Practices

- *Use random variables to represent uncertain outcomes.* Associating a random variable with a random process forces you to become specific about what you think might happen. You have to specify the possibilities and the chances of each.
- *Draw the random variable.* Pictures provide important summaries of data, and the same applies to random variables. The plot of the probability distribution of a random variable is analogous to the histogram of data. We can guess the mean and the standard deviation from the probability distribution. If you draw the probability distribution, you are less likely to make a careless mistake calculating its mean or its standard deviation.

- *Recognize that random variables represent models.* The stock market is a lot more complicated and subtle than flipping coins. We'll never know every probability. Models are useful, but they should not be confused with reality. Question probabilities as you would data. Just because you have written down a random variable does not mean that the phenomenon (such as the gains of a stock) actually behaves that way.
- *Keep track of the units of the random variables.* As with data, the units of random variables are important. It does not make sense to compare data that have different units, and the same applies to random variables.

Pitfalls

- *Do not confuse $\bar{x}$ with μ or s with σ.* Greek letters highlight properties of random variables. These come from weighting the outcomes by probabilities rather than computing them from data. Similarly, do not mistake a histogram for a probability distribution. They are similar, but recognize that the probability distribution of the random variable describes a model for what *will* happen, but it's only a model.

- *Do not mix up X with x.* Uppercase letters denote random variables, and the corresponding lowercase letters indicate a generic possible value of the random variable.
- *Do not forget to square constants in variances.* Remember that $\text{Var}(cX) = c^2\text{Var}(X)$. Variances have squared units.

CHAPTER SUMMARY

A **random variable** represents the uncertain outcome of a random process. The **probability distribution** of a random variable X identifies the possible outcomes and assigns a probability to each. The probability distribution is analogous to the histogram of numerical data. An **expected value** is a weighted sum of the possible values of an expression that includes a random variable. The expected value of a random variable itself is the **mean** $\mu = \text{E}(X)$ of the random variable. The **variance** of a random variable $\sigma^2 = \text{Var}(X)$ is the expected value of the squared deviation from the mean. The **standard deviation** is the square root of the variance. The **Sharpe ratio** of an investment is the ratio of its net expected gain to its standard deviation.

Key Terms

expected value, 201
mean μ of a random
 variable, 200
parameter, 200
probability distribution, 198

random variable, 197
 continuous, 198
 discrete, 198
Sharpe ratio, 208
standard deviation, 203
statistical model, 199

symbol
 μ (mean), 200
 σ (std. dev.), 203
 σ^2 (variance), 203
variance, 203

■ Formulas

The letters a, b, and c stand for constants in the following formulas, and subscripted lowercase symbols (such as x_1) represent possible values of the random variable X.

Mean or expected value of a discrete random variable X

Assuming that the random variable can take on any of the k possible values $x_1, x_2, \ldots, x_k$, then its expected value is

$$\mu = E(X)$$
$$= x_1 p(x_1) + x_2 p(x_2) + \cdots + x_k p(x_k)$$
$$= \sum_{i=1}^{k} x_i p(x_i)$$

Variance of a discrete random variable

Assuming that the random variable can take on any of the k possible values $x_1, x_2, \ldots, x_k$, then its variance is

$$\sigma^2 = \mathrm{Var}(X)$$
$$= E(X - \mu)^2$$
$$= (x_1 - \mu)^2 p(x_1) + (x_2 - \mu)^2 p(x_2) + \cdots + (x_k - \mu)^2 p(x_k)$$
$$= \sum_{i=1}^{k} (x_i - \mu)^2 p(x_i)$$

A short-cut formula simplifies the calculations (avoiding the need to subtract μ before squaring) and is useful if you have to do these calculations by hand:

$$\sigma^2 = E(X^2) - \mu^2$$
$$= x_1^2 p(x_1) + x_2^2 p(x_2) + \cdots + x_k^2 p(x_k) - \mu^2$$

Standard deviation of a random variable X is the square root of its variance.

$$\sigma = \sqrt{\mathrm{Var}(X)} = \sqrt{\sigma^2}$$

Adding a constant to a random variable

$$E(X \pm c) = E(X) \pm c$$
$$SD(X \pm c) = SD(X), \ \mathrm{Var}(X \pm c) = \mathrm{Var}(X)$$

Multiplying a random variable by a constant

$$E(cX) = cE(X)$$
$$SD(cX) = |c|SD(X), \ \mathrm{Var}(cX) = c^2\mathrm{Var}(X)$$

Adding a constant and multiplying by a constant

$$E(a + bX) = a + bE(X)$$
$$\mathrm{Var}(a + bX) = b^2\mathrm{Var}(X)$$

Sharpe ratio of the random variable X with mean μ and variance σ^2

$$S(X) = \frac{\mu - r_f}{\sigma}$$

The symbol r_f is the risk-free rate of interest, the rate of interest paid on savings that are not at risk of loss or default.

■ About the Data

The random variable in this chapter that is used to model changes in the value of stock in IBM has three possible values ($-\$5$, 0, and $\$5$). Even so, this model is a better match to reality than you might expect. The mean and SD of X are close to those of the daily percentage change in this stock. Percentages are relevant because we set the value of the stock to $\$100$; that way, a change of $\$5$ is a 5% change.

The histogram in Figure 9.4 summarizes percentage changes in the price of IBM stock for 2,519 trading days from January 1994 through the end of 2003.

Figure 9.4 includes a boxplot to convince you that there's a reason for the axes to extend so far from zero. Some data fall near the edges of the plot. IBM's stock rose or fell by close to 15% on several days during these 10 years.

The mean and standard deviation of X approximate the mean and standard deviation of these data. The average percentage change in IBM over these 10 years is 0.103% per day. The mean of the random variable X is $\mu = 0.100\%$. The variation is also similar. The SD of these daily returns is $s = 2.25\%$, close to $\sigma = 2.23\%$ in the example.

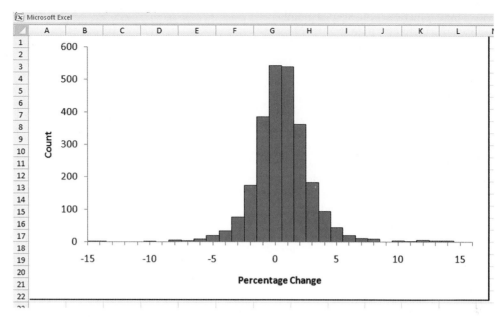

FIGURE 9.4 Daily percentage changes in IBM stock.

EXERCISES

Mix and Match

Match the descriptive phrases to the symbols on the right. In these exercises, X denotes a random variable.

1. Expected value of X	(a) $E(X - \mu)$
2. Variance of X	(b) $10X$
3. Standard deviation of X	(c) $(X - 0.04)/\sigma$
4. Shorthand notation for $P(X = x)$	(d) $X + 10$
5. Has 10 times the standard deviation of X	(e) $E(X - \mu)^2$
6. Is always equal to zero	(f) $\sqrt{\operatorname{Var}(X)}$
7. Increases the mean of X by 10	(g) μ
8. Has standard deviation 1	(h) $p(x)$

True/False

Mark each statement True or False. If you believe that a statement is false, briefly say why you think it is false.

Exercises 9–14. A cable television provider is planning to change the way that it bills customers. The new billing would depend on the number of computers and televisions that customers connect to its network. The provider has asked managers to participate in the planning by describing a random variable X that represents the number of connections used by a randomly selected current customer and a variable Y that represents the number of connections used by a randomly selected customer *after* the change in billing. All customers must have at least one connection.

9. The function defined as $p(1) = 1/2$, $p(2) = 1/4$, $p(3) = 1/8$, $p(4) = 1/8$, and $p(x) = 0$ otherwise is a probability distribution that could be used to model X.

10. The function defined as $p(1) = 1/2$, $p(2) = 1/4$, $p(3) = 1/2$, $p(4) = -1/4$, and $p(x) = 0$ otherwise is a probability distribution that could be used to model X.

11. If the new style of billing increases the average number of connections used by customers, then $E(X) > E(Y)$.

12. If a customer can have from 1 to 10 possible connections in the current setup, then $E(X)$ must also be in the range from 1 to 10.

13. The mean $E(Y)$ must be an integer because the random variable can only be an integer.

14. If the cable network counts the number of connections used by 100 customers, then the average count of these customers is the same as $E(X)$.

Exercises 15–20. An insurance company uses a random variable X to describe the actual cost of an accident incurred by a female driver who is 20 to 30 years old. A second random variable Y models the actual cost of an accident by a male driver in the same range of ages. Both random variables are measured in dollars.

15. If the mean of Y is \$3,200, then the variance of Y must be larger than \$3,200.

16. If $E(X) = \$2,300$, then $P(X \le 2,300) = 1/2$.

17. The units of both the mean and standard deviation of Y are dollars.

18. If the policy limits the coverage of an accident to \$500,000, then both $E(X)$ and $E(Y)$ must be less than \$500,000.

19. If costs associated with accidents rise next year by 5%, then the mean of the random variable X should also increase by 5%.

20. If costs associated with accidents rise next year by 5%, then the standard deviation of the random variable Y should also increase by 5%.

Think About It

The following plots show the probability functions of four random variables: X, Y, Z, and W. Use these pictures to answer Exercises 21–30. Estimate the needed probabilities from the information shown in the plots.

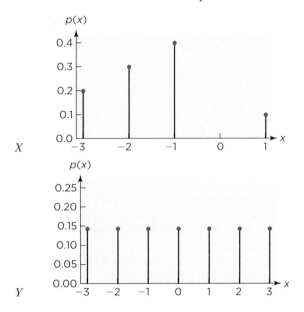

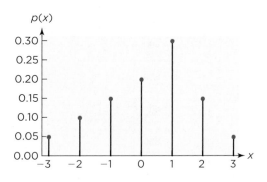

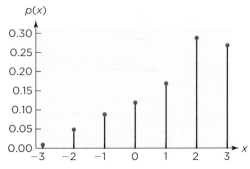

21. What is the probability $X = -2$?

22. What is the probability that $Y \ge 2.5$?

23. What is the probability that $Z \le -3$ or $Z = 3$?

24. Which is larger, $P(W > 0)$ or $P(Y > 0)$?

25. Is the $E(Z)$ positive, negative, or close to zero?

26. Which random variable has the largest man?

27. Which random variable has the largest SD? (You do not need to compute the SD from the shown probabilities to answer this question.)

28. Pick the SD of W: 0.15, 1.5, or 15. (You do not need to compute the SD from the shown probabilities to select from these three choices.)

29. The median of the random variable Y is a value m such that $P(Y \le m) = 1/2$. What is the median of Y?

30. The lower quartile of a random variable X is the smallest q so that $P(X \le q) \ge 0.25$. (This is analogous to the lower edge of the box in the boxplot.) What's the lower quartile of X?

31. A game involving chance is said to be a fair game if the expected amount won or lost is zero. Consider the following arcade game. A player pays \$1 and chooses a number from 1 to 10. A spinning wheel then randomly selects a number from 1 to 10. If the numbers match, the player wins \$5. Otherwise the player loses the \$1 entry fee.
    (a) Define a random variable W that is the amount won by the player and draw its probability distribution. (Capital letters other than X can be used for random variables.) Use negative values for losses, and positive values for winnings.
    (b) Find the mean of W. Is this a fair game?

32. Many state governments use lotteries to raise money for public programs. In a common type of lottery, a customer buys a ticket with a three-digit number

from 000 to 999. A machine (such as one with bouncing balls numbered 0 to 9) then selects a number in this range at random. Each ticket bought by a customer costs $1, whether the customer wins or loses. Customers with winning tickets are paid $500 for each winning ticket.

(a) Sketch the probability distribution of the random variable X that denotes the net amount won by a customer. (Notice that each customer pays $1 regardless of whether he or she wins or loses.)

(b) Is this a fair game? (See Exercise 31 for the definition of a fair game.) Does the state want a fair game?

(c) Interpret the expected value of X for a person who plays the lottery.

33. Another way to define a fair game is that a player's probability of winning must be equal to the player's share of the pot of money awarded to the winner. All money is put into a pot at the start of the game, and the winner claims the entire amount that is in the pot. (Compare this definition to that in Exercise 31.)

(a) The player and the host each put $20 into the pot. The player rolls a die and wins the pot if the die produces an even number. Is this game fair to the player?

(b) A player at a casino puts $1 into the pot and names a card from a standard deck (e.g., ace of diamonds). The casino puts $99 into the pot. If a randomly selected card matches the choice of the player, the player wins the pot. Is this game fair to the player?

34. Two companies are competing to define a new standard for digital recording. Only one standard will be adopted commercially, and it is unclear which standard the consumer market will choose. Company A invests $10 million to promote its standard, and Company B invests $20 million to promote its. If the outcome is to be viewed as a fair game (in the sense defined in Exercise 33), what is the chance that the market accepts the proposal of Company A?

You Do It

Bold names shown with the question identify the data table for the problem.

35. Given that the random variable X has mean $\mu = 120$ and SD $\sigma = 15$, find the mean and SD of each of these random variables that are defined from X:

(a) $X/3$
(b) $2X - 100$
(c) $X + 2$
(d) $X - X$

36. Given that the random variable Y has mean $\mu = 25$ and SD $\sigma = 10$, find the mean and SD of each of these random variables that are defined from Y:

(a) $2Y + 20$
(b) $3Y$
(c) $Y/2 + 0.5$
(d) $6 - Y$

37. An investor buys the stock of two companies, investing $10,000 in each. The stock of each company either goes up by 80% after a month (rising to $18,000) with probability 1/2 or drops by 60% (falling to $4,000) with

probability 1/2. Assume that the changes in each are independent. Let the random variable X denote the value of the amount invested after one month.

(a) Find the probability distribution of X.
(b) Find the mean value of X.
(c) Does the mean value represent the experience of the typical investor?

38. Imagine that the investor in Exercise 37 invests $10,000 for one month in a company whose stock either goes up by 80% after a month with probability 1/2 or drops 60% with probability 1/2. After one month, the investor sells this stock and uses the proceeds to buy stock in a second company. Let the random variable Y denote the value of the $10,000 investment after two months.[10]

(a) Find the probability distribution of Y.
(b) Find the mean value of Y.
(c) Does the mean value represent the experience of the typical investor?

39. A construction firm bids on a contract. It anticipates a profit of $50,000 if it gets the contract for the full project, and it estimates a profit of $20,000 on a shared project. The company estimates there's a 20% chance it will get the larger contract and a 75% chance it will get the smaller contract; otherwise, it gets nothing.

(a) Define a random variable to model the outcome of the bid.
(b) What is the expected profit earned on these contracts? Report units with your answer.
(c) What is the standard deviation of the profits?

40. A law firm takes cases on a contingent fee basis. If the case goes to trial, the firm expects to earn $25,000 as part of the settlement if it wins and nothing if it does not. The firm wins one-third of the cases that go to trial. If the case does not go to trial, the firm earns nothing. Half of the cases do not go to trial. Be sure you add the appropriate units.

(a) Define a random variable to model the earning of taking a case of this type.
(b) What is the expected value of such a case to the firm?
(c) What is the standard deviation of the earnings?

41. Even with email, offices use a lot of paper. The manager of an office has estimated how many reams (packs of 500 sheets) are used each day and has assigned these probabilities.

X = # of reams of paper used	0	1	2	3	4	5
$P(X = x)$	0.05	0.25	0.35	0.15	0.15	0.05

(a) How many reams should the manager expect to use each day?
(b) What is the standard deviation of the number of reams?
(c) If the storeroom has 20 reams at the start of the day, how many should the manager expect to find at the end of the day?

[10] See *A Mathematician Plays the Stock Market* by J. A. Paulos for similar problems.

(d) Find the standard deviation of the number of reams found at the end of the day.

(e) Find the mean and standard deviation of the number of pages of paper used each day (assuming that each ream is completely used).

42. The maintenance staff of a large office building regularly replaces fluorescent ceiling lights that have gone out. During a visit to a typical floor, the staff may have to replace several lights. The manager of this staff has given the following probabilities to the number of lights (identified by the random variable Y) that need to be replaced on a floor:

Y	0	1	2	3	4
$P(Y = y)$	0.2	0.15	0.2	0.3	0.15

(a) How many lights should the manager expect that the staff will replace on a floor?

(b) What is the standard deviation of the number of lights on a floor that are replaced?

(c) If a crew takes six lights to a floor, how many should it expect to have left after replacing those that are out?

(d) Find the standard deviation of the number of lights that remain after replacing those needed on a floor.

(e) If it takes 10 minutes to replace each light, how long should the manager expect the crew to take when replacing the lights on a floor?

43. Companies based in the United States that operate in foreign nations must convert currencies in order to pay obligations denoted in those currencies. Consider a company that has a contract that requires it to pay 1,000,000 Venezuelan bolivars for an oil shipment in six months. Because of economic uncertainties, managers at the company are unsure of the future exchange rate. Local contacts indicate that the exchange rate in six months could be 2.15 bolivars per dollar (with probability 0.6) or might rise dramatically to 5 bolivars per dollar (with probability 0.4).

(a) If managers accept these estimates, what is the expected value today, in dollars, of this contract?

(b) The expected value of the exchange rate is $2.15(0.6) + 5(0.4) = 3.29$ bolivars/dollar. If we divide 1,000,000 by 3.29, do we get the answer found in part (a)? Explain why or why not.

44. During the 1960s and into the 1970s, the Mexican government pegged the value of the Mexican peso to the U.S. dollar at 12 pesos per dollar. Because interest rates in Mexico were higher than those in the United States, many investors (including banks) bought bonds in Mexico to earn higher returns than were available in the United States. The benefits of the higher interest rates, however, masked the possibility that the peso would be allowed to float and lose substantial value compared to the dollar. Suppose you are an investor and believe that the probability that the exchange rate for the next year remains at 12 pesos per dollar is 0.9, but that the rate could soar to 24 per dollar with probability 0.1.

(a) Consider two investments: Deposit $1,000 today in a U.S. savings account that pays 8% annual interest (rates were high in the 1970s), or deposit $1,000 in a Mexican account that pays 16% interest. The latter requires converting the dollars into pesos at the current rate of 12 pesos/dollar, and then after a year converting the pesos back into dollars at whatever rate then applies. Which choice has the higher expected value in one year? (In fact, the peso fell in value in 1976 by nearly 50%, catching some investors by surprise.)

(b) Now suppose you are a Mexican with 12,000 pesos to invest. You can convert these pesos to dollars, collect 8% interest, and then convert them back at the end of the year, or you can get 16% from your local Mexican investment. Compare the expected value in pesos of each of these investments. Which looks better?

(c) Explain the difference in strategies that obtain the higher expected value.

45. A manufacturer of inexpensive printers offers a model that retails for $150. Each sale of one of these models earns the manufacturer $60 in profits. The manufacturer is considering offering a $30 mail-in rebate. Assume that a randomly selected customer has probability p of purchasing this model of printer.

(a) Assume that the availability of a rebate increases the probability of purchase from p to p^*. How much does the chance of purchase need to increase in order for the expected net profits to increase? (Assume that the manufacturer has plenty of printers to sell and that the cost to the manufacturer for the rebate is the stated $30.)

(b) Not all rebates that are offered are used, with the chance of the rebate being used hovering around 40%. If this is so, how much does the chance of purchase need to increase in order for the expected net profits to increase? Identify any assumptions you make.

46. With homeowners buying valuable appliances such as plasma TVs and electronic appliances, there's more concern about the cost of being hit by lightning.[11] Suppose that you've just bought a $4,000 flat-screen TV. Should you also buy the $50 surge protector that guarantees to protect your TV from electric surges caused by lightning?

(a) Let p denote the probability that your home is hit by lightning during the time that you own this TV, say five years. In order for the purchase of the surge protector to have positive long-term value, what must the chance for being hit by lightning be?

(b) There are about 100 million households in the United States, and fewer than 10,000 get hit by lightning each year. Do you think the surge protector is a good deal on average? Be sure to note any assumptions you make.

47. An insurance salesman visits up to three clients each day, hoping to sell a new policy. He stops for the day once he makes a sale. Each client independently decides whether to buy a policy; 10% of clients purchase the policy.

---

[11] See "For Homeowners, Lightning Threat Is Growing Worry," *Wall Street Journal*, November 21, 2006.

(a) Create a probability model for the number of clients the salesman visits each day.

(b) Find the expected number of clients.

(c) If the salesman spends about $2\frac{1}{2}$ hours with each client, then how many hours should he expect to be busy each day?

(d) If the salesman earns $3,000 per policy sold, how much can he expect to make per day?

48. An office machine costs $10,000 to replace unless a mysterious, hard-to-find problem can be found and fixed. You need this machine to keep the office moving. Repair calls from any service technician cost $500 each, and you're willing to spend up to $2,000 to get this machine fixed. You estimate that a repair technician has a 25% chance of fixing it.

(a) Create a probability model for the number of visits needed to fix the machine or exhaust your budget of $2,000. (That is, there can be at most four visits.)

(b) Find the expected number of service technicians you'll call in.

(c) Find the expected amount spent on this machine. (You must spend $10,000 for a new one if you do not find a vendor that repairs yours.)

49. The ATM at a local convenience store allows customers to make withdrawals of $10, $20, $50, or $100. Let X denote a random variable that indicates the amount withdrawn by the customer who just entered the store. The probability distribution of X is

$$p(10) = 0.2$$
$$p(20) = 0.5$$
$$p(50) = 0.2$$
$$p(100) = 0.1$$

(a) Draw the probability distribution of X.

(b) What is the probability that a customer withdraws more than $20?

(c) What is the expected amount of money withdrawn by a customer?

(d) The expected value is not a possible value of the amount withdrawn. Interpret the expected value for a manager.

(e) Find the variance and standard deviation of X.

50. A company orders components from Japan for its game player. The prices for the items that it orders are in Japanese yen. When the products are delivered, it must convert dollars into yen and then make the transfer to the Japanese producer. When its next order is delivered (and must be paid for), it believes that the exchange rate of dollars to yen will take on the following values with the shown probabilities.

Conversion Rate (yen per dollar)	Probability
100	0.1
110	0.2
120	0.4
130	0.2
140	0.1

(a) Draw the probability distribution of a random variable X that denotes the exchange rate on the day of the delivery. Describe the shape of this distribution.

(b) If the current exchange rate is 120 ¥/$, what is the probability that the exchange rate will be higher on the delivery date?

(c) What is the expected value of the exchange rate on the date of the delivery?

(d) If the company exchanges $100,000 on the delivery date, how many yen can it expect to receive?

(e) What is the expected value in dollars of ¥10,000 on the delivery date?

51. **Iverson** While playing basketball for the Philadelphia 76ers, Allen Iverson had the following results during the 2005–2006 NBA season (as of February 5, 2005).[12] This table lists the number of baskets made of several types along with the number of attempts.

Type of Shot	Made	Attempts
Free throw (1 point)	382	481
Field goal (2 points)	462	990
3-point goal	47	133

Use the implied percentages for each type of shot as probabilities to answer the following questions:

(a) Which opportunity has the highest expected value for Iverson: two free throw attempts, one field goal attempt, or one attempted 3-point shot?

(b) Which has larger variation in outcome for Iverson, attempt a 2-point basket or attempt a 3-point basket?

(c) If the 76ers are losing by 1 point near the end of a game, which of the following offers the bigger chance for a win (i.e., scoring 2 points) by the 76ers: Iverson attempts a field goal or Iverson attempts two free throws? Are any assumptions needed for this comparison?

52. **Akers** In seven professional football seasons from 1999 to 2005, kicker David Akers of the Philadelphia Eagles recorded the following results for field goal attempts of various distances.

Distance	Made	Attempts
0–19	2	2
20–29	40	40
30–39	55	61
40–49	48	69
50+	10	17

Use the implied percentages as probabilities to answer the following questions:

(a) A made field goal scores 3 points. What is the expected value (in points scored) if Akers attempts a 45-yard kick?

[12] From http://www.usatoday.com/sports/basketball.

(b) During a game, Akers attempts kicks of 27, 38, and 51 yards. What is the expected contribution of these kicks to the Eagles' score in the game? What assumptions do you need?

(c) The Eagles have the ball and can allow Akers to attempt a 37-yard kick. They can also go for a first down. If they go for a first down and fail, they score nothing. If they make the first down, half of the time Akers will get a chance to kick from less than 30 yards and half of the time they score a touchdown (worth 7 points). They have a 35% chance to make the first down. What should they do?

53. **Daily Stocks** These data are daily percentage changes of several stocks. Use these data to compute the Sharpe ratio for Exxon. Here are the steps.

(a) Describe the shape and any key features of the histogram.

(b) Find the mean and standard deviation of the percentage changes. Use the properties of these data to define a random variable that is the percentage change in Exxon stock in a future day.

(c) If the interest cost for a daily investment in Exxon is 2 basis points (i.e., 0.02% per day), find the Sharpe ratio for a daily investment of $1,000 in this stock.

(d) Find the Sharpe ratio for a daily investment of $10,000 in this stock.

(e) How does the Sharpe ratio for Exxon compare to that for owning IBM? (Use the mean and SD for IBM from the text example.)

(f) If an investor compares the Sharpe ratio of Exxon to that for IBM to decide which investment is better, then what assumptions about future returns are being made?

(g) Is it okay to summarize the data by collecting them into a histogram, or does the histogram mask important patterns?

54. Use the data from Exercise 53 to compute the Sharpe ratio for Disney. Here are the steps.

(a) Describe the shape and any key features of the histogram.

(b) Find the mean and standard deviation of the percentage changes. Use the properties of these data to define a random variable that is the percentage change in Disney stock in a future day.

(c) If the interest cost for a daily investment in Disney is 2 basis points (i.e., 0.02% per day), find the Sharpe ratio for a daily investment of $1,000 in this stock.

(d) Does the Sharpe ratio indicate whether the investor is better off with more money or less money invested in Disney? That is, can an investor use the Sharpe ratio to decide whether to invest $1,000, $5,000, or $10,000 in Disney?

(e) How does the Sharpe ratio for Disney compare to that for IBM? (Use the mean and SD for IBM from the text.)

(f) If an investor compares the Sharpe ratio for Disney to that for IBM to decide which investment is better, then what assumptions about future returns are being made?

(g) Does this sequence of percentage changes show a pattern? Is it okay to summarize the data by collecting them into a histogram, or does this summary hide important patterns?

4M Project Management

A construction company has two large building projects that it manages. Both require about 50 employees with comparable skill levels. At a meeting, the site managers from the two projects got together to estimate the labor needs of their projects during the coming winter season. At the end of the meeting, each manager provided a table to summarize estimates for labor needs as determined by the weather.

Site 1:

	Winter Conditions			
	Mild	**Typical**	**Cold**	**Severe**
Number of labor employees	100	70	40	20

Site 2:

	Winter Conditions			
	Mild	**Typical**	**Cold**	**Severe**
Number of labor employees	80	50	40	30

A subsequent discussion with experts at the National Weather Service (NWS) indicated the following probabilities for the various types of winter weather.

	Winter Conditions			
	Mild	**Typical**	**Cold**	**Severe**
Probability	0.30	0.40	0.20	0.10

Motivation

a) Why should the company consult with these site managers about the labor needs of their project for the coming winter season?

b) Is it more useful to have a probability distribution for the weather, or should management base its decisions on only the most likely form of winter weather (i.e., typical weather)?

Method

c) Identify the key random variable for planning the labor needs of these two sites.

Mechanics

d) Find the probability distribution for the number of labor employees needed at both sites.

e) Find the expected total number of labor employees needed for both sites.

f) Find the standard deviation of the total number of labor employees needed at both sites.

Message

g) Summarize your findings for management. How many laborers do you think will be needed during the coming season? Are you sure about your calculation? Why or why not?

4M Credit Scores

Credit scores rate the quality of a borrower, suggesting the chances that a borrower will repay a loan. Borrowers with higher scores are more likely to repay loans on time. Borrowers with lower scores are more likely to default. For example, those with scores below 600 have missed 2.5 payments on average in the last six months, compared to almost no late payments for those with scores above 700. A credit score below 650 is often considered subprime. Borrowers with lower ratings have to pay a higher rate of interest to obtain a loan.

Credit scores were once only used to determine the interest rate on loans and decide whether a borrower was qualified for a $200,000 mortgage, for example. Recently, other businesses have begun to use them as well. Some auto insurance policies are more expensive for drivers with lower credit scores because the companies have found that drivers with lower scores are more risky when it comes to accidents as well.

An insurance agent has just opened an office in a community. Conversations with the local bank produced the probabilities of scores shown in this table. The table also shows how much insurance costs customers.

Credit Score of Borrower	Annual Interest Rate	Annual Premium for Car Insurance	Probability of Score
700–850	6%	$500	0.60
650–699	7%	$600	0.10
600–649	8%	$800	0.15
500–599	10%	$1,100	0.10
Below 500	14%	$1,500	0.05

Motivation

a) The insurance agent is paid by commission, earning 10% of the annual premium. Why should he care about the expected value of a random variable?

b) Why should he also care about the variance and standard deviation?

Method

c) Identify the random variable described in this table that is most relevant to the agent.

d) How is this random variable related to his commission?

Mechanics

e) Graph the probability distribution of the random variable of interest to the agent.

f) Find the expected commission earned by the agent for writing one policy.

g) Find the variance and standard deviation of the commission earned by the agent for writing one policy.

Message

h) Summarize your results for the agent in language he can appreciate.

i) Do you have any advice for the insurance company based on these calculations?

Association between Random Variables

THREE STOCK EXCHANGES DOMINATE IN THE UNITED STATES: the New York Stock Exchange (NYSE), the American Stock Exchange (AMEX), and the NASDAQ.

The NYSE is the largest of these exchanges. It lists stock in 2,500 companies worth *trillions* of dollars. The NYSE reaches outside of the United States and includes stock in 450 foreign companies. Electronic trading has become more popular on the NYSE, but you can still find traders yelling out orders on the floor of the exchange as in the "good old days."

The AMEX lists generally smaller companies than the NYSE and offers more trading in options, contracts that allow the holder to buy or sell a stock at a future point in time. The NASDAQ is an electronic stock exchange. The NASDAQ became well known in the 1990s with the surge in technology stocks. Technology companies listed on the NASDAQ include Amazon, Apple, eBay, Google, Intel, Microsoft, and Yahoo from the United States as well as Infosys (India), Research in Motion (Canada), and Ericsson (Sweden).

Together, these three exchanges offer investors thousands of choices. Most investors take advantage of the variety and diversify. They spread their money over many stocks, forming a portfolio of investments. If the price of one stock decreases, the hope is that others in the portfolio will make up for the loss.

To understand the properties of portfolios, we need to manipulate several random variables at once. That's the topic of this chapter. If the random variables were independent of each other, this would be simple to do, but stocks tend to rise and fall together. Independence is not a good model for stocks or many other situations. For instance, the success of one business or product line often presages the success—or failure—of another. Consumers often mimic one another. Random variables for these events must capture this dependence. Several important ideas in this chapter echo properties of data covered in Chapters 5 and 6. For instance, covariance and correlation summarize the dependence of random variables in a manner analogous to the dependence in data.

10.1 | PORTFOLIOS AND RANDOM VARIABLES

Investors choose stocks on the basis of what they believe will happen. The decision of how to allocate money among the several stocks that form a portfolio depends on how one thinks the stocks will perform in the future. We can model these possibilities as random variables, with one random variable representing each stock in the portfolio. We concentrate on portfolios of two stocks and the role of dependence.

Two Random Variables

Let's offer the day trader from Chapter 9 the opportunity to buy shares in two companies, IBM and Microsoft. For this example, we assume that a share of IBM or Microsoft stock costs $100 today. Two random variables describe how the values of these stocks change from day to day. The random variable X denotes the change in the value of stock in IBM; it is the same random variable used in Chapter 9. The random variable Y represents the change in Microsoft. Table 10.1 shows the probability distributions for X and for Y.

TABLE 10.1 Probability distributions for two stocks.

	IBM Stock		Microsoft Stock	
	x	$P(X = x)$	y	$P(Y = y)$
Increases	$5	0.11	$4	0.18
No change	0	0.80	0	0.67
Decreases	−$5	0.09	−$4	0.15

As in Chapter 9, we've limited each random variable to three outcomes so that we can illustrate several calculations.

In these calculations, we need to distinguish the mean and variance of X from those of Y. In Chapter 9, we considered one random variable at a time, so the symbol μ was unambiguous; it was the mean of that random variable. With two (or more) random variables, we have to distinguish which mean goes with which variable. We will use subscripts and write μ_X for the mean of X and μ_Y for the mean of Y. Similarly, σ_X is the standard deviation of X and σ_Y is the standard deviation of Y. For the probability distributions in Table 10.1, $p(x)$ denotes the probability distribution of X and $p(y)$ denotes the distribution of Y. Because you cannot tell whether $p(2)$ means $P(X = 2)$ or $P(Y = 2)$, we'll add a subscript when needed. For example, $p_X(2) = P(X = 2)$.

Comparisons and the Sharpe Ratio

Assume that the day trader can invest $200 in one of three ways. She can put all $200 into IBM or into Microsoft (buying two shares of one or two shares of the other), or she can divide her investment equally between the two (buying one share of each). She is going to borrow the $200 that she invests. Her savings account earns 0.015% for the day (about 4% per year), so the lost interest amounts to 3 cents. Which portfolio should she choose? One of the many ways to compare portfolios is to find the Sharpe ratio of each. We introduced the Sharpe ratio in Chapter 9. The numerator of the Sharpe ratio holds the expected gains after subtracting lost interest. The denominator is the standard deviation. The investment with the largest Sharpe ratio offers the greatest reward (higher expected gain) relative to the risk (measured by the standard deviation). The Sharpe ratio is typically computed from returns. Since the price of shares in both IBM and Microsoft is equal to $100, dollar changes are percentage changes.

Let's review the calculation of the Sharpe ratio. The mean and SD for investing $100 in IBM for a day are $\mu_X = \$0.10$ with $\sigma_X = \$2.23$. Hence, the Sharpe ratio for X is

$$S(X) = \frac{\mu_X - \$.015}{\sigma_X} = \frac{\$.10 - \$.015}{\$2.23} \approx 0.038$$

The constant $\$.015$ in the numerator is the interest sacrificed by investing $100 in IBM instead of leaving it in the bank, 1.5 cents. The Sharpe ratio has no units; those in the numerator cancel those in the denominator.

The Sharpe ratio remains the same if she invests all $200 in IBM. If she doubles her investment to $200, her earnings are $2X$, twice what she gets if she invests $100. She either earns $10, breaks even, or loses $10. Using the rules from Chapter 9 (repeated in the margin), the mean and SD of $2X$ are

$$E(2X) = 2\mu_X \quad \text{and} \quad SD(2X) = 2\sigma_X$$

The Sharpe ratio of investing $200 in IBM for a day is then

$$S(2X) = \frac{2(\mu_X - 0.015)}{2\sigma_X} = \frac{\mu_X - 0.015}{\sigma_X} = S(X)$$

The common factor 2 cancels in the numerator and denominator; the Sharpe ratio is the same regardless of how much she invests in IBM.

Let's compare this to the Sharpe ratio for Microsoft. Again, the Sharpe ratio is the same regardless of how much she invests in Microsoft. The Sharpe ratio for investing in Microsoft is

$$S(Y) = \frac{\mu_Y - 0.015}{\sigma_Y}$$

For practice, we'll find μ_Y directly from the probability distribution $p(y)$. The mean μ_Y is the weighted average of the three outcomes.

$$\mu_Y = 4(0.18) + 0(0.67) - 4(0.15)$$
$$= \$.12$$

The calculation of the variance is similar, though more tedious because we have to subtract the mean and then square each deviation.

$$\sigma_Y^2 = (4 - 0.12)^2(0.18) + (0 - 0.12)^2(0.67) + (-4 - 0.12)^2(0.15)$$
$$\approx 5.27$$

Even though changes in the value of Microsoft are closer to zero than those of IBM ($4 rather than $5), the probability distribution of Y places more weight on the nonzero outcomes. Consequently, the variance of Y is larger than the variance of X even though its range is smaller. Plugging the values of μ_Y and σ_Y^2 into the formula for $S(Y)$ gives the Sharpe ratio for investing in Microsoft, which is larger than the Sharpe ratio for investing in IBM.

$$S(Y) = \frac{\mu_Y - \$.015}{\sigma_Y} = \frac{\$.12 - \$.015}{\$\sqrt{5.27}} \approx 0.046$$

multiplying a random variable by a constant

$$E(cX) = cE(X)$$
$$SD(cX) = cSD(X)$$
$$Var(cX) = c^2Var(X)$$

Table 10.2 summarizes the two single-stock portfolios.

TABLE 10.2 Means, variances, standard deviations, and Sharpe ratios for X and Y.

		Mean	Variance	SD	Sharpe Ratio
IBM	X	$.10	4.99	$2.23	0.038
Microsoft	Y	$.12	5.27	$2.30	0.046

If the day trader invests in one stock, the Sharpe ratio favors Microsoft. The larger mean of Microsoft compensates for its greater risk (larger SD). The next section considers if she can do better by diversifying her investment.

10.2 | JOINT PROBABILITY DISTRIBUTION

To complete the comparison, we need the Sharpe ratio for the portfolio that divides $200 equally between IBM and Microsoft. That will take more effort because this portfolio combines two different random variables.

To find the Sharpe ratio for this portfolio, we need the expected value and SD of $X + Y$. We can find the mean of $X + Y$ from first principles. The mean of $X + Y$ is an expected value, so we need the probability distribution of the sum. For example, what's the probability $P(X + Y = 9)$? That event occurs only when $X = 5$ and $Y = 4$, implying that

$$P(X + Y = 9) = P(X = 5 \text{ and } Y = 4)$$

We need the probability that two events happen simultaneously—the probability of an intersection. The distributions $p(x)$ and $p(y)$ do not answer such questions unless we assume independence. If the random variables are independent, then the probability is the product

$$P(X = 5 \text{ and } Y = 4) = P(X = 5) \times P(Y = 4)$$

Stocks are seldom independent, however. More often, shares tend to rise and fall together like boats rising and falling with the tides.

Unless we know that events are independent, we need the joint probability distribution of X and Y. The **joint probability distribution** of X and Y, labeled $p(x, y)$, gives the probability of events of the form $(X = x \text{ and } Y = y)$. Such events describe the simultaneous outcomes of *both* random variables. The joint probability distribution resembles the percentages shown in a contingency table, but with probabilities in place of relative frequencies. (If we believe the percentages observed in past data also describe the chances in the future, we can use these percentages to estimate the joint probability distribution.) Table 10.3 gives the joint probability distribution of X and Y.

joint probability distribution The probability of two (or more) random variables taking on specified values simultaneously, denoted as $P(X = x \text{ and } Y = y) = p(x, y)$.

TABLE 10.3 Joint probability distribution of X and Y.

		X			
		$x = -5$	$x = 0$	$x = 5$	$p(y)$
Y	$y = 4$	0.00	0.11	0.07	0.18
	$y = 0$	0.03	0.62	0.02	0.67
	$y = -4$	0.06	0.07	0.02	0.15
	$p(x)$	0.09	0.80	0.11	1

The marginal distributions along the bottom and right of Table 10.3 are consistent with those in Table 10.1. This is no accident. Once you have the joint probability distribution, the probability distribution of each random variable

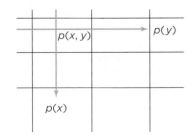

is the sum of the probabilities in the rows or the columns. The calculations are precisely those used to find the margins of a contingency table (Chapter 5). For example, $P(X = 0)$ is

$$P(X = 0) = P(X = 0 \text{ and } Y = 4) + P(X = 0 \text{ and } Y = 0)$$
$$+ P(X = 0 \text{ and } Y = -4)$$
$$= 0.11 + 0.62 + 0.07 = 0.80$$

The probabilities add because the events are disjoint. Only one of these pairs can happen tomorrow.

Independent Random Variables

The joint probability distribution determines whether X and Y are independent. In order for random variables to be independent, the joint probability must match the product of the marginal probabilities.

Independent random variables   Two random variables are independent if (and only if) the joint probability distribution is the product of the marginal probability distributions.

X and Y are independent $\iff$ $p(x, y) = p(x) \times p(y)$ for all x, y

---

The definition of independent random variables resembles that for independent events. Two events **A** and **B** are *independent* if $P(\mathbf{A} \text{ and } \mathbf{B}) = P(\mathbf{A}) \times P(\mathbf{B})$. When you hear *independent*, think "multiply." An important consequence of independence is that expected values of products of independent random variables are products of expected values.

Multiplication Rule for the Expected Value of a Product of Independent Random Variables   *The expected value of a product of **independent** random variables is the product of their expected values.*

$$E(XY) = E(X)E(Y)$$

EXCHANGE RATES

MOTIVATION | STATE THE QUESTION

Companies in the United States that do business internationally worry about currency risk. If a customer in Europe pays in euros, a company in the United States has to convert those euros into dollars to pay its employees. Exchange rates between currencies fluctuate, so €100 today won't convert to the same number of dollars tomorrow. Whatever the exchange rate is today, it will probably be different a week from now. The timeplot on the next page shows how the rate varied in 2008 and early 2009.

Consider the situation faced by a firm whose sales in Europe average €10 million each month. The firm must convert these euros into dollars. The current exchange rate faced by the company is 1.40 $/€, but it could go lower or higher. We may not know the direction, but we can anticipate that the exchange rate will change in the future. What should this firm expect for the dollar value of European sales next month?

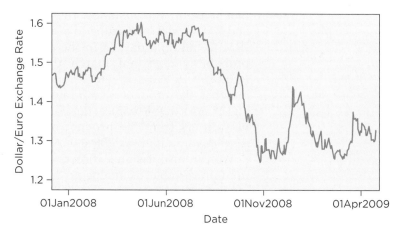

DESCRIBE THE DATA AND SELECT AN APPROACH

The language of random variables lets us describe the question and form our answer concisely. This particular problem involves three random variables. One random variable, call it S, stands for sales next month in euros. The random variable R stands for the exchange rate next month when the conversion takes place. The random variable D is the value of sales in dollars. These are related by the equation $D = S \times R$.

To find $E(D)$, we will assume that sales S is independent of the exchange rate R. This assumption simplifies the calculations because independence implies that the expected value of a product is the product of the expected values.

DO THE ANALYSIS

Unless we have an insight into the future of exchange rates, a commonsense estimate of the expected value of the exchange rate is the current rate, $E(R) = 1.40$ \$/€. Because S and R are (assumed to be) independent,

$$E(D) = E(S \times R) = E(S) \times E(R) = €10{,}000{,}000 \times 1.40 = \$14 \text{ million}$$

SUMMARIZE THE RESULTS

On the basis of typical sales of €10 million and the current exchange rate, European sales of this company next month convert to $14 million, on average. This calculation does, however, require two important assumptions. To obtain this estimate, we assumed that sales next month were on average the same as in the past and that changes in sales are independent of changes in the exchange rate. *Be sure to note when your answer requires a major assumption.*

caution It is much more difficult to find $E(D)$ in the currency example if the random variables are dependent. Assuming that they are independent simplifies the calculations, but that does not make it true. There's a good chance they might be dependent. Exchange rates often track perceptions of economic strength. For example, if the European economy is growing faster than the economy in the United States, you might see the number of dollars per euro increase. At the same time, European sales might also grow along with the economy. That combination would make the exchange rate and sales dependent: A higher exchange rate would come with larger sales—a double win. Of course, both could go the other way. If the variables are dependent, you need the joint distribution to find the expected value of the product.

Dependent Random Variables

The joint probability distribution in Table 10.3 implies that X and Y are not independent. For instance, the joint distribution indicates that $P(X = -5$ and $Y = 4) = 0$. According to $p(x, y)$, it's not possible for the price of stock in IBM to fall on a day when the price of stock in Microsoft rises. This joint probability does not match the product of the marginal probabilities $(0.09 \times 0.18 \neq 0)$. In fact, none of the joint probabilities in Table 10.3 matches the product of the marginal probabilities. For instance, $P(X = 5$ and $Y = 4) \neq P(X = 5) \times P(Y = 4)$. Because the joint probability distribution differs from the product of the marginal distributions, X and Y are *dependent* random variables.

We can also assign a direction to the dependence. Dependence between random variables is analogous to association in data. Recall the description of association in scatterplots (Chapter 6). The dependence between changes in the values of IBM and Microsoft is *positive*. Both random variables are more likely to move in the same direction rather than in opposite directions. There's a much higher probability that both will rise (or that both will fall) than would be the case were these random variables independent.

10.3 | SUMS OF RANDOM VARIABLES

Products of random variables are common, but sums of random variables are everywhere. It's a good thing, too, that sums occur more often. The expected value of a product is the product of the expected values *only* when the random variables are independent. Sums are another matter. The expected value of a sum of random variables is the sum of the expected values, always. We don't have to restrict this rule to independent random variables. It *always* works.

Addition Rule for the Expected Value of a Sum of Random Variables
The expected value of a sum of random variables is the sum of their expected values.

$$E(X + Y) = E(X) + E(Y)$$

This rule says, "The average of a sum is the sum of the averages." The Addition Rule extends to sums of more than two random variables. For example,

$$E(X + Y + W + Z) = E(X) + E(Y) + E(W) + E(Z)$$

This rule makes it easy to find the mean of the portfolio that mixes IBM and Microsoft. We know $\mu_X = 0.10$ and $\mu_Y = 0.12$, so the mean of the sum is

$$E(X + Y) = \mu_X + \mu_Y = 0.10 + 0.12 = 0.22$$

(p. 235) Without this rule, finding $E(X + Y)$ requires a tedious calculation (shown in Behind the Math: Life without the Addition Rule).

The same good fortune does *not* apply to the variance. In general, the variance of a sum is *not* the sum of the variances. The sum of the variances in this example is

$$\sigma_X^2 + \sigma_Y^2 \approx 4.99 + 5.27 = 10.26$$

(p. 235) The variance of the sum is in fact much larger, $\text{Var}(X + Y) = 14.64$ (worked out in Behind the Math: The Variance of a Sum). The value of the portfolio with both stocks is more variable than the sum of the variances because of the positive dependence between X and Y. Positive dependence increases the variance of the sum because large values of X tend to happen with large values of Y, and small values of X with small values of Y. By increasing the chance that the largest and smallest pair up, positive dependence increases the chance of big deviations from $\mu_X + \mu_Y$ and inflates the variance.

Although its value is more variable than we might have guessed, the mixed portfolio has a larger Sharpe ratio than those that invest in IBM or Microsoft alone. The Sharpe ratio of the mixed portfolio that combines IBM with Microsoft is

$$S(X + Y) = \frac{(\mu_X + \mu_Y) - 2r_f}{\sqrt{\text{Var}(X + Y)}} \approx \frac{0.22 - 0.03}{\sqrt{14.64}} \approx 0.050$$

Diversifying, spreading the investment over two stocks, improves the Sharpe ratio, but not as much as adding the variances would suggest. Table 10.4 compares the Sharpe ratios of the different portfolios.

TABLE 10.4 Sharpe ratios of three portfolios based on stock in IBM and Microsoft.

Portfolio	Sharpe Ratio
All IBM	0.038
All Microsoft	0.046
Mix of IBM and Microsoft	**0.050**

What Do You Think?

Shoppers at a convenience store buy either a 12-ounce can or 32-ounce bottle of soda along with either a 4-ounce or 16-ounce bag of chips. The random variable X denotes the amount of soda, and the random variable Y denotes the amount of chips. The means and variances of the marginal distributions are

$$\mu_x = 23 \text{ ounce}, \sigma_x^2 = 99 \qquad \mu_y = 8.8 \text{ ounce}, \sigma_y^2 = 34.56$$

		X	
		12 oz.	**32 oz.**
Y	**4 oz.**	0.30	0.30
	16 oz.	0.15	0.25

a. Find the marginal distributions of X and Y.[1]
b. What is the expected total weight of a purchase? (Assume that liquid ounces weigh an ounce.)[2]
c. Are X and Y positively dependent, negatively dependent, or independent?[3]
d. Is the variance of the total weight $X + Y$ equal to, larger than, or smaller than $\sigma_x^2 + \sigma_y^2$? You don't need to find the variance; just explain your choice.[4]

10.4 | DEPENDENCE BETWEEN RANDOM VARIABLES

We can find the variance of $X + Y$ from the joint probability distribution $p(x, y)$, but that's tedious. It requires a lot of calculation as well as the joint distribution. Think of specifying the joint distribution $p(x, y)$ of the stock values if the random variables X and Y were measured to the penny; we would

---

[1] $p(x) = \{0.45, 0.55\}$ for the two columns; $p(y) = \{0.6, 0.4\}$ for the two rows.
[2] $E(X + Y) = 23 + 8.8 = 31.8$ ounces.
[3] Dependent; for example, $P(X = 12, Y = 4) = 0.3 \neq 0.6 \times 0.45$. There's positive dependence. Someone who wants a lot of chips wants a lot of soda to go with them.
[4] The variance of the sum is larger than the sum of the variances because of positive dependence. In case you're curious, the variance is $0.3(16 - 31.8)^2 + 0.3(36 - 31.8)^2 + 0.15(28 - 31.8)^2 + 0.25(48 - 31.8)^2 = 147.96$.

have to assign hundreds of probabilities. Fortunately, we don't need the joint probability distribution to find the variance of a sum if we know the covariance of the random variables.

Covariance

The covariance between random variables resembles the covariance between columns of data. The covariance between numerical variables x and y is the average product of deviations from the means (Chapter 6).

$$\text{cov}(x, y) = \frac{(x_1 - \bar{x})(y_1 - \bar{y}) + (x_2 - \bar{x})(y_2 - \bar{y}) + \cdots + (x_n - \bar{x})(y_n - \bar{y})}{n - 1}$$

It is almost an average; the divisor $n - 1$ matches the divisor in s^2. The covariance of random variables replaces this average by an expected value.

The **covariance between random variables** is the expected value of the product of deviations from the means.

$$\text{Cov}(X, Y) = E((X - \mu_X)(Y - \mu_Y))$$

$\text{Cov}(X, Y)$ is positive if the joint distribution puts more probability on outcomes with X and Y both larger or both smaller than μ_X and μ_Y, respectively. It is negative if the joint distribution assigns more probability to outcomes in which $X - \mu_X$ and $Y - \mu_Y$ have opposite signs. Table 10.5 repeats the joint distribution for the stock values and highlights the largest probability in each column.

TABLE 10.5 Joint probability distribution with highlighted cells to show the largest column probability.

			X	
		$x = -5$	$x = 0$	$x = 5$
	$y = 4$	0.00	0.11	0.07
Y	$y = 0$	0.03	0.62	0.02
	$y = -4$	0.06	0.07	0.02

(p. 235)

The highlighted cells lie along the positive diagonal of the table, suggesting positive dependence between X and Y. As shown in Behind the Math: Calculating the Covariance, the covariance is indeed positive: $\text{Cov}(X, Y) = 2.19\ \$^2$. The units of the covariance match those of the variance in this example (squared dollars) because both random variables are measured on a dollar scale. In general, the units of $\text{Cov}(X, Y)$ are those of X times those of Y.

Covariance and Sums

Aside from confirming the presence of positive dependence between the values of these two stocks, what's the importance of covariance? Not much, unless you want to know the variance of a sum. For that task, it's perfect.

Addition Rule for Variances   *The variance of the sum of two random variables is the sum of their variances **plus** twice their covariance.*

$$\text{Var}(X + Y) = \text{Var}(X) + \text{Var}(Y) + 2\text{Cov}(X, Y)$$

The covariance measures the effect of dependence on the variance of the sum. If $\text{Cov}(X, Y) > 0$, the variance of the sum of the two random variables is larger than the sum of the variances. That's what happens in the example of the stocks. Using the Addition Rule for Variances,

$$\text{Var}(X + Y) = \text{Var}(X) + \text{Var}(Y) + 2\,\text{Cov}(X, Y)$$
$$= 4.99 + 5.27 + 2 \times 2.19$$
$$= 14.64 \;\$^2$$

This simple calculation—possible if you know the covariance—gives the same variance as that obtained by lengthy direct calculations.

The lower the covariance falls, the smaller the variance of the sum. If $\text{Cov}(X, Y) < 0$, then the variance of $X + Y$ is smaller than the sum of the variances. For the day trader, this means that investing in stocks that have negative covariance produces a better portfolio than investing in stocks with positive covariance (assuming equal means and variances for the stocks). Consider what the Sharpe ratio would have been had the covariance between IBM and Microsoft been negative, say -2.19. The variance of $X + Y$ would be much smaller.

$$\text{Var}(X + Y) = \text{Var}(X) + \text{Var}(Y) + 2\text{Cov}(X, Y) = 4.99 + 5.27 - 2 \times 2.19 = 5.88$$

The actual Sharpe ratio is 0.050 (Table 10.4). With negative dependence, the Sharpe ratio would have been 50% larger.

$$S(X + Y) \approx \frac{0.22 - 0.03}{\sqrt{5.88}} \approx 0.078$$

Correlation

Covariance has measurement units that make it difficult to judge the strength of dependence. The covariance between IBM and Microsoft stock is 2.19 $\$^2$. Is that a lot of dependence? With units like these, it's hard to tell. The correlation produces a more interpretable measure of dependence by removing the measurement units from the covariance. The definition of the correlation between random variables is directly comparable to that of the correlation between numerical variables (Chapter 6).

The **correlation between two random variables** is the covariance divided by the product of standard deviations.

$$\text{Corr}(X, Y) = \frac{\text{Cov}(X, Y)}{\sigma_X \sigma_Y}$$

Since σ_X has the same units as X and σ_Y has the same units as Y, the correlation has no measurement units. It is a scale-free measure of dependence. Because the correlation is a parameter of the joint distribution of X and Y, it too is denoted by a Greek letter. The usual choice is the letter ρ (rho, pronounced "row"). The correlation between the stock values is

$$\rho = \text{Corr}(X, Y) = \frac{\text{Cov}(X, Y)}{\sigma_X \sigma_Y} \approx \frac{2.19}{\sqrt{4.99 \times 5.27}} \approx 0.43$$

Because it removes the units from the covariance, the correlation does not change if we change the units of X or Y. Suppose we measured the change in value of X and Y in cents rather than dollars. For example, X could be -500, 0, or 500 cents. This change in units increases the covariance by 100^2 to 21,900 $\textcent^2$. The correlation remains 0.43 whether we record the values in dollars or cents.

The correlation between random variables shares another property with the correlation in data: It ranges from -1 to $+1$.

$$-1 \le \text{Corr}(X, Y) \le 1$$

For the correlation to reach -1, the joint probability distribution of X and Y would have to look like Table 10.6.

TABLE 10.6 A joint distribution with negative correlation $\rho = -1$.

			X	
		$x = -5$	$x = 0$	$x = 5$
	$y = 4$	p_1	0	0
Y	$y = 0$	0	p_2	0
	$y = -4$	0	0	p_3

All of the probability would have to concentrate in cells on the negative diagonal. The largest value of X pairs with the smallest value of Y and the smallest value of X pairs with the largest value of Y. Any outcome that produces the middle value for X has to produce the middle value for Y. At the other extreme, for the correlation to reach its maximum, 1, the joint probability distribution of X and Y would have to look like Table 10.7.

TABLE 10.7 A joint distribution with positive correlation $\rho = 1$.

			X	
		$x = -5$	$x = 0$	$x = 5$
	$y = 4$	0	0	p_3
Y	$y = 0$	0	p_2	0
	$y = -4$	p_1	0	0

All of the probability concentrates on the positive diagonal. The largest value of X pairs with the largest value of Y, values in the middle pair up, and the smallest value of X pairs with the smallest value of Y. (Compare Table 10.7 to the corresponding scatterplot of data with correlation 1 shown in Figure 6.8 of Chapter 6.)

You can calculate the correlation from the covariance or the covariance from the correlation if the standard deviations are known. If the correlation is known, then the covariance is

$$\text{Cov}(X, Y) = \text{Corr}(X, Y)\sigma_X\sigma_Y$$

The correlation and standard deviations are enough to find the variance of a sum and the Sharpe ratio of the portfolio. We don't need the entire joint distribution.

Covariance, Correlation, and Independence

Independence is a fundamental concept in probability. Events or random variables are independent if whatever happens in one situation has no effect on the other. So, if two random variables are independent, what's their covariance? Zero.

To see that independence implies that the covariance is zero, think back to the definition of *independence*. If random variables are independent, then the expected value of the product is the product of the expected values. The covariance *is* the expected value of a product. If X and Y are independent, the covariance is zero.

$$\begin{aligned}\text{Cov}(X, Y) &= \text{E}((X - \mu_X)(Y - \mu_Y)) \\ &= \text{E}(X - \mu_X) \times \text{E}(Y - \mu_Y) \\ &= 0 \times 0 = 0\end{aligned}$$

uncorrelated Random variables are uncorrelated if the correlation between them is zero; $\rho = 0$.

Since correlation is a multiple of covariance, independence implies that the correlation is also zero. The covariance and correlation between independent random variables X and Y are zero. If the correlation between X and Y is zero, then X and Y are **uncorrelated**.

Because independence implies the covariance is zero, the expression for variance of a sum of independent random variables is simpler than the general result. The covariance is zero and drops out of the expression for the variance.

Addition Rule for Variance of Independent Random Variables   The variance of the sum of *independent* random variables is the sum of their variances.

$$\text{Var}(X + Y) = \text{Var}(X) + \text{Var}(Y)$$

---

caution   This rule extends to sums of more than two independent random variables. For example, if X, Y, Z, and W are independent, then

$$\text{Var}(X + Y + W + Z) = \text{Var}(X) + \text{Var}(Y) + \text{Var}(W) + \text{Var}(Z)$$

Be careful: It's simpler than the Addition Rule for Variances, *but this rule only applies when the covariance is zero.*

independence and correlation

Independence $\Rightarrow$ Corr = 0
Independence $\not\Leftarrow$ Corr = 0

Independence implies that the covariance and correlation are zero. The opposite implication, however, is not true. Knowing that $\text{Corr}(X, Y) = 0$ does not imply that X and Y are independent. Independence is the stronger property. Correlation and covariance measure a special type of dependence, the type of dependence that is related to the variance of sums. Random variables can be dependent but have no correlation. Here's an example. Suppose that X and Y have the joint probability distribution given in Table 10.8.

TABLE 10.8 Dependent but uncorrelated random variables.

		X		
		$x = -10$	$x = 0$	$x = 10$
Y	$y = 10$	0	0.50	0
	$y = -10$	0.25	0	0.25

These random variables are dependent because $p(x, y) \neq p(x)p(y)$. For example, $P(X = 0, Y = 10) = 0.50$, but $P(X = 0)P(Y = 10) = (0.5)(0.5) = 0.25$. The covariance, however, is zero. Both means, μ_X and μ_Y, are zero. For example,

$$\mu_X = (-10) \times 0.25 + (0) \times 0.50 + (10) \times 0.25 = 0$$

Because the means are zero, the covariance simplifies to

$$\text{Cov}(X, Y) = \text{E}((X - \mu_X)(Y - \mu_Y)) = \text{E}(XY)$$

Because only three cells in the joint distribution have positive probability, we can calculate the covariance easily from the joint distribution as a weighted sum.

$$\begin{aligned}
\text{Cov}(X, Y) &= \text{E}(XY) \\
&= (0 \times 10) \times 0.50 + (-10 \times -10) \times 0.25 + (-10 \times 10) \times 0.25 \\
&= (100 - 100) \times 0.25 = 0
\end{aligned}$$

caution   The random variables are dependent, but the covariance and correlation are zero. *Don't confuse the absence of covariance or correlation with independence.* Uncorrelated *does not imply* independent.

10.5 | IID RANDOM VARIABLES

Day traders also need to decide how long to invest. The day trader we've been helping gives up the same amount of interest if she invests $200 for one day or $100 for two days. Is one plan better than the other?

If she invests $200 on Monday in IBM, then the random variable that describes the change in the value of her investment is $2X$. Both shares change the same way; both rise, fall, or stay the same. The mean is $E(2X) = 2E(X) = 0.20$, and the variance is $Var(2X) = 4Var(X) = 4 \times 4.99$. The Sharpe ratio is thus

$$S(2X) = \frac{2\mu_X - 2r_f}{\sqrt{Var(2X)}} = \frac{0.20 - 0.03}{\sqrt{4 \times 4.99}} \approx 0.038$$

That's the same as $S(X)$ because doubling her money has no effect on the Sharpe ratio.

Now let's figure out what happens if she patiently invests $100 for two days, say Monday and Tuesday. The patient approach requires a second random variable that identifies what happens on Tuesday. The use of a *different* random variable for the second day is essential. This second random variable has the same probability distribution as that for Monday, but it isn't the same random variable. Just because the stock goes up on Monday, it doesn't have to go up Tuesday; the outcomes on the two days can differ.

identically distributed
Random variables are identically distributed if they have a common probability distribution.

Random variables with the same probability distribution are **identically distributed**. It is common to distinguish identically distributed random variables with subscripts. For example, we will denote the change on Monday as X_1 and denote the change on Tuesday as X_2. Both X_1 and X_2 have the probability distribution $p(x)$ from Table 10.1, the distribution for the change in value of IBM stock with mean $\mu_X = \$.10$ and standard deviation $\sigma_X = \$2.23$.

The introduction of a second random variable requires that we specify whether it is independent of the first random variable. How is the change in the value of IBM on Monday related to the change on Tuesday? A common model for stocks (known as the random walk model) *assumes* that the change in the value of a stock on one day is independent of the change the next day. This assumption implies X_1 and X_2 are independent.

Random variables that

are independent of each other and

share a common probability distribution

iid Abbreviation for *independent and identically distributed.*

are so common in statistics that the combination of these two properties gets its own abbreviated name. X_1 and X_2 are **iid**, short for **i**ndependent and **i**dentically **d**istributed.

The following rule summarizes the mean and variance for sums of iid random variables. It's a special case of rules that we have already used, but it's worth mentioning because sums of iid random variables are so common.

Addition Rule for iid Random Variables   *If n random variables* $(X_1, X_2, \ldots, X_n)$ *are iid with mean* μ_X *and standard deviation* σ_X, *then*

$$E(X_1 + X_2 + \cdots + X_n) = n\mu_X$$
$$Var(X_1 + X_2 + \cdots + X_n) = n\sigma_X^2$$
$$SD(X_1 + X_2 + \cdots + X_n) = \sqrt{n}\sigma_X$$

Watch out for the standard deviation. The standard deviation of the sum is the square root of the variance, so we get $\sqrt{n}$ rather than n.

Let's work out the Sharpe ratio for the two-day strategy, treating X_1 and X_2 as iid random variables. The mean is easy: $E(X_1 + X_2) = 2\mu_X = \$.20$. For the variance, independence implies that the variances add.

$$\mathrm{Var}(X_1 + X_2) = 2\sigma_X^2 = 2(4.99) = 9.98$$

The Sharpe ratio is

$$S(X_1 + X_2) = \frac{2\mu_X - 2r_f}{\sqrt{2\sigma_X^2}} = \frac{0.20 - 0.03}{\sqrt{9.98}} \approx 0.054$$

That's slightly larger (better) than the Sharpe ratio for mixing IBM and Microsoft (0.050) and much higher than the Sharpe ratio for buying IBM for a day (0.038).

Why is the Sharpe ratio for the two-day strategy larger? The difference is the variation. The mean of $2X$ equals the mean of $X_1 + X_2$, but the variances differ. The variance of loading up with two shares in IBM for one day is twice the variance of buying one share for two days.

$$\mathrm{Var}(X_1 + X_2) = 2\sigma_X^2 \quad \text{but} \quad \mathrm{Var}(2X) = 4\sigma_X^2$$

Spreading the investment over two days retains the mean but reduces the variance. The smaller variance produces a larger Sharpe ratio, thanks to independence.

IID Data

There's a strong link between iid random variables and data that do not show patterns, what we called simple data in Chapter 6. The absence of patterns is to data what iid is to random variables. We've modeled changes in the value of IBM stock using iid random variables. To see how this compares to reality, the timeplot in Figure 10.1 shows daily percentage changes in the value of IBM stock during 2007.

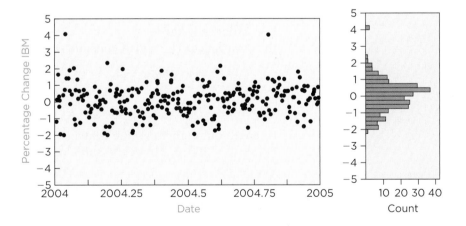

FIGURE 10.1 Daily percentage changes in IBM stock value.

The lack of pattern in the timeplot suggests that we can model changes in IBM stock using iid random variables, such as X_1 and X_2. The histogram beside the timeplot estimates the distribution of the percentage changes. We expect what happens tomorrow to look like a random draw from this histogram that summarizes what's happened previously.

10.6 | WEIGHTED SUMS

We've focused on the variance of sums of random variables. We could have introduced constants in this sum as well. The analysis of real portfolios needs this flexibility. Perhaps the day trader has $1,000 to invest. That would allow her, for instance, to buy two shares of IBM, four of Microsoft, and leave $400 in the bank. In this case, her net worth tomorrow would change by

$$2X + 4Y + 0.06$$

The last 6 cents are the interest earned on $400 in the bank. It would be tedious to repeat all of the previous work to handle this sum.

Fortunately, our last rule for this chapter covers expected values and variances for weighted sums of random variables.

Addition Rule for Weighted Sums   *The expected value of a weighted sum of random variables is the weighted sum of the expected values:*

$$E(aX + bY + c) = aE(X) + bE(Y) + c$$

The variance of a weighted sum is

$$Var(aX + bY + c) = a^2\,Var(X) + b^2\,Var(Y) + 2ab\,Cov(X, Y)$$

The constant c drops out of the variance. Adding (or subtracting) a constant does not change the variation. As for the other constants, if you remember the formula $(a + b)^2 = a^2 + b^2 + 2ab$, then you will have an easier time remembering the formula for the variance. The previous rules are special cases of this one. For instance, if $a = b = 1$ and $c = 0$, you get the Addition Rule for Expected Values.

Let's try the extended rule. The expected value of $2X + 4Y + 0.06$ is

$$\begin{aligned}
E(2X + 4Y + 0.06) &= 2E(X) + 4E(Y) + 0.06 \\
&= 2(0.10) + 4(0.12) + 0.06 \\
&= \$.74
\end{aligned}$$

The variance is harder because you have to square the multipliers and include the covariance. The added constant 0.06 does not appear in the variance.

$$\begin{aligned}
Var(2X + 4Y + 0.06) &= 2^2\,Var(X) + 4^2\,Var(Y) + 2 \times (2 \times 4) \times Cov(X, Y) \\
&= 4(4.99) + 16(5.27) + 16(2.19) \\
&= 139.32
\end{aligned}$$

In case you're curious, the Sharpe ratio for this portfolio is $(0.74 - 0.15)/\sqrt{139.32} \approx 0.050$, about the same as that for the balanced mix of IBM and Microsoft.

Variance of Differences

caution   The Addition Rule for Weighted Sums hides a surprise. What's the variance of $X - Y$? The difference between random variables is a weighted sum with $a = 1$, $b = -1$, and $c = 0$. The variance is

$$Var(X - Y) = Var(X) + Var(Y) - 2\,Cov(X, Y)$$

If X and Y are independent, then $Var(X - Y) = Var(X) + Var(Y)$, which is the same as $Var(X + Y)$.

This seems odd, but it begins to make sense if you think about investing for two days. Suppose we want to know the mean and standard deviation of the *difference*, $X_1 - X_2$. Since each day has an expected change of \$.10, the expected difference is $E(X_1 - X_2) = E(X_1) - E(X_2) = 0.10 - 0.10 = 0$. If we subtract the variances, we get 0, and that doesn't make sense. The day trader is not guaranteed the same outcome both days. If the outcomes on the two days are independent, the difference $X_1 - X_2$ ranges from \$5 − (−\$5) = \$10 (up then down) to (−\$5) − \$5 = −\$10 (down then up). The variability of the difference is just as large as the variability of the sum.

4M EXAMPLE 10.2   CONSTRUCTION ESTIMATES

MOTIVATION ▸ STATE THE QUESTION

Construction estimates involve a mixture of costs, combining different types of labor and materials. These costs tend to move together. Projects that require a lot of labor also typically involve higher material costs.

Let's help a family think about the costs of expanding their home. Talking to contractors reveals that the estimates of the cost of a one-room addition vary. On average, the addition will take *about* three weeks for two carpenters (240 hours of carpentry), with a standard deviation of 40 hours to allow for weather. The electrical work on average requires 12 hours with standard deviation 4 hours. Carpentry goes for \$45 per hour (done through a contractor), and electricians cost \$80 per hour. The amount of both types of labor could be higher (or lower), with correlation $\rho = 0.5$. What can we anticipate the labor costs for the addition to be?

METHOD ▸ DESCRIBE THE DATA AND SELECT AN APPROACH

Let's estimate the expected costs for labor and get a sense for the variation in these costs. Begin by identifying the relevant random variables. This problem has three. We will use the random variable X for the number of carpentry hours, Y for the number of electrician hours, and T for the total costs. These are related by the equation (with T measured in dollars) $T = 45X + 80Y$.

MECHANICS ▸ DO THE ANALYSIS

For the expected value, we can use the Addition Rule for Weighted Sums.

$$E(aX + bY) = aE(X) + bE(Y)$$

This rule implies that

$$E(T) = E(45X + 80Y) = 45E(X) + 80E(Y) = 45 \times 240 + 80 \times 12$$
$$= \$11,760$$

For the variance, the second part of the Addition Rule for Weighted Sums says that

$$\text{Var}(aX + bY) = a^2 \text{Var}(X) + b^2 \text{Var}(Y) + 2ab \, \text{Cov}(X, Y)$$

The correlation determines the covariance from the relationship

$$\text{Cov}(X, Y) = \rho\sigma_X\sigma_Y = 0.5 \times 40 \times 4 = 80$$

Plugging in the values, the variance of the total cost is

$$\text{Var}(T) = \text{Var}(45X + 80Y) = 45^2\,\text{Var}(X) + 80^2\,\text{Var}(Y) + 2(45)(80)\text{Cov}(X, Y)$$

$$= 45^2 \times 40^2 + 80^2 \times 4^2 + 2(45)(80) \times 80$$

$$= 3{,}240{,}000 + 102{,}400 + 576{,}000$$

$$= 3{,}918{,}400$$

The standard deviation converts the variance to the scale of dollars,

$$\text{SD}(T) = \sqrt{\text{Var}(T)} = \sqrt{3{,}918{,}400} \approx \$1{,}979$$

MESSAGE | SUMMARIZE THE RESULTS

tip

Rounding often helps in the presentation of numerical results. Given the uncertainty implied by the standard deviation, it makes sense to round the total costs to $12,000 rather than present them as $11,760. We'd say that the expected cost for labor for adding the addition is around $12,000. This figure is not guaranteed, and the owner should not be too surprised if costs are higher (or lower) by $2,000. (If we believe that the distribution of costs is bell shaped, then the Empirical Rule implies that there's about a 2/3 chance that costs are within SD(T) of the mean.)

Best Practices

- *Consider the possibility of dependence.* Independent random variables are simpler to think about, but often unrealistic unless we're rolling dice or tossing coins. Ignoring dependence can produce poor estimates of variation. Make sure that it makes sense for random variables to be independent before you treat them as independent.
- *Only add variances for random variables that are uncorrelated.* Check for dependence before adding variances. Expected values add for *any* random variables, but variances only add when the random variables are uncorrelated. Independence implies that the random variables are uncorrelated.

- *Use several random variables to capture different features of a problem.* Isolate each random event as a separate random variable so that you can think about the different components of the problem, one at a time. Then think about how much dependence is needed between random variables.
- *Use new symbols for each random variable.* Make sure to represent each outcome of a random process as a different random variable. Just because each random variable describes a similar situation doesn't mean that each random outcome will be the same. Write $X_1 + X_2 + X_3$ rather than $X + X + X$.

Pitfalls

- *Do not think that uncorrelated random variables are independent.* Independence implies that the correlation is zero, but not the other way around. Correlation only captures one type of

dependence. Even if the correlation is zero, the random variables can be very dependent.
- *Don't forget the covariance when finding the variance of a sum.* The variance of a sum is the

sum of the variances *plus* twice the covariance. The covariance adjusts for dependence between the items in the sum.

- *Never add standard deviations of random variables.* If X and Y are uncorrelated, $\text{Var}(X + Y) = \text{Var}(X) + \text{Var}(Y)$. The same does not hold for standard deviations. Variances add for independent random variables; standard deviations do not.

- *Don't mistake Var(X − Y) for Var(X) − Var(Y).* Variances of sums can be confusing even if the random variables are independent. You'll know you made a mistake if you calculate a negative variance.

 BEHIND the Math

Life without the Addition Rule

Like many conveniences, the Addition Rule for the Expected Value of a Sum of Random Variables is difficult to appreciate until you see what you'd be doing without it. As you skim this section, appreciate that you don't need to do these calculations if you remember the Addition Rule.

To calculate the expected value of a sum of random variables *without* the Addition Rule, we have to return to the definition of an expected value. We need to list all of the possible values of the sum along with their probabilities. The expected value is then the weighted sum.

The calculation is tedious because this sum has *two* subscripts, one for each random variable. For the simple random variables in this chapter, the sum runs over nine pairs of outcomes of X and Y, one for each cell in the joint probability distribution in Table 10.3. Here's the calculation, laid out to match the summands to the probabilities in Table 10.3. (For example, $X = -5$ and $Y = 4$ in the first row and column of Table 10.3.)

$$
\begin{aligned}
E(X + Y) = {} & (-5 + 4)(0.00) + (0 + 4)(0.11) \\
& + (5 + 4)(0.07) + (-5 + 0)(0.03) \\
& + (0 + 0)(0.62) + (5 + 0)(0.02) \\
& + (-5 - 4)(0.06) + (0 - 4)(0.07) \\
& + (5 - 4)(0.02) \\
= {} & 0.22
\end{aligned}
$$

Which do you prefer, adding $\mu_X + \mu_Y$ by the Addition Rule or direct calculation?

The Variance of a Sum

Direct calculation of the variance of $X + Y$ uses all nine probabilities from the joint probability distribution in Table 10.3. It is more tedious than calculating $E(X + Y)$ because we need to subtract the mean and square the deviation *before* multiplying by the probabilities. Here's the formula:

$$\text{Var}(X + Y) = \sum_{x,y}[(x + y) - (\mu_X + \mu_Y)]^2 \, p(x, y)$$

Each summand is the product of a squared deviation of $x + y$ from $\mu_X + \mu_Y$ times the joint probability. Each summand shown below comes from a cell in the joint distribution in Table 10.3.

$$
\begin{aligned}
& \text{Var}(X + Y) \\
& = (-1 - 0.22)^2(0.00) + (4 - 0.22)^2(0.11) \\
& \quad + (9 - 0.22)^2(0.07) + (-5 - 0.22)^2(0.03) \\
& \quad + (0 - 0.22)^2(0.62) + (5 - 0.22)^2(0.02) \\
& \quad + (-9 - 0.22)^2(0.06) + (-4 - 0.22)^2(0.07) \\
& \quad + (1 - 0.22)^2(0.02) \\
& \approx 14.64
\end{aligned}
$$

The variance of the sum is 40% bigger than the sum of the variances. The difference is the result of the dependence (positive covariance) between the two random variables. The value is rounded to 14.64 in order to match the result given by the covariance, shown in the text.

Calculating the Covariance

Because it's an expected value, the covariance is the sum of all nine possible outcomes $(x - \mu_X)(y - \mu_Y)$ weighted by the probabilities $p(x, y)$.

$$
\begin{aligned}
\text{Cov}(X, Y) = {} & (-5 - 0.10)(4 - 0.12) \\
& \times 0.00 + (0 - 0.10)(4 - 0.12) \\
& \times 0.11 + (5 - 0.10)(4 - 0.12) \\
& \times 0.07 + (-5 - 0.10)(0 - 0.12) \\
& \times 0.03 + (0 - 0.10)(0 - 0.12) \\
& \times 0.62 + (5 - 0.10)(0 - 0.12) \\
& \times 0.02 + (-5 - 0.10)(-4 - 0.12) \\
& \times 0.06 + (0 - 0.10)(-4 - 0.12) \\
& \times 0.07 + (5 - 0.10)(-4 - 0.12) \times 0.02 \\
= {} & 2.19 \, \$^2
\end{aligned}
$$

Each of the summands shows $x - \mu_X$ and $y - \mu_Y$ as well as $p(x, y)$. The units of the covariance are those of X times those of Y, which works out to squared dollars in this example since both X and Y are measured in dollars.

CHAPTER SUMMARY

The **joint probability distribution** $p(x, y)$ of two random variables gives the probability for simultaneous outcomes, $p(x, y) = P(X = x, Y = y)$. The joint probability distribution of **independent random variables** factors into the product of the marginal distributions, $p(x, y) = p(x)p(y)$. Sums of random variables can be used to model a portfolio of investments. The expected value of a **weighted sum of random variables** is the weighted sum of the expected values. The variance of a weighted sum of random variables depends on the **covariance**, a measure of dependence between random variables. For independent random variables, the covariance is zero. The **correlation** between random variables is a scale-free measure of dependence. Random variables that are **iid** are independent and identically distributed. Observations of iid random variables produce simple data, data without patterns.

▉ Key Terms

Addition Rule for the Expected Value of a Sum of Random Variables, 224
Addition Rule for iid Random Variables, 230
Addition Rule for Variance of Independent Random Variables, 229
Addition Rule for Variances, 226

Addition Rule for Weighted Sums, 232
correlation between random variables, 227
covariance between random variables, 226
identically distributed, 230
iid, 230
independent random variables, 222

joint probability distribution, 221
Multiplication Rule for the Expected Value of a Product of Independent Random Variables, 222
symbol
  ρ (correlation), 227
uncorrelated, 229

▉ Formulas

Joint probability distribution of two random variables

$$p(x, y) = P(X = x \text{ and } Y = y)$$
$$= P(X = x) \times P(Y = y) \quad \text{if } X \text{ and } Y \text{ are independent}$$

Multiplication Rule for Expected Values.
If the random variables X and Y are independent, then

$$E(XY) = E(X)E(Y)$$

If the random variables are dependent, then the expected value of the product must be computed from the joint distribution directly.

Addition Rule for Expected Values of Sums

$$E(X + Y) = E(X) + E(Y)$$

For constants a, b, and c,

$$E(aX + bY + c) = aE(X) + bE(Y) + c$$

Addition Rule for Variances of Sums

$$\text{Var}(X + Y) = \text{Var}(X) + \text{Var}(Y) + 2\,\text{Cov}(X, Y)$$

The formula requires twice the covariance because variance is the expected *squared* deviation. When we expand the square, we get $a^2 + b^2 + 2ab$. In general, for constants a, b, and c,

$$\text{Var}(aX + bY + c)$$
$$= a^2\,\text{Var}(X) + b^2\,\text{Var}(Y) + 2ab\,\text{Cov}(X, Y)$$

If the random variables are independent, then

$$\text{Var}(X + Y) = \text{Var}(X) + \text{Var}(Y)$$

Covariance between random variables

$$\text{Cov}(X, Y) = E((X - \mu_X)(Y - \mu_Y))$$

A shortcut formula for the covariance avoids finding the deviations from the mean first.

$$\text{Cov}(X, Y) = E(XY) - \mu_X\mu_Y$$

If the random variables are independent, then

$$\text{Cov}(X, Y) = 0$$

Correlation between random variables

$$\text{Corr}(X, Y) = \rho = \frac{\text{Cov}(X, Y)}{\sigma_X\sigma_Y}$$

$$\text{Cov}(X, Y) = \rho\sigma_X\sigma_Y$$

About the Data

The random variables in this chapter are simple models of real assets. Though limited to three outcomes, these random variables mimic important attributes of the daily percentage changes in the values of stock of IBM and Microsoft. For example, the mean of the 2,519 percentage changes in Microsoft stock over the 10 years, 1994 to 2003, is 0.124% with standard deviation 2.41%. For the random variable Y, $\mu_Y = \$.12$ and $\sigma_Y = \$2.30$. Because we set the price of the stock to $100, dollar changes in Y are percentage changes.

The joint distribution of X and Y in Table 10.3 reproduces the correlation between percentage changes in IBM and Microsoft, as well as the means and standard deviations of both X and Y. The scatterplot in Figure 10.2 graphs the daily percentage change in Microsoft (y-axis) versus the daily percentage change in IBM (x-axis). There's quite a bit of scatter about a positive linear trend; the correlation in the data is $r = 0.438$. The correlation between the random variables is nearly the same, $\rho = 0.43$.

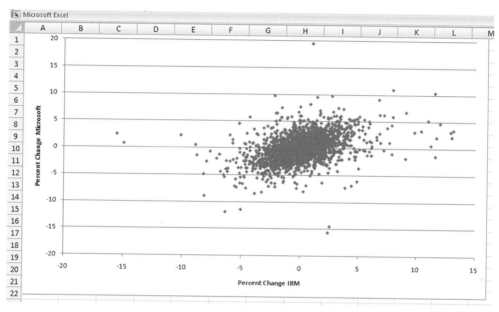

FIGURE 10.2 Scatterplot of percentage changes in Microsoft versus IBM.

EXERCISES

Mix and Match

Match the concept described in words in the left column with the correct symbols or notation in the right column.

1. Consequence of positive covariance

2. Covariance expressed using correlation

3. True for uncorrelated random variables

4. Weighted sum of two random variables

5. Sharpe ratio of a random variable

6. Defines independent random variables

7. Implies that X and Y are identically distributed

8. Symbol for the correlation between random variables

9. Symbol for a joint probability distribution

10. Sequence of iid random variables

(a) $p(x, y)$

(b) $\dot{\rho}$

(c) $p(x, y) = p(x)p(y)$

(d) $\rho\sigma_x\sigma_y$

(e) $\mathrm{Var}(X + Y) > \mathrm{Var}(X) + \mathrm{Var}(Y)$

(f) $p(x) = p(y)$

(g) X_1, X_2, X_3

(h) $\mathrm{Var}(X + Y) = \mathrm{Var}(X) + \mathrm{Var}(Y)$

(i) $S(Y)$

(j) $3X - 2Y$

True/False

Mark each statement True or False. If you believe that a statement is false, briefly say why you think it is false.

Exercises 11–16. An office complex leases space to various companies. These leases include energy costs associated with heating during the winter. To anticipate costs in the coming year, the managers developed two random variables X and Y to describe costs for equivalent amounts of heating oil (X) and natural gas (Y) in the coming year. Both X and Y are measured in dollars per Btu of heat produced. The complex uses both fuels for heating, with $\mu_X = \mu_Y$ and $\sigma_X = \sigma_Y$.

11. If managers believe that costs for both fuels tend to rise and fall together, then they should treat X and Y as independent.

12. Because the complex uses both fuels for heating, a negative covariance between X and Y would increase the uncertainty about future costs.

13. Because the means and SDs of these random variables are the same, the random variables X and Y are identically distributed.

14. Because the means and SDs of these random variables are the same, the random variables X and Y are dependent.

15. If told that the costs of heating oil and natural gas are uncorrelated, an analyst should then treat the joint distribution as $p(x, y) = p(x)p(y)$.

16. By generating half of its heating using heating oil and half using natural gas in the coming year, the complex decreases the variation in costs if X and Y are independent.

Exercises 17–22. As a baseline when planning future advertising, retail executives treat the dollar values of sales on consecutive weekends as iid random variables. The amounts sold on two consecutive weekends (call these X_1 and X_2) are independent and identically distributed random variables with mean μ and standard deviation σ.

17. On average, retailers expect to sell the same amount on the first weekend and the second weekend.

18. If retail sales are exceptionally low on the first weekend, retailers expect lower than average sales on the following weekend.

19. The standard deviation of total sales over consecutive weekends is twice σ.

20. The difference between the amount sold on the first weekend X_1 and the amount sold on the second weekend X_2 is less variable than the total amount sold when both are combined.

21. If a promotion were to introduce negative dependence between the amounts sold on two weekends, then we would need the correlation in order to find $E(X_2 - X_1)$.

22. If a promotion were to introduce negative dependence between the amounts sold on two weekends, then we would need the correlation or covariance in order to find $Var(X_2 - X_1)$.

Think About It

23. If investors want portfolios with small risk (variance), should they look for investments that have positive covariance, have negative covariance, or are uncorrelated?

24. Does a portfolio formed from the mix of three investments have more risk (variance) than a portfolio formed from two?

25. What is the covariance between the random variable Y and itself, $Cov(Y, Y)$? What is the correlation between a random variable and itself?

26. If an investor decides on the all-Microsoft portfolio (in the text example), does it make sense to use the Sharpe ratio to decide whether to invest $2,000 or $4,000? Why or why not?

27. If the covariance between the prices of two investments is 100,000, does this tell you that the correlation between the two is close to 1?

28. If the correlation between the price of Microsoft stock and the price of an Xbox is 1, could you predict the price of an Xbox from knowing the price of Microsoft stock? How well?

29. Would it be reasonable to model the daily sales of a downtown restaurant as a sequence of iid random variables?

30. If percentage changes in the value of a stock are iid with mean 0, then how should we predict the change tomorrow if the change today was a 3% increase? How would the prediction change if the value today decreased by 2%?

31. Kitchen remodeling is a popular way to improve the value of a home.
    (a) If X denotes the amount spent for labor and Y the cost for new appliances, do you think these would be positively correlated, negatively correlated, or independent?
    (b) If a family is on a strict budget that limits the amount spent on remodeling to $25,000, does this change your impression of the dependence between X and Y?

32. An insurance company has studied the costs from covering auto accidents. Its policies offer separate coverage for personal injury and damage to property. It has found that the amount paid in these two categories is highly positively correlated. What is the impact of this dependence on the amounts paid for accident claims?

You Do It

33. Independent random variables X and Y have the means and standard deviations as given in the following table. Use these parameters to find the expected value and SD of the following random variables that are derived from X and Y:
    (a) $2X - 100$
    (b) $0.5Y$

(c) $X + Y$

(d) $X - Y$

	Mean	SD
X	1,000	200
Y	2,000	600

34. The independent random variables X and Y have the means and standard deviations as given in the following table. Use these parameters to find the expected value and SD of the following random variables that are derived from X and Y:
    (a) $8X$
    (b) $3Y - 2$
    (c) $(X + Y)/2$
    (d) $X - 2Y$

	Mean	SD
X	100	16
Y	-50	25

35. Repeat the calculations of Exercise 33. Rather than treat the random variables X and Y as independent, assume that $Cov(X, Y) = 12,500$.

36. Repeat the calculations of Exercise 34. Rather than treat the random variables X and Y as independent, assume that $Corr(X, Y) = -0.5$.

37. If the variance of the sum is $Var(X + Y) = 8$ and $Var(X) = Var(Y) = 5$, then what is the correlation between X and Y?

38. What's the covariance between a random variable X and a constant?

39. A student budgets $60 weekly for gas and a few quick meals off campus. Let X denote the amount spent for gas and Y the amount spent for quick meals in a typical week. Assume this student is very disciplined and sticks to the budget, spending $60 on these two things each week.
    (a) Can we model X and Y as independent random variables? Explain.
    (b) Suppose we assume X and Y are dependent. What is the effect of this dependence on the variance of $X + Y$?

40. A pharmaceutical company has developed a new drug that helps insomniacs sleep. In tests of the drug, it records the daily number of hours asleep and awake. Let X denote the number of hours awake and let Y denote the number of hours asleep for a typical patient.
    (a) Explain why the company must model X and Y as dependent random variables.
    (b) If the company considers the difference between the number of hours awake and the number asleep, how will the dependence affect the SD of this comparison?

41. Drivers for a freight company have a varying number of delivery stops. The mean number of stops is 6 with standard deviation 2. Two drivers operate independently of one another.
    (a) Identify the two random variables and summarize your assumptions.
    (b) What is the mean and standard deviation of the number of stops made in a day by these two drivers?
    (c) If each stop by one driver takes 1 hour and each stop by the other takes 1.5 hours, how many hours do you expect these two drivers to spend making deliveries?
    (d) Find the standard deviation of the amount of time needed for deliveries in part (b).
    (e) It is more likely the case that the driver who spends more time making deliveries also has fewer to make, and conversely that the other driver has more deliveries to make. Does this suggest that the two random variables may not meet the assumptions of the problem?

42. Two classmates both enjoy playing online poker. They both claim to win $300 on average when they play for an evening, even though they play at different sites on the Web. They do not always win the same amounts, and the SD of the amounts won is $100.
    (a) Identify the two random variables and summarize your assumptions.
    (b) What is the mean and standard deviation of the total winnings made in a day by the two classmates?
    (c) Find the mean and standard deviation of the difference in the classmates' winnings.
    (d) The classmates have decided to play in a tournament and are seated at the same virtual game table. How will this affect your assumptions about the random variables?

43. Customers at a fast-food restaurant buy both sandwiches and drinks. The following joint distribution summarizes the numbers of sandwiches (X) and drinks (Y) purchased by customers.

		X	
		1 sandwich	2 sandwiches
	1 drink	0.40	0.20
Y	2 drinks	0.10	0.25
	3 drinks	0	0.05

(a) Find the expected value and variance of the number of sandwiches.
(b) Find the expected value and variance of the number of drinks.
(c) Find the correlation between X and Y. (*Hint:* You might find the calculations easier if you use the alternative expression for the covariance shown in the Formulas section at the end of the chapter.)

(d) Interpret the size of the correlation for the manager of the restaurant.

(e) If the profit earned from selling a sandwich is $1.50 and from a drink is $1.00, what is the expected value and standard deviation of the profit made from each customer?

(f) Find the expected value of the ratio of drinks to sandwiches. Is it the same as μ_Y/μ_X?

44. An insurance agent likes to sell two types of policies to his clients, selling them both life insurance and auto insurance. The following joint distribution summarizes the properties of the number of life insurance policies sold to an individual (X) and the number of auto policies (Y).

		X	
		0	**1 life policy**
Y	**0**	0.1	0.25
	1 auto policy	0.25	0.4

(a) Find the expected value and variance of the number of life insurance policies.

(b) Find the expected value and variance of the number of auto insurance policies.

(c) Find and interpret E(XY) in the context of this problem.

(d) Find the correlation between X and Y. (*Hint:* You can use the alternative formula for the covariance shown in the Formulas section at the end of the chapter.)

(e) Interpret the size of the correlation for the agent.

(f) The agent earns $750 from selling a life insurance policy and $300 from selling an auto policy. What is the expected value and standard deviation of the earnings of this agent from policies sold to a client?

45. During the 2004–2005 NBA season, LeBron James of the Cleveland Cavaliers attempted 308 three-point baskets and made 108. He also attempted 1,376 two-point baskets, making 687 of these. Use these counts to determine probabilities for the following questions.

(a) Let a random variable X denote the result of a two-point attempt. X is either 0 or 2, depending on whether the basket is made. Find the expected value and variance of X.

(b) Let a second random variable Y denote the result of a three-point attempt. Find the expected value and variance of Y.

(c) In a game, LeBron attempts 5 three-point baskets and 20 two-point baskets. How many points do you expect him to score? (*Hint:* Use a collection of iid random variables, some distributed like X and some like Y.)

(d) During this season, LeBron averaged 21.2 points from two- and three-point baskets. (He scored an average of 6 more points per game from free throws.) In the game described in part (c), he made 2 three-pointers and 10 two-point baskets. Does this seem typical for his total? Be sure to identify any further assumptions that you make.

46. Professional football kicker Jay Feely of the New York Giants attempted 42 field goals during the 2005 season, making 35. Of the kicks, he made 24 out of 27 that were for less than 40 yards, and he made 11 out of 15 longer kicks. Use these counts to form probabilities to answer these questions.

(a) Let the random variable X denote the points scored when Feely attempts a field goal of less than 40 yards. If he makes it, the Giants score 3 points. If he misses, they get none. Find the mean and variance of X.

(b) Similarly, let Y denote the points scored when Feely attempts a field goal of more than 40 yards. Find the mean and variance of Y.

(c) During a game, Feely attempts three kicks of less than 40 yards and two of more than 40 yards. How many points do you expect him to contribute to the total for the Giants?

(d) During the 2005 season, Feely scored 105 points on field goals, about 6.6 per game. Does the game in part (c) appear typical, or does it seem like a high- or low-scoring game for Feely?

47. Homes in a community served by a utility company annually use an expected 12,000 kilowatt-hours (kWh) of electricity. The same utility supplies natural gas to these homes, with an annual expected use of 85 thousand cubic feet (85 MCF). The homes vary in size and patterns of use, and so the use of utilities varies as well. The standard deviation of electricity use is 2,000 kWh and the SD of gas use is 15 MCF.

(a) Let the random variable X denote the electricity use of a home in kWh, and Y denote the use of natural gas in MCF. Do you think that the utility should model X and Y as independent?

(b) It makes little sense to add X and Y because these variables have different units. We can convert them both to dollars, then add the dollar amounts. If energy costs $.09 per kilowatt-hour of electricity and $10 per thousand cubic feet of natural gas, then how much on average does a household in the community spend for utilities?

(c) If the correlation between electric use and gas use is 0.35, find the SD of the total utility bill paid by these households.

(d) Total expenditures for electricity and gas in the United States averaged $1,640 per household in 2001. Do the expenditures of the homes served by this utility seem unusually high, unusually low, or typical?

48. A direct sales company has a large sales force that visits customers. To improve sales, it designs a new training program. To see if the program works, it puts some new employees through this program and others through the standard training. It then assigns two salespeople to a district, one trained in the standard way and the other by the new program. Assume that salespeople trained using the new method sell on average $52,000 per month and those trained by the old method sell on average $45,000 per month. The standard deviation of both amounts is $6,000.

(a) If the dollar amounts sold by the two representatives in a district are independent, then do you expect the salesperson trained by the new program to sell more than the salesperson trained by the old program?

(b) Because the sales representatives operate in the same district, the company hopes that the sales will be positively correlated. Explain why the company prefers positive dependence using an example with $\rho = 0.8$.

49. A construction firm places bids on various portions of the work in building a new office tower. The key bids it submits are for electrical work and for plumbing. The bid for the electrical work estimates 64 weeks of labor (e.g., 1 electrician for 64 weeks or 8 for 8 weeks). The bid for the plumbing estimates 120 weeks of labor. Standard procedures indicate that the standard deviations for these estimates are 6 weeks for electrical work and 15 weeks for plumbing.

(a) Find the expected number of weeks of labor provided by the company if it wins both bids.

(b) Would you expect to find positive, negative, or no correlation between the numbers of weeks of the two types of labor needed for the construction?

(c) Find the standard deviation of the total number of weeks of work if the correlation between the weeks of labor for electrical work and plumbing is $\rho = 0.7$.

(d) What is the effect of the dependence between the number of weeks of electrical and plumbing labor? In particular, when preparing bids, would the firm prefer more or less dependence?

(e) The firm earns a profit of $200 per week of electrical work and a profit of $300 per week of plumbing work. What are the expected profits from this contract and the standard deviation of those profits? Assume $\rho = 0.7$ as in part (c).

(f) Do you think that the firm will make more than $60,000 on this contract?

50. A retail company operates two types of clothing stores in shopping malls. One type specializes in clothes for men, and the other in clothes for women. The sales from a men's clothing store average $800,000 annually and those from a women's clothing store average $675,000 annually. The standard deviation of sales among men's stores is $100,000 annually, and among women's stores is $125,000. In a typical mall, the company operates one men's store and one women's store (under different names).

(a) What are the expected annual sales for the stores owned by this company in one shopping mall?

(b) Would you expect to find that the sales in the stores are positive or negatively correlated, or do you expect to find that the sales are uncorrelated?

(c) If the company operated two women's stores rather than one of each type, would you expect the dependence between the sales at the two women's clothing stores to be positive or negative or near zero?

(d) Find the standard deviation of the total sales if the correlation between the sales at the men's store and women's store is $\rho = 0.4$.

(e) The rent for the space in the shopping mall costs $30 per square foot. Both types of stores occupy 2,500 square feet. What is the expected value and standard deviation of total sales in excess of the rent costs?

(f) If labor and other expenses (such as the cost of clothing that is sold) to operate the two stores at the mall costs the company $750,000 annually, do you think there is a good chance that the company might lose money?

4M Real Money

The relevance of variance and covariance developed for random variables in this chapter carries over to the analysis of returns on real stocks. The data file "daily_stocks" includes the daily percentage changes in the value of stock in Citigroup and stock in Exxon.

Motivation

a) An investor is considering buying stock in Citigroup and Exxon. Explain to the investor, in nontechnical language, why it makes sense to invest some of the money in both companies rather than putting the entire amount into one or into the other.

Method

b) Why is the Sharpe ratio useful for comparing the performance of two stocks whose values don't have the same variances?

c) Based on the observed sample correlation in these data, does it appear sensible to model percentage changes in the values of these stocks as independent?

Mechanics

d) Find the Sharpe ratio for a daily investment concentrated in Citigroup, concentrated in Exxon, or split between Citigroup and Exxon, with half of the money invested in each. Use the same interest rate as in the text (0.015). Does this mix look as good as the mix of IBM and Microsoft in this chapter?

e) You can reproduce the calculations for the mix obtained in part (d) by manipulating the underlying data that provide the values of the means and variances. Form a new column in the data that is the average of the columns with the percentage changes for Citigroup and Exxon. Find the mean and SD of this new column and compare these to the values found in part (d).

f) Find the Sharpe ratio for an investment that puts three-quarters of the money into Citigroup and one-quarter into Exxon. Compare this combination to the mix considered in part (d). Which looks better?

Message

g) Explain your result to the investor. Does it matter how the investor divides the investment between the two stocks?

4M Planning Operating Costs

Management of a chain of retail stores has the opportunity to lock in prices for electricity and natural gas, the two energy sources used in the stores. A typical store in this chain uses electricity for lighting and air conditioning. In the winter, natural gas supplies heat. Managers at a recent meeting settled on the following estimates of typical annual use of electricity and natural gas by the stores. They estimated the chances for varying levels of use based on their own experiences operating stores and their expectation for the coming long-term weather patterns.

Thousand kilowatt-hours	200	300	400	500
Chances	5%	25%	40%	30%

Thousand cubic feet	600	800	1,000	1,200
Chances	5%	25%	40%	30%

The cost of electricity is roughly $100 per thousand kilowatt-hours, and the cost of natural gas is about $12 per thousand cubic feet.

Motivation

a) Does the company know exactly how much it will spend on energy costs to operate a store in the coming year?

Method

b) Identify random variables for the amount of electricity that is used (X) and the amount of natural gas that is used (Y). What are the marginal probability distributions for these random variables?

c) Define a third random variable T that combines these two random variables to determine the annual energy operating costs.

d) We don't have the joint distribution for X and Y. Do you think that it is appropriate to model the two random variables X and Y as independent?

Mechanics

e) Find the expected value and variance of the amount of electricity used (in thousands of kilowatt-hours).

f) Find the expected value and variance of the amount of natural gas that is used (in thousands of cubic feet).

g) The correlation between X and Y is believed to be $\rho = 0.4$. Using this value, find the mean and variance of T.

h) How would your answer to part (g) change if the correlation were 0?

Message

i) Use the properties of T to summarize typical energy operating costs for management of this retail firm.

j) What assumptions underlie your analysis? Are there any items that you've treated like constants that might more accurately be treated as random variables (if you had enough additional information do so)?

Probability Models for Counts

PHARMACEUTICAL ADVERTISING APPEARS ON TELEVISION, in magazines, and even on racecars, but much is done the old-fashioned way—in person. Pharmaceutical sales representatives, known as detail reps, visit doctors. Pharmaceutical companies such as Merck and Pfizer together employ 90,000 detail reps, whose compensation makes up a large fraction of the billions of dollars spent annually on promotion in the pharmaceutical industry. You may have noticed some of the trinkets they leave at your doctor's office: magnets, pens, notepads, and even picture frames adorned with the names of medical products. If you look around the waiting room the next time you visit your doctor, you may see one of these couriers. You can recognize detail reps because they don't look sick, plus they're generally nicely dressed.

The goal of every detail rep is to meet face-to-face with the doctor. In a few minutes, a detail rep hands the doctor literature pulled from professional journals and talks up one or two of her or his company's latest drugs. Detail reps also leave samples that the doctor can give to patients. Sampling introduces doctors and patients to new drugs. And, along the way, the reps leave those reminders of their visit.

What does it take to be a good detail rep? Certainly enthusiasm, persistence, and a willingness to overcome obstacles help. Rumor has it that only 40% of detail visits ever reach the doctor, and even then for just a moment. Detail reps typically visit the offices of 10 doctors each day. Visiting the office, however, is not the same as meeting the doctor. There's a random process at work. The doctor may be away or unavailable. How many doctors should those who manage promotion expect a detail rep to meet in a day? Is a detail rep who meets 8 or more doctors in a day doing exceptionally well?

This chapter introduces several discrete random variables designed to answer questions like this one. These random variables model counting problems and provide a systematic method to find the needed probabilities. Your task is to recognize which random variable to use and verify that the necessary assumptions hold. Once you recognize the right random variable, you can take advantage of the known properties of these common random variables. We will show formulas for calculating various probabilities, but software can handle most of these details.

11.1 | RANDOM VARIABLES FOR COUNTS

Let's restate the questions posed in the introduction using a random variable. In this chapter, the random variable Y denotes the number of doctors a detail rep meets during 10 office visits. The question, "How many doctors should a rep be expected to meet?" asks for the mean of Y, $E(Y)$. The second question, "Is a rep who meets 8 or more doctors doing exceptionally well?" requires $P(Y \geq 8)$. To find these properties of the random variable, we'll identify a pattern to the randomness, one that's common to many counting problems.

Bernoulli Random Variable

We will think of a visit to a doctor's office as a chance event with two possible outcomes: The detail rep either meets the doctor or not. The success or failure of a visit resembles a familiar random process: tossing a coin. Ideally three characteristics of these visits to the doctor resemble those associated with tossing a coin. Random events with these three characteristics are known as **Bernoulli trials**.

Bernoulli trials Random events with three characteristics:

1. Two outcomes
2. Fixed probability
3. Independence

- *Each visit produces one of two possible outcomes*, often generically called *success* and *failure*. For detailing, a visit is a success if the rep meets the doctor or a failure if not.
- *The probability of success, denoted p, is the same for every visit.* The chance that a detail rep gets past the receptionist to see the doctor is $p = 0.4$. For tossing a fair coin, $p = 0.5$.
- *The results of the visits are independent.* Independence seems plausible when tossing a coin; the coin does not remember the last toss. Independence is debatable for detailing. Does success in meeting the doctor during the first detail visit affect what happens next? Independence says not.

Bernoulli random variable
A random variable with two possible values, 0 and 1 (often called failure and success), determined in a Bernoulli trial.

Each Bernoulli trial defines a **Bernoulli random variable**. A Bernoulli random variable B has two possible values: $B = 1$ if the trial is a success and $B = 0$ if the trial fails.

$$B = \begin{cases} 1 & \text{if the trial is a success} \\ 0 & \text{if the trial is a failure} \end{cases}$$

Bernoulli random variables are also called *indicators* because they indicate whether or not success occurs.

We can find the mean and variance of B from first principles. The expected value of B is the probability for success p.

$$E(B) = 0 \times P(B = 0) + 1 \times P(B = 1)$$
$$= 0 \times (1 - p) + 1 \times p = p$$

The variance of B is the probability of success times the probability of failure.

$$\begin{aligned} \text{Var}(B) &= (0 - p)^2\, P(B = 0) + (1 - p)^2\, P(B = 1) \\ &= p^2(1 - p) + (1 - p)^2 p \\ &= p(1 - p) \end{aligned}$$

The largest variance occurs when $p = 1/2$, when success and failure are equally likely. The most uncertain Bernoulli trials, those with the largest variance, resemble tosses of a fair coin.

A collection of Bernoulli trials defines iid Bernoulli random variables, one for each trial. As in Chapter 10, we use subscripts to identify the iid random variables. For example, to denote the outcomes of 10 detail rep visits, start by letting B_1 denote the outcome of the first visit. Then $B_1 = 1$ if the rep meets the doctor during the first visit, and $B_1 = 0$ otherwise.

Similarly, let $B_2, B_3, \ldots, B_{10}$ denote the success or failure of the rest of the visits. If the detailing visits are indeed Bernoulli trials, then these random variables are iid because Bernoulli trials are independent with a common probability of success.

Counting Successes

Most random variables that express counts are sums of Bernoulli random variables. Whether we're counting the number of doctors visited by a detail rep, employees absent from work, or defective machine parts produced, Bernoulli random variables are used to model the random process. For example, the total number of successful visits of a detail rep is a sum of the Bernoulli random variables that are defined by the 10 trials.

$$Y = B_1 + B_2 + B_3 + B_4 + B_5 + B_6 + B_7 + B_8 + B_9 + B_{10}$$

What Do You Think?

Do you think that the counts in the following situations are sums of Bernoulli trials? Explain briefly.

a. The number of employees who sign up for retirement payroll deductions[1]
b. The number of household insurance claims filed by homeowners who live in a residential subdivision[2]
c. The number of patients who have a positive response in a test of a new type of medication[3]

---

binomial random variable
A random variable that is the sum of n Bernoulli variables, each having probability p of success.

The random variable Y that counts the total number of successes is a binomial random variable. Every **binomial random variable** is the sum of a given number of iid Bernoulli random variables. Two parameters identify a binomial random variable:

n, the number of Bernoulli trials and
p, the probability of success for each trial

The values of n and p depend on the problem. For detailing 10 offices with a 40% chance of success at each, $n = 10$ and $p = 0.4$. Alternatively, imagine a

---

[1] The employees likely interact. Those conversations suggest that the trials, whether an employee elects the payroll deduction, would be dependent. Friends, for example, might decide to enroll (or not) as a group.
[2] Perhaps, but not if a storm (think tornado or hurricane) hit the area. Then all might need to file a damage claim. Barring an event that affects the whole area, the claims are likely to be independent with roughly equal chances.
[3] Yes, assuming that the patients are treated separately and have similar health status at the start of the experiment.

manufacturing process that has a 10% chance of making defective items. If we sample 100 items from the production, then $n = 100$ and $p = 0.1$. As shorthand, we'll identify a binomial random variable with parameters n and p like this: $Y \sim \text{Bi}(n, p)$. Read the squiggle ~ as "is a random variable with probabilities given by."

11.2 | BINOMIAL MODEL

Chapter 9 describes the use of random variables as *models* of real phenomena. In that chapter, we proposed a random variable as a model for changes in the value of a stock. Though simple, the random variable allows us to think carefully about the trade-offs of investing. In this chapter, we use a binomial random variable to assess the performance of detail reps. If the 10 visits of a detail rep to offices are Bernoulli trials, then there's no question that we've got the right model. The total count Y is a binomial random variable.

Assumptions

| caution | Just because we decide to treat the office visits of a detail rep as Bernoulli trials does not mean that they are. Real trials don't have to be independent with equal chance for success. In fact, unless we work at it, it's unlikely for trials to be independent with equal chance for success. |

binomial model The use of a binomial random variable to describe a real phenomenon.

When we use binomial random variables to describe real phenomena, we will say that we're using a **binomial model**. A binomial model uses a binomial random variable to describe a real process, not some idealized experiment. The success of this model depends on how well the random variable describes the phenomenon.

Let's think about the office visits of the detail rep. Is it reasonable to treat visits as Bernoulli trials? Which do you think is more likely?

a. A detail rep randomly picks offices from anywhere in the assigned territory to visit each day.

or b. A detail rep goes to offices that are located near each other, such as those of doctors who work together in a medical center.

Unless a detail rep wants to spend all day in the car, chances are that the rep follows plan b. The results are likely to be dependent. Doctors in the same practice are likely to respond similarly.

Though it can be unpleasant and costly, we need to be proactive to get Bernoulli trials. To make the visits closer to coin tosses, we could provide a randomized schedule of visits for each detail rep to follow. By controlling the schedule, we can reduce the chances for one outcome to affect others. Of course, detail reps resent this oversight. They have better ways to spend their time than criss-crossing their territory according to a schedule. To them, it's foolish not to visit the doctor next door just because someone wants independent trials.

| tip | **Whenever you use a probability model, think about the assumptions.** Unless your application meets these assumptions, treat the results of the model with caution. Models seldom match real processes perfectly, but we need for them to be close. If not, the results can be misleading. |

Finite Populations

One reason for the mismatch between Bernoulli trials and situations encountered in practice is the finite size of real populations. In the example of detailing, suppose that 40% of all receptionists in the territory are willing to let the detail rep get to the doctor. Once the rep succeeds, there's one less willing receptionist. Each success makes it more likely that the next receptionist blocks the path. The trials are consequently dependent.

The most familiar illustration of this type of dependence occurs when dealing cards. If you deal cards from a shuffled, 52-card deck, what's the probability that the second card is an ace if the first card is an ace? The probability of an ace on the first draw is 4/52. The probability of an ace on the second, given an ace on the first, is smaller (3/51) since fewer aces remain. These trials are dependent: The probability that the second card is an ace depends on what happens with the first. Anytime we build trials based on randomly selecting from a finite collection, the sequence of trials is dependent.

Fortunately this type of dependence is irrelevant unless we're dealing with a small collection of possible sites. If the territory covered by the detail rep has 2,500 offices, the effect is negligible. If the territory has 25 offices, it's a problem.

10% Condition   If trials are selected at random from a finite collection, it is okay to ignore dependence *caused by sampling a finite population* if the selected trials make up less than 10% of the collection.

If you violate the 10% condition, you'll need to learn about another random variable (called the hypergeometric).

Other sources of dependence affect probabilities *regardless* of the size of the population. These are a more serious concern in most applications. For instance, consider surveying the opinions of 10 customers randomly selected at a supermarket on a Saturday morning. Because these customers come from a large population (the surrounding population), the 10% condition says we don't need to worry about the dependence implied by selecting from a finite population. Suppose, however, that the customers are interviewed together. One individual might dominate the interview. Would the outcome represent 10 individual voices or the voice of a single outspoken fan or critic?

tip   Rather than worry about a finite population, spend your energy thinking about ways to get independent responses in each trial.

11.3 | PROPERTIES OF BINOMIAL RANDOM VARIABLES

Mean and Variance

Because a binomial random variable such as Y is a sum of iid random variables, we can use the rules for expected values and variances from Chapter 10 to find its mean and variance. To find the mean of Y, the Addition Rule for Expected Values tells us that

$$\begin{aligned} E(Y) &= E(B_1) + E(B_2) + \cdots + E(B_n) \\ &= p + p + \cdots + p \\ &= np \end{aligned}$$

The mean of a binomial random variable is n times the probability of success p. We expect a fraction p of the trials to result in a success.

For the variance, use the Addition Rule for Variances. Because Bernoulli random variables are independent of one another, the variance of a binomial random variable is the sum of the variances.

$$\text{Var}(Y) = \text{Var}(B_1) + \text{Var}(B_2) + \cdots + \text{Var}(B_n)$$
$$= p(1-p) + p(1-p) + \cdots + p(1-p)$$
$$= np(1-p)$$

The variance of a binomial random variable is n times the variance of one trial. Without the assumption of Bernoulli trials, we could not do this calculation so easily.

For the example of detailing, the number of successful visits out of 10 trials is $Y \sim \text{Bi}(n = 10, p = 0.4)$. We expect the rep to see, on average, $np = 10 \times 0.4 = 4$ doctors in 10 visits. The variance of Y is $np(1-p) = 10 \times 0.4 \times (1 - 0.4) = 2.4$ with standard deviation $\sqrt{2.4} \approx 1.55$ doctors. A detail rep who successfully meets 8 doctors in 10 visits has performed

$$(8 - \text{E}(Y))/\text{SD}(Y) = (8 - 4)/1.55 \approx 2.6$$

standard deviations above the mean. Is that much better than average? To answer this question, we need the probability distribution of Y.

Binomial Probabilities

(p. 256)

Finding the probability distribution of a binomial random variable requires some expertise with methods for counting the number of ways to mix two types of objects. Those methods are not useful to us more generally, so we have relegated them to the section Behind the Math: Binomial Counting. Two simple examples, however, illustrate the ideas and remind us once more of the importance of Bernoulli trials.

Start with an easy case. What is the probability that a detail rep does not see a doctor in 10 visits? That is, if $Y \sim \text{Bi}(10, 0.4)$, what is $P(Y = 0)$? To have no successes, every Bernoulli random variable must be zero. Because these are independent, the probabilities multiply.

$$P(Y = 0) = P(B_1 = 0 \, and \, B_2 = 0 \, and \ldots and \, B_{10} = 0)$$
$$= P(B_1 = 0)P(B_2 = 0) \cdots P(B_{10} = 0)$$
$$= (1 - p)^{10} = 0.6^{10} \approx 0.006$$

There is not much chance for a rep to strike out in 10 visits if $p = 0.4$.

What about $P(Y = 1)$, the chance that a rep sees any one of the 10 doctors? For this to happen, *exactly* one of the B's has to be 1 and the rest must all equal 0. The probability that the first visit is a success and the rest are failures is

$$P(B_1 = 1)P(B_2 = 0)P(B_3 = 0) \cdots P(B_{10} = 0) = p(1 - p)^9$$

Similarly, the probability that the second visit succeeds and the rest are failures is

$$P(B_1 = 0, B_2 = 1, B_3 = 0, \ldots, B_{10} = 0) = p(1 - p)^9$$

Regardless of which visit succeeds, the probability of a specific success in 10 trials is $p(1 - p)^9$. Because there are 10 possible visits that might succeed, the probability for one success is

$$P(Y = 1) = 10p(1 - p)^9 = 10(0.4)(0.6)^9 \approx 0.040$$

In general, each binomial probability has two parts.

1. The probability of a specific sequence of Bernoulli trials with y successes in n attempts
2. The number of sequences that have y successes in n attempts

binomial coefficient The number of arrangements of y successes in n trials, denoted $_nC_y$ and sometimes written (p. 256) as $\binom{n}{y}$.

The probability is $p^y(1-p)^{n-y}$ because the trials are independent. A formula gives the number of ways to label y out of n trials as successful. The formula is known as the **binomial coefficient**, written $_nC_y$. (Pronounce this as "n choose y." Your calculator probably has a button for doing this calculation. Behind the Math: Binomial Counting shows how to find $_nC_y$ if you need to do the calculation directly.)

Combining the two parts, the binomial probability for y successes among n trials is

$$P(Y = y) = {_nC_y}\,p^y(1-p)^{n-y}$$

For 10 detail rep visits, $n = 10$ and $p = 0.4$. Hence, the probability distribution of the binomial random variable Y is

$$P(Y = y) = {_{10}C_y}(0.4)^y(0.6)^{10-y}$$

Hence, the probability for $y = 8$ successful detail visits is about 1%.

$$P(Y = 8) = {_{10}C_8}(0.4)^8(0.6)^2 = 45(0.4)^8(0.6)^2 \approx 0.011$$

Now that we have a formula for $p(y)$, we can graph the probability distribution as shown in Figure 11.1.

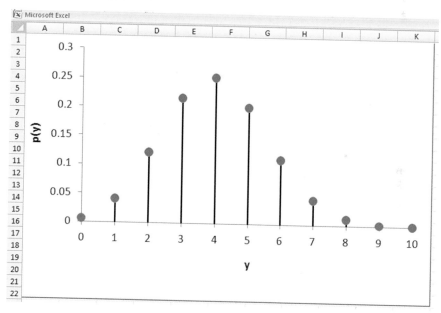

FIGURE 11.1 The probability distribution of $Y \sim \text{Bi}(10, 0.4)$.

The maximum probability occurs at the mean $E(Y) = 4$. The mode (location with largest probability) is always near the mean for binomial random variables.

To find the probability that a detail rep meets 8 or more doctors in a day, we have to add the last three probabilities shown in Figure 11.1. These represent three disjoint events, all with small probability.

$$P(Y \geq 8) = P(Y = 8) + P(Y = 9) + P(Y = 10)$$
$$\approx 0.01062 + 0.00157 + 0.00010 = 0.01229$$

There's slightly more than a 1% chance for a detail rep to meet 8 or more doctors during 10 office visits. We'd expect fewer, about 4 per day. A rep that regularly sees 8 doctors in 10 visits looks like a star!

Summary

A binomial random variable counts the number of successes in n Bernoulli trials. Two parameters identify every binomial random variable: the number of Bernoulli trials n and the probability of success p.

Binomial random variable, $Y \sim \text{Bi}(n, p)$

n = number of trials

p = probability of success

y = number of successes in n Bernoulli trials

$$P(Y = y) = {}_nC_y\, p^y(1 - p)^{n-y}, \quad \text{where } {}_nC_y = \frac{n!}{y!(n - y)!}$$

Mean: $E(Y) = np$

Variance: $\text{Var}(Y) = np(1 - p)$

What Do You Think?

a. Which of the following situations seem suited to a binomial random variable?[4]

The total cost of items purchased by five customers

The number of items purchased by five customers

The number of employees choosing a new health plan

b. If a public radio station solicits contributions during a late-night program, how many listeners should it expect to call in if $n = 1{,}000$ are listening and $p = 0.15$? How large is the SD of the number who call in?[5]

4M EXAMPLE 11.1   FOCUS ON SALES

MOTIVATION   STATE THE QUESTION

During the early stages of product development, some companies use focus groups to get the reaction of customers to a new product or design. The particular focus group in this example has nine randomly selected customers. After being shown a prototype, they were asked, "Would you buy this product if it were sold for $99.95?"

The questionnaire allowed only two answers: yes or no. When the answers were totaled, six of the nine customers in the group answered yes. Prior to the focus group, the development team claimed that 80% of customers would want to buy the product at this price. If the developers are right, what should we expect to happen at the focus group? Is it likely for six out of nine customers in the group to say they'd buy it?

METHOD   DESCRIBE THE DATA AND SELECT AN APPROACH

This application calls for counting a collection of yes/no responses, the perfect setting for Bernoulli trials and a binomial random variable. On the basis of the claims of the developers, the random variable $X \sim \text{Bi}(n = 9, p = 0.8)$

---

[4] Only the number of employees choosing the new health plan, so long as employess are not swayed by the opinion of one outspoken employee. The total cost and number of items are not sums of outcomes of Bernoulli trials; you can buy more than one item, for instance.

[5] We expect 150 to call, with SD $= \sqrt{np(1 - p)} \approx \sqrt{1000(0.15)(0.85)} \approx 11.3$.

represents the possibilities in a focus group of nine customers. An important caveat is that Bernoulli trials are also independent of one another; this may be true of a focus group.

DO THE ANALYSIS

The expected value of X is $np = 9(0.8) = 7.2$, higher than the observed number of responses. The standard deviation of X is $\sqrt{np(1 - p)} = \sqrt{9(0.2)(0.8)} = 1.2$.

This plot shows the probability distribution for the random variable. The probability of exactly six saying yes: $P(X = 6) = {}_9C_6(0.8)^6(0.2)^3 \approx 0.18$. That's not the most likely count, but it is common.

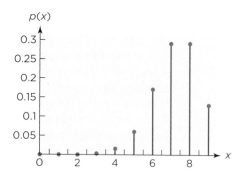

SUMMARIZE THE RESULTS

The results of this focus group are in line with what we'd expect to see if the development team is right. If the team is right, about one-sixth of focus groups of nine customers will result in six who say yes to the purchase question. It is important that those who run the focus group allow each member of the group to fill in his or her questionnaire without being overly influenced by others in the group.

Let's return to the assumption of independence. It's easy for dependence to creep into a focus group. If an outspoken person dominates the group, we've effectively got fewer trials. As a result, the results will be more variable than anticipated by a binomial model. For example, the binomial model in Example 11.1 implies that it is unlikely for the group to be unanimous. The probability of all saying yes is $P(X = 9) = {}_9C_9p^9(1 - p)^0 = (0.8)^9 \approx 0.134$. If, however, one person can persuade the others to follow her or his opinion, there's an 80% chance that they will all say yes.

11.4 | POISSON MODEL

Bernoulli trials are hard to find for many types of counting problems. For some of these, a second type of discrete random variable is helpful when the underlying random process behaves in a continuous manner. As examples, consider these situations.

- The number of imperfections per square meter of glass panel used to make plasma televisions
- The number of robot malfunctions per day on an automobile assembly line
- The number of telephone calls arriving at the help desk of a software company during a 10-minute period

Each situation produces a count, whether it's the count of flaws in a manufactured product or the number of telephone calls. Every case, however, requires some imagination to find Bernoulli trials. For the help desk, for instance, we could slice the 10 minutes into 600 one-second intervals. If the intervals are short enough, at most one call lands in each. We've got success-failure trials and a binomial model, but it takes a bit of imagination to find the trials.

Poisson Random Variable

Poisson random variable

A random variable describing a count of the number of events in a random process that occurs at a steady rate denoted by λ.

There's a better way to model counts such as these. Notice that each of these three situations has a rate. The manager of the help desk, for instance, would be dumbfounded if you asked for the probability of a call arriving during the next second, but he could easily quote you the average number of calls per hour. Similarly, we have a rate for defects per square meter or malfunctions per day.

A **Poisson random variable** describes the number of events determined by a random process during an interval of time or space. A Poisson random variable has one parameter. This parameter λ (spelled lambda, but pronounced "lam-da") is the rate of events, or arrivals, within *disjoint* intervals. If you were told "We typically receive 100 calls per hour at this call center," then $\lambda = 100$ calls/hour. Keep track of the measurement units on λ.

If X denotes a Poisson random variable with parameter λ [abbreviated $X \sim \text{Poisson}(\lambda)$], then its probability distribution is

$$P(X = x) = e^{-\lambda}\frac{\lambda^x}{x!}, \qquad x = 0, 1, 2, \ldots$$

Unlike those of a binomial random variable, the possible values of this random variable keep going and going. Conceptually, there's no limit on the size of a Poisson random variable. [In this expression, the symbol $x!$ denotes the factorial function $x! = x(x - 1)\cdots(2)(1)$, with $0! = 1$. The letter e stands for the base of natural logs, $e \approx 2.71828$, as appears in the exponential function $\exp(x) = e^x$.]

To illustrate the calculations, suppose that calls arrive at the help desk at a rate of 90 calls per hour. We're interested in the chance that no call arrives in the next minute. In order to use the Poisson probability distribution, we have to adjust λ to suit the time interval of interest. The appropriate rate is $\lambda = 90/60 \approx 1.5$ calls *per minute*. If X is a Poisson random variable with $\lambda = 1.5$, then the probability of no calls during the next minute is

$$P(X = 0) = e^{-\lambda}\lambda^0/0! = e^{-1.5} \approx 0.223$$

The probability of one call during the next minute is 1.5 times larger

$$P(X = 1) = e^{-\lambda}\lambda^1/1! = 1.5e^{-1.5} \approx 0.335$$

and the probability of two calls is

$$P(X = 2) = e^{-\lambda}\lambda^2/2! = (1.5^2/2)e^{-1.5} \approx 0.251$$

Figure 11.2 graphs the probability distribution $P(X = x)$, for x ranging from 0 to 7. The probability distribution keeps on going and going, but the probabilities approach zero beyond those shown here.

If the number of calls is a Poisson random variable with rate $\lambda = 1.5$ calls per minute, what would you guess is the mean number of calls per minute? It's $\lambda = 1.5$. It is not obvious, but λ is also the variance of a Poisson random variable.

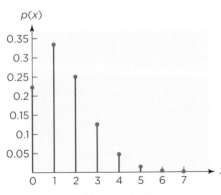

FIGURE 11.2 Poisson probability distribution with $\lambda = 1.5$.

Poisson random variable, $X \sim$ Poisson(λ)

λ = expected count of events over a time interval or a region

X = number of events during an interval or in the region

$$P(X = x) = e^{-\lambda}\frac{\lambda^x}{x!}, \quad x = 0, 1, 2, \ldots$$

Expected value: $E(X) = \lambda$

Variance: $Var(X) = \lambda$

Poisson model A model in which a Poisson random variable is used to describe counts of real data.

Poisson Model

The **Poisson model** refers to using a Poisson random variable to describe a process in the real world. The necessary assumptions resemble those needed in order to use the binomial model. Whereas the binomial model requires independent trials, the Poisson model assumes that events in separate intervals are independent. The binomial model assumes that the chance for success p is the same for every trial; the Poisson model assumes that the rate of events stays the same.

Let's continue with telephone calls arriving at a help desk. The help desk handles 90 calls per hour, on average. Before we use a Poisson random variable to compute probabilities, we should check the assumptions. First, arrivals in disjoint intervals should be independent. Hence, if the help desk gets more calls than usual in an hour, it should not expect the rate of calls to drop below average during the next hour. (That would imply negative dependence.) Second, the rate of arrivals should be constant. This assumption makes sense during the regular business hours. However, the rate would probably decrease at night or on weekends.

Deciding between a binomial model and a Poisson model is straightforward. Use a binomial model when you recognize distinct Bernoulli trials. Use a Poisson model when events happen at a constant rate over time or space.

4M EXAMPLE 11.2 | DEFECTS IN SEMICONDUCTORS

MOTIVATION | STATE THE QUESTION

Computer chips are made from slices of silicon known as wafers. A complex process (known as photolithography) etches lines in the wafer, and these lines become the circuits that enable the chip to process data. With feature sizes of 45 nanometers (about 0.0000018 inch) and smaller, defects in the wafer cause real problems.

Before companies like AMD and Intel buy wafers from a supplier, they want to be sure that each wafer does not have too many defects. No one expects perfection, but the number of defects must be small. In this example, a supplier claims that its wafers have 1 defect per 400 square centimeters (cm²). Each wafer is 20 centimeters in diameter, so its area is $\pi r^2 = \pi(10)^2 \approx 314\,cm^2$.

What is the probability that a wafer from this supplier has no defects? What is the mean number of defects and the SD? (The SD shows how consistent the number of defects will be from wafer to wafer.)

METHOD | DESCRIBE THE DATA AND SELECT AN APPROACH

The relevant random variable is the number of defects on a randomly selected wafer. The type of defect should not carry over from wafer to wafer, so it seems reasonable that these counts should be independent. A Poisson model suits this situation.

MECHANICS | DO THE ANALYSIS

The supplier claims a defect rate of 1 per $400\,\text{cm}^2$. Since a wafer has $314\,\text{cm}^2$, we model the number of defects on a randomly chosen wafer as the random variable $X \sim \text{Poisson}(\lambda = 314/400)$.

The mean number of defects is $\lambda = 314/400 = 0.785$. The probability of no defect on a wafer is then $P(X = 0) = e^{-0.785} \approx 0.456$. The SD for the Poisson model is the square root of the mean, so $\sigma = 0.886$ defects/wafer.

MESSAGE | SUMMARIZE THE RESULTS

If the claims are accurate, the chip maker can expect about 0.9 defects per wafer. Because the variation in the number of defects is small, there will not be much variation in the number of defects from wafer to wafer. About 46% of the wafers should be free of defects.

Best Practices

- *Ensure that you have Bernoulli trials if you are going to use the binomial model.* Bernoulli trials allow only one of two outcomes, with a constant probability of success and independent outcomes. If the conditions of the trials change, so might the chance for success. Results of one trial must not influence the outcomes of others.
- *Use the binomial model to simplify the analysis of counts.* When the random process that you are interested in produces counts from discrete events, use the binomial model rather than figuring out the details from the start.
- *Use the Poisson model when the count accumulates during an interval.* For counts associated with rates, such as events per hour or per square mile, use the Poisson model. The intervals or regions must be disjoint so that the counts are independent. You also need to be careful that the intervals or regions have the same size.
- *Check the assumptions of a model.* Whichever model you use, think about whether the assumptions of independence and stable

behavior make sense. Dependence can arise in many ways unless you've been proactive in designing trials that are independent.
- *Use a Poisson model to simplify counts of rare events.* If you have a large number of Bernoulli trials with small probability p, a Poisson model is simpler for calculations. For example, if $Y \sim \text{Bi}(n = 125, p = 0.01)$, then

$$P(Y \leq 1)$$
$$= P(Y = 0) + P(Y = 1)$$
$$= {}_{125}C_0\, p^0 (1 - p)^{125} + {}_{125}C_1\, p^1 (1 - p)^{124}$$
$$\approx 0.285 + 0.359 = 0.644$$

The corresponding Poisson model sets λ to the mean of the binomial random variable, $\lambda = np = 1.25$. If $X \sim \text{Poisson}(1.25)$, then we obtain almost the same probability.

$$P(X \leq 1) = P(X = 0) + P(X = 1)$$
$$= e^{-1.25} + 1.25 e^{-1.25}$$
$$\approx 0.287 + 0.358 = 0.645$$

Pitfalls

- *Do not presume independence without checking.* Just because you need independent events to use a binomial or Poisson model does not mean that your events are independent. You have to work to get independent events.
- *Do not assume stable conditions routinely.* A binomial model requires the probability p be the same for all trials. Similarly, a Poisson model presumes the rate λ stays the same. Most trials or processes are observed over time, however; this means that conditions may change and alter the characteristics.

Software Hints

Statistical software includes the binomial, Poisson, and other discrete random variables. Software can compute the probabilities of any simple event. For example, if X is a binomial random variable with parameters $n = 34$ and $p = 0.45$, software can calculate $P(X = 15)$ or the cumulative probability $P(X \le 15)$ instantly. All you need to do is identify the random variable and specify its parameters.

Software can also simulate iid data from various distributions. If you'd like to see what data that follow a Poisson distribution look like, you can generate thousands of cases and have a look.

EXCEL

Excel includes formulas for computing probabilities associated with commonly used random variables. For binomial random variables, the function

BINOMDIST(x, n, p, FALSE)

gives the probability $P(X = x)$ for X, a binomial random variable with parameters n and p. For example,

BINOMDIST(3, 10, .5, FALSE)

gives the probability of getting 3 heads in 10 tosses of a fair coin. The third argument TRUE or FALSE, controls whether the formula gives the probability of a specific value or the cumulative probability up to the value. Hence, the formula

BINOMDIST(x, n, p, TRUE)

gives the cumulative probability $P(X \le x)$.

The formula POISSON(y, λ, FALSE) computes the probability $P(Y = y)$ for a Poisson random variable with mean λ, and the formula POISSON(y, λ, TRUE) computes the cumulative probability $P(Y \le y)$.

MINITAB

Following the menu items

Calc > Probability Distributions > Binomial

opens a dialog that allows you to obtain probabilities associated with a binomial distribution. Suppose that X denotes a binomial random variable with parameters n and p. In the dialog, you can choose between the probability $P(X = x)$ and the cumulative probability $P(X \le x)$. All you need to do is pick the one you want and supply values for n (the number of trials) and p (the probability of success).

The value of x used in the probability calculation must be in a column of the spreadsheet, so you'll probably want to use this feature of Minitab with a calculation spreadsheet rather than a data spreadsheet. For example, if the column C_1 has the values 1, 2, and 3 in the first three rows, then specifying $n = 10$, $p = 0.25$, and C_1 as the input column produces the output

0	0.056314
1	0.187712
2	0.281568

For example, $P(X = 2) = 0.281568$.

The menu sequence

Calc > Probability Distributions > Poisson

leads to a similar dialog that specifies a Poisson distribution. In place of n and p, the dialog requests λ, the mean of the distribution.

JMP

To obtain probabilities from JMP, define a formula for a column using the formula calculator. (For example, right click on the header at the top of the column in the spreadsheet view.) In the functions view of the calculator, pick the Probability collection of functions. If you pick the item Binomial Probability, the formula becomes

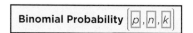

Fill in the boxes for p, n, and k with either numbers that you type or the names of columns. The column is filled with $P(X = k)$ for a binomial random variable X with parameters n and p. Using a column of values for k allows you to get several probabilities at once. Selecting the formula Binomial Distribution computes the cumulative probabilities.

The similar formulas Poisson Probability

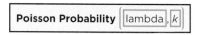

and Poisson Distribution compute the probability $P(Y = k)$ and the cumulative probability $P(Y \le k)$ for a Poisson random variable Y with mean λ.

◼ BEHIND the Math

Binomial Counting

Consider the number of ways for a detail rep to have 2 successful office visits among 10 attempts. These could be the first 2 visits, the first and third, or any pair. No matter which 2 visits succeed, however, the probability of a specific sequence of visits with 2 successes and $10 - 2 = 8$ failures is $p^2(1 - p)^8$. To find the binomial probability, we need to count the number of ways for a rep to have 2 successes in 10 visits.

Let's start by picking which visits succeed. We can pick any of 10 to be the first, and any of the remaining 9 to be second. Hence we can pick out two visits in any of $10 \times 9 = 90$ *ordered* ways. We say that these are ordered because this count distinguishes picking 8 and then 3 from picking 3 and then 8. Since we only care about the number of successful visits, not the order, we have to divide by 2 to correct for double counting. That leaves $90/2 = 45$ ways to have 2 successes among 10 visits. Pulling this together, the probability for 2 successes among the 10 trials is

$$P(Y = 2) = \frac{10 \times 9}{2} p^2 (1 - p)^8$$
$$= 45(0.4)^2(0.6)^8$$
$$\approx 0.121$$

The binomial coefficient $_nC_y$ determines the number of ways to label y out of n trials as successes. The binomial coefficient uses the factorial function. For a positive integer $k > 0$, the factorial function is the product of integers from k down to 1.

$$k! = k \times k - 1 \times \cdots \times 1$$

Define $0! = 1$. Once you have factorials, it's easy to write a compact expression for the number of ways to label k successes among n trials.

$$_nC_y = \frac{n!}{y!(n - y)!}$$

This formula is compact, but don't calculate all of those factorials if you do this by hand. For example, for 2 successes in 10 trials,

$$_{10}C_2 = \frac{10!}{2!(10 - 2)!}$$
$$= \frac{10 \times 9 \times 8 \times 7 \times 6 \times 5 \times 4 \times 3 \times 2 \times 1}{(2 \times 1) \times (8 \times 7 \times 6 \times 5 \times 4 \times 3 \times 2 \times 1)}$$
$$= \frac{10 \times 9}{2 \times 1} = 45$$

That's the same count that we got before, but with lots of redundant terms that cancel in the numerator and denominator. For picking out 3 from 10, there are 120 sequences of 10 trials with exactly 3 successes.

$$_{10}C_3 = \frac{10!}{3!(10 - 3)!}$$
$$= \frac{10 \times 9 \times 8 \times 7 \times 6 \times 5 \times 4 \times 3 \times 2 \times 1}{(3 \times 2 \times 1) \times (7 \times 6 \times 5 \times 4 \times 3 \times 2 \times 1)}$$
$$= \frac{10 \times 9 \times 8}{3 \times 2 \times 1} = 120$$

They Add Up!

Do the probabilities given by the binomial probability distribution add up to 1? It's certainly not obvious that

$$\sum_{y=0}^{n} {_nC_y} p^y (1 - p)^{n-y} = 1$$

It's an amazing fact that these sum to 1, though not a recent discovery.

The key to adding up these probabilities is a more general result called the Binomial Theorem, which gives the random variable its name. The Binomial Theorem says that for any two numbers a and b, we can expand the nth power of the sum as follows:

$$(a + b)^n = \sum_{y=0}^{n} {_nC_y} a^y b^{n-y}$$

If we pick $a = p$ and $b = 1 - p$, then clearly $a + b = 1$. The sum on the right side is the sum of the binomial probabilities.

You can figure out the mean of a binomial random variable this way, but we don't recommend it. It's a fact that

$$\sum_{y=0}^{n} y {_nC_y} p^y (1 - p)^{n-y} = np$$

but it's much easier—and more useful for applications—to think of binomial random variables as sums.

CHAPTER SUMMARY

The number of successes in n **Bernoulli trials** produces a **binomial random variable**. The parameters of a binomial random variable are n (the number of trials) and p (the chance for success). The **binomial model** uses a binomial random variable to describe counts of successes observed for a real phenomenon. The use of a probability model requires assumptions about the real phenomenon, most often stability (constant probability of success) and independence. Counts based on events in disjoint intervals of time or space produce a **Poisson random variable**. A Poisson random variable has one parameter, its mean λ. The **Poisson model** uses a Poisson random variable to describe counts in data.

▧ Key Terms

10% condition, 247
Bernoulli random variable, 244
Bernoulli trial, 244
binomial coefficient, 249

binomial model, 246
binomial random variable, 245
Poisson model, 253
Poisson random variable, 252

symbols
  $x!$ (factorial), 252
  λ (Poisson rate), 252
  e (exponential), 252

▧ Formulas

Binomial random variable, $Y \sim \text{Bi}(n, p)$

Parameters: n (number of trials), p (chance for success)

Assumptions: independent trials, equal probability p

$$\text{E}(Y) = np$$
$$\text{Var}(Y) = np(1 - p)$$
$$P(Y = y) = {}_nC_y\, p^y(1 - p)^{n-y}, \quad \text{where}$$
$${}_nC_y = \frac{n!}{y!(n - y)!}, \quad y = 0, 1, 2, \ldots, n$$

Poisson random variable, $X \sim \text{Poisson}(\lambda)$

Parameter: λ (rate)

Assumptions: independence in disjoint sets, constant rate

$$\text{E}(X) = \lambda$$
$$\text{Var}(X) = \lambda \text{ (same as the mean)}$$
$$P(X = x) = e^{-\lambda}\frac{\lambda^x}{x!}, \quad x = 0, 1, 2, \ldots$$

EXERCISES

Mix and Match

For these matching exercises, Y is a binomial random variable with parameters n and p, X is a Poisson random variable with parameter λ, and B_1 and B_2 are independent Bernoulli trials with probability of success p.

1. Expression for the variance of Y	(a) np
2. Symbol denoting the chance of failure	(b) $e^{-\lambda}$
3. Expression for the mean of X	(c) $e^{-\lambda}\lambda^2/2$
4. Probability that Y is zero	(d) p
5. Probability that X is 2	(e) 0
6. Covariance between B_1 and B_2	(f) p^n
7. Expression for the expected value of Y	(g) $1 - p$
8. The expected value of B_1	(h) $(1 - p)^n$
9. Probability that Y is n	(i) $np(1 - p)$
10. Probability that X is zero	(j) λ

True/False

Mark each statement True or False. If you believe that a statement is false, briefly say why you think it is false.

Exercises 11–16. An auditor inspects 25 transactions processed by the business office of a company. The auditor selects these transactions at random from the thousands that were processed in the most recent three months. In the past, the auditor has found 10% of transactions at this type of company to have been processed incorrectly (including such errors as wrong amount, incorrect budget code, incomplete records).

11. If the selected 25 transactions are treated as Bernoulli trials, the resulting binomial calculations will be in error because of sampling from a finite population.

12. The binomial model for this problem sets $n = 25$ and $p = 0.10$.

13. Assuming Bernoulli trials, there is a 1/1,000 chance for at least one error among the first three transactions that the auditor checks.

14. Assuming Bernoulli trials, the auditor should expect to find more than 3 errors among these transactions.

15. A binomial model would be more appropriate for this problem if the auditor picked the first 25 transactions during the three-month period.

16. It would be unlikely for the auditor to discover more than 10 errors among these transactions because such an event lies more than 4 SDs above the mean.

Exercises 17–22. A textile mill produces fabric used in the production of dresses. Each dress is assembled from patterns that require 2 square yards of the material. Quality monitoring typically finds 12 defects (snagged threads, color splotching) per 100 square yards when testing sections of this fabric.

17. A Poisson model is well suited to this application so long as the defects come in small clusters of two or three together.

18. To use a Poisson model for the number of defects, we must assume that defects occur at an increasing rate throughout the manufacturing of the fabric.

19. An increase in the amount of fabric from 2 to 4 square yards doubles the mean of a Poisson model for the number of defects in the fabric used to make a dress.

20. An increase in the amount of fabric from 2 to 4 square yards doubles the SD of a Poisson model for the number of defects in the fabric used to make a dress.

21. A Poisson model implies that the probability of a defect in the material used to make a dress is $2 \times 12/100$.

22. A Poisson model implies that doubling the amount of fabric does not affect the chance of finding a defect in the fabric.

Think About It

23. Is the binomial model suited to these applications?
    (a) The next five cars that enter a gasoline filling station get a fill-up.

(b) A poll of the 15 members of the board of directors of a company indicates that 6 are in favor of a proposal to change the salary of the CEO.

(c) A company realizes that about 10% of its packages are not being sealed properly. When examining a case of 24, it counts the number that are unsealed.

24. Is the Poisson model suited to these applications?
    (a) The number of customers who enter a store each hour
    (b) The number of auto accidents reported to a claims office each day
    (c) The number of broken items in a shipment of glass vials

25. If the office visits of a detail rep are Bernoulli trials, then which of these two sequences of six office visits is more likely (S stands for success and the rep sees the doctor, and F for failure): SSSFFF or SFFSFS?

26. If the underlying trials are positively dependent (meaning that a success, say, is more likely followed by another success), would you expect the variance of the total number of successes to be larger or smaller than the binomial expression $np(1 - p)$? (*Hint:* Imagine huge dependence, so large that every trial matches the first. What does that do to the chance for all successes or all failures?)

27. Four customers in a taste test sample two types of beverage (called A and B here) and are asked which is preferred. If customers cannot tell the two beverages apart, are we more likely to see the response AABB or AAAA?

28. In the taste test of Exercise 27, are we more likely to find that the four customers prefer the same brand or to find their opinions evenly divided between the two brands?

29. A software project requires 10 components within the next three months. Each component is produced by a programming team that operates independently of the other teams. Each team has a 50% chance of finishing on time within the next three months. Is there a 50% chance that the project will be finished on time?

30. A manager supervises five experimental projects. Each has a 50-50 chance for producing an innovative, successful new product. The teams that are developing each project work separately, so the success of one project is independent of the success of other projects. A successful project returns 10 times the amount invested. Which should the manager choose: invest her full budget in one project, or split it evenly across all 5?
    (a) Compare her options in terms of expected values.
    (b) Compare her options in terms of risk, in the sense of the chances that she will have nothing to show for investing her budget in some or all of these projects.

31. If a type of car can be ordered in three body styles and six colors, how many distinct versions can be ordered?

32. If a manager randomly pairs his staff of 10 into two groups of 5 each to prepare a project, how many ways are there for the staff to be split into these two groups?

You Do It

33. A manager randomly selected 25 large cash transactions at a bank that were made in January. The manager then carefully tracked down the details of these transactions to see that the correct procedures for reporting these large transactions were followed. This bank typically makes 1,500 such transactions monthly.
    (a) Is it appropriate to use a binomial model in this situation?
    (b) If the chance for a procedural error is 10%, is it likely that the manager finds more than two such transactions? For this component, argue informally, without computing the probability, from the characteristics of the binomial model.
    (c) Find the probability in part (b).

34. A commuter airline deliberately overbooks flights, figuring that only 75% of passengers with reservations show up for its flights. It flies small propeller planes that carry 20 passengers.
    (a) In order to use a binomial model in this problem, what assumptions are necessary? Are these reasonable?
    (b) If the airline allows 25 passengers to book reservations for a flight, what is the probability that the flight will be oversold? Do not use the probability distribution and instead think about the mean and shape of the binomial distribution.
    (c) Find the probability of overbooking. Did the rough approach taken in part (b) get close enough to the right answer?

35. Every now and then even a good diamond cutter has a problem and the diamond breaks. For one cutter, the rate of breaks is 0.1%.
    (a) What probability model seems well suited to this problem? Why?
    (b) If this cutter works on 75 stones, what is the probability that he breaks 2 or more?

36. A manufacturer of LCD panels used in computer displays has decided to ship only panels with at most one defective pixel. Defective pixels are the result of randomly located flaws in the glass used to make the panel. To be profitable, the manufacturer must ship 85% of its panels. If an LCD panel has $1,024 \times 768 = 786,432$ pixels, how small must the chance for a defective pixel be in order for the probability of shipping a panel to be 0.95?

37. An insurance company offers two policies that cover the cost of repairs from driving accidents. A policy costing $1,000 annually has a $1,000 deductible (meaning the driver is responsible for paying the first $1,000 in damages before the insurance kicks in), whereas a policy that costs $3,000 annually has a $250 deductible. Assume that accidents that get reported to the insurance company require more than $1,000 to repair the car. If we model the number of accidents as a Poisson random variable with mean μ, when is the policy with the $1,000 deductible cheaper for the driver on average?

38. The Food and Drug Administration has veto power over the choice of drug names. In 2004, it used this power regularly, rejecting 36% of the names proposed by companies for reasons such as sounding too much like another product (and causing confusion at the pharmacy). Suppose that a company spends $500,000 developing each proposed name, but there's a 50-50 chance of a name being rejected.[5]
    (a) If the review of names occurs independently, what is the probability that the company will spend more than $1 million developing a name?
    (b) What is the expected cost of developing a name? [*Hint:* For any value x and $0 \le p < 1$, $p + 2p^2 + 3p^3 + \cdots = p/(1 - p)^2$.]

39. A dairy farmer accidentally allowed some of his cows to graze in a pasture containing weeds that would contaminate the flavor of the milk from this herd. The farmer estimates that there's about a 10% chance of a cow grazing on some of the flavorful weeds.
    (a) Under these conditions, what is the probability that none of the 12 animals in this herd ate the tasty weeds?
    (b) Does the Poisson model give a good estimate of the probability that no animal ate the weed?

40. Historically a bank expects about 5% of its borrowers to default (i.e., not repay). The bank currently has 250 loans outstanding.
    (a) In order to use a binomial model to compute the probabilities associated with defaults, what must the bank assume about the behavior of these borrowers?
    (b) Do the necessary assumptions listed in part (a) appear reasonable in the context of this problem?
    (c) The bank has reserves on hand to cover losses if 25 of these loans were to default. Will these reserves will be enough?

41. A basketball player attempts 20 shots from the field during a game. This player generally hits about 35% of these shots.
    (a) In order to use a binomial model for the number of made baskets, what assumptions are needed in this example? Are they reasonable?
    (b) How many baskets would you expect this player to make in the game?
    (c) If the player hits more than 11 shots (12, 13, 14,..., or 20), would you be surprised?
    (d) How many points would you expect the player to score if all of these are 2-point shots?
    (e) If this player randomly takes half of the shots from 3-point range and half from 2-point range and makes both with 35% chance, how many points would you expect the player to score?

42. The kicker in a football game scores 3 points for a made field goal. A kicker hits 60% of the attempts taken in a game.
    (a) What assumptions are needed in order to model the number of made kicks in a game as a binomial random variable?

(b) If the kicker tries five attempts during a game, how many points would you expect him to contribute to the team's score on field goals?

(c) What is the standard deviation of the number of points scored by the kicker?

(d) Why would the SD of the number of points be useful to the coach?

4M Market Survey

A marketing research firm interviewed visitors at a car show. The show featured new cars with various types of environmentally oriented enhancements, such as hybrid engines, alternative types of fuels (such as biofuels), and aerodynamic styling. The interviews of 25 visitors who indicated they were interested in buying a hybrid included the question: What appeals most to you about a hybrid car? The question offered three possible answers: (a) savings on fuel expenses, (b) concern about global warming, and (c) desire for cutting-edge technology.

Motivation

a) Why would a manufacturer be interested in the opinions of those who have already stated they want to buy a hybrid? How would such information be useful?

Method

b) The question offered three choices, and customers could select more than one. If a manufacturer of hybrids is interested in the desire for cutting-edge technology, can we obtain Bernoulli trials from these responses?

c) What random variable describes the number of visitors who indicated that cutting-edge technology appeals to them in a hybrid car? Be sure to think about the necessary assumptions for this random variable.

Mechanics

d) Past shows have found that 30% of those interested in a hybrid car are drawn to the technology of the car. If that is still the case, how many of the 25 interviewed visitors at this show would you expect to express interest in the technology?

e) If five at this show were drawn to the cutting-edge technology, would this lead you to think that the appeal of technology had changed from 30% found at prior shows?

Message

f) Summarize the implications of the interviews for management, assuming that 5 of the 25 visitors expressed a desire for cutting-edge technology.

4M Safety Monitoring

Companies that sell food and other consumer products watch for problems. Occasional complaints from dissatisfied customers always occur, but an increase in the level of complaints may signal a seious problem that is better handled sooner than later. With food safety, that's a particular concern. Getting out ahead of the bad news and actively working to fix a problem can save the reputation of a company.

The company in this exercise sells frozen, prepackaged dinners. It generally receives one or two calls to its problem center each month from consumers who say "Your food made me sick." Usually, it's not the food that is to blame, but some other cause like the flu. If the number of calls rises to a rate approaching six calls per month, however, then there's a real problem. To be specific, let's say that the normal rate of calls is 1.5 per month and that the rate of calls when there is a serious problem is 6 per month.

Motivation

a) It is important for a company to react quickly when it recognizes that a product threatens public health. Why is it also important for a company not to overreact, acting as if there is a problem when in fact everything is operating normally?

b) What are the trade-offs between reacting quickly to a real problem and overreacting when in fact there is no real problem?

Method

c) What type of random variable seems suited to modeling the number of calls during a normal period? During a problem period?

d) What assumptions are necessary for the use of the random variable chosen in part (c)? Do these seem reasonable in the context of this problem?

Mechanics

e) During normal months, would it be surprising to receive more than 3 calls to the problem center?

f) During problem months, would it be surprising to receive more than 3 calls to the problem center?

g) The company seldom has problems, and management believes, without feedback from the problem center, that there is a 5% chance of a problem. If the company receives more than 3 calls to the problem center in a month, then what is the probability that there is in fact a problem? (*Hint:* Think about Bayes' Rule and reversing conditional probabilities.)

Message

h) More than 3 calls arrived at the problem center in the past month. Explain for management the chances that there is a problem.

The Normal Probability Model

PRICES OF STOCK PLUMMETED IN OCTOBER 1987. At the end of September, stocks in the United States were worth $3.3 trillion. A month later, their value had fallen to $2.5 trillion. The timeplot in Figure 12.1 tracks the value of the market day by day in October. (Horizontal gaps in the plot are weekends.)

October began with a slight rise. By Friday, October 16, however, the market had slumped to $2.9 trillion. That drop was only a prelude. On the following Monday, the total value sank $500 billion; half a trillion dollars evaporated in one day. It's no wonder they call this day "Black Monday" on Wall Street.

The sudden collapse left investors stressing over the possibility of another big drop. Many decided to do something familiar: buy insurance. If you drive, you have insurance in case you have an accident. If you invest, you have insurance in case the market has an "accident." Insurance for the stock market resembles car insurance. You select and pay for various types of coverage up front. If you have an accident, the policy pays for covered damages. If you don't have an accident, the insurance company

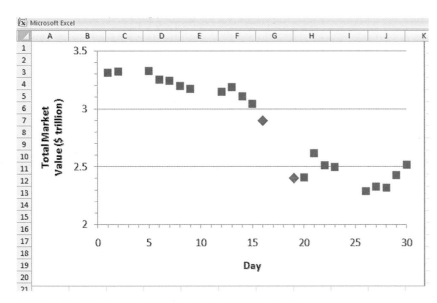

FIGURE 12.1 Total value of stocks during October 1987.

keeps your money. Insurance for stocks works in a similar way. You can buy coverage that pays you if the market stumbles.

How much should an investor expect to pay for this insurance? If you've ever had an accident, you know that your driving record affects your premium. The same logic applies to insurance for stocks. If there's a good chance that the market will take a dive, then insurance costs more.

Models for counts are not suited to data like stock indices that can take on basically any possible value. Stock prices, insurance costs, process yields, and many other business situations call for random variables that are able to represent a continuum of values. That's where the models in this chapter come into the picture. As always, though, we have to make sure that the model fits the application. The models of this chapter are based on the normal distribution, the theory that underlies the Empirical Rule for bell-shaped distributions.

12.1 | NORMAL RANDOM VARIABLE

Advertisements often quote a price, such as offering a one-carat diamond for $4,000. All one-carat diamonds are not the same, and some cost considerably more and others less. The histogram in Figure 12.2 shows prices of 97 one-carat diamonds of similar cut and clarity.

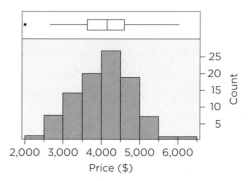

FIGURE 12.2 Histogram of the prices of 97 one-carat diamonds.

One of these diamonds costs a bit more than $2,000 whereas others cost more than $5,000. On average, these diamonds cost $4,066, but the large standard deviation ($738) is a reminder of the variation of prices.

Suppose we're looking at one-carat diamonds. If we pick out a diamond, how much might we have to pay? This histogram provides a point of reference. Do we need to remember everything about this figure, or can we summarize it more compactly? The range from the minimum to maximum price, about $2,000 to $6,000, is a start, but that's a wide interval and does not hint at the way values pile up near the mean.

Models for counts presented in Chapter 11 are poorly suited to this task. Costs can take on virtually any value, even if we round prices to the nearest dollar. We need a random variable whose probability distribution can represent an almost continuous range of prices, as well as reproduce the bell shape evident in Figure 12.2.

This bell shape appears in the histogram of many types of data, both in business and other domains. For example, the histogram on the left of Figure 12.3 shows monthly percentage changes in the U.S. stock market from 1970 through the end of 2007, and the histogram on the right shows X-ray measurements of the density of hip bones.

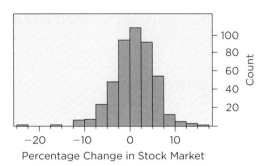
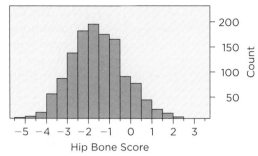

FIGURE 12.3 Monthly percentage changes in the U.S. stock market from 1970 to 1987 and X-ray measurements of bone density.

Black Monday (October 19, 1987, discussed in the introduction) produced the extreme outlier on the left of the histogram of market returns. Otherwise, the bell shape of this histogram resembles that of the prices of one-carat diamonds. The histogram on the right of Figure 12.3 shows X-ray measurements of bone density of 1,102 women in a study of osteoporosis. It too has a bell shape. Models for counts cannot handle a continuous range of values or a mix of positive and negative numbers.

The underlying data in these three histograms are rather different: prices of diamonds, monthly percentage changes in the stock market, and X-ray measurements of bone density. Nonetheless, all have bell-shaped histograms. This similarity allows us to model them all using **normal random variables**. The probability distribution of a normal random variable is *the* bell curve.

normal random variable A random variable whose probability distribution defines a standard bell-shaped curve.

Central Limit Theorem

It's no accident that many types of data have bell-shaped histograms. The probability distribution of *any* random variable that is the sum of *enough* independent random variables is bell shaped. If the random variables to be summed have a normal distribution to begin with, then the sum of them has a normal distribution: Sums of normally distributed random variables are normally distributed. What is remarkable, however, is that sums of just about any random variables are *eventually* normally distributed.

The bell shape emerges even in tossing a coin. If we toss a coin once, we have a single Bernoulli trial. There's nothing bell shaped about its distribution. As we count the results from more and more tosses, though, the bell-shaped mound appears, as shown in Figure 12.4.

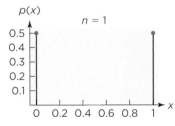
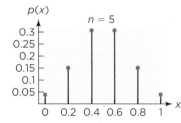
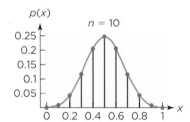

FIGURE 12.4 The Central Limit Theorem is evident in the proportion of heads in *n* tosses of a coin.

The *x*-axis in this figure shows the proportion of heads in *n* tosses of a fair coin. The red curve in the right frame is a normal probability distribution. The normal distribution nearly matches the binomial probabilities when adding up the number of heads in only $n = 10$ tosses of a coin. This result holds more generally than coin tossing and is usually called the Central Limit Theorem.

Central Limit Theorem The probability distribution of a sum of independent random variables of comparable variance tends to a normal distribution as the number of summed random variables increases.

The Central Limit Theorem explains why bell-shaped distributions are so common: Observed data often represent the accumulation of many small factors. Consider the stock market, for instance. The change in the value of the market from month to month is the result of vast numbers of investors, each contributing a small component to the movement in prices.

Normal Probability Distribution

standard normal distribution The probability distribution of a normal random variable with mean 0 and SD 1.

Two parameters identify each normal distribution: the mean μ and variance σ^2. The graph in Figure 12.5 shows the probability distribution of a normal random variable with mean $\mu = 0$ and variance $\sigma^2 = 1$. This particular distribution is called the **standard normal distribution**. A normal probability distribution is a positive function that traces out a smooth, bell-shaped curve. The curve is unimodal and symmetric around its center at $\mu = 0$, as shown in Figure 12.5.

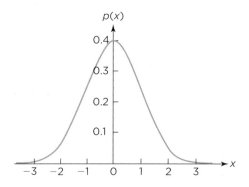

FIGURE 12.5 Standard normal probability distribution.

continuous random variable A random variable that can conceptually assume any value in an interval.

A normal random variable is an example of a **continuous random variable**. A continuous random variable can take on *any* value in an interval rather than just those that we can enumerate in a list. Rather than assign probability to specific values, the probability distribution of a continuous random variable assigns probabilities to intervals. The probability of an interval is the area under the probability distribution over that interval. For example, the shaded area in Figure 12.6 shows $P(0 \leq X \leq 1)$.

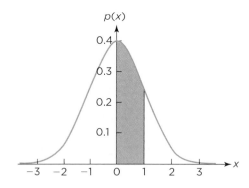

FIGURE 12.6 Probabilities are areas under the curve.

The total area between the bell curve and the x-axis is 1, matching the total probability of a sample space. The shaded portion of that area shown here is $P(0 \leq X \leq 1)$.

Shifts and Scales

By changing μ and σ^2, we can shift the location of a normal distribution and alter its scale. Whatever the values for μ and σ^2, however, the distribution always has the same bell-shaped mound around the mean. In general, the notation $X \sim N(\mu, \sigma^2)$ identifies X as a normal random variable with mean μ and variance σ^2. The Greek letters remind us that μ and σ^2 are parameters that specify a random variable.

Convention uses the variance as the second parameter rather than the standard deviation. In examples that assign numerical values to μ and σ^2, we often show the variance as the square of the standard deviation—for example, $N(10, 5^2)$ for a normal distribution with $\mu = 10$ and $\sigma = 5$. This notation makes it easier to recognize σ, the standard deviation, which is generally more useful in applications than σ^2. This notation also reminds us that the second attribute within the parentheses is the variance, not the standard deviation.

The peak of a normal distribution (its mode) lies at the mean μ. By changing μ, we can slide the distribution up and down the x-axis. If we increase μ from 0 to 3, for instance, the whole distribution shifts to the right as in Figure 12.7.

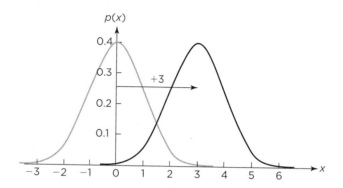

FIGURE 12.7 The mean μ locates the center of the normal distribution.

The shifted curve in Figure 12.7 is the probability distribution of a normal random variable with mean $\mu = 3$ and $\sigma = 1$. Changes to the mean have no impact on the standard deviation of a normal random variable (unlike the binomial and Poisson random variables in Chapter 11).

The variance σ^2 controls the spread of the curve. If σ^2 is small, the curve is tall and thin, concentrated around μ. If σ^2 is large, the distribution spreads out from μ, as illustrated in Figure 12.8.

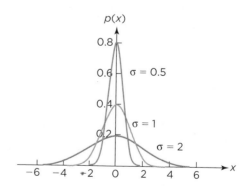

FIGURE 12.8 The variance σ^2 controls the scale of the normal distribution.

12.2 | THE NORMAL MODEL

normal model A model in which a normal random variable is used to describe an observable random process with μ set to the mean of the data and σ set to s.

We match a normal distribution to data by choosing values for μ and σ to match the mean and SD of the data. We call this use of a normal random variable to describe data a **normal model**. Once we pick μ and σ, we can use the normal model to find any probability. That's part of the appeal of a normal model: Once you name the mean and standard deviation, the area under the bell curve determines any probability.

Figure 12.9 superimposes a normal distribution over the histogram of prices of one-carat diamonds shown in Figure 12.2.

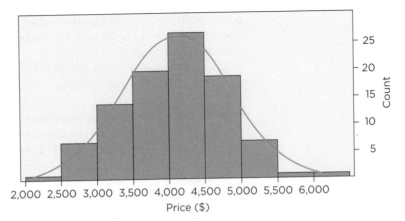

FIGURE 12.9 Normal model for diamond prices superimposed on the histogram.

To align the normal distribution with the histogram, we set μ to $\bar{x} = \$4,066$ and set σ to $s = \$738$. It's not a perfect match, but it appears close. The normal curve (in red) follows the heights of the bars in the histogram. (We adjusted the vertical scale of the normal distribution so that the area under the normal curve in Figure 12.9 is the same as the area under the histogram.)

A normal distribution is also a good description of the histograms of changes in the stock market and X-ray measurements (Figure 12.3). All we need to do is match the mean and variance of the normal distribution to the corresponding properties of the data. To align the normal curve to the histogram of market changes, we set $\mu = 0.972\%$ and $\sigma = 4.49\%$, matching the average and standard deviation of percentage changes. For the measurements of bone density, we set $\mu = -1.53$ and $\sigma = 1.3$.

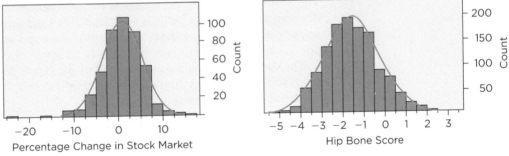

FIGURE 12.10 Normal models for changes in stock prices and bone density measurements.

A normal model is a particularly good match to the distribution of measurements of bone density on the right of Figure 12.10.

What Do You Think?

Which of the following situations appear suited to the use of a normal model? If there's a problem, explain briefly what it is.

a. Heights of adult male customers who shop at a clothing store[1]
b. Value (in dollars) of purchases at the same store[2]
c. Number of items purchased by a customer at one time[3]

Standardizing

The method for obtaining a probability using a normal model resembles the procedure used with the Empirical Rule. We start by converting the measured

---

[1] An okay match. Height can take on any value (conceptually) and concentrates around the mean.
[2] Less likely to be normal. Probably more relatively small purchases and few very large purchases (skewed).
[3] Not well suited to normal. This would produce discrete, small counts and would be better modeled using the methods of Chapter 11.

z-score Number of standard deviations from the mean, computed as $z = \dfrac{x - \mu}{\sigma}$

values into z-scores. A **z-score** measures the number of standard deviations that separate a value from the mean. The z-score of the value x is

$$z = \frac{x - \mu}{\sigma}$$

Let's find the probability that a randomly chosen one-carat diamond costs more than $5,000. Let X denote the random variable in our normal model for the prices of one-carat diamonds; we want to know $P(X > \$5,000)$. We start by converting $5,000 into a z-score. Substituting $\mu = \$4,066$ and $\sigma = \$738$, a cost of $5,000 becomes the z-score

$$z = \frac{5{,}000 - \mu}{\sigma} = \frac{5{,}000 - 4{,}066}{738} \approx 1.27$$

A $5,000 diamond lies 1.27 SDs above the mean cost of one-carat diamonds.

standard normal random variable A normal random variable with mean 0 and SD equal to 1.

The steps that convert the cost of a diamond into a z-score also convert the normal random variable $X \sim N(\mu, \sigma^2)$ into a **standard normal random variable**. A standard normal random variable has parameters $\mu = 0$ and $\sigma = 1$. Figure 12.5 shows the probability distribution of a standard normal random variable. The usual symbol for a standard normal random variable is Z.

$$Z = \frac{X - \mu}{\sigma}$$

The mean of Z is 0.

$$E(Z) = E\left(\frac{X - \mu}{\sigma}\right)$$
$$= \frac{1}{\sigma}(EX - \mu)$$
$$= 0$$

The SD and variance of Z are both 1.

$$Var(Z) = Var\left(\frac{X - \mu}{\sigma}\right)$$
$$= \frac{1}{\sigma^2} Var(X - \mu)$$
$$= \frac{\sigma^2}{\sigma^2} = 1$$

Once standardized, we only need the distribution of Z rather than the specific normal distribution that matches the data. Taken together, the standardization works like this.

$$P(X > \$5{,}000) = P\left(\frac{X - \mu}{\sigma} > \frac{5{,}000 - \mu}{\sigma}\right) = P\left(Z > \frac{5{,}000 - 4{,}066}{738} \approx 1.27\right)$$

Be sure to do the same thing to both sides of the inequality. If you subtract μ and divide by σ on one side, do the same to the other. All that's left is to find the probability associated with a standard normal random variable.

The Empirical Rule, Revisited

We can now reveal the source of those fractions 2/3 and 19/20 in the Empirical Rule. The Empirical Rule pretends that data obey a normal model. If $X \sim N(\mu, \sigma^2)$, then

$$P(\mu - \sigma \le X \le \mu + \sigma) = P(-1 \le Z \le 1) \approx 0.6827 \approx 2/3$$
$$P(\mu - 2\sigma \le X \le \mu + 2\sigma) = P(-2 \le Z \le 2) \approx 0.9545 \approx 19/20$$
$$P(\mu - 3\sigma \le X \le \mu + 3\sigma) = P(-3 \le Z \le 3) \approx 0.9973 \approx almost\ all$$

These probabilities appear in the last column of the table inside the back cover of this book. Figure 12.11 repeats the relevant rows used in the Empirical Rule.

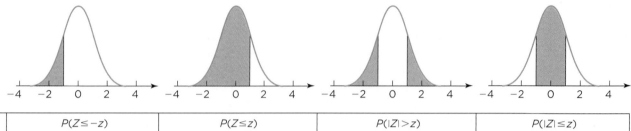

z	$P(Z \leq -z)$	$P(Z \leq z)$	$P(|Z| > z)$	$P(|Z| \leq z)$
1	0.1587	0.8413	0.3173	**0.6827**
2	0.02275	0.97725	0.04550	**0.95450**
3	0.00135	0.99865	0.00270	**0.99730**

FIGURE 12.11 Origins of the Empirical Rule.

The first column of the table in Figure 12.11 lists the z-values. The next two columns give probabilities to the left of $-z$ and to the left of z. These columns add to 1 because the total area under curve is 1 and the curve is symmetric. For instance, $P(Z \leq -1) = 1 - P(Z \leq 1)$. The fourth column gives the probability in the two tails, $P(|Z| > z)$, and the fifth column gives probability around zero, $P(|Z| \leq z)$. The last two columns also add to 1.

Because normal random variables are continuous, they do not assign a probability to a specific value. Hence, for example, $P(Z < 1) = P(Z \leq 1)$ and $P(Z > 2) = P(Z \geq 2)$. (You have to distinguish $<$ from $\leq$ or $>$ from $\geq$ with discrete random variables.)

4M EXAMPLE 12.1 | SATS AND NORMALITY

MOTIVATION | STATE THE QUESTION

Educational Testing Services (ETS) designs the math component of the SAT so that scores should have a distribution that is normally distributed, with a national mean of 500 and a standard deviation of 100. A company is looking for new employees who have good quantitative skills, as indicated by a score of 600 or more on the math SAT. What percentage of test-takers meets this threshold?

METHOD | DESCRIBE THE DATA AND SELECT AN APPROACH

The random variable for this problem is the math score of a randomly selected individual who took the SAT. From what we are given, we will model this score as a normally distributed random variable with mean 500 and SD 100, $X \sim N(\mu = 500, \sigma^2 = 100^2)$.

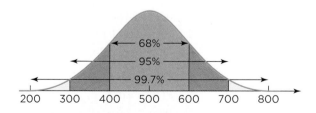

MECHANICS | DO THE ANALYSIS

A score of 600 is $z = 1$ standard deviation above the mean. Precisely, 68.3% of the normal distribution falls within 1 SD of the mean. That implies (Complement Rule) that 31.7% of test-takers score more than 1 SD from the mean. Half of these do better, so we expect that $(100 − 68.3)/2 = 31.7/2 = 15.85\%$ score higher than 600.

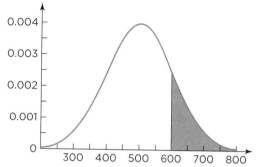

> **tip** It's often helpful to sketch the distribution of the random variable, particularly when dealing with calculations for normally distributed random variables.

The sketch helps you make sure that you have not made a careless error. This sketch shows the normal distribution, and the shaded area matches $P(X \geq 600)$.

MESSAGE | SUMMARIZE THE RESULTS

About one-sixth of those who take the math SAT score more than 600. Recruiters should be aware that, although a score of more than 600 is not that common, they should expect to find more than enough candidates who exceed this requirement.

Using Normal Tables

An SAT score of 600 falls *exactly* 1 standard deviation above the mean. What if the score is 680? After conversion, the z-score is $(680 − 500)/100 = 1.80$, between 1 and 2 standard deviations above the mean. From the Empirical Rule, about one-sixth of people score better than 600 (half of 1/3) and only 1 in 40 (half of 1/20) score better than 700. Can we be more specific than "between 1/6 and 1/40"? We have the same problem finding the probability of picking a one-carat diamond that costs more than $5,000. This z-score is 1.27.

When the z-score does not fall exactly 1, 2, or 3 standard deviations from the mean, we can find the needed probabilities from software or from a normal table. The table inside the back cover of this book shows probabilities for z-scores to one decimal. Figure 12.12 extracts the relevant portion of the table for finding normal probabilities when $z = 1.8$.

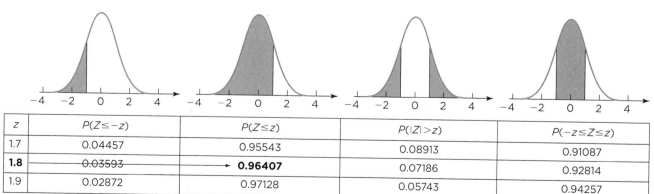

z	$P(Z \leq -z)$	$P(Z \leq z)$	$P(|Z| > z)$	$P(-z \leq Z \leq z)$
1.7	0.04457	0.95543	0.08913	0.91087
1.8	0.03593	**0.96407**	0.07186	0.92814
1.9	0.02872	0.97128	0.05743	0.94257

FIGURE 12.12 Using a normal table to find $P(Z \leq 1.8)$.

The probability of scoring less than 680 is the probability in the third column labeled $P(Z \leq z)$, 0.96407. A score of 680 puts you in the 96th percentile; 96% of those taking the test are expected to score at this level or lower.

Let's use the table to find the probability of picking out a one-carat diamond that costs more than \$5,000, which is $P(Z > 1.27)$.

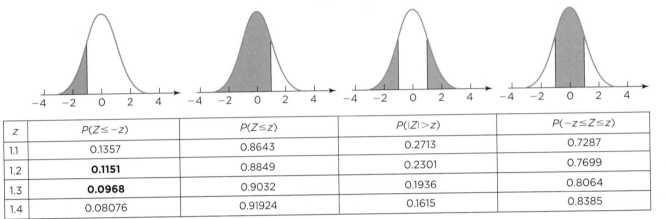

z	$P(Z \leq -z)$	$P(Z \leq z)$	$P(|Z| > z)$	$P(-z \leq Z \leq z)$
1.1	0.1357	0.8643	0.2713	0.7287
1.2	**0.1151**	0.8849	0.2301	0.7699
1.3	**0.0968**	0.9032	0.1936	0.8064
1.4	0.08076	0.91924	0.1615	0.8385

FIGURE 12.13 More normal probabilities.

In Figure 12.13 the probability lies between 0.1151 and 0.0968 because $P(Z > z) = P(Z < -z)$. That's about 0.10 if we interpolate and is probably more than accurate enough. To find precise probabilities for z-scores that are not listed in the table, such as $z = 1.27$, use a computer or calculator. (Excel and statistics packages have functions to find normal probabilities; see the Software Hints at the end of the chapter.) Using software gives $P(Z > 1.27) = 0.10204$. It's also good to use software for finding these probabilities because you often need several. For example, what's $P(-0.5 \leq Z \leq 1)$? Because most tables of the normal distribution tell you $P(Z \leq z)$ for any value of z, the easiest way to find this probability is to write it as the difference:

$$P(-0.5 \leq Z \leq 1) = P(Z \leq 1) - P(Z \leq -0.5) = 0.8413 - 0.3085 = 0.5328$$

Figure 12.14 illustrates this calculation visually, using the shaded areas to represent the probabilities.

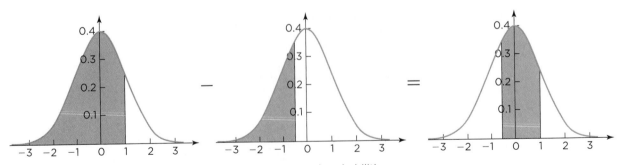

FIGURE 12.14 Calculating normal probabilities.

Don't hesitate to draw lots of pictures like those in Figure 12.14. The picture will help you avoid a mistake.

12.3 | PERCENTILES

The example that we have considered uses a normal model to find a probability, the chance of choosing a diamond that costs more than $5,000. Such problems ask us to find the probability associated with being larger or smaller than some value. Other problems require working in the opposite direction that starts with the probability. Here's an illustration.

Many types of packaging include a weight on the label. For example, a box of cereal might list the weight as 16 ounces. That's seldom the exact weight. Automated packaging systems fill the boxes so quickly that there's no time to weigh each one precisely. Consequently, there's variation in the weights. Most consumers are not going to complain if the box of cereal has more than 16 ounces, but they won't be too happy if the box has less.

Assume that the packaging system fills boxes to an average weight of, say, 16.3 ounces with standard deviation 0.2 ounce and that the weights are normally distributed. Define $X \sim N(\mu = 16.3, \sigma^2 = 0.2^2)$. The probability of this system producing an underweight box is then

$$P(X < 16) = P\left(\frac{X - 16.3}{0.2} < \frac{16 - 16.3}{0.2}\right) = P(Z < -1.5) \approx 0.067$$

Let's assume that we can change the mean but not the SD of the process that fills the boxes. Where should we set the mean in order to reduce the chance of an underweight box to 1/2 of 1%? To find this mean, we have to find the value z such that $P(Z < z) = 0.005$.

quantile Another name for a percentile of a distribution. Quantiles of the standard normal distribution are denoted z.

The sought value z is a percentile, or **quantile**, of the standard normal distribution. The pth quantile of a probability distribution is that value x such that $P(X \leq x) = p$. A quantile function starts with a probability and gives you back a value, just the opposite of the tables we've used so far. That's the point of the table of the standard normal quantile function in the back of the book. Figure 12.15 gives the portion of that table we need to use in this problem.

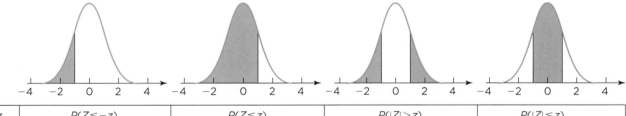

z	$P(Z \leq -z)$	$P(Z \leq z)$	$P(\lvert Z \rvert > z)$	$P(\lvert Z \rvert \leq z)$
2.3263	0.01	0.99	0.02	0.98
2.5758	**0.005**	0.995	0.01	0.99
2.8070	0.0025	0.9975	0.005	0.995

FIGURE 12.15 Portion of the table for the standard normal quantile function.

Setting $z = -2.5758$ cuts off the bottom 0.005 of the standard normal distribution. To finish the question, we need to find the value for μ that gives this z-score.

$$\frac{16 - \mu}{0.2} = -2.5758 \quad \Rightarrow \quad \mu = 16 + 0.2(2.5758) \approx 16.52$$

On average, boxes need roughly 1/2 ounce more cereal than the labeled quantity in order to produce only 0.5% packages that are underfilled.

4M EXAMPLE 12.2	VALUE AT RISK

MOTIVATION ▸ STATE THE QUESTION

In finance, the acronym VaR does not stand for variance; it means "value at risk." J.P. Morgan introduced value at risk in the 1980s as a way to answer a common question asked by investors: "How much might I lose?" VaR works by using a probability model to rule out the 5% worst things that might happen over some time horizon, such as the next year. The question becomes, "How low could my investment go if I exclude the worst 5% of outcomes?" You can use any probability model to find the VaR, but the normal model is common.

Imagine that you manage the $1 million portfolio of a wealthy investor. The portfolio is expected to average 10% growth over the next year with standard deviation 30%. Let's convey the risk of owning this investment in everyday language using the VaR that excludes the worst 5% of the scenarios. (VaR is computed using any of various probabilities, such as 5%, 1%, or even 0.1%.)

METHOD ▸ DESCRIBE THE DATA AND SELECT AN APPROACH

VaR calculations require a random variable to model the possibilities for the investment. For this example, we'll model the percentage change next year in the portfolio as $N(10, 30^2)$, a 10% increase on average with a 30% SD.

MECHANICS ▸ DO THE ANALYSIS

From the normal table, the probability of a standard normal being less than -1.645 is 0.05, $P(Z \leq -1.645) = 0.05$. With $\mu = 10\%$ and $\sigma = 30\%$, this works out to a change in value of $\mu - 1.645\sigma = 10 - 1.645(30) = -39.3\%$. That's a loss of $393,000. The picture of the distribution is particularly helpful in this application. This figure shows the distribution of percentage changes during the next year and the worst 5% of scenarios that we have excluded.

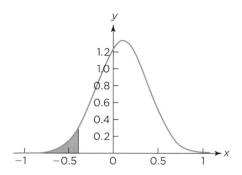

MESSAGE ▸ SUMMARIZE THE RESULTS

The annual value at risk for this portfolio is $393,000 at 5% (eliminating the worst 5% of situations). An investment adviser better be prepared to offer some other investments if this seems too risky for the client.

12.4 | DEPARTURES FROM NORMALITY

Departures from normality typically come in several forms:

- **Multimodality.** Multiple modes suggest that the data may come from two or more distinct groups. Your data may be a mix of samples from different populations. If so, try to identify the groups. For example, one mode may be associated with men in the sample and a second mode with women.

- **Skewness.** A lack of symmetry often results when data are confined to positive values or have a limiting threshold. For instance, salaries don't go below zero and become more rare as the amount increases. You can expect skewness.
- **Outliers.** Examine outliers carefully. Outliers may indicate that a data-entry error was made or that some data are fundamentally different from others. If you find outliers, you may need to set them aside or, better yet, do the analysis both with and without them. Anytime that data are excluded, you must report them. Discovering and understanding outliers may be the most informative part of your analysis.

tip

We use two approaches to check whether data have a normal distribution, one visual and one based on summary statistics.

Normal Quantile Plot

Many histograms are bell shaped, but that does not guarantee that we should use a normal model. Inspecting the histogram is a good start at checking whether a normal model is a good match, but it's only a start. Histograms highlight the center of the distribution, but deviations from normality most often occur in the extremes. These deviations are hard to see in a histogram because the counts are small in the tails of the histogram.

A **normal quantile plot** helps you decide whether a normal model is appropriate. A normal quantile plot is a type of scatterplot. Rather than showing pairs of data, however, it takes only one coordinate from the data. The other comes from a normal model. We put the data on the y-axis. The normal model defines the x-axis. If the normal model is a good description of the data, the points in the scatterplot should concentrate along the diagonal of the plot, as sketched in Figure 12.16.

A normal quantile plot compares the spacing of the observations to the spacing that is expected if the data are normally distributed. According to a normal model, data should pack relatively closely together in the center (near $\bar{x}$) and gradually spread out in the extremes. The horizontal coordinates in the normal quantile plot capture this notion. These coordinates are predicted z-scores assuming a normal model. We will label these $\hat{z}$. Because these are z-scores, 0 on the x-axis denotes the mean and 1, for instance, denotes 1 SD above the mean. As an example, Figure 12.17 shows the normal quantile plot of the prices of the one-carat diamonds.

normal quantile plot A diagnostic scatterplot; if the data track the diagonal line, the data are normally distributed.

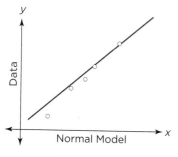

FIGURE 12.16 Schematic diagram for a normal quantile plot.

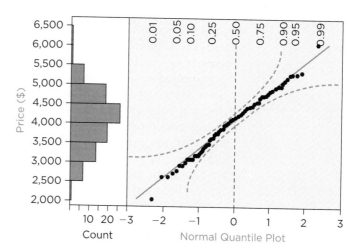

FIGURE 12.17 Normal quantile plot of diamond prices.

The predicted z-scores pack densely near 0, with wider gaps farther from μ. If the gaps between adjacent observations match those predicted by normality, then the points in the normal quantile plot track along the red diagonal reference line. The equation for this line is

$$y = \bar{x} + s\hat{z}$$

Deviations from the reference line indicate deviations from normality. For example, Figure 12.18 zooms in on the lower left-hand corner of the normal quantile plot in Figure 12.17.

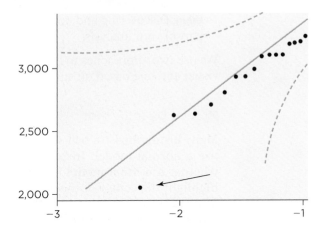

FIGURE 12.18 Magnified view of the lower left corner of the normal quantile plot.

Let's concentrate on the point identified by the arrow in Figure 12.18. This is the cheapest of the 97 diamonds. We can read its price from the vertical axis; it costs slightly more than $2,000 ($2,027, to be exact). The horizontal axis tells us that the smallest z-score in 97 observations from a normal distribution should be about $\hat{z} = -2.32$, about 2.32 standard deviations below the mean. Hence, the predicted price on the diagonal line is $\bar{x} - 2.32\,s = 4{,}066 - 2.32 \times 738 \approx$ $2,354. At $2,027, this diamond costs $327 less than normality predicts. The quantile plot highlights that this diamond is cheaper than predicted by normality by plotting the point below the reference line.

Even data that *are* normally distributed will not lie *exactly* along the diagonal line. The dashed curves in the normal quantile plot in Figure 12.17 tell you if the data are close enough for the normal model to be a reasonable description. In this case, the data stay close enough to the reference line because all of the points lie inside the dashed curves. A normal model seems well suited to these data, even in the tails of the distribution. (For details on the normal quantile plot, see Behind the Math: The Normal Quantile Plot at the end of this chapter.)

(p. 278)

What does a quantile plot look like when the data are *not* normally distributed? Figure 12.19 shows an example. These data are prices of a broader selection of diamonds. Rather than limit the data to diamonds of comparable quality, we included all 686 one-carat diamonds offered by a jeweler. Some are quite special and others have flaws. The distribution is skewed because there are relatively more diamonds that cost less. The evident skewness produces a crescent shape in the normal quantile plot. Diamonds with high prices cost more than expected by normality; the outliers in the boxplot lie above the reference line in the quantile plot. On the low side, inexpensive diamonds are not so cheap as normality predicts. Hence, points in the normal quantile plot representing the lower side of the distribution also lie above the reference line.

caution   Take your time with the normal quantile plot. Until you've seen a few of these, they are peculiar. First, check whether the data track along the diagonal reference line. If they do, the distribution of the data resembles a normal distribution. If you find that the data drift away from the reference line, look at the accompanying histogram to see what's going on.

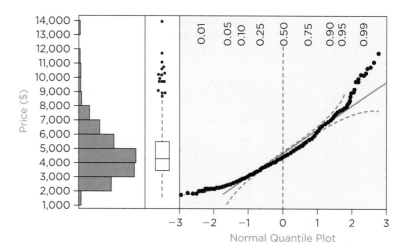

FIGURE 12.19 Quantile plot of a mixture of one-carat diamonds of varying quality.

Skewness and Kurtosis

Two statistics summarize departures from normality. Both are near zero when the data are normally distributed. We will use these statistics in later chapters as measures of abnormality. The calculation of these summaries starts with the standardized observations

$$z = \frac{x - \bar{x}}{s}$$

If we know the mean and variance of the population, we use those to define the z-scores instead ($z = (x - \mu)/\sigma$).

skewness Lack of symmetry of a distribution; average of z^3. $K_3 \approx 0$ for normal data.

Skewness measures the lack of symmetry. It is the average of the z-scores raised to the third power.

$$K_3 = \frac{z_1^3 + z_2^3 + \cdots + z_n^3}{n}$$

where z_1 is the z-score of the first observation, z_2 is the z-score of the second observation, and so forth. If the data are right skewed as on the right in Figure 12.20, then $K_3 > 0$. If the data are left skewed, then $K_3 < 0$. If the data are symmetric, then $K_3 \approx 0$ (left frame of Figure 12.20). For example, the skewness of the distribution of one-carat diamonds of like quality is $K_3 \approx -0.2$, indicating a nearly symmetric distribution. If we mix diamonds of varying quality, the skewness grows to $K_3 \approx 1.1$. If $K_3 > 1$, there's quite a bit of skewness and the data are not normally distributed.

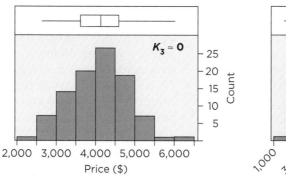

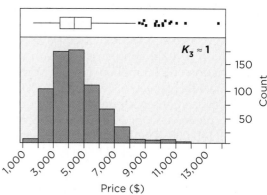

FIGURE 12.20 The distribution that mixes diamonds of varying quality is skewed to the right.

kurtosis Prevalence of outliers in a distribution; average of the z^4. $K_4 \approx 0$ for normal data.

Kurtosis is another measure that captures a deviation from normality. Kurtosis measures the prevalence of outliers. The kurtosis is the average z-score raised to the fourth power, minus 3.

$$K_4 = \frac{z_1^4 + z_2^4 + \cdots + z_n^4}{n} - 3$$

Subtracting 3 gives $K_4 \approx 0$ for normally distributed data. Distributions with a uniform shape don't have tails and produce negative kurtosis; distributions with long tails (many outliers) have positive kurtosis. The kurtosis of the prices of the diamonds of similar quality is $K_4 \approx -0.1$, compared to $K_4 \approx 2.4$ for the diamonds of mixed quality. The outliers in the distribution of the prices of the mixed-quality diamonds on the right-hand side of Figure 12.20 produce the positive kurtosis.

Best Practices

- *Recognize that models approximate what will happen.* Statistical models use properties of data to anticipate what might happen next. Even if a normal model provides a great summary of your data, that does not mean future data will also be normally distributed.
- *Inspect the histogram and normal quantile plot before using a normal model.* Pay attention to how well the normal model describes the data in the extremes. Models help by simplifying the world, but at the end of the day, it's being correct that matters.
- *Use z-scores when working with normal distributions.* In the end, there's only one normal distribution, the standard normal with mean 0 and standard deviation 1. All calculations with normal distributions come down to figuring out how many standard deviations separate a value from μ.
- *Estimate normal probabilities using a sketch and the Empirical Rule.* It's easy to make a careless mistake (like missing a minus sign) when finding a normal probability. If you first estimate the probability with a sketch and the Empirical Rule, you'll avoid most mistakes.
- *Be careful not to confuse the notation for the standard deviation and variance.* The standard notation for a normal random variable is $X \sim N(\mu, \sigma^2)$. Hence, $X \sim N(7, 25)$ implies that $\mu = 7$ and $\sigma = 7$. To make it clear that the second argument is the variance, we write $X \sim N(7, 5^2)$.

Pitfalls

- *Do not use a normal model without checking the distribution of the data.* You can always match a normal model to your data by setting μ to $\bar{x}$ and setting σ to s. A sample of CEOs averages $2.8 million in total compensation with a standard deviation of $8.3 million. Using a normal model, we would expect 68% of the CEOs to have compensations within 1 standard deviation of the mean, a range of from $-\$5,500,000$ to $\$11,140,000$. The lower limit is negative. What went wrong? The distribution is *skewed*, not symmetric, and the normal model is far off target.
- *Do not think that a normal quantile plot can prove that data are normally distributed.* The normal quantile plot only indicates whether data plausibly follow a normal distribution. If you've got a small sample, however, almost any data might appear normally distributed.
- *Do not confuse standardizing with normality.* Subtracting the mean and dividing by the standard deviation produces a new random variable with mean zero and SD 1. It's only a normal random variable if you began with a normal random variable.

Software Hints

Software can find any normal probability quickly and accurately. For $X \sim N(\mu, \sigma^2)$, you can find the cumulative probability $P(X \leq x)$ for a column of values given x, or you can go in the other direction. The inverse of the normal distribution, the quantile function, finds the quantile x that cuts off a fraction p of the distribution. For example, to find the quantile z such that $P(Z \leq z) = 0.25$, apply the inverse distribution to 0.25.

EXCEL

Excel includes several functions for normal probabilities. The function

$$\text{NORMDIST}(x, \mu, \sigma, \text{TRUE})$$

computes $P(X \leq x)$ for X, a normal random variable with mean μ and standard deviation σ. If the fourth argument is FALSE, Excel computes the height of the normal probability distribution (the bell curve) at the point x. You can use that formula to plot the normal curve. The formula

$$\text{NORMINV}(p, \mu, \sigma)$$

finds the inverse of the cumulative normal distribution, returning the quantile x such that $P(X \leq x) = p$. The related formulas NORMSDIST and NORMSINV assume $\mu = 0$ and $\sigma = 1$ (standardized).

Excel does not produce normal quantile plots. You can build your own using several of the built-in functions, but that requires more steps than we can show here. DDXL will draw them, but it calls these normal probability plots. Start by selecting the columns in the Excel spreadsheet that have the data, then follow the menu commands

DDXL > Charts and Plots . . .

Next choose the item Normal Probability Plot from the Click-Me menu and fill in the name of the column of data.

MINITAB

Following the menu items

Calc > Probability Distributions > Normal

opens a dialog that allows you to obtain various normal probabilities. Suppose that X denotes a normal random variable with parameters μ and σ. In the dialog, fill in the values for the mean and standard deviation, then choose between the probability density (which gives the height of the bell curve),

the cumulative probability, and the inverse of the cumulative probability.

You can enter the value of x used in the calculation of the cumulative probability (or p for the inverse of the cumulative probability) in this dialog. You can also use a calculation spreadsheet rather than a data spreadsheet. For example, if the column C_1 has the values 0, 1, and 2 in the first three rows, then specifying $\mu = 0$, $\sigma = 1$, and C_1 as the input column produces the output

0 0.500000
1 0.841345
2 0.977250

For example, $P(X \leq 2) = 0.977250$; about 97.725% of a standard normal distribution lies to the left of 2.

In place of the normal quantile plot, Minitab supplies a normal probability plot with data along the x-axis and percentages along the vertical axis. If the data are a sample from a normal population, the data should fall close to a diagonal reference line. To obtain this plot, follow the menu items

Graph > Probability Plot

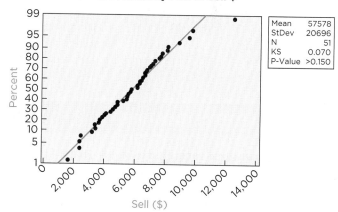

Normal Probability Plot of Sell $

Mean	57578
StDev	20696
N	51
KS	0.070
P-Value	>0.150

JMP

To obtain normal probabilities, define a formula using the formula calculator. (For example, right click on the header at the top of the column in the spreadsheet view.) In the functions list of the calculator, pick the Probability item. Then pick the item Normal Distribution so that the formula becomes

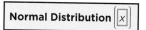

If you put the insertion marker on the box for x and then click the insert button (shown as a caret ^), the formula becomes

Now fill in the boxes for $x, \mu,$ and σ with either numbers that you type or the names of columns. (By default, the formula assumes that $\mu = 0$ and $\sigma = 1$.) After you click the OK button, the formula column is filled with $P(X \leq x)$ for a normal random variable X with mean μ and standard deviation σ. Using a column of values for x allows you to get several probabilities at once. Selecting the formula Normal Density computes the height of the bell curve and Normal Quantile computes the inverse of the cumulative normal distribution.

To obtain a normal quantile plot, use the menu sequence

Analyze > Distribution

and select the variable from the dialog. After you click OK and see the histogram of the variable, click the red triangle beside the name of the variable in the output window and choose the item Normal Quantile Plot from the pop-up menu. If you orient the histogram vertically, you'll get a normal quantile plot like those shown in this chapter.

 BEHIND the Math

The Normal Quantile Plot

Where do the expected locations $\hat{z}$ along the x-axis in a normal quantile plot come from? Here's an outline of the procedure.

The data used in the normal quantile plot shown in Figure 12.17 include $n = 97$ observations. To a good approximation, the normal model predicts the smallest z-score for a sample of this size to be the z-score associated with probability $1/(n + 1) = 1/98$. Using software, $P(Z \leq -2.32) \approx 1/98$. That means that the smallest of 97 observations should have z-score $z = -2.32$ SDs *if* the data are normally distributed. Plugging this z-score into the formula for the reference line, $y = \bar{x} + s\hat{z}$, the normal model predicts the

smallest value to be $4{,}066 + 738(-2.32) = \$2{,}354$. The minimum cost in the data, $\$2{,}027$, is smaller. This deviation appears in the quantile plot as a point far below the reference line.

Let's do one more. The normal model underpredicts the second smallest observation. Because $P(Z \leq -2.05) \approx 2/98$, the predicted z-score for the second smallest observation is $\hat{z} = -2.05$. Hence, the predicted second smallest cost is $4{,}066 + 738(-2.05) \approx \$2{,}553$. The second cheapest diamond costs $\$2{,}606$; the quantile plot graphs this point slightly above the reference line (see Figure 12.18). This process continues for the rest of the data.

CHAPTER SUMMARY

A **normal random variable** has a bell-shaped probability distribution. The two parameters of a normal random variable are its mean μ and its standard deviation σ. A **standardized normal random variable**, often denoted by Z, has mean 0 and standard deviation 1. The **Central Limit Theorem** offers one reason why this distribution appears so often in practice. The Empirical Rule relies on probabilities from the

normal distribution. A **normal model** for data matches the parameters of a normal random variable to the mean and standard deviation of the data. The **normal quantile plot** allows us to check how well a normal model describes the variation in data. The **skewness** and **kurtosis** of a distribution are summary statistics that measure the symmetry and presence of outliers relative to a normal distribution.

■ Key Terms

▇ Formulas

The **probability distribution** of a normal random variable with mean μ and variance σ^2 $X \sim N(\mu, \sigma^2)$ is

$$p(x) = \frac{\exp\left(-\frac{1}{2}\left(\frac{x - \mu}{\sigma}\right)^2\right)}{\sigma\sqrt{2\pi}}$$

Skewness is the average of the observed z-scores raised to the third power.

$$K_3 = \frac{z_1^3 + z_2^3 + \cdots + z_n^3}{n}$$

Kurtosis is the average of the observed z-scores raised to the fourth power minus 3.

$$K_4 = \frac{z_1^4 + z_2^4 + \cdots + z_n^4}{n} - 3.$$

▇ About the Data

The prices of diamonds come from the website of McGivern Diamonds. Data for the stock market are the monthly percentage changes in the value-weighted index from the Center for Research in Security Prices (CRSP). This index includes all of the stocks listed in the New York Stock Exchange, the American Stock Exchange, and the NASDAQ. Like the S&P 500, the value-weighted index describes the earnings of an investor who buys stocks in proportion to the total market value of the firms rather than investing equally. The data on osteoporosis are from a clinical trial that established the benefits of a pharmaceutical for the improvement of bone density in women.[4]

[4] E. Lydick, M. Melton, K. Cook, J. Turpin, M. Melton, R. A. Stine, and C. Byrnes (1998), "Development and Validation of a Simple Questionnaire to Facilitate Identification of Women Likely to Have Low Bone Density." *American Journal of Managed Care*, 4, 37–48.

EXERCISES

Mix and Match

Exercises 1–10. Match each statement on the left to the notation on the right. X denotes a normally distributed random variable: $X \sim N(\mu, \sigma^2)$. A googol, the namesake of Google, is 10^{100}. The random variable Z denotes a standard normal random variable, $Z \sim N(0, 1)$.

1. Mean of X	(a) 1/2
2. Variance of X	(b) $P(Z < 1)$
3. Probability of X being less than its mean	(c) 0.05
4. Probability of X being less than 1 SD above its mean	(d) 2/3
5. Standard deviation of Z	(e) 1/(1 googol)
6. Probability that a z-score based on X is less than 1 in magnitude	(f) μ
7. Proportion of a normal distribution that is more than 20σ from μ	(g) σ^2
8. Difference between value of $P(Z < -x)$ and value of $P(Z > x)$	(h) 1
9. Distribution of the random variable $\mu + \sigma Z$	(i) 0
10. Probability that $Z > 1.96$ plus the probability that $Z < -1.96$	(j) $N(\mu, \sigma^2)$

True/False

Mark each statement True or False. If you believe that a statement is false, briefly explain why you think it is false.

Exercises 11–16. The current age (in years) of 400 clerical employees at an insurance claims processing center is normally distributed with mean 38 and SD 6.

11. More employees at the processing center are older than 44 than between 38 and 44.

12. Most of the employees at this center are older than 25.

13. If the company were to convert these ages from years to days, then the ages in days would also be normally distributed.

14. If none of these employees leaves the firm and no new hires are made, then the distribution a year from now will be normal with mean 39 and SD 7.

15. A training program for employees under the age of 30 at the center would be expected to attract about 36 employees.

16. If the ages of these employees were mixed with those of other types of employees (management, sales, accounting, information systems), then the distribution would also be normal.

Exercises 17–22. The number of packages handled by a freight carrier daily is normally distributed. On average, 8,000 packages are shipped each day, with standard deviation 600. Assume the package counts are independent from one day to the next.

17. The total number of packages shipped over five days is normally distributed.

18. The difference between the numbers of packages shipped on any two consecutive days is normally distributed.

19. The difference between the number of packages shipped today and the number shipped tomorrow is zero.

20. If each shipped package earns the carrier $10, then the amount earned per day is not normally distributed.

21. The probability that more packages are handled tomorrow than today is 1/2.

22. To verify the use of a normal model, the shipper should use the normal quantile plot of the amounts shipped on a series of days.

Think About It

23. How large can a single observation from a normal distribution be?

24. If used to model data that have limits (such as the age of a customer using a credit card or amount spent by a customer), what problem does a normal model have?

25. Would the skewness K_3 and kurtosis K_4 of the 100 observations shown in the following histogram both be close to zero, or can you see that one or the other would differ from zero? You don't need to give a precise value, just indicate the direction away from zero.

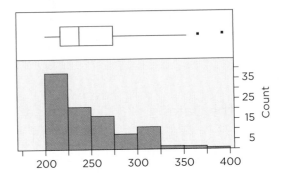

26. Would the skewness K_3 and kurtosis K_4 of the 100 observations shown in the following histogram both be close to zero, or can you see that one or the other would differ from zero? You don't need to give a precise value, just indicate the direction away from zero.

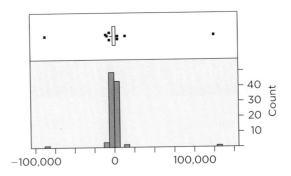

27. If $X_1 \sim N(\mu, \sigma)$ and $X_2 \sim N(\mu, \sigma)$ are iid normal random variables, then what's the difference between $2X_1$ and $X_1 + X_2$? They both have normal distributions.

28. If $X_1 \sim N(\mu, \sigma)$ and $X_2 \sim N(\mu, \sigma)$ are iid, then what is the distribution of $(X_1 - X_2)/(\sqrt{2}\sigma)$?

29. Which one of the following normal quantile plots indicates that the data...
    (a) Are nearly normal?
    (b) Have a bimodal distribution? (One way to recognize a bimodal shape is a gap in the spacing of adjacent data values.)
    (c) Are skewed?
    (d) Have outliers to both sides of the center?

I.

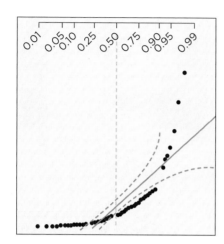

II.

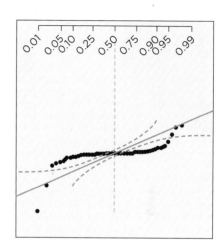

III.

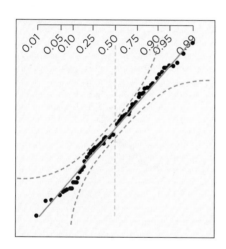

IV.

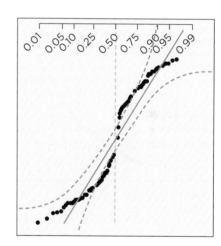

30. Match each histogram to a normal quantile plot.

(a)
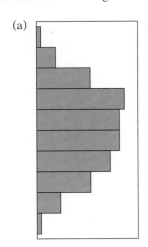

(b)
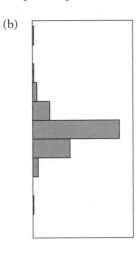

(c)
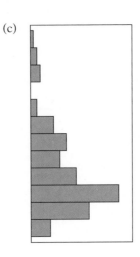

(d)
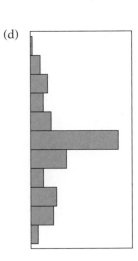

I.
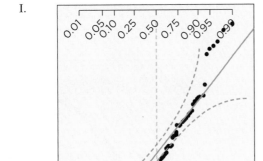

II.
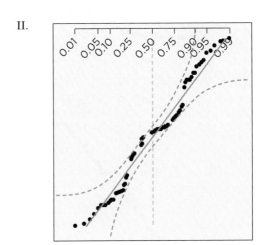

III.
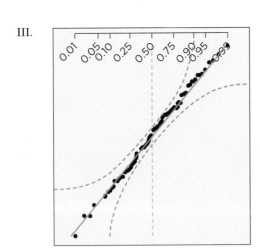

IV.
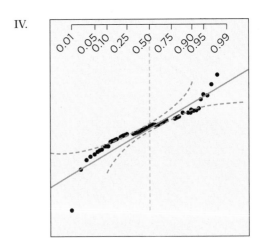

31. One of these pairings of histogram with normal quantile plot shows data that have been rounded. The other shows the same data without rounding. The rounding in the example is similar to the rounding in SAT scores.
    (a) Which is which? How can you tell?
    (b) Why do the histograms appear so similar if the data are different?

I.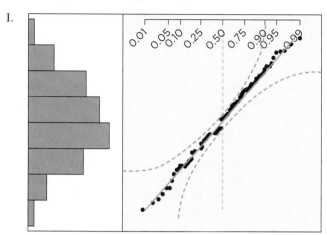
II.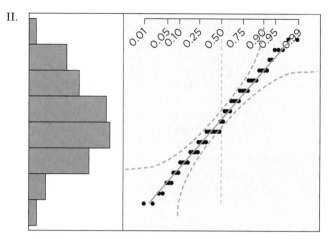

32. The normal quantile plot of daily stock returns for General Motors during 1992–1993 (507 trading days) shows an anomaly, a flat spot at zero.
    (a) What's happened?
    (b) Why does the anomaly not appear in the histogram?

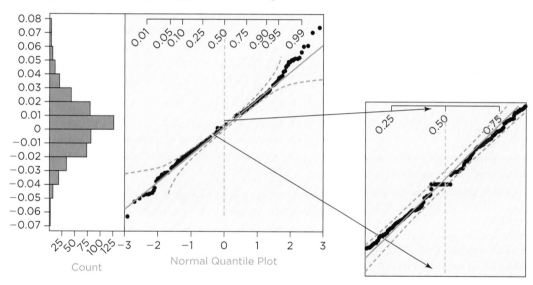

33. The amount of electricity supplied by a utility to residences in a community during the summer depends on the local weather as well as the habits of the people in the community.
    (a) Would a normal model be reasonable to describe the amount of electricity used from one day to the next? Explain your answer briefly.
    (b) Would it be reasonable to treat a sequence of daily electricity consumption values as iid normal random variables? Again, explain your answer briefly.

34. Dress shoes of a specific style for women come in sizes ranging from 4 to 12. A retailer believes that this range of sizes is adequate for all but 5% of its female customers.
    (a) What normal model for sizes is consistent with the opinion of the retailer?
    (b) If the retailer uses a normal model, then should it order more shoes of sizes 4 and 5 or more shoes of sizes 11 and 12?
    (c) Can a normal model possibly be the right model for the proportion of shoe sizes of the retailer's female customers?

35. The weekly salary paid to employees of a small company that supplies part-time laborers averages $700 with a standard deviation of $400.
    (a) If the weekly salaries are normally distributed, estimate the fraction of employees that make more than $300 per week.
    (b) If every employee receives a year-end bonus that adds $100 to the paycheck in the final week, how does this change the normal model for that week?

(c) If every employee receives a 5% salary increase for the next year, how does the normal model change?

(d) If the lowest salary is $300 and the median salary is $500, does a normal model appear appropriate?

36. A specialty foods company mails out "gourmet steaks" to customers willing to pay a gourmet price. The steaks vary in size, with a mean uncooked weight of 1.2 pounds and standard deviation of 0.10 pound. Use a normal model for the weight of a steak.

(a) Estimate the proportion of steaks that weigh less than 1 pound.

(b) If a standard order has 5 steaks, then what model can be used to describe the total weight of the steaks in a standard order? (Do you need to make any assumptions?)

(c) If the weights are converted to ounces (16 ounces = 1 pound), then what normal model would be appropriate for the weight of an individual steak?

(d) The lower quartile, median, and upper quartile of the weights of the steaks are 1.125 pounds, 1.2 pounds, and 1.275 pounds, respectively. Do these summaries suggest that the use of a normal model is appropriate?

37. A contractor built 30 similar homes in a suburban development. The homes have comparable size and amenities, but each has been sold with features that customize the appearance, landscape, and interior. The contractor expects the homes to sell for about $450,000. He expects that one-third of the homes will sell either for less than $400,000 or more than $500,000.

(a) Would a normal model be appropriate to describe the distribution of sale prices?

(b) What data would help you decide if a normal model is appropriate? (You cannot use the prices of these 30 homes; the model is to describe the prices of *as-yet-unsold* homes.)

(c) What normal model has properties that are consistent with the intuition of the contractor?

38. An accounting firm assists small businesses file annual tax forms. It assigns each new client to a CPA at the firm who specializes in companies of that type and size. For instance, one CPA specializes in boutique clothing retailers with annual sales of about $2 million. To speed the process, each business submits a preliminary tax form to the accounting firm.

(a) Would a normal model be useful to describe the total size of adjustments when a CPA reviews the preliminary tax forms? (For example, suppose the preliminary form claims that the business owes taxes of $40,000. If form completed by the CPA says the tax obligation is $35,000, then the adjustment is −$5,000.)

(b) What data would help you decide if a normal model is appropriate? (You cannot use data from the current year; those data are not yet available.)

(c) If the average adjustment obtained by the CPA who specializes in clothing retailers is −$7,000 (i.e., $7,000 less than indicated on the preliminary form), then what SD implies that all but a few business end up with lower taxes after the work of these accountants? (Assume a normal model for this question.)

(d) Would a normal model be useful to describe the total size of adjustments for all of the CPAs at this accounting firm?

You Do It

39. Find these probabilities for a standard normal random variable Z. Be sure to draw a picture to check your calculations. Use the normal table in the back of the book or software.

(a) $P(Z < 1.5)$
(b) $P(Z > -1)$
(c) $P(|Z| < 1.2)$
(d) $P(|Z| > 0.5)$
(e) $P(-1 \leq Z \leq 1.5)$

40. Find these probabilities for a standard normal random variable Z. Be sure to draw a picture for each to check your calculations. Use the normal table in the back of the book or software.

(a) $P(Z \geq 0.6)$
(b) $P(Z < -2.3)$
(c) $P(-0.8 \leq Z < 0.8)$
(d) $P(|Z| > 1.5)$
(e) $P(0.3 \leq Z \leq 2.2)$

41. Using the normal table in the back of the book or software, find the value of z that makes the following probabilities true. You might find it helpful to draw a picture to check your answers.

(a) $P(Z < z) = 0.20$
(b) $P(Z \leq z) = 0.50$
(c) $P(-z \leq Z \leq z) = 0.50$
(d) $P(|Z| > z) = 0.01$
(e) $P(|Z| < z) = 0.90$

42. Using the normal table in the back of the book or software, find the value of z that makes the following probabilities hold. Again, you might find it helpful to draw a picture to check your answers.

(a) $P(Z < z) = 0.35$
(b) $P(Z \geq z) = 0.60$
(c) $P(-z \leq Z \leq z) = 0.40$
(d) $P(|Z| > z) = 0.005$
(e) $P(|Z| < z) = 0.99$

43. (a) Find the value at risk (VaR) for an investment of $100,000 at 2%. (That is, find out how low the value of this investment could be if we rule out the worst 2% of outcomes.) The investment is expected to grow during the year by 8% with SD 20%. Assume a normal model for the change in value.

(b) To reduce the VaR to $20,000, how much more expected growth would be necessary? Assume that the SD of the growth remains 20%.

(c) If the VaR of an investment is $20,000 for a one-year holding period, does that mean that the VaR for a two-year holding period is $40,000?

44. (a) Find the VaR for an investment of $500,000 at 1%. (That is, find out how low the value of this investment could be if we rule out the worst 1% of outcomes.) The investment is expected to grow during the year by 10% with SD 35%. Assume a normal model for the change in value.

(b) How much smaller would the annual SD of the investment need to become in order to reduce the VaR to $200,000? Assume that the expected growth remains 10%.

(c) If this performance is expected to continue for two years, what is the VaR (at 1%) if the investment is held for two years rather than one? Use the features of the investment in part (a).

45. An insurance company found that 2.5% of male drivers between the ages of 18 and 25 are involved in serious accidents annually. To simplify the analysis, assume that every such accident costs the insurance company $65,000 and that a driver can only have one of these accidents in a year.

(a) If the company charges $2,500 for such coverage, what is the chance that it loses money on a single policy?

(b) Suppose that the company writes 1,000 such policies to a collection of drivers. What is the probability that the company makes a profit on these policies? Assume that the drivers don't run into each other and behave independently.

(c) Does the difference between the probabilities of parts (a) and (b) explain how insurance companies stay in business? Large auto insurers are certainly profitable. One report, for example, claims that Allstate pays out less than $0.50 in accident claims for every dollar collected in premiums. (*Business Week*, 5/1/2006)

46. A tire manufacturer warranties its tires to last at least 20,000 miles or "you get a new set of tires." In its experience, a set of these tires lasts on average 26,000 miles with SD 5,000 miles. Assume that the wear is normally distributed. The manufacturer profits $200 on each set sold, and replacing a set costs the manufacturer $400.

(a) What is the probability that a set of tires wears out before 20,000 miles?

(b) What is the probability that the manufacturer turns a profit on selling a set to one customer?

(c) If the manufacturer sells 500 sets of tires, what is the probability that it earns a profit after paying for any replacements? Assume that the purchases are made around the country and that the drivers experience independent amounts of wear.

47. A hurricane bond pays the holder a face amount, say $1 million, if a hurricane causes major damage in the United States. Suppose that the chance for such a storm is 5% per year.

(a) If a financial firm sells these bonds for $60,000, what is the chance that the firm loses money if it only sells one of these?

(b) If the firm sells 1,000 of these policies, each for $60,000, what is the probability that it loses money?

(c) How does the difference between the probabilities of parts (a) and (b) compare to the situation of a life insurance company that writes coverage to numerous patients that live or die independently of one another?

48. **Hedge Funds** The data are returns of 533 hedge funds reported for the month of December 2005. The returns are computed as the change in value of assets managed by the fund during the month divided by the value of the assets at the start of the month.

(a) Describe the histogram and boxplot of the returns. Is the histogram symmetric and unimodal? Would you say that this histogram is bell shaped?

(b) Are there any outliers? Which fund had the highest return? The lowest?

(c) What normal model describes the distribution of returns of these hedge funds?

(d) Is the normal model in part (c) a good description of the returns? Explain how you decided whether it's a good match or not.

49. **Pharma Promotion** The data give the market share (0 to 100%) of a brand-name pharmaceutical product in 224 metropolitan locations. The market share indicates the percentage of all prescriptions for this type of medication that are written for this brand.

(a) Describe the histogram and boxplot of the market shares. Is the histogram symmetric and unimodal? Would you say that this histogram is bell shaped?

(b) Are there any outliers? In what location does this brand have the highest market share? The lowest?

(c) The market share can only be as large as 100 and cannot be negative. Since a normal model allows for any value, is it possible for a normal model to describe these shares? Would a normal model be reasonable if the average share were closer to 0 or 100%?

(d) What normal model describes the distribution of returns of these shares?

(e) Is the normal model in part (d) a good description of these shares? Explain how you decided whether it's a good match or not.

4M Normality of Stock Returns

Percentage changes (or returns) on the stock market follow a normal distribution—if we don't reach too far into the tails of the distribution. Are returns on stocks in individual companies also roughly normally distributed? This example uses monthly returns on McDonald's stock from 1980 through the end of 2005 (312 months).

Motivation

a) If the returns on stock in McDonald's during this period are normally distributed, then how can the performance of this investment be summarized?

b) Should the analysis be performed using the stock price, the returns on the stock, or the percentage changes? Does it matter?

Method

c) Before summarizing the data with a histogram, what assumption needs to be checked?

d) What plot should be used to check for normality if the data can be summarized with a histogram?

Mechanics

e) If the data are summarized with a histogram, what features are lost, and are these too important to conceal?

f) Does a normal model offer a good description of the percentage changes in the price of this stock?

g) Using a normal model, estimate the chance that the price of the stock will increase by 10% or more in at least one month during the next 12 months.

h) Count the historical data rather than using the normal model to estimate the probability of the price of the stock going up by 10% or more in a coming month. Does the estimate differ from that given by a normal model?

Message

i) Describe these data using a normal model, pointing out strengths and any important weaknesses or limitations.

4M Normality and Transformation

We usually do not think of the distribution of income as being normally distributed. Most histograms show the data to be skewed, with a long right tail reaching out toward Bill Gates. This sort of skewness is so common, but normality so useful, that the *lognormal model* has become popular. The lognormal model says that the logarithm of the data follows a normal distribution. (It does not matter which log you use because one log is just a constant multiple of another.)

For this exercise, we'll use a sample of incomes from the 2000 Census. These gives the total household income for a sample of 258 residences in San Antonio, Texas.

Motivation

a) What advantage would there be in using a normal model for the logs rather than a model that described the skewness directly?

b) If poverty in this area is defined as having a household income less than $20,000, how can you use a lognormal model to find the percentage of households in poverty?

Method

c) These data are reported to be a sample of households in the San Antonio area. If the data are from one cluster in a cluster sample, would that cause problems in using these data?

d) How do you plan to check whether the lognormal model is appropriate for these incomes?

Mechanics

e) Does a normal model offer a good description of the household incomes? Explain.

f) Does a normal model offer a good description of the logarithm of household incomes? Explain.

g) Using the lognormal model with parameters set to match this sample, find the probability of finding a household with income less than $20,000.

h) Is the lognormal model suitable for determining this probability?

Message

i) Describe these data using a lognormal model, pointing out strengths and any important weaknesses or limitations.

Case: Managing Financial Risk

INVESTMENT RISK

THE DICE GAME

UNDERSTANDING WHAT HAPPENS

CASE SUMMARY

Financial markets can feel overwhelming. Markets offer so many choices with different promises. How can you make sense of it all and come away feeling that you've made good choices? The right place to start is by understanding the role of random variation. Luck and chance play a large role in deciding which investment wins and which investment loses. This case uses a simulation with dice to follow the performance of several investments. Even though the roll of a die offers only six outcomes, dice provide enough randomness to simulate important features of real investments.

INVESTMENT RISK

When it comes to investments, **risk** is the variance of the percentage changes in value (returns) over time. The two histograms in Figure 1 summarize monthly percentage changes of two investments.

The histogram on the top shows monthly percentage changes on the stock market from 1926 through 2005; the histogram on the bottom shows US Treasury Bills (basically, a short-term savings bond). There's almost no month-to-month variation (on this scale) in Treasuries. The stock market is much riskier. Sometimes stocks go up a lot (the largest monthly increase is 38%), and other times go down. This variation in what happens from month to month is risk.

If stocks are so risky, why does anyone buy them? The answer is simple: stocks offer the potential for higher gains as compensation for taking risks. Stocks over these years grew on average 12% annually, compared to 4% for Treasuries. Is this enough reward for the risk? If those were your retirement dollars, would you put the money into stocks?

It's not an easy question to answer. The following simulation shows you how to understand the relationship between risk and reward. Several questions

---

risk   The variance of the returns on an investment.

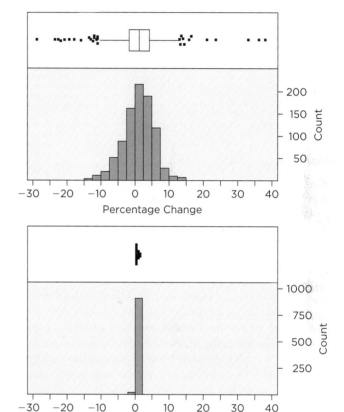

FIGURE 1 Percentage changes of stocks (top) and Treasury Bills (bottom).

along the way get you to think before you simulate, and then contrast what you expected to happen with the observed outcome.

THE DICE GAME

To learn about risks, we are going to simulate a small financial market comprised of three investments. We can do the simulation with a computer, but it's more interesting to watch the investments evolve as we roll dice that determine how the investments perform. Three dice simulate three investments. Each roll of all three determines what happens in a "year" in this market. Doing the simulation as a team works well if you have three people. One person rolls the dice, another keeps track of the dice and reads off their values, and a third records the outcomes.

The investments represented by the dice are rather different. One investment meets our intuitive definition of a risky investment like stocks whereas another resembles Treasuries. A third lies between these extremes. These investments have the characteristics shown in Table 1. The labeling by colors matches colors of widely available dice.

TABLE 1 Properties of three investments.

Investment	Expected Annual Percentage Change	SD of Annual Percentage Change
Green	8.3%	20%
Red	71%	132%
White	0.8%	4%

Here's how to interpret Table 1. Suppose we start with $1,000 in the investment called *Red*. Table 1 says to expect the value of our investment to be 71% larger at the end of a year, growing on average to $1,710. Similarly, after one year, $1,000 in *Green* grows on average to $1,083 and $1,008 in *White*.

Red is the best choice if we only care about what happens on average. We've learned, however, that variances are important as well. The standard deviation of *Red* is the largest of the three, 132%. Not only does *Red* offer the largest average growth, but it promises large gains and losses. How should we balance the average growth for *Red* versus its standard deviation? The average return on *Red* is 8.5 times that for *Green*, but its SD is 6.6 times larger than the SD for *Green*. Is this a good trade-off?

Question 1 Which of the three investments summarized in Table 1 is the most attractive to you? Why?

Dice Investments

Percentage changes are familiar, but there are better choices for simulating investments. We'll switch to **gross returns**. The gross return on an investment over a time period is the ratio of its final value to its initial value. If an investment has value V_t at the end of year t, then the gross return during year t is the ratio $R_t = V_t/V_{t-1}$. Gross returns simplify the

---

gross return The ratio of the value of an asset at the current time to its value at a previous time.

---

calculations and avoid mistakes that happen with percentage changes. If we multiply the value at the start of the year by the gross return, we get the value at the end of the year.

Dice determine the gross returns for each investment. Table 2 shows the correspondence between die outcomes and gross returns.

TABLE 2 Gross returns for the dice simulation.

Outcome	Green	Red	White
1	0.8	0.05	0.95
2	0.9	0.2	1
3	1.1	1	1
4	1.1	3	1
5	1.2	3	1
6	1.4	3	1.1

For example, if the green die rolls 1, then the gross return on *Green* is 0.8. Each dollar invested in *Green* falls to $0.80 (a 20% drop). If the red die rolls 1, its gross return is 0.05 (a 95% drop).

Let's work through an example for two years. Each investment in the simulation begins with $1,000, as indicated in the first row of the data collection form in Table 3. A roll of all three dice simulates a year in this market. Suppose that first roll of the dice shows these outcomes:

(Green 2)   (Red 5)   (White 3)

The value 2 for the green die tells us to use the gross return 0.9 from the second row of Table 2 for *Green*; *Green's* value falls to $1000 \times 0.9 = \$900$. The values after the first year are:

Green $\$1,000 \times 0.9 = \$900$
Red $\$1,000 \times 3 = \$3,000$
White $\$1,000 \times 1 = \$1,000$

A second roll of all three dice determines how these values change in the second year. For example, the second roll of the dice might give

(Green 4)   (Red 2)   (White 6)

The gross return for *Green* in this year is 1.1 (from the fourth row of Table 2); *Green* increases by 10%. After two rounds, the investments are worth

Green $\$900 \times 1.1 = \$990$
Red $\$3,000 \times 0.2 = \$600$
White $\$1,000 \times 1.1 = \$1,100$

TABLE 3 Illustrative calculations for two years.

Round	Value			Multiplier			
	Green	**Red**	**White**	**Green**	**Red**	**White**	
Start	$1,000	$1,000	$1,000	0.9	3	1	
1	900	3,000	1,000	1.1	0.2	1.1	
2	990	600	1,100				

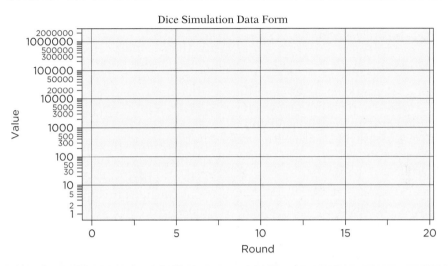

Dice Simulation Data Form

Round	Value			Multiplier			
	Green	**Red**	**White**	**Green**	**Red**	**White**	
Start	$1000	$1000	$1000				
1							
2							
3							
4							
5							
6							
7							
8							
9							
10							
11							
12							
13							
14							
15							
16							
17							
18							
19							
20							

Table 3 keeps track of the results after two "years." *White* has the largest value, followed by *Green*, and then *Red*. Use the larger version of this table (shown below Table 3) to hold the results of your simulation and to plot the values of the investments on the accompanying grid.

TABLE 4 Expanded data form that includes calculations for Pink.

	Value			Multiplier				Value Pink
Round	Green	Red	White	Green	Red	White	Pink	
Start	$1,000	$1,000	$1,000	0.9	3	1	2	$1,000
1	900	3,000	1,000	1.1	0.2	1.1	0.65	2,000
2	990	600	1,100					1,300

Simulation

Run a simulation for 20 years, keeping track of the results as in Table 3. Be sure to plot the values in the chart as the simulation progresses. Roll all three dice to find the outcome for each year. The gross returns in Table 2 determine the value of each investment as the dice are rolled.

Question 2 Which of the three investments has the largest value after 20 years in your simulated market?

Did you pick this investment initially? Having watched the simulation happen, try to explain why this investment won. Was it chance, or something that you expect to hold up over the long term?

A Two-Investment Portfolio

Few investors buy just one stock. Putting all of your money into one company means that if this company stumbles, then you could lose it all. Rather than bet on one company, most investors spread the *risk* by purchasing portfolios. A **portfolio** reduces the risk by investing in a variety of stocks rather than just one. Putting everything into one stock seems very *risky*. But how can we measure risk?

A fourth investment for the dice simulation is a portfolio that mixes *Red* and *White*, so we'll call it *Pink*. The dice can be put away. The previously recorded gross returns for *Red* and *White* determine the returns for *Pink*. It's easiest to describe what to do with an example.

Using the outcomes from the previous example, the gross return for *Pink* in the first year is the average of those for *Red* and *White* (3 and 1):

$$Pink: \quad \$1,000 \times \frac{3 + 1}{2} = \$2,000$$

---

portfolio An investment formed by spreading the invested money among several stocks, bonds, or accounts.

In the second year, *Red* has gross return 0.2 and *White* has gross return 1.1. The compounded value for *Pink* becomes

$$Pink: \quad \$2,000 \times \frac{0.2 + 1.1}{2} = \$1,300$$

Table 4 shows the sample data after adding *Pink* in the first two years.

Before figuring out how *Pink* does in the simulation, consider this question. Since *Pink* mixes *Red* and *White*, where should it be at the end of the simulation? Think carefully about this question.

Question 3 How does your group expect *Pink* to perform? Will it be better or worse than the others?

Go ahead and fill in the returns and values for *Pink*. *Remember*, you don't need to roll the dice again. Use the same gross returns that you already have for *Red* and *White*. Don't average the final values for *Red* and *White* either: that provides the wrong answer. You need to average the returns in each period. When you're done, we have a final question.

Question 4 Did *Pink* perform as well as you expected? How would you explain the performance of *Pink*?

UNDERSTANDING WHAT HAPPENS

If the results for the simulation are like most, *Pink* finishes with the highest amount. *Red* is exciting, but has a bad habit of crashing. *Green* does okay, usually in second place behind *Pink*, but it's not as interesting as *Pink* or *Red*. *White* is boring. To see how we could have anticipated these results, we will define random variables associated with these investments. Two properties of these random variables, namely means and variances, explain what happens in the simulation.

Random Variables

Table 2 defines three random variables, one for each investment simulated by the dice. Each random variable defines a correspondence between the possible outcomes of a random experiment (rolling the dice) and returns. Assuming you've got fair dice, the six outcomes are equally likely, so each row in Table 2 happens with probability 1/6.

Let's identify three random variables by the first letters of the colors of the dice: G, R, and W. We'll focus on G, the random variable associated with the gross returns determined by the green die. Figure 2 shows the probability distribution of G.

Because G is a discrete random variable (one with a finite list of possible values), we've shown the probabilities as individual points tied to the outcomes along the x-axis. Figure 2 resembles a histogram with very narrow intervals at {0.8, 0.9, 1.1, 1.1, 1.2, 1.4}.

Chapter 9 defines the expected value of a random variable to be the weighted average of the possible outcomes, weighted by probabilities. Because 1/6th of the gross returns for *Green* are at 0.8, 1/6th at 0.9, and so forth,

$$E(G) = \frac{0.8 + 0.9 + 2(1.1) + 1.2 + 1.4}{6} = 1.0833$$

Table 1 expresses the properties as percentage changes. For example, the average percentage change is 8.3% for *Green*. The conversion is easy: to get a percentage change, subtract 1 from the gross return and multiply by 100.

Random variables also have variances. The variance of G is the expected squared deviation from its mean 1.0833,

$$\text{Var}(G) = \frac{(0.8 - 1.083)^2 + (0.9 - 1.083)^2 + 2(1.1 - 1.083)^2 + (1.2 - 1.083)^2 + (1.4 - 1.083)}{6}$$

$$\approx 0.0381$$

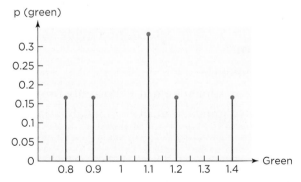

FIGURE 2 Probability distribution of Green.

The standard deviation of G is the square root of its variance, $\text{SD}(G) = \sqrt{0.0381} \approx 0.195$. Let's connect this to Table 1. The conversion from returns to percentage changes requires that we subtract 1 (shifts do not change the SD) and multiply by 100 (which increases the SD by 100). After rounding to 2 digits, the SD of percentage changes in *Green* is 20%. If we work out the means and SDs for R and W as we have done for G, we will get the other values in Table 1.

Properties of a Portfolio

The fourth random variable represents *Pink*. We won't use "*P*" to denote the random variable for *Pink* because P stands for probability. Let's use K for *Pink*.

Let's start with the expected gross return on *Pink*. In the dice simulation, the gross return for *Pink* in each period is the average of the gross returns for *Red* and *White*. If that's the way we manipulate the outcomes, then that's the way we manipulate the random variables. So, we can write $K = (R + W)/2$. Now it's easy to see that the expected value of *Pink* is the average of the expected values of *Red* and *White*:

$$E(K) = E\left(\frac{R + W}{2}\right)$$
$$= \frac{E(R) + E(W)}{2}$$
$$= \frac{1.71 + 1.008}{2} = 1.36$$

Pink gives up half of the return on *Red*. Now let's find its variance.

We get the variance of K from those of R and W. Constants factor out of variances of random variables just as they do for data: We square them. When we pull out the 1/2, we have to square it.

$$\text{Var}(K) = \text{Var}\left(\frac{R + W}{2}\right) = \frac{1}{4}\text{Var}(R + W)$$

The returns on *Red* and *White* are independent because the outcome of the white die does not affect the outcome of the red die. As shown in Chapter 10, the variance of a sum of independent random variables is the sum of their variances.

$$\text{Var}(R + W) = \text{Var}(R) + \text{Var}(W)$$

Now all we have to do is plug in the variances of *Red* and *White* from Table 1:

$$\text{Var}(K) = \frac{1}{4}(\text{Var}(R) + \text{Var}(W))$$

$$= \frac{1.32^2 + 0.04^2}{4}$$

$$= 0.436$$

The standard deviation for *Pink* is $\text{SD}(K) = \sqrt{0.436} \approx 0.66$.

For *Pink* to have this smaller variance, we have to maintain the balance between *Red* and *White*. Not only do we need *Red* and *White*, we have to rebalance the mix of assets. **Rebalancing** a portfolio means periodically dividing the total value among the components.

Consider the values of *Pink* shown in Table 4. In the first round, *Red* goes up by 3 and *White* stays fixed. Starting with $500 (half of $1,000) in each, *Pink* finishes the first round valued at

$$3(500) + 1(500) = \$2,000$$

Before the second round, this wealth has to be rebalanced. That means that we split the $2,000 so that we have $1,000 in each of *Red* and *White*. In the second round, *Red* loses 80% of its value and *White* goes up by 10%. At the end of the second round, the value of *Pink* is

$$0.2(1,000) + 1.1(1,000) = \$1,300$$

Notice what would have happened had we not rebalanced. The factor 0.2 would be applied to all $1,500 in *Red*. Continual rebalancing protects half of the wealth by investing in *White* to avoid the volatility of *Red*. Rebalancing saves past earnings on *Red*, protecting gains from the volatility of *Red*.

Investors in stocks often lose sight of the need to rebalance. During the run-up of prices in the stock market during the dot-com surge in the late 1990s, investors made huge profits. Rather than put some aside, they stayed on board and faced large losses when the bubble burst in 2000–2001.

Table 5 summarizes the properties of the four random variables. This table holds the key to the success of *Pink*. Mixing *Red* with *White* gives up half of the expected return of Red. At the same time, however, *Pink* has only one-fourth of the variance of *Red*. We'll see that this trade-off is enough to make *Pink* usually come out on top.

TABLE 5 Means and SDs for the four random variables in the dice simulation.

Color	Expected Annual Return	Standard Deviation
Green	8.3%	20%
Red	71%	132%
White	0.8%	4%
Pink	35.5%	66%

We say "usually" because some groups occasionally do very well with *Red*. You might be one of those. Because of independence, an initial investment of $1,000 in *Red* grows on average to an astonishing $1,000 \times (1.71)^{20} = \$45,700,000$. It is possible for *Red* to reach huge values if we're lucky, but that's not likely. Most find *Red* on the way out by the end of the simulation. In fact, if we keep rolling, *Red* eventually *must* fall to zero.

Volatility Drag

Why does reducing the variance make *Pink* a winner? How can we tell that *Red* eventually crashes?

To build some intuition, think about this example. Suppose your starting salary after graduation is $100,000 (to keep the math easy). Things go well, and you get a 10% raise after a year. Your salary jumps to $110,000. The business does not do so well in the following year and you take a 10% cut. At the end of the second year, your salary is $99,000. Even though the average percentage change is zero (up 10%, down 10%), the variability in the changes leaves you making 99% of your starting salary. You've lost money even though the average percentage change is zero.

Because variance reduces gains and is often called volatility, the effect of variation in the returns on the value of an investment is called the volatility drag. More precisely, the **volatility drag** is half of the variance of the returns. Volatility drag reduces the average return, producing the **long-run return**.

Long-Run Return =

= Expected Annual Return − Volatility Drag

= Expected Annual Return − (Annual Variance)/2

---

rebalancing   Holding the value of the components of a portfolio to a specified mix.

volatility drag   Reduction of the long-run value of an investment caused by variation in its returns; computed as one-half the variance of the return.

The long run return describes how an investment performs when held over many time periods. Table 6 gives the long-run return on the investments in the dice simulation.

Pink has by far the largest long-run return, growing at a long-run rate of about 13.7% when we allow for its combination of mean and variance. *Red* and *White* don't look as good. The long-run return for *Red* is negative 16.1%; over the long haul, *Red* loses 16.1% of its value per period. Eventually, investors in *Red* lose everything. *White* gains little. Although intuition suggests that mixing two lemons is still a lemon, that's not the case with these investments. *Pink* is a great investment.

TABLE 6 Long-run returns of the four investments in the dice simulation.

Color Die	E Return	Variance	Long-Run Return
Green	0.083	$0.20^2 = 0.0400$	$0.083 - 0.0400/2 = 0.063$
Red	0.710	$1.32^2 = 1.7424$	$0.71 - 1.7424/2 = -0.1612$
White	0.008	$0.04^2 = 0.0016$	$0.008 - .0016/2 = 0.0072$
Pink	0.355	$0.66^2 = 0.4356$	$0.355 - 0.4356/2 = 0.1372$

REAL INVESTMENTS

We made up *Red*, but you can buy *Green* and *White* today. The timeplot in Figure 3 shows monthly gross returns for the whole U.S. stock market and Treasuries (minus inflation). The month-to-month variation of returns on Treasuries is much smaller than the variation in returns on stocks. The returns on Treasuries seem constant compared to the gyrations of the stock market.

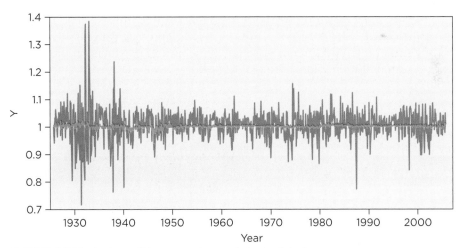

FIGURE 3 Historical monthly gross returns of stocks (blue) and Treasury Bills (red).

Several memorable events are apparent in Figure 3. The Great Depression occurred in the late 1920s and through the 1930s. The coming of the Second World War brought calm to the stock market. We can also see the isolated drop in October 1987.

The returns look like independent observations without a pattern over time, so we can summarize them in histograms (though this hides the clusters of volatile returns). The histograms in Figure 4 share a common scale from 0.7 to 1.4 (a 30% drop to a 40% gain). Gross returns on the stock market need this range; returns on Treasuries never venture far from 1.

Table 7 gives the summary statistics of the monthly gross returns. On average the stock market grew $100 \times 12(.0069) \approx 8.28\%$ annually (above inflation) from 1926 through 2005. Returns on Treasuries were flat, just keeping pace with inflation. The average annual percentage change above inflation was $100 \times 12(.00052) \approx 0.62\%$. The two investments have very different variation. The month-to-month

---

long-run return The expected return minus the volatility drag.

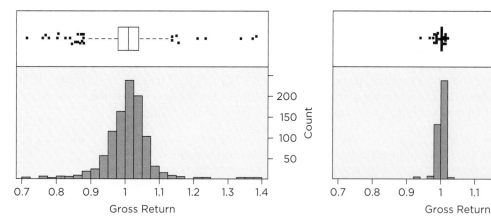

FIGURE 4 Histograms of gross returns on stocks (left) and Treasuries (right).

TABLE 7 Summary of monthly gross returns on stocks and Treasury Bills.

	Stock	**T-Bill**
Mean	1.0069	1.00052
SD	0.0548	0.00543
Variance	0.0030	0.00003
n	960	960

SD for stocks is 0.0548. Assuming independence over time, this implies an annual standard deviation of about 0.19—about 20%.

If we compare the properties of stocks and Treasuries to Table 1, we see that *Green* models the gross returns on the stock market and *White* models the returns on Treasuries.

Broader Implications

The implications for investing from this simulation carry over to other aspects of corporate finance as well. Just as individuals need to find the right mix of investments for themselves, companies must choose the best mix of investments.

Every project, ranging from research on the drawing board to products on the shelves, forms the portfolio of a company's investments. Unless a company carefully accounts for the risks of these projects, it can end up as overextended as many borrowers are today.

CASE SUMMARY

The **risk** of owning an investment is quantified by the variance of the returns on the asset. **Volatility drag** shows how the risk of an asset such as stock reduces the **long-run return** on the investment. By forming a **portfolio** of several assets, investors can trade lower returns for less volatility and obtain a higher long-run performance so long as they **rebalance** the portfolio.

■ Key Terms

gross returns, 288
long-run return, 293

portfolio, 290
rebalancing, 292

risk, 287
volatility drag, 292

Questions for Thought

1. If you repeat the dice game several times, or perhaps in a class, you might find that *Red* occasionally comes out the winner. Seeing that all of the others lose with *Red*, how might it be that some team comes out ahead with *Red*?[1]

2. Is *Pink* the best mix of *Red* and *White*? There's no rule that says a portfolio has to be an even mix of two investments. How might you find a better mixture?

3. *Red* is artificial; we don't know an investment that would work so well if paired with returns on Treasuries (*White*). What other attribute of the simulation is artificial? (*Hint:* This attribute makes it easy to find the variance of *Pink*.)

4. The role of randomness in investing has parallels in gambling, such as in the following game. Consider a slot machine. When you put in x dollars and pull the handle, it randomly pays out either $0.7x$, $0.8x$, $0.9x$, $1.1x$, $1.2x$, or $1.5x$. The payouts are equally likely. Here are two ways to play with the machine.

   (a) Put in \$1, pull the handle, and keep what you get. Repeat.

   (b) First, put in \$1 and pull the handle. Then put back into the machine whatever the machine pays. Repeat.

   (c) Can you win money with either approach? Which is the better way to play? (We suggest using dice to simulate these methods.)

---

[1] For more on this perspective, see our paper D. P. Foster and R. A. Stine (2006). " Warren Buffett: A Classroom Simulation of Risk and Wealth when Investing in the Stock Market." *The American Statistician*, **60**, 54–61.

Case: Modeling Sampling Variation

The normal curve is mathematical and abstract, but it shows up again and again in data and statistics. You can discover this phenomenon first-hand in a variety of places, even product packaging. This hands-on experiment introduces two important statistical concepts: the presence of differences between samples (what we call sampling variation) as well as the role of averaging in producing normal distributions. We'll then use the normal distribution for counting the number of items in an automatically filled package.

A SAMPLING EXPERIMENT

This experiment works well in a class, with some friends, or—if you really like chocolate—all by yourself. It requires two packages of milk chocolate M&M's® if you are doing it alone, or one for each person for groups of two or more. We use milk chocolate M&M's, but you can use any variety. Make sure that both bags are the same variety. You should also visit the company Web site to see if it shows the underlying proportions of colors for your variety; the distribution of colors varies among different types. We used packages labeled 1.69 ounces; these have about 60 pieces.

Milk chocolate M&M's currently come in six colors: brown, yellow, red, blue, orange, and green. The colors are not equally likely. Brown, once the only color, is less common. Blue, a recent addition to the palette, is more common and usually accounts for more than one-sixth of plain M&M's.

Open a bag of M&M's and count the number of candies in each color. Figure out the percentage of blue candies in your bag and write it down.

Question 1 Are more than one-sixth of the M&M's in your bag blue?

Now open the second bag and find the percentage of blue M&M's in this second bag. Write down this percentage as well.

Question 2 Are more than one-sixth of the candies in your second bag blue? Does the percentage of blue pieces in the second bag match the percentage in the first?

A Model for Counts

Mars, the manufacturer of M&M's, is privately held and secretive about its operations, but we can guess that elves don't count the M&M's that go into every bag. It's more likely that the bags are filled from a large "bowl" that mixes the colors. The packaging process basically scoops the contents of each bag from this bowl of differently colored pieces. How can we model the colors in a bag?

The binomial model (Chapter 11) is a reasonable model to describe the number of blue M&Ms in a bag. Every binomial model has two characteristics, or parameters, that identify the model. These parameters are n, the total number of counted items, and p, the probability of a "success." Let's define a success in this case as finding a blue M&M. The total count n is the number of candies in the bag; n for your bags should be *about* 58.

All manufacturing processes have variation, and sometimes it is a surprise to consumers. You might think that every package of M&M's has the same number of candies; after all, they are labeled to contain the same weight. Automated packaging systems, however, are not so consistent, and the number of pieces varies from bag to bag. We got the counts shown in the histogram in Figure 1 by counting the pieces in a case of 48 bags, each labeled to weigh 1.69 ounce.

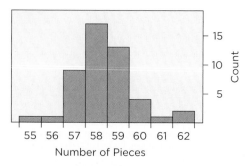

FIGURE 1 Counts of M&M's in 48 bags.

Using the Model

A binomial model tells us how many blue M&M's to expect in a bag if we know n and p. The formula is not surprising. The expected number of successes is np, the number of items times the chance for a success. With $n = 58$ and $p = 1/6$, we expect about $58/6 = 9.67 \approx 10$ blue pieces.

Does your package have 10 blue M&M's? Probably not. The product np is the expected value of the number of blue M&M's. The name "expected value" is a little confusing. It does *not* mean that you'll should expect to find 10 blue M&M's in every bag. The expected value is the number of blue pieces that we should find *on average* when looking at many bags. If we were to open thousands of bags of M&M's and count the number of blue pieces in each, on average we'd get 9.67 blue pieces per bag if the binomial model holds with $n = 58$ and $p = 1/6$.

We must also consider the variation. All manufacturing processes have variation, and packaging M&M's is no exception. A binomial model also says how much variation there should be in the count of blue M&M's. According to this model, the variance among the number of blue pieces is $np(1 - p)$. For $n = 58$ and $p = 1/6$, the variance is 8.06 *pieces*^{2}. Variance has a squared scale, so we take its square root and use the standard deviation. The SD of the number blue is $\sqrt{8.06} \approx 2.84$ pieces.

If the mean count of blue M&M's is 9.67 pieces with SD = 2.84 pieces, what more can be said about the number of blue pieces in other bags? Imagine looking at a histogram of data with mean 9.67 and SD = 2.84. If the histogram is bell shaped, can we say more about how the counts concentrate around the mean?

If the histogram is bell shaped, then the empirical rule, a normal model, suggests that the number of blue M&M's in 95% of the bags (19 out of 20) should be within 2 SDs of the mean. That gives a range of 9.67 ± 5.68, roughly 4 to 15 blue M&M's in a bag with 58 pieces. Did your two bags of M&M's have between 4 and 15 blue M&M's? Of the 48 packages we opened to produce Figure 1, the number of blue M&M's is between 4 and 15 for all but 2. That works out to 95.83% of the bags.

THE CENTRAL LIMIT THEOREM

Why should a normal model work for counts? Let's use a binomial model to see the answer. We can use this model to find

$$p(x) = P(x \text{ blue M\&M's in a bag})$$

for any choice of x from 0 to 58. It's as if we opened many thousands of bags and counted the number of blue M&M's in each. The model gives us a probability, the eventual relative frequency of the counts of blue candies in many bags.

Figure 2 shows the probabilities of different counts of blue M&M's in bags of 58 pieces with $n = 58$ and $p = 1/6$.

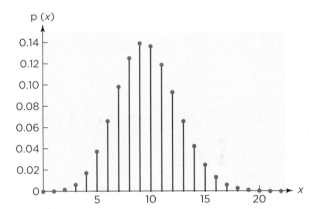

FIGURE 2 Probability distribution of the number of blue M&M's in bags with 58 pieces.

Look at the shape. It's no wonder that a normal model works for counts of blue M&M's. Had we opened thousands of bags, we would have gotten a bell-shaped distribution after all. But why should counting blue M&M's produce a bell-shaped distribution?

Counting Possibilities

Imagine that a bag of M&M's contains only 1 M&M. We either get 1 blue M&M or none. If we open many bags with one piece in each and draw the histogram of the relative frequency of blue pieces, it would eventually look like Figure 3 if $p = 1/6$.

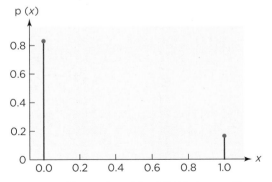

FIGURE 3 Chance for a blue M&M in bags with 1 piece.

There's nothing bell-shaped about this distribution. Only two values are possible, with 5/6 at zero and 1/6 at 1.

What about a bag with 2 pieces? If the bag has 2 pieces, we could get 0, 1, or 2 blue pieces. In most cases, we'd find no blue pieces. The chance for no blue pieces is $P(not\ blue) \times P(not\ blue) = (5/6)^2 = 0.6944$, if we believe that the packaging independently samples pieces from the big bowl with $p = 1/6$. Similarly, the chance for 2 blue pieces is $(1/6)^2 = 0.0278$. That leaves the chance for 1 blue piece. We could get the blue piece either first or second, so the chance for 1 blue piece is $2(1/6)(5/6) = 0.2778$. As a check, these sum to $1 : 0.6944 + 0.0278 + 0.2778 = 1$. Figure 4 shows the graph of the probabilities:

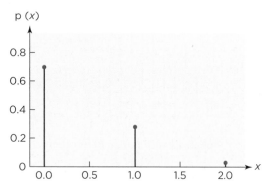

FIGURE 4 Probability distribution of the number of blue pieces in bags with 2 pieces.

It's not very bell shaped either. Let's see what happens for bags with 6 pieces. Figure 5 shows the probability distribution.

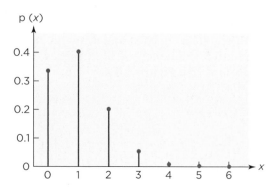

FIGURE 5. Probability distribution of the number of blue pieces in bags with 6 pieces.

We can see the beginnings of a bell-shaped distribution. The distribution is skewed, but there's less chance of bags with either all or no blue M&M's.

The reason that the bell-shaped distribution emerges from these counts has less to do with the chance of a blue piece and more to do with the number of ways to get a bag with *some* blue M&M's. Look at the diagram in Figure 6.

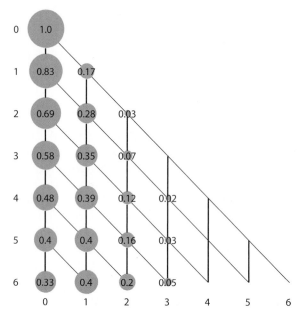

FIGURE 6 Probability tree for finding blue M&M's in bags of varying numbers of pieces.

Each row in Figure 6 identifies the number of M&M's in a bag. The top row has none, and the bottom row represents the probabilities associated with a bag that has $n = 6$ pieces. The nodes within each row give the probabilities for the number of blue M&M's found in a bag. The column tells how many blue M&M's were found. The size of each node is proportional to the probability. Labels show probabilities larger than 0.01.

The top row with one node is the starting point. If a bag has no pieces, then we don't have any blue M&M's. We're certain of that, so this node has probability one. The second row describes the probabilities for the number of blue pieces in a bag with one M&M. Most likely, we won't find a blue M&M, so we still have zero. With probability 1/6, we might find a blue piece and travel down the narrow path to the right. These probabilities match those in Figure 2.

As we move down the rows in Figure 6, the probabilities change. Gradually, but inevitably, the probability moves away from the left column and spreads out. By the time we reach the last row, a bell curve begins to emerge and the probabilities match those in Figure 5.

What does this have to do with a bell-shaped normal distribution? Consider the nodes labeled "0" and "6" in the bottom row of Figure 6. How many paths down the tree lead to these positions?

Only one path leads to each of these nodes. Because there is only one path that reaches the extremes, less and less probability remains at the edges as we consider larger and larger bags.

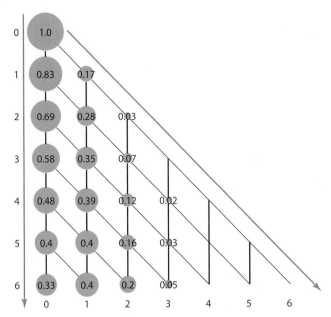

FIGURE 7 Only one path in the tree leads to the nodes representing bags of 0 or 8 blue pieces.

Now consider a centrally positioned node such as the one representing 2 blue pieces in the bottom row. Imagine all of the paths that lead to this node. Figure 8 highlights one of them.

If we count all of the paths from the top to this node, we will find $15 = {}_6C_2$ distinct paths! The proliferation of paths leading to nodes away from the edges causes probability to accumulate near the mean of the distribution.

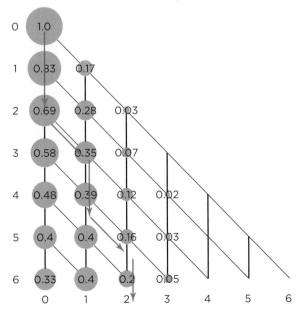

FIGURE 8 Many paths lead to nodes near the expected value.

The formal name for the math behind the emerging bell shape is the Central Limit Theorem. The CLT tells us that sums of random variables tend to follow a bell-shaped distribution as the number of added terms grows. Regardless of the probabilities, we'll eventually end up with a bell-shaped histogram.

USING A NORMAL MODEL

The implications of normality reach deep into decision making. The fact that a normal model can describe the distribution of virtually any sum allows us to solve problems that would otherwise be intractable.

Weighing in Place of Counting

Manufacturing has become widely diversified. It's rare to find a factory that makes all of the parts from raw materials and assembles the final product. The auto industry, for example, makes heavy use of distributed manufacturing: Various suppliers ship parts designed to specifications to the manufacturer for assembly. Many common parts are packaged by weight. If a shipment of parts to an assembly plant is supposed to include the 200 bolts needed to assemble 10 cars, chances are that no one counted them. More likely, they were packaged by weight instead.

If every bolt weighs exactly the same, say 1 ounce, it would be easy to fill a box with 200 of them. Weigh the box, tare the scale, and then add bolts until the scale shows 200 ounces. Unless it's a cheap bathroom scale, the box holds 200 bolts. That's good, because for want of a bolt, one of the cars will not be fully assembled.

The problem with counting by weighing is that most things do not have the same weight. Not bolts, not even M&M's. We used a scale accurate to 0.01 grams to weigh 72 M&M's sampled from several bags. On average, these M&M's weigh 0.86 grams with standard deviation 0.04 grams. Figure 9 shows the histogram.

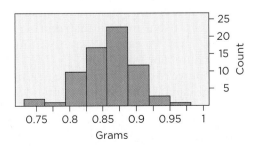

FIGURE 9 Weights of individual milk chocolate M&M's.

The variation in the weights of M&M's is common in manufacturing and complicates counting by weighing. If there's a lot of variation relative to the average weight, the count could be off by several pieces. That might not matter when counting M&M's, but if it brings an assembly line to a halt for want of a bolt, the costs quickly exceed that of including a few extra bolts in the shipment.

A key characteristic of manufactured parts is the coefficient of variation, the ratio of the standard deviation to the mean (Chapter 4). Smaller values of the coefficient of variation indicate more consistent items. For the M&M's shown in Figure 9, the coefficient of variation is $0.04/0.86 \approx 5\%$. Values of the coefficient of variation near 5% are typical for many bulk items.

Let's recognize the presence of variation in the weights of bolts. Assume that bolts weigh 1 ounce, on average, with a coefficient of variation $\sigma/\mu = 0.10$. Here's the question:

Question 4 How much should a box of bolts weigh in order to be 99.99% sure that it includes 200 bolts?

We want to be *very* sure that there's not a shortage at the factory. If we keep adding bolts until the weight reaches 200 ounces, there's a good chance that we included a few heavy bolts and the package will come up short with fewer than 200. We need to add bolts until the weight is larger than 200 ounces. But how much larger? We need to set a threshold, call it x, for the weight.

Take a guess. How large does the threshold x need to be to virtually guarantee that the package has enough bolts? Do you think that $x = 210$ ounces is enough? How about 250?

Using Normality

To find the threshold (which is smaller than you might think), we're going to exploit the CLT. The total weight of a box of bolts adds the weights of many bolts, so we can use a normal distribution to find probabilities. First we need a way to connect the weight to the number of bolts.

Imagine that we're adding bolts one at a time and watching the scale. Think about the following two events.

$\mathbf{A}_k = \{\text{weight of } k - 1 \text{ bolts is less than } x \text{ ounces}\}$

$\mathbf{B}_k = \{\text{at least } k \text{ bolts are needed}$
$\qquad \text{for the weight to exceed } x \text{ ounces}\}$

These two events are equivalent; one cannot happen without the other. If 189 bolts weigh less than 200 ounces, then at least 190 are needed to make the weight exceed 200 ounces. Because the events are equivalent, they occur with the same probability, $P(\mathbf{A}_k) = P(\mathbf{B}_k)$.

The event $\mathbf{B}_{200}$ means that a package that weighs x ounces has at least 200 bolts. To make sure that the package has at least 200 bolts with 99.99% probability, we need to choose the threshold x so that

$$P(\mathbf{B}_{200}) = P(\mathbf{A}_{200}) > 0.9999$$

Because of this equivalence, we can use the CLT to find the threshold x. The CLT says that the weight of 200 bolts should be close to a normal distribution. All we need are the mean and SD of the package weight; then we can find $P(\mathbf{A}_{200})$.

Let's first summarize what we know about one bolt. The average weight of a bolt is $\mu = 1$ ounce. We can find the standard deviation from the coefficient of variation. The ratio $\sigma/\mu = 0.10$, so $\sigma = 0.10 \times \mu = 0.10$. We don't know anything else about the bolts. We haven't even looked at a histogram of these. That's okay, too, because the CLT works regardless of the distribution of the weights.[1]

The properties of a randomly chosen bolt determine the mean and SD of the total weight. The mean of the sum is easy because the expected value of a sum is equal to the sum of the expected values. The mean weight of n bolts is $n \times \mu = n$ ounces. To find the variance of the total weight, we need an assumption. Namely, that the weights of the bolts vary independently of one another. If the bolts we're weighing are randomly mixed, this might be sensible. If the bolts arrive in production order, however, we ought to check whether there's some sort of dependence in the weights. For example, wear and tear on the cutting tools that shape the bolts might cause the weights to gradually trend up or down. Let's presume independence and push on.

If the weights of the bolts are independent, the variance of the sum of the weights is the sum of the variances. If W_i is a random variable that represents the weight of the ith bolt, then the variance of the total weight of n bolts is

$$\begin{aligned} \text{Var}(W_1 + W_2 + \ldots + W_n) \\ = \text{Var}(W_1) + \text{Var}(W_2) + \ldots + \text{Var}(W_n) \\ = n\,\sigma^2 = 0.01\,n \end{aligned}$$

Consequently, the SD of the total weight is $0.1 \sqrt{n}$. $\sqrt{n}$ is critical. If T_n stands for the total weight of

---

[1] The CLT eventually works for any distribution, producing a bell-shaped distribution. You might need to add a huge number of things together, however, for the bell shape to emerge. The closer to normal you start with, the better off you are in the end.

n bolts, $T_n = W_1 + W_2 + \ldots + W_n$, then T_n becomes more normally distributed as n increases, with mean n ounces and SD equal to $0.1\sqrt{n}$ ounces.

We get the threshold x by using the normal table found inside the back cover of the book or software. We can be "99.99% sure" that the package has 200 bolts if we choose x so that

$P(\mathbf{A}_{200})$
$\quad = P\{\text{weight of 199 bolts is less than } x \text{ ounces}\}$
$\quad = 0.9999.$

That is, we need to find x so that $P(T_{199} < x) = 0.9999$. To find x using the normal table, standardize T_{199} by subtracting its mean and dividing by its standard deviation:

$$P(T_{199} < x) = P\left(\frac{T_{199} - 199}{.1\sqrt{199}} < \frac{x - 199}{.1\sqrt{199}}\right)$$
$$= P\left(Z < \frac{x - 199}{1.411}\right)$$
$$= 0.9999$$

(As usual, Z stands for a standard normal random variable with mean 0 and SD 1.) From the normal table, $P(Z < 3.719) = 0.9999$, meaning that

$$\frac{x - 199}{1.411} = 3.719$$

Solving for x, the threshold is $x = 199 + 3.719(1.411) = 204.25$.

We're guessing that this threshold is *a lot* smaller than you guessed. Considering that we want to have enough bolts with probability 0.9999, it may seem like the box should weigh a lot more. The explanation is that the variability of the total weight is smaller than you might expect. That's the importance of that $\sqrt{n}$ in the SD for the total weight. The SD of the sum increases more slowly than the average.

Does It Matter?

Why go to so much trouble? How much could a few bolts cost anyway? Even if it's a special bolt made to unusual specifications, perhaps it costs $2. Why not toss in a few more?

Perhaps your company outsources the manufacture of the bolts to China. You buy them for $2 and include them in the shipment of parts to the auto assembler for $2.25 each. (If you charge much more, you can be sure someone else will become the supplier. After all, it's not as if you made the bolts.) If you add 10% extra bolts, the shipment of 220 bolts sells for $200 \times \$2.25 = \$450$. (The customer only pays for the 200 ordered.) The 220 bolts cost you $440. You net a slim $10. Following the procedure with $x \approx 204$, most shipments would have about 204 bolts. You'd still get paid $450, but now your costs drop to $408 and the profit rises to $42 per shipment for the bolts alone.

The profits multiply as you apply these ideas to other parts and over many shipments. In 2005, 16 million cars were assembled in North America.

The histogram in Figure 10 shows the distribution of the weights (in grams) of the 48 packages of M&M's that we sampled in this chapter.

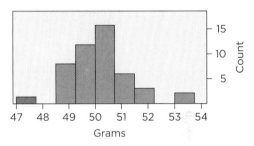

FIGURE 10 Package weights of M&M's.

The packages are labeled to weigh 1.69 ounces, which converts to 47.9 grams. All but one weigh more than the labeled package weight. That's extra candy that would not need to be there if the manufacturing process were variation free. The average weight of these packages is 50.2 grams. That's about 5% more than what would be needed if every package had been filled to the labeled weight. Mars could reduce its chocolate costs by 5% if it could find a way to remove the variation in the packaging. It must not be that easy.

CASE SUMMARY

Modern packaging produces variation among individual items. With a few assumptions, we can model this variation using a normal model that allows us to anticipate differences from item to item. The normal model works because sums of independent random variables, whether counts or measured values to many digits of accuracy, eventually have a normal distribution. The bell curve appears because of the variety of ways to get sums that are near the expected value of the total. We can exploit normality for many things, including monitoring modern automated machinery.

■ Questions for Thought

1. M&M's weigh 0.86 grams on average with SD = 0.04 grams, so the coefficient of variation is $0.04/0.86 \approx 0.047$. Suppose that we decide to label packages by count rather than weight. The system adds candy to a package until the weight of the package exceeds a threshold. How large would we have to set the threshold weight to be 99.5% sure that a package has 60 pieces?

2. Suppose the same system is used (packaging by count), but this time we only want to have 10 pieces in a package. Where is the target weight? (Again, we want to be 99.5% sure that the package has at least 20 pieces.)

3. Comparing your answers to parts a and b, is the method of counting by weight more suited to large or small counts?

Inference

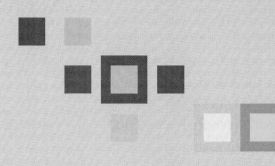

13 CHAPTER

Samples and Surveys

YOU MAY HAVE SEEN THE CLAIM "RANKED TOPS IN INITIAL QUALITY" IN ADVERTISEMENTS FOR NEW CARS. Have you ever wondered who decides which model wins and how the winner is determined?

J. D. Power and Associates annually sends a questionnaire to owners of new cars. Did they find defects? Do they enjoy the way the car performs? Would they buy the same model again? Each year, the Initial Quality Study contacts more than 50,000 purchasers and lessees. Their responses rank the models and determine which wins the prize.

The Initial Quality Study sounds big until you consider that millions of cars and light trucks get sold or leased in the United States every year. J. D. Power contacts less than 1/2 of 1% of those getting a new vehicle; most never get a questionnaire. Plus, J. D. Power isn't getting thousands of responses for each model. These customers bought or leased all sorts of vehicles, from Honda Accords to Ford trucks to Toyota hybrids. When spread over hundreds of models, a survey of 50,000 buyers begins to look small. What can be learned from a relatively tiny proportion of customers?

It's a common problem. Decisions often require knowing a characteristic of a great number of people or things. For example,

- A retailer wants to know the national market share of a brand before deciding to stock the item on its shelves.
- The foreman of a warehouse will not accept a shipment of electronic components unless he's sure that the vast majority of the components in the shipment operate correctly.
- Managers in the human resources department determine the salary for new employees based on wages paid for such work around the country.

The direct, comprehensive approach to each decision requires a huge amount of data. For instance, managers in the human resources (HR) department would like to know the starting salaries offered by every employer for this type of work around the country. That's not going to happen. Instead, they have to rely on what can be learned from a small subset. Such a subset is known as a sample.

***

This chapter introduces the concepts that motivate and guide the collection of samples. By focusing on a subset of the whole group, sampling converts an impossible task into a manageable problem. The key to successful sampling is the random selection of items from the larger group. The use of a sample in place of the whole, however, introduces sampling variation: The results from one sample will differ from those provided by another sample. Much of statistics deals with making decisions in the presence of this sampling variation.

13.1 | TWO SURPRISING PROPERTIES OF SAMPLING

survey Posing questions to a subset to learn about the larger group.

population The entire collection of interest.

sample The subset queried in a survey.

representative Reflecting the mix in the entire population.

bias Systematic error in choosing the sample.

It's hard to go a day without hearing the latest opinion poll when an election approaches. Newspapers regularly report the results of the latest surveys. A **survey** asks questions of a subset of people who belong to a much larger group called the **population**. The subset of the population that is included in the survey is called a **sample**. When done correctly, a survey suggests the opinions of the population, without the need to contact every member of the population. A sample that presents a snapshot of the population is said to be **representative**. Samples that distort the population, such as one that systematically omits a portion of the population, are said to have **bias**.

A key message of this chapter is that it's hard to get a representative sample. To avoid bias, you might be tempted to handpick the sample carefully, but that would usually be a mistake. Well-intentioned rules designed to make a sample representative often fail. Suppose a national retailer chooses a sample by picking one customer from every county of every state. The resulting sample will include shoppers from all around the country, but it probably won't be representative of the retailer's customers. It's likely that most of its customers come from densely populated areas, and these customers would be underrepresented in the sample.

Rather than add more rules to fix problems introduced by previous rules, a simpler method—random selection—works better for getting a representative sample. That's the first surprise about sampling. When it comes to sampling, the right way is to choose the sample from the population at random.

There's also a second surprise. Larger populations don't require larger samples. We do not set out to gather a fixed percentage, say 5% or 10%, of the population. For instance, suppose we'd like to know the average amount spent by customers who visit shopping malls. Consider two samples, each with 100 shoppers. One sample has shoppers who frequent a local mall, and the other has shoppers from malls around the country. The population of shoppers at the local mall is much smaller than for malls around the country. The surprising fact is that we learn as much about the mean amount spent in the national population from the national sample as we learn about the mean of the local population from the local sample.

To recap, the two surprises are as follows:

1. The best way to get a representative sample is to pick members of the population at random.
2. Larger populations don't require larger samples.

We will introduce the reasoning behind these surprises in this chapter, and in later chapters fill in more details.

Following convention, **n** identifies the size of the sample. If it's known, **N** stands for the size of the population. The methods that we use to determine n presume that n is much smaller than N, and our interest lies in problems in which the population is too large to reach every member.

n, N
n = sample size
N = population size

Randomization

The best strategy for choosing a sample is to select cases randomly. The deliberate introduction of randomness when selecting a sample from a population provides the foundation for inference, or drawing conclusions about the population from the sample. The statistical techniques in later chapters all begin with the premise that our data are a random sample. The following analogy suggests the reasoning.

Suppose you've decided to change the color of the living room in your apartment. So, you head down to Home Depot for paint. Once you choose a color from the palette of choices, a salesperson mixes a gallon for you on the spot. After a computer adds pigments to a universal base, the salesperson puts the can in a machine that shakes it for several minutes.

To confirm the color, the salesperson dips a gloved finger into the can and smears a dab of paint on the lid. That smear is just a few drops, but it's enough to tell whether you got the color you wanted. Because the paint has been thoroughly mixed by the shaker, you trust that the color of the smear *represents* the color of the entire can. This test relies on a small sample, the few drops in that smear, to represent the entire population, all of the paint in the can.

You can imagine what would happen if the salesperson were to sample the paint before mixing. Custom paints typically use a white base, so the smear would probably give the impression that the paint was pure white. If the salesperson happened to touch the added pigment, you'd get the misleading impression that the paint was a rich, dark color. Shaking the paint before sampling mixes the pigments with the base and allows a finger-sized sample to be *representative* of the whole can.

randomization Selecting a subset of the population at random.

Shaking works for paint, but how do we "shake" a population of people? The answer is to select them at random. **Randomization**, selecting a subset of the population at random, has essentially the same result as shaking a can of paint. A randomly selected sample is representative of the whole population. Randomization ensures that *on average* a sample mimics the population.

Here's an example. We drew two samples, each with 8,000 people, at random from a database of 3.5 million customers who have accounts at a bank. The entire database of customers is the population. Table 13.1 shows how well the means and percentages match up for two samples and the population.

TABLE 13.1 Comparison of two random samples from a database of 3.5 million.

	Age (yr)	Female (%)	Number of Children	Income (1–7)	Homeowner? (% yes)
Sample 1	45.12	31.54	1.91	5.29	61.4
Sample 2	44.44	31.51	1.88	5.33	61.2
Population	**44.88**	**31.51**	**1.87**	**5.27**	**61.1**

We didn't consider variables such as age or income category when we drew these two samples. Even so, the average ages in the samples, 45.12 and 44.44 years old, are both close to the average age (44.88 years) in the population.

Similarly, averages of the number of children in the samples, 1.91 and 1.88, are close to the average number of children in the population, 1.87. Randomization produces samples whose averages resemble those in the population. It does this without requiring us to construct a sample whose characteristics match key attributes of the population. That's good, because in order to match characteristics of the sample to those of the population we'd have to know the properties of the population!

Because randomizing avoids bias, randomizing enables us to infer characteristics of the population from a sample. Such inferences are among the most powerful things we do with statistics. Large samples with bias are much less useful than small samples that are representative. The accompanying story of the *Literary Digest* illustrates the dangers in bias.

The Literary Digest It can be instructive to see a dismal failure. The following failure in survey sampling is infamous. In the early twentieth century, newspapers asked readers to "vote" by returning sample ballots on a variety of topics. Internet surveys are modern versions of the same idea. A magazine called the *Literary Digest* was known for its ballots. From 1916 through 1932, its mock voting had correctly forecast the winner in presidential elections.

The 1936 presidential campaign pitted Alf Landon versus Franklin Roosevelt. For this election, the *Literary Digest* sent out 10 million ballots. Respondents returned 2.4 million. The results were clear. Alf Landon would be the next president by a landslide: 57% to 43%. In fact, Landon carried only two states. Roosevelt won 62% to 37%. Landon wasn't the only loser: The *Digest* itself went bankrupt. What went wrong?

The *Digest* made a critical blunder: Where do you think the *Digest* got a list of 10 million names and addresses in 1936? You might think of phone numbers, and that's one of the lists the *Digest* used. In 1936, during the Great Depression, telephones were luxuries. A list of phone numbers included far more rich than poor people. The other lists available to the *Digest* were even less representative—drivers' registrations, its own list of subscribers, and memberships in exclusive organizations such as country clubs.

The main campaign issue in 1936 was the economy. Roosevelt's core supporters tended to be poor and were underrepresented in the *Digest*'s sample. The results of the *Digest*'s survey did not reflect the opinions of the overall population.

Someone else, however, did predict the outcome. Using a sample of 50,000, a young pollster named George Gallup predicted that Roosevelt would win 56% of the vote to Landon's 44%. Though smaller, his sample *was representative* of the actual voters. The Gallup Organization went on to become one of the leading polling companies.

Sample Size

How large a sample do we need? Common sense tells you that bigger is better. What is surprising is that how much bigger has nothing to do with the size of the population. The design of a survey does not require that we sample a fixed percentage of the population. For large populations, only the number of cases in the sample matters. Unless the population is small, the size of the population doesn't affect our ability to use a sample to infer properties of the larger population.

The Financial Mood

STOCKS

The view that stocks are a safe investment has rebounded slightly, but fewer people think the stock market will go up.

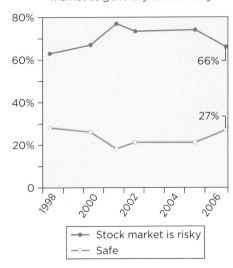

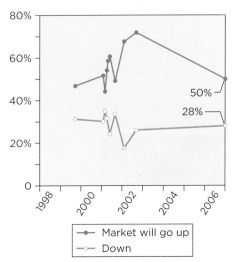

More than 100,000,000 voters cast ballots in U.S. presidential elections, but polls typically query fewer than 1,200 people. For instance, the survey described in the preceding charts summarizes opinions on the economy of 922 American adults in January 2007.[1] A sample of this size is an almost infinitesimal portion of the population. Even so, this survey reveals the attitudes of the entire population to within about ±3%. We'll begin to explain how we can make such a claim in Chapter 14. First, however, we have to get a representative sample. Without a representative sample, we won't be able to draw appropriate conclusions about the larger population.

How can we learn so much about the population from a sample? To build some intuition, let's open that can of paint again. If you're painting a whole house rather than a single room, you might buy one of those large, 5-gallon buckets of paint. Do you expect the salesperson to make a bigger smear to convince you that it's the right color? No, not as long as it's well mixed. The same, small smear is enough to make a decision about the entire batch, no matter how large. The fraction of the population that you've sampled doesn't matter so long as it's well mixed.

Simple Random Sample

simple random sample (SRS) A sample of *n* items chosen by a method that has equal chance of picking any sample of size *n* from the population.

A procedure that makes every sample of size *n* from the population equally likely produces a **simple random sample**, abbreviated **SRS**. An SRS is the standard against which we compare other sampling methods. Virtually all of the theory of statistics presumes simple random samples.

How do you get a simple random sample? Methods that give everyone in the population an equal chance to be in the sample don't necessarily produce a representative sample. Instead, the procedure must assign an equal chance to every *combination* of *n* members of the population. Consider, for instance, a clothier that has equal numbers of male and female customers. We could

---

[1] "Investors Greet New Year with Ambivalence," *New York Times*, Janary 2, 2007.

sample these customers this way: Flip a coin; if it lands heads, select 100 women at random. If it lands tails, select 100 men at random. Every customer has an equal chance of being selected, but every sample is of a single sex—hardly representative.

In order to obtain a simple random sample, we first need the sampling frame. The **sampling frame** is a list of items (voters, computer chips, shoppers, etc.) from which to draw the sample. Ideally, the sampling frame is a list of the members of the population. (The next subsection describes situations in which it's not possible to have a sampling frame.) Once we have the sampling frame, a spreadsheet makes it easy to obtain a simple random sample.

sampling frame A list of items from which to select a random sample.

- Place the list of the N items in the sampling frame in the first column of a spreadsheet.
- Add a second column of random numbers. Most spreadsheet programs have a function for generating random numbers.
- Sort the rows of the spreadsheet using the random numbers in the second column.
- The items in the first n rows of the resulting spreadsheet identify a simple random sample.

You can also obtain a simple random sample by picking items from the sampling frame systematically. For example, you might interview every 10th person from an alphabetical list of customers. For this procedure to produce an SRS, you must start the systematic selection at a randomly selected position in the list. When there is no reason to believe that the order of the list could be associated with the responses, **systematic sampling** gives a representative sample. If you use a systematic sample, you must justify the assumption that the systematic sampling is not associated with any of the measured variables. The ease of sampling using random numbers on a computer makes the use of systematic samples unnecessary.

systematic sampling Selecting items from the sampling frame following a regular pattern.

Identifying the Sampling Frame

The hard part of getting an SRS is to obtain the sampling frame that lists every member of the population of interest. The list that you have often differs from the list that you want. Election polls again provide an intuitive (and important) example. The population for a poll consists of people who *will* vote in the coming election. The typical sampling frame, however, lists registered voters from public records. The sampling frame identifies people who *can* vote, not those who will. Those who actually vote seldom form a random subset of registered voters.

Hypothetical populations complicate identifying the sampling frame. The population of voters is real; we can imagine a list with the name of everyone who will vote. Some populations are less tangible. Consider a biotech company that has developed a new type of fruit, a disease-resistant orange that possesses a higher concentration of vitamin C. Horticulturists grew 300 of these hybrid oranges by grafting buds onto existing trees. Scientists measured their weight and nutritional content. Are these 300 oranges the population or a sample? After all, these are the only 300 oranges of this variety ever grown.

Venture capitalists who invest in the company do not like the answer that these 300 oranges are the population. If these 300 oranges are the population, scientists cannot infer anything about oranges grown from this hybrid in the future. Without some claims about more oranges than these, it's going to be difficult to justify any claims for the advantages of this hybrid. Such claims force us to think of these oranges as a sample, even though there's no list or sampling frame in the usual sense. If these 300 oranges offer a snapshot of the population of all possible oranges that might be grown, then these 300 do indeed form a sample. Of course, the sample is representative only if later growers raise their crops as carefully as the horticulturists who grew these.

Similar concerns arise when sampling from a manufacturing or sales operation. Data that monitor production, sales, orders, and other business activities are often most naturally thought of as sampling a process. There's no fixed population of outcomes, no simple sampling frame. Rather, we have to justify that our sampling procedure produces data that are representative of the process at large. For example, consider sampling the production of a factory that produces LCD panels used in HD television screens. The panels are cut from larger sheets. For example, a factory might divide a large sheet into nine panels like this.

If inspectors always sample the panel at the lower right corner, then the sampling would not be representative and would miss problems that occur elsewhere in the sheet. It would be better to choose the panel from a sheet randomly rather than inspecting one position.

What Do You Think?

In each case, answer yes or no and think about how you would explain your answer to someone.

a. If asked to take blood from a sample of 20 cattle from a large herd at a ranch, do you think a cowhand would be inclined to take a simple random sample?[2]

b. When marketers collect opinions from shoppers who are willing to stop and fill out a form in a market, do you think that they get a simple random sample of shoppers?[3]

13.2 | VARIATION

The results obtained in a survey depend on which members of the population happen to be included. The percentage of consumers who say that they prefer Coke™ to Pepsi™, for instance, varies depending on the composition of the sample. If the results vary from one sample to the next, what can we conclude about the population?

Estimating Parameters

Supermarket chains worry about competition from Wal-Mart. Responding to their concerns, the Food Marketing Institute reported that 72% of shoppers say that a "supermarket is their primary food store." What does that mean? Does this statement mean that Wal-Mart isn't a threat to traditional grocers?

---

[2] Most likely, the cowhand would collect blood from the first 20 cows that he could get to rather than look for some that might be more aggressive or harder to find.

[3] No. The sample will be representative of customers who are willing to stop and fill out a form, but many shoppers are in a hurry and will not want to take the time to complete a form.

This statement evidently refers to the opinions of a sample. The Food Marketing Institute can't possibly know the proportion of *all* shoppers who visit supermarkets. Reality is too complex. These must be results from a sample, even though that detail is not mentioned in the headline. If the results refer to a sample, then what *does* the 72% mean? What should we infer about the population?

Most surveys report a mean and perhaps a standard deviation. Readers often interpret these as population characteristics, but they are not. **To make this distinction clear, statistics uses different names and symbols to distinguish characteristics of the population from those of a sample.** Characteristics of the population, such as its mean and variance, are **population parameters**. Characteristics of a sample are known as **sample statistics**. Sample statistics are used to **estimate** the corresponding parameters of the population. For instance, we might use a sample mean to estimate the population mean.

Statistics differ from parameters, and statistical notation distinguishes one from the other. Traditionally, Greek letters denote population parameters, as in the notation for random variables. Random variables and populations are similar with regard to sampling. A random variable stands for an idealized distribution of possible outcomes, and a population collects all possible items that we might see in a sample. **We often use a random variable to represent a random choice from the population.** Often, the letter that stands for a statistic corresponds to the parameter in an obvious way. For instance, the standard deviation of the data is s, and the population standard deviation is σ (sigma, Greek for s). The letter r denotes the correlation in data, and the correlation in the population is ρ (rho). For the slope of the line associated with the correlation, b is the statistic whereas β (beta) identifies the parameter of the population.

Alas, the pattern is irregular. The mean of a population is μ (because μ is the Greek letter for m). Rather than use m for the sample mean, longstanding convention puts a bar over anything when we average it, so we write $\bar{x}$ or $\bar{y}$ for sample means. Proportions are also irregular. In this book, p denotes the population proportion whereas $\hat{p}$ is the sample proportion. In the study of grocery marketing, the Food Marketing Institute claims that $\hat{p} = 0.72$. The issue becomes whether $\hat{p}$ and p are close to each other.

Table 13.2 summarizes the correspondence among several sample statistics and population parameters.

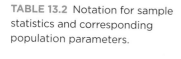

tip

population parameter A characteristic, usually unknown, of the population.

sample statistic An observed characteristic of a sample.

estimate The use of a sample characteristic (statistic) to guess or approximate a population characteristic (parameter).

tip

TABLE 13.2 Notation for sample statistics and corresponding population parameters.

Name	Statistic	Parameter
Mean	$\bar{y}$	μ (mu, pronounced "mew")
Standard deviation	s	σ (sigma)
Correlation	r	ρ (rho, pronounced "row")
Slope of line	b	β (beta)
Proportion	$\hat{p}$	p

Sampling Variation

There's one population, and it has a fixed set of parameters. There's a population mean μ and a population variance σ^2. These are two numbers, and if our data table listed the entire population, we could calculate these values. Each sample, however, has its own characteristics. Although two SRSs are picked from the same population, the mean of one is not likely to match the mean of the second.

As an example, consider a database of several million purchases made by holders of a major credit card. Each transaction records the total amount of the purchase, such as a charge of $95.54 for clothing. The average value of a purchase in this population is approximately $78. Let's consider what happens if we sample 100 transactions from this population and find the average amount. The histogram in Figure 13.1 shows the averages from many such samples, all from this one population. (We took 10,000 samples.) For each sample of 100 transactions, we calculated its average and added that value to the histogram.

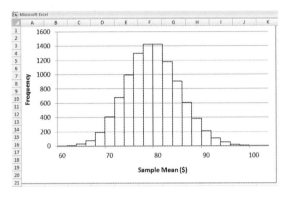

FIGURE 13.1 Each sample has its own mean, as shown in this histogram of sample averages.

The averages of these samples cluster around the population mean $\mu = 78 (the red line in the histogram), but some averages are considerably different from μ. At the far left, the average amount in one random sample is about $60, whereas averages in other samples reach toward $100 at the far right. A manager who fails to appreciate the differences among samples might not appreciate the size of credit card transactions if she happened to get the sample with the small average. Using samples in decision making requires an understanding of the variation among samples shown in Figure 13.1.

This variability in the value of a statistic from sample to sample is known as **sampling variation**. To emphasize this variation, the symbol $\overline{X}$ (written with a capital letter) stands for the collection of all possible mean values defined by the different possible samples. That is, $\overline{X}$ is a random variable that represents the mean of a randomly chosen sample. Once we take our sample, we can calculate the mean $\overline{x}$ of this specific sample. The symbol $\overline{x}$ (lowercase) stands for the mean of the observed sample, the number we calculate from the observed data. A major task in the following chapters is to describe the distribution of $\overline{X}$ without knowing the properties of the population.

sampling variation Variation of a statistic from one sample to the next caused by selecting random subsets of the population.

Sampling variation occurs for all types of statistics, not just the average. For example, the sample proportion varies from one sample to another as well. As an illustration, again from politics, Figure 13.2 tracks approval polls for President George W. Bush from 2001 to 2006. Each point in the figure is the proportion in a sample who said that they approved of President Bush.

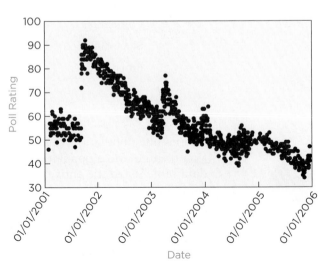

FIGURE 13.2 Different polls give different results.

Not only does this figure suggest that public opinion changes over time, but this figure also shows that the survey results depend upon the sample. Polls taken at the same time don't give the same ratings. There's visible sampling variation in the wide range of results that occur at any point in time.

Sampling variation is the price we pay for working with a sample rather than the population. The sample reveals its average, and as these figures show, that average won't necessarily be close to the population mean. Fortunately, statistics allows us to quantify the effects of sampling variation and reach conclusions about the population if the sample is a random sample.

4M EXAMPLE 13.1	**EXIT SURVEYS**

MOTIVATION

STATE THE QUESTION

Businesses use a variety of methods to keep up with regular customers, such as registration cards included with electronics purchases or loyalty programs at supermarkets. These tell the business about the customers who buy things, but omit those who don't.

Consider the questions of a clothing retailer located in a busy mall. The owner has data about the customers who frequently make purchases, but she knows nothing about those who leave without making a purchase. Do they leave because they did not find what they were looking for, because the sizes or colors were wrong, or because the prices were too high?

 Every survey should have a clear objective. A precise way to state the objective is to identify the population and the parameter of interest. In this case, the objective is to learn the percentage of customers who left without making a purchase for each of the reasons listed above.

METHOD

DESCRIBE THE DATA AND SELECT AN APPROACH

A survey is necessary. It is not possible to speak with everyone who leaves the store or have all of them answer even a few questions. On the basis of the volume of business, the owner wants to get a sample of about 50 weekend customers.

 The hard part of designing a survey is to describe the population and how to sample it. Even if you don't have a list, identify the ideal sampling frame. Then you can compare the actual survey to this ideal. The ideal sampling frame in this example would list every shopper over the weekend who did not make a purchase. That list does not exist, but the store can try to sample shoppers as if it did. Someone will have to try to interview shoppers who do not make a purchase as they leave.

MECHANICS

DO THE ANALYSIS

The store is open from 9 A.M. to 9 P.M., with about 25 shoppers walking through per hour (300 per day). Most don't make a purchase. If a surveyor interviews every 10th departing shopper on both Saturday and Sunday, then the sample size will be $n = 60$. If a shopper refuses to participate, the surveyor will ask the next customer and make a note. To be reliable, a survey needs a record of nonresponses. If there are too many of these, the sample may not be representative.

SUMMARIZE THE RESULTS

On the basis of the survey, the owner will be able to find out why shoppers are leaving without buying. Perhaps she's not changing inventory fast enough or has stocked the wrong sizes.

13.3 | ALTERNATIVE SAMPLING METHODS

Stratified and Cluster Samples

Simple random samples are easy to use. In later chapters we assume that our data are an SRS. Most commercial polls and surveys like those you see on the news, however, are more complex. There's a reason for the complexity: to save time and money. Though more complex, all sampling designs share the idea that random chance, rather than personal preference, determines which members of the sampling frame appear in the sample.

Surveys that sample large populations are typically more complicated than simple random samples. A **stratified random sample** divides the sampling frame into homogeneous groups, called **strata**, before the sample is selected. Simple random sampling is used to pick items for the sample within each stratum.

Suppose the manager of a large hotel wants to survey customers' opinions about the quality of service. Most customers (90%) are tourists. The remaining 10% travel for business. The manager suspects that the two groups of guests have different views on service at the hotel. He would like a survey that reveals properties of both groups. If he selects 100 guests at random, he might get a sample with 95 tourists and only 5 business travelers. He's not going to learn much about business travelers from 5 responses. To learn more about business travelers, he can instead divide the population of customers into two strata, tourists and business travelers, and sample within each. For example, he could sample, say, 75 tourists and 25 business travelers. Now he'll learn something about both groups. The catch is that this procedure deliberately overrepresents business travelers. The manager will have to adjust for that if he wants to describe the population of all customers. (Statistics packages adjust for stratifying by introducing sampling weights into the calculations.)

Cluster sampling is a type of stratified sampling that is natural in situations that cover a wide geographic area. In cluster sampling, we first select a simple random sample from a list of geographic units (clusters), such as census tracts that have comparable population. Given the selection of the clusters, we randomly sample within each cluster. These geographic clusters form the strata. For instance, a national survey of homeowners on remodeling would be prohibitively expensive unless the surveyor can visit several homes in each locale.

stratified random sample A sample derived from random sampling within subsets of similar items that are known as strata.

strata Subsets of similar items in the population.

cluster sampling A type of stratified sampling in which the strata are determined geographically.

4M EXAMPLE 13.2　ESTIMATING THE RISE OF PRICES

STATE THE QUESTION

Businesses, consumers, and the government are all concerned about inflation, the rise in costs for goods and services. Inflation pressures businesses to increase salaries and prices, leads consumers to cut back on what they purchase, and compels governments to pay more for entitlement programs such as Social Security. But what is the level of inflation? No one knows the price paid in every consumer transaction. Let's consider how the Bureau of Labor Statistics (BLS) estimates inflation. What goes into the consumer price index (CPI), the official measure of inflation in the United States?

METHOD

DESCRIBE THE DATA AND SELECT AN APPROACH

The BLS uses a survey to estimate inflation. The survey is done monthly. We'll focus on the urban consumer price index that measures inflation in 38 urban areas. The target population consists of the costs of *every* consumer transaction in these urban areas during a specific month.

MECHANICS

DO THE ANALYSIS

The BLS has the list of the urban areas and a list of people who live in each area, but it does not have a list of every sales transaction. To get a handle on the vast scope of the economy, the BLS divides the items sold into 211 categories and estimates the change in price for each category in every location ($211 \times 38 = 8{,}018$ price indices). To compute each index, the BLS sends data collectors into the field, where they price a sample of items in selected stores in every location. Because the sample only includes transactions from some stores and not others, this is a clustered sample. Also, the definition of 211 types of transactions adds another type of clustering to the survey. The choice of items and stores changes over time to reflect changes in the economy. Personal computers are now included, but were not a major category until recently. Once current prices have been collected, these are compared to prices measured last month, category by category. (To learn more about the CPI, visit the BLS online and check out Chapter 17 in the *Handbook of Methods* from the BLS Web site.)

MESSAGE

SUMMARIZE THE RESULTS

The urban consumer price index is an estimate of inflation based on a complex, clustered sample in selected metropolitan areas. If you live outside the covered areas or perhaps spend your money differently from the items covered in the survey, your impression of inflation may be considerably different from the CPI.

Census

Every survey must balance accuracy versus cost. The cost goes up with every additional respondent. Ultimately, the choice of the sample size n depends on what you want to learn. Larger surveys reveal more, but you have to weigh those gains against rising costs.

Consider a survey of customers at a newly renovated retail store. Management would like to know how customers react to the new design in the various departments. The sample must be large enough for the survey to include customers who visit each portion of the store. If the sample does not include shoppers who visit the hardware department, the survey will not have much to say about reactions to this department, other than indicate that customers perhaps could not find this area!

Wouldn't it be better to include everyone? A comprehensive survey of the entire population is called a **census**. Though a census gives a definitive answer, these are rare in practice. Cost is the overriding concern, but other complications arise: practicality and change over time.

census A comprehensive survey of the entire population.

Another complication is the difficulty of listing and contacting the entire population. For instance, the U.S. Census undercounts the homeless and illegal immigrants; they're hard to find and reluctant to fill out official forms. Some individuals don't want to be found! On the other hand, the U.S. Census ends up with too many college students. Many are included by their families at home and then counted a second time at school.

For manufacturing, a comprehensive census that includes every item is impractical. If you were a taste tester for the Hostess Company, you probably wouldn't want to taste *every* Twinkie™ on the production line. Aside from the fact that you couldn't eat every one, Hostess wouldn't have any left to sell. This is a common attribute of testing manufacturing processes. Many testing procedures are destructive; they test an item by seeing what it takes to break it.

The time that it takes to complete a census also can defeat the purpose. In the time that it takes to finish a census, the process or population may have changed. We can see that happening in the approval polls shown in Figure 13.2; events shift opinions. When the population is changing (and most do), it makes more sense to collect a sequence of smaller samples in order to detect changes in the population.

Voluntary Response

voluntary response sample
A sample consisting of individuals who volunteer when given the opportunity to participate in a survey.

Samples can be just as flawed as a poorly done census. One of the least reliable samples is a **voluntary response sample**. In a voluntary response sample, a group of individuals is invited to respond, and those who do respond are counted. Though flawed, you see voluntary response samples all the time: call-in polls for the local news, 800 numbers, and Internet polls.

Voluntary response samples are usually biased toward those with strong opinions. Which survey would *you* respond to—one that asked,

> "Should the minimum age to drive a car be raised to 30?" or one that asked, "Should women's and men's shoe sizes correspond?"

Experience suggests that people with negative opinions tend to respond more often than those with equally strong positive opinions. How often do customers write to managers when they're happy with the product that they just bought? Even though every individual has the chance to respond, these samples are not representative. The resulting *voluntary response bias* invalidates the survey.

Convenience Samples

convenience sampling
A sampling method that selects individuals who are readily available.

Another sampling method that usually fails is convenience sampling. **Convenience sampling** surveys individuals who are readily available. Though easy to contact, these individuals may not be representative. A farmer asked to pick out a sample of cows to check the health of the herd isn't likely to choose animals that seem unruly or energetically run away. Surveys conducted at shopping malls suffer from this problem. Interviewers tend to select individuals who look easy to interview.

In spite of the problems, convenience sampling is widespread. When a company wants to learn reactions to its products or services, whom does it survey? The easiest people to sample are its current customers. After all, the company has a list of them with addresses and phone numbers, at least those who sent in registration cards. No matter how it selects a sample from these customers, the sample remains a convenience sample. Unless the company reaches beyond this convenient list, it will never learn how the rest of the market feels about its products.

What Do You Think?   What problems do you foresee if Ford uses the following methods to sample customers who purchased an Expedition (a large SUV) during the last model year?[4]

    a. Have each dealer send in a list with the names of 5% of its customers who bought an Expedition.
    b. Start at the top of the list of all purchasers of Ford vehicles and stop after finding 200 who bought an Expedition.
    c. Randomly choose 200 customers from those who voluntarily mailed in customer registration forms.

13.4 | CHECKLIST FOR SURVEYS

Unless surveys are done correctly, the sample data will be flawed. If you start with a biased sample, it won't matter how well you do the subsequent analysis. Your conclusions are questionable.

Most businesses don't conduct their own surveys. They rely on data that someone else collects. To get the most out of surveys, whether you analyze the data yourself or read someone else's analysis, make sure that you can answer a few questions about the origins of the sample. Most summaries of surveys omit these issues, so you might have to ask for more details. We've talked about the first two:

- *What was the sampling frame? Does it match the population?*
- *Is the sample a simple random sample? If not, what is the sampling design?*

Though fundamental, you often won't find answers to these questions until you ask. In addition, there are a few more questions that we have not yet discussed but are worth answering before you rely on the results of a sample.

- *What is the rate of nonresponse?* The design of an SRS includes a list of individuals from the sampling frame that the survey *intends* to contact. You can be sure that some individuals are less willing to participate than others.

caution The problem with nonresponse is that those who don't respond may differ from those who do. It's usually impossible to tell what the nonrespondents would have said had they participated. Rather than sending out a large number of surveys for which the response rate will be low, it is usually better to select a smaller sample for which you have the resources to ensure a high response rate. Response rates of surveys vary, but seldom reach 100%. If the response rate is low, say 20%, you've got to wonder if the people who participated resemble those who declined. An SRS with a low response rate resembles a voluntary response sample rather than a randomly chosen subset of the population.

- *How was the question worded?* The wording of questions can have a dramatic effect on the nature of the responses. Asking a question with a leading statement is a good way to bias the response. Many surveys, especially those conducted by special interest groups, present one side of an issue before the question itself. For example, the poll in Figure 13.3 contacted 1,229 adults and asked about telephone wiretapping. Take a close look at the third and fourth questions.[5]

---

[4] Each method has a problem. Dealers might choose customers that are sure to give them a high rating, and large dealers will be sending in more names than small dealers. The first 200 on the list may be eager buyers and different from those who buy later in the model year. Those who return supplemental information *voluntarily* may have more strongly held opinions.
[5] *New York Times*, January 27, 2006.

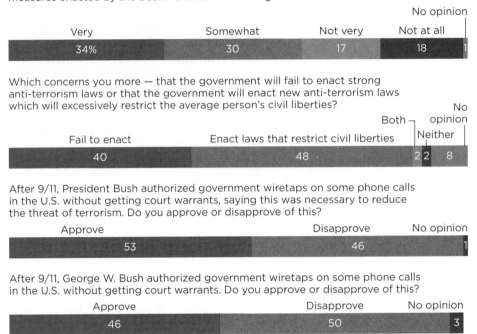

Public Sentiment on Eavesdropping

How concerned are you about losing some of your civil liberties as a result of the measures enacted by the Bush Administration to fight terrorism?

Very	Somewhat	Not very	Not at all	No opinion
34%	30	17	18	1

Which concerns you more — that the government will fail to enact strong anti-terrorism laws or that the government will enact new anti-terrorism laws which will excessively restrict the average person's civil liberties?

Fail to enact	Enact laws that restrict civil liberties	Both	Neither	No opinion
40	48	2	2	8

After 9/11, President Bush authorized government wiretaps on some phone calls in the U.S. without getting court warrants, saying this was necessary to reduce the threat of terrorism. Do you approve or disapprove of this?

Approve	Disapprove	No opinion
53	46	1

After 9/11, George W. Bush authorized government wiretaps on some phone calls in the U.S. without getting court warrants. Do you approve or disapprove of this?

Approve	Disapprove	No opinion
46	50	3

FIGURE 13.3 Changing the wording of questions affects the results in this opinion poll.

The wording of the third question refers to President Bush and mentions the threat of terrorism. Responding to this wording, a majority approves the use of wiretaps without warrants. In the fourth question (the questions were not presented one right after another in the survey), the wording refers to George W. Bush rather than President Bush and omits the link to terrorism. With this wording, fewer than half approve.

Placement of items in a survey affects responses as well. Let's consider one of the most important surveys around: an election. Studies have found that candidates whose names are at the top of the ballot get about 2% more votes on average than they would have had their name been positioned elsewhere on the ballot. It's a lot like putting items on the grocery shelf at eye level; like shoppers, undecided voters gravitate to the first things that get their attention. Two percent is small, but in a tight election, that could swing the outcome.[6]

- *Did the interviewer affect the results?* The interviewer has less effect in automated polls, but many detailed surveys are done in person. If there's chemistry—good or bad—between the interviewer and the respondent, the answer may reflect more about their interaction than the response that you're trying to measure. In general, respondents tend to answer questions in a way that they believe will please the interviewer, either consciously or unconsciously. The sex, race, attire, or behavior of the interviewer can influence responses by providing subtle indications that certain answers are more desirable than others.

- *Does survivor bias affect the survey?* **Survivor bias** occurs when certain long-lived items are more likely to be selected for a sample than others. The name "survivor bias" comes from medical studies. Physicians noticed that studies of certain therapies looked too good because the subjects were

survivor bias Bias in a sample that arises from selecting items that are present in the sampling frame for a longer period of time.

[6] "In the Voting Booth, Bias Starts at the Top," *New York Times*, November 11, 2006.

never those who had severe cases. Those patients did not survive long enough to enter a study.

Survivor bias is easiest to explain in an example that illustrates how common this problem is in business. Many investors have money in hedge funds. To learn about the fees charged by these funds, an analyst developed a list of hedge funds and sampled 50 of these for detailed analysis. Even though the analyst used an accurate sampling frame (he had a list of all of the active hedge funds) and took a random sample from the list, his results suffer from survivor bias. His results overrepresent successful funds, those that survive longer. Because they are largely unregulated, hedge funds choose whether to report their performance. You might suspect that some report only when they've done well and conceal their results otherwise. Those that have collapsed don't report anything.

caution Survivor bias is also a problem with analyses of the stock market. A random sample of companies that are currently listed on the major stock exchanges is more likely to contain successful companies than those which have failed or lost substantial value.

Best Practices

- *Randomize.* Avoid complicated sampling plans and devote your energy to asking the right questions. Spreadsheets make it easy to obtain a random sample from the sampling frame.
- *Plan carefully.* Once bias creeps into a survey, the results become unreliable. Not much can be done to patch up a botched survey, so spend the time up front to design a good survey. A smaller, careful survey is more informative than a larger, biased analysis.
- *Match the sampling frame to the target population.* A survey begins with the sampling frame. The sampling frame defines the population for a survey. If the sampling frame fails to match the population, your sample will not reflect the population that you intended.
- *Keep focused.* Surveys that are too long are more likely to be refused, reducing the response rate and biasing the results. When designing a survey, remember the purpose. For each question you include, ask yourself, "What would I do if I knew the answer to this question?" If you don't have a use for the answer, then don't ask the question.
- *Reduce the amount of nonresponse.* Nonresponse converts a simple random sample into a voluntary response or convenience sample. Those who decline to participate or make themselves hard to reach often differ in many ways from those who are cooperative and easy to find. Some telephone polls call a respondent a second time in case the first call arrived at an inconvenient moment. Others offer a slight reward for participating. For example, the retailer in Example 13.1 might offer departing shoppers a beverage or a free sample of perfume in order to get them to answer a question or two. Keep track of the level of nonresponse so that you can answer questions about participation. Don't hide this aspect of the survey.
- *Pretest your survey.* If at all possible, test the survey in the form that you intend to use it with a small sample drawn from the population (not just friends of yours). Look for misunderstandings, confusing questions, or other sources of bias. The pretest might also suggest why some may refuse to participate. If you fix these problems in advance, the response rate in the survey will be higher. Redesign your survey as needed.

Pitfalls

- *Don't conceal flaws in your sample.* Be honest about the origins of your data. If the sample has problems, such as lack of randomization or voluntary response bias, be up front about these concerns. Let your audience decide how much faith they can attach to the results.
- *Do not lead the witness.* Make sure that the questions in a survey do not influence the response. You want to learn from the sample, not influence the sample. The interaction between interviewer and respondent can also influence the answers.
- *Do not confuse a sample statistic for the population parameter.* The mean of the sample is not the mean of the population. Had we drawn a different sample, we'd have observed a different sample mean. Later chapters consider how close the sample statistic is likely to be to the population parameter.

- *Do not accept results just because they agree with what you expect.* It's easy to question the results of a survey when you discover something that you didn't expect. That's okay, but remember to question the survey methods when the survey results agree with your preconceptions as well.

  For example, a survey done to measure the success of promoting pharmaceutical drugs found that promotion had no effect on the habits of doctors. Stunned, managers at the marketing group dug into the survey and found that software had distorted the linkage between promotion and sales. Doctors were not being accurately tracked. Do you think that the managers would have taken this effort if they had found the results they expected?

Software Hints

EXCEL

The procedure described in the text for constructing a simple random sample is simple to use in Excel. Use the Excel function RAND to generate uniformly distributed random values between 0 and 1. Follow the menu commands Formulas > Insert > Function > Rand. Then sort the rows of the spreadsheet in the order of the generated random numbers.

MINITAB

To use the procedure described in the text to obtain an SRS, follow the sequence of menu commands

> Calc > Random Data > Uniform...

These commands open a dialog that allows you to fill a column of the data table with random numbers. Let's say you put the random numbers in column C_4 and would like to sample values from C_1, C_2, and C_3. Once you have the random numbers, use the command

> SORT   C1 C2 C3   C11 C12 C13   BY C4

to sort the data columns. This command puts the sorted values from columns C_1, C_2, and C_3 into columns C_{11}, C_{12}, and C_{13}. If that seems confusing, then use the command

> Calc > Random Data > Sample from columns...

In the resulting dialog, indicate the columns you'd like to randomly sample and how many rows (cases) you'd like in the sample.

JMP

To use the procedure described in the text to obtain an SRS, use the JMP calculator to insert a column of random numbers into the data table. Right-click on the header of an empty column, pick Formula ..., and insert one of the items in the Random collection of functions. Then follow the menu commands

> Tables > Sort

to use this column to sort the data table. The sorted data appear in a new data table.

Alternatively, you can also use the built-in sampling feature to obtain a random subset of rows of the data table. Follow the menu sequence

> Tables > Subset

In the resulting dialog, indicate the columns you'd like to randomly sample and how many rows you'd like in the sample. You can also indicate a percentage. JMP builds a new data table with a random sample as you've requested.

CHAPTER SUMMARY

A **sample** is a representative subset of a larger **population**. Samples provide **sample statistics** that allow us to estimate **population parameters**. To avoid **bias**, **randomization** is used to select items for the sample from the **sampling frame**, a list of items in the target population. A **simple random sample (SRS)** is chosen in such a way that all possible samples of size n are equally likely. Other types of samples, such as **cluster samples** or **stratified samples**, use special designs to reduce costs without introducing bias. **Sampling variation** occurs because of the differences among randomly selected subsets. A **census** attempts to enumerate every item in the population rather than a subset. **Voluntary response samples** and **convenience samples** are likely to be unrepresentative, and nonresponse and **survivor bias** effects can introduce further biases.

■ Key Terms

bias, 305
census, 315
cluster sampling, 314
convenience sampling, 316
estimate, 311
population, 305
population parameter, 311
randomization, 306
representative, 305

sample, 305
sample statistic, 311
sampling frame, 309
sampling variation, 312
simple random sample
  (SRS), 308
strata, 314
stratified random sample, 314
survey, 305

survivor bias, 318
symbols
  n, 306
  N, 306
  $\bar{X}$, 312
  $\bar{x}$, 312
systematic sampling, 309
voluntary response bias, 316
voluntary response sample, 316

■ About the Data

The approval ratings from polls shown in Figure 13.2 come from Steve Ruggles at the University of Minnesota.

EXERCISES

Mix and Match

Match the concept to the correct description.

1. Sample	(a)	A complete collection of items desired to be studied
2. Census	(b)	A list of items in the population
3. Target population	(c)	A subset of a larger collection of items
4. Statistic	(d)	A sample within homogeneous subsets of the population
5. Parameter	(e)	A characteristic of a sample
6. Sampling frame	(f)	The result if a sampling method distorts a property of the population
7. Simple random sample	(g)	A comprehensive study of every item
8. Stratified sample	(h)	The result if a respondent chooses not to answer questions
9. Bias	(i)	A characteristic of a population
10. Nonresponse	(j)	Sample chosen so that all subsets of size n are equally likely

True/False

Mark each statement True of False. If you believe that a statement is false, briefly explain why you think it is false.

11. Every member of the population is equally likely to be in a simple random sample.

12. The size of the sample for a survey should be a fixed percentage of the population size in order to produce representative results.

13. Bias due to the wording of questions causes different samples to present different impressions of the population.

14. A census offers the most accurate accounting of the characteristics of the target population.

15. The sampling frame is a list of every person who appears in a sample, including those who did not respond to questions.

16. Randomization produces samples that mimic the various characteristics of the population without systematic bias.

17. Voluntary response samples occur when the respondents are not paid for their participation in a survey.

18. Sampling variation occurs when respondents change their answers to questions presented during an interview or change their answers when offered a repeated question.

19. The wording of questions has been shown to have no influence on the responses in surveys.

20. Larger surveys convey a more accurate impression of the population than smaller surveys.

Think About It

Exercises 21–26. List the characteristics of (a)–(f) on the basis of the brief description of the survey that is shown.
   (a) Population
   (b) Parameter of interest
   (c) Sampling frame
   (d) Sample size
   (e) Sampling design, including whether randomization was employed
   (f) Any potential sources of bias or other problems with the survey or sample

21. A business magazine mailed a questionnaire to the human resource directors of all of the *Fortune* 500 companies, and received responses from 23% of them. Among those who responded, 40% reported that they did not find that such surveys intruded significantly on their workday.

22. *PC Magazine* asks all of its readers to participate in a survey of their satisfaction with different brands of computer systems and peripherals. In the 2004 survey, more than 9,000 readers rated the products on a scale from 1 to 10. The magazine reported that the average rating assigned by 225 readers to a Kodak compact digital camera was 7.5.

23. A company packaging snack foods maintains quality control by randomly selecting 10 cases from each day's production. Each case contains 50 bags. An inspector selects and weighs two bags.

24. Inspectors from the food-safety division of the Department of Agriculture visit dairy farms unannounced and take samples of the milk to test for contamination. If the samples are found to contain dirt, antibiotics unsuited for human consumption, or other foreign matter, the milk will be destroyed.

25. A vendor opens a small booth at a supermarket to offer customers a taste of a new beverage. The staff at the booth offer the beverage to adults who pass though the soda aisle of the store near a display of the product. Customers who react favorably receive a discount coupon toward future purchases.

26. The information that comes with a flat-screen television includes a registration card to be returned to the manufacturer. Among questions that identify the purchaser, the registration card asks him or her to identify the cable provider or other source of television programming.

27. A bank with branches in a large metropolitan area is considering opening its offices on Saturday, but it is uncertain whether customers will prefer (1) having walk-in hours on Saturday or (2) having extended branch hours during the week. Listed below are some of the ideas proposed for gathering data. For each, indicate what kind of sampling strategy is involved and what (if any) biases might result.
   (a) Put a big ad in the newspaper asking people to log their opinions on the bank's Web site.
   (b) Randomly select one of the branches and contact every customer at that bank by phone.
   (c) Send a survey to every customer's home, and ask the customers to fill it out and return it.
   (d) Randomly select 20 customers from each branch. Send each a survey, and follow up with a phone call if he or she does not return the survey within a week.

28. Four new sampling strategies have been proposed to help the bank in Exercise 27 determine whether customers favor opening on Saturdays versus keeping branches open longer during the week. For each, indicate what kind of sampling strategy is involved and what (if any) biases might result.
   (a) Sponsor a commercial during a TV program, asking people to dial one of two phone numbers to indicate which option they prefer.
   (b) Hold a meeting at each branch and tally the opinions expressed by those who attend the meetings.
   (c) Randomly select one day at each branch and contact every customer who visits the branch that day.
   (d) Go through the bank's customer records, selecting every 100th customer. Hire a survey research company to interview the people chosen.

29. Two members of a bank's research group have proposed different questions to ask in seeking customers' opinions.

    Question 1: Should United Banks take employees away from their families to open on Saturdays for a few hours?

    Question 2: Should United Banks offer its customers the flexibility of convenient banking over the weekend?

    Do you think responses to these two questions might differ? How?

30. An employer replaced its former paycheck system with a paperless system that directly deposits payments into employee checking or savings accounts. To verify that proper deductions have been made, employees can check their pay stubs online. To investigate whether employees prefer the new system, the employer distributed an email questionnaire that asked

    Do you think that the former payroll system is not inferior to the new paperless system?

    Do you think that this question could be improved? How?

31. Between quarterly audits, a company checks its accounting procedures to detect problems before they become serious. The accounting staff processes payments on about 120 orders each day. The next day a supervisor checks 10 of the transactions to be sure they were processed properly.
    (a) Propose a sampling strategy for the supervisor.
    (b) How would you modify the sampling strategy if the company makes both wholesale and retail sales that require different bookkeeping procedures?

32. A car manufacturer is concerned that dealers conceal unhappy customers by keeping them out of surveys conducted by the manufacturer. The manufacturer suspects that certain dealers enter incorrect addresses for dissatisfied customers so that they do not receive the satisfaction survey that is mailed by the manufacturer. If a survey of 65 current customers at a dealership indicates that 55% rate its service as exceptionally good, can the manufacturer estimate the proportion of all customers at this dealership who feel that its service is exceptionally good? Can it estimate the proportion at other dealerships?

4M Guest Satisfaction

Companies in the competitive hotel industry need to keep up with the tastes of their visitors. If the rooms seem dated and the service is slow, travelers will choose a different destination for their next visit. One means of monitoring customer satisfaction is to put cards in guest rooms, asking for feedback. These are seldom returned, and when they are returned, it is usually because of some unusual event.

To get a more representative sample, a large hotel frequently used by business travelers decided to conduct a survey. It contacted every guest who stayed in the hotel on a randomly chosen weekday during the previous two months

(June and July 2005). On the date chosen (Tuesday, July 19, 2005), the hotel had 437 guests. With several follow-up calls (unless the customer asked to be excluded), the hotel achieved an 85% response rate. The hotel could have mailed questionnaires to guests much more cheaply, at about one-tenth of the cost of the telephone calling.

The key question on the survey was, "Do you plan to stay with us on your next visit to our area?" In the survey, 78% of guests responded Yes to this question.

Motivation

a) What can the company hope to learn from this survey that it could not get from the cards left in guest rooms?

Method

b) The response rate from mailed questionnaires is typically less than 25%. Would it make sense to mail out survey forms to more customers rather than go to the expense of calling?

c) Carefully describe what each observation in this survey represents. What is the population?

d) Does the procedure yield a random sample of guests at the hotel, or does this design make the survey more likely to include some guests than others?

Mechanics

e) In calling the customers, one of the interviewers was a man and one was a woman. Might this difference produce a difference in answers?

f) Some customers were called repeatedly in order to improve the response rate. If the repeated calling annoyed any of these customers, how do you think this might bias the results?

Message

g) In describing the results of the survey, what points about the randomization should be made in order to justify the reliability of the results?

h) How should the hotel deal with the 15% of customers from that day who either did not reply or asked to be excused from the survey and not called again?

4M Tax Audits

The Internal Revenue Service collects personal income and corporate taxes. Most of the job involves processing the paperwork that accompanies taxes. The IRS verifies, for example, the reported Social Security numbers and compares reported earnings from W-2 and 1099 forms to the earnings reported by each taxpayer. Forms with errors or discrepancies lead to follow-up letters and, in some cases, the dreaded IRS audit. The IRS has limited personnel and cannot audit every taxpayer's return. Sampling is a necessity. The IRS temporarily stopped doing audits for a few years, but resumed auditing random samples of tax returns in 2007.[7]

---

[7] "The Next Audit Scare," *Wall Street Journal*, June 13, 2007.

Motivation

a) Why is it important for the IRS to audit a sample of all returns, not just those flagged as having an anomaly?

Method

b) For certain types of audits, an agent visits the residence of the taxpayer. What sort of sampling method is well suited to audits that require a personal visit?

c) If the IRS selects a sample of income tax forms for inspection at random, most will be from individuals with earnings below $100,000. According to the U.S. Census, about 85% of households earn less than $100,000 and half earn less than $50,000. If the IRS would like to have 30% of the audited tax returns cover households that earn more than $100,000 (with the rest chosen at random), how should it choose the sample?

Mechanics

d) If the IRS selects 1,000 personal income tax returns at random from among all submitted for the calendar year 2007, will it obtain a random sample of taxpayers for this year? (*Note:* Married couples typically file a joint return.)

e) An auditor selected a random sample of returns from those submitted on April 15, 2008. Explain why these are not a representative sample of all of the returns for 2007.

Message

f) If the IRS is interested in finding and discouraging tax cheats, what sort of sampling methods should it advertise in press releases?

Sampling Variation and Quality

DELIVERY SERVICES LOVE PORTABLE GPS DEVICES THAT CAN LOCATE YOUR POSITION, ANYPLACE ON THE PLANET. Management can find shipments, estimate arrival times, and decide the best route for a rush package. National sales forces use GPS devices to track sales representatives and record visits to clients.

The heart of a GPS is a chip that talks to satellites. Not only can it figure out its location, the chip is durable. These chips can withstand rough conditions, like baking on the roof of a truck traveling over rough roads in West Texas. Manufacturers of GPS chips work hard to achieve these results. They cannot wait for chips to break in the field. Manufacturers must instead test chips at the factory. A highly accelerated life test (HALT) shakes, shocks, and roasts a chip. The tests get harsher until the tested chip fails or testing ends. Most chips pass early tests, which catch the most serious flaws. Later tests vibrate the chips more, raise the voltage, and boost the temperature. Table 14.1 gives a typical plan for the tests.

14.1 **SAMPLING DISTRIBUTION OF THE MEAN**

14.2 **CONTROL LIMITS**

14.3 **USING A CONTROL CHART**

14.4 **CONTROL CHARTS FOR VARIATION**

CHAPTER SUMMARY

A full HALT sequence takes all day, requires constant oversight, and ruins the chip. The manufacturer is willing to sacrifice a few chips in order to learn whether the process is functioning as designed. In the end, however, it's not the sample of test chips that managers worry about; what matters most to managers is the larger collection of chips that get shipped.

How should operators monitor these tests? This chapter describes one approach: control charts. Control charts determine whether a process is functioning as designed on the basis of properties of a sample of the production. The design of control charts must allow for sampling variation in order to balance two types of errors common to

TABLE 14.1 Design conditions for a HALT procedure.

Test No.	Vibration	Voltage	Temperature
1	Low	Normal	50°C
2	None	+15%	50°C
3	None	Normal	60°C
4	None	Normal	80°C
5	Medium	+25%	50°C
6	None	Normal	100°C
7	Low	+15%	100°C
8	None	Normal	110°C
9	None	+30%	50°C
10	Medium	Normal	60°C
11	None	+15%	80°C
12	Low	+25%	100°C
13	High	+35%	80°C
14	High	+45%	120°C
15	High	+60%	130°C

all statistical decisions. By understanding sampling variation, managers can maintain a delicate balance between one error, incorrectly stopping a functioning process, and another error, failing to detect a malfunctioning process.

14.1 | SAMPLING DISTRIBUTION OF THE MEAN

GPS chips are made in a manufacturing facility called a fab. Every 30 minutes, the fab processes another silicon wafer into a batch of chips. An engineer randomly samples a chip from each wafer to undergo HALT. If the chip fails on the first test, the engineer records a 1. If it fails the second, she records a 2, and so forth. If the chip endures all 15 tests, the engineer records a 16.

Variation among the chips complicates testing. It would be easy to monitor the process if every chip were to pass, say, *exactly* seven tests when the process is working correctly. We could recognize immediately when the process changes. If a single chip fails after 6 or 8 tests, then the process has changed. Real processes, however, are less consistent. Even when the process is functioning properly, there's variation among HALT scores. A monitoring system has to allow for this variation.

To distinguish a change in the process from random variation, we have to know several properties of the variation of the process. These properties tell us what to expect when the process is functioning normally. For the fab in this example, the engineers who designed the production line claim that this process yields chips that pass on average 7 stages of the HALT procedure in Table 14.1. Not every chip is going to pass exactly seven tests: Some will do better, others worse. The designers claim that the standard deviation of the HALT scores should be 4. If the standard deviation is much larger, very good chips will be mixed with poor chips.

The mean and standard deviation provided by the designers describe the manufacturing process, not a handful of chips. These are parameters of a *probability model* that describes the process, the hypothetical population of all chips. Let's distinguish these parameters of the population with Greek letters. On average, chips should pass $\mu = 7$ tests with standard deviation $\sigma = 4$ tests. Let's also add the *assumption* that the observed HALT scores are independent from one chip to another.

Engineers process 20 chips through HALT each day. If we average these 20 scores, what should we find if the process is running as designed? Because we're describing the outcome of an experiment that is subject to chance, we need a random variable. Let's write the average HALT score for the 20 chips measured any day as $\bar{X}$. The capital letter is to remind us that $\bar{X}$ is a random variable that stands for the distribution of the average HALT score over many days, not the average of a specific sample.

Benefits of Averaging

What can we say about the variation of the average HALT score $\bar{X}$ from day to day when the process is operating as designed? To get a feel for the problem, let's look at some data. HALT scores for individual chips have a lumpy, discrete distribution. For example, the histogram in Figure 14.1 shows the scores obtained from testing 400 chips. The first bar counts the number of chips that fail the first test, the second bar counts those that fail the second test, and so forth.

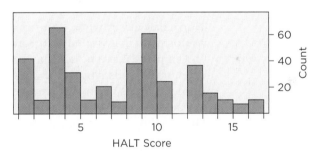

FIGURE 14.1 HALT scores from testing 400 chips.

In contrast, the histogram in Figure 14.2 summarizes the sample-to-sample variation among *average* HALT scores over 54 days. To make it easier to compare the histogram of the averages to the histogram of individual HALT scores, this figure uses the same scale for the *x*-axis as Figure 14.1.

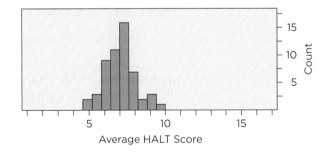

FIGURE 14.2 Histogram of the average HALT scores in 54 samples of 20 chips.

Comparison of these histograms shows that averaging leaves the center of the distribution near 7, but changes two other aspects of the distribution:

1. The sample-to-sample variance among average HALT scores is smaller than the variance among individual HALT scores.
2. The distribution of average HALT scores appears more bell shaped than the distribution of individual HALT scores.

The first effect of averaging seems reasonable: Averaging reduces variation. For instance, consider the variation among the weights of randomly chosen people. If we weigh one person chosen at random, we might weigh a child or adult. If we weigh 20 people, however, the average is unlikely to be so small or so large. If sampling at random, we're unlikely to choose 20 children or 20 football players. The group more likely mixes large and small people. The same thinking applies to the HALT results. It is not uncommon to find a chip that passes fewer than 5 tests or more than 11. It would be surprising, however, to get 20 consecutive chips that perform so poorly or so well. Typically, each day supplies a mix of chips; virtually every average in Figure 14.2 falls between 5 and 9. Consequently, means vary less from one another than the underlying data do. This decrease in variation allows us to design a method able to detect small changes in the underlying process.

The second effect of averaging, the appearance of a bell-shaped distribution, allows us to define what a surprising average is. In order to detect a change in the process, we need a way of distinguishing an unusual event, one that would signal that the manufacturing process has changed, from the random variation that occurs naturally when the process is running as designed. The bell-shaped variation in Figure 14.2 hints that we can use a normal model. Even though HALT scores of individual chips are not normally distributed, a normal model describes the variation among averages.

Normal Models

Normality arises when describing averages for two reasons. Data either begin as normally distributed or the averaging itself produces a normal distribution. In Chapter 12, we noted that sums of normally distributed random variables are normally distributed. If data are observations of normally distributed random variables, then the averages of such data are also normally distributed. We seldom know, however, that the shape of the population is normal; many statisticians believe that data are never *exactly* normally distributed. Instead, normality typically arises through the averaging process because of the Central Limit Theorem. The Central Limit Theorem (also discussed in Chapter 12) shows that the sampling distribution of averages is *approximately* normal even if the underlying population is not normally distributed. The approximation is accurate, provided the sample size is large enough for averaging to smooth away deviations from normality.

To decide whether a sample is large enough to apply the Central Limit Theorem, check the **sample size condition**. The condition requires two statistics from the sample: the skewness (K_3) and kurtosis (K_4) of the data.

Sample Size Condition   Unless it is known that the observed data are a sample from a normally distributed population, check this condition using the skewness K_3 and kurtosis K_4 (Chapter 12) from z-scores of the data.

$$K_3 = \frac{z_1^3 + z_2^3 + \cdots + z_n^3}{n} \qquad K_4 = \frac{z_1^4 + z_2^4 + \cdots + z_n^4}{n} - 3$$

A normal model provides an accurate approximation to the sampling distribution of $\overline{X}$ if the sample size n is larger than 10 times the squared skewness *and* larger than 10 times the absolute value of the kurtosis,

$$n > 10K_3^2 \quad \text{and} \quad n > 10|K_4|$$

In the analysis of control charts, it is important to use data from a period when the process is known to be operating as designed when calculating the skewness and kurtosis. Using data during a period when the chip-making process is operating correctly gives $K_3^2 \approx 0.1$ and $K_4 \approx -1$. For the HALT scores, $n = 20$ chips are tested each day. Because $n = 20$ is larger than both $10K_3^2 = 1$ and $10|K_4| = 10$, these data satisfy the sample size condition. We can use a normal model to describe the sampling variation of the daily HALT averages.

Standard Error of the Mean

We will conclude that the chip-making process is running as designed so long as the daily average HALT score falls within an operating range. Because $\overline{X}$ is approximately normally distributed for this process, percentiles of the normal distribution determine the operating range. We will choose these percentiles so that $\overline{X}$ falls within this range with high probability when the process is operating correctly. All we need are the mean and SD of $\overline{X}$. It's not hard to see that the expected value of $\overline{X}$ is μ, the same as the expected value for each chip. The harder task is to find the standard deviation of $\overline{X}$. The comparison of the histograms in Figures 14.1 and 14.2 indicates that the averages are less variable than the underlying data. The question is, how much less?

The sample-to-sample standard deviation of $\overline{X}$ is known as the **standard error of the mean**. The standard error of the mean *is* a standard deviation, but it measures the sampling variation, which is the variability of the mean from sample to sample.

Standard Error of the Mean   The standard error of the mean of a simple random sample of n measurements from a process or population with standard deviation σ is

$$\text{SD}(\overline{X}) = \text{SE}(\overline{X}) = \frac{\sigma}{\sqrt{n}}$$

The standard error of the average is sometimes written as $\sigma_{\overline{X}}$. The formula for $\text{SE}(\overline{X})$ shows two properties of the standard error of the mean:

1. The standard error is proportional to σ. As data become more variable (σ increases), averages become more variable.
2. The standard error is inversely proportional to the square root of n. The larger the sample size n, the smaller the sampling variation of the average from sample to sample.

tip

The standard error decreases proportionally to changes in the square root of n, not n itself. The fact that the denominator of the expression for the standard error is the square root of n, rather than n itself, is worth remembering.

The reason for introducing a special name, standard error, for the variation of an average is to distinguish this type of variation among samples from the standard deviation of data within a sample (variation from one observation to another). Standard error is visible in a collection of samples, as in Figure 14.2, but in later chapters we will only have one sample from the population. We get to see sampling variation directly in quality control because we observe a sequence of samples. In other situations, we typically have one sample. Sampling variation is still relevant in those cases but more subtle. The methods used to study a portfolio in Chapter 10 can be used to determine the standard error of the mean (see Behind the Math: Properties of Averages).

(p. 343)

Sampling Distribution

The design of the chip-making process indicates that the HALT score of a chip has mean $\mu = 7$ with $\sigma = 4$. Hence, the approximate normal model for the variation among means of sets of $n = 20$ HALT scores is

$$\overline{X} \sim N\left(\mu = 7, \frac{\sigma^2}{n} = \frac{4^2}{20} \approx 0.89^2\right)$$

When the fab runs as designed, the *average* number of tests passed by 20 chips is normally distributed with mean 7 and standard error 0.89. [The abbreviation for a normal distribution is $N(mean, variance)$. Because we often need the standard deviation rather than the variance, we will write the variance as the square of the standard deviation. The number that gets squared, here 0.89, is the standard deviation.]

sampling distribution The probability distribution that describes how a statistic, such as the mean, varies from sample to sample.

This distribution is our model for the **sampling distribution** of the mean of a sample of 20 HALT scores. The sampling distribution of a statistic describes its sampling variation. When we model the mean HALT score as $\overline{X} \sim N(7, 0.89^2)$, we're describing the variation of the average from one sample to the next. The standard error is the standard deviation of this distribution.

The sampling distribution tells us where to place limits to monitor the operation of the chip-making process. To monitor this process, the sampling distribution shows what to expect when the process operates as designed. The average HALT score of a sample of 20 chips *should* look like a random draw from the sampling distribution of the mean, shown in Figure 14.3.

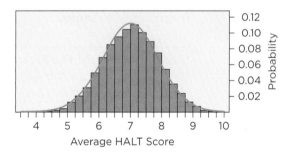

FIGURE 14.3 Normal model for the sampling distribution of the mean HALT score.

This histogram (really, the smooth bell curve) is our model for the sampling distribution and our point of reference.

What Do You Think?

a. To reduce the variation in average HALT scores, an engineer advised doubling the sample size, from 20 to 40. What is the effect of the larger sample size on the sampling distribution of the mean? What stays the same? What changes?[1]

b. The amount spent by customers in the express lane of a supermarket averages $25, with SD = $10. The distribution of amounts is slightly right skewed. What normal model describes the average amount spent by 100 randomly selected customers?[2]

---

[1] The sampling distribution remains normal with mean $\mu = 7$. The SE falls to $4/\sqrt{40} \approx 0.632$, smaller than that with $n = 20$ by a factor of $\sqrt{2} \approx 1.4$. The sample size condition is also easily satisfied because n is larger.

[2] Normal, with mean $\mu = \$25$ and standard error $\sigma/\sqrt{n} = 10/\sqrt{100} = \$1$. A normal model is reasonable because the Central Limit Theorem assures us that the sampling distribution of averages is approximately normal (though we should check).

14.2 | CONTROL LIMITS

Daily averages should resemble random draws from the sampling distribution in Figure 14.3; otherwise, something has gone awry in the chip-making process. Operators should expect the mean HALT score to be near 7. Judging from Figure 14.3, a daily mean HALT score at 6.5 is to be expected, but a mean less than 4 would be a surprise. Such small averages are rare and suggest that production does not meet the claims of the design. The hard decisions come, however, when we see an average HALT score around 5. Such values happen by chance when the process is operating correctly, but not often. Would such a mean HALT score lead you to recommend to shut down the fab and look for a problem?

The supervising manager has to make this call. If the process is not functioning as designed, the supervisor should stop production. But how is a supervisor to distinguish a problem from bad luck? The sampling distribution in Figure 14.3 shows that it is *possible* for the mean of 20 chips to be 5 when the process is operating as designed.

Our approach works like this:

- If the average HALT score $\overline{X}$ is close enough to $\mu = 7$, conclude that the process is functioning as designed. Let it continue to run.
- If the average HALT score is far from μ, conclude that the process is not functioning as designed. Stop production.

The trick is to define the operating range that determines whether $\overline{X}$ is close enough to μ. Our range is the symmetric interval from $\mu - L$ to $\mu + L$. If $\overline{X}$ lies within this operating range,

$$\mu - L \leq \overline{X} \leq \mu + L$$

we'll say that $\overline{X}$ is close enough to μ and let production continue. Otherwise, $\overline{X}$ is far from μ, and we should stop the process in order to find and correct the problem. Because this operating range controls whether we allow the process to continue, $\mu - L$ and $\mu + L$ are called **control limits**. The endpoint $\mu + L$ is the upper control limit (UCL) and $\mu - L$ is the lower control limit (LCL). The process is said to be out of control if a daily average falls outside the control limits. Once the process goes out of control, we can no longer be confident that its output conforms to the design.

Why would a manager stop the process if $\overline{X} > \mu + L$? The chips appear to be doing better than claimed by the design. That's true, but managers need to identify this type of change as well. Why is the process working better than designed? What led to the improvement? If indeed the process has improved, then the supervisor might speed up production. Perhaps the company can negotiate a higher price with customers or advertise that its chips are better than ever. It's a change worth noticing.

control limits Boundaries that determine whether a process is out of control or should be allowed to continue.

Type I and Type II Errors

The remaining task is to choose L. If the supervisor places the control limits too close to $\mu = 7$ (L is small), operators are likely to think that there's a problem even when the process is functioning correctly. On the other hand, wide limits let the process continue too often even though it's off target.

Table 14.2 summarizes the two types of errors that can occur. At the time that the supervisor makes her decision, the production line is either working as designed or needs adjustment. She can make the right choice, or she can make a mistake as laid out in the table.

TABLE 14.2 Correct decisions and errors.

		Supervisor Chooses to	
		Continue	Shut Down
State of process	Working as designed	✔	✗$_1$
	Not working as designed	✗$_2$	✔

Crosses identify the mistakes. The supervisor can shut down the process when it's functioning as designed (✗$_1$), or she can let a failing process continue (✗$_2$).

Let's start with the error marked ✗$_1$. This error indicates that the supervisor shut down the process when it was running as designed. This error happens if $\overline{X}$ lands outside the control limits even though the process is operating as designed. Production managers dread this error because they often earn a substantial portion of their income in the form of bonuses for meeting a quota. Shutting down the line too often puts these bonuses out of reach. This mistake, unnecessarily taking action, is known in statistics as a **Type I error**. The manager shut down the line when she should have left it alone.

Type I error The mistake of taking action when no action is needed.

It's easy to eliminate the chance of a Type I error, but only at the risk of a different mistake. The supervisor can eliminate any chance for a Type I error: *Never* utter the words "Shut down the line." The flaw with this logic is that it ignores the error marked ✗$_2$ in Table 14.2. A supervisor who refuses to stop production regardless of the value of $\overline{X}$ risks allowing the process to continue when it is out of control. This failure to take appropriate action is called a **Type II error**.

Type II error The mistake of failing to take action when needed.

Type II errors are often more expensive than Type I errors, but the expenses are often delayed. If the supervisor fails to act when the process is out of control, the fab may ship faulty chips. That might not matter in the short run—she met her production quota and got a bonus—but eventually these mistakes come back to harm the manufacturer. Perhaps the fab will lose a customer after delivering a shipment of bad chips. Stories of hikers lost in the mountains because their GPS failed generate bad publicity that no amount of advertising can overcome.

Setting the Control Limits

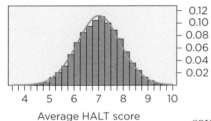

If we take the supervisor's concern over meeting her production quota as paramount, then we need to concentrate on the first error. We should declare the process out of control only if we're reasonably sure that the process is no longer operating according to the design specifications. To see how to set these limits, look again at the sampling distribution of the mean in Figure 14.3, repeated in the margin here. If the supervisor says, "Shut down the line" when the average HALT score is, say, less than 6 or more than 8, what is the chance that she is committing a Type I error? What is the probability that $\overline{X}$ is less than 6 or more than 8 even though the process is running as designed?

To answer this question (and find the chance of a Type I error), we use the normal model for the sampling distribution of $\overline{X}$. If the process is working as designed, then $\overline{X} \sim N(7, 0.89^2)$. The chance for a Type I error is the probability that $\overline{X} < 6$ or $\overline{X} > 8$:

$$P(\overline{X} < 6 \ or \ \overline{X} > 8) = 1 - P(6 \le \overline{X} \le 8)$$
$$= 1 - P\left(\frac{6 - 7}{0.89} \le \frac{\overline{X} - 7}{0.89} \le \frac{8 - 7}{0.89}\right)$$
$$= 1 - P(-1.1 \le Z \le 1.1)$$
$$= 0.27$$

As in Chapter 12, Z denotes a standard normal random variable, $Z \sim N(0, 1)$. The probability for a Type I error is commonly denoted α (alpha, first letter in the Greek alphabet). Control limits at 6 and 8 mean that $\alpha = 0.27$.

Most production managers would find a 27% chance of unnecessarily shutting down production too large. It is too common to get an average outside of the control limits 6 and 8 when the process is operating as designed. To reduce the chance for a Type I error, the production supervisor must move the limits farther out from μ, increasing the distance of $\overline{X}$ from μ. Suppose the supervisor prefers a 5% chance for a Type I error. From tables of the normal distribution, we know that $P(Z < -1.96 \text{ or } Z > 1.96) = 0.05$ if $Z \sim N(0, 1)$. Hence, to have a 5% chance of a Type I error, the supervisor should stop the process if

$$\overline{X} < \mu - z_{0.025}\text{SE}(\overline{X}) \qquad \text{or} \qquad \overline{X} > \mu + z_{0.025}\text{SE}(\overline{X})$$
$$< 7 - 1.96(0.89) \approx 5.26 \qquad \qquad > 7 + 1.96(0.89) \approx 8.74$$

She can reduce α further to 1% by setting the limits at ($z_{0.005} = 2.58$)

$$\overline{X} < 7 - 2.58(0.89) \approx 4.70 \qquad \text{or} \qquad \overline{X} > 7 + 2.58(0.89) \approx 9.30$$

tip **It is important to notice that the data do not determine the control limits.** The control limits for monitoring $\overline{X}$ come from

1. The parameters of the process
2. Our willingness to accept a chance for a Type I error

The only way we are able to detect a change in the process is to risk being wrong.

Balancing Type I and Type II Errors

A production manager cannot simultaneously reduce the chances for *both* Type I and Type II errors by moving the control limits. Wide control limits for $\overline{X}$ reduce the chance for a Type I error but at the same time increase the chance for a Type II error. Narrow control limits reduce the chance of a Type II error but increase the chance of a Type I error. You cannot simultaneously reduce the chances for both Type I and Type II errors by moving the limits.

More data can help. Larger sample sizes reduce $\text{SE}(\overline{X})$ so that we can move the limits closer to μ and keep the same chance for a Type I error. If the fab were to test $n = 40$ chips each day, for example, then $\text{SE}(\overline{X}) = \sigma/\sqrt{40} = 4/\sqrt{40} \approx 0.632$. If the supervisor keeps $\alpha = 0.05$, then the limits move to

$$\overline{X} < 7 - 1.96(0.632) \approx 5.76 \qquad \text{or} \qquad \overline{X} > 7 + 1.96(0.632) \approx 8.24$$

These limits are closer to μ and allow the supervisor to detect a smaller change in the process (smaller chance for Type II error), but the chance for a Type I error remains 5%.

Statistics traditionally focuses on Type I errors ($\boldsymbol{\times}_1$). This emphasis suits the production supervisor. She's concerned about keeping the line running. The rationale within statistics, however, is mathematical rather than substantive. Because a Type I error occurs when the process is functioning as designed, we can determine α because the design specifies μ and σ. Hence we know the sampling distribution of $\overline{X}$ and can find control limits that obtain a given Type I error rate. We seldom, however, have enough information to find the chance of a Type II error ($\boldsymbol{\times}_2$). Doing so requires that we know μ and σ when the process is *not* functioning as designed. It's hard enough to know μ and σ when the process *is* functioning correctly. These parameters might be just about anything when the process is out of control.

caution   In the rest of this book, we'll follow convention and focus on Type I error. Don't lose sight, however, of the possible consequences of a Type II error. A procedure with a 1% chance of a Type I error may have a very large chance for missing a problem (a large chance for a Type II error).

What Do You Think?   A national chain of home improvement centers tracks weekly sales at its locations. Typically, sales average $500,000 per week at each store, with the standard deviation among stores at $75,000.

a. Assuming a normal model is appropriate for the sampling distribution, which model describes the sampling distribution of the average of weekly sales at a sample of 25 stores?[3]

b. Should management interpret average sales of $425,000 in a sample of 25 stores as a large deviation from the baseline of $500,000?[4]

c. A new manager overreacted to a weekly sales report and ordered a new advertising program, even though nothing really changed. What sort of error (Type I or Type II) has this manager committed?[5]

14.3 | USING A CONTROL CHART

Technicians recorded HALT scores for a sample of 20 chips each day during a trial period when the fab was known to be operating as designed. During this period, we know that $\mu = 7$ and $\sigma = 4$. Because the fab is operating correctly during this period, we can use these data to see how control limits perform. If $\overline{X}$ lands outside the control limits during this trial period, we know that we've found a Type I error because we know the process is operating as designed. Each point in the timeplot in Figure 14.4 shows the HALT score of one or more individual chips.

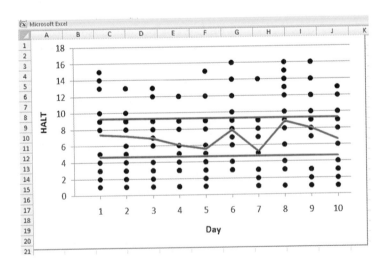

FIGURE 14.4 HALT scores of chips sampled for 10 days.

Because HALT scores are integers, you do not see 20 points each day; some scores are achieved by more than one chip. Some chips held up well (passing

[3] The normal model for the sampling distribution of the mean is
$N(\mu = 500{,}000, \sigma^2/n = 75{,}000^2/25 = (\$15{,}000)^2)$.
[4] Definitely. Though it is not surprising for one store to have sales of $425,000, for 25 stores to have &425,000 as the average of their sales is 5 standard errors below the expected mean.
[5] Type I. The manager took an action (launching new advertising) when nothing had changed.

all 15 tests), whereas a few failed the first test (scoring 1). The jagged blue line in Figure 14.4 connects the daily means. The parallel, horizontal red lines are the 99% control limits; these are located at

$$\mu \pm z_{0.005}\text{SE}(\overline{X}) = \mu \pm z_{0.005}\frac{\sigma}{\sqrt{n}}$$

$$= 7 \pm 2.58\frac{4}{\sqrt{20}}$$

$$\approx 7 \pm 2.58(0.89) \approx [4.70 \text{ to } 9.30]$$

X-bar chart Name of a control chart that tracks sample averages.

A chart such as that in Figure 14.4 that tracks the mean of a process with control limits is known as an **X-bar chart**.

In our example, because the process is known to be operating as designed during these 10 days, there is a 1% chance that the average HALT score on any given day lies outside these limits (Type I error). Because the means in Figure 14.4 remain within these control limits, a supervisor using this chart would not make a Type I error by incorrectly shutting down the process.

Suppose instead, however, that we move the control limits closer to μ by allowing a larger, 5% chance for a Type I error. Tighter limits increase the chance of a Type I error. The 5% limits are

$$\mu \pm z_{0.025}\text{SE}(\overline{X}) = 7 \pm 1.96\frac{4}{\sqrt{20}} \approx 7 \pm 1.96(0.89) \approx [5.26 \text{ to } 8.74]$$

Figure 14.5 shows the data with these 95% control limits rather than the 99% limits in the prior figure.

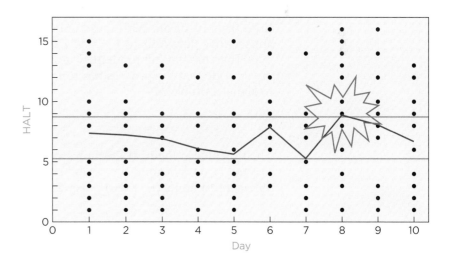

FIGURE 14.5 HALT scores with 95% control limits.

The 95% limits in this chart incorrectly signal that the process went out of control on the eighth day; the average HALT score on the eighth day is larger than the upper control limit. This is a Type I error because we know that the process is working properly during this period.

Repeated Testing

There's a larger chance for a Type I error during these 10 days than you might think. Suppose the supervisor sets $\alpha = 0.05$. When the production line is running normally, there's a 95% chance that the mean stays within the control limits each day. That does not imply, however, that the chance for a Type I error during a 10-day period is 5%. For 10 consecutive days, the probability that *all* 10 averages stay within the control limits is (assuming independence)

$$P(\text{within limits for 10 consecutive days})$$
$$= P(\text{within on day 1}) \times \cdots \times P(\text{within on day 10})$$
$$= 0.95^{10} \approx 0.60$$

tip

The chance of a Type I error during several days is larger than the chance on any one day. A 5% chance for a Type I error on any *one* day implies there is a 100% − 60% = 40% chance for a Type I error during 10 days. **Repeated testing eventually signals a problem.** Testing the process for 25 consecutive days increases the chance that a Type I error will occur during this period to $1 - 0.95^{25} \approx 0.72$.

Smaller values of α reduce the chance that a Type I error occurs, but at a cost. Reducing α requires moving the control limits farther from μ, increasing the chance of missing a problem. By reducing α to 0.01, the supervisor lowers the chance for a Type I error during 25 days to $1 - 0.99^{25} \approx 0.22$. She can reduce the chance further by moving the control limits yet farther from μ, but she then has to accept a greater chance of a Type II error. Too much concern for Type I errors defeats the purpose: The whole point of monitoring a process is to detect a problem when it happens.

As a compromise, many references and software packages by default place control limits for the mean at

$$\mu - 3\frac{\sigma}{\sqrt{n}} \quad \text{and} \quad \mu + 3\frac{\sigma}{\sqrt{n}}$$

These limits reduce the chance for a Type I error each day to 0.0027, the probability of a normal random variable falling more than three standard deviations from its mean. With these limits, the chance of a Type I error during 25 consecutive days falls to $1 - 0.9973^{25} \approx 0.065$. The downside is the risk of a Type II error—failing to identify a problem when it does happen. The limits are so wide that the process has to fail badly before $\overline{X}$ gets far enough from μ to signal a problem.

Recognizing a Problem

The timeplot in Figure 14.6 shows what happens as testing at the fab continued for 11 days beyond the period shown in Figure 14.4. This plot shows the 99% control limits ($\alpha = 0.01$) over 21 days.

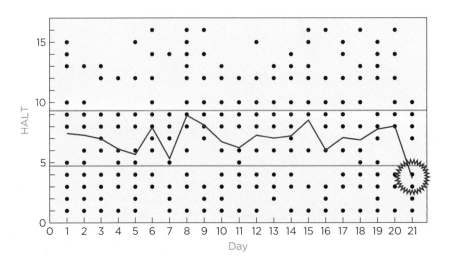

FIGURE 14.6 Continued control chart for mean HALT scores with $\alpha = 0.01$.

The daily mean remained within the control limits for 10 more days. Then, on Day 21, the average HALT score dropped below the lower control limit, and the supervisor shut down production. Technicians discovered that the etching process used in making the chips had become contaminated. In this case, an average outside the limits correctly identified a problem. The circled mean in Figure 14.6 is not an error; it's the right answer. The process *was* broken.

How do we know? We know because the failure in the process was identified when technicians inspected the equipment; we cannot tell from Figure 14.6 alone. All we can see in Figure 14.6 is an average outside the control limits. An average outside the control limits implies one of two things has happened:

1. Bad luck (a Type I error occurred).
2. A problem has been correctly identified.

Which is it? As managers, we'll act as though we've identified a problem. We assume that a mean outside the control limits implies that the process is broken. Hopefully, operators can find the problem and resume production.

Control Limits for the *X*-Bar Chart   The $100(1 - \alpha)\%$ control limits for monitoring averages of a sample of n measurements from a process with mean μ and standard deviation σ are

$$\mu - z_{\alpha/2}\frac{\sigma}{\sqrt{n}} \quad \text{and} \quad \mu + z_{\alpha/2}\frac{\sigma}{\sqrt{n}}$$

The multiplier $z_{\alpha/2}$ controls α, the chance of a Type I error. The z_α is the upper $100(1 - \alpha)$ percentile of a standard normal distribution,

$$P(Z > z_\alpha) = \alpha \quad \text{for } Z \sim N(0, 1)$$

For example, $z_{0.025} = 1.96$ and $z_{0.005} = 2.58$. The process is under control so long as $\overline{X}$ remains within this operating range.

14.4 | CONTROL CHARTS FOR VARIATION

Control charts can be used to monitor any sample statistic by comparing it to the corresponding parameter of the process. In each case, the control limits have to allow for sampling variation in the statistic. For a manufacturing process, it is also important to monitor the variability of a process. The mean of the process might stay close to its target while the variation grows larger than it should.

Here's an example. A food processor weighs samples of 15 packages of frozen food from each day's production. The packages are labeled to weigh 25 ounces. To accommodate random variation, the packaging system is designed to put 26.5 ounces *on average* into each package. The timeplot in Figure 14.7 tracks the weights of packages over a 15-day period, with a horizontal line at the labeled package weight.

When operating correctly, the system has little chance of underfilling packages. The mean is $\mu = 26.5$ ounces and $\sigma = 0.5$ ounce. Because the labeled package weight is 3σ below the mean, the chance of an underweight package is only $P(Z < -3) = 0.00135$ (if the weights are normally distributed). Figure 14.7 suggests that the process operated as designed for the first few days. Something evidently happened between Days 5 and 10. The mean remains near 26.5 ounces, but the weights become more variable.

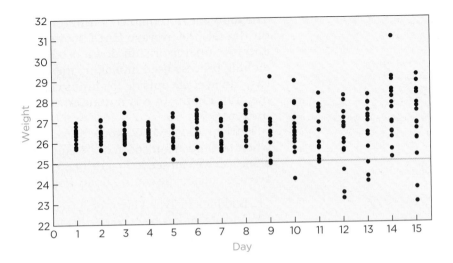

FIGURE 14.7 Package weights.

An *X*-bar chart is slow to detect this problem. Figure 14.8 shows 99% control limits for the mean weight of $n = 15$ packages as two red horizontal lines. The lower control limit is

$$\mu - z_{0.005}(\sigma/\sqrt{n}) = 26.5 - 2.58(0.5/\sqrt{15}) \approx 26.17 \text{ ounces}$$

and the upper control limit is

$$\mu + z_{0.005}(\sigma/\sqrt{n}) = 26.5 + 2.58(0.5/\sqrt{15}) \approx 26.83 \text{ ounces}$$

The plot shown in Figure 14.8 joins the daily averages with a blue line.

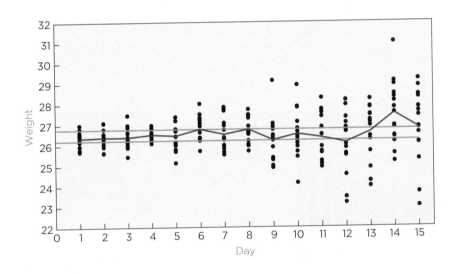

FIGURE 14.8 Control limits for the mean.

This chart does not identify a problem until Day 12, even though Figure 14.7 suggests the problem began earlier. Excessive variation eventually affects averages, but monitoring the average is an inefficient way to detect a change in the variation. It is better to track *both* the mean *and* a measure of the variation.

An **S-chart** tracks the standard deviation *s* from sample to sample. As long as *s* remains within its control limits, the process is functioning within its design parameters. Once *s* falls outside these limits, there's evidence that the process is out of control. A similar chart, called an **R-chart**, tracks the range rather than the SD.

S-chart Name of a control chart that tracks sample standard deviations.

R-chart Name of a control chart that tracks sample ranges.

The control limits in the *S*-chart require a formula for the standard error of the standard deviation. The details of this formula are not important (see Behind the Math: Limits for the *S*-Chart), and we will rely on software. Qualitatively, the standard error of *s* behaves like the standard error of the mean. In particular, the more variable the data become, the more variable the sample standard deviation *s* becomes. The larger the samples become, the more consistent *s* is from sample to sample.

(p. 343)

For the food packages, we set $\alpha = 0.01$ with $\mu = 26.5$ and $\sigma = 0.5$. Figure 14.9 shows the resulting control charts. The *X*-bar chart tracks the daily averages (left), and the *S*-chart tracks the daily standard deviations (right).

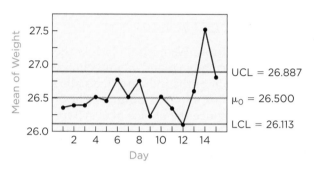
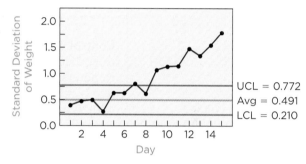

FIGURE 14.9 Control charts for package weights.

This figure shows only the summary statistics, not the individual observations within each sample. The first two points in the *X*-bar chart are 26.35 and 26.39; these are the average weights (in ounces) on the first two days. The first two values in the *S*-chart are 0.39 and 0.47; these are standard deviations (in ounces) for the 15 packages on the first day and the 15 packages on the second. Figure 14.9 also adds horizontal lines that locate the target parameter. The *S*-chart detects the problem more quickly than the *X*-bar chart. The standard deviation of the packages (0.81) exceeds the upper control limit on Day 7, five days before the *X*-bar chart detects a problem. (The *S*-chart is not centered on $\sigma = 0.5$; it's centered at 0.491, slightly lower. That's because the sample standard deviation *s* is a biased estimate of σ. On average, the sample standard deviation is slightly less than σ.)

4M EXAMPLE 14.1 | **MONITORING A CALL CENTER**

MOTIVATION | **STATE THE QUESTION**

Control charts have become common in many business sectors, even the service industry. Many companies operate hotlines that customers may call for help. By monitoring the volume and nature of the calls, a company can learn how customers are reacting to its products.

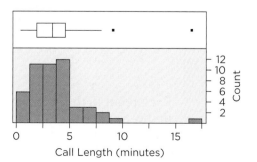

In this example, a bank is tracking calls related to its Internet bill-paying service. Problems generally occur when customers register. During the previous year and a half, questions regarding registration took 4 minutes on average to resolve, with standard deviation 3 minutes. The histogram on the preceding page shows the times for 50 calls during one day. Most calls take 2 or 3 minutes, but one lasted more than 15 minutes.

The bank wants a system to monitor the length of calls. If calls are getting longer or shorter, then changes to the service may have affected the customers who use it. The bank is willing to monitor 50 calls a day, but does not want to intrude further on system resources.

METHOD

DESCRIBE THE DATA AND SELECT AN APPROACH

Control charts are an obvious solution. We need to specify the parameters of the process when it is working normally (μ and σ) and choose α. For the parameters, records from the last 18 months indicate that calls should average $\mu = 4$ minutes with standard deviation $\sigma = 3$ minutes. The costs associated with Type I and Type II errors influence the choice of α. A Type I error (thinking calls have changed when they are in fact the same) will lead to expensive software testing. To reduce the chance for this error, we will follow convention by setting $\alpha = 0.0027$ and placing the control limits three standard errors from the parameter.

tip | Check the conditions before you get into the details. You will need to use a different method if the conditions are not met. The distribution of call lengths is skewed, but the sample size condition for using a normal model is satisfied. Using data from the first few days (while everything appears okay), $K_3^2 \approx 3.2$ and $K_4 \approx 5$. We can use a normal model for the distribution of $\overline{X}$ but only just barely. The limits in the S-chart are less reliable when data are skewed, so we will interpret the S-chart informally.

MECHANICS

DO THE ANALYSIS

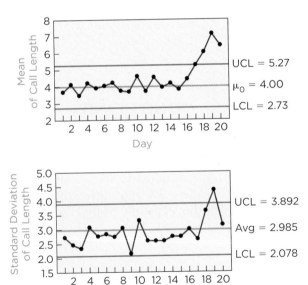

Each point in the *X*-bar chart is an average of the lengths of 50 calls; points in the *S*-chart are standard deviations. The choice $\alpha = 0.0027$ puts the control limits at ± 3 standard errors from the process target. For the mean, $SE(\overline{X}) = \sigma/\sqrt{n} = 3/\sqrt{50} \approx 0.42$ minute, so the control limits for tracking the mean are about $4 \pm 3 \times 0.42$. Software figures out the limits for the S-chart [about $3 \pm 3 \times \text{SE}(s)$].

Both charts indicate that the length of calls to the help center changes. The mean crosses the upper control limit on Day 17. On Day 19, both the mean and the SD are outside of the control limits. The length of the calls has increased on average, as has the variability of the length.

MESSAGE | SUMMARIZE THE RESULTS

Interpret the results in language that avoids technical terms that the audience won't understand. The length of time required for calls to this help line has changed. The average length has increased and the lengths have become more variable. Either the bank's bill-paying service has gotten harder to use or perhaps it is attracting new customers who are less familiar with the Internet and require more help.

Best Practices

- *Think hard about which attribute of the process to monitor.* Software will draw the pictures, but it's up to managers to pick the characteristic. In addition to monitoring HALT scores, the production supervisor might also monitor the number of chips produced every hour to be sure that production is keeping pace with orders.

- *Use both X-bar charts and S-charts charts to monitor a process.* A process might be fine on average while its variability increases. We won't know unless we chart both $\bar{x}$ and s.

- *Set the control limits from process characteristics, not data.* The underlying process characteristics determine appropriate values for μ and σ. If using data to choose these, use data from a period when the process is known to be operating as designed.

- *Set the control limits before looking at the data.* Once set, don't go back in time and pretend the limits are set differently. It distorts the analysis to

move the limits just to signal a problem or keep the process from crossing the limits. The control limits have to be set before the process runs.

- *Carefully check before applying control limits to small samples.* The standard control limits come from using a normal model to describe the sampling variation of $\bar{X}$. This model is usually only an approximation to the distribution. The resulting control limits are fine for monitoring the mean when the sample size condition is met. The limits in the S-chart are more sensitive to outliers and skewness than those in the X-bar chart.

- *Recognize that control charts eventually signal a problem.* Unless we set $\alpha = 0$ and never indicate a problem, repeated testing will eventually show a statistic outside the control limits even when the process is in control. Accept that these false signals will happen; it's the cost of being able to detect problems.

Pitfalls

- *Do not concentrate on one error while ignoring the other.* A small value for α means little chance for a Type I error, but conversely may mean a large chance for a Type II error. In many situations, the cost of a Type II error—letting a malfunctioning process continue to run—may ultimately be more expensive than the cost of stopping production. Managers have to keep both costs in mind.

- *Do not assume that the process has failed if a value appears outside the control limits.* If $\bar{x}$ or s goes outside the control limits, halt the process and look for the problem. Recognize, though, that Type I errors happen. It's no one's fault. It's the price that we pay for being able to detect a problem when it happens.

- *Avoid confusing Type I and Type II errors.* It is easy to confuse these labels. To help distinguish them, some fields, such as medicine, rename these kinds of errors. For instance, a medical diagnosis that incorrectly indicates an illness in a healthy person is called a false positive error.

We'd call it a Type I error. Similarly, a test that fails to detect an illness commits a false negative error. That's a Type II error.

Software Hints

Minitab and JMP provide X-bar and S-charts like those in this chapter. In addition, both packages offer other control charts, including runs charts (to check the assumption that the data taken from the process over time are independent) and cusum (<u>cumulative</u> <u>sum</u>) charts, which have useful properties and some advantages over the basic control charts shown here. If using a package to obtain the kurtosis, ensure that the software subtracts 3. (The sample kurtosis should be near zero for a sample from a normal distribution.)

EXCEL

Excel does not have built-in functions for control charts, so we have to draw our own using the formulas in this chapter. For example, suppose that we obtain 10 measurements per day and put these in one row using the 10 columns B through K, with column A holding the date. In column L, use the formula AVERAGE to find the mean of the 10 daily values, and in column M put the formula STDEV for the standard deviation. The function SKEW computes the skewness and KURT computes the kurtosis.

If the DDXL package is installed, then use it to make X-bar and S-charts (as well as R-charts). After selecting the portion of the spreadsheet with the data table, follow the menu commands

DDXL > Process control

and choose the item labeled "Mu, R, S Control Charts" from the Click me menu in the dialog. The selected data have to include at least one column used to group the data (such as identifying the day by a number in the HALT scores example) and a numerical column that provides the observed measurements.

MINITAB

Following the menu items

Stat > Control Charts > Variables Charts for Subgroups > X-bar S

opens a dialog for creating a pair of control charts, one for the mean (X-bar chart) and one for the standard deviation (S-chart). Pick the variable that is to be shown; it should be in a single column of the data table. Then specify how the data are grouped. The data can be grouped 10 consecutive rows at a time, for example, or another column can be used to specify the groups. Options in the dialog allow us to specify parameters of the underlying process (μ and σ). Without these, the software by default uses the mean and SD of the data. Click the OK button to generate the plots.

JMP

Following the menu items

Graph > Control Chart > X-bar

opens a dialog for creating both X-bar and S-charts. In the dialog, check the boxes at the left for the X-bar and S-charts. In the parameters section of the dialog, use the default setting for getting control limit at ±3 standard errors.

Next, pick the variable that is to be shown and enter its name in the Process field. The data to be charted should be stacked in a single column of the data table. Then specify how the data are grouped. The data can be grouped into batches of fixed size, such as 10 consecutive rows at a time (set the sample size to be the constant 10) or grouped by the values in another column (specify a variable in the Sample Label field.) To specify parameters of the underlying process (μ and σ), click the Specify Stats button and fill in values for the process mean (called mean measure) and sigma. Click the OK button when ready to see the plots. The summary stats in the control charts are linked back to the underlying data; clicking on a sample mean, for instance, highlights the rows of the data table that have been averaged.

To obtain the skewness and kurtosis, use the menu commands

Analyze > Distribution

For JMP's summary of a variable. Then access the Display Options menu by right-clicking on the summary and choosing the option for More Moments.

▣▶ BEHIND the Math

Properties of Averages

Methods used for finding means and variances of sums of random variables in Chapter 10 produce the sampling distribution of $\overline{X}$. First, write $\overline{X}$ in terms of the underlying individual measurements.

$$\overline{X} = \frac{X_1 + X_2 + \cdots + X_n}{n}$$

The averaged values are *assumed to be iid* random variables with common mean μ and SD σ. Because the expected value of a sum is equal to the sum of the expected values, the mean of $\overline{X}$ is also μ.

$$\mathrm{E}(\overline{X}) = \mathrm{E}\left(\frac{X_1 + X_2 + \cdots + X_n}{n}\right)$$
$$= \frac{1}{n}(\mathrm{E}(X_1) + \mathrm{E}(X_2) + \cdots + \mathrm{E}(X_n))$$
$$= \frac{n\mu}{n} = \mu$$

We might say "The average of the averages is the average," but that's confusing.

A similar calculation produces the variance of $\overline{X}$. To get the variance of $\overline{X}$, we use another assumption: independence. Variances of sums of independent random variables are sums of variances.

$$\mathrm{Var}(\overline{X}) = \mathrm{Var}\left(\frac{X_1 + X_2 + \cdots + X_n}{n}\right)$$
$$= \frac{1}{n^2}\big(\mathrm{Var}[X_1] + \mathrm{Var}[X_2] + \cdots + \mathrm{Var}[X_n]\big)$$
$$= \frac{n\sigma^2}{n^2} = \frac{\sigma^2}{n}$$

The normal model for the sampling distribution of $\overline{X}$ has mean $\mathrm{E}(\overline{X}) = \mu$ and variance $\mathrm{Var}(\overline{X}) = \sigma^2/n$.

Limits for the *S*-Chart

These details are esoteric and involved. The expressions for the limits used in the *S*-chart presume that the data are a sample of n cases from a normal population with variance σ^2. Given that the data are normally distributed, the standard error of the standard deviation is $\mathrm{SE}(S) = \sqrt{2}\sigma^2/\sqrt{n}$ (the standard deviation of a standard deviation). The upper and lower control limits for the standard deviation are then located at $c\sigma \pm z_a\mathrm{SE}(S)$, where the constant $c < 1$ and approaches 1 as n increases.

CHAPTER SUMMARY

The **sampling distribution** of the mean of a sample of n items describes the variation among averages from sample to sample. When the **sample size condition** is satisfied, the sampling distribution of the mean can be modeled as a normal distribution, $\overline{X} \sim N(\mu, \sigma^2/n)$. The **standard error of the mean** is the standard deviation of the sampling distribution, $\mathrm{SE}(\overline{X}) = \sigma/\sqrt{n}$. A **Type I error** occurs when we take unnecessary action (halting production when all is fine); a **Type II error** occurs when we fail to take appropriate action to remedy a problem (not stopping a faulty production process). The upper and lower **control limits** determine the probability of a Type I error, denoted by α. An **X-bar chart** monitors averages of samples. An **S-chart** monitors the standard deviations, and an **R-chart** monitors the ranges.

▣ Key Terms

■ Formulas

Sampling distribution of the mean if data are normally distributed is also a normal distribution with the population mean μ and and standard error given by σ over the square root of the sample size.

$$\overline{X} \sim N(\mu, \sigma^2/n)$$

Standard error of the mean

$$SE(\overline{X}) = \mu/\sqrt{n}$$

Sample size condition.
To model $\overline{X}$ as having a normal distribution when the population is not known to be normally distributed, the sample size must be larger than 10 times the squared skewness K_3^2 and larger than 10 times the absolute value of the kurtosis K_4.

$$n > 10K_3^2 \quad \text{and} \quad n > 10|K_4|$$

Software packages such as Excel may use a more refined estimate for the skewness and the kurtosis, computing these via the formulas

$$K_3 = \frac{n}{(n-1)(n-2)} \sum_{i=1}^{n} z_i^3$$

and

$$K_4 = \frac{n(n+1)}{(n-1)(n-2)(n-3)} \sum_{i=1}^{n} z_i^4 - \frac{3(n-1)^2}{(n-2)(n-3)}$$

These refinements make little difference from the expressions in the text unless n is small.

Upper and lower control limits for an average (*X*-bar chart) are

$$\mu \pm z_{\alpha/2}\sigma/\sqrt{n}$$

where z_α satisfies $P(Z > z_\alpha) = \alpha$.

■ About the Data

The HALT data are from publications that summarize quality control procedures in the semiconductor industry. The data on packaging frozen food and monitoring the duration of telephone calls were obtained from consulting work performed by colleagues at the Department of Statistics at the Wharton School.

EXERCISES

Mix and Match

1. Which of the following *X*-bar charts indicates a process that is out of control?

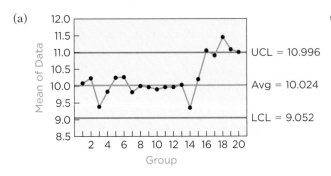

(a)

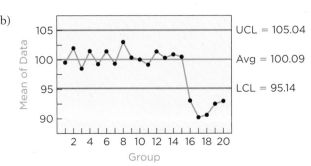

(b)

(c)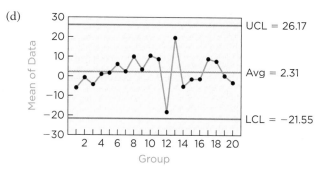

(d)

2. Which of the following *X*-bar charts show that a process went out of control?

(a)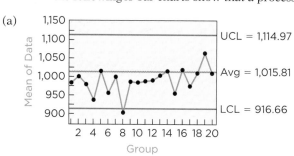

(b)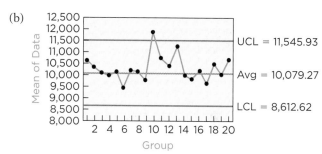

(c)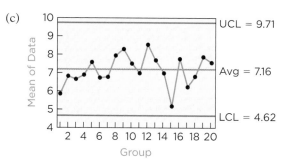

(d)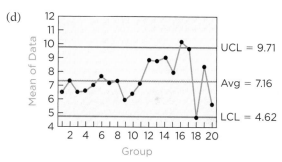

3. Which, if any, of these combinations of an *X*-bar chart and an *S*-chart suggest a problem? If there's a problem, did you find it in the *X*-bar chart, the *S*-chart, or both?

(a)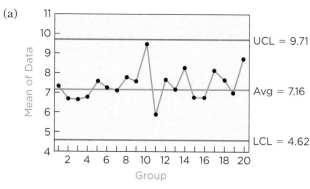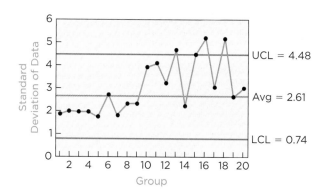

(b)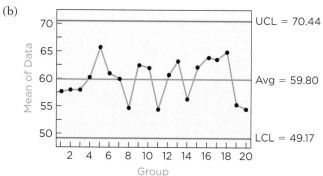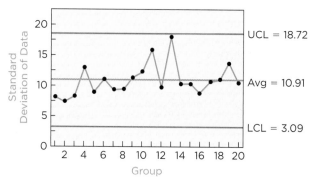

(c)

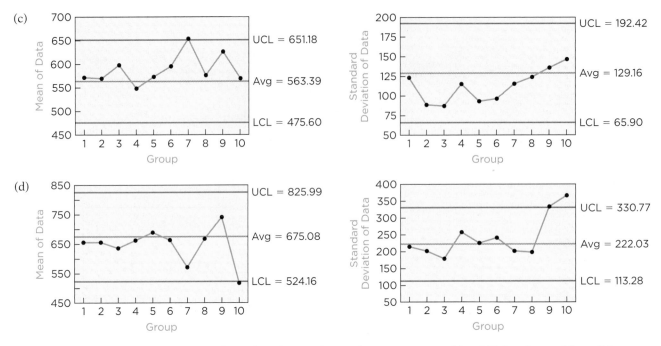

4. Which, if any, of these combinations of an X-bar chart and an S-chart suggest a problem? If there's a problem, did you find it in the X-bar chart, the S-chart, or both?

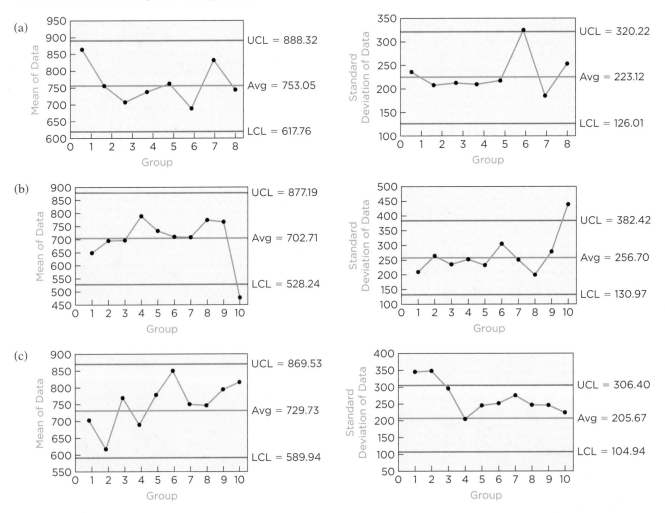

(d)

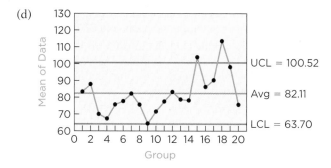

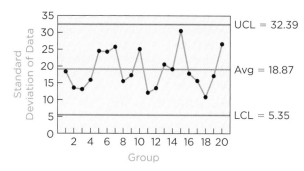

True/False

Mark each statement True or False. If you believe that a statement is false, briefly explain why you think it is false.

Exercises 5–12. The manager of a warehouse monitors the volume of shipments made by the delivery team. The automated tracking system tracks every package as it moves through the facility. A sample of 25 packages is selected and weighed every day. On the basis of contracts with customers, the mean weight should be $\mu = 22$ pounds with $\sigma = 5$ pounds.

5. A sampling distribution describes the variability among average weights from day to day.

6. Before using a normal model for the sampling distribution of the average package weights, the manager must confirm that weights of individual packages are normally distributed.

7. A Type I error occurs if the mean weight μ and standard deviation σ do not change.

8. If the average weight of packages increases and the manager does not recognize the change, then the manager has committed a Type II error.

9. The standard error of the daily average is $SE(\overline{X}) = 1$.

10. An X-bar chart with control limits at 12 pounds and 32 pounds has a 5% chance of a Type I error.

11. To have a small chance for a Type II error, the manager of the warehouse should locate the control limits in the X-bar chart at 22 ± 3 pounds.

12. By expanding the control limits in the X-bar chart from 22 ± 2 to 22 ± 4 pounds, the manager reduces the chance of a Type I error by 50%.

Exercises 13–20. Auditors at a bank randomly sample 100 withdrawal transactions made at ATM machines each day and use video records to verify that authorized users of the accounts made the transactions. The system records the amounts withdrawn. The average withdrawal is typically $50 with SD $40. Deposits are handled separately.

13. A histogram of the average withdrawal amounts made daily over the span of a month should cluster around $50.

14. A histogram of the daily standard deviations of the withdrawal amounts over the span of a month should cluster around $40.

15. It would be surprising to discover a transaction for more than $100.

16. The monitoring system will have a higher chance of detecting a change in transaction behavior if the control limits in the X-bar chart are moved closer to $50.

17. If the daily average withdrawal exceeds the upper control limit in the X-bar chart, then the procedure has committed a Type I error.

18. An upcoming holiday weekend increases the size of most withdrawals. If the average on such a day remains inside the control limits, then a Type I error has occurred.

19. By sampling 200 transactions daily, auditors can move the control limits in the X-bar chart closer to $50 without increasing the chance of a Type I error.

20. Control limits in an X-bar chart at $50 \pm 3 \times 40/\sqrt{n}$ based on samples of size $n = 100$ are more likely to produce a Type I error than control limits at $50 \pm 3 \times 40/\sqrt{n}$ based on samples of size $n = 200$.

Think About It

21. Rather than stop the production when a mean crosses the control limits in the X-bar chart, a manager has decided to wait until two consecutive means lie outside the control limits before stopping the process. The control limits are at $\mu \pm 2\sigma/\sqrt{n}$.
    (a) By waiting for two consecutive sample means to lie outside the control limits, has the manager increased or decreased the chance for a Type I error?
    (b) What is the probability of a Type I error if this procedure is used?

22. Suppose that a manager monitors both the X-bar and S-charts. The manager will stop the process if either a sample mean lies outside the control limits or a sample standard deviation lies outside the control limits. The control limits are set at ± 3 standard error bounds.
    (a) By tracking both X-bar and S-charts, what is the probability of a Type I error?
    (b) What assumption do you need in order to answer part a?

23. Which of the following processes would you expect to be under control, and which would you expect not to be under control? Explain briefly why or why not.
    (a) Daily sales at each checkout line in a supermarket
    (b) Number of weekday calls to a telephone help line
    (c) Monthly volume of shipments of video game software
    (d) Dollar value of profits of a new startup company

24. Do you think the following data would represent processes that were under control or out of control? Explain your thinking.
    (a) Monthly shipments of snow skis to retail stores
    (b) Number of daily transactions at the service counter in a local bank
    (c) Attendance at NFL games during the 16-week season
    (d) Number of hourly hits on a corporate Web site

25. A monitoring system records the production of an assembly line each hour. The design calls for output to be four units per hour. Current policy calls for stopping the line to check for flaws if fewer than two items are produced in an hour. Use what you learned about counting random variables in Chapter 11.
    (a) What random variable seems suited to modeling the number of items produced each hour? What assumptions would you need to verify?
    (b) What is the chance of a Type I error in this procedure? Use the random variable chosen in part (a).
    (c) What is the chance of a Type II error in this procedure if production falls to three units per hour?

26. A bottler carefully weighs bottles coming off its production line to check that the system is filling the bottles with the correct amount of beverage. By design, the system is set to slightly overfill the bottles to allow for random variation in the packaging. The content weight is designed to be 1,020 g with standard deviation 8 g. Assume that the weights are normally distributed. (Use tables or software for the normal distribution.)
    (a) A proposed system shuts down the facility if the contents of a bottle weigh less than 1,000 g. What is the chance of a Type I error with this system?
    (b) Suppose the mean content level of bottles were to fall to 1,000 g. What is the chance that the proposed system will miss the drop in average level and commit a Type II error?

27. Auto manufacturers buy car components from suppliers who deliver them to the assembly line. A manufacturer can schedule staff time to check a sample of 5 parts a day for defects or a sample of 25 parts over the weekend when the assembly line is slowed.
    (a) If the manufacturer is interested in finding a flaw that shows up rarely, which approach will have a better chance of finding it?
    (b) What if the manufacturer is concerned about a problem that will be evident and persistent?

28. The Web site of a photo processor allows customers to send digital files of picture to be printed on high-quality paper with durable inks. When the inks used by the processor start to run out, the color mix in the pictures gradually degrades.
    (a) How could the facility use a quality control process to identify when it needed to switch ink cartridges? Will it be necessary to check every photo or only a sample of photos?
    (b) If the facility samples photos, would it do better to group the photos into a large batch and then calculate a mean and standard deviation (for the measured color mix) or only wait a little while before calculating the mean and SD?

29. Where should the control limits for an X-bar chart be placed if the design of the process sets $\alpha = 0.0027$ with the following parameters (assume that the sample size condition for control charts has been verified)?
    (a) $\mu = 10$, $\sigma = 5$, and $n = 18$ cases per batch
    (b) $\mu = -4$, $\sigma = 2$, and $n = 12$ cases per batch

30. Where should the control limits in an X-bar chart be placed if the design of the process sets $\alpha = 0.01$ with the following parameters (assume that the sample size condition for control charts has been verified)?
    (a) $\mu = 100$, $\sigma = 20$, and $n = 25$ cases per batch
    (b) $\mu = 2000$, $\sigma = 2000$, and $n = 100$ cases per batch

31. An X-bar control chart monitors the mean of a process by checking that the average stays between $\mu - 3\sigma/\sqrt{n}$ and $\mu + 3\sigma/\sqrt{n}$. When the process is under control,
    (a) What is the probability that five consecutive sample means of n cases stay within these limits?
    (b) What is the probability that *all* of the means in a control chart for 100 days fall within the control limits?

32. Using X-bar and S-charts that set control limits at ± 3 standard errors,
    (a) What is the probability that one or more of the first 10 sample averages from a process that is under control lands *outside* the control limits?
    (b) What is the probability that both the sample average and the sample standard deviation stay within their control limits for one period if the process meets the design conditions? Be sure to state any assumptions you make.
    (c) When the process is under control, are we more likely to signal a problem falsely if we' re using both the X-bar and S-charts or if we only use one of them?

You Do It

Names shown in bold with the exercises identify the associated data file.

33. **Shafts** These data give the diameter (in thousandths of an inch) of motor shafts that will be used in automobile engines. Each day for 80 weekdays, five shafts were sampled from the production line and carefully measured. For these shafts to work properly, the diameter must be about 815 thousandths of an inch. Engineers designed a process that they claim will produce shafts of this diameter, with $\sigma = 1$ thousandth of an inch.[6]
    (a) If the diameters of shafts produced by this process are normally distributed, then what is the probability of finding a shaft whose diameter is more than 2 thousandths of an inch above the target?
    (b) If we measure 80 shafts independently, what is the probability that the diameter of every shaft is

[6] From D. P. Foster, R. A. Stine, and R. Waterman (1996), *Basic Business Statistics: A Casebook* (New York: Springer).

between 813 and 817 thousandths of an inch? Between 812 and 818 thousandths?

(c) Group the data by days and generate *X*-bar and *S*-charts, putting the limits at ± 3 SE. Is the process under control?

(d) Group the data by days and generate *X*-bar and *S*-charts, putting the control limits at ± 2 SE. Is the process under control?

(e) Explain the results of parts (c) and (d). Do these results lead to contradictory conclusions, or can you explain what has happened?

(f) Only five shafts are measured each day. Is it okay to use a normal distribution to set the control limits in the *X*-bar chart? Do you recommend changes in future testing?

34. **Door Seam** A truck manufacturer monitors the width of the door seam as vehicles come off its assembly line. The seam width is the distance between the edge of the door and the truck body, in inches. These data are 62 days of measurements of a passenger door seam, with 10 trucks measured each day. It has been claimed that the process has average width 0.275 with $\sigma = 0.1$.[7]

(a) If the seam widths at this assembly line are normally distributed, then what is the probability of finding a seam wider than 1/2 inch?

(b) If the process is under control, what is the probability of finding the mean of a daily sample of 10 widths more than 3 standard errors away from $\mu = 0.275$?

(c) Group the data by days and generate *X*-bar and *S*-charts, putting the limits at ± 3 SE. Is the process under control?

(d) If the process is under control, how does looking at *both* the *X*-bar and *S*-charts affect the chance for reaching an incorrect decision that the process is not in control, compared to looking at just the *X*-bar chart?

(e) Ten measurements are averaged each day. Is this a large enough sample size to justify using a normal model to set the limits in the *X*-bar chart? Do you recommend changes in future testing?

35. **Insulator** One stage in the manufacture of semiconductor chips applies an insulator on the chips. This process must coat the chip evenly to the desired thickness of 250 microns or the chip will not be able to run at the desired speed. If the coating is too thin, the chip will leak voltage and not function reliably; if the coating is too thick, the chip will overheat. The process has been designed so that 95% of the chips have a coating thickness between 247 and 253 microns. Twelve chips were measured daily for 40 days, a total of 480 chips.[8]

(a) Do the data meet the sample size condition if we look at samples taken each day?

(b) Group the data by days and generate *X*-bar and *S*-charts with control limits at ± 3 SE. Is the process under control?

(c) Describe the nature of the problem found in the control charts. Is the problem an isolated incident (which might be just a chance event), or does there appear to be a more systematic failure?

36. **Web Hits** Most commercial Internet sites monitor site traffic so that they can recognize attacks by hackers, such as a distributed denial of service attack (DDoS). In this attack, the hacker hijacks a collection of the computers connected to the Internet and has them all attempt to access the commercial site. The sudden increase in activity overloads the server and makes it impossible for legitimate users to log in. The systems engineer says that during typical use, the site gets 20 hits per minute, on average. He didn't have any idea when asked about a standard deviation. These data show the number of hits per minute over three hours one day, from 7:30 to 10:30 A.M.

(a) Group the data into 15-minute periods and generate *X*-bar and *S*-charts with control limits at ± 3 SE. Is the process under control?

(b) Describe the nature of the problem found in the control charts. Do the data suggest that a DDoS attack is happening?

4M Monitoring an Email System

A firm monitors the use of its email system. A sudden change in activity might indicate a virus spreading in the system, and a lull in activity might indicate problems on the network. When the system and office are operating normally, about 16.5 messages move through the system on average every minute, with a standard deviation near 8.

The data for this exercise count the number of messages sent every minute, with 60 values for each hour and eight hours of data for four days (1,920 rows). The data cover the period from 9 A.M. to 5 P.M. The number of users on the system is reasonably consistent during this time period.

Motivation

a) Explain why the firm needs to allow for variation in the underlying volume. Why not simply send engineers in search of the problem whenever e-mail use exceeds a rate of, say, 1,000 messages?

b) Explain why it is important to monitor both the mean and the variance of volume of email on this system.

Method

c) Because the computer support team is well-staffed, there is minimal cost (aggravation aside) in having someone check for a problem. On the other hand, failing to identify a problem could be serious because it would allow the problem to grow in magnitude. What value do you recommend for α, the chance of a Type I error?

d) To form a control chart, we will accumulate the counts into blocks of 15 minutes rather than use the raw counts. What are the advantages and disadvantages of computing averages and SDs over a 15-minute period compared to using the data for 1-minute intervals?

---

[7] Ibid.
[8] Ibid.

Mechanics

e) Build the X-bar and S-charts for these data with $\alpha = 0.0027$ (i.e., using control limits at ± 3 SE). Do these charts indicate that the process is out of control?

f) What is the probability that the control charts in part (e) signal a problem even if the system remains under control over these four days?

g) Repeat part (e), but with the control limits set according to your choice of α. (*Hint:* If you used $\alpha = 0.0027$, think harder about part (c).) Do you reach a different conclusion?

Message

h) Interpret the result from your control charts (using your choice of α) using nontechnical language. Does a value outside the control limits guarantee that there's a problem?

4M Quality Control and Finance (Dell Stock Prices)

Control charts are also useful when you look at financial data. These charts help you recognize that some processes are not very simple: Their features change over time. X-bar and S-charts are convenient to plot means and standard deviations, and the control limits supply a point of reference for deciding whether something unusual is going on.

For this exercise, consider the daily prices of stock in Dell Computers during 2002–2003.

Motivation

a) Explain why a sequence of prices of a stock will not appear to be in control. Confirm your explanation by looking at the timeplot of the prices of Dell stock during these years.

b) Explain whether it would be useful to monitor both the mean and the variance of the sequence of daily returns on a stock.

Method

c) If you observe a period of stable behavior, suggest how you can use these data to set the mean and SD for monitoring the future returns.

d) How many days should you collect data in order to compute a mean and standard deviation for tracking in the control chart? In particular, if you use 10 days (about two weeks) will you be able to rely on the Central Limit Theorem? How can you tell?

Mechanics

e) Use the data for 2002 to choose values for μ and σ. Does the series of returns form a simple time series?

f) Do the returns in 2002 appear to be normally distributed?

g) Generate the control chart for 2003 (10 days at a time) using values for μ and σ chosen in part (e). Are the returns under control?

Message

h) Explain your findings in language that an investor will understand. Avoid statistical jargon (e.g., Type I error, α, and the like).

Confidence Intervals

THIS CHART FROM THE FEDERAL RESERVE SHOWS THAT U.S. HOUSEHOLDS HAVE ACCUMULATED CLOSE TO $15 *TRILLION* IN OUTSTANDING DEBT. That's enough to buy *everything* produced in the European Union this year on credit. To get a share of the revenue generated by this borrowing, universities, museums, and zoos ally with major lenders to sponsor an affinity credit card. Before agreeing to a deal, the sponsor and lender, such as Chase or Bank of America, want to be sure it will be profitable. It costs money to entice customers to sign up, process the application, and manage the account.

Two characteristics determine the initial success of these credit cards. First, a substantial proportion of those who receive an offer have to accept. If only one person signs up, the originating bank won't make a profit. Second, those who accept the offer have to use

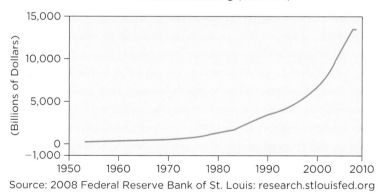

Household Sector: Liabilities: Household Credit Market Debt Outstanding (CMDEBT)

Source: 2008 Federal Reserve Bank of St. Louis: research.stlouisfed.org the card. The big profits lie in the interest earned on outstanding balances, typically about 18% annually. Unless enough customers accept a card and carry a balance, the offering will not generate the profits that justify the start-up costs.

How many customers will accept an offer of a new credit card? How large a balance will they carry? Until it can answer these questions, a bank cannot make an informed decision on whether to launch a new credit card.

This chapter introduces confidence intervals that can be used to answer such questions on the basis of the information in a sample. Given a sample, a confidence interval provides a range that estimates an unknown parameter of the sampled population, such as a mean or proportion. The length of the confidence interval directly conveys the precision of the estimate. When the sample reveals a great deal about the population, the resulting narrow confidence intervals suggest precise answers. With less information, we get wide confidence intervals that indicate that we know less about the population.

15.1 | RANGES FOR PARAMETERS

To introduce confidence intervals, let's evaluate the launch of a new affinity credit card. The contemplated launch proposes sending preapproved applications to 100,000 alumni of a large university. That's the population. Two parameters of this population determine whether the card will be profitable:

- p, the proportion who *will* return the application
- μ, the average monthly balance that those who accept the card *will* carry

To estimate these parameters, the credit card issuer sent preapproved applications to a sample of 1,000 alumni. Of these, 140 accepted the offer (14%) and received a card. The histogram in Figure 15.1 shows the average revolving balance for these customers in the 3 months after receiving the card.

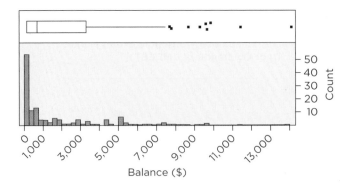

FIGURE 15.1 Balances of a sample of credit card accounts.

The distribution of balances is right skewed, with a large number at zero. An account with zero balance identifies either a transactor (who pays the statement balance in full each month to avoid finance charges) or someone who carries this card only to use "just in case." Table 15.1 summarizes the key statistics of the trial launch of the affinity card.

TABLE 15.1 Summary statistics in the test of a new affinity credit card.

Number of offers	1,000
Number accepted	140
Proportion who accepted	$\hat{p} = 0.14$
Average (balance)	$\bar{x} = \$1,990.50$
SD (balance)	$s = \$2,833.33$
Skewness K_3	1.750
Kurtosis K_4	2.975

Among those who received a card, the average monthly revolving balance is $1,990.50 with standard deviation of $2,833.33.

Confidence Interval for the Proportion

What should we conclude about p, the proportion in the population of 100,000 alumni who will accept the offer if the card is launched on a wider scale? To answer this question, we will construct a confidence interval. A **confidence interval** is a range of plausible values for a parameter based on a sample from the population. Confidence intervals provide an intuitive measure of the influence of sampling variation. We know that the value of a statistic such as $\hat{p}$ depends on the sample we observe. Had we gotten a different sample of alumni in this example, it's unlikely that the proportion accepting the card in that sample would also be 0.14. How different from 0.14 might the proportion in this second sample be? Once we recognize that different samples have different proportions, what can we conclude about p, the proportion in the population? Is $\hat{p}$ close to p?

The procedure for constructing a confidence interval resembles the development of control limits discussed in the previous chapter. Both control limits and confidence intervals rely on the sampling distribution of a statistic. The difference arises in the roles played by the sample and population. Control charts presume a value for a process parameter, then provide control limits that indicate where the mean of a sample should be found. The process parameters determine these limits. Confidence intervals work in the other direction. A sample determines a range for the population parameter.

The key tool used in building a confidence interval is the sampling distribution of a statistic. The sampling distribution of the sample proportion describes how $\hat{p}$ varies from sample to sample. If there's little variation from sample to sample, then not only are the sample proportions similar to one another, they are also close to the population proportion p. Because proportions are averages (see Behind the Math: Proportions Are Averages), the Central Limit Theorem implies a normal model for the sampling distribution of $\hat{p}$ if the sample size n is large enough.

(p. 370)

$$\hat{p} \sim N\left(p, \frac{p(1-p)}{n}\right)$$

The mean and variance of this normal model come from properties of binomial random variables (Chapter 11). The expected value of $\hat{p}$ is the population proportion p, so p lies at the center of the sampling distribution. The variance of $\hat{p}$ resembles that of $\bar{X}$, but with $p(1-p)$ in place of σ^2. The standard error of $\hat{p}$ is

$$SE(\hat{p}) = \sqrt{\frac{p(1-p)}{n}}$$

confidence interval A range of values for a parameter that is compatible with the data in a sample.

Since the sample-to-sample variation of $\hat{p}$ is normally distributed if n is large enough, proportions in 95% of samples are within about two standard errors of the mean, p. More precisely, if we use the percentile of the normal distribution, $z_{0.025} = 1.96$, then it is true that

$$P(-1.96 \; \mathrm{SE}(\hat{p}) \leq \hat{p} - p \leq 1.96 \; \mathrm{SE}(\hat{p})) = 0.95$$

In words, $\hat{p}$ lies within 1.96 standard errors of p in 95% of samples. Figure 15.2 shows the normal model for the sampling distribution of $\hat{p}$ and shades the region within 1.96 standard errors of p.

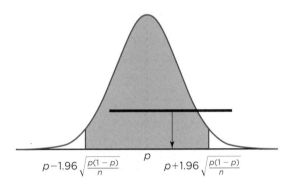

FIGURE 15.2 Normal model of the sampling distribution of the sample proportion.

Because we choose one of these samples at random (an SRS), the proportion in our sample is a random draw from the sampling distribution in Figure 15.2. There is a 95% chance that $\hat{p}$ lies in the shaded region of the sampling distribution, so $\hat{p}$ is within 1.96 standard errors of p. **It is worth repeating this property of random samples: The sample statistic in 95% of samples lies within about two standard errors of the population parameter.** The arrow in Figure 15.2 indicates $\hat{p}$ in one such sample. For any of these samples, the interval formed by reaching 1.96 standard errors to the left and right of $\hat{p}$ holds p, illustrated by the heavy line centered at the arrow in Figure 15.2. It follows that when we draw a random sample, there is a 95% chance that we will get one of these samples for which this interval includes p.

$$P(\hat{p} - 1.96 \; \mathrm{SE}(\hat{p}) \leq p \leq \hat{p} + 1.96 \; \mathrm{SE}(\hat{p})) \approx 0.95$$

If the standard error is small, then $\hat{p}$ is probably close to p.

　　We almost have a useful procedure, but there's one more issue to resolve. Because we do not know p, we do not know the standard error of $\hat{p}$. The plug-in estimated standard error

SE vs. se
SE uses population parameters. **se** estimates SE by substituting sample statistics for population parameters.

$$\mathrm{se}(\hat{p}) = \sqrt{\frac{\hat{p}(1 - \hat{p})}{n}}$$

that uses $\hat{p}$ in place of p, however, works quite well. It remains approximately true that $\hat{p}$ lies within $1.96\,\mathrm{se}(\hat{p})$ of p in 95% of samples.

$$P(\hat{p} - 1.96 \; \mathrm{se}(\hat{p}) \leq p \leq \hat{p} + 1.96 \; \mathrm{se}(\hat{p})) \approx 0.95$$

The interval from $\hat{p} - 1.96\,\mathrm{se}(\hat{p})$ to $\hat{p} + 1.96\,\mathrm{se}(\hat{p})$ is the 95% **confidence interval for p**. It is often called a **z-interval** because it uses the percentile from a normal distribution ($z_{0.025} = 1.96$) to set the number of multiples of the standard error.

z-interval A confidence interval based on using a normal model for the sampling distribution.

　　A confidence interval is said to cover its parameter if the parameter lies within the interval. If

$$\hat{p} - 1.96\,\mathrm{se}(\hat{p}) \leq p \leq \hat{p} + 1.96\,\mathrm{se}(\hat{p})$$

then the interval covers p. The value 95%, or 0.95, is the **coverage**, or **confidence level**, of the interval. To alter the coverage, change the number of multiples of the standard error by using a different percentile of the normal distribution. For instance, the interval with $z_{0.005} = 2.58$ has 99% coverage because

$$P(\hat{p} - 2.58\,\text{se}(\hat{p}) \leq p \leq \hat{p} + 2.58\,\text{se}(\hat{p})) = 0.99$$

Notice that increasing the coverage requires increasing the width of the confidence interval.

Confidence Interval for p The $100(1 - \alpha)\%$ z-interval for p is the interval from

$$\hat{p} - z_{\alpha/2}\sqrt{\hat{p}(1 - \hat{p})/n} \qquad \text{to} \qquad \hat{p} + z_{\alpha/2}\sqrt{\hat{p}(1 - \hat{p})/n}$$

where $P(Z > z_\alpha) = \alpha$ for $Z \sim N(0, 1)$. For a 95% confidence interval, use $z_{\alpha/2} = 1.96$.

Checklist

✓ **SRS condition.** The observed sample is a simple random sample from the relevant population. If taken from a finite population, the sample comprises less than 10% of the population.

✓ **Sample size condition (for proportion).** Both $n\hat{p}$ and $n(1 - \hat{p})$ are larger than 10. These bounds control skewness and kurtosis.

Assumptions and Conditions

The confidence interval for the proportion requires two assumptions. We never know whether these *assumptions* are true, but we can check the relevant *conditions*.

The first condition is critical. Don't let the details of building a confidence interval disguise the importance of beginning with an SRS. **A confidence interval depends on beginning with a simple random sample from the population.** Without an SRS, you won't have a meaningful confidence interval. Before starting calculations, confirm that your data are representative. For credit cards, that would mean checking that this test market is representative of the success in other communities: Alumni elsewhere may be more or less receptive. The confidence interval would not adjust for this bias in the sample.

The second condition ensures the sample is large enough to justify using a normal model to approximate the sampling distribution of $\hat{p}$. If both $n\hat{p}$ (the number of successes) and $n(1 - \hat{p})$ (the number of failures) are larger than 10, then the Central Limit Theorem produces a good approximation to the sampling distribution. (Chapter 17 describes confidence intervals that can be used for proportions if the sample size condition is not met.)

For the credit card experiment, $n = 1{,}000$ and $\hat{p} = 0.14$. First, we check the conditions. The data are a simple random sample and constitute less than 10% of the local population of alumni, so these data meet the SRS condition. The data also meet the sample size condition because we observe more than 10 successes and 10 failures. We can calculate the estimated standard error:

$$\text{se}(\hat{p}) = \sqrt{\frac{\hat{p}(1 - \hat{p})}{n}} = \sqrt{\frac{0.14(1 - 0.14)}{1000}} \approx 0.01097$$

Hence the 95% confidence interval for p is the range

$$\hat{p} - 1.96\,\text{se}(\hat{p}) \qquad \text{to} \qquad \hat{p} + 1.96\,\text{se}(\hat{p})$$

$$0.14 - 1.96 \times 0.01097 \quad \text{to} \quad 0.14 + 1.96 \times 0.01097 \approx [0.1185 \text{ to } 0.1615]$$

We often write a confidence interval with its endpoints enclosed in square brackets, [... to ...]. We can also write the interval in the preceding equation as [11.85% to 16.15%]; percentage symbols clarify that the interval describes a proportion. With 95% confidence, the population proportion that will accept this offer is between about 12% and 16%. If the bank decides to launch the card, might 20% accept the offer? It's possible but not likely given the information in our sample; 20% is outside the 95% confidence interval for p. The bank is more likely to be successful in its planning if it anticipates a response rate within the interval.

In practice, most confidence intervals are 95% intervals. The choice of 95% coverage is traditional in statistics but is not required. We could offer planners a wider 99% interval or a narrower 80% or even 50% interval. It's a question of how we balance the width of the interval with the chance that the interval does not include the parameter. For instance, the 99% interval in this example is $0.14 \pm 2.58(0.01097) \approx [11\% \text{ to } 17\%]$; this interval reduces the chance of missing p from 0.05 down to 0.01. On the other hand, it conveys less precision to the decision maker. The 80% interval, $0.14 \pm 1.28(0.01097) \approx [13\% \text{ to } 15\%]$, looks precise, but is produced by a method that is more likely to omit p.

Ideally, we would like a narrow 99.99% interval. Such an interval almost certainly contains the parameter *and* conveys that we have a very good idea of its value; however, we generally cannot have both a narrow interval and exceptionally high coverage. Over time, statisticians have settled on 95% coverage as a compromise. Lower coverage leads to more intervals that do not contain the parameter; higher coverage leads to unnecessarily long intervals. The convention of using 95% coverage has become so strong that readers may become suspicious when the coverage differs from this benchmark.

Sample Size

Confidence intervals show the benefit of having a large sample. The larger the sample grows, the smaller the standard error. The smaller the standard error becomes, the narrower the 95% confidence interval. A narrow interval conveys that we have a precise estimate of the parameter.

In the example, the width of the confidence interval for p with $n = 1,000$ is about 4.3%, $0.1615 - 0.1185 = 0.043$. Had the sample been one-fourth as large, with $n = 250$ and $\hat{p} = 0.14$, the standard error would have been twice as big. With $n = 250$,

$$\text{se}(\hat{p}) = \sqrt{\frac{0.14(1 - 0.14)}{250}} \approx 0.02195$$

As a result, the confidence interval would be twice as wide.

$$[0.14 - 1.96(0.02195) \text{ to } 0.14 + 1.96(0.02195)] \approx [0.09698 \text{ to } 0.1830]$$

That confidence intervals narrow as we learn more about the population mimics how people behave. When asked a question about a topic about which we know little, it's common to reply vaguely, giving a wide range. For instance, what's the GDP of Brazil? Unless you've been studying South American economies or come from Brazil, your best guess is likely to come with a wide range. Someone who has studied South American economies, however, might reply "$1.6 to $1.7 trillion."[1] Confidence intervals suggest precision in this way as well.

---

[1] According to its online fact book, the CIA estimated the GDP of Brazil to have been $1.655 trillion in 2006.

| **What Do You Think?** | An auditor checks a sample of 225 randomly chosen transactions from among the thousands processed in an office. Thirty-five contain errors in crediting or debiting the appropriate account. |

a. Does this situation meet the conditions required for a z-interval for the proportion?[2]
b. Find the 95% confidence interval for p, the proportion of all transactions processed in this office that have these small errors.[3]
c. Managers claim that the proportion of errors is about 10%. Does that seem reasonable?[4]

15.2 | CONFIDENCE INTERVAL FOR THE MEAN

A similar procedure produces a confidence interval for μ, the mean of a population. The similarity comes from using a normal model for the sampling distribution of the statistic,

$$\overline{X} \sim N\left(\mu, \frac{\sigma^2}{n}\right)$$

This sampling distribution implies that

$$P(\overline{X} - 1.96\sigma/\sqrt{n} \le \mu \le \overline{X} + 1.96\sigma/\sqrt{n}) = 0.95$$

The average of 95% of samples lies within $1.96\sigma/\sqrt{n}$ of μ. Once again, *the sample statistic lies within about two standard errors of the corresponding population parameter in 95% of samples.* As in Chapters 13 and 14, $\overline{X}$ is a random variable that represents the mean of a randomly chosen sample, and $\overline{x}$ is the mean of the observed sample. (We should do the same for $\hat{p}$, but $\hat{P}$ looks too much like the P that stands for probability.)

As when finding the confidence interval for p, we do not know the standard error of $\overline{X}$. The solution is to plug in a sample statistic in place of the unknown population parameter. In this case, we replace σ by the standard deviation s of the sample. As before, we label the estimated standard error using lowercase letters.

$$\text{SE}(\overline{X}) = \frac{\sigma}{\sqrt{n}} \qquad \text{se}(\overline{X}) = \frac{s}{\sqrt{n}}$$

When dealing with averages, we also make a second adjustment that accounts for the use of s in place of σ.

Student's *t*-Distribution

Student's *t*-distribution A model for the sampling distribution that adjusts for the use of an estimate of σ.

Student's *t*-distribution compensates for substituting s for σ in the standard error. Provided that the data are a sample from a population that is normally distributed, Student's t-distribution is the *exact* sampling distribution of the random variable

$$T_{n-1} = \frac{\overline{X} - \mu}{S/\sqrt{n}}$$

[2] Yes; $\hat{p} = 35/225 \approx 0.156$ and $n\hat{p}$ and $n(1 - \hat{p})$ are larger than 10. Presumably, this is an SRS, and the sample is less than 10% of the population.
[3] $\hat{p} \pm 1.96\sqrt{\hat{p}(1 - \hat{p})/n} = 0.156 \pm 1.96 \times 0.0242 = 0.156 \pm 0.047$, or [0.109 to 0.203].
[4] No, 10% seems too low (at 95% confidence). It appears that the proportion is larger.

degrees of freedom (df) The denominator of an estimate of σ^2, $n - 1$ for testing the mean.

(p. 370)

The sampling distribution of T_{n-1} is not normal because the divisor varies from sample to sample. (The capital S in the denominator is a reminder that the sample standard deviation is also a random variable that changes from sample to sample.)

A parameter known as the **degrees of freedom (df)** controls the shape of Student's t-distribution. The degrees of freedom is the divisor in S^2, $n - 1$. The larger n, the better S^2 estimates σ^2 and the more a t-distribution resembles a standard normal distribution. Both distributions are symmetric and bell shaped. The important difference lies in the extremes: A t-distribution spreads its area farther from the center. The tails cover samples in which S is smaller than σ, inflating $|T_{n-1}|$ (see Behind the Math: Degrees of Freedom).

The graphs in Figure 15.3 show the standard normal distributions (solid black) and t distributions (dashed red) with 3 or 9 degrees of freedom.

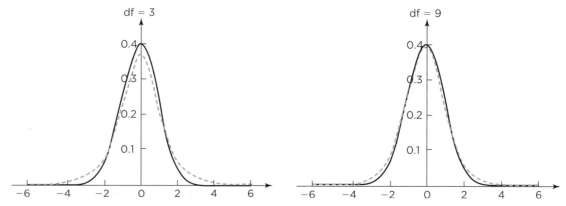

FIGURE 15.3 Standard normal (black) and Student's t-distributions (red).

To show the differences more clearly, the graphs in Figure 15.4 magnify the right tails of the distributions and show that the t-distribution puts much more area farther from 0 than the normal distribution does.

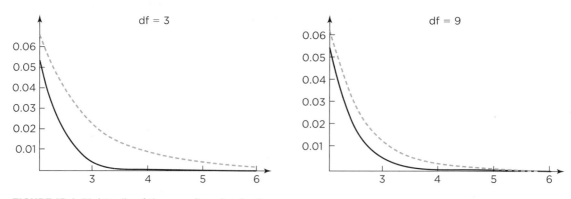

FIGURE 15.4 Right tails of the sampling distributions.

The introduction of degrees of freedom often obscures an important aspect of using a t-distribution: It presumes that the population is normally distributed. The Central Limit Theorem shows that a normal model approximates the distribution of the average when sampling *any* population. To obtain the exact adjustment for the use of S, however, the t-distribution assumes that the sample comes from a *normal* population. Fortunately, research has shown that the t-distribution works well even when the population is not exactly normally distributed.

t-Interval for the Mean

Percentiles from the t-distribution with $n - 1$ degrees of freedom determine the number of standard errors needed for 95% coverage.

$$P(-t_{0.025,n-1}S/\sqrt{n} \leq \overline{X} - \mu \leq t_{0.025,n-1}S/\sqrt{n}) \approx 0.95$$

Percentiles from the t-distribution are slightly larger than those from the normal distribution (for example, $t_{0.025,n-1} > z_{0.025}$), making the t-interval wider than the corresponding z-interval. The differences become small once n reaches 30 or more.

> **Confidence Interval for μ** The $100(1 - \alpha)\%$ confidence t-interval for μ is
>
> $$\overline{x} - t_{\alpha/2,n-1}s/\sqrt{n} \quad \text{to} \quad \overline{x} + t_{\alpha/2,n-1}s/\sqrt{n}$$
>
> where $P(T_{n-1} > t_{\alpha/2,n-1}) = \alpha/2$ for T_{n-1} distributed as a student's t-random variable with $n - 1$ degrees of freedom.
>
> **Checklist**
>
> ✓ **SRS condition.** The observed sample is a simple random sample from the relevant population. If taken from a finite population, the sample comprises less than 10% of the population.
> ✓ **Sample size condition.** The sample size is larger than 10 times the squared skewness and 10 times the absolute value of the kurtosis, $n > 10K_3^2$ and $n > 10|K_4|$ (see Chapters 12 and 14).

In the credit card example, $n = 140$ customers accepted the offer and carry an average monthly balance $\overline{x} = \$1,990.50$ with $s = \$2,833.33$. The squared skewness and absolute value of the kurtosis of the balances are $K_3^2 \approx 3.1$ and $|K_4| \approx 3.0$ (see Table 15.1), so these data satisfy the sample size condition ($n > 10K_3^2$ and $n > 10|K_4|$).

To obtain the 95% t-interval for μ, first notice that we have $n - 1 = 140 - 1 = 139$ degrees of freedom. To find the percentile $t_{0.025,139}$, use software or the table in the Appendix of this book. Tables that give percentiles $t_{\alpha/2,n-1}$ of the t-distribution differ from those for the standard normal because the t-distribution depends on n. Table 15.2 shows how the information is presented. Each row in the table is labeled by the degrees of freedom. The columns give percentiles $t_{\alpha,df}$.

TABLE 15.2 Percentiles of the t-distribution (a larger table is in the Appendix).

df	0.10	0.05	0.025	0.01	0.005
			α		
1	3.078	6.314	12.706	31.821	63.657
2	1.886	2.920	4.303	6.965	9.925
⋮	⋮	⋮	⋮	⋮	⋮
100	1.290	1.660	1.984	2.364	2.626
150	1.287	1.655	1.976	2.351	2.609
Normal	1.282	1.645	1.960	2.327	2.576

The distribution of T_{n-1} approaches a standard normal distribution as n increases. The bottom row of Table 15.2 gives the corresponding percentiles for a normal distribution. With *enough* degrees of freedom, the t-distribution matches the standard normal distribution.

In this example, the percentile from the t-distribution is $t_{0.025,139} \approx 1.98$, which is slightly larger than $z_{0.025} = 1.96$. The 95% confidence interval for the average balance carried by alumni who accept this credit card is

$$\bar{x} \pm t_{0.025,139}\, s/\sqrt{140}$$
$$= -1,990.50 - 1.98\,(2,833.33/\sqrt{140}) \quad \text{to} \quad 1,990.50 + 1.98\,(2,833.33/\sqrt{140})$$
$$\approx \$1,516.369 \quad \text{to} \quad \$2,464.631$$

We are 95% confident that μ lies between $\$1,516.36$ and $\$2,464.63$. Might μ be $\$2,000$? Yes, $\$2,000$ lies within the confidence interval. Might μ be $\$1,250$? It could be, but that's outside the confidence interval and *not* compatible with our sample at a 95% level of confidence.

What Do You Think?

Office administrators claim that the average amount on a purchase order is $\$6,000$. A SRS of 49 purchase orders averages $\bar{x} = \$4,200$ with $s = \$3,500$. Assume the data meet the sample size condition.

a. What is the relevant sampling distribution?[5]
b. Find the 95% confidence interval for μ, the mean of purchase orders handled by this office during the sampled period.[6]
c. Do you think the administrators' claim is reasonable?[7]

15.3 | INTERPRETING CONFIDENCE INTERVALS

In the preceding section, we wrote, "We are 95% confident that μ lies in the range $\$1,516.369$ to $\$2,464.631$." Let's figure out what that means.

tip   **The first step in interpreting a confidence interval is to round the endpoints to something sensible.** Although we need to carry out the calculations to full precision, it makes little sense to present endpoints to the nearest $\$0.001$ if the interval includes values of μ that range from $\$1,700$ to $\$2,200$. The number of digits to show depends on the sample size, but most intervals should be rounded to two or three digits. For example, we'd summarize the interval for μ in the credit-card example after rounding as $[\$1,500$ to $\$2,500]$ or $[\$1,520$ to $\$2,460]$. Whatever the choice, do the intermediate calculations to full accuracy; rounding comes only at the last step. (After rounding, we typically obtain the same interval whether we use $\bar{x} \pm t_{0.025,n-1}s/\sqrt{n}$ or $\bar{x} \pm 2s/\sqrt{n}$. The latter interval, the estimate ± 2 standard errors, is a handy back-of-the-envelope 95% confidence interval.)

Now that we've rounded the endpoints to something easier to communicate, let's think about the claim "We are 95% confident...." Imagine placing a bet with a friend *before* seeing the sample. The outcome of the bet depends on whether the confidence interval covers μ. We wager $\$95$ that the interval $\bar{X} \pm t_{0.025,n-1}S/\sqrt{n}$ will cover μ; our friend bets $\$5$ that it won't. Whoever is right gets all $\$100$. Now, let's add the fact that our friend knows μ, but not which sample will be observed. Even though our friend knows μ, this remains a fair bet because our share of the total wager ($\$95$ out of $\$100$) matches our chance for winning the pot.

---

[5] The sampling distribution of the average is approximately $N(\mu, s^2/49)$.
[6] $4,200 \pm 2.01[3,500/\sqrt{49}] = [\$3,195$ to $\$5,205](t_{0.025,48} = 2.01)$.
[7] No, $\$6,000$ lies far above the confidence interval and is not compatible with these data.

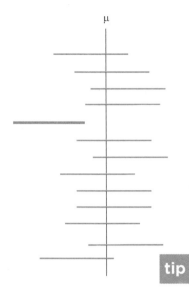

Compare that situation to the same bet *after* calculating the interval. Do we really want to bet in this case? Our friend knows μ, so she knows whether μ lies inside our confidence interval. If she takes this bet, don't expect to win!

The difference lies in the timing. The formula $\bar{X} \pm t_{0.025,n-1}S/\sqrt{n}$ produces a range that covers μ in 95% of samples. The observed sample is either one of the good ones, in which the confidence interval covers μ, or it's not. When we say "We are 95% confident," we're describing a *procedure* that works for 95% of samples. If we lined up the intervals from many, many samples, 95% of these intervals would contain μ and 5% would not. Our sample and confidence interval are a random draw from these. We either got one of the 95% of samples for which the confidence interval covers μ or not.

Common Confusions

To avoid common mistakes, let's consider several *incorrect* interpretations of a confidence interval. **Most errors of interpretation can be spotted by remembering that a confidence interval offers a range for a population parameter.** Confidence intervals describe the population parameter, not the data in this sample or another sample.

1. *"Ninety-five percent of all customers keep a balance of $1,520 to $2,460."* This error is so common that it's worth looking again at the histogram of the 140 customer balances shown in Figure 15.5.

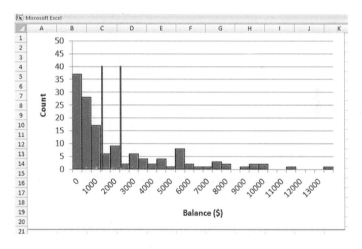

FIGURE 15.5 Histogram of account balances showing the confidence interval.

The confidence interval (identified by the two vertical red lines) doesn't contain many data, much less 95% of the balances. The confidence interval gives a range for the population mean μ, not the balance of an individual.
2. *"The mean balance of 95% of samples of 140 accounts will fall between $1,520 and $2,460."* The confidence interval describes μ, an unknown constant, not the means of other samples.
3. *"The mean balance μ is between $1,520 and $2,460."* Closer, but still incorrect. The average balance in the population does not *have* to fall between $1,520 and $2,460. This is a 95% confidence interval. It might not contain μ.

Here's the way we prefer to state a confidence interval when presenting results to a nontechnical audience:

"I am 95% confident that the mean monthly balance for the population of customers who accept an application lies between $1,520 and $2,460."

The phrase "95% confident" hides a lot. It's our way of saying that we're using a procedure that produces an interval that covers μ in 95% of samples.

Follow the Money The profitability of a new credit card depends on whether income from those who get a card is enough to pay for the expense of launching the card. Promotional costs in this example are $300,000 for advertising plus $5 to send each of the 100,000 mailed offers, a total of $800,000. For those who accept the offer, the bank spends another $50 to set up an account. For profits, the bank earns 10% of the average balance. Let's work through two examples.

> **profitable scenario** 10% accept, $2,500 balance; bank profits $1.2 million.

If $\hat{p}$ = 10% of those who receive the offer accept, it costs $500,000 to set up 10,000 accounts at $50 each; the total cost of the launch is $800,000 + $500,000 = $1.3 million. If the average monthly balance of these 10,000 accounts is $\bar{x}$ = $2,500, the bank earns $2.5 million. After subtracting the costs, the bank nets $1.2 million. Here's the formula.

$$\text{Profit} = 100,000 \times \hat{p} \times (\bar{x}/10 - \$50) - \$800,000$$
$$= 100,000 \times 0.10 \times (\$2,500/10 - \$50) - \$800,000$$
$$= \$1,200,000$$

Consider what happens if, however, 5% return the application and carry a smaller average balance of $1,500. The bank still has $800,000 in promotional costs, plus $250,000 to set up the accounts, totaling $1,050,000. The interest from 5,000 customers who carry a $1,500 balance is $750,000, and so the bank loses $300,000.

> **unprofitable scenario** 5% accept, $1,500 balance; bank loses $.3 million.

$$\text{Profit} = 100,000 \times 0.05 \times (\$1,500/10 - \$50) - \$800,000$$
$$= -\$300,000$$

15.4 | MANIPULATING CONFIDENCE INTERVALS

We can manipulate a confidence interval to obtain ranges for related quantities. For instance, suppose that federal regulators require a bank to maintain cash reserves equivalent to 10% of the outstanding balances. How much cash is that on average per customer? Because the 95% confidence interval for μ is [$1,520 to $2,460], the 95% confidence interval for 10% of μ (0.1 $\times$ μ) is [$152 to $246].

If [L to U] is a 100(1 − α)% confidence interval for μ, then

$$[c \times L \text{ to } c \times U]$$

is a 100(1 − α)% confidence interval for $c \times \mu$ and

$$[c + L \text{ to } c + U]$$

is a 100(1 − α)% confidence interval for $c + \mu$.

The same rule applies if the parameter is p rather than μ. More generally, you can substitute a confidence interval for a parameter in many types of mathematical expressions. Functions like logs, square roots, and reciprocals are monotone; they keep going in the same direction.

If [L to U] is a 95% confidence interval for μ and f is a monotone increasing function, then

$$[f(L) \text{ to } f(U)]$$

is a 95% confidence interval for $f(\mu)$.

If *f* is a monotone decreasing function, then

$$[f(U) \text{ to } f(L)]$$

is a 95% confidence interval for $f(\mu)$.

For example, if $[1.30 \text{ \$/€ to } 1.70 \text{ \$/€}]$ is a 95% confidence interval for the expected value of the dollar/euro exchange rate, then

$$[1/1.70 = 0.588 \text{ €/\$ to } 1/1.30 = 0.769 \text{ €/\$}]$$

is a 95% confidence interval for the reciprocal, the expected euro/dollar exchange rate.

Combining Confidence Intervals

With $p = 0.14$ and $\mu = \$1,990.50$, Table 15.3 shows a solid profit for the bank if it rolls out this affinity card.

TABLE 15.3 Profitability if sample statistics match parameters of the population.

	$300,000	Up-front cost of promotion
	$500,000	Mailing costs (100,000 @ $5)
	$700,000	Cost to set up accounts (14,000 @ $50)
Total cost	$1,500,000	
Income	$2,786,700	Interest (14,000 @ 0.1 × $1,990.50)
Net profit	$1,286,700	

The calculation of the profit depends on several statistics.

$$
\begin{aligned}
\text{Profit} &= 100,000 \times \hat{p} \times (\bar{x}/10 - 50) - 800,000 \\
&= 100,000 \times 0.14 \times (1,990.50/10 - 50) - 800,000 \\
&= \$1,286,700
\end{aligned}
$$

It is unrealistic, however, to expect p and μ to match $\hat{p}$ and $\bar{x}$. Just because 14% of *these* alumni return the application does not mean that 14% *of the population* will. An average balance of $1,990.50 in this sample does not guarantee $\mu = \$1,990.50$.

Confidence intervals are handy in this situation: Substitute intervals for $\hat{p}$ and $\bar{x}$ in place of the observed values and propagate the uncertainty. Rather than a number, the result is a range that expresses the uncertainty. For the product of two intervals, we combine the intervals by multiplying their limits: *lower × lower* and *upper × upper*. (We round the intervals to avoid cluttering the calculations with extra digits.)

$$
\begin{aligned}
\text{Profit} &= 100,000 \times [0.12 \text{ to } 0.16] \times ([1,520 \text{ to } 2,460]/10 - 50) - 800,000 \\
&= 100,000([0.12 \text{ to } 0.16] \times [102 \text{ to } 196]) - 800,000 \\
&= [1,224,000 \text{ to } 3,136,000] - 800,000 \\
&= [\$424,000 \text{ to } \$2,336,000]
\end{aligned}
$$

Even at the low side of the interval, the bank makes a profit, and the potential profits could reach more than $2.3 million.

This approach works well when exploring scenarios. These calculations separate $\hat{p}$ (usually determined by marketing) from the interest earnings represented by $\bar{x}/10$. The final expression shows how uncertainty in estimates of p and μ lead to uncertainty in the profits. That transparency provides talking points for discussions with colleagues in marketing and finance.

The weakness of this approach is that we end up with a confidence interval with unknown coverage. About all we can conclude is that the coverage of the interval [$424,000 to $2,336,000] is more than 90%. (See Behind the Math: Combining Confidence Intervals.)

(p. 371)

Changing the Problem

We can often avoid combining confidence intervals by creating a new variable. In this example, let's work directly with the profit earned from each customer rather than indirectly constructing the profit from $\hat{p}$ and $\bar{x}$.

Each customer who does not accept the card costs the bank $8: the sum of $5 for sending the application plus $3 for promotion. Each customer who accepts the card costs the bank $58 ($8 plus $50 for setting up the account), but the bank earns 10% of the revolving balance. The following variable measures the profit, in dollars, earned from a customer:

$$y_i = \begin{cases} -8, & \text{if offer not accepted} \\ \dfrac{\text{Balance}}{10} - 58, & \text{if offer accepted} \end{cases}$$

We form this variable and calculate the summary in Table 15.4.

TABLE 15.4 Summary statistics for profit earned from each customer.

$\bar{y}$	$12.867
s	$117.674
n	1,000
$s/\sqrt{n}$	$3.721
95% t-interval	$5.565 to $20.169
Skewness	6.175
Kurtosis	44.061

The constructed variable is not normally distributed: 86% of the y's are -8 because 86% of the customers declined the offer. All of these -8s contribute to the skewness and kurtosis. Nonetheless, because of the very large sample, these data meet the sample size condition. We can use the t-interval because $n = 1,000$ is larger than $10K_3^2 \approx 381$ and $10|K_4| \approx 441$. The t-interval for the average profit per offer is $12.867 \pm t_{0.025,999}(3.721) \approx [$5.57 to $20.17]$. For 100,000 offers, the interval extends from $557,000 to $2,017,000.

This confidence interval has two advantages over the previous interval for the mean profit. First, it's narrower (about $1.5 million versus $2.0 million). It lies entirely inside the previous interval. Second, this interval *is* a 95% confidence interval. Narrower with known coverage is a good combination. (By combining everything into one variable, however, we no longer distinguish the role of the acceptance rate p from that of the average balance μ.)

15.5 | MARGIN OF ERROR

An informal, back-of-the-envelope 95% confidence interval for μ replaces the percentile from the t-distribution with 2.

$$[\bar{x} - 2s/\sqrt{n} \text{ to } \bar{x} + 2s/\sqrt{n}]$$

margin of error $2\dfrac{s}{\sqrt{n}}$

Though informal, the extent of this interval to either side of $\bar{X}$ (or similarly around $\hat{p}$) is known as the **margin of error (ME)**.

$$\text{Margin of Error} = 2\frac{s}{\sqrt{n}}$$

A *precise* confidence interval has a small margin of error. Three factors determine the margin of error.

1. *Level of confidence.* The multiplier $2 \approx t_{0.025,n-1}$ comes from wanting 95% coverage and using a t-distribution (or normal model) to describe the sampling distribution.
2. *Variation of the data.* The smaller the standard deviation s, the narrower the confidence interval becomes. It's usually not possible to reduce the standard deviation without changing the population.
3. *Number of observations.* The larger the sample, the shorter the interval because the standard error $s/\sqrt{n}$ gets smaller as n increases.

> **caution** The reduction in the margin of error achieved with a larger sample is not proportional to the increase in the sample size. The margin of error decreases proportionally to increases in $\sqrt{n}$. As a result, to cut the margin of error in half, you need four times as many cases. Costs, however, typically rise in proportion to n. So, you'd spend four times as much to slice the margin of error in half.

Determining Sample Size

How large a sample do we need? The usual answer is larger, but data collection costs time and money. How much is enough? Before collecting data, it's a good idea to know whether the sample we can afford is adequate for what we want to learn.

If we know the needed margin of error, we can estimate the necessary sample size. Use this formula to determine the sample size.

$$\text{Margin of Error} = \frac{2\sigma}{\sqrt{n}} \quad \Rightarrow \quad n = \frac{4\sigma^2}{(\text{Margin of Error})^2}$$

The necessary sample size depends on σ. Sample size calculations are rarely exact, so we won't find out if the margin of error is as small as we want until we get the sample. Complicating matters, the sample SD s is not available to estimate σ because we have to choose n *before* collecting the sample. If we have no idea of σ, a common remedy is to gather a small sample to estimate σ. A **pilot sample** is a small sample of, say, 10 to 30 cases used to estimate σ.

pilot sample A small, preliminary sample used to obtain an estimate of σ.

For example, a company is planning a line of frozen organic dinners for women. It would like to design the meals to suit the daily calorie intake of customers. Its nutritionists want to know the average calorie intake of female customers to within ± 50 calories with 95% confidence. A pilot sample of 25 women gives a standard deviation $s = 430$ calories. To obtain the sought margin of error (50 calories) requires a sample size of

$$n = \frac{4s^2}{\text{ME}^2} = \frac{4(430^2)}{50^2} \approx 295.8$$

The company thus plans to survey 300 customers.

In the case of proportions, we don't need a pilot sample. Suppose we're doing a survey of brand awareness. We need the survey to have a margin of error of 3% or less. Whatever the estimate for $\hat{p}$, we want to claim that $\hat{p}$ lies within ± 0.03 of p (with 95% confidence). For this to happen, the margin of error must be no more than 0.03.

$$\text{Margin of Error} = \frac{2\sigma}{\sqrt{n}} \leq 0.03$$

For proportions, $\sigma = \sqrt{p(1-p)} \leq 1/2$ no matter what the value of p.[8] If we choose n so that the margin of error is 0.03 when $\sigma = 1/2$, then the margin of error cannot be larger than 0.03 when we get our sample. If $\sigma = 1/2$, the margin of error is

$$\text{Margin of Error} = \frac{2\sigma}{\sqrt{n}} \leq \frac{2(1/2)}{\sqrt{n}} \leq 0.03$$

The necessary sample size is then

$$n = \frac{1}{(\text{Margin of Error})^2} = \frac{1}{0.03^2} \approx 1{,}111.1$$

A survey with 1,112 or more customers guarantees the margin of error is 0.03 or smaller.

This formula explains why we do not see surveys with a 2% margin of error. To guarantee a 2% margin of error, the survey would need $n = 1/0.02^2 = 2{,}500$ respondents. Evidently, the increased precision isn't perceived to be worth more than doubling the cost. Table 15.5 shows the sample sizes needed for several choices of the margin of error.

TABLE 15.5 Sample sizes needed for various margins of error for a survey (95% coverage).

n	Margin of Error
100	10%
205	7%
400	5%
625	4%
1,112	3%
2,500	2%
10,000	1%

Keep in mind that n is the number of respondents, not the number of questionnaires mailed out. A low response rate turns a sample into an unreliable voluntary response sample. As discussed in Chapter 13, it is better to spend resources on increasing the response rate than on surveying a larger group.

4M EXAMPLE 15.1 | PROPERTY TAXES

MOTIVATION | **STATE THE QUESTION**

A Midwestern city faces a budget crunch. To close the gap, the mayor is considering a tax on businesses that is proportional to the amount spent to lease property in the city. How much revenue would a 1% tax generate?

---

[8] The maximum of $f(x) = x(1-x)$ occurs at $x = 1/2$. This can be proved with calculus or by drawing the graph of $f(x)$.

<table>
<tr><td align="right">METHOD</td><td>DESCRIBE THE DATA AND SELECT AN APPROACH</td></tr>
</table>

Determine parameter.

Identify population.

Describe data.

Choose interval.

Check conditions.

DESCRIBE THE DATA AND SELECT AN APPROACH

If we have a confidence interval for μ, the average cost of a lease, we can obtain a confidence interval for the amount raised by the tax. The city has 4,500 businesses that lease properties; these are the population. If we multiply a 95% confidence interval for μ by 1% of 4,500, we'll have a 95% confidence interval for the total revenue.

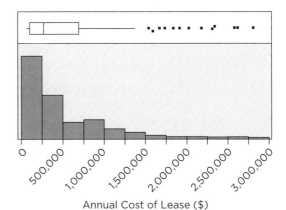

Annual Cost of Lease ($)

The data are the costs of a random sample of 223 recent leases. The histogram of the lease costs is skewed. One lease costs nearly $3,000,000 per year, whereas most are far smaller.

We will use a 95% t-interval interval for μ.

Checking the conditions, we find that both are satisfied.

✓ **SRS condition.** The sample consists of less than 10% of the population of leases, randomly chosen from the correct population.

✓ **Sample size condition.** The sample size $n = 223$ cases. Since $10K_3^2 \approx 38$ and $10|K_4| \approx 41$ (see the summary table in Mechanics), we have enough data to meet this condition.

MECHANICS | DO THE ANALYSIS

This table summarizes the key summary statistics of the lease costs, including the skewness K_3 and kurtosis K_4. The confidence interval is $\bar{x} \pm (t_{0.025,223} \approx 1.97)s/\sqrt{n}$.

$\bar{x}$	$478,603.48
s	$535,342.56
n	223
$s/\sqrt{n}$	$35,849.19
95% interval	$407,981 to $549,226
Skewness K_3	1.953
Kurtosis K_4	4.138

MESSAGE | SUMMARIZE THE RESULTS

We are 95% confident that the average cost of a lease is between $410,000 and $550,000. (Rounding to tens of thousands seems more than enough digits for such a long interval.) On average, we can be 95% confident that the tax will raise between $4,100 and $5,500 per business, and thus between $18,450,000 and $24,750,000 citywide (multiplying by 4,500, the number of business leases in the city).

EXAMPLE 15.2 │ A POLITICAL POLL

MOTIVATION STATE THE QUESTION

The mayor was so happy with the amount raised by the business tax that he's decided to run for reelection. Only 40% of registered voters in a survey done by the local newspaper ($n = 400$), however, think that he's doing a good job. What does this indicate about attitudes among all registered voters?

METHOD DESCRIBE THE DATA AND SELECT AN APPROACH

Determine parameter.
Identify population.
Describe data.
Choose interval.
Check conditions.

The parameter of interest is the proportion in the population of registered voters who think that the mayor is doing a good job. The data reported in the news are a sample (allegedly) from this population. We'll use a 95% z-interval for p to summarize what we can reasonably conclude about p from this sample. The conditions for using this interval are satisfied.

✓ **SRS condition.** The newspaper hires a reputable firm to conduct its polls, so we'll assume that the pollsters got a simple random sample. Also, n is much less than 10% of the population. We'd like to see the precise question and find out the rate of nonresponse, but let's give the pollsters the benefit of the doubt.

✓ **Sample size condition.** Both $n\hat{p}$ and $n(1 - \hat{p})$ are larger than 10.

MECHANICS DO THE ANALYSIS

The estimated standard error is

$$\text{se}(\hat{p}) = \sqrt{\hat{p}(1 - \hat{p})/n} = \sqrt{0.4 \times 0.6/400} \approx 0.0245$$

We use a z-interval for proportions even though we have estimated the standard error. The 95% z-interval for p is

$$[0.40 - 1.96(0.0245) \text{ to } 0.40 + 1.96(0.0245)] \approx [0.352 \text{ to } 0.448]$$

MESSAGE SUMMARIZE THE RESULTS

We can tell the mayor that he can be 95% confident that between 35% and 45% of the registered voters think that he is doing a good job. Fewer than half appear pleased. Perhaps he needs to convince more voters that the business tax will be good for the city or remind them that it's not them but businesses that will pay this tax! (Of course, businesses might then pass the tax on to customers.)

Best Practices

- *Be sure that the data are an SRS from the population.* If we don't start with a representative sample, it won't matter what we do. A questionnaire that finds that 85% of people enjoy filling out surveys suffers from nonresponse bias even if we put confidence intervals around this (biased) estimate.

- *Stick to 95% confidence intervals.* Unless there is a compelling reason for an interval to have larger or smaller coverage, use 95%. The most common alternatives are 90% and 99%. Have an explanation ready when choosing an interval whose coverage is not 95%.

- *Round the endpoints of intervals when presenting the results.* Software produces many digits of accuracy, but these aren't helpful when presenting our results. By rounding, we avoid littering our summary with superfluous digits that hide the important differences.

- *Use full precision for intermediate calculations.* Store the intermediate results in the calculator or write down the full answer. Round only at the final step.

Pitfalls

- *Do not claim that a 95% confidence interval holds μ.* The name *95% confidence interval* means that the interval could be wrong. We have not seen the population, only a sample. Don't claim or imply that we know the parameter lies in this range.
- *Do not use a confidence interval to describe other samples.* Confidence intervals describe a feature of the population, not a feature of another sample. A confidence interval doesn't describe individual responses or statistics in samples; it's a statement about a population parameter.

- *Do not manipulate the sampling to obtain a particular confidence interval.* A confidence interval has the indicated coverage properties when formed from a simple random sample. If we were instead to compute a sequence of intervals while sampling until the interval had certain properties (such as one that excludes zero), those inferential properties would no longer hold. We'd still have an interval, but not a confidence interval. Gather the sample, then compute the confidence interval.

Software Hints

A *z*-interval uses a percentile from the standard normal distribution in place of the *t*-percentile.

EXCEL

Use the function CONFIDENCE to obtain a *z*-interval. This function returns the half-length of the interval,

$$\text{CONFIDENCE}(\alpha, \sigma, n) = z_{\alpha/2}\sigma/\sqrt{n}$$

Hence, the formula AVERAGE(*data range*) − CONFIDENCE(0.05, α, n) produces the lower limit for a 95% confidence interval.

The function TINV(α, d) returns the $100(1 - \alpha/2)$ percentile of a *t*-distribution with d degrees of freedom. To obtain the 95% *t*-interval from data in a range (*range* = A1 ... A25), the formula

$$\text{AVERAGE}(range) - \text{TINV}(0.05, \text{ROWS}(range) - 1)$$
$$\text{STDEV}(range)/\sqrt{\text{ROWS}(range)}$$

produces the lower endpoint of the interval, and the formula

$$\text{AVERAGE}(range) + \text{TINV}(0.05, \text{ROWS}(range) - 1)$$
$$\text{STDEV}(range)/\sqrt{(\text{ROWS}(range)}$$

produces the upper endpoint.

DDXL simplifies this task. Having selected the data in the spreadsheet, follow the menu items DDXL > Confidence Intervals and then choose the type of interval from the Click me menu. The option 1 Var(iable) Prop(ortion) Interval implements the *z*-interval for a proportion. The options 1 Var(iable) t Interval and 1 Var(iable) z Interval cover the two types of intervals for μ.

MINITAB

The menu sequence

Stat > Basic Statistics > 1-Sample *Z* ...

opens a dialog that computes the z-interval for the mean. Select the column that holds the data, enter the standard deviation σ, and optionally specify a level of confidence. The default is 95%. The output includes $\bar{x}, s$, and the standard error of the mean $s/\sqrt{n}$. If the data consist of 0s and 1s, this will generate the z-interval for p. (See Behind the Math: Proportions Are Averages.) The sequence

Stat > Basic Statistics > 1-Sample t ...

opens a dialog that computes the t-interval for the mean.

JMP

The menu sequence

Analyze > Distribution

opens a dialog that allows us to pick the column that holds the data for the test. Specify this column, click OK, and JMP produces the histogram and boxplot. Below the histogram, JMP shows $\bar{x}, s$, the standard error, and the endpoints of the 95% confidence t-interval for μ.

To use a different level of confidence and obtain the z-interval, click on the red triangle in the output window beside the name of the variable above the histogram. Choose the item Confidence Interval, specify the interval of confidence, and insert a value for σ.

▶ BEHIND the Math

Proportions Are Averages

A sample proportion is related to a binomial random variable (Chapter 11). The connection is apparent if we use a Bernoulli random variable to indicate whether a customer accepts the offered credit card. A column of 0/1 indicators of an event is called a *dummy variable*. In the example of credit cards, the dummy variable indicates whether the customer accepts:

$$X_i = \begin{cases} 1 & \text{if accepts} \\ 0 & \text{otherwise} \end{cases}$$

The proportion in the sample is the average of these indicators:

$$\hat{p} = \frac{X_1 + X_2 + \cdots + X_{1,000}}{1,000} = \frac{M}{1,000}$$

M is a random variable that stands for the number who accept. Because the sample of customers is an SRS from a large population, the indicators X_i for the 1,000 customers are independent with common probability p. Hence, M is a binomial random variable with parameters $n = 1,000$ and p; $M \sim \text{Bi}(n = 1,000, p)$.

M is approximately normally distributed because of the Central Limit Theorem (Chapter 12). All we need are its mean and variance. The mean of M is the number of customers in the sample times the probability of acceptance, $\text{E}(M) = np$. The variance of M is $\text{Var}(M) = np(1 - p)$. Hence,

$$\text{E}(\hat{p}) = \text{E}(M)/n = p$$
$$\text{Var}(\hat{p}) = \text{Var}(M)/n^2 = p(1 - p)/n$$

To compare estimated standard errors, notice that the sum of squared deviations about $\bar{x}$ simplifies:

$$\sum_{i=1}^{n} (x_i - \bar{x})^2$$

$$= \sum_{i=1}^{n} (x_i - \hat{p})^2 = (n - n_1)(0 - \hat{p})^2 + n_1(1 - \hat{p})^2$$

n_1 is the number of 1s, so $\hat{p} = n_1/n$. If we plug in $n\hat{p}$ for n_1, then

$$\sum_{i=1}^{n} (x_i - \bar{x})^2 = n\hat{p}^2(1 - \hat{p}) + n\hat{p}(1 - \hat{p})^2$$

$$= n\hat{p}(1 - \hat{p})(\hat{p} + [1 - \hat{p}]) = n\hat{p}(1 - \hat{p})$$

If we divide the left side by $n - 1$, s^2 is

$$s^2 = \hat{p}(1 - \hat{p})\left(\frac{n}{n - 1}\right)$$

We might have guessed most of this formula from the population version, $\sigma^2 = p(1 - p)$.

Degrees of Freedom

Degrees of freedom arises because we need to estimate μ to get s^2. The use of $\bar{x}$ in place of μ in the formula for s^2 has a surprising effect. The sum of the squared deviations is too small:

$$\sum (x_i - \bar{x})^2 \leq \sum (x_i - \mu)^2$$

We can prove this with calculus, but it makes sense intuitively: $\bar{x}$ is the average of *this* sample whereas

μ is the average of the population. Consequently, $\overline{X}$ is closer to its sample than μ is.

Dividing by $n - 1$ rather than n in the expression for s^2 makes up for this effect. In the language of random variables, it can be shown that the expected value of s^2 is σ^2.

$$E(s^2) = E\left(\frac{\sum_{i=1}^{n}(X_i - \overline{X})^2}{n - 1}\right)$$

$$= E\left(\frac{\sum_{i=1}^{n}(X_i - \mu)^2}{n}\right) = \sigma^2$$

Because it's right on average, s^2 is an *unbiased estimate* of σ^2. The degrees of freedom is the denominator of the unbiased estimate.

To appreciate the name *degrees of freedom*, consider this example. When we use $\bar{x}$ in place of μ to estimate σ^2, it's as though we fixed one of the sample values, leaving the other $n - 1$ free to vary. For example, suppose that $n = 5$ and

$$x_1 = 1, \quad x_2 = 2, \quad x_3 = 3, \quad x_4 = 4, \quad \text{and} \quad x_5 = ?$$

If we know that $\bar{x} = 3$, then we can easily figure out that $\bar{x}_5 = 5$. Knowing $\bar{x}$ and any four of the sampled values is enough to tell us the fifth. Given $\bar{x}$, only $n - 1 = 4$ of the sample values are *free to vary*. So there are 4 degrees of freedom.

Combining Confidence Intervals

Write the z-interval for p as $[L(p) \text{ to } U(p)]$ and the z-interval for μ as $[L(\mu) \text{ to } U(\mu)]$. Because each is a 95% interval,

$$P(L[p] \le p \le U[p]) = P(L[\mu] \le \mu \le U[\mu]) = 0.95$$

What can we conclude about $P(L[p]L[\mu] \le p\mu \le U[p]U[\mu])$? Not as much as we'd like. If we observe a sample in which both intervals cover their parameters, then the product of the endpoints covers $p\mu$. That is, if the events $\mathbf{A} = L(p) \le p \le U(p)$ and $\mathbf{B} = L(\mu) \le \mu \le U(\mu)$ both occur, then $L(p)L(\mu) \le p\mu \le U(p)U(\mu)$. (This is true so long as both endpoints are positive, as in this example.) If these events are independent, then the coverage of the combined interval is at least

$$P\big(L[p]L[\mu] \le p\mu \le L[p]L[\mu]\big) \ge P(\mathbf{A} \text{ and } \mathbf{B})$$
$$= P(\mathbf{A})P(\mathbf{B})$$
$$= 0.95^2 = 0.9025$$

There's no guarantee, however, that the two intervals cover their parameters independently.

CHAPTER SUMMARY

Confidence intervals provide a range for a parameter. The **coverage** (or **confidence level**) of a confidence interval is the probability of getting a sample in which the interval includes the parameter. Most often, confidence intervals have coverage 0.95 and are known as 95% confidence intervals. The **margin of error** is the half-length of the 95% confidence interval. A **z-interval** uses percentiles from a normal distribution, whereas a t-interval uses (slightly larger) percentiles from a **Student's t-distribution**. Use a z-interval for proportions even with an estimated standard error; use a t-interval for the mean when the standard deviation is estimated from the data.

■ Key Terms

■ Formulas

z-interval for the proportion
The $100(1 - \alpha)$% confidence interval for the population proportion p is

$$\hat{p} \pm z_{\alpha/2}\text{se}(\hat{p}) = \hat{p} \pm z_{\alpha/2}\sqrt{\frac{\hat{p}(1 - \hat{p})}{n}}$$

For the typical 95% interval, set $\alpha = 0.05$ and $z_{0.025} = 1.96 \approx 2$.

t-interval for the mean

The $100(1 - \alpha)\%$ confidence interval for μ when using an estimate of the standard error of $\bar{x}$ with $n - 1$ degrees of freedom is

$$\bar{x} \pm t_{\alpha/2\,n-1}\,\text{se}(\bar{x}) = \bar{x} \pm t_{\alpha/2\,n-1}\frac{s}{\sqrt{n}}$$

Margin of error

Half of the length of the approximate 95% *z*-interval:

$$\text{Margin of Error} = \frac{2\sigma}{\sqrt{n}} \quad \text{or} \quad \frac{2\sqrt{p(1 - p)}}{\sqrt{n}}$$

▇ About the Data

The banking data for this chapter's example come from a research project that studied the performance of consumer loans done at the Wharton Financial Institutions Center. In order to simplify the analysis, we have ignored another important variable that determines the profitability of credit cards: default rates. For more information on the practice of cooperative offers for affinity cards, see the article "The College Credit-Card Hustle" in *Business Week* (July 17, 2008).

The data on the value of business leases come from an analysis of real estate prices conducted by several enterprising MBA students. We've changed the data to keep the prices in line with current rates. The figure at the start of the chapter is from the online Fred II system of the Federal Reserve Bank of St. Louis.

EXERCISES

Mix and Match

Match each item on the left with its correct description on the right.

1. $\bar{y} \pm 2\text{se}(\bar{y})$	(a) Sampling distribution of $\bar{X}$
2. $\hat{p} \pm \text{se}(\hat{p})$	(b) Margin of error
3. $2\text{se}(\bar{X})$	(c) 100% confidence interval for p
4. $N(\mu, \sigma^2/n)$	(d) Estimated standard error of $\bar{Y}$
5. $s/\sqrt{n}$	(e) Estimated standard error of $\hat{p}$
6. $\sigma/\sqrt{n}$	(f) An interval with about 95% coverage
7. $1/(0.05)^2$	(g) Actual standard error of $\bar{Y}$
8. $[0, 1]$	(h) About 2 for moderate sample sizes
9. $\sqrt{\hat{p}(1 - \hat{p})/n}$	(i) An interval with 68% coverage
10. $t_{\alpha/2,n-1}$	(j) Sample size needed for 0.05 margin of error

True/False

Mark each statement True or False. If you believe that a statement is false, briefly explain why you think it is false.

11. All other things the same, a 90% confidence interval is shorter than a 95% confidence interval.

12. Ninety-five percent of *z*-intervals have the form of a statistic plus or minus more than 3 standard errors of the statistic.

13. By increasing the sample size from $n = 100$ to $n = 400$, we can reduce the margin of error by 50%.

14. If we double the sample size from $n = 50$ to $n = 100$, the length of the confidence interval is reduced by half.

15. If the 95% confidence interval for the average purchase of customers at a department store is $50 to $110, then $100 is a plausible value for the population mean at this level of confidence.

16. If the 95% confidence interval for the number of movie-goers who purchase from the concession stand is 30% to 45%, then fewer than half of all moviegoers do not purchase from the concession stand.

17. If zero lies inside the 95% confidence interval for μ, then zero is also inside the 99% confidence interval for μ.

18. There is a 95% chance that the mean $\bar{y}$ of a second sample from the same population is in the range $\bar{x} \pm 1.96\sigma/\sqrt{n}$.

19. To guarantee a margin of error of 0.05 for the population proportion p, a survey needs to have at least 500 respondents.

20. The 95% t-interval for μ only applies if the sample data are nearly normally distributed.

Think About It

21. Convert these confidence intervals.
    (a) [11 pounds to 45 pounds] to kilograms (1 pound = 0.453 kilogram)
    (b) [$2,300 to $4,400] to yen (Use the exchange rate of $1 = ¥116.3.)
    (c) [$79.50 to $101.44] minus a fixed cost of $25
    (d) [$465,000 to $729,000] for total revenue of 25 retail stores to a per-store revenue

22. Convert these confidence intervals.
    (a) [14.3 liters to 19.4 liters] to gallons (1 gallon = 3.785 liters)
    (b) [€234 to €520] to dollars (Use the exchange rate 1 dollar = 0.821 euro.)
    (c) 5% of [$23,564 to $45,637]
    (d) 250 items at a profit of [$23.4 to $32.8] each (Give the interval for the total.)

23. What are the chances that $\bar{X} > \mu$?

24. What is the coverage of the confidence interval $[\hat{p}$ to $1]$?

25. Which is shorter, a 95% z-interval for μ or a 95% t-interval for μ? Is one of these always shorter, or does the outcome depend on the sample?

26. Which is more likely to contain μ, the z-interval $\bar{X} \pm 1.96\sigma/\sqrt{n}$ or the t-interval $\bar{X} \pm t_{0.025,n-1}S/\sqrt{n}$?

27. The clothing buyer for a department store wants to be sure the store orders the right mix of sizes. As part of a survey, she measured the height (in inches) of men who bought suits at this store. Her software reported the following confidence interval:

    With 95.00% confidence, $70.8876 < \mu < 74.4970$

    (a) Explain carefully what the software output means.
    (b) What's the margin of error for this interval?
    (c) How should the buyer round the endpoints of the interval to summarize the result in a report for store managers?
    (d) If the researcher had calculated a 99% confidence interval, would the output have shown a longer or shorter interval?

28. Data collected by the human resources group at a company produced this confidence interval for the average age of MBAs hired during the past recruiting season.

    With 95.00% confidence, $26.202 < \mu < 28.844$

    (a) Explain carefully what the software output means.
    (b) What is the margin of error for this interval?
    (c) Round the endpoints of the interval for a scale appropriate for a summary report to management.
    (d) If the researcher had calculated a 90% confidence interval, would the interval be longer or shorter?

29. A summary of sales made in the quarterly report of a department store says that the average retail purchase was $125 with a margin of error equal to $15. What does the margin of error mean in this context?

30. A news report summarizes a poll of voters and then adds that the margin of error is plus or minus 4%. Explain what that means.

31. To prepare a report on the economy, analysts need to estimate the percentage of businesses that plan to hire additional employees in the next 60 days.
    (a) How many randomly selected employers must you contact in order to guarantee a margin of error of no more than 4% (at 95% confidence)?
    (b) Suppose you want the margin of error to be based on a 98% confidence interval. What sample size must you use now?

32. A political candidate is anxious about the outcome of the election.
    (a) To have his next survey result produce a 95% confidence interval with margin of error of no more than 0.025, how many eligible voters are needed in the sample?
    (b) If the candidate fears that he's way behind and needs to save money, how many voters are needed if he expects that his percentage among voters p is near 0.25?

33. The Basel II standards for banking specify procedures for estimating the exposure to risk. In particular, Basel II specifies how much cash banks must keep on hand to cover bad loans. One element of these standards is the following formula, which expresses the expected amount lost when a borrower defaults on a loan:

    $$\text{Expected Loss} = \text{PD} \times \text{EAD} \times \text{LGD}$$

    where PD is the probability of default on the loan, EAD is the exposure at default (the face value of the loan), and LGD is the loss given default (expressed as a percentage of the loan).

    For a certain class of mortgages, 6% of the borrowers are expected to default. The face value of these mortgages averages $250,000. On average, the bank recovers 80% of the mortgaged amount if the borrower defaults by selling the property.
    (a) What is the expected loss on a mortgage?
    (b) Each stated characteristic is a sample estimate. The 95% confidence intervals are [0.05 to 0.07] for PD, [$220,000 to $290,000] for EAD, and [0.18 to 0.23] for LGD. What effect does this uncertainty have on the expected loss?

(c) What can be said about the coverage of the range implied by combining these intervals?

34. Catalog sales companies such as L.L. Bean encourage customers to place orders by mailing seasonal catalogs to prior customers. The expected profit from each mailed catalog can be expressed as the product

$$\text{Expected Profit} = p \times D \times S$$

where p is the probability that the customer places an order, D is the dollar amount of the order, and S is the percentage profit earned on the total value of an order. Typically 10% of customers who receive a catalog place orders that average $125, and 20% of that amount is profit.
  (a) What is the expected profit under these conditions?
  (b) The response rates and amounts are sample estimates. If it costs the company $2.00 to mail each catalog, how accurate does the estimate of p need to be in order to convince you that the next mailing is going to be profitable?

You Do It

35. Find the appropriate percentile from a t-distribution for constructing the following confidence intervals:
  (a) 90% t-interval with $n = 12$
  (b) 95% t-interval with $n = 6$
  (c) 99% t-interval with $n = 15$

36. Find the appropriate for the following:
  (a) 90% t-interval with $n = 5$
  (b) 95% t-interval with $n = 20$
  (c) 99% t-interval with $n = 2$

37. Consider each situation described below. Identify the population and the sample, explain what the parameter p or μ represents, and tell whether the methods of this chapter can be used to create a confidence interval. If so, find the interval.
  (a) The service department at a new car dealer checks for small dents in cars brought in for scheduled maintenance. It finds that 22 of 87 cars have a dent that can be removed easily. The service department wants to estimate the percentage of all cars with these easily repaired dents.
  (b) A survey of customers at a supermarket asks whether they found shopping at this market more pleasing than at a nearby store. Of the 2,500 forms distributed to customers, 325 were filled in and 250 of these said that the experience was more pleasing.
  (c) A poll asks visitors to a Web site for the number of hours spent Web surfing daily. The poll gets 223 responses one day. The average response is three hours per day with $s = 1.5$.
  (d) A sample of 1,000 customers given loans during the past two years contains 2 who have defaulted.

38. Consider each situation. Identify the population and the sample, explain what p or μ represents, and tell whether the methods of this chapter can be used to create a confidence interval. If so, indicate what the interval would say about the parameter.

(a) A consumer group surveys 195 people who recently bought new kitchen appliances. Fifteen percent of them expressed dissatisfaction with the salesperson.
(b) A catalog mail order firm finds that the number of days between orders for a sample of 250 customers averages 105 days with $s = 55$.
(c) A questionnaire given to customers at a warehouse store finds that only 2 of the 50 who return the questionnaire live more than 25 miles away.
(d) A factory is considering requiring employees to wear uniforms. In a survey of all 1,245 employees, 380 forms are returned, with 228 employees in favor of the change.

39. Hoping to lure more shoppers downtown, a city builds a new public parking garage in the central business district. The city plans to pay for the structure through parking fees. During a two-month period (44 weekdays), daily fees collected averaged $1,264 with a standard deviation of $150.
  (a) What assumptions must you make in order to use these statistics for inference?
  (b) Write a 90% confidence interval for the mean daily income this parking garage will generate, rounded appropriately.
  (c) The consultant who advised the city on this project predicted that parking revenues would average $1,300 per day. On the basis of your confidence interval, do you think the consultant was correct? Why or why not?
  (d) Give a 90% confidence interval for the total revenue earned during five weekdays.

40. Suppose that for budget planning purposes the city in Exercise 39 needs a better estimate of the mean daily income from parking fees.
  (a) Someone suggests that the city use its data to create a 95% confidence interval instead of the 90% interval first created. Would this interval be better for the city planners? (You need not actually create the new interval.)
  (b) How would the 95% interval be worse for the city planners?
  (c) How could city planners achieve an interval estimate that would better serve their planning needs?
  (d) How many days' worth of data must planners collect to have 95% confidence of estimating the true mean to within $10? Does this seem like a reasonable objective? (Use a z-interval to simplify the calculations.)

41. A sample of 150 calls to a customer help line during one week found that callers were kept waiting on average for 16 minutes with $s = 8$.
  (a) Find the margin of error for this result if we use a 95% confidence interval for the length of time all customers during this period are kept waiting.
  (b) Interpret for management the margin of error.
  (c) If we only need to be 90% confident, does the margin of error become larger or smaller?
  (d) Find the margin of error for a 90% confidence interval.

42. A sample of 300 orders for take-out food at a local pizzeria found that the average cost of an order was $23 with $s = \$15$.
    (a) Find the margin of error for this result if you want a 95% confidence interval for the average cost of an order.
    (b) Interpret for management the margin of error.
    (c) If we need to be 99% confident, does the margin of error become larger or smaller?
    (d) Find the margin of error for a 99% confidence interval.

43. A book publisher monitors the size of shipments of its textbooks to university bookstores. For a sample of texts used at various schools, the 95% confidence interval for the size of the shipment was 250 ± 45 books. Which, if any, of the following interpretations of this interval is/are correct?
    (a) All shipments are between 205 and 295 books.
    (b) 95% of shipments are between 160 and 340 books.
    (c) The procedure that produced this interval generates ranges that hold the population mean for 95% of samples.
    (d) If we get another sample, then we can be 95% sure that the mean of this second sample is between 160 and 340.
    (e) We can be 95% confident that the range 160 to 340 holds the population mean.

44. A catalog sales company promises to deliver orders placed on the Internet within three days. Follow-up calls to randomly selected customers show that a 95% confidence interval for the proportion of all orders that arrive on time is $88\% \pm 6\%$. What does this mean? Are the following conclusions correct? Explain.
    (a) Between 82% and 94% of all orders arrive on time.
    (b) 95% of all random samples of customers will show that 88% of orders arrived on time.
    (c) 95% of all random samples of customers will show that 82% to 94% of orders arrived on time.
    (d) We are 95% sure that between 82% and 94% of the orders placed by the customers in this sample arrived on time.
    (e) On a randomly chosen day, we can be 95% confident that between 82% and 94% of the large volume of orders will arrive on time.

45. Find the 95% z-interval or t-interval for the indicated parameter.
    (a) μ $\bar{x} = 152, s = 35, n = 60$
    (b) μ $\bar{x} = 8, s = 75, n = 25$
    (c) p $\hat{p} = 0.5, n = 75$
    (d) p $\hat{p} = 0.3, n = 23$

46. Show the 95% z-interval or t-interval for the indicated parameter.
    (a) μ $\bar{x} = -45, s = 80, n = 33$
    (b) μ $\bar{x} = 255, s = 16, n = 21$
    (c) p $\hat{p} = 0.25, n = 48$
    (d) p $\hat{p} = 0.9, n = 52$

47. Direct mail advertisers send solicitations (junk mail) to thousands of potential customers hoping that some will buy the product. The response rate is usually quite low. Suppose a company wants to test the response to a new flyer and sends it to 1,000 randomly selected people. The company gets orders from 123 of the recipients and decides to do a mass mailing to everyone on its mailing list of over 200,000. Create a 95% confidence interval for the percentage of those people who will order something.

48. Not all junk mail comes from businesses. Internet lore is full of familiar scams, such as the desperate foreigner who needs your help to transfer a large amount of money. The scammer sends out 100,000 messages and gets 15 replies. Can the scam artist make a 95% confidence interval for the proportion of victims out there, or is there a problem with the methods of this chapter?

49. A package of light bulbs promises an average life of more than 750 hours per bulb. A consumer group did not believe the claim and tested a sample of 40 bulbs. The average lifetime of these 40 bulbs was 740 hours with $s = 30$ hours. The manufacturer responded that its claim was based on testing hundreds of bulbs.
    (a) If the consumer group and manufacturer both make 95% confidence intervals for the population's average lifetime, whose will probably be shorter? Can you tell for certain?
    (b) Given the usual sampling assumptions, is there a 95% probability that 740 lies in the 95% confidence interval of the manufacturer?
    (c) Is the manufacturer's confidence interval more likely to contain the population mean because it is based on a larger sample?

50. In January 2005, a company that monitors Internet traffic (WebSideStory) reported that its sampling revealed that the Mozilla browser launched in 2004 had grabbed a 4.6% share of the market.[9]
    (a) If the sample was based on 2,000 users, could Microsoft conclude that Mozilla has a 5% share of the market?
    (b) WebSideStory claims that its sample includes 30,000,000 daily Internet users. If that's the case, then what can Microsoft conclude about Mozilla's share of the market?

51. In its 2004 survey of employees, Watson-Wyatt reported that 51% had confidence in the actions of senior management.[10] To be 95% confident that at least half of all employees have confidence in senior management, how many would have to be in the survey sample?

52. Fireman's Fund commissioned an online survey in 2004 of 1,154 wealthy homeowners to find out what they knew about their insurance coverage.[11]
    (a) When asked whether they knew the replacement value of their home, 63% replied yes. In this case, should Fireman's Fund conclude that more than half of wealthy homeowners know the value of their homes?
    (b) Round the interval into a form suitable for presentation.

[9] "The Gnat Nipping at Microsoft," *Business Week*, 1/24/2005, p. 78. Mozilla evolved into Firefox.
[10] http://www.watsonwyatt.com/news/press.asp?ID=14047. The sample size is 12,703.
[11] http://www.firemansfund.com/servlet/dcms?c=about&rkey=107.

(c) The results of the survey were accompanied by the statement, "In theory, with probability samples of this size, one could say with 95 percent certainty that the results have a statistical precision of plus or minus 3 percentage points. This online sample was not a probability sample." What is the point of this comment?

53. Click fraud has become a major concern as more and more companies advertise on the Internet. When Google places an ad for a company with its search results, the company pays a fee to Google each time someone clicks on the link. That's fine when it's a person who's interested in buying a product or service, but not so good when it's a computer program pretending to be a customer. An analysis of 1,200 clicks coming into a company's site during a week identified 175 of these clicks as fraudulent.[4]

(a) Under what conditions does it make sense to treat these 1,200 clicks as a sample? What would be the population?

(b) Show the 95% confidence interval for the population proportion of fraudulent clicks in a form suitable for sharing with a nontechnical audience.

(c) If a company pays Google $4.50 for each click, give a confidence interval (again, to presentation precision) for the mean cost due to fraud *per click*.

54. Philanthropic organizations, such as the Gates Foundation, care about the return on their investment, too. For example, if a foundation spends $7,000 for preschool education of children who are at risk, does society get a reasonable return on this investment? A key component of the social benefit is the reduction in spending for crimes later in life. In the High/Scope Perry Preschool Project, 128 children were randomly assigned to a regular education or to an intensive preschool program that would have cost $10,600 per pupil in 2005. By age 27, those in the preschool program averaged 2.3 fewer crimes (with standard error 0.85) than those who were not in the preschool program.[5]

(a) Explain why it is important that the children in this study were randomly assigned to the preschool program.

(b) Does the 95% confidence interval for the reduction in crime in the preschool group include zero? Is this important?

(c) If the total cost per crime averages $25,000 (in costs for police, courts, and prisons), give a confidence interval for the savings produced by this preschool program.

4M Promotion Response

A phone company launched an advertising program designed to increase the number of minutes of long-distance calls made by customers. To get a sense of the benefits of the program, it ran a small test of the promotion. It first selected a sample of 100 customers of the type being targeted by the promotion. This sample of 100 customers used an average of 185 minutes per month of long-distance service. The company then included a special flyer in its monthly statement to those customers for the next two billing cycles. After receiving the promotion, these same customers were using 215 minutes per month. Did the promotion work?

Motivation

a) Explain why it makes sense for the company to experiment with a sample of customers before rolling this program out to all of its subscribers.

b) What is the advantage of measuring the response using the same customers? What is a weakness?

Method

Let X_1 denote the number of minutes used by a customer before the promotion, and let X_2 denote the number of minutes after the promotion. Use μ_1 for the mean of X_1 (before the promotion) and μ_2 for the mean of X_2 (after).

c) The data on the number of minutes used by a customer during a given month are rather skewed. How does this affect the use of confidence intervals?

d) Form a new variable, say $Y = X_2 - X_1$, that measures the change in use. How can you use the 95% confidence interval for the mean of Y?

Mechanics

e) Form 95% confidence intervals for the mean μ_1 of X_1 and the mean μ_2 of X_2.

f) Form the 95% confidence interval for the mean of the difference Y.

g) Which of the intervals in parts (e) and (f) is shortest? Is this a good thing? Explain why this interval is far shorter than the others.

Message

h) Interpret, after appropriate rounding, the confidence interval for Y in the context of this problem. Are there any caveats worth mentioning?

i) If the advertising program is rolled out to 10,000 subscribers in this region, what sort of increase in phone usage would you anticipate?

4M Leasing

An auto manufacturer leases cars to small businesses for use in visiting clients and other business travel. The contracted lease does not specify a mileage limit and instead includes a depreciation fee of $0.30 per mile. The contract includes other origination, maintenance, and damage fees in addition to the fee that covers the mileage. These leases run for one year.

---

[4] "Click Fraud," *Business Week*, October 2, 2006.
[5] "Going Beyond Head Start," *Business Week*, October 23, 2006. Additional data were derived from *The Economics of Investing in Universal Preschool Education in California*, L. A. Karoly and J. H. Bigelow (2005), Rand.

A sample of 150 cars (all were a particular model of four-door sedan) returned to their dealers early in this program averaged 21,714 miles, with standard deviation $s = 2,352$ miles. Currently, this manufacturer has leased approximately 10,000 of these vehicles. When the program was launched, the planning budget projected that the company would earn (in depreciation fees) $6,500 on average per car.

Motivation

a) Should the manufacturer assume that if it were to check every leased car, the average would be 21,714 miles driven?
b) Can the manufacturer use a confidence interval to check on the claim of $6,500 earnings in depreciation fees?

Method

c) Are the conditions for using a 95% confidence interval for the mean number of miles driven per year satisfied?
d) Does the method of sampling raise any concerns?

e) Can the manufacturer estimate, with a range, the amount it can expect to earn in depreciation fees per leased vehicle, on average?

Mechanics

f) Construct the 95% confidence interval for the number of miles driven per year on average for leased cars of this type.
g) Construct the 95% confidence interval for the earnings over the one-year lease, in a form suitable for presentation.

Message

h) Interpret the 95% confidence interval for the number of miles driven over the one-year period of the lease.
i) Interpret the 95% confidence interval for the average amount earned per vehicle. What is the implication for the budget claim?
j) Communicate a range for the total earnings of this program, assuming 10,000 vehicles.

CHAPTER 16

Statistical Tests

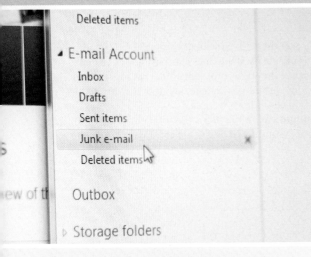

SPAM IS ANNOYING. NOT ONLY DOES JUNK EMAIL TAKE TIME TO DELETE, YOU SOMETIMES MISTAKENLY THROW OUT MESSAGES YOU WANT TO SAVE ALONG WITH THE JUNK. Beyond annoyance, spam costs businesses billions of dollars in lost productivity and the expense of lost messages. Studies at businesses show that 40% or more of the arriving email is spam, and reports in the press claim that nearly 95% of all email is spam.[1]

To contain the flood, companies buy software that filters out junk mail before it reaches desktops. An office with a staff of 36 employees currently uses a free filtering system that reduced the amount of spam annoying its employees to 24% of the incoming messages. To filter out more, the office manager is evaluating a commercial filtering system. The vendor that licenses this product claims it works better than the free system and will not lose any valid messages. To demonstrate how well its software works, the vendor applied its filtering system to email arriving at the office. After passing through the filter, a sample of 100 messages contained only 11% spam and no valid messages were removed.

This performance comes at a price. The vendor licenses this software for $15,000 a year. At this price, the accounting department says the software will pay for itself by improving productivity if it reduces the level of spam to less than 20%. A reduction in spam from 24% to 20% saves the company $15,000, which covers the cost of the license. Further reductions add profit.

The demonstration of the vendor's software suggests that it will be profitable, but does its performance on one sample *prove* it is worth the cost? Should management conclude that buying the software is a good move?

---

[1] "Spam Back to 94% of All E-Mail," *New York Times*, March 31, 2009.

To answer this question, we will use a statistical test. Like confidence intervals, statistical tests infer characteristics of the population from a sample. Rather than provide a range for a population parameter, a statistical test considers the plausibility of a specific claim about the population. Claims about the population are called hypotheses. Each hypothesis implies an action, and taking the wrong action leads to costs. We'll use a test to find the action that lowers our costs.

16.1 | CONCEPTS OF STATISTICAL TESTS

Statistical tests have much in common with control charts and confidence intervals. All three methods help managers make decisions from data while balancing two types of errors. Statistical tests focus attention on the costs and chances for these errors when we make a specific decision.

Null and Alternative Hypotheses

Think about the email that would pass through the filtering system described in the introduction over the next year if it were installed. That's the population of interest. Some email that slips past the filter will be spam; call this proportion p. Purchasing the filtering system will be profitable if $p < 0.20$. If $p \geq 0.20$, the software will not remove enough spam to compensate for the licensing fee. These claims about p are mutually exclusive; only one of them can be true.

statistical hypothesis Claim about a parameter of a population.

These competing claims about the proportion of spam that will get through the filtering system are hypotheses. A **statistical hypothesis** asserts that a population has a certain property. The hypotheses we consider specify a range or value for a parameter of the population, such as its mean. In this example, the population is the collection of all email messages that will pass through the filtering software *if* the office manager licenses this software. The parameter is p, the proportion of spam in this population.

Hypotheses imply actions. The hypothesis that $p \geq 0.20$ implies that the software will not be cost effective; too much spam will slip through. If the office manager believes this hypothesis, then there's no need to license the software. If the manager rejects this claim in favor of the contradictory hypothesis that $p < 0.20$, then the software should be purchased.

null hypothesis (H_0) The hypothesis that specifies a default course of action unless contradicted by data.

Statistical tests use data to evaluate these competing claims about the population. The approach requires that we designate one hypothesis as the null hypothesis. The **null hypothesis (H_0)** is associated with preserving the status quo, taking no action. The null hypothesis expresses the default belief that we accept in the absence of data. For control charts, the null hypothesis asserts that the process is under control and should be allowed to continue operating. In scientific problems, the null hypothesis describes the current state of knowledge. For the test of the filtering system, the default choice is to leave the computing system as is; don't purchase the software unless data offer compelling evidence to the contrary. The null hypothesis is thus

$$H_0: p \geq 0.20$$

alternative hypothesis (H_a) A hypothesis that contradicts the assertion of the null hypothesis.

The contradictory hypothesis is known as the **alternative hypothesis (H_a, or H_1 in some books) and is as follows:

$$H_a: p < 0.20$$

one-sided test A test in which the null hypothesis allows any value of a parameter larger (or smaller) than a specified value.

These hypotheses lead to a **one-sided test**. The test is one sided because it rejects H_0 only if $\hat{p}$ is small. It does not reject H_0 for large values of $\hat{p}$ which agree with H_0. One-sided tests are common in business. The expense of the proposed action needs to be justified, and a break-even analysis indicates what needs to be achieved. If the test shows that a proposal will generate profits, management rejects H_0 and implements the innovation. Otherwise, most managers are not interested in how much the proposal might lose.

two-sided test A test in which the null hypothesis asserts a specific value for the population parameter.

We will later see **two-sided tests**. A two-sided test resembles a control chart. An X-bar chart, for instance, signals that a process is out of control if the mean of a sample exceeds the UCL *or* falls below the LCL. A deviation in *either* direction from the design parameter indicates a need for action. Two-sided tests are useful when comparing two samples (Chapter 18) and in the analysis of regression models (Part IV), but are equivalent to confidence intervals. We find confidence intervals more natural and use them unless the decision is one sided.

To convince the office manager to reject H_0 and license its product, the vendor must demonstrate beyond reasonable doubt that its software will reduce the level of spam below 20%. As this language suggests, the null hypothesis is analogous to the presumption of innocence in the law. Legal systems derived from English common law presume the accused is not guilty unless evidence convinces the court otherwise. Unless the prosecution can prove that a defendant is guilty "beyond a reasonable doubt," the jury declares the defendant not guilty. For the filtering software, it's up to the vendor to prove that its software works.

Retaining the null hypothesis does not mean that H_0 is true. Rather, retaining H_0 means that we take the action associated with the null hypothesis. If a jury decides that the evidence fails to convict the defendant, this lack of evidence does not prove that the defendant is innocent. Juries do not return a verdict of innocent; they say "not guilty." If data are consistent with the null hypothesis, they lend support to H_0. In that case, we *retain* H_0 and continue the current practice. Lending support is not the same as proving H_0. When data are inconsistent with the null hypothesis, we *reject* it. This doesn't prove that the alternative hypothesis is true either.

> **Follow the Money** Here's the calculation used to arrive at the break-even point for the purchase of the filtering software. Currently, 36 employees at the office are affected by spam. These employees earn $40 per hour on average in wages and benefits. If dealing with 24% spam wastes an estimated 15 minutes per employee per day, then a reduction from 24% to 20% saves 2.5 minutes each day. That works out to $1.67 per employee per day, or $60 per day for the staff of 36. Over 250 workdays annually, the total comes to $15,000. If the commercial software reduces spam from 24% to 20%, the purchase breaks even. If the software does better, the office comes out ahead.

Type I and II Errors

Either $p \geq 0.20$ or it's not. The only way to know for sure is to install the software, check *every* message that comes through over the next year, and count the spam. That's not going to happen. The office manager has to decide now whether to license the software. Sampling is necessary, and it is up to the manager to find a way to sample the type of email traffic he expects to arrive over the next year. Sampling also introduces a second concern: A sample reveals its own proportion of spam, $\hat{p}$, not the proportion in the population.

The errors in this situation are analogous to those for control charts (Table 14.2). If H_0 is true and the manager retains H_0, then he's made the right decision. However, the sample proportion $\hat{p}$ could be small even though H_0 is true. That situation may lead the manager to reject H_0 incorrectly, licensing

TABLE 16.1 Decisions about
the null hypothesis.

		Manager's Decision	
		Retain H_0	Reject H_0
Population	$H_0: p \geq 0.20$	✓	✗$_1$
	$H_a: p < 0.20$	✗$_2$	✓

software that will not be cost effective. That's the Type I error ✗$_1$ in Table 16.1. Similarly, it's possible for sampling variation to result in a large proportion $\hat{p}$ even if H_a is true. That may cause the office manager to miss an opportunity to reduce costs. That's the Type II error ✗$_2$.

Other Tests

We did not call them hypothesis tests, but we already used this idea in previous chapters. The visual test for association, normal quantile plots, and control charts all use tests of hypotheses.

The first example of a hypothesis test that we encountered is the visual test for association (Chapter 6). The visual test for association asks whether we can pick out the original scatterplot of y on x when this graph is mixed into a collection of scatterplots that show scrambled versions of the data. If we recognize the original, there's association. If not, the pattern is too weak for us to say that y and x are associated.

Let's use the language of hypothesis testing to describe an example of the visual test for association. The null hypothesis for the visual test for association is that there's no association between two variables shown in a scatterplot. For example, the scatterplot in Figure 16.1 graphs properties of 30 countries in the Organization for Economic Cooperation and Development (OECD). This figure plots the gross domestic product (GDP, in dollars per capita) versus the percentage of GDP collected in taxes in each country in 2005. Should we reject H_0 and conclude that GDP is associated with the tax rate?

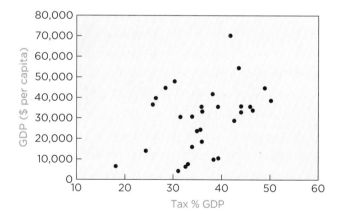

FIGURE 16.1 Scatterplot of gross domestic product (per capita) versus taxes as a percentage of GDP for 30 nations in the OECD.

To test H_0, Figure 16.2 mixes the original plot in Figure 16.1 with nine other frames that scramble the pairs of (x, y) coordinates to remove the association. Do you recognize the original without having to refer back and forth to Figure 16.1?

Suppose there is no association. You might still pick the original plot just by guessing. If the null hypothesis of no association is true, then there's 1 chance in 10 that you pick the original plot from the 10 frames in Figure 16.2. That's the chance of a Type I error; you picked the original even though there

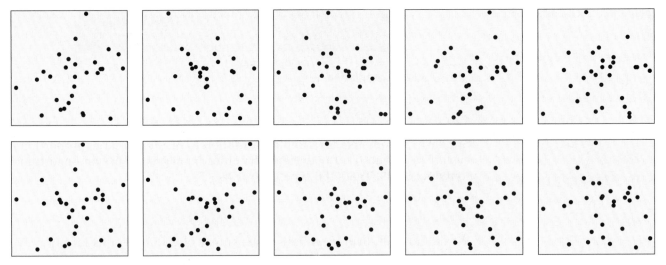

FIGURE 16.2 Visual test for association showing the original and nine scrambled plots.

is no association. If there is association and you miss it, then a Type II error occurred.[2]

The boundaries that surround the diagonal reference line in a normal quantile plot also define a statistical test. The null hypothesis tested by a normal quantile plot is

H_0: Data are a sample from a normally distributed population.

The normal quantile plot shown in Figure 16.3 summarizes the per capita GDP amounts graphed in Figure 16.1.

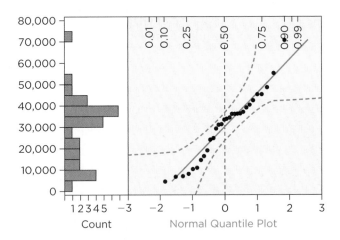

FIGURE 16.3 The dashed bands in a normal quantile plot define a test for normality.

Although the histogram hints at a bimodal shape, the data remain close enough to the diagonal reference line and we do not reject H_0. Had any point fallen outside these bands, then we would have rejected H_0. The bands are drawn so that if H_0 holds, then there is only a 5% chance of any point lying outside these limits. Hence, there is a 5% chance of a Type I error. With a small sample, such as $n = 30$ as shown, these bands are rather wide. The data have to get far from the reference line in order to reject H_0; hence, there is a good chance for a Type II error. That may be the case in this example: The population is not normally distributed, but the data remain inside the dashed bands.

---

[2] The original is the fourth plot in the first row. It's not so easy to identify, confirming the weak association.

Control charts (Chapter 14) define a sequence of two-sided tests. Each day we inspect the production and compare the mean for that day to the control limits. The null hypothesis H_0 is that the process is in control and should be allowed to run. The alternative hypothesis H_a is that the process is out of control and should be halted. The default action is to allow the process to continue running unless a sample mean falls outside the control limits. That happens rarely with the default control limits; the chance for a Type I error is only 0.00027 when the process is operating correctly. In other words, even if the process is operating normally, there's a slight chance that the average of a sample will land outside the limits. A Type II error occurs when the process changes but the average stays within the control limits.

Sampling Distribution

Statistical tests rely on the sampling distribution of the statistic that estimates the parameter specified in the null and alternative hypotheses. In the main example of this chapter, $\hat{p}$ denotes the proportion of spam in a sample; $\hat{p}$ estimates p. If the sample contains a lot of spam, then $\hat{p}$ lies within the region specified by H_0 and we retain H_0. The data agree with the null hypothesis and no action is warranted.

The interesting cases occur when the sample statistic lies outside the region specified by H_0. That's the situation in this example: The sample of $n = 100$ messages contained only 11 that were spam. Hence, $\hat{p} = 0.11$ is outside the region specified by H_0. In these cases, we have to measure how far the statistic lies from the region specified by H_0. In particular, the key question to answer is

> "What is the chance of getting a sample that differs from H_0 by as much as this one if H_0 is true?"

The null hypothesis claims that $p \geq 0.20$, but the sample estimate is $\hat{p} = 0.11$. Is that surprising? Is the sample statistic so far from the region specified by H_0 that we should reject H_0? Or might this be a fluke? Perhaps we just happened to get an unusual sample.

We answer this and similar questions by using a normal model for the sampling distribution of $\hat{p}$. Chapter 15 introduced this sampling distribution to find the confidence interval for $\hat{p}$. The normal model for the sampling distribution of $\hat{p}$ is

$$\hat{p} \sim N\left(p, \frac{p(1-p)}{n}\right)$$

What Do You Think?

A retailer sent a sample of 85 salespeople to a training program that emphasized self-image, respect for others, and manners. Management decided that if the rate of complaints fell below 10% per month after the training (fewer than 10 complaints per 100 employees), it would judge the program a success. Following the program, complaints were received about 6% of the 85 employees who participated.

a. State the null hypothesis and the alternative hypothesis.[3]
b. Identify the relevant sample statistic.[4]
c. Describe a model for the sampling distribution of the test statistic if, in the full population, 10% of employees generate complaints.[5]

---

[3] The null hypothesis H_0 is that the program is not warranted, that p (the population proportion who generate a complaint after attending the training program) is 10% or more. H_a: $p < 0.10$.
[4] $\hat{p}$, the proportion of the 85 in the test group who generate a complaint.
[5] The count is binomial, with $n = 85$ and $p = 0.10$ under H_0. We approximate the sampling distribution of $\hat{p}$ as normal with mean $p = 0.10$ and standard error $\sqrt{0.1 \times 0.9/85} \approx 0.033$.

16.2 | TESTING THE PROPORTION

If the office manager rejects H_0 on the basis of observing $\hat{p} = 0.11$, what is the chance that he has made a Type I error? The sampling distribution of $\hat{p}$ provides the answer. The sampling distribution of $\hat{p}$ describes the frequencies of $\hat{p}$ among *all possible* samples of size n from the population. If H_0 holds with $p = 0.20$, the sampling distribution of $\hat{p}$ has mean $p = 0.20$ and standard error $\text{SE}(\hat{p}) = 0.04 = \sqrt{0.2(1 - 0.2)/100}$, shown (in black) in the plot in Figure 16.4.

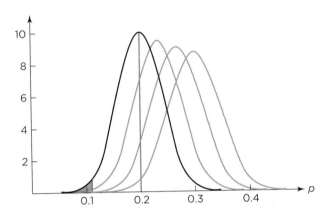

FIGURE 16.4 Possible sampling distributions of the sample proportion if H_0 holds.

It's as though we calculated $\hat{p}$ for every possible sample from this population and summarized them with a histogram. By choosing a random sample for testing the software, we've chosen *one* $\hat{p}$ at random from the sampling distribution.

z-Test and p-Value

If the office manager rejects H_0 based on $\hat{p} = 0.11$, the probability of incorrectly rejecting H_0 (a Type I error) could be as large as the shaded area in Figure 16.4. The observed sample could be one of those for which $\hat{p} = 0.11$ or less even though $p = 0.20$. The sampling distribution with the smallest value for p allowed by H_0 ($p = 0.20$) makes the probability of a Type I error as large as possible. The gray curves in Figure 16.4 show other sampling distributions allowed by H_0, but these do not assign as much probability to the left of $\hat{p}$. The maximum chance of a Type I error if we reject H_0 based on the observed statistic is called the **p-value** of the test. (The name p-value has nothing to do with the parameter p. The name is short for prob value, but no one calls it that.)

p-value The largest chance of a Type I error if H_0 is rejected based on the observed test statistic.

To calculate the p-value, we count the number of standard errors that separate $\hat{p}$ from the region specified by H_0. This count of the number of standard errors is called the **z-statistic** and is sometimes called a test statistic. If p_0 is the boundary of the null hypothesis, then the z-statistic for the test of a null hypothesis of a proportion is

z-statistic Number of standard errors that separate the test statistic from the region specified by H_0.

$$z = \frac{\text{Deviation of sample statistic from } H_0}{\text{Standard error of sample statistic}} = \frac{\hat{p} - p_0}{\sqrt{p_0(1 - p_0)/n}}$$

Plugging in $\hat{p} = 0.11$, $p_0 = 0.20$, and $n = 100$, the z-statistic is

$$z = \frac{\hat{p} - p_0}{\sqrt{p_0(1 - p_0)/n}} = \frac{0.11 - 0.20}{\sqrt{0.20(1 - 0.20)/100}} = -2.25$$

The proportion of spam in the observed sample of 100 email messages is 2.25 standard errors below the boundary of the null hypothesis. Because our test of H_0 is based on a z-statistic with a known standard error (known from the null hypothesis), it is often called a **z-test**.

z-test Test of H_0 based on a count of standard errors separating H_0 from test statistic.

The normal model for the sampling distribution converts the z-statistic into a p-value. What is the probability of getting a sample in which $\hat{p}$ lies this far (or farther) from H_0 if in fact H_0 is true? For a normal distribution, values more than 2.25 standard deviations below the mean are rare and happen in slightly more than 1% of samples.

$$P(Z \leq z) = P(Z \leq -2.25) \approx 0.012$$

tip

This probability is the p-value of the test. **Interpret the p-value as a weight of evidence against H_0 or as the plausibility of H_0. Small p-values mean that H_0 is not plausible.** The chance of getting a sample with 11% or less spam if H_0 holds is at most 1.2%. Samples of $n = 100$ messages in which $\hat{p} \leq 0.11$ do happen when H_0 is true, but not very often. For most managers, getting such a sample would be enough evidence to reject H_0 and license the software.

α-Level

α-level Threshold that sets the maximum tolerance for a Type I error.

statistically significant Statistically significant data contradict the null hypothesis and lead us to reject H_0 (p-value $< \alpha$).

Common practice is to reject H_0 if the p-value is less than a preset threshold. This allowance for a Type I error is known as the **α-level** of the test. The most common choice in statistics for the α-level is $\alpha = 0.05$. If the p-value of the test is less than the α-level, we reject H_0 in favor of H_a and declare a **statistically significant** outcome. Otherwise, we retain H_0. In this example, we reject the null hypothesis that claims that the software is not cost effective at level $\alpha = 0.05$; the test is statistically significant because the p-value is less than 0.05. In control charts we say that the process is out of control. With tests, we declare a statistically significant difference from the null hypothesis.

For situations in which the cost of incorrectly rejecting H_0 (committing a Type I error) is high, we might reduce the α-level to 1% or less. If the cost is small, then a larger choice for α is appropriate. Consider these examples:

- A researcher claims that the proportion of employees who live within walking distance of the office is higher than 10 years ago. You might be willing to reject the null hypothesis of no change with $\alpha = 0.1$ or more.
- A business analyst claims that customers complain about rude salespeople at your store more often than those of a competitor. Rejecting if the p-value is less than 1% protects you from overreacting to the claims of the analyst ($\alpha = 0.01$).

Why not require α to be smaller than 0.05 all the time? We could, but doing so increases the chance for the other error. The lower we set the α-level, the greater the chance for a Type II error. Small values of α make it hard to reject the null hypothesis, even when it is false and should be rejected. Consider this example: Had the office manager set $\alpha = 0.01$, for instance, then the test of the filtering software would *not* reject H_0 because the p-value of the test (0.012) is larger than 0.01.

Type II Error

Let's change our perspective for a moment and look at this test from the vendor's point of view. If we were the software vendor trying to make the sale, we should worry that the office manager might make a Type II error: fail to license the vendor's software even though it works.

To analyze the situation, the vendor must have a good idea of how well its software performs. Suppose that, on the basis of experiences with similar customers, the vendor believes that its software removes all but 15% of spam. The vendor believes that $H_0: p \geq 0.20$ is false and that its software will be profitable for the office to license. Is it likely that the office manager will recognize the benefits in a test of $n = 100$ messages, or might the office manager fail to appreciate the value of the software?

The sampling distribution has the answer. Assume that the office manager sets $\alpha = 0.05$. How far from H_0 must $\hat{p}$ be in order to reject H_0? From tables or software, $P(Z < -1.645) = 0.05$ for $Z \sim N(0, 1)$. Hence, to reject H_0, $\hat{p}$ must be at least 1.645 standard errors below H_0. This number of standard errors requires $\hat{p}$ to be less than

$$p_0 - z_{0.05} \, \text{SE}(\hat{p}) = 0.20 - 1.645 \sqrt{0.2(1 - 0.2)/100} \approx 0.134$$

Figure 16.5 shows the sampling distribution of $\hat{p}$ if the vendor's beliefs about the software are correct ($p = 0.15$).

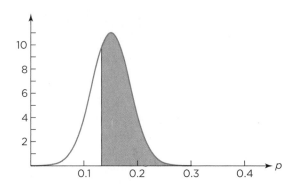

FIGURE 16.5 Probability of a Type II error if $p = 0.15$.

If $p = 0.15$, the normal model for the sampling distribution in Figure 16.5 has mean $p = 0.15$ and variance $p(1 - p)/n = (0.15)(0.85)/100 \approx 0.036^2$.

$$\hat{p} \sim N(0.15, 0.036^2)$$

The shaded area in Figure 16.5 is the probability of a Type II error. For these samples, $\hat{p}$ is too large to reject H_0, even though H_0 is false. The probability that $\hat{p}$ lies above the office manager's threshold (0.134) even though the software is cost effective is

$$P\left(Z \geq \frac{0.134 - 0.15}{\sqrt{0.15(1 - 0.15)/100}}\right) \approx P(Z \geq -0.448) \approx 0.67$$

This calculation shows that about two-thirds of samples fail to demonstrate that the software is cost effective, even though $p = 0.15$. Only one sample in three leads to the correct decision to reject H_0. The large chance for a Type II error implies that this test has little power. The **power** of a test is the probability that it rejects H_0. By design, the power of a test when H_0 is true is α. We don't want much power when H_0 is true; we want large power when H_0 is false. If the test has little power when H_0 is false, it is likely to miss meaningful deviations from the null hypothesis and produce a Type II error.

power The probability that a test can reject H_0.

If the vendor is correct about the ability of its software to filter spam, it should never have agreed to a test with a small sample of only 100 messages. This test lacks power. There's too much chance for its software to fail to convince the office manager of its value. The vendor should have insisted on a larger sample of email. A test with a larger sample would reduce the chance of a Type II error while keeping the α-level fixed. (See Exercise 38.)

Summary

In the following table, the null hypothesis specifies $H_0: p \geq p_0$ (as in the test of the email filter) or $H_0: p \leq p_0$ (illustrated in the following 4M example). **The equal sign always stays with the null hypothesis.** The symbol z_α stands for the $100(1 - \alpha)$ percentile of the standard normal distribution, $P(Z > z_\alpha) = \alpha$. For example, $z_{0.05} = 1.645$.

Population parameter	p	
Sample statistic	$\hat{p}$	
Standard error	$\text{SE}(\hat{p}) = \sqrt{p_0(1 - p_0)/n}$	
Null hypothesis	$H_0: p \leq p_0$	$H_0: p \geq p_0$
Alternative hypothesis	$H_a: p > p_0$	$H_a: p < p_0$
Reject H_0 if or	$p\text{-value} < \alpha$ $(\hat{p} - p_0)/\text{SE}(\hat{p}) > z_\alpha$	$p\text{-value} < \alpha$ $(\hat{p} - p_0)/\text{SE}(\hat{p}) < -z_\alpha$

The two rejection rules in each column are equivalent; one uses the p-value and the other uses the z-statistic.

Like confidence intervals, hypothesis tests require assumptions. Check the following conditions before using this test. The test of a proportion requires the same two assumptions as the confidence interval for the proportion. Both the test and the confidence interval rely on the use of a normal model for the sampling distribution of $\hat{p}$. Both of these conditions are met in the spam filtering example.

Checklist

✓ **SRS condition.** The observed sample is a simple random sample from the relevant population. If sampling is from a finite population, the sample comprises less than 10% of the population.

✓ **Sample size condition (for proportion).** Both np_0 and $n(1 - p_0)$ are larger than 10.

4M EXAMPLE 16.1 | DO ENOUGH HOUSEHOLDS WATCH?

MOTIVATION | STATE THE QUESTION

Digital video recorders allow us to skip commercials when watching TV. Some allow advertisers to see which ads are watched. A study used this technology in a sample of 2,500 homes in Omaha, Nebraska. MediaCheck monitored viewers' reactions to a Burger King ad that featured the band Coq Roq. The ad won critical acclaim, but MediaCheck found that only 6% of households saw the ad. An analysis indicates that an ad has to be viewed by 5% or more households to be cost effective. On the basis of this test, should the local sponsor run this ad?

METHOD | DESCRIBE THE DATA AND SELECT AN APPROACH

Let's relate this question to a hypothesis test. To do that, we need to identify the population and sample, list the hypotheses, and choose α.

List hypotheses, α.

Identify population.

Describe data.

Choose test.

Check conditions.

The break-even point is 5%. Because of the expense of running the ad, managers want to be assured enough viewers will see it. The test needs to prove beyond a reasonable doubt that the proportion p who will watch this ad if run in the general market exceeds this threshold. Hence the null hypothesis is $H_0: p \leq p_0 = 0.05$ and the alternative is $H_a: p > 0.05$. Let's use $\alpha = 0.05$. The population consists of all households in Omaha who watch TV (or perhaps elsewhere too if we think their viewing habits are similar). The sample consists of the 2,500 households surveyed by MediaCheck.

Choose the appropriate test. The test in this case is a one-sided z-test of a proportion. Before getting caught up in details, verify the items in the checklist. You may need to use a different method if a condition is not met.

✓ **SRS condition.** Let's assume that MediaCheck worked hard to get a representative sample. It might be useful to check, however, whether those who declined the offer to participate have different viewing habits.

✓ **Sample size condition.** Both $2{,}500 \times 0.05$ and $2{,}500 \times 0.95$ are larger than 10. A normal model provides an accurate approximation to the sampling distribution.

MECHANICS   DO THE ANALYSIS

This plot shows the sampling distribution of $\hat{p}$ if $p = 0.05$. The observed sample proportion seems unusually large. To find the p-value, we first calculate the z-statistic. The standard error of $\hat{p}$ is $SE(\hat{p}) = \sqrt{p_0(1 - p_0)/n} = \sqrt{0.05(1 - 0.05)/2{,}500} \approx 0.0044$. Hence, the observed proportion $\hat{p} = 0.06$ lies $z = (0.06 - 0.05)/0.0044 \approx 2.3$ standard errors above the largest proportion allowed by H_0. From tables of the normal distribution, the p-value is about $0.011 < \alpha$, so we reject H_0. The shaded area in this graph shows the p-value.

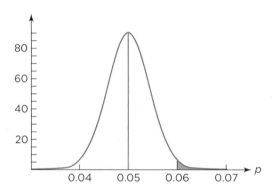

MESSAGE   SUMMARIZE THE RESULTS

Statistically significantly more than 5% of households watch this commercial. Hence, the sponsor should conclude that the ad draws more than a break-even viewing rate (albeit by a slight margin). The Coq Roq ad has proven itself cost effective in a statistical test and demonstrated that it should be run.

16.3 | TESTING THE MEAN

Tests of a mean resemble tests for proportions because we can think of a proportion as an average (see Chapter 15). The test of a hypothesis about μ replaces $\hat{p}$ and $SE(\hat{p})$ with $\overline{X}$ and $SE(\overline{X}) = \sigma/\sqrt{n}$. As when forming confidence intervals, a normal model approximates the sampling distribution if some assumptions hold. In practice, however, we seldom know σ and cannot use $SE(\overline{X})$. Unlike in the case of a proportion, the null hypothesis about μ does not tell us what to use for σ. Instead, we have to estimate σ from the sample.

To illustrate testing a mean, consider the decision faced by a firm that manages rental properties. The firm is assessing an expansion into the Denver metropolitan area. To cover its costs, the firm needs rents in this area to average

more than \$500 per month. Are rents in Denver high enough to justify the expansion?

To frame this question as a test, a conservative firm would assert as the null hypothesis that the expansion into the Denver rental market will not be profitable. Data need to prove otherwise. (A rapidly growing firm that thrives on expansion might reverse the hypotheses; for it, the norm is growth.) Because of the expense of expanding into a new area, managers at the firm have chosen a 1% chance for a Type I error (an unprofitable expansion). The population is the collection of all rental properties in the Denver metropolitan area, and the parameter is the mean rental μ. The one-sided hypotheses are

$$H_0: \mu \le \mu_0 = \$500$$
$$H_a: \mu > \mu_0 = \$500$$

where μ_0 denotes the break-even value of the mean specified by H_0.

To test H_0, the firm obtained rents for a sample of $n = 45$ rental units in the Denver area. Among these, the average rent is $\bar{x} = \$647$ with sample standard deviation $s = \$299$. As shown in the histogram in Figure 16.6, many units rent for less than \$500 in Denver. Does this mean H_0 is true?

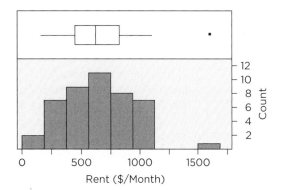

FIGURE 16.6 Summary of rents for a sample of 45 properties in Denver.

Although many individual properties rent for less than \$500, the firm is interested in the average rent. This histogram shows rents for *individual* properties. That's not the right scale for judging whether $\bar{x}$ is far from μ_0. It's the sampling variation of $\overline{X}$ that matters. We need to know the chance of getting a sample whose *mean* lies $\$647 - \$500 = \$147$ from μ_0 if H_0 holds.

t-Statistic

If we knew σ, we could use $Z = (\overline{X} - 500)/(\sigma/\sqrt{n})$ to test H_0. The null hypothesis, however, specifies μ_0 but says nothing about σ. For tests of a proportion, p_0 determines both the mean and scale of the sampling distribution $\mathrm{SE}(\hat{p}) = \sqrt{p_0(1 - p_0)/n}$. A hypothesis that specifies μ_0 does not imply a value for σ.

The obvious solution is to repeat what is done when building a confidence interval for μ: Replace σ by the standard deviation s of the data. As in Chapter 15, we label the estimated standard error using lowercase letters to distinguish it from the standard error obtained when σ is known:

$$\mathrm{SE}(\overline{X}) = \frac{\sigma}{\sqrt{n}} \quad \mathrm{se}(\overline{X}) = \frac{s}{\sqrt{n}}$$

Because we have replaced σ by the estimate s, we use a t-statistic to measure the distance that separates the observed statistic from the region given by H_0.

$$t = \frac{\text{Deviation of sample statistic from } H_0}{\text{Estimated standard error of sample statistic}}$$

$$= \frac{\bar{x} - \mu_0}{s/\sqrt{n}}$$

$$= \frac{647 - 500}{299/\sqrt{45}} = 3.298$$

***t*-statistic** The number of estimated standard errors from $\bar{x}$ to μ_0.

$$t = \frac{\bar{x} - \mu_0}{s/\sqrt{n}}$$

Student's *t*-distribution A model for the sampling distribution that adjusts for the use of an estimate of σ.

A ***t*-statistic** (or t-ratio) is $\bar{x} - \mu_0$ divided by the *estimated* standard error. It counts the number of estimated standard errors that separate $\bar{x}$ from μ_0. In this example, $\bar{x}$ lies about 3.3 estimated standard errors above μ_0. To judge whether that's far enough, we use **Student's *t*-distribution** to account for the use of an estimate of σ^2.

t-Test and *p*-Value

For the sample of $n = 45$ rents in the Denver area, the observed t-statistic is $t = 3.298$. The observed mean lies about 3.3 estimated standard errors from H_0. In order to test H_0: $\mu \le \mu_0 = \$500$, we need a p-value for this test statistic. Because we use the t-statistic as the test statistic, this procedure is called a ***t*-test** for the mean of one sample.

***t*-test** A test that uses a t-statistic as the test statistic.

You get the p-value from software or from a table of the t-distribution, such as the one found inside the back cover. A shorter version is shown in Table 16.2. Here's the procedure for using the table. Because $n = 45$ in this example, there are $n - 1 = 44$ degrees of freedom, which is a value for the degrees of freedom that does not appear in the table. In such cases, use the closest row in the table with fewer degrees of freedom. In this case, that means use the row with 40 degrees of freedom. The observed t-statistic 3.298 in this example is larger than any value in this row. That means that the p-value is less than 0.005. In fact, our software computed the p-value to be 0.00097.

TABLE 16.2 Percentiles of the t-distribution (a larger table is inside the back cover).

	α				
df	**0.10**	**0.05**	**0.025**	**0.01**	**0.005**
1	3.078	6.314	12.706	31.821	63.657
⋮	⋮	⋮	⋮	⋮	⋮
40	1.303	1.684	2.021	2.423	2.704
50	1.299	1.676	2.009	2.403	2.678
Normal	**1.282**	**1.645**	**1.960**	**2.327**	**2.576**

Because the p-value is less than the chosen level $\alpha = 0.01$, we reject H_0 and conclude that average rents in Denver exceed the break-even amount \$500. If H_0 is true, there's little chance (less than 1/1,000) of getting a sample of 45 properties with average rent \$647 or more per month.

Summary

Because we estimate σ using the sample standard deviation s, use a t-statistic to test H_0: $\mu \le \mu_0$ (or H_0: $\mu \ge \mu_0$). In the following table, symbol $t_{\alpha,n-1}$ denotes the $100(1 - \alpha)$ percentile of a t-distribution with $n - 1$ degrees of freedom.

Population parameter	μ	
Sample statistic	$\bar{x}$	
Standard error (estimate)	$se(\overline{X}) = s/\sqrt{n}$	
Null hypothesis	$H_0: \mu \leq \mu_0$	$H_0: \mu \geq \mu_0$
Alternative hypothesis	$H_a: \mu > \mu_0$	$H_a: \mu < \mu_0$
Reject H_0 if or	$p\text{-value} < \alpha$ $(\bar{x} - \mu_0)/se(\overline{X}) > t_{\alpha,n-1}$	$p\text{-value} < \alpha$ $(\bar{x} - \mu_0)/se(\overline{X}) < -t_{\alpha,n-1}$

This test requires two conditions that are similar to those in the z-test of a proportion. We checked both of these conditions when using the 95% t-interval for the mean in Chapter 15.

Checklist

✓ **SRS condition.** The observed sample is a simple random sample from the relevant population. If sampling is from a finite population, the sample comprises less than 10% of the population.

✓ **Sample size condition.** Unless it is known that the population is normally distributed, a normal model can be used to approximate the sampling distribution of $\overline{X}$ if n is larger than 10 times *both* the squared skewness and the absolute value of the kurtosis, $n > 10K_3^2$ and $n > 10|K_4|$.

For the analysis of rents in Denver, we are given that these data are a representative sample. The sample size condition also holds since $10\,K_3^2 \approx 4$ and $10|K_4| \approx 10$ are both less than $n = 45$.

4M EXAMPLE 16.2 | COMPARING RETURNS ON INVESTMENTS

MOTIVATION | STATE THE QUESTION

Hypothesis testing needs a point of reference that defines the null hypothesis. Previous examples use break-even analyses. When it comes to evaluating an investment, common sense supplies the null hypothesis. In Chapters 9 and 10, we studied the fortunes of a day trader who invests in IBM stock. She could instead invest in something that's less risky, such as U.S. Treasury Bills. From 1980 through 2005, T-Bills returned 0.5% each month. Does stock in IBM return more on average than T-Bills?

METHOD | DESCRIBE THE DATA AND SELECT AN APPROACH

List hypotheses, α.
Identify population.
Describe data.
Choose test.
Check conditions.

Let's set this up as a hypothesis test. The null hypothesis is $H_0: \mu \leq 0.005$, versus $H_a: \mu > 0.005$, where μ is the mean monthly return on the stock. We'll use $\alpha = 0.05$. The term μ is the mean of the population of all possible *future* monthly returns. We have to think of the observed returns as a part of the ongoing process that produces returns on IBM.

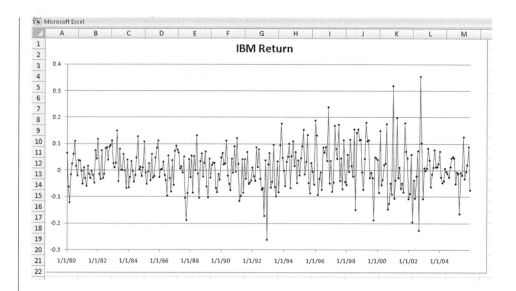

The sample consists of the returns on IBM in the 312 months from January 1980 through December 2005 that are shown in the timeplot above. There are no patterns in these returns, and the histogram (not shown) is bell-shaped like a normal distribution. Because σ is estimated from the sample, we'll use a t-test of the mean. Checking the assumptions, we need to think hard about the first one.

✓ **SRS condition.** In order to think of these observations as a sample from the population of future returns, we have to assume that the phenomenon that produced these returns in the past will continue into the future. This assumption underlies much of financial analysis, so we will accept it here.

✓ **Sample size condition.** The skewness $K_3 \approx 0.3$ and the kurtosis $K_4 \approx 1.6$. This condition is met since $n = 312$. The kurtosis is relatively large for these data because of the occasional large or small return. Financial returns generally show some kurtosis.

MECHANICS | ### DO THE ANALYSIS

The mean return on IBM stock during this period is $\bar{x} = 0.0106$ with $s = 0.0805$. The estimated standard error is $se(\bar{X}) = s/\sqrt{n} = 0.0805/\sqrt{312} \approx 0.0046$. The t-statistic is

$$t = (\bar{x} - \mu_0)/se(\bar{X}) = (0.0106 - 0.0050)/0.0046 \approx 1.22$$

The average return on IBM stock is about 1.22 standard errors above the return on T-Bills. That's not far on the scale of sampling variation, so the p-value is larger than 0.05. The p-value determined by a normal model is $P(Z > 1.22) \approx 0.111$. The p-value from a t-distribution with $n - 1 = 311$ degrees of freedom is nearly identical: $P(T_{311} > 1.22) \approx 0.112$.

MESSAGE | ### SUMMARIZE THE RESULTS

tip

When summarizing results, avoid terms like "null hypothesis," "α-level," and "p-value" unless those who are listening understand them. Monthly IBM returns from 1980 through 2005 do not bring statistically significantly higher earnings than comparable investments in U.S. Treasury Bills during this period. Stock in IBM averages a higher return (about 1.1% versus 0.5%), but the advantage can be explained by chance. If average returns on IBM stock and Treasuries in the long term are equivalent, the observed advantage for IBM on average would happen in about 11% (the *p*-value) of samples of this duration.

16.4 | OTHER PROPERTIES OF TESTS

Significance versus Importance

caution Statistical significance does not imply that you've made an important or meaningful discovery, even though it may sound that way. A test produces a statistically significant result whenever we reject H_0 at the chosen α-level. A tiny *p*-value does not mean that the difference discovered is dramatically profitable; it means that the null hypothesis is inconsistent with the data.

In the example of rentals in Denver, suppose we had observed a much larger sample with $n = 2,000$. In that case, the estimated standard error of the sample mean is $s/\sqrt{n} = \$6.69$. We reject H_0 using $\alpha = 0.05$ for any sample which has an average rental more than about $1.645 \times \$6.69 = \$11.00$ above $500 per month. If, for example, the average rent in Denver had been $515 per month (rather than $647), the *p*-value would have been 0.01. We'd still reject H_0, but this small difference does not presage much in the way of profits. Remember, μ_0 is the break-even rent. At $500, the income matches the cost. At $515, the firm is making a profit, but not much.

The size of the sample affects the *p*-value of a test. With enough data, a trivial difference from H_0 leads to a statistically significant outcome. Here's another example. Prior to a large, national advertising campaign, customers assigned an average rating of 5.8 (out of 7 on a Likert scale as defined in Chapter 2) to the quality of service at a chain of hotels. After the advertising campaign, which extolled the service at these hotels, the average rating grew to 5.86. That's an almost meaningless change in the ratings, particularly when we learn that the standard deviation of these ratings is 2.05. The follow-up survey, however, was huge, collecting 5,594 rating cards from customers as they checked out. If we treat 5.8 as the mean rating preadvertising, then this tiny shift produces a *t*-statistic equal to

$$t = \frac{\bar{x} - \mu}{s/\sqrt{n}} = \frac{5.86 - 5.8}{2.05/\sqrt{5,594}} \approx 2.19$$

If the chain tests $H_0: \mu \leq 5.8$ versus $H_a: \mu > 5.8$ with these data, then it can reject H_0 with *p*-value 0.014. The test is statistically significant at $\alpha = 0.05$ even though the size of the effect appears trivial. Statistical significance means that the difference from H_0 is large compared to the variation from sample to sample, not large when costs or profits are taken into account.

Confidence Interval or Test?

In general, anything done with a test can also be done with a confidence interval (CI) and vice versa. Confidence intervals provide a range of plausible values for the population parameter whereas tests give a detailed evaluation of a specific proposal for the population.

We have found that casual readers understand the implications of confidence intervals more readily than tests. Confidence intervals make positive statements about the population, such as offering a range for a parameter. Tests provide negative statements. Rejecting H_0 does not tell us what μ is; rather, rejecting H_0 tells us what μ isn't. A test indicates whether data contradict a specific claim about the population; you may not find that claim very sensible. Confidence intervals let the reader decide which values make an interesting null hypothesis, and the width of the interval immediately conveys the precision of the results. For instance, the hotel exit survey produces this 95% confidence interval for the average rating.

$$\bar{x} \pm t_{0.025,n-1}\, s/\sqrt{n} = 5.86 \pm 1.96(2.05)/\sqrt{5{,}594} \approx [5.806 \text{ to } 5.914]$$

The narrow confidence interval immediately conveys that we have a precise estimate of μ, and it also shows how close this range is to the prior rating, 5.8.

Notice that the null value specified in the hypothesis test ($\mu_0 = 5.8$) lies outside of this interval. That's always the case: If μ_0 is outside the 95% confidence interval, then we can reject a null hypothesis H_0 that claims $\mu = \mu_0$ at $\alpha = 1 - 0.95 = 0.05$.

Let's take a closer look at the relationship between tests and confidence intervals. The launch of the credit card discussed in Chapter 15 provides all of the ingredients needed for a hypothesis test. Consider the variable Y that tracks the profit earned per customer. The null hypothesis is that the card is not profitable; average profits are equal to or less than 0.

$$H_0: \mu_y \leq \$0$$

The alternative hypothesis $H_a: \mu_y > \$0$ claims that the launch will be profitable. Using summary statistics from Table 15.4, the t-statistic for testing H_0 is

$$t = \frac{\bar{y} - \mu_0}{\text{se}(\bar{y})} = \frac{12.867 - 0}{117.674/\sqrt{1{,}000}} \approx 3.46$$

The p-value is about 0.0003, much less than any usual α-level. We reject H_0 and conclude that the test of the launch has shown beyond reasonable doubt that the credit card will be profitable. The t-statistic also tells us that 0 is not inside the 95% confidence interval. The t-statistic indicates that $\bar{y}$ lies 3.46 standard errors away from $\mu_0 = 0$. Since the 95% confidence interval holds values that are within about two standard errors of the hypothesized mean, 0 must lie outside of the 95% confidence interval. It's not compatible with the data at a 95% level of confidence.

Though clearly related, the test of $H_0: \mu_y \leq 0$ and the confidence interval for μ_y answer different questions. The one-sided test tells you whether a proposal has proven itself profitable (with a 5% chance for a Type I error). To reject H_0, the profits have to be statistically significantly *more* than $\$0$ per person. Confidence intervals provide different information. What if someone asks how high the profits might go? A one-sided test does not answer this question. It simply concludes that average profits are positive without putting a limit on how high. The confidence interval, on the other hand, gives a range for the profits per person, about $\$5.60$ to $\$20$. Since 0 lies outside this interval, we're 95% confident that the program is profitable *and* we have an upper limit for the profitability.

You don't get this extra information from a confidence interval for free, however. Consider an example in which the outcome is less clear. Suppose that the average profit $\bar{y}$ per customer had been smaller; for example, $\bar{y} = \$7$. Then the 95% confidence interval would have been about $\$7 \pm 2(3.7) = [-0.40 \text{ to } \$14.40]$; zero lies just inside this interval. Even so, the one-sided test still rejects $H_0: \mu \leq 0$. The t-statistic for testing H_0 is $t = (7 - 0)/3.7 \approx 1.89$ with p-value $= 0.029$. The test rejects H_0 because the p-value $0.029 < 0.05$ even though $\$0$ lies inside the confidence interval.

A CI provides . . . a range of parameter values that are compatible with the observed data.

A test provides . . . a precise analysis of a specific, hypothesized value for a parameter.

That's the drawback of using a confidence interval. Because the 95% confidence interval tells us *both* an upper *and* a lower limit for μ_y, it is less sensitive than the one-sided test when setting a lower limit for the average profitability μ_y. In other words, the one-sided test is more powerful than the confidence interval in this situation. Just remember, however, that the one-sided test only rejects H_0 without indicating the possible size of profits.[6]

Best Practices

- *Pick the hypotheses before looking at the data.* If you peek into the data to determine a value for μ_0 far from the data, it makes little sense to use this same sample to test H_0. After seeing the mean of the data, for example, it's easy to pick a null hypothesis that we can reject, but that's not a real hypothesis test and the *p*-value has no inferential interpretation.
- *Choose the null hypothesis on the basis of profitability.* Retaining H_0 means business as usual. Reject H_0 when convincing evidence shows that a new procedure offers a profitable alternative. Formulating hypotheses that have economic meaning is fundamental to using hypothesis testing to make decisions.
- *Pick the α-level first, taking account of both errors.* You might later change your mind, but you ought to be able to decide on the tolerable chance for a Type I error before running the test. If the cost of a Type I error is large compared to the cost of failing to act, choose a relatively small value for α (0.01 or smaller). If the costs appear similar, choose a larger value for α (perhaps 0.10 or larger).
- *Think about whether α = 0.05 is appropriate for each test.* It's the common choice (and the typical default used by software), but a better approach is to assess the costs of the errors. As with confidence intervals, the default choice α = 0.05 is common practice; have an explanation ready if you change the value of α.
- *Make sure to have an SRS from the right population.* If the data are not a representative sample of the relevant population, the rest of the calculations are unreliable. Don't pretend to have an SRS.
- *Use a one-sided test.* In most situations, managers want to know if the data indicate that a new project will be profitable. The manager isn't interested in an approach that loses money. If the decision is two sided, like a control chart, then use a confidence interval.
- *Report a p-value to summarize the outcome of a test.* Don't just report that the test was statistically significant. Instead, provide a summary that shows the size of the effect (such as the distance from H_0) and give a *p*-value. Providing a *p*-value allows others to decide whether they believe the evidence against H_0 is strong enough to reject H_0.

Pitfalls

- *Do not confuse statistical significance with substantive importance.* If the *p*-value is less than α, the data show a statistically significant deviation from H_0. This may not represent a meaningful difference. Large samples produce small standard errors, declaring trivial differences from H_0 statistically significant while being economically meaningless.
- *Do not think that the p-value is the probability that the null hypothesis is true.* The *p*-value assumes H_0 is true. Given that H_0 is true, the *p*-value is the probability of rejecting H_0

---

[6] A method known as a one-sided confidence interval gives a range for μ of the form $[-\infty, x]$ or $[x, +\infty]$. Such intervals correspond to one-sided tests, but in this book we stick to two-sided intervals.

incorrectly on the basis of results as far from H_0 as those observed in the sample.

- *Avoid cluttering a test summary with jargon.* Hypothesis testing introduces a lot of terminology that most listeners are not going to understand. Summarize results in everyday language.

Software Hints

EXCEL

Excel does the z-test, but we'll need to do more with formulas to perform the t-test. The function ZTEST performs a test using the mean of a sample. The formula

$$\text{ZTEST}(\textit{data range}, \mu_0, \sigma)$$

tests the one-sided null hypothesis $H_0: \mu \leq \mu_0$ versus $H_a: \mu > \mu_0$ and returns a p-value. For example,

$$\text{ZTEST(A1, A200, 15, 3)}$$

tests $H_0: \mu \leq 15$ versus $H_a: \mu > 15$ using data in rows 1 through 200 of column A with $\sigma = 3$. If μ_0 is omitted, then it's assumed to be zero. If σ is omitted, it's estimated from the data using s, but a t-distribution is not used to find the p-value. The Excel Help for this function describes how to get p-values for other null hypotheses.

The like-named function TTEST does not perform the one-sample t-test; it does a two-sample test as used in Chapter 18. To compute a t-statistic, use the formula

SQRT(ROWS(*data range*))

(AVERAGE(*data* range) − μ_0)/

STDEV(*data range*)

The function TDIST(t, df, 1) returns the p-value using a t-distribution with df degrees of freedom. Just plug in the absolute value of the t-statistic.

DDXL simplifies these tasks. After selecting the column that holds the data for the test, follow the menu items DDXL > Hypothesis Tests. From the Click Me menu, select either 1 Var(iable) z Test or 1 Var(iable) t Test. After you indicate the column in the dialog, DDXL produces a summary with the test and p-value. We can set various values for μ_0 and α and determine whether the test is one or two sided.

MINITAB

The sequence of menu items

Stat > Basic Statistics > 1 − Sample Z...

opens a dialog that computes the one-sample test of a mean. Pick the column that holds the data, enter the population standard deviation, check the box for doing a test, and specify the value μ_0 for the null hypothesis.

By default, the output shows the z-statistic and p-value for the two-sided test of $H_0: \mu = \mu_0$. To obtain the one-sided p-value, specify a one-sided alternative hypothesis via the options button in the dialog. Be sure to specify the direction of the alternative hypothesis, not the null! The output shows $\bar{x}$, s, and the standard error of the mean $s/\sqrt{n}$. The similar command 1 − Sample t... performs the t-test.

Minitab also includes a specialized test for proportions. The menu sequence Stat > Basic Statistics > 1 Proportion ... produces a p-value that is similar to that from the z-test if n is large but differs in small samples. See Chapter 17 for further discussion of the effects of small samples.

JMP

The sequence of menu items

Analyze > Distribution

opens a dialog that allows us to pick the column that holds the data for the test. Specify this column, click OK, and JMP produces the histogram and boxplot of the data (assuming numerical data). In the output window, click on the red triangle beside the name of the variable above the histogram. In the resulting pop-up menu, choose the item Test Mean. A dialog appears that allows us to specify the hypothesized mean μ_0 and optionally enter σ (otherwise the software uses s in place of σ and returns the t-interval). Click the OK button in this dialog. The output window now includes another section labeled Test Mean = value. In addition to showing the sample mean $\bar{x}$ and standard deviation s, the output shows the z-statistic or t-statistic and several p-values. Only one of these p-values is relevant for our test.

For example, in the following output, $\mu_0 = 2.5$ and $\bar{x} = 2.7$. For the one-sided test of $H_0: \mu \leq 2.5$ versus $H_a: \mu > 2.5$, the relevant p-value is circled (0.1653). We gave the software σ, so this output is a z-test. Had the software estimated s from the data, it would have been a t-test.

Test Mean = value	
Hypothesized Value	2.5
Actual Estimate	2.7
df	9
Std Dev	0.67495
Sigma given	0.65
	z Test
Test Statistic	0.9730
Prob > $\|z\|$	0.3305
Prob > z	0.1653
Prob < z	0.8347

The other p-values are for a two-sided test ($H_0: \mu = 2.5$) and for the one-sided test in the opposite direction ($H_0: \mu \geq 2.5$). For one-sided tests, match the direction of the inequality, labeling the output with the direction of the inequality in H_a.

If the data are categorical and a proportion is tested, JMP produces a specialized p-value designed for such data. The p-value resembles that from the z-test if n is large, but can be considerably different if n is small. See Chapter 17 for further discussion of the effects of small samples.

CHAPTER SUMMARY

A **statistical hypothesis** asserts that a population has a certain property. The **null hypothesis** H_0 asserts a claim about a parameter of a population. The **alternative hypothesis** asserts that H_0 is false. The choice of the null hypothesis typically depends on a break-even economic analysis. This situation produces a **one-sided test** that rejects H_0 either for large or for small values of a test statistic (not both). The test statistic counts the number of standard errors between the statistic and the region specified by H_0. If the standard error is known, the test statistic is a **z-statistic (z-test)**; it is a **t-statistic (or t-ratio; t-test)** if the standard error is estimated. If the **p-value** is less than a specified threshold, the **α-level** of the test, the observed sample statistic has a **statistically significant** difference from H_0, and the test rejects H_0 in favor of the alternative hypothesis. A normal distribution determines the p-value of a z-test; **Student's t-distribution** determines the p-value for a t-test. A test that incorrectly rejects H_0 when it is true results in a Type I error. A test that fails to reject H_0 when it is false results in a Type II error. A test that is unable to reject H_0 when it is false lacks **power**.

■ Key Terms

α-level, 385
alternative hypothesis (H_a), 379
null hypothesis (H_0), 379
one-sided test, 379
power, 386

p-value, 384
statistical hypothesis, 379
statistically significant, 385
Student's t-distribution, 390
t-statistic (or t-ratio), 390

t-test, 390
two-sided test, 380
z-statistic, 384
z-test, 384

■ Formulas

Reject H_0 if the p-value is less than α, the chosen threshold for Type I errors (false positive errors). The p-value is computed from the z-statistic or t-statistic by reference to a standard normal distribution.

One-sided z-test of a proportion

Compute $z = (\hat{p} - p_0)/\text{SE}(\hat{p})$ with

$$\text{SE}(\hat{p}) = \sqrt{\frac{p_0(1 - p_0)}{n}}$$

To test $H_0: p \leq p_0$ vs. $H_a: p > p_0$, use
$$p\text{-value} = P(Z > z)$$

To test $H_0: p \geq p_0$ vs. $H_a: p < p_0$, use
$$p\text{-value} = P(Z < z)$$

One-sided t-test of a mean

Compute $t = (\bar{x} - \mu_0)/\text{se}(\overline{X})$ with
$$\text{se}(\overline{X}) = s/\sqrt{n}$$

T_{n-1} denotes a t-statistic with $n - 1$ degrees of freedom.

To test $H_0: \mu \leq \mu_0$ vs. $H_a: \mu > \mu_0$ use
$$p\text{-value} = P(T_{n-1} > t)$$

To test $H_0: \mu \geq \mu_0$ vs. $H_a: \mu < \mu_0$, use
$$p\text{-value} = P(T_{n-1} < t)$$

About the Data

The example of spam filtering comes from our work on classification models to identify spam. We modified the percentages so that the calculations would be easier to describe. Numerous articles in the business press describe the costs of dealing with email, such as "Postage Due, With Special Delivery, for Companies Sending E-Mail to AOL and Yahoo," *New York Times* (February 5, 2006).

We obtained the sample of rents in the Denver metropolitan area from the 1-percent public use

microdata sample (PUMS) available from the 2000 Census. These are available from the U.S. Bureau of the Census on the Web.

The example on advertising penetration comes from the article "Counting the Eyeballs," *Business Week* (January 16, 2006). The monthly returns on IBM stock are from the Center for Research in Security Prices (CRSP), accessed through the Wharton Research Data System.

EXERCISES

Mix and Match

Match the description of each concept with the correct symbol or term.

1. One-sided null hypothesis	(a) t-statistic
2. Common symbols for the alternative hypothesis	(b) μ_0
3. Maximum tolerance for incorrectly rejecting H_0	(c) p-value
4. Counts the number of standard errors that separate an observed statistic from the boundary of H_0	(d) p-value $< \alpha$
5. Counts the number of estimated standard errors that separate an observed statistic from the boundary of H_0	(e) Type I error
6. Largest value of the α-level for which a test can reject the null hypothesis	(f) z-statistic
7. Occurs if the p-value is less than the α-level when H_0 is true	(g) Type II error
8. Occurs if the p-value is larger than the α-level when H_0 is false	(h) α-level
9. Symbol for the largest or smallest population mean specified by the null hypothesis	(i) $H_0: \mu \geq 0$
10. Indicates that a test obtains a statistically significant result	(j) H_a, H_1

True/False

Mark each statement True or False. If you believe that a statement is false, briefly explain why you think it is false.

Exercises 11–18. A retailer maintains a Web site that it uses to attract shoppers. The average purchase amount is $80. The retailer is evaluating a new Web site that would, it hopes, encourage shoppers to spend more. Let μ represent the average amount spent per customer at its redesigned Web site.

11. The appropriate null hypothesis for testing the profitability of the new design sets $\mu_0 = \$80$.

12. The appropriate null hypothesis for testing the profitability of the new design is $H_0: \mu \leq \mu_0$.

13. If the α-level of the test is $\alpha = 0.05$, then there is at most a 5% chance of incorrectly rejecting H_0.

14. If the p-value of the test of H_0 is less than α, then the test has produced a Type II error.

15. If the test used by the retailer rejects H_0 with the α-level set to $\alpha = 0.05$, then it would also reject H_0 with $\alpha = 0.01$.

16. The larger the sample size used to evaluate the new design, the larger the chance for a Type II error.

17. If the standard deviation is estimated from the data, then a z-statistic determines the p-value.

18. If the t-statistic rejects H_0, then we would also reject H_0 had we obtained the p-value using a normal distribution rather than a t-distribution.

Exercises 19–26. An accounting firm is considering offering investment advice in addition to its current focus on tax planning. Its analysis of the costs and benefits of adding this service indicates that it will be profitable if 40% or more of its current customer base use it. The firm plans to survey its customers. Let p denote the proportion of its customers who will use this service if offered, and let $\hat{p}$ denote the proportion who say in a survey that they will use this service. The firm does not want to invest in this expansion unless data show that it will be profitable.

19. If H_0 holds, then $\hat{p}$ in the sample will be less than 0.4.

20. By setting a small α-level, the accounting firm reduces the chance of a test indicating that it should add this new service even though it is not profitable.

21. If $\hat{p}$ is larger than 0.4, a test will reject the appropriate null hypothesis for this context.

22. The larger the absolute value of the z-statistic that compares $\hat{p}$ to p, the smaller the p-value.

23. The mean of the sampling distribution of $\hat{p}$ that is used to determine whether a statistically significant result has been obtained in this example is 0.4.

24. The standard error of the sampling distribution of $\hat{p}$ depends on an estimate determined from the survey results.

25. The p-value of the test of the null hypothesis in this example is the probability that the firm should add the investment service.

26. Larger samples are more likely than smaller samples to produce a test that incorrectly rejects a true null hypothesis.

Think About It

27. A pharmaceutical company is testing a newly developed therapy. If the therapy lowers the blood pressure of a patient by more than 10 mm, it is deemed effective.[7] What are the natural hypotheses to test in a clinical study of this new therapy?

28. A chemical firm has been accused of polluting the local river system. State laws require the accuser to prove the polluting by a statistical analysis of water samples. Is the chemical firm worried about a Type I or a Type II error?

29. The research labs of a corporation occasionally produce breakthroughs that can lead to multibillion-dollar blockbuster products. Should the managers of the labs be more worried about Type I or Type II errors?

30. Modern combinatorial chemistry allows drug researchers to explore millions of products when searching for the next big pharmaceutical drug. An analysis can consider literally hundreds of thousands of compounds. Suppose that none of 100,000 compounds

in reality produces beneficial results. How many would a compound-by-compound testing procedure with $\alpha = 0.05$ nonetheless indicate were effective? Would this cause any problems?

31. Consider the following test of whether a coin is fair. Toss the coin three times. If the coin lands either all heads or all tails, reject H_0: $p = 1/2$. (The p denotes the chance for the coin to land on heads.)
    (a) What is the probability of a Type I error for this procedure?
    (b) If $p = 3/4$, what is the probability of a Type II error for this procedure?

32. A consumer interest group buys a brand-name kitchen appliance. During its first month of use, the appliance breaks. The manufacturer claims that 99% of its appliances work for a year without a problem.
    (a) State the appropriate null hypothesis that will facilitate a hypothesis test to analyze this claim.
    (b) Do the data supply enough information to reject the null hypothesis? If so, at what α-level?

33. A jury of 12 begins with the premise that the accused is innocent. Assume that these 12 jurors were chosen from a large population, such as voters. Unless the jury votes unanimously for conviction, the accused is set free.
    (a) Evidence in the trial of an innocent suspect is enough to convince half of all jurors in the population that the suspect is guilty. What is the probability that a jury convicts an innocent suspect?
    (b) What type of error (Type I or Type II) is committed by the jury in part (a)?
    (c) Evidence in the trial of a guilty suspect is enough to convince 95% of all jurors in the population that the suspect is guilty. What is the probability that a jury fails to convict the guilty suspect?
    (d) What type of error is committed by the jury in part (c)?

34. To demonstrate that a planned commercial will be cost effective, at least 60% of those watching the programming need to see the commercial (rather than switching stations or using a digital recorder to skip the commercial). Typically, half of viewers watch the commercials. It is possible to measure the behavior of viewers in 1,000,000 households, but is such a large sample needed?

35. The biostatistician who designed a study for investigating the efficacy of a new medication was fired after the study. The tested null hypothesis states that the drug is no better than a placebo. The t-statistic that was obtained in the study of 400 subjects was $t = 20$, rejecting the null hypothesis and showing that the drug has a beneficial effect. Why was management upset?

36. A consultant claims that by following his advice, an insurance-processing center can reduce the time required to process a claim for storm damage from the current average length of 6.5 to 5 workdays. He claims that his method obviously works so H_0: $\mu \leq 6.5$. He says he can save the insurer further money because he can test this hypothesis by measuring the time to process a single claim. What's wrong with this approach?

[7] Historically, blood pressure was measured using a device that monitored the height of a column of mercury in a tube. That's not used anymore, but the scale remains.

37. The Human Resources (HR) group gives job applicants at a firm a personality test to assess how well they will fit into the firm and get along with colleagues. Historically, test scores have been normally distributed with mean μ and standard deviation $\sigma = 25$. The HR group wants to hire applicants whose true personality rating μ is greater than 200 points. (Test scores are an imperfect measure of true personality.)
    (a) Before seeing test results, should the HR group assert as the null hypothesis that μ for an applicant is greater than 200 or less than 200?
    (b) If the HR group chooses $H_0: \mu \leq 200$, then for what test scores (approximately) will the HR group reject H_0 if $\alpha = 2.5\%$?
    (c) What is the chance of a Type II error using the procedure in part (b) if the true score of an applicant is 225?

38. A test of filtering software examined a sample of $n = 100$ messages. If the filtering software reduces the level of spam to 15%, this test only has a 33% chance of correctly rejecting $H_0: p \geq 0.20$. Suppose instead of using 100 messages, the test were to use $n = 400$.
    (a) In order to obtain a p-value of 0.05, what must be the percentage of spam that gets through the filtering software? (*Hint:* The z-statistic must be -1.645.)
    (b) With the larger sample size, does $\hat{p}$ need to be as far below $p_0 = 0.20$ as when $n = 100$? Explain what threshold moves closer to p_0.
    (c) If in fact $p = 0.15$, what is the probability that the test with $n = 400$ correctly rejects H_0?

You Do It

39. An appliance manufacturer stockpiles washers and dryers in a large warehouse for shipment to retail stores. Some appliances get damaged in handling. The long-term goal has been to keep the level of damaged machines below 2%. In a recent test, an inspector randomly checked 60 washers and discovered that 5 of them had scratches or dents. Test the null hypothesis $H_0: p \leq 0.02$ in which p represents the probability of a damaged washer.
    (a) Do these data supply enough evidence to reject H_0? Use a binomial model from Chapter 11 to obtain the p-value.
    (b) What assumption is necessary in order to use the binomial model for the count of the number of damaged washers?
    (c) Test H_0 by using a normal model for the sampling distribution of $\hat{p}$. Does this test reject H_0?
    (d) Which test procedure should be used to test H_0? Explain your choice.

40. The electronic components used to assemble a cellular phone have been exceptionally reliable, with *more than* 99.9% working correctly. The head of procurement believes that the current supplier meets this standard, but he tests components just the same. A test of a sample of 100 components yielded no defects. Do these data prove beyond reasonable doubt that the components continue to exceed the 99.9% target?

(a) Identify the relevant population parameter and null hypothesis.
(b) Explain why it is not appropriate to use a normal approximation for the sampling distribution of the proportion in this situation.
(c) Determine if the data supply enough evidence to reject the null hypothesis. Identify any assumptions you need for the calculation.

41. A company that stocks shelves in supermarkets is considering expanding the supply that it delivers. Items that are not sold must be discarded at the end of the day, so it only wants to schedule additional deliveries if stores regularly sell out. A break-even analysis indicates that an additional delivery cycle will be profitable if items are selling out in more than 60% of markets. A survey during the last week in 45 markets found the shelves bare in 35.
    (a) State the null and alternative hypotheses.
    (b) Describe a Type I error and a Type II error in this context.
    (c) Find the p-value of the test. Do the data supply enough evidence to reject the null hypothesis if the α-level is 0.05?

42. Field tests of a low-calorie sport drink found that 80 of the 100 who tasted the beverage preferred it to the regular higher-calorie drink. A break-even analysis indicates that the launch of this product will be profitable if the beverage is preferred by more than 75% of all customers.
    (a) State the null and alternative hypotheses.
    (b) Describe a Type I error and a Type II error in this context.
    (c) Find the p-value for a test of the null hypothesis. If $\alpha = 0.10$, does the test reject H_0?

43. The management of a chain of hotels avoids intervening in the local management of its franchises unless problems become far too common to ignore. Management believes that solving the problems is better left to the local staff unless the measure of satisfaction drops below 33%. A survey of 80 guests who recently stayed in the franchise in St. Louis found that only 20% of the guests indicated that they would return to that hotel when next visiting the city. Should management intervene in the franchise in St. Louis?
    (a) State the null and alternative hypotheses.
    (b) Describe Type I and Type II errors in this context.
    (c) Find the p-value of the test. Do the data supply enough evidence to reject the null hypothesis if $\alpha = 0.025$?

44. An importer of electronic goods is considering packaging a new, easy-to-read instruction booklet with DVD players. It wants to package this booklet only if it helps customers more than the current booklet. Previous tests found that only 30% of customers were able to program their DVD player. An experiment with the new booklet found that 16 out of 60 customers were able to program their DVD player.
    (a) State the null and alternative hypotheses.
    (b) Describe Type I and Type II errors in this context.

(c) Find the *p*-value of the test. Do the data supply enough evidence to reject the null hypothesis if $\alpha = 0.05$?

45. A variety of stores offer loyalty programs. Participating shoppers swipe a bar-coded tag at the register when checking out and receive discounts on certain purchases. Stores benefit by gleaning information about shopping habits and hope to encourage shoppers to spend more. A typical Saturday morning shopper who does not participate in this program spends $120 on her or his order. In a sample of 80 shoppers participating in the loyalty program, each shopper spent $130 on average during a recent Saturday, with standard deviation $s = \$40$. Is this statistical proof that the shoppers participating in the loyalty program spend more on average than typical shoppers? (Assume that the data meet the sample size condition.)
    (a) State the null and alternative hypotheses. Describe the parameters.
    (b) Describe the Type I and Type II errors.
    (c) How large could the kurtosis be without violating the CLT condition?
    (d) Find the *p*-value of the test. Do the data supply enough evidence to reject the null hypothesis if $\alpha = 0.05$?

46. Brand managers become concerned if they discover that customers are aging and gradually moving out of the high-spending age groups. For example, the average Cadillac buyer is older than 60, past the prime middle years that typically are associated with more spending. Part of the importance to Cadillac of the success of the Escalade model has been its ability to draw in younger customers. If a sample of 50 Escalade purchasers has average age 45 (with standard deviation 25), is this compelling evidence that Escalade buyers are younger on average than the typical Cadillac buyer? (Assume that the data meet the sample size condition.)
    (a) State the null and alternative hypotheses. Describe the parameters.
    (b) Identify the Type I (false positive) and Type II (false negative) errors.
    (c) Find the *p*-value of the test using a normal model for the sampling distribution. Do the data supply enough evidence to reject the null hypothesis if $\alpha = 0.025$?

47. Refer to the analysis of shoppers in Exercise 45.
    (a) If several of those participating in the loyalty program are members of the same family, would this cause you to question the assumptions that underlie the test in this question?
    (b) Several outliers were observed in the data for the loyalty program. Should these high-volume purchases be excluded?

48. Refer to the analysis of car buyers in Exercise 46.
    (a) Suppose the distribution of the ages of the buyers of the Cadillac Escalade is skewed. Does this affect the use of a normal model as the sampling distribution? How skewed can the data become without preventing use of the *t*-test?
    (b) The distribution of the ages of these buyers is in fact bimodal, with one group hovering near 30 and

the other at 60+. How does this observation affect the test?

49. Refer to the analysis of shoppers in Exercise 45. If the population mean spending amount for shoppers in the loyalty program is $135 (with $\sigma = \$40$), then what is the probability that the test *procedure* used in this question will fail to reject H_0?

50. Refer to the analysis of car buyers in Exercise 46. If the population mean age for purchasers of the Cadillac Escalade is 50 years (with $\sigma = 25$ years), then what is the probability that the test *procedure* used in this question will fail to reject H_0?

51. Banks frequently compete by adding special services that distinguish them from rivals. These services can be expensive to provide. The bank hopes to retain customers who keep high balances in accounts that do not pay large interest rates. Typical customers at this bank keep an average balance of $3,500 in savings accounts that pay 2% interest annually. The bank loans this money to other customers at an average rate of 6%, earning 4% profit on the balance. A sample of 65 customers was offered a special personalized account. After three months, the average balance in savings for these customers was $5,000 ($s = \$3,000$). If the service costs the bank $50 per customer per year, is this going to be profitable to roll out on a larger scale?
    (a) State the null and alternative hypotheses. Describe the parameters.
    (b) Describe Type I and Type II errors in this context.
    (c) What is necessary for the sample size to be adequate for using a *t*-test?
    (d) Find the *p*-value of the test. Do the data supply enough evidence to reject the null hypothesis if $\alpha = 0.05$? (Assume that the data meet the sample size condition.)

52. Headhunters locate candidates to fill vacant senior positions in companies. These placement companies are typically paid a percentage of the salary of the filled position. A placement company that specializes in biostatistics is considering a move into information technology (IT). It earns a fee of 15% of the starting salary for each person it places. Its numerous placements in biostatistics had an average starting salary of $125,000. Its first 50 placements in IT had an average starting salary of $140,000 ($s = \$20,000$) but produced higher costs at the agency. If each placement in IT has cost the placement company $1,200 more than each placement in biostatistics, should the firm continue its push into the IT industry?
    (a) State the null and alternative hypotheses. Describe the parameters.
    (b) Describe Type I and Type II errors in this context.
    (c) Find the *p*-value of the test. Do the data supply enough evidence to reject the null hypothesis if $\alpha = 0.10$? (Assume that the data meet the sample size condition.)

4M Direct Mail Advertising

Performance Tires plans to engage in direct mail advertising. It is currently in negotiations to purchase a mailing list of the names of people who bought sports cars within the last three years. The owner of the mailing list claims that sales generated by contacting names on the list will more than pay for the cost of using the list. (Typically, a company will not sell its list of contacts, but rather provides the mailing services. For example, the owner of the list would handle addressing and mailing catalogs.)

Before it is willing to pay the asking price of $3 per name, the company obtains a sample of 225 names and addresses from the list in order to run a small experiment. It sends a promotional mailing to each of these customers. The data for this exercise show the gross dollar value of the orders produced by this experimental mailing. The company makes a profit of 20% of the gross dollar value of a sale. For example, an order for $100 produces $20 in profit.

Should the company agree to the asking price?

Motivation

a) Why would a company want to run an experiment? Why not just buy the list and see what happens?

b) Why would the holder of the list agree to allow the potential purchaser to run an experiment?

Method

c) Why is a hypothesis test relevant in this situation?

d) Describe the appropriate hypotheses and type of test to use. Choose (and justify) an α-level for the test. Be sure to check that this choice satisfies the necessary conditions.

Mechanics

e) Summarize the results of the sample mailing. Use a histogram and appropriate numerical statistics.

f) Test the null hypothesis and report a p-value. If examination of the data suggests problems with the proposed plan for testing the null hypothesis, revise the analysis appropriately.

Message

g) Summarize the results of the test. Make a recommendation to the management of Performance Tires (avoid statistical jargon, but report the results of the test).

4M Reducing Turnover Rates

A specialist in the Human Resources department of a national hotel chain is looking for ways to improve retention among hotel staff. The problem is particularly acute among those who maintain rooms, work in the hotel restaurant, and greet guests. Within this chain, among those who greet and register guests at the front desk, the annual percentage who quit is 52%.[8] Among the employees who work the front desk, more than half are expected to quit during the next

year. The specialist in HR has estimated that the turnover rate costs $20,000 per quitter, with the cost attributed to factors such as

- The time a supervisor spends to orient and train a new employee
- The effort to recruit and interview replacement workers
- The loss of efficiencies with a new employee rather than one who is more experienced and takes less time to complete tasks
- Administrative time both to add the new employee to the payroll and to remove the prior employee

To increase retention by lowering the quit rate, the specialist has formulated a benefits program targeted at employees who staff the front desk. The cost of offering these benefits averages $2,000 per employee. The chain operates 225 hotels, each with 16 front-desk employees. As a test, the specialist has proposed extending improved benefits to 320 employees who work the front desk in 20 hotels.

Motivation

a) Why would it be important to test the effect of the employee benefits program before offering it to all front-desk employees at the hotel chain?

b) If the benefits program is to be tested, how would you recommend choosing the hotels? How long will the test take to run? (There is no best answer to this question; do your best to articulate the relevant issues.)

Method

c) An analyst proposed testing the null hypothesis H_0: $p \leq 0.52$, where p is the annual quit rate for employees who work the main desk *if* the new program is implemented. Explain why this is *not* the right null hypothesis.

d) Another analyst proposed the null hypothesis H_0: $p \geq 0.52$. While better than the choice in part (c), what key issue does this choice of H_0 ignore? What is needed in order to improve this null hypothesis?

Mechanics

e) If the chosen null hypothesis is H_0: $p \geq 0.42$, what percentage of these 320 must stay on (not quit) in order to reject H_0 if $\alpha = 0.05$?

f) Assume the chosen null hypothesis is H_0: $p \geq 0.42$. Suppose that the actual quit rate among employees who receive these new benefits is 35%. What is the chance that the test of H_0 will correctly reject H_0?

Message

g) Do you think that the owners of this hotel chain should run the test of the proposed benefits plan? Explain your conclusion without using technical language.

---

[8] The Bureau of Labor Statistics (BLS) estimates that the annual quit rate within the leisure industry in the United States in 2007 was more that 52%. See the report "Openings and Labor Turnover: January 2008." This report is available online from the BLS. Search for the acronym JOLTS that statidenties the survey used to collect this information.

Alternative Approaches to Inference

BUSINESSES HAVE TO SOLVE THE PRINCIPAL-AGENT PROBLEM. Ideally, employees of a company (the agents) share the goals of the business's management (the principal). The principal-agent problem refers to situations in which these goals no longer coincide; factors encourage employees to behave in a way that does not benefit the company as a whole. Principal-agent problems often arise in setting compensation. Salaries should motivate employees to work in a manner that's beneficial to the company as a whole rather than the individual alone. For instance, a manufacturer might pay employees a piece rate rather than an hourly wage. The more items produced, the more the employee is paid and the more goods the manufacturer has to sell.

The principal-agent problem becomes harder to solve when employees operate in less easily watched, far-flung locations. As an example, consider the situation faced by an automobile insurance company.

Agents of the company operate out of small offices around the country. Think about what can happen if agents are paid a percentage of the up-front price of the insurance policies that they sell. Unless the company carefully reviews the records of the insured drivers, agents might be encouraged to sell policies to risky drivers. The agent gets paid right away, but the insurance company faces losses from subsequent accidents. (Some believe a similar problem contributed to the meltdown in the U.S. housing market and banking industry that began in 2007.)

One way to avoid a misalignment of incentives is to change the compensation structure. The insurer could pay employees incrementally over the life of the policy rather than at the time of the initial sale. Alternatively, the company can monitor the claims produced by different agents, comparing each agent to a standard. For example, it is known that annual claims on auto insurance average near $3,200 nationwide, with the median claim equal to about $2,000. These differ because the distribution of the sizes of claims is highly skewed due to the occasional large amount.

We will show how a company can take the second approach in this chapter, using methods of inference that do not rely on a normal model for the sampling distribution of the statistic. The confidence intervals and tests developed in Chapters 15 and 16 use a normal model for the sampling distribution, but normal models do not provide the best approach in every situation. What should we do, for instance, when data are skewed and fail to meet the sample size condition? This chapter offers alternatives, emphasizing how to draw inferences when data are not normally distributed. This chapter also introduces a different type of statistical interval that is designed to hold future observations rather than bracket a parameter.

17.1 | A CONFIDENCE INTERVAL FOR THE MEDIAN

The histogram in Figure 17.1 shows the distribution of claims on 42 auto insurance policies that were sold by a local agency. Most of the claims are reasonably small, but the distribution is very skewed. The diamond in the boxplot is the 95% confidence interval for the mean.

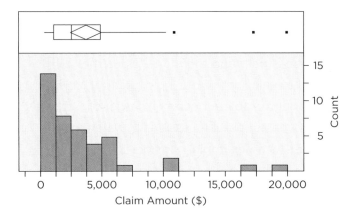

FIGURE 17.1 Claims on auto insurance policies are highly skewed.

As noted in Chapter 4, the median is smaller than $\bar{x}$ when the data are right skewed. For this sample, the average claim is $3,632 with s = $4,254 whereas the median claim is $2,456. Half of the claims are for $2,456 or less, but these account for only 14% of the total cost of the claims. A few expensive claims generate most of the costs. Is the average of this sample compatible with the nationwide average $3,200?

To answer this question, we can use a 95% confidence interval for mean cost, treating these 42 claims as a sample from a population with mean μ. If the local agency is operating differently from what is typical around the country, the overall average cost of its claims μ may be higher or lower than $3,200. If the confidence interval for μ includes $3,200, there is not a significant difference from the national standard. If $3,200 lies outside the interval (particularly if costs are higher than the national average), then there is evidence that policies sold by this agency generate more expensive claims than the insurer expects.

Before we should come to a decision based on the t-interval for μ, we need to check the appropriate conditions. The data are a random sample of claims from this agency, so the SRS condition is met. Checking the sample size condition, however, we find that the skewness in this sample is $K_3 = 2.407$, implying that the sample size must be at least $10K_3^2 = 10(2.407)^2 \approx 58$ in order to rely on a normal model for the sampling distribution. It isn't. Also, the absolute value of the kurtosis $|K_4| = 6.473$, requiring $n > 65$. We need another method for constructing a confidence interval.

For the sake of comparison, we're going to calculate the nominal 95% t-interval for μ that is shown in Figure 17.1. The word *nominal* suggests that this is a 95% confidence interval in name alone; we have little faith that this interval has 95% coverage. The interval is

$$\bar{x} \pm t_{0.025,41}s/\sqrt{n} = \$3{,}632 \pm 2.02 \times \$4{,}254/\sqrt{42} = [\$2{,}306 \text{ to } \$4{,}958]$$

The national average \$3,200 lies well inside this range. Were we to accept this interval, then there is no evidence to suggest that policies written by this agency generate statistically significantly higher claims on average.

Nonparametric Statistics

There are two reasons to consider an alternative to the t-interval for the mean in this example:

- First, the sample does not satisfy the sample size condition. The data are skewed and the sample is too small for us to use a method that relies on a normal model for the sampling distribution.
- The second consequence of the skewness is that the mean is highly variable from sample to sample. Skewed distributions produce many outliers that cause the mean to vary considerably from sample to sample.

These two reasons suggest that a confidence interval based on $\bar{x}$ is unreliable, with unknown coverage. We can do better by combining a different summary statistic with an alternative approach to finding the confidence interval.

nonparametric statistics
Statistical methods that make few assumptions about the shape of the population.

Nonparametric statistics avoid making assumptions about the shape of the population. In spite of their name, nonparametric statistics involve parameters of the distribution of the population, and inferences based on nonparametric statistics do require assumptions. Rather than make assumptions about the shape of the population or sampling distribution, however, nonparametric statistics typically require only that the data be a random sample from the population of interest.

Nonparametric statistics often rely on sorting the data. Sorting is tedious to do by hand, and consequently nonparametric methods were slow to gain popularity when calculations were done by hand. Computers removed this barrier. The connection to sorting also affects the choice of parameters. In place of the mean, nonparametric methods are more suited to parameters such as the population median. We use the Greek letter θ (theta) for the population median. Half of the population is smaller than θ and half is larger, analogous to how the median of a sample divides the data in half. If the random variable X denotes a random draw from a population with median θ, then $P(X \leq \theta) = 1/2$.

If the distribution of the population is symmetric, then the mean is equal to the median ($\mu = \theta$). In that case, inferences about the median are inferences about the mean. If the distribution is right skewed, as in this example of insurance claims, the mean and median are quite different. For these claims, the median is smaller than the mean, $\theta < \mu$. The average auto insurance claim in the nation is about $3,200 whereas the national median claim is near $2,000.

Nonparametric Confidence Interval

The procedure for finding a confidence interval for the median differs from that used to find a confidence interval for the mean. In place of summary statistics and sampling distributions, we sort the data into increasing order and count. The end result is a confidence interval for θ that takes into account sampling variation. The justification of the interval is also different: Rather than appeal to the Central Limit Theorem, we only need basic concepts from probability.

The first step in finding a confidence interval for θ is to sort the observed data. To show the calculations (which are usually done by computer), we will work with the first 10 claims in this sample, as if $n = 10$. After illustrating the calculations, we'll build the interval for the full sample with all 42 claims. The first 10 claims in the sample in order are (to the nearest dollar)

$617 $732 $1,298 $2,450   **$3,168**   **$3,545** $4,498 $4,539 $4,808 $5,754

The least expensive claim out of these 10 is $617, and the most expensive claim is $5,754. Because $n = 10$ is even, the sample median m is the average of the two middle claims, $m = (3,168 + 3,545)/2 = \$3,356.50$.

In place of a normal model or t-distribution, the nonparametric confidence interval for the median relies on a simple but clever count to obtain a confidence interval. Let $X_1, X_2, \ldots, X_{10}$ be random variables that stand for the costs of 10 randomly chosen claims. Because of the importance of sorting, it is convenient to have a notation that denotes the ordered values. The ordered values are known as **order statistics** and are usually written as

order statistics The sample values in ascending order.

$$X_{(1)} < X_{(2)} < \cdots < X_{(10)}$$

Parentheses around the subscripts signal that these are ordered random variables. In keeping with previous chapters, we use lowercase letters to distinguish observed values from random variables. For example, the two smallest values in this sample are $x_{(1)} = \$617$ and $x_{(2)} = \$732$.

Like all methods of inference, nonparametric intervals require assumptions. Every method of inference that we use requires that the observed data be a random sample from the population of interest; nonparametric methods are no different. Unlike the t-interval, which adds conditions to justify a normal model, nonparametric intervals are satisfied with having a random sample. If the data are a random sample from a population with median θ, then we know that

1. The probability that a random draw from the population is less than or equal to θ is 1/2.
2. The observations in the random sample are independent.

These two properties of a sample are enough to find the confidence interval for θ.

Since θ is the median of the population, it must be the case that

$$P(X_1 \le \theta) = 1/2, \quad P(X_2 \le \theta) = 1/2, \quad \ldots, \quad P(X_{10} \le \theta) = 1/2$$

Because the data are a sample, the random variables $X_1, X_2, \ldots, X_{10}$ are independent. Hence, the chance of getting a sample of 10 claims in which *every* claim is less than θ is the product

$$P(X_1 \le \theta \text{ and } X_2 \le \theta \text{ and} \ldots \text{ and } X_{10} \le \theta) = (1/2)^{10}$$

The probability of a sample in which all 10 observations are less than or equal to θ is very small, about 1 in 1,000: $(1/2)^{10} = 1/2^{10} \approx 0.001$.

A confidence interval emerges from looking at this and similar statements from a different point of view. If every observation is less than θ, then the largest observation, the maximum, is also less than θ. Hence,

$$P(X_{(10)} \leq \theta) = (1/2)^{10}$$

That is, the only way for the maximum to be less than θ is for every item in the sample to be less than θ.

Let's do another calculation, then find a pattern. What is the probability that exactly 9 of the 10 claims are less than or equal to θ? That's the same as the probability of getting 9 heads in 10 tosses of a fair coin. The counting methods introduced in Chapter 11 imply that

$$P(9 \text{ out of } 10 \text{ claims} \leq \theta) = {}_{10}C_9(1/2)^{10} = 10(1/2)^{10}$$

Using order statistics, the event in which exactly 9 out of 10 claims are less than θ occurs if and only if θ lies between the two largest values, so

$$P(X_{(9)} \leq \theta < X_{(10)}) = 10(1/2)^{10} \approx 0.01$$

Continuing in this way, this type of counting assigns probabilities to the intervals between the ordered observations. The probability that θ lies between two adjacent order statistics is

$$P(X_{(i)} \leq \theta < X_{(i+1)}) = {}_{10}C_i(1/2)^{10}$$

If we define $x_{(0)} = -\infty$ and $x_{(11)} = +\infty$, this formula works for the 11 disjoint intervals defined by the 10 values of the sample. The first interval is everything below the smallest value, and the last interval is everything above the largest value. Table 17.1 summarizes the probabilities, which are also the probability distribution for a binomial random variable with $n = 10$ and $p = 1/2$ (Chapter 11).

$${}_nC_k = \frac{n!}{k!\,(n-k)!}$$

TABLE 17.1 Probabilities that the population median lies between the ordered observations.

i	Interval	Probability
0	$\theta < 617$	0.00098
1	$617 \leq \theta < 732$	0.0098
2	$\mathbf{732} \leq \theta < 1{,}298$	0.044
3	$1{,}298 \leq \theta < 2{,}450$	0.117
4	$2{,}450 \leq \theta < 3{,}168$	0.205
5	$3{,}168 \leq \theta < 3{,}545$	0.246
6	$3{,}545 \leq \theta < 4{,}498$	0.205
7	$4{,}498 \leq \theta < 4{,}539$	0.117
8	$4{,}539 \leq \theta < \mathbf{4{,}808}$	0.044
9	$4{,}808 \leq \theta < 5{,}754$	0.0098
10	$5{,}754 \leq \theta$	0.00098

To form a confidence interval for θ, we combine several segments. Because the events that define the segments are disjoint, the probabilities add. Using two segments to illustrate the calculations, we have

$$P(X_{(9)} \leq \theta) = P(X_{(9)} \leq \theta < X_{(10)} \text{ or } X_{(10)} \leq \theta)$$
$$= P(X_{(9)} \leq \theta < X_{(10)}) + P(X_{(10)} \leq \theta)$$
$$= 0.0098 + 0.00098 = 0.01078$$

To get a confidence interval for θ, we choose the segments that have the most probability. For example, the interval $[x_{(2)}$ to $x_{(8)}] = [\$732$ to $\$4,808]$ is a confidence interval for the median with coverage $0.044 + 0.117 + \cdots + 0.044 \approx 0.978$. Its coverage is the sum of the probabilities of the included segments. (These segments are bracketed in Table 17.1.) As is the case in general, we cannot get a nonparametric interval for the median whose coverage is exactly 0.95. For instance, the shorter confidence interval $[\$1,298$ to $\$4,808]$ has coverage $0.978 - 0.044 = 0.934$ because it omits the segment from $\$732$ to $\$1,298$, which has probability 0.044.

To summarize, using a sample of 10 claims, we obtain a 97.8% confidence interval for θ. The interval $[\$732$ to $\$4,808]$ *is* a 97.8% confidence interval for the median so long as the data are a random sample from a population with median θ. This interval suggests that the claims from this agency are compatible with the insurance company's experience. The company expects $\theta \approx \$2,000$, which lies well within the interval. (The hypothesis test associated with this confidence interval is known as the sign test.)

What Do You Think?

The salaries of three senior executives listed in the annual report of a publicly traded firm are $\$310,000$, $\$350,000$, and $\$425,000$. Let θ denote the median salary earned by this type of executive, and assume these three are a random sample from the population of senior executives at firms like this one.

a. Which value is $x_{(2)}$?[1]
b. What is the probability that all three of these executives earn more than θ?[2]
c. What is the probability that the interval $\$310,000$ to $\$350,000$ holds θ?[3]
d. As a confidence interval for θ, what is the coverage of the interval $\$310,000$ to $\$425,000$?[4]

Parametric versus Nonparametric

Let's compare the nonparametric interval for the median, using all 42 cases, to the t-interval for the mean. As in the illustration with $n = 10$, we begin by sorting the 42 claims into ascending order. The probability contribution of each segment defined by the order statistics is then

$$P(X_{(i)} \le \theta < X_{(i+1)}) = {}_{42}C_i(1/2)^{42}$$

A spreadsheet that uses a binomial distribution is useful for this calculation. Adding up the contributions of segments as in Table 17.1, we find that

$$P(X_{(14)} \le \theta < X_{(27)}) = 0.946$$

which is close to the usual 95% coverage. The ordered values from the data are $x_{(14)} = \$1,217$ and $x_{(27)} = \$3,168$, so the 94.6% confidence interval for the median claim is $[\$1,217$ to $\$3,168]$.

Let's compare the nonparametric interval for θ to the nominal t-interval with the same coverage. The nominal 94.6% coverage t-interval for μ has larger endpoints and is longer, running from $\$2,330$ to $\$4,934$. The t-interval is longer in order to accommodate the variability of the mean from sample to sample, and because the data fail to meet the sample size condition, we cannot be sure of its coverage.

---

[1] The middle value, $\$350,000$.
[2] P(all 3 larger than θ) $= (1/2)^3 = 1/8$.
[3] $P(X_{(1)} < \theta \le X_{(2)}) = {}_3C_1(1/2)^3 = 3/8$.
[4] The coverage of the range of this small sample is $3/8 + 3/8 = 3/4$.

Nonparametric methods are not without limitations, however. The confidence interval for the median illustrates two weaknesses common to nonparametric methods:

1. The coverage is limited to certain values determined by sums of binomial probabilities. In general, we cannot exactly obtain a 95% interval that is the *de facto* standard in many applications.
2. The parameter of the population is related to percentiles rather than the mean. The median is not equal to the mean if the distribution of the population is skewed.

tip

The second concern is the more serious limitation. We often need a confidence interval for μ; an interval for θ just won't do. The most important example concerns confidence intervals for total costs or profits. **If we care about the total, we care about the mean, not the median.** By multiplying a confidence interval for μ by n, we obtain a confidence interval for the total amount associated with n items. That's not possible to do with the median.

To illustrate this issue, consider the situation faced by a manager in charge of health benefits. Her company provides health care benefits for its 60 employees. Rather than pay the $5,000 charged annually per employee for health insurance, the company is considering self-insuring. By self-insuring, the company would put money into an account monthly to be used to pay future claims. If it puts $25,000 into this account each month, the total cost annually will match the annual total cost of health insurance ($300,000), but the company gets to retain control of these funds and earn interest. To estimate the amount that will be withdrawn monthly, the company collected health expenses from its employees. The histogram in Figure 17.2 shows the amounts paid in a recent month.

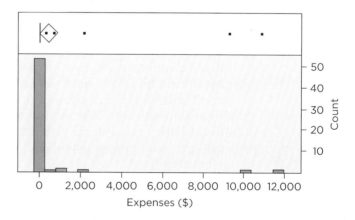

FIGURE 17.2 Health care expenses during a recent month for 60 employees.

The average amount in this month is $\bar{x} = \$410$ with $s = \$1,846$. The median is zero, and the nonparametric interval for θ with coverage 0.948 is from zero to zero! Most employees do not see a doctor or make a claim during a typical month, so the median payment is zero. The confidence interval for the median is not very helpful for anticipating future costs. The confidence interval for the mean cost per employee

$$\bar{x} \pm t_{0.025,59}s/\sqrt{60} = [-\$67 \text{ to } \$887]$$

is more helpful. Clearly, we cannot trust that this is a 95% confidence interval: The skewness ($K_3^2 = 26.7$) is so extreme that we would need $n = 10K_3^2 = 267$ in order to use a normal sampling distribution for $\bar{X}$. Plus, the lower limit of the interval is negative. Even so, the upper limit provides an important warning about how high total costs might climb: $60 \times \$887 = \$53,220$. This upper limit is not so unreasonable if we think about it. Figure 17.2 shows that 2 out of 60 employees required expensive care this month, with large payments of about

$10,000 each. At this rate, it would only take 5 out of 60 to produce $50,000 in costs, near the upper limit of the confidence interval for μ. Perhaps this company ought to stick to buying health insurance rather than risk the expense that would come if several employees required an extensive stay in a hospital.

17.2 | TRANSFORMATIONS

When dealing with positive data that are right skewed like the insurance claims in the histograms of Figures 17.1 and 17.2, it is often recommended to transform the data to obtain a more symmetric distribution. As an illustration, the histogram in Figure 17.3 summarizes base 10 logs of the 42 auto insurance claims from the auto claims example. Each value of this new variable is the common logarithm of the amount of a claim. Since these are base 10 logs, we can associate 2 with $100, 3 with $1,000, and 4 with $10,000 to make plots easier to interpret.

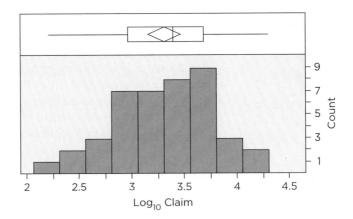

FIGURE 17.3 Base 10 logarithm of the value of insurance claims.

If you compare this histogram to the original histogram in Figure 17.1, the log transformation has removed most of the skewness from the sample. If we check the normal quantile plot (Chapter 12), the log of the claims could be a sample from a normally distributed population. All of the data remain within the dashed region around the diagonal reference line in Figure 17.4. We cannot reject the null hypothesis that these data are a sample from a normal population.

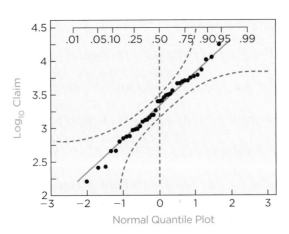

FIGURE 17.4 Normal quantile plot of the log of the claims.

The skewness and kurtosis of the logarithms of the claims are close to zero. After taking logs, $K_3 = -0.166$ and $K_4 = -0.370$. Hence, these data satisfy the sample size condition, and we can use a t-interval for the mean of the population on a log scale. If we use the symbol y to denote the log of the claims ($y = \log_{10} x$), then the average of the logarithm of the claims is $\bar{y} = 3.312$ with $s_y = 0.493$. The 95% confidence t-interval for μ_y is then

$$3.312 \pm 2.02 \times 0.493/\sqrt{42} = [3.16 \text{ to } 3.47]$$

This interval satisfies the conditions for using a t-interval; we can rely on the coverage being 95%. But now we run into a problem: How are we to interpret a confidence interval for the mean on the $\log_{10}$ scale? **Unless we can interpret a confidence interval, it is not of much value.**

tip

caution The difficulty with interpreting this interval is that the log of an average is not the same as the average of the logs. The average claim is $\bar{x} = \$3{,}632$ and $\log_{10}(\bar{x}) \approx 3.560$. That is larger than the average of the data on the log scale ($\bar{y} = 3.312$).

$$\bar{y} = \text{average(log)} < \text{log(average)} = \log(\bar{x})$$

Similarly, if we "unlog" $\bar{y}$ (raise 10 to the power indicated by the mean), we get $10^{3.312} = \$2{,}051$. That's nowhere near the average claim $\bar{x}$. If we transform the endpoints of the confidence interval on the log scale back to the scale of dollars, we have similar problems. When expressed as dollars, the confidence interval based on the logarithm of the claims is

$$[10^{3.16} \text{ to } 10^{3.47}] = [\$1{,}445 \text{ to } \$2{,}951]$$

On the scale of dollars, the 95% confidence t-interval computed from the logs does not include $\bar{x}$, the observed average claim!

The explanation is that the t-interval obtained from the logs is similar to the nonparametric confidence interval for the median.

t-interval obtained from logs	$[\$1{,}445 \text{ to } \$2{,}951]$
Nonparametric interval for median	$[\$1{,}217 \text{ to } \$3{,}168]$

Transforming to a log scale does remove the skewness and allows us to use a t-interval. The use of this transformation, however, changes the meaning of the confidence interval. A t-interval computed on a log scale and converted back to the scale of dollars resembles a confidence interval for the median rather than a confidence interval for μ.

That's not a problem if we want a confidence interval for the median θ and would rather not deal with binomial probabilities and nonparametric methods. (Transforming to logs and forming a confidence interval is easy with most software, but the nonparametric interval for θ is less common.) It is a problem, however, if we mistakenly believe that we have found a confidence interval for μ. If we want a confidence interval for μ, we are stuck with the t-interval for the mean of the observed sample and all of the attendant limitations (the uncertain coverage and large width).

17.3 | PREDICTION INTERVALS

prediction interval Interval that holds a future draw from the population with chosen probability.

Prediction intervals are statistical intervals that sometimes, with great misfortune, get confused with confidence intervals. A **prediction interval** (or the related type of interval, a tolerance interval) is a range that contains a chosen percentage of the population. Rather than hold a parameter such as μ with a given level of confidence, a prediction interval holds a future draw from a population. A confidence interval for the mean insurance claim, for instance, asserts a range for μ, the average in the population. A prediction interval anticipates the size of the next claim, allowing for the random variation associated with an individual.

Prediction Intervals for the Normal Model

If a population is normally distributed with mean μ and standard deviation σ, then 95% of the population lies between $\mu - 1.96\sigma$ and $\mu + 1.96\sigma$. That's the 95% prediction interval. This range is called a prediction interval because we can think of μ as a prediction of a random draw from the population. The limits at $\mu \pm 1.96\sigma$ bound the size of the probable error that comes with this prediction,

$$P(-1.96\sigma \leq X - \mu \leq 1.96\sigma) = 0.95$$

Ninety-five percent of prediction errors $X - \mu$ are less than 1.96σ in magnitude. In general, the $100(1 - \alpha)\%$ prediction interval for a normal population is $\mu - z_{\alpha/2}\sigma$ to $\mu + z_{\alpha/2}\sigma$. Notice that the sample size n does not affect this interval; the interval is the same regardless of n because we assume that μ and σ are known rather than estimated from the data. Also, the width of the interval is determined by σ, not the standard error.

Two aspects of this prediction interval change if μ and σ are estimated from the data. First, percentiles $t_{\alpha/2,n-1}$ of the t-distribution replace the normal percentile $z_{\alpha/2}$. Second, the interval widens to compensate for using the estimate $\bar{x}$ in place of μ. The use of an estimate of μ increases the likely size of a prediction error. The variance of the prediction error grows from σ^2 if μ is known to $(1 + 1/n)\sigma^2$ if the predictor is $\bar{x}$.

If $X \sim N(\mu, \sigma)$ is an independent draw from the same population as the sample that produces $\overline{X}$ and the standard deviation is S, then

$$P\left(\overline{X} - t_{\alpha/2,n-1}\, S\sqrt{1 + \frac{1}{n}} \leq X \leq \overline{X} + t_{\alpha/2,n-1}\, S\sqrt{1 + \frac{1}{n}}\right) = 0.95$$

Both adjustments (replacing z by t and inserting the factor times S) make this interval wider than the prediction interval based on μ and σ. It is reasonable to require a wider interval when working with estimates since we know less about the population.

Prediction Interval for Normal Population The $100(1 - \alpha)\%$ prediction interval for an independent draw from a normal population is the interval

$$\bar{x} - t_{\alpha/2,n-1}\, s\sqrt{1 + \frac{1}{n}} \quad \text{to} \quad \bar{x} + t_{\alpha/2,n-1}\, s\sqrt{1 + \frac{1}{n}}$$

where $\bar{x}$ and s estimate μ and σ of the normal population on the basis of a sample of n observations. This interval is only applicable when sampling a normally distributed population.

Nonparametric Prediction Interval

A simple nonparametric procedure produces a prediction interval that does not require assumptions about the shape of the population. This nonparametric interval is also easy to compute. Nonparametric methods are particularly useful for prediction intervals. Because a prediction interval must anticipate the next value, we cannot rely on the Central Limit Theorem to smooth away deviations from normality.

The nonparametric prediction interval relies on a fundamental property of the order statistics $X_{(1)} < X_{(2)} < \cdots < X_{(n)}$ of a sample. So long as our data lack ties (each value is unique, at least conceptually), then

$$P(X_{(i)} \leq X \leq X_{(i+1)}) = 1/(n + 1)$$

That is, every gap between adjacent order statistics has equal probability $1/(n + 1)$ of holding the next observation. The interval below the minimum $X_{(1)}$ and the interval above the maximum $X_{(n)}$ also have probability $1/(n + 1)$.

$$P(X \leq X_{(1)}) = 1/(n + 1) \quad \text{and} \quad P(X_{(n)} \leq X) = 1/(n + 1)$$

These properties of the order statistics hold regardless of the shape of the population.

For example, the smallest two insurance claims among the 42 in the auto claims example are $x_{(1)} = \$158$ and $x_{(2)} = \$255$. Hence, the probability that an independently selected claim from the sampled population is less than $\$158$ is $1/(n + 1) = 1/(42 + 1) \approx 0.023$. Similarly, the probability that the next claim is between $\$158$ and $\$255$ is also 0.023.

As with the nonparametric interval for the median, we can combine several segments to increase the coverage. Each segment that is included increases the coverage of the prediction interval by $1/(n + 1)$. In general, the coverage of the interval between the ith largest value $X_{(i)}$ and the jth largest value $X_{(j)}$ is $(j - i)/(n + 1)$ so long as $j > i$. Using the claims data,

$$P(x_{(2)} \leq X \leq x_{(41)}) = P(\$255 \leq X \leq \$17{,}305) = (41 - 2)/43 \approx 0.91$$

There is a 91% probability that the next claim is between $\$255$ and $\$17,305$.

What Do You Think?

The executive salaries listed in the prior What Do You Think? are $\$310,000$, $\$350,000$, and $\$425,000$. Assume these three are a random sample.

a. What is the probability that the next draw from this population is larger than $\$310,000$?[5]
b. Find a 50% prediction interval for the next random draw from this population.[6]
c. Is this the only possible 50% interval, or are there others?[7]

4M EXAMPLE 17.1 | **EXECUTIVE SALARIES**

MOTIVATION | STATE THE QUESTION

Fees earned by an executive placement service are 5% of the starting annual total compensation package. A current client is looking to move into the telecom industry as CEO. How much can the firm expect to earn by placing this client as a CEO?

METHOD | DESCRIBE THE DATA AND SELECT AN APPROACH

In this situation, directors of the placement service want a range for the likely earning from placing this candidate. They are not interested in what happens

---

[5] 0.75, because there is a 25% chance for each of the four subintervals.
[6] The middle interval, from $\$310,000$ to $\$425,000$.
[7] There are six 50% intervals if we include the intervals below the smallest and above the largest observations; pick any pair of the four subintervals defined by the data.

on average (μ or θ), but rather want a range that accommodates the variability among individuals; they want a prediction interval.

The relevant population is hypothetical. From public records, the placement service has the total compensation (in millions of dollars) earned by all 23 CEOs in the telecom industry. That's the population of current CEOs. We can, however, think of these 23 as a sample from the collection of all compensation packages that are available within this industry. These 23 amounts represent a sample of what is possible.

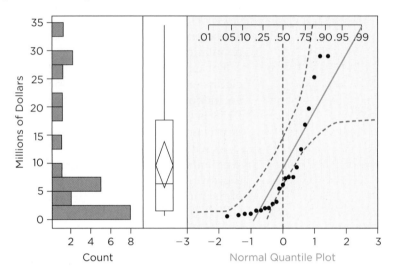

It is clear from the histogram of these data that the distribution is not normal. Consequently, we will use a nonparametric prediction interval. We'll use the 75% coverage prediction interval obtained by trimming off the top three and bottom three subintervals. The coverage of the prediction interval between $X_{(3)}$ and $X_{(21)}$ is $(21 - 3)/24 = 0.75$.

✓ **SRS condition.** If we view these compensation figures as a sample from the negotiation process used in the telecom industry to set salaries, then it is reasonable to call these data a sample. It is hard to believe, though, that the data are independent. These executives know how much each other earns. We will note this source of possible dependence in our message.

MECHANICS

DO THE ANALYSIS

The following table shows a portion of the sorted observations, in dollars:

$x_{(1)}$	$200,000
$x_{(2)}$	$421,338
$x_{(3)}$	**$743,801**
$x_{(4)}$	$806,060
$x_{(5)}$	$1,627,999
⋮	⋮
$x_{(19)}$	$20,064,332
$x_{(20)}$	$25,719,144
$x_{(21)}$	**$29,863,393**
$x_{(22)}$	$29,914,396
$x_{(23)}$	$34,606,343

The interval $x_{(3)}$ to $x_{(21)}$ is \$743,801 $\approx$ \$750,000 to \$29,863,393 $\approx$ \$30,000,000 and is a 75% prediction interval.

MESSAGE | SUMMARIZE THE RESULTS

The compensation package of three out of four placements in this industry can be predicted to be in the range from about 3/4 to 30 million dollars. The implied fees thus range from a relatively paltry \$37,500 all the way up to \$1,500,000. This range is wide because of the massive variation in pay packages, from \$200,000 to more than \$30 million. If we allow for possible dependencies in the data, the range may be larger still. The best recommendation might be to suggest that the placement agency needs to learn why some salaries are so much higher than others. (We will come to methods for exploiting such explanations to obtain better prediction limits beginning in Chapter 19.)

17.4 | PROPORTIONS BASED ON SMALL SAMPLES

Student's t-distribution provides a confidence interval for means of small samples from a normal population. The nonparametric interval for the median handles inference for the center of a population that is not normal. Neither applies to finding a confidence interval for a proportion when the sample size is small. An example with a small sample shows the problems that we run into if we use a method designed for other situations.

Here are the results of a small survey. We observe seven shoppers at a supermarket. The data record whether the customer used a credit card to pay for his or her purchase:

Yes, No, No, Yes, No, No, No

The sample proportion $\hat{p} = 2/7 \approx 0.286$. How should we compute a 95% confidence interval for p, the population proportion that pay with a credit card? Nonparametric methods don't help because the data have only two distinct values. Nonparametric methods rely on sorting distinct values. Because of the matching values, sorting the data is not informative.

We cannot use a t-interval either. These Yes and No data (or 1s and 0s if we represent the data as a dummy variable) are not a sample from a normal population. If we had a large enough sample, the Central Limit Theorem would take care of us. But $n = 7$ is too small. If we ignore those cautions and compute the confidence interval, the lower endpoint of the z-interval shown in Table 17.2 confirms a problem.

TABLE 17.2 Summary of a small survey.

$\hat{p}$	0.2857
n	7
$\sqrt{\hat{p}(1 - \hat{p})/n}$	0.1707
95% z-interval	−0.0489 to 0.6204

The 95% z-interval is −4.9% to 62%. The confidence interval for the proportion p includes *negative* values. That doesn't make sense because we know that $p \geq 0$. The normal model extends the lower limit of the confidence interval below zero into a range that is not possible. A t-distribution would do the same. The sample size condition for proportions (Chapters 15 and 16) prevents this problem by requiring both $n\hat{p} \geq 10$ and $n(1 - \hat{p}) \geq 10$. In order to use the

z-test or z-interval for a proportion, this condition requires at least 10 successes and 10 failures. This small sample includes just 2 successes and 5 failures.

One remedy for this problem is to work directly with a binomial distribution. Another that is easier to use moves the sampling distribution away from 0 and 1. The way this is done is remarkably easy: Add four artificial cases to the data, 2 successes and 2 failures. The interval is centered on the adjusted proportion $\widetilde{p} = (\#\text{successes} + 2)/(n + 4)$ and uses $\widetilde{p}$ to determine the standard error in the revised z-interval.

> **Wilson's Interval for a Proportion**[8] Add 2 successes and 2 failures to the data and define
>
> $$\widetilde{p} = \frac{\#\text{ successes} + 2}{n + 4}, \qquad \widetilde{n} = n + 4$$
>
> Then compute the usual z interval from the augmented data.
>
> $$\left[\widetilde{p} - z_{\alpha/2}\sqrt{\frac{\widetilde{p}(1 - \widetilde{p})}{\widetilde{n}}} \quad \text{to} \quad \widetilde{p} + z_{\alpha/2}\sqrt{\frac{\widetilde{p}(1 - \widetilde{p})}{\widetilde{n}}} \right]$$
>
> **Checklist**
>
> ✓ **SRS condition.** The data are a simple random sample from the relevant population. In this case, the data are Bernoulli trials.

This simple adjustment moves the sampling distribution closer to 1/2, away from the troublesome boundaries at 0 and 1. This adjustment also packs the sampling distribution closer to $\widetilde{p}$. We interpret the adjusted interval as if it were the original confidence interval for p. (Some software packages offer another confidence interval for p known as the score interval. It gives similar results but is more challenging to compute.)

In the example of customers checking out from a supermarket, adding 2 successes and 2 failures increases $n = 7$ to $\widetilde{n} = 11$ and gives $\widetilde{p} = (2 + 2)/(7 + 4) \approx 0.364$. The revised 95% z-interval for p is

$$\widetilde{p} \pm 1.96\sqrt{\widetilde{p}(1 - \widetilde{p})/\widetilde{n}} = 0.364 \pm 1.96\sqrt{0.364(1 - 0.364)/11}$$
$$= [0.080, 0.648]$$

With 2 successes and 2 failures added to the data, Wilson's z-interval for the proportion no longer includes negative values.

4M EXAMPLE 17.2 | **DRUG TESTING**

MOTIVATION | STATE THE QUESTION

Surgery can remove a tumor, but that does not mean the cancer won't return. To prevent a relapse, pharmaceutical companies develop drugs that delay or eliminate the chance for a relapse.

To assess the impact of the medication, we need a baseline for comparison. If patients routinely survive for a long time without the drug, it is difficult to make the case that health insurance should pay thousands of dollars for the treatment. That said, studies designed to establish the baseline are expensive and take years to complete. The expense can exceed $30,000 to $40,000 per subject if surgery or hospitalization is involved.

---

[8] This interval is studied in A. Agresti and B. A. Coull (1998), "Approximate Is Better Than 'Exact' for Interval Estimation of Binomial Proportions," *American Statistician*, *52*, 119–126. Wilson also proposed a more elaborate interval, but this one does quite well and is easy to compute.

The company in this example is developing a drug to prolong the time before a relapse of cancer. For the drug to appeal to physicians and insurers, management believes that the drug must cut the rate of relapses in half. That's the goal, but scientists need to know if that's possible. As a start, they must know the current time to relapse.

METHOD	## DESCRIBE THE DATA AND SELECT AN APPROACH

Identify parameter.

Identify population.

Describe data.

Pick an interval.

Check conditions.

The parameter of interest is p, the probability of relapse over a relevant time horizon. In this case, the horizon is two years after surgery. The population is the collection of adults who are older than 50 and have this operation. The data in this case are 19 patients who were observed for 24 months after surgery to remove a single tumor. Among these, 9 suffered a relapse: Within two years, doctors found another tumor in 9 of the 19 patients.

We will provide the scientists a 95% confidence interval for p. The sample is small, with 9 who relapse and 10 who do not. Before we go further, let's check the relevant conditions.

✓ **SRS condition.** These data come from a medical study sponsored by the National Institutes of Health (NIH) and published in a leading journal. Protocols used by NIH to identify subjects should produce a sample.

✗ **Sample size condition.** We do not quite have enough data to satisfy this condition. It requires at least 10 successes and 10 failures.

These data are a random sample from the population, but we lack enough data to satisfy the sample size condition. To make the 95% confidence interval more reliable, we will add 2 successes and 2 failures and use Wilson's interval.

MECHANICS | ## DO THE ANALYSIS

This table shows the standard output of a statistics package without any adjustments. Although this interval does not include negative values as in the prior example, we cannot rely on it being a 95% confidence interval for p since the sample is too small.

$\hat{p}$	0.4737
n	19
$\sqrt{\hat{p}(1-\hat{p})/n}$	0.1145
95% z-interval	0.2492 to 0.6982

Wilson's interval is similar, but shifted toward 0.5 and narrower. By adding two successes and two failures, $\hat{p} = (9 + 2)/(19 + 4) \approx 0.478$ and the interval is

$$0.478 \pm 1.96 \sqrt{0.478(1 - 0.478)/(19 + 4)} \approx [0.27 \text{ to } 0.68]$$

MESSAGE | ## SUMMARIZE THE RESULTS

We are 95% confident that the proportion of patients under these conditions who survive at least 24 months without remission lies in the range of 27% to 68%. For the proposed drug to cut this proportion in half, it must be able to reduce the rate of relapses to somewhere between 13% and 34%.

Best Practices

- *Check the assumptions carefully when dealing with small samples.* With large samples, you can typically rely on the Central Limit Theorem to justify using a normal model for the sampling distribution. You can't do that with small samples. For these, you must either make strong assumptions, such as assume that the population is normally distributed, or find methods that avoid such assumptions (the nonparametric confidence interval for the median).

- *Consider a nonparametric alternative if you suspect non-normal data.* If data are a sample from a normal population, the *t*-interval is the best alternative. If the data aren't normally distributed, a *t*-interval may look foolish (such as including negative values in a confidence interval for a parameter that you know has to be positive).

- *Use the adjustment procedure for proportions from small samples.* It's easy to add 2 successes and 2 failures, and the resulting interval has much better coverage properties.

- *Verify that your data are a simple random sample.* We talked a lot about random samples in Chapter 13, but this is the most important condition of all. Every statistical method requires that the data be a random sample from the appropriate population. The alternative approaches in this chapter avoid the assumption of normality, but all still assume that you begin with a random sample from the population of interest.

Pitfalls

- *Avoid assuming that populations are normally distributed in order to use a t-interval for the mean.* A *t*-interval may be appropriate, but only if you are quite sure that the population you're sampling has a normal distribution. You won't know unless you check.

- *Do not use confidence intervals based on normality just because they are narrower than a nonparametric interval.* In some examples, a confidence interval that assumes normality may be narrower than an interval that does not make this assumption. That's a good reason for using the usual normality-based interval if you're sure that normality holds, but otherwise it's misleading. That 95% *t*-interval may have substantially less than 95% coverage.

- *Do not think that you can prove normality using a normal quantile plot.* The normal quantile plot can show that data are not a sample from a normal population. If the data remain close to the diagonal reference in this display, however, that's no proof that the population is normal. It only shows that the data *may* be a sample from a normal population. The bounds around the diagonal reference line are so wide when dealing with small samples that almost any sample looks as if it may be normally distributed; that does not mean that it is. You cannot prove the null hypothesis of normality.

- *Do not rely on software to know which procedure to use.* Software will only do so much for you. Most packages do not check to see that your sample size is adequate for the method that you've used. For all the software knows, you may be a teacher preparing an example to illustrate what can go wrong. Just because your software can compute a *t*-confidence interval for μ does not mean that you should.

- *Do not use a confidence interval when you need a prediction interval.* If you want a range that anticipates the value of the next observation, you need a prediction interval. Most 95% confidence intervals contain only a very small proportion of the data; they are designed to hold the population mean with the chosen confidence level, not to contain data values.

Software Hints

Modern statistics packages remove the need to know many of the details of the calculations discussed in this chapter. Most packages include so-called nonparametric methods that implement the methods included in this chapter or provide suitable alternatives.

EXCEL

Excel has all of the tools needed to compute the non-parametric intervals for the median and prediction intervals. For prediction intervals, sort the data into ascending order. For the confidence interval for the median, begin by sorting the data. Once the data are sorted, use the built-in function BinomDist to obtain the binomial probabilities. For example, consider the probabilities given in the last column of Table 17.1. In this example, $n = 10$. Add a column that identifies the rows of the table, running from 0 to 10. Then use the formula BinomDist $(i, 10, 0.5, \text{False})$ for $i = 0, 1, 2, \ldots, 10$ to compute the probabilities in the last column.

DDXL includes several nonparametric tests, emphasizing methods for comparing two samples. The included sign test is equivalent to the nonparametric interval for the median described in this chapter. The sign test rejects $H_0: \theta = 10$, for example, at $\alpha = 0.05$, if 10 is not inside the 95% confidence interval. The output does not include the associated confidence interval for θ, however.

MINITAB

Minitab includes a variety of nonparametric statistics, primarily related to comparison of two or more samples. The options are found by following the menu commands

Stat > Nonparametrics > ...

The sign test and Mood's median test provide confidence intervals for the median. The method shown in the text is equivalent to the one-sample sign test. The output of these tests includes the associated confidence interval for the median θ.

Minitab will also compute the binomial probabilities needed for the nonparametric interval for the median. The procedure for computing this interval resembles that for Excel. Use the commands

Calc > Probability Distributions > Binomial ...

Fill in the dialog with the appropriate values for n and p.

Minitab includes a specialized confidence interval designed for proportions. Follow the menu commands

Stat > Basic Statistics > 1 Proportion ...

to use this approach. (Minitab computes an interval based on the likelihood ratio test rather than the simpler method discussed in this chapter.)

JMP

JMP includes nonparametric methods for comparison of two or more samples, but not the nonparametric interval for the median described in this chapter. The procedure for computing this interval resembles that for Excel. For example, consider the calculations illustrated in Table 17.1. Begin by entering a column of values from 0 to $n = 10$. Then define a column using the built-in formula for the binomial probabilities (open the formula calculator and select the probability formula Binomial Probability). Fill in the boxes in the formula so that it looks like this:

Binomial Probability(0.5, 10, Column)

Column identifies the column holding the values 0 to 10 in the data table. After you click the apply button, JMP fills the column with the needed binomial probabilities as found in the last column of Table 17.1.

JMP includes the score interval for the proportion mentioned in the text. The menu item

Analyze > Distribution

opens a dialog that allows for picking the column that holds the data for the test. If the variable is categorical, the results show a frequency table. Click the red triangle above the bar chart and select the item Confidence Interval and choose the level of the interval. The output window expands to show the score confidence interval for the proportion in each category.

CHAPTER SUMMARY

Not all data satisfy the conditions for a z-interval (proportions) or a t-interval (means). Most often, the failure is due to a small sample or distribution that is far from the normal. Using **nonparametric statistics** for the confidence interval for the median is an alternative to using the t-interval for the mean. The nonparametric interval uses the sorted data, known as **order statistics**. By adding four artificial observations, **Wilson's interval for a proportion** produces a more reliable z-interval for the proportion p. **Prediction intervals** are statistical intervals designed to hold a chosen percentage of the population. These intervals do not benefit from averaging and hence require either a nonparametric approach or strong assumptions about the form of the population.

Key Terms

nonparametric statistics, 405
order statistics, 406

prediction interval, 411
Wilson's interval for a proportion, 416

Formulas

Order statistics

The data are put into ascending order, identified by parentheses in the subscripts:

$$\text{Minimum} = X_{(1)} < X_{(2)} < \cdots < X_{(n)} = \text{Maximum}$$

Nonparametric confidence interval for the median

Combine subintervals defined by the ordered data with probabilities assigned to the subintervals from the binomial distribution with parameters $n + 1$ and $1/2$. The probability attached to the subinterval

between the ith largest value and the $(i + 1)$st largest value is

$$P(X_{(i)} < \theta \leq X_{(i+1)}) = {_n}C_i(1/2)^n$$

Nonparametric prediction interval

Combine subintervals defined by the ordered data, with equal probability $1/(n + 1)$ assigned to each. If X denotes an independent random draw from the population, then

$$P(X_{(i)} < X \leq X_{(i+1)}) = 1/(n + 1)$$

About the Data

The data for the cancer trial are from the Web site of the National Center for Health Statistics (NCHS). Data on the cost of claims made to auto insurers come from the Insurance Information Institute.

Both sources are available online. Salaries of telecom executives come from company listings in the telecom industry in the 2003 Execucomp database.

EXERCISES

Mix and Match

Match each description on the left to the appropriate method or situation on the right.

1. Describes statistical methods that do not make assumptions about the shape of the population

2. Proportion of the population distribution less than the median

3. Symbols denoting the ordered observations in a sample of size n

4. Probability that the population median lies between $X_{(i)}$ and $X_{(i+1)}$

5. Probability that the population median is larger than the next three observations

6. Transformation that reduces the amount of right skewness

7. A statistical interval designed to hold a population parameter in a chosen fraction of samples

8. A statistical interval designed to hold a fraction of the population

9. A confidence interval for the proportion that adds artificial observations

10. Chance that the next observation is smaller than the minimum of a sample of size 3

(a) Logarithm

(b) $X_{(1)}, X_{(2)}, \ldots, X_{(n)}$

(c) Wilson's interval

(d) $1/2$

(e) Prediction interval

(f) $1/8$

(g) Nonparametric

(h) $1/4$

(i) $P(Y = i)$ if $Y \sim \text{Binomial}(n, 1/2)$

(j) Confidence interval

True/False

Mark each statement True or False. If you believe that a statement is false, briefly explain why you think it is false.

Exercises 11–16. Four employees were recently hired. Their ages at the time of hiring are

$$x_1 = 20, \quad x_2 = 34, \quad x_3 = 25, \quad x_4 = 22$$

11. The Central Limit Theorem assures us that we can use a t-interval for the mean of the population from which these employees were hired.

12. If it were known that the population of employee ages from which these four are a sample was normally distributed, then a 95% t-interval for μ, the mean age, would have coverage exactly 0.95.

13. Before we can find a confidence interval for the population median age, we need to check the skewness and kurtosis of this small sample.

14. The nonparametric prediction interval for the median does not require or make any assumptions about the data at hand.

15. There is a 1 in 5 chance that the population median is larger than 34.

16. There is a 1 in 5 chance that the age of the next employee will be less than 20.

17. The small-sample interval for the proportion is more accurate than the z-interval for a proportion if we add 4 successes or 4 failures to the observed data.

18. The small-sample interval for the proportion is unnecessary if the data are a sample from a normal population.

19. The population median is the same as the population mean if the population is symmetric.

20. Given a sample from a normally distributed population, the 95% Student's t-interval for the mean is shorter, on average, than the 95% confidence interval for the median.

Think About It

21. In looking over sales generated by 15 offices in his district, a manager noticed that the distribution of sales was right skewed. On a $\log_{10}$ scale the data were symmetric, so he computed the 95% t-interval for the mean on the log scale. He then raised 10 to the power of the endpoints of the interval to get back to dollars. He claims that this is a 95% confidence interval for mean sales. What do you think?

22. An investor scanned over the daily summary of the stock market in the *Wall Street Journal*. She found 12 stocks whose return exceeded the return on the market that day. The one-sample 95% t-interval based on these data does not include zero. She concluded, "It's easy to find a stock that beats the returns on the market." Do you agree with her analysis or conclusion? Why or why not?

23. Does either of the samples summarized by the following normal quantile plots appear normally distributed? If not, how does the shape of the sample deviate from normality? Both samples have $n = 30$ cases.

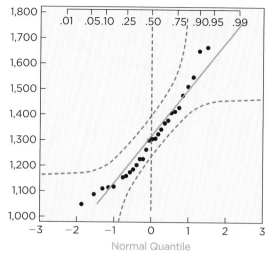

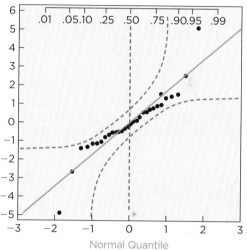

24. The following two normal quantile plots show two more samples. Both samples come from the same populations as those in Exercise 23, but this time we have larger samples with $n = 250$. Does either sample appear normally distributed? Does either plot suggest a different conclusion than the corresponding plot in Exercise 23? Why might that happen?

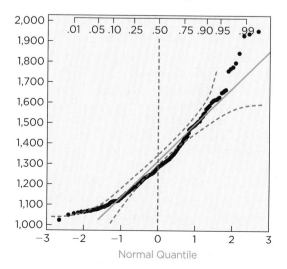

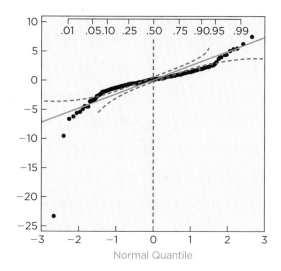

25. The adjusted 95% interval for the proportion based on a small sample is shorter than the usual 95% z-interval. Does that mean that the coverage of the adjusted interval is less than 95%?

26. An analyst decided to use either the 95% t-interval for the mean or the corresponding 95% interval for the median, whichever was shorter. His sample was symmetric, so the mean and median are roughly the same in his population. What do you think of this approach?

27. You have a very small sample, with $n = 1$. What is the chance that the median in the population is larger than your single value? Smaller?

28. This time you have a larger sample, with $n = 2$.
    (a) What is a 50% confidence interval for the population median?
    (b) What do you think of the use of the one-sample t-interval for the mean in this situation, and why?

29. A taste test has customers taste two types of cola drink and asks which is preferred. An analyst then finds the sample proportion $\hat{p}$ who prefer Brand 1. The analyst will conclude that customers prefer one brand to the other if a 95% confidence interval for the population proportion p does *not* include 1/2. Which interval is more likely to include 1/2, the small sample interval discussed in this chapter or the z-interval for the proportion?

30. Why wouldn't you use Student's t-interval for a proportion based on a small sample?

31. A retailer has data on sales in its stores yesterday. It would like a range that has a 50% chance of covering the sales of a randomly selected store today. Which does the retailer need, a prediction interval or a confidence interval?

32. A tax auditor wants to identify tax returns that report unusually low or unusually high deductions. Does the auditor need a prediction interval or a confidence interval?

33. A printout in the back of a market report summarized sales in retail outlets with two intervals but did not label which was a confidence interval for the mean and which was a prediction interval. If both are 95% intervals, can you tell which is which?

34. The manager of the HR group needs to hire a new technical analyst and wants to offer the next candidate a very attractive salary. Which is more useful to the manager, a prediction interval for the salary of a new technical analyst or a confidence interval for the median salary of a technical analyst?

35. A 95% confidence interval for the mean of a sample of positive numbers can include negative values if the sample size is small and the data are highly skewed. Is it possible for the nonparametric confidence interval for the median based on a sample of positive values to include negative numbers in its range?

36. Does taking logarithms to remove skewness improve the sampling properties of a nonparametric interval for the median?

You Do It

37. The manager of a convenience store tracked total sales during a small sample of eight recent daytime shifts (an eight-hour period of the day). The total sales during these periods were

$2,243 $2,014 $1,964 $1,889 $2,502 $2,685 $2,146 $4,592

    (a) Does this sample appear to be normally distributed on the basis of a normal quantile plot?
    (b) On the basis of the CLT condition, do these data seem appropriate as the basis for a t-interval for the mean?
    (c) Assuming that the data are a sample from a normally distributed population, find the 95% one-sample t-interval for the mean, both with and without the outlier. How does the 95% interval change if the last data point (the largest) is excluded?
    (d) Explain why the lower endpoint of the 95% confidence t-interval is larger without the outlier than with the outlier.

38. Most of the stock trades at an office that handles over-the-counter retail stock sales are fairly small, but sometimes a big trade comes through. These data show the dollar value of a sample of 11 trades made during the previous day.

    $1,500, $1,000, $1,000, $500, $750, $850, $1,800, $1,250, $15,500, $1,750, $2,250

    (a) Using a normal quantile plot, determine whether these data might be a sample from a normally distributed population.
    (b) On the basis of the skewness and kurtosis of this sample, how large a sample does the CLT condition require in order to rely on a t-interval for the mean?
    (c) Assuming that the data are nearly normal, find the 95% one-sample t-interval for the mean. What does it mean that the interval includes zero?
    (d) How does the 95% interval change if the largest data point is excluded?

(e) Explain why the lower endpoint of the 95% confidence interval is positive after the outlier is removed although it is negative with the outlier.

39. The interval from the smallest to largest amounts sold during the eight shifts in Exercise 37 is $1,889 to $4,592. If this range is treated as a nonparametric confidence interval for the median, what is the coverage probability of this interval?

40. The interval $750 to $2,250 excludes the largest and smallest trades executed at the office described in Exercise 38. If we treat this interval as a confidence interval for the associated population median, what's the coverage?

41. A manager conducted a small survey of office employees to find out how many play the office pool that selects the winning teams during the NCAA basketball tournament. Of the 14 asked, 10 said that they do. If these data are a random sample of office employees around the country, do you think that more than half of office employees participate in such pools during March Madness?

4M Stopping Distances

Better brakes make safer cars, and manufacturers compete to have the best. To keep this competition honest, the Department of Transportation (DOT) occasionally tests cars to see that safety features perform as advertised. Rather than rely on testing done by manufacturers or Consumers Union, DOT buys cars such as the Toyota Camry described in this example and tests them itself.

For this experiment, DOT measured the stopping distance for a small sample of 10 Camrys. All 10 stops were done under carefully monitored, nearly identical conditions. Each car was going 50 miles per hour. Each time the driver stopped on a patch of *wet* pavement. The same driver

performed every test, and technicians checked the road to ensure that the conditions remained the same. On average, the stops required 176 feet. With 450,000 Camrys sold each year, these 10 are a small sample.

Motivation

a) A competitor's model takes 180 feet under these conditions to stop from 50 miles per hour. How can Toyota use these results to its advantage?

b) As a driver, would you rather have a 95% confidence interval for the mean stopping distance or a 95% prediction interval for the actual distance in your next emergency stop?

Method

c) What is the consequence of having the same driver at the wheel in the DOT testing?

d) Do these data meet the conditions necessary for a t-interval for the mean stopping distance?

e) On the basis of the distribution of these stopping distances, will inferences about the mean resemble inferences about the median in the population?

Mechanics

f) Compute the 97.8% confidence interval for the mean and compare it to the corresponding 97.8% confidence interval for the median.

g) Test the null hypothesis $H_0: \mu \geq 180$, as appropriate for these data.

Mechanics

h) On the basis of these data, can Toyota advertise that its cars stop in less than 180 feet under these conditions?

CHAPTER 18 Comparison

DIETS ARE BIG BUSINESS. A diet that helps people lose weight also helps its discoverer (or promoter) earn a lot of money. In 1972 Robert Atkins published *Dr. Atkins' New Diet Revolution*. His diet bucked the prevailing wisdom by recommending a diet rich in fats rather than carbohydrates, and his book sold more than 10 million copies.

Fitness centers have noticed the interest in losing weight as well. The promise of a trim, chiseled figure attracts and keeps members. If the fitness center helps them control their weight, members keep paying their fees. Members who don't lose weight are more likely to quit. The chain of fitness centers that we'll advise in this chapter credits dietary counseling with retaining 5,000 subscribers nationwide who would have otherwise dropped out. A better diet would keep more.

Management is considering licensing a proprietary diet to include in an upcoming promotion. If it adopts the proprietary approach, centers will benefit from name recognition, but the chain will have to pay a $200,000 licensing fee in order to use the name. By comparison, fitness centers can advertise a government-recommended food pyramid without paying more than the cost of a few posters.

Is one diet more effective than the other? Name recognition aside, should the chain expect members on one diet to lose more weight than those on the other? For the example in this chapter, we'll use the Atkins diet to represent a trendy program. For the adoption of the Atkins diet to be more profitable than a conventional diet, it's not enough for members to lose weight. A trendy diet has to retain enough additional members to cover the $200,000 up-front cost for the license to use the name.

Which choice would you recommend? Which statistics help you decide?

This chapter introduces inferential statistics that test for differences between two populations. In particular, our interest concerns the difference between the means of two populations. Two samples, one from each population, supply the data. The methods in this chapter include both confidence intervals and tests. Tests support a break-even analysis whereas confidence intervals provide informative upper and lower bounds for the difference. The concepts and formulas that define two-sample methods resemble those designed for one sample, with adjustments that accommodate a second sample. (Chapter 26 shows how to compare more than two populations.)

The presence of a second sample introduces an important substantive concern that is not an issue when the data come from one population. Are differences between the samples *caused* by the variable that labels the groups or is there a different, lurking explanation? The possibility of lurking variables demands more than passing familiarity with the data. It won't matter how the statistics are done if the data don't justify the comparison.

18.1 | DATA FOR COMPARISONS

Let's frame the comparison between two diets as a test of the difference between the means of two populations. Chapter 16 introduced hypothesis tests as a method to demonstrate profitability beyond a reasonable doubt. Rejecting the null hypothesis H_0 of no change or status quo implies that the alternative has demonstrated itself to be more profitable than an economic threshold at the chosen α level, usually $\alpha = 0.05$. The same framework applies to comparisons of two populations.

In order to state the hypotheses, we need to add some notation. Because we have two populations, symbols for parameters and statistics such as μ and $\bar{x}$ must identify the population. For this example, the subscript A identifies parameters and statistics associated with the Atkins diet, and the subscript C identifies those for the conventional diet. For instance, μ_A denotes the mean number of pounds lost in the population if members go on the Atkins diet, and μ_C denotes the mean number of pounds lost on the conventional diet. The difference, $\mu_A - \mu_C$, measures the extra weight lost on average by those in the Atkins plan. To demonstrate profitability, an economic analysis (see the Follow the Money section that follows) shows that the Atkins diet has to win by more than 2 pounds, on average. The null hypothesis

$$H_0: \mu_A - \mu_C \leq 2$$

implies that the Atkins diet falls short of this threshold. If data reject H_0 in favor of the alternative, $H_a: \mu_A - \mu_C > 2$, then the Atkins plan looks like a winner. (Generically, the subscripts 1 and 2 identify the two populations. In specific examples, however, we prefer identifiable letters.)

The first step in comparing two samples requires us to think carefully about the data that will lead us either to reject H_0 or not. The data used to compare two groups typically arise in one of three ways:

1. *Run an experiment that isolates a specific cause.* This is the best way to get data for comparing two populations, and its advantages are summarized in the next section.
2. *Obtain random samples from two populations.* This is often the best we can do; experiments often are simply impossible to do.
3. *Compare two sets of observations.* This common approach often leads to problems that we must guard against.

The first two usually lead to reliable conclusions. The third often leads to a serious mistake in the interpretation of the results.

Follow the Money Management of the fitness chain has experience with dieting programs. It believes that each pound of weight lost on average when participating in a sponsored diet retains 1,000 members nationally who would otherwise not renew their membership. Each retained member adds $100 in profits, so an additional pound taken off earns the chain $1,000 \times \$100 = \$100,000$. If the Atkins diet reduces weight by 2 pounds *more* than the conventional diet, then the resulting increase in membership profits will pay for the $200,000 licensing fee. That's the break-even point: The Atkins diet needs to generate at least 2 more pounds in weight loss to make it the more profitable choice. Otherwise licensing fees will cost more than the gain in membership dues.

Experiments

> **experiment** Procedure that uses randomization to produce data that reveal causation.
>
> **factor** A variable manipulated by the experimenter in order to judge its effect on subjects in an experiment.
>
> **treatment** A level of a factor chosen by the experimenter in order to see the effect on subjects.
>
> **randomization** Random assignment of subjects to treatments.

An **experiment** produces data that reveal causal relationships. In the most common experiment, the experimenter manipulates a variable called a **factor** to discover its effect on a second variable, the response. In the ideal experiment, the experimenter

1. *Selects a random sample* from a population
2. *Assigns* subjects at random to **treatments** defined by the factor
3. *Compares* the responses of subjects between the treatments

The second step that uses a random procedure to assign the subjects is called **randomization**. This particular experimental procedure is known as a completely randomized experiment in one factor. The name *treatment* for a level of a factor originates from the use of randomized experiments in medical tests. The factor in medical experiments is a drug or therapy, and the treatments consist, for instance, of varying medications. The factor in our comparison of diets is the diet offered to a member, with two levels: Atkins or conventional. The response is the amount of weight that is lost, measured in pounds.

An experiment compares hypothetical populations. There's only one "real" population; for the fitness centers, the population consists of current and possible members of these fitness centers. The two populations that we compare result from actions of the experimenter. One population represents the weight lost *if* members follow the Atkins diet, whereas the second population represents the weight lost *if* members follow the conventional diet.

Experiments are common in marketing because they justify conclusions of cause and effect. For instance, product placement is believed to influence the brand shoppers choose in a supermarket. Suppose Post Raisin Bran is positioned at eye level in Store A, with Kellogg's Raisin Bran positioned on the lowest shelf. In Store B, the positions are reversed, with Kellogg's at eye level and Post on the bottom shelf. If we compare the amount of each brand purchased per customer at each store, we have a problem. Are differences in sales between the stores caused by the difference in shelf placement, or are differences due to other variables that distinguish the stores? Absent randomization, we'll never be sure whether Post Raisin Bran sells more per customer than Kellogg's at Store A than Store B *because* of the shelf placement or because of other differences between the stores.

Confounding

The importance of randomization may be hard to accept. Physicians involved in medical studies would rather assign patients to the therapy that they think is best for that individual. For the results of an experiment to be valid, however, randomization is necessary. Without randomization, we cannot know if observed differences are due to treatments or to choices made by physicians.

If we let customers choose a diet, are the results caused by the diet or something about the kind of people who select each diet? Randomization eliminates the ambiguity.

In the comparison of the Atkins diet to the conventional diet, we cannot be sure whether differences in weight loss are due to the diet or something else unless participants are randomly assigned. Suppose that we were to conduct the study in two locations, with one site placing customers on the conventional diet and the other placing them on the Atkins diet. Nothing in the data would enable us to distinguish the effects of diet from location. The population near one location might be younger or have a different mix of men and women. When the levels of one factor (diet) are associated with the levels of another factor (location), we say that these two factors are confounded. Randomization eliminates **confounding**, simplifying the interpretation of statistical results. These issues of interpretation resemble Simpson's paradox (Chapter 5). Confounding is another name for the same confusion.

confounding Mixing the effects of two or more factors when comparing treatments.

Confounding matters because incorrect beliefs about cause and effect lead to poor decisions. If managers interpret the confounded comparison as meaning one diet causes more weight loss than the other, they may pick the wrong diet. They may also get lucky and pick the right diet for the wrong reason. In either case, the data are likely to generate poor estimates of the relative advantages. The advantages of randomization are so strong in medicine that randomized studies are generally required by the FDA before a pharmaceutical is approved for use.

In spite of the benefits, it is not always possible to randomize. Any study that compares men to women runs into this problem. Often, the best that the experimenter can do is sample *independently* from two populations, in this case, from the population of men and from the population of women. Comparisons of the salaries of men and women run into arguments about confounding all the time. Are observed differences in income caused by gender or something else?

Confounding also contaminates comparisons over time. Suppose a company showed one advertisement last quarter and a different advertisement in the current quarter. The company then compares same-store sales in the two periods. Other variables that change over time are confounded with the type of advertisement. Was it the ad or some other factor that changed (a surge in unemployment or perhaps gasoline prices) that caused a change in sales?

The third source of data—comparing two sets of observations—usually produces confounding. In this situation, we have a data table that includes a numerical variable and a categorical variable with two levels. The categorical variable is the factor and the numerical variable is the response. We can *pretend* that we ran an experiment to assign the cases to the groups or *pretend* that we independently sampled two populations. Chances are that we did neither.

18.2 | TWO-SAMPLE *t*-TEST

The null hypothesis specifies that the difference between the population means is less than or equal to a predetermined constant, labeled D_0,

$$H_0: \mu_1 - \mu_2 \leq D_0$$

The constant D_0 usually comes from an economic analysis of the problem. For the experiment that compares diets, the null hypothesis asserts that $\mu_A - \mu_C \leq 2$ so that $D_0 = 2$ pounds. The sample means $\bar{x}_1$ and $\bar{x}_2$ estimate μ_1 and μ_2.

To test H_0, we have to decide if the observed difference $\bar{x}_1 - \bar{x}_2$ is far enough from the region specified by H_0. If $\bar{x}_1 - \bar{x}_2 \leq D_0$, the data agree with H_0 and there's no reason to reject it. If $\bar{x}_1 - \bar{x}_2 > D_0$, the data contradict H_0.

Even so, we have to be sure at the chosen level of confidence that this result did not happen by chance alone. For that, we use a t-statistic as when testing the mean of a single sample.

All t-statistics used in testing hypotheses share a common structure, so it's useful to review the one-sample t-test. The one-sample t-test in Chapter 16 compares the sample mean to a value μ_0 specified by a null hypothesis. It does this by counting the number of standard errors that separate $\bar{x}$ from μ_0. The estimated standard error of the average is $s/\sqrt{n}$, so the t-statistic (known formally as the one-sample t-statistic for the mean) is

$$t = \frac{\bar{x} - \mu_0}{\text{se}(\bar{x})}$$

If $H_0: \mu \leq \mu_0$ is true, there's less than 1 chance in 20 of getting a sample whose average is more than $t_{0.05, n-1} \approx 1.7$ standard errors above μ_0. Equivalently, we reject H_0 if the p-value of the test is less than $\alpha = 0.05$.

The t-statistic for testing the difference between two means shares this structure but relies on different estimates. It too computes the number of standard errors that separate an estimate from the null hypothesis. Rather than use a single mean, however, the estimate is the difference between the means. The two-sample test compares $\bar{x}_1 - \bar{x}_2$ to D_0, measured on the scale of the standard error of the difference,

$$t = \frac{(\bar{x}_1 - \bar{x}_2) - D_0}{\text{se}(\bar{x}_1 - \bar{x}_2)}$$

two-sample t-statistic
Test statistic for testing $H_0: \mu_1 - \mu_2 \leq D_0$.

$$t = \frac{(\bar{x}_1 - \bar{x}_2) - D_0}{\text{se}(\bar{x}_1 - \bar{x}_2)}$$

two-sample t-test Test of $H_0: \mu_1 = \mu_2$ using the two-sample t-statistic.

The full name for this t-statistic is the *two-sample t-statistic for the difference between means*. Let's shorten that to **two-sample t-statistic** and call the test the **two-sample t-test**.

The two-sample t-statistic requires the standard error of the difference between two sample averages, shown below. Most software provides the estimated standard error with the output of the test. The calculation of the degrees of freedom for the two-sample t-statistic is messy, and we defer this task to software. The complication arises because the sampling distribution of the two-sample t-statistic is not exactly a t-distribution; an elaborate calculation of the degrees of freedom produces a close match. The expression for the degrees of freedom appears at the end of the chapter. (For more background on the standard error, see Behind the Math: Standard Errors for Comparing Means, which also describes an alternative estimate of the standard error, known as the *pooled estimate*, that is easier to do by hand but requires an additional assumption.)

(p. 443)

The null hypothesis in this summary is $\mu_1 - \mu_2 \leq D_0$. If the situation asserts $\mu_2 - \mu_1 \leq D_0$, swap the labels on the two groups. When arranged as shown in the table, the test rejects H_0 when the t-statistic is large and positive.

Population parameters	μ_1, μ_2
Null hypothesis	$H_0: \mu_1 - \mu_2 \leq D_0$
Alternative hypothesis	$H_a: \mu_1 - \mu_2 > D_0$
Sample statistic	$\bar{x}_1 - \bar{x}_2$
Estimated standard error	$\text{se}(\bar{x}_1 - \bar{x}_2) = \sqrt{\dfrac{s_1^2}{n_1} + \dfrac{s_2^2}{n_2}}$
Test statistic	$t = (\bar{x}_1 - \bar{x}_2 - D_0)/\text{se}(\bar{x}_1 - \bar{x}_2)$
Reject H_0 if or	p-value $< \alpha$ $\qquad t > t_{\alpha/2}$

Consult your software for the degrees of freedom of the *t*-statistic. The two-sample *t*-test requires that we check several conditions. The checklist is longer than for previous inferential methods because two samples are involved. The first two conditions in the checklist are the most important. Both hold for experiments.

Checklist

✓ **No obvious lurking variables.** The only systematic difference between the samples is due to the factor that defines the groups. A randomized experiment achieves this. Otherwise, consider whether there might be another variable, a lurking factor, that explains the observed difference.

✓ **SRS condition.** Ideally, the observed samples should be the result of randomly assigning treatments to a sample from the population. Otherwise, the two samples should be independent random samples from two populations. Each sample constitutes less than 10% of its population.

✓ **Similar variances.** The test allows the two populations to have different variances, but we ought to notice whether the variances are similar. They don't have to match, but it is foolish to ignore a visible difference in variation. If one diet produces more consistent results, for instance, that would be important to know.

✓ **Sample size condition.** Each of the two samples, evaluated separately, must satisfy the sample size condition used in the one-sample *t*-test. The number of observations in *each* sample must exceed 10 times the squared skewness K_3^2 and 10 times the absolute value of the kurtosis $|K_4|$ in each sample.

What Do You Think?

A marketing team designed a promotional Web page to increase online sales. Visitors to www.name-of-this-company.com were randomly directed to the old page or the new page. During a day of tests, 300 visitors to the site were randomly assigned. The 169 visitors who were directed to the old page spent $\bar{x}_{old} = \$253$ on average ($s_{old} = \$130$); those directed to the new page spent $\bar{x}_{new} = \$328$ on average ($s_{new} = \$161$). Software computed the standard error of the difference to be $17.30 with about 295 degrees of freedom.

a. Should we be concerned with the possibility of confounding?[1]
b. Does the new page generate statistically significantly higher sales than the old page? State the null hypothesis and whether it's rejected. (Assume that these samples are large enough to satisfy the sample size condition.)[2]
c. A manager claims that all of the past data should be used to estimate μ_{old} rather than limit the comparison to the small sample observed during one day. Is the manager's proposal a good idea?[3]

---

[1] It is always good to be concerned about confounding, but the randomization removes the concern.
[2] Yes, at the usual 5% tolerance for an error, $t = (328 - 253)/(17.3) \approx 4.3$. The *p*-value is much less than 0.05.
[3] No, while the suggestion to use all of the data for the existing method is useful, these data are not under the same conditions (different times, for example) and so do not provide a fair comparison of the success of the pages.

COMPARING TWO DIETS

STATE THE QUESTION

Rather than conduct an experiment of its own to compare the Atkins diet to a conventional diet, the chain of fitness centers did some research and discovered that someone had done the work for them. In this experiment, scientists at the University of Pennsylvania selected a sample of 63 subjects from the local population of obese adults. From these, experimenters randomly assigned 33 to the Atkins diet and 30 to the conventional diet. Does this experiment show, with $\alpha = 0.05$, that the Atkins diet is worth licensing?

DESCRIBE THE DATA AND SELECT AN APPROACH

List hypotheses, α level.

Identify population.

Describe data.

Choose test.

Check conditions.

The null hypothesis is $H_0: \mu_A - \mu_C \leq 2$ versus $H_a: \mu_A - \mu_C > 2$. The sample was drawn from the population of obese adults in the Philadelphia area. For each subject, the researchers measured the number of pounds lost during the six-month study. To test H_0, we'll use the two-sample t-test with $\alpha = 0.05$ and not require the variances to be identical. To check the last two conditions, we need a display of the data.

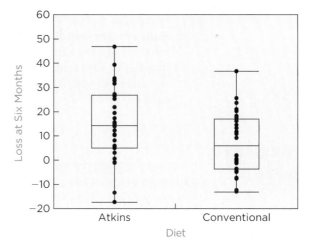

Side-by-side comparison boxplots of the weight lost in the two groups are handy for this task. The dots within each boxplot show weights of individual subjects in that group. The column of dots summarized by the boxplot on the left represents the 33 subjects on the Atkins diet; the column of dots on the right represent the 30 subjects on the conventional diet.

Neither diet outperforms the other for everyone. The overlap of these boxplots shows that some participants did better on one diet, and some did better on the other. One person on the Atkins diet, for instance, lost nearly 50 pounds whereas several others on the same diet gained weight. The median weight lost on the Atkins diet is several pounds higher than for the conventional diet. (The horizontal line in the middle of each boxplot shows the median for that group.)

Let's check the conditions for the two-sample t-test.

✓ **No obvious lurking variables.** The randomization that was used to assign subjects to the two diets implies that confounding factors are not a problem.

✓ **SRS condition.** The experiment began with a random sample from the local population of obese adults. These 63 adults definitely represent less than 10% of the population of Philadelphia.

✓ **Similar variances.** The interquartile ranges of the boxplots appear similar, which is all that is needed for the two-sample *t*-test.

✓ **Sample size condition.** Both samples meet the requirements of the sample size condition; the skewness and kurtosis are both less than 1 in both samples.

MECHANICS

DO THE ANALYSIS

This table summarizes the means, standard deviations, skewness, and kurtosis of the two samples.

	Number	Mean	Std. Dev.	Skewness	Kurtosis
Atkins	$n_A = 33$	$\bar{x}_A = 15.42$	$s_A = 14.37$	−0.052	0.100
Conventional	$n_C = 30$	$\bar{x}_C = 7.00$	$s_C = 12.36$	0.342	−0.565

Dieters on the Atkins plan lost an average of $\bar{x}_A = 15.42$ pounds after six months, compared to $\bar{x}_C = 7.00$ pounds on the conventional diet. The observed difference $\bar{x}_A - \bar{x}_C = 8.42$ is larger than the break-even 2-pound advantage. Before concluding that the Atkins program should be profitable (at level $\alpha = 0.05$), we need to rule out the possibility that this experiment was a fluke.

To test the null hypothesis, the two-sample *t*-statistic is

$$t = \frac{(\bar{x}_A - \bar{x}_C) - 2}{\text{se}(\bar{x}_A - \bar{x}_C)} = \frac{(15.42 - 7.00) - 2}{\sqrt{\dfrac{14.37^2}{33} + \dfrac{12.36^2}{30}}} \approx 1.91$$

The difference between the sample averages is 1.91 standard errors away from the difference $D_0 = 2$ specified by the null hypothesis. Our software computed the degrees of freedom to be 60.8255 with *p*-value = 0.0308. Because the *p*-value is less than $\alpha = 0.05$, we reject H_0. (The calculation of degrees of freedom does not usually produce an integer.)

 (p. 443)

A common variation on the two-sample *t*-test adds the assumption $\sigma_1^2 = \sigma_2^2$. This assumption permits the use of an alternative standard error (called the pooled estimate in Behind the Math: Standard Errors for Comparing Means). Because the variances in the two samples are similar, the results agree. With the added assumption, there are $n_1 + n_2 - 2 = 61$ degrees of freedom. The *p*-value is $P(T_{61} \geq 1.91) = 0.0304$, almost matching the *p*-value obtained by the test that does not presume equal variances.

MESSAGE

SUMMARIZE THE RESULTS

This experiment shows that the average weight loss of obese dieters on the Atkins diet exceeds the average weight loss on the conventional diet by more than 2 pounds over a six-month period. That's proof (with a 5% chance of making a false claim) that the gains from licensing the Atkins diet will compensate for the initial licensing fee. *Be sure to disclose all important caveats so others aren't missing potentially vital information.* Managers of the fitness centers need to be warned, that this study sampled an obese population in the Philadelphia area. Unless the membership of the fitness center resembles this population, these results may not apply to its population. For instance, we might question whether this difference will also apply to people who are not obese in the first place.

18.3 | CONFIDENCE INTERVAL FOR THE DIFFERENCE

The two-sample *t*-test determines whether the average of one sample is statistically significantly larger than the other. Some situations, however, lack the financial information required for a break-even analysis that determines the boundary D_0 in H_0. Perhaps someone else handles the finances and only wants input on the difference between the methods. Or, we might prefer a range for the difference rather than a yes or no demonstration of profitability. Confidence intervals are more appropriate in these situations.

An intuitive comparison relies on confidence intervals for the mean of each sample. We will cover a better method, but this approach may be necessary when we only have access to summary results. Table 18.1 shows the means, standard deviations, and 95% confidence intervals for each sample in the diet comparison in Example 18.1.

TABLE 18.1 Comparison statistics for two diets.

Atkins		Conventional
15.424242	Mean	7.006667
14.371065	Std Dev	12.360754
2.501681	Std Err Mean	2.256754
20.520001	Upper 95% Mean	11.622248
10.328484	Lower 95% Mean	2.391085
33	*n*	30

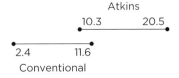

The two 95% confidence intervals in Table 18.1 overlap. The upper endpoint of the confidence interval for the mean weight loss on the conventional diet is larger than the lower endpoint of the confidence interval for the mean loss on the Atkins plan.

The mean for the Atkins diet and the mean for the conventional diet might both be, say, 11 pounds. Had the intervals been disjoint, we could have concluded that the mean loss in the population for the Atkins diet is statistically significantly greater than for the conventional diet. Disjoint intervals imply a statistically significant difference. Overlapping intervals as seen in this example are inconclusive.

A better, definitive approach uses the confidence interval for the difference between the means. Rather than compare two intervals, we summarize what we know about $\mu_1 - \mu_2$ in a single interval. There's a simple principle at work here. **Make a confidence interval for the relevant parameter or combination of parameters.** If we want to make an inference about $\mu_1 - \mu_2$, then make a confidence interval for $\mu_1 - \mu_2$ rather compare than two separate intervals.

The same ingredients that produce a hypothesis test determine the confidence interval. Without assuming $\sigma_1^2 = \sigma_2^2$, the sampling distribution is again a *t*-like distribution, including the elaborate calculation of degrees of freedom. The $100(1 - \alpha)\%$ **two-sample confidence interval** for the difference between sample means is

$$\bar{x}_1 - \bar{x}_2 - t_{\alpha/2}\,\text{se}(\bar{x}_1 - \bar{x}_2) \quad \text{to} \quad \bar{x}_1 - \bar{x}_2 + t_{\alpha/2}\,\text{se}(\bar{x}_1 - \bar{x}_2)$$

We use software to handle the details of this calculation to save time for thinking about conditions. The conditions for the two-sample confidence interval are the same as those for the two-sample *t*-test. As we have seen, these data satisfy the conditions.

The confidence interval for the difference in weight lost on the Atkins and conventional diets is approximately 1.7 to 15.2 pounds. Unlike the informal comparison of two separate intervals, the two-sample interval excludes zero. (The break-even value 2 lies inside this confidence interval even though the one-sided test rejects H_0: $\mu_1 - \mu_2 \leq 2$. As we have seen in previous chapters, confidence intervals provide both upper and lower limits whereas the break-even analysis concerns only a lower limit. Because of its focus on only the lower limit, the one-sided test is able to reject 2 as a plausible value for $\mu_1 - \mu_2$.)

Interpreting the Confidence Interval

The confidence interval in Table 18.2 is easy to interpret because these data come from an experiment. The subjects were randomly assigned to the two diets. The confidence interval implies that people in the sampled population who go on the Atkins diet on average lose between 1.7 and 15.2 pounds more than they would on a conventional diet. That's a wide range, but it *excludes* 0. The difference between the averages in the two samples is too large to explain as sampling variation. We cannot say exactly how much more an individual would lose on the Atkins plan, but we can be 95% confident that the average difference lies between 1.7 and 15.2 pounds. The difference $\mu_1 - \mu_2$ might plausibly be 0 or any other value inside the interval.

TABLE 18.2 Numerical summary of a two-sample *t*-interval.

Difference	8.4176
Std Err Difference	3.3692
Degrees of Freedom	60.8255
95% Confidence Interval	1.6801 to 15.1551

statistically significantly different If the 95% confidence interval for the difference does not include zero, the means are statistically significantly different.

When the 95% confidence interval for $\mu_1 - \mu_2$ does not include zero, we say that the two sample means are **statistically significantly different** from each other. Zero is not a plausible value for $\mu_1 - \mu_2$. Not only that, we know which is larger. We would say, "Dieters on the Atkins plan lost statistically significantly *more* weight than those on the regular diet." The comparison of separate intervals lacks power and does not detect this difference.

The word *statistically* often gets dropped from *statistically significant* and leads to confusion. Someone who is 100 pounds overweight may not view losing 2 more pounds as significant, confusing the statistical sense of a significant difference with the colloquial sense. Also, confidence intervals are not always right; the interval might omit $\mu_1 - \mu_2$. With a 95% confidence interval, such an error happens for 5% of samples.

Finally, what should we conclude if the confidence interval for $\mu_1 - \mu_2$ does include 0? Finding zero inside the interval means that the samples *could* have come from populations with the same mean. It does not mean that they *do*. When the confidence interval for the difference includes 0, the lower endpoint is negative and the upper endpoint is positive. Such a confidence interval for $\mu_1 - \mu_2$ means that we don't know whether $\mu_1 - \mu_2$ is positive or negative. Either μ_1 is larger than μ_2, or μ_2 is larger than μ_1. The confidence interval bounds the magnitude of the difference, but does not identify which population has the larger mean.

> **Confidence Interval for $\mu_1 - \mu_2$** The $100(1 - \alpha)\%$ confidence *t*-interval for $\mu_1 - \mu_2$ is
>
> $$\bar{x}_1 - \bar{x}_2 - t_{\alpha/2}\,\text{se}(\bar{x}_1 - \bar{x}_2) \quad \text{to} \quad \bar{x}_1 - \bar{x}_2 + t_{\alpha/2}\,\text{se}(\bar{x}_1 - \bar{x}_2)$$
>
> where
>
> $$\text{se}(\bar{x}_1 - \bar{x}_2) = \sqrt{\frac{s_1^2}{n_1} + \frac{s_2^2}{n_2}}$$

The degrees of freedom are approximately $n_1 + n_2 - 2$, depending on the variances in the two samples.

Checklist (see the two-sample *t*-test)

✓ **No obvious lurking variables**
✓ **SRS condition**
✓ **Similar variances**
✓ **Sample size condition**

| 4M **EXAMPLE 18.2** | **EVALUATING A PROMOTION** |

MOTIVATION

STATE THE QUESTION

In the early days of overnight shipping, offices used specialized couriers rather than an overnight delivery service such as FedEx. Many thought that FedEx delivered only letters and did not think to use FedEx for shipping large packages. To raise awareness, FedEx developed promotions to advertise its capabilities.

To assess the possible benefit of a promotion, an overnight service pulled shipping records for a random sample of 50 offices that received the promotion and another random sample of 75 that did not. The offices were not randomly assigned to the two groups; rather, we sampled independently from two populations.

METHOD

Identify parameters.

Identify population.

Describe data.

Choose method.

Check conditions.

DESCRIBE THE DATA AND SELECT AN APPROACH

μ_{yes} is the mean number of packages shipped by companies that received the promotion, and μ_{no} is the mean for those that did not receive the promotion. These parameters are means of the population of offices, one group having received the promotion and the other not. The data are random samples from these two populations at the same point in time. Unfortunately, these were not randomly assigned. To compare the means, we'll use the 95% two-sample *t*-interval.

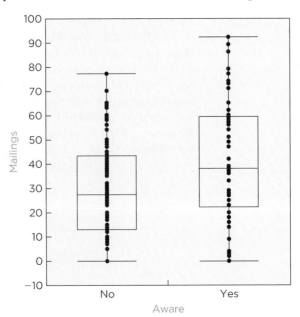

Let's check the conditions for this interval. We use a plot and summary statistics to check the last two.

? **No obvious lurking variables.** There could be confounding because we don't know how the delivery service decided which offices should receive the promotion. For example, it could be the case that only larger offices received the promotion. Our comparison would then mix the effect of the promotion with the size of the office.

✓ **SRS condition.** Each sample was randomly drawn from the relevant population, so each is an SRS. These are small samples relative to the size of the population.

✓ **Similar variances.** The interquartile ranges in the figure are similar. Plus, we will use the test that does not require the variances to be equal in the two populations.

✓ **Sample size condition.** Both samples meet this condition. For example, the skewness and kurtosis in the group that did not receive the promotion are $K_3 \approx 0.4$ and $K_4 \approx -0.7$. Neither sample appears skewed or contaminated by outliers.

MECHANICS

DO THE ANALYSIS

Our software computed the following summary:

$\bar{x}_{yes} - \bar{x}_{no}$	12.3067
$se(\bar{x}_{yes} - \bar{x}_{no})$	4.2633
Degrees of Freedom	85.166
95% CI	3.8303 to 20.7830

The difference between sample averages is about 12.3 packages, and the 95% confidence interval spans the wide range of 3.83 to 20.78 more packages. As in prior examples, the formula gives fractional degrees of freedom.

MESSAGE

SUMMARIZE THE RESULTS

There is a statistically significant difference between the average number of packages shipped by offices that received the promotion and by those that did not. Offices that received the promotion used the overnight delivery service to ship from 4 to 21 more packages on average than those that did not receive the promotion.

Include important caveats in your summary. Without an economic analysis of the costs of the promotion, we should not conclude that the promotion is cost effective. Also, because these data are not from an experiment, other factors could influence the outcome. If the delivery service was promoted only to legal firms, for example, then this analysis compares legal firms that received the promotion to other types of offices that did not. Is it the promotion or the type of office that affects the use of services?

18.4 | OTHER COMPARISONS

Comparison arises in many situations, and not all of these involve comparing the means in two independent samples. We may need to compare other attributes of the samples, such as proportions. We may also have data that provide two measurements for each subject: paired data. The following sections consider comparing proportions and comparisons that use paired data. In each case, we limit ourselves to confidence intervals. Though the situations differ, 95% confidence intervals retain the familiar, approximate form,

$$\text{Estimated Difference} \pm 2 \times \text{Estimated Standard Error of the Difference}$$

Comparing Proportions

The comparison of sample proportions proceeds as the comparison of sample averages. After all, proportions are averages. The 95% confidence interval for $p_1 - p_2$ is the estimate $\hat{p}_1 - \hat{p}_2$ plus or minus about 2 standard errors.

Differences from the two-sample confidence interval for $\mu_1 - \mu_2$ concern the sample size condition and use of a normal percentile. The sample size condition for proportions requires each sample to have at least 10 successes and 10 failures. Hence, each sample must have at least 20 observations. The percentile from the normal distribution replaces the percentile from the t-distribution. The following summarizes the two-sample confidence interval for proportions:

> **Confidence Interval for $p_1 - p_2$** The $100(1 - \alpha)\%$ confidence z-interval for $p_1 - p_2$ is
>
> $$\hat{p}_1 - \hat{p}_2 - z_{\alpha/2}\, se(\hat{p}_1 - \hat{p}_2) \quad \text{to} \quad \hat{p}_1 - \hat{p}_2 + z_{\alpha/2}\, se(\hat{p}_1 - \hat{p}_2)$$
>
> where the estimated standard error of the difference is
>
> $$se(\hat{p}_1 - \hat{p}_2) = \sqrt{\frac{\hat{p}_1(1 - \hat{p}_1)}{n_1} + \frac{\hat{p}_2(1 - \hat{p}_2)}{n_2}}$$
>
> **Checklist**
>
> ✓ **No obvious lurking variables**
> ✓ **SRS condition**
> ✓ **Sample size condition (for proportion).** Both $n_1\hat{p}_1$ and $n_1(1 - \hat{p}_1)$ are greater than 10, and both $n_2\hat{p}_2$ and $n_2(1 - \hat{p}_2)$ are greater than 10.

4M EXAMPLE 18.3 | COLOR PREFERENCES

MOTIVATION ▸ STATE THE QUESTION

Fashion buyers for department stores try to anticipate which colors and styles will be popular during the next season. Getting the reaction of a sample of consumers can help. In this example, a department store has sampled customers from its operations in the East. For logistical reasons, it has a separate sample of customers from its western operations. Each customer was shown designs for the coming fall season. One design features red fabrics, the other features violet. If customers in the two regions have similar preferences, buyers can order the same line of clothing for both districts. Otherwise, the buyers will have to do a special order for each district.

METHOD ▸ DESCRIBE THE DATA AND SELECT AN APPROACH

Buyers want to know whether the proportion of customers that prefer each design differs in the two locations. There are two populations of customers, one for each region (eastern and western districts). The data are samples from these, with a sample of $n_e = 60$ from the East and $n_w = 72$ from the West. Each customer picked a favorite design (red or violet).

Identify parameters.

Identify population.

Describe data.

Choose method.

Check conditions.

We will use a 95% two-sample confidence interval for the difference in the proportions. Let's check the conditions.

? **No obvious lurking variables.** The data are samples from two populations, so we have to be concerned about a possible lurking variable. For example, if the sample in the West has younger customers than the sample in the East, we have confounded location with age. We cannot rule this out from the description and we need more information.

✓ **SRS condition.** The description of the data indicates that this holds for each location.

✓ **Sample size condition for proportions.** The data table shows that more than 10 customers in each location chose each of the designs. The data meet this condition.

MECHANICS

DO THE ANALYSIS

The following mosaic plot and contingency table summarize the results.

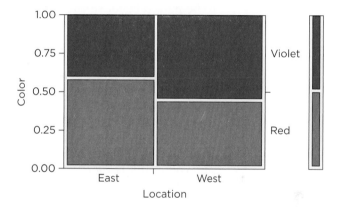

Count Column %		District		
		East	**West**	**Total**
Design	**Red**	35 58.33	32 44.44	67
	Violet	25 41.67	40 55.56	65
	Total	60	72	132

Customers in the eastern district prefer red (58% to 42%), whereas those in the western district prefer violet (56% to 44%). The difference in the proportions is $\hat{p}_E - \hat{p}_W = 0.5833 - 0.4444 = 0.1389$. The two-sample confidence interval is then

$$0.1389 \pm 1.96\sqrt{\frac{0.5833(1 - 0.5833)}{60} + \frac{0.4444(1 - 0.4444)}{72}}$$

$$\approx 0.1389 \pm 1.96(0.08645)$$

$$\approx -0.031 \text{ to } 0.308$$

Zero lies just inside the 95% confidence interval.

MESSAGE │ SUMMARIZE THE RESULTS

There is not a statistically significant difference between these samples in preference for the two designs. That said, customers in the East favor the red lineup (58%) more often than to those in the West (44%). With 95% confidence, the difference lies in the range −3% to 30%. The population preferences may be the same, but it looks as if there's a preference for red in the East. Before we order the same lineup for both districts, we should (1) consider gathering larger samples so that the confidence interval would be narrower and (2) check whether other differences between the customers used in this comparison might explain the differences in preference.

Paired Comparisons

One of the best ways to compare strategies is to put them in a head-to-head contest. For example, rather than have one group of people rate the taste of Pepsi and a different group rate the taste of Coke, we obtain a better idea of preferences by having the *same* people taste and rate both colas. This head-to-head approach is called a **paired comparison** of the treatments. It is said to be paired because each rater tries both drinks. Paired data look different from two-sample data for comparisons in a data table. Paired comparisons produce two variables (two columns of data) for each observation. Each row is a subject, and the two columns record their reactions to the treatments.

paired comparison A comparison of two treatments using dependent samples designed to be similar.

Pairing isolates the effect of the treatment, reducing the random variation that can hide a difference. Some of the difference in ratings between two samples comes from the composition of the samples rather than from the treatments, Coke and Pepsi. Sampling variation may mean that one sample has a larger proportion of cola lovers than the other. The average of this group might be high due to the members of the sample, not the treatment. This sample would highly rate any cola. By having each person rate both, we remove this source of random variation. The resulting comparison is also simpler to interpret because it rules out other sources of confounding. Reports on retailing, for example, give changes in same-store sales rather than mix new locations with old locations. Medical studies use a similar design. Rather than compare a new pharmaceutical drug to a standard therapy on two samples, doctors treat each subject with both. Paired comparisons nonetheless require care. For example, misinterpreted taste tests are blamed for the failed launch of New Coke.[4] (For a more technical explanation of the advantage of pairing, see Behind the Math: Standard Errors for Comparing Means. This section shows that pairing reduces the variability of the difference $\bar{x}_1 - \bar{x}_2$ if the measurements are positively correlated.)

(p. 443) ▶

Randomization remains relevant in a paired comparison. We should vary the order in which the subjects are exposed to the treatments. Rather than have each person taste Coke first and then Pepsi, we should randomly order the presentation. The same phenomenon applies to a comparison of drugs. One drug might have some residual effect that would influence how the other subsequently performs. To rule out the effect of the ordering of the treatments, each should be presented first to a randomly selected half of the sample.

Paired comparisons can be very effective, but are often impossible. For example, an experiment to test the value of additives in concrete increases the load until it destroys the sample. We cannot destroy it twice. We can, however, come closer to a paired comparison by thinking harder about our choice of subjects. For instance, if we are interested in how men and women respond to

---

[4] Malcolm Gladwell (2005), *Blink: The Power of Thinking Without Thinking* (New York: Little, Brown).

a commercial, we cannot change the sex of a person. But we can find a man and a woman who have similar incomes, jobs, ages, and so forth. By randomly choosing pairs of similar subjects, we obtain many of the benefits of pairing. The difficulty is that the sampling procedure becomes more complex. We need random samples of pairs of similar subjects rather than a random sample from each of two populations.

Once we have the paired data, the statistical analysis is simple. We start by forming the difference within each pair. If x_i is the ith observation in one group and y_i is the paired item in the second group, we work with the differences $d_i = x_i - y_i$. In effect, pairing converts a two-sample analysis into a one-sample analysis. Our data are a sample of the possible differences between the two groups.

> **Paired Confidence Interval for $\mu_1 - \mu_2$** Calculate the n differences between the pairs of measurements, $d_i = x_i - y_i$. Let $\bar{d}$ denote the mean of the differences and let s_d denote the standard deviation of the differences. The $100(1 - \alpha)\%$ confidence paired t-interval for $\mu_d = \mu_1 - \mu_2$ is
>
> $$\bar{d} - t_{\alpha/2,n-1}\frac{s_d}{\sqrt{n}} \quad \text{to} \quad \bar{d} + t_{\alpha/2,n-1}\frac{s_d}{\sqrt{n}}$$
>
> where $P(T_{n-1} > t_{\alpha/2,n-1}) = \alpha/2$ and T_{n-1} is a t-random variable with $n - 1$ degrees of freedom.
>
> **Checklist**
>
> ✓ **No obvious lurking variables.** Pairing does not remove every possible lurking factor. At a minimum, we must make sure that if two treatments are applied to the same subject the order of the treatments has been randomized.
> ✓ **SRS condition.** The observed sample of differences $x_i - y_i$ is a simple random sample from the collection of all possible differences.
> ✓ **Sample size condition.** The skewness and kurtosis of the sample of n differences $d_i = x_i - y_i$ must be such that $n > 10K_3^2$ and $n > |K_4|$.

4M EXAMPLE 18.4	SALES FORCE COMPARISON

MOTIVATION ▸ STATE THE QUESTION

The profit in merging two companies often lies in eliminating redundant staff. The merger of two pharmaceutical companies (call them A and B) allows senior management to eliminate one of their sales forces. Which one should the merged company retain?

METHOD ▸ DESCRIBE THE DATA AND SELECT AN APPROACH

Identify parameters.

Identify population.

Describe data.

Choose test.

Check conditions.

To help reach a decision, we can compare data on the performance of the two sales forces. Let's compare the average level of sales obtained during a recent period by the two sales forces. We'd like to sample from the future performance of these groups, but we'll have to settle for using data for a recent period and think of these data as a random sample from the distribution of future sales for each group. As with any process over time (recall control charts), we have to think of the observed data as a sample from a continuing process.

Rather than have two independent samples, we have paired samples. Both sales forces market similar products and were organized into 20 comparable geographical districts. For each district, the data give the average dollar sales per representative per day in that district. Because each district has its own mix of population, cities, and cultures, it makes the most sense to directly compare the sales forces in each district. Some districts have higher sales than others because of the makeup of the district. We will use the difference obtained by subtracting sales for Division B from sales of Division A in each district, getting a confidence interval for $\mu_A - \mu_B$.

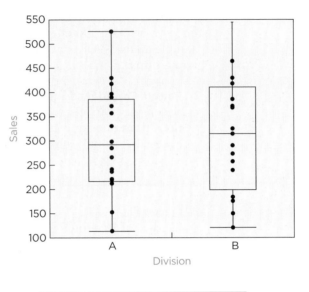

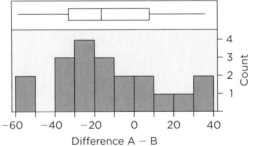

The data appear similar when viewed as two samples. The comparison box-plots are virtually identical. The variation in sales across the districts obscures the differences in performance within each district. The differences that are present *within* districts (ranging from −60 to 40) are small relative to the variation *between* districts (100 to 550).

To check the conditions for the paired analysis, we need to inspect the differences. The histogram of the differences shows more negative than positive values, indicating that in a district-by-district comparison, sales of Division A appear smaller than those in Division B.

✓ **No obvious lurking variables.** By matching up the sales in comparable districts, we have avoided comparing sales of one division in a poor district to sales of the other in a good district. The comparisons should be fair.
✓ **SRS condition.** We will treat the differences $d_i = x_i - y_i$ as a simple random sample from the collection of all possible differences. We have to question the assumption of independence, however, because the districts are geographically adjacent. If sales in one district go down, it could lead to changes in the sales in a neighboring district.
✓ **Sample size condition.** The skewness and kurtosis of the differences are both less than 1, so the data meet the sample size condition.

| MECHANICS | DO THE ANALYSIS |

Our software reports this summary of the differences

$\bar{d}$	−13.5
s_d	26.7474
n	20
$s_d/\sqrt{n}$	5.9809
95% CI	−26.0181 to −0.9819

The mean difference is \$13.5 per representative per day, with an estimated standard error $s_d/\sqrt{n}$ of about \$6 per representative per day. The 95% t-interval for the mean differences is about −\$26.02 to −\$0.98. This interval excludes zero, so we have found a statistically significant difference.

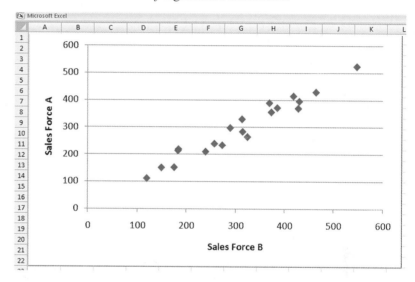

This scatterplot confirms the benefit of a paired comparison. It shows that sales in these districts are highly correlated ($r = 0.97$). Both sales forces do well in some districts and poorly in others. Had the samples been independent, we would have found little correlation between these two sets of measurements. The large correlation improves the comparison because as a result we compare sales of both groups in comparable regions.

| MESSAGE | SUMMARIZE THE RESULTS |

On average, Sales Force B sells more per day than Sales Force A. By comparing sales per representative head to head in each district, we have isolated a statistically significant difference in performance.

Best Practices

- *Use experiments to discover causal relationships.* A statistically significant result means that the two samples evidently come from populations with different means. That's all. A statistically significant difference between the means of two samples does not imply that we've found the cause of the difference. Unless an experiment produced the data, the differences could be due to other things, those pesky lurking factors.

- *Plot your data.* Formulas make it easy to forget the importance of looking at your data. We spend time with the formulas so that you can see how they work, but don't forget the importance of looking at the data. Skewness, outliers, and other anomalies are just as important, maybe more so. You won't know there's a problem unless you look.
- *Use a break-even analysis to formulate the null hypothesis.* If you're going to use a test, use a financial analysis to decide how much better one mean needs to be than the other. Otherwise, use the two-sample confidence interval to gauge the size of the difference between the means in the two populations.
- *Use one confidence interval for comparisons.* If you want to compare two means, then use a single confidence interval for the difference. Two confidence intervals, one for each mean, are fine if you want to talk about the means separately. But if you're comparing them, use the confidence interval for the difference.
- *Compare the variances in the two samples.* Although the test and interval of this chapter concentrate on the difference between the means, notice whether the variances are similar. It is often useful to know that one procedure produces less consistent results (more variation) than the other.
- *Take advantage of paired comparisons.* By spending the time to pair corresponding items in the two groups, the comparison of the treatments focuses on the treatments rather than other distinctions between the two groups.

Pitfalls

- *Don't forget confounding.* Confounding affects how we interpret the results. On seeing a statistically significant result, it is tempting to believe that the difference is caused by the factor that labels the groups: male versus female, new versus old, automatic versus manual. That's fine if the data come from an experiment. If not, other differences between the groups—differences attributed to lurking variables—may cause the differences.
- *Do not assume that a confidence interval that includes zero means that the difference is zero.* A confidence interval for $\mu_1 - \mu_2$ that includes zero indicates that $\mu_1 - \mu_2$ might plausibly equal zero, at the chosen level of confidence. That interval also indicates that $\mu_1 - \mu_2$ might be a lot of other things as well, some positive and some negative.
- *Don't confuse a two-sample comparison with a paired comparison.* When using the two-sample test or interval, make sure there is no relationship between the two groups. If there's a natural pairing, then use the paired standard error. A common mistake happens when data measure before-and-after characteristics. You should not, for instance, use two-sample methods to compare same-store sales from this year to last year. This type of data requires a paired comparison. The responses are not independent because the data have repeated measurements of the same stores.
- *Don't think that equal sample sizes imply paired data.* Make sure that there is an association between each observation in one sample and the corresponding observation in the other. Pairing improves the analysis only when there is a strong association between the two sets of measurements. The data should resemble two measurements taken on the same object, such as the same person, store, or location.

Software Hints

For all of these packages, the simplest way to perform a paired analysis is to subtract the paired items and then use the sample of differences as in the one-sample methods of Chapters 15 and 16.

EXCEL

The built-in function TTEST computes the two-sample *t*-test. The expression TTEST(*range1, range2,* 1, 3) uses *range1* to identify the data in one sample and *range2* to identify those in the other. The third argument, 1 indicates a one-sided test of $H_0: \mu_1 - \mu_2 \leq 0$, and the fourth argument allows different variances in the two samples. The formula returns the *p*-value of the test. Before using this formula, check whether the sample averages agree with H_0. If they do, there's no need to continue; you cannot reject H_0. If the sample means contradict H_0, then run the test.

To set $D_0 \neq 0$ (the default in Excel), adjust the data by D_0. For example, in the text, we test $H_0: \mu_A - \mu_C \leq 2$. Do this with Excel by adding 2 to the data in the range for method c. In effect, we're testing $H_0: \mu_A - (\mu_C + 2) \leq 0$. By changing the fourth argument from 3 to 1, you obtain a paired test.

DDXL provides more alternatives and point-and-click options. The two samples must be in separate columns of the spreadsheet. In Excel, follow the menu command

DDXL > Hypothesis tests

The option 2 Var t-Test provided by the "click me" button offers the two-sample t-test; the dialog for this test allows for picking the α level, choosing H_0, and selecting whether to use the default or pooled standard error. The option 2 Var Prop Test produces the two-sample test of proportions. The menu commands

DDXL > Confidence intervals

provide access to the corresponding confidence intervals.

MINITAB

The menu sequence

Stat > Basic Statistics > 2 − Sample t ...

opens a dialog for performing a two-sample t-test. If the data for both samples are in one column with a second column identifying the group, select the Samples in one column option (the default). If the samples are in different columns, then select the Samples in different columns option. Clicking the OK button produces the two-sided test of $H_0: \mu_1 = \mu_2$ (output shows the t-statistic and p-value) and the 95%

confidence interval for $\mu_1 - \mu_2$. Options provide comparison boxplots of the two samples, a one-sided test, and change in the level of confidence for the confidence interval.

JMP

For a two sample comparison, the data for both samples must be in a single column, with the groups identified by a categorical variable in a separate column. Following the menu sequence

Analyze > Fit Y by X

opens a dialog to identify the column that contains the measurements for both samples (enter this one in the response field) and the column that identifies the groups (the categorical variable). After you click OK, JMP produces a plot with the measurements on the y-axis and groups named on the x-axis.

To obtain the two-sample test and confidence interval, click on the red triangle in the header of the output window that reads "Oneway Analysis ...". Select the item t-test. The output expands to show the two-sample comparison; the summary shows the t-statistic, the standard error, the 95% confidence interval for $\mu_1 - \mu_2$, and the p-values. To specify a different α [and get a $100(1 - \alpha)\%$ confidence interval], use the option Set α level in the pop-up menu obtained by clicking on the red triangle. The menu option Means/Anova/Pooled t computes the t-test using the pooled variance.

In order to obtain skewness and kurtosis estimates for the separate samples, use the Analyze > Distribution command to obtain a dialog, and use the experimental factor as the By variable. The By variable does a separate analysis for each group identified by the By variable.

 Behind the Math

Standard Errors for Comparing Means

The standard error for comparing the averages of two independent samples is simple, if we remember the properties of a portfolio from Chapter 10. We need the special property that variances of sums or differences add.

> The variance of the sum or difference of two uncorrelated random variables is the sum of their variances.

For standard deviations, we can write symbolically what we've just said like this:

$$SD(X - Y) = \sqrt{\mathrm{Var}(X - Y)} = \sqrt{\mathrm{Var}(X) + \mathrm{Var}(Y)}$$

This formula applies *only* when X and Y are uncorrelated. Means of *independent* random samples are

uncorrelated. Otherwise, we need to add the covariance. The standard error of the mean of *one* sample of n observations is

$$SE(\overline{X}) = \frac{\sigma}{\sqrt{n}}$$

The standard error of the difference between two sample means, viewed as the random variables $\overline{X}_1$ and $\overline{X}_2$, is then

$$
\begin{aligned}
SE(\overline{X}_1 - \overline{X}_2) &= SD(\overline{X}_1 - \overline{X}_2) \\
&= \sqrt{\mathrm{Var}(\overline{X}_1) + \mathrm{Var}(\overline{X}_2)} \\
&= \sqrt{\frac{\sigma_1^2}{n_1} + \frac{\sigma_2^2}{n_2}}
\end{aligned}
$$

Because we don't know the variances, we substitute estimates. With sample standard deviations in place of σ_1 and σ_2, the estimated standard error for the difference between two sample means is

$$\text{se}(\overline{x}_1 - \overline{x}_2) = \sqrt{\frac{s_1^2}{n_1} + \frac{s_2^2}{n_2}}$$

Other situations lead to alternative estimates of the sample-to-sample variation of the difference. One alternative is known as the pooled estimate. If we know that the variances of the populations are the same ($\sigma_1^2 = \sigma_2^2 = \sigma^2$), then the common variance factors out like this:

$$\text{SE}(\overline{X}_1 - \overline{X}_2) = \sqrt{\frac{\sigma^2}{n_1} + \frac{\sigma^2}{n_2}} = \sigma\sqrt{\frac{1}{n_1} + \frac{1}{n_2}}$$

If this assumption holds, then we don't need two separate estimates s_1^2 and s_2^2 of one variance σ^2. Instead, it is better to combine the samples and form the **pooled variance estimator**.

$$s_{\text{pool}}^2 = \frac{(n_1 - 1)s_1^2 + (n_2 - 1)s_2^2}{n_1 + n_2 - 2}$$

The pooled estimator of the standard error of the difference between the means uses s_{pool} in place of both s_1 and s_2 in the formula for $\text{se}(\overline{x}_1 - \overline{x}_2)$.

Assuming equal variances, the pooled standard error is

$$\text{se}_{\text{pool}}(\overline{x}_1 - \overline{x}_2) = s_{\text{pool}}\sqrt{\frac{1}{n_1} + \frac{1}{n_2}}$$

The assumption of equal variances also leads to another simplification to the procedure. The degrees of freedom for the two-sample t-statistic that uses the pooled standard error are the sum of the degrees of freedom in the two samples. If $\sigma_1^2 = \sigma_2^2$, the degrees of freedom is $(n_1 - 1) + (n_2 - 1) = n_1 + n_2 - 2$.

Paired samples produce a different standard error that accounts for the dependence between the two samples. This dependence produces a covariance between the averages $\overline{X}_1$ and $\overline{X}_2$. As in Chapter 10, the covariance captures this dependence (notice also that $n_1 = n_2 = n$).

$$\begin{aligned} \text{SE}(\overline{X}_1 &- \overline{X}_2) \\ &= \text{SD}(\overline{X}_1 - \overline{X}_2) \\ &= \sqrt{\text{Var}(\overline{X}_1) + \text{Var}(\overline{X}_2) - 2\,\text{Cov}(\overline{X}_1, \overline{X}_2)} \\ &= \sqrt{\frac{\sigma_1^2}{n} + \frac{\sigma_2^2}{n} - 2\rho\frac{\sigma_1\sigma_2}{n}} \end{aligned}$$

If the covariance is positive (as should happen with pooling), the dependence reduces standard error of the difference between the means.

CHAPTER SUMMARY

Two-sample t-tests and **confidence intervals** for the difference between means allow us to compare results obtained from two samples. For hand calculations, **pooled variances estimators** simplify the calculations but require adding the assumption of equal variances in the populations. Two-sample confidence intervals for proportions proceed in a similar way. **Experiments** provide the ideal data for such comparisons. In an experiment, subjects from a sample are randomly assigned to **treatments** defined by levels of an experimental **factor**. This **randomization** avoids **confounding**, which introduces the possibility of lurking variables. **Paired comparisons** focus on the differences between corresponding subjects by ruling out other differences between the samples.

▇ Key Terms

confounding, 427
experiment, 426
factor, 426
paired comparison, 438
pooled variance
 estimator, 444

randomization, 426
statistically significantly
 different, 433
treatment, 426

two-sample confidence interval, 432
two-sample t-statistic, 428
two-sample t-test, 428

■ Formulas

Standard error of the difference between two sample means

The subscripts 1 and 2 identify the two samples. When estimated from the standard deviations in the two samples without assuming equal variances, the standard error is

$$se(\bar{x}_1 - \bar{x}_2) = \sqrt{\frac{s_1^2}{n_1} + \frac{s_2^2}{n_2}}$$

If the samples are known to come from populations with equal variances, then the standard error uses a pooled estimate of the common variance.

$$se(\bar{x}_1 - \bar{x}_2) = s_{\text{pool}}\sqrt{\frac{1}{n_1} + \frac{1}{n_2}} \quad \text{with}$$

$$s_{\text{pool}}^2 = \frac{(n_1 - 1)s_1^2 + (n_2 - 1)s_2^2}{n_1 + n_2 - 2}$$

Two-sample t-test for the difference in means

Test the null hypothesis $H_0: \mu_1 - \mu_2 \leq D_0$ by comparing the p-value produced by the t-statistic,

$$t = \frac{(\bar{x}_1 - \bar{x}_2) - D_0}{se(\bar{x}_1 - \bar{x}_2)}$$

to the chosen α-level (usually $\alpha = 0.05$). The degrees of freedom is determined by the formula (allowing unequal variances)

$$df = \frac{\left(\frac{s_1^2}{n_1} + \frac{s_1^2}{n_2}\right)^2}{\frac{1}{n_1 - 1}\left(\frac{s_1^2}{n_1}\right)^2 + \frac{1}{n_2 - 1}\left(\frac{s_2^2}{n_2}\right)^2}$$

The degrees of freedom for the pooled test is $n_1 + n_2 - 2$.

Two-sample confidence interval for the difference in means

The $100(1 - 2)\%$ is confidence interval (not requiring equal variances) is

$$\bar{x}_1 - \bar{x}_2 - t_{\alpha/2,df}\, se(\bar{x}_1 - \bar{x}_2) \quad \text{to}$$
$$\bar{x}_1 - \bar{x}_2 + t_{a/2,df}\, se(\bar{x}_1 - \bar{x}_2)$$

■ About the Data

The diet comparison is based on research summarized in G. D. Foster, H. R. Wyatt, J. O. Hill, B. G. McGuckin, C. Brill, B. S. Mohammed, P. O. Szapary, D. J. Rader, J. S. Edman, and S. Klein (2003), A Randomized Trial of a Low-Carbohydrate Diet for Obesity, *New England Journal of Medicine, 348*, 2082–2090. To avoid complications caused by people who drop out of studies and issues of privacy, the data shown here reproduce the summary comparisons of the original research, but are not identical. Data on the success of promoting a delivery service come from a case developed for the MBA course on statistics at Wharton. The data on comparing two sales forces is from D. P. Foster, R. A. Stine, and R. Waterman (1998), *Basic Business Statistics: A Casebook* (New York: Springer).

EXERCISES

Mix and Match

Match each definition on the left with its mathematical expression on the right.

1. Plot used for visual comparison of results in two (or more) groups	(a)	$t = -4.6$
2. Difference between the averages in two samples	(b)	$t = 1.3$
3. Difference between the averages in two populations	(c)	$\mu_1 - \mu_2$
4. Name given to the variable that specifies the treatments in an experiment	(d)	$n_1 + n_2 - 2$
5. Estimate of the standard error of the difference between two sample means	(e)	$n - 1$
6. Avoids confounding in a two-sample comparison	(f)	$\bar{x}_1 - \bar{x}_2$
7. Test statistic indicating a statistically significant result if α is 0.05 and $H_0: \mu \geq 0$	(g)	$\sqrt{\dfrac{s_1^2}{n_1} + \dfrac{s_2^2}{n_2}}$
8. Test statistic indicating that a mean difference is not statistically significant if $\alpha = 0.05$	(h)	Randomization
9. The number of degrees of freedom in a paired t-test	(i)	Factor
10. Approximates the number of degrees of freedom in a two-sample t-test if the sample variances are similar	(j)	Comparison boxplots

True/False

Mark each statement True or False. If you believe that a statement is false, briefly explain why you think it is false.

11. The null hypothesis from a break-even analysis should imply that the two methods are equally profitable.

12. A one-sided test of $H_0: \mu_1 - \mu_2 \leq 10$ rejects H_0 if 10 lies outside of the 95% confidence interval for $\mu_1 - \mu_2$.

13. If the standard two-sample t-test rejects $H_0: \mu_1 - \mu_2 \leq$ \$100, then we know for sure that μ_1 is more than \$100 larger than μ_2.

14. If we double the size of both samples, we reduce the chance that we falsely reject the null hypothesis when using the two-sample t-test.

15. If the two-sample confidence interval for $\mu_1 - \mu_2$ includes 0, then we should conclude that $\mu_1 = \mu_2$.

16. The value of the t-statistic in a two-sample test does not depend on the units of the comparison. (We could, for example, measure the data in dollars or cents.)

17. If the boxplots of the data for the two groups overlap, then the two means are not significantly different.

18. If the confidence interval for μ_1 does not overlap the confidence interval for μ_2, then the two means are statistically significantly different.

19. If the two sample standard deviations are essentially the same ($s_1^2 \approx s_2^2$), then the pooled two-sample interval will agree with the regular two-sample interval for the difference in the means.

20. Pooling the two samples to estimate a common variance σ^2 avoids complications due to confounding.

Think About It

21. A harried commuter is concerned about the time she spends in traffic getting to the office. She times the drive for a couple of weeks and finds that it averages 40 minutes. The next day, she tries public transit and it takes 45 minutes. The next day, she's back on the roads,

convinced that driving is quicker. Does her decision make sense?

22. If the harried commuter in Exercise 21 decides to do more testing, how should she decide on the mode of transportation? Should she, for example, drive for a week and then take public transit for a week? What advice would you offer?

23. An insurance office offers its staff free membership in a local fitness center. The intention is to give employees a chance to exercise and hopefully feel better about their jobs (which are rather sedentary). Rather than ask the employees if they like the service, the company developed a measure of productivity for each employee based on the number of claims handled. To assess the program, managers measured productivity of staff members both before and after the introduction of this program. How should these data be analyzed in order to judge the effect of the exercise program?

24. Doctors tested a new type of contact lens. Volunteers who normally wear contact lenses were given a standard type of lens for one eye and a lens made of the new material for the other. After a month of wear, the volunteers rated the level of perceived comfort for each eye.
    (a) Should the new lens be used for the left or right eye for every patient?
    (b) How should the data on comfort be analyzed?

25. Members of a sales force were randomly assigned to two management groups. Each group employed a different technique for motivating and supporting the sales team. Let's label these groups A and B, and let μ_A and μ_B denote the mean weekly sales generated by members of the two groups. The 95% confidence interval for $\mu_A - \mu_B$ was found to be [$500, $2,200].
    (a) If profits are 40% of sales, what's the 95% confidence interval for the difference in profits generated by the two methods?
    (b) Assuming the usual conditions, should we conclude that the approach taken by Group A sells more than that taken by Group B, or can we dismiss the observed difference as due to chance?
    (c) The manager responsible for Group B complained that his group had been assigned the "poor performers" and that the results were not his fault. How would you respond?

26. Management of a company divided its sales force into two groups. The situation is precisely that of the previous exercise, only now a manager decided the assignment of the groups. The 95% confidence interval for $\mu_A - \mu_B$ was found to be [$500, $2,200].
    (a) Does the lack of randomized assignment of the two groups suggest that we should be more careful in checking the condition of similar variances?
    (b) The manager responsible for Group B complained that her group had been assigned the poor performers and that the results were not her fault. How would you respond?

27. Many advocates for daylight savings time claim that it saves money by reducing energy consumption. Generally it would be hard to run an experiment to test this claim, but as fortune would have it, counties in Indiana provided an opportunity. Until 2006, only 15 of the 92 counties in Indiana used daylight savings time. The rest had remained on standard time. Now all are on daylight savings time. The local energy provider has access to utility records in counties that recently moved to daylight savings time and counties that did not. How can it exploit these data to test the benefits of daylight savings time?

28. Managers of a national retail chain want to test a new advertising program intended to increase the total sales in each store. The new advertising requires moving some display items and making changes in lighting and decoration. The changes can be done overnight in a store. How would you recommend they choose the stores in which to place the advertising? Would data from stores that did not get the new advertising be useful as well?

You Do It

29. **Wine** The recommendations of respected wine critics such as Robert M. Parker, Jr. have a substantial effect on the price of wine. Vintages that earn higher ratings command higher prices and spark surges in demand. These data are a sample of ratings of wines selected from an online Web site from the 2000 and 2001 vintages.
    (a) Do the ratings meet the conditions needed for a two-sample confidence interval?
    (b) Find the 95% confidence interval for the difference in average ratings for these two vintages. Does one of the years look better than the other?
    (c) Suggest one or two possible sources of confounding that might distort the confidence interval reported in part (b).
    (d) Round the endpoints of the interval as needed for presenting it as part of a message.
    (e) Given a choice between a 2000 and 2001 vintage at the same price, which would you choose? Does it matter?

30. **Dexterity** A factory hiring people for tasks on its assembly line gives applicants a test of manual dexterity. This test counts how many oddly shaped parts the applicant can install on a model engine in a one-minute period. Assume that these tested applicants represent simple random samples of men and women who apply for these jobs.
    (a) Find 95% confidence intervals for the expected number of parts that men and women can install during a one-minute period.
    (b) These data are counts, and hence cannot be negative or fractions. How can we use the normal model in this situation?
    (c) Your intervals in part (a) should overlap. What does it mean that the intervals overlap?
    (d) Find the 95% confidence interval for the difference $\mu_{men} - \mu_{women}$.
    (e) Does the interval found in part (d) suggest a different conclusion about $\mu_{men} - \mu_{women}$ than the use of two separate intervals?

(f) Which procedure is the right one to use if we're interested in making an inference about $\mu_{men} - \mu_{women}$?

31. **Used Cars** These data indicate the prices of 156 used BMW cars. Some have four-wheel drive (the model identified by the Xi type) and others two-wheel drive (the model denoted simply by the letter i).

(a) If we treat the data as samples of the typical selling prices of these models, what do you conclude? Do four-wheel drive models command a higher price as used cars, or are differences in average price between these samples typical of sampling variation?

(b) These cars were not randomized to the two groups. We also know that newer cars sell for more than older cars. Has this effect distorted through confounding the confidence interval in part (a)?

32. **Retail Sales** These data give the sales volume (in dollars per square foot) for 37 retail outlets specializing in women's clothing in 2006 and 2007. Do you think that sales changed by a statistically significant amount from 2006 to 2007?

4M Losing Weight

Anyone who's ever watched late-night TV knows how many people want to lose weight the easy way. On the basis of recent medical studies, there may be such a thing. Glaxo Smith Kline bought the drug Orlistat with the hopes of turning it into a modern miracle. Results in prior clinical trials had found that when Orlistat was combined with a diet, subjects taking Orlistat lost more weight. After about a year, a randomized experiment found that obese subjects taking this medication had lost about 11 more pounds, on average, than others who were also dieting, but taking a placebo.[5] Further studies showed similar results.

This drug had moderate success, but Glaxo Smith Kline wanted to take it before the Food and Drug Administration (FDA) and ask the agency to approve it for over-the-counter sales. They wanted it to be available without a prescription. To assemble more information, the company needs help in designing a study to test the drug for college students. The company needs to know how many students it must enroll in order to see if the drug works.

Motivation

Worldwide sales were about $500 million in 2004, but projected to soar to more than $1.5 billion annually with the greater access of over-the-counter sales.

a) Do you think it would be damaging to Glaxo Smith Kline's case if the proposed study did not find the drug to be helpful?

b) If the study costs $5,000 per subject to run but the potential upside is $1 billion, why should the size of the study matter? Why not recruit thousands?

---

[5] J. S. Torgerson, J. Hauptman, M. N. Boldrin, and L. Sjostrom (2004), "XENical in the Prevention of Diabetes in Obese Subjects (XENDOS) Study," *Diabetes Care, 27*, 155–161.

Method

To get FDA recognition, this study requires a randomized experiment. Let's assume half of the subjects receive a placebo and half receive Orlistat.

c) If the subjects are healthy students who take half the dose taken by the obese people in prior studies, would you expect the amount of weight loss to be as large as in previous studies?

d) Is a two-sample confidence interval of the difference in means appropriate for this analysis?

e) Will the presence of lurking factors lead to confusion over the interpretation of the difference in means?

f) What would you recommend if the variation in weight loss is different in the two groups, with those taking Orlistat showing much more variation than those taking the placebo?

Mechanics

For these calculations, let's assume that Glaxo Smith Kline expects those taking Orlistat to have lost 6 pounds more than the controls after six months. Pilot studies indicate that the SD for the amount of weight lost by an individual is $\sigma \approx 5$.

g) If the study enrolls 25 students in each group (a total of 50), is it likely that the difference between the two sample means will be far from zero?

h) Repeat part (g), with 100 students enrolled in each group.

i) Do you think it's likely that the confidence interval with 25 or 100 in each group will include zero if in fact $\mu_1 - \mu_2 = 6$?

j) If the claimed value for σ is too small, what will be the likely consequence for the choice of a sample size?

Message

k) Which sample size would you recommend, 25 in each group or 100?

l) This drug has some embarrassing side effects (e.g., incontinence). Would this affect your recommendation of how many subjects to enroll? Why?

4M Sex Discrimination in the Workplace

Statistical analyses are often featured in lawsuits that allege discrimination. Two-sample methods are particularly common because they can be used to quantify differences between the average salaries of, for example, men and women.

A lawsuit filed against Wal-Mart in 2003 alleged that the retailer discriminated against women. As part of their argument, lawyers for the plaintiffs observed that men who managed Wal-Mart stores in 2001 made an average of $105,682 compared to $89,280 for women who were managers, a difference of $16,402 annually. At the higher level of district manager, men in 2001 made an average of $239,519 compared to $177,149 for women. Of the 508 district managers in 2003, 50 were women (9.8%).

The data used to obtain these numbers are private to the litigation, but we can guess reasonable numbers. Let's focus on the smaller group, the district managers. All we need are standard deviations for the two groups. You'll frequently find yourself in a situation where you only get to see the summary numbers as in this example. Let's use $50,000 for the standard deviation of the pay to women and $60,000 for men. Using the Empirical Rule as a guide, we're guessing that about two-thirds of the female district managers make between $125,000 and $225,000 and two-thirds of the male district managers make $180,000 to $300,000.

Issues of guessing the variation aside, such comparisons have to deal with issues of confounding.

Motivation

a) If a statistical analysis finds that Wal-Mart pays women statistically significantly less than men in the same position, how do you expect a jury to react to this finding?

b) Do you think that Wal-Mart or the plaintiffs have more to gain from doing a statistical analysis that compares these salaries?

Method

c) We don't observe the salaries of either male or female managers. If the actual distribution of these salaries is not normal, does that mean that we cannot use t-tests or intervals?

d) Explain why a confidence interval for the difference in salaries is the more natural technique to use to quantify the statistical significance of the difference in salaries between male and female district managers.

e) We don't have a sample; these are the average salaries of *all* district managers. Doesn't that mean we know μ_M and μ_F? If so, why is a confidence interval useful?

f) These positions at Wal-Mart are relatively high ranking. One suspects that most of these managers know each other. Does that suggest a problem with the usual analysis?

g) Can you think of any lurking factors that might distort the comparison between the means of these two groups?

Mechanics

h) Estimate the standard error of $\bar{x}_M - \bar{x}_F$.

i) Find the 95% two-sample confidence interval. Estimate the degrees of freedom as $n_M + n_F - 2$ and round the endpoints of the confidence interval as needed to present in your message.

j) If the two guessed values for the sample standard deviations are off by a factor of 2 (so that the standard deviation for women is $100,000 and for men is $120,000), what happens to the confidence interval?

k) One way to avoid some types of lurking factors is to restrict the comparison to cases that are more similar, such as district managers who work in the same region. If restricting the managers to one region reduces the sample size to 100 (rather than 508), what effect will this have on the confidence interval?

Message

l) Write a one- or two-sentence summary that interprets for the court the 95% confidence interval.

m) What important caveats should be mentioned along with this interval?

Case: Rare Events

How do we build a confidence interval when every observation in the sample is the same? It may sound unusual at first, but it's common to observe a sample of results in which every case produces the same outcome in medicine, finance, retail, and manufacturing. For instance, we might observe that all of the patients are healthy, none of the customers made a purchase, or that all of the tested components checked out fine. Just because every item in the sample came out the same, however, doesn't mean that the proportion in the population is 0. Can we be sure that no one in the population is sick? The answer is "no," and there's an easy-to-use technique called the Rule of Three that provides a 95% confidence interval for the unknown population proportion.

RARE EVENTS

Statistical analysis typically involves reaching a decision or making an inference in spite of the variation in data. In most problems, this variation conceals important characteristics of the population. The more variation in the data, the harder it becomes to make precise statements about the population. On the other hand, as the sample variance s^2 gets smaller and smaller, the 95% confidence intervals for μ close in tighter and tighter and we learn where μ is. As s^2 approaches 0, the confidence interval collapses to a single point.

There are important cases, however, when $s^2 = 0$ but we nonetheless don't believe that we know the population parameter. Even though there's no variation in the sample, we can ask whether there's variation in the population. The following 3 examples illustrate such cases.

Clinical Trials

The Food and Drug Administration (FDA) requires that companies wishing to sell a pharmaceutical drug in the United States demonstrate the efficacy and safety of the drug in a randomized clinical trial. These trials often have hundreds, if not thousands, of patients. For drugs designed to treat rare illnesses, however, the counts are smaller and the inferences more subtle.

The contingency table shown in Table 1 summarizes a portion of the safety results from one such clinical trial. The trial was a study of a medication for the treatment of severe gout. These data are counts of severe cardiac adverse events among the 212 subjects enrolled in the trial.

TABLE 1 Severe cardiovascular adverse events observed during a clinical trial.

		Treatment ($n = 169$)	Placebo ($n = 43$)
Type of Event	Heart attack	3	0
	Angina	1	0
	Arrhythmia	1	0
	Other	2	0
	Total	8	0

During this trial, investigators identified 8 cases of severe cardiac adverse events among the 169 patients treated with the drug. No severe cardiac events occurred among the 43 patients treated with placebo. A naïve look at Table 1 suggests we have a smoking gun here. None of these adverse events happen in the placebo group compared to 8 under treatment. Even if 1 event occurred in the placebo group, the treatment would still have twice as many heart attacks, adjusting for sample sizes. It looks like this drug is too dangerous to approve for wider use.

In fact these data do *not* indicate a statistically significant difference. The rate of adverse events in the population for placebo may be just as large as that observed in the treated sample. At first, this conclusion may sound like a failure of statistical methodology. We want to show how to think about this example to make this statistical outcome obvious rather than mysterious.

Defaults on Corporate Bonds

Corporations often raise money, such as to pay for a one-time expansion, through the sale of bonds to investors. Instead of offering stock (which is owner-ship in the company), the company sells bonds. Corporate bonds obligate the company to repay investors the principal amount plus interest according to a fixed time schedule. The riskier that investors perceive the company (the greater the chance that it will default on the loan), the higher the rate of interest they demand in order to make a loan to the company. A company might go out of business before it repays its debts, including repaying those who own its bonds.

The rate of interest on a corporate bond depends on expectations of the company and the wider economy. The rate of interest paid by the US government on loans sets a floor for the rate of interest on corporate bonds. If the US government has to pay, say, 4% annual interest on its bonds, then companies pay at least this much. The question is: how much more?

The gap between the rate paid by corporate bonds and the rate paid by Treasury Bonds is known as the default risk premium or credit spread. Large credit spreads (rates that are much higher than those on Treasury Bonds) imply a risky investment. For help in gauging the risk of corporate bonds, investors frequently turn to credit rating agencies. In the United States, the three key agencies are Standard and Poor's, Moody's, and Fitch. These agencies rate the creditworthiness of corporations and assign a letter grade. The rating scale used by Moody's, for instance, assigns a rating of Aaa to companies with the smallest perceived chance of default. Companies with progressively higher chances of default are rated Aa, A, Baa, Ba, B, and then C. Bonds from companies rated Aaa, Aa, A, and Baa are called investment grade. Those with lower ratings are euphemistically called speculative grade or junk bonds. Table 2 summarizes Moody's ratings and the performance of 5,580 US corporate bonds in 2008.

Overall, the ratings in Table 2 are predictive of the risk of default. Default rates are generally lower for bonds issued by companies with better ratings. In

TABLE 2 Number of defaults within Moody's bond rating categories in 2008.

Bond Rating	Number of Issues	Number of Defaults	Percentage Defaulting
Aaa	182	0	0%
Aa	795	4	0.50%
A	1240	4	0.32%
Baa	1138	5	0.44%
Ba	590	9	1.05%
B	1210	24	1.98%
C	425	62	14.6%

2008, the upheaval in the finance industry produced an anomaly: the default rate was higher among Aa rated bonds than among A or Baa rated bonds (0.50% versus 0.32% and 0.44%, respectively).

In 2008, there were no defaults among bonds issued by companies rated Aaa. In fact, if you go back to 1920, you wouldn't see any defaults in this category. (A company might, however, slide out of the group rated Aaa and subsequently default. Such "fallen angels" do happen, but the transition from Aaa to default takes more than a year.) Should we infer that the probability of default by a company rated Aaa is zero?

Electronic Components

When companies ship electronic components to each other, it's common for the purchaser to reserve payment until testing verifies that the order meets the necessary specifications. Often the terms of the sale dictate that the percentage of defective items within the shipment is less than some threshold, say 1%. This threshold is related to the price of the components. Shipments that guarantee a lower rate of 0.1% would command a higher price per unit.

The customer in this example has just received a shipment of 100,000 electronic components used in computer assembly. The customer cannot test them all, and instead tests a random sample of $n = 400$ components from the shipment.

All 400 of these tested components meet the required specifications. There's not a problem in the group. What should the customer conclude about the presence of defective components in the entire batch? Every one of the tested items is okay, but that does not mean that every item in the much larger population is okay, too. How large might the rate of defective components be? Is this enough evidence to convince the customer to accept the shipment?

INFERENCE FOR RARE EVENTS

Whatever the circumstance, we'll describe the results using the success/failure terminology of binomial random variables and inference for proportions. A success might be a good thing (a customer who makes a purchase on a web site) or a bad thing (a component fails a diagnostic test), but we'll call them both successes. The probability of a success in the population is p.

In each of the three examples, we're faced with data that have no sample variation.

- Not one of the 43 patients on the placebo therapy in the clinical trial had a cardiac-related adverse event.

- None of the 182 Aaa-rated issuers of corporate debt defaulted in 2008.
- All of the 400 tested electronic components satisfied the acceptance test.

None of these samples exhibits variation, but most of us wouldn't interpret that to imply that there's no chance of an event in the future. We cannot fall back on standard methods of inference for p to help us. The sample size condition for inference for proportions excludes samples with very low (or very high) rates. The sample size condition requires that we observe at least 10 successes and 10 failures. These samples don't have any successes. (If you try to use the usual confidence interval for a proportion, you'll find that se$(\hat{p}) = 0$ and that the confidence interval for p collapses to a point. That's wrong and leads to incorrect inferences.)

Rule of Three

To get a 95% confidence interval, we have to return to the definition of what it means to have a confidence interval. A confidence interval is a range of values for a population parameter that is compatible with our observed sample. We should not be surprised to learn that p lies inside the 95% confidence interval.

Let's build a 95% confidence interval of the form $[0, p^*]$ with the lower limit set to zero. Since $\hat{p} = 0$, it seems reasonable to allow that p might be zero. Our concern is how large p might be. The only way to see the role of variation is to think in terms of what is possible, not just what was seen. We'll have to think in terms of probabilities.

To find the upper limit p^*, ask yourself: How surprising is it to observe a sample with no successes if p is some value larger than 0? As in the usual test of a proportion, we need for the data to satisfy conditions that allow us to compute probabilities. In particular we have to assume that we observe Bernoulli trials (Chapter 11). That is, the trials have to share a common probability of success p and be independent of one another.

Given these assumptions, the probability that we observe a sample of n observations with $\hat{p} = 0$ is

$$P(\text{no successes in } n \text{ trials}) = p^0(1 - p)^n = (1 - p)^n$$

Each of the n trials has to fail, which happens with probability $1 - p$. Since the trials are independent, the probabilities multiply.

In the example of acceptance sampling, we observe no successes in $n = 400$ trials. Would this be likely to happen if $p = 0.5$? Is it likely to get a random sample of $n = 400$ items with $\hat{p} = 0$ from a pop-

ulation with $p = 0.5$? The answer is no. If $p = 0.5$, then the probability of no successes in a sample of 400 is incredibly small,

$$(1 - p)^n = (1 - 0.5)^{400} \approx 3.87 \times 10^{-121}$$

Clearly, p has to be less than 0.5. How about $p = 0.1$? If $p = 0.1$, then the probability of no successes in a sample of 400 is larger, but still rare

$$(1 - p)^n = (1 - 0.1)^{400} \approx 4.98 \times 10^{-19}$$

To have no successes in so many trials, p must be smaller. Let's try $p = 0.01$. If $p = 0.01$, then the probability of no successes is

$$(1 - p)^n = (1 - 0.01)^{400} \approx 0.018$$

Compared to our usual rule for tests, that's still too surprising for us to accept 0.01 as a plausible value for p. A statistical test with $\alpha = 0.05$ rejects values of the parameter if the probability of the sample is less than α.

We can keep trying different values, but it's easier to solve for p^* that makes the probability of a sample with no successes equal to 0.05. If we solve the equation

$$(1 - p^*)^{400} = 0.05$$

then we get $p^* \approx 0.0074614$ if we round the answer to five significant digits. If $p < p^*$, then the chance of seeing a sample of 400 trials with no successes is larger than 0.05. For example,

If $p = 0.0050, (1 - p)^n = (1 - 0.0050)^{400} \approx 0.135$
If $p = 0.0025, (1 - p)^n = (1 - 0.0025)^{400} \approx 0.367$

The procedure we've just discovered is commonly known as the **Rule of Three**. The Rule of Three defines the 95% confidence interval for p to be $[0, p^*]$ when we observe a sample in which there are no successes. The name "Rule of Three" comes from a simple approximation to p^* that makes the procedure very easy to use. The approximation is $p^* \approx 3/n$. (See Behind the Math: Continuous Compounding). For example, with $n = 400, p^* \approx 0.0074614$. The quick approximation matches p^* if we round to two significant digits, $3/n = 3/400 = 0.0075$. (p.454)

Rule of Three　If a sample of n Bernoulli trials has no successes ($\hat{p} = 0$), then the 95% confidence interval for p is approximately $[0, 3/n]$.

✓ The events are Bernoulli trials (independent with equal chance p of successes.
✓ The sample yielding these trials makes up no more than 10% of the population of possible trials.

Notice that the upper limit of the 95% confidence interval gets smaller as the sample gets larger. No matter what the sample size, the occurrence of 3 successes defines the upper limit. These 3 successes convert to a smaller and smaller upper limit for p as n grows.

Using the Rule of Three

The Rule of Three provides quick answers in two of the three examples that we've used to introduce rare events. Consider the example with adverse events in a clinical trial. In the trial, investigators observed 8 events in the treatment sample ($n = 169$) but none in the placebo sample ($n = 43$). Does this suggest that the drug is a safety risk?

The Rule of Three says "no." According to the rule, we would not be surprised to see another sample from the placebo population with 3 serious cardiac adverse events. Seeing that the treatment population is about 4 times larger than the placebo population, the rate of events in the placebo sample might be larger than that in the treatment sample. The observed difference, stark as it is, is not statistically significant. The presence of adverse events in the treatment group does indicate the need for close follow-up on the wider use of this medication, but the sample sizes are too small to be conclusive.

The Rule of Three similarly quickly answers the question posed in the acceptance sampling application. We want assurances that the defective rate in the full shipment is less than 1%. We observed none in a sample of $n = 400$. Using the Rule of Three, we conclude that p in the full shipment might be as large as $3/n = 3/400 = 0.0075$, less than 1%. Accept the sample.

Determining n

The use of the Rule of Three is straightforward in the previous two examples because it's clear that we have n independent trials. That's the whole point of a randomized clinical trial and acceptance sampling. You're assured of having Bernoulli trials.

For the possibility that Aaa corporate bonds default next year, however, we have to be more careful. Let's think about why a company might default. We can loosely divide the causes of default into two broad categories: idiosyncratic blunders and systematic risks. An idiosyncratic blunder means that the management of a company made a colossal mistake—such as launching a dangerous product resulting in huge liability claims or committing a gross error in accounting for costs—and that the blunder was so large as to bring down the firm. The fact that no previous Aaa company has ever defaulted tells us

right away that the chance for such big mistakes is quite small. The idiosyncratic nature of blunders also suggests independent events to which we can apply the Rule of Three. The relevant count would be all of the previous Aaa bonds that never defaulted over the past years, adding up to about 150 companies for 90 years, or 13,500 bond-years. The Rule of Three bounds the idiosyncratic contribution to the probability of default at $3/13,500 \approx 0.02\%$ which is quite small compared to the default rates observed in Table 2.

The larger source of default risk is the systematic component. Bonds do not default independently of one another. All of these Aaa businesses operate in the same country and are buffeted by the same economic stresses. Chances are that if the economic stress becomes so large that one of these companies defaults, there's a good chance that others will default as well. A major recession could spread into Aaa bonds.

What are the chances of that happening next year? The Rule of Three gives a bound. We've never seen a serious meltdown hit the Aaa market in the past 90 years, so treating these years as independent (as in our analysis of stock returns), the Rule of Three puts the upper limit of the 95% confidence interval for the probability of a meltdown of Aaa bonds at $3/90 \approx 3.33\%$.

Considerations in Finance

To see whether the Rule of Three gives a reasonable bound for defaults on Aaa bonds, we need to compare the 3.33% rate to what we can observe in data. That's harder than you might think. The difficulty is not with the statistics, but rather with the subtleties of the market for corporate bonds. Fear of a meltdown is only one component that affects the yield on a corporate bond.

Here are four additional financial considerations. We have to form reasoned, but subjective, answers to these questions when speculating on things that have not happened.

- First, are investors in bonds satisfied with a 95% confidence interval? Perhaps they'd like a higher level of coverage. More coverage is appealing when you consider that you're guarding against an event in which most other investments are also going to be losing money. For example, increasing the coverage from 0.95 to $0.9975(= 1 - 0.05^2)$ doubles the upper confidence limit from $3/n$ to $6/n$.
- Second, do we expect every Aaa bond to default? A serious financial shock is unlikely to cause every Aaa bond to default. During the Great Depression, the percentage of defaults within the bottom-rated

junk bonds reached 25% in only one year, 1933. Rates were much lower in other categories of bonds.

- Third, what's the value at liquidation of a defaulting company? Bondholders stand at the head of the line to claim the remaining assets of a company that defaults. The historical amount recovered is about 50%, but we might expect less if the economy is in turmoil at the time of default.

- Fourth, what's the effect of prepayment risk? Most bonds allow the issuing company to pay off investors if interest rates fall. The possibility of prepayment hurts investors because they won't be able to turn bond payments into profits if interest rates drop, but they'll have to live with the payments if rates rise.

Combined with the risk of a meltdown, five factors determine the credit spread on a bond. The credit spread (the difference between the rate paid by corporate bonds above that paid by comparable Treasury Bonds) is the product of the

- probability of systematic failure in Aaa bonds
- adjustment for the level of confidence
- share of Aaa companies that fail
- percentage of bond value lost to recovery
- adjustment for prepayment risk

For example, if we use 95% coverage, expect 50% of companies to default, anticipate 80% to be lost, and don't make any adjustment for prepayment risk, we get

$$0.0333 \times 0.50 \times 0.80 = 0.0133, \text{ or } 1.33\%$$

Is this the right upper limit? Based on the history of market prices, it is in the right ballpark. Figure 1 tracks the offered yield on long-term corporate bonds monthly from 1950 through the end of 2008.

The timeplot follows two series: the rate paid by Treasury Bonds (blue, based on yields of 20- and 30-year Treasury Bonds) and the rate paid by Aaa bonds (red). The gap between these rates is the credit spread shown in Figure 2. This time series subtracts the rate paid by Treasury Bonds in Figure 1 from the rate of interest paid on the Aaa bonds.

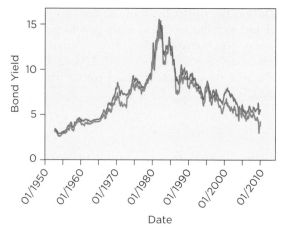

FIGURE 1 Annual percentage yield on Aaa corporate bonds (red) and long-term Treasury Bonds (blue).

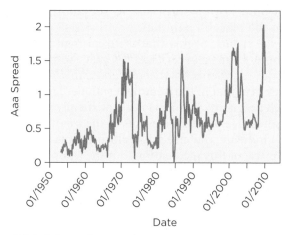

FIGURE 2 Credit spreads on Aaa (red) and Baa (green) corporate bonds.

The credit spread on Aaa bonds has averaged 0.63% since 1953 (when our data on Treasury Bonds begins). Lately, the recession caused rates to increase and even exceed the upper limit of our interval. With so many other choices for the financial adjustments, we might come up with a very different endpoint for the 95% confidence interval. The Rule of Three handles its part of the calculation, but it's up to experts in finance to nail down the other factors that determine the credit spread.

 BEHIND the Math

Continuous Compounding

The Rule of Three comes from an approximation to the solution of the equation $(1 - p^*)^n = 0.05$. You may have seen this approximation used in finance courses to compute continuously compounded interest.

Here's a quick review of continuously compounded interest. Suppose we put $100 into a certificate of deposit (CD) for a year at 5% interest. If the interest is paid at end of the year, the CD would then be worth $100(1 + 0.05) = \$105$. Suppose instead the bank credits interest to the CD each month. Then the CD would earn interest on the interest and become more valuable by the end of the year, growing slightly more to

$$100(1 + 0.05/12)^{12} \approx \$105.1162$$

That's not much more, but suppose the bank were to compound the interest daily (for 365 days). Then the CD would grow to

$$100(1 + 0.05/365)^{365} \approx \$105.1267$$

Continuously compounded interest emerges as we let the number of time periods get larger. If we divide the year into k time periods, then the value of the CD at the end of the year is $100(1 + 0.05/k)^k$.

Calculus shows that in the limit as k increases, the multiplier in this expression grows to

$$\lim_{k \to \infty} (1 + 0.05/k)^k = e^{0.05} \approx 1.051271$$

With continuous compounding, the CD is worth $105.1271. The limit works in general for other rates of interest because

$$\lim_{k \to \infty} (1 + x/k)^k = e^x$$

This last expression is the source of the Rule of Three. In order to solve the equation $(1 - p^*)^n = 0.05$ for p^*, write $p^* = x/n$. Now we need to solve $(1 - x/n)^n \approx e^{-x} = 0.05$ for x. That's easy: $x = -\log_e 0.05 \approx 3.00$.

CASE SUMMARY

The Rule of Three provides a 95% confidence interval for the population proportion p when observing a sample with all failures (or all successes). The interval for p is the range $[0, 3/n]$. The interval requires Bernoulli trials: the data must define independent events with only two possible outcomes and a constant probability.

Key Terms

Rule of Three

Questions for Thought

1. Suppose you observe a sample with all successes. How could you get a 95% confidence interval for p?
2. The Rule of Three generates a 95% confidence interval. What rule would you recommend if you wanted to have a 99.75% confidence interval?
3. Does the Rule of Three work with small samples as well? In particular, if $n = 20$ does the argument leading to the 95% interval $[0, 3/n]$ still apply?

About the Data

The data on adverse events are from an application made to the FDA in 2009 for the use of the medication pegloticase for the treatment of severe gout. Data on corporate bonds and Treasury Bonds is from the US Federal Reserve series H-15. The corporate bonds are classified by Moody's as seasoned bonds with a long time (at least 20 years) to maturity. The counts of default and the default rates in Table were converted from hazard function estimates published by Moody's in the report "Corporate Default and Recovery Rates, 1920–2008." The data on acceptance sampling is from a student-initiated independent summer research project.

Case: Testing Association

MANAGING INVENTORIES

DATA MINING

CASE SUMMARY

If we examine hundreds of contingency tables, how will we separate those that indicate real association from those that happen to show association by chance? Contingency tables of data typically show some association between two categorical variables, even those that are independent. To separate meaningful association from sampling variation, we'll use a test to identify statistically significant differences from independence. The test revisits the chi-squared statistic χ^2 that was introduced as a descriptive measure of association in Chapter 5. By attaching a p-value to this statistic, we obtain a procedure that works in wider generality.

MANAGING INVENTORIES

National retailers, such as Wal-Mart and Target, stock an amazing variety of products in every store: clothing, food, hardware, electronics, health, cleaning, and even garden supplies. As if that's not enough, many items are packaged and sold in a bewildering array of sizes and configurations. Do you want the single serving, the economy size, the family pack, or the bulk case? Are you looking for two batteries, a bubble pack of four or eight, or a contractor's box with three dozen? Do you want a travel size of toothpaste, a medium tube, or the economy three-pack?

Keeping track of so many products is difficult but essential. Retailers carry products in so many sizes and styles because that's what shoppers want. If shoppers everywhere had the same preferences, managers' lives would be simpler. A retailer could stock the same mix of packages at every location. Shoppers aren't all the same, though, and chains that operate in different venues stock their stores differently. It's not enough to choose the brands that customers want; managers have to decide what style of packaging will sell the best in different locations. A suburban customer who drives an SUV to the store can buy and take home larger packages than

an urban shopper who walks in and has to carry her purchases home.

Looking for Association

Let's consider the situation faced by regional managers who control the mix of packaging. We will keep the problem manageable by assuming that the chain operates retail stores at three types of locations: urban, suburban, and rural. Each store stocks many thousands of items. From these, regional managers want to find the right packaging mix to offer for 650 different products. All 650 of these products are packaged in more than one way.

As an example, let's consider the packaging options for a nationally advertised brand of hand soap. The soap is sold as a single bar, a twin pack with two bars, a bundle of eight bars, and recently as a liquid in a pump dispenser. To get a sense of which configurations sell the best at different locations, managers sampled records from the retailer's database of transactions. From these records, the managers identified 200 purchases of this soap at each of the three locations. Table 1 shows the contingency table of counts.

TABLE 1 Sales of hand soap at three types of locations operated by a chain of retail stores.

		Package Type				Total
		Single	Twin	Bundle	Liquid	
Location	Rural	63	76	45	16	200
	Suburban	32	45	71	52	200
	Urban	78	44	24	54	200
	Total	173	165	140	122	600

If the categorical variables *Location* and *Package Type* are independent, then the data should indicate that there's no need to customize the mix of packages. Independence would mean that rural, suburban, and urban customers share the same preferences in packaging. There would not be a need to customize the mix of packages at different locations. The chain should stock the same blend of sizes at every location, namely the mix that matches the marginal distribution of *Package Type* (about $173/600 \approx 29\%$ single bars, 28% twin packs, 23% eight bar bundles, and 20% liquid dispensers). In the absence of association, this blend is suitable for all three types of locations.

The data in Table 1 instead appear dependent. Were *Location* and *Package Type* are independent, then we ought to observe roughly equal counts of sales within the columns of the table. Overall, for example, 29% of the sales are for a single bar. If the two variables were independent, then about 29% of the 200 sales in each row should be for a single bar. Instead, we find fewer sales of single bars in suburban locations and more sales in urban locations. Are these deviations from the expected frequency in Table 1 large, or could random variation produce the observed differences in the counts?

This question requires a statistical test. We'll base our test on the chi-squared statistic. Chapter 5 introduced the chi-squared statistic χ^2 as a measure of dependence. Let's review the definition of χ^2 by calculating this statistic for Table 1. Chi-squared measures dependence by comparing the counts in a contingency table such as Table 1 to the counts in an artificial table in which the two variables are independent. If *Location* and *Package Type* are independent and we hold constant the observed marginal counts, then we expect to find the counts shown in Table 2.

TABLE 2 Expected cell counts under the assumption of independence.

		Package Type				
		Single	Twin	Bundle	Liquid	Total
	Rural	57.667	55.000	46.667	40.667	200
Location	Suburban	57.667	55.000	46.667	40.667	200
	Urban	57.667	55.000	46.667	40.667	200
	Total	173	165	140	122	600

For example, consider the expected count of single bars in rural stores. One third of the sales occur at each type of store because of the way the managers sampled the database, and the marginal distribution puts $173/600 \approx 28.8\%$ of the share on sales of single bars. Hence, out of the 200 sales at each type of location, we expect sales of $200(0.288) = 57.667$ single bars at every location (the first column of Table 2).

The chi-squared statistic compares the hypothetical counts in Table 2 to the actual counts in Table 1. The comparison requires two steps: first square the difference between the expected count from the

observed count. Then divide that squared deviation by the expected count. The chi-squared statistic is the sum of these normalized, squared deviations:

$$\chi^2 = \sum_{cells} \frac{(\text{observed count} - \text{expected count})^2}{\text{expected count}}$$

For these data, $\chi^2 \approx 77.372$. Is that statistically significant association?

Chi-squared Test of Association

Rather than convert the chi-squared statistic into Cramer's V as in Chapter 5, we use it to test the null hypothesis that *Location* and *Package Type* are independent. The *p*-value of that test determines whether there's statistically significant association.

We can compute χ^2 for any table, but to make an inference to the wider population, we require two conditions of the data. These are similar to those needed when testing a proportion (Chapter 15). First, the sample must be an SRS from the population, and second, the sample must be large enough so that the expected counts in the cells of the table are all at least 10. (Some texts weaken this condition to the requirement for at least five.)

The SRS condition is met in the sales example because of the randomized method used to select transactions from the database. This is a large retailer, and 600 transactions for any of these products is a small fraction of the data for each. The second condition also holds. Managers sampled 200 sales from each location in order to be sure that the data would meet the conditions of this test.

Under the null hypothesis that the two categorical variables are independent, χ^2 has a known sampling distribution called by the same name as this statistic, the chi-squared distribution. We generally let software compute the *p*-value for this test, but it is useful to see how the distribution of this statistic depends on the size of the contingency table. The statistic depends on the number of rows and columns in the contingency table, not the total number of cases.

Table 3 lists upper percentiles of the chi-squared distribution; if χ^2 exceeds the percentile in the column labeled 0.05 in Table 3, for example, then we can reject the null hypothesis of independence with $\alpha = 0.05$. (The Appendix has a larger version of this table.)

The trick to using this table is to know which row applies. Table 3 resembles the table of the *t*-distribution (Chapter 15). In particular, the sampling distribution of χ^2 under the null hypothesis of no association depends on the number of degrees of freedom associated with a contingency table.

The degrees of freedom for the χ^2 test is the product of the number of rows in the contingency table

TABLE 3 Upper percentiles of the distribution of the chi-squared statistic.

df		α					
		0.1	**0.05**	**0.02**	**0.01**	**0.005**	**0.001**
	1	2.71	3.84	5.41	6.63	7.88	10.83
	2	4.61	5.99	7.82	9.21	10.60	13.82
	3	6.25	7.81	9.84	11.34	12.84	16.27
	4	7.78	9.49	11.67	13.28	14.86	18.47
	5	9.24	11.07	13.39	15.09	16.75	20.52
	6	10.64	12.59	15.03	16.81	18.55	22.46
	7	12.02	14.07	16.62	18.48	20.28	24.32
	8	13.36	15.51	18.17	20.09	21.95	26.12
	9	14.68	16.92	19.68	21.67	23.59	27.88
	10	15.99	18.31	21.16	23.21	25.19	29.59
	20	28.41	31.41	35.02	37.57	40.00	45.31
	30	40.26	43.77	47.96	50.89	53.67	59.70
	40	51.81	55.76	60.44	63.69	66.77	73.40
	50	63.17	67.50	72.61	76.15	79.49	86.66

minus 1 times the number of columns in the contingency table minus 1. For the contingency table of soap sales (Table 1), χ^2 has $(3 - 1)(4 - 1) = 6$ degrees of freedom. If we look within the row of Table 3 identified by 6 degrees of freedom (df), the observed statistic $\chi^2 = 77.372$ exceeds every percentile. Hence, the p-value of the test of independence is less than 0.001. We reject H_0 and conclude that these data are associated. The choice of the packaging mix for a store depends on its location.

Chi-Squared Test of Association. To test the null hypothesis that two categorical variables are independent in the population, compute the χ^2 statistic for the contingency table formed from observations of the two variables. Compare χ^2 to the upper $100(1 - \alpha)\%$ percentile of a chi-squared distribution with degrees of freedom determined by the size of the contingency table, df = (#rows − 1)(#cols − 1).

✓ **SRS Condition.** The observed sample is a simple random sample from the relevant population. If sampling from a finite population, the sample comprises less than 10% of the population.

✓ **Sample Size Condition.** The expected count in every cell of the hypothetical table of counts is at least 10.

Counting Degrees of Freedom

A heuristic explanation of the formula for the number of degrees of freedom explains this unusual name and why the formula for degrees of freedom is different for a table than for the t-test of a mean.

The degrees of freedom of the t-statistic introduced in Chapter 15 is $n - 1$, where n is the sample size. For the t-statistic, the calculation of s^2 determines the degrees of freedom. As explained in Chapter 15, the calculation of s^2 requires that we first calculate $\bar{x}$ and use this value to determine the variation in the data. That use of the sample mean constrains the data: given $\bar{x}$ and any $n - 1$ of the cases, we can figure out the value of the last case. Hence, there are $n - 1$ degrees of freedom.

A similar argument applies to tables. The marginal counts in Table 1 are uninformative for testing the association between *Location* and *Package Type*. Marginal counts do not reveal whether the variables are independent or highly associated. Association depends on what happens inside the contingency table, not along the margins. Instead, we use the observed margins to compute the table of counts assuming independence (Table 2). Hence, these marginal counts play a role analogous to that of $\bar{x}$ in the t-statistic. They don't tell us the association, but we need them to construct the test statistic.

Given the margins of the table, how many cell counts inside the table are "free to vary?" For example, once we fill in the six cells of Table 4 marked with an ×, we can fill in the remaining cells by using the margins. Counts in the other cells are determined by our choices for these six cells and the margins of the table.

TABLE 4 The size of the table determines the number of degrees of freedom, the count of cells with an *x*.

		Package Type				
		Single	**Twin**	**Bundle**	**Liquid**	**Total**
Location	**Rural**	×	×	×		200
	Suburban	×	×	×		200
	Urban					200
	Total	173	165	140	122	600

Because the sum of the cells in a row or column has to match the associated margin, only 6 degrees of freedom remain in Table 4 given the marginal totals. If we vary the size of the table, we see that once we know the margins, we can fill in the rest of the cells in the table once we fill in a block of (#rows − 1) (#columns − 1) cells.

Equality of Several Proportions

Let's review the chi-squared test of independence by looking at the distribution of sales for another product, paper towels.

TABLE 5 Sales of paper towel packages.

		Package Type		
		Two rolls	**Dozen rolls**	**Total**
Location	**Rural**	89	111	200
	Suburban	75	125	200
	Urban	82	118	200
	Total	246	354	600

This brand of paper towels is sold in two types of packaging: either as a pair of rolls or in a large bundle with 12 rolls. Table 6 shows the counts that we expect if *Location* and *Package Type* are independent for paper towels.

The expected counts for paper towels are closer to the observed counts than for the hand soap. The percentage of two-roll packages appears similar over the three locations. The chi-squared statistic is $\chi^2 \approx 2.03$ with $(3 - 1)(2 - 1) = 2$ degrees of freedom for the test of the null hypothesis of independence. Our software calculates the *p*-value to be 0.36; the test does not indicate the presence of statistically significant dependence. These data could be samples

TABLE 6 Expected counts if location and package type are independent.

		Package Type		
		Two rolls	**Dozen rolls**	**Total**
Location	**Rural**	82	118	200
	Suburban	82	118	200
	Urban	82	118	200
	Total	246	354	600

from populations that have the same distribution of package types for the three locations.

The chi-squared test in this example is closely related to the two-sample test of proportions (Chapter 18). Let p_{rural} denote the population proportion of sales of two-roll packages in rural locations. Similarly let $p_{suburban}$ and p_{urban} denote this proportion for suburban and urban locations, respectively. Since this product comes in only two types of packaging, the test of independence is equivalent to testing the null hypothesis

$$H_0: p_{rural} = p_{suburban} = p_{urban}$$

Instead of comparing two proportions, the chi-squared test allows us to compare proportions in three (or more) populations. If we were to test the difference only between urban and rural locations, then the chi-squared test would be equivalent to the two-sample test of proportions shown in Chapter 18.

DATA MINING

Data mining refers to extracting useful knowledge from what may otherwise appear to be an overwhelming amount of noisy data. Data mining is often associated with elaborate, specialized methods, but we can also data mine by scaling up the way in which we use simpler statistics. The data we're considering here consists of 600 transactions (200 each at rural, suburban, and rural locations) for 650 products that come in multiple types of packaging. That's 390,000 data records (which seems like a lot until you realize these are a small sample of all customer transactions). To make sense of so much data, we will convert the table of counts for each product into a chi-squared statistic. With a little more work needed to make the results comparable across the products, we'll be able to quickly identify products whose sales by package type vary most from location to location.

---

data mining Searching for patterns in large amount of data.

Regional managers can take advantage of this testing procedure to locate products that have to be managed differently for the three types of locations. Consider the products shown in Table 1 and Table 5. The chi-squared test finds statistically significant association for the hand soap, but not paper towels. The test implies that customers at different locations have different preferences in packaging for the soap, but not for the paper towels. These results suggest that managers can centrally allocate the same mix of paper towels to the various locations, but have to customize the mix of soap.

For data mining, we scale up this procedure to the full set of products. Rather than look at products one-by-one, managers can compute χ^2 for each of the 650 products that the chain offers in multiple packaging styles. (There's no need to do this for the products that come in only one packaging style.) By converting the collection of chi-squared statistics into p-values, managers will be able to quickly find the products that have the most association.

Standardizing with p-values

For each product, we constructed a contingency table like that in Table 1 or Table 5. We then computed the value of χ^2 for each of these tables. The histogram and boxplot in Figure 1 summarize the values of χ^2 for all 650 products.

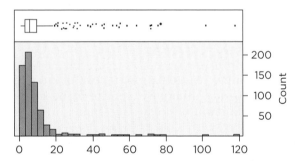

FIGURE 1 Histogram of chi-square statistics for 650 products.

It would be hasty to assume that the largest χ^2 identifies the product with the greatest association between *Location* and *Package Type*. That would be the case if every product had the same number of packaging types, but that's not true. Some products are packaged 2 ways whereas others come in 5 packages. The bar chart in Figure 2 shows the frequencies of the number of packaging types for these products.

Most products come in 3 types of packaging, but some of these have 4 or 5 and others have as few as 2. (Only products with 2 or more packaging types were extracted from the retailer's database.)

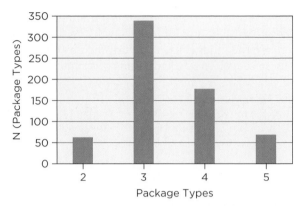

FIGURE 2 Frequencies of the number of products offering different packaging types.

Since the number of package types varies from one product to another, the histogram in Figure 1 mixes observations of different types of random variables. Suppose that the null hypothesis of independence between *Location* and *Package Type* holds for most products. If every product offered four packaging types, then the histogram of chi-squared statistics would estimate of the sampling distribution of the test statistic χ^2. Instead, the number of packaging types varies from one product to the next. We need to separate the values of χ^2 into those that have 4, 6, 8 or 10 degrees of freedom. (Every product is offered in three locations, so the number of degrees of freedom is twice the number of packaging types minus one.)

Grouping the products by the number of packaging types makes it easier for managers to spot the products with the most association. A quick glance at the percentiles of the chi-squared distribution in Table 3 shows that the value of χ^2 increases with the number of degrees of freedom. Bigger tables tend to produce larger values of χ^2. It could be that the products that generate the largest values of χ^2 in Figure 1 also have more packaging types. The side-by-side boxplots in Figure 3 show the chi-squared statistics

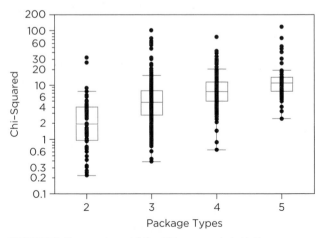

FIGURE 3 Comparison of the chi-squared statistics.

grouped by the number of packaging types. These boxplots share a common scale, and you can see that values of χ^2 that have more degrees of freedom tend to be larger. (The plot uses a log scale so that skewness does not conceal the details.)

To put the test results on a common scale that adjusts for the number of packaging types, we convert each χ^2 into a p-value. This conversion distinguishes chi-squared statistics that indicate statistically significant association from those that are large simply because the contingency table is large.

Now that we've accounted for the size of the contingency table, we can combine these p-values into one histogram. Products with the smallest p-value show the most association; these are the products that require oversight to get the right blend of packages in different locations. The histogram in Figure 4 shows all 650 p-values, one for each product.

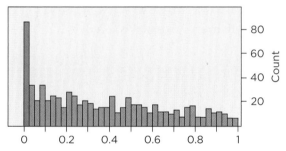

FIGURE 4 Histogram of p-values of the 650 chi-squared tests.

If the null hypothesis of independence held for every product, then we'd expect to find a uniform distribution in this histogram. For example, there's a 5% chance of a p-value being less than 0.05 even though

H_0 holds. The histogram in Figure 4 looks rather flat, except for the prominent bar at the left side.

The tall bar at the left of Figure 4 identifies products with the most statistically significant values of χ^2. Each of the intervals that define the histogram in Figure 4 has length 0.025, so we'd expect about $0.025 \times 650 \approx 16$ in each. The interval from 0 to 0.025 has 84. These products are the most sensitive to location. Managers responsible for inventory control would be well served to start with them. For instance, a manager could begin with the product with the smallest p-value and work in order.

Related Methods

The approach consisting of

1. Calculate a lot of test statistics
2. Study those with the most statistically significant outcome

is a common paradigm in modern quantitative biology. Scientists use this approach to identify genes that may be responsible for an illness. Genetic tests are performed on two samples of individuals, one that is healthy (the control group) and the other that has an illness that is suspected of having a genetic connection (such as various types of cancers). For each subject, a genetic analysis determines the presence or absence of 10,000 or more genes. That's a data table with 10,000 columns for each subject. To reduce the volume of data, scientists use a basic test that compares two groups, like the two-sample t-test, to contrast the amount of each gene present in the control group to the amount present in the studied group. Next, each test is converted into a p-value as we did with χ^2. The most significant tests indicate genes for further, more extensive study.

CASE SUMMARY

The **chi-squared test** of the null hypothesis of independence compares observed frequencies in a contingency table to those expected if the underlying random variables are independent. As a special case we can use this test to compare the equality of several proportions, extending the two-sample comparison in Chapter 18. When used to search for patterns in **data mining**, we need to assign p-values to the test statistic to adjust for the effects of sample sizes and the size of the contingency table.

▮ Key Terms

chi-squared test, 458   data mining, 459

Questions for Thought

1. The data used in the chi-squared analysis has 200 cases for each location. Is it necessary to have the same number of observations from each location for every product?

2. The histogram of *p*-values (Figure 4) shows that 84 products have *p*-value less than 0.025. Does this mean that if we were to examine all of the transactions for these products that we would find *Location* and *Package Type* associated for all 84 of them?

3. Explain how the analysis of packaging types could be used to manage the mix of colors or sizes of apparel in clothing stores that operate in different parts of the United States.

About the Data

The packaging data is based on an analysis developed by several students participating in Wharton's Executive MBA program.

PART IV

Regression Models

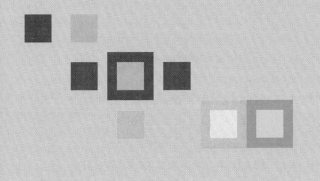

19 Linear Patterns

MANY FACTORS AFFECT THE PRICE OF A COMMODITY. These factors fall into two broad categories: fixed costs and variable costs. Fixed costs are present and of constant size regardless of the quantity; variable costs increase with the quantity. As an example, let's consider the price charged by a jewelry merchant for a diamond.

The variable cost of a diamond depends on its size. A variable cost is expressed as the product of an amount times the marginal cost of the item. For a diamond, the variable cost is the product of its weight (in carats) times the marginal cost (in dollars per caret). (A carat is a unit of weight used for gems; one carat is 0.2 gram.)

Fixed costs are present regardless of the size of the diamond. Fixed costs include overhead expenses, such as the cost of maintaining the store where diamonds are shown or hosting a Web site to advertise the gems online. The ratio of the cost of a diamond to its weight mixes fixed and variable costs together. A one-carat diamond might cost, say, $2,500. That's not the variable cost unless this jeweler has no fixed costs.

Fixed and variable costs can be separated by comparing the prices of diamonds of varying sizes. The relationship between the price and weight in a collection of diamonds of varying weights allows us to separate these costs and come to a better understanding of what determines the final cost of a gem.

The technique that we will use to estimate fixed and variable costs is known as regression analysis. Regression analysis produces an equation that summarizes the association between two variables. In this chapter, we focus on linear association that we can summarize with a line. An estimated slope and intercept identify each line. The interpretation of these estimates is essential when using regression in business applications; for instance, the intercept and slope estimate the fixed and variable costs in the example of pricing diamonds. The fitted line is also used for predicting new observations. We choose the line that provides the best fit to the data, using a criterion known as least squares.

19.1 | FITTING A LINE TO DATA

Consider two questions about the cost of diamonds:

1. What's the average price of diamonds that weigh 0.4 carat?
2. How much more do diamonds that weigh 0.5 carat cost?

One way to answer these questions would be to get two samples of diamonds. The diamonds in one sample weigh 0.4 carat, and those in the other weigh 0.5 carat. The average price in the sample at 0.4 carat answers the first question. The difference in average prices between the two samples answers the second. Going further, we could build confidence intervals or test hypotheses about the means.

Suppose, however, the questions were to change slightly:

1′. What's the average price of diamonds that weigh 0.35 carat?
2′. How much more do diamonds that weigh 0.45 carat cost?

Answering these questions in the same manner requires two more samples of diamonds that weigh either 0.35 or 0.45 carat. The first two samples wouldn't help.

Regression analysis answers these questions differently. Regression analysis produces an equation that, in this context, relates weight to price. Rather than require diamonds of specific weights, regression analysis builds an equation from a sample of diamonds of various weights. We are not limited to diamonds of certain weights; we can use them all. The resulting equation allows us to predict prices at any weight, easily answering questions such as (1) or (1′). Properties of the regression equation also allow measurement of rates of change, such as the increase in average price associated with increased weight needed to answer (2) and (2′).

Equation of a Line

In regression analysis, the variable that we are trying to describe or predict defines the *y*-axis and is called the **response**. The associated variable goes on the *x*-axis and is called the **explanatory variable** or **predictor**. Explanatory variables have many names: factors, covariates, or even independent variables. (The last name is traditional but introduces the word *independent* in a way that has nothing to do with probability and so we avoid this use.) In regression analysis, the symbol *y* denotes the response and the symbol *x* denotes the explanatory variable. To answer the questions about diamonds, we want to understand how the price depends on weight. Hence, price is the response and weight is the explanatory variable.

Regression analysis can produce various types of equations. The most common equation defines a line. Before we fit a line, however, we need to inspect the scatterplot of the data to see that the association between the variables is linear. If the association between *x* and *y* is linear, we can summarize the relationship with a line. As an example, the scatterplot in Figure 19.1 graphs the price in dollars versus the weight in carats for a sample of 320 emerald-cut diamonds.

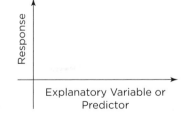

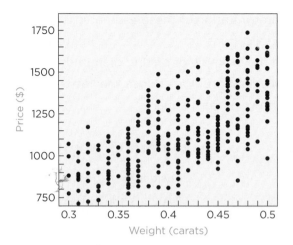

FIGURE 19.1 Prices versus weights for emerald-cut diamonds.

In the terminology of Chapter 6, the scatterplot shows positive, linear association. The correlation between weight and price is $r = 0.66$. The association is evident, but the data show considerable variation around the positive trend. The data do not lie on a single line; diamonds of a given weight do not cost the same. Other characteristics aside from weight influence price, such as the clarity or color of the stone.

We identify the line fit to data by an intercept b_0 and a slope b_1. The equation of the resulting line would usually be written $y = b_0 + b_1x$ in algebra. Because we associate the symbol y with the observed response in the data (and the data do not all lie on a line), we'll write the equation as

$$\hat{y} = b_0 + b_1x$$

fitted value Estimate of response based on fitting a line to data.

The "hat" or caret over y identifies $\hat{y}$ as a **fitted value**, an estimate of y based on an estimated equation. It's there as a reminder that the data vary around the line. The line omits details in order to capture the overall trend. Using the names of the variables rather than x and y, we write the equation of this line as

$$\textit{Estimated } \text{Price} = b_0 + b_1 \text{ Weight}$$

Least Squares

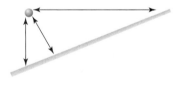

It remains to choose b_0 and b_1. Unless every point lies on a single line (in which case the correlation $r = 1$ or $r = -1$), we have to decide on a criterion for picking the best line. Clearly, we want the fitted line to be close to the data, but there are several ways to measure the distance from a point to a line. We could use the horizontal distance, the perpendicular (shortest) distance, or the vertical distance, as sketched here.

We choose to measure the distance of each data point to the line vertically because we use the fitted line to predict the value of y from x. The vertical distance is the error of this prediction. The vertical deviations from the data points to the line are called **residuals**. The best-fitting line collectively makes the squares of the residuals as small as possible. Each observation defines a residual as follows:

residual Vertical distance of a point from a line,

$$e = y - (b_0 + b_1x)$$

or

$$e = y - \hat{y}$$

$$e = y - \hat{y} = y - b_0 - b_1x$$

Because residuals are vertical deviations, the units of residuals match those of the response. In this example, the residuals are measured in dollars. The arrows in Figure 19.2 illustrate two residuals.

The sign of a residual tells you whether the point is above or below the fitted line. Points above the line produce positive residuals (y_1 in Figure 19.2), and points below the line produce negative residuals (y_2).

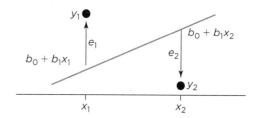

FIGURE 19.2 A residual is the vertical deviation (positive or negative) from the line.

least squares regression Regression analysis that picks the line that minimizes the sum of the squared residuals.

To keep negative and positive residuals from canceling when we add up the residuals, we square them first. We then choose b_0 and b_1 to minimize the sum of the *squared* residuals. This line is called the **least squares regression** line. Even though we rely on software for the calculations, the following expressions reveal that these estimates are related to familiar statistics:

$$b_1 = r\frac{s_y}{s_x} \quad \text{and} \quad b_0 = \bar{y} - b_1\bar{x}$$

The first formula shows that the slope b_1 is the product of the correlation r between x and y and the ratio of the standard deviations. The least squares line *is* the line associated with the correlation introduced in Chapter 6. Hence, if $r = 0$, then $b_1 = 0$, too. The formula for b_0 shows that the fitted line always goes through the point $(\bar{x}, \bar{y})$. (Other formulas for b_0 and b_1 are discussed in Behind the Math: The Least Squares Line.)

(p. 480)

19.2 | INTERPRETING THE FITTED LINE

The slope of the least squares regression line of price on weight for the data in Figure 19.1 is $b_1 = 2{,}670$ and the intercept is $b_0 = 43$. Hence, the equation of the fitted line is

$$\textit{Estimated}\ \text{Price} = 43 + 2{,}670\ \text{Weight}$$

The first step in interpreting a fitted line is to look at a plot that shows the line with the data. Figure 19.3 adds the least squares line to the scatterplot of price on weight.

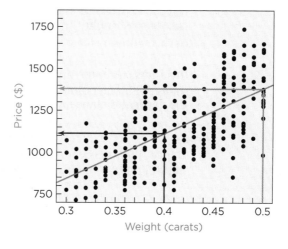

FIGURE 19.3 Estimating prices using the least squares line.

The equation of the fitted line makes it easy to answer questions like those posed at the start of this chapter. The fitted line estimates the average price of a diamond of *any* weight. For example, if the weight is $x = 0.4$ carat, then the estimated average price is (follow the blue arrows in Figure 19.3)

$$\hat{y} = 43 + 2{,}670 \times 0.4 = \$1{,}111$$

The equation of the line also describes how average prices change with weight. To estimate the average price of a $\frac{1}{2}$-carat diamond, set $x = 0.5$ carat. The estimated price is $267 higher than the price of a 0.4-carat diamond (follow the green arrows in Figure 19.3):

$$\hat{y} = 43 + 2{,}670 \times 0.5 = \$1{,}378$$

The slope of the line determines the difference in estimated prices. The difference between these two fitted values is $2{,}670 \times 0.1 = \$267$.

We can also see the residuals in this plot. The price of one of the diamonds that weighs 0.4 carat is $1,009. This diamond is priced *less* than the fitted value. Hence, the residual for this diamond is negative:

$$e = y - \hat{y} = \$1{,}009 - \$1{,}111 = -\$102$$

Another diamond that weighs 0.4 carat is priced at $1,251. It costs more than the line predicts, so the residual for this diamond is positive,

$$e = y - \hat{y} = \$1{,}251 - \$1{,}111 = \$140$$

Interpreting the Intercept

Interpretation of the intercept and slope of the least squares regression line is an essential part of regression modeling. These estimates are more easily interpreted if you are familiar with the data and pay attention to the measurement units of the explanatory variable and the response. **Attaching measurement units to b_0 and b_1 is a key initial step in interpreting these estimates.**

The intercept has measurement units of y. Because the response in this example measures price in dollars, the estimated intercept is not just $b_0 = 43$, it's $b_0 = \$43$. There are two ways to interpret the intercept b_0:

1. It is the portion of y that is present for all values of x.
2. It estimates the average of the response when $x = 0$.

The first interprets b_0 as telling us how much of the response is present, regardless of x. The estimate b_0 is part of the fitted value $\hat{y} = b_0 + b_1 x$ for every choice of x. In this example, a jeweler has costs regardless of the size of the gem: storage, labor, and other costs of running the business. The intercept represents the portion of the price that is present regardless of the weight: *fixed costs*. In this example b_0 estimates that fixed costs make up $43 of the selling price of every diamond.

The second interpretation of the intercept often reveals that the data have little to say about this constant component of the response. Graphically, the intercept is the point where the least squares line crosses the y-axis. If we plug $x = 0$ into the equation for the line, we are left with b_0:

$$\hat{y} = b_0 + b_1 0 = b_0$$

Hence, the intercept is the estimated average value of y when $x = 0$. If we set the weight of a diamond to 0, then

$$\textit{Estimated Price} = 43 + 2{,}670 \times 0 = \$43$$

Interpreted literally, this equation estimates the cost of a weightless diamond to be $43. The peculiarity of this estimate suggests a problem. To see the problem, we need to extend the scatterplot of the data to show b_0. The x-axis in Figure 19.3 ranges from 0.3 carat up to 0.5 carat, matching the range of observed weights. To show the intercept, we need to extend the x-axis to include $x = 0$. The odd appearance of the next scatterplot (Figure 19.4) shows why software generally does not do this: too much white space.

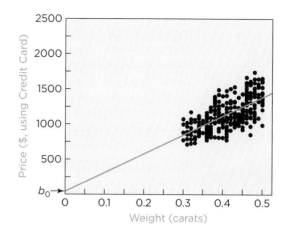

FIGURE 19.4 The intercept is the height of the point where the line and *y*-axis meet.

extrapolation An estimate outside the range of experience provided in the data.

caution The intercept in this example lies far from the data. That's often the case. Unless the range of the explanatory variable includes zero, b_0 lies outside the range of the observations and is an extrapolation. An **extrapolation** is an estimate based on extending an equation beyond conditions observed in the data. Equations become less reliable when extrapolated beyond the data. Saying anything about "weightless diamonds" lies outside what these data tell us, suggesting that our estimate of fixed costs is not well determined. (We quantify this uncertainty in Chapter 21 using confidence intervals.) Unless the range of the explanatory variable includes 0, be careful interpreting b_0.

Interpreting the Slope

The interpretation of the slope typically offers more insights than interpretation of the intercept because the slope describes how differences in the explanatory variable associate with differences in the response. Let's start with the units of b_1. The units of the slope are those of *y* divided by those of *x*. In this example, the slope converts carats to dollars: b_1 = 2,670 dollars per carat. **Once you attach units to b_1, its meaning should be clear.** The slope in this example estimates the marginal cost which is used to find the variable cost. The slope does not mean that a 1-carat diamond costs \$2,670, even on average. The slope is not a prediction of the response. Instead, the slope concerns comparison.

tip

The slope b_1 = \$2,670/carat estimates the difference in average prices of diamonds that differ in weight by 1 carat. Based on the fitted line, the estimated average price of, say, 2-carat diamonds is \$2,670 more than the estimated average price of 1-carat diamonds. That's another extrapolation. All of these diamonds weigh less than half of a carat; none differ in weight by a carat. It is more sensible to interpret the slope in the context of the data. For example, the difference in estimated average price between diamonds that differ in weight by 1/10th of a carat is 1/10th of the slope:

$$b_1(1/10) = \$2,670/10 = \$267$$

Similarly, the difference in estimated price is \$26.7 per point (a point is 1/100th of a carat). Notice that only the *difference* in weight matters because fixed costs affect both prices by the same amount.

caution It is tempting, but incorrect, to describe the slope as "the change in *y caused* by changing *x*." For instance, we might say "The estimated average price of diamonds increases \$26.70 for each increase in weight of one point." That statement puts into words the previous calculation using the fitted line.

Remember, however, that this statement describes the fitted line rather than diamonds. We cannot literally change the weight of a diamond to see how its price changes. Instead, the line summarizes prices of diamonds of varying weights. These diamonds can differ in other ways as well. Suppose it were the case that heavier diamonds had fewer flaws. Such a lurking variable would mean that some of the price increase that we had attributed to weight is due to better quality. Confounding, confusing the effects of explanatory variables, happens in regression analysis as well as two-sample comparisons (Chapter 18).

4M EXAMPLE 19.1 | ESTIMATING CONSUMPTION

MOTIVATION | STATE THE QUESTION

Utility companies in many older communities still rely on "meter readers" who visit homes to read meters that measure consumption of electricity and gas. Unless someone is home to let the meter reader inside, the utility company has to estimate the amount of energy used.

The utility company in this example sells natural gas to homes in the Philadelphia area. Many of these are older homes that have the gas meter in the basement. We can estimate the use of gas in these homes with a regression equation.

METHOD | DESCRIBE THE DATA AND SELECT AN APPROACH

Identify x and y.

Link b_0 and b_1 to problem.

Describe data.

Check for linear association.

The explanatory variable is the average number of degrees below 65° during the billing period, and the response is the number of hundred cubic feet of natural gas (CCF) consumed during the billing period (about a month). The explanatory variable is set to 0 if the average temperature is above 65° (assuming a homeowner won't need heating in this case).

The intercept estimates the amount of gas consumed for activity unrelated to temperature (such as cooking). The slope estimates the average amount of gas used per 1° decrease in temperature. For this experiment, the local utility has 4 years of data ($n = 48$ months) for an owner-occupied, detached home. Based on the scatterplot, the association appears linear.

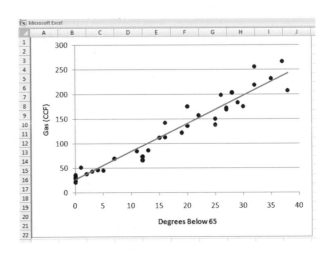

MECHANICS | DO THE ANALYSIS

The fitted least squares line in the scatterplot tracks the pattern in the data very closely (the correlation $r = 0.98$). The equation of the estimated least squares line is

$$\textit{Estimated } \text{Gas (CCF)} = 26.7 + 5.7 \text{ Degrees Below 65}$$

The intercept $b_0 = 26.7$ CCF estimates the amount of gas used for things unrelated to outside temperature. The slope b_1 implies that the estimated average use of gas for heating increases by about 5.7 CCF per 1° drop in temperature. There's relatively little variation around the fitted line.

MESSAGE | SUMMARIZE THE RESULTS

The utility can accurately predict the amount of natural gas used for this home—and perhaps similar homes in this area—without reading the meter by using the temperature during a billing period. During the summer, the home uses about 26.7 hundred cubic feet of gas in a billing period. As the weather gets colder, the estimated average amount of gas rises by 5.7 hundred cubic feet for each additional degree below 65°. For instance, during a billing period with temperature 55° we expect this home to use 26.7 + 5.7 × 10 = 83.7 CCF of gas.

What Do You Think?

A manufacturing plant receives orders for customized mechanical parts. The orders vary in size, from about 30 to 130 units. After configuring the production line, a supervisor oversees the production. The scatterplot in Figure 19.5 plots the production time (in hours) versus the number of units for 45 orders.

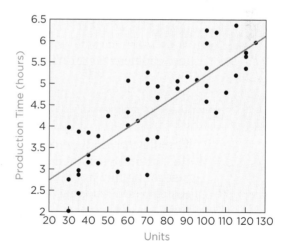

FIGURE 19.5 Scatterplot of production time on order size.

The least squares regression line in the scatterplot (Figure 19.5) is

$$\textit{Estimated } \text{Production Time (Hours)} = 2.1 + 0.031 \text{ Number of Units}$$

a. Interpret the intercept of the estimated line. Is the intercept visible in Figure 19.5?[1]

---

[1] The intercept (2.1 hours) is best interpreted as the estimated time for any orders, regardless of size (e.g., to set up production). It is not visible; it lies farther to the left outside the range of the data and is thus an extrapolation.

b. Interpret the slope of the estimated line.[2]
c. Using the fitted line, estimate the amount of time needed for an order with 100 units. Is this estimate an extrapolation?[3]
d. Based on the fitted line, how much more time does an order with 100 units require over an order with 50 units?[4]

19.3 | PROPERTIES OF RESIDUALS

The least squares line summarizes the association between y and x. The residuals, the deviations from the fitted line, show variation that remains after we account for this relationship. If a regression equation works well, it should capture the underlying pattern. **Only simple random variation that can be summarized in a histogram should remain in the residuals.**

To see what is left after fitting the line, it is essential to plot the residuals. A separate plot of the residuals zooms in on these deviations, making it easier to spot problems. The explanatory variable remains on the x-axis, but the residuals replace the response on the y-axis.

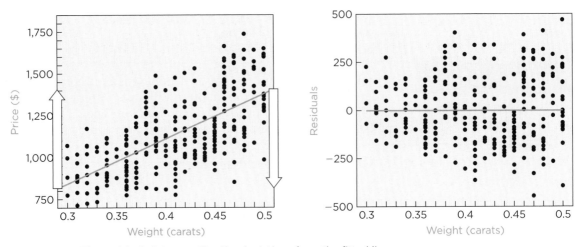

FIGURE 19.6 The residual plot magnifies the deviations from the fitted line.

Visually, the residual plot flattens the line in the initial scatterplot, pulling up the left and pushing down the right until the line becomes horizontal. The horizontal line in the residual plot corresponds to the regression line in the scatterplot. Flattening the line changes the scale of the plot to focus on deviations from the line.

Is there a pattern in the residual plot shown in Figure 19.6? If the least squares line captures the association between x and y, then a scatterplot of residuals versus x should have no pattern. It should stretch out horizontally, with consistent vertical scatter throughout. It should ideally show neither bends nor outliers.

Check for an absence of pattern in the residuals using the visual test for association introduced in Chapter 6. One of the scatterplots in Figure 19.7 is the residual plot from Figure 19.6. The other three scramble the data so that residu-

---

[2] Once it is running, the estimated time for an order is 0.031 hour per unit (or $60 \times 0.031 = 1.9$ minutes).
[3] The estimated time for an order of 100 units is $2.1 + 0.031 \times 100 = 5.2$ hours. This is not an extrapolation because we have orders for less and more than 100 units.
[4] Fifty more units would need about $0.031 \times 50 = 1.55$ additional hours.

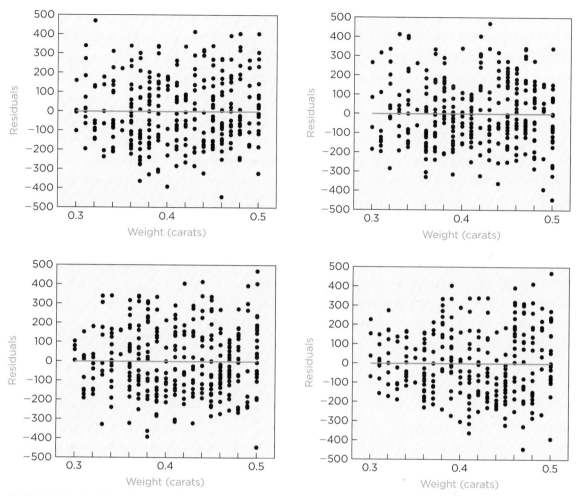

FIGURE 19.7 Applying the visual test for association to residuals.

als are randomly paired with the weights. If all of these plots look the same, then there's no apparent pattern in the residual plot. Which is the original?

The plots in Figure 19.7 appear similar, but we recognize the original as the bottom right. There is a pattern, but the pattern is subtle: The residuals become more variable as the diamonds become larger. Smaller diamonds have more consistent prices than larger diamonds. This pattern is well hidden in the initial scatterplot. Just as a microscope reveals unseen organisms, a residual plot can reveal subtleties invisible to the naked eye. (We'll deal with such problems in Chapter 23.)

Standard Deviation of the Residuals

A regression equation should capture the pattern and leave behind only simple random variation. If the residuals are "simple enough" to be treated as a sample from a population, we can summarize them in a histogram.

The histogram of the diamond residuals appears reasonably symmetric around 0 and bell shaped, with a hint of skewness (Figure 19.8). If the residuals are nearly normal, we can summarize the residual variation with a mean and standard deviation. Because we fit this line using least squares, the mean of the residuals *must be zero*. The standard deviation of the residuals measures how much the residuals vary around the fitted line and goes by many names, such as the **standard error of the regression** or the **root mean squared error (RMSE)**. The formula used to compute the standard deviation

standard error of the regression or **root mean squared error (RMSE)** Alternative names given to the SD of the residuals in a regression.

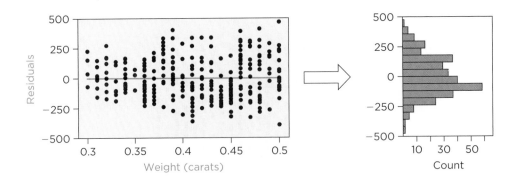

FIGURE 19.8 Summarizing residuals in a histogram.

of the residuals is *almost* the same as the formula used to calculate other standard deviations:

$$s_e = \sqrt{\frac{e_1^2 + e_2^2 + \cdots + e_n^2}{n - 2}}$$

Integer subscripts identify residuals associated with different observations; e_1 denotes the residual for the first diamond, e_2 denotes the residual for the second, and so forth. The mean residual $\bar{e}$ does not appear in this expression because $\bar{e} = 0$. The denominator is $n - 2$ rather than $n - 1$ because the regression line requires two estimates, b_0 and b_1, to calculate each residual that contributes to s_e. (See Chapter 15, Behind the Math: Student's t and Degrees of Freedom.)

If all of the data were exactly on a line, then s_e would be zero. Least squares makes s_e as small as possible, but it seldom can reduce it to zero. For the diamonds, the standard deviation of the residuals is $s_e = \$169$. The units of s_e match those of the response (dollars). Because these residuals have a bell-shaped distribution around the fitted line, the Empirical Rule implies that the prices of about two-thirds of the diamonds are within $169 of the regression line and about 95% of the prices are within $338 of the regression. If a jeweler quotes a price for a 0.5-carat diamond that is $400 more than this line predicts, we ought to be surprised.

19.4 | EXPLAINING VARIATION

A regression line splits the response into two parts, a fitted value and a residual,

$$y = \hat{y} + e$$

The fitted value $\hat{y}$ represents the portion of y that is associated with x, and the residual e represents variation due to other factors. As a summary of the fitted line, it is common to say how much of the variation of y belongs with each of these components.

The histograms in Figure 19.9 suggest how we will answer this question. The histogram shown along the vertical axis in the scatterplot on the left-hand side of Figure 19.9 summarizes the prices, and the histogram on the right summarizes the residuals. The axes share a common scale. The residuals clearly have less variation.

The correlation between x and y determines the reduction in variation. The sample correlation r is confined to the range $-1 \leq r \leq 1$. If we *square* the correlation, we get a value between 0 and 1. The sign does not matter. The squared correlation r^2, or **r-squared**, determines the fraction of the variation accounted for by the least squares regression line. The expression $1 - r^2$ is the fraction of variation that is left in the residuals. If $r^2 = 0$, the regression line

r-squared (r^2) Square of the correlation between x and y, the percentage of "explained" variation.

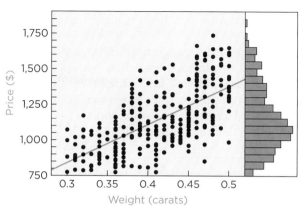 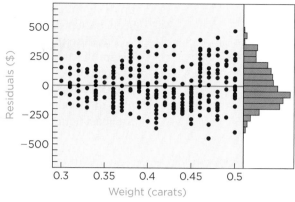

FIGURE 19.9 The prices among diamonds (left) vary more than the residuals (right).

describes none of the variation in the data. In this case, the slope $b_1 = r(s_y/s_x)$ is also zero and the fitted line is flat. If $r^2 = 1$, the line represents all of the variation and $s_e = 0$.

For the diamonds, $r^2 = 0.66^2 = 0.434$ and $1 - r^2 = 0.566$. Because $0 \le r^2 \le 1$, this summary is often described on a percentage scale. For example, we might say "The fitted line explains 43.4% of the variation in price."

Summarizing the Fit of a Line

The term r^2 is a popular summary of a regression because of its intuitive interpretation as a percentage. This lack of units makes it incomplete, however, as a description of the fitted line. For example, r^2 alone does not indicate the size of the typical residual. The standard deviation of the residuals s_e is useful along with r^2 because s_e conveys these units. Because $s_e = \$169$, we know that fitted prices frequently differ from actual prices by hundreds of dollars. For example, in Example 19.1 (household gas use), $r^2 = 0.955$ and $s_e = 16$ CCF. The size of r^2 tells us that the data stick close to the fitted line, but only in a relative sense. The term s_e tells us that, appealing to the Empirical Rule, about two-thirds of the data lie within about 16 CCF of the fitted line. **Along with the slope and intercept, always report both r^2 and s_e so that others can judge how well the equation describes the data.**

There's no hard-and-fast rule for how large r^2 must be. The typical size of r^2 varies across applications. In macroeconomics, regression lines frequently have r^2 larger than 90%. With medical studies and marketing surveys, on the other hand, an r^2 of 30% or less may still indicate an important discovery.

19.5 | CONDITIONS FOR SIMPLE REGRESSION

We must check three conditions before we summarize the association between two variables with a line:

✓ linear
✓ random residual variation
✓ no obvious lurking variable

Two of these conditions are related to scatterplots. Because these conditions are easily verified from plots, regression analysis with one explanatory variable is often called **simple regression**. Once we see the plots, we immediately know whether a line is a good summary of the association. (Regression

simple regression Regression analysis with one explanatory variable.

analysis that uses several explanatory variables is called multiple regression; see Chapter 23.)

linear condition Data in scatterplot have a linear pattern.

The **linear condition** is met if the pattern or association in the scatterplot resembles a line. If the pattern in the scatterplot of y on x does not appear to be straight, stop. If the relationship appears to bend, for example, an alternative equation is needed (Chapter 20). Do not rely on r^2 as a measure of the degree of association without looking at the scatterplot. For example, $r^2 = 0$ for the least squares regression line fit to the data in Figure 19.10.

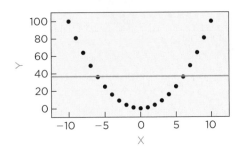

FIGURE 19.10 The r^2 for this regression is zero even though the data have a pattern.

The equation $y = x^2$ perfectly predicts y from x in Figure 19.10, but the pattern is not linear—it bends. As a summary statistic, r^2 measures only the degree of linear association.

The second condition relies on another plot. Because residuals magnify deviations from the fitted line, it's important to examine the scatterplot of the residuals versus x as well as that of y versus x. The residuals must meet the **random residual variation condition**. The plot of the residuals on x should have no pattern, allowing us to summarize them with a histogram. Outliers, evident patterns, or isolated clusters indicate problems worth investigating. We may change our minds about the linear condition after inspecting the residuals. Notice that we have to fit the regression equation and obtain the residuals to check this condition.

random residual variation condition Residuals from a regression show no patterns in plots.

no obvious lurking variable condition No other explanatory variable offers a better explanation of the association between x and y.

The third condition requires thinking rather than plotting. The **no obvious lurking variable condition** is met if we cannot think of another variable that explains the pattern in the scatterplot of y on x. We mentioned lurking variables when interpreting the fitted line in the scatterplot of the prices of the diamonds versus their weights (Figure 19.3). If larger diamonds systematically differ from smaller diamonds in factors other than weight, then these other differences might offer a better explanation of the increase in price. The presence of a lurking variable produces confounding in two-sample tests of the difference in means (Chapter 18). The following example illustrates a relationship that is affected by a lurking variable.

4M EXAMPLE 19.2 | **LEASE COSTS**

MOTIVATION ▶ STATE THE QUESTION

When auto dealers lease cars, they include the cost of depreciation. They want to be sure that the price of the lease plus the resale value of the returned car yields a profit. How can a dealer anticipate the effect of age on the value of a used car?

A manufacturer who leases thousands of cars can group those of similar ages and see how the average price drops with age. Small dealers that need to account for local conditions won't have enough data. They need regression analysis. Let's help a BMW

dealer in the Philadelphia area determine the cost due to depreciation. The dealer currently estimates that $4,000 is enough to cover the depreciation per year.

Identify x and y.

Link b_0 and b_1 to problem.

Describe data.

Check linear condition and lurking variable condition.

DESCRIBE THE DATA AND SELECT AN APPROACH

We will fit a least squares regression with a linear equation to see how the age of a car is related to its resale value. Age in years is the explanatory variable and resale value in dollars is the response. The slope b_1 estimates how the estimated resale price changes per year; we expect a negative slope since resale value falls as the car ages. The intercept b_0 estimates the value of a just-sold car, one with age 0.

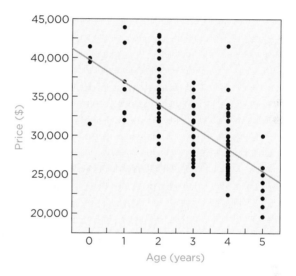

We obtained prices and ages of 218 used BMWs in the 3-series from Web sites advertising certified used BMWs in 2006 in the Philadelphia region. (These data also appear in an example in Chapter 1.)

✓ **Linear.** Seems okay. The price for one car at age 0 seems unusually low, but we don't have many cars in that model year.

✗ **No obvious lurking variable.** This analysis ignores the mileage of the car. Older cars are likely to have been driven farther than newer models. This means that the fitted regression line of price on age mixes the effect of age on price with the effect of mileage on price.

DO THE ANALYSIS

The scatterplot shows linear association. There are no extreme outliers or isolated clusters. The fitted equation is

$$Estimated \ \text{Price} = 39,851.72 - 2,905.53 \ \text{Age}$$

Interpret the slope and intercept within the context of the problem, assigning units to both. Don't proceed without being comfortable with the interpretations. The slope is the annual decrease in the estimated resale value, $2,905.53 per year. The intercept estimates the price of used cars from the current model year to be $39,851.72. The average selling price of new cars like these is around $43,000, so the intercept suggests that a car depreciates about $43,000 − $39,852 = $3,148 as it is driven off the lot.

Use r^2 and s_e to summarize the fit of the equation. The regression shows that $r^2 = 45\%$ of the variation in prices is associated with variation in age, leaving 55% to other factors (such as different options and miles driven). The residual standard deviation $s_e = \$3,367$ is substantial, but we have not taken into account other factors that affect resale value (such as mileage).

Check the residuals. Residual plots are an essential part of any regression analysis.

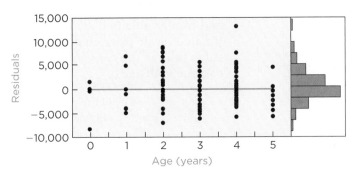

✓ **Random residual variation.** The residuals cluster into groups for cars of each model year. There should be roughly equal scatter at each age. These seem okay. It is typical to see more compact clusters for groups with fewer cases; the range can only get wider as the number of points increases.

MESSAGE | SUMMARIZE THE RESULTS

Our results estimate that used BMW cars (in the 3-series) decline in resale value by $2,900 per year. This estimate combines the effect of a car getting older with the effects of other variables, such as mileage and damages that accumulate as a car ages. Thus, the current lease pricing that charges $4,000 per year appears profitable. We should confirm that fees at the time the lease is signed are adequate to cover the estimated $3,150 depreciation that occurs when the lessee drives the car off the lot.

State the limitations of our regression line. The fitted line leaves more than half of the variation (55%) in prices unexplained. Consequently our estimate of resale value could be off by thousands of dollars ($s_e = \$3,367$). Also, our estimates of the depreciation should only be used for short-term leases. So long as leases last 5 years or less, the estimates should be fine. Longer leases would require extrapolation outside the data used to build our model.

Best Practices

- *Always look at the scatterplot.* Regression is reliable if the scatterplot of the response versus the explanatory variable shows a linear pattern. It also helps to add the fitted line to this plot. Plot the residuals versus *x* to magnify deviations from the regression line.
- *Know the substantive context of the model.* Otherwise, there's no way to decide whether the slope and intercept make sense. If the slope and intercept cannot be interpreted, the model may have serious flaws. Perhaps there's an important lurking factor or extreme outlier. A plot will show an outlier, but we have to know the context to identify a lurking variable.

- *Describe the intercept and slope using units of the data.* Both the intercept b_0 and slope b_1 have units, and these scales aid the interpretation of the estimates. Use these to relate the estimates to the context of the problem, such as estimating fixed and variable costs.
- *Limit predictions to the range of observed conditions.* When extrapolating outside the data, we assume the equation keeps going and going. Without data, we won't know if those predictions make sense. A linear equation often does a reasonable job over the range of observed *x*-values but fails to describe the relationship beyond this range.

Pitfalls

- *Do not assume that changing x causes changes in y.* A linear equation is closely related to a correlation, and correlation is not causation. For example, consider a regression of monthly sales of a company on monthly advertising. A large r^2 indicates substantial association, but it would be a mistake to conclude that changes in advertising cause changes in sales. Perhaps every time this company increased its advertising, it also lowered prices. Was it price, advertising, or something else that determined sales? Another way to see that a large r^2 does not prove that changes in x cause changes in y is to recognize that we could just as easily fit the regression with the roles of the variables reversed. The regression of x on y has the same r^2 as the regression of y on x because the $\text{corr}(x, y) = \text{corr}(y, x)$.

- *Do not forget lurking variables.* With a little imagination, we can think of alternative explanations for why heavier diamonds cost more than smaller diamonds. Perhaps heavier diamonds also have more desirable colors or better, more precise cuts. That's what it means to have a lurking variable: Perhaps it's the lurking variable that produces the higher costs, not our choice of the explanatory variable.

- *Don't trust summaries like r^2 without looking at plots.* Although r^2 measures the *strength* of a linear equation's ability to describe the response, a

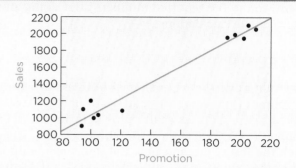

high r^2 does not demonstrate the *appropriateness* of the linear equation. A single outlier or data that separate into two groups rather than a single cloud of points can produce a large r^2 when, in fact, the linear equation is inappropriate. Conversely, a low r^2 value may be due to a single outlier. It may be that most of the data fall roughly along a straight line with the exception of an outlier.

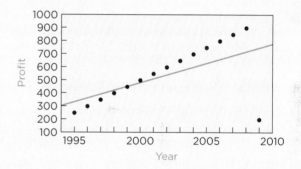

Software Hints

All software packages that fit a regression summarize the fitted equation with several tables. These tables are laid out differently from one package to another, but all contain essentially the same information. The output for simple regression includes a section that resembles the tables that follow. For this chapter, we need to be able to find the estimated coefficients b_0 and b_1 as well as s_e and r^2. These estimates are circled in the table. (This package labels s_e the "root mean square error," or RMSE.)

Summary of Fit: Response = Price	
RSquare	0.434303
Root Mean Square Error	(168.634)
Mean of Response	1146.653
Observations	320

The slope and intercept coefficients are given in a table with four or more columns. Each row of the table is usually labeled with the name of the x-variable, with the intercept labeled "Intercept" or "Constant." The rest of the columns give the estimates along with several properties of the estimates needed for inference. We will use these other columns in Chapter 21. The regression equation summarized in this table is

Estimated Price $= 43.419237 + 2669.8544$ Weight

Parameter Estimates				
Term	Estimate	Std Error	t Ratio	Prob > |t|
Intercept	(43.419237)	71.23376	0.61	0.5426
Weight (carats)	(2669.8544)	170.8713	15.62	<0.0001

Statistics packages like this one often show more digits of the estimates b_0 and b_1 than needed. Ordinarily, round the reported numbers after doing any intermediate calculations.

EXCEL

The right way to do regression with Excel starts with a picture of the data: the scatterplot of Y on X. We talked about how to scatterplot data in Excel in Chapter 6. First select the two columns of data. Then click the insert tab and pick the items to obtain a chart that shows x-y scatterplots. Next, select the scatterplot and follow the menu commands

Chart > Add Trendline...

Pick the option for adding a line, and Excel will add the least squares line to the plot. Double-click the line in the chart and Excel will optionally show the equation and r^2 for the model. These options do not include s_e. Formulas in Excel can find the least squares regression. The formula LINEST does most of the work.

DDXL provides a more complete summary. Select the region of the spreadsheet with the data (including the variable names in the first row). Then follow the menu sequence

DDXL > Regression > ***Click me ***

> Simple Regression

to get to the regression dialog. Fill in the names of the response and explanatory variable and click the OK button. DDXL shows a graph with the fitted line next to a table with the estimates. To plot the residuals, click the button labeled "Check the Residuals."

MINITAB

The procedure resembles constructing a scatterplot of the data. Follow the menu sequence

Graph > Scatterplot...

and choose the icon that adds a line to the plot. After picking variables for the response and explanatory variable, click OK. Minitab then draws the scatterplot with the least squares line added. The command

Stat > Regression > Regression...

opens a dialog in which we can pick the response and explanatory variable. Then click OK. Minitab then prints a summary of the regression in the scrolling window of results.

JMP

Following the menu sequence

Analyze > Fit Y by X

opens the dialog that we used to construct a scatterplot in Chapter 6. Fill in variables for the response and explanatory variable; then click OK. In the window that shows the scatterplot, click on the red triangle above the scatterplot (near the words "Bivariate Fit of ..."). In the pop-up menu, choose the item Fit Line. JMP adds the least squares line to the plot and appends a tabular summary of the model to the output window below the scatterplot.

 BEHIND the Math

The Least Squares Line

The least squares line minimizes the sum of squared residuals,

$$S(b_0, b_1) = \sum_{i=1}^{n} (y_i - b_0 - b_1 x_i)^2$$

Two equations, known as the normal equations, lead to formulas for the optimal values for b_0 and b_1 that determine the least squares line:

$$\sum_{i=1}^{n} (y_i - b_0 - b_1 x_i) = 0 \quad \sum_{i=1}^{n} (y_i - b_0 - b_1 x_i) x_i = 0$$

(These equations have nothing to do with a normal distribution. "Normal" in this sense means perpendicular.) After a bit of algebra, the normal equations give these formulas for the slope and intercept:

$$b_1 = \frac{\sum_{i=1}^{n} (y_i - \bar{y})(x_i - \bar{x})}{\sum_{i=1}^{n} (x_i - \bar{x})^2} = \frac{\text{cov}(x, y)}{\text{var}(x)} = r \frac{s_y}{s_x} \quad \text{and}$$

$$b_0 = \bar{y} - b_1 \bar{x}$$

The least squares line matches the line defined by the sample correlation r in Chapter 6. If you remember that the units of the slope are those of y divided by those of x then it's easy to recall $b_1 = r\, s_y/s_x$.

The normal equations say two things about the residuals from a least squares regression. With $\hat{y} = b_0 + b_1 x$, the normal equations are

$$\sum_{i=1}^{n}(y_i - \hat{y}_i) = 0, \qquad \sum_{i=1}^{n}(y_i - \hat{y}_i)x_i = 0$$

The normal equations tell us that

1. *The mean of the residuals is zero.* The deviation $y - \hat{y}$ is the residual. Because the sum of the residuals is zero, the average residual is zero as well.
2. *The residuals are uncorrelated with the explanatory variable.* The second equation is the covariance between x and the residuals. Because the covariance is zero, so is the correlation.

CHAPTER SUMMARY

The **least-squares regression** line summarizes how the average value of the **response** (y) depends upon the value of an **explanatory variable** or **predictor** (x). The **fitted value** of the response is $\hat{y} = b_0 + b_1 x$. The value b_0 is the intercept and b_1 is the slope. The intercept has the same units as the response, and the slope has the units of the response divided by the units of the explanatory variable. The vertical deviations $e = y - \hat{y}$ from the line are **residuals**. The least squares criterion provides formulas for b_0 and b_1 that minimize the sum of the squared residuals. The r^2 statistic tells the percentage of variation in the response that is described by the equation, and the residual standard deviation s_e gives the scale of the unexplained variation. The linear equation should satisfy the **linear**, **no obvious lurking variables**, and **simple random variation** conditions.

■ Key Terms

condition
  linear, 476
  no obvious lurking
    variable, 476
random residual
  variation, 476
explanatory variable, 465

extrapolation, 469
fitted value, 466
least squares
  regression, 467
predictor, 465
residual, 466
response, 465

root mean squared error
  (RMSE), 473
r-squared (r^2), 474
simple regression, 475
standard error of the
  regression, 473

■ Formulas

Linear equation

$$\hat{y} = b_0 + b_1 x$$

b_0 is the intercept and b_1 is the slope. This equation for a line is sometimes called "slope-intercept form."

Slope
If $r = \text{corr}(y, x)$ is the correlation between the response and the explanatory variable, s_x is the standard deviation of x, and s_y is the standard deviation of y, then

$$b_1 = r\frac{s_y}{s_x}$$

Intercept
The intercept is most easily computed by using the means of the response and explanatory variable and the estimated slope,

$$b_0 = \bar{y} - b_1\bar{x}$$

Fitted value

$$\hat{y} = b_0 + b_1 x$$

Residual

$$e = y - \hat{y} = y - b_0 - b_1 x$$

Standard deviation of the residuals

$$s_e = \sqrt{s_e^2}, \quad s_e^2 = \frac{\sum_{i=1}^{n}(y_i - b_0 - b_1 x_i)^2}{n-2} = \frac{\sum_{i=1}^{n}e_i^2}{n-2}$$

r-squared
The fraction of the variation in y that has been captured or explained by the fitted equation,

$$r^2 = \text{corr}(y, x)^2$$

▨ About the Data

We obtained the data for emerald-cut diamonds from the Web site of McGivern Diamonds in the fall of 2004. The data on prices of used BMW cars were similarly extracted from listings of used cars available within 100 miles of Philadelphia during the fall of 2006.

EXERCISES

Mix and Match

Match each description on the left with its mathematical expression on the right.

1. Symbol for the explanatory variable in a regression	(a) r^2
2. Symbol for the response in a regression	(b) b_0
3. Fitted value from an estimated regression equation	(c) $\bar{y}$
4. Residual from an estimated regression equation	(d) b_1
5. Identifies the intercept in a fitted line	(e) x
6. Identifies the slope in a fitted line	(f) $\hat{y}$
7. Percentage variation described by a fitted line	(g) $b_0 + b_1$
8. Symbol for the standard deviation of the residuals	(h) y
9. Prediction from a fitted line if $x = \bar{x}$	(i) $y - \hat{y}$
10. Prediction from a fitted line if $x = 1$	(j) s_e

True/False

Mark each statment True or False. If you believe that a statement is false, briefly explain why you think it is false.

11. In a scatterplot, the response is shown on the horizontal axis with the explanatory variable on the vertical axis.

12. Regression analysis requires several values of the response for each value of the predictor so that we can calculate averages for each x.

13. If all of the data lie along a single line with nonzero slope, then the r^2 of the regression is 1. (Assume the values of the explanatory variable are not identical.)

14. If the correlation between the explanatory variable and the response is zero, then the slope will also be zero.

15. The use of a linear equation to describe the association between price and sales implies that we expect equal differences in sales when comparing periods with prices $10 and $11 and periods with prices $20 and $21.

16. The linear equation (estimated from a sequence of daily observations)

   Estimated Shipments = b_0 + 0.9 Orders Processed

   implies that we expect twice as many shipments when the number of orders processed doubles only if $b_0 = 0$.

17. The intercept estimates how much the response changes on average with changes in the predictor.

18. The estimated value $\hat{y} = b_0 + b_1 x$ approximates the average value of the response when the explanatory variable equals x.

19. The horizontal distance between y and $\hat{y}$ is known as the residual and so takes its scale from the predictor.

20. The sum of the fitted value $\hat{y}$ plus the residual e is equal to the original data value y.

21. The plot of the residuals on the predictor should show a linear pattern, with the data packed along a diagonal line.

22. Regression predictions become less reliable as we extrapolate farther from the observed data.

Think About It

23. If the correlation between x and y is $r = 0.5$, do x and y share half of their variation in common?

24. The value of $r^2 = 1$ if data lie along a single line. Is it possible to fit a linear regression for which r^2 is exactly equal to zero?

25. In general, is the linear least squares regression equation of y on x the same as the equation for regressing x on y?

26. In what special case does the regression of y on x match the regression of x on y, so that if two fitted lines were drawn on the same plot, the lines would coincide?

27. From looking at this plot of the residuals from a linear equation, estimate the value of s_e.

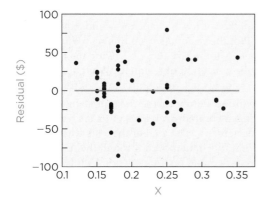

28. This histogram summarizes residuals from a fit that regresses the number of items produced by 50 employees during a shift on the number of years with the company. Estimate s_e from this plot.

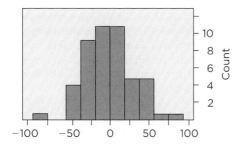

29. A package delivery service uses a regression equation to estimate the fuel costs of its trucks based on the number of miles driven. The equation is of the form *Estimated* Dollars = $b_0 + b_1$ Miles. If gasoline prices go up, how would you expect the fit of this equation to change?

30. A customized milling operation uses the equation $200 plus $150 per hour to give price estimates to customers. If it pays a fixed fee to ship these orders, how should it change this equation if the cost of shipping goes up?

31. If the standard deviation of x matches the standard deviation of y, then what is the relationship between the slope in a least squares regression of y on x and the correlation between x and y?

32. If the correlation between x and y is 0.8 and the slope in the regression of y on x is 1.5, then which of x or y has larger variation?

33. Shoppers at a local supermarket spend, on average, $85 during each shopping trip. Imagine the scatterplot that shows the total amount spent each day in the store on the y-axis versus the number of shoppers each day on the x-axis.
    (a) Would you expect a linear equation to describe these data?
    (b) What would you expect for the intercept of the linear model?

(c) What would you expect for the slope?
(d) Do you expect patterns in the variation around the equation?

34. Costs for building a new elementary school in the United States average about $100 per square foot. In a review of school construction projects in Arizona, the head of the Department of Education examined a scatterplot of the cost of recently completed schools (y) versus the size of the school (in square feet, x).
    (a) Would you expect a linear equation to describe these data?
    (b) What would you expect for the intercept of the linear model?
    (c) What would you expect for the slope?
    (d) Do you expect patterns in the variation around the equation?

35. A division of a multinational retail company prepared a presentation to give at the home office in Paris, France. The presentation includes a scatterplot that shows the relationship between square footage and annual sales in retail outlets owned by the chain. The units in the plot show the size in thousands of square feet and the response in thousands of dollars. A fitted line in the plot is $y = 47 + 650x$.
    (a) Interpret the slope and intercept in the fitted line.
    (b) To present the model in Europe, the plot must be shown with sales denominated in euros rather than dollars (use the exchange rate $1 = €0.82$) and size given in square meters rather than square feet (1 square foot = 0.093 square meter). Find the slope and intercept in these new units.
    (c) Would the r^2 summary attached to the regression model change along with the slope and intercept when the data are changed into euros and meters?
    (d) Would s_e change with the new scales?

36. An assembly plant tracks the daily productivity of the workers on the line. Each day, for every employee, the plant records the number of hours put in (Hours) and the number of completed packages assembled by the employee (Count). A scatterplot of the data for one day shows a linear trend. A fitted line with the equation *Estimated* Count = $-2 + 15$ summarizes this trend.
    (a) Interpret the slope and intercept in the fitted line.
    (b) A carton holds 12 packages. A working day at this plant has 8 hours. Describe the regression line if the data were converted to cartons produced (y) and days (or fraction of a day) worked.
    (c) Would the r^2 summary attached to the regression model change along with the slope and intercept when the data are converted to cartons and days?
    (d) What about the value of s_e? Would it change with the new scales?

You Do It

The name shown with each question identifies the data table to be used for the problem.

37. **Diamond Rings** This data table contains the listed prices and weights of the diamonds in 48 rings offered for sale in *The Singapore Times*. The prices are in Singapore dollars, with the weights in carats.

(a) Scatterplot the listed prices of the rings on the weights of the rings. Does the trend in the average price seem linear?

(b) Estimate the linear equation using least squares. Interpret the fitted intercept and slope. Be sure to include their units. Note if either estimate represents a large extrapolation and is consequently not reliable.

(c) Interpret the summary values r^2 and s_e associated with the fitted equation. Attach units to these summary statistics as appropriate.

(d) What is the estimated difference in price (on average) between diamond rings with diamonds that weigh 0.25 and 0.35 carat?

(e) The slope in this regression is a lot larger than the slope for the emerald diamonds discussed in this chapter. Or is it? Notice that one Singapore dollar is currently worth about $0.65 and convert the slope to an analysis in U.S. dollars.

(f) These are rings, not diamonds. How would you expect the cost of the setting to affect the linear equation between weight and price?

(g) A ring with a 0.18-carat diamond lists for $325 Singapore. Is this a bargain?

(h) Plot the residuals from this regression. If appropriate, summarize these by giving the mean and standard deviation of the collection of residuals. What does the standard deviation of the residuals tell you about the fit of this equation?

38. **Convenience Shopping** It's rare that you'll find a gas station these days that only sells gas. It's become more common to find a convenience store that also sells gas. These data describe the sales over time at a franchise outlet of a major U.S. oil company. Each row summarizes sales for one day. This particular station sells gas, and it also has a convenience store and a car wash. The column labeled *Sales* gives the dollar sales of the convenience store, and the column *Volume* gives the number of gallons of gas sold.

(a) Scatterplot Sales on Volume. Does there appear to be a linear pattern that relates these two sequences?

(b) Estimate the linear equation using least squares. Interpret the fitted intercept and slope. Be sure to include their units. Note if either estimate represents a large extrapolation and is consequently not reliable.

(c) Interpret the summary values r^2 and s_e associated with the fitted equation. Attach units to these summary statistics as appropriate.

(d) Estimate the difference in sales at the convenience store (on average) between a day with 3,500 gallons sold and a day with 4,000 gallons sold.

(e) This company also operates franchises in Canada. At those operations, gas sales are tracked in liters and sales in Canadian dollars. What would your equation look like if measured in these other units? (Note: 1 gallon = 3.7854 liters, and use the exchange rate $1 = $1.1 Canadian.) Include r^2 and s_e as well as the slope and intercept.

(f) The form of the equation suggests that selling more gas produces increases in sales at the associated

store. Does this mean that customers come to the station to buy gas and then happen to buy something at the convenience store, or might the causation work in the other direction?

(g) On one day, the station sold 4,165 gallons of gas and had sales of $1,744 at the attached convenience store. Find the residual for this case. Are these sales higher or lower than you would expect?

(h) Plot the residuals from this regression. If appropriate, summarize these by giving the mean and SD of the collection of residuals. What does the SD of the residuals tell you about the fit of this equation?

39. **Download** Before taking the plunge into videoconferencing, a company ran tests of its current internal computer network. The goal of the tests was to measure how rapidly data moved through the network given the current demand on the network. Eighty files ranging in size from 20 to 100 megabytes (MB) were transmitted over the network at various times of day, and the time to send the files (in seconds) recorded.

(a) Create a scatterplot of Transfer Time on File Size. Does a line seem to you to be a good summary of the association between these variables?

(b) Estimate the least squares linear equation for Transfer Time on File Size. Interpret the fitted intercept and slope. Be sure to include their units. Note if either estimate represents a large extrapolation and is consequently not reliable.

(c) Interpret the summary values r^2 and s_e associated with the fitted equation. Attach units to these summary statistics as appropriate.

(d) To make the system look more impressive (i.e., have smaller slope and intercept), a colleague changed the units of y to minutes and the units of x to kilobytes (1 MB = 1,024 kilobytes). What does the new equation look like? Does it fit the data any better than the equation obtained in part (b)?

(e) Plot the residuals from the regression fit in part (b) on the sizes of the files. Does this plot suggest that the residuals reveal patterns in the residual variation?

(f) Given a goal of getting data transferred in no more than 15 seconds, how much data do you think can typically be transmitted in this length of time? Would the equation provided in part (b) be useful, or can you offer a better approach?

40. **Production Costs** A manufacturer produces custom metal blanks that are used by its customers for computer-aided machining. The customer sends a design via computer (a 3-D blueprint), and the manufacturer comes up with an estimated cost per unit, which is then used to determine a price for the customer. This analysis considers the factors that affect the cost to manufacture these blanks. The data for the analysis were sampled from the accounting records of 195 previous orders that were filled during the last 3 months.

(a) Create a scatterplot for the average cost per item on the material cost per item. Do you find a linear pattern?

(b) Estimate the linear equation using least squares. Interpret the fitted intercept and slope. Be sure to include their units. Note if either estimate represents a large extrapolation and is consequently not reliable.

(c) Interpret the summary values r^2 and s_e associated with the fitted equation. Attach units to these summary statistics as appropriate.

(d) What is the estimated increase in the total cost per finished item if the material cost per unit goes up by $3?

(e) One can argue that the slope in this regression should be 1, but it's not. Explain the difference.

(f) The total cost of an order in these data was $61.16 per unit with material costs of $4.18 per unit. Is this a relatively expensive order given the material costs?

(g) Plot the residuals from this regression. If appropriate, summarize these by giving the mean and standard deviation of the collection of residuals. What does the standard deviation of the residuals tell about the fit of this equation?

41. **Seattle Homes** This data table contains the listed prices (in thousands of dollars) and the number of square feet for 28 homes listed by a realtor in the Seattle area.

(a) Create a scatterplot for the price of the home on the number of square feet. Does the trend in the average price seem linear?

(b) Estimate the linear equation using least squares. Interpret the fitted intercept and slope. Be sure to include their units. Note if either estimate represents a large extrapolation and is consequently not reliable.

(c) Interpret the summary values r^2 and s_e associated with the fitted equation. Attach units to these summary statistics as appropriate.

(d) If a homeowner adds an extra room with 500 square feet to her home, can we use this model to estimate the increase in the value of the home?

(e) A home with 2,690 square feet lists for $625,000. What is the residual for this case? Is it a good deal?

(f) Do the residuals from this regression show patterns? Does it make sense to interpret s_e as the standard deviation of the errors of the fit? Use the plot of the residuals on the predictor to help decide.

42. **Leases** This data table includes the annual prices of 223 commercial leases. All of these leases provide office space in a Midwestern city in the United States.

(a) Create a scatterplot for the annual cost of the leases on the number of square feet of leased space. Does the pattern in the plot seem linear? Does the white space rule[5] hint of possible problems?

(b) Estimate the linear equation using least squares. Interpret the fitted intercept and slope. Be sure to include their units. Note if either estimate represents a large extrapolation and is consequently not reliable.

(c) Interpret the summary values r^2 and s_e associated with the fitted equation. Attach units to these summary statistics as appropriate.

(d) If a business decides to expand and wants to lease an additional 1,000 square feet beyond its current lease, explain how it can use the equation obtained in part (b) to estimate the increment in the cost of its lease. Would this estimate be reliable?

(e) A row in the data table describes a lease for 32,303 square feet. The annual cost for this lease is $496,409. What is the residual for this case? Is it a good deal?

(f) Do the residuals from this regression show any patterns? Does it make sense to interpret s_e as the standard deviation of the errors of the fit? Use the plot of the residuals on the number of square feet to decide.

43. **R&D Expenses** This data file contains a variety of accounting and financial values that describe 493 companies operating in several technology industries in 2004: software, systems design, and semiconductor manufacturing. One column gives the expenses on research and development (R&D), and another gives the total assets of the companies. Both columns are reported in millions of dollars.

(a) Scatterplot R&D Expense on Assets. Does a line seem to you to be a good summary of the relationship between these two variables? Describe the outlying companies.

(b) Estimate the least squares linear equation for R&D Expense on Assets. Interpret the fitted intercept and slope. Be sure to include their units. Note if either estimate represents a large extrapolation and is consequently not reliable.

(c) Interpret the summary values r^2 and s_e associated with the fitted equation. Attach units to these summary statistics as appropriate. Does the value of r^2 seem fair to you as a characterization of how well the equation summarizes the association?

(d) Inspect the histograms of the x- and y-variables in this regression. Do the shapes of these histograms anticipate some aspects of the scatterplot and the linear relationship between these variables?

(e) Plot the residuals from this regression. Does this plot reveal patterns in the residuals? Does s_e provide an adequate summary of the residual variation?

44. **Cars** The cases that make up this dataset are types of cars. For each of 223 types of cars sold in the United States during the 2003 and 2004 model years, we have the base price and the horsepower of the engine (HP).

(a) Create a scatterplot for the price of the car on the horsepower of the car. Does the trend in the average price seem linear?

(b) Estimate the linear equation using least squares. Interpret the fitted intercept and slope. Be sure to include their units. Note if either estimate represents a large extrapolation and is consequently not reliable.

(c) Interpret r^2 and s_e associated with the fitted equation. Attach units to these summary statistics as appropriate.

(d) If a manufacturer raises the engine output of a car by 5HP, should it use $5b_1$ to get a sense of how to increase price?

---

[5] The white space rule was covered in Chapter 4. A plot that is mostly white space doesn't reveal much about the data. A good plot uses its space to show data, not empty space.

(e) A car with 250 HP among these lists for $34,725. What is the residual for this case? Does this car lie above or below the fitted line?

(f) If the price of a car has a positive residual, is it a good deal? What if the residual is negative?

(g) Do you find patterns in the residuals from this regression? Does it make sense to interpret s_e as the standard deviation of the errors of the fit? Use the plot of the residuals on the predictor to help decide.

45. **OECD** The Organization for Economic Cooperation and Development (OECD) tracks various summary statistics of the member economies. The countries lie in Europe, parts of Asia, and North America. Two variables of interest are GDP (gross domestic product per capita, a measure of the overall production in an economy per citizen) and trade balances (measured as a percentage of GDP). Exporting countries tend to have large positive trade balances. Importers have negative balances. These data are from the 2005 report of the OECD.

(a) Describe the association in the scatterplot of GDP on Trade Balance. Does the association in this plot move in the right direction? Does the association appear linear?

(b) Estimate the least squares linear equation for GDP on Trade Balance. Interpret the fitted intercept and slope. Be sure to include their units. Note if either estimate represents a large extrapolation and is consequently not reliable.

(c) Interpret r^2 and s_e associated with the fitted equation. Attach units to these summary statistics as appropriate.

(d) Plot the residuals from this regression. After considering this plot, does s_e provide an adequate summary of the residual variation?

(e) Which country has the largest values of both variables? Is it the country that you expected?

(f) Locate the United States in the scatterplot and find the residual for the United States. Interpret the value of the residual for the United States.

46. **Hiring** A firm that operates a large, direct-to-consumer sales force would like to implement a system to monitor the progress of new agents. A key task for agents is to open new accounts; an account is a new customer to the business. The goal is to identify "superstar agents" as rapidly as possible, offer them incentives, and keep them with the company. To build such a system, the firm has been monitoring sales of new agents over the past 2 years. The response of interest is the profit to the firm (in dollars) of contracts sold by agents over their first year. Among the possible predictors of this performance is the number of new accounts developed by the agent during the first 3 months of work.

(a) Create a scatterplot for Profit from Sales on Number of Accounts. Does a line seem to be a good summary of the association between these variables?

(b) Estimate the least squares linear equation for Profit from Sales on Number of Accounts. Interpret the fitted intercept and slope; be sure to include their

units. Note if either estimate represents a large extrapolation and is consequently not reliable.

(c) Interpret r^2 and s_e associated with the fitted equation. Attach units to these summary statistics as appropriate.

(d) Based on the equation fit in part (b), what is the gain in profit to the firm of getting agents to open 100 additional accounts in the first 3 months? Do you think that this is a reasonable estimate?

(e) Plot the residuals from the regression fit in part (b) on the sizes of the files. Does this plot show random variation?

(f) Exclude the data for agents who open 75 or fewer accounts in the first 3 months. Does the fit of the least squares line change much? Should it?

47. **Promotion** These data describe spending by a major pharmaceutical company for promoting a cholesterol-lowering drug. The data cover 39 consecutive weeks and isolate the area around Boston. The variables in this collection are shares. Marketing research often uses the notion of voice to describe the level of promotion for a product. In place of the absolute spending for advertising, *voice* is the share of a type of advertising devoted to a specific product. Voice puts this spending in context; $10 million might seem like a lot for advertising unless everyone else is spending $200 million.

The column *Market Share* is sales of this product divided by total sales for such drugs in the Boston area. The column *Detail Voice* is the ratio of detailing for this drug to the amount of detailing for all cholesterol-lowering drugs in Boston. Detailing counts the number of promotional visits made by representatives of a pharmaceutical company to doctors' offices.

(a) Do timeplots of Market Share and Detail Voice suggest an association between these series? Does either series show simple variation?

(b) Create a scatterplot for Market Share on Detail Voice. Are the variables associated? Does a line summarize any association?

(c) Estimate the least squares linear equation for the regression of Market Share on Detail Voice. Interpret the intercept and slope. Be sure to include the units for each. Note if either estimate represents a large extrapolation and is consequently not reliable.

(d) Interpret r^2 and s_e associated with the fitted equation. Attach units to these summary statistics as appropriate.

(e) According to this equation, how does the average share for this product change if the detail voice rises from 0.04 to 0.14 (4% to 14%)?

(f) Plot the residuals from the regression fit in part (c) on the sizes of the files. Does this plot suggest that the residuals possess simple variation?

48. **Apple** This dataset tracks the monthly performance of stock in Apple Computer since its inception back in 1980. The data include 300 monthly returns on Apple Computer, as well as returns on the entire stock market, returns on Treasury Bills (short-term, 30-day loans to

Uncle Sam), and inflation. (The column *Market Return* is the return on a value-weighted portfolio that purchases stock in proportion to the size of the company rather than one from each company.)

(a) Create a scatterplot for Apple Return on Market Return. Does a line seem to be a good summary of the association between these variables?

(b) Estimate the least squares linear equation for Apple Return on Market Return. Interpret the fitted intercept and slope. Be sure to include their units. Note if either estimate represents a large extrapolation and is consequently not reliable.

(c) Interpret r^2 and s_e associated with the fitted equation. Attach units to these summary statistics as appropriate.

(d) If months in which the market went down by 2% were compared to months in which the market went up by 2%, how would this equation suggest Apple stock would differ between these periods?

(e) Plot the residuals from the regression fit in part (b) on the market returns. Does this plot suggest that the residuals possess simple variation? Do you recognize the dates of any of the outliers?

(f) Careful analyses of stock prices often subtract the so-called risk-free rate from the returns on the stock. After the risk-free rate has been subtracted, the returns are sometimes called "excess returns" to distinguish them. The risk-free rate is the interest rate returned by a very safe investment, one with no (or at least almost no) chance of default. The return on short-term Treasury Bills is typically used as the risk-free rate. Subtract the risk-free rate from returns on Apple stock and the market, and then refit the equation using these excess returns. Does the equation change from the previous estimate? Explain why it's similar or different.

4M Credit Cards

Banks monitor the use of credit cards to see whether promotions that encourage customers to charge more have been successful. Banks also monitor balances to seek out fraud or cases of stolen credit card numbers. For these methods to work, we need to be able to anticipate the balance on a credit card next month from things that we know today. A key indicator of the balance next month is the balance this month. Those balances that roll over earn the bank high rates of interest, often in the neighborhood of 12% to 24% annually.

The data table for this analysis shows a sample of balances for 923 customers of a national issuer of credit cards,

such as Visa and Mastercard. The four columns in the data table are balances over four consecutive months. For this analysis, we'll focus on the relationship between the balance in the third and fourth months.

Motivation

a) The bank would like to predict the balance of a customer in month 4 from the balance in month 3 to within 10% of the actual value. How well must a linear equation describe the data in order to meet this goal?

b) Explain in management terms how an equation that anticipates the balance next month based on the current balance could be useful in evaluating the success of a marketing program intended to increase customer account balances.

Method

c) Form the appropriate scatterplot of the balances in months 3 and 4. Does a linear model seem like a decent way to describe the association?

d) The scatterplot reveals two types of outliers that deviate from the general pattern. What is the explanation for these outliers?

Mechanics

e) Fit the linear equation using all of the cases. Briefly summarize the estimated slope and intercept as well as the overall summary statistics, R^2 and s_e.

f) Exclude the cases that have a near-zero balance in either month 3 or month 4 (or both). Interpret "near zero" to mean an account with balance $25 or less. Refit the equation and compare the results to the equation obtained in part (e). Do the results change in a meaningful way?

g) Inspect the residuals from the equation fit in part (f). Do these suggest simple variation?

Message

h) Summarize the fit of your equation for management. Explain in clear terms any relevant symbols. (That is, don't just give a value for r^2 and expect that the reader knows regression analysis.)

i) Is the goal of predicting the balance next month within 10% possible? Indicate, for management, why or why not.

Curved Patterns

CHAPTER

THE PRICE OF GASOLINE CAN BE A PAINFUL REMINDER OF THE LAWS OF SUPPLY AND DEMAND. The first big increase in domestic prices struck in 1973–1974 when the Organization of Petroleum Exporting Countries (OPEC) introduced production quotas. Gas prices had varied so little that accurate records had not been kept. The data in Figure 20.1 begin in 1975, in time to capture a second jump in 1979. After selling for 60 to 70 cents per gallon during the 1970s, the average price rose above $1.40 per gallon in 1981. That jump seems small, though, compared to recent increases.

In response to rising prices, Congress passed the Energy Policy and Conservation Act of 1975. This Act established the corporate average fuel economy (CAFE) standards. The CAFE standards set mileage requirements on the cars sold in the United States. The current standard for cars is 27.5 miles per gallon and is slated to increase to 30.2 miles per gallon for 2011 models.

One way to improve fuel mileage is to reduce the weight of the car. Lighter materials, however, cost more than heavier materials of comparable strength. Aluminum and

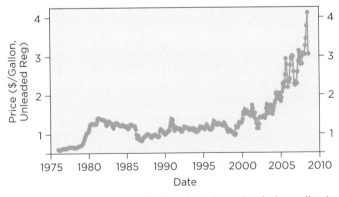

FIGURE 20.1 Average retail price of regular unleaded gasoline in the United States, in dollars per gallon.

composites are more expensive than steel. Before investing in exotic materials, manufacturers want evidence of the benefit. The costs—and benefits—add up when you make millions. What sort of improvements in mileage should a manufacturer expect from reducing the weight of a car by, for example, 200 pounds?

To answer this question, we'll use regression. Regression models require data that meet the linear condition: The association between the response and the explanatory variable needs to be linear. The association between gasoline mileage and weight, however, is not linear. Instead, a curved pattern links mileage and weight; a line is a poor summary of the dependence. In this chapter, we show how to use transformations of the data to describe such curved patterns. The slope and intercept in an equation using transformed variables summarize the pattern, and the interpretation of this equation depends on the transformation.

20.1 | DETECTING NONLINEAR PATTERNS

Linear patterns are a good place to begin when modeling dependence, but they don't work in every situation. There are two ways to recognize problems with linear equations: before looking at the data and while looking at a scatterplot.

At the start of the modeling process before looking at the data, ask this question:

> Should changes in the explanatory variable of a given size come with equal changes in the estimated response, regardless of the value of the explanatory variable?

Linear association means that equally sized changes in x associate with equal changes in y. For example, Chapter 19 develops a linear equation that describes the association between the weight and price of diamonds—relatively small diamonds that weigh less than 1/2 carat. An increase of one-tenth of a carat increases the estimated price by $267, regardless of the size of the diamond. A quick visit to a jeweler will convince you, however, that the association changes as diamonds get larger. The average difference in price between 1.0-carat diamonds and 1.1-carat diamonds is more than $267. As diamonds get larger, they become scarcer, and increments in size command ever-larger increments in price.

The same line of thinking applies to fuel consumption in cars. Do we expect the effect of weight on mileage to be the same for cars of all sizes? Does trimming 200 pounds from a big SUV have the same effect on mileage as trimming 200 pounds from a small compact? If we describe the relationship between mileage (y) and weight (x) using a line, then our answer is "yes." A fitted line $\hat{y} = b_0 + b_1 x$ has one slope, regardless of x.

Scatterplots

The second opportunity to recognize a problem with a linear equation comes when we see plots. The scatterplot in Figure 20.2 graphs mileage (in miles per gallon) versus weight (in thousands of pounds) for 230 passenger vehicles produced in the 2003 and 2004 model years. The line in Figure 20.2 is the least squares fit that regresses mileage on weight.

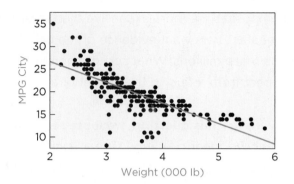

FIGURE 20.2 Mileage versus weight for 230 passenger vehicles sold in the United States.

The scatterplot and fitted line show negative association. Heavier cars on average get fewer miles per gallon. The linear pattern attributes more than half of the variation in mileage to differences in weight ($r^2 = 57\%$), and the standard deviation of the residuals is $s_e = 2.9$ MPG.

Let's interpret the estimated intercept and slope. It would be naïve to interpret b_0 as telling us that a "weightless car" would get 35.6 MPG on average; we can instead think of b_0 as the mileage attainable before accounting for moving the mass of a car. The value of b_0 is a reminder that cars burn fuel regardless of their weight to power the air conditioning and electronics. No matter how we interpret it, however, b_0 represents an extrapolation beyond the range of these data. All of these cars weigh more than 2,000 pounds, putting b_0 far to the left of the data in Figure 20.2. To interpret the slope, the estimate $b_1 = -4.52$ MPG/(1,000 lb) implies that estimated mileage drops by 4.52 miles per gallon for each additional 1,000 pounds of weight. (Weight is measured in *thousands* of pounds.) Returning to the motivating question, the linear equation estimates that mileage would increase on average by $0.2 \times 4.52 = 0.904$ mile per gallon were a car 200 pounds lighter.

Residual Plots

Even though the least squares regression line explains more than half of the variation in mileage ($r^2 = 0.568$), these data fail the linear condition. You have to look closely at Figure 20.2 to see the problem. The fitted line passes below most of the points on the left that represent lightweight cars, above most of the points in the middle, and again below the points on the right that represent heavy cars.

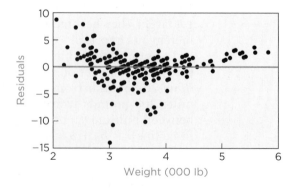

FIGURE 20.3 Residual plot from fitting a line to mileage versus weight shows a curved pattern.

The lack of fit becomes more apparent in the residual plot. After the linear pattern is removed, the curved shape is more evident. The residuals are generally positive on the left (light cars), negative in the middle, and positive on the right (heavy cars). The plot seems to smile at us.

To confirm the presence of a pattern in the residuals, we can use the visual test for association (Chapter 6). Build scatterplots that scramble the weights so that each residual gets matched to a randomly chosen weight. If there's no pattern, then it should not matter which weight goes with which residual. Figure 20.4 shows two scatterplots that scramble the residuals with the weights.

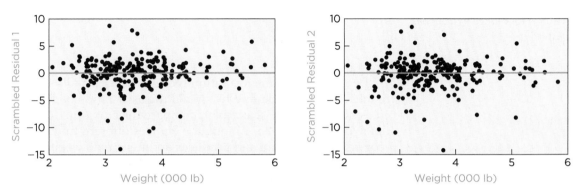

FIGURE 20.4 Checking for random variation in the residuals.

The residual plot in Figure 20.3 is visually distinct from the plots in Figure 20.4. The scales shown on the axes are the same, but the distinction is readily apparent. This distinction implies that there is a pattern in the residuals. The association in the data is not straight enough to be summarized with a line.

20.2 | TRANSFORMATIONS

transformation Re-expression of a variable by applying a function to each observation.

Transformations allow us to use regression analysis to find an equation that describes a curved pattern. A transformation defines a new variable by applying a function to each of the values of an existing variable. In many cases, we can find a transformation of x or y (or both) so that association between the transformed variables is linear, even though the association between the original variables is not.

Two nonlinear transformations are useful in business applications: reciprocals and logarithms. The reciprocal transformation converts the observed data d into $1/d$. The reciprocal transformation is often useful when dealing with variables, such as miles per gallon, that are already in the form of a ratio. The logarithm transformation converts d into $\log d$. Logs are useful when we believe that the association between variables is more meaningful on a percentage scale. For instance, if we believe that percentage increases in price are linearly associated with percentage changes in sales, then log transformations will be useful. (An example of this use of logs appears later in this chapter.) Other transformations such as squares may be needed to obtain a linear relationship.

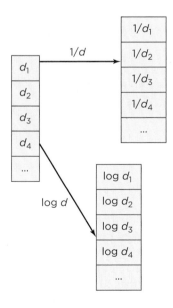

The process of choosing an appropriate transformation is usually iterative: Try a transformation to see whether it makes the association more linear. Figure 20.5 (which depicts Tukey's bulging rule) suggests when to use logs, reciprocals, and squares to convert a bending pattern into a linear pattern. Simply match the pattern in your scatterplot of y on x to one of the shapes in the figure.

For example, the data in Figure 20.2 have a curved pattern that resembles the downward bending curve in the lower left corner of Figure 20.5. We have several choices available to us, but the reciprocal of miles per gallon is easily interpreted and, as we will see, produces a linear pattern. We could also have tried to use the log of miles per gallon, the log of weight, or even both of these. The bulging rule is suggestive; it is up to us to decide which is best for our data.

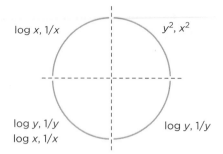

FIGURE 20.5 Possible transformations for modeling a curved pattern.

Deciding on a transformation requires several skills. First, think about the context of the problem: Why should the association be linear? Then, once you see curvature in the scatterplot, compare the curvature to the bending patterns shown in Figure 20.5. Among the choices offered, find one that captures the curvature of the data and produces an interpretable equation. Above all, don't be afraid to try several. **Picking a transformation requires practice, and you may need to try several to find one that is interpretable and captures the pattern in the data.**

tip

20.3 | RECIPROCAL TRANSFORMATION

The reciprocal transformation converts the curved pattern between MPG and Weight in Figure 20.2 into linear association. As part of the transformation, we also multiply by 100 to help with the interpretation. The overall transformation is 100/MPG, the number of gallons that it takes to go 100 miles. Europeans measure fuel consumption on a similar scale, but in liters per 100 kilometers. The scatterplot shown in Figure 20.6 graphs the transformed variable versus weight; the transformation produces linear association and reveals a collection of outliers that we had not noticed previously.

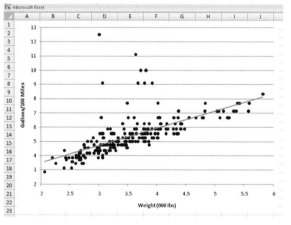

Estimated Gallons/Hundred Miles
$= 1.11 + 1.21$ Weight

$r^2 = 0.412 \quad s_e = 1.04$

FIGURE 20.6 The scatterplot of fuel consumption (gallons/ 100 miles) versus weight has a linear pattern.

The association between these variables is positive (rather than negative as in Figure 20.2) because the transformed response measures fuel consumption rather than fuel efficiency. Figure 20.6 includes the least squares line. Heavier cars consume more gas, so the slope is positive. The pattern looks linear but for the scattering of outliers. After seeing the outliers in Figure 20.6, you can find these same cars in the initial scatterplot in Figure 20.2. These cars are outliers in both views of the data, but the transformation has made them more prominent. You can probably guess which cars have very high fuel consumption for their weights. The plot shown in Figure 20.7 shows the residuals from the regression of gallons per 100 miles on weight; the outliers are even more apparent than in Figure 20.6.

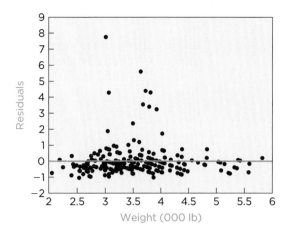

FIGURE 20.7 Residuals from the regression of gallons per hundred miles on weight.

The outliers are exotic sports cars, including a Ferrari, Lamborghini, Maserati, and Aston Martin. The Ferrari Enzo has the largest positive residual; it needs 8 *more* gallons to go 100 miles than expected for cars of its weight. It also packs 660 horsepower. An engine that generates that kind of power burns a lot of gas regardless of the weight of the car.

The transformation of miles per gallon into gallons per 100 miles produces linear association between fuel consumption and weight,

$$\textit{Estimated} \text{ Gallons/Hundred Miles} = 1.11 + 1.21 \text{ Weight}$$

The transformation also changes the interpretation of the slope and intercept. The slope b_1 estimates that the amount of gas needed to drive 100 miles grows by 1.21 gallons on average for each additional 1,000 pounds of weight. The intercept again naïvely speaks of weightless cars, but in terms of gallons per 100 miles. We prefer to interpret b_0 as the fuel burned regardless of the weight. These cars use about 1.11 gallons when driving 100 miles to run air conditioning and other conveniences, regardless of the weight. The value of b_0 remains an extrapolation.

caution Although we are concentrating on the linear condition in this chapter, we ought not forget the issue of a lurking variable. The 230 cars shown in these plots differ in many other ways that might affect fuel consumption aside from weight, such as having different engines and transmissions. Any effect that we attribute to weight could be associated with these lurking variables as well.

Comparing Linear and Nonlinear Equations

Which equation is better: the equation that regresses mileage on weight, or the equation that regresses fuel consumption on weight?

It is tempting to choose the fit with the larger r^2. The r^2 of the initial equation is higher, 57% to 41%. Before concluding this is a meaningful comparison, notice that the equations have different responses. Explaining 57% of the variation in mileage is not comparable to explaining 41% of the variation in fuel consumption. **tip** **Only compare r^2 between regression equations that use the same observations and same response.** The same goes for comparisons of residual standard deviations. These two standard deviations are on different scales. With MPG as the response, $s_e = 2.9$ miles per gallon. With the transformed response, $s_e = 1.04$ gallons/100 miles.

To obtain a meaningful comparison of these equations, show the fits implied by both equations in *one* scatterplot. Then decide which offers the better description of the pattern in the data. While doing this, think about which equation provides a more sensible answer to the motivating question. In this case, which equation provides a better answer to the question "What is the impact on mileage of reducing the weight of a car by 200 pounds?"

Visual Comparisons

A visual comparison requires showing both equations in one plot. We will show both in the scatterplot of MPG on Weight since we're more familiar with MPG. The fitted line from Figure 20.2 is

$$\textit{Estimated } \text{MPG} = 35.6 - 4.52 \text{ Weight}$$

Each increase of Weight by 1 (1,000 pounds) reduces estimated MPG by 4.52 miles per gallon. The equation using the reciprocal of MPG is

$$\textit{Estimated } \frac{\text{Gallons}}{\text{Hundred Miles}} = 1.11 + 1.21 \text{ Weight}$$

When drawn in the scatterplot of *MPG* on Weight, the reciprocal equation produces a curve. To draw the curve, let's work out a few estimated values. For a small car that weighs 2,000 pounds (Weight = 2), the estimated gallons per 100 miles is

$$\textit{Estimated } \frac{\text{Gallons}}{\text{Hundred Miles}} = 1.11 + 1.21 \times 2 = 3.53$$

The estimated mileage is then (100 miles)/(3.53 gallons) = 28.3 miles per gallon. Table 20.1 computes estimates at several more weights.

TABLE 20.1 Comparison of estimated MPG from two regression equations.

Weight (000 lb)	Line	Reciprocal	
	Estimated MPG	Estimated gal/(100 mi)	Estimated MPG
2	35.6 − 4.52 × 2 = 26.56	1.11 + 1.21 × 2 = 3.53	100/3.53 ≈ 28.3
3	35.6 − 4.52 × 3 = 22.04	1.11 + 1.21 × 3 = 4.74	100/4.74 ≈ 21.1
4	17.52	5.95	100/5.95 ≈ 16.8
5	13.00	7.16	100/7.16 ≈ 14.0
6	8.48	8.37	100/8.37 ≈ 11.9

The scatterplot shown in Figure 20.8 shows the linear fit of MPG on Weight (red) along with the curve produced by the reciprocal equation (green).

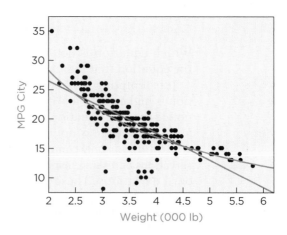

FIGURE 20.8 Comparing predictions from two models.

The curve produced by transforming MPG captures the bend that we saw in the original plot.

The linear and reciprocal equations provide very different estimates of mileage for heavy cars. Consider the estimated mileage of a Hummer H2. Because of its weight (6,400 lb), federal mileage standards exclude the Hummer;

it lacks an official mileage. *Car and Driver* magazine estimates it gets about 10 MPG. Plugging into the linear equation, the estimated mileage is

$$\hat{y} = 35.6 - 4.52 \times 6.4 \approx 6.7 \text{ MPG}$$

The reciprocal equation that estimates its fuel consumption is

$$\hat{y} = 1.11 + 1.21 \times 6.4 \approx 8.85 \text{ gallons/100 miles, or } 11.3 \text{ MPG}$$

That's closer to what *Car and Driver* reports, and more sensible than the very low estimate of 6.7 miles per gallon given by the linear equation.

Substantive Comparison

Let's complete our comparison of these equations by thinking about what each has to say about the association between weight and mileage. The linear equation fixes the slope at −4.52 MPG per 1,000 pounds. An increase of 1,000 pounds reduces estimated mileage by 4.52 MPG. The reciprocal equation treats changes in weight differently. Differences in weight matter less as cars get heavier. We can see this in Figure 20.8; the blue curve gets flatter as the weight increases.

To see why this happens, write the equation for the curve differently. The reciprocal equation is

$$Estimated \ \frac{\text{Gallons}}{\text{Hundred Miles}} = 1.11 + 1.21 \times \text{Weight}$$

If we take the reciprocal of both sides, we get

$$Estimated \ \frac{\text{Hundred Miles}}{\text{Gallons}} = \frac{1}{1.11 + 1.21 \times \text{Weight}}$$

This equation estimates the hundreds of miles per gallon. For example, it estimates that a car that weighs 3,000 pounds gets $1/(1.11 + 1.21 \times 3) \approx 0.21$ hundred miles per gallon, or 21 miles per gallon. In miles per gallon, the equation is

$$Estimated \ \frac{\text{Miles}}{\text{Gallon}} = \frac{100}{1.11 + 1.21 \times \text{Weight}}$$

Let's compare the effects on estimated mileage of several differences in weight. As shown in the margin, the curve is steeper for small cars and flatter for large cars: Differences in weight matter more for small cars. The difference in estimated MPG between cars that weigh 2,000 pounds and cars that weigh 3,000 pounds is relatively large,

$$\frac{100}{1.11 + 1.21 \times 2} - \frac{100}{1.11 + 1.21 \times 3} = 28.3 - 21.1$$
$$= 7.2 \text{ miles per gallon}$$

That's bigger than the effect on mileage implied by the linear model. For cars that weigh 3,000 and 4,000 pounds, the effect falls to

$$\frac{100}{1.11 + 1.21 \times 3} - \frac{100}{1.11 + 1.21 \times 4} = 21.1 - 16.8$$
$$= 4.3 \text{ miles per gallon}$$

That's close to the slope in the linear equation. For larger cars, the effect falls again to

$$\frac{100}{1.11 + 1.21 \times 4} - \frac{100}{1.11 + 1.21 \times 5} = 16.8 - 14.0 = 2.8 \text{ miles per gallon}$$

The diminishing effect of changes in weight makes more sense than a constant decrease. After all, mileage can only go down to zero.

Does the difference in how we model the effect of weight matter? It does to an automotive engineer charged with improving mileage. Suppose that engineers estimate the benefits of reducing weight by 200 pounds using the linear equation. They would estimate that reducing the weight by 200 pounds would improve mileage by $0.2 \times 4.52 \approx 0.9$ MPG, about one more mile per gallon. The reciprocal equation tells a different story. Shaving 200 pounds from a 3,000-pound car improves the mileage by

$$\frac{100}{1.11 + 1.21 \times 2.8} - \frac{100}{1.11 + 1.21 \times 3} = 22.2 - 21.1 = 1.1 \text{ miles per gallon}$$

For a 5,000-pound SUV, however, the improvement is much less:

$$\frac{100}{1.11 + 1.21 \times 4.8} - \frac{100}{1.11 + 1.21 \times 5} = 14.5 - 14.0 = 0.5 \text{ mile per gallon}$$

The reciprocal shows the engineers where to focus their efforts to make the most difference. Although it might be easy to trim 200 pounds from a heavy SUV, it's not going to have much effect on mileage.

Finally, we should confess that we did not pick the reciprocal transformation out of the blue: Many advocate switching from miles per gallon to gallons per mile. An example explains the preference. Consider a manager whose salespeople drive a fleet of cars, each driving 25,000 miles per year. What would have more impact on fuel costs: switching from cars that average 25 MPG to cars that achieve 35 MPG, or switching from limos that average 12 MPG to limos that average 14 MPG? Although the improvement in car mileage looks impressive, swapping those guzzlers that average 12 MPG for alternatives with slightly better MPG has the bigger impact: about 298 gallons saved per limo compared to 286 gallons per car!

If that seems surprising, you're not alone. Most people have the same trouble when estimating fuel savings.[1] If compared using gallons per 100 miles, the savings advantage of swapping the limos is clear. The two cars use either 4 or 2.86 gallons per 100 miles, a difference of 1.14. The limos use either 8.33 or 7.14 gallons per 100 miles, a larger difference of 1.19 gallons per 100 miles.

What Do You Think?   A retail chain operates 65 franchise stores. Each store covers about 12,000 square feet. The winter heating costs are substantial, as shown in the scatterplot in Figure 20.9. The response is the cost per month for heating, and the explanatory variable is the average local temperature during the month.

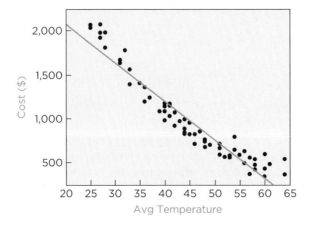

Estimated Cost = 2330 − 29.5 Avg Temperature

FIGURE 20.9 Monthly heating costs versus temperature.

---

[1] R. P. Larrick and J. B. Soll (2008), "The MPG Illusion," *Science*, 320, 1593–1594. To reproduce the results shown, divide the annual mileage by MPG. Cars that get 25 MPG need 1,000 gallons to drive 25,000 miles; at 35 MPG, the consumption drops to 714.3 gallons. Limos that get 12 MPG need 2,083.3 gallons; at 14 MPG, the consumption falls to 1,785.7 gallons.

a. Explain how to tell that these data do not satisfy the linear condition. What plot would probably show the problem more clearly?[2]

b. Which change in temperature appears to have the larger effect on heating costs: a drop from 35 to 25 degrees, or from 60 to 50 degrees?[3]

c. Sketch a bending pattern that captures the relationship between temperature and cost better than the fitted line.[4]

20.4 | LOGARITHM TRANSFORMATION

How much should a retailer charge? Each sale at a high price brings a large profit, but economics suggests that fewer items will sell. Lower prices generate more volume, but each sale generates less profit.

Let's use data to determine the best price for a chain of supermarkets. The two time series shown in Figure 20.10 track weekly sales of a national brand of pet food (number of cans sold) and the average selling price over two years ($n = 104$ weeks).

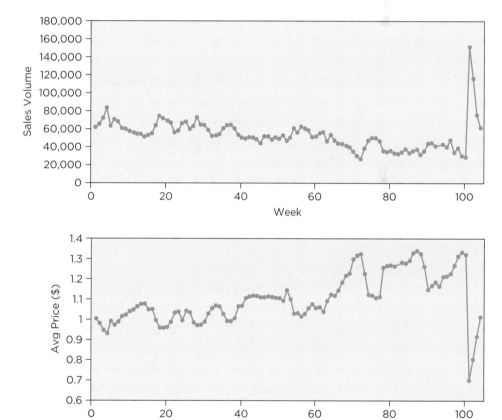

FIGURE 20.10 Timeplots of sales and price of pet food over two years.

The second timeplot shows that the price generally increased over these weeks until a large price reduction near the end of the 104 weeks. The average selling price dropped from $1.32 in week 100 to $0.70 in week 101. Shoppers noticed: Sales volume soared from 29,000 cans to 151,000.

---

[2] The fitted line is too low at the left and right of the plot; costs are higher than estimated by the line when the temperature is low or relatively high. A residual plot would help.

[3] The drop from 35 to 25. The pattern seems steeper on the left than on the right.

[4] A smooth curve that starts in the upper left corner, follows the data below the line and then flattens out at the right. The plot should resemble the blue curve in Figure 20.8.

tip   **Timeplots are great for finding trends and identifying seasonal effects, but they are less useful for seeing the relationship between two series.** For that, a scatterplot works better. But which variable is the response and which is the predictor? To decide, identify the variable that the retailer controls. Retailers set the price, not the volume. That means that price is the explanatory variable and the quantity is the response. (Economic textbooks often draw demand curves the other way around. Economists view both price and quantity as dependent and put price on the *y*-axis out of tradition.)

Before we look at the plot, let's think about what we expect to see. For a commodity like pet foods, we expect larger quantities to sell at lower prices—negative dependence. Should the relationship be linear? Does an increase of, say, $0.10 have the same effect on sales when the price is $0.70 as when the price is $1.20? It does if the relationship is linear.

Scatterplots and Residual Plots

The scatterplot shown in Figure 20.11 graphs the number of cans sold on the average selling price. Rather than show the data as separate timeplots (Figure 20.10), this scatterplot graphs one series versus the other at the same point in time. The association has the expected direction: negative. Lower prices associate with higher quantities sold.

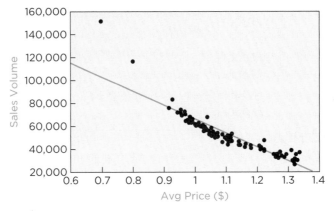

Estimated Sales Volume = 190,480 − 125,190 Price

FIGURE 20.11 These stores sell larger quantities at lower prices.

The two outliers at the left show a big increase in sales at low prices. These outliers are weeks 101 and 102, when stores drastically lowered the price. The scatterplot includes the least squares line.

Let's interpret the estimated slope and intercept. The slope estimates that the chain would sell 125,190 more cans per week on average if the price were reduced by $1 per can. This interpretation requires a large extrapolation. Prices range from $0.70 to $1.30, a range of only $0.60. A better interpretation of the slope uses a smaller price difference in keeping with the range of the explanatory variable. The fitted line estimates that in weeks in which the price differs by $0.10, the chain on average sells 12,519 more cans in those with the lower price. Considering that prices range from about $0.70 to $1.30, the intercept, $b_0 = 190,480$ cans, is a huge extrapolation. We should not pretend to know what would happen if the food were basically free.

The linear fit in Figure 20.11 explains much of the variation in sales ($r^2 = 0.83$), but the relationship between sales volume and price fails the linear condition. The white space rule suggests that we might be missing something in Figure 20.11. The residuals provide a closer look and clarify the problem with the linear equation.

As in the previous example, the plot of the residuals shown in Figure 20.12 makes it easier to spot the curved pattern. The residuals are positive at both low

and high prices and negative in the middle. The fitted line underpredicts sales at both low and high prices. In between, it overpredicts sales. Analogous to a stopped clock, it accurately estimates sales volume at two prices but otherwise misses the target.

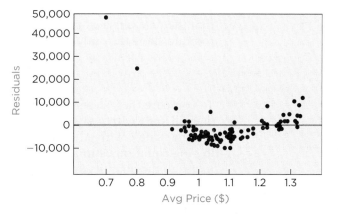

FIGURE 20.12 The nonlinear pattern is more distinct in the residual plot than in the scatterplot in Figure 20.11.

To improve the fit to the data, notice that the curvature in the scatterplot in Figure 20.11 matches the bend in the lower left corner of Tukey's bulging rule (Figure 20.5). This time, for reasons to be explained shortly, we'll use log transformations to capture the curvature. The reciprocal transformation worked well for mileage, but logs provide a more natural interpretation and capture the curvature in this example. The scatterplot shown in Figure 20.13 graphs the natural logarithm (log to base e) of sales volume versus the natural log of price. This relationship appears linear.

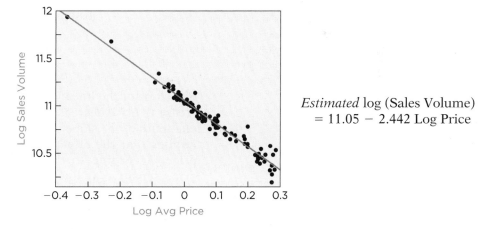

FIGURE 20.13 Quantity and price transformed to log scales.

Estimated log (Sales Volume)
    = 11.05 − 2.442 Log Price

With the data packed so tightly around the line ($r^2 = 0.955$), the white space rule suggests that we should look at the residuals.

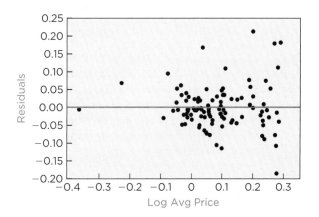

FIGURE 20.14 Residuals on the log scale show no remaining pattern.

The plot of the residuals in Figure 20.14 confirms that the log-log equation captures the curvature that is missed by the linear equation. Even during periods of very low prices, the residuals track near the fitted line. These residuals do not show a systematic departure from the fitted pattern, particularly when compared to those from the linear equation (Figure 20.12). It is okay that the data cluster on the right-hand side of the plot; this clustering makes room to show the outlying weeks on the left when prices were very low.

Comparing Equations

To compare these descriptions of the relationship between price and quantity, we will start with a visual comparison. Table 20.2 shows the details for estimating sales with the linear equation and the log-log equation. The log-log equation estimates the log of the sales volume. To convert these estimates to the quantity scale, we have to take the exponential of the fitted value using the function $\exp(x) = e^x$.

TABLE 20.2 Predicting sales volume with the linear equation and the log-log equation.

Average Price ($)	Linear Equation Estimated Sales	Log-Log Equation	
		Estimated Sales, Log scale	Estimated Sales
0.80	$190{,}480 - 125{,}190 \times 0.80 \approx 90{,}328$	$11.05 - 2.442 \times \log(0.8) \approx 11.594$	$\exp(11.594) \approx 108{,}445$
0.90	$190{,}480 - 125{,}190 \times 0.90 \approx 77{,}809$	$11.05 - 2.442 \times \log(0.8) \approx 11.307$	$\exp(11.307) \approx 81{,}389$
1.00	65,290	11.050	$\exp(11.050) \approx 62{,}944$
1.10	52,771	10.817	$\exp(10.817) \approx 49{,}861$
1.20	40,252	10.605	$\exp(10.605) \approx 40{,}336$

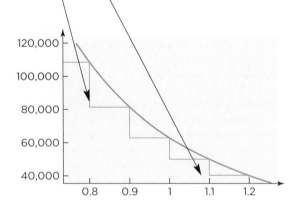

The slope of the linear equation implies that each $0.10 increase in price decreases estimated sales volume by 12,519, one-tenth of the slope of the fitted line. The log-log equation implies that the effect of changing price depends on the price. According to the log-log equation, customers are price sensitive at low prices. An increase from $0.80 to $0.90 leads to a drop of more than 27,000 in the estimated volume. In contrast, customers who are willing to buy at $1.10 are less sensitive to price than those who buy at $0.70. A $0.10 increase from $1.10 to $1.20 leads to a smaller drop of about 9,500 cans in the estimated volume.

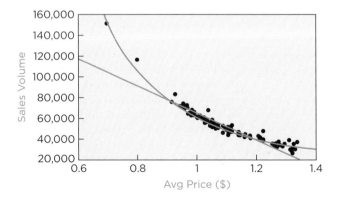

FIGURE 20.15 Comparing models of sales versus price. (The green curve represents a log-log equation.)

The two curves in the scatterplot of sales volume versus price (Figure 20.15) contrast estimated sales from the two equations. The log-log equation (green) captures the bend that is missed by the fitted line (red). This graph also makes it clear that the slope of the log-log equation changes. The curve gets flatter as the price increases. Changes in price have a larger effect on estimated sales volume when the price is low (where the green curve is steep) than when the price is high (where it flattens). We can describe this effect nicely using percentages. A $0.10 increase in price is a larger percentage change at lower prices. At $0.80, a $0.10 increase represents a 12.5% increase in the price. At $1.20, that same $0.10 increase represents an 8.3% increase in the price and hence has a smaller impact on sales.

Elasticity

Most economists wouldn't wait to see a residual plot when building an equation to relate price and quantity. They would have begun with *both* variables on a log scale. There's an important reason: Logs capture how percentage changes relate price and quantity. Logs change how we think about variation. Variation on a log scale amounts to variation among percentage differences.

The slope in an equation that transforms both x and y using logs is known as the **elasticity** of y with respect to x. The elasticity describes how small percentage changes in x are associated with small percentage changes in y (see Behind the Math: Logs and Percentage Changes). In the pet food example, the slope in the log-log model $b_1 = -2.44$ tells us that a 1% increase in price estimates a 2.44% decrease in sales. The intercept in a log-log model is important for calculating the fitted value $\hat{y}$, but not very interpretable substantively.

(p. 505)

elasticity Measure that relates % change in x to % change in y; slope in a log-log regression equation.

What should we tell the retailer? Using the log-log equation, we estimate that a 1% increase in price reduces sales by 2.44%. Is that a good trade-off? These cans usually sell for about $1, with typical sales volume near 60,000 cans per week. If the retailer raises the price by 5% to $1.05, then the log-log equation estimates sales volume to fall by about $5 \times -2.44\% = -12.2\%$. The retailer would sell about $0.122 \times 60,000 \approx 7,320$ fewer cans, but those that sell would bring in $0.05 more. We can find the best price by using the elasticity in a formula that includes the cost to the grocery. If the store spends $0.60 to stock and sell each can, then the price that maximizes profits is as follows (see Behind the Math: Optimal Pricing):

(p. 506)

$$\text{Optimal Price} = \text{Cost}\left(\frac{\text{Elasticity}}{\text{Elasticity} + 1}\right)$$
$$= 0.60\left(\frac{-2.44}{-2.44 + 1}\right)$$
$$\approx 0.60(1.694)$$
$$= \$1.017$$

Price	Estimated Sales	Estimated Profit
$0.80	108,445	$21,689
$0.90	81,389	$24,417
$1.00	62,944	$25,178
$1.10	49,861	$24,931
$1.20	40,336	$24,202

The ratio of the elasticity to 1 plus the elasticity gives the markup as a multiplier to apply to the cost. Most of the time, the retailer has been charging about $1 per can. At $1.02, estimated sales are 59,973; each sale earns $0.42 and estimated profits are about $25,188. At the lower price offered in the discount weeks, estimated sales almost double, 108,445 at $0.80 (Table 20.2). At $0.20 profit per can, however, that nets only $21,689. Big sales move a lot of inventory, but they sacrifice profits.

4M EXAMPLE 20.1   OPTIMAL PRICING

MOTIVATION

STATE THE QUESTION

How much should a convenience store charge for a half-gallon of orange juice? Currently, stores charge $3.00 each. Economics tells us that if orange juice costs c dollars per carton and γ is the elasticity of sales with respect to price, then the optimal price to charge is $c\gamma/(1 + \gamma)$. For this store, each half-gallon of orange juice costs $c = \$1$ to purchase and stock. All we need in order to find the best price is the elasticity. We can estimate the elasticity from a regression of the log of sales on the log of price.

METHOD

Identify x and y.

Link b_0 and b_1 to problem.

Describe data.

Check linear condition and variable condition.

DESCRIBE THE DATA AND SELECT AN APPROACH

The explanatory variable will be the price charged per half-gallon, and the response will be the sales volume. Both are transformed using logs. The slope in this equation will estimate the elasticity of sales with respect to price.

The chain collected data on sales of orange juice over a weekend at 50 locations. The stores are in similar neighborhoods and have comparable levels of business. We have to make sure that these stores sell juice at different prices. If all of the prices are the same, we cannot fit a line or curve. Management made sure that the price varied among stores, with all selling orange juice for between $1.25 and $4.25 per half-gallon.

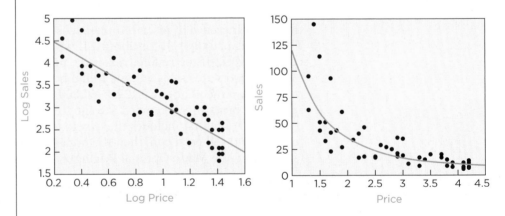

✓ **Linear.** The scatterplot of log of sales volume versus log of price appears linear. The relationship is clearly not linear on the original scale.

✓ **No obvious lurking variable.** The data were collected at the same time in stores that operate in similar areas. It would also be helpful to know that the stores face similar competition.

MECHANICS

DO THE ANALYSIS

The least squares regression of log sales on log price is

$$\textit{Estimated} \text{ log Sales} = 4.81 - 1.75 \text{ log Price}$$

✓ **Random residual variation.** The residual plot from the log-log regression shows no pattern.

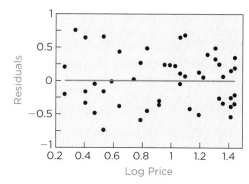

The orange juice costs the chain $1 to stock and sell. The estimated elasticity from the least squares equation is −1.75. If we substitute the estimated elasticity into the optimal pricing formula, the estimated optimal price is

$$c\gamma/(1 + \gamma) = \$1(-1.75)/(1 - 1.75) = \$1(2.33) = \$2.33$$

At $3, selling the predicted 18 cartons earns 18($2) = $36 profit. At $2.33, say, predicted sales grow to about 28 cartons and earn a bit more profit, 28 × 1.33 = $37.24 per store.

MESSAGE | ## SUMMARIZE THE RESULTS

Some people will buy orange juice at a high price. Others only buy orange juice if the price is low enough. Sales would grow if the chain were to lower the price. The chain would make higher profits by decreasing the price from the current $3 to near $2.33. Even though it makes less on each, the increased volume would generate more profit. The typical store sells 18 cartons at $3, giving profits of 18 × ($3 − $1) = $36 per store. At $2.33, a model of the relationship between price and quantity predicts sales of about 28 cartons at a profit of 28 × 1.33 = $37.24 per store. This increase in profits is small, however, and may not be feasible if stocking more orange juice will lead to greater storage costs.

Use words or terminology that your audience understands. Logs probably are not a good choice. Elasticity might not be a good choice either, but it's easy to explain. The elasticity of sales with respect to price is −1.75. This means that each 1% increase in price, on average, reduces sales by about 1.75%.

Best Practices

- *Anticipate whether the association between y and x is linear.* It's useful to think about what to expect before starting the data analysis. In many situations, such as the relationship between promotion and sales or price and quantity, you should expect to find diminishing effects of increases in the explanatory variable. In these situations, finding linear association would be surprising.

- *Check that a line summarizes the relationship between the explanatory variable and the response both visually and substantively.* Lines are common, easy to interpret, and may even obtain a large r^2, but that does not mean a line is the best summary of the dependence. A linear pattern implies that changes in the explanatory variable have the same expected effect on the response, regardless of the size of the predictor variable. Often, particularly in economic problems, the effect of the predictor depends on its size.

- *Stick to models you can understand and interpret.* Reciprocals are a natural transformation if your data are already a ratio. Logs are harder to motivate but are common in economic modeling, such as in demand curves that relate price and quantity. Recognize that you need to be able to interpret the components of your model. If you use elaborate transformations like those used by NASA to study the space shuttle (Chapter 1),

$$p = \frac{0.0195(L/d)^{0.45}(d)\rho_F^{0.27}(V - V^*)^{2/3}}{S_T^{1/4}\rho_T^{1/6}}$$

you might have trouble interpreting the intercept and slope! Unless you can make sense of your model in terms that others appreciate, you're not going to find curves useful.

■ *Interpret the slope carefully.* Slopes in models with transformations mean different things than slopes in a linear model. In particular, the slope in a log-log model is an elasticity, telling you how percentage changes in x are associated with percentage changes in y.

■ *Graph your model in the original units.* If you show most audiences a scatterplot after a transformation, such as logs, few will realize that it means that sales of this product are price sensitive. Look back at the two plots of the fitted log-log model in Example 20.1. Most hasty readers will look at the plot on the left and think that price has a constant effect on sales. They don't notice logs. When shown on the original scales, it's immediately apparent that the effects of price increases diminish at higher prices.

Pitfalls

■ *Don't think that regression only fits lines.* Because regression models always include a slope and intercept, it is easy to think that regression only fits straight lines. Once you use the right transformation, you do fit straight lines. To see the curvature, you have to return to the original scales.

■ *Don't forget to look for curves, even in models with high values of r^2.* A large r^2 does not mean that the pattern is linear, only that there's a lot of association. You won't discover curvature unless you look for it. Examine the residuals, and pay special attention to the most extreme residuals because they may have something to add to the story told by the linear model. It's often easier to see bending patterns in the residual plot than in the scatterplot of y on x.

■ *Don't forget lurking variables.* A consequence of not understanding your data is forgetting that regression equations describe association, not causation. It is easy to get caught up in figuring out the right transformation and forget other factors. Unless we can rule out other factors, we don't know that changing the price of a product caused the quantity sold to change, for example. Perhaps every time the retailer raised prices, so did the competition. Consumers always saw the same difference. Unless that matching continues the next time the price changes, the response may be very different.

■ *Don't compare r^2 between models with different responses.* You should only compare r^2 between models with the same response. It makes no sense, for instance, to compare the proportion of variation in the log of quantity to the proportion of variation in quantity itself.

Software Hints

There are two general approaches to fitting curves. Some software can show the nonlinear curve with the fitted line in a scatterplot of the data. Rather than transform both x and y to a log scale, for instance, add the curve that summarizes the fit of $\log y$ on $\log x$ directly to the scatterplot of y on x, as in Figure 20.8. Most software, however, requires us to transform the data. That is, in this example, we need to build two new columns, one holding the logs of y and the other holding logs of x. To fit the log-log model, fit the linear regression of $\log y$ on $\log x$ as done in Chapter 19.

EXCEL

To decide if a transformation is necessary, start with the scatterplot of y on x. With the chart selected, the menu sequence

Chart > Add Trendline…

allows you to fit several curves to the data (as well as a line). Try several and decide which fit provides the best summary of the data.

If you believe that the association between y and x bends, use built-in functions to compute columns of transformed data (for example, make a column that is equal to the log or reciprocal of another). Then follow the methods in Chapter 19 to fit a line to the transformed data.

MINITAB

If the scatterplot of y on x shows a bending pattern, use the calculator (menu sequence Calc > Calculator ...) to construct columns of transformed data (i.e., build new variables such as $1/x$ or $\log y$). Then follow the methods in Chapter 19 to fit a line to the transformed data.

JMP

The menu sequence

$$\text{Analyze} > \text{Fit Y by X}$$

opens the dialog used to construct a scatterplot. Fill in variables for the response and the explanatory variable, then click OK. In the window that shows the scatterplot, click on the red triangle above the scatterplot (near the words "Bivariate Fit..."). In the pop-up menu, choose the item Fit Special to fit curves to the shown data. Select a transformation for y or x and JMP will estimate the chosen curve by least squares, show the fit in the scatterplot, and add a summary of the equation to the output below the plot.

If you want to see the linear fit to the transformed data, you'll need to use the JMP formula calculator to build new columns (new variables). For example, right-clicking in the header of an empty column in the spreadsheet allows you to define the column using a formula such as log (X). Then fit a regression using the transformed variables as in Chapter 19.

 BEHIND the Math

Logs and Percentage Changes

The slope in a linear equation indicates how the estimate of y changes when x grows by 1. Denote the estimate of y at x by $\hat{y}(x) = b_0 + b_1 x$, and the estimate of y at $x + 1$ by $\hat{y}(x + 1) = b_0 + b_1(x + 1)$. The slope is the difference between these estimates:

$$\hat{y}(x + 1) - \hat{y}(x) = (b_0 + b_1[x + 1]) - (b_0 + b_1 x)$$
$$= b_1$$

To interpret a log-log equation, think in terms of percentages. Instead of looking for what to expect when x grows by 1, think in terms what happens when x grows by 1%. The predicted value for the log of y in the log-log equation at x is

$$\log \hat{y}(x) = b_0 + b_1 \log x$$

Rather than increase x by *adding* 1, multiply x by 1.01 to increase it by 1%. If x increases by 1%, the predicted value becomes

$$\log \hat{y}(1.01x) = b_0 + b_1 \log (1.01x)$$
$$= b_0 + b_1 \log x + b_1 \log 1.01$$
$$= \log \hat{y}(x) + b_1 \log 1.01$$
$$\approx \log \hat{y}(x) + 0.01b_1$$

The last step uses a simple but useful approximation:

$$\log (1 + \text{little bit}) \approx \text{little bit}$$

This approximation is accurate so long as the little bit is less than 0.10. In this example, the approximation

means $\log (1.01) = \log (1 + 0.01) \approx 0.01$. (Try this on your calculator using $\log_e$.) If we rearrange the previous equation, the estimated change that comes with a 1% increase in x is

$$0.01b_1 = \log \hat{y}(1.01x) - \log \hat{y}(x)$$

Since $\log a - \log b = \log (a/b)$, we can write the right-hand side as

$$0.01b_1 = \log \frac{\hat{y}(1.01x)}{\hat{y}(x)}$$
$$= \log \frac{\hat{y}(x) + (\hat{y}(1.01x) - \hat{y}(x))}{\hat{y}(x)}$$
$$= \log \left(1 + \frac{\hat{y}(1.01x) - \hat{y}(x)}{\hat{y}(x)} \right)$$
$$\approx \frac{\hat{y}(1.01x) - \hat{y}(x)}{\hat{y}(x)}$$

little bit

Hence, if we change x by 1%, then the fitted value changes by

$$b_1 \approx 100 \frac{\hat{y}(1.01x) - \hat{y}(x)}{\hat{y}(x)} = \text{Percentage Change in } \hat{y}$$

The slope b_1 in a log-log model tells you how percentage changes in x (up to about $\pm 10\%$) translate into percentage changes in y.

Optimal Pricing

Choosing the price that earns the most profit requires a bit of economic analysis. In the so-called monopolist situation, economics describes how the quantity sold Q depends on the price p according to an equation like this one, $Q(p) = mp^\gamma$. The exponent is constrained to be less than -1 ($\gamma < -1$) or the seller is in a market that allows the seller to charge an infinite price. The symbol m stands for a constant that depends, for example, on whether we're measuring sales in cans or cases. The exponent γ is the **elasticity** of the quantity sold with respect to price.

We don't want to maximize sales—it's profit that matters. Suppose each item costs the seller c dollars. Then the profit at the selling price p is

$$\text{Profit}(p) = Q(p)(p - c)$$

Calculus shows that the price p^* that maximizes the profit is

$$p^* = c\gamma/(1 + \gamma)$$

For example, if each item costs $c = \$1$ and the elasticity $\gamma = -2$, then the optimal price is $c\gamma/(1 + \gamma) = \$1 \times -2/(1 - 2) = \$2$. If customers are more price sensitive and $\gamma = -3$, the optimal price is $\$1.50$. As the elasticity approaches -1, the optimal price soars (tempting governments to regulate prices).

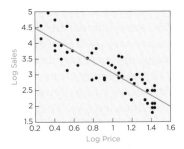

CHAPTER SUMMARY

Regression equations use **transformations** of the original variables to describe patterns that curve. Equations with reciprocals and logs are useful in many business applications and are readily interpretable. Nonlinear equations that use these transformations allow the slope to change, implying that the estimated effects of changes in the explanatory variable on the average response depend on the value of the explanatory variable. The slope in a regression equation that uses logs of both the response and the explanatory variable is the **elasticity** of the response with respect to price.

◼ Key Terms

elasticity, 501

transformation, 491

◼ About the Data

The data in Figure 20.1 that track gasoline prices come from the Energy Information Administration, as reported in the *Monthly Energy Review*. This source also reports other time series of interest, such as petroleum production, natural gas prices and supply, and oil inventories.

A former Wharton MBA student worked in the auto industry and provided the data in this chapter that describe characteristics of cars. The mileage data are the standard EPA estimates for urban driving. Highway mileage produces similar results.

The pricing data come from an MBA student who worked with a retailer during her summer internship. We modified the data slightly to protect the source of the data.

EXERCISES

Mix and Match

Match each description on the left with its mathematical expression on the right.

1. Reciprocal transformation

(a) $\log y = b_0 + b_1 \log x$

2. Opposite of log transformation

(b) $1.01x$

3. Slope estimates elasticity

(c) $\exp(x)$

4. 1% increase in x

(d) $1/y$

5. Approaches b_0 as x gets large

(e) $y = b_0 + b_1 1/x$

6. Random variation

(f)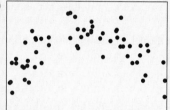

7. Diminishing marginal return

(g)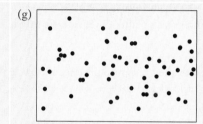

8. Pattern unsuited to log or reciprocal transformations

(h)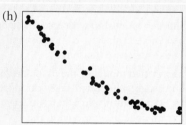

9. Negative, nonlinear association

(i)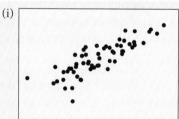

10. Linear pattern

(j)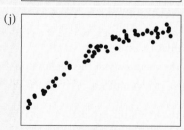

True/False

Mark each statement True or False. If you believe that a statement is false, briefly explain why you think it is false.

11. Regression equations only represent linear trends in data.

12. If the correlation between x and y is larger than 0.5, then a linear equation is appropriate to describe the association.

13. To identify the presence of curvature, it can be helpful to begin by fitting a line and plotting the residuals from the linear equation.

14. The residual standard deviation s_e when fitting the model $1/y = b_0 + b_1 x$ has the same units as $1/y$.

15. The reciprocal transformation is commonly applied to rates, such as miles per gallon or dollars per square foot.

16. If the equation in a model has a log transformation, the interpretation of the slope should be avoided.

17. Transformations such as logs or reciprocals are determined by the position of outliers in a scatterplot.

18. The slope of a regression model which uses $\log x$ as a predictor is known as an elasticity.

19. Transformations in regression affect the r^2 of the fitted model.

20. When returned to the original scale, the fitted values of a model with a transformed response produce a curved set of predictions.

Think About It

21. If diamonds have a linear relationship with essentially no fixed costs, which costs more: a single 1-carat diamond or two $\frac{1}{2}$-carat diamonds? Are these the same?

22. According to the transformed model of the association between weight and mileage, which will save you more gas: getting rid of the 50 pounds of junk that you leave in the trunk of your compact car or removing the 50-pound extra seat from an SUV? Are these the same?

23. Can you think of any lurking factors behind the relationship between weight and fuel consumption among car models?

24. In discussing the estimate of an elasticity, the text mentions changes in the prices of rival brands as a lurking factor. Can you think of another?

25. If the elasticity of quantity with respect to price is close to zero, how are the price and quantity related?

26. If quantity sold increases with price, would the elasticity be positive, negative, or zero?

27. If an equation uses the reciprocal of the explanatory variable, as in $\hat{y} = b_0 + b_1 1/x$, then what does the intercept b_0 tell you?

28. If an equation uses the log of the explanatory variable, as in $\hat{y} = b_0 + b_1 \log x$, then what does the intercept b_0 tell you?

29. This plot shows the residuals from the linear equation relating mileage to weight. What's causing the visible diagonal stripes that are parallel to the shown arrow?

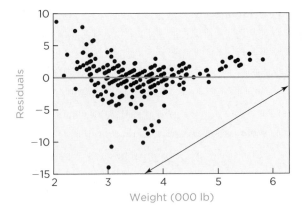

30. If the prices in the equation between price and quantity are expressed in a different currency (such as euros at 1.2 per dollar), how does the elasticity change?

You Do It

31. **Wal-Mart** These data contain quarterly operating income (in millions of dollars) of Wal-Mart from 1990 through the end of 2005 (64 quarters). Operating income is the difference between net sales and the cost of items sold, including both the merchandise and the labor to make the sale.
    (a) Create a scatterplot for operating income on Date. Is the pattern linear?
    (b) Fit a linear trend to the operating income. If you accept the fit of this equation, interpret the slope and intercept.
    (c) Plot the residuals from this linear trend on Date. Do the residuals show random variation, or do you see patterns in the mean or variation of the residuals?
    (d) Create a scatterplot for operating income on date, but use a log scale for the y-axis. Compare this view of the data to the initial plot of operating income versus date.
    (e) Fit the equation $\log(\text{Operating Income}) = b_0 + b_1$ Date. Interpret the slope in the fit of this equation.
    (f) What pattern remains in the residuals from the fit of the log of operating income on date? Can you explain what's happening?
    (g) Which equation offers the better summary of the trend in operating income at Wal-Mart? What's the basis for your choice?

32. **Target** These data contain quarterly operating income (in millions of dollars) of Target stores from 1990 through the end of 2005 (64 quarters). Operating income is the difference between net sales and the cost of things sold, including both the merchandise and the labor to make the sale.
    (a) Create a scatterplot for operating income on Date. Is the pattern linear?

(b) Fit a linear trend to the operating income. If you accept the fit of this equation, interpret the slope and intercept.

(c) Plot the residuals from this linear trend on Date. Do the residuals show random variation, or do you see patterns in the mean or variation of the residuals?

(d) Create a scatterplot for operating income on Date, but use a log scale for the y-axis. Compare this view of the data to the initial plot of operating income versus date.

(e) Fit the equation log (Operating Income) $= b_0 + b_1$ Date. Interpret the slope in the fit of this equation.

(f) What pattern remains in the residuals from the fit of the log of operating income on date? Can you explain what's happening?

(g) Which equation offers the better summary of the trend in operating income at Target? What's the basis for your choice?

33. **Wine** Influential wine critics such as Robert Parker publish their personal ratings of wines, and many consumers pay close attention. Do these ratings affect the price? The data in this exercise are a sample of ratings and prices found online at the Web site of an Internet wine merchant.

(a) Does the scatterplot of the price of wine on the rating suggest a linear or nonlinear relationship?

(b) Fit a linear regression equation to the data, regressing price on the rating. Does this fitted model make substantive sense?

(c) Create a scatterplot for the log of the price on the rating. Does the relationship seem more suited to regression?

(d) Fit a regression of the log of price on the rating. Does this model provide a better description of the pattern in the data?

(e) Compare the fit of the two models to the data. Can you rely on summary statistics like r^2 and s_e?

34. **Display Space** Initial levels of advertising often bring a larger response in the market than later spending. In this example, advertising comes in the form of devoting more shelf space to the indicated product. The level of sales is the weekly total sales of this product at several outlets of a chain of markets. The display space gives the number of shelf feet used to display the item. The data include sales at 48 stores.

(a) Create a scatterplot for the level of sales on the number of shelf feet. Does the relationship appear linear? Do you think that it ought to be linear?

(b) Fit a linear regression equation to the data, regressing sales on the number of shelf feet. Does this fitted model make substantive sense?

(c) Consider a scatterplot that shows sales on the log of the number of shelf feet. Does the relationship seem more linear than in part (a)?

(d) Fit a regression of sales on the log of the number of shelf feet. Does this model provide a better description of the pattern in the data? What do the slope and intercept tell you?

(e) Compare the fit of the two models to the data. Can you rely on summary statistics like r^2 and s_e?

35. **Used Accords** Cars depreciate over time. These data show the prices of Honda Accords listed for sale by individuals in *The Philadelphia Inquirer* in an issue during 2005. One column gives the asking price (in thousands of dollars) and a second column gives the age (in years).

(a) Do you expect the resale value of a car to drop by a fixed amount each year?

(b) Fit a linear equation with price as the response and age as the explanatory variable. What do the slope and intercept tell you, if you accept this equation's description of the pattern in the data?

(c) Plot the residuals from the linear equation on Age. Do the residuals suggest a problem with the linear equation?

(d) Fit the equation

$$\textit{Estimated Price} = b_0 + b_1 \log (\text{Age})$$

Do the residuals from this fit "fix" the problem found in part (c)?

(e) Compare the fitted values from this equation with those from the linear model. Show both in the same scatterplot. In particular, compare what this graph has to say about the effects of increasing age on resale value.

(f) Compare the values of r^2 and s_e between these two equations. Give units where appropriate. Does this comparison agree with your impression of the better model? Should these summary statistics be compared?

(g) Interpret the intercept and slope in this equation.

(h) Compare the change in asking price for cars that are 1 and 2 years old to that for cars that are 11 and 12 years old. Use the equation with the log of age as the explanatory variable. Is the difference the same or different?

36. **Used Camrys** Cars depreciate over time, whether made by Honda or, in this case, Toyota. These data show the prices of Toyota Camrys listed for sale by individuals in *The Philadelphia Inquirer* in an issue during 2005. One column gives the asking price (in thousands of dollars) and a second column gives the age (in years).

(a) Do you expect the resale value of a car to drop by a fixed amount each year?

(b) Fit a linear equation with price as the response and age as the explanatory variable. What do the slope and intercept tell you, if you accept this equation's description of the pattern in the data?

(c) Plot the residuals from the linear equation on Age. Do the residuals suggest a problem with the linear equation?

(d) Fit the equation

$$\textit{Estimated Price} = b_0 + b_1 \log (\text{Age})$$

Do the residuals from this fit "fix" the problem found in part (c)?

(e) Compare the fitted values from this equation with those from the linear model. Show both in the same scatterplot. In particular, compare what this graph has to say about the effects of increasing age on resale value.

(f) Compare the values of r^2 and s_e between these two equations. Give units where appropriate. Does this comparison agree with your impression of the better model? Should these summary statistics be compared?

(g) Interpret the intercept and slope in this equation.

(h) Compare the difference in asking price for cars that are 1 and 2 years old to that for cars that are 11 and 12 years old. Use the equation with the log of age as the explanatory variable. Is the difference the same or different?

37. **Cellular Phones in the United States** Cellular (or mobile) phones are everywhere these days, but it has not always been so. These data from CTIA, an organization representing the wireless communications industry, track the number of cellular subscribers in the United States. The data are semiannual, from 1985 through mid-2006.

(a) From what you have observed about the use of cellular telephones, what do you expect the trend in the number of subscribers to look like?

(b) Create a scatterplot for the number of subscribers on the date of the measurement. Does the trend look as you would have expected?

(c) Fit a linear equation with the number of subscribers as the response and the date as the explanatory variable. What do the slope and intercept tell you, if you accept this equation's description of the pattern in the data?

(d) Create a scatterplot for the same data shown in the scatterplot done in part (b), but for this plot, put the response on a log scale. Does the scatterplot suggest that a curve of the form

Estimated log (Number of Subscribers) = $b_0 + b_1$ Date

is a good summary?

(e) Create a scatterplot for the percentage change in the number of subscribers versus the year minus 1984. (That's like treating 1984 as the start of the cellular industry in the United States.) Does this plot suggest any problem with the use of the equation of the log of the number of subscribers on the date? What should this plot look like if a log equation of the form in part (d) is going to be a good summary?

(f) Summarize the curve in this scatterplot using a curve of the form

Estimated Percentage Growth = $b_0 + b_1$ 1/(Date − 1984)

Does this curve appear to be a better summary of the pattern of growth in the domestic cellular industry?

(g) What's the interpretation of the estimated intercept b_0 in the curve fit in part (f)?

(h) Use the equation from part (f) to predict the number of subscribers in the next period. Do you think this will be a better estimate than that offered by the linear equation or logarithmic curve?

38. **Cellular Phones in Africa** Mobile phones (as cellular phones are often called outside the United States) have replaced traditional landlines in parts of the developing world where it has been impractical to build the infrastructure needed for landlines. These data from the ITU (International Telecommunication Union) estimate the number of mobile and landline subscribers (in thousands) in Sub-Saharan Africa outside of South Africa. The data are annual, from 1995 through 2005.

(a) On the same axes, show timeplots of the two types of subscribers versus year. Does either curve appear linear?

(b) Create a scatterplot for the number of landline subscribers on the year of the count and then fit a linear equation with the number of subscribers as the response and the year as the explanatory variable. Does the r^2 of this fitted equation mean that it's a good summary?

(c) What do the slope and intercept tell you, if you accept this equation's description of the pattern in the data?

(d) Do the residuals from the linear equation confirm your impression of the fit of the model?

(e) Does a curve of the form

Estimated log (Number of Subscribers) = $b_0 + b_1$ Year

provide a better summary of the growth of the use of landlines? Use the residuals to help you decide.

(f) Interpret the slope in the previous equation that uses the log of the number of subscribers. What does it tell you about the growth of this sector?

(g) Fit a similar logarithmic equation to the growth in the number of mobile phone users. The curve is not such a nice fit, but allows some comparison of the rates of growth. How do the approximate rates of growth compare?

39. **Pet Foods, Revisited** This exercise uses base 10 logs instead of natural logs.

(a) Using the pet food data of the text example, transform the price and volume data using natural logs and then using base 10 logs. Then plot the natural log of volume on the natural log of price, and the base 10 log of volume on the base 10 log of price. What's the difference in your plots?

(b) Fit the linear equation of the log of volume on the log of price in both scales. What differences do you find between the fitted slopes and intercepts?

(c) Do the summary statistics r^2 and s_e differ between the two fitted equations?

(d) Create a scatterplot for the $\log_e$ of volume on the $\log_{10}$ of volume, and then fit the least squares line. Describe how this relationship explains the similarities and differences.

(e) Which log should be used to estimate an elasticity, or does it not matter?

40. **Movies** These data describe the box office success of 407 movies released during the years 1998 through

2001. For this analysis, you're interested in the relationship between initial success at the movie theater and subsequent sales for pay-per-view services, such as those offered by cable television. Both columns in the data table are in millions of dollars. Base 10 logs are more useful than natural logs when dealing with large monetary quantities since they give you the number of digits (minus 1).

(a) Create a scatterplot for subsequent sales on the box office sales, and scatterplot the $\log_{10}$ of the subsequent sales on the $\log_{10}$ of the box office sales. Can you describe the pattern in either plot using a linear equation?

(b) Estimate the elasticity of subsequent sales with respect to box office sales using the least squares equation with y given by the $\log_{10}$ of box office sales and x given by the $\log_{10}$ of subsequent sales. (Note: $\log_e x = 2.30262 \log_{10} x$.)

(c) Interpret the elasticity in the context of these data.

(d) How would the fitted equation in part (b) differ had you used natural logs (base e) rather than common logs (base 10)? Would the equation fit the data any better?

(e) Several successful movies that grossed more than $10 million at the box office have unusually large, negative residuals relative to the linear equation of $\log_{10}$ of subsequent sales on $\log_{10}$ of gross. Which movies are these and do you think that they have anything in common?

4M Cars in 1989

In order to have its cars meet the corporate average fuel economy (CAFE) standard of 27.5 MPG, a manufacturer needs to improve the mileage of two of its cars. One design, a small sports car, weighs 2,500 pounds. The other model, a four-door family sedan, weighs 4,000 pounds. If it can improve the mileage of either design by two more miles per gallon, its cars will meet the federal standards.

Use the data for cars from the 1989 model year to answer these questions. The weight of the cars is measured in thousands of pounds, and the city mileage is expressed in miles per gallon.

Motivation

a) Which of these two models should the manufacturer modify? In particular, if the manufacturer needs to reduce the weight of a car to improve its mileage, how can an equation that relates weight to mileage help?

Method

b) Based on the analysis in this chapter for modern cars, what sort of relationship do you expect to find between weight and mileage (city driving) for cars from the 1989 model year—linear or curved?

c) In order to choose the equation to describe the relationship between weight and mileage, will you be able to use summary measures like r^2, or will you have to rely on other methods to pick the equation?

Mechanics

d) Create a scatterplot for mileage on weight. Describe the association between these variables.

e) Fit an equation using least squares that captures the pattern seen in these data. Why have you chosen this equation?

f) Do the residuals from your fitted equation show random variation? Do any outliers stand out?

g) Compare the fit of this equation to that used in this chapter to describe the relationship between weight and mileage for more recent cars. Include in your comparison the slope and intercept of the fitted equation as well as the two summary measures, r^2 and s_e.

Message

h) Summarize the equation developed in your modeling for the manufacturer's management, using words instead of algebra.

i) Provide a recommendation for management on the best approach to use to attain the needed improvement in fuel efficiency.

4M Crime and Housing Prices in Philadelphia

Modern housing areas often seek to obtain, or at least convey, low rates of crimes. Homeowners and families like to live in safe areas and presumably are willing to pay a premium for the opportunity to have a safe home.

These data from *Philadelphia Magazine* summarize crime rates and housing prices in communities near and including Philadelphia. The housing prices for each community are the median selling prices for homes sold in the prior year. The crime rate variable measures the number of reported crimes per 100,000 people living in the community.

Exclude the data for Center City, Philadelphia from this analysis. It's a predominantly commercial area that includes clusters of residential housing. The amount of commercial activity produces a very large crime rate relative to the number of residents.

Motivation

a) How could local political and business leaders use an equation that relates crime rates to housing values to advocate higher expenditures for police?

b) Would an equation from these data produce a causal statement relating crime rates to housing prices? Explain.

Method

c) For modeling the association between crime rates and housing prices, explain why a community leader should consider crime rates the explanatory variable.

d) Do you anticipate differences in the level of crime to be linearly related to differences in the housing prices? Explain in the context of your answer the underlying implication of a linear relationship.

Mechanics

e) Create a scatterplot for the housing prices on crime rates. Describe the association. Is it strong? What is the direction?

f) Fit the linear equation of housing prices on crime rates. Interpret the slope, intercept, and summary statistics (r^2 and s_e).

g) As an alternative to the linear model, fit an equation that uses housing prices as the response with the reciprocal ($1/x$) of the crime rate as the explanatory variable. What is the natural interpretation of the reciprocal of the crime rate?

Message

h) Which model do you think offers the better summary of the association between crime rates and housing prices? Use residual plots, summary statistics, and substantive interpretation to make your case.

i) Interpret the equation that you think best summarizes the relationship between crime rates and housing prices.

j) Does an increment in the crime rate from 1 to 2 per 100,000 have the same impact (on average) on housing prices as the change from 11 to 12 per 100,000?

The Simple Regression Model

THE CAPITAL ASSET PRICING MODEL (CAPM) DESCRIBES THE RELATIONSHIP BETWEEN RETURNS ON A SPECULATIVE ASSET AND RETURNS ON THE WHOLE STOCK MARKET. According to the underlying theory, the market rewards investors for taking unavoidable risks, collectively known as *market risk*. Because investors cannot escape market risk, the market pays them for taking these risks. Other risks, collectively called *idiosyncratic risk*, are avoidable, and the CAPM promises no compensation for these. For instance, if you buy stock in a company that is pursuing an uncertain biotechnology strategy, you cannot expect the market to compensate you for those unique risks.

We can formulate the CAPM as a simple regression. The response R_t is the percentage change in the value of an asset during some time period, identified by the subscript t. Most often, R_t denotes the return on a stock during a month, say. The explanatory variable M_t is the contemporaneous percentage change in the value of the whole stock market. The equation associated with the CAPM is

$$R_t = \beta_0 + \beta_1 M_t + \varepsilon_t$$

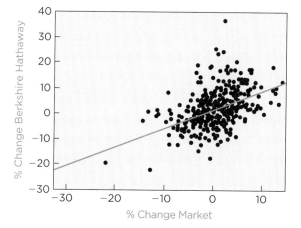

Estimated % Change Berkshire Hathaway = 1.40 + 0.722 % Change Market

FIGURE 21.1 Estimating the CAPM for Berkshire Hathaway.

The added term ε_t represents the effect of everything else on R_t. According to the CAPM, the mean of ε_t is zero and $\beta_0 = 0$. On average, if the return on the market is zero, we expect the return on a stock to be zero as well.

We can use regression analysis to test this theory. The scatterplot in Figure 21.1 shows monthly percentage changes in the price of stock in Berkshire Hathaway, the company managed by the famous investor Warren Buffett, versus those for the entire stock market. These data span October 1976 (the start of public trading in Berkshire Hathaway) through December 2007. The red line is the least squares line.

Previous chapters have shown how to summarize dependence with a line. Now comes inference. For example, is the estimated intercept $b_0 = 1.40$ in this equation large enough to suggest that $\beta_0 \neq 0$, contradicting the CAPM? To answer this question, we need tools for inference: standard errors, confidence intervals, and hypothesis tests. We develop these methods for regression analysis in this chapter.

21.1 | THE SIMPLE REGRESSION MODEL

Chapters 19 and 20 showed how we can use an equation to describe the association between an explanatory variable and the response. Our approach has been descriptive. This chapter turns to inference. To answer a question such as that posed in the introduction about the CAPM, we have to think of our data as a sample from a population. As in previous chapters, the statistical model for the population in regression analysis has random variables and parameters. For example, when making an inference about the mean of the population, we often model data as independent observations of a normally distributed random variable $Y \sim N(\mu, \sigma^2)$.

simple regression model (SRM) Model for the association in the population between an explanatory variable x and response y.

The **simple regression model (SRM)** describes the population in regression analysis with one explanatory variable. This model is more complicated than the models in Part 3 of this book because it must describe the relationship between two variables rather than the variation in one. To do this, the SRM combines an equation that captures the association between the response and the explanatory variable with a description of the remaining variation.

Linear on Average

The equation in the simple regression model specifies how the explanatory variable is related to the *mean* of the response. Because the model describes a population, we use random variables and Greek letters for the parameters of their distributions. The notation for these parameters requires subscripts that identify the random variables, since several random variables appear in the model.

The equation of the **SRM** describes how the conditional mean of the response Y depends upon the value of the explanatory variable X. The **conditional mean** of Y given X, written as $\mu_{y|x} = \mathrm{E}(Y|X = x)$, is the mean of Y given that the explanatory variable X takes on the value x. The conditional mean is analogous to the average within a row or column of a table. For instance, in the CAPM, $\mathrm{E}(Y|X = 5)$ denotes the mean percentage change in the value of stock in Berkshire Hathaway during months in which the market

conditional mean The average of one variable given that another variable takes on a specific value.

$$\mu_{y|x} = \mathrm{E}(Y|X = x)$$

increases by 5%. Similarly, $E(Y|X = 2)$ is the mean percentage change in Berkshire Hathaway during months in which the market increases by 2%. The SRM states that these averages align on a line with intercept β_0 and slope β_1:

$$\mu_{y|x} = E(Y|X = x) = \beta_0 + \beta_1 x$$

Finance analysts traditionally denote β_0 by α and β_1 by β in the CAPM.

> **caution** Don't form the hasty impression that the simple regression model describes only linear patterns. The variables *X* and *Y* may involve transformations such as logs or reciprocals as shown in Chapter 20. The variable *Y* in the SRM might be, for instance, the log of sales rather than sales itself.

Deviations from the Mean

error ε Deviation from the conditional mean specified by the SRM.

The equation of the simple regression model describes what happens on average. The deviations of observed responses around the conditional means $\mu_{y|x}$ are called **errors**. These aren't mistakes, just random variation around the line of conditional means. The usual symbol for an error is another Greek letter, ε, defined by the equation $\varepsilon = y - \mu_{y|x}$. The Greek letter ε (epsilon) is a reminder that we do not observe the errors; these are deviations from the *population* conditional mean, not an average in the data.

Errors can be positive or negative, depending on whether data lie above (positive) or below the conditional means (negative). Because $\mu_{y|x}$ is the conditional mean of Y given X in the population, the expected value of an error is zero, $E(\varepsilon) = 0$. The average deviation from the line is zero.

Because the errors are not observed, the SRM makes three assumptions about them:

- **Independent.** The error for one observation is independent of the error for any other observation.
- **Equal variance.** All errors have the same variance, $Var(\varepsilon) = \sigma_\varepsilon^2$.
- **Normal.** The errors are normally distributed.

If these assumptions hold, then the collection of all possible errors forms a normal population with mean 0 and variance σ_ε^2, abbreviated $\varepsilon \sim N(0, \sigma_\varepsilon^2)$.

Data Generating Process

When a statistical model for the population describes two or more random variables, it is helpful to provide a sequence of steps that show how we arrive at a sample from the population. This sequence is often called a data generating process. The data generating process is particularly important in regression analysis because of our asymmetric view of the roles of the explanatory variable and response. We want to estimate the conditional average of the response given the explanatory variable, not the other way around. As an illustration, let Y denote monthly sales of a company and let X denote its spending on advertising (both scaled in thousands of dollars). To specify the SRM, we assign values to the three parameters β_0, β_1, and σ_ε. Suppose that $\beta_0 = 500$ and $\beta_1 = 2$; if the company spends x thousand dollars on advertising, then the conditional mean of sales in thousands of dollars is

$$\mu_{y|x} = 500 + 2x$$

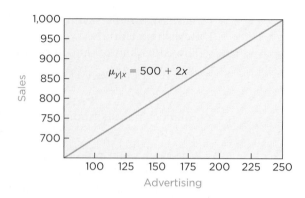

Without advertising ($x = 0$), β_0 indicates that sales average \$500,000. Every dollar spent on advertising increases expected sales by \$2. We further set $\sigma_\varepsilon = 45$. Collected together, the model specifies that sales have a normal distribution with mean $500 + 2x$ and standard deviation 45. These normal distributions line up as in Figure 21.2.

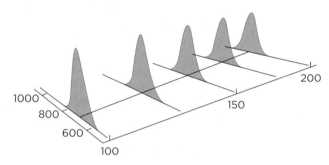

FIGURE 21.2 The simple regression model assumes a normal distribution at each x.

The data generating process defined by the SRM in Figure 21.2 allows the company to choose a value for the explanatory variable. The SRM does not specify how this is done; the company is free to decide how much it wants to advertise. **The SRM does not make assumptions about the values of the explanatory variable.**

tip

Suppose the company spends $x_1 = 150$ (thousand dollars) on advertising in the first month. According to our SRM, the expected value of sales given this advertising is

$$\mu_{y|150} = \beta_0 + \beta_1(150) = 500 + 2(150) = 800, \text{ or } \$800,000$$

Other factors collectively produce a deviation, or error, from $\mu_{y|150}$ that looks like a random draw from a normal distribution with mean 0 and SD σ_ε. Let's denote the first error as ε_1 and imagine that $\varepsilon_1 = -20$. The observed sales during the first month is then

$$y_1 = \mu_{y|150} + \varepsilon_1 = 800 + (-20) = 780, \text{ or } \$780,000$$

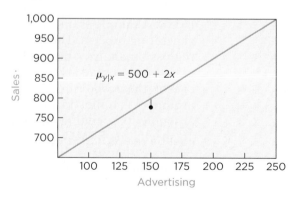

That's the dot in the figure above the dot is below the line because $\varepsilon_1 < 0$. If the company spends \$150,000 for advertising month after month, average sales will eventually settle down to $\mu_{y|150} = \$800,000$.

Let's follow the data generating process for a second month. In the next month, the company spends $x_2 = 100$ (thousand dollars) on advertising. The *same* line determines the average response; the equation for $\mu_{y|x}$ remains the same. Hence, the expected level of sales is

$$\mu_{y|100} = \beta_0 + \beta_1(100) = 500 + 2(100) = 700, \text{ or } \$700{,}000$$

Because the model assumes that the errors are independent of one another, we ignore ε_1 and independently draw a second error, say $\varepsilon_2 = 50$, from the *same* normal distribution with mean 0 and SD equal to σ_ε. The observed sales in the second month is then

$$y_2 = \mu_{y|100} + \varepsilon_2 = 700 + 50 = 750, \text{ or } \$750{,}000$$

This data point lies above the line because ε_2 is positive; it's a bit farther from the line than y_1 since $|\varepsilon_2| > |\varepsilon_1|$.

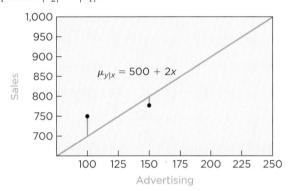

This process repeats each month. The company sets the amount of advertising, and the data generating process defined by the SRM determines sales. The company eventually observes data like those shown in Figure 21.3.

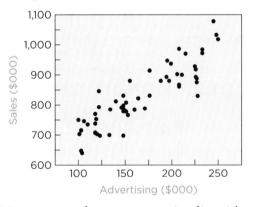

FIGURE 21.3 The observed data do not show the population regression line.

The line is gone. We never see the true regression line; it's a characteristic of the population, not our observed data. Recognize also that the simple regression model offers a simplified view of reality. The SRM is a *model*, not the real thing. Nonetheless the SRM is often a reasonable description of data. The closer data conform to this ideal data generating process, the more reliable inferences become.

Simple Regression Model (SRM)   Observed values of the response Y are linearly related to values of the explanatory variable X by the equation

$$y = \beta_0 + \beta_1 x + \varepsilon, \qquad \varepsilon \sim N(0, \sigma_\varepsilon^2)$$

The observations

1. are *independent* of one another,
2. have *equal variance* σ_ε^2 around the regression line, and
3. are *normally distributed* around the regression line.

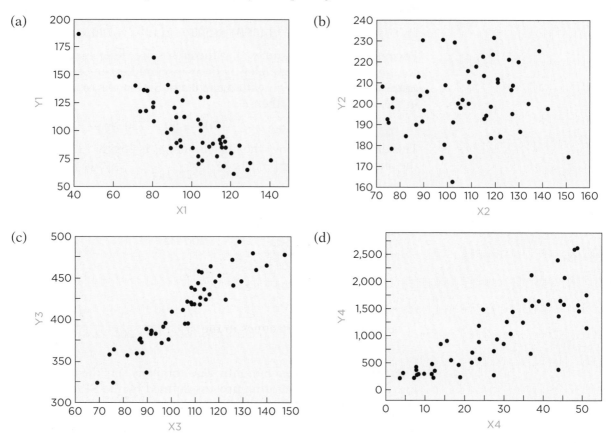

| **What Do You Think?** | Which scatterplots show data that are consistent with the data generating process defined by a simple regression model?[1] |

21.2 | CONDITIONS FOR THE SRM

We never know for certain whether the SRM describes the population. We only observe a sample and the fitted least squares regression line. The best we can do is to check several conditions. We checked these informally in Chapters 19 and 20, but we can now refine our conditions. Instead of checking for random residual variation, we have three specific conditions. If the answer to every question is "yes," then the data match the SRM well enough to move on to inference.

> **Checklist for the Simple Regression Model**
>
> ✓ Is the association between y and x linear?
> ✓ Have we ruled out obvious lurking variables?
> ✓ Are the errors evidently independent?
> ✓ Are the variances of the residuals similar? } Errors appear to be a sample from a normal population.
> ✓ Are the residuals nearly normal?

Let's check these conditions for the regression of percentage changes on Berkshire Hathaway stock versus the market (Figure 21.1).

---

[1] All but (d). In (a), the data track along a line with negative slope. In (b), there's little evidence of a line, but that just implies the slope is near zero. (c) is an example of a linear pattern with little error variation (strong association). (d) fails because the error variation grows with the mean.

✓ **Linear.** The pattern in the scatterplot of y on x in Figure 21.1 seems linear. The association is weak ($r^2 = 0.24$), but what there is seems linear. We confirm this by plotting the residuals versus the explanatory variable. This plot should look like a random swarm of bees, buzzing above and below the horizontal line at zero (Figure 21.4).

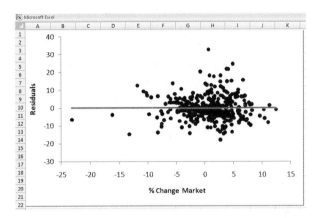

FIGURE 21.4 Residuals from the regression of percentage changes.

There's no evident pattern in Figure 21.4. The data are shifted to the right of the plot to accommodate outliers at the left (two months with large, well-known declines in the market: October 1987 and August 1998). This skewness in the distribution of x does not indicate a problem; the SRM makes no assumption about the distribution of the explanatory variable.

✓ **No obvious lurking variable.** According to the CAPM, there aren't any lurking variables. Some finance experts question the CAPM for just this reason, claiming that other variables predict stock performance.

We have three specific tasks when inspecting the residuals: check for independence, equal variance, and normality.

✓ **Evidently independent.** If the data are time series, as in this example, we can check for the presence of dependence by plotting the residuals over time. The timeplot in Figure 21.5 shows that the residuals vary around zero consistently over time, with no drifts. These appear independent.

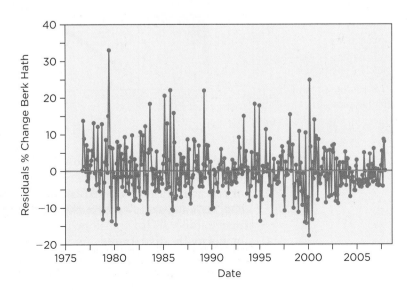

FIGURE 21.5 Timeplot of residuals from the CAPM regression for Berkshire Hathaway.

✓ **Similar variances.** To check this condition, plot the residuals versus x as in Figure 21.4. With the fitted line removed, it is easier to see changes in variation. In this case, the spread of the residuals around the horizontal line at zero appears constant. Although the range of the residuals is largest near $x = 0$ in Figure 21.4, this is the result of having more data when $x = 0$. The range can only get larger as we add more data. Be alert for a fan-shaped pattern or tendency for the variability to grow or shrink.

Notice a subtle problem in this example related to the variation in the residuals? The timeplot of the residuals in Figure 21.5 shows periods in which the residuals appear more and less variable. For instance, the residuals vary less after 2001 than in 2000. These gradual changes in variation do not cause problems in inferences about β_0 or β_1. This aspect of financial data (patterns in the variation) is an area of active research in quantitative finance.

✗ **Nearly normal.** A normal model is often a good description for the unexplained variation. The errors around the fitted line represent the combined effect of other variables on the response. Since sums of random effects tend to be normally distributed (the Central Limit Theorem), a normal model is a good place to start when describing the error variation. Because we don't observe the errors, we substitute the residuals from the least squares regression. To check that the residuals are nearly normal, inspect a histogram and normal quantile plot as in Chapter 12.

The histogram and normal quantile plot shown in Figure 21.6 summarize the distribution of the residuals from the regression of percentage changes in Berkshire Hathaway on those of the market.

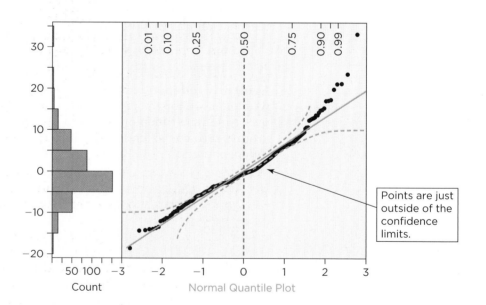

FIGURE 21.6 Histogram and normal quantile plot of residuals.

The residuals track the diagonal reference line except near the arrow. Here they drift too far from the reference line. The distribution of the residuals also has a long right tail. These residuals are not nearly normal. **Fortunately, inferences about β_0 and β_1 work well even if the data are not normally distributed.** Confidence intervals for the slope and intercept are reliable even if the errors are not normally distributed. As when finding confidence intervals for μ, this claim relies on the Central Limit Theorem.

If the residuals are not normally distributed (as in this example in Figure 21.6), check the sample size condition for the residuals (as in Chapter 14). The sample size should be larger than 10 times the larger of the skewness K_3 and the absolute value of the kurtosis K_4 of the residuals. In this example, $K_3 = 0.8$ and $K_4 = 2.5$, so we can rely on the CLT for making inferences about β_0 and β_1.

Modeling Process

There's a lot going on when building and checking a regression model, so let's collect the steps in one place. *Before* we look at plots, think about these two questions:

1. Does a linear relationship make sense?
2. Is the relationship free of obvious lurking variables?

If our answer to either question is "no," then we must find a remedy for the problem. We may need to transform a variable or find more data. *If* our answer to both questions is "yes," then we must start working with the data. We'll need to look at three or perhaps four plots when building a regression model:

1. The scatterplot of y on x,
2. The scatterplot of the residuals e on x,
3. The timeplot of the residuals if the data are time series, and
4. The histogram and normal quantile plot of the residuals.

Use this outline to fit and check the model:

- Plot y versus x and verify that the association appears linear. Don't worry about properties of the residuals until you get an equation that captures the pattern in the scatterplot.
- If the pattern is linear, fit the least squares regression line and obtain the residuals, e.
- Plot the residuals e versus the explanatory variable x. This plot should have no pattern. Curvature suggests that the association between y and x is not linear after all, and any "thickening" indicates different variances in the errors. Note the presence of outliers as well.
- If the data are measured over time, inspect a timeplot of the residuals for signs of dependence.
- Inspect the histogram and normal quantile plot of the residuals to check the nearly normal condition. If the residuals are not nearly normal, check the skewness and kurtosis.

It's a good idea to proceed in this order. For example, if we skip the initial check for linear association, we may find something unusual in the residuals. If the association is not linear, it will often happen that the residuals from the fitted line are not normally distributed. We may conclude, "Ah-ha, the errors are not normally distributed." That would be right, but we would not know how to remedy the problem. Detecting a problem when inspecting the distribution of the residuals offers little advice for how to fix it. **By following the suggested outline, you will detect the problem at a point in the modeling process where it can be fixed.** We'll have more to say about fixing common problems in Chapter 22.

tip

21.3 | INFERENCE IN REGRESSION

A model that survives this gauntlet of checks provides the foundation for statistical inference. Three parameters, β_0, β_1, and σ_ε, identify the population described by the simple regression model. The least squares regression line provides the estimates: b_0 estimates β_0, b_1 estimates β_1, and s_e estimates σ_ε.

Data		SRM
b_0	Intercept	β_0
b_1	Slope	β_1
$\hat{y}$	Line	$\mu_{y\|x}$
e	Deviation	ε
s_e	SD(ε)	σ_ε

Confidence intervals and hypothesis tests work as in inferences for the mean of a population:

- The 95% confidence intervals for β_0 and β_1 identify values for these parameters within about two standard errors of the estimates b_0 and b_1.
- The data supply enough evidence to reject a null hypothesis (such as H_0: $\beta_1 = 0$) if the estimate of the parameters lies too many standard errors from the hypothesized value.

Student's t-distribution determines precisely how many standard errors are necessary for statistical significance and confidence intervals. All we need are standard errors to go with b_0 and b_1.

Standard Errors

Standard errors describe the sample-to-sample variability of b_0 and b_1. Different samples from the population described by the SRM lead to different estimates. How different? That's the job of the standard error: estimate the sample-to-sample variation. If the standard error of b_1 is small, for instance, then not only are estimates from different samples similar to each other, but they are also close to β_1.

The formula for the standard error of the slope b_1 in a least squares regression resembles the formula for the standard error of an average. Both substitute sample standard deviations in place of unknown population parameters, and both take account of the sample size. Recall that the standard error of the average of a sample of n observations $y_1, y_2, \ldots, y_n$ is (Chapter 13)

$$\text{SE}(\overline{Y}) = \frac{\sigma_y}{\sqrt{n}}$$

Because σ_y is an unknown parameter, we plug in the sample standard deviation of the response s_y and use the estimated standard error given by

$$\text{se}(\overline{Y}) = \frac{s_y}{\sqrt{n}}$$

The square root of the sample size n is in the denominator; as the size of the sample increases, the sample-to-sample variation of $\overline{Y}$ decreases.

Formulas for the standard errors of b_0 and b_1 take the same approach. Both require an estimate of a population standard deviation, and both take account of the sample size. We'll focus here on the standard error of b_1; the formula for the standard error of b_0 appears at the end of this chapter in the list of formulas.

The formula for the standard error of the slope b_1 resembles the formula for the standard error of an average, with an extra multiplicative factor. The standard error of the least squares slope is

$$\text{SE}(b_1) = \frac{\sigma_\varepsilon}{\sqrt{n-1}} \times \frac{1}{s_x}$$

The **estimated standard error of b_1** substitutes the sample standard deviation of the residuals s_e for the standard deviation of the errors σ_ε,

$$\text{se}(b_1) = \frac{s_e}{\sqrt{n-1}} \times \frac{1}{s_x} \approx \frac{s_e}{\sqrt{n}} \times \frac{1}{s_x}$$

From this expression, we can see that three attributes of the data influence the standard error of the slope:

- Standard deviation of the residuals, s_e
- Sample size, n
- Standard deviation of the explanatory variable, s_x

The residual standard deviation sits in the numerator of the expression for $se(b_1)$. Since the regression line estimates the conditional mean of Y given X, it is the residuals that measure the variation around the mean in regression analysis. Hence s_e replaces s_y in the formula for the standard error. More variation around the line *increases* the standard error of b_1; the more the data spread out around the regression line, the less precise the estimate of the slope. The sample size n is again in the denominator. Larger samples *decrease* the standard error. The larger the sample grows, the more information we have and the more precise the estimate of the slope becomes.

To appreciate why the standard deviation s_x of the explanatory variable affects the standard error of the slope, compare the data in the two scatterplots shown in Figure 21.7. Each scatterplot shows a sample of 25 observations from the *same* regression model. Which sample provides a better estimate of the slope in a regression of y on x?

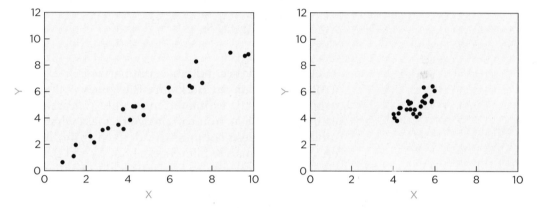

FIGURE 21.7 Which sample reveals more about the slope?

The difference between the samples is that the points in the scatterplot on the left spread out more along the x-axis; s_x is larger for the data in the left-hand scatterplot. This sample produces a more precise estimate of β_1.

The lines sketched in Figure 21.8 fit the data within each sample equally well. The lines on the left are very similar; those on the right have larger differences among slopes and intercepts.

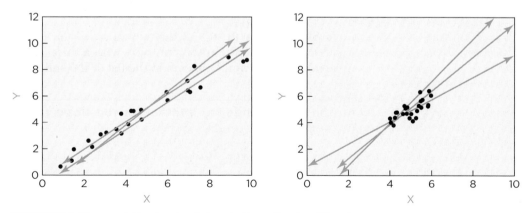

FIGURE 21.8 More variation in x leads to a better estimate of the slope.

Because the data on the left spread out along the x-axis, any line that comes close to these data has a similar slope. Hence, the sample on the left provides more information about β_1. The presence of s_x in the denominator of the formula for $se(b_1)$ captures this aspect of the sample. The larger the variation in the explanatory variable, the smaller the standard error for b_1 becomes.

Role of Software

Statistics packages typically handle the details of calculating both the least squares estimates and their standard errors. Software routinely presents a summary of a least squares regression and the estimates b_0 and b_1 in one or two tables like those shown in Table 21.1.

TABLE 21.1 Estimates for the CAPM regression for Berkshire Hathaway.

r^2	0.187872
s_e	6.517409
n	375

Term		Estimate	Std Error	t-Statistic	p-Value
Intercept	b_0	1.3962046	0.339682	4.11	<.0001
Market	b_1	0.7223495	0.077763	9.29	<.0001

Each row in the second table summarizes the estimate of a parameter in the CAPM regression for Berkshire Hathaway. Values in the row labeled "Intercept" describe the estimated intercept b_0; those in the next row describe the estimated slope b_1. You can probably guess what the last two columns of this table show. These columns are derived from the estimate and its standard error and summarize inferences about β_0 and β_1. The t-statistics and p-values show that both estimates are statistically significantly larger than zero.

Confidence Intervals

The sampling distribution of b_1 is centered at β_1 with standard deviation estimated by $se(b_1)$. If the errors ε are normally distributed or satisfy the CLT condition, then the sampling distribution of b_1 is approximately normal. Since we substitute s_e for σ_ε to calculate the standard error, we use a t-distribution for inference. These conditions imply that the sampling distribution of the ratio

$$T = \frac{b_1 - \beta_1}{se(b_1)}$$

is Student's t with $n - 2$ degrees of freedom, the divisor in s_e. Analogous results apply to the intercept, only with a different formula for its standard error. (That formula appears in the list at the end of this chapter.)

The **95% confidence interval for the slope β_1** in the simple regression model is the interval

$$[b_1 - t_{0.025,n-2} \times se(b_1), b_1 + t_{0.025,n-2} \times se(b_1)]$$

The **95% confidence interval for the intercept β_0** is

$$[b_0 - t_{0.025,n-2} \times se(b_0), b_0 + t_{0.025,n-2} \times se(b_0)]$$

Let's compute these confidence intervals for the CAPM regression summarized in Table 21.1. The sample size $n = 375$, leaving $n - 2 = 373$ degrees of freedom. Hence, the 95% confidence interval for β_1 (beta for Berkshire Hathaway) is

$$b_1 \pm t_{0.025,373} \times se(b_1) = 0.7223495 \pm 1.97 \times 0.077763 \approx [0.569 \text{ to } 0.876]$$

A stock with $\beta_1 = 1$ moves up and down on average with the market. If the market grows by 1%, such a stock grows by 1% on average as well. Because 1 lies outside the confidence interval for β_1, we conclude that β_1 for Berkshire Hathaway is statistically significantly less than 1. Returns on Berkshire Hathaway attenuate returns on the market, reducing the volatility of this asset.

The intercept is the more interesting parameter in the CAPM regression. The 95% confidence interval for β_0 (alpha for Berkshire Hathaway) is

$$b_0 \pm t_{0.025,373} \times \text{se}(b_0) = 1.3962046 \pm 1.97 \times 0.339682 \approx [0.727 \text{ to } 2.065]$$

Since zero lies well outside this interval, alpha for Berkshire Hathaway is statistically significantly larger than zero. Buffett's stock averaged statistically significantly higher returns than predicted by the CAPM: He beat the market by a statistically significant margin over these years.

Hypothesis Tests

Summaries of a least squares regression such as Table 21.1 provide several redundant ways to test the two-sided hypotheses H_0: $\beta_0 = 0$ and H_0: $\beta_1 = 0$. Each t-statistic in Table 21.1 is the ratio of an estimate to its standard error, counting the number of standard errors that separate the estimate from zero. For example, the t-statistic for the intercept in Table 21.1 is

$$b_0/\text{se}(b_0) = 1.3962046/0.339682 \approx 4.11$$

The estimated intercept b_0 lies about 4.11 standard errors above zero. The p-value converts the t-statistic into a probability, as when testing a hypothesis about μ. For the intercept, the p-value shown in Table 21.1 is smaller than 0.0001, far less than the common threshold 0.05. We can reject H_0 if we accept a 5% chance of a Type I error. This test agrees with the corresponding 95% confidence interval: The test rejects H_0: $\beta_0 = 0$ if and only if 0 lies *outside* the 95% confidence interval. Both the 95% confidence interval and the test of H_0: $\beta_0 = 0$ tell us that β_0 (alpha for Berkshire Hathaway) is statistically significantly larger than zero.

The t-statistic and p-value in the second row of Table 21.1 test the null hypothesis H_0: $\beta_1 = 0$; this test considers the population slope rather than the intercept. The t-statistic in this row shows that b_1 lies $b_1/\text{se}(b_1) = 9.29$ standard errors above zero. That's unlikely to happen by chance if the null hypothesis H_0: $\beta_1 = 0$ is true, so the p-value is tiny (less than 0.0001). In plain language, the t-statistic tells us that beta for Berkshire Hathaway is not zero. That's not a surprise; few financial experts would expect the beta for any stock to be zero.

The reason that software automatically tests H_0: $\beta_1 = 0$ is simple: If this hypothesis is true, then the distribution of Y is the same regardless of the value of X—there is no linear association between Y and X. In that case, Y has the same mean regardless of the value of x. For the regression of stock returns in the CAPM, the null hypothesis H_0: $\beta_1 = 0$ implies that returns on Berkshire Hathaway are unrelated to returns on the market.

Special language is often used when the data supply enough evidence to reject H_0: $\beta_1 = 0$. You'll sometimes hear that "the model explains statistically significant variation in the response" or "the slope is significantly different from zero." The first phrase comes from the connection between β_1 and the correlation between X and Y ($\rho = \text{Corr}(X, Y)$). If $\beta_1 = 0$, then $\rho = 0$. By rejecting H_0, we've said that $\rho \neq 0$. Hence $\rho^2 > 0$, too. **Remember, however, that failing to reject H_0: $\beta_1 = 0$ does not prove that $\beta_1 = 0$. Failure to reject H_0 shows that β_1 might be zero, not that it is zero.**

tip

caution | **The output of statistics packages by default always summarizes tests of H_0: $\beta_0 = 0$ and H_0: $\beta_1 = 0$. Be careful: These may not be the hypotheses of interest to us. Just because the data supply enough evidence to reject H_0: $\beta_1 = 0$**

does not imply that the data also reject the hypothesis that interests us. In the CAPM regression, for instance, it is more interesting to test the null hypothesis $H_0: \beta_1 = 1$. (This hypothesis says that average returns on Berkshire Hathaway move up and down with the market.) The appropriate t-statistic for testing this hypothesis is *not* the t-statistic in Table 21.1; that t-statistic compares the slope to 0. The correct t-statistic to test $H_0: \beta_1 = 1$ subtracts the hypothesized value 1 from the estimate:

$$t = (b_1 - 1)/se(b_1) = (0.7223495 - 1)/0.077763 = -3.57047$$

The estimate b_1 lies about 3.6 standard errors below 1. The p-value for this test is 0.0002 and we reject H_0. (This result is consistent with the 95% confidence interval for β_1; 1 is outside of the 95% confidence interval.)

Equivalent Inferences for the SRM We reject the claim that a parameter in the SRM (β_0 or β_1) equals zero with 95% confidence (or a 5% chance of a Type I error) if

a. Zero lies outside the 95% confidence interval for the parameter;
b. The absolute value of the associated t-statistic is larger than $t_{0.025,n-2} \approx 2$; or
c. The p-value reported with the t-statistic is less than 0.05.

What Do You Think?

The scatterplots shown in Figure 21.9 and the following tables summarize the relationship between total compensation of CEOs (y) and net sales (x) at 201 finance companies. Both variables are on a $\log_{10}$ scale. Hence, the slope is the elasticity of compensation with respect to sales, the percentage change in salary associated with a 1% increase in sales. (See Chapter 20.)

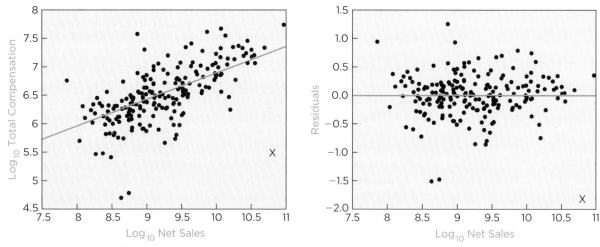

FIGURE 21.9 Log of CEO compensation versus log of net sales in the finance industry.

r^2	0.403728
s_e	0.388446

Term		Estimate	Std Error	t-Statistic	p-Value
Intercept	b_0	1.8653351	0.400834	4.65	<.0001
$\log_{10}$ Net Sales	b_1	0.5028293	0.043318	11.61	<.0001

a. Based on what is shown, check the conditions for the SRM:
   1. linear association
   2. no obvious lurking variable
   3. evidently independent
   4. similar variances
   5. nearly normal[2]
b. What does it mean that the t-statistic for b_1 is bigger than 10?[3]
c. Find the 95% confidence interval for the elasticity of compensation (the slope in this model) with respect to net sales.[4]
d. A CEO claimed to her Board of Directors that the elasticity of compensation with respect to net sales (β_1 in this context) is 1/2. Does this model agree?[5]
e. The outlier marked with an x at the right in the plots in Figure 21.9 is Warren Buffett. Why is this observation an outlier?[6]

4M EXAMPLE 21.1   LOCATING A FRANCHISE OUTLET

MOTIVATION ▸ STATE THE QUESTION

A common saying in real estate is that three things determine the value of a property: "location, location, and location." The same goes for commercial property. Consider where to locate a gas station. Many drivers choose where to buy gas out of convenience rather than loyalty to a particular brand. If a gas station is on their way, they will pull in and fill up. Otherwise, few are going to drive out of their way in order to get gas—especially if they're running low.

Does traffic volume affect sales? We want to predict sales volume at two sites. One site is located on a highway that averages 40,000 "drive-bys" a day; the other gets 32,000. How much more gas can we expect to sell at the busier location?

METHOD ▸ DESCRIBE THE DATA AND SELECT AN APPROACH

Identify x and y.

Link b_0 and b_1 to problem.

Describe data.

Check linear condition and lurking variable condition.

We won't find many stations that have 32,000 and 40,000 drive-bys a day, but we can find sales at stations with different levels of traffic volume. We can use these data in a regression, with Y equal to the average sales of gas per day (thousands of gallons) and X equal to the average daily traffic volume (in thousands of cars). The data used to fit the line in this example were obtained from sales during a recent month at $n = 80$ franchise outlets. All charge roughly the same price and are located in similar communities.

The intercept β_0 sets a baseline of gas sales that occur regardless of traffic intensity (perhaps due to loyal customers). The slope β_1 measures the sales per passing car. The 95% confidence interval for 8,000 times the estimated slope will indicate how much more gas to expect to sell at the busier location.

---

[2] The relationship seems linear (with logs). This variation seems consistent. It's hard to judge normality without a quantile plot, but these residuals probably meet the CLT condition.
[3] The t-statistic (11.61) indicates that b_1 is more than 11 standard errors away from 0, and hence statistically significant. The t-statistic also anticipates that 0 is not in the 95% confidence interval.
[4] The confidence interval for the slope is $0.503 \pm 1.97(0.0433) \approx [0.42 \text{ to } 0.59]$.
[5] 0.5 lies inside the confidence interval for β_1; 1/2 is a plausible value for the elasticity.
[6] Buffett earns a small salary (he makes his money in stock investments).

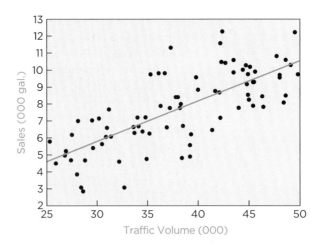

✓ **Linear.** The scatterplot suggests that the association is linear.

✓ **No obvious lurking variable.** These stations are in similar areas with comparable prices. We should also check that they face similar competition.

MECHANICS | DO THE ANALYSIS

The following tables summarize the least squares regression equation.

r^2	0.548596
s_e	1.505407
n	80

Term		Estimate	Std Error	t-Stat	p-Value
Intercept	b_0	−1.338097	0.945844	−1.41	0.1611
Traffic Volume (000)	b_1	0.236729	0.024314	9.74	<.0001

The following scatterplot graphs the residuals versus the explanatory variable, and the histogram and normal quantile plot summarize the distribution of the residuals.

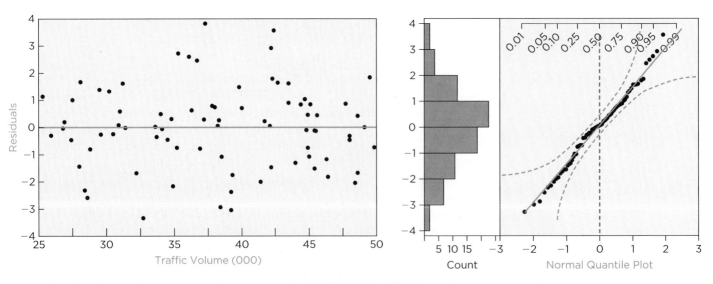

✓ **Evidently independent.** Nothing in the plots of the data suggests a problem with the assumption of independence.

✓ **Similar variances.** This is confirmed in the plot of the residuals on traffic volume. We might have expected more variation at the busier stations, but since these data are averages, that effect is not evident.

✓ **Nearly normal.** The histogram of the residuals is reasonably bell shaped with no large outliers. The points in the normal quantile plot stay near the diagonal.

Since the conditions for the SRM are met, we can proceed to inference. The 95% confidence interval for β_1 is

$$b_1 \pm t_{0.025,78}\,se(b_1) = 0.236729 \pm 1.99 \times 0.024314$$
$$\approx [0.188 \text{ to } 0.285 \text{ gallons/car}]$$

Hence, a difference of 8,000 cars in daily traffic volume implies a difference in average daily sales of

$$8,000 \times [0.188 \text{ to } 0.285 \text{ gallons per car}] \approx [1,507 \text{ to } 2,281 \text{ more gallons per day}]$$

MESSAGE | SUMMARIZE THE RESULTS

Based on a sample of 80 stations in similar communities, we expect (with 95% confidence) that a station located at a site with 40,000 drive-bys will sell on average from 1,507 to 2,281 more gallons of gas daily than a location with 32,000 drive-bys. *To make this range more memorable, it is often helpful to round the interval to emphasize the leading digits.* For example, when presenting this interval, it would be useful to report it as 1,500 to 2,300 more gallons daily.

It is also useful to mention possible lurking variables. We should supplement this analysis by checking that these stations face similar levels of competition and verify that they charge comparable prices. If, for example, stations at busy locations were charging higher prices, this equation may underestimate the benefit of the busier location: Estimates from this model would mix the effect on sales of increasing traffic (positive effect) with increasing price (negative effect).

21.4 | PREDICTION INTERVALS

Regression is often used to predict the response for new, unobserved cases. Suppose that we know that the value of the explanatory variable for a case is x_{new}, but we have yet to observe the corresponding response y_{new}. For instance, we know how much a company plans to spend on advertising next month (x_{new} is known), but we don't yet know the level of sales this advertising will produce (y_{new}). The SRM provides a framework that predicts y_{new} and anticipates the accuracy of this prediction.

$x_{\text{new}}, y_{\text{new}}$ Values of the explanatory variable and the response we'd like to predict.

Leveraging the SRM

The SRM model implies that y_{new} is determined by the equation

$$y_{\text{new}} = \beta_0 + \beta_1 x_{\text{new}} + \varepsilon_{\text{new}}, \qquad \varepsilon_{\text{new}} \sim N(0, \sigma_{\varepsilon}^2)$$

The summand ε_{new} is a random error term that represents the influence of other factors on the new observation. According to the SRM, ε_{new} is normally distributed with mean 0 and the same standard deviation σ_{ε} as the errors in the observed sample. This error term is also assumed to be independent of the observed data.

To predict y_{new}, we plug in the least squares estimates of β_0 and β_1, and we estimate ε_{new} by its mean, 0. The resulting prediction is $\hat{y}_{\text{new}} = b_0 + b_1 x_{\text{new}}$. This is the same formula that is used to obtain the fitted values, only now we're using it to predict a case for which we don't have the response.

The SRM also describes the accuracy of $\hat{y}_{\text{new}}$. The SRM states that the response is normally distributed around the line $\beta_0 + \beta_1 x$ at every value of x.

Hence, there's a 95% chance that y_{new} will fall within $1.96\sigma_\varepsilon$ of $\beta_0 + \beta_1 x_{new}$. We can write this as the following probability statement:

$$P(\beta_0 + \beta_1 x_{new} - 1.96\sigma_\varepsilon \leq y_{new} \leq \beta_0 + \beta_1 x_{new} + 1.96\sigma_\varepsilon) = 0.95$$

The resulting 95% ideal **prediction interval**

prediction interval An interval designed to hold a fraction (usually 95%) of the values of the response for a given value *x* of the explanatory variable in a regression.

$$[\beta_0 + \beta_1 x_{new} - 1.96\sigma_\varepsilon,\ \beta_0 + \beta_1 x_{new} + 1.96\sigma_\varepsilon]$$

holds 95% of all responses at x_{new}. Even if we knew β_0 and β_1, we would still not know y_{new} because of the other factors that affect the response. We call this range a *prediction* interval rather than a *confidence* interval because it makes a statement about the location of a new observation rather than a parameter of the population. This prediction interval, however, is not useful in practice because we don't know β_0, β_1, or σ_ε.

To get a prediction interval we can use, we substitute $\hat{y}_{new}$ for $\beta_0 + \beta_1 x_{new}$ and s_e for σ_ε. These changes require a different estimate of scale called the standard error of $\hat{y}_{new}$, or $se(\hat{y}_{new})$. There is a 95% chance that y_{new} will fall within $t_{0.025, n-2}se(\hat{y}_{new})$ of the estimated line:

$$P(\hat{y}_{new} - t_{0.025, n-2}se(\hat{y}_{new}) \leq y_{new} \leq \hat{y}_{new} + t_{0.025, n-2}se(\hat{y}_{new})) = 0.95$$

The **95% prediction interval** for the response y_{new} in the simple regression model is the interval

$$[\hat{y}_{new} - t_{0.025, n-2}se(\hat{y}_{new}),\ \hat{y}_{new} + t_{0.025, n-2}se(\hat{y}_{new})]$$

where $\hat{y}_{new} = b_0 + b_1 x_{new}$ and

$$se(\hat{y}_{new}) = s_e\sqrt{1 + \frac{1}{n} + \frac{(x_{new} - \bar{x})^2}{(n-1)s_x^2}}$$

The standard error of a prediction $se(\hat{y}_{new})$ is tedious to calculate because it adjusts for the position of x_{new} relative to the observed data. The farther x_{new} is from $\bar{x}$, the longer the prediction interval becomes.

In many situations, we can avoid this calculation by resorting to a simple "back-of-the-envelope" approximation. **So long as we're not extrapolating far from $\bar{x}$ and have a moderately sized sample, then $se(\hat{y}_{new}) \approx s_e$ and $t_{0.025, n-2} \approx 2$.** These produce the approximate 95% prediction interval

$$[\hat{y} - 2s_e,\ \hat{y} + 2s_e]$$

tip

Reliability of Prediction Intervals

Prediction intervals are reliable within the range of observed data, the region in which the approximate interval $[\hat{y} - 2s_e,\ \hat{y} + 2s_e]$ holds. Predictions that extrapolate beyond the data rely heavily on the SRM. Often, a relationship that is linear over the range of the observed data changes outside that range.

As an example of the problems of extrapolation, the scatterplots in Figure 21.10 at the top of the next page show prices for diamonds that weigh less than 1/2 carat (left, as in Chapter 19) and for a larger collection (right).

The line in both frames is fit only to diamonds that weigh up to 1/2 carat. The shaded area denotes the 95% prediction intervals for price. Although this line offers a good fit for smaller diamonds in the left frame, it underpredicts prices for larger gems. The price of a diamond grows faster than the fitted line anticipates; the prices of most of the larger diamonds lie above the extrapolated 95% prediction intervals shown in the right frame of Figure 21.10.

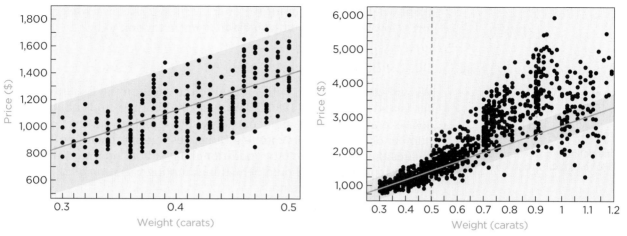

FIGURE 21.10 Prediction intervals fail when the SRM does not hold.

> **caution** Prediction intervals are also sensitive to the assumptions of constant variance and normality. A fan-shaped pattern that indicates increasing variation is a serious issue when using prediction intervals. If the variance of the errors around the regression line increases with the size of the prediction (as happens on the right in Figure 21.10), then the prediction intervals will be too narrow for large items. We will deal with this issue in Chapter 22. Similarly, if the errors are not normally distributed, the *t*-percentiles used to set the endpoints of the interval can be off target. Prediction intervals depend on normality of each observation, not normality produced by averaging. We cannot rely on the averaging effects of the CLT to justify prediction intervals because prediction intervals describe variability of individual observations. Before using prediction intervals, verify that residuals of the regression are nearly normal.

4M EXAMPLE 21.2   MANAGING NATURAL RESOURCES

MOTIVATION ▸ **STATE THE QUESTION**

Many commercial fish species have been virtually wiped out by overfishing. For instance, the stock of Mediterranean blue-fin tuna is nearly depleted and the large cod fishery in the northwest Atlantic is threatened. Worldwide, fishing has only been able to produce a steady yield of 85 million tons of fish per year over the last decade in spite of rising demand, higher prices, and better technology.

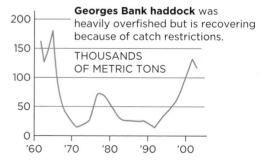

Georges Bank haddock was heavily overfished but is recovering because of catch restrictions.

THOUSANDS OF METRIC TONS

To protect native species from overfishing, many governments regulate fishing. For example, the Canadian government limits the number of days that fishing boats trapping Dungeness crabs can operate in its waters. To manage the commercial fleet requires understanding how the level of effort (number of boat-days) influences the size of the catch.

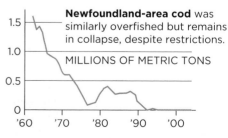

Newfoundland-area cod was similarly overfished but remains in collapse, despite restrictions.

MILLIONS OF METRIC TONS

What would you predict for the crab catch in a season with 7,500 days of effort? How accurate would you claim your prediction to be?

DESCRIBE THE DATA AND SELECT AN APPROACH

We will use regression, with Y equal to the catch in the fishery near Vancouver Island in 28 years (1980–2007, measured in thousands of pounds of Dungeness crabs that were landed) and with X equal to the level of effort (total number of days by boats catching Dungeness crabs). The slope in the SRM indicates the average catch per additional day a boat operates; the intercept is an extrapolation, but should be within sampling variation of zero if the model is a good match to the data.

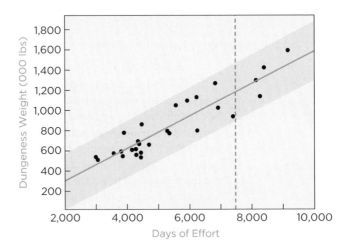

✓ **Linear.** The scatterplot suggests that the association between harvest weight and effort is strong and linear.

? **No obvious lurking variable.** These data are time series, so we need to check for possible lurking variables by considering the time plot of the residuals. Perhaps the industry has steadily improved the type of trap and is able to catch more crabs in a day of effort than in the past.

DO THE ANALYSIS

The following tables summarize the fit of the least squares regression equation.

r^2	0.840
s_e	123.33
n	28

Term		Estimate	Std Error	t-Stat	p-Value
Intercept	b_0	−32.83512	77.73172	−0.42	0.6762
Days of Effort	b_1	0.16081	0.01377	11.68	<0.0001

These data form a time series, so we begin with a timeplot of the residuals. The following plot shows the residuals from the model, plotted versus year. Although there may be short-term trends, the data do not show a clear pattern, suggesting both independence of the underlying model errors and a lack of an obvious lurking variable. (We'll cover other methods to check residuals like these in Chapter 22.)

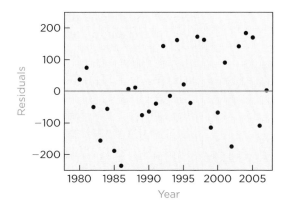

The following scatterplot graphs the residuals versus the explanatory variable, and the normal quantile plot summarizes the distribution of the residuals. *Because prediction intervals must account for the variation due to other factors (the error variation), it is essential that the model meet the conditions of the SRM.*

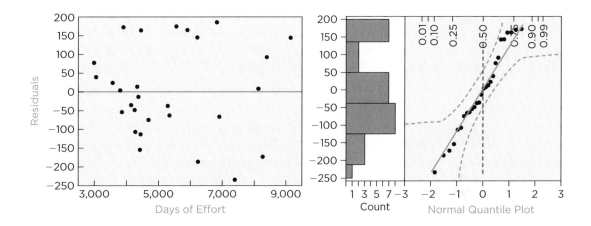

✓ **Evidently independent.** Nothing in these plots suggests dependence.

✓ **Similar variances.** Confirmed in the plot of the residuals on days of effort.

✓ **Nearly normal.** The histogram of the residuals is not very bell shaped, but with such a small sample these data could be normally distributed.

Since the conditions for the SRM are met, we can proceed to inference and prediction. The 95% confidence interval for β_1 is

$$b_1 \pm t_{0.025,26} \, se(b_1) = 0.16081 \pm 2.056 \times 0.01377$$

$$\approx [0.132 \text{ to } 0.189 \text{ thousand pound/boat-day}]$$

This interval does not include zero (b_1 is statistically significantly different from 0; we can reject H_0: $\beta_1 = 0$). Because the t-statistic for b_0 is small (the estimate is only 0.42 standard error below zero), the intercept is not statistically significantly different from zero; the 95% confidence interval for β_0 includes zero.

Because we reject H_0: $\beta_1 = 0$, the regression of weight on days of effort provides a better prediction of the catch than we can generate without knowing this variable. Had we not rejected H_0, then knowing the days of effort would not be useful since we could not tell whether β_1 was positive, negative, or zero. **Explanatory variables that do not explain statistically significant variation in the response don't produce more accurate predictions.**

tip

Now let's find the prediction and prediction interval. From the summary of the regression in the preceding table, the fitted line is

$$\hat{y} = b_0 + b_1 x = -32.83512 + 0.16081x$$

The predicted catch in a year with $x = 7{,}500$ days of effort is

$$\hat{y} = b_0 + b_1(7{,}500) = -32.83512 + 0.16081(7{,}500)$$
$$= 1{,}173.24 \text{ thousand pounds}$$

(or 1,173,240 pounds). Since we're not extrapolating, we might consider using the "back-of-the-envelope" approximate 95% prediction interval, namely

$$\hat{y} \pm 2\,s_e = 1{,}173.24 \pm 2 \times 123.33 = [926.58 \text{ to } 1{,}419.9 \text{ thousand pounds}]$$

Because this is a small sample, we refer to intervals computed by software. The exact 95% prediction interval is wider, from 908.44 to 1,438.11 thousand pounds. We'll use this interval in our message.

MESSAGE | SUMMARIZE THE RESULTS

Our analysis of the relationship between days of effort and total catch finds statistically significant linear association between these variables. On average, each additional day of effort (per boat) increases the harvest by about 160 pounds (that's 1,000 times $b_1 = 0.16081$, rounded to the nearest 10). In a season with 7,500 days of effort, we anticipate a total harvest of about 1,173,240 pounds. Because of the influence of other, unanticipated factors (this is the error variation in the model), there is a 95% probability that the catch will be between 908,440 and 1,438,110 pounds.

tip **Rounding often produces estimates that your audience will remember, particularly as the numbers get large.** This prediction interval indicates that we'd expect a total harvest between 910,000 and 1,440,000 pounds (or perhaps with more rounding, a range of 0.9 to 1.4 million pounds).

Best Practices

- *Verify that your model makes sense, both visually and substantively.* If you cannot interpret the slope, then what's the point of fitting a line? If the relationship between x and y isn't linear, there's no sense in summarizing it with a line.
- *Consider other possible explanatory variables.* The single explanatory variable may not be the only important influence on the response. If you can think of several other variables that affect the response, you may need to use multiple regression (Chapter 23).
- *Check the conditions, in the listed order.* The farther up the chain you find a problem, the more likely you can fix it. If you model data using a line when a curve is needed, you'll get residuals with all sorts of problems. It's a consequence of using

the wrong equation for the conditional average value of Y. Take a look at the two plots shown in Figure 21.11. In the scatterplot on the left, the pattern clearly bends. We fit a line anyway and saved the residuals. The resulting residuals are not nearly normal, but the normal quantile plot does not reveal the origin of the problem.

- *Use confidence intervals to express what you know about the slope and intercept.* Confidence intervals convey uncertainty and show that we don't learn things like the exact beta of a stock from data. Rounding helps in this context. Nothing makes you look more out of touch than claiming "the cost per carat is $2,333.6732 to $3,006.0355." The cost might be $2,500 or $2,900, and you're worried about $0.0032? Round the values!

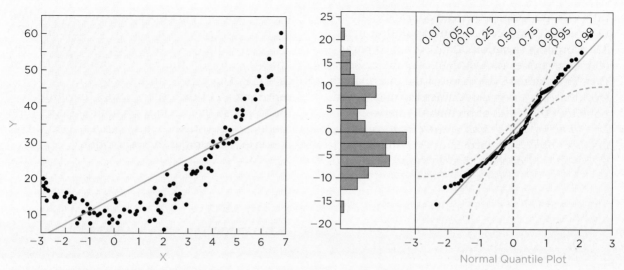

FIGURE 21.11 Fitting a line to a nonlinear pattern often produces residuals that are not nearly normal.

- *Check the assumptions of the SRM carefully before using prediction intervals.* Many inferences work well even if the data are not normally distributed. The Central Limit Theorem often produces normally distributed estimates of parameters even though the data are far from normally distributed. Prediction intervals, however, rely on the normal distribution to set a range for a single new value, and the CLT offers no help.

- *Be careful when extrapolating.* It's tempting to think that because you have prediction *intervals*, they'll take care of all of your uncertainty so you don't have to worry about extrapolating. Wrong: The interval is only as good as the model. Prediction intervals presume that the SRM holds, both within the range of the data *and* outside.

Pitfalls

- *Don't overreact to residual plots.* If you stare at a residual plot long enough, you'll see a pattern. Use the visual test for association if you're not sure whether there's a pattern. Even samples from normal distributions have outliers and irregularities every now and then.

- *Do not mistake varying amounts of data for unequal variances.* If the data have more observations at some values of the explanatory variable than others, it may seem as though the variance is larger where you have more data. That's because you see the range of the residuals when you look at a scatterplot, not the SD. The range can only grow with more data. The errors that are estimated by the residuals in the plot shown in Figure 21.12 have equal variance, but there are only 3 residuals in one location and 100 in the other.

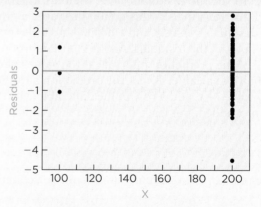

FIGURE 21.12 Residuals appear to have higher variances where there are many.

- *Do not confuse confidence intervals with prediction intervals.* Prediction intervals are ranges for single, as yet unobserved values. Confidence intervals are ranges for fixed characteristics of the population, such as the population mean. The sampling variation of the estimated parameters determines the width of confidence intervals. The standard deviation of the unexplained (residual) variation determines the major component of the width of prediction intervals.

- *Do not expect that r^2 and s_e must improve with a larger sample.* Standard errors get smaller as n grows, but r^2 doesn't head to 1 and s_e to zero.

Both r^2 and s_e reflect the variation of the data around the regression line. More data provide a better estimate of this line, but even if you know β_0 and β_1, there's still variation around the line. The errors around the line $\beta_0 + \beta_1 x$ represent all of the other factors that influence the response that are not accounted for by your model. Those other factors remain even if you increase the number of observations.

Software Hints

It's best to begin a regression analysis with a scatterplot of y on x. For checking the SRM, we need to go further and examine plots of the residuals.

EXCEL

Once we start checking regression models, we find it much easier to use DDXL rather than the commands built into Excel. After selecting the region with the data, follow the menu sequence

> DDXL > Regression > ***Click me***
> > Simple Regression

to get to the regression dialog. Fill in the names of the response and explanatory variable and click the OK button. The DDXL window shows a plot of the regression with summaries of the estimates and standard errors. To see the residuals, click the button labeled "Check the Residuals." This action produces a graph of the residuals versus x and a normal quantile (probability) plot of the residuals. Another button produces a table of the prediction intervals at the values of our explanatory variable. To get the prediction summary at any value of x, add a row to the data table with this value of x added, but leave the value for y empty.

MINITAB

To see plots of the residuals, use the menu sequence

> Stat > Regression . . . > Regression

and select the response and explanatory variable. Click the **Graphs** button and pick the 4-in-1 option to see all of the residual plots together. The

collection includes a scatterplot of the residuals on the fitted values and a normal probability plot of the residuals. The output that summarizes the regression itself appears in the scrolling session window.

JMP

Follow the menu sequence

> Analyze > Fit Y by X

to construct a scatterplot and add a regression line. (Click on the red triangle above the scatterplot near the words "Bivariate Fit of . . . .") Once you add the least squares fitted line, a button labeled Linear Fit will appear below the scatterplot.

Click on the red triangle in this field and choose the item Plot Residuals to see a plot of the residuals on the explanatory variable. To obtain a normal quantile plot of the residuals, save them first as a column in the data table. Click on the red triangle in the Linear Fit button and choose the item Save Residuals. To get the normal quantile plot, follow the menu sequence

> Analyze > Distribution

and choose the residuals to go into the histogram. Once the histogram is visible, click on the red triangle immediately above the histogram and choose the item Normal Quantile Plot.

The **simple regression model** (SRM) provides an idealized description of the association between two numerical variables. This model has two components. The equation of the SRM describes the association between the explanatory variable x and the response y. This equation states that the **conditional mean** of Y given $X = x$ is a line, $\mu_{y|x} = \beta_0 + \beta_1 x$. Both variables X and Y may require transformations. The second component of the SRM describes the random variation around this pattern as a sample of independent, normally distributed **errors** ε with constant variance. Before using this model we check the evidently independent, similar variances, and nearly normal conditions.

Confidence intervals for the parameters β_0 and β_1 of the linear equation are centered on the least squares estimates b_0 and b_1. The 95% confidence intervals are $b_0 \pm t \times \text{se}(b_0)$ and $b_1 \pm t \times \text{se}(b_1)$. The standard summary of a regression includes a t-statistic and p-value for testing $H_0: \beta_0 = 0$ and $H_0: \beta_1 = 0$. A **prediction interval** measures the accuracy of predictions of new observations. Provided the SRM holds, the approximate 95% prediction interval for an observation at x is $\hat{y}_{\text{new}} \pm 2s_e$.

■ Key Terms

condition, 519
   evidently independent, 519
   nearly normal, 520
   similar variances, 520

conditional mean, 514
error ε, 515
prediction interval, 530
simple regression model (SRM), 514

■ Formulas

Simple regression model

The conditional mean of the response given that the value of the explanatory variable is x is

$$\mu_{y|x} = \text{E}(y|x) = \beta_0 + \beta_1 x$$

Individual values of the response are

$$y_i = \beta_0 + \beta_1 x_i + \varepsilon_i$$

In this equation error terms ε_i are assumed

1. be independent of each other,
2. have equal standard deviation σ_ε, and
3. be normally distributed.

Checklist of conditions for the simple regression model

✓ Linear association
✓ No obvious lurking variable
✓ Evidently independent observations
✓ Similar residual variances
✓ Nearly normal residuals

Standard error of the slope

$$\text{se}(b_1) = s_e \sqrt{\frac{1}{(n-1)s_x^2}} \approx \frac{s_e}{\sqrt{n}} \times \frac{1}{s_x}$$

where s_x is the standard deviation of the explanatory variable,

$$s_x^2 = \frac{(x_1 - \bar{x})^2 + (x_2 - \bar{x})^2 + \cdots + (x_n - \bar{x})^2}{n - 1}$$

Standard error of the intercept

$$\text{se}(b_0) = s_e \sqrt{\frac{1}{n} + \frac{\bar{x}^2}{(n-1)s_x^2}} \approx \frac{s_e}{\sqrt{n}} \times \sqrt{1 + \frac{\bar{x}^2}{s_x^2}}$$

where s_e is the standard deviation of the residuals

$$s_e^2 = \frac{e_1^2 + e_2^2 + \cdots + e_n^2}{n - 2}$$

If $\bar{x} = 0$, the formula for $\text{se}(b_0)$ reduces to $s_e/\sqrt{n}$. The farther $\bar{x}$ is from 0, the larger the standard error of b_0 becomes. Large values of $\text{se}(b_0)$ warn that the intercept may be an extrapolation.

Standard error of prediction

When a simple regression model is used to predict the value of an independent observation for which $x = x_{\text{new}}$,

$$\text{se}(\hat{y}_{\text{new}}) \approx s_e \sqrt{1 + \frac{1}{n} + \frac{(x_{\text{new}} - \bar{x})^2}{(n-1)s_x^2}} \approx s_e$$

The approximation by $\text{se}(\hat{y}_{\text{new}}) \approx s_e$ is accurate so long as x_{new} lies within the range of the observed data and n is moderately large (on the order of 40 or more).

About the Data

To keep the discussion from bogging down, we avoided some details about the stock returns. To be precise, CAPM requires "excess returns" rather than simple returns. Simple returns, or percentage changes, ignore the interest that could have been earned on the money that's invested. Suppose you put $1,000 in a bank that pays 6% annual interest (1/2% per month). At the end of a month, you have 1,000 × 1.005 = $1,005, guaranteed. If you buy stock, the value of your investment depends on what happens in the market. If the stock goes up by more than 1/2%, you come out ahead. If the stock rises by 5% that month, for example, you could sell the stock for $1,050, put the money back in the bank, and come out $45 ahead. That $45 is the excess gain earned on the stock beyond what you could have made without risking a loss. That's really what the CAPM is about: the reward for taking risks. The standard approach to calculating excess returns subtracts the interest paid on U.S. government bonds, the so-called "risk-free rate." The shown percentage changes earned by Berkshire Hathaway and the market are 100 times the excess returns on each. If Berkshire Hathaway returned 3% in a month, and the risk-free rate is 1/2%, then the excess return is 2.5%. For the whole stock market, we use returns on the so-called value-weighted stock market index; interest rates on 30-day Treasury Bills provide the risk-free rate in the examples. The stock returns and risk-free returns are from the Center for Research in Security Prices (CRSP).

The data on harvests of Dungeness crabs are from tables that appear in several reports of Fisheries and Oceans Canada. The data on sales at gasoline stations are from a consulting project performed for a national oil company that operates gasoline stations around the United States. The salaries of CEOs in the finance industry are from the Compustat database for 2003.

EXERCISES

Mix and Match

Match each description on the left with the correct graph or equation on the right, taking into account all the items on pages 539 and 540 also.

1. Straight enough to fit a line

2. Not straight enough to fit a line

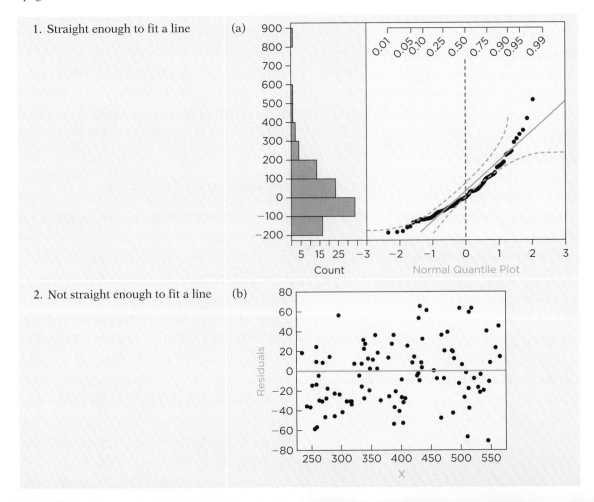

3. Similar variances

(c)

4. Not similar variances

(d)

5. Nearly normal

(e)

6. Not nearly normal

(f)

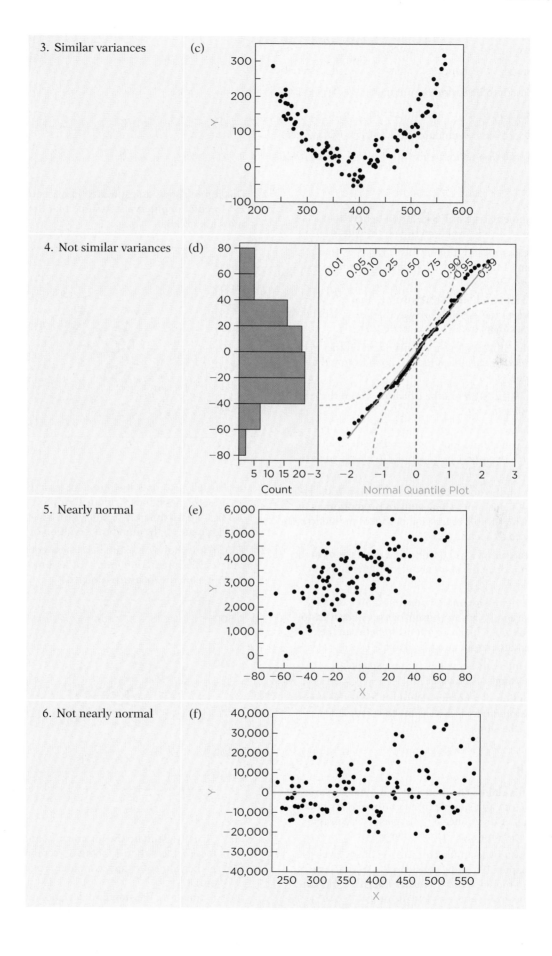

7. Average of y given x in the SRM	(g) $s_e/(\sqrt{n}\, s_x)$
8. Standard deviation of errors	(h) t-statistic
9. Standard deviation of residuals	(i) $\hat{y} \pm 2s_e$
10. Ratio of b_1 to its standard error	(j) σ_ε
11. Approximate standard error of b_1	(k) s_e
12. Approximate prediction interval	(l) $\mu_{y|x} = \beta_0 + \beta_1 x$

True/False

Mark each statement True or False. If you believe that a statement is false, briefly explain why you think it is false.

13. The simple regression model (SRM) requires that a histogram of the response look like a normal distribution.

14. In the ideal situation, the SRM assumes that observations of the explanatory variable are independent of one another.

15. Errors in the SRM represent the net effect on the response of other variables that are not accounted for by the model.

16. The errors in the SRM are the deviations from the least squares regression line.

17. To estimate the effect of advertising on sales using regression, you should look for periods with steady levels of advertising rather than periods in which advertising varies.

18. An increase in the observed sample size from 100 customers to 400 customers in a study of promotion response using simple regression produces predictions that are twice as accurate.

19. Doubling the sample size used to fit a regression can be expected to reduce the standard error of the slope by about 30%.

20. The simple regression model presumes, for example, that you have appropriately used logs or other transformation to obtain a linear relationship between the response and the explanatory variable.

21. Prediction intervals get wider as you extrapolate outside the range of the data.

22. The assumption of a normal distribution for the errors in a regression model is critical for the confidence interval for the slope.

Think About It

23. Consider this claim: The model for data that says each observation of y is a sample from a normal distribution with mean μ and variance σ^2 is a special case of the simple regression model. Do you agree?

24. Explain why a two-sample comparison of averages as done in Chapter 18 can be considered a special case of a regression model. (*Hint:* Consider assigning the value $x = 0$ to one group and $x = 1$ to the other.)

25. This figure shows an estimated linear equation with 95% prediction intervals. Assume that the units of x and y are dollars.

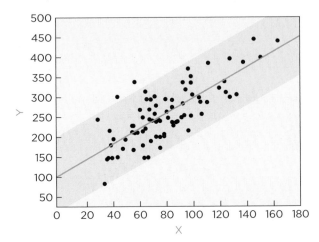

(a) Visually estimate the intercept b_0 and the slope b_1.
(b) Is se(b_1) less than 1, about 1, or more than 1?
(c) Is r^2 approximately 0.3, 0.5, or 0.8?
(d) Estimate s_e, the standard deviation of the residuals.

26. This figure shows an estimated linear equation along with its 95% prediction intervals. Assume that x counts customers and y measures dollars.

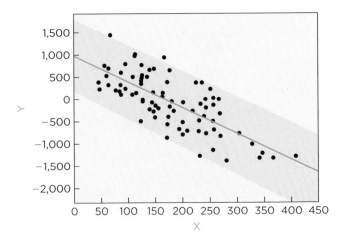

(a) Visually estimate the intercept b_0 and the slope b_1.
(b) Is se(b_1) less than 3, about 3, or more than 3?
(c) Is the value of r^2 approximately equal to 0.2, 0.5, or 0.8?
(d) Estimate s_e, the standard deviation of the residuals.

27. You suspect that the pattern relating sales (y) to levels of advertising (x) is not linear (perhaps a log transformation is needed to represent diminishing marginal returns[7]). Explain how you can still use the SRM as a framework for modeling the variation in the data.

28. Suppose that large diamonds (more than 1.5 carats) sold at retail tend to be of very mixed quality, whereas small diamonds have consistent quality (more uniform color, clarity, and attractive cut). If this is the case, can we use the SRM to describe the relationship between price and weight among diamonds of widely varying size?

29. A company tracks the level of sales at retail outlets weekly for 36 weeks. During the first 12 weeks, a fixed level of advertising was used each week to draw in customers. During the second 12 weeks, the level of advertising changed. During the last 12 weeks, a third level of advertising was used. What does the SRM have to say about the average level of sales during these three periods? (Treat sales as y and advertising as x and think of the data as 36 weeks of information.)

30. Referring to the previous scenario, suppose that during the first 12 weeks, this company was the only clothing retailer in a busy mall. During the second 12 weeks, a rival company opened. Then, during the third period, a second rival opened. Would the SRM still be a useful description of the relationship between sales and advertising?

31. The collection of emerald diamonds contains replications, several diamonds at each weight. Had we fit the linear equation using average price at each weight rather than individual diamonds, how would the fitted equation and summary measures (r^2 and s_e) change? (You don't have to create it—just describe in general terms how the equation and related measures would change.)

32. A marketing research company showed a sample of 100 male customers a new type of power tool. Each customer was shown the tool and given the chance to use it for several minutes. The customer was then asked to rate the quality of the tool (from 0 to 100) and told the tool would cost one of 10 prices, in $10 increments from $60 to $150. The tool is the same; only the stated price differs. In summarizing the results, the research company reported a linear equation summarizing the fit of the average quality rating versus the offered price:

    Estimated Average Rating = 115 − 0.5 Offered Price, with $r^2 = 0.67$ and $s_e = 15$

    How would the estimated equation and summary statistics likely differ had the company used individual ratings rather than averages?

33. Sometimes changes in stock prices are recorded as percentages, other times as returns. The difference is a factor of 100: Percentage changes are 100 times the returns. How does this choice affect the estimated value for β_1 in a regression that appears in the CAPM? What about β_0 in the CAPM?

34. In the CAPM regression, can you determine the risk of owning a stock (defined as the variation in returns) from the value of β_1 alone, or do you need to know more? Explain why β_1 is or is not enough, and if it's not enough, describe what else you need.

35. The following output summarizes the results of fitting a least squares regression to simulated data. Because we constructed these data using a computer, we know the SRM holds and we know the parameters of the model. We chose $\beta_0 = 7$, $\beta_1 = 0.5$, and $\sigma_\varepsilon = 1.5$. The fit is to a sample of $n = 50$ cases.

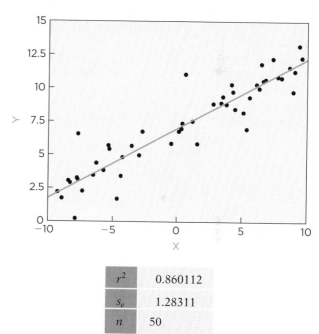

r^2	0.860112
s_e	1.28311
n	50

Term		Estimate	Std Error
Intercept	b_0	6.993459	0.181933
X	b_1	0.5134397	0.029887

(a) We know that β_1 is 1/2. Why isn't $b_1 = 1/2$?
(b) Do the 95% confidence intervals for β_0 and β_1 "work" in this example? (By "work," we mean that these intervals contain β_0 and β_1, respectively.)
(c) What's going to change in this summary if we increase the sample size from $n = 50$ to $n = 5,000$?

36. Representing a large auto dealer, a buyer attends car auctions. To help with the bidding, the buyer built a regression equation to predict the resale value of cars purchased at the auction. The equation is

    Estimated Resale Price ($) = 22,000 − 2,150 Age (years), with $r^2 = 0.48$ and $s_e = \$3,100$

---

[7] What are diminishing marginal returns? Diminishing marginal returns are common in advertising. You get a lot of bang for the dollar when the product is initially promoted. Those first ads have a large effect. Later on, as the market is saturated, the money spent for ads has less effect. The benefit of advertising diminishes as the level of advertising grows.

(a) Which is more predictable: the resale value of one five-year-old car, or the resale value of a collection of 25 cars, all of which are five years old?

(b) According to the buyer's equation, what is the estimated resale value of a five-year-old car? The average resale value of a collection of 25 cars, each five years old?

(c) Could the prediction from this equation overestimate or underestimate the resale price of a car by more than $2,500?

You Do It

For each data table, before answering the questions, determine whether the simple regression model is a reasonable description of the association between the two indicated variables. In particular, consider the conditions needed for the reliable use of the SRM. Then answer the listed questions. The data tables for these exercises are the same as those used for the like-numbered exercises in Chapter 19.

37. **Diamond Rings** This data table contains the listed prices and weights of the diamonds in 48 rings offered for sale in *The Singapore Times*. The prices are in Singapore dollars, with the weights in carats. Formulate the regression model with price as the response and weight as the explanatory variable.

(a) Could these data be a sample from a population in which the population intercept is zero? Should $\beta_0 = 0$?

(b) Is $800 an unusually high price for a ring with a diamond that weighs 0.25 carat?

38. **Convenience Shopping** It has become common to find a convenience store that also sells gas. These data describe the sales over time at a franchise outlet of a major U.S. oil company. Each row summarizes sales for one day. This particular station sells gas, and it also has a convenience store and a car wash. The column labeled *Sales* gives the dollar sales of the convenience store, and the column *Volume* gives the number of gallons of gas sold. Formulate the regression model with dollar sales as the response and number of gallons sold as the predictor. When checking the conditions for using the SRM, remember that these data are sequential.

(a) Give a confidence interval for the difference in average sales in the convenience store between days on which the station sells 2,000 gallons of gas and those on which the station sells 3,000 gallons.

(b) Is $1,800 an unusually low level of sales in the convenience store on a day that the pumps record having sold 3,000 gallons?

39. **Download** Before purchasing videoconferencing equipment, a company ran tests of its current internal computer network. The goal of the tests was to measure how rapidly data moved through the network given the current demand on the network. Eighty files ranging in size from 20 to 100 megabytes (MB) were transmitted over the network at various times of day, and the time to send the files recorded. Formulate the SRM with y given by Transfer Time and x given by File Size.

(a) Is the correlation between file size and transfer time significantly different from zero?

(b) Estimate the average "setup" time to start the file transfer. This time is used to synchronize the computers making the transfer and is unrelated to file size. Provide an interval suitable for presenting.

(c) To speed transfers, the company introduced compression software that halves the size of a file when sent over the network. On average, what is the improvement in time when sending a 50-MB file? State your answer as a range.

40. **Production Costs** A manufacturer produces custom metal blanks that are used by its customers for computer-aided machining. The customer sends a design via computer (a 3-D blueprint), and the manufacturer comes up with an estimated cost per unit, which is then used to determine a price for the customer. This analysis considers the factors that affect the cost to manufacture these blanks. The data for the analysis were sampled from the accounting records of 195 previous orders that were filled during the last 3 months. Formulate the regression model with y as the cost per unit and x as the material cost per unit.

(a) Does the material cost per unit explain statistically significant variation in the average cost of an order?

(b) A customer has called back with a revised order. Rather than use materials that cost $2 per unit, the customer prefers to use a cheaper material that reduces the cost to $1.60 per unit. Based on the fit of the indicated model, how should this reduction affect the average cost per unit ordered? Give your answer as a 95% confidence interval.

(c) Do you have any qualms about presenting this interval as a 95% confidence interval?

41. **Seattle Homes** This dataset contains the listed prices (in thousands of dollars) and the number of square feet for 28 homes in the Seattle area. The data come from the Web site of a Seattle realtor offering homes in the area for sale. For the SRM to work, we need to formulate the model in terms of cost per square foot. Use the selling price per square foot as y and the reciprocal of the number of square feet as x. (*Note:* If you keep track of the dimensions for the slope and intercept, you'll see that one represents fixed costs and the other, marginal costs.)

(a) Give a 95% confidence interval for the fixed cost (the portion of the cost that does not change with the size of the home) associated with these home prices, along with a brief interpretation.

(b) Give a 95% confidence interval for the marginal costs (the cost that is related to number of square feet), along with a brief interpretation.

(c) How much might a buyer pay, per square foot, for a specific home with 3,000 square feet in the Seattle area? Give a range, rounded appropriately, to show the buyer.

(d) How much in *total* might a buyer pay for a 3,000-square-foot home in the Seattle area? Give a range, rounded appropriately, to show the buyer.

42. **Leases** This dataset includes the annual prices of 223 commercial leases. All of these leases provide office space in a Midwestern city in the United States. For the response, use the cost of the lease per square foot. As the explanatory variable, use the reciprocal of the number of square feet.
    (a) Give a 95% confidence interval for the fixed cost (the portion of the cost that does not change with the size of the lease) of these leases, along with a brief interpretation.
    (b) Give a 95% confidence interval for the variable cost (the cost determined by the number of square feet) of these leases, along with a brief interpretation.
    (c) How much might a company pay, per square foot, for a specific lease with 15,000 square feet in this metropolitan area? Give a range, rounded appropriately, to show to management.
    (d) How much in *total* might the company pay for a 15,000-square-foot lease? Give a range, rounded appropriately, to show to management.

43. **R&D Expenses** This data file contains a variety of accounting and financial values that describe 493 companies operating in several technology industries in 2004: software, systems design, and semiconductor manufacturing. One column gives the expenses on research and development (R&D), and another gives the total assets of the companies. Both of these columns are reported in millions of dollars. These data need to be expressed on a log scale; otherwise, two outlying companies (Intel and Microsoft) dominate the analysis. Use the natural logs of both variables rather than the original variables in the data table. (Note that the variables are recorded in millions, so 1,000 = 1 billion.)
    (a) What difference in R&D spending (as a percentage) is associated with a 1% increase in the assets of a firm? Give your answer as a range, in presentation precision.
    (b) Revise your model to use base 10 logs of assets and R&D expenses. Does using a different base for both log transformations affect your answer to part (a)?
    (c) Estimate, with a range, the R&D expenses of a firm with $1 billion in assets. Be sure to express your range on a dollar scale.

44. **Cars** The cases that make up this dataset are types of cars. For each of 223 types of cars sold in the United States during the 2003 and 2004 model years, we have the base price and the horsepower of the engine (HP). Use the SRM of the price of the car on the horsepower to answer these questions.
    (a) A manufacturer has developed an improved, high-performance engine that boosts the engine HP by 20 HP, from 180 to 200 HP. It plans to offer this engine as an option on its current model. On average, how much does this model suggest that the company increase the price of the car to reflect the greater engine power? Give your answer as a 95% confidence interval.
    (b) Do you have any qualms about presenting this interval as an appropriate 95% range?
    (c) Based on the fit of this regression model, what would you expect to pay for a car with a 200 HP

engine? Give your answer as a 95% prediction interval. Do you think that standard prediction interval is reasonable? Explain.

45. **OECD** The Organization for Economic Cooperation and Development (OECD) tracks various summary statistics of its member economies. The countries lie in Europe, parts of Asia, and North America. Two variables of interest are GDP (gross domestic product per capita, a measure of the overall production in an economy per citizen) and trade balances (measured as a percentage of GDP). Exporting countries tend to have large positive trade balances. Importers have negative balances. These data are from the 2005 report of the OECD. Formulate the SRM with GDP as the response and Trade Balance as the explanatory variable.
    (a) On average, what is the per capita GDP for countries with balanced imports and exports (i.e., with trade balance zero)? Give your answer as a range, suitable for presentation.
    (b) The foreign minister of Krakozia has claimed that by increasing the trade surplus of her country by 2%, she expects to raise GDP per capita by $4,000. Is this claim plausible given this model?
    (c) Suppose that OECD uses this model to predict the GDP for a country with balanced trade. Give the 95% prediction interval.
    (d) Do your answers for parts (a) and (c) differ from each other? Should they?

46. **Hiring** A firm that operates a large, direct-to-consumer sales force would like to put in place a system to monitor the progress of new agents. A key task for agents is to open new accounts; an account is a new customer to the business. The goal is to identify "superstar agents" as rapidly as possible, offer them incentives, and keep them with the company. To build such a system, the firm has been monitoring sales of new agents over the past two years. The response of interest is the profit to the firm (in dollars) of contracts sold by agents over their first year. Among the possible predictors of this performance is the number of new accounts developed by the agent during the first three months of work. Formulate the SRM with y given by the natural log of Profit from Sales and x given by the natural log of Number of Accounts.
    (a) Is the elasticity of profit with respect to the number of accounts statistically significantly more than or less than 0.30 or equal to 0.30?
    (b) If the firm could train these sales representatives to increase the number of accounts created by 5%, would this be enough to improve the overall level of profits by 1%?
    (c) If a sales representative opens 150 accounts, what level of sales would you predict for this individual? Express your answer as an interval.
    (d) Would you expect the sales of about half the sales representatives with this many accounts to be less than the center of the interval formed to answer part (c), and half to be more?

47. **Promotion** These data describe spending by a major pharmaceutical company for promoting a cholesterol-lowering drug. The data cover 39 consecutive weeks and

isolate the area around Boston. The variables in this collection are shares. Marketing research often describes the level of promotion in terms of voice. In place of the level of spending, *voice* is the share of advertising devoted to a specific product. Voice puts spending in context; $10 million might seem like a lot for advertising unless everyone else is spending $200 million.

The column *Market Share* is sales of this product divided by total sales for such drugs in the Boston area. The column *Detail Voice* is the ratio of detailing for this drug to the amount of detailing for all cholesterol-lowering drugs in Boston. Detailing counts the number of promotional visits made by representatives of a pharmaceutical company to doctors' offices. Formulate the SRM with y given by the Market Share and x given by the Detail Voice.

(a) Is there statistically significant linear association between market share and the share of detailing?

(b) An advocate of reducing spending for detailing has argued that this product would have more than 20% of the market even if there were no detailing. Does this model support this claim?

(c) Accounting calculations show that at the current level of effort, the profit earned by gaining 1% more market share is 6 times the cost of acquiring 1% more detail voice. Should the company increase detailing?

(d) This regression uses share for both y and x. Had the analysis been done in percentages, would any of the previous answers change? If so, how?

48. **Apple** This dataset tracks the monthly performance of stock in Apple Computer since its inception in 1980. The data include 300 monthly returns on Apple, Inc., as well as returns on the entire stock market, Treasury Bills, and inflation. (The column labeled *Market Return* in the data table is the return on a value-weighted portfolio that purchases stock in proportion to the size of the company rather than one of each stock.) Formulate the SRM with Apple Return as the response and Market Return as the predictor.

(a) Is there statistically significant linear association between returns on stock in Apple and returns on the market?

(b) Is the estimate of the alpha for this stock (β_0) significantly different from zero?

(c) Is the estimate of the beta for this stock (β_1) significantly different from one?

(d) This regression uses returns. Had the analysis been done in percentages instead (the percentage changes are 100 times the returns), would any of the previous answers change? If so, how?

4M High-Frequency Data in Finance

This chapter introduced the capital asset pricing model to motivate several problems that regression analysis solves nicely. The slope and intercept in this model have particularly important meaning, with β_0 (known as alpha in finance) denoting the mean return of the idiosyncratic risk

over a time period (such as a month) and β_1 (known simply as beta) measuring the association with the market.

The importance of β_0 and β_1 in finance leads analysts to want better estimates of these quantities. We've used monthly data in these examples, but this exercise introduces daily returns. Here's the question: Can we get better estimates of β_0 and β_1 from using daily data? To put the estimates of β_0 on common scales, let's annualize the estimated rate of return. For monthly data, annualize the estimate of β_0 by multiplying by 12. For daily data, multiply the estimate by 250. (A typical year has 250 trading days.) Estimates of β_1 require no adjustment.

The two data tables in this example describe returns on stock in Apple Computer. One data table, introduced in the exercises of Chapter 19, uses 300 monthly returns on Apple from 1981 through 2005. The second data table, new for this example, provides 6,323 daily returns on stock in Apple over this same period.

Motivation

a) What is the importance of knowing β_0 more accurately? People in finance care a lot about this quantity, but why?

b) Why would you expect that using daily data rather than monthly data over a given time period would produce more accurate estimates of β_0 and β_1?

Method

c) Verify that the SRM that sets y to be returns on stock in Apple and sets x to be returns on the market meets the needed conditions. Be sure to consider both sampling rates (daily and monthly) and note any outliers or anomalies.

d) How can you use confidence intervals based on the estimates of β_0 and β_1 from the two fitted regression equations? Will they be identical? How close is close enough?

Mechanics

e) Fit the two simple regression models, one using daily data and the other monthly. Interpret the estimates of β_0 and β_1 in each case.

f) Form interval estimates of the annualized mean return on the intrinsic risk for the daily and monthly data. Are the intervals compatible, in the sense of both center and length?

g) Compare the estimates of beta from the daily and monthly data. Are the intervals compatible, in the sense of both center and length?

Message

h) Give an interval estimate of the annualized mean return on the intrinsic risk and β_1 for Apple. As part of your message, make sure that you've rounded appropriately.

i) If you could get data on a finer time scale, say every 30 minutes, would you be able to get better estimates of β_0 or β_1? (Tic data track the price of a stock with every trade, producing returns on a very fine time scale for actively traded investments.)

Regression Diagnostics

THE INTERNET HAS CHANGED MANY BUSINESS MODELS, GENERATING NEW APPROACHES TO RETAIL MARKETING, MEDICAL CARE, AND SERVICE INDUSTRIES. You can buy whatever you want on your computer, be it a video, a music player, or a car. You can even buy a house.

Realtors now offer "virtual tours" of homes. No need to spend the day driving from house to house. You can explore them from the comfort of your chair. Though convenient, this efficient approach to selling real estate can make it hard to appreciate a home. Without visiting the property and walking the area, how are you going to decide if the property is worth the price that's being asked?

22.1 **PROBLEM 1: CHANGING VARIATION**

22.2 **PROBLEM 2: LEVERAGED OUTLIERS**

22.3 **PROBLEM 3: DEPENDENT ERRORS AND TIME SERIES**

CHAPTER SUMMARY

A long-standing practice has been to find the selling prices of comparable homes that sold recently in the area. If the house next door sold for $450,000 last week, then you'd feel odd about paying $600,000 for this one that looks just like it, right? What makes this home so much better that the price is $150,000 higher? Perhaps the homes are not so similar after all—although it would take some very nice appliances in the kitchen to add up to $150,000!

To estimate the price of a home, you can use the average price of many similar homes. But what if you cannot find many similar properties? How can you estimate the price if, for instance, no other home of the same size has sold recently? We'll use regression analysis to handle this problem. Regression analysis allows for the use of prices of homes of varying size to estimate the price of a home of a specific size.

The problem is that as homes get larger, the prices tend to vary more as well. That's one of the complications of regression analysis addressed in this chapter. This chapter describes three problems that often affect regression models: changing variation in the data, outliers, and dependence among the observations. For each problem, we begin with an example in which one or more problems affect the interpretation and use of a fitted regression equation. Each example shows how to

- recognize that there is a problem,
- identify the consequences of the problem, and then
- fix the problem when possible.

The fix is often easy, but we first have to recognize that there's a problem.

22.1 | PROBLEM 1: CHANGING VARIATION

The scatterplot in Figure 22.1 graphs the offered prices of 94 houses in Seattle, Washington, versus their sizes in square feet. All of these are detached, single-family houses within a particular section of the city, and all share similar construction. The houses differ, however, in size. The line in the scatterplot is the least squares line.

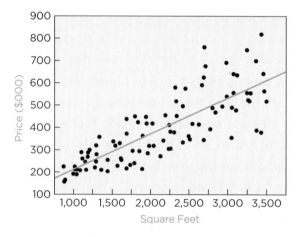

FIGURE 22.1 Prices of larger homes are higher and more variable than prices of smaller homes.

The pattern is clear: Bigger houses cost more, but only on average. Price does not always increase with size. One house with 1,300 square feet, for instance, commands a higher price than a larger house with 2,100 square feet. Evidently, other attributes of the smaller house more than compensate for its smaller size. In addition to the upward trend, there's another pattern in the association between size and price. The variation increases with size. Both the average and standard deviation of price increase with the size of the house.

Table 22.1 summarizes the least squares line in Figure 22.1.

TABLE 22.1 Simple regression of price on square feet.

r^2	0.6545
s_e	91.05
n	94

Term		Estimate	Std Error	t-Statistic	p-Value
Intercept	b_0	50.598687	27.41166	1.85	0.0681
Size	b_1	0.1594259	0.012076	13.20	<0.0001

Let's review the interpretation of this summary, starting with the overall goodness of fit. The value of r^2 shows that the fitted line $\hat{y} = b_0 + b_1 x$ represents almost two-thirds (65.45%) of the variation in prices. The line describes most of the variation among prices, but it leaves a lot of room for error. Size is an important factor in real estate, but not the only one. The standard deviation of the residuals around the fit is s_e = 91.05, or \$91,050. If we were to predict the price of a house using this model, we would often be off by \$100,000 or more in either direction.

Fixed Costs, Marginal Costs, and Variable Costs

The intercept and slope of the line summarized in Table 22.1 represent fixed costs and marginal costs. These concepts appear in previous chapters but are important enough to review here.

The intercept b_0 = 50.598687 naïvely estimates the cost of a house with no square feet. That's an extrapolation. We are better served by interpreting the intercept as the fixed cost of a home purchase. A fixed cost is present regardless of the size of the house. All houses, regardless of size, have a kitchen and appliances. Because the houses in this analysis are located near each other, they also share the climate and access to the same public schools. The combination of these attributes determines the fixed cost.

Referring to Table 22.1, the stated 95% confidence interval for the intercept is

$$b_0 \pm t_{0.025,92} \, \text{se}(b_0) = 50.598687 \pm 1.99 \times 27.41166 \approx [-3.951 \text{ to } 105.148]$$

Rounding to multiples of \$1,000 looks reasonable once you've seen this interval: The expected fixed costs lie between −\$4,000 and \$105,000 (with 95% confidence). This interval includes zero, meaning that we're not sure of the sign. Fixed costs could be very large, or they could be virtually nothing or slightly negative. This range implies that the fitted line does not yield a precise estimate of fixed costs.

The slope b_1 = 0.1594259 thousand dollars per square foot estimates the marginal cost of an additional square foot of space. The marginal cost determines how the total price increases with size. The marginal cost times the number of square feet gives the variable cost of the house. A homebuyer can reduce the variable cost by choosing a smaller house. Fixed costs, in contrast, remain regardless. For these houses, each additional square foot adds about \$160 to the estimated price. When comparing houses, for example, we expect a home with 3,500 square feet to be priced on average about \$160,000 more than another with 2,500 square feet. We'd be smart to add a range to this estimate. The 95% confidence interval for the slope given by the fitted model is

$$b_1 \pm t_{0.025,92} \, \text{se}(b_1) = 0.1594259 \pm 1.99 \times 0.012076 \approx [0.1354 \text{ to } 0.1835]$$

This confidence interval implies a range of \$135,400 to \$183,500 for the average difference in price associated with 1,000 square feet.

Detecting Differences in Variation

This sample of homes reveals a lot about the market for real estate in this section of Seattle. The regression estimates fixed costs in the 95% confidence interval

Fixed costs: −\$4,000 to \$105,000

The estimated marginal costs are in the interval

Marginal costs: \$135 to \$184 per square foot

These 95% confidence intervals rely on the simple regression model. In our haste to interpret the fitted line, we have not carefully checked the conditions for the SRM:

✓ Linear association
✓ No obvious lurking variables
✓ Evidently independent observations
✓ Similar residual variances
✓ Nearly normal residuals

Judging from Figure 22.1, the pattern seems linear. Also, the description of the data removes some concerns about lurking variables. For example, if larger homes have a scenic view that smaller homes lack, then some of the association that we're attributing to size might be due to the better view. That's not the case for these data (all of the homes are in the same neighborhood), but that's the sort of thing that can go wrong with regression if we don't understand the data.

These data do not satisfy the similar variances condition. To identify this problem, use the scatterplot of the residuals on x. **The scatterplot of the residuals on the explanatory variable is useful for detecting deviations from linearity and changes in the variation around the trend.** These problems are more apparent in this plot (Figure 22.2) than in the scatterplot of y on x (Figure 22.1).

tip

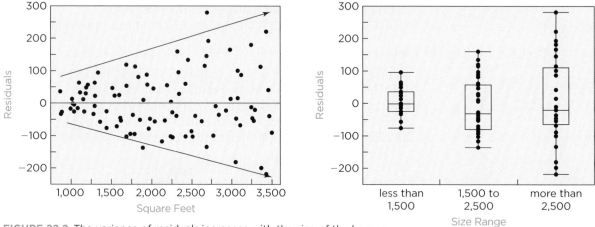

FIGURE 22.2 The variance of residuals increases with the size of the homes.

The scatterplot of e versus x has a fan-shaped appearance. As houses get larger, the residuals become more variable. Prices of houses with about 1,000 square feet stay close to the fitted line; residuals at the left of the scatterplot are within about $50,000 of the line. Prices of houses with 3,000 or more square feet vary farther from the line, reaching out more than plus or minus $200,000. The side-by-side boxplots on the right of Figure 22.2 confirm this pattern. These boxplots summarize the residuals grouped into three bins: houses with less than 1,500 square feet, those with 1,500 to 2,500 square feet, and those that are larger. The boxplots show more and more variation as the size increases.

Although the variation of the residuals grows with size, the fitted model summarizes them with *one* standard deviation ($s_e = \$91,050$). The SRM gives *one* estimate for the variation of the residuals when in fact the error variation changes. For small homes, s_e is too large. For large homes, s_e is too small. The technical name for this situation is to say that the errors are heteroscedastic. **Heteroscedastic** errors have different variances. The SRM assumes that the errors are **homoscedastic**, having equal variation.

heteroscedastic Having different amounts of variation.

homoscedastic Having equal amounts of variation.

Consequences of Different Variation

The presence of changing error variation has two consequences for a regression analysis. First, imagine what would happen if we were to use the regression summarized in Table 22.1 to predict the price of a small house. The predicted price, in thousands of dollars, for a house with 1,000 square feet is

$$\hat{y} = b_0 + b_1 \times 1,000 = 50.598687 + 0.1594259 \times 1,000$$
$$= 210.0246, \text{ or } \$210,025$$

Rounded to thousands of dollars, the approximate 95% prediction interal is

$$\hat{y} \pm 2s_e \approx \$210{,}000 \pm 2 \times \$91{,}000 = \$28{,}000 \text{ to } \$392{,}000$$

This interval is much wider than it needs to be.

At the same time, the fitted model underestimates the variation in prices for large houses. For example, the predicted price of a house with 3,000 square feet is

$$\hat{y} = b_0 + b_1 \times 3{,}000 = 50.598687 + 0.1594259 \times 3{,}000$$
$$= 528.8767, \text{ or } \$528{,}876$$

The approximate 95% prediction interval for the price is

$$\hat{y} \pm 2s_e \approx \$529{,}000 \pm \$182{,}000 = \$347{,}000 \text{ to } \$711{,}000$$

Although this prediction interval is quite wide, it is nonetheless too narrow to obtain 95% coverage.

The scatterplot in Figure 22.3 summarizes the problem. The shaded region around the fitted line shows the 95% prediction intervals given by the SRM (as given by the formulas in Chapter 21) for the least squares regression of price on size. All of the points representing smaller homes lie well inside the shaded region, whereas five points representing larger homes (points marked with ×) lie outside the prediction intervals.

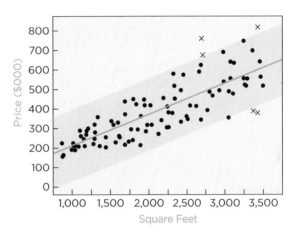

FIGURE 22.3 The 95% prediction intervals are too wide for small homes and too narrow for large homes.

Overall, these prediction intervals have done their job: About 95% of the data (89 out of 94) lie inside the prediction intervals. The problem is that all of the points outside the intervals represent large houses, and points for small houses are far inside the bounds. The prediction intervals are too wide at the left of the plot (small houses) and too narrow at the right (large houses).

The second consequence of heteroscedasticity is more subtle. It concerns the standard errors and confidence intervals that express the precision of the estimated intercept and slope, here our estimates of fixed and marginal costs. Inferences based on these estimates rely on the simple regression model. Since the SRM fails to hold for these data, the confidence intervals computed in the previous section are unreliable. The stated interval for marginal costs, $135 to $184 per square foot, *is probably not* a 95% confidence interval. We calculated this interval by using the formula for a 95% confidence interval, but these formulas are not reliable if the data fail to conform to the SRM.

Consequences of Changing Variation (Heteroscedasticity)

1. Prediction intervals are too narrow or too wide.
2. Confidence intervals for slope and intercept are not reliable.
3. Hypothesis tests regarding β_0 and β_1 are not reliable.

What Do You Think?

The scatterplot in Figure 22.4 shows the regression of annual sales of hospital supplies versus the number of sales representatives who work for an equipment manufacturer.

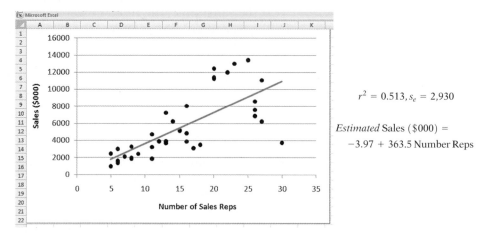

$r^2 = 0.513, s_e = 2{,}930$

Estimated Sales ($000) = $-3.97 + 363.5$ Number Reps

FIGURE 22.4 Scatterplot and regression of sales versus number of representatives in territories.

Each point shows sales (in thousands of dollars) in 37 territories; territories are distinct regions located around the United States. This manufacturer puts more representatives in districts with more sales leads, so the number of representatives varies from territory to territory.

a. Explain why we should expect to find heteroscedasticity in this regression.[1]
b. What additional plot would be most useful to check for a problem?[2]
c. Construct the approximate 95% prediction interval for sales in a territory with five representatives from this regression. Do you see a problem with the interval?[3]

Fixing the Problem: Revise the Model

Most regression modeling is iterative. No one gets it right the first time, every time. We try our best, see how it works, and then refine the model as we learn from the data.

Let's think in terms of fixed costs and variable costs. If we let F represent fixed costs and M Marginal costs, the equation of the SRM becomes

$$\text{Price} = F + M \times \text{Sq Ft} + \varepsilon$$

The problem in this example is that the errors ε vary more for large homes than small homes. The local market keeps the price of a small home above a certain level, but even a great small home is still small. Larger homes, on the other hand, allow more opportunities for enhancements that add value than smaller homes. A plain large home has a low price, but a very attractive large home commands a huge price.

 Differences in size often produce heteroscedastic data, so it's good that a simple remedy usually fixes the problem. To explain the remedy, we need a concept from Chapter 9. If Y is a random variable and c is a constant, then the

---

[1] A territory with five really good (or poor) sales representatives can only generate a limited range of sales. Each rep can only do so much. An office with 30 could have 30 really good reps (or 30 weak reps), producing a wide range in possible sales.
[2] The scatterplot of the residuals on the explanatory variable is useful for seeing heteroscedasticity.
[3] Expected sales in a territory with 5 reps is about $-3.97 + 363.5(5) \approx \$1{,}814$ (thousand). The lower bound of the approximate 95% prediction interval is $\hat{y} - 2s_e = 1{,}814 - 2(2{,}630) < 0$. Negative sales are not possible. The interval is too wide.

standard deviation of Y divided by c is $\text{SD}(Y/c) = \text{SD}(Y)/c$. To use this result in regression, notice that errors for large homes have large variation and errors for small homes have small variation. If we divide the errors for large homes by a large value and divide the errors for small homes by a small value, we end up with data that have more similar variances. (A technique known as *weighted least squares* is often used in this situation. Our approach shows why and how well it works.)

You may be wondering how we're going to divide the errors by anything since we never observe ε. It turns out to be surprisingly easy. Let's go back to the previous equation for home prices. Divide *both* sides of the equation by the number of square feet and simplify:

$$\frac{\text{Price}}{\text{Sq Ft}} = \frac{F + M \times \text{Sq Ft} + \varepsilon}{\text{Sq Ft}}$$

$$= \frac{F}{\text{Sq Ft}} + M + \frac{\varepsilon}{\text{Sq Ft}}$$

$$= M + F \times \frac{1}{\text{Sq Ft}} + \varepsilon'$$

Dividing price by square feet has the right effect: The error term ε is divided by a large value for large houses and a small value for small houses.

The revised equation has a different response and a different explanatory variable. The response becomes the price per square foot and the explanatory variable becomes the reciprocal of the number of square feet. The errors in the revised equation (labeled ε') are those in the prior equation divided by the number of square feet. (This use of the reciprocal differs from its use in Chapter 20. In Chapter 20, we used the reciprocal to obtain a linear pattern. In this example, we started with a linear pattern. The reciprocal arises here because we manipulated the equation to equalize the variance of the errors.)

Another aspect of the model changes when we divide by the number of square feet. This change can be confusing: The intercept and slope swap roles. In the revised equation, the marginal cost M is the intercept. The slope is F, the fixed costs. If we keep track of the units of the intercept and slope, we'll have no problem recognizing the change of roles. In the revised equation, the units of the response are \$/Sq Ft, matching those of a marginal cost. The units on the slope are \$, appropriate for a fixed cost.

The scatterplots in Figure 22.5 summarize the fit of the revised equation, graphing both price per square foot and the residuals versus 1/Sq Ft.

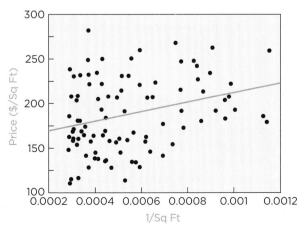
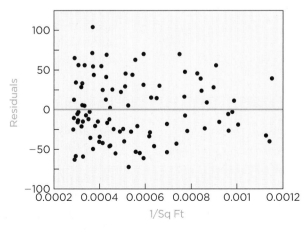

FIGURE 22.5 Cost per square foot versus the reciprocal of square feet (data left and residuals right).

The residuals seem to become less variable as 1/Sq Ft increases. It looks that way, but it's not the case. The appearance is most likely an artifact of having less data on the right-hand side of the plots. As noted in Chapter 21, the range gets larger as we add more data. Most of the residuals cluster on the left of these graphs because the distribution of 1/Sq Ft is skewed. This skewness does not violate an assumption since the SRM makes no assumption about the distribution of the explanatory variable.

tip **To avoid the illusion of different variances, use boxplots to compare the variation.** The boxplots in Figure 22.6 show subsets of the residuals from the regression of price per square foot on the reciprocal of size. The groups are the same as those shown previously in Figure 22.2 (houses with less than 1,500 square feet, from 1,500 to 2,500 square feet, and more than 2,500 square feet).

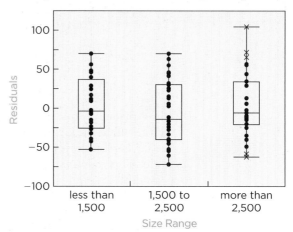

FIGURE 22.6 The variance of the residuals is similar in the two groups.

The residuals have similar variances. Small differences in the interquartile ranges are not enough to indicate a problem; each IQR is about $55 to $60 per square foot. In contrast, the interquartile ranges of the boxplots of the initial residuals (shown in Figure 22.2) steadily increase from about $60 for small homes to more than $170 for large homes.

We'll summarize this analysis with a 4M presentation.

4M EXAMPLE 22.1 | ESTIMATING HOME PRICES

MOTIVATION ▸ STATE THE QUESTION

A company is relocating several managers to the Seattle area. The company needs to budget for the cost of helping these employees move to Seattle. We will break down prices into fixed and variable costs to help management prepare to negotiate with realtors.

METHOD ▸ DESCRIBE THE DATA AND SELECT AN APPROACH

Identify x and y.

Link b_0 and b_1 to problem.

Describe data.

Check for linear association.

The analysis uses simple regression. The explanatory variable is the reciprocal of the size of the home (1 over the number of square feet) and the response is the price per square foot. The intercept in this regression estimates the expected price per square foot (marginal cost) and the slope estimates the fixed costs. We will use the sample of 94 homes for sale in Seattle.

✓ **Linear.** The scatterplot (Figure 22.5) shows linear association.

✓ **No obvious lurking variable.** Nothing other than size distinguishes the larger homes from the smaller homes within this sample.

MECHANICS | DO THE ANALYSIS

Start by verifying other conditions. If these check out, proceed to the details of the fitted equation.

✓ **Evidently independent.** This seems okay because these homes were randomly sampled from available listings within a neighborhood of Seattle.

✓ **Similar variances.** The variation in the price per square foot is stable (compared to the changing variation in prices).

✓ **Nearly normal.** Now that we have similar variances, we can check for normality. The histogram appears somewhat bimodal. Even so, the residuals stay within the limits around the diagonal reference line. These are nearly normal.

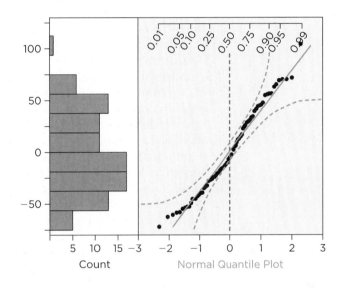

r^2	0.09119
s_e	39.41

Term	Estimate	Std Error	*t*-Statistic	*p*-Value
b_0	157.753	10.52	14.99	<0.0001
b_1	53,887	17,736	3.04	0.003

The fitted equation is

$$\textit{Estimated } \$/\text{Sq Ft} = 157.753 + 53{,}887/\text{Sq Ft}$$

The intercept is the typical marginal cost per square foot in Seattle, about $158. The 95% confidence interval for the cost per square foot is

$$b_0 \pm t_{0.025,92}\, se(b_0) = 157.753 \pm 1.99(10.52) = [136.8182 \text{ to } 178.6878]$$

which rounds to, say, $140 to $180 per square foot. The slope estimates the fixed costs at around $54,000 regardless of home size. The 95% confidence interval is

$$b_1 \pm t_{0.025,92}\, se(b_1) = 53{,}887 \pm 1.99(17{,}736) = [18{,}592.36 \text{ to } 89{,}181.64]$$

which rounds to $19,000 to $89,000 (or perhaps $20,000 to $90,000).

MESSAGE | SUMMARIZE THE RESULTS

Based on prices of a sample of 94 homes listed for sale in Seattle, we estimate that prices of homes in this neighborhood run about $140 to $180 per square foot, on average. The variable cost on a 2,000 square foot home, for example, averages from $280,000 to $360,000. Average fixed costs associated with the purchase are in the range $19,000 to $89,000, with 95% confidence.

These results rely on a small sample of homes. We can expect that other factors aside from size, notably location, lot size, and style of construction, also affect prices. We can model those effects when we get more data.

Comparing Models with Different Responses

It may be hard to accept that the revised regression is better than the initial regression of price on size. The modified regression of price per square foot on the reciprocal of square feet has $r^2 = 0.09$, far less than the $r^2 = 0.65$ of the initial model.

How can we claim that a model with such a smaller r^2 is better? That's easy: The model with price per square foot is a better match to the SRM. Without the SRM, we cannot make inferences about fixed and variable costs or form reliable prediction intervals. We should not judge which is the better model by comparing the r^2 statistics. The first model explains variation in the total price. That's easy: Large homes cost more. The second model explains variation in the price per square foot. That's hard, and as a result, the model has a smaller r^2.

Superficial comparisons based on r^2 distract from meaningful comparisons that reflect on how we plan to use our model. We're not just trying to get a large r^2; we want to infer fixed and variable costs and predict prices. Even though the revised model has a smaller r^2, it provides more reliable *and narrower* confidence intervals for fixed and variable costs. That's right: The model with the smaller r^2 in this case produces shorter confidence intervals. Table 22.2 compares the estimates of fixed costs from the two regressions. (We shaded the confidence interval from the initial model in orange to flag it as unreliable due to the failure of the similar-variances condition.)

TABLE 22.2 Comparision of estimated fixed costs from two regression models.

Response	Similar Variances?	Estimated Fixed Cost	95% Confidence Interval	
			Lower	Upper
Price	No	$50,599	−$4,000	$105,000
Price/Sq Ft	Yes	$53,887	$18,000	$89,000

A comparison of confidence intervals for the marginal cost is similar (Table 22.3). The revised model that adjusts for heteroscedasticity gives a shorter 95% confidence interval.

TABLE 22.3 Comparision of estimated marginal cost.

Response	Similar Variances?	Estimated Marginal Cost	95% Confidence Interval	
			Lower	Upper
Price	No	$159/Sq Ft	$135/Sq Ft	$184/Sq Ft
Price/Sq Ft	Yes	$158/Sq Ft	$137/Sq Ft	$179/Sq Ft

The superiority of the second model becomes clearer if we consider prediction. Consider predicting the prices of two houses, a small house with 1,000 square feet and a large house with 3,000 square feet. Table 22.4 shows the results, rounded to thousands of dollars.

TABLE 22.4 Comparison of approximate 95% prediction intervals for the prices of two homes.

Size (sq ft)	Response	Similar Variances?	95% Prediction Interval		
			Lower	Upper	Length
1,000	Price	No	$28,000	$392,000	$364,000
	Price/Sq Ft	Yes	$133,000	$290,000	$157,000
3,000	Price	No	$347,000	$711,000	$364,000
	Price/Sq Ft	Yes	$291,000	$765,000	$474,000

The regression that uses price per square foot as the response gives more sensible prediction intervals. For the house with 1,000 square feet, its 95% prediction interval is less than half as long as that from the initial model, with a plausible lower bound. For the house with 3,000 square feet, the revised regression produces a wider range that accommodates the larger variation in prices among big houses.

To obtain prediction intervals for price from the model with price per square foot as the response, start by finding the prediction interval for the price per square foot. The predicted price per square foot for a house with 1,000 square feet is

$$\hat{y} = 157.753 + 53,887/1,000 = \$211.640/\text{Sq Ft}$$

The approximate 95% prediction interval for the price per square foot is

$$\hat{y} \pm 2 \times s_e = 211.640 \pm 2 \times 39.41 \approx [\$133/\text{Sq Ft to } \$290/\text{Sq Ft}]$$

Hence, a house with 1,000 square feet is predicted to cost between about $133,000 and $290,000.

For a house with 3,000 square feet, the predicted price per square foot is

$$\hat{y} = 157.753 + 53,887/3,000 \approx \$175.715/\text{Sq Ft}$$

The approximate 95% prediction interval is then

$$\hat{y} \pm 2 \times s_e = 175.715 \pm 2 \times 39.41 \approx [\$97/\text{Sq Ft to } \$255/\text{Sq Ft}]$$

Hence, the prediction interval for the price of a house with 3,000 square feet is $291,000 to $765,000 (i.e., 3,000 times the shown range).

After adjusting for heteroscedasticity, regression gives a *longer* prediction interval for the price of a larger home than for that of a smaller home. The revised model recognizes that the variation in the price increases and builds this aspect of the data into prediction intervals. It correctly claims more accurate predictions of the prices of smaller houses and less accurate predictions of the prices of larger houses.

22.2 | PROBLEM 2: LEVERAGED OUTLIERS

Outliers are another common problem in regression. Outliers are observations that stand away from the rest of the data and appear distinct in a plot. The following example illustrates how a particular type of outlier affects regression and considers how we decide whether to retain the point or exclude it from the modeling.

Let's stay in the housing market. Rather than finding the carpenters, electricians, and plumbers needed for a home renovation, most homeowners hire a contractor. The contractor finds the right people for each task, manages the schedule, and checks that the work is done properly. When picking a contractor, homeowners usually solicit several bids, asking each contractor to estimate the cost of completing the work. Contractors who offer high bids make a nice profit—if the homeowner accepts their bid. A contractor wants to offer the highest bid that gets accepted *and* covers costs. If he bids too low, he loses money on the project. If he bids too high, he does not get the job.

Let's take the contractor's point of view. This contractor is bidding on a project to construct an 875-square-foot addition to a house. He's kept data that record his costs for $n = 30$ similar projects. All but one of these projects, however, are smaller than 875 square feet. His lone project of this size is an outlier at 900 square feet. The scatterplot in Figure 22.7 graphs his costs versus the size of the additions and shows two fitted lines.

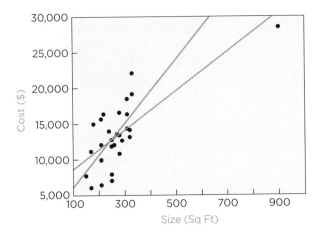

FIGURE 22.7 Contractor costs versus size of project.

The red line is the least squares fit to all of the data; the green line is the least squares fit excluding the outlier. If the contractor includes the outlier, the fitted line is flatter, passing near this point. Data such as this outlier that are near the minimum or maximum of the explanatory variable are said to be **leveraged**. The name *leverage* refers to the ability of these points to tilt the regression line in their direction. If the contractor excludes the leveraged outlier, the line becomes steeper and produces a fit with a larger slope.

leveraged A leveraged observation in regression has a small or large value of the explanatory variable.

Consequences of an Outlier

Table 22.5 contrasts the least squares fits, both with and without the outlier.

TABLE 22.5 Comparison of two fits, with and without a leveraged outlier.

Including Outlier		Excluding Outlier
0.558567	r^2	0.376464
3,196.80	s_e	3,093.18
30	n	29

Estimate	Std Error	Term	Estimate	Std Error
5,887.74	1,400.02	b_0	1,558.17	2,877.88
27.44	4.61	b_1	44.74	11.08

The standard errors provide a scale to measure the differences between the fitted lines. Estimates of the slope and intercept that differ by fractions of a standard error aren't so far apart; such differences can be attributed to sampling variation. Estimates that differ by several standard errors, however, are quite far apart.

We can interpret the intercepts in these equations as estimates of the contractor's fixed costs, the money spent to get a project going and maintain the business, regardless of the size of the project. With the outlier included, the least squares estimate of fixed costs is $5,887.74. Without the outlier, the estimate of fixed costs falls to $1,558.17. **Use the standard error from the regression without the outlier to compare the estimates.** Dropping one observation out of 30 shifts the estimated fixed costs by

$$\frac{5,887.74 - 1,558.17}{2,877.88} \approx 1.50 \text{ standard errors}$$

The outlier shifts the estimate of fixed costs to near the upper endpoint of the 95% confidence interval obtained when the outlier is excluded. That's a substantial change. The addition of one observation shifts estimated fixed costs nearly outside the confidence interval defined by the other 29.

Comparison of the slopes produces a similar conclusion. The slopes in both models estimate marginal costs that determine how the total cost depends on the size of the project. The fit with the outlier estimates the marginal cost at $27.44 per square foot, compared to $44.74 per square foot without the outlier. Using the standard error from the fit *without* the outlier, this change is large relative to the precision of the estimate. The estimated slope in the regression with the outlier lies

$$\frac{27.44 - 44.74}{11.08} \approx -1.56 \text{ standard errors}$$

below the estimate obtained from the other 29 cases. As with the intercept, one case dramatically changes our estimate.

What Do You Think?

A single point in each of the three plots in Figure 22.8 is highlighted with an × symbol. Each scatterplot shows the least squares regression estimated *without* this point.

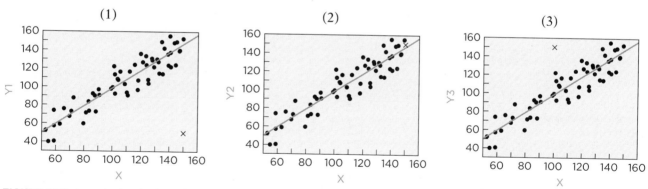

FIGURE 22.8 A marked point is excluded from the regression in each scatterplot.

a. Indicate whether the marked point is leveraged in each case.[4]
b. Will the r^2 of the regression increase, decrease, or stay about the same if the marked point is added to a regression?[5]
c. How will adding the marked point to the regression affect the estimated slope and intercept?[6]

---

[4] The marked point is leveraged in (1) and (2), but not (3).
[5] R^2 will decrease in (1), stay about the same (a bit higher) in (2), and be a bit lower in (3).
[6] The slope will decrease and the intercept increase in (1), both will stay about the same in (2), and the intercept will increase a little and the slope stay about the same in (3).

Extrapolating Prediction Intervals

The presence of a leveraged outlier also affects prediction. Consider predicting the cost to the contractor of building the 875-square-foot addition. With the outlier, regression predicts the cost to the contractor to be

$$\hat{y} = b_0 + b_1 \times 875 = 5{,}887.74 + 27.44 \times 875 = \$29{,}897.74$$

The approximate 95% prediction interval, rounded to the nearest dollar, is

$$\hat{y} \pm 2s_e = 29{,}898 \pm 2 \times 3{,}197 \approx [\,\$23{,}504 \text{ to } \$36{,}292\,]$$

The approximate prediction interval is reliable so long as we have not extrapolated. When predicting under conditions outside the range of the explanatory variable, we must use the prediction interval that uses the formula for $se(\hat{y})$ given in Chapter 21. This adjustment increases the length of the prediction interval to account for extrapolation. Accounting for extrapolation produces a 95% prediction interval that is considerably wider than the approximate interval:

$$\hat{y} \pm t_{0.025,28}\, se(\hat{y}) = [\,\$21{,}166 \text{ to } \$38{,}638\,]$$

An 875-square-foot addition lies at the edge of these data if the outlier is included. Without the outlier, this prediction represents a huge extrapolation from the rest of the data.

The scatterplots in Figure 22.9 show the effect on prediction intervals of removing the outlier. The shaded region in the scatterplot on the left shows 95% prediction intervals based on the regression fit with the outlier. These intervals gradually widen as the size of the project increases.

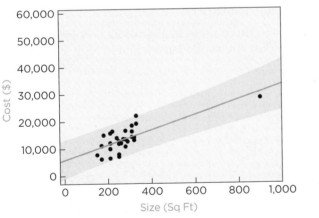 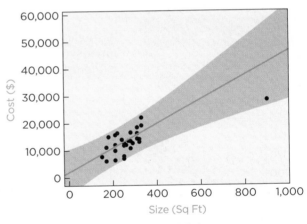

FIGURE 22.9 Prediction intervals from a model that uses the outlier (left) and omits the outlier (right).

The right panel of Figure 22.9 shows prediction intervals produced by the regression that excludes the outlying project. The 95% prediction interval for the total cost of an 875-square-foot addition runs from $25,191 to $56,221. Without knowing what happened with the previous big project, a prediction of the cost of this addition is a huge extrapolation and comes with a very wide 95% prediction interval.

With the outlier, the 95% prediction interval for total costs is $21,166 to $38,638. Without the outlier, the interval shifts upward and widens to $25,191 to $56,221. The fact that the low ends of these intervals are similar is little consolation. If the contractor bids at the low end of either interval, his costs are almost surely going to be larger than his bid. Suppose then that he bids toward the upper end, say at $50,000. If he gets the work at this price, then the fit with the outlier estimates that he's likely to make a handsome profit; look at the difference between $50,000 (his bid) and $38,638 (upper limit of his cost). The fit without the outlier suggests, however, that he might lose money. If the contractor reaches for a safe, high bid (say $60,000), someone else is likely to get the work.

Fixing the Problem: More Information

The issue with leveraged observations in regression is whether to use them to estimate the regression model or not. Leveraged observations have a great deal of impact on the fitted model. With the outlier in this example, the standard error of the estimated slope is $4.61/Sq Ft. Without the outlier, the standard error of b_1 grows to $11.08. We have a much more precise estimate of the marginal cost (the slope) if we include the outlier. This precision translates into more accurate predictions of variable costs and total costs as well.

If we can determine that the outlier describes what we can expect to happen the *next* time we experience these conditions, then we should use it. One observation is better than none. On the other hand, if we are not assured of getting the same kind of results, we are better off excluding the outlier and accepting less precise estimates. Clearly, if we discover that the outlier is the result of entering the wrong data, then we should exclude the outlier until we can correct the data.

We need more information to decide whether to keep the outlier in this example. These regressions show what happens and inform the contractor how to bid in each case, but the fits alone do not tell us which model is better. We need to find out whether this next large project resembles the previous large project. If the large projects are similar, the contractor should use this experience with big projects to bid on the next one. Some experience is better than none. On the other hand, if the new project is different, perhaps involving a different design or type of material, then it's best to exclude this outlier and use a higher bid. The contractor might not get this work, but by bidding conservatively, he won't lose money either.

22.3 | PROBLEM 3: DEPENDENT ERRORS AND TIME SERIES

The only way to guarantee that errors in regression are independent is to collect random samples from the population for each value of the explanatory variable. In general, plots do not show the presence of dependence among the errors, with the exception of data collected over time. Time series data provide another variable, namely time itself, that we can use to detect dependence.

Detecting Dependence Using the Durbin-Watson Statistic

The key diagnostic plot for the simple regression model graphs the residuals on the explanatory variable. By removing the pattern that relates y to x, the scatterplot of e on x focuses on the residual variation. In order to see dependence among the residuals, we need to graph the residuals versus another variable aside from x. If we only have y and x, we have no other plots to use. With time series data, though, we've always got another variable: time.

A pattern in the timeplot of the residuals indicates dependence in the errors of the model. One source of dependence is the presence of lurking variables. The errors ε accumulate everything else that affects the response, aside from the single explanatory variable in the model. If there's another variable that affects y, we may be able to see its influence in the timeplot of the residuals.

Durbin-Watson statistic
Statistic used to detect sequential dependence in residuals from regression.

To quantify the amount of dependence in the residuals, we use the Durbin-Watson statistic. The **Durbin-Watson statistic** tests for correlation between adjacent residuals. If $e_1, e_2, \ldots, e_n$ denote the residuals in time order, then the Durbin-Watson statistic D is defined as

$$D = \frac{(e_2 - e_1)^2 + (e_3 - e_2)^2 + \cdots + (e_n - e_{n-1})^2}{e_1^2 + e_2^2 + \cdots + e_n^2}$$

The formula for the Durbin-Watson statistic is straightforward but tedious to compute, particularly when n is large. Most software packages can compute it.

The type of correlation measured by the Durbin-Watson statistic between adjacent observations is known as **autocorrelation**. When the data used in a regression are measured over time, we are particularly interested in the autocorrelation between adjacent errors in the SRM, $\rho_\varepsilon = \text{Corr}(\varepsilon_t, \varepsilon_{t-1})$. (The symbol ρ is used to denote correlation between two random variables. See Chapter 10 to review these ideas.) The Durbin-Watson statistic tests the null hypothesis that the autocorrelation between adjacent errors is 0,

autocorrelation Correlation between consecutive observations in a sequence.

$$H_0: \rho_\varepsilon = 0$$

(p. 565)

To get a sense for how the test works, it can be shown that the Durbin-Watson statistic is related to the autocorrelation in the residuals: $D \approx 2(1 - \text{corr}(e_t, e_{t-1}))$. If adjacent residuals are uncorrelated, then $\text{corr}(e_t, e_{t-1}) \approx 0$ and $D \approx 2$. (See Behind the Math: The Durbin-Watson Statistic for more details.) As with any test, failing to reject H_0 does not prove it is true.

If $D \approx 2$, we have not proven that $\rho_\varepsilon = 0$ or that the errors are independent; we have only shown that whatever autocorrelation is present is too small to indicate a problem.

Modern software provides a p-value for the Durbin-Watson test of $H_0: \rho_\varepsilon = 0$. If the p-value is less than 0.05, reject H_0 and deal with the dependent errors. If the software does not provide a p-value, then use Table 22.6, which lists critical values for D in simple regression (A more extensive table appears in the Appendix). If D is less than the lower limit or higher than the upper limit, the residuals indicate dependent errors. Roughly, the test rejects H_0 when $D < 1.5$ or $D > 2.5$.

TABLE 22.6 Critical values at alpha–level 0.05 of the Durbin-Watson statistic D.[7]

n	**Reject $H_0: \rho_\varepsilon = 0$ if**	
	D is less than	**D is greater than**
15	1.36	2.64
20	1.41	2.59
30	1.49	2.51
40	1.54	2.46
50	1.59	2.41
75	1.65	2.35
100	1.69	2.31

Consequences of Dependence

Positive autocorrelation leads to a serious problem in regression analysis. If $\rho_\varepsilon > 0$, then the standard errors reported in the summary of the fitted model are too small. They ought to be larger. The estimated slope and intercept are correct on average, but these estimates are less precise than the output suggests.

[7] From the paper J. Durbin and G. S. Watson (1951), "Testing for Serial Correlation in Least Squares Regression. II" *Biometrika* **38**, 159–177. Table 22.6 lists d_u, the threshold at which the Durbin-Watson test signals a potential problem. The more extensive table in the Appendix gives d_U and d_L, the definitive threshold.

As a consequence, we mistakenly think that our estimates are more precise than they really are. Because the shown standard errors are too small, the nominal 95% confidence intervals implied from the regression are too short. Their coverage may be much less than 0.95. Similarly, p-values for the usual tests of $H_0: \beta_0 = 0$ and $H_0: \beta_1 = 0$ are too small.

The best remedy for autocorrelation is to incorporate the dependence between adjacent observations into the regression model. Doing that requires using more than one explanatory variable in the regression analysis. We describe that approach in Chapter 27. For now, we'll content ourselves with recognizing the presence of autocorrelation.

4M EXAMPLE 22.2 — CELL PHONE SUBSCRIBERS

MOTIVATION ▸ STATE THE QUESTION

The task in this example is to predict the market for cellular telephone services. These predictions can be used by managers in this industry to plan future construction, anticipate capacity shortages, and hire employees. Predictions also provide a baseline to measure the effect of promotions.

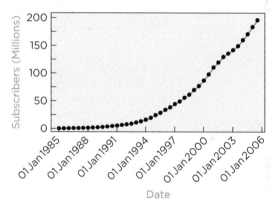

The timeplot shown above tracks the rapid growth of cellular services in the United States. The observations, taken at six-month intervals, record the number of subscribers to cellular telephone services from December 1984 through June 2005 ($n = 42$).

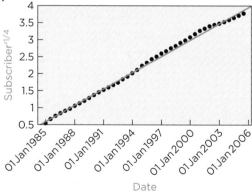

The growth in the number of subscribers seems exponential, but in fact it's slower. Curiously, the rate of growth is captured by taking the $\frac{1}{4}$ power of the number of subscribers. We do not have an explanation for this transformation, other than to say that the growth becomes remarkably linear after using this transformation.[8]

---

[8] See D. P. Foster, R. A. Stine, and R. Waterman (1998), *Business Statistics Using Regression* (New York: Springer), for further discussion of the choice of this transformation.

METHOD | DESCRIBE THE DATA AND SELECT AN APPROACH

We will use simple regression to predict the future number of subscribers. Because of the nonlinear pattern in the rate of growth, the quarter power of the number of subscribers, in millions, is the response. The explanatory variable is the date (1985 for December 1984, 1985.5 for June 1985, and so forth). We can use the estimated line to extrapolate the observed data and assign a range using prediction intervals—if the model meets the needed conditions. Because of the transformation, the slope and intercept lack intuitive interpretations.

✓ **Linear.** Seems okay from the scatterplot, with a very high r^2. The white space rule, however, suggests caution: We cannot see the deviations around the fitted line. The data concentrate in a small portion of the plot.

? **No obvious lurking variable.** It is naïve to think that this industry grew without being propelled by technology and marketing. This model has only the passage of time to explain the growth. Certainly these other variables are changing as well and would be related to the response.

MECHANICS | DO THE ANALYSIS

The timeplot below graphs residuals from the least squares regression. After removing the linear trend, we can see a meandering pattern in the residuals. These do not look like independent observations.

✗ **Evidently independent.** These residuals clearly meander over time. Positive and negative residuals come in clusters. The Durbin-Watson statistic confirms this impression: $D = 0.11$. That's near the minimum possible value of D and far less than the critical value 1.54 in Table 22.6 (use the line for $n = 40$, the closest sample size less than the observed sample size). We reject $H_0: \rho_\varepsilon = 0$.

✗ **Similar variances.** In addition to autocorrelation, we see a variation in the residuals that appears to increase in later years, with wider excursions from the fitted line.

? **Nearly normal.** The residuals are evidently not a sample (the data appear dependent with changing variation) and should not be summarized in a histogram.

The least squares fit is *Estimated* Subscribers$^{1/4} = -317.4 + 0.16$ Date. Because of the failure to meet the conditions of the SRM, we cannot interpret the standard errors. They are not meaningful even though the regression has a very high r^2.

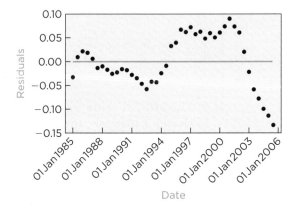

MESSAGE | SUMMARIZE THE RESULTS

There's a strong upward trend in the number of subscribers. Using a novel transformation, we can summarize the historical trend as

$$\textit{Estimated Subscribers}^{1/4} = -317 + 0.16 \, \text{Date}$$

Though the model is a good fit, we cannot rely on statistical inferences from regression to quantify the uncertainty of predictions because these data fail to meet the conditions of the SRM. To improve the model, we need to identify explanatory variables that measure progress in the industry, such as locations of cellular networks, technological progress, and prices. For that, we need multiple regression, the topic of Chapter 23.

Best Practices

- *Make sure that your model makes sense.* If you cannot interpret the key parts of your model, including the intercept, slope, and residual standard deviation, you're likely to blunder. If the equation is linear, be sure to think about why you believe that the effect of the predictor is consistent, up and down the x-axis. You can often anticipate a problem by recognizing a flaw in the interpretation.
- *Plan to change your model if it does not match the data.* If the data just don't fit your initial model, you might not have the right variables. Be flexible and willing to learn from the data

to improve the model. Most regression modeling is iterative. No one gets it right the first time, every time. Try your best, see how it works, and then refine the model as you learn from the data. For many problems, particularly those related to outliers, you may need to ask more questions about the data before you can go on.
- *Report the presence and how you handle any outliers.* Others might not agree with your choices, so it's best to keep a record of any outliers you find, particularly if you decide to exclude them from the analysis.

Pitfalls

- *Do not rely on summary statistics like r^2 to pick the best model.* If you look back at the examples of this chapter, we hardly considered r^2. We certainly did not use it to decide whether we had a good model. You want to have a model that explains the variation in the data, but ensure that the model makes sense and doesn't just separate an outlier from the rest of the cases.
- *Don't compare r^2 between regression models unless the response is the same.* Explaining variation in price per square foot is not the same as explaining variation in price itself. Make sure that the response in both regressions is the same (including the same cases) before you contrast the values of r^2.

- *Do not check for normality until you get the right equation.* A problem in the regression equation often produces residuals that do not resemble a sample from a normal population. To see how to fix the problem, you need to find it before you get to the point of looking at the histogram and normal quantile plot of the residuals.
- *Don't think that your data are independent if the Durbin-Watson statistic is close to 2.* The Durbin-Watson statistic tests for autocorrelation between adjacent residuals. Dependence among the errors comes in many forms that may not show up as autocorrelation between adjacent errors. For example, seasonal

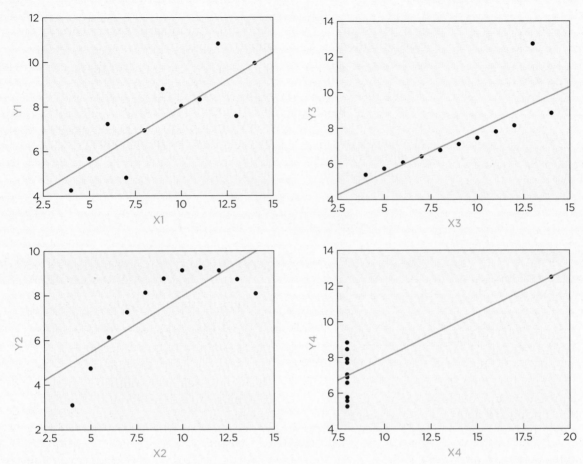

FIGURE 22.10 Anscombe's quartet: different data, but identical fits.

dependence occurs in monthly data when errors this year are related to those during the same season a year earlier. Errors that are separated by 12 months rather than adjacent are dependent.

- *Never forget to look at plots of the data and model.* Plots are the best tools when checking for problems in a regression. You won't see a problem unless you plot the data. A model with a large r^2 may not be a good fit to the data. If you don't believe us, look at the four scatterplots on shown above.[9]

The fit of the least squares line is identical in all four cases! Not just similar, identical: same r^2, same s_e, same b_0 and b_1, and even the same confidence intervals.

TABLE 22.7 Summary of the fit for every dataset in Anscombe's quartet.

r^2	0.667
s_e	1.237
n	11

Term	Estimate	Std Error	t-Stat	p-Value
b_0	3.0000	1.1247	2.67	0.0257
b_1	0.5000	0.1179	4.24	0.0022

Always plot your data! If all you see is Table 22.7, then your data might look like any of those plots above. A plot instantly reveals whether curvature, outliers, or other problems exist.

[9] F. J. Anscombe (1973), "Graphs in Statistical Analysis," *The American Statistician* **27**, 17–21.

Software Hints

Plots are the most important diagnostics. We covered those that show the data and residuals in Chapter 21. The normal quantile plot (or similar normal probability plot) helps quantify the size of outliers. Transformations are sometimes used in correcting problems due to heteroscedasticity; Chapter 20 covers transformations in regression. If there is a problem using the Durbin-Watson test, methods shown in Chapter 27 can remedy the problem. The fix requires adding more explanatory variables to the regression model, so we cover that after introducing multiple regression in the following chapters.

EXCEL

To obtain the Durbin-Watson statistic, write a small formula. First, calculate the residuals from the regression. That's a simple formula once we have b_0 and b_1. (Excel does this automatically if you use the regression command from the Data Analysis menu.) If the residuals are in the range A1...A100, for example, then the following expression computes the Durbin-Watson statistic:

$$= \text{sumxmy2(A1:A99, A2:A100)/sumsq(A1:A100)}$$

If the value of this formula is near 2, you're OK. The hard part is that the importance of a deviation from 2 depends on n and the explanatory variable. Table 22.6 gives a few critical values to determine whether D is far from 2. Roughly, $D < 1.5$ (or $D > 2.5$) requires action.

MINITAB

In addition to plots, the regression command

$$\text{Stat} > \text{Regression} > \text{Regression} \dots$$

optionally produces a list of unusual observations and the Durbin-Watson statistic. To obtain these, you have to select these options in the dialog box used to pick the columns for the model. To obtain a list of unusual observations, click on the button labeled "Results . . ." and select this option. To obtain the Durbin-Watson statistic, click on the button labeled "Options . . ." and check the box next to Durbin-Watson statistic. Use Table 22.6 to interpret the size of the Durbin-Watson statistic.

JMP

Follow the menu sequence

$$\text{Analyze} > \text{Fit Model}$$

to construct the regression. (This tool offers more options than the Analyze > Fit Y by X command.) Pick the response and explanatory variable. (The single explanatory variable goes in the large section of the dialog box labeled "Construct model effects.") Click the Run Model button to obtain a summary of the least squares regression. The summary window combines the now-familiar numerical summary statistics as well as several plots.

Click the red triangle at the top of the output window and select the item Row Diagnostics > Durbin Watson Test. JMP adds a section to the window that summarizes the regression. To test for the presence of statistically significant dependence (autocorrelation), click the red triangle in the header of the Durbin-Watson section. Pick the offered item and ignore any messages about a lengthy calculation. JMP computes the p-value for testing H_0: data are independent (no autocorrelation).

BEHIND the Math

The Durbin-Watson Statistic

D is near 2 if the model errors are uncorrelated. To see that this is the case, expand the square in the numerator of the formula for D to obtain

$$D = \frac{\sum_{t=2}^{n} \left(e_t - e_{t-1}\right)^2}{\sum_{t=1}^{n} e_t^2} = \frac{\sum_{t=2}^{n} \left(e_t^2 + e_{t-1}^2 - 2e_{t-1}e_t\right)}{\sum_{t=1}^{n} e_t^2}$$

Dealing with slight differences in the limits of the sums gives

$$D = \frac{\sum_{t=2}^{n} e_t^2 + \sum_{t=1}^{n-1} e_t^2 - 2\sum_{t=2}^{n} e_t e_{t-1}}{\sum_{t=1}^{n} e_t^2}$$

$$= \frac{2\left(\sum_{t=1}^{n} e_t^2 - \sum_{t=2}^{n} e_t e_{t-1}\right) - e_1^2 - e_n^2}{\sum_{t=1}^{n} e_t^2}$$

$$\approx \frac{2\left(\sum_{t=1}^{n} e_t^2 - \sum_{t=2}^{n} e_t e_{t-1}\right)}{\sum_{t=1}^{n} e t_t^2}$$

$$= 2\left(1 - \frac{\sum_{t=2}^{n} e_t e_{t-1}}{\sum_{t=1}^{n} e_t^2}\right)$$

$$= 2(1 - r)$$

In the final line, r is the ratio of the two sums. You might recognize it: r is *almost* the autocorrelation between the residual at time t and the residual at time $t - 1$. If the errors are independent, then $r \approx 0$ and $D \approx 2$. As r approaches 1, D approaches 0.

The calculation of a p-value from D is hard to carry out, and some software won't show a p-value. The difficulty is easy to appreciate: We observe the residuals, not the errors.

CHAPTER SUMMARY

Reliable use of the simple regression model requires that we check for several common problems: heteroscedasticity, outliers, and dependence. **Heteroscedasticity** occurs when the underlying errors lack constant variance. **Homoscedasticity** occurs when random variables have equal variation. Transformations may correct this problem. Outliers, particularly **leveraged** outliers at the boundaries of the explanatory variable, can exert a strong pull on the model. Comparisons of the fit with and without an outlier are useful. Dependent residuals indicate that the model errors are dependent, violating a key assumption of the SRM. For time series, the **Durbin-Watson statistic** tests for a particular type of dependence known as **autocorrelation**.

■ Key Terms

autocorrelation, 560
Durbin-Watson statistic, 560

heteroscedastic, 548
homoscedastic, 548

leveraged, 556

■ Formulas

Durbin-Watson statistic

If $e_1, e_2, \ldots, e_n$ are the sequence of residuals from a regression, then the Durbin-Watson statistic is

$$D = \frac{\sum_{t=2}^{n} (e_t - e_{t-1})^2}{\sum_{t=1}^{n} e_t^2}$$

The numerator sums the squares of the $n - 1$ differences between adjacent residuals; the denominator sums the squares of all n residuals.

■ About the Data

We collected the home prices illustrating heteroscedasticity from the Web site of a realtor operating in Seattle in 2005–2006. We sampled homes listed for sale over a range of sizes. The data on contractor prices are discussed in the casebook *Business Statistics Using Regression* (Springer, 1998) by Foster, Stine, and Waterman. The data on cellular telephones comes from a biannual survey of the industry conducted by the Cellular Telecommunications & Internet Association (CTIA). CTIA reports the data in its *Semiannual Wireless Industry Survey*.

EXERCISES

Mix and Match

Match each description on the left with its mathematical expression on the right.

1. Use this plot to check the linear enough condition.
2. Use this plot to check for dependence in data over time.
3. Use this plot to check the similar variances condition.
4. Use this plot to check the nearly normal condition.
5. Term that describes data with unequal error variation
6. Another name for the assumption that the SRM makes about the variance of errors
7. An observation in a regression model with an unusually large or small value of x
8. Statistic used to detect dependence in sequences of residuals
9. An observation that deviates from the pattern in the rest of the data
10. The phrase *independent observations* describes data collected in this manner.

(a) Normal quantile plot of residuals
(b) Heteroscedasticity
(c) Time plot of residuals
(d) Durbin-Watson statistic
(e) Leveraged
(f) Plot of residuals on x
(g) Random sample from a population
(h) Homoscedasticity
(i) Scatterplot of y on x
(j) Outlier

True/False

Mark each statement True or False. If you believe that a statement is false, briefly explain why you think it is false.

11. If the SRM is used to model data that do not have constant variance, then 95% prediction intervals produced by this model are longer than needed.

12. When data do not satisfy the similar variances condition, the regression predictions tend to be too high on average, overpredicting most observations.

13. A common cause of dependent error terms is the presence of a lurking variable.

14. The Durbin-Watson test quantifies deviations from a normal population that are seen in the normal quantile plot.

15. A leveraged outlier has an unusually large or small value of the explanatory variable.

16. The presence of an outlier in the data used to fit a regression causes the estimated model to have a lower r^2 than it should.

17. Because residuals represent the net effects of many other factors, it is rare to find a group of residuals from a simple regression that is normally distributed.

18. The nearly normal condition is critical when using prediction intervals.

19. We should exclude from the estimation of the regression equation any case for which the residual is more than $3s_e$ away from the fitted line.

20. If the Durbin-Watson statistic is near zero, we can conclude that the fitted model meets the no lurking variable condition, at least for time series data.

21. The best plot for checking the similar variances condition is the scatterplot of y on x.

22. In regression modeling, it is more important to fit a model that meets the conditions of the SRM than to maximize the value of r^2.

Think About It

23. Data on sales have been collected from a chain of convenience stores. Some of the stores are considerably larger (more square feet of display space) than others. In a regression of sales on square feet, can you anticipate any problems?

24. As part of locating a new factory, a company investigated the education and income of the local population. To keep costs low, the size of the survey of prospective employees was proportional to the size of the community. What possible problems for the SRM would you expect to find in a scatterplot of average income versus average education for communities of varying size?

25. A hasty analyst looked at the normal quantile plot for the residuals from the regression shown and concluded that the model could not be used because the residuals were not normally distributed. What do you think of this analysis of the problem?

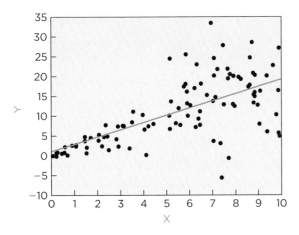

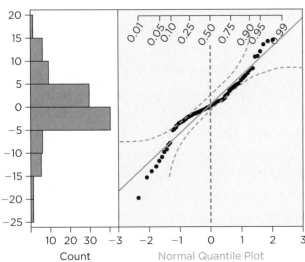

26. A second analyst looked at the same data as in Exercise 25 and concluded that the use of the SRM for prediction was fine, on average, because the fitted line clearly tracks the mean of y as the value of x increases. What do you think of this analysis?

27. (a) If the observation marked with an "x" in the following plot is removed, how will the slope of the least squares line change?

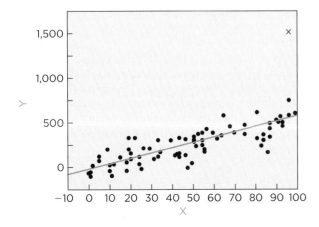

   (b) What will happen to r^2 and s_e?
   (c) Is this observation leveraged?

28. (a) If the observation marked with an × in the following plot is removed, how will the slope of the least squares line change?

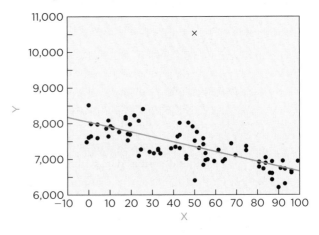

   (b) What will happen to r^2 and s_e?
   (c) Is this observation leveraged?

29. (a) If the observation marked with an × in the following plot is removed, how will the slope of the least squares line change?

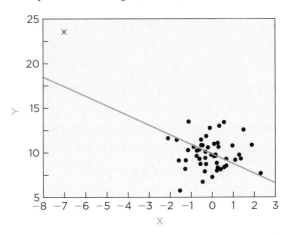

   (b) What will happen to r^2 and s_e?
   (c) Is this observation leveraged?

30. (a) If the observation marked with an × in the following plot is removed, how will the slope of the least squares line change?

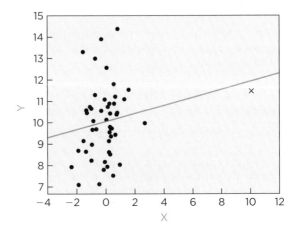

(b) What will happen to r^2 and s_e?

(c) Is this observation leveraged?

31. Management of a retail chain has been tracking the growth of sales, regressing the company's sales versus the number of outlets. Their data are weekly, spanning the last 65 weeks, since the chain opened its first outlets. What lurking variable might introduce dependence into the errors of the SRM?

32. Supervisors of an assembly line track the output of the plant. One tool that they use is a simple regression of the count of packages shipped each day versus the number of employees who were active on the assembly line during that day, which varies from 35 to about 50. Identify a lurking variable that might violate one of the assumptions of the SRM.

33. If the Durbin-Watson statistic is near 2 for the fit of a SRM to monthly data, have we proven that the errors are independent and meet the assumption of the SRM?

34. The Durbin-Watson statistic for the fit of the least squares regression in this figure is 0.34. Should you interpret the value of D as indicating dependence, or is it really an artifact of a different problem? (*Note:* The value of the explanatory variable is getting larger with the sequence order of the data.)

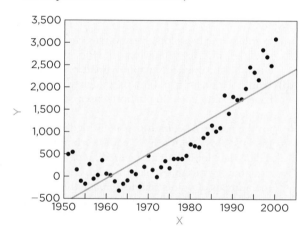

You Do It

35. **Diamond Rings** This data table contains the listed prices and weights of the diamonds in 48 rings offered for sale in *The Singapore Times*. The prices are in Singapore dollars, with the weights in carats. Use price as the response and weight as the explanatory variable. These rings hold relatively small diamonds, with weights less than $\frac{1}{2}$ carat. The Hope Diamond weighs in at 45.52 carats. Its history and fame make it impossible to assign a price, and smaller stones of its quality have gone for $600,000 per carat. Let's say 45.52 carats $\times$ $750,000/carat = $34,140,000 and call it $35 million. For the exchange rate, assume that 1 U.S. dollar is worth about 1.6 Singapore dollars.

(a) Add an imaginary ring with the weight and this price of the Hope Diamond (in Singapore dollars) to the data set as a 49th case. How does the addition of the Hope Diamond to these other rings

change the appearance of the plot? How many points can you see?

(b) How does the fitted equation of the SRM to this data change with the addition of this one case?

(c) Explain how it can be that *both* R^2 and s_e increase with the addition of this point.

(d) Why does the addition of one point, making up only 2% of the data, have so much influence on the fitted model?

36. **Convenience Shopping** These data describe the sales over time at a franchise outlet of a major U.S. oil company. This particular station sells gas, and it also has a convenience store and a car wash. Each row summarizes sales for one day at this location. The column labeled *Sales* gives the dollar sales of the convenience store, and the column *Volume* gives the number of gallons of gas sold. Formulate the regression model with dollar sales as the response and number of gallons sold as the predictor.

(a) These data are a time series, with five or six measurements per week. (The initial data collection did not monitor sales on Saturday.) Does the sequence plot of residuals from the fitted equation indicate the presence of dependence?

(b) Calculate the Durbin-Watson statistic D. (Ignore the fact that the data over the weekend are not adjacent.) Does the value of D indicate the presence of dependence? Does it agree with your impression in part (a)?

(c) The residual for row 14 is rather large and positive. How does this outlier affect the fit of the regression of sales on gallons?

(d) Should the outlier be removed from the fit?

37. **Download** Before plunging into videoconferencing, a company tested the speed of its internal computer network. The tests were designed to measure how rapidly data moved through the network under a typical load. Eighty files ranging in size from 20 to 100 megabytes (MB) were transmitted over the network at various times of day, and the time to send the files (in seconds) recorded. Formulate the SRM with y given by Transfer Time and x given by File Size.

(a) Plot the residuals versus the x-variable and in time order, as indicated by the column labeled *Hours past 8*. Does either plot suggest a problem with the SRM?

(b) Compute the Durbin-Watson D statistic. Does it indicate a problem? For a series of this length, any value of D that is more than $\frac{1}{2}$ away from 2 is statistically significant with p-value smaller than 0.01.

(c) Explain or interpret the result of the Durbin-Watson test.

38. **Production Costs** A manufacturer produces custom metal blanks that are used by its customers for computer-aided machining. The customer sends a design via computer (a 3-D blueprint), and the manufacturer comes up with an estimated price per unit, which is then used to determine a price for the customer. The data for the analysis were sampled from

the accounting records of 195 orders that were filled during the previous three months. Formulate the regression model with y as the average cost per unit and x as the material cost per unit.

(a) Does the scatterplot of y on x or the plot of the residuals on x indicate a problem with the fitted equation?

(b) Use the context of this regression to suggest any possible lurking variables. Recognize that the response reflects the cost of all inputs to the manufacturing task, not just the materials that are used.

(c) Consider a scatterplot of the residuals from this regression on the number of labor hours. Does this plot suggest that the labor input is a lurking variable?

39. **Seattle Homes** This data table contains the listed prices (in thousands of dollars) and the number of square feet for 28 homes in or near Seattle. The data come from the Web site of a realtor offering homes in the area. Use the selling price per square foot as y and the reciprocal of the number of square feet as x.

(a) The data used previously for this analysis exclude a home with 2,500 square feet that costs $1.5 million ($1,500 thousand) and is on a lot with 871,000 square feet. Add this case to the data table and refit the indicated model.

(b) Compare the fit of the model with this large home to the fit without this home. Does the slope or intercept differ by very much between the two cases? Use one estimated model as your point of reference.

(c) Which is more affected by the outlier: the estimated fixed costs or the estimated marginal cost?

(d) Outliers often shout "There's a reason for me being different!" Consider the nonmissing values in the column labeled *Lot Size*. These give the number of square feet for the size of the lot that comes with the home. Does this column help explain the outlier and suggest a lurking variable?

40. **Leases** This data table gives annual costs of 223 commercial leases. All of these leases provide office space in a Midwestern city in the United States. For the response, use the cost of the lease per square foot. As the explanatory variable, use the reciprocal of the number of square feet.

(a) Identify the leases whose values lie outside the 95% prediction intervals for leases of their size. Does the location of these data indicate a problem with the fitted model? (*Hint:* Are all of these residuals on the same side, positive or negative, of the regression?)

(b) Given the context of the problem (costs of leasing commercial property), list several possible lurking variables that might be responsible for the size and position of leases with large residual costs.

(c) The leases with the four largest residuals have something in common. What is it, and does it help you identify a lurking variable?

41. **R&D Expenses** This table contains accounting and financial data that describe 493 companies operating in technology industries: software, systems design, and semiconductor manufacturing. One column gives the expenses on research and development (R&D), and another gives the total assets of the companies. Both columns are reported in millions of dollars. Use the logs of both variables rather than the originals. (That is, set y to the natural log of R&D expenses, and set x to the natural log of assets. Note that the variables are recorded in millions, so $1,000 = 1$ billion.)

(a) What problem with the use of the SRM is evident in the scatterplot of y on x as well as in the plot of the residuals from the fitted equation on x?

(b) If the residuals are nearly normal, of the values that lie outside the 95% prediction intervals, what proportion should be above the fitted equation?

(c) Based on the property of residuals identified in part (b), can you anticipate that these residuals are not nearly normal—without needing the normal quantile plot?

42. **Cars** The cases that make up this dataset are types of cars. For each of 223 types of cars sold in the United States during the 2003 and 2004 model years, these data include the base price (in dollars) and the horsepower of the engine. Use the price of the car as the response and the horsepower as the explanatory variable.

(a) Compare the two plots: Price versus Horsepower and $\log_{10}$ of Price versus $\log_{10}$ of Horsepower. Does either seem suited, even approximately, to the SRM?

(b) Would the linear equation, particularly the slope, have the same meaning in both cases considered in part (a)?

(c) Fit the preferred equation as identified in part (a); then use this rough test for equal variance. First, divide the plot in half at the median of the explanatory variable in your model. Second, count the number of points that lie outside the 95% prediction limits in each half of the data. Are these randomly distributed between the two halves? Are the error terms homoscedastic?

43. **OECD** The Organization for Economic Cooperation and Development (OECD) tracks summary statistics of the member economies. The countries are located in Europe, parts of Asia, and North America. Two variables of interest are GDP (gross domestic product per capita, a measure of the overall production in an economy per citizen) and trade balance (measured as a percentage of GDP). Exporting countries have positive trade balances; importers have negative trade balances. These data are from the 2005 report of the OECD. Formulate the SRM with GDP as the response and Trade Balance as the explanatory variable.

(a) In 2005, Luxembourg reported the highest positive balance of trade, 21.5% of GDP and per capita GDP equal to $70,000. Fit the least squares equation both with and without Luxembourg and compare the results. Does the fitted slope change by very much?

(b) Explain any differences between r^2 and s_e for the two fits considered in part (a).

(c) Luxembourg also has the second smallest population among the countries. Does this explain the size of the difference between the two equations in part (a)? Explain.

44. **Hiring** A firm that operates a large, direct-to-consumer sales force would like to build a system to monitor the progress of new agents. The goal is to identify "super-star agents" as rapidly as possible, offer them incentives, and keep them with the company. A key task for agents is to open new accounts; an account is a new customer to the business. To build the system, the firm has monitored activities of new agents over the past two years. The response of interest is the profit to the firm (in dollars) of contracts sold by agents over their first year. Among the possible expla-nations of this performance is the number of new accounts developed by the agent during the first three months of work. Formulate the SRM with y given by the natural log of Profit from Sales and x given by the natural log of Number of Accounts.

(a) Locate the most negative residual in the data. Which case is this?

(b) Explain some characteristics that distinguish this employee from the others. (*Hint:* Consider the data in the *Early Commission* and *Early Selling* columns. Both are measured in dollars and measure the qual-ity of business developed in the first three months of working for this firm.)

(c) How does the fit change if this point is set aside, excluded from the original regression? Compare the fitted model both with and without this employee.

(d) Explain the magnitude of the change in the fit. Why does the fit change by so much or so little?

45. **Promotion** These data describe spending by a major pharmaceutical company for promoting a cholesterol-lowering drug. The data cover 39 consecutive weeks and isolate the area around Boston. The variables in this collection are shares. Marketing research often de-scribes the level of promotion in terms of voice. In place of the level of spending, *voice is* the share of advertising devoted to a specific product.

The column *Market Share* is sales of this product divided by total sales for such drugs in the Boston area. The column *Detail Voice* is the ratio of detailing for this drug to the amount of detailing for all cholesterol-low-ering drugs in Boston. Detailing counts the number of promotional visits made by representatives of a pharmaceutical company to doctors' offices. Formulate the SRM with y given by the Market Share and x given by the Detail Voice.

(a) Identify the week associated with the outlying value highlighted in the figure below. (The figure shows the least squares fitted line.) Does this week have unusually large sales given the level of promotion, or unusually low levels of promotion? Take a look at the timeplots to help you decide.

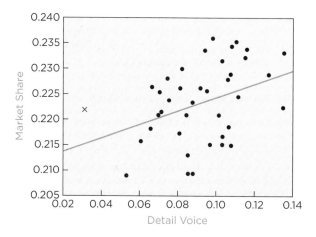

(b) How does the fitted regression equation change if this week is excluded from the analysis? Are these large changes?

(c) The r^2 of the fit gets larger and s_e gets smaller with-out this week; however, the standard error for b_1 in-creases. Why?

(d) These are time series data. Do other diagnostics suggest a violation of the assumptions of the SRM?

46. **Apple** This dataset tracks monthly performance of stock in Apple Computer since its inception in 1980. The data include 300 monthly returns on Apple, Inc., as well as returns on the entire stock market, Treasury Bills (short-term, 30-day loans to the government), and inflation. (The column *Market Return* is the return on a value-weighted portfolio that purchases stock in proportion to the size of the company rather than one of each stock.) Formulate the SRM with Apple Return as the response and Market Return as the predictor.

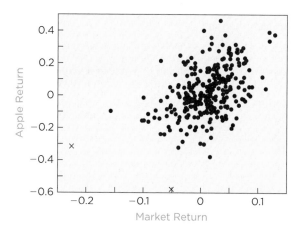

(a) Identify the time period associated with each of the two outliers highlighted in this scatterplot. What's special, if anything, about these two months?

(b) Which observation is more important to the statisti-cal significance of the fit of the least squares equa-tion? That is, if we want to keep the absolute value of the t-statistic as large as possible, which month should be retained?

(c) Explain why the month that keeps the *t*-statistic large is more important than the other month.

(d) Explain why removing either observation has little effect on the least squares fit.

4M Do Fences Make Good Neighbors?

For this exercise, you're a real-estate developer. You're planning a suburban housing development outside Philadelphia. The design calls for 25 homes that you expect to sell for about $450,000 each. If all goes as planned, you'll make a profit of $50,000 per house, or $1.25 million overall.

If you add a security wall around the development, you might be able to sell each home for more. Gates convey safety and low crime rates to potential homebuyers. The crime rate in the area where you are building the development is already low, about 15 incidents per 1,000 residents. A security consultant claims a gate and fence would reduce this further to 10 per 1,000.

If this consultant is right, is it worth adding the gate and wall? The builders say that it will cost you about $875,000 ($35,000 more per house) to add the gate and fence to the development—if you do it now while construction is starting. If you wait until people move in, the costs will rise sharply.

You have some data to help you decide. The data include the median selling price of homes in communities in the Philadelphia area. The data also include the crime rate in these communities, expressed in incidents per 1,000 residents. This analysis will use the reciprocal of this rate. These data appeared in the April 1996 issue of *Philadelphia Magazine*. Because of values of homes have increased a lot since 1996, let's assume that prices have *doubled* since these data were measured. (These data also appear in an exercise in Chapter 20. That exercise focuses on the transformation. For this exercise, we will focus on the use of the model with the transformation.)

Motivation

a) Assume that the addition of a gate and wall have the effect of convincing potential buyers that the crime rate of this development will "feel" like 10 crimes per 1,000 rather than 15. How much does this have to increase the value of these homes (on average) in order for building the security fence to be cost effective?

b) If the regression model identifies a statistically significant association between the price of housing and the number of people per crime (the reciprocal of the crime rate), will this prove that lowering the crime rate will pay for the cost of constructing the security wall?

Method

c) Plot the selling prices of homes in these communities versus 1,000 divided by the crime rate. Does the plot seem straight enough to continue? (The variable created by 1,000 divided by the crime rate is the number of residents per crime.)

d) Fit the linear equation to the scatterplot in part (c). If you accept the fit of this equation, what do you think about building the wall? Be sure to take the doubling of home prices into account.

Mechanics

e) Which communities are leveraged in this analysis? What distinguishes these communities from the others?

f) Which communities are outliers with unusually positive or negative residuals? Identify these in the plot of the residuals on the explanatory variable.

g) Does this model meet the conditions needed for using the SRM for inference about the parameters? What about prediction intervals?

h) If we ignore any problems noted in the form of this model, would the usual inferences lead us to tell the developer to build the wall? (Again, remember to take account of the doubling of prices since 1996 into account.)

Message

i) How would you answer the question for the developer? Should the developer proceed with the wall?

j) What could you do to improve the analysis? State your suggestions in a form that the developer would understand.

Multiple Regression

EXPANDING BUSINESSES MUST DECIDE WHERE TO LOCATE NEW OUTLETS. Consider the choices available to a chain of restaurants. Characteristics of the community that surrounds a site, such as the size and affluence of the local population, influence the success of a new restaurant. It would not make sense to locate an expensive restaurant that caters to small business meals in a neighborhood of large, working-class families.

It's not hard to list other factors that can influence the success of a business. For a fast-food restaurant, convenient access to a busy highway is essential. Few are willing to drive a mile off of their preferred commute to grab a meal if there are more convenient alternatives. A location that looks good to one business, however, most likely appeals to competitors as well. It's rare to find a McDonald's restaurant that is not near a Burger King, Wendy's, or other fast-food outlet.

Most sites require some sort of trade-off. It's common to find locations that mix attractive attributes with poor attributes. Locations near an upscale shopping mall are likely to have many competitors as well. Which is better: to be far from the competition or to be in a more affluent area? How is a manager to separate the advantages of one attribute from the disadvantages associated with another?

23.1 THE MULTIPLE REGRESSION MODEL
23.2 INTERPRETING MULTIPLE REGRESSION
23.3 CHECKING CONDITIONS
23.4 INFERENCE IN MULTIPLE REGRESSION
23.5 STEPS IN FITTING A MULTIPLE REGRESSION
CHAPTER SUMMARY

That's the challenge for multiple regression analysis, the modeling technique introduced in this chapter. Rather than describe the association of a single explanatory variable with the response, multiple regression describes the relationship between several explanatory variables and the response. By examining the explanatory variables simultaneously, multiple regression separates their effects on the response and reveals which variables really matter. The inclusion of several explanatory variables also allows multiple regression to obtain more accurate predictions than is possible with simple regression. These benefits come with costs, however. Multiple regression offers deeper insights than simple regression, but it also requires more care in the analysis of data and interpretation of the results. This chapter emphasizes models with two explanatory variables; later chapters use more.

23.1 | THE MULTIPLE REGRESSION MODEL

multiple regression model (MRM) Model for the association in the population between multiple explanatory variables and a response.

The **multiple regression model (MRM)** resembles the simple regression model (SRM). The difference lies in the equation; the equation of the MRM allows several explanatory variables. For example, the regression of y on two explanatory variables x_1 and x_2 describes how the response y depends on the values of x_1, x_2, and a random error term ε:

$$y = \beta_0 + \beta_1 x_1 + \beta_2 x_2 + \varepsilon$$

This equation implies a stronger version of linear association than found in simple regression. In this case, equal changes in x_1 are associated with equal changes in the mean of y, regardless of the value of x_2 and vice versa. For instance, the following equation describes how sales of a product (y) depend on the amount spent on advertising (x_1) and the difference in price between the product and its major competitor (x_2):

$$\text{Sales} = \beta_0 + \beta_1 \text{ Advertising Spending} + \beta_2 \text{ Price Difference} + \varepsilon$$

According to this equation, the benefits of advertising are boundless and unaffected by price differences. If $\beta_1 > 0$, then higher advertising means higher sales, on average. That's a problem if the level of advertising reaches a point of diminishing return. (A transformation might be called for as in Chapter 20.) The equation also implies that the impact of advertising is the same regardless of the price difference. That would not be sensible if the advertising exploits a price advantage.

Though it makes strong assumptions about the association between the response y and the explanatory variables, multiple regression does not model the explanatory variables themselves. Managers and competitors determine what to spend for advertising and the difference in prices; the model does not assume how these values are determined.

In addition to the equation, the MRM includes assumptions about the error term ε. The assumptions about these errors match those in the SRM: The errors in the MRM are independent observations sampled from a normal distribution with mean 0 and equal variance σ_ε^2.

k The number of explanatory variables in the multiple regression. ($k = 1$ in simple regression.)

Multiple Regression Model The observed response y is linearly related to k explanatory variables $x_1, x_2, \ldots,$ and x_k by the equation

$$y = \beta_0 + \beta_1 x_1 + \beta_2 x_2 + \cdots + \beta_k x_k + \varepsilon, \qquad \varepsilon \sim N(0, \sigma_\varepsilon^2)$$

The unobserved errors ε in the model

1. are *independent* of one another,
2. have *equal variance* σ_ε^2, and
3. are *normally distributed* around the regression equation.

This model provides another way to think about the errors in regression. The SRM says that one explanatory variable x is associated with y through a linear equation,

$$y = \mu_{y|x} + \varepsilon \quad \text{with} \quad \mu_{y|x} = \beta_0 + \beta_1 x$$

Only x systematically affects the response. Other variables have such diffuse effects that we lump them together and call them "random error."

That often seems too simple. Advertising and price influence sales of a product, but other variables could affect sales as well. For example, the income and tastes of consumers, store locations, delays in shipments, or even bad weather might also have an effect. The "real" equation looks more like

$$y = \beta_0 + \beta_1 x_1 + \beta_2 x_2 + \beta_3 x_3 + \beta_4 x_4 + \beta_5 x_5 + \cdots$$

with many explanatory variables affecting the response. We either are unaware of all of these explanatory variables, don't observe them, or believe them to have little effect. The SRM bundles all but x_1 into the error,

$$y = \beta_0 + \beta_1 x_1 + \underbrace{\beta_2 x_2 + \beta_3 x_3 + \beta_4 x_4 + \beta_5 x_5 + \cdots}_{\varepsilon}$$

$$= \beta_0 + \beta_1 x_1 + \varepsilon$$

Multiple regression allows us to include more of these variables in the equation of our model rather than leave them as part of the error term.

This way of thinking about the errors, as the sum of the effects of all of the omitted variables, suggests why the errors are often nearly normal. If none of the omitted variables stands out, then the Central Limit Theorem tells us that the sum of their effects is roughly normally distributed. As a result, it's not too surprising that we often find normally distributed residuals. If we omit an important variable whose effect stands out (a lurking variable), however, the residuals probably will not be normally distributed. **Deviations of the residuals from normality may suggest an important omitted variable.**

tip

What Do You Think?

A simple regression that describes the prices of new cars offered for sale at a large auto dealer regresses Price (y) on Engine Power (x).

a. What other explanatory variables probably affect the price of a car?[1]
b. What happens to the effects of these other explanatory variables if they are not used in the regression model?[2]

23.2 | INTERPRETING MULTIPLE REGRESSION

Let's consider an example like that described in the introduction to this chapter. We want to determine how two variables influence sales at stores in a chain of women's apparel stores (annually, in dollars per square foot of retail space). Each of the 65 stores in our data is located in a shopping mall and occupies 3,000 square feet. One explanatory variable is the median household income of the area served by the mall (in thousands of dollars); the other is the number of competing apparel stores in the same mall. We expect outlets in malls that serve more affluent neighborhoods will have higher sales, whereas we expect outlets that face more competition will have lower sales. A regression that describes these sales would be useful for choosing new locations for stores as well as evaluating current franchises.

Scatterplot Matrix

Simple regression begins with the scatterplot of y on x. Multiple regression begins with a plot as well, but with more than two variables, we have more

---

[1] Others include the model of the car and options (e.g., a sunroof or navigation system).
[2] Variables that affect y that are not explicitly used in the model become part of the error term.

scatterplot matrix A table of scatterplots arranged as in a correlation matrix.

choices for what to plot. To organize the plots we use a **scatterplot matrix**. A scatterplot matrix is a graphical version of the correlation matrix introduced in Chapter 6. A scatterplot matrix is a table of scatterplots, with each cell showing a scatterplot of the variable that labels the column versus the variable that labels the row. Figure 23.1 shows the scatterplot matrix for the three variables in this analysis with the corresponding correlation matrix.

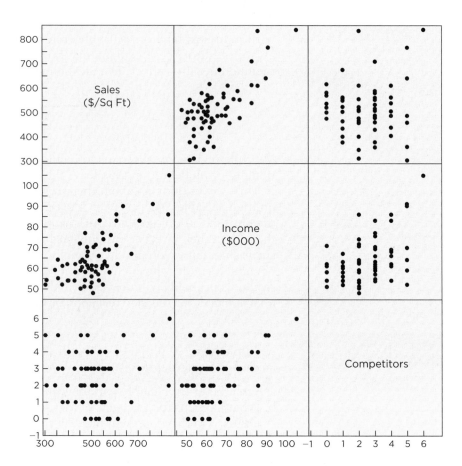

FIGURE 23.1 Scatterplot matrix of sales, income, and competitors.

	Sales	**Income**	**Competitors**
Sales	1.0000	0.7080	0.0666
Income	0.7080	1.0000	0.4743
Competitors	0.0666	0.4743	1.0000

Rather than distill the association down to a number as in the correlation matrix, a scatterplot matrix shows the scatterplot of each pair of variables. Like a correlation matrix, a scatterplot matrix also has a lot of duplication: The scatterplot matrix includes the scatterplot of y on x as well as x on y. Because of the redundancies, we concentrate on the upper right triangle of the scatterplot matrix. The two scatterplots in the first row of this scatterplot matrix graph sales per square foot on local income and on the number of competitors, the two explanatory variables. The scatterplot of income on the number of competitors (in the third cell of the second row) shows the relationship between the explanatory variables. As you will learn in this chapter, the association between the explanatory variables plays an important role in multiple regression.

Figure 23.1 confirms the positive association we expected between sales and income and shows that this association is linear. The association between sales and the number of competitors is weak at best; because we have seen the plot, we know that this weak association is not the result of bending patterns or outliers. Bending patterns between the response and explanatory variables would suggest transformations (Chapter 20) and the presence of outliers might reveal data errors. **Spending a few moments to become familiar with your data in a scatterplot matrix can save you hours figuring out a multiple regression.** Because none of these scatterplots indicates a problem, we're ready to fit the multiple regression.

tip

What Do You Think?

The scatterplot in Figure 23.2 shows three variables: y, x_1, and x_2. Assume that we are planning to fit a multiple regression of y on x_1 and x_2.

a. Which pair of variables has the largest positive correlation?[3]
b. What is the maximum value, approximately, of x_1?[4]

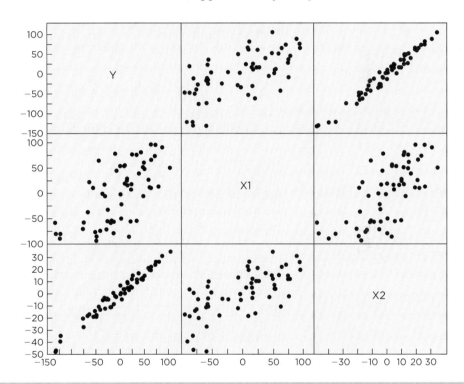

FIGURE 23.2 Scatterplot matrix of three variables.

R-squared and s_e

Table 23.1 summarizes the estimated multiple regression of sales on *both* income and the number of competitors. This summary includes familiar terms such as s_e and b_0 that appear in simple regression. Several new items are present, however, including R^2 and $k = 2$ estimated slopes.

We grayed out numbers in the summary related to inference; we need to check the conditions of the MRM before trusting these items. Using the estimates from Table 23.1, the equation of the fitted model is

$$\textit{Estimated Sales} = \hat{y} = 60.3587 + 7.966 \text{ Income} - 24.165 \text{ Competitors}$$

[3] The largest is between y and x_2 in the upper right corner.
[4] About 100. Read this from the scale for x_1 along either axis.

TABLE 23.1 Summary of the
multiple regression of sales per
square foot on two explanatory
variables.

R^2	0.5947
$\overline{R}^2$	0.5816
s_e	68.0306
n	65

Term		Estimate	Std Error	*t*-Statistic	*p*-Value
Intercept	b_0	60.3587	49.2902	1.22	0.2254
Income	b_1	7.9660	0.8383	9.50	<.0001
Competitors	b_2	−24.1650	6.3899	−3.78	0.0004

As in simple regression, least squares determines the intercept b_0 and slopes b_1 and b_2 to minimize the sum of squared residuals. We will rely on software for the calculations, but the end of this chapter gives formulas for b_0, b_1, and b_2. The residuals, denoted e, are again deviations between observed sales y and estimated sales $\hat{y}$, $e = y - \hat{y}$.

The summary of the regression in Table 23.1 begins with two statistics that summarize how well the model describes the response. In place of r^2 in simple regression, the summary shows $R^2 = 0.5947$. The "big-R" version of r-squared has the same interpretation as the "little-r" version: $R^2 = 0.5947$ indicates that the fitted equation "explains" 59.47% of the store-to-store variation in sales per square foot. Let's compare that to how much variation can be explained separately by the explanatory variables. Recall that r^2 in simple regression is the square of the correlation between y and x. From the correlations between sales per square foot and the explanatory variables in Figure 23.1, we can see that R^2 is larger than either r^2. The simple regression of Sales on Income has $r^2 = 0.708^2 \approx 0.5013$, and the regression of Sales on Competitors has $r^2 = 0.0666^2 \approx 0.0044$. Not only is R^2 larger than either of these, it is larger than their sum. The combination of explanatory variables explains more variation in sales than the sum of the parts. (This will not always be the case, however.)

R^2 is also the square of a correlation, but not between y and either explanatory variable. Instead, R^2 is the square of the correlation between y and $\hat{y}$. The larger R^2 becomes, the larger the correlation between the response and the fitted values. R^2 typically grows as explanatory variables are added to the regression (see Behind the Math: Why R^2 Increases). Because R^2 never decreases when an explanatory variable is added to a regression, it is common to see a related statistic known as **adjusted R-squared (denoted $\overline{R}^2$)** in the summary of a multiple regression. Adjusted R^2 takes account of the number of explanatory variables k and the number of observations n. Adjusted R^2 is computed from R^2 as follows:

(p. 593) ▢▶

adjusted R-squared ($\overline{R}^2$)
Adjusts R^2 in regression to account for sample size n and model size k.

$$\overline{R}^2 = 1 - (1 - R^2)\left(\frac{n - 1}{n - k - 1}\right)$$

Adjusted R^2 is always smaller than R^2, $\overline{R}^2 < R^2$. In the regression of Sales on Income and Competition, $\overline{R}^2 = 0.5816$ is slightly smaller than $R^2 = 0.5947$ because the ratio $(n - 1)/(n - k - 1) = 0.969$ is close to 1.

To appreciate the definition of $\overline{R}^2$, we need to consider the other summary statistic s_e in Table 23.1. As in simple regression, the standard deviation of the residuals $s_e = \$68.03$ per square foot estimates the standard deviation of the errors σ_ε. The formula for s_e is almost the same as in simple regression:

$$s_e = \sqrt{\frac{e_1^2 + e_2^2 + \cdots + e_n^2}{n - k - 1}} = \sqrt{\frac{e_1^2 + e_2^2 + \cdots + e_n^2}{n - 3}} = \$68.03$$

residual degrees of freedom
The sample size minus the number of estimated coefficients in the regression equation; $n - k - 1$.

The difference lies in the divisor: We subtract 3 from n in the denominator to account for the three estimates (b_0, b_1, and b_2). The divisor in s_e, called the **residual degrees of freedom**, is $n - k - 1$ in a multiple regression with k explanatory variables. The residual degrees of freedom also appears in the expression for $\overline{R}^2$. As a result, s_e and $\overline{R}^2$ move in opposite directions when an explanatory variable is added to a regression: $\overline{R}^2$ gets larger whenever s_e gets smaller.

Marginal and Partial Slopes

The summary of the multiple regression in Table 23.1 gives the estimated intercept b_0 and slopes b_1 and b_2. The layout corresponds to that of a simple regression. Each estimate comes with a standard error that determines its t-statistic (the ratio of the estimate to its standard error) and p-value. Before we conduct tests of the coefficients, we must consider how to interpret these estimates and then check the conditions of the multiple regression model. We'll first handle the interpretation. The interpretation of the intercept b_0 proceeds as in simple regression: $b_0 = 60.3587$ estimates the average sales to be about $60 per square foot when Income and Competition are zero. As in many other examples, b_0 in this equation is an extrapolation. The estimated slopes are more interesting.

Let's start with the slope for Income. Consider two stores, one located in a mall with median household income at $60,000 and the other in a mall with median income at $70,000. Rather than compare average sales of these locations directly, multiple regression takes account of the other explanatory variable, the number of competitors. This variable is associated with Income. The scatterplot matrix and correlation matrix in Figure 23.1 show that locations with higher incomes tend to face more competitors. The correlation between Income and Competitors is $r = 0.47$.

The slope for Income in the multiple regression adjusts for the typical presence of more competitors in wealthier locations. Rather than compare sales of stores in locations with *different* incomes (say, $60,000 versus $70,000) and various numbers of competitors, multiple regression estimates what happens if the stores face the *same* number of competitors. The slope $b_1 = 7.966$ for Income (Table 23.1) estimates that a store in a location with $10,000 more income would sell on average $10b_1 = \$79.66$ more per square foot than a store in the less affluent location *with the same number of competitors*. Because the slope in a multiple regression separates the effects of the explanatory variables, it is known as a **partial slope**. (The name is analogous to the term *partial derivative* in calculus.)

partial slope Slope of an explanatory variable in a multiple regression that statistically excludes the effects of other explanatory variables.

A similar interpretation applies to the slope of Competitors in the multiple regression. The statistic $b_2 = -24.165$ implies that, among stores in equally affluent locations, each additional competitor lowers average sales by $24.165 per square foot. Notice that b_2 is negative whereas the correlation between sales and number of competitors is positive. The explanation for this change in the sign is the role of income. The partial slope b_2 takes account of changes in income that typically come with changes in competition. The correlation between sales and number of competitors does not: The correlation is positive because we tend to find more competitors in affluent locations where average sales are higher.

Let's compare these partial slopes to the slopes in simple regressions of Sales on Income and Sales on Competitors shown in Table 23.2 (on the next page). Because the slopes in a simple regression average over excluded explanatory variables, we call them **marginal slopes**. The associated scatterplots are in the top row of the scatterplot matrix in Figure 23.1.

marginal slope Slope of an explanatory variable in a simple regression.

Neither the intercept nor slopes of these simple regressions match the corresponding estimates in the multiple regression. **Partial and marginal slopes only agree when the explanatory variables are uncorrelated.** (See the formulas at the end of this chapter.) The marginal slope (6.4625) in the regression of Sales on Income estimates that stores in locations with median income $70,000, say sell

tip

TABLE 23.2 Simple regressions of sales per square foot on income (left) and competitors (right).

	Income		Competitors
	0.501258	r^2	0.004438
	74.86883	s_e	105.7784

Estimate	Std Error	t-Statistic	Term	Estimate	Std Error	t-Statistic
97.0724	53.1821	1.83	b_0	502.2016	25.4437	19.74
6.4625	0.8122	7.96	b_1	4.6352	8.7469	0.53

for $10\,b_1 = \$64.625$ more per square foot on average than stores in locations with \$60,000 incomes. That's less than the estimate (\$79.66) from the multiple regression because the simple regression of Sales on Income compares

stores in locations with *higher* incomes and *more* competitors

to

stores in locations with *lower* incomes and *fewer* competitors.

The marginal slope mixes the effect of income (positive) with the effect of competition (negative). The different number of competitors suppresses the benefit of higher incomes revealed by the multiple regression.

The marginal slope for Competitors has the opposite sign from the partial slope. The marginal slope $b_1 = 4.6352$ finds that stores with more competitors have higher average sales per square foot. The marginal slope is positive because locations with higher incomes tend to draw more competitors, making it appear as if getting more competitors increases average sales. Multiple regression separates the effects of these explanatory variables and conveys what we ought to anticipate with the arrival of another competitor: On average, each additional competitor reduces sales per square foot by about \$24 if local incomes remain the same.

Path Diagram

path diagram Schematic drawing of the relationships among the explanatory variables and the response.

A **path diagram** summarizes the relationship among several explanatory variables and the response. In this example, we expect higher levels of income to associate with both more competitors and higher sales per square foot. We also anticipate that more competition reduces sales per square foot. A path diagram represents these associations as shown in Figure 23.2.

FIGURE 23.3 Path diagram for the multiple regression of Sales on Income and Competition.

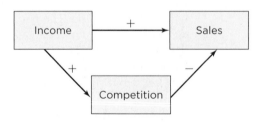

Consider the implications of this path diagram for the simple regression of sales on income. Income has a positive direct effect on sales; if no other variable changes, higher incomes increase sales. Differences in income, however, also have an indirect effect on sales via the number of competitors. Higher incomes attract more competitors, and the presence of more competitors lowers sales. This indirect effect of income lowers sales. The simple regression of Sales on Income mixes the *increase* in sales associated with higher levels of income with the *decrease* in sales associated with more competitors that accompany higher levels of income. (Behind the Math: Path Diagrams shows how to quantify the effects in the path diagram by attaching numerical values to the arrows.)

(p. 593)

The correlation among the explanatory variables also explains why you cannot add the r^2-statistics from simple regressions to obtain the R^2 of a multiple regression. If we add r^2s from the simple regressions, we either double-count the variation that is explained by both explanatory variables or miss the suppressing effect of an explanatory variable. If the explanatory variables are very correlated, we cannot separate their effects. The presence of large correlations among the explanatory variables in regression is known as **collinearity** (or multicollinearity) and is discussed further in Chapter 24.

collinearity Correlation among the explanatory variables in a multiple regression that makes the estimates uninterpretable.

What Do You Think?

a. Suppose that several high-profile technology businesses move their corporate headquarters to locations near a mall, boosting typical incomes in that area by \$5,000. If the collection of retailers in the mall remains the same, how much would you expect sales to increase at an outlet of the type considered in the preceding discussion? Should you use the marginal or partial slope for income?[5]

b. A contractor replaces windows and siding in suburban homes. He records material costs for these jobs (costs of replacement windows and siding). The homes vary in size; repairs to larger homes usually require more windows and siding. He fits two regressions: a simple regression of material costs on the number of windows and a multiple regression of costs on the number of windows *and* square feet of siding. Which should be larger: the marginal slope for the number of windows or the partial slope for the number of windows in the multiple regression?[6]

23.3 | CHECKING CONDITIONS

Before proceeding to inference, we need to check the conditions of the MRM. Graphical analysis using the scatterplot matrix provides an initial check: This plot reveals gross outliers and bending patterns. To complete our check of the model, we have to estimate the equation and use the residuals from the fit. The residuals allow us to check that the errors in the model

✓ are independent
✓ have equal variation
✓ follow a normal distribution

As in simple regression, we rely on plots to verify the associated conditions.

Calibration Plot

Two scatterplots summarize the overall fit of a multiple regression. These plots are analogous to the scatterplot of y on x and the scatterplot of e on x used in simple regression. Indeed, most diagnostic plots used to check a multiple regression convert it into a simple regression in one way or another.

calibration plot Scatterplot of the response y on the fitted values $\hat{y}$.

A **calibration plot** is analogous to the scatterplot of y on x. Rather than plot y on one explanatory variable, the calibration plot graphs y versus the combination of explanatory variables that is most correlated with the response. Which is

---

[5] Use the partial slope because the increase in income did not come with an increase in competition. The \$5,000 increase would be expected to boost sales by $5 \times 7.966 = \$39.83$ per square foot.

[6] The marginal slope is larger. Bigger homes that require replacing more windows also require more siding; homes with more windows are bigger. Hence the marginal slope combines the cost of windows and the cost of more siding. Multiple regression separates these costs. In the language of path diagrams, the marginal slope combines a positive direct effect with a positive indirect effect.

that? The fitted value $\hat{y} = b_0 + b_1 x_1 + b_2 x_2$. Figure 23.4 shows the calibration plot of the multiple regression of Sales on Income and Competition.

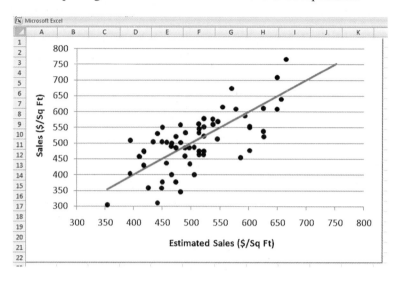

FIGURE 23.4 Calibration plot for the regression of sales on income and and the number of competitors.

Whereas the scatterplot of y on x shows r^2, the calibration plot shows R^2 because R^2 is the square of the correlation between y and $\hat{y}$. The tighter the data cluster along the diagonal in the calibration plot, the larger R^2.

Residual Plots

The plot of the residuals $e = y - \hat{y}$ on x is useful in simple regression because the residuals zoom in on the deviations from the fitted line. The analogous plot of e on $\hat{y}$ is useful in multiple regression. The scatterplot in Figure 23.5 graphs the residuals on the fitted values for the multiple regression of sales per square foot on income and number of competitors.

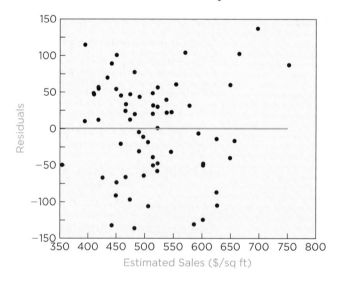

FIGURE 23.5 Scatterplot of residuals on fitted values.

From this plot, we can estimate that s_e is less than $75 per square foot; all of the residuals lie within $150 of the horizontal line at zero, the average of the residuals. (The Empirical Rule suggests about 95% of the residuals are within $\pm 2 s_e$ of zero.) In fact, $s_e = \$67.49$. As in simple regression, the residuals should suggest a swarm of bees buzzing around the horizontal line at zero, with no evident pattern.

The most common uses of the scatterplot of e on $\hat{y}$ in multiple regression are to identify outliers and to check the similar variances condition. Data often become more variable as they get larger. Since $\hat{y}$ tracks the estimated response, this plot is the natural place to check for changes in variation. If a pattern appears, either a trend in the residuals or changing variation (typically increasing from left to right), the data do not meet the conditions of the MRM. For the sales data, we might expect more variation among stores with higher sales per square foot, but that does not seem to be the case. The variation among stores with larger estimated sales per square foot (points on the right side of Figure 23.5) is similar to the variation among the more numerous stores with lower sales per square foot.

It is also useful to plot the residuals versus each explanatory variable. In general, outliers, skewness, and changing variation are revealed in either the calibration plot or the plot of e on $\hat{y}$. Separate plots of the residuals versus individual explanatory variables, like those shown in Figure 23.6, are useful to verify that the relationships are linear.

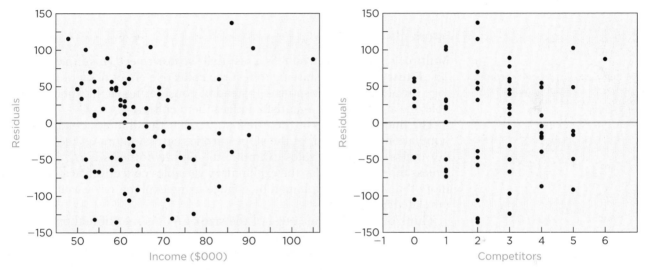

FIGURE 23.6 Scatterplots of the residuals versus each explanatory variable.

A consequence of the least squares fitting procedure is that the residuals have mean zero and are uncorrelated with each explanatory variable. Hence, these scatterplots of the residuals should show no pattern. With only 65 observations, we should only react to very clear patterns. **Don't overreact to trivial anomalies; if you stare at residual plots long enough, you may think that there's a pattern even though there is not.** Don't forget those faces in the clouds discussed in Chapter 1.

The last condition to check is the nearly normal condition. As in simple regression, we inspect the normal quantile plot of the residuals (Figure 23.7).

| tip |

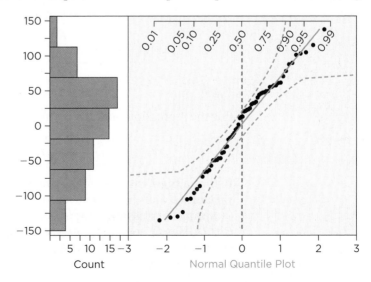

FIGURE 23.7 Normal quantile plot of residuals from the multiple regression.

The distribution of the residuals is slightly skewed, but the bounds in the normal quantile plot show that this small sample could be a sample from a normal population. These data are nearly normal.

Since the data pass the conditions of the MRM, we're set for inference.

23.4 | INFERENCE IN MULTIPLE REGRESSION

Inference in multiple regression typically proceeds in two stages. First, we test whether the overall model explains more variation than we would expect by chance alone. If it does, we test the contributions of individual explanatory variables. These two stages are equivalent in simple regression because there's only one explanatory variable.

Inference for the Model: F-test

F-test Test of the explanatory power of the model as a whole.

Multiple regression introduces a test that we don't need in simple regression. The **F-test** uses the F-statistic to test the collective effect of *all* of the explanatory variables on the response. Instead of considering each explanatory variable one at a time, the F-statistic judges them collectively.

The null hypothesis of the F-test states that the data are a sample from a population in which *all* of the slopes in the regression are 0. The intercept is not included. For the sales example, the null hypothesis is $H_0: \beta_1 = \beta_2 = 0$. Unless the data reject H_0, the explanatory variables collectively explain nothing more than could be explained by chance using explanatory variables that are unrelated to y.

Think of the F-test as a test of the size of R^2. As mentioned previously, R^2 cannot decrease whenever we add an explanatory variable to a regression. In fact, if we add enough explanatory variables, we can make R^2 as large as we want. Before being impressed by a regression with a large R^2, ask these two questions: "How many explanatory variables are in the model?" and "What is the sample size?" A model with $R^2 = 98\%$ using $k = 2$ explanatory variables with $n = 1,000$ is impressive. A model that obtains $R^2 = 98\%$ using $k = 45$ explanatory variables fit to $n = 50$ observations is not.

F-statistic Ratio of the sample variance of the fitted values to the variance in the residuals.

The F-statistic makes this adjustment for us. Like $\overline{R}^2$, the F-statistic "charges" for each explanatory variable. The F-statistic is the ratio of the sample variance of the fitted values to the variance of the residuals. For a multiple regression with k explanatory variables, the **F-statistic** is

$$F = \frac{\text{Estimated variance of fitted values}}{\text{Estimated variance of residuals}}$$

$$= \frac{\dfrac{R^2}{k}}{\dfrac{1 - R^2}{n - k - 1}} = \frac{R^2}{1 - R^2} \frac{n - k - 1}{k}$$

For the regression of Sales on Income and Competitors, $n = 65$ with $k = 2$, so

$$F = \frac{0.5947}{1 - 0.5947} \frac{65 - 2 - 1}{2} = 45.49$$

Is that a big F-statistic? The answer depends on the number of explanatory variables k and the sample size n. If the model has $k = 1$ explanatory variable (a simple regression), then F is the square of the t-statistic for the slope, $F = t^2$. Hence, for moderate sample sizes, values of F above 4 are statistically significant.

If $k \geq 2$, we need to consult a p-value. Software that builds a regression routinely includes a table in the output called the Analysis of Variance that resembles Table 23.3 for the regression of Sales on Income and Competitors.

TABLE 23.3 The analysis of variance summary includes a p-value for the F-test.

Source	Sum of Squares	df	Mean Square	F-Statistic	p-Value
Model	421,107.76	2	210,554	45.4940	<.0001
Error	286,946.30	62	4,628		
Total	708,054.06	64			

(p. 594)

This table includes quite a few details about the regression, but we only need the F-statistic and p-value in the last column of Table 23.3. (Behind the Math: The ANOVA Table and the F-Statistic describes the other components of this table and their relationship to R^2 and s_e.) The associated p-value is less than 0.0001, so we reject $H_0: \beta_1 = \beta_2 = 0$. Assuming H_0 and the MRM, it is rare to find R^2 as large as 59.47% with 2 explanatory variables and $n = 65$. Hence, these explanatory variables together *explain statistically significant variation* in sales, and we next consider contributions of individual explanatory variables.

caution If the overall F-test is not statistically significant, be wary of tests of individual slopes. It is possible that a subsequent test of a specific coefficient will appear statistically significant by chance alone. This caution is particularly relevant when exploring models with many explanatory variables. If we bypassed the F-test and cherry-picked the most significant individual slope, chances are that we'd be fooled by randomness. We will revisit the F-test in Chapter 24.

What Do You Think?

The contractor in the previous "What Do You Think" exercise got excited about multiple regression. He particularly liked the way R^2 went up as he added another explanatory variable. By the time he was done, his model reached $R^2 = 0.892$. He has data on costs at $n = 49$ homes, and he used $k = 26$ explanatory variables. Does his model impress you?[7]

Inference for One Coefficient

Standard error remains the key to inference for the intercept and individual slopes in multiple regression. As in simple regression, each row in Table 23.4 reports an estimate, its standard error, and the derived t-statistic and p-value.

TABLE 23.4 Estimated coefficients from the multiple regression of sales on income and competitors.

Term		Estimate	Std Error	t-Statistic	p-Value
Intercept	b_0	60.3587	49.2902	1.22	0.2254
Income	b_1	7.9660	0.8383	9.50	<.0001
Competitors	b_2	−24.1650	6.3899	−3.78	0.0004

[7] The overall F-statistic is $F = R^2/(1 - R^2) \times (n - k - 1)/k = 0.892/(1 - 0.892) \times (49 - 26 - 1)/26 \approx 7$. That's statistically significant. His model does explain more than random variation, but he's going to have a hard time sorting out what all of those slopes mean.

The procedure for building confidence intervals for these estimates is the same as in simple regression. We only need to adjust the degrees of freedom that determine the t-percentile. For a regression with k explanatory variables, the 95% confidence intervals for the coefficients have the form

$$\text{estimate} \pm t_{0.025,n-k-1}\text{se}(\text{estimate})$$

In this example, $n = 65$ and $k = 2$, leaving $n - k - 1 = 62$ residual degrees of freedom. The t-percentile is $t_{0.025,62} = 1.999$. Table 23.5 gives the confidence interval for each estimate.

TABLE 23.5 Confidence intervals for the coefficients in the multiple regression.

Term		95% Confidence Interval
Intercept	b_0	-38.17 to 158.89 \$/Sq Ft
Income (\$000)	b_1	6.29 to 9.64 (\$/Sq Ft)/\$1,000 income
Competitors	b_2	-36.94 to -11.39 (\$/Sq Ft)/competitor

We interpret these confidence intervals as in simple regression. For example, consider the confidence interval for the slope of the number of competitors. On average, the partial slope for number of competitors indicates that annual sales fall on average by

$$[24.165 \pm 2 \times 6.389] \approx \$11.39 \text{ to } \$36.94 \text{ per square foot}$$

with the arrival of another competitor, assuming that the level of income at the location remains the same.

The t-statistic that accompanies each estimate in Table 23.4 tests the null hypothesis that the associated parameter is zero. If the slope of an explanatory variable is zero in the population, then this variable does not affect the response once we take account of the other explanatory variables. We'll write this generic null hypothesis as $H_0: \beta_j = 0$. Remember that the implications of this null hypothesis depend on which explanatory variables are in the model.

The mechanics of this test are those of simple regression. Each t-statistic is the ratio of an estimate to its standard error, counting the number of standard errors that separate the estimate from zero:

$$t_j = \frac{b_j - 0}{\text{se}(b_j)}$$

Values of $|t_j| > t_{0.025,n-k-1} \approx 2$ are "far" from zero. Roughly, if $|t_j| > 2$, then the p-value < 0.05. The t-distribution with $n - k - 1$ degrees of freedom determines the precise p-value.

The t-statistics and p-values in Table 23.4 indicate that both slopes are significantly different from zero. For example, by rejecting $H_0: \beta_1 = 0$, we are claiming that average sales increase with the income of the surrounding community at stores facing equal numbers of competitors. We can also reject $H_0: \beta_2 = 0$. Both inferences are consistent with confidence intervals; neither 95% confidence interval in Table 23.5 includes 0.

The test of a slope in a multiple regression also has an *incremental* interpretation related to the predictive accuracy of the fitted model. By rejecting $H_0: \beta_2 = 0$, for instance, we can conclude that the regression that includes Competitors explains statistically significantly more variation in sales than a regression with Income alone. The addition of Competitors improves the fit of a regression that uses Income alone. Adding Competitors to the simple regression of Sales on Income, increases R^2 by a statistically significant amount. Similarly, by rejecting $H_0: \beta_1 = 0$, we conclude that adding Income to the simple regression of Sales on Competitors statistically significantly improves the of the fit.

> **tip** The order of the variables in the table doesn't matter; a *t*-statistic adjusts for **all** of the other explanatory variables.

If we reject $H_0: \beta_j = 0$ at level $\alpha = 0.05$, then it follows that

1. Zero is not in the 95% confidence interval for β_j;
2. The regression model with x_j explains statistically significantly more variation than the regression without x_j.

Prediction Intervals

Prediction intervals in multiple regression resemble their counterparts in simple regression. When predicting cases that have the characteristics of those in the observed sample, we can approximate the 95% prediction intervals as in simple regression: Use the predicted value from the estimated equation plus or minus 2 times s_e: $\hat{y} \pm 2s_e$.

As an example, the approximate 95% prediction interval for sales per square foot at a location with median income $70,000 and 3 competitors (a typical combination; see Figure 23.1) is

$$\hat{y} \pm 2s_e = (b_0 + b_1x_1 + b_2x_2) \pm 2s_e$$
$$= (60.3587 + 7.966(70) - 24.165(3)) \pm 2 \times 68.031$$
$$\approx \$545.48 \pm \$136.06 \text{ per square foot}$$

The exact interval computed by software is nearly identical:

$$\hat{y} \pm t_{0.025,n-k-1}s_e(\hat{y}) = 545.48 \pm 1.999 \times 68.68 = \$545.48 \pm \$137.29$$

It is easy, however, to extrapolate inadvertently, and we recommend letting software handle these calculations. Consider the following example. These data include stores that have 1 competitor and stores in communities with median incomes of $100,000 or more. The scatterplot matrix in Figure 23.1, however, shows that no location combines these attributes. If we predict sales in a location with Income = $100,000 and Competitors = 1, we have extrapolated; our data do not have this combination of the explanatory variables. The approximate prediction interval with no adjustment for extrapolation is

$$\hat{y} \pm 2s_e = (b_0 + b_1x_1 + b_2x_2) \pm 2s_e$$
$$= (60.3587 + 7.966(100) - 24.165(1)) \pm 2 \times 68.031$$
$$\approx \$832.79 \pm \$136.06 \text{ per square foot}$$

Because this combination is an extrapolation, the prediction interval should be wider. The exact 95% prediction interval is

$$832.79 \pm t_{0.025,62} \times 77.11 = \$832.79 \pm \$144.22 \text{ per square foot}$$

The moral of the story is simple: Let software compute the width of prediction intervals in multiple regression.

23.5 | STEPS IN FITTING A MULTIPLE REGRESSION

Let's summarize the steps that we used in fitting this multiple regression. As in simple regression, it pays to start with the big picture.

1. What problem are you trying to solve? Do these data help you? Until you know enough about the data and problem to answer these questions, there's no need to fit a complicated model.
2. Check the scatterplots of the response versus each explanatory variable and look for relationships among the explanatory variables, as in a scatterplot matrix. (Chapter 24 further discusses the role of correlation

among the explanatory variables.) Understand the measurement units of the variables, identify outliers, and look for bending patterns.

3. If the scatterplots of y on the explanatory variables appear straight enough, fit the multiple regression. Otherwise, find a transformation to straighten out a relationship that bends as in Chapter 20.

4. Obtain the residuals and fitted values from your regression.

5. Make scatterplots that show the overall model (y on $\hat{y}$ and e versus $\hat{y}$). The plot of e versus $\hat{y}$ is the best place to check the similar variances condition.

6. Scatterplot the residuals versus individual explanatory variables. Patterns in these plots indicate a problem.

7. Check whether the residuals are nearly normal. If not, be wary of prediction intervals.

8. Use the F-statistic to test the null hypothesis that the collection of explanatory variables has no effect on the response.

9. If the F-statistic is statistically significant, test and interpret individual partial slopes. If not, proceed with caution to tests of individual coefficients.

4M EXAMPLE 23.1 | SUBPRIME MORTGAGES

MOTIVATION ▸ STATE THE QUESTION

Subprime mortgages were prominent business news in 2007 and 2008 during the meltdown in financial markets. A subprime mortgage is a home loan made to a risky borrower. Banks and hedge funds plunged into the subprime housing market because these loans earn higher interest payments—so long as the borrower keeps paying. Defaults on loans derived from subprime mortgages brought down several financial institutions in 2008 (Lehman Brothers, Bear Sterns, and AIG) and led to federal bailouts totaling hundreds of billions of dollars.

For this analysis, a banking regulator would like to verify how lenders are using credit scores to determine the rate of interest paid by subprime borrowers. A credit score around 500 indicates a risky subprime borrower, one who might not repay the loan. A score near 800 indicates a low-risk borrower. The regulator wants to isolate the effect of credit score from other variables that might affect the interest rate. For example, the loan-to-value ratio (LTV) captures the exposure of the lender to default. As an illustration, if LTV = 0.80 (80%), then the mortgage covers 80% of the value of the property. The higher the LTV, the more risk the lender faces if the borrower defaults. The data also include the stated income of the borrower and value of the home, both in thousands of dollars.

METHOD ▸ DESCRIBE THE DATA AND SELECT AN APPROACH

We use multiple regression. The explanatory variables are the LTV, the credit score and income of the borrower, and home value. The response is the annual percentage rate of interest on the loan (APR).

The data describe $n = 372$ mortgages obtained from a credit bureau. These loans are a random sample of mortgages within the geographic territory of this regulator. Scatterplots of APR on the explanatory variables seem linear, albeit with varying levels of association. Among the explanatory variables, only the score and LTV are moderately correlated with each other.

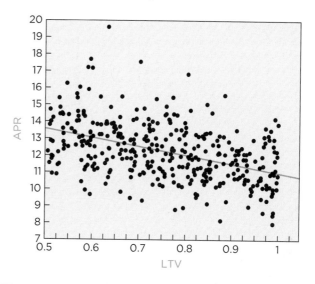

Correlations	APR	LTV	Score	Income	Home
APR	1.0000	−0.4265	−0.6696	−0.0046	−0.1043
LTV	−0.4265	1.0000	0.4853	−0.0282	0.0280
Score	−0.6696	0.4853	1.0000	0.0092	0.1088
Income	−0.0046	−0.0282	0.0092	1.0000	0.2096
Home Value	−0.1043	0.0280	0.1088	0.2096	1.0000

✓ **Linear.** Scatterplots such as this one (APR versus LTV) indicate linear relationships. There's no evident bending. Other plots indicate moderate association, no big outliers, and no bending patterns.

✓ **No obvious lurking variable.** We can imagine other variables that might be relevant, such as the type of housing or the employment status of the borrower. Some aspects of the borrower (age, race) should not matter unless something illegal is going on.

MECHANICS | DO THE ANALYSIS

✓ **Evidently independent.** These data are an SRS from the relevant population.

✓ **Similar variances.** This plot of residuals e on fitted values $\hat{y}$ at the top of the next page shows consistent variation over the range of fitted values. (The positive outlier and skewness in the residuals indicate deviations from normality rather than a lack of constant variation.)

✗ **Nearly normal.** The normal quantile plot shown on the following page confirms the skewness of the residuals. The regression underpredicts APR by up to 7%, but overpredicts by no more than 2.5%. The Central Limit Theorem nonetheless justifies using normal sampling distributions for the coefficients. The averaging that produces normal distributions for the coefficients, however, does not help when predicting individual cases. Prediction intervals from this model are not reliable. (See Chapter 22.)

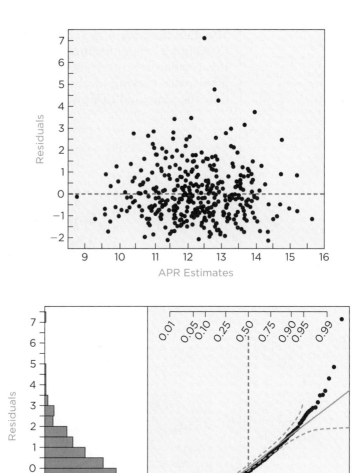

The fitted model has $R^2 = 0.4631$ with $s_e = 1.244$. The regression explains about 46% of the variation in interest rates. The overall F-test shows that this performance is highly statistically significant. The F-statistic is

$$F = R^2/(1 - R^2) \times (n - 4 - 1)/4 = 0.4631/(1 - 0.4631) \times (372 - 5)/4 \approx 79$$

with p-value < 0.0001. We reject H_0 that all four slopes are 0. These four explanatory variables together explain statistically significant variation in APR.

Term		Estimate	SE	*t*-Statistic	*p*-Value
Intercept	b_0	23.7254	0.6859	34.59	<0.0001
LTV	b_1	−1.5888	0.5197	−3.06	0.0024
Credit Score	b_2	−0.0184	0.0014	−13.65	<0.0001
Income	b_3	0.0004	0.0033	0.12	0.9036
Home Value	b_4	−0.0008	0.0008	−0.92	0.3589

Using the estimates from this summary, the fitted equation is

Estimated APR = 23.7254 − 1.589 LTV − 0.0184 Score + 0.0004 Income − 0.0008 Home Value

The first two slopes are significantly different from zero. The slope for LTV has a wide confidence interval of $-1.588 \pm 1.97(0.520)$, about -2.6 to -0.6. (The t-percentile for a 95% confidence interval with $n - 5 = 372 - 5 = 367$ degrees of freedom is $1.9664 \approx 1.97$.) The effect of the borrower's credit score is stronger, with a larger t-statistic and tighter confidence interval $-0.0184 \pm 1.97(0.0014)$, about -0.021 to -0.016. The coefficients of income and home value are not statistically significant; confidence intervals for these include 0. The addition of each of them has not improved the fit.

MESSAGE | SUMMARIZE THE RESULTS

Regression analysis shows that characteristics of the borrower (credit score) and loan LTV affect interest rates in this market. These two factors together describe almost half of the variation in interest rates. Neither income of the borrower nor the home value improves a model with these two variables.

When describing the effects of the explanatory variables, convey that these are partial, not marginal, effects. Consider the APR paid by two borrowers who are similar but have different credit scores. The borrowers have comparable income and are seeking loans for equal percentages of homes of the same value. Subprime borrowers with credit score 500 pay on average interest rates between 3.2% and 4.2% higher than comparable borrowers with credit score 700. (*It is helpful to convert the interval to a more useful scale for presenting the results. This range is 200 times the interval for the slope of the credit score.*) Loans with higher loan-to-value ratios have lower interest rates; a possible explanation for this counterintuitive result is that borrowers who take on high loan-to-value mortgages pose better risks than are captured by the credit score, income, or value of the home. This aspect of the analysis requires further study.

Don't conceal flaws, such as the skewness. This analysis relies on a sample of 372 out of the thousands of mortgages in this area. The data suggest that some loans have much higher interest rates than what we would expect from this basic analysis. Evidently, some borrowers are riskier than these characteristics indicate and other variables that measure risk should be identified.

Best Practices

- *Know the context of your model.* It's important in simple regression, and even more important in multiple regression. How are you supposed to guess what factors are missing from the model or interpret the slopes unless you know something about the problem and the data?

- *Examine plots of the overall model and individual explanatory variables before interpreting the output.* You did it in simple regression, and you need to do it in multiple regression. It can be tempting to dive into the output rather than look at the plots, but you'll make better decisions by being patient.

- *Check the overall F-statistic before looking at the t-statistics.* If you look at enough t-statistics, you'll eventually find explanatory variables that are statistically significant. If you check the

F-statistic first, you'll avoid the worst of these problems.

- *Distinguish marginal from partial slopes.* A marginal slope combines the direct and indirect effects. A partial slope separates the effects of the variables in the model. Some would say that the partial slope "holds the other variables fixed" but that's only true in a certain mathematical sense; we didn't hold anything fixed in the sales example, we just compared sales in stores in different locations and used regression to adjust for the member of competition.

- *Let your software compute prediction intervals in multiple regression.* Extrapolation is hard to recognize in multiple regression. The approximate interval $\hat{y} \pm 2s_e$ only applies when not extrapolating. Let software do these calculations.

Pitfalls

- *Don't confuse a multiple regression with several simple regressions.* It's really quite different. The slopes in simple regressions match the partial slopes in multiple regression only when the explanatory variables are uncorrelated.
- *Do not become impatient.* Multiple regression takes time to learn and do. If you skip the initial plots, you may fool yourself into thinking you've figured it all out, only to discover that your big *t*-statistic is the result of a single outlier.
- *Don't believe that you have all of the important variables.* Even though we added a second variable to our model and it made a lot of sense, that does not mean that two is enough. In most applications, it is impossible to know whether you've found all of the relevant explanatory variables.
- *Do not think that you've found causal effects.* Unless you got your data by running an experiment (this occasionally happens), regression does not imply causation, no matter how many explanatory variables appear in the model.
- *Do not interpret an insignificant t-statistic to mean that an explanatory variable has no effect.* If we do not reject $H_0: \beta_1 = 0$, this slope might be zero. It also might be close to zero; the confidence interval tells us how close. If the confidence interval includes zero, the partial slope might be positive or might be negative. Just because we don't know the direction (or sign) of the slope doesn't mean it's 0.
- *Don't think that the order of the explanatory variables in a regression matters.* The partial slopes and *t*-statistics adjust for *all* of the other explanatory variables. The adjustments that produce the partial slopes are not performed sequentially.

Software Hints

EXCEL

Excel and DDXL produce scatterplots and correlations one at a time for each pair of variables; neither offers a scatterplot matrix. To fit a multiple regression, follow the menu commands

Data > Data Analysis ... > Regression

(If this option doesn't appear in the Data menu, you need to add these commands. See the Software Hints in Chapter 19.) Selecting a range with several columns as explanatory variables produces a multiple regression analysis. With DDXL, follow the commands DDXL > Regression. Pick the Multiple Regression item from the list offered in the Click Me menu. The DDXL output window shows a summary of the fitted model. Click on the Check Residuals button to get a plot of e on $\hat{y}$ and a normal quantile plot of the residuals.

MINITAB

To obtain a correlation matrix, use the commands

Stat > Basic Statistics > Correlation ...

to open a dialog box. Pick the numerical variables for the correlation matrix and click the OK button. To get a scatterplot matrix, use the command

Graph > Matrix Plot ... > Simple

to open a dialog box for a matrix of plots; then pick the variables, starting with the response.

The menu sequence

Stat > Regression ... > Regression

opens a dialog box that defines a multiple regression. Pick the response variable; then choose the explanatory variables. Click the Results button and select the second option to get the familiar summary of the estimated coefficients and analysis of variance table. The Graphs button of this dialog box allows you to view several displays of the residuals as part of the output. We recommend the "Four in one" option.

JMP

To obtain a scatterplot matrix and a correlation matrix, use the menu commands

Analyze > Multivariate Methods > Multivariate

Click the button labeled "Y, Columns" to insert the response first into the list of variables, followed by the explanatory variables. If you don't see both the correlation and the scatterplot matrix, click the red triangle at the top of the output window to change the options.

The menu sequence Analyze > Fit Model constructs a multiple regression if two or more variables are entered as explanatory variables. Click the Run Model button to obtain a summary of the least squares regression. The summary window combines numerical summary statistics with several optional plots, including the calibration plot of y on $\hat{y}$ and e on $\hat{y}$. Click the red triangle at the top of the output window to access a pop-up menu that offers several plots of the regression and various diagnostic options.

 BEHIND the Math

Path Diagrams

A path diagram shows a regression model as a graph of nodes and edges (see Figure 23.3). Nodes represent variables and edges identify slopes. A path diagram joins correlated explanatory variables to each other with a double-headed arrow that symbolizes their association. Numbers attached to the paths that quantify these associations come from several regression equations.

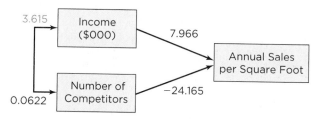

FIGURE 23.8 Path diagram of the regression of sales per square foot on income and number of competitors.

Two simple regressions determine the values attached to the double-headed edge between the explanatory variables. One regresses Competitors on Income, and the other regresses Income on Competitors:

Estimated Competitors = −1.519 + 0.0622 Income

Estimated Income = 55.46 + 3.615 Competitors

The arrow from Income to Sales indicates that an increase of $1,000 in local incomes is associated, on average, with an increase of $7.966 per square foot, assuming equal numbers of competitors. That's the "direct effect" of a wealthier location, holding fixed the effects on competition: Customers have more to spend and sales are higher, on average.

Higher incomes also have an "indirect effect": Wealthier locations attract more competitors. Following the edge from Income to Competitors, stores in wealthier locations draw an average of 0.0622 more competitors per $1,000 than less prosperous locations. More competitors reduce sales: The path from number of competitors to sales shows that 0.0622 more competitors reduces average sales by

$$0.0622 \times -24.165 \approx -1.503 \ \$/\text{Sq Ft}$$

If we add this indirect effect to the direct effect, stores in a more affluent location sell

$$7.966 - 1.503 = 6.463 \ \$/\text{Sq Ft}$$

more on average than those in the locations with incomes that are $1,000 less. Look back at the summary of the simple regression of sales on income (Table 23.2). The marginal slope is 6.463 $/Sq Ft, exactly what we've calculated from the path diagram.

$$\underset{\text{direct effect}}{7.966} \quad + \quad \underset{\text{indirect effect}}{0.0622 \times (-24.165)} \quad \underset{=}{\approx} \quad \underset{\text{marginal effect}}{6.463 \ \$/\text{Sq Ft}}$$

Path diagrams help us to think carefully about the slope in a simple regression. The marginal slope blends the direct effect of an explanatory variable with its indirect effects. It's appropriate for the marginal slope to add indirect effects, so long as we do not forget them and interpret the marginal slope as though it represented the direct effect.

Path diagrams help us *see* something else, too. When are the marginal and partial slopes the same? They match if there are no indirect effects. This happens only when the explanatory variables are uncorrelated, breaking the pathway for the indirect effect.

Why R^2 Increases

The least squares estimates in a multiple regression minimize the sum of the squared residuals. With one explanatory variable, least squares minimizes

$$\min_{b_0, b_1} \sum_{i=1}^{n} (y_i - b_0 - b_1 x_{1,i})^2$$

by picking estimates for b_0 and b_1. In a way, this expression includes the second explanatory variable x_2 but with its slope constrained to be zero:

$$\min_{b_0, b_1} \sum_{i=1}^{n} (y_i - b_0 - b_1 x_{1,i} - 0 x_{2,i})^2$$

A multiple regression using x_2 removes this constraint, and the least squares process is free to pick a slope for x_2. It solves this problem instead:

$$\min_{b_0, b_1} \sum_{i=1}^{n}(y_i - b_0 - b_1 x_{1,i} - b_2 x_{2,i})^2$$

Now that the least squares process is free to change b_2, it can find a smaller residual sum of squares. That's why R^2 goes up. A multiple regression with two explanatory variables has more choices. This flexibility allows the fitting procedure to explain more variation and increase R^2.

The ANOVA Table and the *F*-Statistic

The *p*-value for the *F*-statistic is typically located in the analysis of variance table, or ANOVA table. This table summarizes the overall fit of the regression. Table 23.6 shows the ANOVA table for the two-predictor model in this chapter from Table 23.3.

TABLE 23.6 ANOVA for the two-predictor regression.

Source	Sum of Squares	df	Mean Square	*F*-Statistic	*p*-Value
Regression	421,107.76	2	210,553.88	45.4940	<.0001
Residual	286,946.30	62	4,628.16		
Total	708,054.06	64			

This table determines R^2 and s_e^2. Something that isn't found elsewhere is the *p*-value of the *F*-statistic.

The ANOVA table gives a detailed accounting of the variation in *y*. The ANOVA table starts with the sum of squared deviations around $\bar{y}$, typically called the total sum-of-squares (SS). The table then divides this variation into two components, one determined by the fit of the regression and the other determined by the residuals. Here,

$$\text{Total SS} = \text{Regression SS} + \text{Residual SS}$$
$$708,054.06 = 421,107.76 \quad + 286,946.30$$

Here are the definitions of these sums of squares, in symbols:

$$\sum_{i=1}^{n}(y_i - \bar{y})^2 = \sum_{i=1}^{n}(\hat{y}_i - \bar{y})^2 + \sum_{i=1}^{n}(y_i - \hat{y}_i)^2$$

This equality is not obvious, but it's true. (It's a generalization of the Pythagorean theorem for right triangles: $c^2 = a^2 + b^2$.) The Total SS splits into the Regression SS plus the sum of squared residuals (Residual SS). R^2 is the ratio

$$R^2 = \frac{\text{Regression SS}}{\text{Total SS}} = \frac{421,107.76}{708,054.06} \approx 0.5947$$

The *F*-statistic adjusts R^2 for the sample size n and number of explanatory variables k. Mean squares in Table 23.6 are sums of squares divided by constants labeled df (for degrees of freedom). The degrees of freedom for the regression is the number of explanatory variables, k. The residual degrees of freedom is n minus the number of estimated parameters, $n - k - 1$. The mean square for the residuals is what we have been calling s_e^2. The *F*-statistic is the ratio of these mean squares:

$$F = \frac{\text{Mean Square Regression}}{\text{Mean Square Residual}}$$
$$= \frac{\text{Regression SS}/k}{\text{Residual SS}/(n - k - 1)}$$
$$= \frac{210,554}{4,628} = 45.4940$$

where

$$\text{Sum of squares} = \text{SS}$$
$$\text{Regression SS} = \text{Model SS}$$
$$= \text{Explained SS}$$
$$\text{Residual SS} = \text{Error SS}$$
$$= \text{Unexplained SS}$$

CHAPTER SUMMARY

The **multiple regression model (MRM)** extends the simple regression model by incorporating other explanatory variables in its equation. A **scatterplot matrix** graphs the response versus the explanatory variables and includes plots of the association among the explanatory variables. The statistics R^2 and $\bar{R}^2$ measure the proportion of variation explained by the fitted model. **Partial slopes** in the equation of an

MRM adjust for the other explanatory variables in the model and typically differ from **marginal slopes** in simple regressions. Correlation among the explanatory variables causes the differences. A **path diagram** is a useful drawing of the association that distinguishes the direct and indirect effects of explanatory variables. A **calibration plot** of y on $\hat{y}$ shows the overall fit of the model, visualizing R^2. The plot of residuals e on $\hat{y}$ allows a check for outliers and similar variances. The ***F*-statistic** and ***F*-test** measure the overall significance of the fitted model. Individual t-statistics for each partial slope test the incremental improvement in R^2 obtained by adding that variable to a regression that contains the other explanatory variables.

■ Key Terms

adjusted R-squared $(\overline{R}^2)$, 578
calibration plot, 581
collinearity, 581
F-statistic, 584
F-test, 584

multiple regression model
  (MRM), 574
path diagram, 580
residual degrees of freedom, 579
scatterplot matrix, 576

slope, 579
  marginal, 579
  partial, 579
symbol k, 574

■ Formulas

In each formula, k denotes the number of explanatory variables.

Estimates of $b_0, b_1,$ and b_2

For computing slopes by hand, here are the formulas for a multiple regression with 2 explanatory variables.

$$b_1 = \frac{s_y}{s_{x_1}} \frac{\text{corr}(y, x_1) - \text{corr}(y, x_2)\text{corr}(x_1, x_2)}{1 - \text{corr}(x_1, x_2)^2}$$

$$b_2 = \frac{s_y}{s_{x_2}} \frac{\text{corr}(y, x_2) - \text{corr}(y, x_1)\text{corr}(x_1, x_2)}{1 - \text{corr}(x_1, x_2)^2}$$

$$b_0 = \bar{y} - b_1\bar{x}_1 - b_2\bar{x}_2$$

The ratio of standard deviations attaches units to the slopes. The product of correlations in the numerator and squared correlation in the denominator of these expressions adjust for correlation between the two explanatory variables. If the x_1 and x_2 are uncorrelated, the slopes reduce to marginal slopes of y on x_1 or x_2. If the correlation between x_1 and x_2 is ± 1, these formulas fail because zero appears in the denominator. Multiple regression is not defined if $\text{corr}(x_1, x_2) = \pm 1$ because you don't really have two explanatory variables, you have one.

R^2 is the square of the correlation between the response and the fitted values, y and $\hat{y}$. It is also the proportion of the variation in the response captured by the fitted values. As such, R^2 is the ratio of the sum of squared deviations of the fitted values around $\bar{y}$ to the sum of squared deviations of the data around $\bar{y}$,

$$R^2 = \frac{(\hat{y}_1 - \bar{y})^2 + (\hat{y}_2 - \bar{y})^2 + \cdots + (\hat{y}_n - \bar{y})^2}{(y_1 - \bar{y})^2 + (y_2 - \bar{y})^2 + \cdots + (y_n - \bar{y})^2}$$

Adjusted R^2

$$\overline{R}^2 = 1 - (1 - R^2)\left(\frac{n - 1}{n - k - 1}\right)$$

F-statistic

$$F = \frac{\dfrac{R^2}{k}}{\dfrac{1 - R^2}{n - k - 1}} = \frac{R^2}{1 - R^2} \frac{n - k - 1}{k}$$

Divide the sum of squared residuals by n minus the number of estimated coefficients, including the intercept. For a multiple regression with $k = 2$ explanatory variables,

$$s_e^2 = \frac{\sum\limits_{i=1}^{n} e_i^2}{n - k - 1} = \frac{\sum\limits_{i=1}^{n} (y_i - b_0 - b_1 x_{i,1} - b_2 x_{i,2})^2}{n - 2 - 1}$$

■ About the Data

The data on store sales are based on a consulting project aiding a firm to locate stores in good locations. The study of subprime mortgages is based on an analysis of loans described in the paper "Mortgage Brokers and the Subprime Mortgage Market" written by A. El Anshasy, G. Elliehausen, and Y. Shimazaki of George Washington University and The Credit Research Center of Georgetown University.

EXERCISES

Mix and Match

Match the plot or test in the first column with the condition that can be inspected, test statistic, or estimate from the second column.

1. Scatterplot of y on x_1	(a) Similar variances
2. Scatterplot of y on x_2	(b) F-statistic
3. Scatterplot of x_2 on x_1	(c) Collinearity
4. Direct effect of x_2	(d) Nearly normal errors
5. Indirect effect of x_2	(e) Nonlinear effect
6. Scatterplot of y on $\hat{y}$	(f) Marginal slope of x_1 on x_2
7. Scatterplot of e on $\hat{y}$	(g) Partial slope for x_2
8. Normal quantile plot	(h) Leveraged outlier
9. Test $H_0: \beta_1 = \beta_2 = 0$	(i) t-statistic
10. Test $H_0: \beta_2 = 0$	(j) Calibration plot

True/False

Mark each statement True or False. If you believe that a statement is false, briefly explain why you think it is false.

11. Adjusted R^2 is less than regular R^2.

12. The statistic s_e falls when an explanatory variable is added to a regression model.

13. A slope in a simple regression is known as a partial slope because it ignores the effects of other explanatory variables.

14. A partial slope estimates differences between average values of observations y that match on the other explanatory variables.

15. The partial slope for an explanatory variable has to be smaller in absolute value than its marginal slope.

16. If the confidence interval for the marginal slope of x_1 includes zero, then the confidence interval for its partial slope includes zero as well.

17. The partial slope corresponds to the direct effect in a path diagram.

18. The indirect effect of an explanatory variable is the difference between the marginal and partial slopes.

19. If we reject $H_0: \beta_1 = \beta_2 = 0$ using the F-test, then we should conclude that both slopes are different from zero.

20. If we reject $H_0: \beta_2 = 0$, then we can conclude that the increase in R^2 obtained by adding x_2 to the regression is statistically significant.

21. The main use for a residual plot is finding nonlinear effects in multiple regression.

22. A calibration plot summarizes the overall fit of a regression model.

Think About It

23. An analyst became puzzled when analyzing the performance of franchises operated by a fast-food chain. The correlation between sales and the number of competitors within 3 miles was positive. When she regressed sales on the number of competitors and population density, however, she got a negative slope for the number of competitors. How do you explain this?

24. In evaluating the performance of new hires, the human resources division found that candidates with higher scores on its qualifying exam performed better. In a multiple regression that also used the education of the new hire as an explanatory variable, the slope for test score was near zero. Explain this paradox for the manager of the human resources division.

25. The human resources department at a firm developed a multiple regression to predict the success of candidates for available positions. Drawing records of new hires from five years ago, analysts regressed current annual salary on age at the time of hire and score on a personality test given to new hires. The path diagram on the next page summarizes the fitted model. Age is coded in

years, the test is scored from 1 to 20, and annual salary is in thousands of dollars ($M).

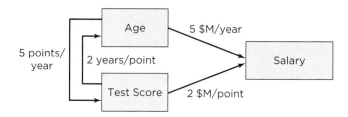

(a) Write down the equation for the multiple regression model.

(b) Which is larger: the direct or indirect effect of test score?

(c) Find the marginal slope of salary on test score.

(d) If you were a new applicant and could take a special course that in a week's time could raise your test score by 5 points, would the course be worth the $25,000 being charged? Which slope is relevant: marginal or partial?

26. A marketing research analysis considered how two customer characteristics affect their customers' stated desire for their product. Potential customers in a focus group were shown prototypes of a new convenience product and asked to indicate how much they would like to buy such a product. Scores were obtained on a 0 to 100 rating scale. The marketing group measured the age (in years) and income (in thousands of dollars, $M). The following path diagram summarizes the estimates in the associated multiple regression.

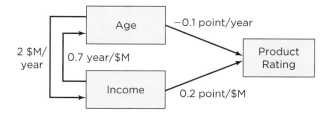

(a) Does the path diagram provide enough information to compute the fitted values from the multiple regression?

(b) What is the sign of the marginal correlation between age and product rating?

(c) Describe the ideal customer for this product, at least as indicated by the summarized model.

27. The following correlation matrix shows the pairwise correlations among three variables: two explanatory variables x_1 and x_2 and the response denoted by y. For example, $\text{corr}(y, x_2) = 0.2359$.

	y	x_1	x_2
y	1.0000	0.8667	0.2359
x_1	0.8667	1.0000	0.0000
x_2	0.2359	0.0000	1.0000

(a) Why does it make sense to put 1s down the diagonal of this table?

(b) Find the slope of the simple regression of y on x_1, if you can. If you cannot, indicate what's missing.

(c) Do you think that the marginal slope for y on x_2 will be similar to the partial slope for x_2 in the multiple regression of y on x_1 and x_2?

28. The following correlation matrix shows the pairwise correlations among three variables. The variables are the "expert" ratings assigned to wines by well-known connoisseurs (from 0 to 100), the year of the vintage (year in which the grapes were harvested), and the listed price on a Web site. For example, corr(year, price) = 0.3222.

	Rating	Year	Price
Rating	1.0000	0.0966	0.7408
Year	0.0966	1.0000	0.3222
Price	0.7408	0.3222	1.0000

(a) Would a multiple regression of the ratings on year and price explain more than half of the variation in the ratings?

(b) If we regressed the standardized value of the rating (that is, subtract the mean rating and divide by the SD of the rating) on the standardized price (subtract the mean price and divide by the SD of the price), what would be the slope?

(c) Are the partial and marginal slopes for price identical? Explain.

29. The following correlation matrix and the scatterplot matrix on the next page summarize the same data, only we scrambled the order of the variables in the two views. If the labels X, Y, Z, and T are as given in the scatterplot matrix, label the rows and columns in the correlation matrix.

	1.0000	0.8618	−0.1718	−0.3987
	0.8618	1.0000	−0.1992	−0.4589
	−0.1718	−0.1992	1.0000	−0.0108
	−0.3987	−0.4589	−0.0108	1.0000

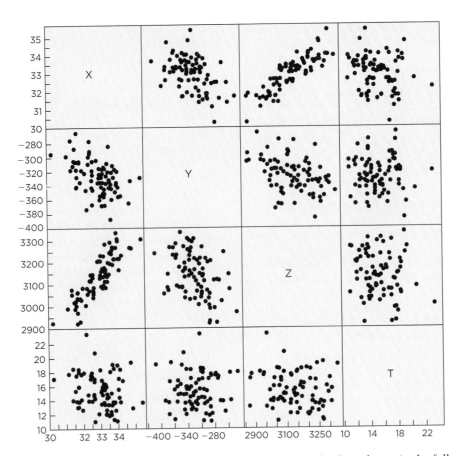

30. Identify the variable by matching the description on the following page to the data shown in the following scatterplot matrix. The plot shows 80 observations.

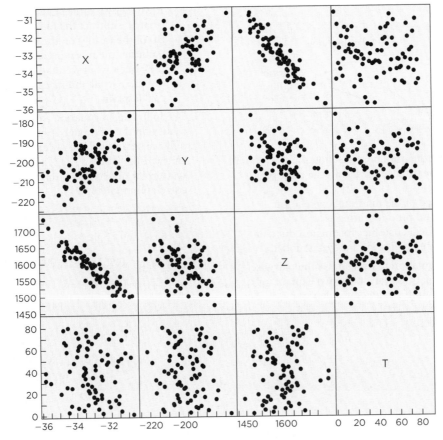

(a) The sequence 1, 2, 3, ..., 80
(b) Has mean -200
(c) Most highly *positively* correlated pair of variables
(d) Uncorrelated with Y
(e) Identify any outliers in these data. If you don't find any, then say so.

31. An airline developed a regression model to predict revenue from flights that connect "feeder" cities to its hub airport. The response in the model is the revenue generated by flights operating to the feeder cities (in thousands of dollars per month), and the two explanatory variables are the air distance between the hub and the feeder city (Distance, in miles) and the population of the feeder city (in thousands). The least squares regression equation based on data for 37 feeder locations last month is

$$Estimated\ \text{Revenue} = 87 + 0.3\,\text{Distance} + 1.5\,\text{Population}$$

with $R^2 = 0.74$ and $s_e = 32.7$.
(a) The airline plans to expand its operations to add an additional feeder city. One possible city has population 100,000 and is 250 miles from the hub. A second possible city has population 75,000 and is 200 miles from the hub. Which would you recommend if the airline wants to increase total revenue?
(b) What is the interpretation of the intercept in this equation?
(c) What is the interpretation of the partial slope for Distance?
(d) What is the interpretation of the partial slope for Population?

32. A national motel chain has a model for the operating margin of its franchises. The operating margin is defined to be the ratio of net profit to total revenue (as a percentage). The company plans to use this model to help it identify profitable sites to locate new hotels. The response in the model is the operating margin, and the explanatory variables are the number of available hotel rooms currently within 3 miles of the site (Rooms) and the square feet of office space (Office, in thousands of square feet) near the site. The estimated regression based on sites of 100 motels operated by this chain is

$$Estimated\ \text{Margin} = 54 - 0.0073\,\text{Rooms} + 0.0216\,\text{Office}$$

with $R^2 = 45\%$ and $s_e = 8.4$.
(a) Two possible sites are similar, except that one is near an office complex with 400,000 square feet, whereas the other is near 50,000 square feet of offices. Within a mile of the location near the office complex, a competing hotel has 2,250 rooms, whereas competitors near the other location offer 300 rooms. Which site would you expect to generate a higher operating margin? How much higher?
(b) What is the interpretation of the intercept in this equation?
(c) What does it mean that the partial slope for Rooms is negative?
(d) Interpret the partial slope for Office.

33. This table gives further details of the multiple regression estimated in Exercise 31. Assume that the MRM satisfies the conditions for using this model for inference.

	Estimate	SE	*t*-Statistic	*p*-Value
Intercept	87.3543	55.0459		
Distance	0.3428	0.0925		
Population	1.4789	0.2515		

(a) Fill in the *t*-statistics.
(b) Estimate the *p*-values using the Empirical Rule. Only rough estimates are needed.
(c) Does the addition of Distance to a simple regression using only population as an explanatory variable produce a statistically significant increase in R^2?
(d) If the population of a city were to grow by 10,000, on average what effect would this growth have on revenue? Give a range, rounded to presentation precision.

34. This table gives further details of the multiple regression estimated in Exercise 32. Assume that the MRM satisfies the conditions for using this model for inference.

	Estimate	SE	*t*-Statistic	*p*-Value
Intercept	53.9826	5.1777		
Rooms	-0.0073	0.0013		
Office	0.0216	0.0176		

(a) Fill in the column of *t*-statistics.
(b) Estimate the column of *p*-values using the Empirical Rule. Only rough estimates are needed.
(c) Does the addition of Office to a simple regression using only Rooms as an explanatory variable produce a statistically significant increase in R^2?
(d) Does the addition of Rooms to a simple regression using only Office as an explanatory variable produce a statistically significant increase in R^2?

35. Refer to the context of the airline in Exercise 31 part (c). Assume that the estimated model meets the conditions for using the MRM for inference.
(a) Does the estimated multiple regression equation explain statistically significant variation in revenue among these feeder cities?
(b) If this model is used to predict revenue for a feeder city, how accurate would you expect those predictions to be?

36. Refer to the context of the motel chain in Exercise 32. Assume that the estimated model meets the conditions for using the MRM for inference.
(a) Does the estimated multiple regression equation explain statistically significant variation in operating margins among these hotels?

(b) If this model is used to predict the operating margin for a site, how accurate would you expect the prediction to be?

You Do It

37. **Gold Chains** These data give the prices (in dollars) for gold link chains at the Web site of a discount jeweler. The data include the length of the chain (in inches) and its width (in millimeters). All of the chains are 14-carat gold in a similar link style. Use the price as the response.
    (a) Examine the scatterplots of the response versus the two explanatory variables as well as the scatterplot between the responses. Do you notice any unusual features in the data? Do the relevant plots appear straight enough for multiple regression?
    (b) Find the correlation between each pair of variables. Which correlation is largest? Explain why this correlation is so much larger than the others.
    (c) Fit the multiple regression of price on length and width. Show a summary of the fitted model. (Save the diagnostics for part (d).)
    (d) Even though the equation fit in part (c) has a large R^2 and both slopes are significantly different from zero, the estimated regression does **not** meet the conditions of the MRM. Explain why.
    (e) You can obtain a better model by combining the two explanatory variables in a way that captures an important omitted variable. Do this, and see if the model improves. (*Hint:* Concentrate on identifying the obvious missing variable from this model. You can build a very good proxy for this variable using the given columns.)
    (f) Summarize the fit of your improved model.

38. **Convenience Shopping (introduced in Chapter 19)** These data describe the sales over time at a franchise outlet of a major U.S. oil company. Each row summarizes sales for one day. This particular station sells gas, and it also has a convenience store and a car wash. The response Sales gives the dollar sales of the convenience store. The explanatory variable Volume gives the number of gallons of gas sold, and Washes gives the number of car washes sold at the station.
    (a) Examine scatterplots of the response versus the two explanatory variables as well as the scatterplot between the responses. Do you notice any unusual features in the data? Do the relevant plots appear straight enough for multiple regression?
    (b) Find the correlation between each pair of variables. Which correlation is largest? Explain why this correlation is so much larger than the others.
    (c) Fit the multiple regression of sales on volume and the number of car washes. Show a summary of the fitted model. (Save the diagnostics for part (d).)
    (d) Does the fitted model meet the conditions for using the MRM for inference?
    (e) Assume that the model meets the conditions for the MRM. Interpret carefully the estimated

slope for the number of car washes. In your interpretation, include a range for the effect of this variable.

39. **Download (introduced in Chapter 19)** Before purchasing videoconferencing equipment, a company tested its current internal computer network. The tests measured how rapidly data moved through its network given the current demand on the network. Eighty files ranging in size from 20 to 100 megabytes (MB) were transmitted over the network at various times of day, and the time to send the files (in seconds) recorded. The time is given as the number of hours past 8 A.M. on the day of the test. Use the transfer time as the response, with the file size and time of day as explanatory variables.
    (a) Examine scatterplots of the response versus the two explanatory variables as well as the scatterplot between the responses. Do you notice any unusual features in the data? Do the relevant plots appear straight enough for multiple regression?
    (b) Do you think, before fitting the multiple regression, that the partial slope for the file size will be the same as its marginal slope? Explain.
    (c) Fit the multiple regression of the transfer time on the file size and the time of day. Summarize the estimates obtained for the fitted model.
    (d) Does the fit of this model meet the conditions of the MRM?
    (e) Compare the sizes of the t-statistics of the fitted model to the overall F-statistic. Do these tests agree with each other?
    (f) Compare the confidence interval for the marginal slope for file size to the confidence interval for the partial slope for file size. How are these different?
    (g) Does the path diagram for the multiple regression offer a suggestion for the differences noticed in the previous questions?

40. **Production Costs (introduced in Chapter 19)** A manufacturer produces custom metal blanks that are used by its customers for computer-aided machining. The customer sends a design via computer, and the manufacturer comes up with an estimated cost per unit, which is then used to determine a price for the customer. The data for the analysis were sampled from the accounting records of 195 orders filled during the previous three months. Formulate the regression model with y as the average dollar cost per unit and x_1 as the material cost per unit and x_2 as the labor hours per unit.
    (a) Examine scatterplots of the response versus the two explanatory variables as well as the scatterplot between the responses. Do you notice any unusual features in the data? Do the relevant plots appear straight enough for multiple regression?
    (b) Fit the indicated multiple regression and show a summary of the estimated features of the model.
    (c) Does the estimated model appear to meet the conditions for the use of the MRM?
    (d) Has the addition of labor hours per unit resulted in a model that explains statistically significantly more variation in the average cost per unit?

(e) Interpret the estimated slope for labor hours. Include in your answer the 95% confidence interval for the estimate.

(f) Does this model promise accurate predictions of the cost to fill orders based on the material cost and labor hours?

41. **Home Prices** In order to help clients determine the price at which their house is likely to sell, a realtor gathered a sample of 150 purchase transactions in her area during a recent three-month period. For the response in the model, use the price of the home (in thousands of dollars). As explanatory variables, use the number of square feet (also in thousands) and the number of bathrooms.

(a) Examine scatterplots of the response versus the two explanatory variables as well as the scatterplot between the responses. Do you notice any unusual features in the data? Do the relevant plots appear straight enough for multiple regression?

(b) Fit the indicated multiple regression and show a summary of the estimated features of the model.

(c) Does the estimated model appear to meet the conditions for the use of the MRM?

(d) Does this estimated model explain statistically significant variation in the prices of homes?

(e) Compare the marginal slope for the number of bathrooms to the partial slope. Explain why these are so different, and show a confidence interval for each.

(f) A homeowner asked the realtor if she should spend $40,000 to convert a walk-in closet into a small bathroom in order to increase the sale price of her home. What does your analysis indicate?

42. **Leases (introduced in Chapter 19)** This data table gives annual costs of 223 commercial leases. All of these leases provide office space in a Midwestern city in the United States. For the response, use the cost of the lease (in dollars per square foot). As explanatory variables, use the reciprocal of the number of square feet and the age of the property in which the office space is located (denoted as Age, in years).

(a) Examine scatterplots of the response versus the two explanatory variables as well as the scatterplot between the responses. Do you notice any unusual features in the data? Do the relevant plots appear straight enough for multiple regression?

(b) Fit the indicated multiple regression and show a summary of the estimated features of the model.

(c) Does the estimated model appear to meet the conditions for the use of the MRM?

(d) Does this estimated model explain statistically significant variation in the costs per square foot of the leases?

(e) Interpret the slope for the age of the building, including in your answer the confidence interval for this estimate.

(f) Can you identify a lurking variable? Could this lurking variable affect the coefficients of the explanatory variables?

43. **R&D Expenses (introduced in Chapter 19)** This data table contains accounting and financial data that describe 504 companies operating in technology industries: software, systems design, and semiconductor manufacturing. The variables include the expenses on research and development (R&D), total assets of the company, and net sales. All columns are reported in millions of dollars; the variables are recorded in millions, so 1,000 = 1 billion. Use the natural logs of all variables rather than the originals.

(a) Examine scatterplots of the response versus the two explanatory variables as well as the scatterplot between the responses. Do you notice any unusual features in the data? Do the relevant plots appear straight enough for multiple regression?

(b) Fit the indicated multiple regression and show a summary of the estimated features of the model.

(c) Does the estimated model appear to meet the conditions for the use of the MRM?

(d) Does the fit of this model explain statistically significantly more variation in the log of spending on R&D than a model that uses the log of assets alone?

The multiple regression in part (b) has all variables on a natural log scale. To interpret the equation, note that the sum of natural logs is the log of the product, $\log_e x + \log_e y = \log_e (xy)$ and that $b \log_e x = \log_e x^b$. Hence, the equation

$$\log_e y = b_0 + b_1 \log_e x_1 + b_2 \log_e x_2$$

is equivalent to

$$y = e^{b_0} x_1^{b_1} x_2^{b_2}$$

The slopes in the log-log regression are exponents in an equation that describes y as the product of the explanatory variables raised to different powers. These powers are the partial elasticities of the response with respect to the predictors. (See Chapter 20 for a discussion of elasticities.)

(e) Interpret the slope for net sales in the equation estimated by the fitted model in part (b). Include the confidence interval in your calculation.

(f) The marginal elasticity of R&D spending with respect to net sales is about 0.79. Why is the partial elasticity in the multiple regression for net sales so much smaller? Is it really that much smaller?

44. **Cars (introduced in Chapter 19)** This data table gives various characteristics of 223 types of cars sold in the United States during the 2003 and 2004 model years. Use the base price as the response and the horsepower of the engine (HP) and the weight of the car (given in thousands of pounds) as explanatory variables.

(a) Examine the calibration plot and the plot of the residuals e on the fitted values $\hat{y}$ for the multiple regression of base price on HP and weight. Do these plots reveal any problems in the fit of this model?

(b) Revise the variables in the model so all are on the scale defined by $\log_{10}$. Has this common transformation fixed the problems identified in part (a)? To answer this question, refit the model on the log scale and consider the calibration and residual plots for the revised model.

This multiple regression has all variables on a log scale as in Exercise 43, but using base 10 logs. To interpret this model, recall that the sum of logs is the log of the product, $\log_{10} x + \log_{10} y = \log_{10}(xy)$ and that $b \log_{10} x = \log_{10} x^b$. Hence, an equation of the form

$$\log_{10} y = b_0 + b_1 \log_{10} x_1 + b_2 \log_{10} x_2$$

is equivalent to the product

$$y = 10^{b_0} x_1^{b_1} x_2^{b_2}$$

The slopes in the log-log regression are exponents in a model that estimates y as the product of the explanatory variables raised to different powers. These powers are the partial elasticities of the response with respect to the predictors. (See Chapter 20 for a discussion of elasticities.)

(c) Is the partial elasticity for weight equal to zero? Estimate the partial elasticity from the multiple regression of $\log_{10}$ price on $\log_{10}$ HP and $\log_{10}$ weight.

(d) Compare the partial elasticity for weight (the slope for $\log_{10}$ weight in the multiple regression) to the marginal elasticity for price with respect to weight (the slope for $\log_{10}$ weight in a simple regression of $\log_{10}$ price on $\log_{10}$ weight). Are these estimates very different? Use confidence intervals to measure the size of any differences.

(e) Does the path diagram for this model offer an explanation for the differences in the confidence intervals found in part (d)? Explain.

(f) Based on your analysis, describe the effect of weight on price. Does it have an effect? Are heavier cars more expensive, on average?

45. **OECD (introduced in Chapter 19)** An analyst at the United Nations is developing a model that describes GDP (gross domestic product per capita, a measure of the overall production in an economy per citizen) among developed countries. She is using national data for 29 countries from the 2005 report of the Organization for Economic Cooperation and Development (OECD). She started with the equation (estimated by least squares)

Estimated per capita GDP = $26,714 + $1,441 Trade Balance

The trade balance is measured as a percentage of GDP. Exporting countries tend have large positive trade balances. Importers have negative balances. This equation explains only 37% of the variation in per capita GDP, so she added a second explanatory variable, the number of kilograms of municipal waste per person.

(a) Examine scatterplots of the response versus the two explanatory variables as well as the scatterplot between the responses. Do you notice any unusual features in the data? Do the relevant plots appear straight enough for multiple regression?

(b) Do you think, before fitting the multiple regression, that the partial slope for trade balance will be the same as in the equation shown? Explain.

(c) Fit the multiple regression that expands the one-predictor equation by adding the second explanatory variable to the model. Summarize the estimates obtained for the fitted model.

(d) Does the fit of this model meet the conditions of the MRM?

(e) Draw the path diagram for this estimated model. Use it to explain why the estimated slope for the trade balance has become smaller than in the simple regression shown.

(f) Give a confidence interval, to presentation precision, for the slope of the municipal waste variable. Does this interval imply that countries can increase their GDP by encouraging residents to produce more municipal waste?

46. **Hiring (introduced in Chapter 19)** A firm that operates a large, direct-to-consumer sales force would like to build a system to monitor the progress of new agents. The goal is to identify "superstar agents" as rapidly as possible, offer them incentives, and keep them with the company. A key task for agents is to open new accounts; an account is a new customer to the business. The response of interest is the profit to the firm (in dollars) of contracts sold by agents over their first year. These data summarize the early performance of 464 agents. Among the possible explanations of this performance are the number of new accounts developed by the agent during the first 3 months of work and the commission earned on early sales activity. An analyst at the firm began the modeling by estimating the simple regression

Estimated Log Profit = 8.98 + 0.28 Log Accounts

This equation explains 18% of the variation in the log of profits. Formulate the MRM with y given by the natural log of Profit from Sales and the natural log of Number of Accounts and the natural log of Early Commissions. (For a discussion of models in logs, see Exercises 43 and 44.) Notice that some cases have value 0 for early commission. Rather than drop these (you cannot take the log of zero), replace the zeros with a small positive constant. Here we'll use $1 and continue by taking the log with the value added on.

(a) Examine scatterplots of the response versus the two explanatory variables as well as the scatterplot between the responses. Be sure to keep all of the variables on the scale of natural logs. Do you notice any unusual features in the data? Do the relevant plots appear straight enough for multiple regression?

(b) Do you think, before fitting the multiple regression, that the partial elasticity for number of accounts will be the same as the marginal elasticity?

(c) Fit the multiple regression that expands the one-predictor equation by adding the second explanatory variable to the model. Summarize the estimates obtained for the fitted model.

(d) Does the fit of this model meet the conditions of the MRM?

(e) Does the confidence interval for the partial elasticity for the number of accounts indicate a large shift from the marginal elasticity?

(f) Use a path diagram to illustrate why the marginal elasticity and partial elasticity are either similar or different.

(g) Which would likely be more successful in raising the performance of new hires: a training program that increased the number of accounts by 5% but did not change early selling, or a program that raised both by 2.5%? Can you answer this question from the estimated model?

47. **Promotion (introduced in Chapter 19)** These data describe promotional spending by a pharmaceutical company for a cholesterol-lowering drug. The data cover 39 consecutive weeks and isolate the area around Boston. The variables in this collection are shares. Marketing research often describes the level of promotion in terms of voice. In place of the level of spending, *voice* is the share of advertising devoted to a specific product.

The column *Market Share* is sales of this product divided by total sales for such drugs in the Boston area. The column *Detail Voice* is the ratio of detailing for this drug to the amount of detailing for all cholesterol-lowering drugs in Boston. Detailing counts the number of promotional visits made by representatives of a pharmaceutical company to doctors' offices. Similarly, *Sample Voice* is the share of samples in this market that are from this manufacturer. Formulate the MRM with y given by the Market Share and x given by Detail Voice and Sample Voice.

(a) Examine scatterplots of the response versus the two explanatory variables as well as the scatterplot between the responses. Do you notice any unusual features in the data? Do the relevant plots appear straight enough for multiple regression?

(b) Fit the indicated multiple regression and show a summary of the estimated features of the model.

(c) Does the estimated model appear to meet the conditions for the use of the MRM?

(d) Does this estimated model explain statistically significant variation in the market share?

(e) At a fixed level of sampling, do periods with increased detailing show significantly larger market share?

(f) Does the fit of the multiple regression imply that the pharmaceutical company should no longer invest in detailing and only rely on sampling? Discuss briefly.

48. **Apple (introduced in Chapter 19)** This data table tracks monthly performance of stock in Apple Computer since its inception in 1980. The data include 300 monthly returns on Apple Computer, as well as returns on the entire stock market, Treasury Bills (short-term, 30-day loans to the government), and inflation. (The column *Market Return* is the return on a value-weighted portfolio that purchases stock in proportion to the size of the company rather than one of each stock.) Formulate the model with Apple Return as the response and Market Return and IBM Return as explanatory variables.

(a) Examine scatterplots of the response versus the two explanatory variables as well as the scatterplot between the responses. Do you notice any unusual features in the data? Do the relevant plots appear straight enough for multiple regression?

(b) Fit the indicated multiple regression and show a summary of the estimated features of the model.

(c) Does the estimated model appear to meet the conditions for the use of the MRM?

(d) The regression of returns on Apple on market returns estimates β for this stock to be about 1.5. Does the multiple regression suggest a different slope for the market?

(e) Give a confidence interval for the slope of IBM returns and carefully interpret this estimate.

(f) Does the addition of IBM returns improve the fit of the model with just market returns by a statistically significant amount? Does this imply that we've found an improved trading scheme?

4M Residual Car Values

When car dealers lease a car, how do they decide what to charge? One answer, if you've got a lot of unpopular cars to move, is to charge whatever it takes to get the cars off the lot. A different answer considers the so-called residual value of the car at the end of the lease. The residual value of a leased car is the value of this car in the used-car market, also known as the previously owned car market.

How should we estimate the residual value of a car? The residual value depends on how much the car was worth originally, such as the manufacturer's list price. Let's take this off the table by limiting our attention to a particular type of car. Let's also assume that we're looking at cars that have not been damaged in an accident.

What else matters? Certainly the age of the car affects its residual value. Customers expect to pay less for older cars, on average. Older cars have smaller residual value. The term of the lease, say 2 or 3 years, has to cover the cost of the ageing of the car. Another factor that affects the residual value is the type of use. An older car that's in great condition might be worth more than a newer car that's been heavily driven. It seems as though the cost of a lease ought to take both duration and use into account.

We'll use data on the resale prices of used BMW cars gathered from the World Wide Web. We introduced these data in an example in Chapter 1. These 218 cars are late-model BMW cars in BMW's popular 3-series.

Motivation

a) Why does a manufacturer need to estimate the residual value of a leased car in order to determine annual payments on the lease?

Method

b) Explain why we should use a multiple regression to estimate the effects of age and mileage simultaneously, rather than use two simple regression models.

c) The cars are all late-model cars in the same series. Explain why this similarity avoids some of the problems that we have when looking at a cross-section of different types of cars (such as nonlinear patterns or different variances).

d) Check scatterplots of the variables. Do the relationships appear straight enough to permit using multiple regression with these variables?

Mechanics

e) Fit the appropriate multiple regression model.

f) Does this model meet the conditions for using the MRM as the basis for inference?

g) Build confidence intervals for the partial effects of age and mileage.

Message

h) Summarize the results of your model. Recommend terms for leases that cover the costs of aging and mileage.

i) Do you have any caveats that should be pointed out with your recommended terms? For example, are there any evident lurking variables?

Building Regression Models

WHAT EXPLANATORY VARIABLES BELONG IN A REGRESSION MODEL FOR STOCK RETURNS? The capital asset pricing model (CAPM, introduced in Chapter 21) has an answer. The regression ought to have *one* explanatory variable: the percentage change in the value of the market.

% Change in Stock $= \beta_0 + \beta_1$ % Change in the Market $+ \varepsilon$

In addition, the CAPM says that the intercept β_0 is 0. In finance, β_0 is called alpha and β_1 is called simply beta.

The capital asset pricing model continues to attract supporters, but numerous rivals have appeared. These rivals specify a regression equation with additional explanatory variables. Whereas the CAPM says that it's good enough to use returns on the whole stock market, competitors claim that two, three, or more explanatory variables determine the movements in the value of a stock.[1] Which explanatory variables belong in this regression?

This question is hard to answer because the explanatory variables used to model stock returns are related to each other. For instance, some rivals to the CAPM add variables that distinguish the performance of stock in small and large companies. Though different, the performance of both small and large companies depends on overall economic conditions, making these variables correlated. If the explanatory variables are highly correlated with each other, regression has a hard time separating their effects. If the correlation among the explanatory variables is high enough, it may appear as if none of the explanatory variables is useful even though the model as a whole explains most of the variation in the response. It's frustrating to have a model that predicts well but offers no explanation of how or why.

---

[1] Fama and French led the assault on beta and the CAPM in the 1990s with their three-factor model. See, for example, E. F. Fama and K. R. French (1995), "Size and Book-to-Market Factors in Earnings and Returns," *Journal of Finance*, **50**, 131–155, and more recently J. L. Davis, E. F. Fama, and K. R. French (2000), "Characteristics, Covariances and Average Returns: 1929 to 1997," *Journal of Finance*, **55**, 389–406.

This chapter introduces the model-building process, with an emphasis on the difficulties of working with correlated explanatory variables. Regression modeling starts with an initial model motivated by theory, such as the CAPM, or experience. Once we have that initial model, we usually seek other explanatory variables that improve the fit and produce better predictions. Collinearity, correlation among explanatory variables, complicates this search. In looking for a model that explains more variation, we can expect to find redundant, correlated explanatory variables. High correlations may conceal variables that in fact improve the model. This chapter also shows how to deal with collinearity and recognize better models.

24.1 | IDENTIFYING EXPLANATORY VARIABLES

Regression modeling is iterative. We start with a model that makes substantive sense, and then refine and improve its fit to data. These improvements could be transformations, but more often come in the form of added explanatory variables. We discover these variables by exploring the data and by learning more about the problem. At each step, we make sure that the model meets the conditions of the multiple regression model (MRM) before relying on statistical tests. Often, patterns in the residuals of early models in this process suggest other variables to add to the model.

The Initial Model

To illustrate the modeling process, we'll build a model that describes returns on stock in Sony Corporation, known for electronics, videogames, and movies. The CAPM provides us with a theoretically motivated starting point, a simple regression of returns on stock in Sony on returns on the whole stock market. In a different field, the initial model might be a multiple regression. We then consider adding explanatory variables that improve the fit of the initial regression. Much research has been devoted to finding explanatory variables that improve the CAPM regression. We'll consider some of those in this chapter.

Figure 24.1 summarizes the fit of the CAPM regression for stock in Sony. The scatterplot graphs monthly percentage changes in the value of Sony stock versus those in the stock market during 1995–2008 ($n = 168$). Two outliers are highlighted. The timeplot tracks the residuals from the simple regression and locates these outliers in time (December 1999 and April 2003).

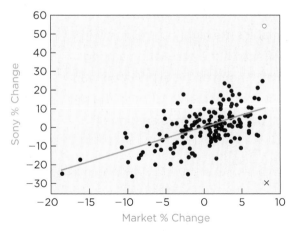
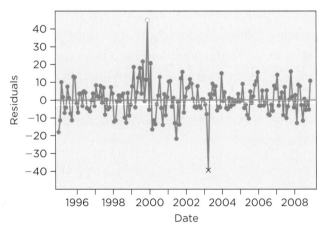

FIGURE 24.1 Percentage change in the value of stock in Sony Corporation versus the market.

As in Chapter 21, the value-weighted index represents the whole stock market. The value-weighted index assumes that you invest more in big companies than in small companies. If stock in Google, say, makes up 5% of the value of stocks in the market, then the value-weighted index assumes that 5% of your money is invested in Google as well. Table 24.1 summarizes the fit and estimates of this simple regression.

TABLE 24.1 CAPM regression for returns on Sony.

r^2	0.2793
s_e	9.2197
n	168

Term	Estimate	Std Error	t-Statistic	p-Value
Intercept	−0.2789	0.7134	−0.39	0.6964
Market % Change	1.2945	0.1558	8.02	<.0001

Let's check the conditions for the SRM. The scatterplot in Figure 24.1 shows linear association, and the timeplot of residuals shows no evidence of dependence over time. The scatterplot of the residuals versus the percentage change in the market in Figure 24.2 confirms that the residuals have similar variances.

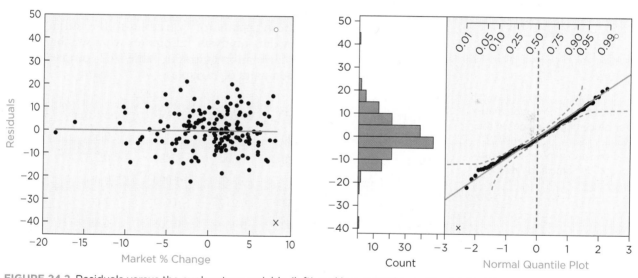

FIGURE 24.2 Residuals versus the explanatory variable (left) and in a normal quantile plot (right).

Conditions?
1. Linear
2. No lurking variable
3. Evidently independent
4. Similar variances
5. Nearly normal

Except for the two outliers (identified in Figure 24.1), the residuals conform to the SRM. The outliers suggest that the model may occasionally be off target by more than s_e and normality would suggest, and the sample kurtosis of the residuals is rather large ($K_4 = 4.2$). Aside from these two months, however, the variance of the residuals is constant and nearly normal. For inferences regarding the slope and intercept, we can rely on the averaging effects of the CLT since $10|K_4| < n = 168$. (About the Data at the end of this chapter discusses the events that produced these outliers.)

Having verified the conditions of the SRM, we can proceed to inference. The estimates in Table 24.1 are consistent with the CAPM; in particular, the estimated intercept $b_0 = -0.2789$ is not significantly different from 0 ($t = -0.39$). Deviations of this magnitude (or larger) happen by chance in 70% of samples of this size if $\beta_0 = 0$ in the population (p-value = 0.6964). The 95% confidence interval for β_0 is 0.2789 ± 1.97 × 0.7134 ≈ [−1.13 to 1.68]. β_0 (alpha for Sony) might plausibly be -1, $+1$, or any other value inside the confidence interval, including 0. The estimate of β_1 is 1.29 and highly statistically significant ($t = 8.02$ with p-value less than 0.0001).

Identifying Other Variables

Picking explanatory variables to add to a regression relies on substantive insights and data analysis. Substantive insights are key. We cannot use data we don't have, and we won't know which data to collect for our analysis unless we know enough about the problem.

A naïve approach to improving this regression is to include "more of the same" explanatory variables. The whole market index alone explains $r^2 = 28\%$ of the variation in percentage changes in Sony stock. Perhaps we can explain more variation with another stock index. For instance, the Dow Jones Industrial Average (DJIA) tracks stocks in 30 large companies rather than the whole market. It also assumes that we invest differently than the whole market index. Rather than invest in proportion to size, the DJIA assumes that we divide our investment equally among these companies.

Research in finance suggests other variables. Papers by Fama and French[2] recommend variables that contrast how different segments of the market perform. Two of the variables they propose are the differences in performance between small and large companies (Small-Big) and between growth and value stocks (High-Low).[3] Chances are that we wouldn't have thought of these without learning quite a bit of finance.

Before we add any of these explanatory variables to the initial simple regression, some preliminary data analysis anticipates what will happen when we expand the model. The correlation matrix in Table 24.2 compactly summarizes the association between pairs of variables.

TABLE 24.2 Correlation matrix for percentage changes in the value of Sony, two market indices, and two variables proposed by Fama and French.

Correlations	Sony	Market	Dow	Small-Big	High-Low
Sony % Change	1.000	0.528	0.417	0.351	−0.382
Market % Change	0.528	1.000	0.887	0.217	−0.439
Dow % Change	0.417	0.887	1.000	−0.064	−0.205
Small-Big	0.351	0.217	−0.064	1.000	−0.452
High-Low	−0.382	−0.439	−0.205	−0.452	1.000

We put the response in the first row so that this row shows the correlation between it and each possible explanatory variable. The possible explanatory variables have correlations with magnitudes in the range of 0.3 to 0.4 with the percentage change in Sony. The correlation matrix also shows that the DJIA and the whole market index are highly correlated. The square of the correlation between these indices is $r^2 = 0.887^2 \approx 0.79$, so about 79% of the variation in the two indices is shared. **The square of any correlation can be interpreted as the r^2 of the regression of one variable on the other.**

The correlation matrix compactly summarizes the association between variables, but we cannot judge whether the association is linear or whether outliers affect the correlations. To answer those questions, we use the scatterplot matrix shown in Figure 24.3.

[2] See the footnote in the introduction.
[3] Growth stocks have high market value compared to assets on the books of the company, whereas value stocks have low market value compared to the book value.

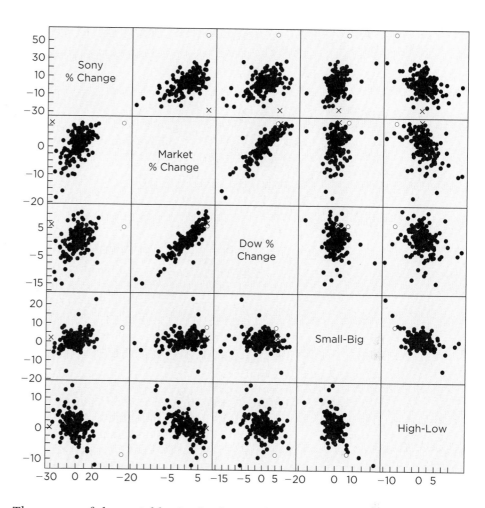

FIGURE 24.3 Scatterplot matrix of variables for the regression model.

The names of the variables in the diagonal cells label the axes. For example, "Sony % Change" labels the *y*-axis of the four plots in the first row as well as the *x*-axis of the four plots below the label in the first column. Similarly, "Market % Change" labels the *y*-axis of the four plots in the second row and the *x*-axis of the four plots in the second column.

The first row of scatterplots indicates that the association between the response and these variables appears linear. A few more outliers appear at the fringes of some scatterplots, but these tend to fall in line with the association in the rest of the data. The scatterplot matrix in Figure 24.3 also draws our attention to the high correlation between the DJIA and the whole market index. Observations of highly correlated variables concentrate along a diagonal, and a scatterplot matrix makes this clustering easy to spot. A narrow cluster of points along the diagonal draws our attention better than a number in the correlation matrix.

Adding Explanatory Variables

We have 168 observations and four candidate explanatory variables. With so many observations relative to the number of variables, we find it easiest to build a regression with all four variables included rather than adding one at a time. We'd take a different approach if we had 40 possible explanatory variables. With only a handful, we add them all. Table 24.3 summarizes the multiple regression of percentage changes in Sony stock on four variables: percentage changes in the whole market index and the Dow Jones Industrial Average as well as the two Fama-French variables.

TABLE 24.3 Summary of the regression of Sony returns versus returns on two market indices and the Fama-French variables.

R^2	0.3463
$\overline{R}^2$	0.3303
s_e	8.8609
n	168

Source	DF	Sum of Squares	Mean Square	F-Statistic	p-Value
Model	4	6,779.817	1,694.95	21.5875	<0.0001
Error	163	12,798.042	78.52		
Total	167	19,577.858			

Term	Estimate	Std Error	t-Statistic	p-Value
Intercept	−0.1865	0.6977	−0.27	0.7896
Market % Change	0.6976	0.4408	1.58	0.1154
Dow % Change	0.3689	0.4331	0.85	0.3956
Small-Big	0.6859	0.2248	3.05	0.0027
High-Low	−0.3418	0.2477	−1.38	0.1695

Before we investigate the coefficients in this model, we check the conditions of the MRM. For example, the scatterplot in Figure 24.4 graphs the residuals of the multiple regression versus the fitted values.

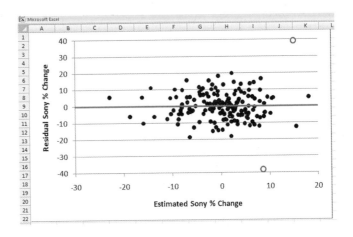

FIGURE 24.4 Scatterplot of residuals versus fitted values from the multiple regression.

The outliers remain unexplained, but this and other residual plots confirm that the estimated model meets the conditions of the MRM.

Let's move on to the test of the whole model. The F-statistic in Table 24.3 ($F = 21.59$) is highly statistically significant because its p-value is less than 0.0001; we can reject $H_0: \beta_1 = \beta_2 = \beta_3 = \beta_4 = 0$. This multiple regression explains statistically significant variation in percentage changes in the value of Sony stock. It would be very unlikely for a regression with $k = 4$ explanatory variables to explain 34.6% of the variation in 168 observations if all of the slopes in the population were 0. By rejecting H_0, the F-test shows that something is going on; this *collection* of explanatory variables explains statistically significant variation in the response. The F-test does not, however, indicate a specific variable as being important.

We can also conclude from the summary in Table 24.3 that this multiple regression explains statistically significantly more variation ($R^2 \approx 35\%$) than the

CAPM regression ($r^2 \approx 28\%$, Table 24.1). **To judge the statistical significance of an increase in R^2, we cannot rely on R^2 alone; R^2 always increases when an explanatory variable is added to a regression.** To see if an added variable statistically significantly improves the fit of a model, we need to check its t-statistic. Recall from Chapter 23 that the t-statistic for each explanatory variable in multiple regression tests whether the addition of that variable statistically significantly improves the fit of the model when compared to a regression without that explanatory variable.

The t-statistics and p-values show that *only* the Fama-French variable Small-Big improves a regression that contains all of the others. The t-statistic for Small-Big is 3.05 (p-value = 0.0027). Since this p-value is less than 0.05, we conclude that Small-Big statistically significantly improves the fit of a model that uses the other three.

Adding these explanatory variables to the initial simple regression also dramatically alters the slope of Market % Change. When considered alone in the initial regression, Market % Change has slope 1.2495 with t-statistic 8.02 (Table 24.1). It alone explains 28% of the variation in percentage changes in the value of Sony stock. After these variables are added, the slope of Market % Change falls to $b_1 = 0.6976$ with t-statistic 1.58 (Table 24.3). This once important explanatory variable no longer contributes statistically significant variation to the multiple regression. We need to understand why these changes occur.

24.2 | COLLINEARITY

We can understand the loss of statistical significance from a substantive perspective. The marginal slope of Market % Change in the simple regression measures how average prices of Sony stock change when the market changes. The fitted model from Table 24.1 is (to one decimal)

$$\textit{Estimated } \text{Sony}\,\%\,\text{Change} = -0.3 + 1.2\,\text{Market}\,\%\,\text{Change}$$

If the market grows, stock in Sony tends to grow. If the market drops, stock in Sony tends to drop as well. To be specific, consider months in which the market either does not change or grows by 1%. The marginal slope indicates that stock in Sony grows 1.2% more on average in months in which the market grows than in months in which the market does not.

The partial slope for the market index makes a similar comparison, but under restrictive conditions. The fitted model from Table 24.3 is

$$\textit{Estimated } \text{Sony}\,\%\,\text{Change} = 0.1 + 0.7\,\text{Market}\,\%\,\text{Change} + 0.4\,\text{Dow}\,\%\,\text{Change} + 0.7\,\text{Small-Big} - 0.3\,\text{High-Low}$$

Again, let's compare returns on Sony in months with no change in the market and months with 1% growth. To interpret the partial slope of Market % Change, all of these months must have the *same* values of Dow % Change, Small-Big, and High-Low. The partial slope indicates that, on average, Sony stock is 0.7% higher in months in which the market is up 1% compared to months in which the market is steady, assuming equal levels of the other variables. The partial slope compares returns between months with *different* percentage changes in the market but the *same* percentage change in the DJIA and the same levels of the Fama-French variables. The high correlation between Market % Change and Dow % Change ($r = 0.89$) shows that the data offer little information about what happens when one index moves and the other stays the same. Both usually move together. As a result, the estimate b_1 is not precise. **The lack of information due to collinearity produces imprecise estimates of the partial slopes when the explanatory variables are highly correlated.**

Variance Inflation Factor

Although it complicates the interpretation of regression models, collinearity does not violate an assumption of the MRM. Inferences implied by hypothesis tests and confidence intervals remain valid. Even so, the resulting loss of precision may produce such wide confidence intervals that the estimates are not useful. Consider the standard errors of the marginal and partial slopes of Market % Change. The standard error of the marginal slope of Market % Change in the simple regression is $se(b_1) \approx 0.16$. The standard error of the partial slope of Market % Change in the multiple regression is more than 2.8 times as large, about 0.44. The multiple regression explains more variation in the response, yet the slope is less precisely determined.

This loss of precision is regression's way of indicating that the data offer little information about the partial slope. Recall the approximate formula for the standard error of the marginal slope (Chapter 21):

$$se(b_1) \approx \frac{s_e}{\sqrt{n}} \times \frac{1}{s_x}$$

The greater the variation of the explanatory variable becomes (s_x is the standard deviation of the explanatory variable), the smaller the standard error. It is easier to detect a change in y if the values of x are more diverse. In the CAPM regression, percentage changes in the market have standard deviation 4.58% and range from less than -15% to nearly 10% in a month. The simple regression of Sony % Change on Market % Change in Figure 24.1 uses *all* of this variation to estimate the slope. Multiple regression has to use less.

Multiple regression estimates each partial slope using only variation that is *unique* to each explanatory variable. To separate the influences of several explanatory variables, multiple regression requires that each explanatory variable have unique variation that is not reproduced by the other explanatory variables. To estimate the partial slope of Market % Change, the data must contain months in which the market moves in a way that we cannot anticipate from the other explanatory variables.

The **variance inflation factor (VIF)** quantifies the amount of unique variation in each explanatory variable and uses this to summarize the effect of collinearity. The standard error for the partial slope of an explanatory variable X_1 in a multiple regression resembles the standard error in simple regression, but with an extra term. This extra term is the square root of the variance inflation factor:

$$se(b_1) = \left(\begin{array}{c} \text{standard error if } x\text{'s} \\ \text{are uncorrelated} \end{array} \right) \times \left(\begin{array}{c} \text{adjustment} \\ \text{for collinearity} \end{array} \right)$$

$$\approx \frac{s_e}{\sqrt{n}\, s_x} \times \sqrt{\text{VIF}(x_1)}$$

The larger the VIF becomes, the larger the standard error.

Collinearity among the explanatory variables determines the VIF. In a regression with two explanatory variables x_1 and x_2, the square of the correlation between x_1 and x_2 determines the variance inflation factor:

$$\text{VIF}(x_1) = \text{VIF}(x_2) = \frac{1}{1 - r^2}, \qquad r = \text{corr}(x_1, x_2)$$

The denominator $1 - r^2$ measures the proportion of unique variation left in x_1 after regressing out the effects of x_2 and vice versa. The larger the amount of unique variation that remains, the smaller the standard error. If x_1 and x_2 are uncorrelated, then VIF = 1; there's no collinearity and the formula for the standard error of the slope is the same as that used in simple regression. If the explanatory variables are perfectly correlated ($r = 1$), then no unique

variance inflation factor (VIF) Measure of the effect of collinearity on the precision of a partial slope.

variation remains and the standard error becomes infinitely large; multiple regression cannot provide distinct slopes for two variables that are in essence the same.

For larger models, the variance inflation factor uses the R^2-statistic from a multiple regression among the explanatory variables. In a model with k explanatory variables, the VIF for x_j is

$$\text{VIF}(x_j) = \frac{1}{1 - R_j^2}$$

where R_j^2 is the R^2-statistic in the regression of x_j on *all* of the other explanatory variables.

For example, if we regress Market % Change on the three other explanatory variables, we find that $R_1^2 = 0.885$; only 11.5% of the variation in Market % Change is unique ($1 - R_1^2 = 1 - 0.885 = 0.115$). Consequently, the variance inflation factor for Market % Change is

$$\text{VIF}(x_1) = \frac{1}{1 - R_1^2} = \frac{1}{1 - 0.885} \approx 8.7$$

As a result, collinearity roughly triples the standard error of this slope because $\sqrt{\text{VIF}(x_1)} \approx 2.9$. Table 24.4 adds VIFs to the summary of the slopes in the multiple regression. (The VIF does not apply to the intercept.)

TABLE 24.4 Summary of multiple regression showing variance inflation factors.

Term	Estimate	Std Error	t-Statistic	p-Value	VIF
Intercept	−0.1865	0.6977	−0.27	0.7896	—
Market % Change	0.6976	0.4408	1.58	0.1154	8.66
Dow % Change	0.3689	0.4331	0.85	0.3956	7.54
Small-Big	0.6859	0.2248	3.05	0.0027	1.62
High-Low	−0.3418	0.2477	−1.38	0.1695	1.61

Variance inflation factors are handy in regression models with several explanatory variables. First, they quantify the effects of collinearity, summarizing the impact of redundancies among the explanatory variables. Second, they save us a lot of time. Consider the Fama-French variable High-Low. Is High-Low not statistically significant in Table 24.4 because it is redundant (as is the case for Market % Change) or is it simply unrelated to the response? To decide, look at the VIF. Because the VIF of High-Low is near 1, collinearity has little effect on this explanatory variable. High-Low is not related to returns on Sony stock once we've taken account of the other explanatory variables.

There's no definitive rule for what constitutes a large VIF. We could have VIF > 10 and still get a narrow confidence interval and statistically significant estimate. Generally, though, finding VIF > 5 or 10 suggests our variables are highly redundant and that the partial slopes will be difficult to interpret.

Signs of Collinearity

This list summarizes several things that happen if we add an explanatory variable that is highly correlated with others in a regression model.

1. *R^2 increases less than we'd expect.* R^2 increases by only about 0.07 when we expand the model, even though the correlation between Dow % Change and Sony % Change is 0.42.
2. *Slopes of correlated explanatory variables in the model change dramatically.* The marginal slope of Market % Change is 1.25. After we add Dow % Change

and the Fama-French variables, the slope for the market falls and is not statistically significant.

3. *The F-statistic is more impressive than individual t-statistics.* The p-value for the F-statistic is much less than 0.0001 (Table 24.3), but only one explanatory variable appears so significant.
4. *Standard errors for partial slopes are larger than those for marginal slopes.* This happens even though the multiple regression explains more variation than the simple regression.
5. *Variance inflation factors increase.*

These effects occur to some extent whenever you add to a regression *any* explanatory variable that is correlated with other explanatory variables; it's a question of degree. The only situation in which the slopes don't change occurs if the added explanatory variable is uncorrelated with those already in the regression. In this special case, the slopes of the other variables remain as they were.

What Do You Think?

a. Is it possible for collinearity to reduce the standard error of a slope estimate, or must it increase the standard error?[4]
b. In a multiple regression with 2 explanatory variables, the correlation between X_1 and X_2 is 0.84. What is the VIF for X_1? For X_2?[5]
c. The following table summarizes the regression of Y on X_1 and X_2. Does it appear that collinearity is responsible for the insignificant effect of X_1?[6]

Term	Estimate	Std Error	t-Statistic	p-Value
Intercept	2.4404306	3.057573	0.80	0.4307
X_1	0.0092054	0.133619	0.07	0.9455
X_2	3.2182753	0.107313	29.99	<0.0001

Remedies for Collinearity

Collinearity not only makes it hard to interpret a multiple regression, it also blurs the estimates because it increases standard errors. You end up with wide confidence intervals and t-statistics that move toward to 0. What should you do about collinearity? There's a range of actions.

Remove redundant explanatory variables. The catch is deciding which to remove. If the t-statistic of an explanatory variable is close to zero, then that variable is not contributing to the model. The small t-statistic is the result of collinearity if the explanatory variable has a large marginal correlation with the response, but no effect in the multiple regression (like the market in our example). A large VIF distinguishes these explanatory variables.

Re-express your explanatory variables. If you know the context, you can often think of a way to combine several explanatory variables. For example, the Consumer Price Index (CPI) blends prices for a collection of products. This one variable replaces several prices that are highly correlated. Similarly, models in the social sciences frequently use a variable called socioeconomic

[4] Collinearity always increases the SE. VIF = 1 when there is no correlation and can only increase.
[5] The VIF for both explanatory variables is $1/(1 - corr(X_1, X_2)^2) = 1/(1 - 0.84^2) \approx 3.4$.
[6] No. Find the SE assuming the two explanatory variables are uncorrelated. If X_1 and X_2 were uncorrelated (with VIF = 1), the SE for X_1 would be $0.133/\sqrt{3.4} \approx 0.072$. The estimate b_1 would still be much less than 1 standard error from zero. It's just not going to contribute to the fit.

status. This variable combines income, education, and employment status into one variable so that a modeler does not need to include several collinear explanatory variables.

Nothing. Almost every multiple regression has correlated explanatory variables. If the estimated model has statistically significant explanatory variables with sensible estimates, we may not need to do anything about collinearity.

If we plan to use regression to predict new cases that resemble the observed data, collinearity may not be a large concern. We can have a very predictive model with a large R^2 and substantial collinearity. Be cautious, however: It is easy to accidentally extrapolate from collinear data. (See the discussion of extrapolation in Chapter 23.)

To handle the collinearity between Market % Change and Dow % Change, a simple re-expression removes most of the association. The similarity of these indices (both have the same units) suggests that the average of the two might make a better explanatory variable than either one alone. At the same time, the Dow Jones Industrial Average and the whole market index are different, so let's include the difference between the two as an explanatory variable.

Table 24.5 summarizes the multiple regression with these two variables in place of the original collinear pair.

TABLE 24.5 Multiple regression with rearranged explanatory variables.

R^2	0.3463
s_e	8.8709

Term	Estimate	Std Error	*t*-Statistic	*p*-Value	VIF
Intercept	−0.1865	0.6977	−0.27	0.7896	—
(Market + Dow)/2 % Chg	1.0665	0.1685	6.33	<0.0001	1.14
Market − Dow % Change	0.1644	0.4288	0.38	0.7019	1.78
Small-Big	0.6859	0.2248	3.05	0.0027	1.62
High-Low	−0.3418	0.2477	−1.38	0.1695	1.61

This regression has the same R^2 and s_e as the regression shown in Table 24.3 because the two equations produce the same fitted values. If we start from the regression summarized in Table 24.5, we can work our way back to the equation given by Table 24.4.

$$
\begin{aligned}
\textit{Estimated } \text{Sony \% Change} &= -0.1865 + 1.0665(\text{Market} + \text{Dow})/2 \\
&\quad + 0.1644(\text{Market} - \text{Dow}) + 0.6859 \text{ Small-Big} - 0.3418 \text{ High-Low} \\
&= -0.1865 + (1.0665/2 + 0.1644) \text{ Market} + (1.0665/2 - 0.1644) \text{ Dow} \\
&\quad + 0.6859 \text{ Small-Big} - 0.3418 \text{ High-Low} \\
&= -0.1865 + 0.6976 \text{ Market} + 0.3689 \text{ Dow} + 0.6859 \text{ Small-Big} \\
&\quad - 0.3418 \text{ High-Low}
\end{aligned}
$$

The advantage of this formulation is that the average and difference of the indices are almost uncorrelated ($r = 0.11$) so there is little collinearity to cloud our interpretation of this regression. The average of the indices is clearly an important explanatory variable; the difference is not. The intercept and other slopes match those in Table 24.3. The disadvantage of re-expression is that our model no longer has the clearly identified market indices as explanatory variables. If we want those, we have to consider removing variables from the regression.

24.3 | REMOVING EXPLANATORY VARIABLES

After several explanatory variables are added to a model, some of those added and some of those originally present may not be statistically significant. For instance, High-Low is not statistically significant in our modeling of the percentage changes in the value of Sony stock. Since this variable is not statistically significant, we could remove it from the model without causing a significant drop in R^2. Nonetheless, we often keep explanatory variables of interest in the regression whether significant or not, so long as they do not introduce substantial collinearity.

The reason for keeping these variables in the regression is simple: If we take High-Low out of this model, we're setting its slope to 0. Not close to 0—exactly 0. By leaving High-Low in the model, we obtain a confidence interval for its effect on the response, given the other explanatory variables. The 95% confidence interval for the partial slope of High-Low is $-0.3418 \pm 1.97(0.2477) \approx$ $[-0.83 \text{ to } 0.15]$. The slope of High-Low in the population could be rather negative, -0.5 or perhaps -0.75, but not so positive as 0.5. The confidence interval puts limits on the probable sizes. Because the confidence interval includes 0, we cannot claim to know the direction of its effect. Perhaps it's negative. Perhaps it's positive. If we were to remove this variable from our model, all we can say is that it does not affect the response.

On the other hand, we cannot leave both Market % Change and Dow % Change together in the regression. When both are in the same regression, neither is statistically significant, whereas either index taken separately is significant. Which should we remove? In this case, both statistics and substance point in the same direction: remove Dow % Change. The statistics point in this direction because Dow % Change adds less to the regression in Table 24.3; its t-statistic is 0.85 compared to 1.58 for Market % Change. Substantively, finance theory says to keep the whole market index as well; it represents a broader picture of returns on all stock as called for by the CAPM.

MARKET SEGMENTATION

MOTIVATION   STATE THE QUESTION

The manufacturer of a new mobile phone is considering advertising in one of two magazines. Both appeal to affluent consumers, but subscribers to one magazine average 20 years older than subscribers to the other. The industry will be watching the launch, so the manufacturer wants a flurry of initial sales. Which magazine offers a more receptive audience?

The manufacturer hired a market research firm to find the answer. This firm obtained a sample of 75 consumers. The firm individually showed each consumer a prototype of the design. After trying out the prototype, each consumer reported the likelihood of purchasing such a phone, on a scale of 1–10 (1 implies little chance of purchase and 10 indicates almost certain purchase). The marketing firm also measured two characteristics of the consumers: age (in years) and reported income (in thousands of dollars).

METHOD   DESCRIBE THE DATA AND SELECT AN APPROACH

We will use a multiple regression of the ratings on age and income to isolate the effects of age from income. The partial slopes for age and income will show how these variables combine to influence the rating. We are particularly interested in the partial slope of age to answer the posed question.

Plots remain important, even though we have several variables. A scatterplot matrix gives a compact summary of the relevant scatterplots, and the correlation matrix summarizes the degree of association.

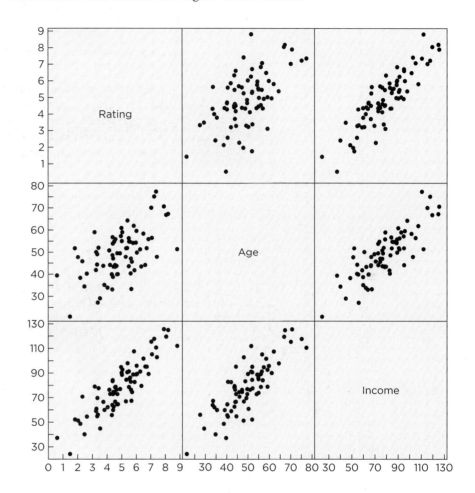

	Rating	**Age**	**Income**
Rating	1.0000	0.5867	0.8845
Age	0.5867	1.0000	0.8286
Income	0.8845	0.8286	1.0000

The scatterplot matrix shows that the association is linear, with no distorting outliers. The correlation between the two explanatory variables, corr(Age, Income) = 0.8286, implies collinearity: $\text{VIF} = 1/(1 - r^2) = 1/(1 - 0.8286^2) \approx 3.2$.

✓ **Linear.** The scatterplots of the response in the scatterplot matrix (top row) look straight enough.

✓ **No obvious lurking variable.** Hopefully, the survey got a good mix of men and women. It would be embarrassing to discover later that all of the younger customers in the survey were women, and all of the older customers were men. To avoid such problems, check that the survey participants were randomly selected from a target consumer profile or use randomization.

MECHANICS | DO THE ANALYSIS

Because there are no apparent violations of the conditions of the MRM, we can summarize the overall fit of the model. We start by checking the conditions of the MRM and verifying that the model explains statistically significant variation.

R^2	0.8507
s_e	0.6444

Term	Estimate	Std Error	t-Statistic	p-Value	VIF
Intercept	0.5177	0.3521	1.47	0.1459	—
Age	−0.0716	0.0125	−5.74	<0.0001	3.2
Income	0.1007	0.0064	15.63	<0.0001	3.2

We next check the conditions on the errors by examining the residuals. Here are the plot of e on $\hat{y}$ and the normal quantile plot of the residuals.

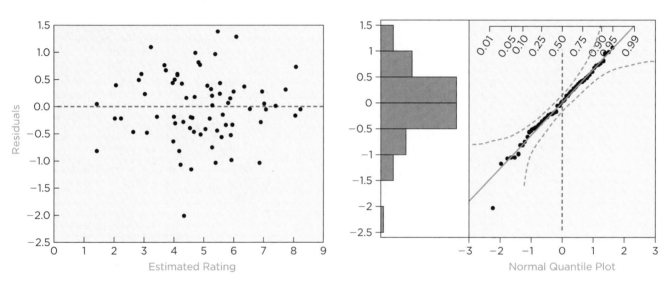

✓ **Evidently independent.** These data come from an experiment with 75 randomly selected consumers. These should be representative. Each consumer saw the device separately to avoid introducing dependence.

✓ **Similar variances.** The residuals resemble a random swarm of points, with no tendency for changing variation or extreme outliers.

✓ **Nearly normal.** The normal quantile plot shows that the residuals are nearly normal in spite of the one straggling negative outlier.

Since the model meets the conditions of the MRM, we can proceed to inference. The F-statistic for testing the overall fit of the model is

$$F = R^2/(1 - R^2) \times (n - k - 1)/k = (0.85/0.15) \times (75 - 3)/2 \approx 204$$

with p-value < 0.0001. We can reject H_0 that both slopes are zero. This model explains statistically significant variation in the ratings. The estimated equation is *Estimated* Rating $\approx 0.52 - 0.072$ Age $+ 0.10$ Income. Notice the change in the direction of the association between Age and Rating. The partial slope for Age is negative even though the correlation between Age and Rating is positive.

Although collinear, both predictors explain significant variation (both p-values are less than 0.05). Hence, both explanatory variables contribute to the regression. The 95% confidence interval for the slope of Age is ($t = 1.99$)

$$-0.0716 \pm 1.99 \times 0.0125 = [-0.0965 \text{ to } -0.0467 \text{ rating point/year}]$$

For consumers with comparable incomes, the 20-year age difference in subscribers of the two magazines implies a shift of $20 \times [-0.965 \text{ to } -0.0467] = [-1.930 \text{ to } -0.935]$ on the average rating.

MESSAGE | SUMMARIZE THE RESULTS

The manufacturer should advertise in the magazine with younger subscribers. We can be 95% confident that a younger, affluent audience assigns on average a rating 1 to 2 points higher (out of 10) than that of an older, affluent audience.

Don't avoid the issue of collinearity. Someone hearing these results may be aware of the positive correlation between age and rating, so it's better to explain your results. Although older customers rate this design higher than younger customers, this tendency is an artifact of income. Older customers typically have larger incomes than younger customers. When comparing two customers with comparable incomes, we can expect the younger customer to rate the phone more highly.

Note any important caveats that are relevant to the question. This analysis presumes that customers who assign a higher rating will indeed be more likely to purchase the phone. Also, this analysis accounts for only the age and income of the consumer. Other attributes, such as sex or level of use of a phone, might affect the rating as well.

In Example 24.1, the slope of *Age* changes sign when adjusted for differences in income. Substantively, this change makes sense because only younger customers with money to spend find the new design attractive. There's also a plot that shows what is going on. Figure 24.5 shows a scatterplot of *Rating* on *Age*, with the least squares line.

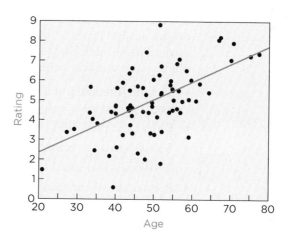

FIGURE 24.5 Scatterplot of rating on age with a fitted line.

The slope of the simple regression is clearly positive: On average, older customers rate the phone more highly. To see the association of age and rating among customers with comparable incomes, as in the multiple regression, let's (a) group consumers with similar levels of income and then (b) fit regressions within these groups. We picked out three groups, described in Table 24.6.

TABLE 24.6 Characteristics of three groups of customers with similar incomes.

Income	Number of Cases	Color
Less than $60,000	14	Blue
$70,000–80,000	17	Green
More than $100,000	11	Red

The scatterplot of *Rating* on *Age* in Figure 24.6 shows a simple regression within each of these groups.

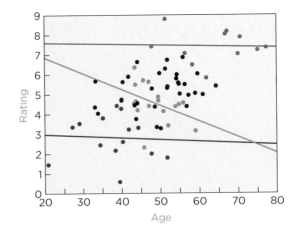

FIGURE 24.6 Slopes of simple regressions within subsets defined by income are negative.

The slope within each is negative even though the correlation between Rating and Age is positive. (With small groups, the slopes vary considerably from one to another.) That's what the multiple regression tells us: Among customers with similar incomes, younger customers are the biggest fans.

4M EXAMPLE 24.2 | RETAIL PROFITS

MOTIVATION | **STATE THE QUESTION**

A chain of pharmacies is looking to expand into a new community. It has data on the annual profits (in dollars) of pharmacies that are located in 110 cities around the United States. For each city, the chain also has the following variables that managers believe are associated with profits:

- Income (median annual salary) and disposable income (median income net of taxes),
- Birth rate (per 1,000 people in the local population),
- Social Security recipients (per 1,000 people in the local population),
- Cardiovascular deaths (per 100,000 people in the local population), and
- Percentage of the local population aged 65 or more.

Managers would like to have a model that would (a) indicate whether and how these variables are related to profits, (b) provide a means to choose new communities for expansion, and (c) predict sales at current locations to identify underperforming sites.

METHOD | **DESCRIBE THE DATA AND SELECT AN APPROACH**

We will build a multiple regression. The definitions of the possible explanatory variables suggest that we should expect collinearity: Two variables measure income and four variables are related to the age of the local population.

The following correlation matrix confirms our suspicions: The correlation between income and disposable income, for example, is 0.777, and the correlation between the number of Social Security recipients and the percentage aged 65 or more is 0.938. (These correlations are shaded in the following table.)

	Profit	Income	Disp Inc	Birth Rate	Soc Sec	CV Death	% 65+
Profit	1.000	0.264	0.474	−0.347	0.668	0.609	0.774
Income	0.264	1.000	0.777	−0.130	−0.130	−0.050	−0.056
Disp Inc	0.474	0.777	1.000	−0.256	0.063	0.056	0.165
Birth Rate	−0.347	−0.091	−0.256	1.000	−0.585	−0.550	−0.554
Soc Sec	0.668	−0.130	0.063	−0.585	1.000	0.853	0.938
CV Death	0.609	−0.050	0.056	−0.550	0.853	1.000	0.867
% 65+	0.774	−0.056	0.165	−0.554	0.938	0.867	1.000

Each of these characteristics of the local community is correlated with profits at the pharmacy. Birth rate is negatively correlated with profits.

The following scatterplot matrix shows the response with four of these variables. (Showing them all makes the plots hard to fit on a page.) Birth rate is negatively associated with the number of Social Security recipients, rate of cardiovascular disease, and proportion 65 and older. Communities with relatively high birth rates, not surprisingly, tend to be younger.

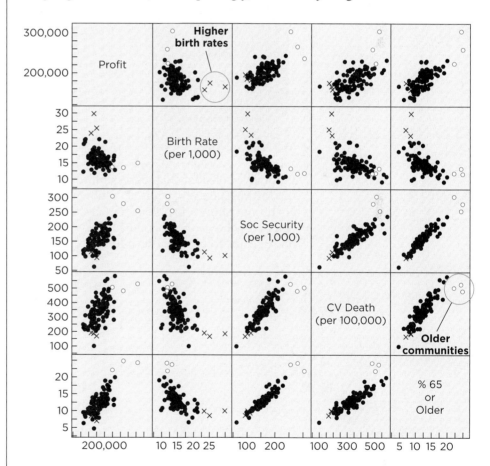

The scatterplot matrix also identifies communities that are distinct from the others. For example, the distribution of birth rates is skewed, with relatively high rates in Texas and Utah ($\times$ in the figures). In contrast, the outliers with high concentrations of people 65 and older are in Florida (o in the figures).

✓ **Linear.** Scatterplots of the response in the scatterplot matrix (top row) appear linear, at least for those that show stronger association.

✓ **No obvious lurking variable.** Several other variables might influence profits, such as the age of the drug store (newer stores might be in better locations) or the foot traffic (more passers-by would lead to heavier business). Other information obtained when these data were collected indicates that these stores are randomly scattered among cities so that omitted variables ought not distort our estimates.

DO THE ANALYSIS

We begin with a regression that includes all six explanatory variables.

R^2	0.7558
$\bar{R}^2$	0.7415
s_e	14019
n	110

Term	Estimate	Std Error	*t*-Statistic	*p*-Value	VIF
Intercept	13,161.34	19,111.39	0.69	0.4926	—
Income	0.60	0.59	1.02	0.3116	2.95
Disposable Income	2.54	0.73	3.46	0.0008	3.30
Birth Rate (per 1,000)	1,703.87	563.67	3.02	0.0032	1.70
Soc Security (per 1,000)	−47.52	110.21	−0.43	0.6673	10.04
CV Death (per 100,000)	−22.68	31.46	−0.72	0.4726	4.71
% 65 or Older	7,713.85	1,316.21	5.86	−0.0001	11.39

Before we consider this regression further, we need to check the conditions of the MRM. The calibration plot shown on the left below and plot of the residuals do not show a problem. The association in the calibration plot is strong (this plot shows the correlation between y and $\hat{y}$, the square root of R^2) and there seems to be no pattern in the residual plot.

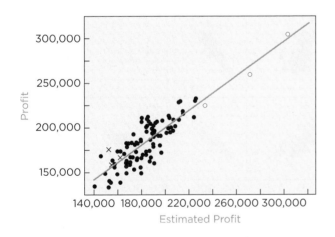
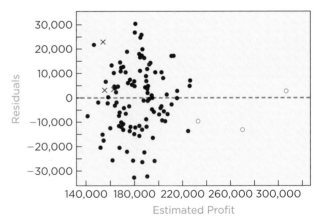

The regression predicts the highest profits in Florida, which has high concentrations of people 65 and older. (We flagged these outliers in the scatterplot matrix.) Further plots of the residuals, such as the following scatterplot of the residuals versus the percentage 65 and older and the normal quantile plot of the residuals, indicate that this model meets the conditions of the MRM. There's no pattern in the plot of the residuals on this explanatory variable, and the residuals appear normally distributed.

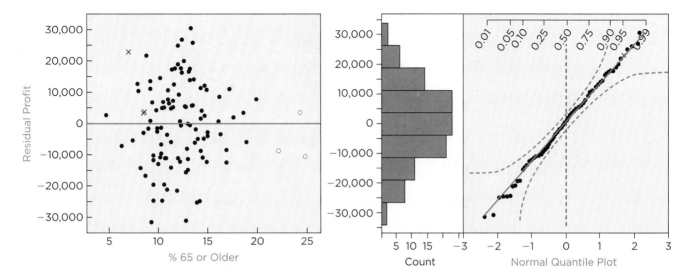

✓ **Evidently independent.** We can think of these data as a random sample of the locations at which stores with pharmacies might be located. Had several stores been in the same city, we would question the assumption of independence.

✓ **Similar variances.** Scatterplots of the residuals versus fitted values and explanatory variables resemble a random swarm of points, with no tendency for changing variation or extreme outliers.

✓ **Nearly normal.** The normal quantile plot shows that the residuals are nearly normal.

Having checked the conditions of the MRM, we proceed to inference. The *F*-test indicates that the model explains statistically significant variation in profits,

$$F = R^2/(1 - R^2) \times (n - k - 1)/k = (0.7558/0.2442) \times (110 - 6 - 1)/6 \approx 53$$

with *p*-value less than 0.0001. We reject $H_0: \beta_1 = \beta_2 = \cdots = \beta_6 = 0$. This collection of explanatory variables explains statistically significant variation in profits.

The model as a whole explains statistically significant variation in profits, but three of the explanatory variables are not statistically significant: income, Social Security recipients, and deaths from cardiovascular disease. The VIFs offer an explanation for the lack of statistical significance of some estimates: collinearity. Income is highly correlated with disposable income (which is significant), and the other two explanatory variables are highly correlated with the percentage above 65 (which is significant). Because these redundant variables are not statistically significant and make it difficult to estimate those that are, we remove them from the model. **Remove variables one at a time to make sure that you don't omit a useful variable that was disguised by collinearity.** The table at the top of the next page shows a summary of the simplified regression model with three explanatory variables.

tip

The collinear explanatory variables contributed little to the explained variation: R^2 falls without these (from 0.7558 to 0.7521) but the adjusted R^2 grows a bit (from 0.7415 to 0.7450). The remaining estimated slopes are statistically significant and more precisely determined. For example, the standard error of the slope for % 65 or Older falls from 1,316 in the six-variable regression to 466 in this model, and its *t*-statistic grows from 5.86 to 14.22. Without the diluting effect of redundant measures this simpler model provides a more precise estimate of the effect of age of the local population.

With the model simplified, we focus on the contrast between the marginal and partial associations. In particular, birth rate has a positive partial slope

R^2	0.7521
$\bar{R}^2$	0.7450
s_e	13,455
n	110

Term	Estimate	Std Error	*t*-statistic	*p*-value	VIF
Intercept	10,044.61	15,344.79	0.65	0.5141	—
Disposable Income	3.24	0.41	7.83	<0.0001	1.07
Birth Rate (per 1,000)	1,874.05	526.50	3.56	0.0006	1.50
% 65 or Older	6,619.21	465.53	14.22	<0.0001	1.44

compared to a negative correlation with profits. Collinearity also explains this change in sign. When considered marginally, cities with low birth rates typically have a larger concentration of older residents who in turn generate profits. Cities with high birth rates tend to lack a large elderly population and generate smaller profits. The multiple regression separates the effects of birth rates from age (and income) and reveals that cities with higher birth rates produce higher average profits when compared to cities with lower birth rates but comparable income and percentage of residents above 65.

In preparation for the message, here are 95% confidence intervals for the slopes of the explanatory variables ($t_{0.025,n-3-1} = 1.98$):

Disposable income: $3.24 \pm 1.98(0.41) = [2.428 \text{ to } 4.051]$
Birth rate: $1,874.05 \pm 1.98(526.5) = [831.58 \text{ to } 2,916.52]$
% 65 and older: $6,619.21 \pm 1.98(465.53) = [5,697.461 \text{ to } 7,540.959]$

The next calculation finds estimated profits for a new location with median disposable income $22,642, 14.4 births per 1,000, and 11.4% who are 65 or more years old:

Estimated Profits $= 10,044.61 + 3.24(22,642) + 1,874.05(14.4) + 6,619.21(11.4)$
$= \$185,850$

Our software calculates $se(\hat{y}) = \$13,577$. The 95% prediction interval for profits for a new location is $\hat{y} \pm t_{0.025,106}se(\hat{y}) = [\$158,968 \text{ to } \$212,733]$.

MESSAGE | SUMMARIZE THE RESULTS

This analysis finds that three characteristics of the local community affect estimated profits of pharmacies located in drug stores: disposable income, age, and birth rates. Increases in each of these characteristics lead to greater profits. On average, pharmacies in locations with more disposable income earn more profits. When compared to profits in locations with the same proportion of residents 65 and older and comparable birth rate, pharmacy profits increase on average $2,400 to $4,100 per $1,000 increase in the median local disposable income. Similarly, comparing sites with different birth rates, we expect $800 to $2,900 more in profits on average for each increase in the birth rate by 1 per 1,000. Also, profits increase on average from $5,700 to $7,500 for each 1% increase in the percentage of the local population above 65 (assuming as before no difference in the other characteristics).

The data show that site selection will have to trade off these characteristics. Our data have few communities with both high birth rates and a large percentage 65 years old and older. Communities tend to have either high birth rates or large percentages of older residents. The equation produced by the fitted model allows us to combine these attributes to anticipate profits at new locations.

As an illustration of the use of this model, we predict profits of a new store in Kansas City, MO. The three relevant local characteristics are

Disposable income:	$22,642
Birth rate:	14.4 per 1,000
Percentage 65+:	11.4

Under these conditions, our model predicts profits in the range $159,000 to $213,000 with 95% probability.

Note any important weakness or limitation of your model. This analysis presumes that the stores used in the analysis are representative of typical operations. In anticipating the sales at new locations, we assume that such stores will operate in the same manner during similar economic conditions.

Best Practices

- *Begin a regression analysis by looking at plots.* We've said it before. Start with plots. A scatterplot matrix makes it easy to skim plots between y and the explanatory variables. These plots also help identify the extent of the collinearity among the explanatory variables and identify important outliers.
- *Use the F-statistic for the overall model and a t-statistic for each explanatory variable.* The overall *F*-statistic tells you about the whole model, not any one explanatory variable. If you have a question about the whole model, look at the *F*-statistic. If you have a question about a single explanatory variable, look at its *t*-statistic.
- *Learn to recognize the presence of collinearity.* When you see the slopes in your model change as you add or remove a variable, recognize that you've got collinearity in the model. Make sure you know why. Variance inflation factors provide a concise numerical summary of the effects of the collinearity.
- *Don't fear collinearity—understand it.* In Example 24.1, we could have gotten rid of the collinearity by removing one of the variables. The simple regression of the rating on the age doesn't show any effects of collinearity—but it also leads to the wrong conclusion. In examples like the stock market illustration, you can see that two variables measure almost the same thing. It makes sense to combine them or perhaps remove one.

Pitfalls

- *Do not remove explanatory variables at the first sign of collinearity.* Collinearity does not violate an assumption of the MRM. The MRM lets you pick values of the explanatory variables however you choose. In many cases, the collinearity is an important part of the model: The partial and marginal slopes are simply different. If an explanatory variable is statistically significant, don't remove it just because it's related to others.
- *Don't remove several explanatory variables from your model at once.* If these variables are collinear (highly correlated), it could be the case that any one of them would be a very distinct, important explanatory variable. By using several of them, you've let collinearity mask the importance of what these variables measure.

If you remove them all from the model, you might miss the fact that any one of them could be useful by itself.

Software Hints

EXCEL

Excel and DDXL do not provide VIFs for a regression. You can compute these using the formulas given in this chapter.

MINITAB

The menu sequence

Stat > Regression ... > Regression

constructs a multiple regression if several variables are chosen as explanatory variables. To add VIFs to the summary of the regression, click the Options button in the dialog box that constructs the model and select the check-box labeled Variance inflation factors.

JMP

The menu sequence

Analyze > Fit Model

constructs a multiple regression if two or more variables are entered as explanatory variables. Click the Run Model button to fit the least squares regression. The summary window combines numerical summary statistics with several plots. To see VIFs for the estimates, right-click on the tabular summary of the regression coefficients. In the pop-up menu, select Columns > VIF.

BEHIND the Math

Variance Inflation Factors in Larger Models

The VIF is tedious to calculate in models with more than two explanatory variables. With two explanatory variables, all you need is the correlation $\text{corr}(x_1, x_2)$. With more than two, you need between them, a multiple regression. In general, to find the VIF for an explanatory variable, regress that explanatory variable on all of the other explanatory variables. For instance, to find the VIF for x_1 in a regression with three other explanatory variables,

regress x_1 on $x_2, x_3,$ and x_4. Denote the R^2 of that regression as R_1^2. The VIF for x_1 is then

$$\text{VIF}(x_1) = \frac{1}{1 - R_1^2}$$

The R^2-statistic from the multiple regression replaces the square of the simple correlation between x_1 and x_2. As R_1^2 increases, less unique variation remains in x_1 and the effects of collinearity (and VIF) grow.

CHAPTER SUMMARY

Large correlations between explanatory variables in a regression model produce collinearity. Collinearity leads to surprising coefficients with unexpected signs, imprecise estimates, and wide confidence intervals. You can detect collinearity before fitting a model in the correlation matrix and scatterplot matrix. Once you have fit a regression model, use the **variance inflation factor** to quantify the extent to which collinearity increases the standard error of a slope. Remedies for collinearity include combining variables and removing one of a highly correlated pair.

Key Term

variance inflation factor (VIF), 612

Formulas

Standard error of a slope in multiple regression

$$se(b_j) = \frac{s_e}{\sqrt{n-1}\, s_x} \times \sqrt{VIF(x_j)}$$

Variance inflation factor (VIF)

In multiple regression with two explanatory variables, the VIF for both x_1 and x_2 is

$$VIF(x_1) = VIF(x_2) = \frac{1}{1 - \left(corr(x_1, x_2)\right)^2}$$

In general, with k explanatory variables, replace the correlation with the R^2-statistic from a regression of x_j on the other explanatory variables:

$$VIF(x_j) = \frac{1}{1 - R_j^2},$$
$$R_j^2 = R^2(x_j \text{ on } x_1, \ldots, x_{j-1}, x_{j+1}, \ldots, x_k)$$

About the Data

As noted in the discussion of the data in Chapter 21, the CAPM requires excess returns that account for the rate of interest paid on risk-free investments. As in Chapter 21, the percentage changes used in the regressions of this chapter are "excess percentage changes" because we have subtracted out the risk-free rate implied by 30-day Treasury Bills. The underlying data come from CRSP, the Center for Research in Security Prices, and were obtained through WRDS, Wharton Research Data Services.

Regarding the outliers in returns on Sony stock, the huge gain in December 1999 capped off a year in which stock in Sony soared in value as Sony reor-

ganized its business units and positioned itself as a leader in Internet technology. During December, Sony announced a two-for-one stock split, producing a surge in buying. By April 2003, the stock had fallen from its peak at the start of 2000 and was approaching a new low. The announcement in April of lower than expected earnings in its fourth quarter (which included holiday sales) led to a sell-off of shares.

The data in Example 24.1 on consumer products come from a case study developed in the MBA program at Wharton. Portions of the data in Example 24.2 come from consulting research; the explanatory variables are from the U.S. Census of Metropolitan Areas.

EXERCISES

Mix and Match

Match each task or property of a regression model in the left-hand column with its description in the right-hand column.

1. Test $H_0: \beta_1 = \beta_2 = 0$	(a) t-statistic for b_1
2. Test $H_0: \beta_2 = 0$	(b) $1 - R^2$
3. Test $H_0: \alpha = 0$	(c) $s_{x_2}^2$
4. Effect of collinearity on $se(b_1)$	(d) $VIF(x_1)$
5. Correlations among variables	(e) t-statistic for b_2
6. Scatterplots among variables	(f) F-statistic
7. Percentage of variation in residuals	(g) Scatterplot matrix
8. Test whether adding x_1 improves fit of model	(h) $s_{x_2}^2 / VIF(x_2)$
9. All variation in x_2	(i) t-statistic for b_0
10. Unique variance in x_2	(j) Correlation matrix

True/False

Mark each statement True or False. If you believe that a statement is false, briefly explain why you think it is false.

11. Excess percentage changes in the value of an investment subtract the cost of borrowing from the percentage changes in the value of the investment.

12. The capital asset pricing model indicates that the estimated intercept in a regression of excess percentage changes in a stock on those in the market is zero.

13. If a multiple regression has a large F-statistic but a small t-statistic for each predictor (i.e., the t-statistics for the slopes are near zero), then collinearity is present in the model.

14. The F-statistic is statistically significant only if some t-statistic for a slope in multiple regression is statistically significant.

15. If the R^2 of a multiple regression with two predictors is larger than 80%, then the regression explains a statistically significant fraction of the variance in y.

16. If the t-statistic for X_2 is larger than 2 in absolute size, then adding X_2 to the simple regression containing X_1 produces a significant improvement in the fit of the model.

17. We can detect outliers by reviewing the summary of the associations in the scatterplot matrix.

18. A correlation matrix summarizes the same information in the data as is given in a scatterplot matrix.

19. In order to calculate the VIF for an explanatory variable, we need to use the values of the response.

20. If $\text{VIF}(X_2) = 1$, then we can be sure that collinearity has not inflated the standard error of the estimated partial slope for X_2.

21. The best remedy for a regression model that has collinear predictors is to remove one of those that are correlated.

22. It is not appropriate to ignore the presence of collinearity because it violates one of the assumptions of the MRM.

Think About It

23. Collinearity is sometimes described as a problem with the data, not the model. Rather than filling the scatterplot of x_1 on x_2, the data concentrate along a diagonal. For example, the following plot shows monthly percentage changes in the whole stock market and the S&P 500 (in excess of the risk-free rate of return). The data span the same period considered in the text, running from 1995 through 2008.

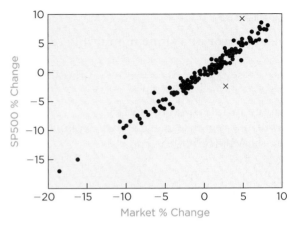

(a) Data for two months (February and March of 2000) deviate from the pattern evident in other months. What makes these months unusual?

(b) If you were to use both returns on the market and those on the S&P 500 as explanatory variables in the same regression, would these two months be leveraged?

(c) Would you want to use these months in the regression or exclude them from the multiple regression?

24. Regression models that describe macroeconomic properties in the United States often have to deal with large amounts of collinearity. For example, suppose we want to use as explanatory variables the disposable income and the amount of household credit debt. Because the economy in the United States continues to grow, both of these variables grow as well. Here's a scatterplot of quarterly data, from 1960 through 2008. Both are measured in billions of dollars. Disposable income is in red and credit debt is in green.

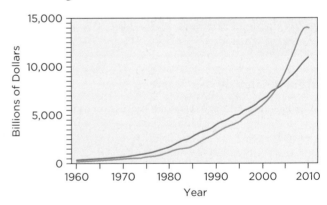

(a) This plot shows timeplots of the two series. Do you think that they are correlated? Estimate the correlation.

(b) If the variables are expressed on a log scale, will the transformation to logs increase, decrease, or not affect the correlation between these series?

(c) If both variables are used as explanatory variables in a multiple regression, will you be able to separate the two?

(d) You're trying to build a model to predict how changes in the macroeconomy will affect consumer demand.

You've got sales of your firm over time as the response. Suggest an approach to using the information in both of these series in a multiple regression that avoids some of the effects of collinearity.

25. The version of the CAPM studied in this chapter specifies a simple regression model as

$$100(S_t - r_t) = \beta_0 + 100\,\beta_1(M_t - r_t) + \varepsilon$$

where M_t are the returns on the market, S_t are the returns on the stock, and r_t are the returns on risk-free investments. Hence, $100(M_t - r_t)$ are the excess percentage changes for the market and $100(S_t - r_t)$ are the excess percentage changes for the stock. What happens if we work with the excess returns themselves rather than the percentage changes? In particular, what happens to the t-statistic for the test of $H_0: \alpha = 0$?

26. The following histograms summarize monthly returns on Sony, the whole stock market, and risk-free assets during 1995–2008. Having seen this comparison, explain why it does not make much difference whether we subtract the risk-free rate from the variables in the CAPM regression.

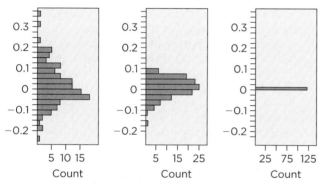

27. In Example 24.1 the data show correlation between the income and age of the customer. This produces collinearity and makes the analysis tricky to interpret. The marketing research group could have removed this collinearity by collecting data with these two factors made uncorrelated. For example, for incomes of $60,000, $70,000, . . . , $120,000, they could have found two customers, say, 25, 35, 45, 55, and 65 years old. That would have given them 70 observations (7 income levels, with 10 in each).
    (a) Explain why Income and Age would be uncorrelated for these data.
    (b) Would the marginal slope be the same as the partial slope when analyzing these data?
    (c) Would the marginal slope for Age when estimated for these data have a positive or negative sign?

28. To find out whether employees are interested in joining a union, a manufacturing company hired an employee relations firm to survey attitudes toward unionization. In addition to a rating of their agreement with the statement "I do **not** think we need a union at this company" (on a 1–7 Likert scale), the firm also recorded the number of years of experience and the salary of the employees. Both of these are typically positively correlated with agreement with the statement.

(a) In building a multiple regression of the agreement variable on years of experience and salary, would you expect to find collinearity? Why?
(b) Would you expect to find the partial slope for salary to be about the same as the marginal slope, or would you expect it to be noticeably larger or smaller?

29. Modern steel mills are very automated and need to monitor their substantial energy costs carefully to be competitive. In making cold-rolled steel (as used in bodies of cars), it is known that temperature during rolling and the amount of expensive additives (expensive metals like manganese and nickel give steel desired properties) affect the number of pits per 20-foot section. A pit is a small flaw in the surface. To save on costs, a manager suggested the following plan for testing the results at various temperatures and amounts of additives.

   90°, 0.5% additive
   95°, 1.0% additive
   100°, 1.5% additive
   105°, 2.0% additive
   110°, 2.5% additive

Multiple sections of steel would be produced for each combination, with the number of pits computed.
(a) If Temperature and Additive are used as predictors together in a multiple regression, will this approach yield useful data?
(b) Would you stick to this plan, or can you offer an alternative that you think is better? What would that approach be?

30. A builder is interested in which types of homes earn a higher price. For a given number of square feet, the builder gathered prices of homes that use the space differently. In addition to price, the homes vary in the number of rooms devoted to personal use (such as bathrooms or bedrooms) and rooms devoted to social use (enclosed decks or game rooms). Because the homes are of roughly equal size (equal numbers of square feet), the more space devoted to private use, the less devoted to social use. The variable Private denotes the number of square feet used for private space and Social the number of square feet for social rooms.
(a) Would you expect to find collinearity in a multiple regression of Price on Private and Social? Explain.
(b) Rather than use both Private and Social as two variables in a multiple regression for Price, suggest an alternative that might in the end be simpler to interpret as well.

You Do It

We investigated the use of the MRM for inference in the following examples in Chapter 23. In answering the following questions, assume unless indicated otherwise that you can use the MRM for inference. If you're concerned that it's not appropriate, see the analyses of these data in Chapter 23. For each dataset, if your software supports it, prepare a scatterplot matrix as a first step in your analysis.

Otherwise, you might want to look back at the individual scatterplots of these data that were used in the exercises of Chapter 23.

31. **Gold Chains (introduced in Chapter 23)** These data give the prices (in dollars) for gold link chains at the Web site of a discount jeweler. The data include the length of the chain (in inches) and its width (in millimeters). All of the chains are 14-carat gold in a similar link style. Use the price as the response. For one explanatory variable, use the width of the chain. For the second, calculate the "volume" of the chain as π times its length times the square of half the width, Volume = π Metric Length $\times$ (Width/2)^2. To make the units of volume mm^3, first convert the length to millimeters (25.4 mm = 1 inch).
    (a) The explanatory variable Volume includes Width. Are these explanatory variables perfectly correlated? Can we use them both in the same multiple regression?
    (b) Fit the multiple regression of Price on Width and Volume. Do both explanatory variables improve the fit of the model?
    (c) Find the variance inflation factor and interpret the value that you obtain.
    (d) What is the interpretation of the coefficient of Volume?
    (e) The marginal correlation between Width and Price is 0.95, but its slope in the multiple regression is negative. How can this be?

32. **Convenience Shopping (introduced in Chapter 19)** These data describe sales over time at a franchise outlet of a major U.S. oil company. (The data file has values for two stations. For this exercise, use only the 283 cases for site 1.) Each row summarizes sales for one day. This particular station sells gas, and it also has a convenience store and a car wash. The response Sales gives the dollar sales of the convenience store. The explanatory variable Volume gives the number of gallons of gas sold, and Washes gives the number of car washes sold at the station.
    (a) Fit the multiple regression of Sales on Volume and Washes. Do both explanatory variables improve the fit of the model?
    (b) Which explanatory variable is more important to the success of sales at the convenience store: gas sales or car washes? Do the slopes of these variables in the multiple regression provide the full answer?
    (c) Find the variance inflation factor and interpret the value that you obtain.
    (d) One of the explanatory variables is just barely statistically significant. Assuming the same estimated value, would a complete lack of collinearity have made this explanatory variable noticeably more statistically significant?

33. **Download (introduced in Chapter 19)** Before purchasing videoconferencing equipment, a company tested its current internal computer network. The tests measured how rapidly data moved through its network given the current demand on the network. Eighty files ranging in size from 20 to 100 megabytes (MB) were transmitted over the network at various times of day, and the time to send the files (in seconds) recorded. The time is given as the number of hours past 8 A.M. on the day of the test.
    (a) Fit the multiple regression of Transfer Time on File Size and Hours past 8. Does the model, taken collectively, explain statistically significant variation in transfer time?
    (b) Does either explanatory variable improve the fit of the model that uses the other? Use a test statistic for each.
    (c) Find the variance inflation factors for both explanatory variables. Interpret the values that you obtain.
    (d) Can collinearity explain the paradoxical results found in parts (a) and (b)?
    (e) Would it have been possible to obtain data in this situation in a manner that would have avoided the effects of collinearity?

34. **Production Costs (introduced in Chapter 19)** A manufacturer produces custom metal blanks that are used by its customers for computer-aided machining. The customer sends a design via computer, and the manufacturer comes up with an estimated cost per unit, which is then used to determine a price for the customer. The data for the analysis were sampled from the accounting records of 195 orders that were filled during the previous three months.
    (a) Fit the multiple regression of Average Cost on Material Cost and Labor Hours. Both explanatory variables are per unit produced. Do both explanatory variables improve the fit of the model that uses the other?
    (b) The estimated slope for labor hours per unit is much larger than the slope for material cost per unit. Does this difference mean that labor costs form a larger proportion of production costs than material costs?
    (c) Find the variance inflation factors for both explanatory variables. Interpret the value that you obtain.
    (d) Suppose that you formulated this regression using total cost of each production run rather than average cost per unit. Would collinearity have been a problem in this model? Explain.

35. **Home Prices (introduced in Chapter 23)** In order to help clients determine the price at which their house is likely to sell, a realtor gathered a sample of 150 purchase transactions in her area during a recent three-month period. The price of the home is measured in thousands of dollars. The number of square feet is also expressed in thousands, and the number of bathrooms is just that. Fit the multiple regression of Price on Square Feet and Bathrooms.
    (a) Thinking marginally for a moment, should there be a correlation between the square feet and the number of bathrooms in a home?
    (b) One of the two explanatory variables in this model does not explain statistically significant variation in the price. Had the two explanatory variables been uncorrelated (and produced these estimates), would this variable have been statistically significant? Use the VIF to find your answer.

(c) We can see the effects of collinearity by constructing a plot that shows the slope of the multiple regression. To do this, we remove the effect of one of the explanatory variables from the other variables. Here's how to make a so-called partial regression leverage plot for these data. First, regress Price on Square Feet and save the residuals. Second, regress Bathrooms on Square Feet and save these residuals. Now, make a scatterplot of the residuals from the regression of Price on Square Feet on the residuals from the regression of Bathrooms on Square Feet. Fit the simple regression for this scatterplot, and compare the slope in this fit to the partial slope for Bathrooms in the multiple regression. Are they different?

(d) Compare the scatterplot of Price on Bathrooms to the partial regression plot constructed in part (c). What has changed?

36. **Leases (introduced in Chapter 19)** This data table gives annual costs of 223 commercial leases. All of these leases provide office space in a Midwestern city in the United States. The cost of the lease is measured in dollars per square foot, per year. The number of square feet is as labeled, and Parking counts the number of parking spots in an adjacent garage that the realtor will build into the cost of the lease. Fit the multiple regression of Cost per Sq Ft on 1/Sq Ft and Parking/Sq Ft. (Recall that the slope of 1/Sq Ft captures the fixed costs of the lease, those present regardless of the number of square feet.)

(a) Thinking marginally for a moment, should there be a correlation between the number of parking spots and the fixed cost of a lease?

(b) Interpret the coefficient of Parking/Sq Ft. Once you figure out the units of the slope, you should be able to get the interpretation.

(c) One of the two explanatory variables explains slightly more than statistically significant variation in the price. Had the two explanatory variables been uncorrelated (and produced these estimates), would the variation have been more clearly statistically significant? Use the VIF to see.

(d) We can see the effects of collinearity by constructing a plot that shows the slope of the multiple regression. To do this, we have to remove the effect of one of the explanatory variables from the other variables. Here's how to make a so-called partial regression leverage plot for these data. First, regress Cost/Sq Ft on Parking/Sq Ft and save the residuals. Second, regress 1/Sq Ft on Parking/Sq Ft and save these residuals. Now, make a scatterplot of the residuals from the regression of Cost/Sq Ft on Parking/Sq Ft on the residuals from the regression of 1/Sq Ft on Parking/Sq Ft. Fit the simple regression for this scatterplot, and compare the slope in this fit to the partial slope for 1/Sq Ft in the multiple regression. Are they different?

(e) Compare the scatterplot of Cost/Sq Ft on 1/Sq Ft to the partial regression plot constructed in part (d). What has changed?

37. **R&D Expenses (introduced in Chapter 19)** This data table contains accounting and financial data that describe 493 companies operating in technology industries: software, systems design, and semiconductor manufacturing. The variables include the expenses on research and development (R&D), total assets of the company, and net sales. All columns are reported in millions of dollars, so 1,000 = $1 billion. Use the natural logs of all variables and fit the regression of Log R&D Expenses on Log Assets and Log Net Sales.

(a) Thinking marginally for a moment, would you expect to find a correlation between the log of the total assets and the log of net sales?

(b) Does the correlation between the explanatory variables change if you work with the data on the original scale rather than on a log scale? In which case is the correlation between the explanatory variables larger?

(c) In which case does correlation provide a more useful summary of the association between the two explanatory variables?

(d) What is the impact of the collinearity on the standard errors in the multiple regression using the variables on a log scale? Does the size of the VIF tell you that the two explanatory variables cannot be statistically significant?

(e) We can see the effects of collinearity by constructing a plot that shows the slope of the multiple regression. To do this, we have to remove the effect of one of the explanatory variables from the other variables. Here's how to make a so-called partial regression leverage plot for these data. First, regress Log R&D Expenses on Log Net Sales and save the residuals. Second, regress Log Assets on Log Net Sales and save these residuals. Now, make a scatterplot of the residuals from the regression of Log R&D Expenses on Log Net Sales on the residuals from the regression of Log Assets on Log Net Sales. Fit the simple regression for this scatterplot, and compare the slope of this simple regression to the partial slope for Log Assets in the multiple regression. Are they different?

(f) Compare the scatterplot of Log R&D Expenses on Log Assets to the partial regression plot constructed in part (e). What has changed?

38. **Cars (introduced in Chapter 19)** This data table gives characteristics of 223 types of cars sold in the United States during the 2003 and 2004 model years. Fit a multiple regression with the $\log_{10}$ of the base price as the response and the $\log_{10}$ of the horsepower of the engine (HP) and the $\log_{10}$ of the weight the car (given in thousands of pounds) as explanatory variables.

(a) Does it seem natural to find correlation between these two explanatory variables, either on a log scale or in the original units?

(b) How will collinearity on the log scale affect the standard errors of the slopes of these predictors in the multiple regression?

(c) One of the explanatory variables in the multiple regression is not statistically significant. Can this be attributed to the effect of collinearity of the standard error of this estimate?

(d) We can see the effects of collinearity by constructing a plot that shows the slope of the multiple regression. To do this, we have to remove the effect of one of the explanatory variables from the other variables. Here's how to make a so-called partial regression leverage plot for these data. First, regress Log_{10} Price on Log_{10} HP and save the residuals. Second, regress Log_{10} Weight on Log_{10} HP and save these residuals. Now, make a scatterplot of the residuals from the regression of Log_{10} Price on Log_{10} HP on the residuals from the regression of Log_{10} Weight on Log_{10} HP. Fit the simple regression for this scatterplot, and compare the slope in this fit to the partial slope for Log_{10} Weight in the multiple regression. Are they different?

(e) Compare the scatterplot of Log_{10} Price on Log_{10} Weight to the partial regression plot constructed in part (e). What has changed?

39. **OECD** An analyst at the United Nations is developing a model that describes GDP (gross domestic product per capita, a measure of the overall production in an economy per citizen) among developed countries. For this analysis, she uses national data for 30 countries from the 2005 report of the Organization for Economic Cooperation and Development (OECD). Her current equation is

Estimated per capita GDP $= \beta_0 + \beta_1$ Trade Balance $+ \beta_2$ Waste per capita

Trade balance is measured as a percentage of GDP. Exporting countries tend to have large positive trade balances. Importers have negative balances. The other explanatory variable is the annual number of kilograms of municipal waste per person.

(a) Is there any natural reason to expect these explanatory variables to be correlated? Suppose the analyst had formulated her model using national totals as

GDP $= \beta_0 + \beta_1$ Net Export (\$) $+ \beta_2$ Total Waste (kg) $+ \varepsilon$

Would this model have more or less collinearity? (You should not need to explicitly form these variables to answer this question.)

(b) One nation is particularly leveraged in the marginal relationship between per capita GDP and trade balance. Which is it?

(c) Does collinearity exert a strong influence on the standard errors of the estimates in the analyst's multiple regression?

(d) Because multiple regression estimates the partial effect of an explanatory variable rather than its marginal effect, we cannot judge the effect of outliers on the partial slope from their position in the scatterplot of y on x. We can, however, see their effect by constructing a plot that shows the partial slope. To do this, we have to remove the effect of

one of the explanatory variables from the other variables. Here's how to make a so-called partial regression leverage plot for these data.

First, regress per capita GDP on per capita Waste and save the residuals. Second, regress Trade Balance on per capita Waste and save these residuals. These regressions remove the effects of waste from the other two variables. Now, make a scatterplot of the residuals from the regression of per capita GDP on per capita Waste on the residuals from the regression of Trade Balance on per capita Waste. Fit the simple regression for this scatterplot, and compare the slope in this fit to the partial slope for Trade Balance in the multiple regression. Are they different?

(e) Which nation, if any, is leveraged in the partial regression leverage plot constructed in part (d)? What would happen to the estimate for this partial slope if the outlier were excluded?

40. **Hiring (introduced in Chapter 19)** A firm operates a large, direct-to-consumer sales force. The firm would like to build a system to monitor the progress of new agents. The goal is to identify "superstar agents" as rapidly as possible, offer them incentives, and keep them with the firm. A key task for agents is to open new accounts; an account is a new customer to the business. The response of interest is the profit to the firm (in dollars) of contracts sold by agents over their first year. These data summarize the early performance of 464 agents. Among the possible explanations of performance are the number of new accounts developed by the agent during the first 3 months of work and the commission earned on early sales activity. An analyst at the firm is using an equation of the form (with natural logs)

Log Profit $= \beta_0 + \beta_1$ Log Accounts $+ \beta_2$ Log Early Commission

For cases having value 0 for early commission, the analyst replaced zero with \$1.

(a) The choice of the analyst to fill in the 0 values of early commission with 1 so as to be able to take the log is a common choice (you cannot take the log of 0). From the scatterplot of Log Profit on Log Early Commission, you can see the effect of what the analyst did. What is the impact of these filled-in values on the marginal association?

(b) Is there much collinearity between the explanatory variables? How does the presence of these filled-in values affect the collinearity?

(c) Using all of the cases, does collinearity exert a strong influence on the standard errors of the estimates in the analyst's multiple regression?

(d) Because multiple regression estimates the partial effect of an explanatory variable rather than its marginal effect, we cannot judge the effect of outliers on the partial slope from their position in the scatterplot of y on x. We can, however, see their effect by constructing a plot that shows the partial slope. To do this, we have to remove the effect of one of the explanatory variables from the other variables. Here's how to make a so-called partial regression leverage plot for these data.

First, regress Log Profit on Log Accounts and save the residuals. Second, regress Log Commission on Log Accounts and save these residuals. These regressions remove the effects of the number of accounts opened from the other two variables. Now, make a scatterplot of the residuals from the regression of Log Profit on Log Accounts on the residuals from the regression of Log Commission on Log Accounts. Fit the simple regression for this scatterplot, and compare the slope in this fit to the partial slope for Log Commission in the multiple regression. Are they different?

(e) Do the filled-in cases remain leveraged in the partial regression leverage plot constructed in part (d)? What does this view of the data suggest would happen to the estimate for this partial slope if these cases were excluded?

(f) What do you think about filling in these cases with 1 so that we can take the log? Should something else be done with them?

41. **Promotion (introduced in Chapter 19)** These data describe promotional spending by a pharmaceutical company for a cholesterol-lowering drug. The data cover 39 consecutive weeks and isolate the area around Boston. The variables in this collection are shares. Marketing research often describes the level of promotion in terms of voice. In place of the level of spending, *voice* is the share of advertising devoted to a specific product.

The column *Market Share* is sales of this product divided by total sales for such drugs in the Boston area. The column *Detail Voice* is the ratio of detailing for this drug to the amount of detailing for all cholesterol-lowering drugs in Boston. Detailing counts the number of promotional visits made by representatives of a pharmaceutical company to doctors' offices. Similarly, *Sample Voice* is the share of samples in this market that are from this manufacturer.

(a) Do any of these variables have linear patterns over time? Use timeplots of each one to see. (A scatterplot matrix becomes particularly useful.) Do any weeks stand out as unusual?

(b) Fit the multiple regression of Market Share on three explanatory variables: Detail Voice, Sample Voice, and Week (which is a simple time trend, numbering the weeks of the study from 1 to 39). Does the multiple regression, taken as a whole, explain statistically significant variation in the response?

(c) Does collinearity affect the estimated effects of these explanatory variables in the estimated equation? In particular, do the partial effects create a different sense of importance from what is suggested by marginal effects?

(d) Which explanatory variable has the largest VIF?

(e) What is your substantive interpretation of the fitted equation? Take into account collinearity and statistical significance.

(f) Should both of the explanatory variables that are not statistically significant be removed from the model at the same time? Explain why doing this would *not* be such a good idea, in general. (*Hint:* Are they collinear?)

42. **Apple (introduced in Chapter 19)** These data track monthly performance of stock in Apple Computer since its inception in 1980. The data include 300 monthly returns on Apple Computer, as well as returns on the entire stock market, the S&P 500 index, stock in IBM, and Treasury Bills (short-term, 30-day loans to the government). (The column *Whole Market Return* is the return on a value-weighted portfolio that purchases stock in the three major U.S. markets in proportion to the size of the company rather than one of each stock.) Formulate the regression with excess returns on Apple as the response and excess returns on the whole market, the S&P 500, and IBM as explanatory variables. (Excess returns are the same as excess percentage changes, only without being multiplied by 100. Just subtract the return on Treasury Bills from each.)

(a) Do any of these excess returns have linear patterns over time? Use timeplots of each one to see. (A scatterplot matrix becomes particularly useful.) Do any months stand out as unusual?

(b) Fit the indicated multiple regression. Does the estimated multiple regression explain statistically significant variation in the excess returns on Apple?

(c) Does collinearity affect the estimated effects of these explanatory variables in the estimated equation? In particular, do the partial effects create a different sense of importance from what is suggested by marginal effects?

(d) Which explanatory variable has the largest VIF?

(e) How would you suggest improving this model, or would you just leave it as is?

(f) Interpret substantively the fit of your model (which might be the one the question starts with).

4M Budget Allocation

Collinearity among the predictors is common in many applications, particularly those that track the growth of a new business over time. The problem is worse when the business has steadily grown or fallen. Because the growth of a business affects many attributes of the business (such as assets, sales, number of employees, and so forth; see Exercise 37), the simultaneous changes that take place make it hard to separate important factors from coincidences.

For this exercise, you're the manager who allocates advertising dollars. You have a fixed total budget for advertising, and you have to decide how to spend it. We've simplified things so that you have two choices: print ads or television.

The past two years have been a time of growth for your company, as you can see from the timeplot of weekly sales during this period. The data are in thousands of dollars, so you can see from the plot that weekly sales have grown from about $2.5 million up to around $4.5 million.

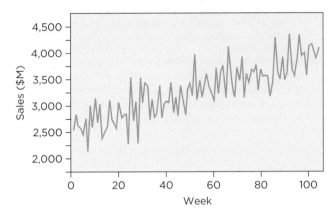

Other things have grown as well, namely your expenditures for TV and print advertising. This timeplot shows the two of them (TV in green, and print ads in red), over the same 104 weeks.

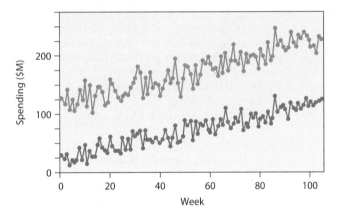

With everything getting larger over time, all three variables are highly correlated.

Motivation

a) How are you going to decide how to allocate your budget between these two types of promotion?

Method

b) Explain how you can use multiple regression to help decide how to allocate the advertising budget between print ads and television ads.

c) Why would it not be enough to work with several, more easily understood simple regression models, such as sales on spending for television ads or sales on spending on print ads?

d) Look at the scatterplot matrix of Sales, the two explanatory variables (TV Adv and Print Adv), and a time trend (Week). Do the relationships between the variables seem straight enough to fit a multiple regression?

e) Do you anticipate that collinearity will affect the estimates and standard errors in the multiple regression? Use the correlation matrix of these variables to help construct your answer.

Mechanics

Fit the multiple regression of Sales on TV Adv and Print Adv.

f) Does the model satisfy the assumptions for the use of the MRM?

g) Assuming that the model satisfies the conditions for the MRM,
   i. Does the model as a whole explain statistically significant variation in Sales?
   ii. Does each individual explanatory variable improve the fit of the model, beyond the variation explained by the other alone?

h) Do the results from the multiple regression suggest a method for allocating your budget? Assume that your budget for the next week is $360,000.

i) Does the fit of this model promise an accurate prediction of sales in the next week, accurate enough for you to think that you have the right allocation?

Message

j) Everyone at the budget meeting knows the information in the plots shown in the introduction to this exercise: Sales and both types of advertising are up. Make a recommendation with enough justification to satisfy their concerns.

k) Identify any important concerns or limitations that you feel should be understood to appreciate your recommendation.

Categorical Explanatory Variables

IN 2001, SIX WOMEN FILED A LAWSUIT AGAINST WAL-MART IN FEDERAL COURT IN SAN FRANCISCO. Their suit (which is still pending as of this writing) charged that Wal-Mart discriminates against female employees. One of the women claimed that she was paid $8.44 per hour, but that men with less experience in the same position earned $9 per hour. The wages of six employees are small amounts of money for a huge business like Wal-Mart. The scope of this litigation changed, however, when this lawsuit gained class-action status in 2004. By gaining class-action status, this suit functions on behalf of all others who are eligible to make the same claim. The case expanded from six women to include *any* woman who works for or worked for Wal-Mart. That's a lot of employees. More than half of the employees who work at the 4,000 Wal-Mart stores in the United States are women, perhaps as many as two million employees.

Statistics play a big role in cases such as this. In granting class-action status, U.S. District Judge Martin Jenkins wrote that the "plaintiffs present largely uncontested *descriptive statistics* which show that women working at Wal-Mart stores are paid less than men in every region, that pay disparities exist in most job categories, that the salary gap widens over time, that women take longer to enter management positions, and that the higher one looks in the organization the lower the percentage of women" (our emphasis).[1]

To go beyond description to inference, each side offered a regression analysis that produced a conclusion in its favor. Regression analysis is relevant because the amounts paid to men and women at Wal-Mart are not the result of a randomized experiment.

---

[1] "Judge Certifies Wal-Mart Class Action Lawsuit," The Associated Press, June 22, 2004. As of late fall 2009, the case remains in litigation.

Without randomization, the men and women being compared may differ systematically in a variety of ways. Perhaps there's an explanation for the gap in hourly pay aside from male versus female.

This chapter introduces the use of categorical variables as explanatory variables in regression. A categorical explanatory variable reveals whether and how membership in a group affects the response, all the while taking account of other explanatory variables. An important use of categorical explanatory variables in regression is to adjust for other differences when comparing the means of the response between groups. Rather than simply compare the average salary of men versus women, for instance, regression analysis can adjust this comparison for other differences among employees. As you will discover in this chapter, the conclusions depend upon how this adjustment is done.

25.1 | TWO-SAMPLE COMPARISONS

We don't have the data from the lawsuit against Wal-Mart, but we do have a sample of salaries of 174 mid-level managers at a large firm. As in the lawsuit against Wal-Mart, the average salaries of male and female employees differ. The salaries of the 115 men in this sample average $144,700 whereas the salaries of the 59 women average $140,000. The histograms in Figure 25.1 contrast the salaries of both groups.

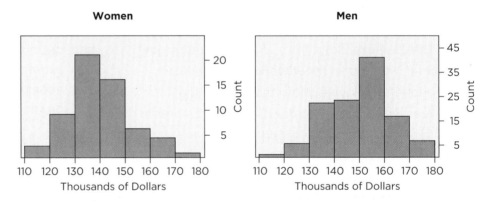

FIGURE 25.1 Histograms of salaries of female and male managers.

The histogram of salaries of male managers appears shifted to the right. Can we attribute this difference to sampling variation, or is this an indication of a systematic pattern that might carry over to the whole firm?

Before performing a two-sample test or building a confidence interval for the difference between the means as in Chapter 18, we need to verify the conditions for inference. The four conditions required for two-sample inferences are as follows:

1. The data are simple random samples (SRS).
2. There's no lurking variable that explains the differences.
3. The groups have similar variances.
4. Each sample is nearly normal.

The only difference from the conditions for regression is the absence of the linear condition. The SRS condition implies independent observations.

If the salaries satisfy these conditions, we can use the confidence interval for the difference in average salaries between men and women to determine whether the observed differences are statistically significant. Table 25.1 gives the confidence interval for the difference between the means.

Difference	**4.670**
Standard Error	1.986
95% Confidence Interval	0.750 to 8.591

With 95% confidence, the difference in population mean salaries lies between $750 and $8,591. Because 0 is not in the confidence interval, the observed difference in average salaries ($4,670, bold in Table 25.1) is statistically significantly different from 0; we reject $H_0: \mu_{men} = \mu_{women}$. Women make statistically significantly less on average at this firm, and the difference is too large to dismiss as a consequence of sampling variation.

Confounding Variables

Three of the conditions for two-sample inference appear to have been met (normal quantile plots, for instance, confirm that the samples are nearly normal). The difficult condition to resolve is the possibility of a lurking variable, also called a confounding variable. Without a randomized experiment, we must be careful that there's not another explanation for the statistically significant difference between the average salaries. For instance, some managers have more experience than others, and experience and salary are correlated. Managers with more experience typically earn higher salaries, as shown in Figure 25.2.

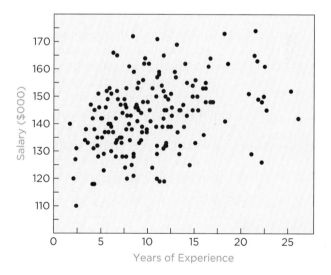

FIGURE 25.2 Salary and years of experience are correlated.

The correlation between salary and years of experience is $r = 0.36$. Because the number of years of experience is correlated with the response, there's a possibility that the level of experience confounds the interpretation of the two-sample t-interval. Perhaps men in this sample have more experience than women, and it's this difference rather than gender that explains the difference in average salary. **A variable such as experience is a confounding variable if it is correlated with the response and the two groups differ with respect to this variable.** Figure 25.2 shows that the number of years of experience is correlated with salary, so it meets the first requirement of a confounding variable. To meet the second requirement, male managers must differ from female managers in years of experience. If male managers have more or less years

tip

of experience, on average, than female managers, then the two-sample comparison of salaries summarized in Table 25.1 isn't a fair comparison.

Subsets and Confounding

If we have enough data, there's an easy way to control for the effects of a confounding variable: Restrict the analysis to cases with matching levels of the confounding variable. If we restrict the comparison of salaries to male and female managers with similar experience, we eliminate the number of years of experience as a source of confounding. For example, our data include 24 managers with about 5 years of experience (from 4 to 6 years). Table 25.2 contrasts the results for this subset (first column) to the results for the whole sample (second column)

TABLE 25.2 Comparison of salaries for managers with about 5 years of experience and all managers.

	5 Years Experience	All Managers
Difference (thousands of dollars)	0.378	4.670
Standard Error of Difference	4.268	1.986
95% Confidence Interval	−8.473 to 9.228	0.750 to 8.591
n	24	174

The 95% confidence interval for the difference in average salaries within this subset includes 0; this *t*-test of the difference in average salaries is not statistically significant. Among managers with about 5 years of experience, women and men appear to earn comparable salaries. The salary of the 15 women in this subset averages $137,733; the average salary of the 9 men is $378 higher at $138,111.

Restricting the comparison to managers with similar experience removes experience from the comparison, but it also omits most of the sample. As a result, the standard error of the difference between the average salaries of men and women in Table 25.2 based on this subset is much larger than when using the entire sample (4.268 versus 1.986). These 24 cases do not produce a precise estimate of the difference, and the confidence interval for the difference becomes much longer, extending from −$8,473 all the way to $9,228. Had we begun with a larger sample, with thousands of managers at each year of experience, we might have had enough data to work with a subset. Because we have a relatively small sample, however, we have to use our data more carefully.

What Do You Think?

A supermarket compared the average value of purchases made by shoppers who use coupons to the average of shoppers who do not. On average those who use coupons spend $32 more.

a. Name several lurking variables that could affect this comparison.[2]
b. How could the supermarket have avoided confounding caused by possible lurking variables?[3]
c. An analyst believes that family size is a possible lurking variable. How could the store, with enough data, control for the possible effects of family size on the comparison of the average purchase amounts?[4]

---

[2] Possible confounders are income, number of children, age, and frequency of visits to the store.
[3] Randomize the distribution of coupons to its customers. That would be hard to do in the United States because of the widespread distribution of coupons via newspapers and other uncontrolled sources.
[4] Restrict the test to customers with a specific size of the family (assuming this is known).

25.2 | ANALYSIS OF COVARIANCE

analysis of covariance A regression that combines categorical and numerical explanatory variables.

The previous analysis compares the average salaries of male and female managers by limiting the comparison to a subset. Although it removes the confounding effect of experience, this subset has only 24 managers with about 5 years of experience. Not only does this approach reduce the size of the sample, it also restricts the comparison to managers with about 5 years of experience. What about the difference between average salaries for managers with 2, 10, or 15 years of experience?

Regression analysis provides another approach. Regression allows us to use all of the data to compare salaries of male and female managers while at the same time accounting for differences in experience. The use of regression in this way is known as the analysis of covariance. An **analysis of covariance** is a regression that combines dummy variables with confounding variables as explanatory variables. The resulting regression adjusts the comparison of means for the effects of the confounding variables.

Regression on Subsets

To appreciate how regression adjusts for a confounding variable, recall that the equation of a regression model estimates the mean of the response conditional on the explanatory variables. In the SRM, the conditional mean of the response changes linearly with the explanatory variable,

$$\mathrm{E}(Y|X = x) = \mu_{y|x} = \beta_0 + \beta_1 x$$

Suppose that two simple regressions describe the relationship between salary (Y) and years of experience (X) for male and female managers. Write the equation for the mean salary of female managers as

$$\mu_{\mathrm{Salary|Years,f}} = \beta_{0,\mathrm{f}} + \beta_{1,\mathrm{f}}\ \mathrm{Years}$$

and the equation for the mean salary of male managers as

$$\mu_{\mathrm{Salary|Years,m}} = \beta_{0,\mathrm{m}} + \beta_{1,\mathrm{m}}\ \mathrm{Years}$$

Subscripts distinguish the slope and intercept for male and female managers. Using this notation, let's revisit what we did by restricting the comparison to managers with about 5 years of experience. The average salary of the men with about 5 years of experience estimates $\mu_{\mathrm{Salary|5,m}}$, and the average salary of the women with 5 years of experience estimates $\mu_{\mathrm{Salary|5,f}}$.

Once you've written down the equations for mean salary in the two groups, it's easy to see that regression offers an alternative way to estimate the difference. Rather than estimate these means using two small samples, we can instead fit regressions of Salary on Years for male managers and for female managers. The difference between the fitted lines at Years = 5 estimates $\mu_{\mathrm{Salary|5,m}} - \mu_{\mathrm{Salary|5,f}}$. Two regression lines allow us to use all of the data to compare salaries at any number of years of experience. The scatterplot of Salary on Years in Figure 25.3 shows two estimated lines. The lines and points are colored blue for men and red for women. **Color-coding is helpful when comparing subsets of your data.**

The red line fit to female managers has a steeper slope: The estimated salary rises faster with experience for women than for men. If the lines were parallel, the salary gap at every grade level would be the same. In this case, the fitted lines cross near 11 years of experience. Because they cross, the estimated salary of women is higher in some ranges of experience, and the estimated salary of men is higher in others. It appears that among managers with more

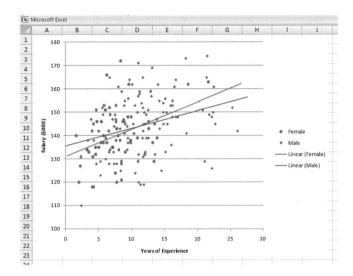

FIGURE 25.3 Simple regressions fit separately to men (blue) and women (red).

Female				**Male**	
0.139		r^2		0.093	
11.225		s_e		12.059	
59		n		115	
Est	**SE**	**Term**	**Est**	**SE**	
130.989	3.324	Intercept	135.600	2.901	
1.176	0.388	Years	0.761	0.223	

than 11 years of experience, women earn higher average salaries than men; the comparison reverses for managers with less than 11 years of experience.

Combining Regressions

In order to judge the statistical significance of differences between the fitted lines, we need standard errors for the differences between the estimated intercepts and slopes. To get these, we combine the two simple regressions shown in Figure 25.3 into one multiple regression. The multiple regression will have a coefficient that estimates the difference between the slopes and another coefficient that estimates the difference between the intercepts. We can read off the standard errors of the differences from the summary of the estimated coefficients.

Joining these equations requires a dummy variable and a second variable that uses the dummy variable. For this example, the dummy variable Group identifies whether a manager is male or female. We code Group as 1 for men and 0 for women.

$$\text{Group} = \begin{cases} 1 & \text{if manager is male} \\ 0 & \text{if manager is female} \end{cases}$$

It does not matter which group you represent as 0 and which you represent as 1, so long as you remember which is which.

The second variable is the product of the dummy variable and the confounding variable, Group $\times$ Years. The product of two explanatory variables in a regression model is known as an **interaction**. For male managers, the

interaction The product of two explanatory variables in a regression model.

interaction variable matches Years. For female managers, the interaction Group × Years is 0. Table 25.3 shows four examples from the data table.

TABLE 25.3 Calculation of the interaction variable for several managers in the data table.

Row	Sex	Group	Years	Interaction
1	male	1	16.7	$1 \times 16.7 = 16.7$
2	male	1	6.7	$1 \times 6.7 = 6.7$
4	female	0	13.9	$0 \times 13.9 = 0$
5	female	0	8.5	$0 \times 8.5 = 0$

We now regress Salary on three explanatory variables: Years, Group, and Group × Years. Table 25.4 summarizes the multiple regression. Some of the estimated coefficients should look familiar.

TABLE 25.4 Summary of the multiple regression that combines two simple regressions.

R^2	0.135
s_e	11.785
n	174

Term	Estimate	Std Error	t-Statistic	p-Value
Intercept	130.989	3.490	37.53	<0.0001
Years	1.176	0.408	2.89	0.0044
Group	4.611	4.497	1.03	0.3066
Group × Years	−0.415	0.462	−0.90	0.3709

Interpreting Coefficients

The intercept and slope of Years in this multiple regression match those of the regression of Salary on Years for female managers. The equation of the simple regression for the salary of female managers is

$$\textit{Estimated } \text{Salary} = 130.989 + 1.176 \text{ Years}$$

Compare these estimates to those in Table 25.4. The intercept in the multiple regression matches the intercept in Table 25.4, and the partial slope for Years in the multiple regression matches the slope in Table 25.4. That's no accident. **The intercept and slope for the group coded as 0 in the dummy variable are the intercept and slope of the explanatory variable in the multiple regression.**

The intercept and slope for the group coded as 1 by the dummy variable (male managers in this example) are embedded in the estimates associated with the dummy variable Group. Let's start with the intercept. The intercept in the regression for women is $b_{0,f} = 130.989$. The intercept in the regression for the men is $b_{0,m} = 135.600$. The difference between these is

$$b_{0,m} - b_{0,f} = 135.600 - 130.989 = 4.611$$

On average, male managers with no experience make \$4,611 more than new female managers. That's also the slope of Group in the multiple regression in Table 25.4. Now let's continue with the slopes. The difference between the slopes in the simple regressions for men and women is

$$b_{1,m} - b_{1,f} = 0.761 - 1.176 = -0.415$$

tip

tip

The estimated average salary increases $415 per year of experience faster for women than for men. This difference between the slopes matches the slope of the interaction in the multiple regression. **The slope of the dummy variable is the difference between estimated intercepts. The slope of the interaction is the difference between estimated slopes in the simple regressions.**

To see why estimates in these three regression equations have these connections, write out the equation of the multiple regression:

$$\mu_{Salary|Years,Group} = \beta_0 + \beta_1 \text{ Years} + \beta_2 \text{ Group} + \beta_3 \text{ Group} \times \text{Years}$$

If we plug in Group $= 0$, we restrict the equation to female managers, and as a result the equation reduces to the simple regression for women. With Group $= 0$, the parts of the equation that involve the dummy variable disappear:

$$\mu_{Salary|Years,0} = \beta_0 + \beta_1 \text{ Years} + \beta_2 \times 0 + \beta_3 \times 0 \times \text{Years}$$
$$= \beta_0 + \beta_1 \text{ Years}$$

tip

The intercept and the slope of Years in the multiple regression are the intercept and slope in the equation for female managers (what we called $\beta_{0,f}$ and $\beta_{1,f}$). The other coefficients, β_3 and β_4, represent differences from this equation. **The equation for the group coded as 0 in the dummy variable forms a baseline for comparison.**

To find the equation for men, plug Group $= 1$ into the equation from the multiple regression. The equation becomes

$$\mu_{Salary|Years,1} = \beta_0 + \beta_1 \text{ Years} + \beta_2 \times 1 + \beta_3 \times 1 \times \text{Years}$$
$$= (\beta_0 + \beta_2) + (\beta_1 + \beta_3) \text{ Years}$$

That's the equation from the simple regression of salary on Years for men. The intercept is $\beta_{0,m} = \beta_{0,f} + \beta_2$, and the slope is $\beta_{1,m} = \beta_{1,f} + \beta_3$. The coefficient of the dummy variable $\beta_2 = \beta_{0,m} - \beta_{0,f}$ is the difference between the intercepts. The coefficient of the interaction β_3 is the difference $\beta_{1,m} - \beta_{1,f}$ between the marginal slopes.

What Do You Think?

In the study of the relationship between couponing and purchase amount (see the previous "What Do You Think?"), an analyst working for the supermarket estimated the following multiple regression for the amount of a sale to a family:

Est. Sales $= 47.23 + 18.04$ Family Size $- 22.38$ Coupon $+ 4.51$ Family Size $\times$ Coupon

Sales are in dollars, Family Size counts the number of family members in the household, and Coupon is a dummy variable coded as 1 if a coupon was used in the purchase and 0 otherwise.

a. Had the analyst fit a simple regression of sales on family size for only those who do not use coupons, what would be the fitted equation?[5]
b. What is the fitted equation for those who do use coupons?[6]

25.3 | CHECKING CONDITIONS

The remaining analysis of this regression requires inference, so we must check the conditions of the MRM before continuing. The use of a dummy variable and an interaction does not change the conditions of the MRM: We

---

[5] Set the variable *Coupon* $= 0$, leaving $47.23 + 18.04$ *Family Size*.
[6] Set *Coupon* $= 1$ and combine the estimates as $(47.23 - 22.38) + (18.04 + 4.51)$ *Family Size*.

need to confirm the conditions for linearity and lurking variables, and then check the conditions on the errors. For this regression the colored scatterplot in Figure 25.3 shows linear, albeit weak, association between Salary and Years for the two groups. As for lurking variables, we can imagine several possible lurking variables when comparing two subsets of managers, such as educational background and business aptitude. Because these might also affect the comparison of salaries, we have to proceed cautiously. (This concern helps you appreciate a randomized experiment; it controls for *all* lurking variables.)

The three remaining conditions are that the data are evidently independent, with similar variances and a normal distribution. Because the cases within each group are samples from the respective populations, we are comfortable with the assumption of independence. The condition of similar variances requires special care in a regression model that compares groups. We allow the averages to be different in the groups, so why should we believe that the variances of the errors around the two lines are the same?

Checking for Similar Variances

Checking the condition of similar variances highlights an important difference between fitting two simple regressions and fitting them together as one multiple regression. Although the multiple regression reproduces the intercepts and slopes of the two simple regressions, it offers only one estimate of the standard deviation of the errors ($s_e = \$11,785$, Table 25.4). According to the assumptions of the MRM, that's all that it needs; all of the observations, both men and women, have equal error variation. In contrast, each simple regression offers a separate estimate of the standard deviation of the errors. In the simple regression for women, $s_e = \$11,225$, whereas for men, the estimate is larger, $s_e = \$12,059$ (Figure 25.3). These estimates will never match exactly (unless someone makes up the data), so how close should they be? There is a hypothesis test of the equality of variances (another type of *F*-test), but this test is very sensitive to the assumption of normality. We recommend plots instead.

For models that do not include dummy variables, we've checked the similar variances condition in previous chapters by plotting the residuals on the fitted values. That remains a useful diagnostic plot because it's possible for the variation to increase with the size of the fitted value. In this example, salaries could become more variable as they grow.

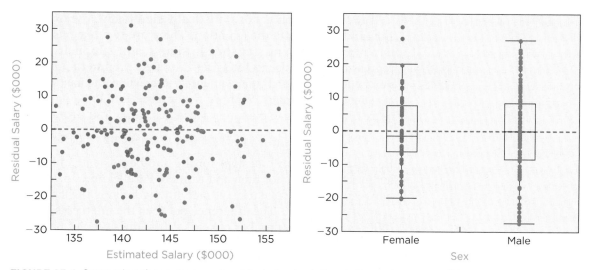

FIGURE 25.4 Comparing the variances of residuals to check the similar variances condition.

The presence of distinct groups in the data motivates a second plot that focuses on the variation of the residuals in the two groups. Comparison (or side-by-side) boxplots of the residuals show whether the residual variation is similar. In Figure 25.4, the scatterplot on the left graphs the residuals versus the fitted values. The plot on the right pulls together the residuals from the groups to make it easier to compare the residual variation in the two groups.

The similar lengths of the boxes show that the two groups have similar variances. We adopt the convention that the two groups have similar variances unless the length of one box, the interquartile range (IQR), is more than twice the length of the other. (If the groups are very small, less than 20 observations each, it becomes very hard to compare the variances of the residuals. The IQR varies so much from sample to sample that this procedure is not conclusive. We'll have to rely on our judgment.)

It is worth remembering that the two-sample t-test in Chapter 18 does not require the assumption of equal variances. Instead, the two-sample t-test uses an elaborate calculation of degrees of freedom. This adjustment is not available when we use regression to adjust for a confounding variable, providing yet another reminder of the value of having data from a randomized experiment. We analyze data from an experiment using a two-sample t-test and do not need the assumption of equal variances.

Finally, the last condition to verify is that the residuals appear nearly normal. As in previous examples, we check this condition by inspecting the normal quantile plot of the residuals. In this example, the normal quantile plot (not shown) shows that the residuals are nearly normal, and we can proceed to inference.

25.4 | INTERACTIONS AND INFERENCE

Once we verify that the fitted model meets the conditions of the MRM, the first test to consider is the overall F-test. If the overall model is statistically significant, then test whether the interaction is statistically significant *before* considering other explanatory variables.

principle of marginality If the interaction $X_1 \times X_2$ is in the regression, keep both X_1 and X_2 as explanatory variables.

If the interaction is statistically significant, retain it in the model. Standard practice, sometimes called the **principle of marginality**, is to keep the components of a statistically significant interaction in the regression regardless of their level of significance. For example, if the interaction $X_1 \times X_2$ is statistically significant, the principle of marginality says to retain both X_1 and X_2 in the regression whether or not X_1 and X_2 are separately statistically significant. (Example 25.1 later in this chapter illustrates this principle.)

If the interaction is not statistically significant, remove it from the regression and re-estimate the equation. There are two reasons for removing an interaction that is not statistically significant: simplicity and collinearity. A regression is simpler to interpret if it lacks the interaction. **The reason for the simpler interpretation is that without an interaction, the multiple regression fits parallel lines to the groups.** Without the interaction, the equation of the multiple regression has two explanatory variables, *Years* and *Group*.

$$\mu_{\text{Salary}|\text{Years},\text{Group}} = \beta_0 + \beta_1 \text{ Years} + \beta_2 \text{ Group}$$

Because there's only one slope for the number of years, this equation fits parallel lines to the data for men and women. Parallel lines mean that the difference between estimated salaries of men and women is the same regardless of the years of experience. The difference between the intercepts (at Years = 0) is the difference at every number of years. If the lines are not parallel, the difference in salary between men and women depends on the years of experience.

By removing an interaction that is not statistically significant, we're acting as though we have proven $H_0: \beta_3 = 0$ is true. We cannot prove that a null hypothesis is true. Even so, this common practice produces more interpretable, useful regression models.

Interactions and Collinearity

The second reason for removing an interaction that is not statistically significant is collinearity. Models with less collinearity are easier to interpret and supply more accurate estimates of the slopes. It's easy to see that an interaction increases collinearity. The interaction Group × Years is correlated with Years since it is Years for men (see Table 25.3). The interaction also matches the dummy variable for women. Table 25.5 shows the VIFs (Chapter 24) in the multiple regression.

TABLE 25.5 The interaction in a multiple regression introduces collinearity.

Term	Estimate	Std Error	t-Statistic	VIF
Intercept	130.989	3.490	37.53	–
Years	1.176	0.408	2.89	5.33
Group	4.611	4.497	1.03	5.68
Group × Years	−0.415	0.462	−0.90	13.07

The large VIF for the interaction confirms that the interaction is highly correlated with the other explanatory variables. With so much collinearity, it is not surprising to find that the coefficient of the interaction is not statistically significantly different from 0 ($t = -0.90$). We cannot reject the null hypothesis that the slopes for men and women in the populations match.

Because the interaction is not statistically significant in this example, we remove it from the regression. The following regression without the interaction has much less collinearity (smaller VIFs for the remaining variables). Removing the interaction has little effect on R^2 (falling from 0.135 to 0.131) because the interaction is not statistically significant. This drop in R^2 is attributable to random variation.

TABLE 25.6 Multiple regression without the interaction.

R^2	0.131
s_e	11.779
n	174

Term	Estimate	Std Error	t-Statistic	p-Value	VIF
Intercept	133.468	2.132	62.62	<0.0001	–
Years	0.854	0.192	4.44	<0.0001	1.19
Group	1.024	2.058	0.50	0.6193	1.19

Without the interaction, the common slope for Years is 0.854 thousand dollars ($854) per year of experience for both male and female managers. This estimate compromises between the slopes for Years in the separate regressions (1.176 for women and 0.761 for men).

Parallel Fits

The slope for Group in Table 25.6 estimates the difference between the intercepts for the male and female managers. Setting Group = 0, the estimated salary for a female manager is

$$Estimated \ Salary = 133.468 + 0.854 \ Years$$

To find the salary for a male manager, substitute Group = 1 into the fit to get the equation

$$Estimated \text{ Salary} = (133.468 + 1.024) + 0.854 \text{ Years}$$

The coefficient of the dummy variable shifts the line up by 1.024 thousand dollars ($1,024). Because the estimated lines have the same slope, the fits are parallel and the difference of $1,024 between the intercepts applies regardless of the number of years of experience. Figure 25.5 shows the nearly identical, parallel fits implied by the multiple regression without the interaction.

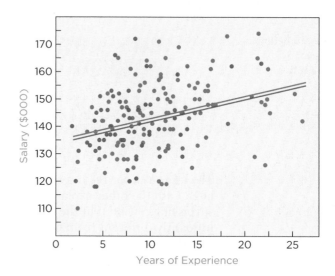

FIGURE 25.5 Regression without an interaction fits parallel lines to the groups.

Statistics confirms the visual comparison: The fits of these parallel lines are not statistically significantly different. The t-statistic for Group in Table 25.6 shows that the difference falls $t = 0.50$ standard error above 0, with a p-value of 0.6193. Assuming that average salaries of men and women grow at the same rate with experience (parallel), the 95% confidence interval for the difference in salary is −$3,030 to $5,078 ($1,000 \times 1.024 \pm 1.97 \times 2.058$). This model finds no statistically significant difference between the average salaries of male and female managers when comparing managers with equal years of experience. (Making an inference in a regression with a statistically significant interaction is more complicated. The difficulties are described in Behind the Math: Comparisons with Interactions.)

(p. 655)

Does this analysis prove that this firm does not discriminate in setting salaries? The analysis of covariance (the multiple regression with the dummy variable) shows that confounding contaminates the initial two-sample comparison. Managers of the company can explain that the initial two-sample t-test, which is statistically significant, is due to the male managers having more experience. Male managers average 12.1 years of experience whereas female managers average 7.2. The difference in salaries in the initial t-test can be explained by this difference in experience rather than whether the manager is male or female. This gap in experience, however, raises another issue: Why is it that women at the firm have less experience?

It is common for different models to produce different conclusions when statistical models appear in legal trials. In a discrimination case such as the class-action suit against Wal-Mart described in the introduction, the conclusion often depends on which explanatory variables are in the model.

What Do You Think?

The following table summarizes the regression for sales related to coupons considered in earlier sections. The data include $n = 135$ cases, of which 60 used coupons.

Term	Estimate	Std Error	t-Statistic	VIF
Intercept	47.236	10.253	4.607	—
Family Size	18.038	7.362	2.450	4.324
Coupon	−22.383	5.251	−4.263	3.250
Coupon × Family Size	5.507	2.207	2.495	7.195

a. What is the interpretation of the confidence interval for the coefficient of Coupon in the fitted model?[7]
b. Should the interaction be removed and the model re-estimated?[8]

PRIMING IN ADVERTISING

MOTIVATION ▸ STATE THE QUESTION

It is often useful to get the word out ahead of the launch of a new product, a practice referred to as *priming*. Then when the product is ready to sell, there's pent-up demand. We have seen this done with previews of movies; it's useful in other situations as well.

When it began in 1973 as Federal Express, FedEx made a name for itself by guaranteeing overnight delivery. Companies couldn't use email back then, and FedEx had great success delivering FedEx express mail. Then they came upon the idea of the Courier Pak. The Courier Pak allowed customers to send up to two pounds. Customers at the time thought of express mail for key correspondence, not sending a manuscript for a book or the transcript of court testimony. A first wave of promotion was devoted to raising awareness through use of clever television ads. The second wave sent sales representatives to existing clients. The data in this application describe how 125 customers who were visited by FedEx employees reacted to this promotion.

The questions that motivate our analysis reflect concerns about the costs and benefits of this promotion. Even if the visits by sales representatives increased the use of Courier Paks, costs made this an expensive promotion. Management has two specific questions:

1. How many shipments were generated by a typical contact hour?
2. Was the promotion more effective for customers who were already aware of FedEx's business?

---

[7] The 95% confidence interval is $-22.383 + 2.00 \times 5.251$, which indicates that coupon users spend from $11.88 to $32.89 on average less, irrespective of family size (difference of intercepts).
[8] No. The effect is statistically significant and the interaction must be retained.

METHOD

Describe data.

Identify x and y.

Link the equation to the problem.

Check linear and lurking variable condition.

DESCRIBE THE DATA AND SELECT AN APPROACH

Describe the data and how the model addresses the motivating problem. The data consist of a sample of 125 customers selected randomly from the database of a regional office. Of these, 50 were aware of Courier Paks prior to conversations with sales representatives, and 75 were not. For each customer, the data give the number of times in the following month that the customer used Courier Paks (the response) and the length of time (in hours) spent with the customer.

We will use a regression model with the number of Courier Paks used after the promotion as the response. The explanatory variables are the amount of time spent with the client by a sales representative and a dummy variable indicating whether or not the customer had been aware of the Courier Paks prior to the visit. The dummy variable is coded

$$\text{Aware} = \begin{cases} 1 & \text{if the client was aware of Courier Paks} \\ 0 & \text{otherwise} \end{cases}$$

We will also use an interaction between Aware and the hours of effort.

Relate this model to the business questions. The interaction indicates whether prior awareness of Courier Paks affects how the personal visit influenced the customer. If there's no interaction, the slope of the dummy variable is the difference in use of Courier Paks in the two groups, regardless of the time spent by the FedEx sales representative.

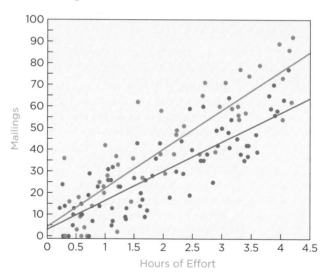

This color-coded scatterplot shows regressions fit separately to the two groups. Those who were aware are shown in green, the others in red. The plot suggests an interaction because the slopes appear different and the data in the two groups separate once the hours of effort exceed 2 or 3.

✓ **Linear.** The association within each group appears linear. There's no evident bending in either group.

✓ **No obvious lurking variable.** We need to think about whether there's another explanation for the use of Courier Paks aside from the promotion and effort. For example, if some customers were offered discounts or were in markets saturated by television advertising, then these conditions might alter the comparison.

MECHANICS

DO THE ANALYSIS

Since there are no violations of the major conditions, we continue with a summary of the overall fit shown on the following page.

R^2	0.7665
s_e	11.1634
n	125

Term	Estimate	SE	*t*-Statistic	*p*-Value
Intercept	2.454	2.502	0.98	0.3286
Hours	13.821	1.088	12.70	<0.0001
Aware	1.707	3.991	0.43	0.6697
Aware × Hours	4.308	1.683	2.56	0.0117

Before inference, check the conditions by examining the residuals. The scatterplot on the left graphs the residuals versus the explanatory variable, and the normal quantile plot below it summarizes the distribution of the residuals.

✓ **Evidently independent.** Nothing in plots of the data suggests a problem with the assumption of independence. Plus, these data are a random sample. Dependence could nonetheless be present if the data include several cases of visits by the same sales representative.

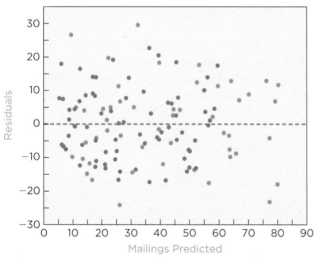

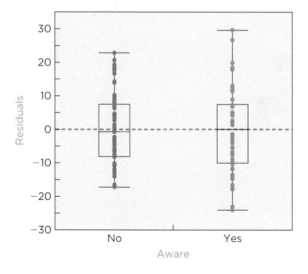

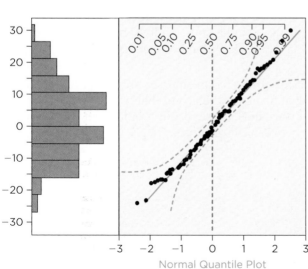

✓ **Similar variances.** This is confirmed in the plot of the residuals on the explanatory variable. The comparison boxplots shown above indicate that the residual variance is similar in the two groups.

✓ **Nearly normal.** The histogram of the residuals is reasonably bell shaped with no large outliers. The points in the normal quantile plot stay near the diagonal.

Since the conditions of the MRM are met, we proceed to inference. The F-statistic is $F = 0.7665/(1 - 0.7665) \times (125 - 4)/3 \approx 132$ with p-value < 0.0001. We reject H_0 that all slopes are zero. This model explains statistically significant variation.

The interaction between awareness and hours of effort is statistically significant; the fits are not parallel. Following the principle of marginality, we retain Aware as an explanatory variable in the regression even though it is not, by itself, statistically significant. (Its t-statistic is $t = 0.43$ with p-value 0.67.)

Show the fits for the two groups. For those not aware (Aware = 0, to one decimal)

$$\textit{Estimated} \text{ Courier Paks} = 2.5 + 13.8 \text{ Hours}$$

and for those who were aware (Aware = 1)

$$\textit{Estimated} \text{ Courier Paks} = (2.5 + 1.7) + (13.8 + 4.3) \text{ Hours}$$
$$= 4.2 + 18.1 \text{ Hours}$$

The interaction implies that the gap between the estimated fits gets wider as the number of hours increases. At Hours = 0 (no visiting by a sales representative), the fits are very similar, 2.5 versus 4.2 Courier Paks. The difference in the use of Courier Paks between those who are aware and those who are unaware is not statistically significant ($t = 0.43$). The gap widens as the number of hours increases. At 3 hours of effort, the model estimates $2.5 + 13.8 \times 3 \approx 43.9$ Paks for those who were not previously aware compared to $4.2 + 18.1 \times 3 \approx 58.5$ for those who were.

(p. 655) Using software (which makes the adjustments for an interaction described in Behind the Math: Comparisons with Interactions), the confidence interval at 3 hours for those who were not aware is 40.5 to 47.3 Courier Paks compared to 54.7 to 62.4 Courier Paks for those who were aware. These confidence intervals do not overlap, implying a statistically significant difference at 3 hours of contact. (The difference increases with the number of hours because of the interaction.)

MESSAGE

SUMMARIZE THE RESULTS

After so much analysis, don't forget to answer the motivating questions. Priming produces a statistically significant increase in the subsequent use of Courier Paks when followed by a visit from a sales representative. With no follow-up by a sales representative, there's no statistically significant difference in the use of Courier Paks. With follow-up, customers who were primed are more receptive. Each additional hour of contact with a sales representative produces about 4.3 more uses of Courier Paks with priming than without priming. At 3 hours of contact, we find a statistically significant difference of 14.6 more Courier Paks (58.5 versus 43.9), on average, for those with priming than those without.

This analysis presumes that the contacted businesses are comparable. If other differences exist (such as legal firms for the group that were aware and manufacturers for the group that were not), these results would need to be adjusted to reflect other differences between the groups.

25.5 | REGRESSION WITH SEVERAL GROUPS

Comparisons often involve more than two groups. For example, a clinical trial of a new medication might compare doses of a new pharmaceutical drug to a rival product and a placebo. Or, a marketing study may compare the sales of a product in several geographic regions or the success of several types of advertising. Unless the data come from a randomized experiment, we typically need to use regression to adjust for confounding variables.

As an example, the data in Figure 25.6 measure sales of stores operated by a retail chain in three markets: urban, suburban, and rural. The response is sales in dollars per square foot. One explanatory variable is the median household income in the surrounding community (in dollars); another is the size of the surrounding population (in thousands). The data include sales at 87 locations. Of these, 27 are rural (red), 27 are suburban (green), and the remaining 33 are urban (blue).

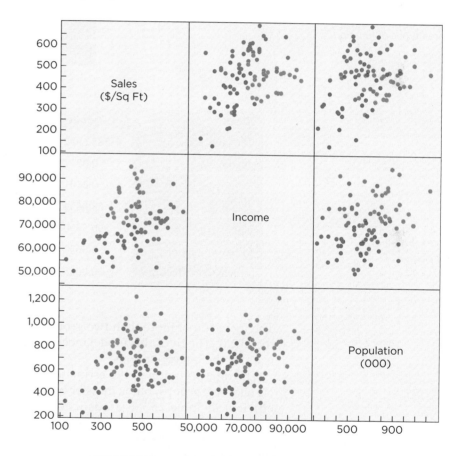

FIGURE 25.6 Scatterplot matrix and correlation matrix of retail sales variables.

	Sales	**Income**	**Pop**
Sales ($/Sq Ft)	1.000	0.385	0.313
Income	0.385	1.000	0.287
Population	0.313	0.287	1.000

The association within each group appears linear, and color-coding shows that the characteristics of the three types of locations overlap.

We would like to compare the sales per square foot in different locations, taking into account the substantial differences in income and population.

A multiple regression with dummy variables can do that, but now we need a second dummy variable. Regression analysis with two groups uses one dummy variable to distinguish two categories. To distinguish three groups requires a second dummy variable. In general, to distinguish J groups requires $J - 1$ dummy variables.

The dummy variables we use in this example are defined as follows:

$$\text{Suburban Dummy} = \begin{cases} 1 & \text{suburban} \\ 0 & \text{otherwise} \end{cases} \quad \text{and} \quad \text{Urban Dummy} = \begin{cases} 1 & \text{urban} \\ 0 & \text{otherwise} \end{cases}$$

Stores in the rural location are identified by having a value of 0 for both dummy variables, so a third dummy variable would be redundant.

Table 25.7 summarizes the regression of sales per square foot on Income, Population, both dummy variables, and the interaction between these and Income.

TABLE 25.7 Multiple regression for retail sales per square foot in three types of location.

R^2	0.7154
s_e	62.6584
n	87

Term	Estimate	Std Error	t-Statistic	p-Value
Intercept	−388.6992	139.6706	−2.78	0.0067
Income	0.0097	0.0022	4.47	<0.0001
Population (000)	0.2401	0.0401	5.99	<0.0001
Urban Dummy	468.8654	161.5639	2.90	0.0048
Suburban Dummy	390.5890	183.3185	2.13	0.0362
Urban × Income	−0.0053	0.0025	−2.14	0.0356
Suburban × Income	−0.0068	0.0026	−2.56	0.0122

The inclusion of more than two groups does not alter any of the conditions of the MRM. The fit of this model meets the necessary conditions, so we concentrate on the interpretations.

The interpretation of the estimates is similar to the interpretation of models with two groups, but is messier because there are more estimates. Each coefficient in Table 25.7 is statistically significantly different from zero. (The interaction of location with Population is not significant and was not used) The equation for the group not represented by a dummy variable, rural locations, omits all of the terms associated with dummy variables and is the baseline for comparison. The fitted model for rural stores is

$$\textit{Estimated Sales (\$/Sq Ft)} = -388.6992 + 0.0097 \text{ Income} + 0.2401 \text{ Population (000)}$$

The negative intercept is a substantial extrapolation from these data. The fit indicates that average sales rise about $0.01 per square foot per dollar of increased local income in rural locations; average sales rise about $0.24 per square foot with an increase of 1,000 in population (assuming this increase does not come with an increase in income).

Coefficients associated with dummy variables reflect differences of stores in other locations from rural stores. The fitted model for an urban location, for instance, adds items associated with Urban Dummy; the slope of Urban

Dummy shifts the intercept and the slope of Urban Dummy $\times$ Income alters the slope for Income. The equation for sales of urban stores is

$$\begin{aligned} Estimated \text{ Sales } (\$/\text{Sq Ft}) &= (-388.6992 + 468.8654) + (0.0097 - 0.0053) \text{ Income} \\ &\quad + 0.2401 \text{ Population } (000) \\ &= 80.1662 + 0.0044 \text{ Income} + 0.2401 \text{ Population } (000) \end{aligned}$$

Hence, sales at a given income are higher in urban than rural stores (this can be seen in Figure 25.6), but do not grow as fast with increases in income. Similarly, sales in suburban locations are estimated to be

$$\begin{aligned} Estimated \text{ Sales } (\$/\text{Sq Ft}) &= (-388.6992 + 390.5890) + (0.0097 - 0.0068) \text{ Income} \\ &\quad + 0.2401 \text{ Population } (000) \\ &= 1.8898 + 0.0029 \text{ Income} + 0.2401 \text{ Population } (000) \end{aligned}$$

Population has the same effect in every location because the model does not have an interaction between Population and the dummy variables for location.

The choice of which group to code with 0s on the dummy variables affects the ease of making comparisons. With the rural group as the baseline (or "left-out") group, the t-statistics associated with dummy variables in Table 25.7 indicate the statistical significance of differences between suburban and rural stores and between urban and rural stores. This formulation of the model does not provide a t-statistic that compares the intercepts and slopes for suburban versus urban stores. The easiest way to get such a comparison would be to replace either Urban Dummy or Suburban Dummy by

$$\text{Rural Dummy} = \begin{cases} 1 & \text{rural} \\ 0 & \text{otherwise} \end{cases}$$

The model specified using Urban Dummy and Rural Dummy would then make it easy to contrast stores in these locations to stores in suburban locations. The overall fit would be the same ($R^2 = 0.7154$, $s_e = 62.6584$), but the coefficients would change to suit the new interpretation.

Best Practices

- *Be thorough in your search for confounding variables.* It might be tempting, but don't stop looking for a confounding variable when you get an answer that you like. The firm in the discrimination example in this chapter might want to stop with a regression that shows it doesn't discriminate, but that does not mean that the analysis has adjusted for all of the relevant differences between the groups.
- *Consider interactions.* Begin with an interaction in the multiple regression that includes dummy variables that represent groups, then remove the interaction *only* if it's not statistically significant, to simplify the interpretation and reduce collinearity. If you don't start with an interaction in the model, you might forget to check for differences in the effects of the explanatory variables.
- *Choose an appropriate baseline group.* The group identified by a value of 0 on the dummy variable provides a baseline for comparison. In a model with dummy variables and interactions, the intercept and slope of the explanatory variable are the intercept and slope for this group, as if a regression had been fit to only the cases in this group. Other coefficients estimate differences from this group.
- *Write out the fits for separate groups.* It's easy to get confused when you introduce categorical information into the regression with dummy variables. If you write out the fits of the model for each group, you will begin to see the differences more clearly. After you've done a lot of modeling with dummy variables, you won't need to do this, but it helps as you're learning the material.
- *Be careful interpreting the coefficient of the dummy variable.* If the regression has an interaction, the slope of the dummy variable estimates the difference between the groups when the numerical variable in the interaction equals 0. That

may be quite an extrapolation from the data and not a meaningful place to compare the means of the groups.
- *Check for comparable variances in the groups.* Multiple regression assumes equal variances for the errors in every group. You should check this assumption carefully using comparison boxplots of the residuals in the two groups, in addition to the regular plot of the residuals versus the fitted values.
- *Use color-coding or different plot symbols to identify subsets of observations in plots.* Colors or symbols work nicely to highlight differences among groups shown together in a scatterplot.

Pitfalls

- *Don't think that you've adjusted for all of the confounding factors.* You might think that one explanatory variable is enough to adjust for differences between two groups, but unless you've done an experiment, there's little reason to think you have all of the lurking factors. You may need to add several explanatory variables to remove the effects of confounding.
- *Don't confuse the different types of slopes.* Slopes that involve a dummy variable compare the model for the category coded as 1 to the baseline category. The slope of the dummy variable itself estimates the difference between the intercepts, and the slope of an interaction estimates the difference between the slopes of the explanatory variable.
- *Don't forget to check the conditions of the MRM.* Regression models become more complicated with the addition of dummy variables, particularly when an interaction is present. In the rush to check the interaction (to see if it can be removed), it can be tempting to go straight to the t-statistics without stopping to check the conditions of the MRM. Don't get hasty; keep checking the conditions before moving on to decisions that require inference.

Software Hints

Some software automatically handles categorical variables as explanatory variables in a regression model. You'll still need to make sure that you know exactly how it was done in order to interpret the fitted equation.

EXCEL

The Excel menu sequence Data > Data Analysis . . . > Regression expects the ranges that specify each variable to be numerical. The same applies to using DDXL. That means you need to construct dummy variables to represent categorical information before attempting to specify the model. Use the IF formula to build these indicators (e.g., = IF(cell = "M",1,0)). Be careful; Excel uses the symbol "=" to check whether two cells are the same as well as to identify a cell that contains a formula.

MINITAB

The dialog produced by the menu sequence

Stat > Regression . . . > Regression

expects each variable to be numerical. Before starting, then, you need to construct indicators. The menu sequence

Calc > Make Indicator Variables . . .

automates this process. Pick the column that identifies the groups (such as a column of labels that identify the groups), specify columns to hold the indicators, and Minitab builds the associated dummy variables. You can then add these to the regression.

JMP

The menu sequence

Analyze > Fit Model

constructs a multiple regression and happily accepts categorical variables as explanatory variables. To build an interaction, select the names of the two variables from the list on the left of the Fit Model dialog and click the Cross button. Coefficients associated with dummy variables in the output are identified by names that include the category field in brackets (such as Sex[Male], Sex[Female]). JMP offers several methods for representing categorical data in regression. To obtain dummy variable estimates as illustrated in this chapter, select the red triangle at the top of the output window. In the pop-up menu, choose the item labeled Estimates > Indicator Parameterization Estimates.

You can also build indicator variables yourself and add these to a regression as numerical data. To build a dummy variable, add a column and define its values using the formula calculator. (Follow the menu commands Cols > New column, and click on the Column Properties button and select the Formula item.) The IF formula is in the Conditional group.

 BEHIND the Math

Comparisons with Interactions

When a regression includes an interaction, comparisons of the average of the response in one group versus that of another require that we compare two estimated coefficients at once. For example, the model in Example 25.1 has the form

$$\text{Estimated Courier Paks} = b_0 + b_1 \text{ Hours} + b_2 \text{ Aware} + b_3 \text{ Hours} \times \text{Aware}$$

For those who were not previously aware, the estimated mean response is

$$\text{Estimated Courier Paks (Not aware)} = b_0 + b_1 \text{ Hours}$$

For the locations that were aware before the visit, the fit is

$$\text{Estimated Courier Paks (Aware)} = (b_0 + b_2) + (b_1 + b_3) \text{ Hour}$$

For locations with x hours of contact, the estimated difference in mean use of Courier Paks is the difference between these fits:

Aware	$(b_0 + b_2) + (b_1 + b_3)x$
Not aware	$- b_0 + b_1x$
Difference	$b_2 + b_3x$

For example, with $x = 3$ hours of contact, the estimated difference is $b_2 + 3b_3$. For tests or confidence intervals, we need the standard error of this weighted sum of two slopes. It is important to recognize that the standard error of $b_2 + 3b_3$ is *not* the sum of the standard error of b_2 plus 3 times the standard error of b_3. The standard error of this weighted sum is the square root of the variance, $\text{Var}(b_2 + 3b_3)$, where we need to think of b_2 and b_3 as random variables. As in Chapter 10, this calculation must account for the *covariance* between the estimates b_2 and b_3:

$$\text{Var}(b_2 + 3b_3) = \text{Var}(b_2) + 3^2\text{Var}(b_3) + 2 \times 3 \text{ Cov}(b_2, b_3)$$

Unless software provides this covariance, we cannot find the standard error of the difference between means for a given number of hours of contact.

To complete this calculation, our software estimates the following covariance matrix (a table of variances and covariances) of the coefficients in the regression (shown to three decimals).

Covariance	Intercept	Hours	Aware	Hours × Aware
Intercept	6.260	−2.333	−6.259	2.333
Hours	−2.333	1.184	2.333	−1.184
Aware	−6.259	2.333	15.928	−5.773
Hours × Aware	2.333	−1.184	−5.773	2.832

For example, the variance of the estimated intercept is 6.260 (the square of its standard error). The covariance between the slopes of Aware and the interaction is −5.773. The estimated variance of $b_2 + 3b_3$ is then

$$15.928 + 9(2.832) + 6(-5.773) = 6.778$$

Notice that the negative covariance implies that the variance of the sum of b_2 plus $3b_3$ is much less than the sum of the variances. The estimated standard error is $\text{se}(b_2 + 3b_3) = \sqrt{6.778} \approx 2.603$.

Referring back to Example 25.1, the estimated average number of Courier Paks at aware locations with 3 hours of contact is 58.5, compared to 43.9 at locations that were not aware. The 95% confidence interval for the difference between these is then $(t_{0.025,121} = 1.98)$

$$(58.5 - 43.9) \pm 1.98(2.603)$$
$$= [9.446 \text{ to } 19.754 \text{ Paks}]$$

Since this interval does not include 0, the difference between the use of Courier Paks in the two situations is statistically significantly different from 0.

CHAPTER SUMMARY

Dummy variables are used to include categorical data as explanatory variables in regression models. When used as explanatory variables, dummy variables indicate group membership. A multiple regression that includes numerical explanatory variables and a dummy variable that identifies group membership is known as an **analysis of covariance**. An analysis of covariance adjusts for possible confounding variables when comparing the means of two or more groups. **Interaction** variables allow the multiple regression to fit different slopes to each group. When comparing groups using regression, it is important to compare the variance of the residuals associated with the separate groups.

■ Key Terms

analysis of covariance, 639

interaction, 640

principle of marginality, 644

■ About the Data

The salary data used in this chapter are from *Business Analysis Using Regression: A Casebook* by Foster, Stine, and Waterman (Springer, 1998). The data are based on an analysis of wage discrimination at a major university, with adjustments to assure confidentiality. The data on the launch of FedEx Courier Paks come from a case study developed for an introductory course in marketing at Wharton. The analysis of retail sales in different locations is from a student project done to satisfy a requirement in an MBA statistics course at Wharton.

EXERCISES

Mix and Match

Match each definition on the left with its mathematical expression on the right. Refer to the following regression equation in this exercise:

$$y = \beta_0 + \beta_1 x + \beta_2 D + \beta_3 xD + \varepsilon$$

where y is the annual salary of an employee (in thousands of dollars) and x denotes the years of experience. D is coded as 1 for college graduates and is coded as 0 for those graduating high school but not college.

1. Intercept for high school graduate	(a) β_1
2. Intercept for college graduate	(b) β_3
3. Has units $thousand/year, high school graduate	(c) $\beta_0 + \beta_2$
4. Has units $thousand/year, college graduate	(d) $\beta_0 + \beta_1 10$
5. Difference in slopes	(e) xD
6. Difference in intercepts	(f) ε
7. Interaction	(g) $\beta_1 + \beta_3$
8. Equal variances	(h) $\beta_0 + \beta_2 + (\beta_1 + \beta_3)10$
9. Average salary for high school graduate with 10 years of experience	(i) β_0
10. Average salary for college graduate with 10 years of experience	(j) β_2

True/False

Mark each statement True or False. If you believe that a statement is false, briefly explain why you think it is false.

11. Confounding arises in a two-sample *t*-test when the groups differ in ways other than the labeling that distinguishes the groups.

12. An analysis of covariance is another name for the use of randomization to avoid confounding.

13. A dummy variable is a numerical encoding that assigns the value 1 to the members of a category and assigns the value 0 to others.

14. To build the interaction between *x* and a dummy variable *D*, we multiply *x* times *D*.

15. If the multiple regression implies parallel fits, the slope of the dummy variable is the difference between the two fitted lines.

16. A multiple regression with a numerical predictor and a dummy variable as two explanatory variables implies parallel fits to the two groups.

17. The purpose of an interaction is to force fits in the groups to be parallel.

18. Interactions introduce collinearity into a multiple regression and should be removed from the model if not statistically significant.

19. If neither the interaction nor the dummy variable is statistically significant in an analysis of covariance, then there's no lurking factor that confounds the results of the related two-sample *t*-test.

20. To be a confounding variable, the variable must be related to *y* and to the dummy variable that indicates group membership.

21. A major assumption of the use of regression with dummy variables is that the size of the two groups must be approximately the same in order to increase the variation of the dummy variable.

22. To check the similar variances condition in models with a dummy variable, use comparison boxplots of *y* versus the categorical variable.

23. To fit a multiple regression that compares the mean values of five groups identified by a categorical variable requires using five dummy variables.

24. An analysis of covariance involving four groups requires that the residuals associated with each group have similar variances.

Think About It

25. The following comparison boxplots show the revenue generated by individual sales representatives who operate in divisions supervised by two different managers. What is the problem with using a two-sample *t*-test to judge the statistical significance of the apparent difference?

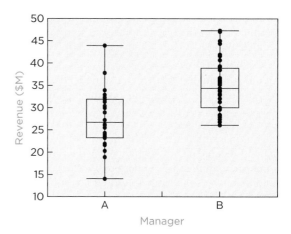

Level	Number	Mean	Std Dev
A	24	27.5625	6.51579
B	37	35.0216	5.65160

26. An auditor collected a random sample of about 100 invoices sent out in the current fiscal year and compared the amounts of these invoices to those of a second random sample of invoices in the prior fiscal year. These boxplots summarize the amounts (in dollars) of the two sets of invoices.

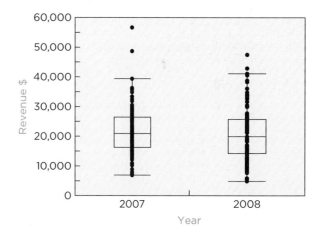

Year	Number	Mean	Std Dev
2007	111	21713.4	8245.42
2008	109	20582.7	8763.01

Would you suggest that the auditor perform a two-sample *t*-test to compare the mean values of these invoices, or can you suggest one (or more) lurking factors that should be taken into account prior to the comparison?

27. When fitting the regression of *y* on *x* for two groups, we can estimate the slope and intercept within each group either by fitting two simple regressions or by fitting one multiple regression. If simple regressions are so much easier to interpret, why combine them into one multiple regression?

28. Multiple regression requires an assumption that the combination of the two simple regressions does not require. What is it, and what condition of the multiple regression does it affect?

29. An industry analyst constructed a model describing the cost of building cars at plants operated by different manufacturers in North America. As a first step, the analyst regressed total production cost (in dollars) on the number of labor hours for a sample of vehicles. The data used came from two plants, one operated by a domestic manufacturer under contract with the labor union known as the United Auto Workers (UAW), and the other operating a nonunionized plant. (UAW members cost more than nonunion labor; *The Wall Street Journal* in May 2006 estimated total costs run $74 per hour if benefits are included.) The analyst included a dummy variable in the regression indicating the plant. Do you think the analyst should also include an interaction (between plant and labor hours)?

30. Matsushita is well known for the efficiency of its automated factories. Facing pressure from developing Asian producers with lower labor costs, the company reconfigured robots in its factory in Saga, Japan. After the modification, it takes 40 minutes to configure the assembly line and start production. Formerly, it took about 20 hours.[9] Once production begins, the plant runs as previously; the robots are the same, only reconfigured to simplify changing tasks. How will this modification change the nature of the equation fitted in order to analyze the association between the time to complete a production run (the response) and the number of units produced? Do you expect the slope, intercept, and error variance all to change? Note the interpretation of these parameters in the context of these data.

31. A two-sample *t*-test has a lot in common with simple regression. This output summarizes the results of fitting a simple regression with only a dummy variable as the explanatory variable. The data are the same salary data used in the text, with salary regressed on Group.

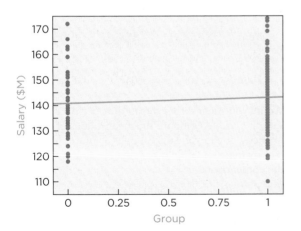

R^2	0.0312
s_e	12.402
n	174

Term	Estimate	Std Error	*t*-Stat	*p*-Value
Intercept	140.0339	1.614561	86.73	<0.0001
Group	4.6705	1.986004	2.35	0.0198

(a) Interpret the estimated intercept and slope in the fitted simple regression.

(b) What is the relationship between the *t*-statistic for the slope in this simple regression and the *t*-statistic for the two-sample *t*-test? (See Table 25.1.)

(c) What assumption is needed in this simple regression approach to a two-sample *t*-test that we did not require previously?

32. This output summarizes a simple regression fit to the data on marketing Courier Paks in Example 25.1.

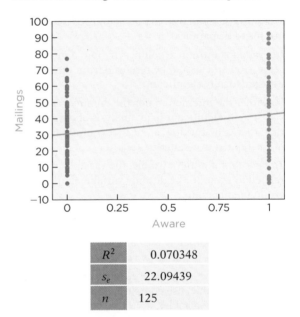

R^2	0.070348
s_e	22.09439
n	125

Term	Estimate	Std Error	*t*-Stat	*p*-Value
Intercept	29.693333	2.551241	11.64	<0.0001
Aware	12.306667	4.033866	3.05	0.0028

(a) Summarize the estimated equation of the simple regression model.

(b) The *t*-statistic for the slope in this model is statistically significant. Assuming the conditions of the SRM hold, what does this tell us?

(c) The variability of the two groups seems somewhat different. Why might that be the case, considering the role of the hours of promotion in this example?

33. The analysis of covariance emphasizes the use of regression to fix a problem with the two-sample *t*-test that has a confounding variable. You can also think of the use of a dummy variable as a way to fix a problem

in the regression of y on x. Take a look at this scatterplot:

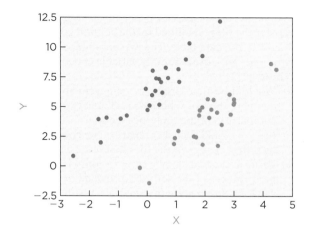

(a) If we fit parallel slopes to these data, with one line for the red and another for the green points, what do you think the slope will be?
(b) What happens if we estimate the slope while ignoring the presence of two clear groups? That is, if we fit a simple regression of y on x using all of the data?

34. After a manufacturer closed an old assembly plant, it re-trained its production employees to use new machines in a more highly automated robotic facility. The automated facility allows the plant to fill small orders of customized parts rather than turn out identical copies. After a weeklong training period, a group of these long-time employees were put to work. Another group of workers were new hires that did not undergo this extensive training. In a study of the value of this training program, an analyst regressed the number of items produced on the time required (in minutes) for completion of the order. The data are shown in this plot; trained employees are colored green.

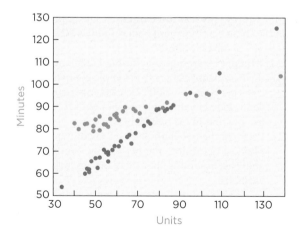

(a) If we fit a separate equation to each group, then what is the interpretation of the intercept in either fit? Include the units as part of your description. What do these intercepts tell you about the training?

(b) What is the interpretation of the slopes in both regressions? Include the units as part of your description and relate these slopes to the training program.
(c) Will an analysis of covariance require an interaction term, or can you skip this step and only fit a dummy variable to distinguish the two groups?

35. The following output summarizes the fit of an analysis of covariance to the data in Exercise 33. The variable D denotes a dummy variable, with $D = 1$ for values colored green and 0 otherwise.

Term	Estimate	Std Error	t-Statistic	p-Value
Intercept	6.0486237	0.208994	28.94	<0.0001
x	2.0349806	0.180043	11.30	<0.0001
D	−5.999216	0.506347	−11.85	<0.0001
Dx	−0.186497	0.264486	−0.71	0.4843

(a) Does the fit of the model suggest parallel equations for the two groups?
(b) How would the output change if the coding of the dummy variable D were reversed (so that 0s became 1s and vice versa)?
(c) What should be the next step in analyzing these data, specifically thinking of the form of the fitted model?

36. The following output summarizes the fit of an analysis of covariance to the data in Exercise 34. The variable D denotes a dummy variable, with $D = 1$ for values colored green and 0 otherwise.

Term	Estimate	Std Error	t-Statistic	p-Value
Intercept	29.437929	1.036176	28.41	<0.0001
Units	0.7083286	0.014987	47.26	<0.0001
D	40.532759	1.436375	28.22	<0.0001
$D \times$ Units	−0.457095	0.020499	−22.30	<0.0001

(a) What is the interpretation of the coefficient of D in the fit of this multiple regression? Use the context of the analysis in your answer.
(b) For what size production run does it appear that the trained employees (shown in green with $D = 1$) appear more productive than the employees who did not receive training (red, $D = 0$)? If one group is always better than the other, say so.

You Do It

37. **Emerald Diamonds** These data are a subset of the diamonds used in Chapter 19. This data table of 144 diamonds includes the price (in dollars), the weight (in carats), and the clarity grade of the diamonds. The diamonds have clarity grade either VS1 or VVS1. VVS1 diamonds are nearly flawless; VS1 diamonds have more visible (but still small) flaws.
(a) Would it be appropriate to use a two-sample t-test to compare the average prices of VS1 and VVS1 diamonds, or is this relationship confounded by the weights of the diamonds?

(b) Perform the two-sample t-test to compare the prices of the two grades of diamonds. Summarize this analysis as if there are no lurking variables. Do you get the sort of difference that would be expected from the definitions of the categories?

(c) Compare the prices of the two types of diamonds using an analysis of covariance. Summarize the comparison of prices based on this analysis. Use a dummy variable coded as 1 for VVS1 diamonds and 0 otherwise. (Assume for the moment that the model meets the conditions for the MRM.)

(d) Compare the results from parts (b) and (c). What can you conclude about the cost of diamonds of these two grades? You should take into account the precision of the estimates and your answer to part (a).

(e) What problem bedevils the multiple regression used for the analysis of covariance that is not present in the two-sample t-test?

38. **Convenience Shopping (introduced in Chapter 19)** These data expand the data table introduced in Chapter 19 by introducing data from a second location. For each of two service stations operated by a national petroleum refiner, we have the daily sales in the convenience store located at the service station. The data for each day give the sales at the store (in dollars) and the number of gallons of gas sold. For Site 1, the data cover 283 days; for Site 2, the data cover 285 days.

(a) Would it be appropriate for management of this chain of service stations to rate the operators of the convenience stores based on a two-sample comparison of the sales of the convenience stores during these two periods, or would such a comparison be confounded by different levels of traffic (as measured by the volume of gas sold)?

(b) Perform the two-sample t-test to compare the sales of the two service stations. Summarize this analysis, assuming that there are no lurking variables.

(c) Compare the sales at the two sites using an analysis of covariance. Summarize the comparison of sales based on this analysis. Use a dummy variable coded as 1 for Site 1 and 0 otherwise. (Assume for the moment that the model meets the conditions for the MRM.)

(d) Compare the results from parts (b) and (c). Do they agree? Explain why they agree or differ. You should take into account the precision of the estimates and your answer to part (a).

(e) Does the estimated multiple regression used in the analysis of covariance meet the similar variances condition?

(f) Suppose an analyst fit the simple regression of sales in the convenience store on gas sales, ignoring the distinction between the two sites. Does this pooling of all the data together affect the relationship between sales in the store and gas sales?

39. **Download (introduced in Chapter 19)** Before purchasing videoconferencing equipment, a company ran tests of its current internal computer network. The goal of the tests was to measure how rapidly data moved through the network given the current demand on the network. Eighty files ranging in size from 20 to 100 megabytes (MB) were transmitted over the network at various times of day, and the time to send the files (in seconds) recorded. Two types of software were used to transfer the files, identified by the column labeled Vendor in the data table. The two possible values are "MS" and "NP"; use a dummy variable coded as 1 when Vendor = "MS."

(a) Would it be appropriate for management to compare the two vendors based on a two-sample comparison of the times needed to transfer the files, or would such a comparison be confounded by different sizes of the files that were sent?

(b) Perform the two-sample t-test to compare the performance of the software provided by the two vendors. Summarize this analysis, assuming that there are no lurking variables.

(c) Compare the sales at the two sites using an analysis of covariance. Summarize the comparison of sales based on this analysis. Use a dummy variable coded as 1 for Site 1 and 0 otherwise. (Assume for the moment that the model meets the conditions for the MRM.)

(d) Compare the results from parts (b) and (c). Do they agree? Explain why they agree or differ. You should take into account the precision of the estimates and your answer to part (a).

(e) Does the estimated multiple regression used in the analysis of covariance meet the similar variances condition?

40. **Production Costs (introduced in Chapter 19)** A manufacturer produces custom metal blanks that are used by its customers for computer-aided machining. The customer sends a design via computer (a 3-D blueprint), and the manufacturer comes up with an estimated price per unit, which is then used to determine a price for the customer. This analysis considers the factors that affect the cost to manufacture these blanks. The data for the analysis were sampled from the accounting records of 195 previous orders filled during the last three months. The data measure performance at two plants, identified as "OLD" and "NEW" in the column *Plant*.

(a) Would it be appropriate for management to compare the two plants using a two-sample comparison of the costs per unit, or would such a comparison be confounded by different requirements for machine use per unit in the two plants?

(b) Perform the two-sample t-test to compare the average cost per unit at the two plants. Summarize this analysis, assuming that there are no lurking variables.

(c) Compare the average cost per unit at the two plants using an analysis of covariance. Summarize the comparison based on this analysis. Represent these categories using a dummy variable coded as 1 if the plant is new. (Assume for the moment that the model meets the conditions for the MRM.)

(d) Compare the results from parts (b) and (c). Do they agree? Explain why they agree or differ. You should take into account the precision of the estimates and your answer to part (a).

(e) Does the estimated multiple regression used in the analysis of covariance meet the similar variances condition?

41. **Seattle Home Prices** This data table expands the data introduced in Chapter 19 on the prices of homes in the Seattle area. One realtor operating in Seattle listed all 28 homes for sale in the original data table. This table includes prices and sizes of 8 more homes listed by a different realtor in Seattle. As previously, we'll look at the price per square foot, using the reciprocal of the number of square feet as the explanatory marginal. In this model, the intercept estimates the variable cost per square foot and the slope of 1/Sq Ft estimates the fixed costs present regardless of the size of the home.

(a) Create a scatterplot of the cost per square foot of the homes on the reciprocal of the size of the homes. Do you see a difference in the relationship between cost per square foot and 1/Sq Ft for the two realtors? Use color-coding or different symbols to distinguish the data for the two realtors.

(b) Based on your visual impression formed in part (a), fit an appropriate regression model that describes the fixed and marginal costs for these realtors. Use a dummy variable coded as 1 for Realtor B to represent the different realtors in the regression.

(c) Does the estimated multiple regression fit in part (b) meet the conditions for the MRM?

(d) Interpret the estimated coefficients from the equation fit in part (b), if it is Okay to do so. If not, indicate why not. What does the fitted model tell you about the properties offered by the realtors?

(e) Would it be appropriate to use the estimated standard errors shown in the output of your regression estimated in part (b) to set confidence intervals for the estimated intercept and slopes? Explain.

42. **Leases (introduced in Chapter 19)** This data table includes the annual prices of 223 commercial leases. All of these leases provide office space in a Midwestern city in the United States. In previous exercises, we estimated the variable costs (costs that increase with the size of the lease) and fixed costs (those present regardless of the size of the property) using a regression of the cost per square foot on the reciprocal of the number of square feet. The intercept estimates the variable costs and the slope estimates the marginal cost. Some of these leases cover space in the downtown area, whereas others are located in the suburbs. The variable Location identifies these two categories.

(a) Create a scatterplot of the cost per square foot of the lease on the reciprocal of the square feet of the lease. Do you see a difference in the relationship between cost per square foot and 1/Sq Ft for the two locations? Use color-coding or different symbols to distinguish the data for the two locations.

(b) Based on your visual impression formed in part (a), fit an appropriate regression model that describes the fixed and variable costs for these leases. Use a dummy marginal coded as 1 for leases in the city and 0 for suburban leases.

(c) Does the estimated multiple regression fit in part (b) meet the conditions for the MRM?

(d) Interpret the estimated coefficients from the equation fit in part (b), if it is OK to do so. If not, indicate why not.

(e) Would it be appropriate to use the estimated standard errors shown in the output of your regression estimated in part (b) to set confidence intervals for the estimated intercept and slopes? Explain.

43. **R&D Expenses** This data file contains a variety of accounting and financial values that describe companies operating in several technology industries: software, systems design, and semiconductor manufacturing. One column gives the expenses on research and development (R&D), and another gives the total assets of the companies. Both of these columns are reported in millions of dollars. This data table expands previous versions (introduced in Chapter 19) by adding data for 2003 to the data for 2004. To estimate regression models, we need to transform both expenses and assets to a log scale.

(a) Plot the log of R&D expenses on the log of assets for 2003 and 2004 together in one scatterplot. Use color-coding or distinct symbols to distinguish the groups. Does it appear that the relationship is different in these two years, or can you capture the association with a single simple regression?

A common question asked when fitting models to subsets is "Do the equations for the two groups differ from each other?" For example, does the equation for 2003 differ from the equation for 2004? We've been answering this question informally, using the *t*-statistics for the slopes of the dummy variable and interaction. There's just one small problem: We're using two tests to answer one question. What's the chance for a false positive error? If you've got one question, better to use one test.

To see if there's any difference, we can use a variation on the *F*-test for R^2. The idea is to test both slopes at once rather than separately. The method uses the change in the size of R^2. If the R^2 of the model increases by a statistically significant amount when we add *both* the dummy variable and interaction to the model, then something changed and the model is different. The form of this incremental, or partial, *F*-test is

$$F = \frac{\text{Change in } R^2/\text{number of added slopes}}{(1 - R^2_{\text{full}})/(n - k_{\text{full}} - 1)}$$

In this formula, k_{full} denotes the number of variables in the model with the extra features, including dummy variables and interactions. R^2_{full} is the R^2 for that model. As usual, a big value for this *F*-statistic is 4.

(b) Add a dummy variable (coded as 0 for 2004 and 1 for 2003) and its interaction with Log Assets to the model. Does the fit of this model meet the conditions for the MRM? Comment on the consequences of any problem that you identify.

(c) Assuming that the model meets the conditions for the MRM, use the incremental *F*-test to assess the size of the change in R^2. Does the test agree with your visual impression? (The value of k_{full} for the model with dummy and interaction is 3, with 2 slopes added. You will need to fit the simple regression of Log R&D Expensed on Log Assets to get the R^2 from this model.)

(d) Summarize the fit of the model that best captures what is happening in these two years.

44. **Cars** The cases that make up this dataset are types of cars. For each of 223 types of cars sold in the United States during the 2003 and 2004 model years, we have the base price and the horsepower of the engine (HP). In previous exercises, we found that a model for the association of price and horsepower required taking logs of both variables. (We used base 10 logs.) The column *Location* denotes the "home country" or continent of the manufacturer. (This is a bit loose, since Ford owns Jaguar and GM owns Saab. We coded these as Europe anyhow. Similarly, we labeled Chrysler as United States even though it was absorbed by Daimler–Mercedes.) Since we have three groups, to simplify the analysis, we'll compare domestic cars to European imports. The data set for this exercise hence excludes cars from Asian manufacturers.

(a) Plot the $\log_{10}$ of price on the $\log_{10}$ of horsepower for cars from both groups of manufacturers in one scatterplot. Use color-coding or distinct symbols to distinguish the groups. Does it appear that the relationship is different in these two years, or can you capture the association with a single simple regression?

(b) Add a dummy variable (coded as 0 for U.S. and 1 for European designs) and its interaction with $\log_{10}$ of HP to the model. Does the fit of this model meet the conditions for the MRM? Comment on the consequences of any problem that you identify.

(c) Assuming that the model meets the conditions for the MRM, use the incremental F-test to assess the size of the change in R^2. (See the discussion of this test in Exercise 43.) Does the test agree with your visual impression? (The value of k_{full} for the model with dummy and interaction is 3, with 2 slopes added. You will need to fit the simple regression of $\log_{10}$ of price on $\log_{10}$ of horsepower to get the R^2 from this model.)

(d) Compare the conclusion of the incremental F-test to those of the tests of the coefficients of the dummy variable and interaction separately. Do these agree? Explain the similarity or difference.

45. **Movies** These data (used also in Chapter 20) describe the box office success of 224 movies released during the years 1998 through 2001. For this analysis, we're interested in the relationship between initial success at the movie theater and subsequent sales for pay-per-view services, such as those offered by cable television. All of these movies are rated either G or PG, with Audience set to "Family," or rated R, with Audience set to "Adult." We dropped movies rated PG-13.

(a) Plot the $\log_{10}$ of subsequent sales on the $\log_{10}$ of the box office gross for movies from both groups in one scatterplot. Use color-coding or distinct

symbols to distinguish the groups. Does it appear that the relationship between box office success and subsequent video sales differs for the two categories, or can you capture the association with a single simple regression?

(b) Add a dummy variable (coded as 1 for adult audiences and 0 for family audiences) and its interaction with $\log_{10}$ of box office gross to the model. Does the fit of this model meet the conditions for the MRM? Comment on the consequences of any problem that you identify.

(c) Assuming that the model meets the conditions for the MRM, use the incremental F-test to assess the size of the change in R^2. (See the discussion of this test in Exercise 43.) Does the test agree with your visual impression? (The value of k_{full} for the model with dummy and interaction is 3, with 2 slopes added. You will need to fit the simple regression to get its R^2 for comparison to the multiple regression.)

(d) Compare the conclusion of the incremental F-test to those of separate tests of the coefficients of the dummy variable and interaction separately. Do these agree? Explain the similarity or difference.

(e) What does your analysis say about the subsequent success of movies? Does the box office gross tell you something different about movies intended for adults versus movies for the family?

46. **Hiring (introduced in Chapter 19)** A firm that operates a large, direct-to-consumer sales force would like to be able to put in place a system to monitor the progress of new agents. A key task for agents is to open new accounts; an account is a new customer to the business. The goal is to identify "superstar agents" as rapidly as possible, offer them incentives, and keep them with the company. To build such a system, the firm has been monitoring sales of new agents over the past two years. The response of interest is the profit to the firm (in dollars) of contracts sold by agents over their first year. Among the possible predictors of this performance is the number of new accounts developed by the agent during the first three months of work. Some of these agents were located in new offices, whereas others joined an existing office (see the column labeled *Office*).

(a) Plot the log of profit on the log of the number of accounts opened for both groups in one scatterplot. Use color-coding or distinct symbols to distinguish the groups. Does the coloring explain an unusual aspect of the "black and white" scatterplot? Does a simple regression that ignores the groups provide a reasonable summary?

(b) Add a dummy variable (coded as 1 for new offices and 0 for existing offices) and its interaction with Log Number of Accounts to the model. Does the fit of this model meet the conditions for the MRM? Comment on the consequences of any problem that you identify.

(c) Assuming that the model meets the conditions for the MRM, use the incremental F-test to assess the size of the change in R^2. (See the discussion of this test in Exercise 43.) Does the test agree with your visual impression? (The value of k_{full} for the model with dummy and interaction is 3, with 2 slopes added. You will need to fit the simple regression to get its R^2 for comparison to the multiple regression.)

(d) Compare the conclusion of the incremental F-test to those of the tests of the coefficients of the dummy variable and interaction separately. Do these agree? Explain the similarity or difference.

(e) What do you think about locating new hires in new or existing offices? Would you recommend locating them in one or the other (assuming it could be done without disrupting the current placement procedures)?

47. **Promotion (introduced in Chapter 19)** These data describe spending by a pharmaceutical company to promote a cholesterol-lowering drug. The data cover 39 consecutive weeks and isolate the metropolitan areas near Boston, Massachusetts, and Portland, Oregon. A subset of this data was introduced in Chapter 19.

The variables in this collection are shares. Marketing research often describes the level of promotion in terms of voice. In place of level of spending, *voice* is the share of advertising devoted to a specific product. Voice puts spending in context; $10 million might seem like a lot for advertising unless everyone else is spending $200 million. The column *Market Share* is sales of this product divided by total sales for such drugs in each metropolitan area. The column *Detail Voice* is the ratio of detailing for this drug to the amount of detailing for all cholesterol-lowering drugs in the metro area. Detailing counts the number of promotional visits made by representatives of a pharmaceutical company to doctors' offices.

(a) A hasty analyst fit the regression of Market Share on Detail Voice with the data from both locations combined. The analyst found a statistically significant slope for Detail Voice, estimated larger than 1. (This finding implies that a 1% increase in the share of detailing would get on average 1% more of the market.) What mistake has the analyst made?

(b) Propose an alternative model and evaluate whether your alternative model meets the conditions of the MRM so that you can do confidence intervals.

(c) What's your interpretation of the relationship between detailing and market share? If you can, offer your impression as a range.

48. **iTunes** The music on an Apple iPod can be stored digitally in several formats. A popular format for Apple is known as AIFF, short for Audio Interchange File Format. Another format is known as AAC, short for Advanced Audio Coding. Files on an iPod can be in either of these formats or both. The 596 songs in this dataset use a mixture of these two formats.

(a) Based on the scatterplot of the amount of space needed on the length of the songs, propose a model for how much space (in megabytes, MB) is needed to store a song of a given number of seconds.

(b) Evaluate whether your model meets the conditions of the MRM so that you can do confidence intervals.

(c) Interpret the estimated slopes in your model.

(d) Construct, if appropriate, a prediction interval for the amount of disk space required to store a song that is 240 seconds long using AAC and then AIFF format. How can you get intervals? (Be imaginative: The obvious approach has some problems.)

4M Home Prices

A national real-estate developer builds luxury homes in three types of locations: urban cities ("city"), suburbs ("suburb"), and rural locations that were previously farm-lands ("rural"). The response variable in this analysis is the *change* in the selling price per square foot from the time the home is listed to the time at which the home sells,

Change = (final selling price − initial listing price)/ number of square feet

The initial listing price is a fixed markup of construction and financing costs. These homes typically sell for about $150 to $200 per square foot. The response is negative if the price falls; positive values indicate an increase such as occurs when more than one buyer wants the home. Other variables that appear in the analysis include the following:

Square Feet	Size of the property, in square feet
Bathrooms	Number of bathrooms in the home
Distance	Distance in miles from the nearest public school

The observations are 120 homes built by this developer that were sold during the last calendar year.

Motivation

a) Explain how a regression model that estimates the change in the value of the home would be useful to the developer?

Method

b) Consider a regression model of the form

$$\text{Change} = \beta_0 + \beta_1\, 1/\text{Square Feet} + \beta_2\, \text{Baths} + \beta_3\, \text{Distance} + \varepsilon$$

Why use the variable 1/Square Feet rather than Square Feet alone? (*Hint:* Recall models with fixed and variable costs from Chapter 22.)

c) The model described in part (b) does not distinguish one location from another. How can the regression model be modified to account for differences among the three types of locations?

d) Why might interactions that account for differences in the three types of locations be useful in the model? Do you expect any of the possible interactions to be important?

Mechanics

e) Use a scatterplot matrix to explore the data. List any important features in the data that are relevant for regression modeling. (Use color-coding if your software allows.)

f) Fit an initial model specified as in part (b) without accounting for location. Then compare the residuals grouped by location using side-by-side boxplots. Does this comparison suggest that location matters? In what way?

g) Extend the model fit in part (b) to account for differences in location. Be sure to follow the appropriate procedure for dealing with interactions and verify whether your model meets the conditions of the MRM.

Message

h) Summarize the fit of your model for the developer, showing three equations for the three locations (with rounded estimates). Help the developer understand the importance of the terms in these equations.

i) Point out any important limitations of your model. In particular, are there other variables that you would like to include in the model but that are not found in the data table?

Analysis of Variance

BY 1870, AMERICANS WERE MOVING OFF FARMS AND INTO CITIES. Of the 12.5 million in the workforce, fewer than half (48%) worked on farms. Thirty years later at the dawn of the 20th century, farming claimed only 36% of the 29 million in the workforce. And by 2005, farming employed just 2.2 million out of the 142 million in the workforce—less than 2%.

Huge strides in productivity made this transition possible. In 1800, it's estimated that it took 56 hours to farm a single acre that produced, on average, 15 bushels of wheat. (One bushel of wheat weighs 60 pounds.) By 1900, that same acre of wheat required only 15 hours of labor, but the yield had fallen to 14 bushels. Today, it takes an hour of labor to raise and harvest an acre of wheat, and the yield has soared to more than 40 bushels.

Where is statistics in all of this? The amount of grain produced per acre only began to improve recently. These gains came from learning how to match the fertilizer, insecticides, irrigation, and seed varieties to the local climate to improve the yield. Agricultural experiment stations in states from Connecticut to California regularly test farming techniques to find the best combination for that region. Deciding what works requires statistics because random variation obscures the answers. Did the yield go up this year because we added more fertilizer, or did it go up because it rained more? Maybe it's the difference in the temperature or the type of seed that was used. Or maybe it's plain luck. To find the best combination requires statistics. In fact, the inferential concepts of statistics originated as methods to answer such questions.

The methods in this chapter draw inferences from the comparison of averages of several groups. Chapter 18 describes two-sample comparisons using a *t*-test. This chapter extends our ability to make comparisons by showing how to use regression to compare the averages of several groups. The two-sample *t*-test is equivalent to a simple regression; a test that compares more than two groups is equivalent to a multiple regression. As in Chapter 25, dummy variables in the regression identify the categories. Interpreting the models in this chapter is a good review of this use of dummy variables. The comparison of several categories also introduces an issue that we have not had to deal with previously, one caused by the many simultaneous inferences that we have available to us.

26.1 | COMPARING SEVERAL GROUPS

We're going to help a farmer in eastern Colorado decide which variety of winter wheat to plant. The methods we use apply equally well to comparisons of the sales of several products, productivity in several factories, or incomes in several locations. The common factor in these comparisons is the presence of several groups. The farmer in Colorado is considering five varieties, named Endurance, Hatcher, NuHills, RonL, and Ripper, a drought-resistant variety. Which of these do we think will produce the most wheat per acre?

Our data come from recent performance trials run by Colorado State University. The performance trials are a large agricultural experiment designed to compare different varieties and planting methods. In these trials, randomly chosen plots of ground in eastern Colorado were planted with these five varieties. Farmers sowed eight randomly chosen plots with each type of seed. Combines harvested the wheat at the end of the season and measured the bushels per acre. By growing each variety in randomly chosen plots, the experiment avoids confounding and produces a fair comparison without requiring us to search for lurking variables as in Chapter 25. Because we have the same number of observations of each variety of wheat, the experiment is said to be **balanced**. This balance simplifies the analysis of the data.

Our analysis of these yields follows the same steps that we have used in previous regression analyses:

- Plot the data to find patterns and surprising features.
- Propose a regression model for the data.
- Check conditions associated with the model.
- Test the appropriate hypothesis and draw a conclusion.

The similarity occurs because we can set up the comparison of averages as a regression. By framing comparison as regression, we can exploit all that we have learned about regression in a new context.

balanced experiment An experiment with an equal number of observations for each treatment.

Comparing Groups in Plots

The side-by-side boxplots shown in Figure 26.1 on the next page summarize the results of the wheat trials. The experiment produces eight observations of the yield per acre for each variety. Each point in Figure 26.1 shows the yield of a variety at one of those plots. Table 26.1 summarizes the mean and standard deviation of the yield for each variety. Subscripts on $\bar{y}$ and s identify the variety.

Let's consider how we might answer the motivating question from looking at these boxplots and summary statistics. Ripper, the drought-resistant variety, typically produces lower yields than Endurance or Hatcher; the boxplot for Ripper is shifted down from those for Endurance and Hatcher. That said, the yields from a few plots sown with Ripper are higher than some of the yields for Endurance or Hatcher; the boxplots overlap. Perhaps the differences

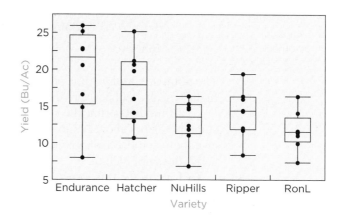

FIGURE 26.1 Yields of winter wheat (bushels per acre) in the Colorado experiment.

TABLE 26.1 Summary statistics from a trial of five varieties of wheat.

Variety	Mean	Std Dev
Endurance	$\bar{y}_1 = 19.58$	$s_1 = 6.08$
Hatcher	$\bar{y}_2 = 17.54$	$s_2 = 4.91$
NuHills	$\bar{y}_3 = 12.90$	$s_3 = 3.07$
Ripper	$\bar{y}_4 = 14.08$	$s_4 = 3.37$
RonL	$\bar{y}_5 = 11.68$	$s_5 = 2.66$

among the varieties in Figure 26.1 result from random variation. Maybe Ripper was unlucky in this experiment. If that's the case, we could expect to find rather different outcomes in another experiment or during next year's growing season. To make a recommendation, we need to determine the statistical significance of the differences among these five varieties.

Relating the *t*-Test to Regression

Before considering all five varieties, let's start with a simpler question: Is there a statistically significant difference between the average yield of Endurance and the average yield of the others? We'll first answer this question using a two-sample *t*-test (Chapter 18), and then show that this test is equivalent to a simple regression. Once we have that connection, it will be easy to see how to use regression to compare five varieties at once.

Let's start by checking the conditions for a two-sample *t*-test. One sample consists of the yields of the 8 plots growing Endurance, and the second sample holds the yields of the 32 plots growing other varieties. The randomized design of the experiment avoids confounding variables, and the randomized selection of plots produces a random sample of locations. The comparison boxplots in Figure 26.1 suggest similar variances, though with few points in each batch, it is hard to be sure. Such small samples from each variety also make it difficult to check for normality. We don't see any big outliers or skewness, so we will treat the yields of each type as nearly normal and proceed to the test. (Chapter 17 discusses methods for small samples.)

Because the variances appear comparable in Figure 26.1, we use the two-sample *t*-test that pools the variances. We would normally use the two-sample test that permits different variances among the groups, but to make the connection to regression, we assume equal variances. The pooled, two-sample *t*-test finds a statistically significant difference between the average yield of Endurance (19.58 bushels per acre) and the average yield of the other four varieties (14.05 bushels per acre). Table 26.2 summarizes the test.

TABLE 26.2 Two sample *t*-test of yields of winter wheat for Endurance versus all others.

Term	Estimate	Std Error	*t*-Statistic	*p*-Value
Difference	5.53	1.79	3.10	0.0037

The *t*-statistic and *p*-value show that Endurance has a statistically significantly higher mean yield per acre than the combination of other varieties. The *t*-statistic $t = 3.10$ indicates that the observed difference of 5.53 bushels per acre is 3.10 standard errors above 0. Consequently, the 95% confidence interval for the difference between the average yields, $5.53 \pm t_{.025,38}1.79 = [1.91$ to 9.14 bushels per acre], does not include 0.

To formulate this *t*-test as a regression, define a dummy variable as in Chapter 25 that identifies plots that grew Endurance.

$$D(\text{Endurance}) = \begin{cases} 1 & \text{if plot is seeded with Endurance} \\ 0 & \text{otherwise} \end{cases}$$

Now fit the simple regression of the yields of all 40 plots on this dummy variable. Because the explanatory variable is a dummy variable that takes on only 2 values (0 and 1), the scatterplot of yield versus $D(\text{Endurance})$ shown in Figure 26.2 presents the data in two columns, one at $x = 0$ (other varieties) and the other at $x = 1$ (Endurance).

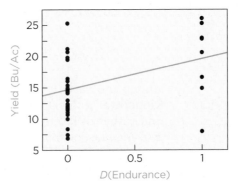

FIGURE 26.2 Scatterplot of yield versus a dummy variable for Endurance.

Table 26.3 summarizes the simple regression in Figure 26.2.

TABLE 26.3 Simple regression of yield versus a dummy variable for Endurance.

r^2	0.202
s_e	4.517
n	40

Term	Estimate	Std Error	*t*-Statistic	*p*-Value
b_0	14.05	0.80	17.60	<0.0001
b_1	5.53	1.79	3.10	0.0037

The means of the two groups determine the estimates in the simple regression. To see this connection, construct the fitted values of the regression when the dummy variable $D(\text{Endurance}) = 0$ and 1.

If $D(\text{Endurance}) = 0$: $\hat{y} = b_0 + b_1 D(\text{Endurance})$
$$= b_0$$
$$= 14.05 \text{ bushels/acre}$$

That's the average yield for other varieties shown in Figure 26.2.

If $D(\text{Endurance}) = 1$: $\hat{y} = b_0 + b_1 D(\text{Endurance})$
$$= b_0 + b_1$$
$$= 14.05 + 5.53 = 19.58 \text{ bushels/acre}$$

That's the average yield for Endurance.

Not only does the slope $b_1 = 5.53$ bushels/acre match the difference between the means, but the standard error, t-statistic, and p-value for the slope also match those of the two-sample test. The summary of the two-sample t-test in Table 26.2 matches the summary of the slope in the simple regression in Table 26.3. **Testing the slope in a simple regression of Y on a dummy variable D is equivalent to a pooled two-sample t-test of the difference between means.**

What Do You Think?

The manager of a call center randomly selected 20 employees to attend training sessions that teach how to resolve a question more quickly. The manager also selected 20 other employees who did not attend these sessions as a comparison group. After the training sessions were completed, the manager used regression to compare the two groups. The response is the average length of time (in minutes) required by the 40 chosen employees to resolve calls during the next week. The single explanatory variable is a dummy variable *Attend* that is coded 1 for those who attended the training sessions and 0 for those who did not. This table summarizes the least squares regression.

Term	Estimate	Std Error	t-Statistic	p-Value
Intercept	5.32	2.21	2.41	0.0210
Attend	−1.45	2.67	0.54	0.5903

a. How long on average do those who did not attend the training spend on a call? What is the average length for those who did attend the training?[1]
b. If one were to compare the average length of calls for those who attended the sessions to that for those who did not using a pooled two-sample t-test, what would be the p-value of that test? Would that test find a statistically significant difference between the means?[2]
c. What is missing from this comparison of the two groups?[3]

Comparing Several Groups Using Regression

The connection between the pooled two-sample t-test and simple regression hints that we can use multiple regression to compare the means of several groups. Simply add more dummy variables.

For the wheat trials, define dummy variables for the other varieties:

$D(\text{Hatcher}) = 1$ if plot grows variety Hatcher and 0 otherwise.
$D(\text{NuHills}) = 1$ if plot grows variety NuHills and 0 otherwise.
$D(\text{Ripper}) = 1$ if plot grows variety Ripper and 0 otherwise.
$D(\text{RonL}) = 1$ if plot grows variety RonL and 0 otherwise.

Each row of the data table that represents a plot sown with Ripper has the values shown in the first row of Table 26.4 at the top of the next page. Each row that represents a plot sown with Endurance has the values in the second row.

---

[1] b_0 is the average length for those who did not receive training, 5.32 minutes. The average for those who did receive training (those for whom the dummy variable is 1) is $b_0 + b_1 = 5.32 - 1.45 = 3.87$ minutes.
[2] The p-value is 0.5903, shown for the slope in the table. The test would not find a statistically significant difference.
[3] Where are the plots? We need to see, at a minimum, boxplots comparing the data in these groups in order to check the conditions for the two-sample comparison.

TABLE 26.4 Dummy variables for plots sown with Ripper (first row) and Endurance (second row).

$D(\text{Endurance})$	$D(\text{Hatcher})$	$D(\text{NuHills})$	$D(\text{Ripper})$	$D(\text{RonL})$
0	0	0	1	0
1	0	0	0	0

We cannot put all five of these dummy variables into a regression at once. Least squares would complain because these variables are perfectly collinear. Because only one of these dummy variables is equal to 1 in any row of the data table, their sum is always equal to 1. Hence, we can calculate any one of these dummy variables from the other four. For example, the value of $D(\text{RonL})$ is

$$D(\text{RonL}) = 1 - (D(\text{Endurance}) + D(\text{Hatcher}) + D(\text{NuHills}) + D(\text{Ripper}))$$

As noted in Chapter 25, we need only $J - 1$ dummy variables to represent J categories. Since the categorical variable *Variety* has $J = 5$ levels, only $J - 1 = 4$ dummy variables are needed to distinguish the categories. We'll use the first four dummy variables, but any four produce the same conclusions.

Table 26.5 summarizes the regression of yield on $D(\text{Endurance})$, $D(\text{Hatcher})$, $D(\text{NuHills})$, and $D(\text{Ripper})$. As in Chapter 25, we refer to the category that is not explicitly represented by a dummy variable in the regression as the baseline category; in this example, the variety RonL is the baseline category. A multiple regression such as this in which all of the explanatory variables are dummy variables produces an **analysis of variance** (also called **ANOVA** or ANOVA regression).

analysis of variance (ANOVA) The comparison of two or more averages obtained by fitting a regression model with dummy variables.

TABLE 26.5 Multiple regression of yield on four dummy variables produces an ANOVA regression.

R^2	0.359
s_e	4.217
n	40

Term	Estimate	Std Error	t-Statistic	p-Value
Intercept	**11.68**	1.49	7.84	<.0001
$D(\text{Endurance})$	7.90	2.11	3.75	0.0006
$D(\text{Hatcher})$	5.86	2.11	2.78	0.0087
$D(\text{NuHills})$	1.22	2.11	0.58	0.5682
$D(\text{Ripper})$	2.40	2.11	1.14	0.2633

Interpreting the Estimates

To interpret the slopes in the multiple regression summarized in Table 26.5, we proceed as in simple regression with one dummy variable. Start with the intercept and then consider the slopes.

The intercept in an ANOVA regression is the average yield for the baseline variety, RonL. If all of the dummy variables in the regression are set to 0,

$$D(\text{Endurance}) = D(\text{Hatcher}) = D(\text{NuHills}) = D(\text{Ripper}) = 0$$

then we are considering a plot sown with RonL. In this case, the fit of the model reduces to the intercept shown in boldface in Table 26.5.

$$\hat{y} = b_0 = 11.68 \text{ bushels/acre}$$

Table 26.1 confirms that the intercept 11.68 is the mean for RonL, $\bar{y}_5$. In general, fitted values are means of categories. The fitted value for every observation in the jth category is the mean of the response for those cases $\bar{y}_j$.

By setting each of the dummy variables in turn to 1, we find that each slope compares the average response of a category to the average of the baseline category. For instance, if we set $D(\text{Endurance}) = 1$ (and set the others to 0 as we must), then the fitted value from the multiple regression is

$$\hat{y} = b_0 + b_1 D(\text{Endurance}) + b_2 D(\text{Hatcher}) + b_3 D(\text{NuHills}) + b_4 D(\text{Ripper})$$
$$= b_0 + b_1$$
$$= 11.68 + 7.90 = 19.58 \text{ bushels/acre}$$

Table 26.1 shows that this is the mean yield $\bar{y}_1$ of Endurance. Since b_0 is the mean yield for RonL, the slope b_1 of $D(\text{Endurance})$ is the *difference* between the average yields of Endurance and RonL. Endurance averages $b_1 = 7.9$ more bushels of wheat per acre than RonL.

As another example, if $D(\text{Hatcher}) = 1$, then the fitted value is

$$\hat{y} = b_0 + b_1 D(\text{Endurance}) + b_2 D(\text{Hatcher}) + b_3 D(\text{NuHills}) + b_4 D(\text{Ripper})$$
$$= b_0 + b_2$$
$$= 11.68 + 5.86 = 17.54 \text{ bushels/acre}$$

That's the mean yield $\bar{y}_2$ of Hatcher in Table 26.1. Hatcher averages $b_2 = 5.86$ more bushels/acre than RonL. Similarly, the average yield of NuHills is $b_3 = 1.22$ more bushels per acre than RonL, and the average yield of Ripper is 2.40 more bushels per acre than RonL.

To recap, when using an ANOVA regression, remember that

- Fitted values are means of the categories.
- The intercept is the average response for the baseline category.
- Each slope is the difference between the average of the response in a specific category and the average of the response in the baseline category.

ANOVA Regression Model

The parameters of the MRM for an ANOVA regression are derived from the population means. Just as the estimated intercept b_0 matches the sample mean $\bar{y}_5$, the parameter β_0 matches μ_5, the mean yield for the baseline category RonL. The slope parameters represent differences between the population mean yields of Endurance, Hatcher, NuHills, and Ripper from RonL. For instance, $b_1 = \bar{y}_1 - \bar{y}_5$ is the difference between the sample means of Endurance and RonL. The corresponding parameter β_1 is the difference between the population means, $\mu_1 - \mu_5$. The other parameters are defined analogously:

$$\beta_2 = \mu_2 - \mu_5, \beta_3 = \mu_3 - \mu_5, \text{ and } \beta_4 = \mu_4 - \mu_5$$

The equation of the MRM can then be written in terms of the population means as

$$y = \beta_0 + \beta_1 D(\text{Endurance}) + \beta_2 D(\text{Hatcher}) + \beta_3 D(\text{NuHills}) + \beta_4 D(\text{Ripper}) + \varepsilon$$
$$= \mu_5 + (\mu_1 - \mu_5)D(\text{Endurance}) + (\mu_2 - \mu_5)D(\text{Hatcher}) + (\mu_3 - \mu_5)D(\text{NuHills})$$
$$+ (\mu_4 - \mu_5)D(\text{Ripper}) + \varepsilon$$

It is also common to express the model for an ANOVA regression in a different style. Imagine that we have arranged the wheat yields in a table with 8 rows (for the plots) and 5 columns (for the varieties of wheat). Let y_{ij} denote the yield for the ith observation in the jth category. The first subscript identifies the row (plot), and the second subscript identifies the column (variety). For instance, y_{52} denotes the yield of the fifth plot sown with the second variety, Hatcher. In a balanced experiment such as this one, the table has the same number of rows in each column.

tabular notation Notation in which y_{ij} denotes the response for the ith case in the jth group. There are a total of j groups, with n_j in the jth group.

This **tabular notation** produces a simpler, but equivalent, equation for the population regression model. The equation states that observations in a category are randomly distributed around the mean of that category:

$$y_{ij} = \mu_j + \varepsilon_{ij}$$

Because we are still using regression, the assumptions about the errors ε are the same as in other regression models: The errors are independent, normally distributed random variables with mean 0 and variance σ_ε^2. An ANOVA regression in which the dummy variables derive from a single categorical variable is often called a **one-way analysis of variance**.

One-Way Analysis of Variance   This regression model compares the averages of the groups defined by the J levels of a single categorical variable. The observations in each group are a sample from the associated population.

Equation

$$y_{ij} = \mu_j + \varepsilon_{ij}$$

Assumptions

The errors ε_{ij}

1. are independent,
2. have equal variance σ_ε^2, and
3. are normally distributed, $\varepsilon_{ij} \sim N(0, \sigma_\varepsilon^2)$.

We can also interpret the model for the one-way analysis of variance as stating that $y_{ij} \sim N(\mu_j, \sigma_\varepsilon^2)$. That is, the data within each category are normally distributed around the category mean with variance σ_ε^2.

What Do You Think?

A recent study describes a randomized experiment that compares three therapies for lower back pain: basic exercise, specialized motor-control exercise, and spinal manipulation.[4] Motor-control exercise requires one-on-one coaching, and spinal manipulation requires treatment by a therapist. Both cost more than basic exercise. The study randomly assigned 240 people, with 80 to each therapy. Researchers then measured the improvement after 8 weeks in how well subjects were able to do things that previously caused pain. Figure 26.3 summarizes the results.

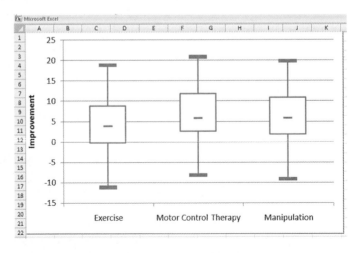

FIGURE 26.3 Side-by-side comparison boxplots show the improvement obtained by three therapies.

[4] M. L. Ferreira, P. H. Ferreira, R. D. Herbert, et al. (2007), "Comparison of General Exercise, Motor Control Exercise, and Spinal Manipulative Therapy for Chronic Low Back Pain: A Randomized Trial," *Pain*, **131**, 31–37.

Therapy	Mean	SD
Exercise	4.09	6.00
Motor Control	6.89	6.03
Manipulation	6.43	6.04

Suppose that we fit a multiple regression of improvement on two dummy variables, one identifying subjects taught specialized exercise and the second identifying those treated with spinal manipulation:

$$\text{\textit{Estimated} Improvement} = b_0 + b_1 D(\text{Specialized}) + b_2 D(\text{Manipulation})$$

a. What would happen if we included a third dummy variable, $D(\text{Exercise})$, in this multiple regression?[5]
b. Based on the summary, what is the value of the intercept b_0?[6]
c. What are the values and interpretations of the estimated slopes b_1 and b_2?[7]

26.2 | INFERENCE IN ANOVA REGRESSION MODELS

Inference in an analysis of variance makes a claim about the population means of the groups represented in the data. Before considering such inferences, we need to verify the conditions of the associated regression model.

Checking Conditions

The conditions for regression that are relevant in ANOVA regression are

~~Linear association~~
✓ No obvious lurking variable
✓ Evidently independent
✓ Similar variances
✓ Nearly normal

We put a line through the linear condition because this condition is automatic for an ANOVA regression. Because the slopes in an ANOVA regression are differences between averages, there is no need for fitting a curve. As for lurking variables, this condition is also automatic—if the data are from a randomized experiment such as the wheat trials. If the data are not from a randomized experiment, we need to investigate possible confounding variables as when using a two-sample t-test. For example, it would be a serious flaw in the analysis of the wheat trials if plots for some varieties received more irrigation or fertilizer than others and we did not adjust for this confounding.

Checking the remaining conditions proceeds as in other regression models. We use the residuals from the fitted model to check these visually. Residuals in an ANOVA regression are deviations from the means of the groups, $e_{ij} = y_{ij} - \bar{y}_j$. To check the condition of similar variances, inspect side-by-side boxplots of the residuals as shown in Figure 26.4.

[5] The three dummy variables are perfectly collinear and redundant. Only two dummy variables are needed.
[6] The intercept $b_0 = 4.09$, the mean of the omitted group, those with exercise.
[7] The slopes are differences in the means of the motor control and manipulation subjects from the mean for the group represented by the intercept. So, $b_1 = 6.89 - 4.09 = 2.8$ and $b_2 = 6.43 - 4.09 = 2.34$.

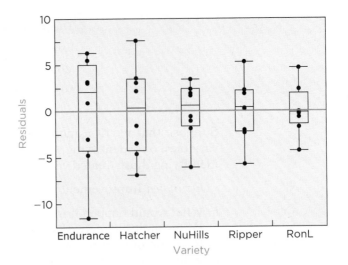

FIGURE 26.4 Side-by-side comparison boxplots of the residuals.

The average of each group of residuals is 0 because the residuals within a group are deviations from the mean of that group. With every group centered at zero, it's easier to compare the variation because the boxes line up. Figure 26.4 suggests that the two varieties with higher yields (Endurance and Hatcher, on the left) are more variable (less consistent yields) than the others. The samples are too small to be definitive, but this might be worth further study.

> **caution** When comparing two groups, we conclude that the data have similar variances so long as the interquartile ranges of the residuals in the two groups are within a factor of 2 (Chapter 25). With more than two groups, we must allow for a wider range in observed variation. So long as the IQRs of the residuals in each group are similar, within a factor of 3 to 1 with up to five groups, this condition is satisfied. Effective comparison of variances, however, requires 20 or more cases in each group. For small samples, such as those from the wheat trial (8 in each group), variance estimates are so imprecise that the comparison of IQRs may not reveal a problem. Fortunately, the p-value that measures statistical significance is reliable when comparing the averages of equally sized (balanced) samples even if the data don't have the identical variances.

Finally, we use a normal quantile plot of the residuals to check for normality (Figure 26.5). These residuals seem nearly normal even allowing for a very negative residual. (The yield from one plot with Endurance is much lower than the others.)

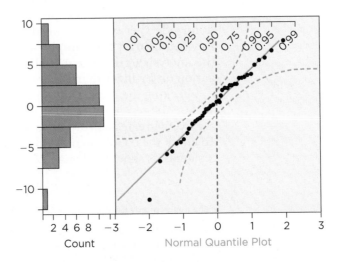

FIGURE 26.5 Normal quantile plot of the residuals.

| **What Do You Think?** | Refer to the study of back pain (Figure 26.3) to answer these questions. |

a. What is the fitted value $\hat{y}$ for the subject in the exercise group who improved by 9 points? What is the residual for this subject?[8]
b. Does it appear that the back pain data meet the assumptions of regression?[9]
c. Estimate the value of s_e, the standard deviation of the residuals in the multiple regression of improvement on D(Specialized) and D(Manipulation).[10]

F-Test for the Difference among Means

Now that we've checked the conditions of the model, we can perform inference. Are the differences among the average yields of the five varieties of wheat statistically significant? If not, the data don't help us make a recommendation to the wheat farmer. If so, which variety should we choose?

As in other multiple regression models, we begin with the overall F-test that was introduced in Chapter 23. The overall F-test tests the null hypothesis that the slopes in a regression are simultaneously 0. For the regression of wheat yields, the F-test considers the null hypothesis that all of the slopes of the dummy variables are zero:

$$H_0: \beta_1 = \beta_2 = \beta_3 = \beta_4 = 0$$

The correspondence $\beta_j = \mu_j - \mu_5$ between slopes and means leads to an important interpretation of H_0 in an ANOVA regression. H_0 implies that all four differences among the means are 0. If all four differences are 0, then all five means are the same. Hence, H_0 implies that all five varieties produce the same average yield:

$$H_0: \mu_1 = \mu_2 = \mu_3 = \mu_4 = \mu_5$$

If H_0 holds, it does not matter which variety the farmer grows. Table 26.6 summarizes the F-test.

TABLE 26.6 ANOVA table for the multiple regression of yield on three dummy variables.

Source	df	Sum of Squares	Mean Square	F	p-Value
Regression	4	348.554	87.139	4.900	0.0030
Residual	35	622.362	17.782		
Total	39	970.916			

You can check that the F-statistic in Table 26.6 matches what we get if we compute F from R^2 (R^2 for the multiple regression is given in Table 26.5):

$$F = R^2/(1 - R^2) \times (n - k - 1)/k = 0.359/(1 - 0.359) \times (40 - 5)/4 \approx 4.90$$

This table also includes a p-value. Because the p-value is less than 0.05, we reject H_0. We conclude that there is a statistically significant difference among

[8] In ANOVA, the fitted value is the mean of the category, in this case the mean improvement for the exercise group. So, $\hat{y} = 4.09$. The residual is the deviation $y - \hat{y} = 9 - 4.09 = 4.91$.
[9] Yes. Randomization means we don't have to worry about confounding variables. The data seem independent and the SDs in the groups are similar. For normality, we need to see the normal quantile plot, but the data show no evident outliers or skewness.
[10] Look at the SDs in the table following Figure 26.3. The residuals are the deviations from the group mean. That's what the SDs of the groups measure. Since these are nearly the same in every group, s_e is about 6.

the mean yields of these five varieties. The yield depends on which variety the farmer chooses. (For another description of terms that appear in Table 26.6, see Behind the Math: Within and Between Sums of Squares.)

(p. 685)

Understanding the *F*-Test

TABLE 26.7 Hypothetical means of four categories.

Category	Mean
a	10
b	4
c	0
d	−2

We can motivate the *F*-test in an ANOVA regression by considering some artificial data. Suppose the averages of the response in four categories are as shown in Table 26.7. Are these averages statistically significantly different, or are these averages so similar as to be the result of random variation?

It is important to recognize that we cannot answer this question from the averages alone. To assess statistical significance, we need to know both the sample sizes and the variation within each group. Figure 26.6 shows two situations. In each, there are four groups, each with 10 observations. The means of the categories are exactly those in Table 26.7. In which case does it appear that the means are statistically significantly different?

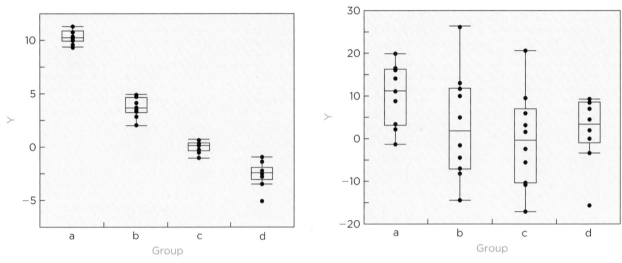

FIGURE 26.6 Side-by-side comparison boxplots for two sets of responses.

The frame on the left is like an image in sharp focus—every detail stands out. The difference $\bar{y}_a - \bar{y}_b = 6$ is large compared to the variation within the groups; the boxplots do not overlap. The frame on the right lacks focus and the details blur. The difference $\bar{y}_a - \bar{y}_b$ is hardly noticeable on the right; the variation within the groups obscures the difference between the means. The *F*-test agrees with this visual impression. It finds statistically significant differences on the left, but not on the right. The data in the left frame of Figure 26.6 produce a large *F*-statistic ($F = 445$, highly statistically significant) whereas the data in the right frame produce a small *F*-statistic ($F = 2.05$, which is not statistically significant).

Confidence Intervals

The one-way analysis of variance summarized in Table 26.6 rejects the null hypothesis that the mean yields of the varieties of wheat are the same. There is a statistically significant difference among the means. Like other regression models, the *F*-test does not indicate which differences are statistically significant. We would like to be able to recommend a specific variety, not just tell the farmer that there are differences among the varieties. The same concern arises in business applications of ANOVA regression. We not only want to know that differences exist; we also want to know which category is best.

Since the overall F-statistic rejects the null hypothesis that the means are the same, we consider the statistical significance of individual slopes. In this model, these slopes estimate the difference in yield between each variety and RonL, the baseline category. Table 26.8 repeats portions of Table 26.5 for convenience.

TABLE 26.8 Estimates from the one-way analysis of variance of yields.

Term	Estimate	Std Error
Intercept	11.68	1.49
D(Endurance)	7.90	2.11
D(Hatcher)	5.86	2.11
D(NuHills)	1.22	2.11
D(Ripper)	2.40	2.11

Let's construct confidence intervals for these estimates. The t-percentile needed for each interval is $t_{.025,n-k+1} = t_{.025,n-J} = t_{.025,35} = 2.03$. For example, the 95% confidence interval for $\mu_1 - \mu_5$ (the difference in mean yield between Endurance and RonL) is

$$b_1 \pm t_{.025,35}\text{se}(b_1) = 7.90 \pm 2.03(2.11) \approx 3.62 \text{ to } 12.18 \text{ bushels/acre}$$

This interval does not include zero and hence indicates that Endurance yields statistically significantly more wheat per acre than RonL. Similarly, the 95% confidence interval for $\mu_2 - \mu_5$ is

$$b_2 \pm t_{.025,35}\text{se}(b_2) = 5.86 \pm 2.03(2.11) \approx 1.58 \text{ to } 10.14 \text{ bushels/acre}$$

This interval also excludes zero, indicating another statistically significant difference.

Notice that both of these confidence intervals have the same length. This happens because every slope in the regression (Table 26.8) has the same standard error. The connection between regression slopes and averages explains why the standard errors match. The slopes in the regression are the differences between averages; for example, $b_1 = \bar{y}_1 - \bar{y}_5$. Hence, the confidence interval for β_1 is also the confidence interval for $\mu_1 - \mu_5$:

$$\bar{y}_1 - \bar{y}_5 \pm t_{0.025,35}\text{se}(\bar{y}_1 - \bar{y}_5)$$

The standard error for each slope is the standard error of the difference between two means. The standard error of the difference between sample means is formed as in the pooled two-sample t-test in Chapter 18:

$$\text{se}(\bar{y}_1 - \bar{y}_5) = s_e\sqrt{1/n_1 + 1/n_5} = 4.217\sqrt{2/8} \approx 2.11 \text{ bushels/acre}$$

Now we can see why the standard errors of the slopes match: The experiment is balanced, with equal numbers of observations in each group.

26.3 | MULTIPLE COMPARISONS

The four confidence intervals or t-statistics associated with the slopes of this regression test the differences between the average yield of RonL, the baseline category, and the other varieties. That's okay if we're interested *only* in comparing RonL to the other varieties, but not if we're interested in every comparison. What about differences among the other varieties? Table 26.9 shows every pairwise comparison. (We are interested in just half of the differences in this table because of symmetry. If the confidence interval for, say, $\mu_1 - \mu_2$ includes 0, then so too does the confidence interval for $\mu_2 - \mu_1$.)

TABLE 26.9 Pairwise differences between mean yields of wheat varieties. Margins show average yields.

$\bar{y}_{\text{row}} - \bar{y}_{\text{column}}$	Endurance 19.58	Hatcher 17.54	NuHills 12.90	Ripper 14.08	RonL 11.68
Endurance 19.58	0	2.04	6.68	5.50	7.90
Hatcher 17.54	−2.04	0	4.64	3.46	5.86
NuHills 12.90	−6.68	−4.64	0	−1.18	1.21
Ripper 14.08	−5.50	−3.46	1.18	0	2.40
RonL 11.68	−7.90	−5.86	−1.21	−2.40	0

multiple comparisons
Inferential procedure composed of numerous separate tests. Also called **multiplicity**.

The differences between RonL and the other varieties given by the regression slopes are highlighted.

The half of Table 26.9 above the diagonal holds 10 differences. In general, for J groups, the counting methods of Chapter 11 show that there are ${}_JC_2 = J(J - 1)/2$ pairwise comparisons if we ignore the sign of the difference. Because we must consider many confidence intervals or t-statistics in order to compare all five varieties, this situation presents **multiple comparisons** (sometimes called **multiplicity** or a *post hoc* comparison).

Multiple comparisons introduce an important issue in hypothesis testing. The purpose of a hypothesis test is to distinguish meaningful differences from those that result from random variation. We pick a null hypothesis and then test whether data deviate enough from H_0 to reject this hypothesis. Each test that we have used controls for the chance of falsely rejecting H_0, declaring that we have found a statistically significant result when in fact H_0 holds (a Type I error). Typically, we allow a 5% chance for falsely rejecting H_0. The issue that arises in multiple comparisons is that we are considering many hypotheses at once. For judging the statistical significance of the differences among these yields, we need to use an inferential procedure that allows us to make 10 comparisons while retaining a 5% chance of a Type I error. If we construct confidence intervals for all 10 differences $\mu_i - \mu_j$, we would like to be assured that there's only a 5% chance that *any* of these intervals omits the population difference.

The usual 95% confidence interval does not offer this guarantee. There's a 5% chance that the t-interval for each difference $\mu_j - \mu_k$ does not contain 0 even though $\mu_j = \mu_k$ (a Type I error). If we form 10 intervals using the usual t-interval, too many mistakes are likely; the overall chance for a Type I error is larger than 5%. We will think we've found statistically significant differences even though the means in the population are the same.

In general, it is difficult to determine the chance for a Type I error among a collection of confidence intervals or tests. About the best we can usually do is put an upper bound on the chance for a Type I error. Let the event E_j indicate that the jth interval does not contain the associated population parameter. The chance for at least one Type I error among M confidence intervals with confidence level $100(1 - \alpha)\%$ is (using Boole's inequality from Chapter 7)

$$P(\text{Type I error}) = P(\text{at least one interval is wrong})$$
$$= P(E_1 \text{ or } E_2 \text{ or} \ldots \text{or } E_M)$$
$$\leq P(E_1) + P(E_2) + \cdots + P(E_M) = \alpha M$$

For the wheat varieties, $M = 10$ and $\alpha = 0.05$, so the chance for at least one Type I error could be as large as $10 \times 0.05 = 1/2$. That's too large. If we persist in doing lots and lots of tests at level $\alpha = 0.05$, we'll eventually reject a null hypothesis purely by chance alone.

There are only two ways to reduce the chance for an error: We have to either make fewer comparisons (make M smaller) or increase the level of confidence (make α smaller). If we must compare the means of all five varieties, we can't change M. We have to make α smaller and use longer confidence intervals.

The problem of multiple comparisons occurs in other situations. Control charts (Chapter 13) deal with this issue by widening the control limits to avoid falsely declaring a changed process; control limits are typically set at ± 3 standard errors rather than ± 2 standard errors. Control charts that use 95% confidence intervals falsely indicate a failing process too often to be acceptable in most situations. The issue of multiple comparisons also creeps into most multiple regression analyses. If we construct several 95% confidence intervals, each has a 95% chance of covering its parameter. The chance that they all cover, however, is less than 95%. Conventional regression analysis ignores this issue since we consider the confidence intervals one at a time. In ANOVA regression, however, we consider all of the comparisons together in order to decide which categories are significantly different. In ANOVA regression, we must adjust the tests. (Stepwise regression, described in Statistics in Action. Automated Modeling automates the search for explanatory variables. Because stepwise regression uses the data to decide empirically which slopes to test, it too requires an adjustment for multiple testing.)

Tukey Confidence Intervals

Tukey confidence intervals Confidence intervals for the pairwise comparison of many means that adjust for multiple comparisons by replacing the *t*-percentile.

Statistics offers several ways to identify significant differences among a collection of averages. We will consider two. For the pairwise comparison of several sample averages, most software offers **Tukey confidence intervals** (called Tukey-Kramer intervals when the experiment is not balanced). These intervals hold the chance for a Type I error to 5% over the *entire* collection of confidence intervals. To guarantee this protection, Tukey confidence intervals replace the percentile $t_{.025,n-J}$ by a larger multiple of the standard error denoted $q_{.025,n,J}$. For example, the 95% Tukey confidence interval for the difference between any of the $J = 5$ average yields replaces $t_{.025,35} = 2.030$ by $q_{.025,40,5} = 2.875$. The adjustment for multiple comparisons increases the length of each confidence interval by about 40%. A table of $q_{.025,n,J}$ appears at the end of the chapter with other formulas.

The 95% Tukey confidence interval for the difference between the two best varieties, Endurance and Hatcher, is

$$2.04 \pm 2.875 \times 2.11 \approx 2.04 \pm 6.07 \text{ bushels/acre}$$

This difference is not statistically significant since the interval includes 0. As with *t*-intervals for the differences between means, the width of each Tukey interval is the same because the data are balanced across varieties. For example, the confidence interval for the difference between Endurance and RonL is

$$7.9 \pm 2.875 \times 2.11 \approx 7.9 \pm 6.07 \text{ bushels/acre}$$

This Tukey interval does not include zero: Endurance yields statistically significantly more wheat per acre than RonL.

The difference in yield between two varieties must be more than 6.07 bushels per acre in order to be statistically significant (otherwise the confidence interval includes 0). Referring to the differences between the means in Table 26.9, Endurance produces statistically significantly higher yields than NuHills and RonL. The other comparisons are not statistically significant. Although Endurance would be our choice, its advantage over Hatcher and Ripper could be due to random chance.

Bonferroni Confidence Intervals

Bonferroni confidence intervals Confidence intervals that adjust for multiple comparison by changing the α-level used in the standard interval to α/M for M intervals.

Bonferroni confidence intervals adjust for multiple comparisons but do not require special tables. The underlying approach is quite general and applies to any situation involving several inferences. Tukey's intervals are designed specifically for comparing many pairs of averages.

The underlying justification for Bonferroni intervals is simple. We found that the chance for a Type I error among M 95% confidence intervals could be as large as M (0.05). If we increase the coverage of each interval by setting $\alpha = 0.05/M$, then the chance for a Type I error among all of the intervals is no more than $M(0.05/M) = 0.05$. By reducing the chance for a Type I error for each interval to $0.05/M$, we guarantee that the chance for a Type I error among all of the intervals is no more than 0.05.

For the comparison among the wheat varieties, Bonferroni confidence intervals reduce $\alpha = .05$ to $\alpha/10 = .005$ and replace $t_{.025,35} = 2.08$ by $t_{.0025,35} = 3.00$. The multiplier used by the Tukey interval (2.875) lies between these. If you're interested in all pairwise comparisons, Tukey intervals are shorter than Bonferroni confidence intervals.

What Do You Think?

Table 26.10 summarizes the F-test for the comparison of therapies for low back pain, introduced previously. The value of s_e^2 is the residual mean square, 36.278.

TABLE 26.10 ANOVA table for the comparison of back therapies.

Source	df	Sum of Squares	Mean Square	F	p-Value
Regression	2	360.475	180.237	4.9682	0.0077
Residual	237	8597.925	36.278		
Total	239	8958.400			

a. What is the standard deviation of the residuals?[11]
b. Interpret the p-value = 0.0077.[12]
c. Construct Bonferroni confidence intervals to compare the therapies. Are any of the differences statistically significant?[13]
d. Find the 95% Tukey confidence intervals for the difference between the best and worst therapy. Are the Tukey intervals longer or shorter than those obtained by using the Bonferroni method?[14]

26.4 | GROUPS OF DIFFERENT SIZE

An ANOVA regression does not require an equal number of cases in each group. The calculation and interpretation of the associated regression and ANOVA table proceed as with balanced data. For instance, the intercept remains the mean of the baseline group, and the F-test compares the means of the J populations.

---

[11] s_e is the square root of the residual mean square, $s_e = \sqrt{36.278} \approx 6.023$.
[12] The F-statistic tests the null hypothesis that the groups have a common mean, $H_0: \mu_1 = \mu_2 = \mu_3$. Because $p < 0.05$, we reject H_0 and conclude that there is a statistically significant difference among the therapies.
[13] The critical t-value at $\alpha = .025/6 = .00416$ is approximately 2.66. Hence Bonferroni confidence intervals are (mean difference) $\pm$ (2.66 × 6.02/$\sqrt{40} \approx 2.53$). The biggest difference is statistically significant.
[14] From Figure 26.3, the largest difference is between motor control and regular exercise, $\bar{y}_2 - \bar{y}_1 = 6.89 - 4.09 = 2.8$. Tukey's 95% confidence interval $2.8 \pm q_{.025,240,3} s_e/\sqrt{2/80} = 2.8 \pm 2.359 \times 6.02/\sqrt{40} \approx 2.8 \pm 2.25$, does not include zero, so the difference is statistically significant. These intervals are shorter than those of Bonferroni. (If using the table at the end of this chapter, use the nearest percentile with a smaller sample size $q_{.025,150,3} = 2.368$.)

The effect of unequal group sizes is that we know more about the means of some categories than others. The standard error of an average of n observations is $\sigma/\sqrt{n}$. If one group has, say, 36 observations whereas another has 9, then the standard error of the mean of the larger group ($\sigma/6$) is half the size of the standard error of the mean of the smaller group ($\sigma/3$).

Because of the different sample sizes, unbalanced data produce confidence intervals of different lengths. The estimated standard error for the difference between $\bar{y}_1$ and $\bar{y}_2$ is, for instance,

$$\text{se}(\bar{y}_1 - \bar{y}_2) = s_e\sqrt{\frac{1}{n_1} + \frac{1}{n_2}}$$

If n_1 is much larger than n_2, then our lack of knowledge about the second sample obscures what we learn from the first. Let's consider an extreme example. If $n_1 = 625$ and $n_2 = 25$, then

$$\text{se}(\bar{y}_1 - \bar{y}_2) = s_e\sqrt{\frac{1}{625} + \frac{1}{25}} \approx 0.204 s_e$$

That's virtually the same as the standard error of $\bar{y}_2\,(0.2s_e)$. Consequently, the lengths of t, Tukey, or Bonferroni confidence intervals depend on which means are being compared. Comparisons between categories with many observations produce narrower intervals than comparisons between categories with fewer observations. The lengths of confidence intervals obtained from balanced data are all the same.

The second consequence of unbalanced group sizes is that Tukey intervals are no longer exact. The percentile $q_{.025,n,J}$ assumes balanced data. Nonetheless, one can still use Tukey's intervals with unbalanced data; the intervals hold the chance of a Type I error to no more than 5%. Bonferroni intervals are not affected by unequal sample sizes.

caution The final consequence of unbalanced data is that tests are more sensitive to a lack of constant variation across the several groups. The overall F-test, for example, is known to provide reliable inferences with balanced samples even if the variances are moderately different across the groups. This robustness to a deviation from the assumed conditions weakens as the sample sizes become more disparate.

The bottom line for unbalanced data, then, is simple: ANOVA regression works, but you need to be more careful than in the analysis of the data from a balanced experiment. The assumptions become more important and the details become more complicated.

4M EXAMPLE 26.1 | JUDGING THE CREDIBILITY OF ADVERTISEMENTS

MOTIVATION | STATE THE QUESTION

To launch its first full-sized pickup truck in the United States, Toyota spent $100 million advertising the Toyota Tundra. The edgy ads featured stunts so dangerous that viewers thought they were computer generated. In one ad, a truck speeds through a barrier and then skids to a stop inches before falling into a canyon. Impressive, but do such ads influence consumers?

For this analysis, advertising executives want to compare four commercials for a retail item that make claims of varying strength. How over-the-top can commercials become before the ads go too far and customers turn away in disbelief? Does the strength of the claim in the advertisement affect consumer reactions?

METHOD

DESCRIBE THE DATA AND SELECT AN APPROACH

The data for this analysis are the reactions of a random sample of 80 customers to the advertising. Each customer was randomly assigned to see one of the four commercials. Each group saw a different ad, with ads having different degrees of claims. The claims in the ads are rated as "Tame," "Plausible," "Stretch," and "Outrageous." These levels define a categorical variable.

After seeing an ad, each subject completed a questionnaire. The questionnaire included several items that asked whether the customer believed the ad. By combining these responses, analysts created a single numerical variable (called Credibility) that is positive if the subject believes the claim and is negative otherwise.

A statistically significant analysis of variance will confirm that the strength of the claim matters in how customers react to these ads. We can then use Tukey's method to identify how their responses differed in the various categories.

✓ **Linear.** This is not an issue in an analysis of variance.

✓ **No obvious lurking variable.** The use of randomization in the design of the study makes this one automatic. Otherwise, we would have to be concerned about whether the customers in the four groups were truly comparable but for having seen different ads.

MECHANICS

DO THE ANALYSIS

The remaining conditions depend on residuals, so we'll cover them after fitting the ANOVA regression. The regression includes three dummy variables that identify the "Plausible," "Stretch," and "Outrageous" categories. The baseline category is "Tame," so the slopes in the regression will be average differences from the mean of this category. Here's the fitted regression of credibility on these three dummy variables.

R^2	0.115
s_e	3.915
n	80

Source	df	Sum of Squares	Mean Square	F	p-Value
Regression	3	151.35	50.4502	3.2914	0.0251
Residual	76	1,164.92	15.3280		
Total	79	1,316.28			

Term	Estimate	se	t-Statistic	p-Value
Intercept	0.66	0.88	0.75	0.4532
D(Plausible)	−1.125	1.24	−0.91	0.3664
D(Stretch)	−0.73	1.24	−0.59	0.5572
D(Outrageous)	−3.655	1.24	−2.95	0.0042

Check the rest of the conditions before considering inference.

✓ **Evidently independent.** Each customer watched an ad in a separate cubicle and filled out the form individually. Had the customers been in the same room or discussed the ads together before filling in the questionnaire, they might have influenced each other.

✓ **Similar variances.** The boxplots of the residuals grouped by which ad was seen suggest comparable variation. The boxes are of similar length.

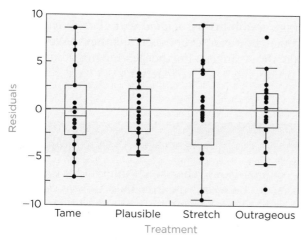

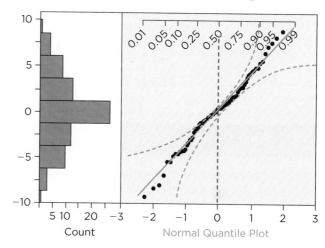

✓ **Nearly normal.** The normal quantile plot looks okay. The residuals stay close enough to the diagonal reference line.

Because the conditions are met, we can make inferences. Start with the F-test. The overall *F*-test has *p*-value 0.0251, which is less than 0.05. Hence, we reject H_0 that the mean credibility of the four ads is equal. *Now interpret the individual estimates, noting the correspondence between slopes and means.* The intercept 0.66 is the mean of credibility for the "Tame" ads. Other slopes estimate the differences between the mean of a group and the "Tame" group. For example, the mean credibility of "Outrageous" ads was 3.655 less than that of "Tame" ads.

Since the F-test is statistically significant, proceed to multiple comparisons using Tukey intervals. (If the F-test is not significant, there are no statistically significant differences to identify.) This table gives the mean credibility ratings within the four groups.

Level	Mean
Tame	0.660
Plausible	−0.465
Stretch	−0.070
Outrageous	−2.995

Because each category has 20 subjects, the estimated standard error for any pairwise comparison of means is $\sqrt{2/20} = 3.915 \sqrt{0.1} \approx 1.238$. Using the table at the end of this chapter, $q_{.025,nJ} = q_{0.025,80,4} = 2.627$. Hence, every pairwise interval is of the form

$$\bar{y}_a - \bar{y}_b \pm (2.627 \times 1.238 \approx 3.25)$$

The only Tukey interval that does not include 0 is the comparison of "Tame" ads to "Outrageous" ads: $\bar{y}_{\text{tame}} - \bar{y}_{\text{out}} \pm 3.25 = (0.66 - (-3)) \pm 3.25 = -3.66 \pm 3.25$.

The Bonferroni percentile is $t_{.025/6,76} = 2.71$ ($_4C_2 = 6$), implying that Bonferroni confidence intervals are slightly longer intervals than those produced by Tukey's procedure. These approaches reach the same conclusion: only the difference between "Tame" and "Outrageous" ads is statistically significant.

MESSAGE | SUMMARIZE THE RESULTS

Results from an experiment with 80 randomly selected consumers suggest avoiding outrageous claims in ads; customers don't believe them. A one-way analysis of variance finds customers place less credibility in ads that make outrageous claims than ads that make tame claims. In general, the stronger the claim, the less the average credibility associated with the ad. The only statistically significant difference, however, is between ads that make the weakest claims and ads that make the strongest. Further testing might identify how far ads can reach without producing negative reactions.

Best Practices

- *Use a randomized experiment to obtain your data.* Spend the effort up front to get experimental data. Comparisons with experimental data are straightforward because you do not have to be concerned with confounding variables as in Chapter 25.
- *Check the assumptions of multiple regression when using ANOVA regression.* An analysis of variance is a regression, so the same assumptions of independence, equal variance, and normality apply. If you've run an experiment, then there's no lurking factor to worry about either.
- *Use Tukey or Bonferroni confidence intervals to identify groups that are statistically significantly different.* These methods allow for multiple comparisons. The usual *t*-interval for the difference between means does not; it assumes you're considering only one confidence interval, not several. Without the adjustment, there's a good chance that you'll declare a difference statistically significant just because you've made lots of comparisons.
- *Recognize the cost of snooping in the data to choose hypotheses.* If your experiment includes many groups, then you'll pay a stiff fee (in the form of longer confidence intervals) because of doing many comparisons. The more intervals you construct when looking for something statistically significant, the wider each must become in order to hold the chance for a Type I error to 5%.

Pitfalls

- *Don't compare the means of several groups using lots of t-tests.* Use a single ANOVA regression instead. A two-sample *t*-test is made for comparing two groups. ANOVA regression is made for comparing the means of two or more groups.
- *Don't forget confounding factors.* Unless your data are from a randomized experiment, there could be other explanations for the differences aside from the labels attached to the groups. No plot is going to show this effect—you have to identify possible lurking variables, get the relevant data, and then look to see whether these explain the differences. Make sure that your data are from an experiment before you treat them as if they were.
- *Never pretend you only have two groups.* If you're comparing men to women or employed to unemployed, then there are two categories. If you run an experiment with 10 treatments and only show the two-sample analysis of the best versus the worst, you're cheating. Don't cherry-pick the two groups that are farthest apart and do a two-sample *t*-test—that's not a legitimate test. You need to do the one-way ANOVA regression, check the *F*-test, and then—if that test is statistically significant—use intervals that adjust for multiplicity.
- *Do not add or subtract standard errors.* The standard error for the difference between two

sample averages is the square root of the sum of their variances. Use the formula

$$\text{se}(\bar{y}_j - \bar{y}_k) = s_e \sqrt{\frac{1}{n_j} + \frac{1}{n_k}}$$

rather than adding or subtracting the standard errors. Variances of independent averages add, not standard errors.

- *Do not use a one-way analysis of variance to analyze data with repeated measurements.* Sometimes there's a reason that the groups have the same number of observations: They're the same observations! If your data have repeated measurements on the same person, place, or thing, you've got what's known as a randomized block experiment. Because the same subjects appear in different groups, such data violate the assumption of independence and require a different type of analysis.

Software Hints

Because the analysis of variance is a regression using categorical variables, you can avoid special tools. To use the software commands from prior chapters, build dummy variables for each group. Alternatively, some software will build an analysis of variance for you without needing you to construct the dummy variables. If you've already got them, that's not such a big deal, but it's helpful if you have several categories.

EXCEL

Use the command

    Data > Data Analysis... > Anova: Single Factor

Excel expects the data for each group to be in a separate column (or row), with an optional label for the group appearing in the first row (or column) of the range. (This is not the standard "Number of rows = n" arrangement we have been using. The groups do not have to be the same size; leave empty cells if the groups are not of the same size.) Excel computes the mean for each group and includes the overall *F*-statistic.

MINITAB

Use the command

    Stat > ANOVA > One-Way...

or

    Stat > ANOVA > One-Way(Unstacked)...

The choice depends on the arrangement of the data. If your data are in two columns with n rows (a numerical Y variable and the categorical variable that defines the groups), pick the first command. The second command accommodates the arrangement required by Excel. The default output includes a table of means and standard deviations along with an overall summary of statistical significance.

JMP

The menu sequence

    Analyze > Fit Y by X

with a categorical numerical variable as Y and a variable as X results in a one-way analysis of variance. Click the red triangle by the header in the output window whose title begins "Oneway Analysis of ..." and choose the option Means/ANOVA/Pooled t (two groups) or Means/ANOVA (if there are more than two groups). JMP adds the locations of the means to the scatterplot of the data and summarizes the group statistics in a table below the plot. (The points in the plot are located along the *y*-axis defined by the response in columns over the labels that identify the groups.)

BEHIND the Math

Within and Between Sums of Squares

The Section "Behind the Math: The ANOVA Table and the *F*-Statistic" in Chapter 23 describes the calculations in the ANOVA table in general terms. Table 26.11 also describes calculations that produce Table 26.6, but specializes these to the context of an ANOVA regression using the tabular notation for the data introduced in this chapter. In this table, J denotes the number of groups, n_j is the number of cases in the *i*th category, and n denotes the total number of observations.

TABLE 26.11 Components of the analysis of variance table for a one-way analysis of variance.

Source	DF	Sum of Squares	Mean Square	F	p-Value
Regression	$J - 1$	$\sum_{j=1}^{J} \sum_{i=1}^{n_j} (\bar{y}_j - \bar{y})^2$	$\dfrac{\text{Regression SS}}{J - 1}$	$\dfrac{\text{Regression Mean Square}}{\text{Residual Mean Square}}$	Test of H_0 that means are equal
Residual	$n - J$	$\sum_{j=1}^{J} \sum_{i=1}^{n_j} (y_{ij} - \bar{y}_j)^2$	$\dfrac{\text{Residual SS}}{n - J} = s_e^2$		
Total	$n - 1$	$\sum_{j=1}^{J} \sum_{i=1}^{n_j} (y_{ij} - \bar{y})^2$			

The mean square is the sum of squares divided by the column labeled df (degrees of freedom). The residual sum of squares is the sum of the squared residuals divided by $n - J$, what we usually call s_e^2. The F-statistic is the ratio of the two mean squares.

$$F = \frac{\text{Regression SS}/(J - 1)}{\text{Residual SS}/(n - J)} = \frac{\text{Regression SS}/(J - 1)}{s_e^2}$$

The denominator is our usual estimate of σ_ε^2, the error variance. In an ANOVA regression, the residuals are deviations of the responses from the means of the categories. Hence, the residual sum of squares measures the variation within the categories; in an ANOVA regression, the residual sum of squares is often called the "within sum of squares." Analogously, the regression sum of squares is often called the "between sum of squares." In an ANOVA regression,

the fitted values are averages of the groups ($\hat{y}_{ij} = \bar{y}_j$). As a result, the sum of the squared deviations of the fitted values around the overall mean for the analysis of wheat yields is

$$\text{Regression SS} = \sum_{j=1}^{5} \sum_{i=1}^{8} (\hat{y}_j - \bar{y})^2$$

$$= 8 \sum_{j=1}^{5} (\bar{y}_j - \bar{y})^2$$

This sum adds up the squared deviations of the group averages from the overall average, earning it the name "between sum of squares." The double sum simplifies because $\hat{y}_{ij} = \bar{y}_j$; the fitted value for every observation in the jth category is the same, $\bar{y}_j$.

CHAPTER SUMMARY

A **balanced** experiment assigns an equal number of subjects to each treatment. Side-by-side boxplots are useful to visually compare the responses in the groups. Experimental data are often arranged in a tabular notation with subscripts to identify observations and categories. To judge the statistical significance of differences among the means of several groups, a **one-way analysis of variance** (**ANOVA**, or ANOVA regression) regresses the response on dummy variables that represent the treatment groups. The regression requires dummy variables

for all but one group. The intercept in the regression is the mean of the baseline category, and each slope estimates the difference between the mean of a group and the baseline group. The analysis of variance table summarizes the statistical significance of the model. In general, t-statistics for individual slopes are inappropriate because of **multiplicity**. If the overall F-test is statistically significant, then **Tukey confidence intervals** or **Bonferroni confidence intervals** are used to identify which pairs of means are significantly different.

Key Terms

Formulas

Notation for data

y_{ij} is the response for the ith observation in the jth group. There are J groups, with n_j observations in the jth group. The total number of observations is

$$n, n = \sum_{j=1}^{J} n_j.$$

Fitted values and residuals

Fitted values are means of the groups, $\hat{y}_{ij} = \bar{y}_j$.

Residuals are the deviations $e_{ij} = y_{ij} - \bar{y}_j$.

Standard error for the difference between two means

$$\text{se}(\bar{y}_j - \bar{y}_k) = s_e \sqrt{\frac{1}{n_j} + \frac{1}{n_k}} \quad \left(= s_e \sqrt{\frac{2}{n_j}} \text{ if } n_j = n_k \right)$$

This is also the standard error for the slope of a dummy variable in a one-way ANOVA because the slope is the difference between two means.

Tukey percentiles $q_{.025,n,J}$

Use percentiles from the following table, $q_{.025,n,J}$, in place of $t_{.025,n-J}$ to adjust for multiplicity when forming a collection of 95% confidence intervals in an ANOVA regression with n_j observations in each of J groups ($n = n_j J$).

If the sample is not balanced, set $n_j = n/J$ and round to the nearest integer.

	$q_{.025,n,J}$					
	Number of Groups (J)					
n_j	**2**	**3**	**4**	**5**	**10**	**20**
2	4.299	4.179	4.071	4.012	3.959	4.040
3	2.776	3.068	3.202	3.291	3.541	3.788
4	2.447	2.792	2.969	3.088	3.411	3.706
5	2.306	2.668	2.861	2.992	3.348	3.665
6	2.228	2.597	2.799	2.937	3.310	3.641
7	2.179	2.552	2.759	2.901	3.285	3.625
8	2.145	2.521	2.730	2.875	3.268	3.613
9	2.120	2.497	2.709	2.856	3.255	3.604
10	2.101	2.479	2.693	2.841	3.244	3.598
20	2.024	2.406	2.627	2.781	3.202	3.569
30	2.002	2.384	2.607	2.762	3.189	3.560
50	1.984	2.368	2.591	2.748	3.178	3.554
100	1.972	2.356	2.580	2.738	3.171	3.549
1,000	1.961	2.345	2.570	2.729	3.164	3.544

About the Data

Data for the civilian workforce described in the introduction to the chapter come from *Historical Statistics of the United States: Colonial Times to 1970* (D-16, D-17, K-445, K-448) and the *Statistical Abstract of the United States* (Table 587). Both are available from the U.S. Census Bureau. Additional figures are from the *Census of Agriculture* produced by the U.S. Department of Agriculture. Be careful about measurements reported in bushels. As a unit of volume, a bushel is equivalent to 35.24 liters. As a weight, a standard bushel of wheat weighs 60 pounds. Bushels of other foods have different weights. A standard bushel of apples weighs 48 pounds but a bushel of spinach weighs 20 pounds. The wheat data are a small subset from the Colorado Winter Wheat Performance Trials in 2007. These experiments annually test 30 or more varieties of wheat at various locations and conditions around the state. Numerous other universities conduct similar agricultural experiments.

The data in Example 26.1 are derived from a paper by M. Goldberg and J. Hartwick (1990), "The Effects of Advertiser Reputation and Extremity of Advertising Claim on Advertising Effectiveness," *Journal of Consumer Research*, **17**, 172–179.

EXERCISES

Mix and Match

Match each term from an ANOVA regression on the left to its symbol on the right. These exercises use the abbreviations SS for sum of squares and MS for mean squares.

1. Observed response		(a) b_0	
2. Number of cases in jth group		(b) $\mu_1 = \mu_2 = \cdots = \mu_j$	
3. Fitted value		(c) y_{ij}	
4. Residual		(d) (Regression MS)/(Residual MS)	
5. Mean of data in omitted category		(e) (Regression SS)/(Total SS)	
6. Difference of two sample means		(f) n_j	
7. Difference of two population means		(g) $\mu_1 - \mu_2$	
8. Null hypothesis of F-test		(h) $\bar{y}_j$	
9. F-statistic		(i) b_1	
10. R^2 in disguise		(j) $y_{ij} - \bar{y}_j$	

True/False

Mark each statement True or False. If you believe that a statement is false, briefly explain why you think it is false.

11. A balanced experiment does not benefit from the use of randomization to assign the treatments to the subjects.

12. The one-way analysis of variance requires balanced data, with an equal number of observations in each group.

13. A two-sample t-test that pools the variances is equivalent to a simple regression of the response on a single dummy variable.

14. The intercept in a regression of y on a dummy variable x is the difference between the mean of y for observations with $x = 0$ and the mean of y for observations with $x = 1$.

15. A regression model that uses only dummy variables as explanatory variables is known as an analysis of variance.

16. A fitted value $\hat{y}$ in a one-way ANOVA is the mean for some group defined by the explanatory dummy variables.

17. The F-test in an ANOVA tests the null hypothesis that all of the groups have equal variance.

18. The average of the residuals within a category used in an ANOVA is zero.

19. The within mean square in an ANOVA is another name for s_e^2, the sample variance of the residuals in the corresponding regression.

20. The F-statistic depends upon which dummy variable defined by a categorical variable is excluded from the regression model.

21. Tukey confidence intervals for the difference between means in an ANOVA are narrower than the corresponding Bonferroni t-intervals.

22. Bonferroni confidence intervals adjust for the effect of multiple comparisons.

Think About It

23. Does the p-value of the two-sample t-test that does not assume equal means match the p-value of the slope on a regression of Y on a dummy variable?

24. Suppose that the subjects in an experiment are reused. For example, each person in a taste test samples every product. Are these data suitable for a one-way ANOVA?

25. A company operates in the United States, Europe, South America, and the Pacific Rim. Management is comparing the costs incurred in its health benefits program by employees across these four regions. It fit an ANOVA regression of the amount spent for samples of 25 workers in each of the four regions. The following table summarizes the estimates.

Analysis of Variance					
Source	df	Sum of Squares	Mean Square	F	p-Value
Regression	3	58,072,149	19,357,383	131.1519	<0.0001
Residual	96	14,169,131	147,595.11	p	
Total	99	72,241,280			

Parameter Estimates				
Term	Estimate	Std Error	t	p
Intercept	1,795.05	76.84	23.36	<0.0001
D(US)	1,238.10	108.66	11.39	<0.0001
D(Europe)	717.84	108.66	6.61	<0.0001
D(Pacific)	−785.58	108.66	−7.23	<0.0001

This analysis was done in dollars. What would change and what would be the same had the analysis been done in euros? (Assume for this exercise that 1 euro = 1.5 U.S. dollars.)

26. The analysis in Exercise 25 uses South America as the omitted category. What would change and what would be the same had the analysis used the United States as the omitted reference category?

27. Consider the data shown in the following plot. Each group has 12 cases. Do the means appear statistically significant? Estimate the p-value: Is it about 0.5, about 0.05, or less than 0.0001?

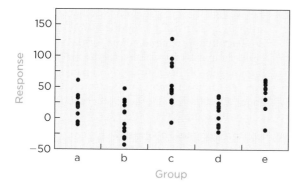

28. Consider the data shown in the following plot. Each group has 12 cases. Do the means appear statistically significant? Estimate the p-value: Is it about 0.5, about 0.05, or less than 0.0001?

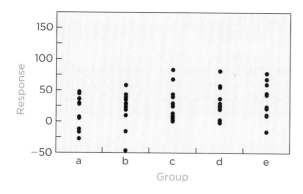

29. It can be shown that for any data $y_1, y_2, \ldots, y_n$ the smallest value for

$$T = \sum_{i=1}^{n} (y_i - M)^2$$

is obtained by setting $M = \bar{y}$. Explain why this implies that the fitted values in ANOVA are the sample averages of the groups.

30. A Web site monitors the number of unique customer visits, producing a total for each day. The following table summarizes the totals by day of the week, averaged over the last 12 weeks. (For example, during this 12-week period, the site averaged 2,350 visitors on Mondays.)

Day	Average Number of Visits
Monday	2,350
Tuesday	2,530
Wednesday	2,190
Thursday	1,940
Friday	2,270
Saturday	3,100
Sunday	2,920

A regression model regressed the number of visits on six dummy variables, representing the days Monday through Saturday (omitting Sunday).
(a) What is the estimated intercept b_0 in the regression?
(b) What is the slope of the dummy variable that represents Monday?
(c) Are the differences among the days statistically significant?

31. Rather than create five dummy variables to represent a categorical variable C with five labels, an analyst defined the variable X by converting the categories to the numbers 1, 2, 3, 4, and 5. Does the regression of Y on X produce the same results as a regression of Y on four dummy variables that represent four of the categories of C?

32. A line of men's shirts was offered in a chain of retail stores at three prices: \$32, \$35, and \$40. Weekly sales were monitored, producing totals at 30 stores in the chain (10 at each price). Which produces a higher R^2: a linear regression of sales on price or an analysis of variance of sales grouped by price (treating price as categorical)?

33. Suppose an ANOVA meets the conditions of the MRM and the F-test rejects the overall null hypothesis that four groups have equal means. Group 1 has the largest sample mean and Group 4 has the smallest. Does the confidence interval for $\mu_1 - \mu_4$ contain 0?

34. Suppose an ANOVA meets the conditions of the MRM and the F-test rejects the overall null hypothesis that five groups have equal means. If the Bonferroni confidence interval (adjusted for pairwise comparisons) for $\mu_1 - \mu_2$ does not include zero, does the Tukey confidence interval for $\mu_1 - \mu_2$ include zero?

35. A research chemist uses the following laboratory procedure. He considers the yield of 12 processes that produce synthetic yarn. He then conducts the two-sample t-test with $\alpha = 0.05$ between the process with the lowest yield and the process with the highest yield. What are his chances for a Type I error?

36. A modeler has constructed a multiple regression with $k = 10$ explanatory variables to predict costs to her firm of providing health care to its employees. To decide which of the 10 explanatory variables is statistically significant, she rejects $H_0: \beta_j = 0$ if the p-value for the t-statistic of a slope is less than 0.05. What is her chance for making a Type I error?

37. The ANOVA in this chapter compares the yield of 5 varieties of wheat. In fact, the Colorado trials involved 54 varieties! Suppose our grower in eastern Colorado was interested in comparisons among all 54 varieties, not just 5. (Assume 8 plots for each variety, so $n = 8 \times 54 = 432$.)
    (a) How many dummy variables would be needed as explanatory variables in the regression?
    (b) Suppose that $s_e = 4.217$ bushels per acre, as in the text analysis of five varieties. With so much more data, are the Tukey and Bonferroni confidence intervals for the differences in average yields longer or shorter than those in the analysis of five varieties?

38. The 95% confidence interval for μ derived from a sample of n observations from a normal population gets smaller as the sample size increases because the standard error of the mean $s/\sqrt{n}$ decreases. Adding more groups to an ANOVA also increases n.
    (a) Does adding more groups improve the accuracy of the estimated mean of the first group?
    (b) Tukey and Bonferroni intervals for the difference between means get longer as more groups are added, even though n increases. Why?

You Do It

39. A marketing research analyst conducted an experiment to see whether customers preferred one type of artificial sweetener over another. In the experiment, the analyst randomly assigned 60 soda drinkers to one of four groups. Each group tried a cola drink flavored with one of four sweeteners and rated the flavor.
    (a) Fill in the missing cells of the ANOVA summary table.

Source	df	Sum of Squares	Mean Square	F	p-Value
Regression		150			0.0077
Residual		800			
Total					

    (b) State and interpret the null hypothesis tested by the F-test in the analysis of this experiment.
    (c) Does the F-test reject this null hypothesis?
    (d) Assuming that the data meet the required conditions, what can we conclude from the F-test about the preferences of consumers?

40. An overnight shipping firm operates major sorting facilities (hubs) in six cities. To compare the performance of the hubs, it tracked the shipping time (in hours) required to process 20 randomly selected priority packages as they passed through each hub.
    (a) Fill in the missing cells of the ANOVA summary table.

Source	df	Sum of Squares	Mean Square	F	p-Value
Regression		600			0.5104
Residual		16,000			
Total					

    (b) State and interpret the null hypothesis tested by the F-test in the analysis of this experiment.
    (c) Does the F-test reject this null hypothesis?
    (d) What is the implication for management of the F-test? Assume that the data meet the required conditions.

41. An insurance company has offices in all 50 states and the District of Columbia. It plans to use an ANOVA to compare the average sales (in dollars) generated per agent among the states and DC. To simplify the analysis, assume that it has an equal number of agents in each state (20 per state).
    (a) If $s_e = \$3,500$, then how different must the average sales per agent in one state be from the average sales per agent in another state in order to be statistically significant? Use the Bonferroni approach to adjust for the effect of multiplicity.
    (b) If managers of the insurance company believe that average sales per agent in most states are within about $2,000 of each other, is there much point in doing this study?

42. A real estate company operates 25 offices scattered around the southeastern United States. Each office employs six agents. Each month, the CEO receives a summary report of the average value of home sales per agent for every office. Because of the high volatility of real estate markets, there's a lot of residual variation ($s_e = \$285,000$). In the current report, the office in Charleston, SC generated the largest average sales at $2,600,000; the office in Macon, GA had the smallest at $2,150,000. Assuming the ANOVA model is reasonable for these data, is this a large range in sales, or should the CEO interpret these differences as the result of random variation?

43. Rather than use a dummy variable (D, coded as 1 for men and 0 for women) as the explanatory variable in a regression of responses of men and women, a model included an explanatory variable X coded as $+1$ for men and -1 for women. (This type of indicator variable is sometimes used rather than a dummy variable.)
    (a) What is the difference between the scatterplot of Y on X and the scatterplot of Y on D?
    (b) Does the regression of Y on D have the same R^2 as the regression of Y on X?
    (c) In the regression of Y on X, what is the fitted value for men? For women?
    (d) Compare b_0 and b_1 in the regression of Y on D to b_0 and b_1 in the regression of Y on X.

44. A department store sampled the purchase amounts (in dollars) of 50 customers during a recent Saturday sale. Half of the customers in the sample used coupons, and the other half did not. The data identify the group using a variable coded as -1 for those who did not use a coupon and $+1$ for those who did use a coupon. The plot at the top of the next page shows the data along with the least squares regression line.

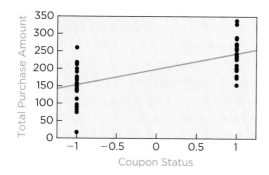

Term	Estimate	Std Error	t-Ratio	p
Intercept	196.57	7.07	27.81	<0.0001
Coupon Status	44.14	7.07	6.24	<0.0001

(a) Interpret the estimated intercept, slope, and value of s_e.

(b) Should managers conclude that customers who use coupons spend statistically significantly more than those who do not?

(c) Suppose the comparison had been done using a pooled two-sample t-test. What would be the value of the t-statistic?

(d) Suppose the comparison had been done using a dummy variable (coded as 1 for coupon users and 0 otherwise) rather than the variable Coupon Status. Give the values of b_0, b_1, and the t-statistic for the estimated slope.

45. **Bread Volume** We can often use an analysis of variance with data that are not well matched to a linear regression, using dummy variables to avoid the need to model a curved relationship. In this example, a bakery ran an experiment to measure the effect of different recipes on the volume of commercial bread loaves (measured in milliliters). Customers generally prefer larger-looking loaves rather than smaller loaves, even if the net weights are the same. In this experiment, the bakery varied the level of potassium bromate from 0 to 4 milligrams.[15]

(a) Consider the linear regression of loaf volume on the amount of potassium bromate. Do the data appear to meet the assumptions required by the simple regression model?

(b) Do the data appear to meet the conditions required for fitting an ANOVA? (Determine this visually, before fitting the ANOVA.)

(c) Fit an ANOVA by treating the amounts of potassium bromate as identifying five categories. Can the bakery affect the volume of its loaves by varying the amount of this ingredient?

(d) If your software provides it, give the 95% Tukey confidence interval for the difference in volume produced with no potassium bromate and 2 milligrams of potassium bromate. Is there a statistically significant difference in volume?

(e) Give the 95% Bonferroni confidence interval for the difference in volume produced with no potassium bromate and 2 milligrams of potassium bromate. Is there a statistically significant difference in volume?

(f) Explain any differences between the conclusions of part (c) and the conclusions of parts (d) and (e).

46. **Display Space** Because an ANOVA does not presume a linear trend, it can be used to check for deviations from the straight-enough condition. The procedure requires replicated observations; we need several values of y at each value of x. The response is the weekly sales of a beverage in 47 stores, and the explanatory variable is the number of feet of shelf space used to display the product.

(a) Fit the linear regression of sales on number of feet of shelf space. Does the relationship meet the straight-enough condition?

(b) Build six dummy variables to represent the values of the explanatory variable (1, 2, 3, ..., 6 with 7 excluded). The dummy variable D_1 identifies stores displaying the product on 1 foot of shelf space, D_2 identifies those with 2 feet, and so forth. Fit the multiple regression of the residuals from the simple regression in part (a) versus the six variables $D_1, D_2, ..., D_6$. Summarize the fit.

(c) Does the regression of the residuals on the dummy variables explain statistically significant amounts of variation in the residuals? Should it?

47. **Stopping Distances** The U.S. Department of Transportation (DOT) tests the braking abilities of various cars. For this experiment, DOT measured the stopping distances on dry pavement at 100 kph (kilometers per hour, about 62 mph) of four models: a Chevrolet Malibu, Toyota Camry, Pontiac Grand Am, and Cadillac DeVille. Each car accerlated to 100 kilometers per hour on a test track; then the brakes were applied. The distance required to come to a full stop was measured, in meters. The test was repeated for each car 10 times under identical conditions.

(a) Plot the data. From your visual inspection of the plot, do you think there are statistically significant differences among these four cars?

(b) Fit a multiple regression of stopping distance on three dummy variables that identify the Malibu, Grand Am, and Cadillac. Interpret the estimated intercept and slopes.

(c) Does a statistical test agree with your visual impression? Test the null hypothesis that the four cars have the same stopping distance.

(d) These stopping distances were recorded in meters. Would the analysis change had the distances been measured in feet instead?

(e) Do these data meet the conditions required for an ANOVA?

(f) Based on these results, what should we conclude about differences in the stopping distance of other cars of these models?

[15] Data used in the classic text *The Analysis of Variance* by H. Scheffé (Wiley, 1959).

48. **Health Costs** The Human Resources (HR) office of a company is responsible for managing the cost of providing health care benefits to employees. HR managers are concerned that health care costs differ among several departments: administration, clerical, manufacturing, information technology (IT), and sales. Staff selected a sample of 15 employees of each type and carefully assembled a summary of health care costs for each.
    (a) Plot the data. From your visual inspection of the plot, do you think there are statistically significant differences among these five departments?
    (b) Fit a multiple regression of health care cost on four dummy variables that identify employees in administration, clerical, IT, and manufacturing. Interpret the estimated intercept and slopes.
    (c) Does a statistical test agree with your visual impression? Test the null hypothesis that costs are the same in the four departments.
    (d) Do these data meet the conditions required for an ANOVA?
    (e) Assuming the data are suitable for ANOVA, is there a statistically significant difference between the costs in the two most costly departments?
    (f) Nationally, the cost of supplying health benefits averages $4,500. Assuming these data are suitable for ANOVA, should the HR managers conclude that any of these departments have particularly high or low costs?

Exercises 49 and 50 illustrate some of the consequences of data with unequal group sizes.

49. **Movie Reviews** Movie studios often release films into selected markets and use the reactions of audiences to plan further promotions. In these data, viewers rate the film on a scale that assigns a score from 0 (dislike) to 100 (great) to the movie. The viewers are located in one of three test markets: urban, rural, and suburban. The groups vary in size.
    (a) Plot the data. Do the data appear suited to ANOVA?
    (b) From your visual inspection, do differences among the average ratings appear large when compared to the within-group variation?
    (c) Fit a multiple regression of rating on two dummy variables that identify the urban and suburban viewers. Interpret the estimated intercept and slopes.
    (d) Are the standard errors of the slopes equal, as in the text example of wheat yields? Explain why or why not.
    (e) Does a statistical test agree with your visual impression of the differences among the groups? Test the null hypothesis that the ratings are the same in the three markets.
    (f) Do these data meet the conditions required for an ANOVA?
    (g) What conclusions should the studio reach regarding the prospects for marketing this movie in the three types of markets?

50. **Song Lengths** This dataset gives the playing time (in seconds) of a sample of 639 songs grouped by genre (as classified by the online seller).

Genre	Number of Songs
Blues	19
Country	96
Folk	63
Jazz	192
Latin	22
Rock	247

(a) From your visual inspection of the song lengths, do differences among the average lengths appear large when compared to the variation within genre?
(b) Fit a multiple regression of rating on five dummy variables that represent the blues, country, folk, jazz, and Latin genres. Interpret the estimated intercept and slopes.
(c) Which estimated slope (exclude the intercept) has the smallest standard error? Explain why this standard error is smaller than those of the other slopes.
(d) Does a statistical test agree with your visual impression of the differences among the groups? Test the null hypothesis that the song lengths are the same in the six genres.
(e) The absolute value of the t-statistics of two estimated slopes in the multiple regression is smaller than 2. Why are these estimates relatively close to zero?
(f) A media company recently acquired a station that was playing rock music. The station can work in more ads if the songs it plays are shorter. By changing to another genre, can the company reduce the typical length of songs? Which one(s)?

4M Credit Risk

Banks often rely on an "originator" that seeks out borrowers whom it then connects with the bank. An originator develops local customers, solicits applications, and manages the review process. By relying on an originator to generate new loans, a bank reduces its need for expensive branch offices, yet retains a large, diverse portfolio of loans. The originator receives a percentage of the face value of the loan at the time the loan begins.

Unless a bank is careful, however, an originator may not be thorough in its review of applications; the loans may be riskier than the bank would prefer. To monitor the origination process, the bank in this analysis has obtained the credit scores of a sample of 215 individuals who applied for loans in the current week. A credit score is a numerical rating assigned to a borrower based on his or her financial history. The higher the score, the less likely it is that the borrower will default on the debt. The bank would like for the credit score of loans in this portfolio to average 650 or more.

This sample was drawn randomly from applications received from five originators identified for confidentiality as "a," "b," "c," "d," and "e." (The bank cannot audit the credit history of every loan; if it did, it would reproduce the work of the originator and incur the associated costs.)

Motivation

a) The bank expects loans from these originators to obtain an average credit score of at least 620. Why is it necessary for the bank to monitor a sample of the credit scores of the loans submitted by originators?

Method

b) How can the bank use this sample to check the objective of an average credit score of at least 620?

c) Even if all of the originators meet the objective of a 620 average credit score, some may be producing better loans than others. How can the bank decide?

Mechanics

d) Summarize the results of the estimated ANOVA model. Include the F-test and estimates from the regression using four dummy variables.

e) Do these data appear suited to modeling with an ANOVA?

f) Assume that the data meet the conditions of the MRM. Do loans from any originator average less than 620 by a statistically significant amount? Be sure to adjust appropriately for multiplicity.

g) Assume that the data meet the conditions of the MRM. Using confidence intervals that have been adjusted for multiplicity, are there any statistically significant differences among the credit scores of these samples?

Message

h) Summarize the results of the ANOVA for the management of the bank. Should the bank avoid some of the originators? Are some better than others?

i) The R^2 for the underlying regression is "only" 9%. Isn't that too low for the model to be useful? Explain why or why not.

27

CHAPTER

Time Series

27.1 **DECOMPOSING A TIME SERIES**

27.2 **REGRESSION MODELS**

27.3 **CHECKING THE MODEL**

CHAPTER SUMMARY

THE VALUE OF GOODS SHIPPED BY A BUSINESS IS AN IMPORTANT MEASURE OF ITS HEALTH, BOTH INSIDE AND OUTSIDE THE COMPANY. If shipments dip unexpectedly, investors may dump their shares and employees grow anxious about job security. Shipments can tell other businesses what to expect as well. Fewer shipments of computers, for instance, mean fewer sales of Microsoft's latest software and fewer service calls for those who install and maintain systems.

Figure 27.1 tracks the value of monthly shipments of computers and electronic products, in billions of dollars, from January 1992 through December 2007. Where do you think that value is headed?

Shipments rose during the 1990s, then peaked at $52.9 billion in December 2000. The dot-com bubble burst in 2001, and shipments fell by $15 billion. In 2003, shipments began to rise again, and the slow growth continued through 2007. Imagine you were back at the end of 2007. Would you predict that slow growth to continue? How much confidence would you attach to your estimate?

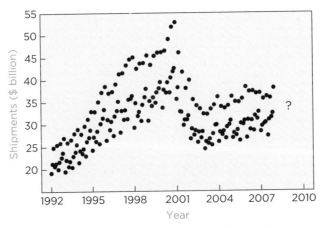

FIGURE 27.1 Value of shipments of computers and electronics in the United States, 1992–2007.

We have used regression to generate predictions and prediction intervals. We'll use the same methods in this chapter, but we'll call these estimates forecasts. A **forecast** is a prediction of future data that anticipates where a time series is going. Distinguishing a prediction from a forecast reminds us that forecasting requires extra precautions. We have learned in prior chapters that predictions from regression are reliable so long as we're not extrapolating, but forecasts are always extrapolations in time, reaching into the future to help businesses plan.

This chapter introduces several methods for modeling time series, emphasizing how to use regression models to forecast time series. Regression models for time series rely on unique properties of time series to build convenient explanatory variables. For example, dummy variables allow a regression to capture seasonal patterns such as those in computer shipments (Figure 27.1). These convenient explanatory variables provide a baseline forecast that is enhanced by incorporating explanatory variables that are unique to each situation.

27.1 | DECOMPOSING A TIME SERIES

forecast A prediction of a future value of a time series that extrapolates historical patterns.

Statistical analysis of a sample of a single numerical variable begins with its histogram. The histogram shows the location, scale, and shape of the data and reveals anomalies such as outliers and skewness that affect subsequent analyses. The analysis of a time series begins differently, commencing with a timeplot such as Figure 27.1 that shows how the data evolve over time. A common framework used to describe the evolution of a time series informally classifies temporal variation into one of three types: trend, seasonal, and irregular:

$$y_t = \underbrace{\text{Trend}_t + \text{Seasonal}_t}_{\text{pattern}} + \underbrace{\text{Irregular}_t}_{\text{random}}$$

The components of this decomposition are as follows:

- *Trend.* The trend is a smooth, slowly meandering pattern. We estimate the trend with averages of adjacent observations or with a regression model that produces a smooth curve.
- *Seasonal.* Seasonal patterns represent cyclical oscillations related to the calendar. The most common seasonal patterns repeat each year, following the seasons, such as the sales of a business that fluctuate with holidays or weather. Sales of snow skis, for instance, drop after the winter, and few consumers shop for a new swimsuit after summer.
- *Irregular.* Irregular variation means just that: random variation that is unpredictable from one time to the next.

Common approaches to forecasting a time series extrapolate the trend and seasonal patterns. The irregular variation suggests how accurate these forecasts will be: **We cannot expect to forecast the future any better than our model describes the past.**

tip

Smoothing

smoothing Removing irregular and seasonal components of a time series to enhance the visibility of the trend.

Smoothing a time series enhances the visibility of the trend by removing the irregular and seasonal components. The most common type of smoothing

moving average A weighted average of adjacent values of a time series.

uses moving averages. **A moving average** is a weighted average of adjacent values in a time series. For example a five-term moving average estimates the trend at time t as the average of y_t and the two observations on either side:

$$\bar{y}_{t,5} = \frac{y_{t-2} + y_{t-1} + y_t + y_{t+1} + y_{t+2}}{5}$$

The more terms that are averaged, the smoother the estimate of the trend becomes. The moving average of enough adjacent terms removes both seasonal and irregular variation. For example, a 13-term moving average $\bar{y}_{t,13}$ of monthly computer shipments produces the smoother green line shown in Figure 27.2.

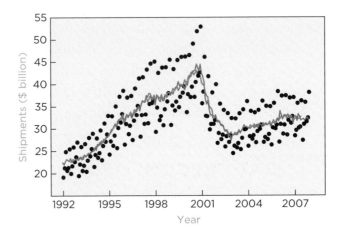

FIGURE 27.2 A 13-term moving average (green) emphasizes the underlying trend in shipments.

The 13-term moving average follows the average level of shipments and removes short-term fluctuations. Because the averaging in this example spans more than a year, $\bar{y}_{t,13}$ smoothes out seasonal variation. The similar, but rougher, red trend in Figure 27.2 is the official "seasonally adjusted" estimate of shipments and is produced by the U.S. Census Bureau.[1]

seasonally adjusted Having had the seasonal component removed.

A **seasonally adjusted** time series is a time series from which the seasonal component has been removed. The difference between the observed time series and the seasonally adjusted series is the seasonal component. Many government-reported time series are released after seasonal adjustment that removes periodic calendar swings. Otherwise, one might mistake a regularly occurring seasonal change as an indication of economic weakness or strength. For instance, raw unemployment rates regularly rise each summer as students enter the workforce. Seasonal adjustment separates this reoccurring change from other trends.

The seasonal component of a time series is periodic. For the time series of computer shipments, a repeating 3-month (quarterly) cycle dominates this component. Figure 27.3 zooms in on the seasonal component for the last three years of data. Shipments peak at the close of each quarter. Evidently, companies ship goods at the end of the quarter to meet projections or obligations. The seasonal component is particularly high in the fourth quarter, reflecting holiday sales.

[1] The U.S. Census Bureau uses a complex algorithm known as X-12 to perform seasonal adjustment. See the paper by D. F. Findley, B. C. Monsell, W. R. Bell, M. C. Otto, and B.-C. Chen (1998). "New Capabilities and Methods of the X-12 Seasonal Adjustment Program," *Journal of Business and Economic Statistics* **16**, 127–176. You can also search online at the Census Bureau for recent updates and reports.

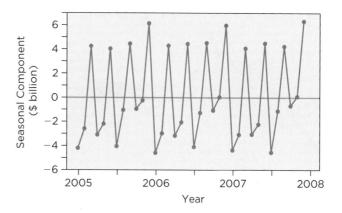

FIGURE 27.3 Seasonal component of computer shipments shows a strong three-month cycle.

Exponential Smoothing

Centered moving averages run into a problem near the beginning and end of a time series. The 13-term moving average $\bar{y}_{t,13}$ shown in Figure 27.2, for instance, requires six values on either side. Missing the first six smoothed values may not be important for forecasting the future of the series, but missing the last six is. We obtain a smooth trend that we can extrapolate if we use a one-sided moving average.

An **exponentially weighted moving average** (EWMA) is a weighted average of current and past data. An EWMA, or exponential smooth, is "exponentially weighted" because the weights applied to the data fall off as a series of powers. The EWMA of y_t is

exponentially weighted moving average A weighted average of past observations with geometrically declining weights.

$$s_t = \frac{y_t + wy_{t-1} + w^2 y_{t-2} + \cdots}{1 + w + w^2 + \cdots} \quad \text{with} \quad 0 \le w < 1$$

The average s_t is a weighted average because the denominator is the sum of the weights applied to y_t and the prior observations. Because $w < 1$, more recent observations receive larger weights than those further in the past. The definition of s_t shows that we only really approximate s_t in data because we don't have the infinite past shown in the sum. It's also clear that a finite sum is a good approximation so long as w is not too close to 1. For example, if $w = 0.75$, then $w^{10} \approx 0.056$ and we can truncate the infinite sum.

A different expression for the EWMA provides a more efficient computational formula and insight into the choice of the weight w. As shown in Behind the Math: Exponential Smoothing, we can also write an EWMA as

(p. 719)

$$s_t = (1 - w)y_t + ws_{t-1}$$

Hence, an EWMA is a weighted average of the current observation y_t and the prior smoothed value s_{t-1}.

This form of the EWMA shows how the choice of w affects s_t. The larger w is, the smoother s_t becomes. If $w = 0$, then $s_t = y_t$ with no smoothing. If w is near 1, then $s_t \approx s_{t-1}$; the smoothed series is nearly constant. To show how the choice of w affects the EWMA, let's start with the seasonally adjusted data. Figure 27.4 compares the exponential smooth of seasonally adjusted computer shipments with $w = 0.5$ (on the left) to the exponential smooth with $w = 0.8$ (on the right).

The EWMA with $w = 0.8$ is smoother than that with $w = 0.5$; it has fewer isolated peaks. This figure also illustrates another consequence of increasing w. Because an EWMA averages past data, the resulting curve trails behind the data. The lagging behavior becomes more noticeable as w increases. The smooth series on the right of Figure 27.4, for instance, takes several months to pick up the large drop in shipments that occurred in 2001.

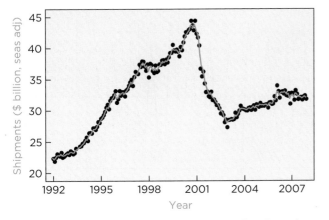
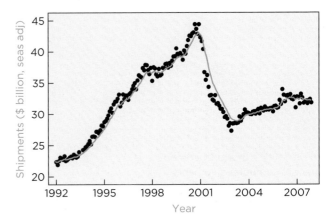

FIGURE 27.4 Exponential smooth of seasonally adjusted computer shipments, with $w = 0.5$ (left) and $w = 0.8$ (right).

Unlike a centered moving average, an exponential smooth produces a forecast of the time series. The forecast of an EWMA is particularly simple. The procedure simply predicts the next observation y_{n+1} to be the last smoothed value s_n. Forecasts further beyond the data stay the same; using an exponential smooth, we forecast all of the future values $y_{n+1}, y_{n+2}, \ldots$ as the last smoothed value s_n.

What Do You Think?   The following timeplots show the monthly value of shipments of nondurable goods (left) and the value of new orders for computers and electronics (right). Both are in billions of dollars.

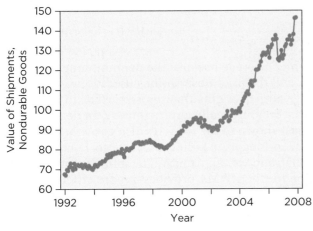
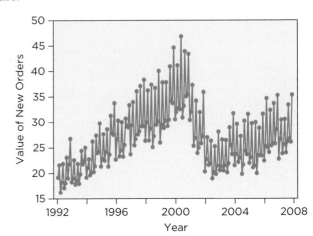

FIGURE 27.5 Value of shipments of nondurable goods (left) and new orders for computers and electronics (right).

a. Which time series clearly shows seasonal variation?[2]
b. Which time series clearly shows the trend component?[3]
c. Which time series will require a longer moving average to show the trend?[4]

27.2 | REGRESSION MODELS

leading indicator An explanatory variable that anticipates coming changes in a time series.

Smoothing helps us see trends in a time series and can even generate a forecast. To get a forecast with a longer horizon, however, we need to identify explanatory variables that anticipate changes in y_t. A **leading indicator** is an explanatory variable that anticipates changes in the time series we are trying to forecast. For

---

[2] New orders, on the right, has a strong seasonal component.
[3] The trend component is clearly visible in the value of shipments on the left. The data hardly seem to need any smoothing for the trend to be visible.
[4] The substantial seasonal variation in new orders will require a longer moving average.

example, a schedule of release dates for new computer chips would be useful for predicting shipments of computers. If Intel plans to release a new computer chip next March, we could anticipate that the introduction of this new chip would influence shipments of assembled products. Leading indicators are hard to find, however, even if you are familiar with an industry. Plus, you have to obtain the data for the leading indicator. (Exercise 39 studies a leading indicator to predict housing construction.)

Since leading indicators are hard to come by, it is good that regression remains useful without one. We can instead build explanatory variables by taking advantage of the sequential order of a time series. One type of explanatory variable is made from the time index t directly, and a second type uses prior values of the response. We call such ad hoc explanatory variables **predictors**; they do not offer much in the way of explanation but nonetheless help us predict the time series.

predictor An ad hoc explanatory variable in a regression model used to forecast a time series.

The resulting regression model for a time series requires the same assumptions as any other: an equation for the mean of y given the explanatory variables combined with the assumptions of independent, normally distributed errors with constant variance. Of these assumptions, independence of the errors is crucial when modeling time series. The assumption of independence means that the model includes predictors that capture the sequential dependence in y.

Polynomial Trends

polynomial trend A regression model for a time series that uses powers of t as the explanatory variables.

A **polynomial trend** is a regression model that uses powers (squares, cubes, etc.) of the time index t as explanatory variables. Typically, the predictors in a polynomial trend include several powers of the time index, such as t, t^2, and t^3:

$$y_t = \beta_0 + \beta_1 t + \beta_2 t^2 + \beta_3 t^3 + \varepsilon_t$$

Such an equation is called a third-degree, or cubic, polynomial.

caution We must be cautious when using a polynomial trend to forecast a time series. Polynomials extrapolate past trends and lead to very poor forecasts should the trends change direction. For example, the plot on the left side of Figure 27.6 shows that computer shipments grew steadily from 1999 through 2000. This plot includes the fit of a simple linear trend, $y_t = b_0 + b_1 t$, along with the 95% prediction interval ($\hat{y}_t \pm 2s_e$) from the simple regression.

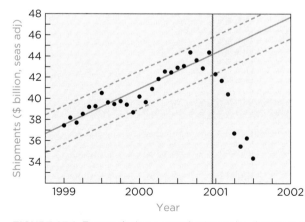
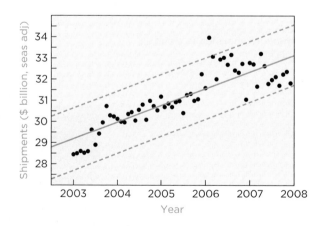

FIGURE 27.6 Extrapolating a trend misses the downturn in 2001. What happens in 2008?

Suppose, at the end of 2000, we had used this simple regression to forecast shipments in 2001. The fitted linear trend generates optimistic, but inaccurate, forecasts of the short-term future. Shipments fell instead of continuing to rise. Now consider the plot on the right side of Figure 27.6, which shows more recent data from 2003 onward. These data again appear to follow a linear trend. Having seen what happened in 2001, however, we ought to be suspicious of the accuracy claimed by the prediction interval shown around this model. That claim is valid only so long as this historical pattern persists.

You cannot rely on a polynomial that fits well during a stable period if you suspect that the stability may be short lived. The object of modeling a time series isn't to get a good fit to the past; rather, it's to have a forecast of what may happen next. A forecast built during an unusually stable period will not represent the actual uncertainty. Unless a model incorporates the causes of changes, it merely extrapolates historical patterns.

tip It's very hard to build a model that reliably predicts the future, but polynomials are notoriously unreliable. **Avoid forecasting with polynomials that have high powers of the time index.** Polynomials with many powers can fit historical data quite well, but seldom generate accurate forecasts. To illustrate the problem, Figure 27.7 shows the fit of a sixth-degree polynomial trend (six powers, $t, t^2, t^3, t^4, t^5,$ and t^6) to the seasonally adjusted shipments from 2000 through 2007.

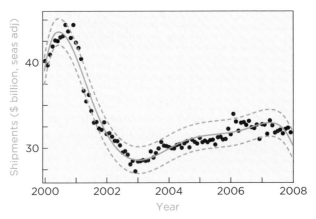

FIGURE 27.7 A sixth-degree polynomial trend for shipments using six terms.

The region around the fitted polynomial trend shows the historical 95% prediction interval. The estimated polynomial shown in Table 27.1 below describes $R^2 = 95.76\%$ of the historical variation, and all of the estimated slopes but one claim to be statistically significant. (We say that the estimates "claim" to be statistically significant because we'll see later in this chapter that this model does not meet the conditions of the MRM.)

TABLE 27.1 Summary of a sixth-degree polynomial trend model.[5]

R^2	0.958
s_e	0.925
n	96

Term	Estimate	Std Error	t-Statistic	p-Value
Intercept	−3850.532	358.6997	−10.73	<0.0001
Date	1.9363	0.1790	10.82	<0.0001
$(\text{Date} - 2003.96)^2$	−0.0711	0.1562	−0.46	0.6501
$(\text{Date} - 2003.96)^3$	−0.6215	0.0440	−14.13	<0.0001
$(\text{Date} - 2003.96)^4$	0.1902	0.0267	7.12	<0.0001
$(\text{Date} - 2003.96)^5$	0.0277	0.0024	11.46	<0.0001
$(\text{Date} - 2003.96)^6$	−0.0105	0.0012	−8.72	<0.0001

[5] To reduce collinearity, the powers of the time index in this model are centered by subtracting the average of the time index (2003.96) before raising to powers. Without this centering, collinearity is much worse.

When extrapolated to 2008, this regression forecasts a huge drop in shipments that year, seen at the right of Figure 27.8.

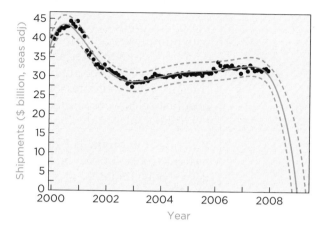

FIGURE 27.8 Polynomial trends have problems when extrapolated outside the data.

According to these forecasts, computer and electronic shipments end by the close of 2009! A recession did begin in 2008, but shipments did not stop. The behavior exhibited in Figure 27.8 is typical: Polynomial trends with high powers wobble wildly outside the data. Beyond the data, however, is precisely where we need a forecast.

4M EXAMPLE 27.1   **PREDICTING SALES OF NEW CARS**

MOTIVATION   STATE THE QUESTION

The U.S. auto industry neared collapse in 2008–2009 with the tightening of credit markets and resulting decline in sales. Was this collapse sudden, or could we have anticipated that sales would drop from historical trends? The regression model that we'll use to find an answer combines a polynomial trend with dummy variables that capture the seasonal pattern.

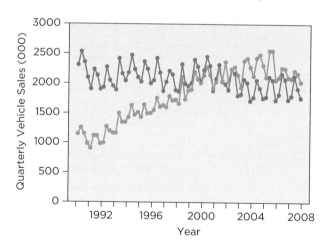

This timeplot follows quarterly sales (in thousands) of cars (blue) and light trucks (red). Light trucks include SUVs and pickups, which fell out of favor with the surge in gasoline prices in 2008. The industry had been surviving on sales of light trucks; sales of cars had been stagnant or falling since 1990.

Car sales were also falling before 2008. How well does a model built using data from 2007 and earlier forecast what happens in 2008?

METHOD

DESCRIBE THE DATA AND SELECT AN APPROACH

The timeplot of car sales shows two patterns: a gradual downward trend along with a regular oscillation, a seasonal pattern. To model the trend as well as the seasonal variation, we'll use regression to combine a polynomial for the trend with dummy variables for the seasons. The dummy variable Q1 is coded 1 for data in the first quarter and 0 otherwise. Similarly, Q2 represents the second quarter and Q3 represents the third quarter. The fourth quarter is the baseline category. (Our estimates exclude 2008.) The slope of the time trend will tell us how fast sales were falling, and the coefficients of the dummy variables will tell us how sales in the four quarters differ from that overall trend.

✓ **Straight enough.** The decline in car sales seems linear. (The trend in sales of light trucks, on the other hand, is not and would need at least a second-degree polynomial.)

? No lurking variable. Our model extends historical trends. Since it does not specify the causes of these trends (such as gasoline prices or a slowing economy), we have to anticipate possible dependence in the unexplained variation due to lurking variables.

MECHANICS

DO THE ANALYSIS

Start with a simple model for the trend, then account for seasonal patterns. A line captures the declining pattern in car sales. (Notice that the seasonal pattern almost "disappears" when the data are pushed together in this plot.) The figure also shows a second-degree polynomial (a quadratic). The quadratic does not offer much improvement in R^2 and has an odd bend. (It's blue in this plot.)

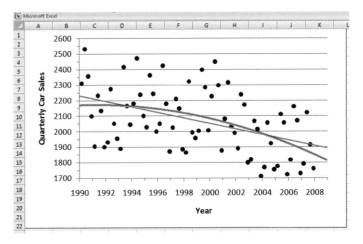

We then add the dummy variables Q1, Q2, and Q3 to capture the seasonal variation. The tables on page 703 summarize the resulting multiple regression. The variable Time is coded on an annual scale as follows: Time = 1990 + 2/12 for the first quarter of 1990, Time = 1990 + 5/12 for the second quarter, and so forth.

The slope for Time indicates that average car sales were falling at a rate of 17,449 cars per year. Since car sales are in the neighborhood of 2 million, that's slow. The slopes of the dummy variables indicate seasonal effects, relative to the baseline category (the fourth quarter). The fourth quarter has the lowest sales relative to the trend because the coefficients of the dummy variables are all positive. Sales in the first quarter, for example, average 53,147 more than in the fourth quarter. The second quarter has the highest sales relative to the trend, averaging 379,448 more than we'd expect in the fourth quarter.

R^2	0.742
s_e	107.479
n	72

Term	Estimate	Std Error	t-Statistic	p-Value
Intercept	36787	4882	7.54	<0.0001
Time	−17.449	2.441	−7.15	<0.0001
Q1	53.147	35.873	1.48	0.1432
Q2	379.448	35.847	10.59	<0.0001
Q3	223.632	35.831	6.24	<0.0001

Check the conditions before considering inference. We cannot continue to inference. This model does not meet the conditions of the MRM.

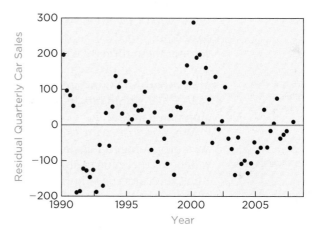

✗ **Evidently independent.** The model does not explain why sales were falling (it only follows the trend) and omits underlying causal variables. The preceding timeplot of the residuals shows a meandering pattern, and the Durbin-Watson statistic is $D = 0.84$. There's dependence in the residuals.

At this point, we needn't check other conditions; the fitted model does not meet the conditions of the multiple regression model. We can use this model to describe trends in car sales, but we cannot form confidence intervals or prediction intervals.

To get rough estimates for 2008, we can plug in values for the predictors. For example, for the first quarter of 2008,

$$\hat{y} = 36787 - 17.449(2008.167) + 53.147 \approx 1{,}799.6 \text{ thousand}$$

For the remaining quarters, the predictions are as follows:

Second: $\hat{y} = 36787 - 17.449(2008.417) + 379.448 \approx 2{,}121.6$ thousand
Third: $\hat{y} = 36787 - 17.449(2008.667) + 223.632 \approx 1{,}961.4$ thousand
Fourth: $\hat{y} = 36787 - 17.449(2008.917) \approx 1{,}733.4$ thousand

MESSAGE | SUMMARIZE THE RESULTS

Sales of cars in early 2008 are in line with historical trends. A regression model with a linear time trend and seasonal factors closely predicts sales of new cars in the first two quarters of 2008, but substantially overpredicts sales in the third and fourth quarters. The following table summarizes the forecasts.

Quarter	Actual Sales	Forecast Sales	Forecast Error
Q1, 2008	1,733,900	1,799,600	−65,700
Q2, 2008	2,123,400	2,121,600	1,800
Q3, 2008	1,734,100	1,961,400	−227,300
Q4, 2008	1,211,800	1,733,400	−521,600

Autoregression

Regression models for time series that use simple trends such as that in Example 27.1 often result in correlated residuals. We shouldn't be surprised to find substantial autocorrelation in the residuals. The regression omits explanatory variables that explain why sales change, and we should expect that these omitted variables are themselves correlated over time. Since these variables are not in the model, they must be in the errors. **Autocorrelated errors in regression most often result from using an incomplete set of explanatory variables.**

Though these omitted explanatory variables are important, we often don't know what they are or we don't have data for them. Fortunately, there's a way to account for their short-term effects in a time series regression without observing them directly. The trick is to use lags of the response variable itself as predictors. An **autoregression** is a regression that uses prior values of y_t as predictors. An autoregression might, for instance, use shipments in January to forecast shipments in February, and use shipments in February to forecast shipments in March.

When we use prior values of y_t to forecast future values, we're using lagged variables. A **lagged variable** is a prior value of the response in a time series. If y_t denotes the value of shipments in month t, then the lagged variable y_{t-1} identifies the value of shipments in the previous month and y_{t-2} is the value of shipments two months back. If we're using a software package that gives a spreadsheet view of the data (and the rows indicate time order), a lagged variable appears as a column that's been pushed down one row, as in Table 27.2.

autoregression A regression that uses prior values of the response as predictors.

lagged variable A prior value of the response in a time series.

TABLE 27.2 Lagging a column shifts the variable down one row in the data table.

Date	Shipments ($B, Seas. Adj.)	Lag Shipments
January 2005	31.184	.
February 2005	30.690	31.184
March 2005	30.853	30.690
April 2005	30.696	30.853
May 2005	30.924	30.696
⋮	⋮	⋮

This visual display makes an important point: Each lag introduces an empty cell at the top of the column. We lose cases as we lag variables.

The simplest autoregression is a simple regression that has one lag, y_{t-1}, as a predictor.

$$y_t = \beta_0 + \beta_1 y_{t-1} + \varepsilon_t$$

This model, summarized in Table 27.3, is called a first-order autoregression, abbreviated AR(1).

Just as we can fit a polynomial with several powers of the time index, we can fit an autoregression with several lags of the response. For example, the equation of a second-order autoregression is

$$y_t = \beta_0 + \beta_1 y_{t-1} + \beta_2 y_{t-2} + \varepsilon_t$$

TABLE 27.3 Summary of the first-order autoregression for shipments.

R^2	0.958
s_e	0.893
n	96

Term	Estimate	Std Error	t-Statistic	p-Value
Intercept	0.9000	0.6964	1.29	0.1994
Lag 1 Shipments	0.9706	0.0209	46.47	<0.0001

The slopes in these models are not very interpretable. Autoregressions are made for short-term forecasting and little else. In a first-order autoregression, the slope is close to the correlation between y_t and y_{t-1}. In a second-order autoregression, the coefficients are less interpretable. Choosing the right number of lags is usually done by trial and error: Fit several and select the model that offers the best combination of goodness of fit (R^2 and s_e) and statistically significant coefficients. A criterion such as adjusted R^2 that does not automatically increase with each added variable is helpful.

Although hard to interpret, autoregressions capture large amounts of dependence. For example, the scatterplot in Figure 27.9 graphs the value of computer shipments on its lag during 2000–2008. This plot takes the place of the usual scatterplot of y on x for an autoregression.

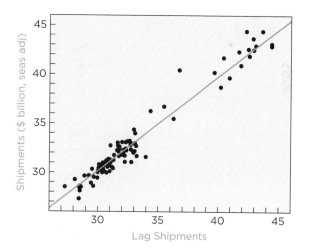

FIGURE 27.9 Scatterplot of shipments on the lag shows very high association.

This simple regression explains as much variation as the sixth-degree polynomial in Figure 27.7, but achieves this with only one predictor, the previous value of the response. Table 27.3 summarizes the fit of this AR(1) model.

Adding more lags improves the fit of the autoregression. Table 27.4 summarizes an autoregression for shipments with four lags, an AR(4) model. [Notice that we don't lose any observations when adding these lags; $n = 96$ for both the AR(1) and AR(4) models. This happens because we used data from 1999 to fill in the lags of shipments, but otherwise exclude these and earlier data from our model.]

The added lags y_{t-2}, y_{t-3}, and y_{t-4} boost R^2 from 95.8% to 96.9%. Though this change is only $96.9 - 95.8 = 1.1\%$, we have to think about changes in R^2 relative to what is possible. The AR(1) model does not explain $100 - 95.8 = 4.2\%$ of the variation in shipments. The AR(4) model explains about 25% of this unexplained variation, $(96.9 - 95.8)/4.2 = 0.262$. Even in the presence of

TABLE 27.4 Summary of the AR(4) model for shipments.

R^2	0.969
s_e	0.783
n	96

Term	Estimate	Std Error	t-Statistic	p-Value	VIF
Intercept	1.0305	0.6125	1.68	0.0959	—
Lag 1 Shipments	0.8173	0.0913	8.95	<0.0001	24.91
Lag 2 Shipments	0.2705	0.1196	2.26	0.0261	43.66
Lag 3 Shipments	0.3661	0.1195	3.06	0.0029	44.56
Lag 4 Shipments	−0.4861	0.0905	−5.37	<0.0001	26.05

substantial collinearity (the VIFs show that the lags are highly correlated with each other), the slopes of all four predictors are statistically significant.

Forecasting an Autoregression

To forecast an autoregression, we extrapolate the time series one period at a time. As an example, let's forecast computer shipments using the AR(1) model summarized in Table 27.3. We start by getting a forecast for January 2008. The fitted equation for the AR(1) model is

$$\hat{y}_t = 0.9000 + 0.9706 y_{t-1}$$

To forecast shipments in January 2008, plug in the value of shipments observed for December 2007. That gives the forecast

$$\hat{y}_{\text{Jan},2008} = 0.9000 + 0.9706 y_{\text{Dec},2007}$$
$$= 0.9000 + 0.9706(31.821) \approx \$31.7385 \text{ billion}$$

We obtain the approximate 95% prediction interval as in other simple regression models.

$$\hat{y}_{\text{Jan},2008} \pm 2s_e = 31.785 \pm 2 \times 0.893$$
$$\approx \$30.0 \text{ billion to } \$33.6 \text{ billion}$$

To forecast February, we need the value of shipments in January:

$$\hat{y}_{\text{Feb},2008} = 0.9000 + 0.9706 y_{\text{Jan},2008}$$

At the end of December 2007, however, we don't know $y_{\text{Jan, 2008}}$—but we do have our forecast. We forecast February by substituting this estimate for the actual value:

$$\hat{y}_{\text{Feb},2008} = 0.9000 + 0.9706 \hat{y}_{\text{Jan},2008}$$
$$= 0.9000 + 0.9706(31.785) \approx \$31.751 \text{ billion}$$

Once we substitute forecasts for data, however, it becomes difficult to determine the prediction interval. The forecast of February relies on the forecast of January, and the uncertainty compounds in a way that is hard to compute. Specialized software that fits time series models, however, usually comes with algorithms that determine prediction intervals for autoregressions. As an example, Figure 27.10 shows the forecasts of shipments produced by the AR(4) model in Table 27.4 along with the exact 95% prediction interval.

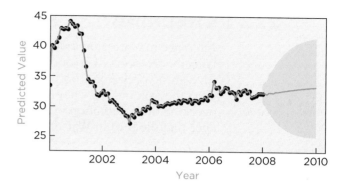

FIGURE 27.10 Fit, forecasts, and prediction interval for the AR(4) model of shipments.

Even though the model explains almost 97% of the variation, the prediction interval rapidly widens as the forecasts reach into 2008 and beyond. Autoregressions provide accurate near-term forecasts but have less precision when extrapolated farther out.

What Do You Think?

Figure 27.11 shows the fit and 95% prediction interval of a sixth-degree polynomial for the value of shipments of nondurable goods (in billions of dollars). All six slopes in the estimated polynomial have p-values much less than 0.05, and $R^2 = 0.98$. The accompanying table summarizes the fit of an AR(1) model (also with $R^2 = 0.98$) for this same time series.

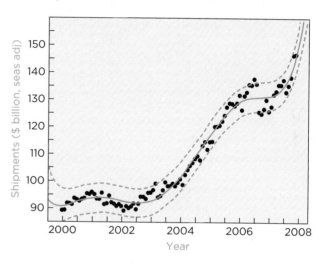

FIGURE 27.11 Fit and forecasts of a sixth-degree polynomial and a tabular summary of an AR(1) model.

AR(1)

Term	Estimate
Intercept	−0.2165
Lag Nondurable	1.0076

a. The polynomial model predicts shipments to grow rapidly—off the charts—in 2008. Do you believe these predictions are reasonable estimates?[6]
b. In December 2007, shipments of nondurable goods were valued at $146 million. What is the forecast from the AR(1) model for February 2008?[7]

---

[6] Not likely. The sudden rise (in this case) is an artifact of fitting a polynomial.
[7] First forecast January, $\hat{y} = -0.2165 + 1.0076 \times 146 = 146.89$. Then use this prediction to forecast February, $-0.2165 + 1.0076 \times 146.89 = \$147.79$ billion. The AR(1) model predicts growth as well, but slower growth.

27.3 | CHECKING THE MODEL

Inference in regression relies on the assumption of independence of the underlying model errors. If errors are positively dependent from one to the next (positive autocorrelation), both s_e and the standard errors of slopes are smaller than they should be. Chapter 22 introduced the Durbin-Watson statistic for checking the assumption of independence. Lagged variables show the dependence that the Durbin-Watson statistic measures.

Autocorrelation and the Durbin-Watson Statistic

If the errors in a time series regression are dependent, then the residual at time t, e_t, is usually correlated with the previous residual e_{t-1}. Negative residuals collect near other negative residuals, and positive residuals occur near other positive residuals. For example, the timeplot on the left of Figure 27.12 shows the residuals from the sixth-degree polynomial trend for computer shipments. The scatterplot on the right side of Figure 27.12 provides another way to see the dependence by plotting e_t versus e_{t-1}.

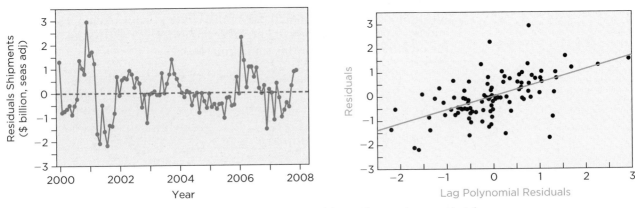

FIGURE 27.12 Residuals from the sixth-degree polynomial model over time and versus their lag.

The scatterplot of e_t on its lag e_{t-1} shows that the sixth-degree polynomial leaves substantial dependence in its residuals. The polynomial has a high R^2 (95.8%), but the remaining variation is dependent. This dependence produces the meandering pattern in the timeplot of the residuals. Meandering patterns are sometimes hard to recognize in a timeplot, so it is useful to see the scatterplot of the adjacent residuals along with the timeplot. In the scatterplot, the correlation between adjacent residuals e_t and e_{t-1} is apparent ($r = 0.56$). This type of correlation between adjacent values in a time series is known as autocorrelation (Chapter 22). The correlation between adjacent values, such as the correlation between e_t and e_{t-1}, is called a first-order autocorrelation and sometimes denoted r_1. The subscript distinguishes the autocorrelation between adjacent values from correlations between values separated further in time. For instance, the correlation between e_t and e_{t-2} is called the second-order autocorrelation and written r_2.

The Durbin-Watson statistic is related to the autocorrelation of the residuals of a regression. The Durbin-Watson statistic for a regression is (Chapter 22)

$$D = \frac{\sum_{t=2}^{n}(e_t - e_{t-1})^2}{\sum_{t=1}^{n}e_t^2} \approx 2(1 - r_1)$$

As shown in Chapter 22, the Durbin-Watson statistic is approximately twice the difference between 1 and the autocorrelation r_1. If the autocorrelation is near zero

($r_1 \approx 0$), then D is near 2. As the autocorrelation increases, D falls toward zero. For example, the first-order autocorrelation for the residuals from the sixth-degree polynomial is $r_1 \approx 0.56$, so $D \approx 2(1 - 0.56) = 0.88$. **If the software provides D, then you can quickly estimate the autocorrelation by using the formula $r_1 \approx 1 - D/2$.**

caution There is an important limitation to the use of the Durbin-Watson statistic: *Do not apply the Durbin-Watson statistic to an autoregression.* The Durbin-Watson statistic is not appropriate for regression models that use a lagged value of y_t as an explanatory variable. Rather, if we suspect there is residual autocorrelation, we should fit an autoregression with more lags. In this example, the statistically significant improvement offered by fitting the AR(4) model shows that the AR(1) model is inadequate and fails to capture all of the autocorrelation.

Residual Plots

Like regular correlations between two variables, an autocorrelation is sensitive to outliers and does not capture nonlinear patterns. To be careful, look at the scatterplot of e_t on e_{t-1} when checking for autocorrelation in the residuals from a time series regression.

Don't forget two other plots of the residuals when fitting a regression model to a time series. One is the familiar plot of the residuals on the explanatory variable (simple regression) or fitted values (multiple regression). The other is a timeplot of the residuals. Because the residuals are plotted along the y-axis in both plots, the points occur in the same vertical positions. The difference lies in the variable on the x-axis, either the fitted values or the times. The two plots in Figure 27.13 show these two views of the residuals from the first-order autoregression of shipments.

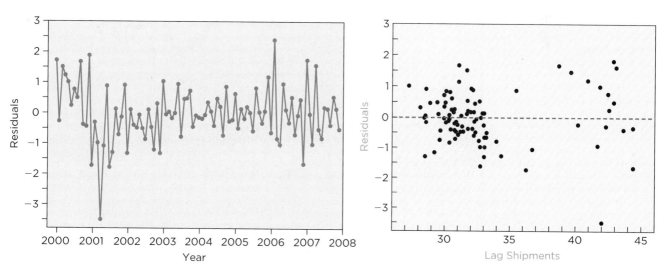

FIGURE 27.13 Diagnostic plots of the residuals for the first-order autoregression of shipments.

Time determines the order on the left; the lag of shipments (the predictor in the regression) sets the order on the right. The timeplot on the left makes it easy to date the residuals. For example, these plots reveal an outlier in April 2001 (which was a bad month for IBM). Shipments were much smaller during that month than you'd expect from shipments in March. The timeplot also shows that the fit of the AR(1) model is not able to keep up with the rapid fall of shipments in 2000–2001. This trend in the residuals suggests that these residuals are dependent—something we have confirmed from the improved fit given by the AR(4) model.

To summarize, inspect these plots of the residuals when fitting a time series regression:

- Timeplot of residuals
- Scatterplot of residuals versus fitted values (e_t versus $\hat{y}_t$)
- Scatterplot of residuals versus lags of the residuals (e_t versus e_{t-1})

4M EXAMPLE 27.2   FORECASTING UNEMPLOYMENT

MOTIVATION   ▸   STATE THE QUESTION

One of the most watched macroeconomic variables in the United States is the unemployment rate, the percentage of the workforce that is unemployed. A variety of federal, state, and local agencies track unemployment. An increase in unemployment anticipates greater demand for services from the government. Companies also watch this rate. Managers at a chain of fast-food restaurants, for instance, worry that rising levels of employment could produce pressure to raise wages, cutting into profits.

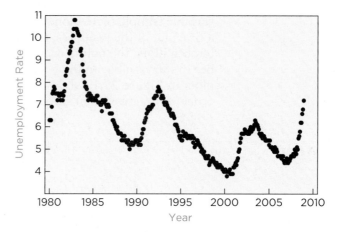

Your analysis of a time series should always start with a timeplot of the data. The preceding timeplot shows monthly civilian unemployment (seasonally adjusted, courtesy of the Bureau of Labor Statistics) from 1980 through 2008. Unemployment generally declined over this period, but a linear time trend misses the oscillations.

We'd like predictions for the first three months of 2009 based on data through the end of 2008. Can a time series regression predict the rapid increase in unemployment that came with the recession in 2009?

METHOD   ▸   DESCRIBE THE DATA AND SELECT AN APPROACH

We will use a multiple regression of the percentage unemployed on lags of unemployment and a time trend. In other words, we'll combine an autoregression with a polynomial trend. The linear trend can capture the gradual descent of unemployment over the last 25 years, and the lags allow the model to capture peaks and troughs around this trend. If the fit meets the conditions of the multiple regression model, we will use it to get a prediction interval for the next month.

To verify the basic conditions, use a scatterplot matrix of y_t and its lags. (We looked at all of them but only show two lags, y_{t-1} and y_{t-6}, here.) We picked these lags after exploring a few preliminary models.

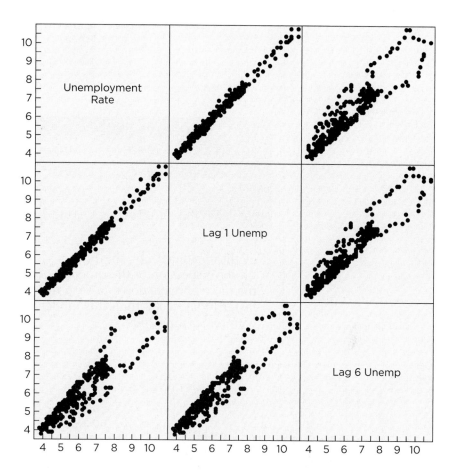

✓ **Straight enough.** The association in the scatterplot matrix of the lags of unemployment appears straight enough, but the tight shapes show considerable collinearity as well (which is to be expected when using lagged variables).

✓ **No lurking variable.** We would like to know the causes of unemployment, but that would take a greater knowledge of economics and psychology. We can be sure that we are leaving out many important variables, but hopefully the lags of unemployment capture most of the effects of omitted variables on the near-term future.

MECHANICS   DO THE ANALYSIS

A linear trend obtained by including the time index Year as a predictor captures the generally downward drift in the unemployment rate, but it is not statistically significant and we do not retain it in our model. The only predictors are three lags of unemployment. We fit several different models before settling on the collection of lags shown in the model summarized at the top of the next page. We started with just one lag, y_{t-1}, and added others as useful, beginning with y_{t-2}. The group shown in the summary of the regression captures all of the autocorrelation. The choice of a six-month lag was based on a hunch regarding the impact of holiday and calendar effects on unemployment. As in other autoregressions, the lagged variables are highly collinear, but nonetheless are statistically significant.

Before inference, check the conditions of the model.

✓ **Evidently independent.** The timeplot of the residuals seems okay. The Durbin-Watson statistic $D = 2.03$ is near 2, but we cannot rely on this statistic since the model uses lagged values of y_t as predictors. Models with other lags offered negligible improvements in the fit. Neither the timeplot

R^2	0.9882
s_e	0.1586
n	348

Term	Estimate	Std Error	*t*-Statistic	*p*-Value	VIF
Intercept	0.086	0.037	2.31	0.0217	—
Lag Unemp	0.958	0.053	18.13	<0.0001	80.94
Lag 2 Unemp	0.192	0.063	3.06	0.0024	114.21
Lag 6 Unemp	−0.164	0.023	−7.27	<0.0001	14.72

of the residuals nor the plot of e_t on e_{t-1} shows a noticeable pattern. Closer inspection of the timeplot of the residuals, however, suggests that the model breaks down toward the end of 2008; unemployment is rising faster in late 2008 than this model can fit. That flow suggests problems with forecasting 2009.

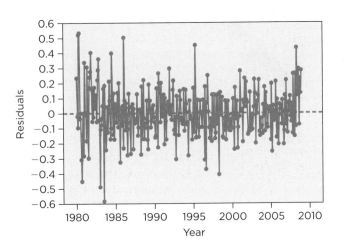

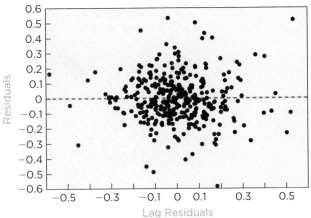

✓ **Similar variances.** The plot of residuals on fitted values has no tendency for changing the variation.

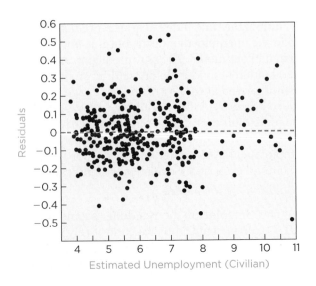

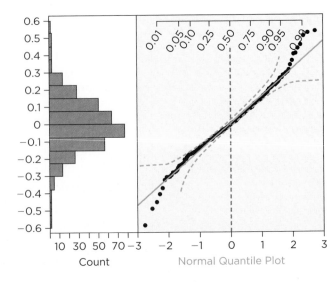

✓ **Nearly normal.** The normal quantile plot of the residuals looks okay, though the residuals appear to have more kurtosis than a normal distribution (fat tails, as is common with financial returns). The effect is modest and does not produce values outside the dashed bands, but it suggests that predictions can be off target by more than a normal model would anticipate.

This regression meets the conditions of the MRM, so we can move to inference. The F-statistic for the fit is huge, $F = 0.9882/(1 - 0.9882) \times (348 - 4)/3 \approx 9602$; we reject H_0 that all slopes are zero. This model explains statistically significant variation. To help with the interpretation, the fitted equation can be written slightly differently as

$$\hat{y}_t = 0.086 + 0.794y_{t-1} + 0.192y_{t-2} + 0.164(y_{t-1} - y_{t-6})$$

The lag at six months shows that the model adjusts for the change from y_{t-1} to y_{t-6}. If unemployment goes up from y_t to y_{t-6}, then this term increases the estimate further.

Present the forecast after inference. We want to predict unemployment in January 2009. Unemployment was 7.2% in December 2008 and 6.8% in November 2008. Six months earlier, in July 2008, it was 5.8%. Filling in the equation for $\hat{y}$ gives the forecast

$$\hat{y}_{\text{Jan,2009}} = 0.086 + 0.794y_{\text{Dec}} + 0.192y_{\text{Nov}} + 0.164(y_{\text{Dec}} - y_{\text{July}})$$
$$= 0.086 + 0.794(7.2) + 0.192(6.8) + 0.164(7.2 - 5.8) \approx 7.34$$

with approximate range $\pm 2s_e$, or about ± 0.32. The 95% prediction interval is 7.02% to 7.66%. The exact interval obtained from software (7.024 to 7.657) matches this range to two decimals.

To get forecasts for February and March 2009, we need to substitute previous predictions for data that we're missing. The forecast for February is

$$\hat{y}_{\text{Feb,2009}} = 0.086 + 0.794\hat{y}_{\text{Jan}} + 0.192y_{\text{Dec}} + 0.164(\hat{y}_{\text{Jan}} - y_{\text{Aug}})$$
$$= 0.086 + 0.794(7.34) + 0.192(7.2) + 0.164(7.34 - 6.2) \approx 7.48$$

The forecast for March uses two previous predictions and climbs to

$$\hat{y}_{\text{Mar,2009}} = 0.086 + 0.794\hat{y}_{\text{Feb}} + 0.192\hat{y}_{\text{Jan}} + 0.164(\hat{y}_{\text{Feb}} - y_{\text{Sep}})$$
$$= 0.086 + 0.794(7.48) + 0.192(7.34) + 0.164(7.48 - 6.2) \approx 7.64$$

(Unemployment was equal to 6.2% in August and September of 2008.)

Since we have to plug in forecasts in order to predict February and March, we cannot use the expression $\hat{y} \pm 2s_e$ to obtain a 95% prediction interval. The use of prior forecasts requires a wider interval and specialized software.

MESSAGE | SUMMARIZE THE RESULTS

A multiple regression fit to monthly unemployment data from 1980 though 2008 predicts that unemployment in January 2009 will be between 7.02% and 7.66%, with 95% probability. Forecasts for February and March call for unemployment to rise further to 7.48% and 7.64%, respectively.

The model that produced these forecasts presumes that historical patterns in unemployment persist into 2009. The model extrapolates these patterns, so one must cautiously interpret the forecasts; an evolving economy could produce unexpected changes to employment that these forecasts do not anticipate.

The economy was different in 2009 than it had been. A recession spread from the financial sector into the wider economy. Even so, the unemployment rate for January 2009 of 7.6% fell inside the 95% prediction interval for that month. The rates for February and March, however, climbed faster than this model anticipated, growing to 8.1% and 8.5%, outside the prediction interval.

4M EXAMPLE 27.3   FORECASTING PROFITS

MOTIVATION

STATE THE QUESTION

Sales of retailers are often seasonal, fluctuating with the holiday shopping season. This timeplot shows gross profits at Best Buy in millions of dollars, quarterly from 1995 through 2007. (Gross profits subtract the cost of goods that were sold from the total sales amount.)

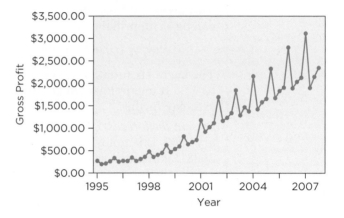

Management has proposed making changes in the way the company is run. To measure the success of those changes, we need a point of reference. We would like to forecast profits in 2008 in order to use forecasts to measure the success of changes. (A 4M exercise at the end of this chapter considers a different approach to forecasting these data that makes use of trends and dummy variables, as in Example 27.1.)

METHOD

DESCRIBE THE DATA AND SELECT AN APPROACH

Our model must account for the pattern in this series. Not only have profits grown nonlinearly (faster and faster), but the growth is also seasonal. Profits soar during the fourth quarter of each year. In addition, the variation of the data appears to be increasing with the level. To simplify the modeling, we will use a transformation that absorbs as much of the pattern as possible, allowing us to use a simple model to describe the remaining variation.

Percentage changes often serve this purpose well. Because of the evident seasonal variation, we will use the percentage change from year to year. Rather than compute the percentage change between adjacent quarters, we will compute the percentage change from one year to the next (such as the percentage change from the fourth quarter of 2006 to the fourth quarter of 2007).

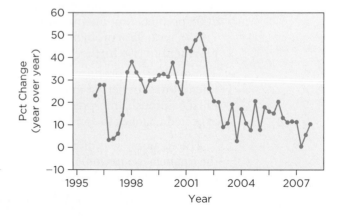

For the rest of this example, y_t denotes these year-over-year percentage changes. The preceding timeplot graphs them. Annual growth was more than 30% in the late 1990s and into 2000, but fell below 20% after the dot-com bubble burst in 2001.

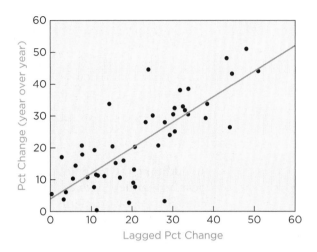

The preceding plot graphs the year-over-year percentage change on its lag.

✓ **Straight enough.** The association between y_t and y_{t-1} appears straight enough, though we should check that the outliers are not all from a particular season.

? **No lurking variable.** Retail sales at Best Buy are likely sensitive to the amount of money that consumers have to spend on non-necessities. Possible causal explanatory variables include the levels of income and unemployment. Without these in our model, we risk having dependent error terms.

MECHANICS   DO THE ANALYSIS

We fit several autoregressive models to the percentage changes. The first lag y_{t-1} captures most of the autocorrelation, but some autocorrelation remains. We capture this using an autoregression with three lags. As we saw in the previous example, a model that uses lagged values of y_t does not need to use all of the intermediate lags. The fitted equation has $R^2 = 71.0\%$ with $s_e = 7.37$ and these coefficients:

Term	Estimate	Std Error	t-Statistic	p-Value
Intercept	3.7909	2.7736	1.37	0.1795
Lag 1 Pct Change	0.8994	0.0981	9.17	<0.0001
Lag 4 Pct Change	−0.4440	0.1492	−2.98	0.0050
Lag 5 Pct Change	0.3798	0.1373	2.77	0.0086

✓ **Evidently independent.** We cannot rely on the Durbin-Watson statistic because this model uses a lag of y_t. We use a scatterplot instead. The plot of e_t on e_{t-1} does not indicate outliers and shows very little autocorrelation. Fits with more lagged variables did not improve the model.

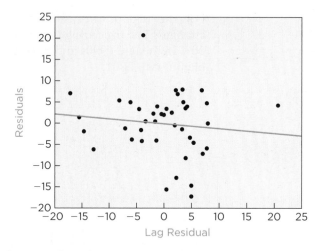

In addition, the timeplot of the residuals shows no patterns and does not suggest further dependence. The plot of the residuals on the fitted values does not indicate major problems but does highlight the one positive outlier from the holiday season of 2001 (which is also evident in the timeplot)

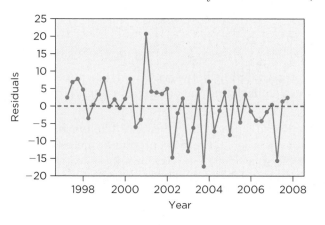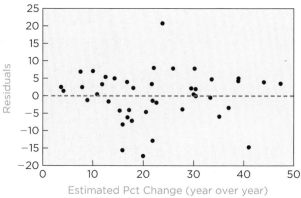

✓ **Similar variances.** The plot of residuals on fitted values has no tendency for changing the variation, though it is hard to recognize changes in variation from so few observations.

✓ **Nearly normal.** The normal quantile plot of the residuals looks fine but for the one large positive outlier noticed previously.

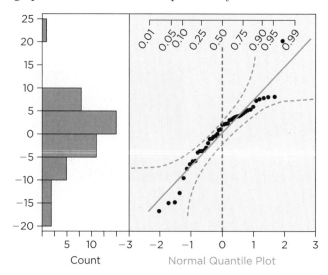

The F-statistic, $F = 0.712/(1 - 0.712) \times (44 - 3 - 1)/3 \approx 32.96$, is statistically significant, so we reject H_0 that all slopes are zero. Individual t-statistics show that each slope is statistically significant.

The prediction interval for the first quarter of 2008 serves as a good reminder that a large R^2 and a statistically significantly fit do not imply precise predictions. Plugging into the estimating equation, we obtain the forecast

$$\hat{y} = 2.971 + 0.911(2.285) - 0.443(0.318) + 0.393(11.282) \approx 9.345\%$$

The standard deviation of the residuals, $s_e = 7.4$, shows that the range of the prediction interval includes zero. We let our software handle the details and obtained a 95% prediction interval of -6.5% to 25%. The uncertainty of the model is so large that we cannot confidently forecast whether sales will rise or fall.

MESSAGE | SUMMARIZE THE RESULTS

A time series regression that describes the year-over-year percentage changes in sales at Best Buy explains statistically significant variation in the series, modeling over 70% of the historical variation. The model predicts profits in the first quarter of 2008 to grow about 9.3% over profits in the first quarter of 2007, rising to $\$1,892 \times 1.093 \approx \$2,068$ million. Although the model explains more than two-thirds of the variation in the percentage changes, considerable variation remains unexplained, and the model does not rule out a substantially larger increase (more than 25%) or some contraction (dropping about 5%).

Because the model lacks causal explanatory variables (such as measures of consumer disposable income), it cannot forecast how future changes in the economy or the actions of competitors will affect sales.

Best Practices

- *Provide a prediction interval with your forecast.* A forecast without a prediction interval is an empty promise. Unless you know the accuracy of a forecast, how can you decide if you need to prepare for what might happen? A sales forecast that predicts 10% growth is one thing, but if the uncertainty is $\pm 15\%$, you'd probably think about this forecast differently.
- *Find a leading indicator.* If you find that one time series reveals changes that later show up in another, use this series as a predictor that explains the patterns in your data. Other models that use time trends or lagged values of the response don't offer any explanation for changes. Leading indicators do.
- *Use lags in plots so that you can see the autocorrelation.* As with any other correlation, look at a scatterplot before you rely on an autocorrelation.
- *Provide a reasonable planning horizon.* A forecast of sales in January that requires data from December is of little value because it does not provide lead time for a business to react. A company has few opportunities if it learns in December that a model forecasts a downturn in

January. A forecast about January made in September, however, offers time to prepare, so long as the forecast gets it right.
- *Enjoy finding dependence in the residuals of a model.* How can finding a problem be a good thing? It's a "glass half full" versus "glass half empty" kind of thing. Dependence in the residuals means that there are other useful predictors that you can build into your model. You just have to find them. Once the model produces normally distributed residuals that are independent with equal variance, there's not much more you can get out of them. You've reached the end of the road. When you're trying to finish an assignment, it's good to get to the end of the road, but finding dependence when you're interested in getting a better forecast is a good thing.
- *Check plots of the residuals.* Don't rely on the Durbin-Watson statistic alone to check for dependence. The Durbin-Watson statistic is handy for finding dependence in the errors of a regression model for a time series. Because this statistic is basically a correlation, you need to

look at the plot of e_t on e_{t-1} to make sure that a correlation is a reasonable summary of the dependence. Outliers and nonlinear patterns cause just as many problems for autocorrelations as they do for the regular types of correlation.

Pitfalls

- *Don't summarize a time series with a histogram unless you're confident that the data don't have a pattern.* Histograms conceal dependence between adjacent values; when you summarize data with a histogram, you're saying that the data are a sample. That's the right thing to do if the data are a random sample of independent observations, but most time series are dependent.

- *Avoid polynomials with high powers.* With enough terms, a polynomial can follow any pattern in data. That doesn't mean that the pattern is going to continue into the future, however. Time changes many things, and the pattern can be one of them.

- *Do not let the high R^2 of a time series regression convince you that predictions from the regression will be accurate.* Autoregressions often have large values of R^2 with very statistically significant slopes. That's fine, but remember that these models don't explain anything. They represent a clever way to capture historical patterns. Sometimes these regularities continue into the future, but they may not. As long as they match, you're okay.

- *Do not include explanatory variables that also have to be forecast.* It's often possible to find an explanatory variable x_t that is highly correlated with y_t. Current inventory levels are related to the size of current shipments, for instance. To use x_t when forecasting y_t, however, we have to forecast x_t itself, putting us back where we started. A leading indicator anticipates changes, so that we can use x_t to forecast, say, y_{t+j} for some $j > 0$.

- *Don't assume more data is better.* Sample size presents a dilemma for modeling time series. Statistics teaches you that big samples are good. Big samples reward you with small standard errors and short confidence intervals. The catch with time series of economic variables is that the further back in time we look, the less likely it is that the data describe the same process.

Software Hints

The software commands for building a multiple regression for a time series are similar to those used for building regression models in general. The nuances have to do with forming lagged variables and polynomials. Some software will handle these details for you so that you do not have to clutter your data table with special columns of powers or lagged variables.

EXCEL

Moving averages and exponential smoothing are easy to do in excel. Follow the Excel commands Data > Data Analysis > Moving Averages and Data > Data Analysis > Exponential Smoothing.

For regression modeling, construct the lagged variables and polynomial powers as extra columns in the data table. Fortunately, that's an easy task in Excel. Watch for missing values, however, that are produced when lagging a variable. You will either need to fill in a value for these from an earlier data source or exclude the row of the data table that has the missing value.

MINITAB

The menu command Graph > Time Series Plot . . . (or Stat > Time Series > Time Series Plot . . .) opens a dialog for building several types of plots of time series. The command Stat > Regression > Fitted Line Plot will produce polynomial regressions with one, two, or three powers of the time index and graph the fit with the data.

In addition, Minitab also offers a set of specialized tools for time series analysis. The menu obtained by following Stat > Time Series includes smoothing (both smoothing with moving averages and several types of exponential smoothing), commands for lagging variables, and algorithms for fitting very complex time series models known as ARIMA models. For exponential smoothing, the software will pick w using an optimization or you can specify the value of w. The analysis given by Stat > Time Series > Trend Analysis will build linear and quadratic polynomials and provide several views of the model and residuals. Stat > Time Series > Decomposition...decomposes a time series into trend, seasonal, and irregular components. We describe the additive version of such a decomposition in this chapter; the software also offers a multiplicative decomposition. The moving average and exponential smoothing commands generate plots of the smoothed data and optionally produce forecasts.

JMP

To get connected timeplots, use the command Graph > Overlay Plot. The Fit Y by X platform will fit polynomials up to sixth-degree. Follow the menu commands that open a scatterplot (Analyze > Fit Y by X) and then use the red triangle pop-up menu to add a polynomial regression to the plot. The software lets you pick the degree of the polynomial; you can contrast several polynomials in the same scatterplot. To use lagged variables in a regression with other explanatory variables, use the formula calculator to form the columns that can be added as predictors. The formula to lag a variable is in the group labeled "Row."

JMP includes an extensive set of models for time series (ARIMA models). You must navigate this collection to find commands for exponential smoothing. Follow the menu commands Analyze > Modeling > Time Series to open a dialog that allows you to pick the response and a time index. Click the OK button and JMP will open a window with a connected time-plot of the series. Click on the red triangle above the timeplot to open a menu of options. Select the item Smoothing Model to get to simple exponential smoothing. (JMP offers several types of exponential smoothing.) When building an exponential smooth, JMP uses an algorithm to pick the value of w. If you want to specify w, you have to build the exponential smooth by defining a formula in a column and using the formula calculator.

BEHIND the Math

Exponential Smoothing

The right way to understand exponential smoothing is as a weighted average of past values,

$$s_t = \frac{y_t + w\,y_{t-1} + w^2 y_{t-2} + \cdots}{1 + w + w^2 + \cdots}$$

This expression, however, would be very slow to compute, even once we truncate the infinite sums. To arrive at the second, more handy expression for s_t, divide the sum in the numerator into two parts and factor out a single w from the numerator of the second summand, like this:

$$s_t = \frac{y_t}{1 + w + w^2 + \cdots} + \frac{w(y_{t-1} + w y_{t-2} + \cdots)}{1 + w + w^2 + \cdots}$$

We can simplify this further by recalling the sum of a geometric series: If $|w| < 1$, then $1 + w + w^2 + \cdots = 1/(1 - w)$. We use this to simplify the first term, and then we get to the desired expression by recognizing that s_{t-1} is in the second summand.

$$s_t = (1 - w)\,y_t + w\,\underbrace{\frac{y_{t-1} + w y_{t-2} + \cdots}{1 + w + w^2 + \cdots}}_{s_{t-1}}$$
$$= (1 - w)\,y_t + w\,s_{t-1}$$

This is a recursive expression: We need s_0 to get s_1, for instance. Typically, software chooses a starting value for s_0 and then uses the recursive expression to rapidly compute $s_1, s_2, \ldots$.

CHAPTER SUMMARY

A time series can be decomposed into trend, seasonal, and irregular components. Statistical methods have been developed to **smooth** and **forecast** time series. **Moving averages** expose the trend by averaging out other components. **Exponential smoothing** extracts the systematic trend from a time series by using past data and gives a forecast.

Polynomial trend models use powers of the time index as explanatory variables, avoiding the need to identify a **leading indicator**, an explanatory variable that anticipates changes in a time series. **Autoregressions** are regression that use previous values of the response, known as **lagged variables**, as predictors.

Key Terms

autoregression, 704
exponentially weighted moving
    average (EWMA), 697
forecast, 695

lagged variable, 704
leading indicator, 698
moving average, 696
polynomial trend, 699

predictor, 699
seasonally adjusted, 696
smoothing, 695

Formulas

Durbin-Watson statistic and autocorrelation
If $r_1 = \text{corr}(e_t, e_{t-1})$, then the Durbin-Watson statistic D is related to the autocorrelation through these equations:

$$D = 2(1 - r_1) \quad \text{and} \quad r_1 = 1 - D/2$$

Exponential smoothing, exponentially weighted moving average (EWMA)
For smoothing weight w, the exponential smooth s_t of a time series y_t is

$$s_t = (1 - w)y_t + w\,s_{t-1}$$

About the Data

The time series tracking shipments is Series 34S in the *Census of Manufacturers*, collected by the U.S. Census Bureau. The data on domestic car sales are from the Bureau of Economic Analysis. We obtained the unemployment data from the FRED system developed at the Federal Reserve Bank of St. Louis.

This time series is also seasonally adjusted to remove periodic variation. The data for profits at Best Buy are from Compustat, a commercial database that tracks quarterly reports at publically listed companies.

EXERCISES

Mix and Match

Match each definition on the left with its mathematical expression on the right.

1. Change in the value of the response	(a)	$y_{10}, y_{11}, y_{12}, y_{13}, y_{14}$
2. Value of the response in the previous time period	(b)	$b_0 + b_1 t$
3. Values averaged in a five-term moving average	(c)	$\beta_0 + \beta_1 y_{t-1} + \beta_2 y_{t-2} + \beta_3 y_{t-3}$
4. Exponentially weighted moving average	(d)	$y_t - y_{t-1}$
5. Equation of a model that fits a linear trend	(e)	$\beta_0 + \beta_1 y_{t-1}$
6. Equation of a fourth-degree polynomial model	(f)	$1 - \text{corr}(e_t, e_{t-1})/2$
7. Equation of a first-order autoregression	(g)	$w s_{t-1} + (1 - w)y_t$
8. Equation of an AR(3) model	(h)	$\text{corr}(e_t, e_{t-1})$
9. Alternative equation for the Durbin-Watson statistic	(i)	y_{t-1}
10. Autocorrelation of adjacent residuals	(j)	$b_0 + b_1 t + b_2 t^2 + b_3 t^3 + b_4 t^4$

True/False

Mark each statement True or False. If you believe that a statement is false, briefly explain why you think it is false.

11. A forecast is the prediction of a future value of a time series.

12. The approximate prediction interval $\hat{y} \pm 2s_e$ works well for forecasts from a polynomial trend model particularly when extrapolating away from the observations.

13. A moving average smoothes a time series y_t by averaging y_t with nearby observations before and after time period t.

14. The seasonal component of a time series consists of a regular, periodic pattern.

15. Outliers produce jumps in the exponential smooth of a time series.

16. With enough powers, a polynomial trend obtains a large value of R^2 for any time series.

17. The reliability of polynomial trend models is highest at the edges of the data (first and last time points) and falls off slowly as we extrapolate.

18. An autoregression is a regression in which the explanatory variables are lagged variables formed from the response.

19. Regression models for time series should use either time trends or lags of y_t as predictors, but never mix the two.

20. The Durbin-Watson statistic can be computed from the autocorrelation of the residuals from the regression.

21. The Durbin-Watson statistic should not be relied upon to evaluate regression models that use lagged values of y_t as explanatory variables.

22. We can only forecast a first-order autoregression one period out because the value needed in the equation for y_{t+1} is not known.

Think About It

23. The exponentially weighted moving average is a one-sided moving average of the time series. The smoothed value s_t is an average of y_t and prior values. The regular moving average is two sided, averaging values on both sides of y_t. For example, a three-term moving average of y_t is

$$\bar{y}_{t,3} = \frac{y_{t-1} + y_t + y_{t+1}}{3}$$

and a five-term moving average is

$$\bar{y}_{t,5} = \frac{y_{t-2} + y_{t-1} + y_t + y_{t+1} + y_{t+2}}{5}$$

(a) Which series will be smoother: a three-term moving average or a five-term moving average? Explain your thinking.
(b) What problem does a two-sided moving average have when the smoothing reaches the last value of the time series?
(c) The most common moving averages have an odd number of terms, such as the three-term and five-term averages in this exercise or the 13-term average used to smooth computer shipments in this chapter. What problem happens if you try to use a moving average with an even number of terms? Suggest a simple remedy for the problem.

24. Smoothing reduces the random variation in data, producing a sequence that reveals the systematic trend in the data. Shouldn't we build models from the smoothed data, which have less random noise, rather than from the original data? Explain why this is, or is not, such a good approach to building a model for forecasting.

25. Many autoregressions are mean-reverting. *Mean-reverting* means that the forecasts eventually tend back to (revert to) the mean of the time series. For example, a manager uses an AR(1) model to predict sales next week using sales in recent weeks. Sales typically run about $250,000 per week. The estimating equation (with sales in thousands of dollars) is

$$\hat{y}_t = 50 + 0.8 y_{t-1}$$

(a) If sales this week are $250,000 (the mean level), what does the equation forecast for next week?
(b) If sales this week are $300,000 ($50,000 above the mean), does the equation forecast sales to increase farther above the mean or to return toward the mean?
(c) In general, when does a first-order autoregression with positive slope ($0 < b_1 < 1$) predict an increase in the time series? A decrease?

26. A chain of photography and electronics stores created a Web site to promote its photography lessons. The number of weekly visitors grew steadily, at a rate of about 300 new visitors each week. This timeplot shows the counts of unique visitors over the last 33 weeks.

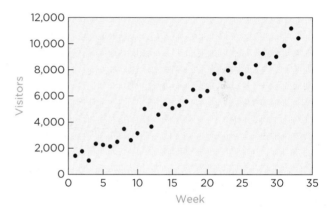

To describe the growth, an analyst used a linear time trend, and estimated the equation to be

$$\hat{y}_t = 666 + 294\, t$$

with $t = 1$ denoting the first week, $t = 2$ the second, and so forth. The company hired a summer employee who had taken some statistics courses, and she suggested using a first-order autoregression [i.e., an AR(1) Model] instead of this time trend.
(a) What do you think the intercept and slope of the AR(1) equation are going to be?
(b) Do you think it's a good idea to use an autoregression in place of the linear time trend in this situation?

27. The two plots in the left-hand column at the top of the next page show exponentially weighted moving averages of the percentage change in the U.S. gross national product (GNP). The time series is quarterly, from 1960 through the first quarter of 2009. On the top for the EWMA is the weight $w = 0.5$, whereas on the bottom $w = 0.9$. The EWMA in both cases is shown as a red line with the surrounding data as black dots. (The Federal Reserve Bank of St. Louis provides these data online.)
(a) Which weight w do you think produces a better summary of the underlying trend in the percentage changes in GNP? Explain your choice.
(b) In the first quarter of 2009 (the last data point available in this series), the GNP was estimated to be $14.238 trillion. What would you forecast,

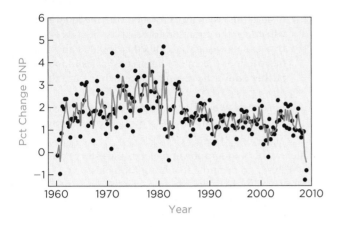

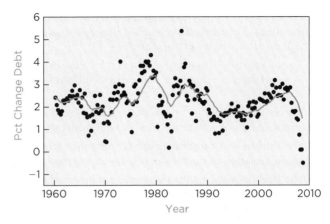

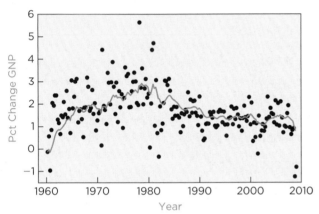

(a) Which weight w do you think produces a better summary of the underlying trend in the percentage change in household debt? Explain your choice.

(b) In the fourth quarter of 2008, household credit debt was estimated to be $13.821 trillion. Having seen these plots, forecast household credit debit in the first quarter of 2009.

(c) Provide a range to accompany your forecast of household credit debt.

29. This timeplot shows the month-to-month changes in the value of shipments of computers and electronics (after seasonal adjustment, as used in the text of this chapter). The change from March to April 2001 was dramatically large and negative, producing the outlier in the timeplot.

having seen these plots, for GNP in the second quarter of 2009?

(c) How accurate would you expect your forecast for GNP to be? From looking at these plots, suggest a range.

28. The following two plots show exponentially weighted moving averages of the percentage change in the amount of household credit market debt. The time series is quarterly, from 1960 through the fourth quarter of 2008. In the first plot the weight for the EWMA $w = 0.5$, whereas in the second plot $w = 0.9$. The EWMA in both cases is shown as a red line with the surrounding data as black dots. (The Federal Reserve Bank of St. Louis provides these data online.)

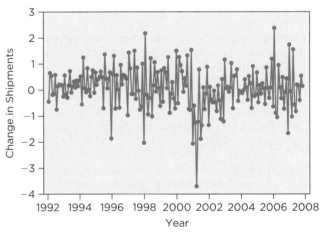

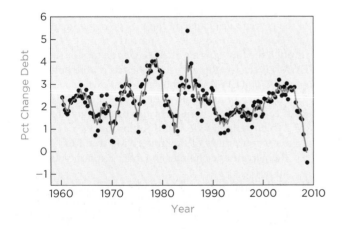

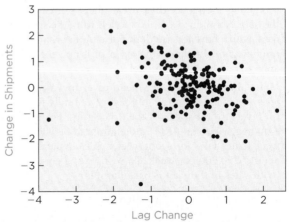

Let y_t denote the change in the value of the shipments. The scatterplot at the bottom of the preceding page shows y_t on its lagged value y_{t-1}.

(a) Does this scatterplot suggest the presence of dependence between adjacent changes in the value of shipments?

(b) Where does the outlier produced by the large drop from March to April appear in the scatterplot of y_t on y_{t-1}?

(c) Do you think that the outlier influences the estimated correlation between y_t and y_{t-1}?

(d) To remove the outlying change in shipments from the estimated correlation between y_t and y_{t-1}, how many data points need to be excluded from the scatterplot?

30. This timeplot charts the value (in millions of dollars) of inventories held by Dell, Incorporated (the computer maker). This time series is quarterly. The scatterplot below the time plot shows the value of inventories on its lag.

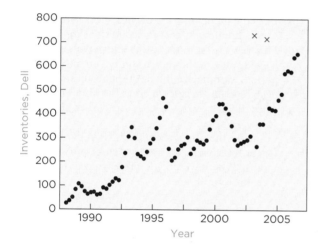

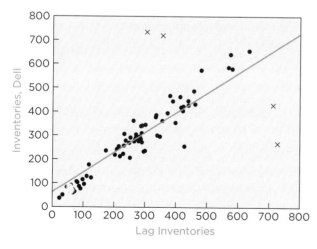

(a) Two outliers are evident in the timeplot (highlighted with ×'s). These occur for the quarters ending January 31, 2003 and January 31, 2004. Why might inventories have been so high at the end of these quarters?

(b) Why do two outliers in the timeplot produce the four outliers seen in the scatterplot of y_t on y_{t-1}?

(c) What is the effect of the outliers on the estimated dependence of inventory levels: Do the outliers create the impression of more or less dependence?

(d) How do you recommend treating these outliers when forecasting future levels of inventory at Dell?

31. This analysis compares the model in the text that has the level of shipments as the response to a model that uses the change in the shipments as the response. (These data are monthly, based on the seasonally adjusted data from 2000 through 2007.) Modeling changes is often more interesting than modeling levels. After all, we often care more about how the future differs from the present than anything else. This scatterplot and table summarize the fit of a regression of the changes, $y_t - y_{t-1}$, in the shipments on the lag of the shipments.

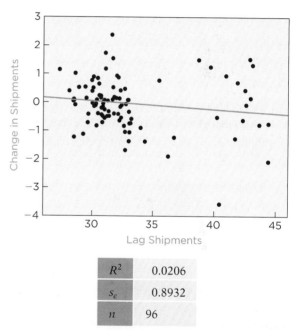

R^2	0.0206
s_e	0.8932
n	96

Term	Estimate	Std Error	t-Statistic	p-Value
Intercept	0.9000	0.6964	1.29	0.1994
Lag 1 Shipments	−0.0294	0.0209	−1.41	0.1626

(a) Why do we have 96 observations for fitting this model, even though we need to use y_t in order to find the change and as the lagged predictor? Shouldn't we lose one observation from lagging the variable?

(b) How do the slope and intercept of this equation differ from those for the first-order autoregression of the level of shipments on its lag (shown in Table 27.3)?

(c) Compare the s_e of this regression to that of the autoregression for the level of shipments. Explain any differences or similarities.

(d) This model has a small R^2 with a slope that is not statistically significant. Why is the fit of this model so poor, whereas that of the AR(1) model is so impressive?

32. The models of this chapter emphasize forecasting the level of a series. For many business decisions, forecasts of change are more important. If we're projecting an increase in sales, then we need to have more items in stock. This timeplot shows sales (in thousands of dollars) over a 25-week period (not including special holiday sales).

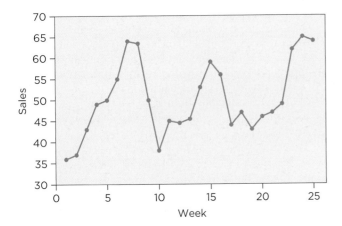

(a) The following output summarizes the fit of an AR(2) model to these data. Assuming that the model meets the usual conditions, is this a good description of the dependence in this series?

R_2	0.5707
s_e	5.5556
n	23

Term	Estimate	Std Error	t-Statistic	p-Value
Intercept	26.7459	8.0932	3.30	0.0035
Lag Sales	0.9746	0.1931	5.05	<.0001
Lag 2 Sales	−0.4960	0.1966	−2.52	0.0202

(b) If we use the same two predictors y_{t-1} and y_{t-2} to describe the changes in sales or use the differences $d_t = y_t - y_{t-1}$, what will be the estimated slopes for the lagged variables?

(c) Explain why the s_e of the regression of y_t on these two lags is the same as the SD of the residuals when regressing the changes on these two lags.

(d) Although the models are equivalent (in the sense that you can get the slope for one from the other), most analysts prefer to fit models to y_t itself rather than to the differences. A partial explanation comes from a comparison of R^2 for the two models. Which model do you think has a larger R^2: the model with y_t as the response, or the model with the differences d_t? (*Hint:* Both models leave the same variance in the residuals, but one has more variance in the response.)

You Do It

33. Exxon Prices of stocks are more appealing than returns in the sense of having a high R^2 because prices suggest many more patterns [see Exercise 32, part (d)]. The time series in this exercise is monthly prices of shares in Exxon from January 1995 through the end of 2005.[8]
   (a) Fit a linear trend model to this time series. The data table includes a column *Month*, which runs 1, 2,..., 132, to use as the time variable t. Does the trend capture the dependence over time, or does it leave substantial dependence in the residuals?
   (b) Add two lags of the price to your model, so that the model includes a linear time trend as well as y_{t-1} and y_{t-2}. Summarize the fit of this model. Do both lags seem necessary?
   (c) Revise the model by removing the second lag y_{t-2} and summarize the fit of the revised model. How has the slope of Month, the time trend, changed from that of the simple linear time trend estimated in part (a)? Explain the reason for the change.
   (d) Does the model in part (c) with a time trend and y_{t-1} as predictors meet the conditions for the MRM?
   (e) Fit a regression model using a linear time trend and lagged value to the returns on Exxon during this period. Summarize the fit of the resulting model. (The returns are calculated as the difference in price divided by the price in the earlier period, $(price_t - price_{t-1}/price_{t-1}.)$
   (f) Compare the use of the models fit in parts (c) and (e) for forecasting the price of stock in Exxon next month. Which would you prefer to use (if either)? Explain your choice.

34. Sears This exercise considers monthly prices of shares in Sears from January 1995 through the end of 2004.[8] The data table includes a column *Month* with values 1, 2, ..., 120 that can be used as the time index t.
   (a) Fit a polynomial trend model to this time series. Try polynomials of various degrees (i.e., try polynomials with t, t^2, and other powers up to t^6). Do any of these, with orders of 6 or less, capture the ups and downs of the prices?
   (b) Compare the ability of a polynomial trend to follow the pattern in this time series to that of two exponentially weighted moving averages (with $w = 0.5$ and $w = 0.9$).
   (c) Fit an autoregression to the time series of prices, using three lags of the price (y_{t-1}, y_{t-2}, and y_{t-3}). Are all of these lagged predictors useful? Remove predictors that appear not to improve the fit of the model, and summarize your final model. (There's a lot of collinearity among these predictors, so only remove predictors one at a time.)
   (d) Does the model you created in part (c) with lags as predictors meet the conditions for the multiple regression model?

---

[8] These prices are derived from CRSP. We adjusted the prices to include the effects of dividends and stock splits during these years.
[9] These prices are derived from CRSP. We adjusted the prices to include the effects of dividends and stock splits during these years.

(e) The data table includes the returns on Sears stock during this period. [The returns are calculated as the difference in price divided by the price in the earlier period, $(\text{price}_t - \text{price}_{t-1})/\text{price}_{t-1}$.] Does the sequence of returns appear simple, or do you find a pattern that can be used to forecast future returns?

(f) Compare the use of the model in part (c) for forecasting the price of Sears stock in the next month to a method that uses the simplicity of the returns. Which would you prefer to use (if either)? Explain your choice.

35. **Compensation** These data measure hourly compensation in the U.S. manufacturing sector from 1987 through the first quarter of 2006. The data are assembled by the Bureau of Labor Statistics from surveys. The value of the index was set to 100 in 1992, so the data indicate relative salaries, not dollar amounts. The data table includes a column *Quarter*, with consecutive values $1, 2, \ldots 77$, for modeling time trends.

(a) Fit a linear trend to summarize the pattern in these data. Would it be correct to describe the residuals as autocorrelated, or is there a better description of the pattern in the residuals from the linear trend?

(b) Form a dummy variable that takes on the value 1 in the 43rd quarter (the 3rd quarter of 1997) and beyond. Add this dummy variable with its interaction with **Quarter** to the simple regression fit in part (a). Interpret the equation of the resulting multiple regression of compensation on Quarter, Dummy and Quarter × Dummy?

(c) Does the multiple regression estimated in part (b) meet the conditions of the MRM, or does it have problems?

(d) Consider a different way to look at the response: its percentage rate of growth. Form the percentage changes in the compensation index and consider the timeplot of these. Does this series appear simple?

(e) Which of these models would you use to forecast this time series? Justify your choice and use it to forecast the index for the second quarter of 2006. Be sure to include an interval with your forecast.

36. **Imports** These data measure net imports into the United States, quarterly from 1981 through the first quarter of 2006 ($n = 101$ quarters). The data are given in billions of dollars, expressed at an annual rate. The data table includes a column named *Quarter*, with consecutive values $1, 2, \ldots 101$, for modeling time trends.

(a) The trend in the value of net imports has a clear bend. Sometimes, one can use a log transformation to "take the bend" out of a time series. Does a timeplot of $\log_{10}$ of the value of the imports produce a more linear trend?

(b) What problems occur if you model the log of imports using a linear trend model? (Fit the simple regression of $\log_{10}$ imports on Quarter.) What conditions of the SRM are not satisfied?

(c) As an alternative to trend models, consider using a method simplifying: Convert the sequence of values of imports into percentage changes. Does the timeplot of the percentage change appear simple?

(d) Forecast the level of imports for the second quarter of 2006, using a model of your choosing. Be sure to include a prediction interval for your forecast.

37. **Inventory** The following regression model forecasts inventory levels at Wal-Mart (in millions of dollars). The predictors are two lags of the aggregate consumer credit debt in the United States (in billions of dollars). Both data series are quarterly from 1980 through 2005 ($n = 104$). The equation of the multiple regression is assumed to be

Estimated $\text{Inventory}_t = b_0 + b_1 \, \text{Debt}_{t-1} + b_2 \, \text{Debt}_{t-3}$

(a) Evaluate the equation by least squares and summarize the estimated equation. What is the interpretation of the estimated intercept and slopes?

(b) An analyst thought it was unusual for the equation of this model to be "missing" the second lag of Debt. Would it improve the fit of this model to add the second lag, or do you recommend leaving it out?

(c) Does the estimated model meet the assumptions of the multiple regression model? Explain why or why not.

(d) Use color codes or different plotting symbols to distinguish the quarter (1, 2, 3, or 4) of the residuals in the timeplot of the residuals. This highlighting is helpful in answering this and part (e). Do the residuals from the estimated equation differ from quarter to quarter? That is, can you distinguish the residuals from certain quarters in the residuals?

(e) If this equation were to be used to forecast inventory levels for the next year (as soon as the consumer debt data were available), would you expect accurate forecasts or would you anticipate larger errors in some quarters than others?

38. **Operating Income** Operating income is the difference between what a company charges for what it sells (net sales) and the cost of the items sold. For example, in the first quarter of 2000, net sales at Lowe's (the chain of hardware megastores) were $4.467 billion. The items that it sold cost Lowe's $3.219 billion, so its operating income for this quarter was $4.467 - 3.219 = \$1.248$ billion. The following simple regression model describes the operating income at Lowe's in terms of loans made for real estate by commercial banks.

Estimated $\text{Operating Income}_t = b_0 + b_1 \, \text{Loans}_t$

The analyst who proposed this equation stated that this relationship holds only since 1990, since the emergence of Lowe's as a major competitor to The Home Depot. The idea is that more home loans mean more busy homeowners shopping for hardware at Lowe's. The data for this example are quarterly and run from 1980 through the end of 2005 (but not all of this is needed in this exercise). The operating income in the data table is given in millions of dollars; the amount of bank loans for real estate is in billions of dollars.

(a) Estimate the Apply least square equation using the 64 quarters from 1990 through 2005. Interpret the estimated intercept slope summarize the estimated equation. What is the interpretation of the estimated intercept and slope?

(b) Does the equation fit in part (a) meet the conditions for the simple regression model? Use color codes or symbols to distinguish the quarter (1, 2, 3, or 4) of the residuals in the plot of the residuals on time. This highlighting is helpful in answering the following questions.

(c) Are the residuals for the different quarters similar, or do you find systematic differences in the different quarters? (Comparison boxplots are helpful, though no substitute for the usual residual plots.)

(d) Regardless of problems from violating conditions, would this equation nonetheless be useful for forecasting the operating income at Lowe's? (*Hint:* Use your common sense. Could you forecast the next quarter using this model?)

(e) Here's an alternative model. If bank loans for real estate are truly related to the operating income at Lowe's in some fundamental way, then the percentage changes in the bank loans ought to be related to percentage changes in Lowe's operating income. Are they? Regress the percentage changes in the operating income at Lowe's on the percentage changes in bank loans for real estate. Summarize the fit of this simple regression. Does this regression meet the conditions for the simple regression model?

(f) What's your conclusion about the relationship between the operating income at Lowe's and the level of lending for real estate?

39. **Housing permits and construction** Construction of new homes is a popular measure of the health of the economy. A slowdown in the construction industry means more unemployment, fewer sales for home stores, and a general economic malaise. These data from the Census Bureau track (seasonally adjusted) sales of new homes and the number of permits issued for constructing new homes, monthly from January 1990 to April 2008. Because builders must apply for permission from the local government to build a house, the number of permits issued today is a possible leading indicator for the number of homes that will be on the market in future months. Let y_t denote the number of single-family homes completed and x_t denote the number of permits issued for single-family homes. (Both counts are in thousands.)

(a) Compare the timeplot of homes completed to the timeplot of permits issued. Do the two time series line up, or does one seem shifted from the other?

(b) Compare the association between y_t and x_t to the association between y_t and x_{t-6}. Which variable, number of current permits or number of permits issued six months earlier, appears to be the better explanatory variable (taken by itself)?

(c) Build a multiple regression designed to predict the number of single-family homes to be completed in July 2008 (three months past the end of this time series). Justify your choice of this model and verify that it meets the conditions required for a multiple regression model.

(d) Use your model to predict housing completions in July 2008, and include a 95% prediction interval. Note any caveats that you would offer with the forecast.

4M Sales at Best Buy

Retailer Best Buy sells computers, software, music, cameras, and other electronic goods. The data for this exercise are the quarterly gross profits of Best Buy, in millions of dollars from 1995 through 2007. The data table includes a column named *Time* that indicates the date of each quarter. Managers at Best Buy expect that there is a substantial increase in profits during the holiday season, but they would like to have a measure of the size of this effect. (These data also appear in Example 27.3; this exercise takes you through an alternative analysis of the same time series.)

Motivation

a) Explain why it would be useful to have an estimate of the size of the seasonal effect on profits.

Method

b) Examine the timeplot of gross profits at Best Buy. Does the magnitude of the seasonal effect appear to change over time?

c) Examine the timeplot of the natural log of gross profits at Best Buy. Does this transformation apparently stabilize the size of the seasonal variation?

d) Let D_1, D_2, and D_3 be dummy variables that identify observations in the first, second, and third quarters. If we add these variables to a polynomial trend model, that uses the gross profits as the response, will the model represent the effects of the seasonal pattern? What if we use the log of gross profits as the response?

Mechanics

e) Fit a third-degree polynomial trend with the three dummy variables to the natural log of gross profits at Best Buy. Do the dummy variables have much effect on the fit of the regression? (See Chapter 20 for hints on working with logs in regression.)

f) Does the model capture all of the dependence in the residuals, or does it appear as though substantial residual autocorrelation remains?

g) Revise the model as you see fit and summarize the statistical significance of the model's estimates.

Message

h) Show a fit of your model with the actual data, on the scale of the original data. (A plot such as this will show how well the model does or does not capture the seasonal variation.)

i) What are the estimated seasonal effects? Interpret (without using esoteric terminology) the seasonal component in this model.

j) Would you recommend forecasting with this model, or limit its use to estimating the seasonal effects?

4M Gasoline Prices

If you operate a refinery and need to buy crude oil (and haven't made a prior arrangement that locked in a lower price), the spot price for crude oil determines what you can expect to pay per barrel (42 gallons). The data for this exercise also include the price of gasoline sold at service stations to retail customers. Both series were compiled by the Department of Energy and represent national averages. The data are weekly from January 1997 through May 2006.

Motivation

a) The rise of gasoline prices along with record profits at Exxon-Mobile produced stories of corporate greed. How could you use a statistical model to explain the rise of prices at the pump?

Method

b) If price rises are immediately passed on to consumers on a 1-for-1 basis, what should you expect for the slope of a simple regression of the price of gas (in dollars per gallon) on the price of crude oil (in dollars per barrel)?

c) Having seen the trends in the prices, do you think it will be easier to work with the prices themselves or with percentage changes? Use plots of the time trends in the percentage changes to help form your answer.

d) Price changes are unlikely to immediately move through the system. It takes a while for the oil that is refined to reach the pump. If the effects of price changes in crude take several weeks to reach the pump, what type of transformations of crude prices can you expect to be useful in explaining the rise in prices at the pump?

Mechanics

e) Fit the simple regression of gasoline prices on contemporaneous prices of crude oil. Is the slope near what you expected?

f) Examine the residuals from the simple regression that you fit in part (e). Do these meet the conditions for fitting the simple regression model? In particular, what flaw common to time series regressions is evident in the residuals from this model?

g) Fit the simple regression of the percentage change in price of gasoline on the percentage change in the price of crude oil in the same period. What does the slope of this regression tell you?

h) Examine the residuals from the regression of the percentage changes in part (g). Identify the two largest outliers and summarize their effects.

i) Add four lags of the percentage change in the price of crude oil to the regression fit in part (g). (The model will then have five predictors: the current percentage change as well as the percentage changes from the prior four weeks.) Does the fit improve? What is the interpretation of the coefficients of the lagged variables?

j) Does collinearity have a strong effect on the fit of the multiple regression in part (h)?

k) Save the residuals from the multiple regression fit in part (h). Form four lags of these residuals and add these lags to the model. Your model should then have a total of nine predictors. Does the addition of these terms improve your fit? What is your interpretation of these coefficients?

l) Does the model you created in part (k) satisfy the conditions of the multiple regression model?

Message

m) Consider the sum of the slopes from the final multiple regression in part (i). What should these add up to? Do they?

n) Interpret the results of your model. What do they indicate about the rate at which price changes move through the system?

Case: Analyzing Experiments

Many businesses have learned that they need to customize their offerings to suit the tastes of local markets. Retailers risk lower sales if they try to stock every store exactly the same,[1] but managing every product differently complicates decisions. An effective way to determine whether customization is necessary is to conduct an experiment. The main objective of the experiment in this case is to determine whether there's an interaction between geographic location and the product choices available to managers. An interaction suggests that decisions may need to be made locally rather than imposed from corporate headquarters. But how do we decide if such an interaction exists?

A PRICING EXPERIMENT

Research has found a variety of ways to influence customer behavior. The study that underlies this project discovered that the way in which a price is presented influences customer perceptions and purchase decisions. The particular example that we'll consider concerns the way a company advertises the prices of automobile repairs. This company operates a national chain of auto repair shops whose services include brake maintenance and repairs. Prices for these services are typically split into the cost of labor and the cost of parts. The brake service offered in this experiment typically costs $245, but is being offered for $198.

The marketing team would like to learn how partitioning of the price affects consumer responses. Following marketing research,[2] the company decided to advertise the brake service with the price partitioned in two ways: $139 for parts and $59 labor and split

evenly as $99 for parts and $99 for labor. The total cost for the repair is the same; all that changes is how the cost is presented to the customer. As a baseline, the ad can list the total price without splitting the cost into parts and labor. In addition, the marketing team is aware that other retailers have found that customizing offerings to suit tastes of local customers helps attract and retain loyal shoppers.[3] Should the company customize its offerings regionally, or can it stick to a common, more easily managed plan?

Here are the two questions that the company would like to answer.

1. Does price partitioning affect sales?
2. If so, does the effect of price partitioning vary by region?

If the effect of price partitioning varies by region, then these two factors interact. If the interaction between region and pricing is strong, then managers will need to use different types of ads in different parts of the country. If there's not much interaction, the company can simplify the advertising by using a single ad everywhere without sacrificing sales.

Balanced Experiment

To learn how price partitioning affects customers in different regions, the company conducted an experiment. Advertisements were prepared that touted the reduced price for brake service. The $198 price was presented in three otherwise identical advertisements

- as a total: $198 without partitioning into costs for parts and labor,
- with small labor costs: $139 for parts with $59 for labor, and
- with small parts costs: $99 for parts with $99 for labor.

The company anticipates that reactions to these prices might differ in various parts of the country. To measure the effect of the advertisements, the company randomly selected 120 franchise outlets from

---

[1] For an example of local preferences in retail, see Statistics in Action: Testing Association (page 456).

[2] R. W. Hamilton and J. Srivastava (2008). When 2+2 is not the same as 1+3: variations in price sensitivity across components of partitioned prices. *Journal of Marketing Research*, **45**, 450–461.

[3] Rosenblum, S. In recession, strategy shifts for big chains. *The New York Times*. June 20, 2009.

around the nation. The franchises were equally divided among four regions, with 30 in the Northeast, South, Midwest, and West regions. Within each region, advertisements were randomly divided among total, small labor, and small parts price partitions. The design of this experiment is balanced (Chapter 26). The data include an equal number of observations in every region, and within each region the data include an equal number of advertisements for each price partition.

The response in this analysis is the average change in daily sales during the promotion from the baseline for each franchise. All of these franchises had been in business for at least two years before the experiment and generated similar total revenue streams.

The allocation of observations to the different conditions in this experiment follows what is known as a **factorial design**. There are two experimental factors, sometimes called "treatments": region and price partition. The data include an equal number of observations for every combination of the two factors. We observe sales in 10 franchises for each combination of region and each price partition. A balanced factorial design produces data that can be analyzed in a planned fashion.

Preliminary Data Analysis

It is convenient to summarize data from a factorial experiment with two factors in a table, with the levels of the two experimental factors labeling the rows and columns. Because this presentation is so natural, this type of experiment is said to have a two-way factorial design. Table 1 shows the average change in daily sales for the 10 franchises for each combination of region and price partition.

Skimming the table suggests what's happening. Overall, daily sales are up from about $100 to more than $600. If we focus within the columns, small labor costs generated the largest increase in the South, whereas small parts costs topped the list in the Midwest. Ads that show the total price seem to be equally well received. If instead we focus within the rows, we can pick out which type of advertising seems best in the different regions. For instance, the price partition with small labor prices generates the highest increase in sales in the Northeast and South, whereas sales increase the most in the Midwest and West when presented with small parts costs.

A plot helps show what is happening. Figure 1 shows a **profile plot** of the mean change in sales for the different regions and types of pricing.

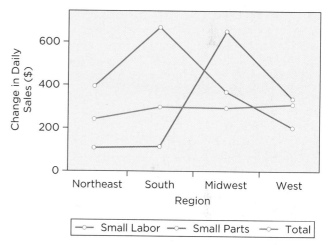

FIGURE 1 Profile plot of the mean sales.

Were there no interaction, or synergy, between the factors, the lines that connect the average values for the different types of pricing would be roughly parallel. Instead, small labor prices produce the largest increase in sales in some regions, and small parts prices produce the largest increase in sales in others. To answer the company's second question, we must determine whether these differences are large compared to sampling variation.

---

factorial design The data include every combination of the experimental factors.

---

profile plot Plot showing means of combinations of factors, joining points associated with a fixed level of one factor.

TABLE 1 Average change in daily sales during the promotion.

		Price Partition			
		Small Labor	**Small Parts**	**Total Price**	**Avg**
	Northeast	$392.20	$106.30	$240.00	$246.17
	South	$665.60	$112.30	$296.60	$358.17
Region	**Midwest**	$366.30	$646.80	$291.10	$434.73
	West	$201.90	$336.20	$308.80	$282.30
	Avg	$406.50	$300.40	$284.13	$330.34

Because Table 1 shows only the average change for each combination, we cannot judge whether these differences are large in comparison to sampling variation. As in a one-way analysis of variance, we need to compare the differences between the means to the variation within the groups. As a first look, the side-by-side boxplots in Figure 2 summarize the marginal distributions of the change in sales for each experimental factor.

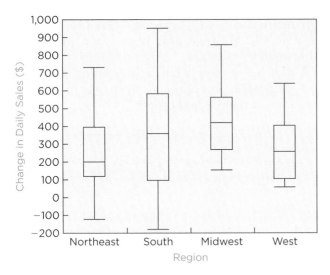

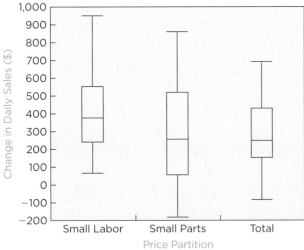

FIGURE 2 Side-by-side boxplots of the change in daily sales.

caution Having looked at the table of means (Table 1), we can tell that these marginal views of the data conceal important information. For example, the boxplots grouped by price partitions suggest that small labor prices produce larger changes. The boxplot for this level is shifted up from the other two in Figure 2. We know from the means in Table 1, however, that small labor costs are favored in the South, but small parts costs are favored in the Midwest. Marginal views of the data such as those in Figure 2 hide the presence of interaction.

TWO-WAY ANALYSIS OF VARIANCE

The marginal views of the data in Figure 2 are fine so long as the two factors don't interact. Experimental factors interact if the effect of the effect of one factor depends on the level of the second factor. That is clearly happening in these data. Customers in the South, for instance, respond differently to price partitioning than customers in the Midwest. The question that remains is whether the differences found in Table 1 and Figure 1 are statistically significant.

In keeping with the tabular layout of Table 1, the standard analysis of data from an experiment with two factors is known as a **two-way analysis of variance**. The method for fitting a two-way analysis of variance resembles that used to fit a one-way analysis of variance in Chapter 26. As in a one-way anova regression, we determine statistical significance by comparing differences among the means to the variation within subsets. The explanatory variables in the regression are dummy variables that represent the categorical variables, only now we need to represent two categorical variables rather than one.

There's another difference that distinguishes a two-way anova regression from the one-way anova regression. We need to include explanatory variables to model interactions. As in other regression models (Chapter 25), interactions in an anova regression are products of explanatory variables. An interaction indicates that the association between an explanatory variable and the response depends on another explanatory variable. We've already seen that this happens in the pricing experiment. The data show that the choice of the most effective price partition depends on the region.

Model

The MRM for a two-way anova gives an equation for the population means associated with Table 1. The equation resembles that in a one-way anova, but adds terms related to the second factor and interaction. First, we'll state this equation without explanatory variables, then add dummy variables to show how the model is estimated with regression.

To write the equation of the MRM, let μ_{ij} denote the mean of the response in the population for observations "treated with" level i of the first factor (rows) and level j of the second factor (columns). We'll call μ_{ij} a cell mean, the mean in the population

two-way analysis of variance Regression model that uses two categorical variables and interactions as explanatory variables.

associated with the cell in row i and column j of Table 1. For example, μ_{12} is the mean population change in sales in the Northeast region (row 1) with small parts pricing (column 2). The equation of a two-way anova decomposes each cell mean into a sum of four components

$$\mu_{ij} = \beta_0 + \beta_i^{\text{row}} + \beta_j^{\text{col}} + \beta_{ij}^{\text{int}}$$

The intercept β_0 is the mean of a baseline cell; the mean of every combination is compared to the mean of this baseline cell. β_i^{row} compares the mean of the baseline cell to the mean of a cell in the same column but different row. Similarly, β_j^{col} compares the mean of the baseline cell to the mean if the column factor changes to level j. The interaction term β_{ij}^{int} captures the synergy produced by combining level i of the row factor with level j of the column factor. Without an interaction, the largest mean for every row (column) is in the same column (row).

Dummy variables allow this description for μ_{ij} to be written as an equation with a collection of explanatory variables. The dummy variables act as "switches" that pick the right coefficients. For example,

$$\mu_{ij} = \beta_0 + \beta_i^{\text{row}}D_i(\text{Row}) + \beta_j^{\text{col}}D_j(\text{Column}) + \beta_{ij}^{\text{int}}D_i(\text{Row})D_j(\text{Column})$$

The βs in this expression are the population regression slopes, and the Ds are dummy variables. For example, $D_1(\text{Row})$ in the example is a dummy variable identifying observations in the Northeast, and $D_2(\text{Column})$ is the dummy variable that identifies observations of the small parts price partition. The interaction is literally a product of separate dummy variables. The equation implies that the mean change in sales in the Northeast with the small parts partition is

$$\mu_{12} = \beta_0 + \beta_1^{\text{row}}D(\text{Northeast}) + \beta_2^{\text{col}}D(\text{Sm Parts}) + \beta_{12}^{\text{int}}D(\text{Northeast})D(\text{Sm Parts})$$

The four regions require three dummy variables, leaving aside the last region (*West*), and the three partitions of price require two more dummy variables (excluding one partition, *Total*). The interactions include all products of these, adding $6 = 3 \times 2$ more dummy variables. Together, we have an intercept plus $3 + 2 + 6 = 11$ dummy variables in the model.

For the example, we've omitted dummy variables for the West and the total price partition, so the mean for this combination defines β_0; the intercept $\beta_0 = \mu_{43}$. The other βs identify how the means for other levels of the factors differ from β_0. It's easier to interpret these once we see the regression output.

Plan for Analysis

The statistical analysis of a two-way anova follows a systematic script. The beauty of a balanced two-way

factorial design is that the analysis follows a straightforward plan. The tests group the coefficients in the MRM, separating coefficients for the factors (β_i^{row} and β_j^{col}) from those for the interaction. Here's the plan:

1. Check the assumptions of the multiple regression model. We start here for any regression model, regardless of the context.
2. Determine the statistical significance of the overall model. As in other regression models, use the p-value from the overall F-statistic. If the overall F-statistic is not statistically significant, there's no meaningful difference among the means for different levels of the experimental factors.
3. If the overall F-statistic is statistically significant, test the significance of the interaction of the experimental factors. If interaction is present, do not proceed to step 4. The effect of one factor depends on the level of the other.
4. If there's not statistically significant interaction, determine the significance of differences among the levels of the experimental factors.

Steps 3 and 4 require a different test of the regression slopes, one that we've not needed in previous analyses.

Checking the Model

Because we're comparing means, not fitting lines, we skip the linear condition (as in a one-way anova regression) and check the conditions on the error terms. The residuals in the two-way anova regression need to meet the usual conditions

✓ Evidently independent
✓ Similar variances
✓ Nearly normal

The random selection of franchises in this experiment suggests that these data are independent observations. We should, however, think about, whether other problems introduce dependence. The problems depend on the context of the analysis. For example, in this analysis, we need to verify that the franchises are independently operated (some franchises share a common management team) and that the advertising for one location does not "leak" into the area served by a different franchise. Both appear to be met in our data.

As in other regressions, the best place to check the similar variances condition is the plot of the residuals on the estimated values from the regression as shown in Figure 3.

The residuals lie in columns above the x-axis because estimated values in an anova regression are the averages of the response for each combination of

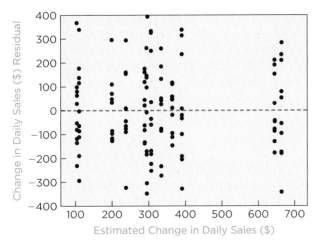

FIGURE 3 Scatterplot of residuals on estimated values of the response.

the levels of the factors. For example, the points at the far right are from the South region with the small labor price partition; this combination has the highest mean response in Table 1 (with estimated response $665.60). The adjacent column of points comes from the Midwest with the small parts price partition; this combination has the second highest mean response ($646.80). The variances of the residuals within these columns of points ought to be similar, and that's the case here. There's no hint that the variation increases with the size of the response, a common problem (Chapter 22).

We use the normal quantile plot of the residuals (Figure 4) to check the nearly normal condition.

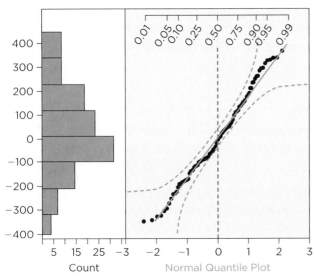

FIGURE 4 Normal quantile plot of residuals from pricing experiment.

These residuals stay within the bands around the diagonal reference line; these data are nearly normal. Had the residuals gone outside these limits, then we

would have to appeal to the Central Limit Theorem. As in other regression models, the sample size n must be larger than 10 times the larger of the squared skewness K_3^2 and the absolute value of the kurtosis $|K_4|$ of the residuals.

Testing the Factors

Testing begins with the overall fit of the model, without trying to isolate which (if any) factor affects the response. The null hypothesis for this initial test is that all of the coefficients of the dummy variables in the model are 0 in the population,

$$H_0^{\text{all}}: \text{all } \beta_i^{\text{row}} = \beta_j^{\text{col}} = \beta_{ij}^{\text{int}} = 0$$

If this hypothesis holds, then the population means for every combination of the factors are the same. As in other multiple regression models, we test H_0^{all} using software to obtain the p-value of the F-statistic in an Analysis of Variance summary table. This table appears along with R^2 and s_e in Table 2.

TABLE 2 Summary statistics of the two-way anova regression for the pricing experiment.

R^2	0.4947
s_e	180.1559
n	120

Analysis of Variance

Source	df	Sum of Squares	Mean Square	F-statistic
Model	11	3432359.7	312033	9.6140
Error	108	3505265.3	32456	**_p_-value**
Total	119	6937625.0		<0.0001

The regression model explains about half of the variation in the sales ($R^2 = 49.47\%$). This performance is highly statistically significant ($F = 9.614$ with p-value < 0.05). We reject H_0^{all}; there are statistically significant differences among the means. That finishes Step #2 of our planned analysis.

We've learned that the multiple regression explains statistically significant variation, so the purpose of Step #3 is to find out why. In particular, Step #3 looks for an interaction between the factors. If there's an interaction, then the effect of pricing depends on region.

Rather than go directly to the coefficient estimates for this test, we stop at a second summary table, one that we have not required before. This table shows tests of three null hypotheses, one for each experimental factor and one for the interaction between these factors (*Region* × *Price Partition*). These null hypotheses group the slopes of the dummy variables. One hypothesis says that there's no interaction, and the other two hypotheses claim that the levels of the factors have no effect on the average change in sales.

H_0^{row} : all regions are the same, $\beta_i^{\text{row}} = 0$

H_0^{col} : all pricing ads are the same, $\beta_j^{\text{col}} = 0$

H_0^{int} : the factors do not interact, $\beta_{ij}^{\text{int}} = 0$

Table 3 summarizes tests of these hypotheses for the two-way anova.

TABLE 3 *F*-tests of the experimental factors and interaction.

Source	df	*F*-statistic	*p*-value
Price Partition	2	5.4428	0.0056
Region	3	6.4904	0.0004
Region × Price Partition	6	12.5662	<0.0001

These tests are called **effect tests** or **partial *F*-tests** because they test whether a subset of the coefficients in the multiple regression are simultaneously 0. The column in Table 3 labeled df (for degrees of freedom) identifies the number of slopes in each test. *Region*, with 4 levels, requires 3 dummy variables, for instance, and so has 3 slopes. The test of the interaction has 6 because it includes a slope for each product of a dummy variable for *Region* with a dummy variable for *Price Partition*.

---

effect test, partial *F*-test Test of a subset of related coefficients in a regression model, such as those defined by one categorical explanatory variable.

Following the analysis plan, we test the interaction hypothesis H_0^{int}. The *F*-statistic for this test in Table 3 is $F = 12.5662$ which has a *p*-value near 0. We reject H_0^{int} and conclude that the effect of the offered pricing depends on the region. Managers have to consider formulating region-specific pricing or lose the benefits of customization.

Regression Coefficients

To understand the effects of the different factors on the response, we need to examine the estimated coefficients. Since we've rejected H_0^{int}, we proceed to the individual coefficient estimates in the regression model. Table 4 shows the estimated coefficients of the dummy variables and highlights the rows that estimate interactions. Notice that the number of estimates in the regression matches the number of means within Table 1. The regression has 12 estimates, an intercept and the slopes of 11 dummy variables. These 12 estimates reproduce the 12 averages in Table 1.

The intercept is the average change in sales for the baseline group, the West with ads that show the total price. To interpret the slopes, review Table 5 (next page). This table subtracts the mean of the baseline group from every cell mean in Table 1.

The last column in Table 5 indicates that ads showing the total price generate less sales outside the West region: $68.80 less in the Northeast, $12.20 less in the South, and $17.70 less in the Midwest. These are exactly the coefficients in the multiple regression for *D*(Northeast), *D*(South), and *D*(Midwest). Similarly, in the West, the last row of Table 5 shows that ads featuring smaller labor prices generated $106.90 less in sales and those with smaller parts prices generated $27.40 more in sales.

TABLE 4 Estimated coefficients of the dummy variables in the two-way anova regression.

Term	Estimate	Std Error	*t*-statistic	*p*-value
Intercept	308.80	56.97	5.42	<0.0001
D(Northeast)	−68.80	80.57	−0.85	0.3950
D(South)	−12.20	80.57	−0.15	0.8799
D(Midwest)	−17.70	80.57	−0.22	0.8265
D(Sm Labor)	−106.90	80.57	−1.33	0.1874
D(Sm Parts)	27.40	80.57	0.34	0.7345
D(Northeast) D(Sm Labor)	259.10	113.94	2.27	0.0249
D(Northeast) D(Sm Parts)	−161.10	113.94	−1.41	0.1603
D(South) D(Sm Labor)	475.90	113.94	4.18	<0.0001
D(South) D(Sm Parts)	−211.70	113.94	−1.86	0.0659
D(Midwest) D(Sm Labor)	182.10	113.94	1.60	0.1129
D(Midwest) D(Sm Parts)	328.30	113.94	2.88	0.0048

TABLE 5 Deviations of cell averages from the average of the reference cell (West, total price).

Region		Price Partition		
		Small Labor	**Small Parts**	**Total Price**
Region	Northeast	83.4	−202.5	−68.8
	South	356.8	−196.5	−12.2
	Midwest	57.5	338	−17.7
	West	−106.9	27.4	0

These values match the coefficients for D(Sm Labor) and D(Sm Price) in Table 4.

Were there no interaction, the rest of the values in Table 5 would be redundant. We would not need them. For example, without an interaction, it would be easy to figure out the cell mean for the Northeast with small labor prices. Without an interaction, the difference between sales in the West and sales in the Northeast would be the same for every type of pricing: $68.80 less in the Northeast than in the West. Similarly, with no interaction, the difference between sales produced by ads showing total price versus ads showing small labor prices would the same in the Northeast as in the West, $106.90 less.

That's not what happens. Sales in the Northeast with ads for smaller labor prices sell $83.40 *more*, not $68.80 + $106.90 = $175.70 *less*. The coefficient of each interaction in the multiple regression is the difference between what we observe and what would happen if there were no interaction. Hence, the coefficient of D(Northeast) D(Sm Labor) is 83.40 + 175.70 = 259.10. Sales in the Northeast produced by the ad featuring small labor costs are $259.10 more than we'd expect if there were no interaction.

Interactions and One-way Anova

The two-way anova regression shows that pricing affects the response to this promotion; the change in sales varies from region to region. The presence of a statistically significant interaction between *Region* and *Price Partition* implies that the effects of the different types of ads vary from one region to another. Price partitioning that features smaller labor costs works best in the Northeast and South, and ads featuring smaller parts costs work best in the Midwest and West.

Now that we have identified the presence of statistically significant interaction, it can be useful to confirm the significance of differences within each region. Examine the profile plot in Figure 1. Are the differences between the three means within each region statistically significant? (We would not need to do this were there no interaction. All we'd need are the overall effects for the row and column factors.)

Because of the interaction, it's simplest to separate the data by region and analyze the differences among the means within a region. That is, use a one-way analysis of variance within each region. (Other methods pool the data, but the advantages of pooling are small and produce a more complicated analysis.) Each one-way anova uses an F-test to compare the three types of ads within a region. If a one-way anova within a region finds a statistically significant difference, we can follow that test with a multiple comparisons procedure (such as Tukey's method from Chapter 26). As can be guessed from Figure 1, the one-way anova in this example finds statistically significant differences due to pricing in the Northeast, South and Midwest, but not the West.

CASE SUMMARY

Data collected for a **two-way analysis of variance** contrasts the effects on the average of the response produced by varying the levels of two categorical variables. A **factorial design** specifies that the data will include every combination of the two variables. Parallel lines in the **profile plot** of the cell means defined by the two factors indicate an absence of interaction between the two factors. The model is estimated by a multiple regression with dummy variables. The **partial F-test** (or **effect test**) of the interaction components of the model indicates whether the presence of interaction is statistically significant.

■ Key Terms

effect test, 733
factorial design, 729

partial F-test, 733
profile plot, 729

two-way analysis of variance, 730

■ Questions for Thought

1. What's the estimated value from the anova regression given by the slopes in Table 4 for the change in sales in the Midwest if ads feature the small labor partition? What's the simple way to get this estimate?

2. The standard errors of the slopes of the dummy variables in Table 4 are all the same (80.57) and less than the common standard error of the interactions (113.94). Why is that?

3. What happens to the profile plot in Figure 1 if we reverse the rows and columns of the data? (Let *Price Partition* define the *x*-axis and connect points in the same region.) Sketch the new plot. Do you still see the presence of interaction?

4. We don't actually need further calculations to compare the effects of the price partitions within regions; everything we need is in Table 1 and Table 2. Using those tables, identify a statistically significant difference among the means in the Northeast. Be sure to adjust for multiple comparisons (Chapter 26).

5. The analysis shown here uses a balanced experiment, with equal numbers of observations for each pairing of the factors. What happens to the estimates and their standard errors if that's not the case? Does your software still work? Try it! Delete a few rows of the data and see what happens.

Case: Automated Modeling

Good regression modeling starts with familiarity with the substance of the problem and a thorough understanding of the response and explanatory variables. When the data include many possible explanatory variables, it can take hours of careful analysis to arrive at a model. Even then, we are seldom sure that we've made complete use of the explanatory variables. The need for an immediate answer or the size of the problem often requires that we resort to an automated search for the explanatory variables that accurately predict the response. Software algorithms that automate the modeling process date back to the first digital computers. For this example, we'll explore the most familiar of these, the algorithm known as stepwise regression.

PREPARATIONS

A change in the ownership of a manufacturing company led to a change in the management of two production facilities. These facilities produce customized metal blocks. Robots owned by customers of this company shape the blocks produced by this manufacturer into detailed parts that go into jet engines and automobiles. When placing an order, customers specify various properties of the blocks that they need, including specific dimensions, materials, and tolerances. Before committing to the order, the manufacturer has to quote a unit price to the customer.

That's become a problem. The turnover in management was not smooth and many details of the operation of these two facilities were lost during the transition. In particular, new managers have found it difficult to estimate the cost of producing blocks that meet the specifications requested by customers. As a result, the company has underbid several recent orders, quoting prices that were ultimately less than what it cost to manufacture the blocks.

Data for Modeling

As a short-term solution, the company hired consultants to build a model that can estimate production costs based on the input specifications. To facilitate the modeling, managers were able to recover information describing 198 prior production runs. For each of these runs (orders for multiple copies of a specific type of block), managers found the average cost per block produced. They also recovered other characteristics of the production. These are listed in Table 1; numbers in the last column give the number of levels of a categorical variable. One of the production facilities is new, located in Arkansas where labor is cheaper. The second is much older, located in New Jersey.

For the rest of this analysis, we'll take on the role of the consultants and build a regression. Even though we have a number of possible explanatory variables already, we'll start by building a few more. Automated searches work better if we can supply more sensible potential explanatory variables. We get these from thinking about the context of the data. Most of the potentially relevant explanatory variables in Table 1 describe variable costs. For example, labor hours and machine hours produce variable costs. (See the discussion of fixed and variable costs in Chapter 22.) To absorb remaining fixed costs, the consultants added the reciprocal of the number of units in the production run. The coefficient of 1/Units in a regression with average cost as the response estimates fixed costs not captured by other explanatory variables. Another explanatory variable added by the consultants is the total cost of metal used, computed as

$$\text{Total Metal Cost} = (\text{Weight Final} + \text{Weight Rem})(\text{Cost Metal/Kg})$$

Figuring that breakdowns introduce a fixed cost (that is, there is a fixed cost per breakdown rather than per unit), they also added the variable Breakdown/Unit. Finally, conversations with employees suggested that work slowed down if a plant got too hot or too cold. So they added the variable

$$\text{Sq Temp Deviation} = (\text{Room Temp} - 75)^2$$

Sq Temp Deviation gets larger the farther room temperature gets from 75 degrees. When combined with

TABLE 1 Variables from company records available for modeling.

Variable	Definition	Levels
Average Cost	in dollars, per block	
Units	number of blocks in the production run	
Precision SD	allowed SD of block dimensions (in mils)	
Weight Final	kilograms of a finished block	
Weight Rem	kg removed during production, per block	
Stamp Ops	number of stamping operations, per block	
Chisel Ops	number of chiseling operations, per block	
Labor Hours	direct labor hours, per block	
Machine Hours	machine hours, per block	
Cost Metal/Kg	dollars/kilogram of input metal	
Room Temp	indoor temperature (F°) during production	
Breakdowns	number of production failures during run	
Detail	whether or not extra detail work is required	2
Rush	whether or not the job was a rush order	2
Manager	supervising production manager	5
Music	type of music played during production	3
Shift	day or night shift	2
Plant	new or old plant	2

the variables listed in Table 1, the data table used by the consultants has the response plus 21 other variables, six of which are categorical.

Preliminary Analysis: Outliers and Collinearity

Though pressed for time, the consultants examined the association among many of the columns in the production data. This preliminary scan revealed some important aspects of the data that influence the subsequent regression modeling. **Time invested** **tip** **in a preliminary examination of the data often greatly simplifies building and interpreting the regression model.** Even with many variables, it is important to look for pronounced outliers and to assess the degree of collinearity. For example, the contingency table of Rush with Detail shown in Table 2 shows that these potential explanatory variables are highly associated.

The association is sensible: jobs produced in a rush seldom require detailed production. The association between Rush and Detail amounts to collinearity between two categorical variables.

There's even stronger association between Plant and Manager. Table 3 shows that managers only work in one plant. The variables Plant and Manager are totally confounded. We won't be able to tell from these data, for instance, whether costs are higher or lower in the new plant than in the old plant because it's new or because of the managers who work there.

We can use a scatterplot matrix to explore the association among several numerical variables at once. Those plots reveal two large outliers. For example, Figure 1 shows the scatterplot of the response (average cost per block) versus labor hours.

TABLE 2 Contingency table of Detail by Rush.

		Rush		
		None	**Rush**	**Total**
Detail	None	5	90	95
	Detail	94	9	103
	Total	99	99	198

TABLE 3 Contingency table of Plant by Manager.

		Manager					
		Jean	**Lee**	**Pat**	**Randy**	**Terry**	**Total**
Plant	New	40	0	0	0	31	71
	Old	0	43	42	42	0	127
	Total	40	43	42	42	31	198

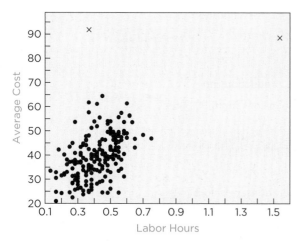

FIGURE 1 The scatterplot of the cost per unit versus labor hours shows two outliers.

Both outliers have extremely high average costs compared to the other runs. A regression model may be able to explain these outliers, but we will exclude them while looking for that model. Otherwise, we might end up with a regression whose variables were chosen to fit two outliers rather than find the pattern in the majority of the data.

With these outliers excluded, the scale of scatterplots focuses on the majority of points. The scatterplot matrix in Figure 2 shows plots between the average cost and five of these variables. We picked these variables to show to show the collinearity among the explanatory variables.

Of these explanatory variables, Labor Hours has the highest correlation with the response ($r = 0.50$). That's smaller than several correlations among the explanatory variables. For example, the correlation between the number of stamping and chiseling operations is 0.77, and the correlation between the number of machine hours and chiseling operations per block is 0.72. These correlations make sense; all three variables describe shaping operations applied to the blocks. These correlations, however, will make it hard to separate the effects of chiseling, stamping, and general machining on the final cost.

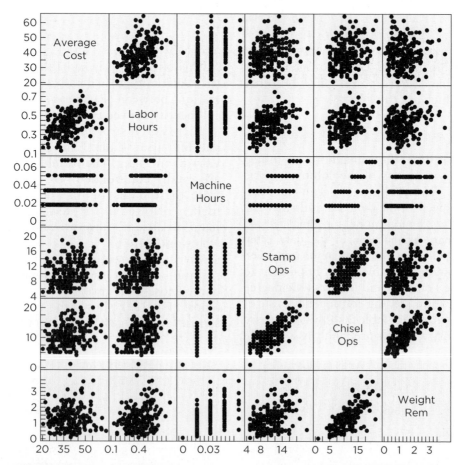

FIGURE 2 Scatterplot matrix of the response (top row) with five possible explanatory variables.

SATURATED MODEL

One approach to building a regression when faced with a large collection of explanatory variables is to use them all. We call this the **saturated model** because we've used every explanatory variable available. We have 21 possible explanatory variables and, after setting aside the two outliers, 196 observations. With more than 9 observations per explanatory variable, we have plenty of data to fit this regression. (Some statisticians suggest fitting the saturated model so long as you have at least 3 observations per explanatory variable.) If the explanatory variables were nearly uncorrelated, we would be able to interpret this model as well. With the substantial collinearity among these, that's not going to happen for these data. The redundancy forces us to remove one variable. Because Plant and Manager are totally confounded, we excluded Plant from the regression. We cannot use Plant and Manager in the same regression since the dummy variables that indicate whether the run occurs in the old or new plant are redundant with those that indicate the manager. For example, Table 3 implies that the dummy variable indicating runs in the new plant is the sum of the dummy variables for runs supervised by Jean and Terry, $D(\text{new}) = D(\text{Jean}) + D(\text{Terry})$.

Table 4 summarizes the fit of the saturated regression. You can check that the residuals from this model meet the usual conditions (evidently independent, similar variances, and nearly normal), so we're justified in examining test statistics.

TABLE 4 Summary of initial regression using all supplied explanatory variables.

R^2	0.5265
Adjusted R^2	0.4568
s_e	6.5909
n	196

Analysis of Variance

Source	df	Sum of Squares	Mean Square	F-statistic
Model	25	8210.895	328.436	7.5608
Error	170	7384.697	43.439	**p-value**
Total	195	15595.591		<.0001

The saturated regression explains $R^2 \approx 53\%$ of the variation in the costs per block among these runs. This regression has 25 explanatory variables,

and the addition of each one pushes R^2 a little higher whether that variable is statistically significant or not. Adjusted R^2, which is close to 46%, is more "honest" about how much variation the model really explains, correcting R^2 for the number of explanatory variables. The sizeable gap between R^2 and adjusted R^2 suggests that many of these explanatory variables are not useful and have only served to make R^2 artificially large.

The equation of the saturated model nonetheless describes statistically significant variation in costs per block. The overall F-statistic, which measures the size of R^2 relative to the number of observations and number of explanatory variables, has p-value much less than 0.05. The model explains real structure, but it's hidden among the slew of explanatory variables.

If we look at the individual coefficient estimates, we find that just 5 out of these 25 estimates have p-values less than 0.05 (Labor Hours, Breakdowns/Unit, Temp Deviation, and the dummy variables for two managers, Jean and Pat). Many of the estimates have unexpected signs. For example, the coefficient of the dummy variable Rush implies that jobs produced in a hurry cost less to produce. (Why does the model have 25 slopes if we only added 20 to the regression? The answer is categorical variables. The categorical variable Manager has 5 levels and so adds 4 dummy variables to the regression. Music has 4 levels and so adds 3 dummy variables.)

To have some hope for interpreting our model, we will build a simpler, more parsimonious model that fits about as well but with fewer explanatory variables.

STEPWISE REGRESSION

Rather than start with every available explanatory variable (which forces us to exclude variables like *Plant* that are redundant), **forward stepwise regression** builds a regression from the ground up, adding one variable to the fit at a time. The algorithm is straightforward.

1. *Initialization.* Stepwise regression begins with an initial model. This model usually has no explanatory variables, but you can force the initial model to include certain explanatory variables. For example, we might force the regression to use the explanatory variable 1/Units to require an estimate of fixed costs.

---

saturated model A regression model that uses all of the available explanatory variables at once.

forward stepwise regression Automatic algorithm that adds variables to a regression.

2. *Search.* The algorithm searches the collection of potential explanatory variables not already included in the regression. It adds the variable that increases the R^2 of the model the most.

3. *Termination.* Forward stepwise regression stops when the variable that improves the fit the most does not meet a minimum threshold. Usually, this condition is expressed by putting a threshold on the *p*-value of the added variable. Unless the *p*-value of the added variable is less than a constant known as the **p-to-enter**, the algorithm halts.

Forward stepwise regression is an example of a **greedy search**. Forward stepwise adds the variable in the search step that improves the current fit the most. It looks one step ahead for the most immediate improvement. Because of collinearity, this variable may not have been the best choice for the next step. For example, consider the first step beginning from an empty initial model. The first variable added to the regression, let's call it X_1, is the explanatory variable which is most correlated with Y. X_1 has the highest correlation with Y, and so generates the simple regression with the largest R^2 statistic. Forward stepwise regression then looks for the explanatory variable that adds the most in addition to X_1. Because of collinearity, it may be the case that the two-variable model using, say, X_2 and X_3 has a larger R^2 than the fit with X_1 and any other variable. Forward stepwise regression won't find that combination because once a variable is in the model, it stays.

Avoiding Over-fitting

The key tuning parameter in running forward stepwise regression is the choice of the *p*-to-enter threshold. The simple rule we recommend is based on the Bonferroni rule discussed in Chapter 26 for multiple comparisons. Because stepwise regression considers many explanatory variables, we have to be concerned about the problem of **over-fitting**. Over-fitting occurs when we add explanatory variables to a regression model that claim to improve the fit but really don't. If a regression model has been over-fit to data, then we'll think that we have a model that

performs well when in fact, it predicts worse than R^2 suggests.

It's easy for over-fitting to occur when using a greedy search like forward stepwise regression. Think again about the first step, beginning from an initial model that has no explanatory variables. Suppose that we have set the *p*-to-enter threshold to our usual threshold for *p*-values, *p*-to-enter = 0.05. At the first step, stepwise regression picks the explanatory variable that is most correlated with the response. Imagine that we have m possible explanatory variables and that in fact none of these is predictive of the response in the population (that is, all of the slopes in the population are 0). By chance alone, 5% of these explanatory variables have a *p*-value less than 0.05. If $m = 100$, we'd expect five *p*-values to be less than 0.05 even though none of the variables is actually related to y. By picking the explanatory variable that has the highest correlation with y in the sample, we've we have tricked ourselves into thinking that we found a useful predictor when in fact it's unrelated to the response in the population.

We recommend setting the *p*-to-enter threshold smaller, taking account of the number of explanatory variables considered. We set the *p*-to-enter = $0.05/m$. This choice holds the probability of adding a variable to the model by chance alone to 5% even if we search through the data to build our model. This threshold makes it likely that if a variable enters the regression, it not only describes the observed sample but will also accurately predict new cases from the population.[1]

Running Stepwise Regression

We ran forward stepwise regression, using it to explore all of the available explanatory variables, starting from an empty initial model. When we use stepwise regression with categorical variables, we allow it to search the dummy variables that identify the groups separately. For example, we do not require that stepwise add all four dummy variables for Manager at once. Rather, it can pick those dummy variables that appear useful. Taking this approach means that we have to count up all of the possible numerical variables plus the dummy variables that represent categorical variables. In this example, we have a total of $m = 26$ variables to search. (There's one more than in the saturated model since we don't have to exclude Plant from the search.) We set *p*-to-enter = $0.05/26 \approx 0.00192$.

---

p-to-enter Forward stepwise regression only adds an explanatory variable to the current model if the *p*-value of the addition is less than this threshold.

greedy search An algorithm that myopically looks for the best immediate improvement without considering several steps ahead.

over-fitting Adding explanatory variables that superficially improve a regression but instead artificially inflate R^2.

---

[1] To make stepwise truly effective requires other adjustments that require modifying the software that builds a stepwise regression. See D. P. Foster and R. A. Stine (2004). Variable selection in data mining: Building a predictive model for bankruptcy. *Journal of the American Statistical Association*, 99, 303–313.

Stepwise regression selects a model with five explanatory variables that reaches $R^2 = 0.4638$, which is about the same as the adjusted R^2 for the original saturated model. Table 5 summarizes the steps of the search, with two steps added beyond where our choice for p-to-enter terminates the search.

TABLE 5 Progress of forward stepwise regression.

Step	Parameter	p-value	R^2
1	Labor Hours	0.0000	0.2460
2	Total Metal Cost	0.0000	0.3371
3	Breakdown/Unit	0.0002	0.3848
4	Temp Deviation	0.0003	0.4254
5	D(New Plant)	0.0003	0.4638
6	1/Units	0.0147	0.4804
7	D(Rush)	0.0356	0.4925

The first variable added to the model is Labor Hours. Recall that this explanatory variable has the highest correlation with average cost per block. The second variable added to the model is Total Metal Cost, which increases R^2 from 24.6% to 33.71%. Among all regressions with Labor Hours and one other explanatory variable, the combination of Labor Hours with Total Metal Cost explains the most variation in the response. Next comes Breakdown/Unit, Temp Deviation, and D(New Plant), a dummy variable indicating runs produced in the new factory.

The column of p-values in Table 5 show the statistical significance of the variable at the time it is added to the model. The forward search stops after adding D(New Plant) because the p-value produced by adding the next variable, ($p = 0.0147$ for 1/Units) is larger than our choice for p-to-enter (0.00192).

Interpreting the Stepwise Model

Stepwise regression searches for a parsimonious model that can predict new cases. The simplicity of the model also avoids the collinearity of the saturated model. As a benefit, slopes in a stepwise regression are often easily interpreted. Table 6 shows the overall summary and table of coefficient estimates.

Overall, this equation explains as much variation in the response as the saturated model if we allow for the large number of explanatory variables in the saturated model (Table 4). R^2 and adjusted R^2 for this model are similar to the adjusted R^2 of the saturated model. The residual SD of the stepwise model $s_e = 6.6344$ almost matches that of the saturated

TABLE 6 Summary of the regression model selected by stepwise regression.

R^2	0.4638
Adjusted R^2	0.4497
s_e	6.6344
n	196

Term	Estimate	Std Error	t-statistic	p-value
Intercept	14.7591	2.1018	7.02	<0.0001
Labor Hours	34.2408	4.0930	8.37	<0.0001
Breakdown/Unit	334.3766	65.8334	5.08	<0.0001
Total Metal Cost	2.2287	0.4056	5.50	<0.0001
Temp Deviation	0.0369	0.0073	5.02	<0.0001
Plant[NEW]	4.5233	1.2263	3.69	0.0003

model (6.5909). Examination of residual plots shows that, like the saturated model, the stepwise model meets the conditions of the MRM.

The slopes in the stepwise model are precisely determined (narrow confidence intervals) and, unlike some in the saturated model, all have simple economic interpretations. The coefficient of Labor Hours, for instance, is $34/hour, estimating the cost per additional hour of labor. To see that these are the units, notice that the response is measured in $/block, and Labor Hours is measured in hours/block. Hence the slope for Labor Hours has units

$$(\$/block)/(hours/block) = \$/hour$$

Similarly, the coefficient of Breakdown/Unit is $334, an estimate of the cost of another production breakdown during the manufacture of an order of blocks. (The units for this slope are the ratio of $/block for the response to breakdown events/block, which reduces to event breakdown $.)

The parsimonious form of the stepwise regression comes in handy when it comes time to use this regression model to predict costs for new production runs. We only need to identify the values of the five variables in this model rather than all of those in the saturated model. Fewer explanatory variables also help to avoid an inadvertent extrapolation when predicting costs of new production runs (see Chapter 23).

RELATED ALGORITHMS

Stepwise regression may be run in various ways. We recommend forward stepwise regression that incrementally adds variables. The alternative version called backward stepwise regression starts with a model having many explanatory variables (usually the saturated regression) and at each step removes the explanatory variable that adds the least to R^2. We prefer a forward search because it avoids conflicts caused by redundant variables (Manager and Plant in this example) and does not require the saturated regression. In some applications, we have more explanatory variables than observations. In that situation, we do not have enough cases to fit the saturated model. A hybrid search combines these two. Mixed stepwise regression runs in the forward direction until collinearity reduces the statistical significance of a variable in the model. When this happens, mixed stepwise removes the variable that is no longer contributing to the fit.

An alternative to stepwise regression known as best subset regression is a global search that finds the explanatory variable with the largest R^2, the two variables that have the highest R^2 together, and so forth. Because there are so many possible models to explore, best subset regression runs very slowly and is only used to fit small models. For example, to find the best combination of 5 variables out of 26 as in our example requires examining 65,780 regressions.

CASE SUMMARY

Forward stepwise regression is a **greedy search** algorithm that identifies explanatory variables to use in modeling the response in a regression. Stepwise regression is capable of identifying a parsimonious model with few explanatory variables that nonetheless describes the data as well as the **saturated model** that includes all of the explanatory variables. The stepwise search picks those explanatory variables that obtain a smaller p-value than given by the **p-to-enter** threshold. We recommend setting this tuning parameter to p-to-enter $= 0.05/m$, where m is the number of potential explanatory variables. This choice reduces the chance for **over-fitting**, adding variables to a model that superficially improve the fit while actually degrading predictive accuracy.

▪ Key Terms

greedy search, 740
over-fitting, 740

p-to-enter, 740
saturated model, 739

forward stepwise regression, 739

▪ Questions for Thought

1. The coefficient of the dummy variable for rushed jobs suggests that rushed jobs cost less to produce. Usually, having to do things in a hurry increases costs. Add the dummy variable $D(\text{Rush})$ to the stepwise model. Does the sign of the slope seem right now? How would you explain what's happening? (*Hint:* look at Table 2.)

2. The stepwise model (Table 6) includes the explanatory variable Temp Deviation, the square of the difference between room temperature and 75 degrees. Why 75? Add Room Temp itself to the regression, and then use the resulting estimates to find the optimal center for this quadratic effect.

3. Suppose you initialize backward stepwise regression from the saturated model. Must the backward search eventually reach the model chosen by the forward search? Try it and see.

4. If you were asked about the costs produced by the number of machine hours, what would you say? The stepwise model omits the explanatory variable Machine Hours, in effect assigning a slope of 0 to this variable. To get a better answer, add this variable to the stepwise model and interpret the estimate and its confidence interval. How do those results compare to the estimate from the saturated model?

Tables

Percentiles of the chi-squared distribution.

Right-tail probability		α				
		0.10	0.05	0.025	0.01	0.005
Values of χ^2_α	df					
	1	2.706	3.841	5.024	6.635	7.879
	2	4.605	5.991	7.378	9.210	10.597
	3	6.251	7.815	9.348	11.345	12.838
	4	7.779	9.488	11.143	13.277	14.860
	5	9.236	11.070	12.833	15.086	16.750
	6	10.645	12.592	14.449	16.812	18.548
	7	12.017	14.067	16.013	18.475	20.278
	8	13.362	15.507	17.535	20.090	21.955
	9	14.684	16.919	19.023	21.666	23.589
	10	15.987	18.307	20.483	23.209	25.188
	11	17.275	19.675	21.920	24.725	26.757
	12	18.549	21.026	23.337	26.217	28.300
	13	19.812	22.362	24.736	27.688	29.819
	14	21.064	23.685	26.119	29.141	31.319
	15	22.307	24.996	27.488	30.578	32.801
	16	23.542	26.296	28.845	32.000	34.267
	17	24.769	27.587	30.191	33.409	35.718
	18	25.989	28.869	31.526	34.805	37.156
	19	27.204	30.143	32.852	36.191	38.582
	20	28.412	31.410	34.170	37.566	39.997
	21	29.615	32.671	35.479	38.932	41.401
	22	30.813	33.924	36.781	40.290	42.796
	23	32.007	35.172	38.076	41.638	44.181
	24	33.196	36.415	39.364	42.980	45.559
	25	34.382	37.653	40.647	44.314	46.928
	26	35.563	38.885	41.923	45.642	48.290
	27	36.741	40.113	43.195	46.963	49.645
	28	37.916	41.337	44.461	48.278	50.994
	29	39.087	42.557	45.722	59.588	52.336
	30	40.256	43.773	46.979	50.892	53.672
	40	51.805	55.759	59.342	63.691	66.767
	50	63.167	67.505	71.420	76.154	79.490
	60	74.397	79.082	83.298	88.381	91.955
	70	85.527	90.531	95.023	100.424	104.213
	80	96.578	101.879	106.628	112.328	116.320
	90	107.565	113.145	118.135	124.115	128.296
	100	118.499	124.343	129.563	135.811	140.177

Percentiles of the F-distribution ($\alpha = 0.01$).

Numerator df

.01	1	2	3	4	5	6	7	8	9	10	11	12	13	14	15	16	17	18	19	20	21	22
1	4052.2	4999.3	5403.5	5624.3	5764.0	5859.0	5928.3	5981.0	6022.4	6055.9	6083.4	6106.7	6125.8	6143.0	6157.0	6170.0	6181.2	6191.4	6200.7	6208.7	6216.1	6223.1
2	98.50	99.00	99.16	99.25	99.30	99.33	99.36	99.38	99.39	99.40	99.41	99.42	99.42	99.43	99.43	99.44	99.44	99.44	99.45	99.45	99.45	99.46
3	34.12	30.82	29.46	28.71	28.24	27.91	27.67	27.49	27.34	27.23	27.13	27.05	26.98	26.92	26.87	26.83	26.79	26.75	26.72	26.69	26.66	26.64
4	21.20	18.00	16.69	15.98	15.52	15.21	14.98	14.80	14.66	14.55	14.45	14.37	14.31	14.25	14.20	14.15	14.11	14.08	14.05	14.02	13.99	13.97
5	16.26	13.27	12.06	11.39	10.97	10.67	10.46	10.29	10.16	10.05	9.96	9.89	9.82	9.77	9.72	9.68	9.64	9.61	9.58	9.55	9.53	9.51
6	13.75	10.92	9.78	9.15	8.75	8.47	8.26	8.10	7.98	7.87	7.79	7.72	7.66	7.60	7.56	7.52	7.48	7.45	7.42	7.40	7.37	7.35
7	12.25	9.55	8.45	7.85	7.46	7.19	6.99	6.84	6.72	6.62	6.54	6.47	6.41	6.36	6.31	6.28	6.24	6.21	6.18	6.16	6.13	6.11
8	11.26	8.65	7.59	7.01	6.63	6.37	6.18	6.03	5.91	5.81	5.73	5.67	5.61	5.56	5.52	5.48	5.44	5.41	5.38	5.36	5.34	5.32
9	10.56	8.02	6.99	6.42	6.06	5.80	5.61	5.47	5.35	5.26	5.18	5.11	5.05	5.01	4.96	4.92	4.89	4.86	4.83	4.81	4.79	4.77
10	10.04	7.56	6.55	5.99	5.64	5.39	5.20	5.06	4.94	4.85	4.77	4.71	4.65	4.60	4.56	4.52	4.49	4.46	4.43	4.41	4.38	4.36
11	9.65	7.21	6.22	5.67	5.32	5.07	4.89	4.74	4.63	4.54	4.46	4.40	4.34	4.29	4.25	4.21	4.18	4.15	4.12	4.10	4.08	4.06
12	9.33	6.93	5.95	5.41	5.06	4.82	4.64	4.50	4.39	4.30	4.22	4.16	4.10	4.05	4.01	3.97	3.94	3.91	3.88	3.86	3.84	3.82
13	9.07	6.70	5.74	5.21	4.86	4.62	4.44	4.30	4.19	4.10	4.02	3.96	3.91	3.86	3.82	3.78	3.75	3.72	3.69	3.66	3.64	3.62
14	8.86	6.51	5.56	5.04	4.69	4.46	4.28	4.14	4.03	3.94	3.86	3.80	3.75	3.70	3.66	3.62	3.59	3.56	3.53	3.51	3.48	3.46
15	8.68	6.36	5.42	4.89	4.56	4.32	4.14	4.00	3.89	3.80	3.73	3.67	3.61	3.56	3.52	3.49	3.45	3.42	3.40	3.37	3.35	3.33
16	8.53	6.23	5.29	4.77	4.44	4.20	4.03	3.89	3.78	3.69	3.62	3.55	3.50	3.45	3.41	3.37	3.34	3.31	3.28	3.26	3.24	3.22
17	8.40	6.11	5.19	4.67	4.34	4.10	3.93	3.79	3.68	3.59	3.52	3.46	3.40	3.35	3.31	3.27	3.24	3.21	3.19	3.16	3.14	3.12
18	8.29	6.01	5.09	4.58	4.25	4.01	3.84	3.71	3.60	3.51	3.43	3.37	3.32	3.27	3.23	3.19	3.16	3.13	3.10	3.08	3.05	3.03
19	8.18	5.93	5.01	4.50	4.17	3.94	3.77	3.63	3.52	3.43	3.36	3.30	3.24	3.19	3.15	3.12	3.08	3.05	3.03	3.00	2.98	2.96
20	8.10	5.85	4.94	4.43	4.10	3.87	3.70	3.56	3.46	3.37	3.29	3.23	3.18	3.13	3.09	3.05	3.02	2.99	2.96	2.94	2.92	2.90
21	8.02	5.78	4.87	4.37	4.04	3.81	3.64	3.51	3.40	3.31	3.24	3.17	3.12	3.07	3.03	2.99	2.96	2.93	2.90	2.88	2.86	2.84
22	7.95	5.72	4.82	4.31	3.99	3.76	3.59	3.45	3.35	3.26	3.18	3.12	3.07	3.02	2.98	2.94	2.91	2.88	2.85	2.83	2.81	2.78
23	7.88	5.66	4.76	4.26	3.94	3.71	3.54	3.41	3.30	3.21	3.14	3.07	3.02	2.97	2.93	2.89	2.86	2.83	2.80	2.78	2.76	2.74
24	7.82	5.61	4.72	4.22	3.90	3.67	3.50	3.36	3.26	3.17	3.09	3.03	2.98	2.93	2.89	2.85	2.82	2.79	2.76	2.74	2.72	2.70
25	7.77	5.57	4.68	4.18	3.85	3.63	3.46	3.32	3.22	3.13	3.06	2.99	2.94	2.89	2.85	2.81	2.78	2.75	2.72	2.70	2.68	2.66
26	7.72	5.53	4.64	4.14	3.82	3.59	3.42	3.29	3.18	3.09	3.02	2.96	2.90	2.86	2.81	2.78	2.75	2.72	2.69	2.66	2.64	2.62
27	7.68	5.49	4.60	4.11	3.78	3.56	3.39	3.26	3.15	3.06	2.99	2.93	2.87	2.82	2.78	2.75	2.71	2.68	2.66	2.63	2.61	2.59
28	7.64	5.45	4.57	4.07	3.75	3.53	3.36	3.23	3.12	3.03	2.96	2.90	2.84	2.79	2.75	2.72	2.68	2.65	2.63	2.60	2.58	2.56
29	7.60	5.42	4.54	4.04	3.73	3.50	3.33	3.20	3.09	3.00	2.93	2.87	2.81	2.77	2.73	2.69	2.66	2.63	2.60	2.57	2.55	2.53
30	7.56	5.39	4.51	4.02	3.70	3.47	3.30	3.17	3.07	2.98	2.91	2.84	2.79	2.74	2.70	2.66	2.63	2.60	2.57	2.55	2.53	2.51
32	7.50	5.34	4.46	3.97	3.65	3.43	3.26	3.13	3.02	2.93	2.86	2.80	2.74	2.70	2.65	2.62	2.58	2.55	2.53	2.50	2.48	2.46
35	7.42	5.27	4.40	3.91	3.59	3.37	3.20	3.07	2.96	2.88	2.80	2.74	2.69	2.64	2.60	2.56	2.53	2.50	2.47	2.44	2.42	2.40
40	7.31	5.18	4.31	3.83	3.51	3.29	3.12	2.99	2.89	2.80	2.73	2.66	2.61	2.56	2.52	2.48	2.45	2.42	2.39	2.37	2.35	2.33
45	7.23	5.11	4.25	3.77	3.45	3.23	3.07	2.94	2.83	2.74	2.67	2.61	2.55	2.51	2.46	2.43	2.39	2.36	2.34	2.31	2.29	2.27
50	7.17	5.06	4.20	3.72	3.41	3.19	3.02	2.89	2.78	2.70	2.63	2.56	2.51	2.46	2.42	2.38	2.35	2.32	2.29	2.27	2.24	2.22
60	7.08	4.98	4.13	3.65	3.34	3.12	2.95	2.82	2.72	2.63	2.56	2.50	2.44	2.39	2.35	2.31	2.28	2.25	2.22	2.20	2.17	2.15
75	6.99	4.90	4.05	3.58	3.27	3.05	2.89	2.76	2.65	2.57	2.49	2.43	2.38	2.33	2.29	2.25	2.22	2.18	2.16	2.13	2.11	2.09
100	6.90	4.82	3.98	3.51	3.21	2.99	2.82	2.69	2.59	2.50	2.43	2.37	2.31	2.27	2.22	2.19	2.15	2.12	2.09	2.07	2.04	2.02
120	6.85	4.79	3.95	3.48	3.17	2.96	2.79	2.66	2.56	2.47	2.40	2.34	2.28	2.23	2.19	2.15	2.12	2.09	2.06	2.03	2.01	1.99
140	6.82	4.76	3.92	3.46	3.15	2.93	2.77	2.64	2.54	2.45	2.38	2.31	2.26	2.21	2.17	2.13	2.10	2.07	2.04	2.01	1.99	1.97
180	6.78	4.73	3.89	3.43	3.12	2.90	2.74	2.61	2.51	2.42	2.35	2.28	2.23	2.18	2.14	2.10	2.07	2.04	2.01	1.98	1.96	1.94
250	6.74	4.69	3.86	3.40	3.09	2.87	2.71	2.58	2.48	2.39	2.32	2.26	2.20	2.15	2.11	2.07	2.04	2.01	1.98	1.95	1.93	1.91
400	6.70	4.66	3.83	3.37	3.06	2.85	2.68	2.56	2.45	2.37	2.29	2.23	2.17	2.13	2.08	2.05	2.01	1.98	1.95	1.92	1.90	1.88
1000	6.66	4.63	3.80	3.34	3.04	2.82	2.66	2.53	2.43	2.34	2.27	2.20	2.15	2.10	2.06	2.02	1.98	1.95	1.92	1.90	1.87	1.85

Denominator df

Percentiles of the F-distribution ($\alpha = 0.01$), continued.

Numerator df

.01	23	24	25	26	27	28	29	30	32	35	40	45	50	60	75	100	120	140	180	250	400	1000
1	6228.7	6234.3	6239.9	6244.5	6249.2	6252.9	6257.1	6260.4	6266.9	6275.3	6286.4	6295.7	6302.3	6313.0	6323.7	6333.9	6339.5	6343.2	6347.9	6353.5	6358.1	6362.8
2	99.46	99.46	99.46	99.46	99.46	99.46	99.46	99.47	99.47	99.47	99.48	99.48	99.48	99.48	99.48	99.49	99.49	99.49	99.49	99.50	99.50	99.50
3	26.62	26.60	26.58	26.56	26.55	26.53	26.52	26.50	26.48	26.45	26.41	26.38	26.35	26.32	26.28	26.24	26.22	26.21	26.19	26.17	26.15	26.14
4	13.95	13.93	13.91	13.89	13.88	13.86	13.85	13.84	13.81	13.79	13.75	13.71	13.69	13.65	13.61	13.58	13.56	13.54	13.53	13.51	13.49	13.47
5	9.49	9.47	9.45	9.43	9.42	9.40	9.39	9.38	9.36	9.33	9.29	9.26	9.24	9.20	9.17	9.13	9.11	9.10	9.08	9.06	9.05	9.03
6	7.33	7.31	7.30	7.28	7.27	7.25	7.24	7.23	7.21	7.18	7.14	7.11	7.09	7.06	7.02	6.99	6.97	6.96	6.94	6.92	6.91	6.89
7	6.09	6.07	6.06	6.04	6.03	6.02	6.00	5.99	5.97	5.94	5.91	5.88	5.86	5.82	5.79	5.75	5.74	5.72	5.71	5.69	5.68	5.66
8	5.30	5.28	5.26	5.25	5.23	5.22	5.21	5.20	5.18	5.15	5.12	5.09	5.07	5.03	5.00	4.96	4.95	4.93	4.92	4.90	4.89	4.87
9	4.75	4.73	4.71	4.70	4.68	4.67	4.66	4.65	4.63	4.60	4.57	4.54	4.52	4.48	4.45	4.41	4.40	4.39	4.37	4.35	4.34	4.32
10	4.34	4.33	4.31	4.30	4.28	4.27	4.26	4.25	4.23	4.20	4.17	4.14	4.12	4.08	4.05	4.01	4.00	3.98	3.97	3.95	3.94	3.92
11	4.04	4.02	4.01	3.99	3.98	3.96	3.95	3.94	3.92	3.89	3.86	3.83	3.81	3.78	3.74	3.71	3.69	3.68	3.66	3.64	3.63	3.61
12	3.80	3.78	3.76	3.75	3.74	3.72	3.71	3.70	3.68	3.65	3.62	3.59	3.57	3.54	3.50	3.47	3.45	3.44	3.42	3.40	3.39	3.37
13	3.60	3.59	3.57	3.56	3.54	3.53	3.52	3.51	3.49	3.46	3.43	3.40	3.38	3.34	3.31	3.27	3.25	3.24	3.23	3.21	3.19	3.18
14	3.44	3.43	3.41	3.40	3.38	3.37	3.36	3.35	3.33	3.30	3.27	3.24	3.22	3.18	3.15	3.11	3.09	3.08	3.06	3.05	3.03	3.02
15	3.31	3.29	3.28	3.26	3.25	3.24	3.23	3.21	3.19	3.17	3.13	3.10	3.08	3.05	3.01	2.98	2.96	2.95	2.93	2.91	2.90	2.88
16	3.20	3.18	3.16	3.15	3.14	3.12	3.11	3.10	3.08	3.05	3.02	2.99	2.97	2.93	2.90	2.86	2.84	2.83	2.81	2.80	2.78	2.76
17	3.10	3.08	3.07	3.05	3.04	3.03	3.01	3.00	2.98	2.96	2.92	2.89	2.87	2.83	2.80	2.76	2.75	2.73	2.72	2.70	2.68	2.66
18	3.02	3.00	2.98	2.97	2.95	2.94	2.93	2.92	2.90	2.87	2.84	2.81	2.78	2.75	2.71	2.68	2.66	2.65	2.63	2.61	2.59	2.58
19	2.94	2.92	2.91	2.89	2.88	2.87	2.86	2.84	2.82	2.80	2.76	2.73	2.71	2.67	2.64	2.60	2.58	2.57	2.55	2.54	2.52	2.50
20	2.88	2.86	2.84	2.83	2.81	2.80	2.79	2.78	2.76	2.73	2.69	2.67	2.64	2.61	2.57	2.54	2.52	2.50	2.49	2.47	2.45	2.43
21	2.82	2.80	2.79	2.77	2.76	2.74	2.73	2.72	2.70	2.67	2.64	2.61	2.58	2.55	2.51	2.48	2.46	2.44	2.43	2.41	2.39	2.37
22	2.77	2.75	2.73	2.72	2.70	2.69	2.68	2.67	2.65	2.62	2.58	2.55	2.53	2.50	2.46	2.42	2.40	2.39	2.37	2.35	2.34	2.32
23	2.72	2.70	2.69	2.67	2.66	2.64	2.63	2.62	2.60	2.57	2.54	2.51	2.48	2.45	2.41	2.37	2.35	2.34	2.32	2.30	2.29	2.27
24	2.68	2.66	2.64	2.63	2.61	2.60	2.59	2.58	2.56	2.53	2.49	2.46	2.44	2.40	2.37	2.33	2.31	2.30	2.28	2.26	2.24	2.22
25	2.64	2.62	2.60	2.59	2.58	2.56	2.55	2.54	2.52	2.49	2.45	2.42	2.40	2.36	2.33	2.29	2.27	2.26	2.24	2.22	2.20	2.18
26	2.60	2.58	2.57	2.55	2.54	2.53	2.51	2.50	2.48	2.45	2.42	2.39	2.36	2.33	2.29	2.25	2.23	2.22	2.20	2.18	2.16	2.14
27	2.57	2.55	2.54	2.52	2.51	2.49	2.48	2.47	2.45	2.42	2.38	2.35	2.33	2.29	2.26	2.22	2.20	2.18	2.17	2.15	2.13	2.11
28	2.54	2.52	2.51	2.49	2.48	2.46	2.45	2.44	2.42	2.39	2.35	2.32	2.30	2.26	2.23	2.19	2.17	2.15	2.13	2.11	2.10	2.08
29	2.51	2.49	2.48	2.46	2.45	2.44	2.42	2.41	2.39	2.36	2.33	2.30	2.27	2.23	2.20	2.16	2.14	2.12	2.10	2.08	2.07	2.05
30	2.49	2.47	2.45	2.44	2.42	2.41	2.40	2.39	2.36	2.34	2.30	2.27	2.25	2.21	2.17	2.13	2.11	2.10	2.08	2.06	2.04	2.02
32	2.44	2.42	2.41	2.39	2.38	2.36	2.35	2.34	2.32	2.29	2.25	2.22	2.20	2.16	2.12	2.08	2.06	2.05	2.03	2.01	1.99	1.97
35	2.38	2.36	2.35	2.33	2.32	2.30	2.29	2.28	2.26	2.23	2.19	2.16	2.14	2.10	2.06	2.02	2.00	1.98	1.96	1.94	1.92	1.90
40	2.31	2.29	2.27	2.26	2.24	2.23	2.22	2.20	2.18	2.15	2.11	2.08	2.06	2.02	1.98	1.94	1.92	1.90	1.88	1.86	1.84	1.82
45	2.25	2.23	2.21	2.20	2.18	2.17	2.16	2.14	2.12	2.09	2.05	2.02	2.00	1.96	1.92	1.88	1.85	1.84	1.82	1.79	1.77	1.75
50	2.20	2.18	2.17	2.15	2.14	2.12	2.11	2.10	2.08	2.05	2.01	1.97	1.95	1.91	1.87	1.82	1.80	1.79	1.76	1.74	1.72	1.70
60	2.13	2.12	2.10	2.08	2.07	2.05	2.04	2.03	2.01	1.98	1.94	1.90	1.88	1.84	1.79	1.75	1.73	1.71	1.69	1.66	1.64	1.62
75	2.07	2.05	2.03	2.02	2.00	1.99	1.97	1.96	1.94	1.91	1.87	1.83	1.81	1.76	1.72	1.67	1.65	1.63	1.61	1.58	1.56	1.53
100	2.00	1.98	1.97	1.95	1.93	1.92	1.91	1.89	1.87	1.84	1.80	1.76	1.74	1.69	1.65	1.60	1.57	1.55	1.53	1.50	1.47	1.45
120	1.97	1.95	1.93	1.92	1.90	1.89	1.87	1.86	1.84	1.81	1.76	1.73	1.70	1.66	1.61	1.56	1.53	1.51	1.49	1.46	1.43	1.40
140	1.95	1.93	1.91	1.89	1.88	1.86	1.85	1.84	1.81	1.78	1.74	1.70	1.67	1.63	1.58	1.53	1.50	1.48	1.46	1.43	1.40	1.37
180	1.92	1.90	1.88	1.86	1.85	1.83	1.82	1.81	1.78	1.75	1.71	1.67	1.64	1.60	1.55	1.49	1.47	1.45	1.42	1.39	1.35	1.32
250	1.89	1.87	1.85	1.83	1.82	1.80	1.79	1.77	1.75	1.72	1.67	1.64	1.61	1.56	1.51	1.46	1.43	1.41	1.38	1.34	1.31	1.27
400	1.86	1.84	1.82	1.80	1.79	1.77	1.76	1.75	1.72	1.69	1.64	1.61	1.58	1.53	1.48	1.42	1.39	1.37	1.33	1.30	1.26	1.22
1000	1.83	1.81	1.79	1.77	1.76	1.74	1.73	1.72	1.69	1.66	1.61	1.58	1.54	1.50	1.44	1.38	1.35	1.33	1.29	1.25	1.21	1.16

Denominator df

Percentiles of the F-distribution (α = 0.05).

Numerator df

.05	1	2	3	4	5	6	7	8	9	10	11	12	13	14	15	16	17	18	19	20	21	22
1	161.4	199.5	215.7	224.6	230.2	234.0	236.8	238.9	240.5	241.9	243.0	243.9	244.7	245.4	245.9	246.5	246.9	247.3	247.7	248.0	248.3	248.6
2	18.51	19.00	19.16	19.25	19.30	19.33	19.35	19.37	19.38	19.40	19.40	19.41	19.42	19.42	19.43	19.43	19.44	19.44	19.44	19.45	19.45	19.45
3	10.13	9.55	9.28	9.12	9.01	8.94	8.89	8.85	8.81	8.79	8.76	8.74	8.73	8.71	8.70	8.69	8.68	8.67	8.67	8.66	8.65	8.65
4	7.71	6.94	6.59	6.39	6.26	6.16	6.09	6.04	6.00	5.96	5.94	5.91	5.89	5.87	5.86	5.84	5.83	5.82	5.81	5.80	5.79	5.79
5	6.61	5.79	5.41	5.19	5.05	4.95	4.88	4.82	4.77	4.74	4.70	4.68	4.66	4.64	4.62	4.60	4.59	4.58	4.57	4.56	4.55	4.54
6	5.99	5.14	4.76	4.53	4.39	4.28	4.21	4.15	4.10	4.06	4.03	4.00	3.98	3.96	3.94	3.92	3.91	3.90	3.88	3.87	3.86	3.86
7	5.59	4.74	4.35	4.12	3.97	3.87	3.79	3.73	3.68	3.64	3.60	3.57	3.55	3.53	3.51	3.49	3.48	3.47	3.46	3.44	3.43	3.43
8	5.32	4.46	4.07	3.84	3.69	3.58	3.50	3.44	3.39	3.35	3.31	3.28	3.26	3.24	3.22	3.20	3.19	3.17	3.16	3.15	3.14	3.13
9	5.12	4.26	3.86	3.63	3.48	3.37	3.29	3.23	3.18	3.14	3.10	3.07	3.05	3.03	3.01	2.99	2.97	2.96	2.95	2.94	2.93	2.92
10	4.96	4.10	3.71	3.48	3.33	3.22	3.14	3.07	3.02	2.98	2.94	2.91	2.89	2.86	2.85	2.83	2.81	2.80	2.79	2.77	2.76	2.75
11	4.84	3.98	3.59	3.36	3.20	3.09	3.01	2.95	2.90	2.85	2.82	2.79	2.76	2.74	2.72	2.70	2.69	2.67	2.66	2.65	2.64	2.63
12	4.75	3.89	3.49	3.26	3.11	3.00	2.91	2.85	2.80	2.75	2.72	2.69	2.66	2.64	2.62	2.60	2.58	2.57	2.56	2.54	2.53	2.52
13	4.67	3.81	3.41	3.18	3.03	2.92	2.83	2.77	2.71	2.67	2.63	2.60	2.58	2.55	2.53	2.51	2.50	2.48	2.47	2.46	2.45	2.44
14	4.60	3.74	3.34	3.11	2.96	2.85	2.76	2.70	2.65	2.60	2.57	2.53	2.51	2.48	2.46	2.44	2.43	2.41	2.40	2.39	2.38	2.37
15	4.54	3.68	3.29	3.06	2.90	2.79	2.71	2.64	2.59	2.54	2.51	2.48	2.45	2.42	2.40	2.38	2.37	2.35	2.34	2.33	2.32	2.31
16	4.49	3.63	3.24	3.01	2.85	2.74	2.66	2.59	2.54	2.49	2.46	2.42	2.40	2.37	2.35	2.33	2.32	2.30	2.29	2.28	2.26	2.25
17	4.45	3.59	3.20	2.96	2.81	2.70	2.61	2.55	2.49	2.45	2.41	2.38	2.35	2.33	2.31	2.29	2.27	2.26	2.24	2.23	2.22	2.21
18	4.41	3.55	3.16	2.93	2.77	2.66	2.58	2.51	2.46	2.41	2.37	2.34	2.31	2.29	2.27	2.25	2.23	2.22	2.20	2.19	2.18	2.17
19	4.38	3.52	3.13	2.90	2.74	2.63	2.54	2.48	2.42	2.38	2.34	2.31	2.28	2.26	2.23	2.21	2.20	2.18	2.17	2.16	2.14	2.13
20	4.35	3.49	3.10	2.87	2.71	2.60	2.51	2.45	2.39	2.35	2.31	2.28	2.25	2.22	2.20	2.18	2.17	2.15	2.14	2.12	2.11	2.10
21	4.32	3.47	3.07	2.84	2.68	2.57	2.49	2.42	2.37	2.32	2.28	2.25	2.22	2.20	2.18	2.16	2.14	2.12	2.11	2.10	2.08	2.07
22	4.30	3.44	3.05	2.82	2.66	2.55	2.46	2.40	2.34	2.30	2.26	2.23	2.20	2.17	2.15	2.13	2.11	2.10	2.08	2.07	2.06	2.05
23	4.28	3.42	3.03	2.80	2.64	2.53	2.44	2.37	2.32	2.27	2.24	2.20	2.18	2.15	2.13	2.11	2.09	2.08	2.06	2.05	2.04	2.02
24	4.26	3.40	3.01	2.78	2.62	2.51	2.42	2.36	2.30	2.25	2.22	2.18	2.15	2.13	2.11	2.09	2.07	2.05	2.04	2.03	2.01	2.00
25	4.24	3.39	2.99	2.76	2.60	2.49	2.40	2.34	2.28	2.24	2.20	2.16	2.14	2.11	2.09	2.07	2.05	2.04	2.02	2.01	2.00	1.98
26	4.23	3.37	2.98	2.74	2.59	2.47	2.39	2.32	2.27	2.22	2.18	2.15	2.12	2.09	2.07	2.05	2.03	2.02	2.00	1.99	1.98	1.97
27	4.21	3.35	2.96	2.73	2.57	2.46	2.37	2.31	2.25	2.20	2.17	2.13	2.10	2.08	2.06	2.04	2.02	2.00	1.99	1.97	1.96	1.95
28	4.20	3.34	2.95	2.71	2.56	2.45	2.36	2.29	2.24	2.19	2.15	2.12	2.09	2.06	2.04	2.02	2.00	1.99	1.97	1.96	1.95	1.93
29	4.18	3.33	2.93	2.70	2.55	2.43	2.35	2.28	2.22	2.18	2.14	2.10	2.08	2.05	2.03	2.01	1.99	1.97	1.96	1.94	1.93	1.92
30	4.17	3.32	2.92	2.69	2.53	2.42	2.33	2.27	2.21	2.16	2.13	2.09	2.06	2.04	2.01	1.99	1.98	1.96	1.95	1.93	1.92	1.91
32	4.15	3.29	2.90	2.67	2.51	2.40	2.31	2.24	2.19	2.14	2.10	2.07	2.04	2.01	1.99	1.97	1.95	1.94	1.92	1.91	1.90	1.88
35	4.12	3.27	2.87	2.64	2.49	2.37	2.29	2.22	2.16	2.11	2.07	2.04	2.01	1.99	1.96	1.94	1.92	1.91	1.89	1.88	1.87	1.85
40	4.08	3.23	2.84	2.61	2.45	2.34	2.25	2.18	2.12	2.08	2.04	2.00	1.97	1.95	1.92	1.90	1.89	1.87	1.85	1.84	1.83	1.81
45	4.06	3.20	2.81	2.58	2.42	2.31	2.22	2.15	2.10	2.05	2.01	1.97	1.94	1.92	1.89	1.87	1.86	1.84	1.82	1.81	1.80	1.78
50	4.03	3.18	2.79	2.56	2.40	2.29	2.20	2.13	2.07	2.03	1.99	1.95	1.92	1.89	1.87	1.85	1.83	1.81	1.80	1.78	1.77	1.76
60	4.00	3.15	2.76	2.53	2.37	2.25	2.17	2.10	2.04	1.99	1.95	1.92	1.89	1.86	1.84	1.82	1.80	1.78	1.76	1.75	1.73	1.72
75	3.97	3.12	2.73	2.49	2.34	2.22	2.13	2.06	2.01	1.96	1.92	1.88	1.85	1.83	1.80	1.78	1.76	1.74	1.73	1.71	1.70	1.69
100	3.94	3.09	2.70	2.46	2.31	2.19	2.10	2.03	1.97	1.93	1.89	1.85	1.82	1.79	1.77	1.75	1.73	1.71	1.69	1.68	1.66	1.65
120	3.92	3.07	2.68	2.45	2.29	2.18	2.09	2.02	1.96	1.91	1.87	1.83	1.80	1.78	1.75	1.73	1.71	1.69	1.67	1.66	1.64	1.63
140	3.91	3.06	2.67	2.44	2.28	2.16	2.08	2.01	1.95	1.90	1.86	1.82	1.79	1.76	1.74	1.72	1.70	1.68	1.66	1.65	1.63	1.62
180	3.89	3.05	2.65	2.42	2.26	2.15	2.06	1.99	1.93	1.88	1.84	1.81	1.77	1.75	1.72	1.70	1.68	1.66	1.64	1.63	1.61	1.60
250	3.88	3.03	2.64	2.41	2.25	2.13	2.05	1.98	1.92	1.87	1.83	1.79	1.76	1.73	1.71	1.68	1.66	1.65	1.63	1.61	1.60	1.58
400	3.86	3.02	2.63	2.39	2.24	2.12	2.03	1.96	1.90	1.85	1.81	1.78	1.74	1.72	1.69	1.67	1.65	1.63	1.61	1.60	1.58	1.57
1000	3.85	3.00	2.61	2.38	2.22	2.11	2.02	1.95	1.89	1.84	1.80	1.76	1.73	1.70	1.68	1.65	1.63	1.61	1.60	1.58	1.57	1.55

Denominator df

Percentiles of the F-distribution ($\alpha = 0.05$), continued.

Numerator df

.05	23	24	25	26	27	28	29	30	32	35	40	45	50	60	75	100	120	140	180	250	400	1000
1	248.8	249.1	249.3	249.5	249.6	249.8	250.0	250.1	250.4	250.7	251.1	251.5	251.8	252.2	252.6	253.0	253.3	253.4	253.6	253.8	254.0	254.2
2	19.45	19.45	19.46	19.46	19.46	19.46	19.46	19.46	19.46	19.47	19.47	19.47	19.48	19.48	19.48	19.49	19.49	19.49	19.49	19.49	19.49	19.49
3	8.64	8.64	8.63	8.63	8.63	8.62	8.62	8.62	8.61	8.60	8.59	8.59	8.58	8.57	8.56	8.55	8.55	8.55	8.54	8.54	8.53	8.53
4	5.78	5.77	5.77	5.76	5.76	5.75	5.75	5.75	5.74	5.73	5.72	5.71	5.70	5.69	5.68	5.66	5.66	5.65	5.65	5.64	5.64	5.63
5	4.53	4.53	4.52	4.52	4.51	4.50	4.50	4.50	4.49	4.48	4.46	4.45	4.44	4.43	4.42	4.41	4.40	4.39	4.39	4.38	4.38	4.37
6	3.85	3.84	3.83	3.83	3.82	3.82	3.81	3.81	3.80	3.79	3.77	3.76	3.75	3.74	3.73	3.71	3.70	3.70	3.69	3.69	3.68	3.67
7	3.42	3.41	3.40	3.40	3.39	3.39	3.38	3.38	3.37	3.36	3.34	3.33	3.32	3.30	3.29	3.27	3.27	3.26	3.25	3.25	3.24	3.23
8	3.12	3.12	3.11	3.10	3.10	3.09	3.08	3.08	3.07	3.06	3.04	3.03	3.02	3.01	2.99	2.97	2.97	2.96	2.95	2.95	2.94	2.93
9	2.91	2.90	2.89	2.89	2.88	2.87	2.87	2.86	2.85	2.84	2.83	2.81	2.80	2.79	2.77	2.76	2.75	2.74	2.73	2.73	2.72	2.71
10	2.75	2.74	2.73	2.72	2.72	2.71	2.70	2.70	2.69	2.68	2.66	2.65	2.64	2.62	2.60	2.59	2.58	2.57	2.57	2.56	2.55	2.54
11	2.62	2.61	2.60	2.59	2.59	2.58	2.58	2.57	2.56	2.55	2.53	2.52	2.51	2.49	2.47	2.46	2.45	2.44	2.43	2.43	2.42	2.41
12	2.51	2.51	2.50	2.49	2.48	2.48	2.47	2.47	2.46	2.44	2.43	2.41	2.40	2.38	2.37	2.35	2.34	2.33	2.33	2.32	2.31	2.30
13	2.43	2.42	2.41	2.41	2.40	2.39	2.39	2.38	2.37	2.36	2.34	2.33	2.31	2.30	2.28	2.26	2.25	2.25	2.24	2.23	2.22	2.21
14	2.36	2.35	2.34	2.33	2.33	2.32	2.31	2.31	2.30	2.28	2.27	2.25	2.24	2.22	2.21	2.19	2.18	2.17	2.16	2.15	2.15	2.14
15	2.30	2.29	2.28	2.27	2.27	2.26	2.25	2.25	2.24	2.22	2.20	2.19	2.18	2.16	2.14	2.12	2.11	2.11	2.10	2.09	2.08	2.07
16	2.24	2.24	2.23	2.22	2.21	2.21	2.20	2.19	2.18	2.17	2.15	2.14	2.12	2.11	2.09	2.07	2.06	2.05	2.04	2.03	2.02	2.02
17	2.20	2.19	2.18	2.17	2.17	2.16	2.15	2.15	2.14	2.12	2.10	2.09	2.08	2.06	2.04	2.02	2.01	2.00	1.99	1.98	1.98	1.97
18	2.16	2.15	2.14	2.13	2.13	2.12	2.11	2.11	2.10	2.08	2.06	2.05	2.04	2.02	2.00	1.98	1.97	1.96	1.95	1.94	1.93	1.92
19	2.12	2.11	2.11	2.10	2.09	2.08	2.08	2.07	2.06	2.05	2.03	2.01	2.00	1.98	1.96	1.94	1.93	1.92	1.91	1.90	1.89	1.88
20	2.09	2.08	2.07	2.07	2.06	2.05	2.05	2.04	2.03	2.01	1.99	1.98	1.97	1.95	1.93	1.91	1.90	1.89	1.88	1.87	1.86	1.85
21	2.06	2.05	2.05	2.04	2.03	2.02	2.02	2.01	2.00	1.98	1.96	1.95	1.94	1.92	1.90	1.88	1.87	1.86	1.85	1.84	1.83	1.82
22	2.04	2.03	2.02	2.01	2.00	2.00	1.99	1.98	1.97	1.96	1.94	1.92	1.91	1.89	1.87	1.85	1.84	1.83	1.82	1.81	1.80	1.79
23	2.01	2.01	2.00	1.99	1.98	1.97	1.97	1.96	1.95	1.93	1.91	1.90	1.88	1.86	1.84	1.82	1.81	1.81	1.79	1.78	1.77	1.76
24	1.99	1.98	1.97	1.97	1.96	1.95	1.95	1.94	1.93	1.91	1.89	1.88	1.86	1.84	1.82	1.80	1.79	1.78	1.77	1.76	1.75	1.74
25	1.97	1.96	1.96	1.95	1.94	1.93	1.93	1.92	1.91	1.89	1.87	1.86	1.84	1.82	1.80	1.78	1.77	1.76	1.75	1.74	1.73	1.72
26	1.96	1.95	1.94	1.93	1.92	1.91	1.91	1.90	1.89	1.87	1.85	1.84	1.82	1.80	1.78	1.76	1.75	1.74	1.73	1.72	1.71	1.70
27	1.94	1.93	1.92	1.91	1.90	1.90	1.89	1.88	1.88	1.86	1.84	1.82	1.81	1.79	1.76	1.74	1.73	1.72	1.71	1.70	1.69	1.68
28	1.92	1.91	1.91	1.90	1.89	1.88	1.88	1.87	1.86	1.84	1.82	1.80	1.79	1.77	1.75	1.73	1.71	1.71	1.69	1.68	1.67	1.66
29	1.91	1.90	1.89	1.88	1.88	1.87	1.86	1.85	1.84	1.83	1.81	1.79	1.77	1.75	1.73	1.71	1.70	1.69	1.68	1.67	1.66	1.65
30	1.90	1.89	1.88	1.87	1.86	1.85	1.85	1.84	1.83	1.81	1.79	1.77	1.76	1.74	1.72	1.70	1.68	1.68	1.66	1.65	1.64	1.63
32	1.87	1.86	1.85	1.85	1.84	1.83	1.82	1.82	1.80	1.79	1.77	1.75	1.74	1.71	1.69	1.67	1.66	1.65	1.64	1.63	1.61	1.60
35	1.84	1.83	1.82	1.82	1.81	1.80	1.79	1.79	1.77	1.76	1.74	1.72	1.70	1.68	1.66	1.63	1.62	1.61	1.60	1.59	1.58	1.57
40	1.80	1.79	1.78	1.77	1.77	1.76	1.75	1.74	1.73	1.72	1.69	1.67	1.66	1.64	1.61	1.59	1.58	1.57	1.55	1.54	1.53	1.52
45	1.77	1.76	1.75	1.74	1.73	1.73	1.72	1.71	1.70	1.68	1.66	1.64	1.63	1.60	1.58	1.55	1.54	1.53	1.52	1.51	1.49	1.48
50	1.75	1.74	1.73	1.72	1.71	1.70	1.69	1.69	1.67	1.66	1.63	1.61	1.60	1.58	1.55	1.52	1.51	1.50	1.49	1.47	1.46	1.45
60	1.71	1.70	1.69	1.68	1.67	1.66	1.66	1.65	1.64	1.62	1.59	1.57	1.56	1.53	1.51	1.48	1.47	1.46	1.44	1.43	1.41	1.40
75	1.67	1.66	1.65	1.64	1.63	1.63	1.62	1.61	1.60	1.58	1.55	1.53	1.52	1.49	1.47	1.44	1.42	1.41	1.40	1.38	1.37	1.35
100	1.64	1.63	1.62	1.61	1.60	1.59	1.58	1.57	1.56	1.54	1.52	1.49	1.48	1.45	1.42	1.39	1.38	1.36	1.35	1.33	1.31	1.30
120	1.62	1.61	1.60	1.59	1.58	1.57	1.56	1.55	1.54	1.52	1.50	1.47	1.46	1.43	1.40	1.37	1.35	1.34	1.32	1.30	1.29	1.27
140	1.61	1.60	1.58	1.57	1.57	1.56	1.55	1.54	1.53	1.51	1.48	1.46	1.44	1.41	1.38	1.35	1.33	1.32	1.30	1.29	1.27	1.25
180	1.59	1.58	1.57	1.56	1.55	1.54	1.53	1.52	1.51	1.49	1.46	1.44	1.42	1.39	1.36	1.33	1.31	1.30	1.28	1.26	1.24	1.22
250	1.57	1.56	1.55	1.54	1.53	1.52	1.51	1.50	1.49	1.47	1.44	1.42	1.40	1.37	1.34	1.31	1.29	1.27	1.25	1.23	1.21	1.18
400	1.56	1.54	1.53	1.52	1.51	1.50	1.50	1.49	1.47	1.45	1.42	1.40	1.38	1.35	1.32	1.28	1.26	1.25	1.23	1.20	1.18	1.15
1000	1.54	1.53	1.52	1.51	1.50	1.49	1.48	1.47	1.46	1.43	1.41	1.38	1.36	1.33	1.30	1.26	1.24	1.22	1.20	1.17	1.14	1.11

Denominator df

Percentiles of the F-distribution ($\alpha = 0.10$).

Numerator df

.1	1	2	3	4	5	6	7	8	9	10	11	12	13	14	15	16	17	18	19	20	21	22
1	39.9	49.5	53.6	55.8	57.2	58.2	58.9	59.4	59.9	60.2	60.5	60.7	60.9	61.1	61.2	61.3	61.5	61.6	61.7	61.7	61.8	61.9
2	8.53	9.00	9.16	9.24	9.29	9.33	9.35	9.37	9.38	9.39	9.40	9.41	9.41	9.42	9.42	9.43	9.43	9.44	9.44	9.44	9.44	9.45
3	5.54	5.46	5.39	5.34	5.31	5.28	5.27	5.25	5.24	5.23	5.22	5.22	5.21	5.20	5.20	5.20	5.19	5.19	5.19	5.18	5.18	5.18
4	4.54	4.32	4.19	4.11	4.05	4.01	3.98	3.95	3.94	3.92	3.91	3.90	3.89	3.88	3.87	3.86	3.86	3.85	3.85	3.84	3.84	3.84
5	4.06	3.78	3.62	3.52	3.45	3.40	3.37	3.34	3.32	3.30	3.28	3.27	3.26	3.25	3.24	3.23	3.22	3.22	3.21	3.21	3.20	3.20
6	3.78	3.46	3.29	3.18	3.11	3.05	3.01	2.98	2.96	2.94	2.92	2.90	2.89	2.88	2.87	2.86	2.85	2.85	2.84	2.84	2.83	2.83
7	3.59	3.26	3.07	2.96	2.88	2.83	2.78	2.75	2.72	2.70	2.68	2.67	2.65	2.64	2.63	2.62	2.61	2.61	2.60	2.59	2.59	2.58
8	3.46	3.11	2.92	2.81	2.73	2.67	2.62	2.59	2.56	2.54	2.52	2.50	2.49	2.48	2.46	2.45	2.45	2.44	2.43	2.42	2.42	2.41
9	3.36	3.01	2.81	2.69	2.61	2.55	2.51	2.47	2.44	2.42	2.40	2.38	2.36	2.35	2.34	2.33	2.32	2.31	2.30	2.30	2.29	2.29
10	3.29	2.92	2.73	2.61	2.52	2.46	2.41	2.38	2.35	2.32	2.30	2.28	2.27	2.26	2.24	2.23	2.22	2.22	2.21	2.20	2.19	2.19
11	3.23	2.86	2.66	2.54	2.45	2.39	2.34	2.30	2.27	2.25	2.23	2.21	2.19	2.18	2.17	2.16	2.15	2.14	2.13	2.12	2.12	2.11
12	3.18	2.81	2.61	2.48	2.39	2.33	2.28	2.24	2.21	2.19	2.17	2.15	2.13	2.12	2.10	2.09	2.08	2.08	2.07	2.06	2.05	2.05
13	3.14	2.76	2.56	2.43	2.35	2.28	2.23	2.20	2.16	2.14	2.12	2.10	2.08	2.07	2.05	2.04	2.03	2.02	2.01	2.01	2.00	1.99
14	3.10	2.73	2.52	2.39	2.31	2.24	2.19	2.15	2.12	2.10	2.07	2.05	2.04	2.02	2.01	2.00	1.99	1.98	1.97	1.96	1.96	1.95
15	3.07	2.70	2.49	2.36	2.27	2.21	2.16	2.12	2.09	2.06	2.04	2.02	2.00	1.99	1.97	1.96	1.95	1.94	1.93	1.92	1.92	1.91
16	3.05	2.67	2.46	2.33	2.24	2.18	2.13	2.09	2.06	2.03	2.01	1.99	1.97	1.95	1.94	1.93	1.92	1.91	1.90	1.89	1.88	1.88
17	3.03	2.64	2.44	2.31	2.22	2.15	2.10	2.06	2.03	2.00	1.98	1.96	1.94	1.93	1.91	1.90	1.89	1.88	1.87	1.86	1.86	1.85
18	3.01	2.62	2.42	2.29	2.20	2.13	2.08	2.04	2.00	1.98	1.95	1.93	1.92	1.90	1.89	1.87	1.86	1.85	1.84	1.84	1.83	1.82
19	2.99	2.61	2.40	2.27	2.18	2.11	2.06	2.02	1.98	1.96	1.93	1.91	1.89	1.88	1.86	1.85	1.84	1.83	1.82	1.81	1.81	1.80
20	2.97	2.59	2.38	2.25	2.16	2.09	2.04	2.00	1.96	1.94	1.91	1.89	1.87	1.86	1.84	1.83	1.82	1.81	1.80	1.79	1.79	1.78
21	2.96	2.57	2.36	2.23	2.14	2.08	2.02	1.98	1.95	1.92	1.90	1.87	1.86	1.84	1.83	1.81	1.80	1.79	1.78	1.78	1.77	1.76
22	2.95	2.56	2.35	2.22	2.13	2.06	2.01	1.97	1.93	1.90	1.88	1.86	1.84	1.83	1.81	1.80	1.79	1.78	1.77	1.76	1.75	1.74
23	2.94	2.55	2.34	2.21	2.11	2.05	1.99	1.95	1.92	1.89	1.87	1.84	1.83	1.81	1.80	1.78	1.77	1.76	1.75	1.74	1.74	1.73
24	2.93	2.54	2.33	2.19	2.10	2.04	1.98	1.94	1.91	1.88	1.85	1.83	1.81	1.80	1.78	1.77	1.76	1.75	1.74	1.73	1.72	1.71
25	2.92	2.53	2.32	2.18	2.09	2.02	1.97	1.93	1.89	1.87	1.84	1.82	1.80	1.79	1.77	1.76	1.75	1.74	1.73	1.72	1.71	1.70
26	2.91	2.52	2.31	2.17	2.08	2.01	1.96	1.92	1.88	1.86	1.83	1.81	1.79	1.77	1.76	1.75	1.73	1.72	1.71	1.71	1.70	1.69
27	2.90	2.51	2.30	2.17	2.07	2.00	1.95	1.91	1.87	1.85	1.82	1.80	1.78	1.76	1.75	1.74	1.72	1.71	1.70	1.70	1.69	1.68
28	2.89	2.50	2.29	2.16	2.06	2.00	1.94	1.90	1.87	1.84	1.81	1.79	1.77	1.75	1.74	1.73	1.71	1.70	1.69	1.69	1.68	1.67
29	2.89	2.50	2.28	2.15	2.06	1.99	1.93	1.89	1.86	1.83	1.80	1.78	1.76	1.75	1.73	1.72	1.71	1.69	1.68	1.68	1.67	1.66
30	2.88	2.49	2.28	2.14	2.05	1.98	1.93	1.88	1.85	1.82	1.79	1.77	1.75	1.74	1.72	1.71	1.70	1.69	1.68	1.67	1.66	1.65
32	2.87	2.48	2.26	2.13	2.04	1.97	1.91	1.87	1.83	1.81	1.78	1.76	1.74	1.72	1.71	1.69	1.68	1.67	1.66	1.65	1.64	1.64
35	2.85	2.46	2.25	2.11	2.02	1.95	1.90	1.85	1.82	1.79	1.76	1.74	1.72	1.70	1.69	1.67	1.66	1.65	1.64	1.63	1.62	1.62
40	2.84	2.44	2.23	2.09	2.00	1.93	1.87	1.83	1.79	1.76	1.74	1.71	1.70	1.68	1.66	1.65	1.64	1.62	1.61	1.61	1.60	1.59
45	2.82	2.42	2.21	2.07	1.98	1.91	1.85	1.81	1.77	1.74	1.72	1.70	1.68	1.66	1.64	1.63	1.62	1.60	1.59	1.58	1.58	1.57
50	2.81	2.41	2.20	2.06	1.97	1.90	1.84	1.80	1.76	1.73	1.70	1.68	1.66	1.64	1.63	1.61	1.60	1.59	1.58	1.57	1.56	1.55
60	2.79	2.39	2.18	2.04	1.95	1.87	1.82	1.77	1.74	1.71	1.68	1.66	1.64	1.62	1.60	1.59	1.58	1.56	1.55	1.54	1.53	1.53
75	2.77	2.37	2.16	2.02	1.93	1.85	1.80	1.75	1.72	1.69	1.66	1.63	1.61	1.60	1.58	1.57	1.55	1.54	1.53	1.52	1.51	1.50
100	2.76	2.36	2.14	2.00	1.91	1.83	1.78	1.73	1.69	1.66	1.64	1.61	1.59	1.57	1.56	1.54	1.53	1.52	1.50	1.49	1.48	1.48
120	2.75	2.35	2.13	1.99	1.90	1.82	1.77	1.72	1.68	1.65	1.63	1.60	1.58	1.56	1.55	1.53	1.52	1.50	1.49	1.48	1.47	1.46
140	2.74	2.34	2.12	1.99	1.89	1.82	1.76	1.71	1.68	1.64	1.62	1.59	1.57	1.55	1.54	1.52	1.51	1.50	1.48	1.47	1.46	1.45
180	2.73	2.33	2.11	1.98	1.88	1.81	1.75	1.70	1.67	1.63	1.61	1.58	1.56	1.54	1.53	1.51	1.50	1.48	1.47	1.46	1.45	1.44
250	2.73	2.32	2.11	1.97	1.87	1.80	1.74	1.69	1.66	1.62	1.60	1.57	1.55	1.53	1.51	1.50	1.49	1.47	1.46	1.45	1.44	1.43
400	2.72	2.32	2.10	1.96	1.86	1.79	1.73	1.69	1.65	1.61	1.59	1.56	1.54	1.52	1.50	1.49	1.47	1.46	1.45	1.44	1.43	1.42
1000	2.71	2.31	2.09	1.95	1.85	1.78	1.72	1.68	1.64	1.61	1.58	1.55	1.53	1.51	1.49	1.48	1.46	1.45	1.44	1.43	1.42	1.41

Denominator df

Percentiles of the *F*-distribution ($\alpha = 0.10$), continued.

Numerator df

.1	23	24	25	26	27	28	29	30	32	35	40	45	50	60	75	100	120	140	180	250	400	1000
1	61.9	62.0	62.1	62.1	62.1	62.2	62.2	62.3	62.3	62.4	62.5	62.6	62.7	62.8	62.9	63.0	63.1	63.1	63.1	63.2	63.2	63.3
2	9.45	9.45	9.45	9.45	9.45	9.46	9.46	9.46	9.46	9.46	9.47	9.47	9.47	9.47	9.48	9.48	9.48	9.48	9.49	9.49	9.49	9.49
3	5.18	5.18	5.17	5.17	5.17	5.17	5.17	5.17	5.17	5.16	5.16	5.16	5.15	5.15	5.15	5.14	5.14	5.14	5.14	5.14	5.14	5.1
4	3.83	3.83	3.83	3.83	3.82	3.82	3.82	3.82	3.81	3.81	3.80	3.80	3.80	3.79	3.78	3.78	3.78	3.77	3.77	3.77	3.77	3.76
5	3.19	3.19	3.19	3.18	3.18	3.18	3.18	3.17	3.17	3.16	3.16	3.15	3.15	3.14	3.13	3.13	3.12	3.12	3.12	3.11	3.11	3.11
6	2.82	2.82	2.81	2.81	2.81	2.81	2.80	2.80	2.80	2.79	2.78	2.77	2.77	2.76	2.75	2.75	2.74	2.74	2.74	2.73	2.73	2.72
7	2.58	2.58	2.57	2.57	2.56	2.56	2.56	2.56	2.55	2.54	2.54	2.53	2.52	2.51	2.51	2.50	2.49	2.49	2.49	2.48	2.48	2.47
8	2.41	2.40	2.40	2.40	2.39	2.39	2.39	2.38	2.38	2.37	2.36	2.35	2.35	2.34	2.33	2.32	2.32	2.31	2.31	2.30	2.30	2.30
9	2.28	2.28	2.27	2.27	2.26	2.26	2.26	2.25	2.25	2.24	2.23	2.22	2.22	2.21	2.20	2.19	2.18	2.18	2.18	2.17	2.17	2.16
10	2.18	2.18	2.17	2.17	2.17	2.16	2.16	2.16	2.15	2.14	2.13	2.12	2.12	2.11	2.10	2.09	2.08	2.08	2.07	2.07	2.06	2.06
11	2.11	2.10	2.10	2.09	2.09	2.08	2.08	2.08	2.07	2.06	2.05	2.04	2.04	2.03	2.02	2.01	2.00	2.00	1.99	1.99	1.98	1.98
12	2.04	2.04	2.03	2.03	2.02	2.02	2.01	2.01	2.01	2.00	1.99	1.98	1.97	1.96	1.95	1.94	1.93	1.93	1.92	1.92	1.91	1.91
13	1.99	1.98	1.98	1.97	1.97	1.96	1.96	1.96	1.95	1.94	1.93	1.92	1.92	1.90	1.89	1.88	1.88	1.87	1.87	1.86	1.86	1.85
14	1.94	1.94	1.93	1.93	1.92	1.92	1.92	1.91	1.91	1.90	1.89	1.88	1.87	1.86	1.85	1.83	1.83	1.82	1.82	1.81	1.81	1.80
15	1.90	1.90	1.89	1.89	1.88	1.88	1.88	1.87	1.87	1.86	1.85	1.84	1.83	1.82	1.80	1.79	1.79	1.78	1.78	1.77	1.76	1.76
16	1.87	1.87	1.86	1.86	1.85	1.85	1.84	1.84	1.83	1.82	1.81	1.80	1.79	1.78	1.77	1.76	1.75	1.75	1.74	1.73	1.73	1.72
17	1.84	1.84	1.83	1.83	1.82	1.82	1.81	1.81	1.80	1.79	1.78	1.77	1.76	1.75	1.74	1.73	1.72	1.71	1.71	1.70	1.70	1.69
18	1.82	1.81	1.80	1.80	1.80	1.79	1.79	1.78	1.78	1.77	1.75	1.74	1.74	1.72	1.71	1.70	1.69	1.69	1.68	1.67	1.67	1.66
19	1.79	1.79	1.78	1.78	1.77	1.77	1.76	1.76	1.75	1.74	1.73	1.72	1.71	1.70	1.69	1.67	1.67	1.66	1.65	1.65	1.64	1.64
20	1.77	1.77	1.76	1.76	1.75	1.75	1.74	1.74	1.73	1.72	1.71	1.70	1.69	1.68	1.66	1.65	1.64	1.64	1.63	1.62	1.62	1.61
21	1.75	1.75	1.74	1.74	1.73	1.73	1.72	1.72	1.71	1.70	1.69	1.68	1.67	1.66	1.64	1.63	1.62	1.62	1.61	1.60	1.60	1.59
22	1.74	1.73	1.73	1.72	1.72	1.71	1.71	1.70	1.69	1.68	1.67	1.66	1.65	1.64	1.63	1.61	1.60	1.60	1.59	1.58	1.58	1.57
23	1.72	1.72	1.71	1.70	1.70	1.69	1.69	1.69	1.68	1.67	1.66	1.64	1.64	1.62	1.61	1.59	1.59	1.58	1.57	1.57	1.56	1.55
24	1.71	1.70	1.70	1.69	1.69	1.68	1.68	1.67	1.66	1.65	1.64	1.63	1.62	1.61	1.59	1.58	1.57	1.57	1.56	1.55	1.54	1.54
25	1.70	1.69	1.68	1.68	1.67	1.67	1.66	1.66	1.65	1.64	1.63	1.62	1.61	1.59	1.58	1.56	1.56	1.55	1.54	1.54	1.53	1.52
26	1.68	1.68	1.67	1.67	1.66	1.66	1.65	1.65	1.64	1.63	1.61	1.60	1.59	1.58	1.57	1.55	1.54	1.54	1.53	1.52	1.52	1.51
27	1.67	1.67	1.66	1.65	1.65	1.64	1.64	1.64	1.63	1.62	1.60	1.59	1.58	1.57	1.55	1.54	1.53	1.53	1.52	1.51	1.50	1.50
28	1.66	1.66	1.65	1.64	1.64	1.63	1.63	1.63	1.62	1.61	1.59	1.58	1.57	1.56	1.54	1.53	1.52	1.51	1.51	1.50	1.49	1.48
29	1.65	1.65	1.64	1.63	1.63	1.62	1.62	1.62	1.61	1.60	1.58	1.57	1.56	1.55	1.53	1.52	1.51	1.50	1.50	1.49	1.48	1.47
30	1.64	1.64	1.63	1.63	1.62	1.62	1.61	1.61	1.60	1.59	1.57	1.56	1.55	1.54	1.52	1.51	1.50	1.49	1.49	1.48	1.47	1.46
32	1.63	1.62	1.62	1.61	1.60	1.60	1.59	1.59	1.58	1.57	1.56	1.54	1.53	1.52	1.50	1.49	1.48	1.47	1.46	1.46	1.45	1.44
35	1.61	1.60	1.60	1.59	1.58	1.58	1.57	1.57	1.56	1.55	1.53	1.52	1.51	1.50	1.48	1.47	1.46	1.45	1.44	1.43	1.43	1.42
40	1.58	1.57	1.57	1.56	1.56	1.55	1.55	1.54	1.53	1.52	1.51	1.49	1.48	1.47	1.45	1.43	1.42	1.42	1.41	1.40	1.39	1.38
45	1.56	1.55	1.55	1.54	1.53	1.53	1.52	1.52	1.51	1.50	1.48	1.47	1.46	1.44	1.43	1.41	1.40	1.39	1.38	1.37	1.37	1.36
50	1.54	1.54	1.53	1.52	1.52	1.51	1.51	1.50	1.49	1.48	1.46	1.45	1.44	1.42	1.41	1.39	1.38	1.37	1.36	1.35	1.34	1.33
60	1.52	1.51	1.50	1.50	1.49	1.49	1.48	1.48	1.47	1.45	1.44	1.42	1.41	1.40	1.38	1.36	1.35	1.34	1.33	1.32	1.31	1.30
75	1.49	1.49	1.48	1.47	1.47	1.46	1.45	1.45	1.44	1.43	1.41	1.40	1.38	1.37	1.35	1.33	1.32	1.31	1.30	1.29	1.27	1.26
100	1.47	1.46	1.45	1.45	1.44	1.43	1.43	1.42	1.41	1.40	1.38	1.37	1.35	1.34	1.32	1.29	1.28	1.27	1.26	1.25	1.24	1.22
120	1.46	1.45	1.44	1.43	1.43	1.42	1.41	1.41	1.40	1.39	1.37	1.35	1.34	1.32	1.30	1.28	1.26	1.26	1.24	1.23	1.22	1.20
140	1.45	1.44	1.43	1.42	1.42	1.41	1.41	1.40	1.39	1.38	1.36	1.34	1.33	1.31	1.29	1.26	1.25	1.24	1.23	1.22	1.20	1.19
180	1.43	1.43	1.42	1.41	1.40	1.40	1.39	1.39	1.38	1.36	1.34	1.33	1.32	1.29	1.27	1.25	1.23	1.22	1.21	1.20	1.18	1.16
250	1.42	1.41	1.41	1.40	1.39	1.39	1.38	1.37	1.36	1.35	1.33	1.31	1.30	1.28	1.26	1.23	1.22	1.21	1.19	1.18	1.16	1.14
400	1.41	1.40	1.39	1.39	1.38	1.37	1.37	1.36	1.35	1.34	1.32	1.30	1.29	1.26	1.24	1.21	1.20	1.19	1.17	1.16	1.14	1.12
1000	1.40	1.39	1.38	1.38	1.37	1.36	1.36	1.35	1.34	1.32	1.30	1.29	1.27	1.25	1.23	1.20	1.18	1.17	1.15	1.13	1.11	1.08

Denominator df

Critical Values d_L and d_U of the Durbin-Watson Statistic D (Critical Values are One-Sided)[a]

	α = 0.05										α = 0.01									
	k = 1		k = 2		k = 3		k = 4		k = 5		k = 1		k = 2		k = 3		k = 4		k = 5	
n	d_L	d_U	d_L	d_U	d_L	d_U	d_L	d_U	d_L	d_U	d_L	d_U	d_L	d_U	d_L	d_U	d_L	d_U	d_L	d_U
15	1.08	1.36	.95	1.54	.82	1.75	.69	1.97	.56	2.21	.81	1.07	.70	1.25	.59	1.46	.49	1.70	.39	1.96
16	1.10	1.37	.98	1.54	.86	1.73	.74	1.93	.62	2.15	.84	1.09	.74	1.25	.63	1.44	.53	1.66	.44	1.90
17	1.13	1.38	1.02	1.54	.90	1.71	.78	1.90	.67	2.10	.87	1.10	.77	1.25	.67	1.43	.57	1.63	.48	1.85
18	1.16	1.39	1.05	1.53	.93	1.69	.82	1.87	.71	2.06	.90	1.12	.80	1.26	.71	1.42	.61	1.60	.52	1.80
19	1.18	1.40	1.08	1.53	.97	1.68	.86	1.85	.75	2.02	.93	1.13	.83	1.26	.74	1.41	.65	1.58	.56	1.77
20	1.20	1.41	1.10	1.54	1.00	1.68	.90	1.83	.79	1.99	.95	1.15	.86	1.27	.77	1.41	.68	1.57	.60	1.74
21	1.22	1.42	1.13	1.54	1.03	1.67	.93	1.81	.83	1.96	.97	1.16	.89	1.27	.80	1.41	.72	1.55	.63	1.71
22	1.24	1.43	1.15	1.54	1.05	1.66	.96	1.80	.86	1.94	1.00	1.17	.91	1.28	.83	1.40	.75	1.54	.66	1.69
23	1.26	1.44	1.17	1.54	1.08	1.66	.99	1.79	.90	1.92	1.02	1.19	.94	1.29	.86	1.40	.77	1.53	.70	1.67
24	1.27	1.45	1.19	1.55	1.10	1.66	1.01	1.78	.93	1.90	1.04	1.20	.96	1.30	.88	1.41	.80	1.53	.72	1.66
25	1.29	1.45	1.21	1.55	1.12	1.66	1.04	1.77	.95	1.89	1.05	1.21	.98	1.30	.90	1.41	.83	1.52	.75	1.65
26	1.30	1.46	1.22	1.55	1.14	1.65	1.06	1.76	.98	1.88	1.07	1.22	1.00	1.31	.93	1.41	.85	1.52	.78	1.64
27	1.32	1.47	1.24	1.56	1.16	1.65	1.08	1.76	1.01	1.86	1.09	1.23	1.02	1.32	.95	1.41	.88	1.51	.81	1.63
28	1.33	1.48	1.26	1.56	1.18	1.65	1.10	1.75	1.03	1.85	1.10	1.24	1.04	1.32	.97	1.41	.90	1.51	.83	1.62
29	1.34	1.48	1.27	1.56	1.20	1.65	1.12	1.74	1.05	1.84	1.12	1.25	1.05	1.33	.99	1.42	.92	1.51	.85	1.61
30	1.35	1.49	1.28	1.57	1.21	1.65	1.14	1.74	1.07	1.83	1.13	1.26	1.07	1.34	1.01	1.42	.94	1.51	.88	1.61
31	1.36	1.50	1.30	1.57	1.23	1.65	1.16	1.74	1.09	1.83	1.15	1.27	1.08	1.34	1.02	1.42	.96	1.51	.90	1.60
32	1.37	1.50	1.31	1.57	1.24	1.65	1.18	1.73	1.11	1.82	1.16	1.28	1.10	1.35	1.04	1.43	.98	1.51	.92	1.60
33	1.38	1.51	1.32	1.58	1.26	1.65	1.19	1.73	1.13	1.81	1.17	1.29	1.11	1.36	1.05	1.43	1.00	1.51	.94	1.59
34	1.39	1.51	1.33	1.58	1.27	1.65	1.21	1.73	1.15	1.81	1.18	1.30	1.13	1.36	1.07	1.43	1.01	1.51	.95	1.59
35	1.40	1.52	1.34	1.58	1.28	1.65	1.22	1.73	1.16	1.80	1.19	1.31	1.14	1.37	1.08	1.44	1.03	1.51	.97	1.59
36	1.41	1.52	1.35	1.59	1.29	1.65	1.24	1.73	1.18	1.80	1.21	1.32	1.15	1.38	1.10	1.44	1.04	1.51	.99	1.59
37	1.42	1.53	1.36	1.59	1.31	1.66	1.25	1.72	1.19	1.80	1.22	1.32	1.16	1.38	1.11	1.45	1.06	1.51	1.00	1.59
38	1.43	1.54	1.37	1.59	1.32	1.66	1.26	1.72	1.21	1.79	1.23	1.33	1.18	1.39	1.12	1.45	1.07	1.52	1.02	1.58
39	1.43	1.54	1.38	1.60	1.33	1.66	1.27	1.72	1.22	1.79	1.24	1.34	1.19	1.39	1.14	1.45	1.09	1.52	1.03	1.58
40	1.44	1.54	1.39	1.60	1.34	1.66	1.29	1.72	1.23	1.79	1.25	1.34	1.20	1.40	1.15	1.46	1.10	1.52	1.05	1.58
45	1.48	1.57	1.43	1.62	1.38	1.67	1.34	1.72	1.29	1.78	1.29	1.38	1.24	1.42	1.20	1.48	1.16	1.53	1.11	1.58
50	1.50	1.59	1.46	1.63	1.42	1.67	1.38	1.72	1.34	1.77	1.32	1.40	1.28	1.45	1.24	1.49	1.20	1.54	1.16	1.59
55	1.53	1.60	1.49	1.64	1.45	1.68	1.41	1.72	1.38	1.77	1.36	1.43	1.32	1.47	1.28	1.51	1.25	1.55	1.21	1.59
60	1.55	1.62	1.51	1.65	1.48	1.69	1.44	1.73	1.41	1.77	1.38	1.45	1.35	1.48	1.32	1.52	1.28	1.56	1.25	1.60
65	1.57	1.63	1.54	1.66	1.50	1.70	1.47	1.73	1.44	1.77	1.41	1.47	1.38	1.50	1.35	1.53	1.31	1.57	1.28	1.61
70	1.58	1.64	1.55	1.67	1.52	1.70	1.49	1.74	1.46	1.77	1.43	1.49	1.40	1.52	1.37	1.55	1.34	1.58	1.31	1.61
75	1.60	1.65	1.57	1.68	1.54	1.71	1.51	1.74	1.49	1.77	1.45	1.50	1.42	1.53	1.39	1.56	1.37	1.59	1.34	1.62
80	1.61	1.66	1.59	1.69	1.56	1.72	1.53	1.74	1.51	1.77	1.47	1.52	1.44	1.54	1.42	1.57	1.39	1.60	1.36	1.62
85	1.62	1.67	1.60	1.70	1.57	1.72	1.55	1.75	1.52	1.77	1.48	1.53	1.46	1.55	1.43	1.58	1.41	1.60	1.39	1.63
90	1.63	1.68	1.61	1.70	1.59	1.73	1.57	1.75	1.54	1.78	1.50	1.54	1.47	1.56	1.45	1.59	1.43	1.61	1.41	1.64
95	1.64	1.69	1.62	1.71	1.60	1.73	1.58	1.75	1.56	1.78	1.51	1.55	1.49	1.57	1.47	1.60	1.45	1.62	1.42	1.64
100	1.65	1.69	1.63	1.72	1.61	1.74	1.59	1.76	1.57	1.78	1.52	1.56	1.50	1.58	1.48	1.60	1.46	1.63	1.44	1.65

[a] n = number of observations; k = number of explanatory variables.

Source: This table is reproduced from Biometrika, 41 (1951): 173 and 175, with the permission of the Biometrika Trustees.

TABLE OF RANDOM DIGITS

Row										
1	96299	07196	98642	20639	23185	56282	69929	14125	38872	94168
2	71622	35940	81807	59225	18192	08710	80777	84395	69563	86280
3	03272	41230	81739	74797	70406	18564	69273	72532	78340	36699
4	46376	58596	14365	63685	56555	42974	72944	96463	63533	24152
5	47352	42853	42903	97504	56655	70355	88606	61406	38757	70657
6	20064	04266	74017	79319	70170	96572	08523	56025	89077	57678
7	73184	95907	05179	51002	83374	52297	07769	99792	78365	93487
8	72753	36216	07230	35793	71907	65571	66784	25548	91861	15725
9	03939	30763	06138	80062	02537	23561	93136	61260	77935	93159
10	75998	37203	07959	38264	78120	77525	86481	54986	33042	70648
11	94435	97441	90998	25104	49761	14967	70724	67030	53887	81293
12	04362	40989	69167	38894	00172	02999	97377	33305	60782	29810
13	89059	43528	10547	40115	82234	86902	04121	83889	76208	31076
14	87736	04666	75145	49175	76754	07884	92564	80793	22573	67902
15	76488	88899	15860	07370	13431	84041	69202	18912	83173	11983
16	36460	53772	66634	25045	79007	78518	73580	14191	50353	32064
17	13205	69237	21820	20952	16635	58867	97650	82983	64865	93298
18	51242	12215	90739	36812	00436	31609	80333	96606	30430	31803
19	67819	00354	91439	91073	49258	15992	41277	75111	67496	68430
20	09875	08990	27656	15871	23637	00952	97818	64234	50199	05715
21	18192	95308	72975	01191	29958	09275	89141	19558	50524	32041
22	02763	33701	66188	50226	35813	72951	11638	01876	93664	37001
23	13349	46328	01856	29935	80563	03742	49470	67749	08578	21956
24	69238	92878	80067	80807	45096	22936	64325	19265	37755	69794
25	92207	63527	59398	29818	24789	94309	88380	57000	50171	17891
26	66679	99100	37072	30593	29665	84286	44458	60180	81451	58273
27	31087	42430	60322	34765	15757	53300	97392	98035	05228	68970
28	84432	04916	52949	78533	31666	62350	20584	56367	19701	60584
29	72042	12287	21081	48426	44321	58765	41760	43304	13399	02043
30	94534	73559	82135	70260	87936	85162	11937	18263	54138	69564
31	63971	97198	40974	45301	60177	35604	21580	68107	25184	42810
32	11227	58474	17272	37619	69517	62964	67962	34510	12607	52255
33	28541	02029	08068	96656	17795	21484	57722	76511	27849	61738
34	11282	43632	49531	78981	81980	08530	08629	32279	29478	50228
35	42907	15137	21918	13248	39129	49559	94540	24070	88151	36782
36	47119	76651	21732	32364	58545	50277	57558	30390	18771	72703
37	11232	99884	05087	76839	65142	19994	91397	29350	83852	04905
38	64725	06719	86262	53356	57999	50193	79936	97230	52073	94467
39	77007	26962	55466	12521	48125	12280	54985	26239	76044	54398
40	18375	19310	59796	89832	59417	18553	17238	05474	33259	50595

Answers

Chapter 2

Mix and Match

1. Brand of car: categorical: drivers
3. Color preference: categorical: consumers in focus group
5. Item size: ordinal: unknown (stocks in stores, purchase amounts)
7. Stock price: numerical: companies
9. Sex: categorical: respondents in survey.

True/False

11. False. Zip codes are numbers, but you cannot perform sensible calculations with them.
13. False. Cases is another name for the rows in a data table.
15. True
17. False. A Likert scale is used for ordinal data.
19. False. Aggregation collapses a table into one with fewer rows.

Think About It

21. a) cross sectional
    b) Whether opened an IRA (categorical), Amount saved (numerical, $)
    c) Did employees respond honestly, particularly when it comes to the amount they report to have saved?
23. a) cross sectional
    b) Service rating (probably ordinal, using a Likert scale)
    c) With only 500 replying, are these representative of the others?
25. a) time series
    b) Exchange rate of U.S. dollar per Canadian dollar (numerical ratio of currencies)
    c) Are the fluctuations in 2005 typical of other years?
27. a) cross sectional
    b) Quality of graphics (categorical, perhaps ordinal from bad to good) Degree of violence (categorical, perhaps ordinal from none to too much)
    c) Did some of the participants influence opinions of others?
29. a) cross sectional (could be converted to a time series)
    b) Name (categorical), Zip code (categorical), Region (categorical), Date of purchase (categorical or numerical, depending on the context), Amount of purchase (numerical, $), Item purchased (categorical)
    c) Presumably recoded the region from the zip code

4M Economic time series

a) Answers will vary, but should resemble the following. By merging the data, we can see how sales of Best Buy move along with the health of the general economy. If sales at Best Buy rise and fall with disposable income, we question the health of this company if the government predicts a drop in the amount of disposable income.
b) A row from FRED2 describes the level of disposable income in a month whereas a row in the company-specific data is quarterly.
c) The national disposable income is in billions and the quarterly sales are in millions.
d) We can aggregate the monthly numbers into a quarterly number, such as an average. Alternatively, we could take the quarterly number and spread it over the months.
e) Name the columns Net Sales ($ billion) and Disp Income ($ trillion) and scale as shown previously. The dates could be recorded in a single column, e.g., 2004:1, 2004:2, etc., or in the style shown in the following table.
f) Here are the merged data for 2004.

Quarter	Net Sales ($ billion)	Disp Income ($ trillion)
Jan–04	$5.48	$8.48
Apr–04	$6.09	$8.58
Jul–04	$6.65	$8.67
Oct–04	$9.23	$8.93

g) Sales at Best Buy rocket up in the fourth quarter (50% higher during the holiday season), but consumers don't have that much more money to spend.

4M Textbooks

a) Various sources report that books cost about $100 per class. In 2003, U.S. Senator Charles E. Schumer of New York released a study showing that the average New York freshman or sophomore pays $922 for textbooks in a year, so reducing the cost 5% would save $46.10 a year and by 10% would save $92.20 a year.
b) Your table should have headings like these. You should use the names of the stores you shopped at if different from these. The first two columns are categorical, with the first identifying the book and the second giving the label. The two columns of prices are both numerical.

Book Title	Type	Price at Amazon	Price at B&N

c) Answers will vary. d) Answers will vary.
e) You should include all of the relevant costs. Some Internet retailers add high shipping costs.
f) Answers will vary. The key point to notice is the value of comparison. Because you've got two prices for the same books, you can compare apples to apples and see whether one retailer is systematically cheaper than the other.

Chapter 3

Mix and Match

In each case, unless noted, bar charts are better to emphasize counts whereas pie charts are better to communicate the relative share of the total amount.

1. Pie chart, bar chart, or Pareto chart
3. Bar chart, Pareto chart or table (3 values)

5. Bar chart, Pareto chart (counts) or pie chart (share)
7. Pareto chart
9. Pie chart (shares) or a table (3 values)
11. Bar chart or table (4 values)

True/False

13. True, but for variables with few categories, a frequency table is often better.
15. False. The frequency is the count of the items.
17. True
19. True
21. True

Think About It

23. Customers tend to stick with manufacturers from the same region. Someone trading in a domestic car tends to get another domestic car whereas someone who trades in an Asian car buys an Asian car. The more subtle message is that those who own Asian cars are more loyal.
25. This is a bar chart comparing the shares of US debt held in these countries.
27. **a)** No, the categories are not mutually exclusive.
   b) Divided bars such as these might work well.

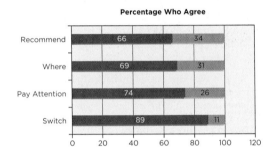

Percentage Who Agree

29. No, the categories are not mutually exclusive.
31. We would have a bar chart with nearly 200 categories.
33. One very long bar (height 900) and five shorter bars of height 20 each.
35. Five bars, each of the same height
37. Frequency table
39. Mode: public. No median since the data are not ordered
41. Modal preference because the median is not defined
43. Ordinal, No, because the responses are ordered

You Do It

45. **a)** It probably accumulates case sales by brand to some degree. Unlikely that every case is represented by a row.
   b) The three smallest brands are combined into an Other category.

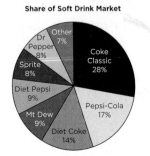

Share of Soft Drink Market

c)

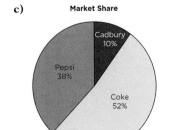

Market Share

d)

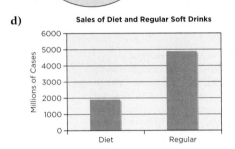

Sales of Diet and Regular Soft Drinks

47. **a)** The addition of this extra row makes up a big part of both pie charts.

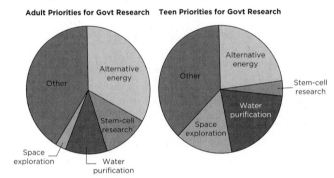

Adult Priorities for Govt Research   Teen Priorities for Govt Research

   b) The side-by-side bar chart works well for this. We no longer need the Other category.

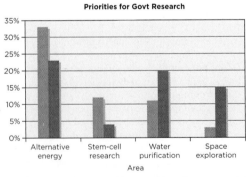

Priorities for Govt Research

   c) No, because the categories would no longer partition the cases into distinct, non-overlapping subsets

49.

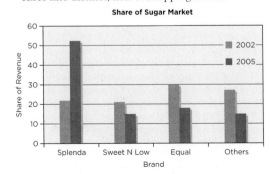

Share of Sugar Market

51. a)

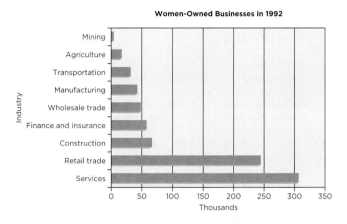

Women-Owned Businesses in 1992

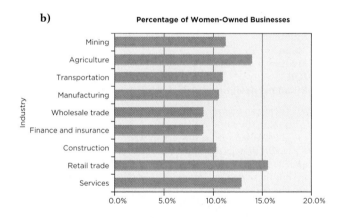

All Businesses in 1995

They are pretty similar, but women-owned business seems to be slightly concentrated in the retail trade.

b)

Percentage of Women-Owned Businesses

No, a pie chart does not show the percentage of women-owned business within each industry.

c) Yes, but slight. Some industries may have grown or shrunk in number during the time gap, but the change would be relatively small due to the broad nature of the categories.

53. a)

Unexpected illness	4,483	15.8%
Planned leave	23.735	84.2%

b)

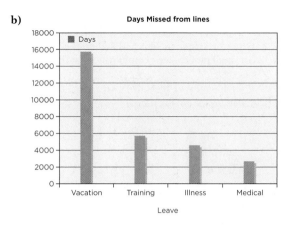

Days Missed from lines

55. a) Yes

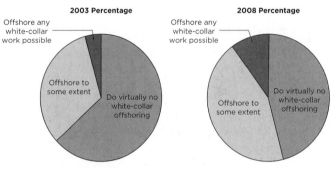

2003 Percentage

2008 Percentage

b) One example:

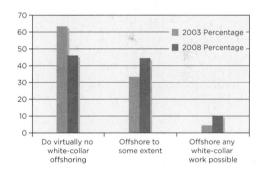

c) The bar chart facilitates comparison and the pie chart makes the relative shares more apparent.

d) Same in 2003, different in 2008

57. a)

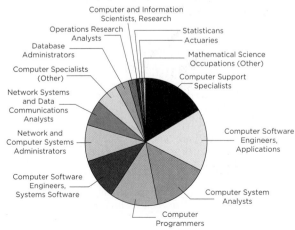

Employment in the Math Sciences

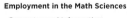

There are so many categories that the dominance of computer-related occupations is hidden unless you look at the labels. Statisticians and Actuaries account for a very small percentage of employment in the math sciences.

b) Computer support specialists

c)

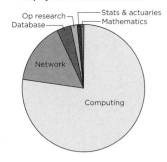

Employment in the Math Sciences

4M Growth Industries

a) It could use the trends suggested by the table to indicate how to shift sales force from declining industries to those that appear stronger and growing.

b) A bar chart makes it simpler to compare the counts within a year; a pie chart would emphasize the share of each industry for each year.

c) A bar chart of the difference between 1980 and 1997 for each industry

d)

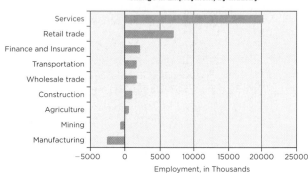

Change in Employment, by Industry

The categories are shown from those with the greatest increase down to those with the largest decline.

e) They have negative bars.

f) Show all nine since there are not that many categories.

g) The growth in service-type industries (including retail) and the decline in basic manufacturing and mining

h) The size of the change relative to the size of the industry is hidden.

Chapter 4

Mix and Match

1. g   **3.** h   **5.** j   **7.** d   **9.** e

True/False

11. False; the box is the median, with lower edge at the 25% point and upper edge at the 75% point.

13. True

15. True

17. False; the Empirical Rule applies only to numerical variables that have a symmetric, bell-shaped distribution.

19. True

21. True

Think About It

23. You cannot tell from the median.

25. It would be very heavily right-skewed.

27. Fixed-rate mortgage

29. Yes

31. Mortgage payments have a larger SD but the coefficient of variation may be larger for the allowances.

33. a) Music only

b) No, the total amount cannot be recovered from the median.

c) No, there is variation in the data which means that the groups may overlap.

35. No. Since 1.2 SDs is not far from the mean, no assumptions are needed.

37. a) Right skewed, with a single peak at zero and trailing off to the right

b) Right skewed with one mode, from moderate prices to very large orders

c) Bell shaped around the target weight

d) Likely bimodal representing male and female students

39. a) 11   **b)** About 5%

c) Median because of the outliers to the far left

d) The rounding of the rates

e) 2

41. a) The mean is about $34,000, the median is about $27,000. The mean is larger because of the skewness.

b) About $30,000

c) The SD is slightly larger due to the skewness.

d) The figure is identical except for the labels on the x-axis.

43. The pricing errors will lead to more variation in prices and a more spread-out histogram.

45. Because the distribution of income is very right skewed, with the upper tail reaching out to very high incomes

47. The shape will be the same as in Figure 4.1 but the labels on the x-axis will change to 60, 120, 180, and so forth. The count axis and bin heights would be the same.

49. a)

Summary	File Size MB	Song Length Sec
Mean	3.8	228
Median	3.5	210
IQR	1.5	90
Standard deviation	1.6	96

b) The mean and median increase by 2 MB, the IQR and SD remain the same.

c) The median would stay about where it is and the IQR would remain the same.

51. a) The mean is about $18,000 and the SD is about $10,000, but these do not capture the bimodal nature of the data.

b) The data are bimodal with cluster centers having means of about $7,500 and $30,000.

c) It's unlikely.

d) The cluster with mean near $7,500 are public; the others are private schools.

You Do It

53. a) The histogram is right skewed. Some of these cars have exceptionally high horsepower.

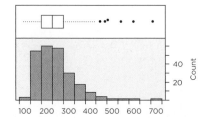

b) The histogram conveys more about the shape of the distribution, whereas the boxplot nails down the position of the median and the IQR. The boxplot also highlights the outlying cars with very high power.

c) The mean is 214 HP with standard deviation 84 HP. The mean is the balance point of the histogram, to the right of the center of the large peak. The SD suggests the spread of the data, but because the distribution is not bell shaped, we should not rely on the Empirical Rule in this case.

d) $c_v \approx 0.39$; s is about 40% of the size of the mean.

e) Several Ferrari models, a Lamborghini, a Rolls Roye, and an Aston Martin. No

f) Typical. Half of the models have more, and half have less.

55. Beatles

a)

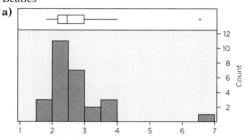

b) The large outlier is the song "Hey Jude," which goes on for 7 minutes and uses 6.6MB.

c)

	Mean	Median
With "Hey Jude"	2.73MB	2.47MB
Without	2.59	2.43

d) Median

e) Mean

57. DP Industry

a) Mean = 529.94, Median = 40, SD = 2131.58. Millions of dollars

b) The histogram is unimodal, but very right skewed. Data occupy little of the plot. There's more to the variation in this column than we can see in this figure.

c) The largest of these is EDS. The next largest is CSC.

d) Sales in the DP industry are concentrated in the few largest companies.

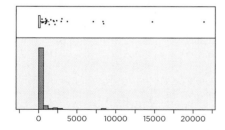

59. Tech stocks

a) Dell   IBM

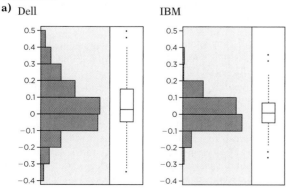

Microsoft

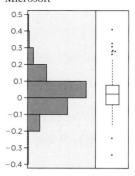

The three histograms are remarkably similar. Each is bell shaped with a few outliers. Dell seems more spread out than IBM or Microsoft, suggesting a larger SD.

b)

	Dell	IBM	Microsoft
Mean	0.0448875	0.0120452	0.0258052
Std Dev	0.1564424	0.091091	0.1046678
Coef of var	3.48521	7.56243	4.05607

Means and SDs are quite useful because each distribution is bell shaped.

c) Returns on IBM are much more variable relative to the mean level.

d) Large positive mean value for the return and small SD for the return; the closer the c_v is to zero, the better.

e) Yes. The company with the largest mean (Dell) also has the largest SD. Similarly, the company with the smallest mean return (IBM) has the smallest SD.

61. Tech Stocks

a)

	Dell	IBM	Microsoft
Mean	0.0448875	0.0120452	0.0258052
Std Dev	0.1564424	0.091091	0.1046678
Sharpe	0.265832664	0.096005094	0.215015506

Dell. Though more volatile, it's earning a much higher rate of return.

b) The mean is the Sharpe ratio for Dell.

c) The Sharpe ratio is used to compare returns on different stocks. Standardizing by forming z-scores is most useful (with the Empirical Rule) for identifying outliers and judging the relative size of different observations of the *same* variable.

4M Financial Ratios

Motivation

a) A single score makes it easier to define a goal rather than having a two-part goal.

b) The ratio measures how much of each dollar in sales is retained toward the net income of the company rather than "lost" as part of the cost of doing business. The ratio is basically a proportion, with no units. (Or, you can multiply by 100 and call it a percentage.)

c) The company could tie compensation, such as the size of a bonus, to the size of this ratio.

d) There are numerous alternative ratios. A similar, more common ratio is the ratio of net income to assets.

Method

e) Histograms of each, perhaps with boxplots in addition

f) If the distribution is very skewed, it may be hard to get a sense of how attainable the goals are.

Mechanics

g)

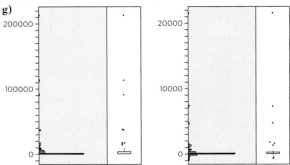

h) Both distributions are very skewed. Although the histogram for net sales (left) runs from 0 to more than $200 billion, the net sales for all but a few is smaller than $10 billion. The median is around $1 billion.

i) The largest 3 outliers (**in both**) are Exxon-Mobil, Chevron, and Conoco.

j)

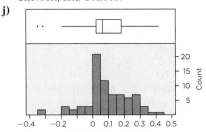

The distribution of the ratio of sales to income is unimodal and much closer to symmetric. The ratio is occasionally negative, with a shape that tails off slowly to the right of zero, sharply to the left). The scale of net income above hides the fact that some companies have negative net income.

k) No, it works in this case because the outliers in both histograms "line up" so that the company with largest income also has largest net sales, the second largest is the same, and so forth. A random ratio could end up even more skewed if the largest in one distribution became paired with the smallest in the other.

Message

l) The ratio of Net Income to Sales in the energy sector ranges from −40% to almost 50%. For most companies, values between 0 to 20% are most common, with about half attaining a ratio of 10% or more.

m) A goal to be in the upper half of the distribution seems reasonable. This requires the ratio to be larger than 10%. A larger goal of 25% is also possible, but much harder to attain. Only about 10% of these companies do so well.

Chapter 5

Mix and Match

1. j **3.** e **5.** a **7.** i **9.** d

True/False

11. True

13. True

15. True

17. False. The value of chi-square is the same in either case.

19. False. Association is not the same as cause and effect. We cannot interpret association as causation because of the possible presence of a lurking variable.

21. False. It is not associated with defective items.

Think About It

23. a) Employed and retired respondents found different rates of satisfaction in resolving the disputed charge.
  b) No. About 70% from each group were satisfied with the outcome of their call.

25. Simpler if they are not associated, because then the choice of the best color does not affect the packaging.

27. a) Administration
  b) Yes, because the composition of the segments differs for the three groups

29. a) Asia Pacific
  b) Latin America and Middle East/Africa
  c) Asia Pacific   **d)** No
  e) Yes. Different manufacturers dominate in different regions.

31. Only if the choice of color is not associated with the style of the vehicle

33. Yes, fewer shoppers with children present would be expected late night.

35. Association is not the same as causation although most scientists now accept the relationship between smoking and cancer as causal.

37. a) 0 **b)** 0
  c) The store should order the same fraction of gloss in each color (4/7 low gloss, 1/7 medium, and 2/7 high gloss).

You Do It

39. a)

	Weekday	Weekend	Total
Premium	126	62	188
Plus	103	28	131
Regular	448	115	563
Total	677	205	882

  b) Premium: 19%; Plus: 15%; Regular: 66%
  c) Weekdays: 67%; Weekend: 33%
  d) No. These conditional distributions are not directly comparable.
  e) Weekends, because there is a greater concentration of premium sales then.

41. a) $x^2 = 12.75, V = 0.12$
  b) V is rather small, indicating weak association between the type of gas and the timing of the purchase.

43. a)

	Question uses "satisfied"	Question uses "dissatisfied"	Total
Very satisfied	139	128	267
Somewhat satisfied	82	69	151
Somewhat dissatisfied	12	20	32
Very dissatisfied	10	23	33
Total	243	240	483

  b) Those surveyed using the word "satisfied"
  c) In a positive sense using the word "satisfied"

45. a) $x^2 = 8.6, V = 0.134$. The association is weak.
  b) $x^2 = 8.1, V = 0.130$. Cramer's V is smaller.

47. a) Yes, the conditional distributions within each row are different.
  b) Some industries have a higher proportion of male employees than female employees.
  c) If $n = 400$ with 100 in each row, $x^2 = 16.5$ and $V = 0.20$. If $n = 1600$ with 400 in each row, chi-squared is four times larger, but V is unchanged.

49. a) The shown proportions differ in the rows of the table, but this is not the type of association that we have studied in this chapter. See part b.

b) This table is *not* a contingency table since the cells show the conditional dist. within each row. The cells are not mutually exclusive.

51. a) The results show clear association, with the appearance that support influences the outcome. Cramer's $V = 0.190$.

  b) The association is so strong as to make it difficult to find a lurking variable that would mitigate what is shown in the table. However, it's possible that the support came after the research, rather than before.

53. a) United, with an on-time arrival rate of 0.811; US Airways's is 0.759.

  b) Yes. The on-time percentage at Denver is 0.824 whereas it's 0.740 in Philadelphia. This is important since most United flights go to Denver, whereas US Airways flies to Philadelphia.

  c) Yes. When compared by destination, US Airways is better: 82.3 vs. 83.4% in Denver and 71.4 vs. 74.3% in Philadelphia.

4M Discrimination in Hiring

Motivation

  a) Yes, namely that the company was laying off a higher proportion of older employees

  b) No, but if association is found, the company might argue that a lurking factor (ability) undermines this analysis.

Method

  c) Within the rows

  d) Cramer's V because of its interpretable range from 0 to 1. It summarizes in a single value the differences in rates of layoffs within the rows of the table.

Mechanics

  e) The value of chi-square is 53.2 and Cramer's V is 0.164, indicating weak association.

Message

  f) The layoffs discriminate against older employees. The overall rate of layoffs is about 3.5%. The rate is less for employees younger than 50 and higher for those who are older. Among employees who are 60 or older, the rate climbs to more than 14%.

  g) It could mean that the association is not due to age, but rather to some other factor that is related to age, such as the ability of the employee to do the work.

4M Picking a Hospital

Motivation

  a) Show the whole table if there is association, but limit the information to the margins if there isn't any.aaa

Method

  b) Conditional

Mechanics

  c)

	Early Stage	Late Stage	
CH	10%	57%	29%
UH	5%	44%	37%

  d) CH: 41.2%; UH: 80.5% **e)** UH. UH **f)** CH

Message

  g) Go to the university hospital regardless of the type of cancer. Even though the university hospital has a higher death rate overall, this is an artifact of its patient mix.

  h) The marginal information is not adequate. To judge the quality of care, information on the mix of patients must be given.

Chapter 6

Mix and Match

1. a) ii **b)** i **c)** iii **d)** iv   **3. a)** iv **b)** iii **c)** i **d)** ii

True/False

5. True

7. False, in general. The pattern could be in a negative direction.

9. True

11. False. The pattern would be linear, with $y \approx 0.1x$, where y denotes revenue and x denotes gross sales.

13. False. The value of the stock would fall along with the economy. We'd rather have one that was negatively related to the overall economy.

15. True

17. True

19. False. The positive association is likely the result of rising gasoline prices in the market overall. Hence a constant volume of customers produces increasing dollar volume.

Think About It

21. a) Response: total cost; explanatory: number of items. Scatterplot: linear, positive direction with lots of variation

  b) Response: items produced; explanatory: hours worked. Scatterplot: positive direction, linear (with perhaps some curvature for long hours), and moderate variation

  c) Response: time; explanatory: weight. Scatterplot: negative direction, probably linear with lots of variation in the times

  d) Response: gallons left; explanatory: number of miles. Scatterplot: negative direction, linear, with small variation around the trend

  e) Response: price change; explanatory: number recommending. Scatterplot: little or no pattern

23. a) Positive association, but weak

  b) Positive, 0.5

  c) It increases the correlation. The correlation would decrease.

  d) No, when the outliers are excluded, the association is too weak to arrive at this conclusion.

25. No change, the correlation is not affected by changes in scale.

27. No, you can add and subtract a constant from a variable without changing the correlation.

29. The slope of the linear relationship measured by the correlation is r. Hence, the predicted z score must be smaller than the observed z score because the absolute value of r is less than 1.

31. a) Yes, they appear to move in opposite directions

  b) No, it also shows a negative association

  c) Approximately -0.70

  d) The timeplot shows the timing of the events. The scatterplot shows the contemporaneous association more clearly and reveals the linear association.

  e) No. We're still only able to consider association, not causation. Other factors in the economy could cause both series to move.

33. The correlation is larger among stones of the same cuts, colors and clarities. These factors add variation around the correlation line. By forcing these to be the same, the pattern is more consistent.

35. Cramer's V measures association between *categorical* variables, the levels of which cannot in general be ordered. Thus, it does not make sense to speak of the *direction* of the association.

37. You have a 1-in-4 chance of guessing the original.

You Do It

39. a) Yes. About 0.5
b) Little or no association

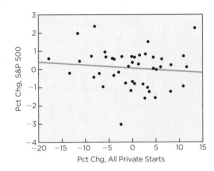

c) $r = -0.115$, slightly negative
d) No, the association is too weak to be useful for prediction.

41. a) Yes
b) Strong, positive, linear association

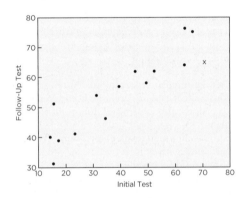

c) The correlation is $r = 0.903$. The correlation is suitable because the association is linear. We'd expect the relative position on the second test to be lower, closer to 1.8.
d) The employee is marked by the x in the figure. The correlation line, with slope 0.9 implies that we expect scores on the second test to be a little closer to the mean than the scores on the first test. The decline for this employee seems consistent with the pattern among the others so we should not judge this employee as becoming less productive.

43. a)

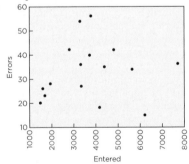

Entered: explanatory; errors: response. Little or no pattern in these data.

b) 0.094 **c)** It would not change.
d) The correlation indicates a very weak association between the number of data values entered and the number of errors. Evidently, those who enter a lot of values are simply faster and more accurate than those who enter fewer
e) There's virtually no association between the number entered and the number of errors.

45. a)

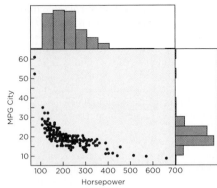

b) Negative pattern, bending with a sharp initial drop and then slowing, with relatively little variation around the curving pattern. The outliers with exceptional mileage are hybrids (Honda Insight and Toyota Prius) and with huge horsepower are exotic sports cars (Lamborghini, Ferrari, etc.).
c) -0.69
d) Moderate negative association. No, because the relationship is not linear.
e) Both marginal distributions are right skewed, with extreme outliers (noted in "b"). Horsepower: mean = 214 HP, SD = 84; Mileage: mean = 19.8, SD 5.5.
f) $z_x = -1.06$. We expect its mileage to be about r SDs = 0.73 above the mean, giving an estimated mileage of $\bar{y} + 0.73 s_y = 23.8$ MPG. Because the relationship is not linear, this estimate is suspect.

47. a)

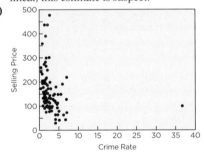

The outlier is Center City, Philadelphia, unusual in terms of crime rate, but not selling price.

b) $r = -0.25$
c)

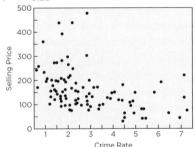

There is a great deal of variation around a weak, negative trend that appears to bend.
d) -0.43
e) No. First, this is aggregated data. Second, correlation measures association, not causation.

49. a) $V \approx 0.23757$
b) 0.24
c) The two are exactly the same here. However, Cramer's V is always positive, whereas the correlation picks up the direction. The squared correlation is always the same as the squared value of V.

51. a) Yes, both show an upward trend (red for compensation, and green for output).

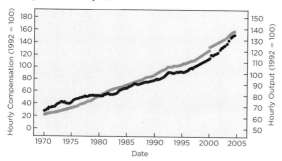

b) It confirms the very high correlation, though the white space rule should make one suspicious.

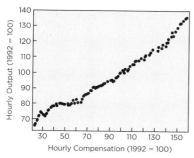

c) 0.988

d) No, correlation does not imply causation. It's more the case that compensation and output are both rising as the economy grows and inflation continues.

53. a) Yes

b) Strong positive linear association, with one pronounced outlier (row 437)

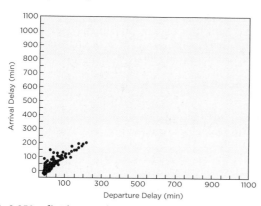

c) 0.958   **d)** The correlation is smaller, 0.907.

e) It would not change.

4M Correlation in the Stock Market

Motivation

a) If the stocks are strongly related, then if one goes up, they all go up and vice versa. Consequently, owning several highly related investments is just like putting all of your money into one of them.

b) Negatively related. This is the motivation behind diversifying one's investments.

Method

c) The investor can measure the association, if it is linear, with three correlations. Only three are necessary since the correlation is symmetric.

d) Look at scatterplots.   **e)** Plot the returns over time.

Mechanics

f)

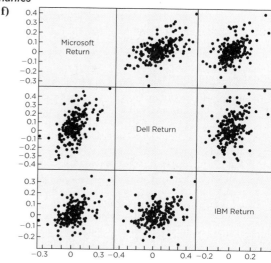

No. Yes, they are linearly associated but not very strongly.

g)

	Correlations		
	Microsoft Return	Dell Return	IBM Return
Microsoft Return	1.0000	0.5500	0.4831
Dell Return	0.5500	1.0000	0.3545
IBM Return	0.4831	0.3545	1.0000

h) They allow us to see whether the correlation is the result of time patterns in the time series.

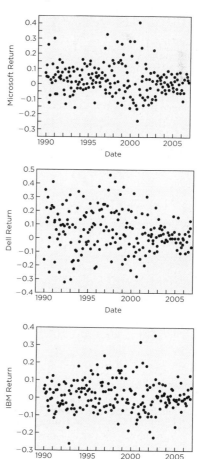

Message

i) The returns on stock in these three companies are positively associated, with considerable variation around the pattern. The returns tend to move up and down together, offering less diversification than an investor might like.

4M Cost Accounting

a) Information about the correlated variable can be used to improve the informal method used to estimate the price of the order. More accurate prices make it possible to offer competitive prices with less chance of losing money on a bid.

b) Cost per unit. Material costs, labor costs, and machining costs

c) For linear associations, correlation identifies which inputs are most associated with the final cost per unit. Strong, positive correlations indicate factors associated with increasing costs, whereas negative correlations indicate factors that are associated with decreasing costs.

d) So that nonlinear patterns or the effects of outliers are not missed

e) The strongest linear association is between the final cost and the number of labor hours. There appear to be four outliers with very high cost.

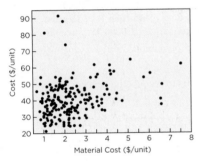

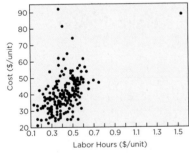

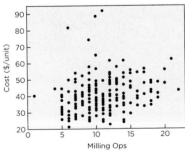

f) Labor hours (0.5086)

	Cost ($/unit)	Material Cost	Labor Hours	Milling Operations
Cost ($/unit)	1.0000	0.2563	0.5086	0.1801
Material Cost	0.2563	1.0000	0.0996	0.3400
Labor Hours	0.5086	0.0996	1.0000	0.3245
Milling Operations	0.1801	0.3400	0.3245	1.0000

g) The four outliers appear to weaken the correlation between the response and material costs, increase the correlation with labor hours, and perhaps weaken the small amount of association between the response and the number of milling operations.

h) Without the four outliers, the correlation with material costs is noticeably higher. The correlation with labor is slightly smaller, and the correlation with milling ops is higher.

	Cost ($/unit)	Material Cost	Labor Hours	Milling Operations
Cost ($/unit)	1.0000	0.3655	0.4960	0.2674
Material Cost	0.3655	1.0000	0.1340	0.3354
Labor Hours	0.4960	0.1340	1.0000	0.4054
Milling Operations	0.2674	0.3354	0.4054	1.0000

i) Amount of labor

j) The size of the correlation implies that much of the variation in cost per unit cannot be attributed to labor alone, and is due to other factors. Management cannot accurately anticipate the cost of an order using just one of these explanatory variables alone.

Chapter 7

Mix and Match

1. j **3.** h **5.** b **7.** d **9.** e

True/False

11. False. The sample space consists of all possible sequences of yes and no that he might record (32 elements total).

13. False. These are not disjoint events.

15. True

17. False. Both events could happen.

19. False. Only if the data lacks patterns does the relative frequency tend to the probability in the long run.

21. False. The intersection **B** and **C** is a subset of the event **A**.

Think About It

23. Items a and b

25. a) {fresh}
  b) {frozen, refrigerated, fresh, deli}
  c) {deli}

27. a) Tall men with a big waist
  b) The choice of waist size is independent of the length of the pant leg.
  c) (**B** *or* **T**)

29. Intersection

31. Item c

33. Not likely. The intensity of traffic would change over the time of day and the day of the week.

35. a) Yes **b)** No

37. Go for the 3-point shot assuming that the outcomes are independent.

39. Pure speculation

41. a) No. The Law of Large Numbers applies to the long-run proportion. Also, accidents could be dependent due to pilot anxiety.
  b) No

You Do It

43. a) 1. S = {blue, orange, green, yellow, red, brown}
  2. 0.37
  3. 0.84

b) 1. **S** = {triples of three colors}
   2. ≈0.0138
   3. 0.013
   4. ≈0.561
45. a) ≈0.238
   b) ≈0.146
   c) 0.438
   d) ≈0.0000628
47. a) ≈0.905
   b) 1 in 10,000
   c) 0.1. The events have such small individual probability that there's not much double-counting.
49. ≈0.262
51. a) 9/16. The necessary assumption of independence is questionable here.
   b) 0. None
53. a) ≈0.782
   b) ≈0.478
   c) ≈0
55. a) 1/8
   b) Yes. The probability of three in a row is ≈ 0.00003.
   c) ≈0.266. Assume that the purchase amount is unrelated to the discount.
57. a) The plot does not indicate anything out of the ordinary.
   b) One explanation involves fouls by the opposing team. Bryant scored 12 points on free throws in the fourth period alone. Misses associated with fouls are counted in the box score as misses just the same. Bryant was fouled on the last three shots (which were misses).

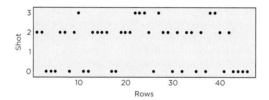

4M Odds: Where's the Rest of the Probability?

a) Odds are useful for figuring out the payouts from wagers, but probabilities are more intuitive if you want to know the relative frequency.
b) 1
c) The complement rule
d) Odds can range from 0 (probability of the event is 0) to arbitrarily large as the probability for the event increases.
e) The probabilities are (in order as shown in the figure)

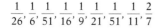

The probabilities are all between 0 and 1, but they don't sum to approximately 0.851, not 1.
f) Where did the rest of the probability go? It goes to the track in the form of money that was wagered, but not paid out. The track takes the rest as profits for running the race and paying its staff.[1]

[1] We're certainly not the first to notice this property of betting odds. Have a look at Brown, L. D., D'Amato, R., and Gertner, R. (1994), "Racetrack betting: Do bettors understand the odds?," *Chance*, 7 (3), 17–23.

4M Auditing a Business

a) By sampling a few transactions and thoroughly investigating these, the auditor is likely to do a better job than by trying to check *every* transaction. It might not even be possible to check every transaction: the firm may generate more transactions in a day than the auditor could check in a day. If the auditor finds no anomalies in, say, 30 or 40 transactions, it's unlikely that there's a lot of fraud going on.
b) At random. Without further input, we need to make sure that we get items from all over the business, without assuming some are more likely to be fraudulent. If, however, managers suspect a problem in certain situations, we should check these first.
c) We'd end up knowing a lot about this one division, but nothing about the others unless we're willing to believe that fraud is equally likely over the entire firm.
d) ≈0.603
e) 0.718
f) Because inventory fraud is more common, we should audit 100 of these transactions. The chance of finding no fraud among these is smaller than the chance of finding a sales fraud. The choice to audit only inventory transactions depends on knowing that there's a higher rate of fraud in these transactions rather than sales transactions. If that knowledge is wrong—or suspect—we could be making the wrong choice. The point, however, is that you should audit the area where you suspect a problem.
g) No. There could be fraud and we missed it. Direct calculations in "e" show that even if there's a 2% chance of sales fraud and a 3% chance of inventory fraud, an evenly divided audit of 50 transactions has a 28% chance of finding only legitimate transactions.

Chapter 8

Mix and Match

1. f **3.** c **5.** b

True/False

7. False. $P(A) > P(A|S)$ could happen, but unlikely.
9. False. $P(A) > P(S)$ does not imply dependence.
11. False. She also needs a joint probability.
13. False. Independence implies that one event does not influence the chances for the other.
15. False. These are disjoint events, so the probability of the intersection must be 0.
17. True
19. True

Think About It

21. Independent. It is unlikely that seeing a Honda, for example, would make us suspect that the next car was also a Honda.
23. Dependent The number of visits today is probably influenced by the same factors that influenced the number of visits yesterday.
25. Most likely independent
27. Independent. If **A** happens, then the chance for **B** remains one-quarter because **B** is one-quarter of the area of **A**.
29. a) Classify everyone as a drug user.
   b) The problem is that the test will not be very specific.

31. No. The statement of the question first gives $P(\mathbf{F}|\mathbf{Y}) = 0.45$ and then $P(\mathbf{Y}|\mathbf{F}) = 0.45$. In order to be independent, $P(\mathbf{F}|\mathbf{Y}) = P(\mathbf{F})$, but we are not given $P(\mathbf{F})$.

33. a) $0.42 = P(\text{working affected grades} | \text{have loan})$. The sample space might be the collection of recent college grads.
b) We cannot tell because we don't know the proportion who worked in college.

You Do It

35. a) A tree is probably easier for two main reasons: the probabilities are given as conditional probabilities and the options for the two types of vehicles are *not* the same.
b) 1/8

37. 0.21

39. ≈0.91

41. We assume that any order of selection of the parts is equally likely.
a) 7/22 **b)** 21/22 **c)** 7/99 **d)** ≈ 0.00884

43. a) 1/12
b) 1/11
c) Because the events are dependent

45. a) Dependent.
b) 21/46
c) ≈0.1439

47. a) 0.543
b) 0.195
c) Education/health care or durable goods.

49. a) 0.82
b) 5/9

51. 0.34

4M Scanner Data

The following table shows the conditional distributions within columns.

	Joint prob Col probability	Number of Dogs Owned: 0	1	2	3	More than 3	Total
Number of dog food items purchased	0	4.87	2.17	0.25	0.02	0.00	7.31
		7.35	7.89	4.51	3.75	0.91	
	1 to 3	16.98	7.34	1.04	0.04	0.02	25.42
		25.61	26.66	18.86	9.58	10.00	
	4 to 6	11.82	5.16	0.93	0.06	0.02	17.99
		17.83	18.73	16.85	12.92	10.00	
	7 to 12	11.60	4.69	1.13	0.12	0.05	17.59
		17.50	17.01	20.54	27.08	25.45	
	More than 12	21.03	8.18	2.16	0.21	0.11	31.69
		31.72	29.71	39.24	46.67	53.64	
	Total	66.30	27.54	5.51	0.45	0.20	100

Motivation

a) No, they are likely dependent. Presumably, shoppers who own more dogs will be inclined to buy more dog food.
b) Most likely it means that the shopper owns a dog, but did not indicate that when applying for the shopper's card or perhaps the dog was acquired after filling out the application. A shopper might also buy the food for a friend, relative, or neighbor. Finally, the self-reported data may not be accurate.

Method

c) Column probabilities because they describe shopping habits given the number of pets owned
d) No. These *joint* probabilities are small because relatively few shoppers report having more than three dogs.
e) Most likely a rounding error or careless computational mistake.

Mechanics

f) The table shown above adds the marginal probabilities in the gray border. In the table, probabilities are shown multiplied by 100 as percentages.
g) $P(\text{more than 3 item} | \text{no dogs}) = 67.03\%$ compared to $P(\text{more than 3 items} | \text{more than 3 dogs}) = 90\%$
h) $P(\text{no dogs} | \text{bought 8}) = \approx 0.659$ so that $P(\text{report own dogs} | \text{bought 8}) = 0.341$

This probability suggests something odd in the data. The most likely number of dogs for someone who buys this much is none! Evidently, the number of dogs seems to be underreported.

Message

i) The scanner data as reported here will not be very useful. The conditional probability of buying more than twelve cans, say, is highest among customers who report owning more than three dogs, but their small number limits sales opportunities. On the other hand, customers who report owning no dogs are the most prevalent group among shoppers who buy large quantities.

Chapter 9

Mix and Match

1. g **3.** f **5.** b **7.** d

True/False

9. True

11. False. The mean of X should be smaller than the mean of Y.

13. False. The mean is the weighted average of possible outcomes; it need not be one of the outcomes.

15. False. The variance measures the spread around the mean; it is not determined by the mean's value.

17. True

19. True

Think About It

21. 0.3

23. 0.10

25. $E(Z) > 0$.

27. Y.

29. 0

31. a) $P(W = 5) = p(5) = 1/10$ and $P(W = -1) = p(-1) = 9/10$
b) −0.4. No, because the mean is not zero

33. a) Yes
b) Better than fair to the player

You Do It

35.

	μ	σ
X/3	40	5
2X − 100	140	30
X + 2	122	15
X − X	0	0

37. a)

Outcome	$P(X)$	X
Both increase 80% to $18,000	0.25	$36,000
One increases	0.5	$22,000
Both fall 60% to $4,000	0.25	$8,000

b) $E(X) = \$22{,}000$
c) Yes, the probability is symmetric around the mean gain of $2,000.

39. a) Let the random variable X denote the earned profits. Then $p(0) = P(X = 0) = 0.05$, $p(20,000) = 0.75$, $p(50,000) = 0.20$.

   b) $25,000

   c) $\sigma \approx \$13,229$

41. a) 2.25 reams

   b) $\sigma \approx 1.26$ reams

   c) 17.75 reams

   d) 1.26 reams

   e) $E(500\,X) = 1125$ pages; $SD(500\,X) \approx 630$ pages

43. a) $359,000

   b) No. Dividing the cost in bolivars by the expected value of the exchange rate gives $1,000,000/3.29 \approx \$304,000$. The error is that $E(1/R) > 1/E(R)$.

45. a) It has to double.

   b) It needs to increase by a factor of approximately 1.25. We assume that whether a customer who is offered a rebate buys a printer and whether they use the rebate are independent.

47. a) Let X denote the number of clients visited each day. Then

$$P(X = 1) = 0.1$$
$$P(X = 2) = 0.09$$
$$P(X = 3) = 0.81$$

   b) 2.71 clients

   c) 6.775 hours

   d) $813

49. a)

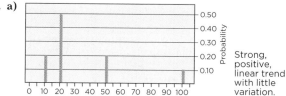

Strong, positive, linear trend with little variation.

   b) 0.3

   c) $32

   d) On average, we expect a customer to withdraw $32. This is the average of the outcomes in the long run, not a value that occurs for any one customer.

   e) Var of $X = 696$; SD of $X = \$26.4$

51. a) Two free throw attempts **b)** 3-point baskets

   c) Two free throws. The calculation for free throws requires independence as well as equal chances for making the shot.

53. a) This histogram is rather bell shaped, centered nearly at zero, with almost all of the daily returns within 5% of zero.

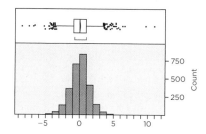

   b) $\bar{x} = 0.0652$, $s = 1.49$ **c)** ≈ 0.030

   d) ≈ 0.030; the Sharpe ratio does not depend on the amount that is invested.

   e) The Sharpe ratio for Exxon is lower, suggesting to an investor that IBM offers a better return for the risks.

   f) That performance of these stocks in the future will resemble the past values

g) A time plot shows that the variation in the returns was higher in the 1990s. This is masked in the histogram.

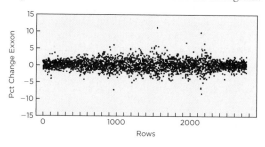

4M Credit Scores

   a) Since he is paid by commission, selling high premium policies is worth more to him than less expensive policies.

   b) These are only probabilities, so his results will vary around the mean. The SD indicates roughly how far from the mean he could be.

   c) Let X = annual premium for a randomly selected customer.

   d) Let the random variable C denote his commission from a policy. Then $C = 0.1 \times X$.

   e) This graph shows the probability distribution of X.

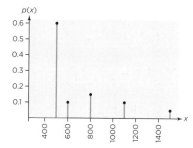

   f) $E(X) = \$665$ so $E(C) = \$66.5$

   g) The SD of $X \approx \$271$ so that $SD(C) \approx \$27$.

   h) The salesman can expect to earn on average about $66 for each policy that he sells but with quite a bit of variation. He can expect 60% of the policies he sells to earn him $50. He can expect 15.% of the policies he sells to pay more than $100.

   i) Because the salesman is paid by commission and hence earns the most by finding the riskiest customers, managers at the insurance company should not be surprised if they find he's writing more than 5% of his policies to the most risky group. He earns $150 for each of these he sells, but only $50 for the group with the best rating.

4M Project Management

   a) The company may need to negotiate with labor unions that represent the employees. It may also need to budget for the costs of the labor force. The amount of labor that it is able to use also affects the time that the project will be completed.

   b) It is more useful to have a probability distribution which conveys that the weather conditions are not known; there's variation in the type of weather that the company can expect.

   c) The total number of labor employees needed at the two sites

   d)

	Winter Conditions			
	Mild	Typical	Cold	Severe
X = total number of labor employees	180	120	80	50
$P(X)$	0.3	0.4	0.2	0.1
$E(X) = 123$				

f) $Var(X) = 1881$, $SD(X) \approx 43.37$

g) On the basis of the projected weather conditions and estimates of labor needs, you estimate the total labor needs of the two projects to be about 125 laborers during the winter. There's a small chance (10%) that only 50 are needed, but a larger chance (30%) that as many as 180 will be needed.

Chapter 10

Mix and Match

1. e **3.** h **5.** i **7.** f **9.** a

True/False

11. False. If the costs move simultaneously, they should be treated as dependent random variables.

13. False. The mean and variance match, but this is not enough to imply that all of the probabilities match and that $p(x) = p(y)$.

15. False. This implies that X and Y have no covariance, but they need not be independent.

17. True

19. False. The SD of the total is sqrt(2) times σ. The variance of the total sales is 2 times σ^2.

21. False. If the effect were simply to introduce dependence (but not otherwise alter the means and SDs), then the expected difference would remain zero.

Think About It

23. Negative covariance.

25. The covariance between Y and itself is its variance. The correlation is 1.

27. No, the covariance depends on the units used to measure the investments.

29. Not likely. Sales during the weekend would probably look rather different than those during the week.

31. a) Positively correlated.

b) A budget constraint may produce negative association or independence.

You Do It

33. a) $E(2X - 100) = 1900$; $SD(2X - 100) = 400$

b) $E(0.5Y) = 1000$; $SD(0.5Y) = 300$

c) $E(X + Y) = 3000$; $SD(X + Y) \approx 632.5$

d) $E(X - Y) = -1000$; $SD(X + Y) \approx 632.5$

35. All of the calculated expected values remain the same. Only the variances and standard deviations change. Unless both X and Y appear, the variance is the same.

a) Unchanged

b) Unchanged

c) $SD(X + Y) \approx 651.9$

d) $SD(X - Y) \approx 612.4$

37. -0.2

39. a) These are dependent, because for example we can write $Y = 60 - X$.

b) Since $X + Y = 60$, the variance is 0.

41. a) Let X_1 and X_2 denote the deliveries for the two. Both drivers are said to operate independently. We assume that the number of deliveries is comparable as well.

b) $E(X_1 + X_2) = 12$ deliveries
$SD(X_1 + X_2) \approx 2.83$ deliveries

c) $E(X_1 + 1.5X_2) = 15$ hours

d) $SD(X_1 + 1.5X_2) \approx 3.61$ hours

e) Yes. It suggests that the counts of the number of deliveries are negatively correlated and not independent, as if the two drivers split a fixed number of deliveries each day.

43. a) $E(X) = 1.5$ sandwiches; $Var(X) = 0.25$ sandwiches^2

b) $E(Y) = 1.45$ drinks; $Var(Y) = 0.3475$ drinks^2

c) $Corr(X, Y) \approx 0.424$

d) Customers who buy more drinks also tend to buy more sandwiches.

e) $E(1.5X + 1Y) = \$3.70$; $SD(1.5X + 1Y) \approx \$1.13$.

f) $E(Y/X) = 1.025$. The ratio of means, $\mu_y/\mu_x \approx 0.97$, is less than 1. These do not agree. In general, for positive random variables, $E(Y/X) \geq E(Y)/E(X)$

45. a) $E(X) \approx 0.999$; $Var(X) \approx 0.999$

b) $E(Y) \approx 1.052$; $Var(Y) \approx 2.049$

c) Let $X_1, X_2 \ldots X_{20}$ denote the 20 attempted two-point shots and let $Y_1, \ldots, Y_5$ denote the three-point attempts. Then the total number of points scored is
$$T = X_1 + X_2 + \ldots + X_{20} + Y_1 + \ldots + Y_5$$
Assuming the X's are identically distributed and the Y's are identically distributed, $E(T) = 25.24$

d) His scoring in this game is within 1 SD of his season performance, so it seems typical from what we are given.

47. a) No. Expect positive dependence.

b) The total cost is $T = 0.09X + 10Y$. $E(T) = \$1,930$.

c) $SD(T) \approx \$272$.

d) The costs for these homes are only about 1.1 SDs above the national figure for 2001. That was a long time ago, so these numbers seem atypical.

49. Let X denote the number of weeks of electrical work, and Y denote the number of weeks of plumbing work.

a) $E(X + Y) = 184$ weeks

b) Positive, because delays or extra time for one type of work suggest similar problems in the other type of work as well.

c) $SD(X + Y) \approx 19.7$ weeks.

d) Less. The larger ρ, the more variable the amount of work, making the bidding process less accurate and harder for the firm to estimate its profits.

e) $E(200X + 300Y) = \$48,800$; $SD(200X + 300Y) \approx \$5,408$.

f) The firm is unlikely to make $60,000 because this amount is more than 2 SDs above the mean.

4M Real Money

1. a) By splitting the investment, the investor obtains a more steady flow of returns than if the money is all placed into one company.

b) It combines the mean and variance of the investment. It takes account not only the growth in most stocks, but penalizes this growth for the cost of borrowing and for the variability of the returns.

c) No

d) The Sharpe ratios for Citigroup and Exxon are $S(\text{Citi}) \approx 0.0407$ and $S(\text{Exxon}) \approx 0.0336$. For the 50-50 split, the ratio is ≈ 0.046. This is similar, but not as high as the Sharpe ratio for the mix of IBM and Microsoft in the chapter.

e) The mean and SD for the new column should match those used in part d. That is, mean = 0.0859 and SD = 1.5291.

f) 0.044

g) By dividing the investment between the two stocks, the investor benefits from the higher average return on stock in Citigroup as well as the low variance on returns earned by Exxon. The combination outperforms, in terms of the Sharpe ratio, either concentrated investment. The precise mixture is less important.

4M Planning Operating Costs

1. a) No. The amounts of the fuels that are used are random.
  b) The marginal distributions are given in the tables of the exercise. For example,

Kilowatt Hours

x	200	300	400	500
$P(X = x) = p(x)$	0.05	0.25	0.40	0.30

  c) $T = 100X + 12Y$ gives the total cost in dollars for energy.
  d) Not likely. A severe winter, for example, would probably mean more electricity for circulating heat in the stores along with the increase in the use of natural gas.
  e) $E(X) = 395$ mkwh; $Var(X) = 7,475$ mkwh^2
  f) $E(Y) = 990$ MCF; $Var(Y) = 29,900$ MCF^2
  g) $E(T) = \$51,380$; $SD(T) \approx \$9665$
  h) The mean $E(T)$ would be the same, but the variance would be smaller by the term added by the covariance, or 14,352,000.
  i) Energy operating costs are expected to be about $50,000 per store. The anticipated uncertainty indicates that costs could be higher (or lower) by about $10,000–$20,000, depending on weather and business activities.
  j) Key assumptions: The chances for the various levels of energy use, the level of dependence between the types of use (correlation), and the costs of the energy (treated here as fixed, though certainly could change).

Chapter 11

Mix and Match

1. i **3.** j **5.** c **7.** a **9.** f

True/False

11. False. These 25 make up a small portion of the total number of transactions
13. False. The chance that the first three are OK is $(999/1000)^3$. Hence, the chance for an error is $1 - (999/1000)^3 \approx 0.003$.
15. False. This clustering would likely introduce dependence.
17. False. A Poisson model requires independent events, not clusters.
19. True
21. False. The rate $\lambda = 0.24$. The chance for no defect is 0.787 and the chance for at least one defect is 0.213.

Think About It

23. a) Yes
  b) No
  c) No, we suspect dependent trials.
25. Equally likely, each with probability $p^3(1 - p)^3$.
27. Equally likely, both with probability $(\frac{1}{2})^4$.
29. No, it's ≈ 0.001.
31. 18

You Do It

33. a) Yes. There may be some dependence, but it is small.
  b) We expect 2.5 such transactions. The binomial model concentrates near its mean, so we would expect about a 50% chance for more than two.
  c) ≈ 0.463
35. a) Binomial or Poisson.
  b) Using the Poisson, $P(X \geq 2) \approx 0.0027$.
37. As long as the driver has 2.67 or fewer accidents a year.
39. a) ≈ 0.282
  b) 0.301

41. a) Assume the shots are all made with 35% accuracy and independently. Somewhat.
  b) 7
  c) Yes. The probability is 0.0196.
  d) 14
  e) 17.5

4M Market Survey

  a) The information could be used to alter the presentation of hybrids in future advertising.
  b) Yes. Assume visitors fill in the forms independently of each other and that the probability a visitor checks a given item remains constant. Consider one item to obtain 0/1 or yes/no response.
  c) Assuming the conditions are met as stated in part b and letting technology be the item of interest, we have a binomial random variable with $n = 25$ and some probability p that must be specified.
  d) 7.5
  e) Assuming $p = 0.3$, $P(X \leq 5) \approx 0.193$. This is not small enough to change the assumed value of p. On note 5 is within 1.1 SDs of the expected value.
  f) The survey found that 20% of visitors at the recent car show with an intention of buying a hybrid car were drawn to the technology of hybrid cars. This is less than the 30% found in other shows. This may indicate a shift in attitudes.

4M Safety Monitoring

  a) Possible reasons: over-reacting may create a perception that the company has a problem even if there was not a real problem; recalls are expensive. Such perceptions cost sales and erode the customer base.
  b) To get ahead of a problem, a firm must detect the problem quickly. The risk is that the problem was a random event that "just happens."
  c) Poisson random variables are well-suited since we have a rate and lack specific Bernoulli trials. $\lambda = 1.5$ in normal times and $\lambda = 5$ during problem periods.
  d) The rate seems independent in the sense that the people that call one day are probably not calling as a result of the calls of others. The problem states two rates; in fact the rates are likely to change gradually rather than dramatically if a problem occurs.
  e) Somewhat surprising, $P(\text{more than 3 calls}) \approx 0.0656$.
  f) No, $P(\text{more than 3 calls}) \approx 0.849$.
  g) ≈ 0.405
  h) It is quite rare to receive so many calls during normal operations, happening in only 6.6% of months. More than 3 calls are, however, quite common when there is a problem. Given that the company receives more than 3 calls at the problem center in a month, there's a 40% chance of a problem.

Chapter 12

Mix and Match

1. f **3.** a **5.** h **7.** e **9.** j

True/False

11. False
  $P(38 < X < 44) = P(0 < Z < 1) \approx 1/3$ whereas $P(X > 44) = P(Z > 1) \approx 1/6$.
13. True
15. True

17. True

19. False. The statement is true on average, but not for specific consecutive days. The random variables have the same distribution, but that does not mean that their outcomes are the same.

21. True

Think About It

23. There is no limit. Even a standard normal with mean 0 and variance 1 can be arbitrarily large, though with tiny probability.

25. The data are skewed to the right, so $K_3 > 0$.

27. Both are normal with mean 2μ, but the variance of the sum is $2\sigma^2$ rather than $4\sigma^2$.

29. A Skewed; B Outliers; C Normal; D Bimodal

31. a) A: original; B: rounding (small gaps and stairsteps in the quantile plot).
b) The amount of rounding is small relative to the size of the bins.

33. a) Yes, so long as the weather was fairly consistent during the time period.
b) No, weather is a dependent process. The amounts might be normally distributed, but not independent.

35. a) 5/6
b) It shifts the mean from $700 to $800.
c) The mean and SD increase by 5% to $735 and $420, respectively.
d) No. It appears that the distribution is skewed to the right.

37. a) Yes. The price of each is the overall average plus various factors that increase and decrease the value of each.
b) Sales data from housing projects recently built by this contractor with similar characteristics.
c) Normal with mean $400,000 and SD $50,000.

You Do It

39. a) 0.93319
b) 0.8413
c) 0.7699
d) 0.6171
e) 0.77449

41. a) −0.84
b) 0
c) 0.6745
d) 2.576
e) 0.90

43. a) $33,000
b) Approximately 21%.
c) No, the value at risk is *smaller* over the longer horizon (primarily because the mean return is so large relative to the SD).

45. a) 0.025
b) The normal approximation gives ≈ 0.00423. A binomial model gives ≈ 0.00514.
c) Yes, to the extent that the company can be profitable by writing many policies whereas it would not if it only sold a few.

47. a) 5%
b) 5%
c) The life insurance firm has independent customers. They don't all die at once. The hurricane bonds do and are dependent. These bonds are more risky than insurance.

49. a) The histogram and boxplot look like a reasonable match for a normal model. The histogram is roughly bell-shaped with some skewness and a sharp cutoff near 30%. The boxplot does not show many outliers.
b) Some minor outliers. The lowest share is 24.9% (in Jersey City, NJ); the highest is 43.7% (in Buffalo, NY).
c) The shares of this product are never very close to zero or 100%, so the data "do not run into the boundary" at the upper and lower limits.
d) $\mu = 33.63\%$ and $\sigma = 3.34$.
e) A normal quantile plot suggests an acceptable match, except for a "kink" in the plot near 30%. This cluster of values is unusual for data that are normally distributed. Otherwise, the normal model describes the distribution of shares nicely, matching well in the tails of the distribution.

4M Normality of Stock Returns

Motivation
a) The mean and standard deviation can be combined with the Empirical Rule to provide a complete description of the performance of the investment.
b) Either returns or percentage changes can be used. The price should not be used since it shows a strong pattern.

Method
c) These data form a time series, so we need to check that independence and stability (no patterns) are reasonable assumptions.
d) Normal quantile plot.

Mechanics
e) Features related to the timing of events. Since the time series of percentage changes does not show patterns, little is lost here by summarizing the data in a histogram.
f) Except for the most extreme values, a normal quantile plot shows that the normal model is a very good description.
g) 0.1018.
h) It increases by 10% or more in 31 out of 312 months, or 9.94% of the time. This is almost identical to the value produced by the normal model.

Message
i) Monthly returns on stock in McDonalds is approximately normally distributed with mean 1.4% and SD 6.7% during the period 1980 through 2005. The use of a normal model to anticipate future events, however, requires that we assume that the variation seen in the past will continue into the future.

4M Normality and Transformation

Motivation
a) Normality is a familiar model for which we have diagnostic plots. It also makes it easy to summarize the data with a mean and a SD.
b) Find the z-score for the log of 20,000 using the parameters set from the logs of this data. Then convert this z-score to a probability using the methods of this chapter.

Method
c) Possibly. The incomes within the cluster may be more similar than those over the whole area resulting in an average income that is possibly too high (or too low) and a SD that is too small.
d) Take logs and examine the normal quantile plot.

Mechanics

   e) No, the data are severely right skewed.

   f) Yes, except in the lower tail. Incomes get lower than the lognormal model predicts.

   g) 0.232

   h) Yes, for incomes of this size and larger. Were we farther into the lower tail, the lognormal would not be a good description of the variation for the very poor.

Message

   i) The log of household incomes in San Antonio is approximately normally distributed with a mean of about 4.6 with standard deviation 0.41. This approximation allows us to use a bell-shaped distribution to describe these otherwise highly skewed data as long as we do not use it for incomes lower than about $15,000 to $20,000.

Chapter 13

Mix and Match

1. c **3.** a **5.** i **7.** j **9.** f

True/False

11. True

13. False. Bias occurs when the sample is not representative.

15. False. The sampling frame is a list of the items in the target population.

17. False. Voluntary response surveys allow self-selection.

19. False

Think About It

21.
Population:	HR directors at Fortune 500 companies
Parameter:	Proportion who find that surveys intrude on workday
Sampling frame:	List of HR directors known to business magazine
Sample size:	115
Sample design:	Voluntary response
Other issues:	Presumably, those who find surveys intrusive lack time to reply.

23.
Population:	Production of the snack foods manufacturer
Parameter:	Mean weight of bags
Sampling frame:	All cartons produced daily
Sample size:	10 cartons × 2 bags each
Sample design:	Cartons are a simple random sample of the production; the two bags sampled within the cartons form a stratified sample. (also a clustered sample)
Other issues:	How are the bags chosen from a carton?

25.
Population:	Adult shoppers
Parameter:	Satisfaction
Sampling frame:	None; the subjects are those who visit the store.
Sample size:	Not given
Sample design:	Convenience sample
Other issues:	This effort is more promotion than survey.

27. a) Voluntary response survey. Are on-line users representative?

   b) Cluster sample/census. Customers at other branches might be different.

   c) Voluntary response, not representative. Only those with strong opinions respond.

   d) Cluster sample. SRS within each branch with plans for follow-up.

29. The wording of Question 1 suggests the bank may not care about the families and is likely to produce a different response than Question 2.

31. a) If payments are numbered and recorded electronically, take an SRS of 10 of these. If on paper, pick a random integer from 1 to 20 and select every 10th payment.

   b) Separate the two types of payments and have different supervisors sample each.

4M Guest Satisfaction

   a) A representative sample allows the company to learn how customers rate its services. Also, results from this survey could be compared to a subsequent survey to determine if changes in practices impact customer experiences.

   b) No, calling reduces non-response bias. Visitors who respond to the initial survey may have different opinions than those who do not.

   c) Unit: "guest day." Population: guests who stay at the hotel during weekdays during the summer season. (This population probably has more business guests than the weekend population.)

   d) Random selection of a weekday implies that the survey over-samples guests who stay longer.

   e) Yes. Female guests might respond differently to questions asked by a woman interviewer; same for male guests. With two interviewers, it would be important to randomize the choice of which guests to contact.

   f) Repeated calls might anger a guest who may never come back or not reply honestly.

   g) The survey samples guests present on a randomly selected day; guests who stay longer are more likely to be present that day and appear in the sample. The random choice of the day is important in order to claim that the survey captured guests on a typical rather than specially selected day that could bias results.

   h) The hotel could thank guests who respond by inviting them to return while reiterating the management's interest in their input. Use the survey to learn opinions as well as as build a satisfied customer base.

4M Tax Audits

   a) A sample of all returns may expose mistakes or cheating that had been unknown. Fear of the random audit may convince taxpayers to comply with tax codes.

   b) Cluster sampling allows the agent to visit several nearby homes quickly rather than spend time traveling from one home to another.

   c) Double the probability of sampling returns for taxpayers earning $100,000 or more. The result is a stratified sample.

   d) No, this will yield a random sample of returns, not taxpayers. Suppose n returns are sampled from all N returns that are filed and n is much smaller than N. The chance of selecting a couple who files jointly is about $1 - (1 - n/N)^n$ whereas the chance that married taxpayers filing separately are both picked is about the square of this probability.

   e) Returns filed on the last date may come from a different type of taxpayer than those filed earlier.

   f) Random sampling since every taxpayer has a chance to be visited by the IRS.

Chapter 14

1. a and b.
3. **a)** S is out of control.
   b) In control.
   c) X-bar is out of control.
   d) Both are out of control.

True/False

5. True
7. False. A Type I error occurs if the system signals a problem but none has occurred.
9. True
11. False. This choice focuses attention on a Type I error and ignores the risk for a Type II error.
13. True
15. False. The mean and SD suggest skewness.
17. False. It may be out of control because the process changed or by chance. Only the latter case produces a Type I error.
19. True

Think About It

21. **a)** Decreased because it becomes more difficult to halt the procedure.
    b) $P(\text{Type I}) = .05^2$
23. **a)** Under control unless there's a special sale or weekend shopping surge.
    b) Under control unless some problem causes a surge in calls.
    c) Out of control with surges during holiday season.
    d) Out of control with upward (or perhaps downward!) trend.
25. **a)** Poisson, with $\lambda = 4$. Verify that counts in disjoint time intervals are independent
    b) $P(Y = 0) + P(Y = 1) \approx 0.092$.
    c) $1 - [P(Y = 0) + P(Y = 1)] \approx 0.80$.
27. **a)** The larger sample.
    b) Check parts daily.
29. **a)** $10 \pm 3 \times 5/\sqrt{18}$
    b) $-4 \pm 3 \times 2/\sqrt{12}$
31. **a)** $0.986573 = 0.9973^5$
    b) $0.763101 = 0.9973^{100}$

You Do It

33. Shafts
    a) By the Empirical Rule, 0.05; $2 \times 0.02275 = .0455$ using the normal table.
    b) $0.95^{80} = 0.0165$. $0.9973^{80} = 0.8055$.
    c) Yes
    d) No
    e) Narrower control limits make one more likely to conclude out of control (and hence a increase the chance of a Type I error).
    f) $K_3 = -0.18$ and $K_4 = 0.15$. Five days is enough since n is larger than 10 times the square of K_3 or $|K_4|$.
35. Insulator
    a) Using the first 5 days, the mean is about 449.88 with $s = 1.14$. The skewness during this period is $K_3 = -0.45$ and $K_4 = 0.81$. Twelve per day is just enough to meet the CLT condition.
    b) The process is out of control in both.
    c) The variation increases beyond that specified in the design. This variation also allows the mean to go out of control.

4M Monitoring an email system

a) The system has inherent variation. If the system had to be checked whenever volume exceeded 1,000

messages, operators would continually hunt phantom problems.

b) A shift in the mean reveals a change in average use, whereas a change in the SD of the process indicates that usage fluctuates.

c) The cost of a Type I error is relatively small compared to the cost of a Type II error. Hence, α should be relatively large, say 0.1.

d) Averaging more data allows use of normal distributions once we have about 6 or more observations per sample. Averages are also able to detect subtle changes more quickly. The key limitation of accumulating more data is that the system requires a minimum sample before signaling a problem.

e) When grouped into 128 15-minute intervals, virtually every mean and SD remains within the limits, but during 1:45 to 2 PM on Sept 25, the SD crosses the UCL and the process is out of control.

f) $1 - 0.9973^{128} \approx 0.293$.

g) If $\alpha = 0.1$, charts frequently indicate that the process is out of control.

h) Control charts signal about 20 "alarms" during these four days. To learn whether an alarm detects a real problem requires close inspection of the system. The design implies each chart will signal about 3 to 4 alarms each day if there is not a problem ($32 \times 0.10 = 3.2$ events/day/chart).

4M Dell Computer Stock

a) Stock prices show trends and are dependent observations. Returns appear more stable with independent variation over time.

b) The mean shows whether the typical level is increasing, and variation measures the risk of a large up or down movement.

Method

c) Use $\bar{x}$ for μ and s for σ.

d) Individual returns in 2002 are nearly normal, but for outliers. Averages of 10 days should be closer to normally distributed. Confirm this by checking the CLT condition during a stable period, such as the second half of 2003. During this period, $K_3 = 0.024$ and $K_4 = -0.27$, so the sample size is adequate to use a normal model for averages of 10 days.

Mechanics

e) The returns lack a trend or pattern; $\mu = 0.00033$, $\sigma = 0.028$.

f) Yes

g) Yes

Message

h) One period may not be indicative of properties of others because the stock market changes over time. Here, the stock market was more volatile in 2002 than in 2003.

Chapter 15

Mix and Match

1. f 3. b 5. d 7. j 9. e

True/False

11. True
13. True
15. True
17. True
19. False. The survey needs $n = 1/0.05^2 = 400$.

Think About It

21. a) [4.983 kg to 20.385 kg]
    b) [¥267,490 to ¥511,720]
    c) [$54.5 to $76.44]
    d) [$18,600 to $29,160] per store
23. $\frac{1}{2}$
25. Cannot tell unless you know σ. If $\sigma = s$, the t-interval is longer because $z_\alpha < t_{\alpha^{n-1}}$. If $s < \sigma$, the t-interval may be shorter.
27. a) We are 95% confident that the mean height of men who visit this store lies between about 70.9 and 74.5 inches.
    b) ≈ 1.8 inches.
    c) Probably to the nearest inch.
    d) Longer.
29. The population average of sales is within $15 of the estimate, with some degree of confidence.
31. a) 625
    b) ≈ 849
33. a) PD $\times$ EAD $\times$ LGD $= \$3,000$
    b) $0.05 \times 220000 \times 0.18 = \$1,980$ to $0.07 \times 290000 \times 0.23 = \$4,669$
    c) If independent, the product has coverage at least $0.95^3 \approx 0.86$. Unknown if dependent.

You Do It

35. a) 1.796
    b) 2.571
    c) 2.977
37. a) Population: cars serviced at this dealership
       Sample: cars serviced during the time of the observations
       Parameter: p – proportion of all cars with these dents
       Statistic: $\hat{p}$ – proportion seen with these dents (25.3%)
       Issues: The sample is small and the context suggests it may not be representative.
       Interval: [0.162, 0.344]
    b) Population: shoppers at supermarket;
       Sample: those who returned the form
       Parameter p: population proportion who find shopping pleasing
       Statistic: $\hat{p}$ – proportion who return the form that find shopping pleasing (250/325);
       Issues: Voluntary response. The sample may be biased.
    c) Population: visitors to web site;
       Sample : those who fill in the questionnaire
       Parameter: μ, average web surfing hours of all visitors
       Statistic: $\bar{y}$, average hours of those who complete survey
       Issues: Relevant population?
       Interval: [2.802, 3.198] hours
    d) Population: – customers given loans during the past 2 years
       Sample: – the 100 customers sampled
       Parameter: p – proportion of customers who default on loan
       Statistic: $\hat{p}$ – proportion of these 1000 who default (0.2%)
       Issue: You cannot use a z-interval because the sample is too small.
39. a) That these data are a SRS from the "population" of revenue streams and that the sample size is sufficient for the CLT condition to be met.
    b) $1,226 to $1,302
    c) Just barely.
    d) $6,130 to $6,510
41. a) 1.291
    b) 95% confident that average waiting time for all callers is within about 1.3 minutes of the average wait of 16 minutes.

c) Smaller.
d) 1.081
43. a) Not correct.
    b) Not correct.
    c) Correct.
    d) Not correct.
    e) Not correct.
45. a) t-interval = [142.963 to 161.037]
    b) t-interval = [−22.9585 to 38.9585]
    c) z-interval = [0.387 to 0.613]
    d) z-interval = [0.113 to 0.487]
47. We are 95% confident that the proportion of people who may buy something is between 10.3% and 14.3%.
49. a) The manufacturer's interval is likely to be shorter because it has a larger sample and smaller standard error.
    b) This is not the correct interpretation.
    c) No, both are 95% intervals.
51. 10,000.
53. a) If this week is typical. All clicks to the company.
    b) [0.126 to 0.166]
    c) [$0.57 to $0.75]

4M Promotion Response

a) The program may have unexpected consequences, such as raising awareness of flaws in the current service. The expense of adding literature and training staff to handle the new program may be considerable.
b) The advantage is that it compares the change in phone use for the same customer, not between different customers. The weakness is that other factors may have occurred during study.
c) The CLT implies that we may nonetheless be able to use a normal/t model for the sampling distribution that underlies the confidence interval.
d) The CI for the mean of Y is a CI for $\mu_2 - \mu_1$.
e, f) This table summarizes the results for X_1, X_2 and their difference.

	Before	After	After–Before
Mean	171.89	202.68	30.75
Std Dev	133.05091	147.10615	68.43612
Std Err Mean	13.305091	14.710615	6.843612
upper 95% Mean	198.29019	231.86905	44.329211
lower 95% Mean	145.48981	173.49095	17.170789

g) The interval for the difference is shorter because of pairing. The data in the before and after columns are dependent, as can be seen in a scatterplot of the after values (X_2) on the before values (X_1). The positive covariance produces a more effective comparison.
h) We estimate the program to increase average use from 17 to 44 minutes per customer. Check that no other large changes occurred in the market during this time period.
i) From 170,000 to 440,000 more minutes.

4M Leasing

a) No, this is true on average for only this small sample.
b) Yes, verify that the confidence interval includes this value.
c) SRS conditions require knowing that the sample is representative of such cars. Verify that the skewness and kurtosis satisfy the CLT condition.
d) Confirm that the pattern of use for returned cars continues in subsequent data.
e) Yes, multiply the interval by 0.30.
f) [21,334.53 to 22,093.47] miles
g) [$6,400 to $6,630] per car.

h) The manufacturer can be 95% confident that leased cars in this fleet on average will be driven between about 21,300 and 22,100 miles.

i) The manufacturer can expect to earn on average, with 95% confidence, between $6,400 and $6,630 on these leases in depreciation fees. Since $6,500 lies in this range, it seems a reasonable estimate.

j) $64 million to $66.3 million.

Chapter 16

Mix and Match

1. i   **3.** h   **5.** a   **7.** e   **9.** b

True/False

Questions 11–18: H_0: $\mu \leq 80$, H_a: $\mu > 80$

11. False

13. True

15. False. The smaller the α-level the less likely to reject H_0.

17. False. When σ is estimated by s, use a t-statistic.

Questions 19–26: H_0: $p \leq 0.4$, H_a: $p > 0.4$

19. False. Not necessarily because of sampling variability.

21. False. A significant result occurs if $\hat{p}$ is large enough to show that more than 40% will use the service.

23. True

25. False. The p-value is the probability of incorrectly rejecting H_0 and adding a service that will not be profitable.

Think About It

27. H_0: $\mu \leq 10$ mm vs H_a: $\mu > 10$.

29. Type II error

31. a) $1/4$.

b) 9/16.

33. a) $1/2^{12} \approx 0.00024$

b) Type I error, false positive

c) $1 - 0.95^{12} \approx 0.46$

d) Type II error, false negative

35. The sample size was too large; a smaller sample would have shown statistical significance.

37. a) H_0: $\mu \leq 200$.

b) More than 249.

c) About 5/6.

You Do It

39. a) Yes, the p-value is $P(X \geq 5) \approx 0.007$ which rejects H_0 for $\alpha = 0.05$.

b) Inspection must produce independent outcomes with constant chance p for finding a defect.

c) $z \approx 3.5 > 1.645$ standard errors above H_0. Reject H_0 for $\alpha = 0.05$.

d) The expected number of events is too small. Use binomial methods.

41. a) H_0: $p \leq 0.6$ versus H_a: $p > 0.6$; p is the proportion of markets selling out.

b) A Type I error implies adding an unnecessary delivery (incorrectly reject H_0). A Type II error occurs if we fail to reject H_0 when it's false, missing an opportunity.

c) Approximately $P(Z > 2,438) = 0.0074$. Yes.

43. a) H_0: $p \geq 0.33$ versus H_a: $p < 0.33$; p is the proportion of all visitors who will indicate a willingness to return.

b) A Type I error implies intervening at the local franchise unnecessarily (incorrectly reject H_0). A Type II error

occurs if we fail to reject H_0 when it's false, missing the opportunity to correct a problem.

c) Approximately $P(Z < -2.53) = 0.0057$. Yes.

45. a) H_0: $\mu \leq \$120$ versus H_a: $\mu > 120$; μ is the average spent by a shopper in the loyalty program.

b) A Type I error implies concluding that loyal shoppers spend more when they do not (incorrectly reject H_0). A Type II error means that loyal shoppers do spend more, but we did not reject H_0.

c) K_4 must be less than 8.

d) About $P(T > 2.236) \approx 0.0140$. Yes.

47. a) Yes. Relatives most likely affect the amount purchased by each other.

b) No. Excluding outliers removes shoppers who add profit.

49. About 4%

51. a) H_0: $\mu \leq \$50$ versus H_a: $\mu > 50$; μ denotes the average increase in interest profit on a savings account when offered this service.

b) A Type I error implies that the bank rolled out the program, but it will not be profitable. A Type II error implies that the bank should have rolled it out (rejected H_0) but did not.

c) K_3^2 must be less than 6.5 and the absolute kurtosis must be less than 6.5.

d) The p-value = $P(T > 0.67) \approx 0.25$. No.

4M Direct Mail Advertising

a) The list is expensive and unnecessary if subsequent marketing will not make enough money to pay for the acquisition and make a profit.

b) To demonstrate that the names are worth the acquisition cost.

c) A break-even analysis requires evidence of profitability beyond the possible consequences of sampling variation.

d) H_0: μ = average earnings from customer $\leq \$3$ vs. H_a: $\mu > 3$. Use a 5% test unless the purchase will wipe out the company if the list does not generate revenue. To avoid the expense, reduce α (e.g., 0.01 or 0.001). The data meets the sample size condition since n exceeds 10 $K_3^2 \approx 114$ and $10|K_4| = 124$. Assume that the vendor provides a random sample of names.

e) The histogram is right-skewed with average $33.66 (including those who do not make a purchase) and $s = \$97.02$.

f) $t = 2.89$ with p-value ≈ 0.002. Reject H_0 for any reasonable α.

g) The test of the mailing list does demonstrate profitability. The average purchase (about $34) is large enough relative to the variability among customers to assure the company that purchasing the list will be profitable.

4M Reducing Turnover Rates

a) At $2,000 each, improved benefits cost $7.2 million. Management should demand proof that it is cost effective.

b) The hotels should be an SRS of those operated by the chain.

c) This H_0 assumes that the program works (lowers the quit rate). H_0 should represent the status quo it does not lower the quit rate.

d) This H_0 has the right direction, but ignores costs. To break-even, the quit rate after implementing the program must be less than 42%, not below 52%.

e) Require $\hat{p} < p_0 - 1.645 \times \sqrt{(p_0 \times (1 - p_0)/n)} = 0.3746$.

f) Power is $P(\hat{p} < 0.3746) \approx 0.82$.

g) Yes, do the test. If the quit rate falls well below the break-even point, say to 35% or less, the savings would be more than $5 million. The chance of detecting such a change is more than 80%.

Chapter 17

Mix and Match

1. g **3.** b **5.** f **7.** j **9.** c

True/False

11. False; CLT assures that a *t*-interval can be used.

13. False; the checks are not needed.

15. False; this is a prediction interval.

17. False; add 4 cases, half success and half failure.

19. True

Think About It

21. The log data are symmetric but should be checked for normality before using the *t*-interval. Also, log(average) ≠ average(log).

23. Both appear nearly normal, though close in the first. With small samples, it is hard to recognize normality.

25. No. Both intervals are based on approximating the sampling distribution of a proportion. The adjusted interval is shorter, but has a different location.

27. $^1/_2$.

29. The small sample interval is more likely to include $^1/_2$ due to shifting $\hat{p}$ toward $^1/_2$. It is not always closer to $^1/_2$ because it is shorter.

31. Prediction interval.

33. The prediction interval is longer.

35. No. The intervals use the observed data that are positive.

You Do It

37. a) In spite of the outlier, yes.

b) No. $K_4 = 5.8$, so we need of $n \geq 58$ to rely on the CLT.

c) With: $1763 to $3245. Without: $1934 to $2478

d) The mean falls from $2504 to $2206, and the SD falls from $886 to $294. The interval's center shifts down but becomes much shorter.

39. $1 - 2/2^8 \approx 0.9922$

41. Use the small-sample interval with $\hat{p} = 12/18 = 2/3$ and standard error ≈ 0.11. The 95% interval $0.667 \pm 2(0.11) = 0.667 \pm 0.22$ includes $^1/_2$ *p* could be $^1/_2$.

4M Stopping Distances

a) A statistically significant advantage could be advertised in commercials that claim the Camry is "safer" than the rival.

b) A prediction interval indicates what is likely to happen the next time we need to stop quickly rather than on average.

c) Using the same driver removes variation produced by different drivers. A professional driver, however, may obtain different performance from the car.

d) Yes, barely.

e) Yes, since the data is relatively symmetric.

f) Mean: $\bar{x} \pm t_{0.011,9} s/\sqrt{10} = 175.67 \pm 4.658767$
Median: $x_{(2)}$ to $x_{(9)} = 169.7$ to 181.7. The intervals are similar, with that for the median being wider.

g) The *t*-statistic is $(\bar{x} - \mu)/(s/\sqrt{10}) = -2.568$ with *p*-value $= 0.0151$.

h) Yes, but add the words "on average". The *p*-value for the test of H_0 is less than $\alpha = 0.05$. Toyota should also note that the results were obtained under ideal test conditions.

Chapter 18

Mix and Match

1. j **3.** c **5.** g **7.** a **9.** e

True/False

11. True

13. False. The probability of a false positive error is typically limited to 5%.

15. False. The difference $\mu_1 - \mu_2$ *might* be zero, but it does not have to be zero.

17. False. The difference between the means may nonetheless be significant.

19. True

Think About It

21. Possibly. Data from one day is not representative of variation in public transit time.

23. Pairing reduces the analysis to a one-sample analysis. Form the differences and compute the confidence interval for the average difference.

25. a) Take 40% of the two endpoints, [$200, $880] profits.

b) Yes, methods used by Group A sell statistically significantly more. Due to the randomization, confounding is not an issue.

c) Randomization makes this bias unlikely.

27. Compare the change in consumption from 2005 to 2006 in homes that were on daylight savings time to the change in homes that moved from standard to daylight savings time.

You Do It

29. Wine

a) The data are numerical and labeled as a sample, but sampling procedure is not identified. Bottlers may only offer for testing exceptionally good choices. Other conditions seem okay, but this flaw in the sampling seems fatal.

b) Software provided the analysis shown below. The interval includes zero, suggesting no statistically significantly difference in means.

2001–2000, Allowing unequal variances	
Difference	0.6880
Std Err Dif	0.7348
95% CI	−0.8094 to 2.1854

c) Type of wine grape, red or white, place of origin could differ.

d) −0.8 to 2.2

e) Wines from the 2001 vintage score higher on average, but the difference is not statistically significant. Even so, unless there's a reason to stay with 2000, pick a 2001 vintage.

31. Used cars

a) The two groups have different variances, but the method does not require equal variances. The 95% CI for the difference is shown in the table below. The Xi model sells for about $600 to $2800 more, on average.

b) No. The average age of the cars in the two groups is identical. Age has not confounded the comparison in "a."

$X_i - I$, allowing unequal variances	
Difference	1698.9
Std Err Dif	546.8
95% CI	618.6 to 2779.3

4M Losing Weight

Motivation

a) Yes. The FDA might find an inconclusive study to be evidence that the drug will not be effective if sold over the counter. Weight-watchers may lose interest.

b) There's no need for so many subjects, especially because of costs. Compliance could also become a problem.

Method

c) Doubtful; subjects are less likely to be over-weight and do not take so much.

d) Yes. This approach is typical.

e) No. The study is a randomized experiment.

f) Use intervals that do not require equal variances and mention in the summary if variances differ.

Mechanics

g) If $\sigma = 5$ and 25 subjects are in each group, then

$$SE(\bar{y}_1 - \bar{y}_2) = \sigma\sqrt{\frac{1}{n_1} + \frac{1}{n_2}} = 5\sqrt{\frac{1}{25} + \frac{1}{25}} = \sqrt{2} \approx 1.4$$

We expect that the mean difference will be 6, with standard error 1.4. Zero is more than 3.5 standard errors away.

h) With 100 in each group, the standard error is

$$SE(\bar{y}_1 - \bar{y}_2) = \sigma\sqrt{\frac{1}{n_1} + \frac{1}{n_2}} = 5\sqrt{\frac{1}{100} + \frac{1}{100}} \approx 0.7$$

The expected difference is more than 7 standard errors from zero.

i) Unlikely in either case.

j) Larger values of σ increase the SE and hence increase the chance that zero will fall into the confidence interval even if the true difference is not zero.

Message

k) Twenty-five seems adequate, but if the study can be run with 100, this larger size might be preferable because of the smaller margin of error.

l) If the factors are embarrassing, a smaller sample size to reduces the possibility of bad publicity during the study.

4M Sex Discrimination in the Workplace

Motivation

a) Many jurors are not likely to distinguish statistically significant from substantively significant. That might lead them into making a large award with substantial punitive damages.

b) Both sides need to anticipate the outcome and decide how to "spin" the results.

Method

c) Because the sample sizes are large, the CLT suggests a normal sampling distribution. Software can find the precise t percentile or we can use 2 from the Empirical Rule.

d) Without a break-even analysis, we are interested in quantifying the difference between μ_1 and μ_2 relative to sampling variation.

e) If these District Managers are the population, no statistics are needed; we observe μ_M and μ_F. We can, however, think of these data as the result of sampling the on-going hiring process at Wal-Mart and treat these as a sample.

f) It is doubtful that managers who know each other negotiate their salaries separately. The data would seem to be dependent.

g) Perhaps women have not worked at Wal-Mart for as long as their male counterparts or are pushing into districts that produce less profit.

Mechanics

h) $se(\bar{x}_M - \bar{x}_F) = \sqrt{60000^2/458 + 50000^2/50} \approx \$7607.$

i) Using the Empirical Rule, we obtain the interval [$47156 to $77584] which rounds to [$47,000 to $78,000].

j) If the two SDs increase by a factor of 2, then the interval would be twice as long, but still not include zero.

k) By reducing the sample size by a factor of about 5, the confidence interval gets longer by $\sqrt{5}$.

Message

l) If we treat the salaries of District Mangers at Wal-Mart as a random sample of pay practices of Wal-Mart, these data show that on average Wal-Mart pays men between $47,000 and $78,000 more than women.

m) We don't have a random sample and so other factors aside from the sex of the manager may influence the results. These other factors include size of the store, profitability, years of experience, work relationships, and so forth.

Chapter 19

Mix and Match

1. e **3.** f **5.** b **7.** a **9.** c

True/False

11. False. The response y appears on the vertical axis, x on the horizontal axis.

13. True

15. True

17. False. The slope indicates change, not the intercept.

19. False. A residual is the *vertical* distance.

21. False. The residual plot should lack patterns.

Think About It

23. No. Square the correlation. These share $1/4$ of their variation.

25. No. Least squares minimizes vertical deviations. If reversed, the deviations are horizontal in the plot of y on x.

27. $s_e = \$32$. If bell-shaped, then $\pm 2\, s_e$ holds about 95% of the residuals.

29. The slope becomes steeper, producing a higher cost per mile driven. The intercept would increase as well.

31. The same.

33. a) Yes. The equation has a constant plus 85 times the number of shoppers.

b) The intercept is about 0, but is likely to be a large extrapolation.

c) The slope is about 85$/shopper.

d) The variation may increase.

35. a) The intercept of $47,000 is likely an extrapolation and not directly interpretable. The slope is $650 per square foot. For every one square foot increase, average annual sales increase by $650.

b) b_0 = : 38.54 thousand. b_1 = : 5,731/sq.meter.

c) r^2 remains the same.

d) Yes. To obtain s_e in euros, multiply the value in dollars by 0.82.

You Do It

37. Diamond rings

a) Yes

b) *Estimated* Price (Singapore dollars) = −259.6259 + 3721.0249 Weight

The intercept is an extrapolation. The slope is 3721 \$S/carat an extrapolation since these gems weight less than $^1/_2$ carat.

c) r^2 = 0.9783, indicating that the fitted equation describes all but about 2% of the variation in prices. The residual SD is s_e = \$S 31.84. The data lie close to the line given the scale of the plot.

d) 372 \$S/ carat.

e) 2420 \$US/carat. The slope for emerald diamonds is slightly higher at 2670 \$US/ carat.

f) The setting adds a fixed cost.

g) Probably, the residual for this ring is −84.78 \$S.

h) The residuals do not show a pattern, suggesting a histogram is a good summary. s_e = \$S 31.84. The Empirical Rule implies about 95% of rings are priced within \$S 64 of the line.

39. Download

a) Linear with substantial residual variation.

b) *Estimated* Transfer Time (sec) = 7.2746633 + 0.3133071 File Size (MB)

b_0 = 7.27 sec estimates "latency" in the network that delays the initial transfer of data. b_1 = 0.3133 seconds per megabyte, transfer rate.

c) r^2 = 0.6246 and s_e = 6.2433. The equation describes about 62% of the variation.

d) The equation fits equally well (same r^2). b_0 becomes 0.1212 minutes; b_1 becomes 0.0000050993 min/kilobyte.

e) The residual variation lacks patterns.

f) Do the regression in reverse, obtaining *Estimated* File Size (MB) = 6.8803225 + 1.993418 Transfer Time (sec). Given 15 seconds, we expect to transmit 36.78 MB.

41. Seattle homes

a) A linear trend with increasing variation.

b) *Estimated* Price (\$000) = 73.938964 + 0.1470966 Square Feet

b_0 (an extrapolation) estimates the average cost of the land. The slope is the cost (in thousands of dollars) per square foot.

c) r^2 = 0.5605; the linear trend describes more than 56% of the home-to-home variation in prices. s_e = \$86.69.

d) Adding 500 square feet might be worth an additional \$73,500, on average. Other factors may lurk behind this association.

e) The residual price for this home is \$155 thousand. It costs more than most of this size.

f) The variation of the residuals increases with the size of the home. A single value cannot not summarize the changing variation.

43. R&D Expenses

a) A strong linear trend, if we accept the two outliers (Microsoft and Intel).

b) *Estimated* R&D Expense = 3.9819 + 0.0869 Assets

b_0 estimates that a company with no assets would spend \$3.75 million on R&D, evidently quite an extrapolation. b_1 estimates average spending of 9 cents out of each additional dollar on R&D.

c) r^2 = 0.9521 with s_e = \$98.2915 million. The large outliers influence the size of r^2.

d) Both are skewed, anticipating the outlier-dominated scatterplot.

e) Most of the data bunch near zero, and the variation increases with assets. A single summary like s_e is inadequate.

45. OECD

a) Positive association of moderate strength, vaguely linear.

b) *Estimated* GDP (per cap) = 26,714.45 + 1,440.51 Trade Bal (%GDP)

b_0 indicates a country with balanced imports and exports averages \$26,714.45 GDP per capita. b_1 indicates countries with one percent higher trade balances average \$1,441 more GDP per capita.

c) r^2 = 0.369; the line describes about one-third of the variation among countries. s_e = \$11,336.78 GDP per capita is the standard deviation of the residuals.

d) The residuals appear random, but the small n makes it hard to detect all but the most extreme patterns.

e) Norway has the largest GDP and Ireland the largest trade balance.

f) The US is a net importer, but has high GDP per capita given its level of imports. GDP in the US is almost \$20,000 (\$19,468) per person higher than anticipated by the fit of this equation.

47. Promotion

a) Timeplots suggest little association.

b) The scatterplot reveals weak association. A line seems to be a reasonable summary of the association.

c) *Estimated* Market Share = 0.211254 + 0.13005 Detail Voice

b_0 estimates that with no detailing, this drug would capture about 21% of the market. The slope indicates that on average, weeks in which promotion increases 1%, share increases 0.13%.

d) r^2 = 0.14178; detail voice describes 14% of the variation in market share over these weeks. s_e = 0.0071.

e) An increase in voice from 4% to 14% increases average share by 1.3%.

f) The residuals appear random, albeit with one straggler (week 6).

4M Credit Cards

a) It is hard to say whether a linear equation can meet this goal. The question asks about relative errors, not absolute errors. The standard deviation of the residuals around the fit measures a constant level of deviation, not a percentage deviation. If s_e = 100 and all of the balances are above \$1,000, the fit would be good enough. r^2 tells us little about the suitability of a linear equation.

b) The equation provides a baseline for comparison. It anticipates the balance of a customer next month assuming business as usual. If these predictions are less than average balances, this gap could be used to measure the gain produced by the marketing program.

c) For most accounts, the pattern is linear. Basically, the balance next month is almost the same as the current balance.

d) The outliers are occasional transactors. (A transactor pays their balance each month to avoid finance charges.) Those along the bottom of the plot paid their balance in full in month 4, but not month 3. Those along the left paid their balance in month 3 but not month 4.

e) *Estimated* Balance M_4 = 141.7698 + 0.9452445 Balance M_3

b_0 = 141.7698, estimates an average a balance of \$142 among those with no balance last month. b_1

estimates that customers with say, $1000 more in balances this month, will carry about $945 more in balances on average next month. $r^2 = 0.9195$ implies that the equation describes more than 90% of the variation using the balance last month. $s_e = \$684$ means that the residual variation is large if we're trying to predict accounts with small balances.

f) The equation for the remaining 627 cases is

Estimated Balance $M4 = 102.7891 + 0.9707906$ Balance $M3$

b_0 is smaller and b_1 is larger. $r^2 = 0.9641$ and $s_e = \$493$.

g) The residual variation has patterns. The rule to remove transactors left "near transactors." These produce a cluster of positive residuals. The rays in the plot suggest that these customers pay a fixed percentage of their balance.

h) Most customers tend to keep their balance from last month. Occasional transactors complicate the analysis because they suddenly pay off their balance.

i) The fit of a linear equation is not up to the challenge on a case-by-case basis. The error of using this equation to predict balances next month can easily be 25% of the typical account balance.

Chapter 20

Mix and Match

1. d   **3.** a   **5.** e   **7.** j   **9.** h

True/False

11. False. Transformations capture curves.

13. True

15. True

17. False. Transformations capture patterns, not make outliers look "more natural."

19. True

Think About It

21. If linear, these would cost the same amount in the absence of fixed costs.

23. Heavier cars typically have more powerful engines that burn more fuel.

25. Unrelated.

27. b_0 is the estimated value when x gets large (the asymptote).

29. The stripes come from the practice of reporting gasoline mileage in whole numbers (rounding).

You Do It

31. Wal-Mart
a) The pattern is not linear; income grows more rapidly in later years.
b) *Estimated* Operating Income $= -570,031 + 286.357$ Date
b_0 is an extrapolation. b_1 indicates that on average operating income grew about $286 million annually at Wal-Mart.
c) The residuals show the positive-negative-positive pattern seen in the text examples. The residuals vary more in the later years.
d) The scatterplot on a log scale is more linear with more consistent variation, but perhaps "bends back" the other way from the initial plot. Logs emphasize the gyration in sales in the late 1990's that is otherwise less apparent.
e) *Estimated* Log Operating Income $= -306.795 + 0.15725$ Date
b_1 estimates the expected annual percentage change in operating income, about 16%.

f) The residuals have a wavy pattern, tracking over time. The 4th quarter (holiday season) has larger positive values than others (higher sales than predicted by the equation).
g) The log transformation reveals details that are otherwise less noticeable (seasonal pattern, dip in sales in the late 90's). The slope also has a nice interpretation as the rate of growth. Do not compare r^2 or s_e because the responses differ.

33. Wine
a) The relationship seems linear, though the floor of prices at zero complicates the analysis.
b) *Estimated* Price $= -1058.588 + 12.18428$ Rating with $r^2 = 0.55$. The linear equation under-predicts poorly and very highly rated rated wines.
c) The plot on the log scale shows a more linear trend.
d) *Estimated* log(Price) $= -19.6725 + 0.25668$ Rating
The log curve does not predict negative prices for lower rated wines.
e) The plot suggests using the log, agreeing with the preference for an equation that does not predict negative prices in common situations. We cannot compare r^2 and s_e because the responses differ.

35. Used Accords
a) Expect larger drops in value in the first few years.
b) *Estimated* Asking price $= 15.463 - 0.94641$ Age
b_0 suggests a new "used" car to be priced near $15,463.
b_1 indicates resale value falls about $946 per year.
c) This equation misses the pattern, underestimating price for very new and very old cars and overestimating price in between.
d) Residuals from the log equation appear more random.
e) The log equation implies price changes more rapidly among newer used cars, then slows as cars age.
f) We can compare these summaries because the response is the same. The log equation fits better has a larger r^2 (0.928 vs. 0.795) and smaller s_e ($1,300 vs. $2,190).
g) b_0 is the estimated asking price for a car that is one year old, $22,993. b_1 estimates that each 1% increase in age reduces average resale price about 0.078 thousand dollars.
h) The estimated asking price drops from $22,993 at age 1 to 17,573 in year 2 (almost $5,500). For older cars, the estimated price drops from $4,243 at year 11 to $3,562 in year 12, only $681.

37. Cellular phones in the US
a) The rapid expansion suggests that we ought to expect nonlinear growth (such as exponential growth).
b) It grows rapidly, with little evident variation around the trend.
c) *Estimated* Subscribers $= -1.96e + 10 + 9,831,282$ Date
b_1 implies average growth near 10 million per year: too fast for the early years and too slow for later years. The negative intercept is a reminder that we cannot extrapolate this equation back to the year zero.
d) On the log scale, the fitted pattern equation bends "the other way" and misses the trend.
e) Percentage changes show a gradual slowing of the rate of growth. Since the rate slows, the logs miss the curvature.
f) *Estimated* Pct Growth $= 1.1793 + 148.013$
$\qquad\qquad 1/(\text{Date}-1984)$
This curve captures the slowing rate of growth.
g) The rate of growth will eventually slow to $b_0 = 1.18$ percent.
h) *Estimated* Pct Growth $\approx 7.6\%$. Applying this equation to the last observation predicts the number of subscribers at 236 million.

39. Pet foods, revisited
  a) The content of the plots are the same; only the axes labels change. Natural logs are larger than base 10 logs by a constant factor.
  b) *Estimated* Log Sales Volume $= 11.0506 - 2.4420$ Log Avg Price
  For base 10 logs, b_1 is the same, but the intercept is smaller

  $$\text{Estimated } \text{Log}_{10} \text{ Volume} = 4.7992 - 2.4420 \, \text{Log}_{10} \text{ Price}$$

  c) r^2 for both is 0.9546; s_e for the log 10 equation is smaller (0.026254 vs. 0.060453) by the same factor that distinguishes the intercepts, 2.30262.
  d) Logs differ by a constant factor, $(2.30262 = \log_e 10)$

  $$\log_e x = 2.30262 \, \log_{10} x$$

  Hence, we can substitute one log for another, as long as we keep the factor 2.30262.
  e) The elasticity is the same regardless of the base of the log.

4M Cars in 1989

  a) The equation will estimate how many pounds of weight reduction are needed. A linear equation indicates that weight can be trimmed from either model with comparable benefit. A nonlinear equation implies it is more advantageous to take the weight from one or the other. (Assuming that weight reductions are equally costly.)
  b) The text example motivates a nonlinear relationship. We expect to find that 1/mileage is linearly related to weight.
  c) We have to rely on the interpretation of the equation and fit to the data. Summary statistics like r^2 and s_e are not useful because the response variable is likely to differ.
  d) There is a strong negative association ($r = -0.86$) between mileage and weight. The pattern bends as in the example of this chapter.
  e) A linear equation misses the pattern. The nonlinear equation captures the tendency for changes in weight to have more benefit on fuel consumption (in gallons per 100 miles) at smaller weights.

  Estimated Gallons/100 Miles $= 0.9432339$
  $+ 1.3615948$ Weight $(000 \, lbs)$

  Also, this equation does not predict mileage to drop below zero as weight increases.
  f) There is a small tendency for larger variation among heavier cars. The diagonal stripes come from rounding the mileage of cars to whole integers. None of the residuals is exceptionally large.
  g) The equation in the text is
  2004: Estimated Gallons/100 miles $= 1.11 + 1.21$ Weight
  compared to
  1989: Estimated Gallons/100 Miles $= 0.94 + 1.36$ Weight
  The intercept has grown and the slope gotten smaller. r^2 for 2004 is 41% with $s_e = 1.04$ gallons per 100 miles. For 1989, explains $r^2 = 76.5\%$ of the variation in consumption with $s_e = 0.42$ gallons per 100 miles.
  h) On average, cars that weigh 3500 pounds on average use 1.4 more gallons to drive 100 miles than cars that weigh 2500 pounds. The effect of weight is not linear; reductions in weight have more effect on fuel consumption for smaller cars than for larger cars.
  i) Based on this equation, suggest taking weight from the smaller car.

4M Crime and Housing in Philadelphia

  a) Leaders could use the equation to estimate the economic payback of increasing police protection.
  b) The equation describes the association between crime rates and housing prices, not causation. Other lurking factors affect housing prices.
  c) A community leader could try to affect the crime rate in order to improve home values. Housing prices are the response affected by changes induced in the crime rate.
  d) Not linear. A linear relationship implies the same average difference in housing prices between communities with crime rates 0 and 1 and communities with crime rates 50 and 51. A nonlinear equation allows incremental effects of the crime rate to diminish as the rate grows.
  e) The direction is negative, with moderate strength ($r = -0.43$).
  f) *Estimated* House Price ($\$$) $= 225,234 - 2,289$ Crime Rate
  b_0 estimates that the average selling price of homes in a community with no crimes is \$225,234.
  b_1 estimates communities that differ by 1 crime per 100,000 differ in average housing prices by \$2,289. r^2 implies that the linear equation describes 18.4% of the variation in prices, and $s_e = \$78,862$ implies that prices vary considerably from this fit.
  g) *Estimated* House Price ($\$$) $= 98,120 + 1,298,243 \, 1/$ (Crime Rate)
  The reciprocal of the crime rate counts multiples of 100,000 people per crime rather than the number of crimes per 100,000 people.
  h) Plots of the residuals are similar. The linear equation has a larger r^2 with smaller s_e. We prefer the reciprocal because it captures the gradually diminishing effect of changes in the crime rate.
  i) Communities with different crime rates also have different housing prices. The connection describes less than 20% of the variation in housing prices among these communities. Changes in small crime rates (such as 1 to 2) have much larger effects on average than differences at higher crime rates (using the reciprocal).
  j) No. The estimated difference in prices goes from \$1,396,363 at 1 crime per 100,000 to \$747,241 at 2. A difference between crime rates of 11 to 12 is associated with an average difference of about \$9,835. A linear equation fixes the slope at \$2,289.

Chapter 21

Mix and Match

1. e **3.** b **5.** d **7.** l **9.** k **11.** g

True/False

13. False. This histogram should resemble a normal distribution.
15. True. (False if you are thinking about curvature.)
17. False. To estimate the slope requires variation in the explanatory variable.
19. True
21. True

Think About It

23. Yes. Let $\beta_1 = 0$ and $\beta_0 = \mu$.
25. a) $b_0 \approx \$100$ and $b_1 \approx 2$.
  b) Less than 1.
  c) 0.5.
  d) About \$50.
27. The response is y and the explanatory variable is x. These can be transformations of the original data.

29. The averages should fall roughly along a line, with comparable variation around each.

31. The slope and intercept ought to be about the same. The value of s_e would shrink since it measures the variation of averages rather than individuals. r^2 would be larger.

33. β_1 remains the same. β_0 carries the units of the response and would be 100 times larger. r^2 remains the same, and s_e changes like the intercept.

35. a) $\beta_1 \neq b_1$ due to sampling variation.
  b) Yes
  c) r^2 will stay about the same. Other estimates probably approach the population parameters: s_e approaches $\sigma_\varepsilon = 1.5$ and β_0 and β_1 approach to 7 and 0.5, respectively. The standard errors will be smaller.

You Do It

37. Diamond rings
  a) β_0 is an extrapolation, and b_0 is about 15 $se(b_0)$ below zero. We expect a *positive* intercept, representing the fixed cost of the ring, though this is an extrapolation.
  b) The prediction interval is [$606.54 to $734.73]. Because $800 lies above this interval, this ring is unusually expensive.

39. Download
  a) Yes, we reject H_0: $\rho = 0$ and H_0: $\beta_1 = 0$.
  b) This is the confidence interval for the intercept, 3.9 to 10.7 sec.
  c) 6.5 to 9.2 seconds saved.

41. Seattle homes
  a) b_1 estimates fixed costs from $-\$13,026$ to $128,871. Fixed costs might be zero, negative, or considerable.
  b) b_0 estimates marginal costs from 111 to 201 \$/SqFt ignore other characteristics.
  c) Use $\pm 2 s_e$ to set the range: 90 to 260 \$/SqFt.
  d) $280,000 to $770,000

43. R&D expenses
  a) The estimated slope is the elasticity. A difference of about 0.755 to 0.825%
  b) No.
  c) $12.5 to $410 million − a wide interval for a model that seems predictive on the log scale.

45. OECD
  a) Use b_0, with confidence interval $22,313 to $31,116
  b) Yes. On average, differences of 1% in trade balances are associated with a difference in GDP of $696 to $2184.
  c) $3,043 to $50,386 per person.
  d) The range in (c) is larger than the confidence interval because it predicts one country, not an average in the population.

47. Promotion
  a) Yes, the t-statistic for the slope 2.47 > 2 with p-value 0.018 < 0.05.
  b) Yes. The confidence interval for β_0 is about 0.0235 to 0.2366.
  c) No, the promotion has not proven itself cost effective. The confidence interval for β_1 includes the 0.16 "break-even" coefficient.
  d) The range for β_0 and the predicted value must to be multiplied by 100. That for the slope remains the same. Conclusions that involve statistical significance are the same.

4M High-frequency Finance

  a) The better we know β_0, the more we are able to find stocks with non-zero mean return on the intrinsic risk. We'd like to buy those with positive mean return.

  b) We expect more accuracy because formulas for the standard errors of b_0 and b_1 have $\sqrt{n}$ in the denominator.
  c) The scatterplot shows much variation around the trend, but the relationship seems straight enough. Residual plots, both versus x and over time, are okay. The timeplot of the residuals shows some periods of higher variance and outliers, but otherwise seems okay.
  d) Use confidence intervals for each regression. If these overlap, the two are consistent. If one is narrower than the other, we obtain more precise estimates.
  e) The fitted model for the daily results. b_0 estimates mean daily return on idiosyncratic risk.

Term	Estimate	Std Error	t Ratio	Prob > |t|
Intercept	0.000252	0.000364	0.69	0.4883
Daily Market Return	1.5294867	0.037537	40.75	0.0000

The slope in the regression with monthly data estimates the same parameter as the slope with daily data since both x and y have been rescaled. b_0 estimates mean monthly idiosyncratic return.

Term	Estimate	Std Error	t Ratio	Prob > |t|
Intercept	0.0049275	0.007911	0.62	0.5338
Market Return	1.5371567	0.175106	8.78	<.0001

  f) The confidence intervals for the intercepts are very similar. From daily data, the confidence interval for the annualized return is −0.11536 to 0.24136. From monthly data, the interval is −0.12760 to 0.24586.
  g) The interval for β_1 given by daily data is more precise. The estimates are similar, but the precision (lengths of the intervals) is different. Daily: 1.455914 to 1.603059, Monthly: 1.192723 to 1.881590
  h) Estimate β_1 from daily data, and use either estimate of β_0. For consistency, take both from the daily equation:

Mean idiosyncratic return	−0.12 to 0.15
beta	1.45 to 1.60

  i) High-frequency data improves the estimate of the beta of a stock, but does not improve the estimate of the mean of the idiosyncratic risk. By using higher frequency data, we can estimate β_1 to very high precision, but this does not improve the estimate of alpha. The higher the frequency of the data, the smaller alpha becomes.

Chapter 22

Mix and Match

1. i or f **3.** f or i **5.** b **7.** e **9.** j

True/False

11. False. When the variance of the errors changes, prediction intervals are likely to be too short in some cases and too long in others.
13. True
15. True
17. False. If omitted variables contribute small, independent deviations, their net affect tends to have bell-shaped variation.
19. False. The decision to exclude data should account for substantive relevance. An unusual case might be the most important data.
21. False. It is easier to see changes in the variation in the plot of residuals on x.

Think About It

23. The data would likely have unequal variation, with more variation among larger stores.

25. The analyst failed to realize an evident lack of constant variation. The variance is smaller on the left than the right.

27. **a)** The slope will become closer to 0.
  b) r^2 would increase, but it is hard to say how much. s_e would be smaller since this observation has the largest residual.
  c) Yes; it is near the right-hand side of the plot.

29. **a)** The slope will increase, moving toward 0.
  b) r^2 will decrease. s_e will stay about the same or be slightly smaller.
  c) Yes, because it lies far below other values of the explanatory variable.

31. Answers will vary. Possibilities include trends in the stock market, interest rates, inflation, etc.

33. No. The Durbin-Watson statistic tests the assumption of uncorrelated. Failure to reject H_0 does not prove it true.

You Do It

35. Diamond rings
  The price of the Hope Diamond comes to S\$ 56 million.
  a) The scaling is such that you can see only 2 points: one for the Hope Diamond, and one for the other 48 diamonds.
  b) The fitted line essentially passes through these 2 points. The slope becomes steeper and intercept becomes more negative. Without the Hope Diamond, the equation is *Estimated* Price (Singapore dollars) $= -260 + 3,721$ Weight (carats)
    With the Hope Diamond, *Estimated* Price (Singapore dollars) $= -251,696 + 1,235,667$ Weight.
  c) r^2 grows from 0.978 to 0.999925 and s_e gets huge, from S\$32 to S\$69,962. r^2 becomes larger because most of the variation in the new response is the difference between rings with small diamonds and this huge gem. s_e is larger because a small error in fitting the Hope Diamond is large when compared to the costs of the other rings.
  d) The Hope Diamond is leveraged. The least squares regression must fit this outlier.

37. Download
  a) No
  b) $D = 2.67$. This is significantly different from 2 with p-value 0.003.
  c) The pattern is one that we have not seen. Rather than meander, these residuals flip sign as in $+, -, +, -, \ldots$

39. Seattle homes
  a) *Estimated* \$/Sq Ft $= 201.018 + 5175.491$/Sq Ft
  b) Use the equation fit without the outlier to set the size of confidence intervals. Estimates with the outlier are not close to those obtained without this home. b_1 nonetheless falls within the range of uncertainty. The intercept is about 2.1 standard errors from the revised estimate.
  c) The intercept represents variable costs and is more affected by the outlier. b_1 changes by more in absolute terms, but falls within random variation. The intercept lies outside the confidence interval if the outlier is included.
  d) Yes, the lot for this home is more than 3 times larger than any other.

41. R&D expenses
  a) Residual variation above the fitted line is more compact than below the line. Negative deviations are more spread out than positive deviations.
  b) 1/2.

c) 24 companies lie outside the exact 95% prediction intervals. Of these, only 2 are positive. We'd expect half, or 12. The SD of a binomial with $n = 24$ and $p = 1/2$ is about 2.45. The observed count is 4.1 SDs below the mean. The residuals are skewed (which can be confirmed from the normal quantile plot).

43. OECD
  a) Visually, the fit does not change much. The fitted equations for all 30 countries: *Estimated* GDP (per cap) $= 26,804 + 1,617$ Trade Bal(%GDP)
    Without Luxembourg: *Estimated* GDP (per cap) $= 26,714 + 1,441$ Trade Bal(%GDP) The CI for β_1 using all of the data is 1009.75 to 2225.19. The slope without Luxembourg lies within this range.
  b) r^2 changes. Without Luxembourg, r^2 is smaller (0.503 to 0.369) whereas s_e is similar (11,298 with compared to 11,336 without). The effect on r^2 is large because this case has the largest response. s_e changes little since the residual at Luxembourg is typical.
  c) No. The regression does not take into account the sizes of the countries.

45. Promotion
  a) Week 6 has unusually low levels of detailing. Voice had been steady at 10% before falling off. Sales remain steady.
  b) The fit with the outlier gives similar estimates. The change in the intercept is 0.5184 (b_0 is larger with the outlier); the slope changes $-.4976$ (b_1 is smaller). Both changes are about 1/2 of a standard error, well within sampling variation.
  c) Week 6 is highly leveraged, so it increases the variation in the explanatory variable. Without this case, we have less variation in x and a larger standard error.
  d) $D = 2.02$ and the timeplot of the residuals shows no pattern. No evidence of a lurking variable over time.

4M Do Fences Make Good Neighbors?

  a) Cost of the security fence is \$35,000 per house, so the value added by reducing the crime rate from 15 to 10 per 1000 must exceed this.
  b) No. First, stat. significant does not equate to cost effective. Second, the model describes association. There could very well be lurking variables.
  c) The plot appears straight enough to proceed. There are several outliers.
  d) *Estimated* House Price ($) $= 97921.6 + 1301.3762 \times 1000$/Crime
    Based on this fit, the average selling price at a crime rate of 1000/15 is about \$369,360 and at 1000/10, the estimated price is \$456,118. The increase seems large enough to cover the \$35,000 per home in costs.
  e) The most leveraged communities are at the the right of the scatterplot, with very low crime rates. The most leveraged is Upper Providence, with Northampton and Solebury close by. Center City Philadelphia has a very high crime rate, but is not leveraged since most of the data are near this side of the plot.
  f) The 4 largest residuals (all positive) are Gladwyn, Villanova, Haverford, and Horsham. The prices in these areas are much larger than expected due to a lurking factor: Location.
  g) No. The residuals are not nearly normal, being skewed. Prediction intervals are thus questionable. The skewness and kurtosis of the residuals are $K_3 = 1.9$ and $K_4 = 4.7$. We have more than 47 cases, so averaging produces normally distributed sampling distributions for the slope and intercept.

h) The estimated difference in average selling price is 66.667 b_1. The estimated change in average value is then about $86,759 with standard error $19,199. The estimated improvement thus lies 2.70 standard errors above the break-even point. Ignore possible dependence due to lurking variables, we can signal the builder to proceed.

i) Build it. The central limit theorem produces a normal sampling distribution for the slope. The confidence interval for the profit earned by improved security is about $87,000 plus or minus about $40,000.

j) Point out as an important caveat that (1) we ought to know about the location of the development and (2) the decision relies on estimates from the security consultant.

Chapter 23

Mix and Match

1. h or e   3. c   5. f   7. a   9. b

True/False

11. True
13. False. A marginal slope includes the effects of other explanatory variables.
15. False. Not necessarily smaller.
17. True
19. False. Perhaps only one differs from zero.
21. False. Its primary use is checking the similar variances condition.

Think About It

23. Collinearity. Busy areas attract more fast food outlets. In densely populated areas, the number of competitors reduces sales.
25. a) Estimated Salary = b_0 + 5 Age + 2 Test Score
    b) Indirect effect.
    c) 12 $M/point.
    d) The partial effect is relevant. It's worth it if you stay longer than 2.5 years since raising the test score by 5 points nets $10,000 annually.
27. a) The correlation of something with itself is 1.
    b) Not without knowing the variance of x_1 and y.
    c) The partial and marginal slopes match because the two x-variables are uncorrelated.
29. The order of the variables in the correlation matrix is Z, X, T, Y (X, Z, T, Y is close).
31. a) City 1. Estimated revenues are: city 1: $312,000/month; city 2: $259,500/month
    b) The intercept, albert an extrapolation, estimates fixed revenue present regardless of distance or population. Perhaps earnings from air freight.
    c) Among comparably populated cities, flights to those that are 100 miles farther away produce $30,000 more revenue per month, on average.
    d) Revenue from flights to cities that are equally distant average $1.5 per 100,000 more to larger cities.
33. a, b) The filled in table is

	Estimate	SE	t-statistic	p-value
Intercept	87.3543	55.0459	1.5869	≈0.10
Distance	0.3428	0.0925	3.7060	<0.01
Population	1.4789	0.2515	5.8803	<0.01

c) Yes, the t-statistic for *Distance* is larger than 2 in absolute value.
d) A confidence interval for 10 times the slope in the population is ($9,700 to $19,800).

35. a) Yes. $F \approx 48.4 \gg 4$ (with p-value $\ll 0.05$).
    b) Assuming a valid model, we ought to be able to predict revenue to within about $65,000 with 95% confidence.

You Do It

37. Gold chains
    a) The data include only several lengths and widths. Width is highly related to price. The two x's are not very correlated. The plots look straight enough.
    b) The largest correlation (0.95) is between price and width.

	Price ($)	Length (Inch)	Width (mm)
Price ($)	1.0000	0.1998	0.9544
Length (Inch)	0.1998	1.0000	0.0355
Width (mm)	0.9544	0.0355	1.0000

c) The fit has $R^2 = 0.94$ and $s_e = 57 with these coefficients . . . .

Term	Estimate	Std Error	t Ratio	Prob > |t|
Intercept	−405.635	62.11863	−6.53	<.0001
Length (Inch)	8.8838083	2.654034	3.35	0.0026
Width (mm)	222.48894	11.64679	19.10	<.0001

d) First, there's a pattern in the residuals. Second, the model is missing an important variable: the amount of gold.
e) Form "volume" as the length (25.4 mm/inch) times the width^2. A pattern remains in the residuals, but less than before. The plot shows previously hidden outliers.
f) With volume, prior explanatory variables lose importance. The model has a much smaller $s_e \approx 17.

R^2	0.994674
s_e	17.0672

Term	Estimate	Std Error	t-statistic	p-value
Intercept	55.118884	34.43198	1.60	0.1225
Length (inch)	0.0451975	0.971144	0.05	0.9633
Width (mm)	−30.59663	16.27885	−1.88	0.0724
Volume (cu mm)	0.0930388	0.005845	15.92	<.0001

39. Download
    a) File sizes increased steadily over the day, meaning that these explanatory variables are associated. The scatterplots of transfer time on file size and time of day seem linear, though their may be some bending for time of day.
    b) The marginal and partial slopes will be very different. File size and time of day are redundant, so the indirect effect of file size will be large.
    c) The multiple regression is

R^2	0.624569
s_e	6.283617

Term	Estimate	Std Error	t-stat	p-value
Intercept	7.1388209	2.885703	2.47	0.0156
File Size (MB)	0.3237435	0.179818	1.80	0.0757
Time (since 8 am)	−0.185726	3.16189	−0.06	0.9533

d) Somewhat. The residual plot suggests slightly more variation at lager files. There is also a slight negative dependence over time. The residuals appear nearly normal with no evidence of trends.
e) No. $F \approx 64$ which is statis. significant. The t-statistics are both less than 2. Thus, we can reject $H_0: \beta_1 = \beta_2 = 0$, but neither $H_0: \beta_1 = 0$ nor $H_0: \beta_2 = 0$.
f) The key difference is the increase in the std. error of the slope. The CI for the partial slope of file size is about −.04 to 0.68 seconds per MB. The marginal slope is about

.26 to .37 seconds per MB. The slopes are about the same, but the range in the multiple regression is much larger.

g) The direct effect of size (multiple regression) is 0.32 sec/MB. The indirect effect is $-.0104532$ sec/MB. The path diagram tells the difference between the indirect and direct effect (slope in the simple and multiple regression), not the changes to standard errors.

41. Home prices

a) Some homes are large and expensive, giving leveraged outliers. The association appear linear. The two explanatory variables are correlated.

b)

$$R^2 \quad 0.533512$$
$$s_e \quad 81.03068$$

Term	Estimate	Std Error	t-stat	p-value
Intercept	107.41869	19.59055	5.48	<.0001
Sq Feet	45.16066	5.78193	7.81	<.0001
Num Bath Rms	14.793861	11.74715	1.26	0.2099

c) Yes, although there is concern over the effect of the leveraged outlier.

d) Yes. $F \approx 84$ which is much larger than needed to assure statistical significance.

e) The 95% CI for the marginal slope is about $63,000 to $101,000 per bathroom. For the partial slope, the CI is about $-9,000$ to 38,000 dollars per bathroom. The ranges are comparable, but the estimates are different. The estimates change because of the correlation between the explanatory variables implies a large indirect effect.

f) Don't do it: She's unlikely to recover the value of the conversion. The value of conversion is expected in the range $-9,000$ to $38,000; $40,000 lies outside this range.

43. R&D expenses

a) The scatterplots (on log scales) show strongly linear trends between y and the explanatory variables as well as between explanatory variables.

b)

$$R^2 \quad 0.80991$$
$$s_e \quad 0.869808$$

Term	Estimate	Std Error	t Ratio	Prob > | t |
Intercept	-1.203173	0.089859	-13.39	<.0001
Log Assets	0.5831633	0.052146	11.18	<.0001
Log Net Sales	0.2284876	0.053194	4.30	<.0001

c) The residuals are skewed, even on the log scale; the residuals are not nearly normal. The model would not be suitable for prediction intervals. The CLT suggests inferences about slopes are okay, but not for predicting individual companies.

d) Yes, the t-statistic indicates this slope is significantly different from zero.

e) Among companies of equal assets, R&D spending averages between 0.12 to 0.33 percent higher among those with 1% higher net sales.

f) Yes, it's smaller. The marginal elasticity is 0.79 ± 0.04, so the confidence intervals do not even overlap. The simple explanation for the difference is that the partial elasticity estimates differences in net sales among companies with equal assets. The marginal elasticity includes the indirect effecs.

45. OECD

a) Scatterplots show strong, linear association between y and the second predictor. This second variable appears more associated with the GDP.

b) The x's are correlated ($r \approx 0.3$). The slope for the trade balance will change because of indirect effects.

c) The estimated model is

$$R^2 \quad 0.772618$$
$$s_e \quad 6934.623$$

Term	Estimate	Std Error	t Ratio	Prob > |t|
Intercept	-4622.225	4796.003	-0.96	0.3440
Trade Bal	959.60593	232.7805	4.12	0.0003
Muni Waste	62.184369	9.153925	6.79	<.0001

d) Yes

e) The direct path from *Trade Balance* to y has coefficient 960 and the path from *Waste* to y has coefficient 62. The path from *Trade Balance* to *Waste* has slope from

Estimated Muni Waste (kg/person) = 503.93174 + 7.7335205 Trade Bal (%GDP)

The path from *Municipal Waste* to *Trade Balance* has slope

Estimated Trade Bal (%GDP) = -4.990754 + 0.0119591 Muni Waste (kg/person)

The indirect effect for *Trade Balance* is about 481, implying its marginal slope is larger than the partial slope. (Countries with larger exports average higher consumption (more trash), and this consumption contributes to GDP.)

f) 43 to 81 $/kg. No, but the associated consumption contributes to GDP.

47. Promotion

a) Scatterplots are vaguely linear, with weak associations between the explanatory variables and response. The largest correlation is between the explanatory variables, so marginal and partial slopes will differ.

b) The estimated model is

$$R^2 \quad 0.2794$$
$$s_e \quad 0.006618$$

Term	Estimate	Std Error	t Ratio	Prob > |t|
Intercept	0.2128209	0.004652	45.74	<.0001*
Detail Voice	0.0166246	0.065259	0.25	0.8004
Sample Voice	0.0219316	0.008364	2.62	0.0127*

c) The residuals look fine, though rather variable. The DW does not find a pattern over time ($D = 2.04$). The residuals are nearly normal.

d) Yes. $F \approx 7 > 4$, so the effect is statistically significant (p-value = 0.0027 < 0.05).

e) No. The partial effect for detailing is not significantly different from zero.

f) No. The model is not causal. The partial slope for detailing is not significantly different from zero. This implies that at a given level of sample share, periods with a higher detailing have not shown gains in market share. Since detailing and sampling tend to come together, it is hard to separate the two.

4M Residual Car Leases

a) Without an estimated residual price, the manufacturer cannot know whether lease price cover costs.

b) We need multiple regression because it is likely that the two factors are related: older cars have been driven further. Marginal estimates double count for the age of the car when estimating the impact of mileage.

c) Most curvature in previous examples with cars arises from combining very different models. Also, nonlinear patterns that come as cars lose value become more evident as cars get older.

d) The plots appear straight, and collinearity is apparent. A few outliers appear.

e)

| | R^2 | 0.510372 |
| | s_e | 3178.879 |

Term	Estimate	Std Error	t Ratio	Prob > |t|
Intercept	40323.937	721.8478	55.86	<.0001
Age	−1853.803	288.8791	−6.42	<.0001
Mileage	−0.124023	0.02375	−5.22	<.0001

f) The residuals have similar variances and are nearly normal. The diagonal stripes come from rounding of the prices. There's an expensive car among these (row 17), but otherwise nothing stands out.

g) For the inverse effect of age on residual value, the 95% CI is $1,280 to $2,430 per year. For mileage, $0.077 to $0.172 per mile. (Both lower residual value)

h) To cover the loss in value of the car over the lease, recommend structuring the lease to cost $2,400 per year with and additional $0.18 per mile. These estimates on average will cover the costs due to aging with 95% confidence.

i) The analysis ignores that these cars cost different amounts at the time of purchase. We do not observed an actual loss, only how time (and mileage) affect the value. Other differences among these cars may be relevant, such valuable options.

Chapter 24

Mix and Match

1. f **3.** i **5.** j **7.** b **9.** c

True/False

11. True

13. True

15. False. Usually true, but not always. If $n = 4$, then $F = 0.8/(2 \times 0.2) = 2$ which is not statistically significant.

17. True

19. False. You need the x's, not y, to compute VIF.

21. False. One of many approaches; the best depends on the context.

Think About It

23. a) In February, the market had positive returns, but S&P was negative. The opposite happened the next month.

b) Very much so. These months are different combinations of the two explanatory variables.

c) Keep them. These points reduce the correlation between the explanatory variables.

25. The t-statistic for testing remains the same. The estimate would be 100 times smaller, but its standard error would also be 100 times smaller.

27. a) The data have 2 cases at every combination of income and age. A scatterplot of age and income is a square grid of dots.

b) Yes, because *Age* and *Income* are uncorrelated.

c) Expect the slope for *Age* to be negative.

29. a) Not as designed. The two explanatory variables are perfectly correlated.

b) The analysis will be simpler if the two variables vary independently over a range of commonly used settings.

You Do It

31. Gold Chains

a) No, the correlation is 0.966. Yes, but with collinearity.

b) Yes

c) $VIF \approx 15$. Collinearity increases the standard errors by about 4 times.

d) For chains of a given width, the retailer charges about 12 cents per additional mm³.

e) Collinearity; the negative slope for width indicates that our estimate of the amount of gold provided by the volume variable is off for the wider chains.

	R^2	0.994674
	s_e	16.72312
	n	28

Term	Estimate	Std Error	t Ratio	Prob > |t|
Intercept	56.521547	16.31446	3.46	0.0019
Width (mm)	−31.01496	13.29882	−2.33	0.0280
Volume (cu mm)	0.1186587	0.005982	19.83	<.0001

33. Download

a) $F \approx 64$, which is statistically significant.

b) No, because the absolute value of neither t is larger than 2.

c) $VIF \approx 42$. Collinearity increases the standard errors of the estimated slopes by about 6.5 times.

d) Yes, the two predictors are highly redundant; transferred files got steadily larger during testing.

e) Yes, they could have randomly chosen the file to send rather than steadily increase the size of the file.

	R^2	0.624569
	s_e	6.283617
	n	80

Term	Estimate	Std Error	t Ratio	Prob > |t|
Intercept	7.1388209	2.885703	2.47	0.0156
File Size (MB)	0.3237435	0.179818	1.80	0.0757
Hours past 8	−0.185726	3.16189	−0.06	0.9533

35. Home prices

a) Yes

b) $VIF \approx 2.2$. Thus, the standard error for the slope of *Bathrooms* would have been smaller by a factor of 1.5, making its t-statistic larger by this same factor. This is not enough to be statistically significant, but closer.

c) The estimated slope in the regression of residuals is the *same* as the partial slope in the multiple regression.

d) There's less variation and less association between residuals. We can see a trend marginally, but only a very weak pattern.

	R^2	0.533512
	s_e	81.03068
	n	150

Term	Estimate	Std Errort	t Ratio	Prob > |t|
Intercept	107.41869	19.59055	5.48	<.0001
Sq Feet (000)	45.16066	5.78193	7.81	<.0001
Bathrooms	14.793861	11.74715	1.26	0.2099

Term	Estimate	Std Error
Intercept	−1.2e − 14	6.593737
Residuals Bathrooms	14.793861	11.7074

37. R&D expenses

a) Yes, large corporations tend to have large values for both assets and sales.

b) On a log scale, $r \approx 0.94$. In dollars, $r \approx 0.95$.

c) On the log scale; the data have a more linear pattern.

d) $VIF \approx 8.6$. Standard errors of the estimated slopes are 2.9 times larger than if we had uncorrelated explanatory variables.

e) The estimated slopes are the same, with a slight difference in the estimates of the standard error.

f) Collinearity reduces the range of *Log Assets*. The slope seems well preserved, just smaller.

R^2	0.80991
s_e	0.869808
n	489

Term	Estimate	Std Error	t Ratio	Prob > | t |
Intercept	−1.203173	0.089859	−13.39	<.0001
Log Assets	0.5831633	0.052146	11.18	<.0001
Log Net Sales	0.2284876	0.053194	4.30	<.0001

Term	Estimate	Std Error
Intercept	$3.1e - 16$	0.039294
Residuals Log Assets	0.5831633	0.052092

39. OECD

a) No. Had the data been formulated as national totals, then most of the variables would indirectly measuring the size of the economies.

b) Luxembourg is leveraged, with a relatively high trade balance.

c) No. $VIF \approx 1.13$. Collinearity increases the standard error slightly.

d) The partial slope matches the slope in the partial regression plot.

e) Luxembourg is more highly leveraged in the regression analysis for the partial slope. Without Luxembourg, the partial slope decreases, and its standard error increases.

Term	Estimate	Std Error
Intercept	$5.8e - 12$	1280.863
Residuals Trade Bal (%GDP)	1157.7626	200.8408

Without Luxembourg ($n = 29$):

Term	Estimate	SE	t-stat	p-val
Intercept	−4622.225	4796.003	−0.96	0.3440
Trade Bal (%GDP)	959.60593	232.7805	4.12	0.0003
Muni Waste (kg/person)	62.184369	9.153925	6.79	<.0001

41. Promotion

a) Yes. There is a strong downward trend in the level of sampling. An outlier in detailing during week 6 is also apparent.

b) Yes, $F \approx 4.86$ with p-value < 0.05.

c) Yes. The marginal slope of detailing is positive. The partial slope is negative (and not statistically significant).

d) Sample voice with $VIF \approx 4.20$.

e) Sampling is the key driver. Sampling is the only one of the 3 variables that contributes statistically significant variation in this fit.

f) Two insignificant variables might be highly correlated with each other (not so in this example).

R^2	0.294436
s_e	0.006632
n	39

Term	Estimate	Std Error	t	p-value	VIF
Intercept	0.2097905	0.005846	35.89	<.0001	·
Detail Voice	−0.011666	0.074983	−0.16	0.8773	2.337
Sample Voice	0.0298901	0.012845	2.33	0.0259	4.193
Week	0.0001239	0.000147	0.84	0.4059	2.436

4M Budget Allocation

a) TV historically got about 2/3 of the budget. I'll maintain some balance, but move toward the allocation that seems most likely to improve sales the most.

b) Multiple regression will show for example, given a fixed level of advertising for print, how much additional TV advertising adds. If one of these is significant but the other is not, then I'll know which of these moves with sales. If both are significant, then the magnitude of the slopes will help me allocate between them.

c) Everything has been rising over time. Marginal slopes reflect this common growth. I need to separate the effects to investigate advertising.

d) Scatterplots look straight-enough. Both forms of advertising are highly correlated with sales and each other.

e) The correlation matrix shows substantial collinearity. The largest correlation is between TV and print advertising. The VIF for these variables is about 9.0. Collinearity increases the estimated standard error by 3 times.

	Sales	Print Adv	TV Adv	Week
Sales	1.0000	0.9135	0.8507	0.8272
Print Adv	0.9135	1.0000	0.9428	0.9294
TV Adv	0.8507	0.9428	1.0000	0.9065
Week	0.8272	0.9294	0.9065	1.0000

R^2	0.835421
s_e	207.0715
n	104

Term	Estimate	Std Error	t-Stat	p-value	VIF
Intercept	2288.985	160.4241	14.27	<.0001	·
TV Adv ($M)	−1.317456	1.681596	−0.78	0.4352	9.00
Print Adv ($M)	16.963068	2.048139	8.28	<.0001	9.00

f) With advertising growing, I am concerned about lurking variables not in the model. The plot of the residuals on the fitted values looks fine, with a consistent level of random variation. The timeplot of the residuals and DW statistic confirm no autocorrelation. The residuals are nearly normal.

g) i. The overall $F \approx 256$ which is statistically significant.
ii. Only one predictor is statistically significant, (*Print Adv*).

h) Recommend shifting from the current emphasis on TV to printed media. With a 50/50 allocation, the estimated sales are about $5,105 thousand rather than $4,008 thousand estimated from the current allocation.

i) s_e indicates that 95% of residuals lie within $414 thousand of the fit. The shift implied by the fit of the model would certainly be visible, if it were to happen. I'm doubtful, however, of the accuracy of the model under such a different allocation of advertising (extrapolation).

j) Spending for printed advertising has a larger association with sales than TV advertising when adjusted for the overall spending. Recommend gradually shifting the budget from the 2/3 TV and 1/3 print toward a more balanced allocation.

k) Because both types of advertising grew over these two years, both series are associated with sales. Once I adjust for the amount of spending on printed ads, spending on TV ads offers little benefit. If I increase my spending on printed ads, this model does not guarantee an increase in sales. A lurking variable, not dollars spent for advertising, may be the key to sales.

Chapter 25

Mix and Match

1. i   **3.** a   **5.** b   **7.** e   **9.** d

True/False

11. True
13. True
15. True

17. False. The purpose of an interaction is to allow the slopes to differ.
19. False. Confounding is always possible without randomization.
21. False. It is helpful if the sizes of the two groups are similar, but not assumed by the model.
23. False. Only 4 are needed to represent 5 groups.

Think About It

25. Is this data from a randomized experiment? If not, do we know whether the sales agents sell comparable products that produce similar revenue streams? Do we know whether the costs for the agents in the two groups are comparable? Without such balance, there are many sources of confounding that could explain the differences that we see in the figure.
27. We combine them in order to compare the intercepts and compare the slopes.
29. Unless you have strong reason to know that the slopes are parallel, one should *always* try an interaction. In this context, the slope is the estimate for the cost per hour of labor and we'd expect it to be higher in the union shop – this is a clear indication that the model needs an interaction.
31. a) The intercept is the mean salary for women ($140,034). The slope is the difference in salaries, with men marginally making $4,671 more than women overall.
    b) They match.
    c) The variances in the two groups are assumed equal.
33. a) About 2.
    b) The slope will be much flatter, closer to zero.
35. a) Yes, since the coefficient of the interaction (Dx) which measures the difference in the slopes of the two groups is not statistically significant.
    b) The coefficients of D and xD would both become positive. The rest of the output would remain as shown.
    c) Remove the interaction term to reduce the collinearity and force the slopes to be parallel.

You Do It

37. **Emerald diamonds**
    a) A two-sample comparison of weight by clarity shows that the average weight is almost the same in the two groups. Weight is unlikely to be a confounding effect in this analysis.

Level	Number	Mean	Std Dev
VS1	90	0.413556	0.054408
VVS1	54	0.408148	0.053661

    b) The two-sample *t*-test finds a statistically significant difference, with VVS1 costing on average about $110 more than VS1 diamonds.

VS1–VVS1, allowing unequal variances

Difference	−112.30	t Ratio	−2.88504
Std Err Dif	38.98	DF	103.4548
Upper CL Dif	−35.11	Prob > $|t|$	0.0048
Lower CL Dif	−189.50	Prob > t	0.9976
Confidence	0.95	Prob < t	0.0024

    c) Because the interaction is not statistically significant, we'll remove it and refit the model.
       Without the interaction, the fits are parallel and the estimated effect for clarity is statistically significant.

$$R^2 \quad 0.497376$$
$$s_e \quad 161.8489$$
$$n \quad 144$$

Term	Estimate	Std Error	t Ratio	Prob > $|t|$
Intercept	−20.44887	105.1595	−0.19	0.8461
Weight (carats)	2785.9054	250.9129	11.10	<.0001
Clarity	127.36823	27.89249	4.57	<.0001

Based on the fit of this multiple regression, we see that for diamonds of comparable weight, those of clarity VVS1 cost on average about $127 more than those of clarity VS1.

    d) From the two-sample comparison, the 95% confidence interval for the mean difference in price is $35 to $190 more for VVS1 diamonds. The estimated mean difference from the multiple regression is $72 to $183 which is shorter because the variation in price due to weight is removed. There's no confounding because the weights are comparable in the two groups. Hence the estimated average differences in price ($112 vs $127) are comparable.
    e) The two groups have similar variances, but the variance increases with the price. Thus, the multiple regression does not meet the similar variances condition.

39. **Download**
    a) The file size is related to the transmission time. However, the file sizes are paired in the two groups. Because of this balance, the file size cannot be a confounding variable. It's the same in both samples.

Level	Number	Mean	Std Dev
MS	40	56.9500	25.7014
NP	40	56.9500	25.7014

    b) The two-sample *t*-test finds a very statistically significant difference in the performance of the software from the two vendors. On average, the software labeled "MS" transfers files in about 5.5 fewer seconds.

MS–NP, allowing unequal variances

Difference	−5.5350	t Ratio	−2.52682
Std Err Dif	2.1905	DF	58.79005
Upper CL Dif	−1.1515	Prob < $|t|$	0.0142
Lower CL Dif	−9.9185	Prob > t	0.9929
Confidence	0.95	Prob < t	0.0071

    c) The interaction in the model is statistically significant, meaning that the two types of software have different rates of transfer (different MB per second).

$$R^2 \quad 0.752229$$
$$s_e \quad 5.138168$$
$$n \quad 80$$

Term	Estimate	Std Error	t Ratio	Prob < $|t|$
Intercept	4.8929786	1.995934	2.45	0.0165
File Size (MB)	0.4037229	0.032012	12.61	<0.0001
Vendor Dummy	4.7633694	2.822677	1.69	0.0956
Vendor Dummy × File Size	−0.180832	0.045272	−3.99	0.0001

The transfer times using MS become progressively less than obtained by the NP software. The small difference in the intercepts occurs because both send small files quickly. The difference emerges only when the files get larger.

    d) The two-sample comparison finds an average difference of 5.5 seconds (range 1 to 10 seconds), with MS transferring files faster. The analysis of covariance also identifies MS as faster, but shows that the gap becomes progressively wider as the file size increases.
    e) No. You can see hints of a problem in the color-coded plot of residuals on fitted values. Similarly, the boxplots of residuals show different variances.

41. Home prices
  a) There's a clear difference, with a much steeper slope (higher fixed costs) for the data for Realtor B.
  b) The model requires both a dummy variable and interaction.

R^2	0.762904
s_e	0.037308
n	36

Term	Estimate	Std Error	t Ratio	Prob < |t|
Intercept	0.155721	0.019713	7.96	<.0001
1/Sq Ft	57.923342	31.2019	1.86	0.0726
Realtor Dummy	−0.176852	0.061062	−2.90	0.0068
Dummy × 1/SqFt	568.5921	110.8419	5.13	<.0001

  c) The data for Realtor B is much less variable around the fitted line than for Realtor A; the residuals do not meet the similar variances condition.
  d) The estimates are fine to interpret even with the evident lack of similar variances, as they reproduce the fitted equations for the separate groups. The intercept, about $156/SqFt is the estimated variable cost for Realtor A homes. The fixed costs for this realtor (slope for 1/SqFt) run about $58,000. For Realtor B, the estimated intercept is near zero (0.158−0.177), suggesting no variable costs! Instead, the prices for this realtor seem to be all fixed costs, with an estimate near $627,000 regardless of the size of the home.
  e) No. The formula for the SE of a regression slope depends on the single estimate se of residual variation, and that is inappropriate in this analysis. We need to have separate estimate of the variance of for the two realtors.

43. R&D expenses
  a) The two look very similar with the colors evenly mixed. A simple regression to both years seems reasonable.

R^2	0.807597
s_e	0.896963
n	985

Term	Estimate	Std Error	t Ratio	Prob < |t|
Intercept	−1.192587	0.062477	−19.09	<.0001
Log Assets	0.7954859	0.012384	64.23	0.0000

  b) The residuals from the multiple regression show some skewness for 2004. Both have comparable variances, however, and share this problem. As you can tell from a normal quantile plot, the combined data are not nearly normal, but since we are working with the slopes (which are averages) we can continue on by using the central limit theorem. However, it is likely that the data are dependent, calling into question any notion of using the usual formulas for standard errors.
  c) Neither added variable is statistically significant and the R^2 has hardly changed from the simple regression.

R^2	0.8077
s_e	0.897636
n	985

Term	Estimate	Std Error	t Ratio	Prob < | t |
Intercept	−1.184021	0.091473	−12.94	<.0001
Log Assets	0.7900453	0.017898	44.14	<.0001
Year Dummy	−0.016828	0.125321	−0.13	0.8932
Dummy × Log Assets	0.0110461	0.024828	0.44	0.6565

$F \approx 0.26$ which is not statistically significant. This agrees with the visual impression conveyed by the original scatterplot: the relationship appears to be the same in both years.
  d) Overall, a common regression model captures the relationship. The elasticity of R&D expenses with respect to assets is about 0.8: on average each 1% increase is assets comes with a 0.8% increase in R&D expenses. There are serious questions, however, about the independence of the residuals in the two years, since there is a pair of measurements on each company.

45. Movies
  a) Adult movies appear to have consistently higher subsequent sales at a given box-office gross than family movies. The fits to the two groups look linear (on this log scale) with a "fringe" of outliers. A common simple regression splits the difference between the two groups.

R^2	0.648668
s_e	0.253298
n	224

Term	Estimate	Std Error	t Ratio	Prob < | t |
Intercept	−1.305742	0.063479	−20.57	<.0001
Log 10 Gross	0.8420019	0.04159	20.25	<.0001

  b) The following results summarize fitting the multiple regression with a dummy variable and interaction.

R^2	0.75236
s_e	0.213623
n	224

Term	Estimate	Std Error	t Ratio	Prob < | t |
Intercept	−1.344678	0.104526	−12.86	<.0001
Log 10 Gross	0.7394797	0.063621	11.62	<.0001
Audience Dummy	−0.070228	0.122375	−0.57	0.5666
Dummy × Log Gross	0.2358524	0.076836	3.07	0.0024

The initial scatterplot appears straight enough within groups, and the plot of residuals on fitted values shows no deviations from the conditions. The comparison box-plots show that the variability is consistent in the two groups. The normal quantile plot confirms that the combined residuals are nearly normal, though a bit skewed.
  c) $F \approx 46$ which is very statistically significant. We'd reject H_0 that the added variables both have slope zero.
  d) The interaction is highly statistically significant, but the slope for the dummy is not. Only one predictor seems useful. The F-test reaches a more impressive view of the value of adding these two predictors because it is not affect by the substantial collinearity between them. The VIFs for these explanatory variables are almost 15, reducing the size of the shown t-statistic for each by about 4.
  e) The estimates show that as the box-office gross increases, movies intended for adult audiences sell statistically significantly better. Each 1% increase in the box-office gross for an adult movie fetches about 0.74% increase in after-market sales. For adult movies, the elasticity jumps to about 0.98%.

47. Promotion
  a) By fitting one line to both groups, rather than within each, the higher sales in Boston inflate the slope.

b) The scatterplot suggests parallel fits in the two locations, with a common slope for detailing. An interaction term is not found to be statistically significant. Residual plots indicate that the following model meets the conditions of the MRM.

Term	Estimate	Std Error	t Ratio	Prob < | t |
Intercept	0.1230925	0.003255	37.81	<.0001
Detail Voice	0.1521676	0.04406	3.45	0.0009
City Dummy	0.0861738	0.002073	41.57	<.0001

c) The effect for detailing has fallen from 1.08 down to 0.15, with a range of 0.06 to 0.24. Rather than get a 1% gain in market share with each 1% increase in detailing voice, the model estimates a far smaller return on this promotion.

4M Home Prices

a) The developer would be able to use the regression model to design houses whose value might be more stable or be increasing. Stable prices would help the developer manage cash flows. The regression might also suggest problems with the initial pricing method used by the developer. The developer might be able to refine its pricing to offer a better initial price.

b) As formulated, β_1 accounts for various "fixed cost" components of the change in value that are present regardless of the size of the home.

c) Add dummy variables for two types of locations, such as rural and suburban, using urban locations as the baseline group. Interactions allow the slopes of the other explanatory variables to vary from one location to another.

d) The interaction of 1/Square Feet and location would indicate different "fixed size" changes in the prices. Similarly, an interaction between *Distance* and location might indicate that distances are not so relevant in cities but very relevant in rural locations.

e) The scatterplot matrix shows curvature (between number of baths and the response) as well as a leveraged outlier (a rural location that is very far from schools).

f) This table summarizes the fit of the model without location effects. The overall R^2 is 0.387.

Term	Estimate	Std Error	t Ratio	Prob < | t |
Intercept	−18.30274	8.33591	−2.20	0.0301*
1/Square Feet	−29167.67	8364.022	−3.49	0.0007*
Number Baths	7.368889	2.059244	3.58	0.0005*
Distance	0.173516	0.669046	0.26	0.7958

Based on the residual boxplots by location, rural locations seem to have larger positive changes on average than predicted by this model, but those differences seem small relative to the variation in these groups. The variation of the residuals is different between locations: the residuals seem more variable in the city and suburb locations.

g) Adding dummy variables for location alone has little effect on the model. Adding interaction terms offer more improvements in the model.

Adding the full set of interactions adds many statistically significantly coefficients to the model. We used the suburban category as the baseline model. Following the principal of marginality, we keep the location dummy variables in the fit since these are involved in statistically significant interactions.

Term	Estimate	SE	t-statistic	p-value
Intercept	−23.42999	14.10251	−1.66	0.0995
1/Square Feet	−23290.43	11704.32	−1.99	0.0491*
Number Baths	12.870567	3.400458	3.78	0.0003*
Distance	−6.391429	2.483663	−2.57	0.0114*
Location[City]	7.0124386	19.14406	0.37	0.7149
Location[Rural]	20.858704	19.58742	1.06	0.2893
Location[City] × 1/Square Feet	44276.312	19285.85	2.30	0.0236*
Location[Rural] × 1/Square Feet	−3450.076	28384.2	−0.12	0.9035
Location[City] × Number Baths	−22.85329	5.763085	−3.97	0.0001*
Location[Rural] × Number Baths	−9.737515	4.723396	−2.06	0.0417*
Location[City] × Distance	7.9910305	3.519161	2.27	0.0251*
Location[Rural] × Distance	6.4266551	2.591713	2.48	0.0147*

The major weakness of the model is the question of other variables that are missing from the model. We consider these in part i. As for the other conditions, the variation in the residuals is more similar after adding interactions and the residuals are nearly normal.

h) Regression analysis finds that location matters, with substantial differences between city, suburban and rural locations. The data indicate the following equations can be used to estimate the change in price per square foot from the initial listing to the final selling price:

Urban (city):
Est Change = −16.42 + 20985.88/sqft − 9.98 Baths + 1.6 Distance

Rural
Est Change = −2.57 − 26740.51/sq ft + 3.13Baths + 0.04 Distance

Suburban
Est Change = −23.43 − 23290.43/sq ft + 12.87 Baths−6.39 Distance

Important features to note:
- Overall prices in the city (adjusted for other features of the homes) rose. Price changes were particularly severe in the suburban market.
- Lots of bathrooms were appealing in the suburbs, but were associated with negative changes in the city (and relatively neutral in the rural areas).
- Distance from a school matters most in the suburbs, and less so elsewhere (particularly in rural areas).

i) The model is not very accurate. For a typical home, the predictions are only within about $25 per square foot with 95% confidence (twice s_e). Almost half of the variation in changes is not explained and attributable to other factors. Another major limitation is the role of the macro economy. Should the economy turn around in the next year, one could not rely on this model to describe home prices.

Chapter 26

Mix and Match

1. c **3.** h **5.** a **7.** g **9.** d

True/False

11. False. Randomization improves any experiment by reducing or eliminating the risk of confounding from a lurking variable.

13. True
15. True
17. False. H_0 claims that the groups have the same mean value.
19. True
21. True

Think About It

23. No. The t-test in general allows different variances in the two groups; it would produce a different standard error for the difference between the sample means.
25. You'd divide by 1.5. The sums of squares in the ANOVA table would be smaller by a factor of 1.5^2, keeping the F-statistic unchanged. All of the estimates in the regression and their standard errors would be divided by 1.5. The t-statistics would remain as shown.
27. The results are highly statistically significant. The p-value is less than 0.0001.
29. The data in two groups determine each slope in a one-way ANOVA, as in the regression view of a two-sample t-test. The least squares regression line has to minimize the deviations away from the line within each group, and that happens if the fit in each group is the mean in that group.
31. No. An analysis of variance fits each group with its own mean, without making assumptions on the relationship among the means. The coding 1, 2, 3, 4, through 5 implies that the means should line up in this order, with equal spacing.
33. We cannot tell; rejecting H_0 does not imply that any specific confidence interval excludes zero.
35. His chance for a Type I error is much larger than 0.05; he should have adjusted for letting the data pick the pair of sample means to compare and set $\alpha = 0.05/(12 \times 11/2) \approx .00076$ as in the Bonferroni interval.
37. a) 53.
    b) Even though n is much larger (8×5 versus 8×54), the Tukey and Bonferroni intervals are quite a bit longer.

You Do It

39. a)

Source	DF	Sum of Squares	Mean Square	F	p-value
Regression	3	150	50.000	3.5	0.0212
Residual	56	800	14.286		
Total	59	1000			

   b) The null hypothesis states that there's no preference for one sweetener over another, $H_0: \mu_1 = \mu_2 = \mu_3 = \mu_4$.
   c) Yes
   d) We can see that there is a statistically significant difference between the average ratings; however, without the means, we cannot tell which is preferred.
41. a) For comparisons between $_{51}C_2 = 1275$ pairs, $\alpha = 0.05/1275 \approx .0000392$. The appropriate t-critical value is $t \approx 4.13$. Hence, to be statistically significant, the difference between two sample averages must be at least

$$4.13 \times s_e \text{ sqrt}(2/20) \approx \$4571$$

   b) No, not unless there is some suspicion that some state is out of line.
43. a) The scale on the x-axis. The x-axis in the plot of Y on D goes from 0 to 1, whereas it goes from -1 to 1 in the plot of Y on X.
   b) Yes
   c) With either explanatory variable, the sample means of the two groups are the fitted values.
   d) The intercept b_0 in the regression of Y on D is the mean for women ($D = 0$) and the slope b_1 is the difference (mean of men minus mean of women). In the regression

of Y on X, the intercept is the overall mean value ($b_0 = \bar{y}$); the slope is half of the difference between the means since the two groups are 2 units apart rather than 1.

45. a) The data do not meet the straight-enough condition. The means in the data appear to first rise then fall with the amount of potassium bromate.
   b) The variation in the data seems to increase slightly with the mean value, but the effect is not substantial. There are no evident outliers nor extreme skewness, though we need to see the residuals from the ANOVA to check the normal quantile plot.
   c) Yes, the F-statistic is barely statistically significant ($p = 0.0404 < 0.05$).

Analysis of Variance

Source	DF	Sum of Squares	Mean Square	F
Regression	4	116537.1	29134.3	2.6289
Residual	80	886582.4	11082.3	p
Total	84	1003119.4		0.0404

   d) The estimated standard error of the difference is

$$s_e \sqrt{(2/n_j)} \approx 36.1 \text{ ml.}$$

   The Tukey interval is

$$\bar{y}_2 - \bar{y}_0 \pm 2.79 \times s_e \sqrt{(2/n_j)} = 91.764 \text{ ml.} \pm 100.72$$

   The difference is not statistically significant.
   e) The Bonferroni interval is

$$\bar{y}_2 - \bar{y}_0 \pm t \times s_e \sqrt{(2/n_j)} = 91.764 \text{ ml.} \pm 104.329$$

   The difference is not statistically significant.
   f) The F-test rejects $H_0: \mu_1 = \mu_2 = \mu_3 = \mu_4 = \mu_5$. The Tukey and Bonferroni intervals, however, find that the largest pairwise difference (between 0 mg and 2 mg) is not statistically significant. Evidently, whatever difference exists among the means is subtle.
47. Stopping distances
   a) There appear to be large, consistent differences among these cars.
   b) The estimated intercept is the mean stopping distance for the omitted category, here the Toyota Camry (48.66 meters). The slope for the Cadillac dummy (-1.02) indicates that the average stopping distance of the Cadillac (47.64 meters) is 1.02 meters less than the average stopping distance for the Camry. The slope for the Malibu dummy (-5.60) indicates that its stops on average were 5.6 meters shorter than those of the Camry.
   c) Yes. $F = 152$ with p-value far less than 0.05.
   d) Changing from meters to feet would increase the estimates in the multiple regression (b_0, b_1, etc.) by a factor of a bit more than 3 since these would be averages in feet rather than meters. The overall F-statistic and p-value would be the same.
   e) Yes. The Cadillac stopped rather quickly once (about 2 meters less than average), producing the most visible outlier. The variation is similar in the groups, with a bit more consistency for the Malibu. The normal quantile plot shows that the data are nearly normal.

   More serious concerns, however, are not revealed in plots. For example, there may be dependence among the measurements. Also, there may be confounding.
   f) The test has a "sample" of one car from each model. We can only conclude, assuming the issues raised in "e" are resolved, that there are differences among these 4 cars – not other cars of the same model.

49. Movie reviews

a) Yes. The variation appears similar in the 3 groups.

b) The side-by-side boxplots suggest that the average urban and suburban customer assign comparable ratings. The rural ratings appear smaller. There is considerable variation within the groups compared to the differences among the centers. That said, these are large groups and the standard error of the average is small for a large sample.

c) The intercept is the average rural rating (49.13). The slope for the dummy variable representing urban views (13.70) is the difference between the average rating of urban views and the average rating of rural viewers. Similarly, the slope for the dummy variable representing suburban views (15.87) is the difference between the average rating of suburban views and the average rating of rural viewers.

d) The standard errors differ because the groups have different sample sizes.

e) The differences among the group means are statistically significant. The p-value for the F-test is 0.0202 < 0.05.

f) The boxplots of the residuals and normal quantile plot of the residuals do not indicate a problem. We might ask about whether the viewers interacted while watching the movie, however, as that might produce dependence.

g) The film gets lower ratings in the rural locations, but there is substantial variation. The only statistically significant difference, using either Tukey or Bonferroni intervals, is between the mean rating in the urban district differs from the mean rating in the rural district.

4M Credit Risk

Motivation

a) As noted, the bank cannot check every application without doing all of the work that the originator is supposed to be doing. A sample will have to do. Because the originator is paid at the time the loan begins, it's in their interest to generate a large volume of loans. That pressure could lead to inferior loans that may pose too large a risk for the bank.

Method

b) $H_0: \mu_j \le 620$. The originator has to prove that it's meeting the objective, not the other way around. Each test needs to be done at level $\alpha = 0.05/5$, not 0.05. Otherwise, there's too much risk of a Type I error.

c) Use simultaneous intervals that adjust for multiplicity.

Mechanics

d) This analysis omits the dummy variable for originator "e". The overall F-test is statistically significant.

Analysis of Variance

Source	DF	Sum of Squares	Mean Square	F
Regression	4	135157.9	33789.5	5.0362
Residual	210	1408968.7	6709.4	*p*
Total	214	1544126.6		0.0007

Parameter Estimates

Term	Estimate	Std Error	t	p
Intercept	668.86	15.21	43.97	<.0001
D(a)	−85.81	24.58	−3.49	0.0006
D(b)	−60.72	19.68	−3.09	0.0023
D(c)	−26.57	19.51	−1.36	0.1745
D(d)	−17.59	17.75	−0.99	0.3230

e) From what we are told, the data should be independent. The data appear to meet the conditions of similar variances and normality.

f) To test whether to reject H_0 for any originator, count the number of *SE*s that separate each mean from 620. Since the groups are of different sizes, the SEs differ. Because we're doing 5 comparisons, we need to use The difference is not statistically significant for any of the originators.

g) We used software to compute the Tukey confidence intervals. These find that originators "c", "d" and "e" are not statistically significantly different. Similarly, originators "a", "b" and "c" are not statistically significantly different. Originators "d" and "e" are different from "a" and "b".

h) Statistical analysis shows that the credit scores of loans originated by "d" and "e" are statistically significantly higher than those originated by "a" and "b". Originator "c" lies in "the middle of the road", but its performance is not statistically significantly different from the others. The average credit ratings of loans originated by all 5 fall within random variation of the target 620 rating. Although the credit ratings of loans originated by "a" and "b" average less than 620, this analysis shows that those differences can be attributed to random variation. A larger sample would be needed to find a statistically significant effect.

i) R^2 is small, but that's an indication of the variation in credit ratings among loans provided by these originators. The standard deviation among the scores originated by a single source is about 82 points. This is about the same as the size of the largest difference in mean ratings. R^2 is a reminder that the differences in mean scores is not large compared to the variation due to other sources. Nonetheless, the differences between means noted in "g" are statistically significant because of the benefits of averaging; averages are less variable than the underlying data.

Chapter 27

Mix and Match

1. d **3.** a **5.** b **7.** e **9.** f

True/False

11. True

13. True

15. True

17. False. The accuracy is low near the edges of the data and can fall off dramatically as we extrapolate.

19. False. See the example of this chapter.

21. True

Think About It

23. a) A 5-term average in general is smoother because it averages more data.

b) We don't know values for the future of the series so there's less averaging that can be done.

c) The time series is no longer centered at the same time points as the data values. For example, suppose that we use a four-term moving average. Should it be defined as

$$\bar{y}_{t,4} = \frac{y_{t-2} + y_{t-1} + y_t + y_{t+1}}{4} \quad \text{or as} \quad \bar{y}_{t,4} = \frac{y_{t-1} + y_t + y_{t+1} + y_{t+2}}{4}?$$

Neither seems very appealing. Perhaps the best remedy is to use the average of the two, which simplifies to

$$\bar{y}_{t,4} = \frac{0.5\, y_{t-2} + y_{t-1} + y_t + y_{t+1} + 0.5\, y_{t+2}}{4}$$

The sum of the weights is 4, but the average is again centered at y_t.

25. a) $250 thousand.
   b) Return toward the mean. The prediction is $290 thousand.
   c) The forecast is for an increase whenever the current value is smaller than the mean, and for a decrease whenever it's greater than the mean.
27. a) The EMWA with more smoothing, $w = 0.9$. The other smoother isn't that smooth and seems to follow random gyrations of the series.
   b) We'd guess growth to be about 1.5 percent in the next quarter, roughly the location of the end of the EWMA. Then GNP would be $\approx \$13.271$ trillion.
   c) From the variation around the EWMA, we'd guess that the percentage change could be as large as about 2.5% or as small as 1%. With these, the range for GDP is $\approx \$13.206$ to $\$13.402$ trillion.
29. a) The estimated correlation is weak and negative at $r = -0.12$.
   b) The one outlier in the time series shows up as the two separated points in the lower left corner of the scatterplot because the scatterplot effectively uses each value of y_t twice.
   c) Yes. The pair of points associated with the outlier produce a "positive" direction to the correlation, whereas the main cluster has a negative slope. Without the pair from the outlier, the correlation falls to -0.26.
   d) One has to exclude two points from the scatterplot.
31. a) We have prior values to find the lags, so we do not lose the first observation.
   b) The slope is smaller by one, and the intercept is the same.
   c) The SD of the residuals s_e is the same because the residuals are the same.
   d) Most of the structure explained by the autoregression in the text is contained in the proximity of y_t to y_{t-1}. By differencing the response, we've removed the easy part of the forecast. This model for the differences has to explain the change in the series, and that's much harder.

You Do It

33. Exxon
   a) Estimated Exxon Price $= 74.803157 + 1.518671$ Month. The model misses considerable pattern in the prices. $D = 0.15$ implies substantial autocorrelation in the residuals.
   b) The t-statistic of the second lag is not far from zero, suggesting we can get a more parsimonious model with less collinearity by removing this term.

RSquare	0.974062
Root Mean Square Error	10.21448
Observations (or Sum Wgts)	130

Term	Estimate	Std Error	t Ratio	Prob > | t |
Intercept	6.40934	3.16132	2.03	0.0447
Month	0.10069	0.05761	1.75	0.0829
Lag 1 Price	0.79758	0.08853	9.01	<.0001
Lag 2 Price	0.13910	0.08916	1.56	0.1212

   c) The slope of the time trend has fallen from near 1.5 in the original simple regression to about 0.1. The change in the slope is due to collinearity. The prices themselves are rising linearly, as seen in the model fit in "a."

RSquare	0.974204
Root Mean Square Error	10.23189
Observations (or Sum Wgts)	131

Term	Estimate	Std Error	t Ratio	Prob > | t |
Intercept	7.0647425	3.105349	2.28	0.0246
Month	0.1171149	0.056729	2.06	0.0410
Lag 1 Price	0.9251095	0.034157	27.08	<.0001

   d) No. Although D is close to 2, the residuals do not have similar variances. The prices become more variable as they get larger. We also have a big outlier, February 2005.
   e) The similarly formulated model using returns finds nothing.

RSquare	0.016168
Root Mean Square Error	0.050128
Observations (or Sum Wgts)	131

Term	Estimate	Std Error	t Ratio	Prob > | t |
Intercept	0.0230028	0.009098	2.53	0.0127
Month	−0.000126	0.000116	−1.09	0.2791
Lag 1 Price	−0.092142	0.088205	−1.04	0.2982

   f) We'd use the returns and guess that the return next month would look like prior returns... averaging about 1.3% increase per month. The distribution of the returns is nearly normal, so we could use the SD of the returns (5% per month) to quantify our uncertainty.
35. Compensation
   a) It's not helpful to describe this structure in the residuals as autocorrelation because that does not help recognize the way to fix the problem: namely, fit a model that captures the trend more completely.
   b)

RSquare	0.995608
Root Mean Square Error	1.919356
Observations (or Sum Wgts)	77

Term	Estimate	Std Error	t Ratio	Prob > | t |
Intercept	79.457816	0.603064	131.76	<.0001
Quarter	0.8109698	0.024434	33.19	<.0001
Late Dummy	−54.40359	2.04544	−26.60	<.0001
Late × Quarter	1.1756546	0.04036	29.13	<.0001

To interpret the fit of the model, start with Dummy = 0, the baseline period (first 42 quarters). For these, the data produce a linear trend

Estimated Hourly Comp $= 79.5 + 0.811$ *Quarter*

In the second period (with Dummy = 1), the fit includes the dummy slope and the interaction (changing the slope) and becomes

$$\begin{aligned} \textit{Estimated Hourly Comp} &= (79.5 - 54.4) \\ &\quad + (0.81 + 1.18)\, \textit{Quarter} \\ &= 25.1 + 1.98\, \textit{Quarter} \end{aligned}$$

Basically, the multiple regression combines two linear regressions, one that fits the early trend, and the second that fits the later linear trend.
   c) No. A sequence plot of the residuals shows that the model leaves substantial pattern in the residuals, and the Durbin-Watson statistic $D = 0.66$ finds statistically significant autocorrelation.
   d) The percentage changes fluctuate around their mean value with no evident pattern or dependence. The variation increased in 2000 (Quarter 53) and appears to have remained large.
   e) We'd use the percentage changes and estimate the variation using the data since 2000. The quarters since 2000 are nearly normal with an SD of about 1.65. The mean seems reasonably stable, so we'd estimate this at the overall mean, about 1 percent per quarter. We'd predict

the next value to be 1% larger than the last observation in the series (176.44 × 1.01), with interval determined by the SD of the percentage changes: 172.407 to 184.054.

37. Inventory

a) The intercept is too far from the data to interpret directly as an estimate at 0. To interpret the slopes, it may be helpful to write the equation as

$$\text{Estimated Inventory} = -6500 - 19\,(\text{Debt}_{t-1} - \text{Debt}_{t-3}) + 5\,\text{Debt}_{t-3}$$

When Debt goes up, the inventory goes down. High levels of consumer debt in the past (3 quarters ago) are associated with larger inventory levels at Wal-Mart.

b) Leave it out. Adding the second lag introduces more collinearity and obfuscates the form of the model without adding to the R^2.

c) No. The residuals have a great deal of structure and are not simple. The data are also autocorrelated within each quarter.

d) Yes. Inventory levels are larger than expected in Q3.

e) The model would look to get Q3 correct (by accident) and over-predict the inventory levels in the other 3 quarters.

39. a) The timeplot shows the dramatic collapse of the housing market associated with problematic home loans in 2007–2008. At the point of the drop, it seems clear that permits leads the way. Other than for this dramatic drop, it is difficult to separate changes in one from the other.

b) The contemporaneous correlation between completions and permits is 0.888 whereas it is larger (0.967) when using the lagged number of permits. Plots confirm that the lagged number is a better predictor.

c) A time trend would have provided a nice fit prior to the collapse of the housing market, but is less useful after the collapse. We found the following model using lags of permits and one lag of housing itself (conveniently 3 months lagged to work with the prediction) to be a good fit.

R^2	0.9497
s_e	51.9587
n	211.0000

Term	Estimate	Std Error	t Ratio	Prob > |t|
Intercept	154.8836	22.1732	6.99	<.0001
Lag 3 Completed	0.1931	0.0636	3.04	0.0027
Lag 3 Permits	0.2940	0.0512	5.74	<.0001
Lag 6 Permits	0.2299	0.0817	2.81	0.0054
Lag 9 Permits	0.1865	0.0660	2.83	0.0052

A timeplot of the residuals shows the presence of a slight amount of autocorrelation in the residuals. The variance of the residuals may be increasing over time, suggesting that the model will be less accurate during the turbulent housing market near the end of the data. Otherwise, the plot of the residuals on the fitted values shows consistent variation. The residuals also appear normally distributed.

d) Substituting the values of the lagged variables into our regression and letting software build the forecast, we obtain an estimate of ≈805,000 homes.

Our software reported the prediction interval as 700,000 to 910,000. Our concern is that the patterns that held in the past may not continue, and the hint that the residuals are becoming more variable supports this theory.

4M Sales at Best Buy

a) An estimate of the seasonal effect would allow managers to anticipate whether recent changes in sales were to be expected from the time of year or from, say, current advertising programs.

b) The data has a strong, nonlinear pattern with strong seasonal variation. The size of the seasonal oscillation is increasing over time.

c) On the log scale, the pattern bends in the other direction (slowing growth rather than increasing growth), and the seasonal variation appears stable.

d) The dummy variables will allow the regression trend to shift up and down depending on the quarter. On the original scale, that will not be so helpful since the size of the shift would need to increase over time. On the log scale, that fixed shift appears reasonable.

e) Two of the dummy variables are statistically significant (though we have not verified the conditions of the regression model).

R^2	0.992751
s_e	0.032532
n	53

Term	Estimate	Std Error	t Ratio	Prob > |t|
Intercept	−211.7551	5.858368	−36.15	<.0001
Time	0.1072808	0.002927	36.65	<.0001
(Time-2001.5)^2	−0.003688	0.000343	−10.76	<.0001
(Time-2001.5)^3	−0.000658	0.000102	−6.44	<.0001
D1	0.1317309	0.012562	10.49	<.0001
D2	−0.035404	0.012799	−2.77	0.0081
D3	−0.011487	0.01277	−0.90	0.3730

f) The Durbin-Watson statistic indicates substantial remaining autocorrelation, also apparent in a plot of the residuals over time. If we were to add a lagged value of y_t to the model, we could absorb this component into the fit.

Durbin-Watson	AutoCorrelation	p-value
0.7170993	0.6232	<.0001

g) We added a two lagged value of the response to the model. The cubic term of the polynomial trend was no longer substantial and was removed. This formulation leaves little structure left in the residuals.

R^2	0.9956
s_e	0.0249
n	51.0000

Term	Estimate	Std Error	t Ratio	Prob > |t|
Intercept	−29.1286	15.9884	−1.82	0.0754
Time	0.0148	0.0081	1.83	0.0742
(Time−2001.75)^2	−0.0012	0.0004	−2.70	0.0100
D1	0.1179	0.0100	11.79	<.0001
D2	−0.1178	0.0194	−6.08	<.0001
D3	−0.0479	0.0275	−1.74	0.0892
Lag Log Gross Profit	0.5149	0.1434	3.59	0.0008
Lag 2 Log Gross Profits	0.3012	0.1395	2.16	0.0364

The timeplot of the residuals shows little of the meandering pattern seen in the residuals from the previous version of the model. The data appear to have constant variance and the residuals are nearly normally distributed as well. The model as a whole is very statistically significant ($F = 1376$). The explanatory variables are statistically significant; we kept time as a predictor even though it is not statistically significant. It's close and by having it in the model we have not anchored the peak of the quadratic to the center of the data. We also kept D3 since we are primarily interested in estimating the seasonal effects.

h) The model follows the up/down seasonal pattern nicely.

i) The key to interpreting the dummy variables is to recognize that all of the comparisons are relative to Q4 and that the effects are multiplicative since the model is on a log scale. We can isolate the role of the season by writing the fit as

Estimated Log Gross Profit = Equation for Q4
+ Seasonal term

When we raise 10 to the power given on both sides (to put the equation back on the scale of the profits, a dollar scale), we have

$$10^{\text{Estimated Log Gross Profit}} = 10^{\text{Equation without dummy}} \times 10^{\text{Seasonal term}}$$

Hence, the effect of Q1 is to raise the fit of profits by about 31% higher than what we'd expect were this Q4. The effect of Q2 is in the opposite direction, about 24% lower than were this Q4. The effect of Q3 is more modest, about 10% lower than in Q4.

j) This model is designed for estimating the seasonal pattern, not forecasting.

4M Gasoline Prices

a) A model can be used to describe how over this time period, prices of the raw crude oil are related to, and determine, the price of gasoline.

b) Immediate price changes would result in a slope of $1/42 \approx 0.024$, the ratio of the size of a barrel of gas to a gallon of fuel.

c) Plots of the prices and percentage changes show that the variation in percentage changes seems simpler, more like iid data than the prices themselves.

d) This lack of immediacy suggests the need for lagged variables so that the price changes do not move instantly through the system, but only move over time.

e) The regression line captures the strong linear trend. The fitted model is Gasoline

($/Gal) = 0.6122189 + 0.0343652 US Spot ($/Bbl)

This slope is larger than our "crude" analysis suggested (about 0.024).

f) The residuals from this model display a strong pattern of dependence. This is confirmed from the Durbin-Watson statistic (0.167).

g) The summary shows a slope much less than one. The percentage changes do not immediately appear in the price of gasoline.

Summary of Fit

RSquare	0.253554
Root Mean Square Error	1.807256
Mean of Response	0.199049
Observations (or Sum Wgts)	487

Parameter Estimates

Term	Estimate	Std Error	t Ratio	Prob > | t |
Intercept	0.1246442	0.082099	1.52	0.1296
%Change US Spot	0.2341702	0.018244	12.84	<.0001

h) The two outliers are from the weeks of September 5, 2005 (this outlier increases the RMSE with little impact on the slope) and March 30, 1998 (leveraged but not influential).

i) The R^2 has improved, and all but one of the added lags of the percentage change are very significant. This model

suggests that percentage changes in the price of crude move through the economy over about 3–4 weeks.

Summary of Fit

RSquare	0.347476
Root Mean Square Error	1.703208
Mean of Response	0.198828
Observations (or Sum Wgts)	483

Parameter Estimates

Term	Estimate	Std Error	t Ratio	Prob > | t |	VIF
Intercept	0.0534869	0.078181	0.68	0.4942	·
%Change US Spot	0.2120529	0.01791	11.84	<.0001	1.08
%Chg US Spot−1	0.0869606	0.018359	4.74	<.0001	1.14
%Chg US Spot−2	0.0627131	0.018281	3.43	0.0007	1.13
%Chg US Spot−3	0.0516171	0.018355	2.81	0.0051	1.14
%Chg US Spot−4	0.0344688	0.017893	1.93	0.0546	1.08

j) No. The slope of the concurrent change remains nearly the same as the value observed in the simple regression (0.212 vs. 0.234).

k) The model offers yet a better fit and captures more of how changes in the price of crude move to the purchase at the filling station.

Summary of Fit

RSquare	0.464844
Root Mean Square Error	1.513707
Mean of Response	0.251296
Observations (or Sum Wgts)	471

Parameter Estimates

Term	Estimate	Std Error	t Ratio	Prob > | t |
Intercept	0.0846268	0.070564	1.20	0.2310
%Change US Spot	0.1842727	0.01639	11.24	<.0001
%Chg US Spot−1	0.0815329	0.016884	4.83	<.0001
%Chg US Spot−2	0.0627179	0.016844	3.72	0.0002
%Chg US Spot−3	0.0582562	0.016861	3.46	0.0006
%Chg US Spot−4	0.0316997	0.016064	1.97	0.0491
Lag Res Pct Chg−1	0.3648019	0.054988	6.63	<.0001
Lag Res Pct Chg−2	0.1298067	0.059844	2.17	0.0306
Lag Res Pct Chg−3	0.0739103	0.059992	1.23	0.2186
Lag Res Pct Chg−4	−0.033269	0.056445	−0.59	0.5559

l) There are still some lingering problems, but the model is now close enough to the multiple regression model. The week of September 5 continues to be a large outlier. The Durbin–Watson statistic is now much closer to 2 and the residuals' autocorrelation is near 0.

Durbin-Watson Number of Obs. AutoCorrelation

Durbin-Watson	Number of Obs.	AutoCorrelation
1.7625995	471	−0.0050

m) The slopes should add to 1. In this form, we can see that the model does. The net effect of these terms (adding the slopes) is 0.954, very close to 1.

n) It seems right in the sense that we expect percentage changes to eventually show up at the pump. The fit of the final regression shows that the full effect happens within about 4 weeks. About 18% of the percentage change shows up almost immediately, followed by about 45% of the change in the next week. The rest occurs over the next 2 to 3 weeks.

Photo Acknowledgments

Chapter 1

Page 2 (Microsoft Zune), Reprinted with permission from Microsoft Corporation.
Page 5 (Satellite images of Hurricane Katrina), National Hurricane Center, National Oceanic and Atmospheric Administration
Page 7 (Katrina track), National Hurricane Center, National Oceanic and Atmospheric Administration
Page 7 (Blue BMW 3551), Gre(BMW of North America, LLC)

Chapter 2

Page 18 (Doctor explaining diagnosis to her patient), Co/Jupiter Unlimited
Page 22 (Student taking a survey), © Richard G. Bingham II/Alamy

Chapter 3

Page 28 (computer mouse), Dave Shutterstock
Page 38 (Old tires), iStockPhoto

Chapter 4

Page 52 (IPod Shuffle), Pearson Education—Addison-Wesley
Page 55 (colorful m&m), Pab\Shutterstock
Page 63 (Pile of Money), iStockPhoto

Chapter 5

Page 77 (Google screen caption), Pearson Education—Addison-Wesley
Page 83 (Broken car window), Luis\istockphoto
Page (Boxes stacked in a warehouse), Jupiter Unlimited
Page 87 (International airport arrivals board), Nuno\Shutterstock
Page 93 (contractor looking at prints), Jupiter Unlimited

Chapter 6

Page 104 (Construction worker on site), Sea\istockphoto.com
Page 119 (Thermostat set at 72 degrees), Sarah Mu//istockphoto.com
Page 119 (Shopping bags), Dreamstime LLC—Royalty Free

Chapter 7

Page 150 (Indian businesspeople wearing headsets), Plush Studios/DH\Alamy Images Royalty Free
Page 156 (Pack of money), Gualberto B\Shutterstock
Page 162 (car industry, automobile), Ricardo/istockphoto.com

Chapter 8

Page 174 (Graduates after the ceremony), Be(Photolibrary.com—Royalty Free
Page 180 (Credit cards), Jupiter Unlimited
Page 182 (Friends watching TV and celebrating), Mo\Alamy Images Royalty Free
Page 188 (Businesswoman sitting in office with laptop on telephone), Monkey Business\Shutterstock

Chapter 9

Page 196 (Maiko Asaba), © Ko Sasaki/The New York Times/Redux
Page 204 (A man operating a fork-lift truck), © Michael Pearcy/Alamy
Page 207 (Indian coins and banknotes), Ajay\Shutterstock
Page 207 (Euros bills: 5 and 50 euros), BigStockPhoto.com

Chapter 10

Page 218 (Giant neon stock ticker on side of Times Square building), istockphoto.com
Page 222 (Currency exchange board showing cross rates between various currencies.), David F\istockphoto.com
Page 233 (Man showing blueprints to couple at construction site), Jupiter Unlimited

Chapter 11

Page 243 (Doctors examining a bottle of pills), Stewart Cohen/Pam\Getty Images, Inc.—Blend Images
Page 250 (Business survey with pen), kare\istockphoto.com

Chapter 12

Page 261 (Black Monday—NY Daily News cover), Pearson Education—Addison-Wesley
Page 268 (Isolated test sheet with a pencil), Jesus J\istockphoto.com
Page 271 (Breakfast cereals on shelf), © FOOD DRINK AND DIET/MARK SYKES/Alamy

Chapter 13

Page 304 (Businessman holding up trophy), Morgan Lane Photography ©Shutterstock;
Page 306 (Color swatches), ©istockphoto.com;
Page 307 (The cover of November 14, 1936 issue of "The Literary Digest"), ©Library of Congress;
Page 311 (Payment on a credit card through the terminal), ©Shutterstock;
Page 313 (Bonding over a Shopping Spree), ©istockphoto.com;
Page 316 (Internet survey), ©Pearson Education—Addison-Wesley.

Chapter 14

Page 325 (GPS navigation system on car dashboard), ©Nick Koudis/Getty Images—Digital Vision;
Page 329 (Microchip testing circuit), ©Chris Knapton/Getty Images, Inc—Stockbyte Royalty Free;
Page 337 (Frozen Foods Aisle), ©Shutterstock.

Chapter 15

Page 351 (Girl filling out an application), ©istockphoto.com;
Page 356 (Mail box), ©Shutterstock;
Page 362 (Open Bank vault), ©istockphoto.com;
Page 366 (Businessman analyzing data), ©Marnie Burkhart/Photolibrary.com—Royalty Free

Chapter 16

Page 378 (LCD macro photo, spam e-mail message concept shot), ©Shutterstock;
Page 389 (A large apartment building with pond and fountain), ©Volodymyr Kyrylyuk/ Shutterstock;
Page 393 (Hotel—business travelers), ©istockphoto.com.

Chapter 17

Page 403 (Car accident on a highway), ©Evgeny Murtola/Shutterstock;
Page 409 (Caucasian, business, adult, talking, male, people), ©Photos.com;
Page 415 (Woman paying for groceries at supermarket checkout), ©Photos.com.

Chapter 18

Page 424 (Happy couple laughing smiling), ©istockphoto.com;
Page 426 (Man shopping in supermarket checking contents of packet), ©Photos.com;
Page 434 (FEDEX express in flight), ©Federal Express Corporation
Page 438 (Two glasses of cola isolated on white), ©Photos.com.

Chapter 19

Page 464 (Jewelry store display case with necklaces and diamond rings), ©istockphoto.com;
Page 467 (Gemologist inspecting diamonds using loupe), ©Getty Images, Inc/Tetra Images Royalty Free;
Page 470 (Gas bill with meter key and coins to illustrate rising energy costs), ©Daisy Daisy/Shutterstock;
Page 476 (Car dealership), ©istockphoto.com.

Chapter 20

Page 488 (New fuel efficient hybrid car design. The future of the industry. Isolated on a white background with a shadow detail drawn in. A pen tool clipping path is included for the car, minus the shadow), ©Michael Shake/ Shutterstock;
Page 498 (Bowl with dog food), ©Photos.com;
Page 502 (Cartons of orange juice on display), ©Russ Du parcq/Shutterstock.

Chapter 21

Page 513 (Company share price information), ©Gal Orbach/Shutterstock;
Page 527 (Gasoline station and convenience store), ©istockphoto.com.

Chapter 22

Page 545 (Sold home for sale sign & house on laptop), ©istockphoto.com;
Page 550 (Home for sale sign in front of new home), ©Shutterstock;
Page 555 (Stock Photo: carpentry), ©Photos.com.

Chapter 23

Page 573 (Beef cheeseburger with potato fries), ©Photos.com;
Page 588 (Signing a contract), istockphoto.com.

Chapter 24

Page 605 (Market analysis), ©istockphoto.com;
Page 611 (LCD television), ©Mike Red/Shutterstock;
Page 616 (Presentation), ©istockphoto.com.

Chapter 25

Page 635 (Employee labeling product in health foods store), ©Photos.com;
Page 639 (Business people in a meeting), ©istockphoto.com;
Page 647 (Receiving Mail), ©istockphoto.com

Chapter 26

Page 665 (Harvesting) ©Orientaly/Shutterstock.
Page 669 (Collection of different wheats isolated on white), ©Beata Becla/Shutterstock;
Page 681 (Sport car) ©Andrejs Pidjass/Shutterstock.

Chapter 27

Page 694 (Pallet truck working at the warehouse), ©Beata Becla/Shutterstock;
Page 694 (Pile of microchips), ©ExaMedia Photography/Shutterstock;
Page 694 (Layoffs and Recession—newspaper headlines documenting deep job cuts), ©JustASC/Shutterstock.

Index

Percentiles of the _t_-distribution, $P(T_{df} \geq t_{\alpha, df}) = \alpha$

df	80%	90%	95%	98%	99%
			α		
1	3.078	6.314	12.706	31.821	63.657
2	1.886	2.920	4.303	6.965	9.925
3	1.638	2.353	3.182	4.541	5.841
4	1.533	2.132	2.776	3.747	4.604
5	1.476	2.015	2.571	3.365	4.032
6	1.440	1.943	2.447	3.143	3.707
7	1.415	1.895	2.365	2.998	3.499
8	1.397	1.860	2.306	2.896	3.355
9	1.383	1.833	2.262	2.821	3.250
10	1.372	1.812	2.228	2.764	3.169
11	1.363	1.796	2.201	2.718	3.106
12	1.356	1.782	2.179	2.681	3.055
13	1.350	1.771	2.160	2.650	3.012
14	1.345	1.761	2.145	2.624	2.977
15	1.341	1.753	2.131	2.602	2.947
16	1.337	1.746	2.120	2.583	2.921
17	1.333	1.740	2.110	2.567	2.898
18	1.330	1.734	2.101	2.552	2.878
19	1.328	1.729	2.093	2.539	2.861
20	1.325	1.725	2.086	2.528	2.845
21	1.323	1.721	2.080	2.518	2.831
22	1.321	1.717	2.074	2.508	2.819
23	1.319	1.714	2.069	2.500	2.807
24	1.318	1.711	2.064	2.492	2.797
25	1.316	1.708	2.060	2.485	2.787
26	1.315	1.706	2.056	2.479	2.779
27	1.314	1.703	2.052	2.473	2.771
28	1.313	1.701	2.048	2.467	2.763
29	1.311	1.699	2.045	2.462	2.756
30	1.310	1.697	2.042	2.457	2.750
31	1.309	1.696	2.040	2.453	2.744
32	1.309	1.694	2.037	2.449	2.738
33	1.308	1.692	2.035	2.445	2.733
34	1.307	1.691	2.032	2.441	2.728
35	1.306	1.690	2.030	2.438	2.724
36	1.306	1.688	2.028	2.434	2.719
37	1.305	1.687	2.026	2.431	2.715
38	1.304	1.686	2.024	2.429	2.712
39	1.304	1.685	2.023	2.426	2.708
40	1.303	1.684	2.021	2.423	2.704
50	1.299	1.676	2.009	2.403	2.678
75	1.292	1.665	1.992	2.377	2.643
100	1.290	1.660	1.984	2.364	2.626
150	1.287	1.655	1.976	2.351	2.609
200	1.286	1.653	1.972	2.345	2.601
Normal	1.282	1.645	1.960	2.327	2.576

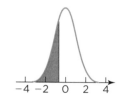 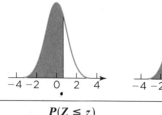 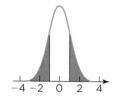 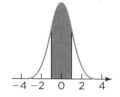

z	$P(Z \le -z)$	$P(Z \le z)$	$P(|Z| > z)$	$P(|Z| \le z)$
0	0.5	0.5	1	0
0.1	0.4602	0.5398	0.9203	0.0797
0.2	0.4207	0.5793	0.8415	0.1585
0.3	0.3821	0.6179	0.7642	0.2358
0.4	0.3446	0.6554	0.6892	0.3108
0.5	0.3085	0.6915	0.6171	0.3829
0.6	0.2743	0.7257	0.5485	0.4515
0.7	0.2420	0.7580	0.4839	0.5161
0.8	0.2119	0.7881	0.4237	0.5763
0.9	0.1841	0.8159	0.3681	0.6319
1	0.1587	0.8413	0.3173	0.6827
1.1	0.1357	0.8643	0.2713	0.7287
1.2	0.1151	0.8849	0.2301	0.7699
1.3	0.0968	0.9032	0.1936	0.8064
1.4	0.08076	0.91924	0.1615	0.8385
1.5	0.06681	0.93319	0.1336	0.8664
1.6	0.05480	0.94520	0.1096	0.8904
1.7	0.04457	0.95543	0.08913	0.91087
1.8	0.03593	0.96407	0.07186	0.92814
1.9	0.02872	0.97128	0.05743	0.94257
2	0.02275	0.97725	0.04550	0.95450
2.1	0.01786	0.98214	0.03573	0.96427
2.2	0.01390	0.98610	0.02781	0.97219
2.3	0.010720	0.989280	0.02145	0.97855
2.4	0.008198	0.991802	0.01640	0.98360
2.5	0.006210	0.993790	0.01242	0.98758
2.6	0.004661	0.995339	0.009322	0.990678
2.7	0.003467	0.996533	0.006934	0.993066
2.8	0.002555	0.997445	0.00511	0.99489
2.9	0.001866	0.998134	0.003732	0.996268
3	0.001350	0.998650	0.002700	0.997300
3.1	0.0009676	0.9990324	0.001935	0.998065
3.2	0.0006871	0.9993129	0.001374	0.998626
3.3	0.0004834	0.9995166	0.0009668	0.9990332
3.4	0.0003369	0.9996631	0.0006739	0.9993261
3.5	0.0002326	0.9997674	0.0004653	0.9995347
3.6	0.0001591	0.9998409	0.0003182	0.9996818
3.7	0.0001078	0.9998922	0.0002156	0.9997844
3.8	0.00007235	0.99992765	0.0001447	0.9998553
3.9	0.00004810	0.99995190	0.00009619	0.99990381
4	0.00003167	0.99996833	0.00006334	0.99993666
4.5	0.000003398	0.999996602	0.000006795	0.999993205
5	0.0000002867	0.9999997133	0.0000005733	0.9999994267
10	7.62×10^{-24}	1	1.52×10^{-23}	1
20	2.75×10^{-89}	1	5.51×10^{-89}	1